Rainer Waser (Ed.)

Nanoelectronics and Information Technology

Advanced Electronic Materials and Novel Devices

Second, corrected edition

WILEY-VCH Verlag GmbH & Co. KGaA

Edited by
Rainer Waser
Department IFF &
CNI – Centre of Nanoelectronic Systems
for Information Technology
Research Centre Jülich, Germany
and
Institut für Werkstoffe der Elektrotechnik
RWTH Aachen, Germany
e-mail: waser@iwe.rwth-aachen.de

All books published by Wiley-VCH are carefully produced. Nevertheless, authors, editors, and publisher do not warrant the information contained in these books, including this book, to be free of errors. Readers are advised to keep in mind that statements, data, illustrations, procedural details or other items may inadvertently be inaccurate.

First edition 2003
Second, corrected edition 2005

Library of Congress Card No.: applied for

A catalogue record for this book is available from the British Library.

Bibliographic information published by Die Deutsche Bibliothek
Die Deutsche Bibliothek lists this publication in the Deutsche Nationalbibliografie; detailed bibliographic data is available in the Internet at ‹http://dnb.ddb.de›.

© 2005 WILEY-VCH Verlag GmbH & Co. KGaA, Weinheim

Printed in the Federal Republic of Germany.
Printed on acid-free paper.

All rights reserved (including those of translation into other languages). No part of this book may be reproduced in any form – by photoprinting, microfilm, or any other means – nor transmitted or translated into a machine language without written permission from the publishers. Registered names, trademarks, etc. used in this book, even when not specifically marked as such, are not to be considered unprotected by law.

Composition: Thomas Pössinger, Aachen
Printing: Druckhaus Darmstadt GmbH, Darmstadt
Bookbinding: Buchbinderei Schaumann, Darmstadt

ISBN-13: 978-3-527-40542-8
ISBN-10: 3-527-40542-9

*To my family,
my friends,
and my colleagues
in Jülich and Aachen.*

Contents

	Contents	4
	Preface	7
	General Introduction	11
I	**Fundamentals**	**25**
1	Dielectrics	31
2	Ferroelectrics	59
3	Electronic Properties and Quantum Effects	79
4	Magnetoelectronics – Magnetism and Magnetotransport in Layered Structures	107
5	Organic Molecules – Electronic Structures, Properties, and Reactions	127
6	Neurons – The Molecular Basis of their Electrical Excitability	145
7	Circuit and System Design	165
II	**Technology and Analysis**	**187**
8	Film Deposition Methods	197
9	Lithography	221
10	Material Removing Techniques – Etching and Chemical Mechanical Polishing	247
11	Analysis by Diffraction and Fluorescence Methods	271
12	Scanning Probe Techniques	295
III	**Logic Devices**	**319**
13	Silicon MOSFETs – Novel Materials and Alternative Concepts	357
14	Ferroelectric Field Effect Transistors	385
15	Quantum Transport Devices Based on Resonant Tunneling	405
16	Single-Electron Devices for Logic Applications	423
17	Superconductor Digital Electronics	443
18	Quantum Computing Using Superconductors	459
19	Carbon Nanotubes for Data Processing	471
20	Molecular Electronics	499
IV	**Random Access Memories**	**525**
21	High-Permittivity Materials for DRAMs	537
22	Ferroelectric Random Access Memories	563
23	Magnetoresistive RAM	589

Contents

V Mass Storage Devices — 605
- 24 Hard Disk Drives — 615
- 25 Magneto-Optical Discs — 631
- 26 Rewritable DVDs Based on Phase Change Materials — 643
- 27 Holographic Data Storage — 657
- 28 AFM-Based Mass Storage – The Millipede Concept — 685

VI Data Transmission and Interfaces — 701
- 29 Transmission on Chip and Board Level — 713
- 30 Photonic Networks — 727
- 31 Microwave Communication Systems – Novel Approaches for Passive Devices — 753
- 32 Neuroelectronic Interfacing: Semiconductor Chips with Ion Channels, Nerve Cells, and Brain — 777

VII Sensor Arrays and Imaging Systems — 807
- 33 Optical 3-D Time-of-Flight Imaging System — 817
- 34 Pyroelectric Detector Arrays for IR Imaging — 829
- 35 Electronic Noses — 847
- 36 2-D Tactile Sensors and Tactile Sensor Arrays — 861

VIII Displays — 877
- 37 Liquid Crystal Displays — 887
- 38 Organic Light Emitting Devices — 911
- 39 Field-Emission and Plasma Displays — 927
- 40 Electronic Paper — 953

Abbreviations — 967
Symbols — 973
Authors — 979
Index — 983

Preface

In 1965, the very early days of integrated circuits, Gordon Moore, the later co-founder of Intel Corp., was asked by the Electronics magazine for their 35th anniversary to report how semiconductor components' integration would develop in future and to make a prediction for the next 10 years. At this time, the Moore group was able to integrate about 60 components on a chip. Using semi log paper and starting from the first planar transistors, he extrapolated to 60,000 components on a chip in 1975. This corresponded to a doubling every year. After the 10 years passed, he was amazed how precisely his prediction had been met. And, with some modifications, the development continued since then. On route, Moore's law has been applied to many developments beyond integration density on chips. For example, it is referred to with respect to the storage density of hard discs, the transmission rate in local and wide area networks, and many other areas. In fact, it seems, it has given name to everything in industry that increases exponentially in performance.

Of course, the atomic structure of matter will some day set limits to the evolution of the integration technology according Moore's law. Historically, limits had been anticipated much earlier – and surpassed. For example, in the early days it was thought that gate oxide thickness could probably not be dropped much below 100 nm because the probability of pinholes increased. Nevertheless, gate oxide thickness now approach 1 nm. The rapidly increasing tunnelling currents represent the next challenge. A solution may be novel dielectric oxides with improved permittivities. Another classic issue has been optical lithography. In the early days, it was believed the limit would be 1 μm. Nevertheless, the principle has been extended to 0.1 μm. And using soft X-rays or, as it is called today, extreme ultraviolet (EUV) lithography, we step towards 0.01 μm in a real big step. These examples demonstrate what this textbook is about: basic concepts and their limits, as well as alternative materials and devices which may circumvent the limits. New opportunities arise on the way towards further miniaturization, such as resonant tunnelling or single electron devices, carbon nanotubes as interconnects or transistors, and eventually complex organic molecules as memory and logic units on the distant horizon, to name just a few. Certainly, if some of these ideas are successful, they probably will not immediately replace Si technology but they will be added onto Si technology for quite some time, using all the benefits which are brought through the (approx.) 100 billion dollar R&D which has been spent on Si technology till today. But these ideas will conserve the validity of Moore's law.

As in the past, there are people asking for the application areas for all that tremendous computational capacity which is ahead of us. And as in the past, some answers to this question are obvious, and others will emerge along the way. Customers certainly would appreciate a truly good speech recognition. As Gordon Moore pointed out in an interview, a system using a real-life speech recognition has to understand context. It has to distinguish between "to" or "two" or "too". If such a system operates in real-time, it requires huge computational capacity. At that level, systems essentially would understand speech. From there, it is a relatively short step towards real-time translation systems, and further to real-time visual recognition of complex daily-life objects e. g. to increase safety in traffic and transportation. Based on this, one may imagine autonomous robots liberating us from undesired daily routine work and assisting us when we get old. Certainly, computational performance on chips has to increase several orders of magnitude before the realization of these ideas becomes feasible, and the developers of software have to bring a similar share.

Scope and Intention of this Textbook

This book has its roots in the 32nd IFF Spring School 2001 at the Research Center Jülich jointly organized with the universities of the region and the industry. The Spring School was dedicated to advanced electronic materials and devices for information technology. Since then the topics have been complemented and many international researchers have been invited to participate in converting the Spring course manuscript book into a real textbook for students. Drafts of this book have been used since 2001 in my Third Year courses at the RWTH Aachen University.

Considering the main title, it is obvious that the area of Nanoelectronics and Information Technology is, of course, much wider than the scope of this book. As the subtitle indicates, the book focuses on materials and devices. It does not cover the field of architecture and design as well as the entire world of software development.

Target readers will be graduate students of physics, electrical engineering and information technology, as well as those studying material science or chemistry and focusing on electronic materials and technologies. The value of the book should lie on its **focus on the underlying principles** of the topics. Contemporary examples are only used to illustrate these principles. Therefore, even if some of the examples are succeeded by more sophisticated ones due to the high dynamics of the field, the basic principles will remain the same.

Preface

This also means that topics and concepts are not only chosen based on the economical potential but by their eductional value. Therefore, the distribution of topics is not primarily balanced with respect to the foreseeable economical development. For example, high permittivity materials such as barium strontium titanate are currently not on the roadmap of the DRAM (Dynamic Random Access Memory) manufacturers. Still, these materials are highly valuable from the physics and technology point of view, because they are ideal to teach the principles of integration of complex oxides onto Si CMOS (Complementary Metal-Oxide-Semiconductor) wafers and the dielectric properties of materials at their stress limits. Superconductor electronics is another example. Although it is currently considered for niche markets only, it shows how logic can work with completely different signals and rules compared to the conventional voltage-level oriented CMOS electronics. This is highly educative if one investigates alternative realizations of logic.

Nevertheless, the topic of the book is not a classical field of science, but rather an extremely fast developing field. Therefore not only the state-of-the-art examples may quickly get outdated, also some of the fundamentals are still under development. Therefore, by nature, the authors were hardly able to give a completely balanced survey of views in the field. Still, they tried to reflect a general consensus among most of the experts in the respective field.

The book tries to provide a link between the short-term topics which will certainly move into production within the next 10 years, such as EUV (Extreme UV) lithography and the aspects described for future Si MOSFETs, and far-reaching topics which are well-ahead or off the main stream, offering high potential on the long term. It is this link which keeps vision focussed on what is necessary for future information technology. This trains students to use their creativity and imagination on the one hand and ensures that they do not get lost in unrealistic speculations, on the other hand. There is a firework of potential future opportunities. Only few will finally make it. But all need to be considered today, to be sure to make the right choice tomorrow.

Organization and Structure of the Textbook

The book is organized in eight *Parts* and fourty topical *Chapters*. Each Part comprises of an area of information technology and starts with an Introduction. It starts with a General Introduction summarizing the basics of the information theory.

Part I - **Fundamentals** - provides at a background on those areas of physics, chemistry, molecular biology, and electrical engineering which are advisable for understanding the succeeding chapters. Part II – **Technology and Analysis** - presents a survey on the current integration techniques used in the main stream Si based CMOS device fabrication and estimations on their extendabilities. As a supplement, alternative fabrication technologies which may become relevant for a future nanoelectronics, as well as the major analytical methods are discussed. Part III - **Logic Devices** comprises new materials which enter into the Si based semiconductor technology to support the further reduction of the size of the devices or to open new functionalities. In addition, the materials and concepts for alternative logic devices are described. Part IV - **Random Access Memories** - demonstrates how high-permittivity oxides, ferroelectrics, and magnetoresistive materials can be utilized to upgrade the high-speed semiconductor memory with respect to density and non-volatility. Part V - **Mass Storage Devices** - resumes the current materials and technologies for state-of-the–art hard discs, magneto-optical devices and phase-change media and looks into completely new concepts for 3-D or nano-scale devices. Part VI - **Data Transmission and Interfaces** - describes the development of interconnects on the chip and board level, optical and microwave data communication, as well as the potential to connect the microelectronic world with biological neural networks. Part VII – **Sensor Arrays** – presents array concepts based on a *spatial* resolution and on a *functional* resolution. Part VIII - **Displays** - summarizes active matrix display technologies based on liquid crystals, organic LEDs, plasma devices, field emitters, and on electronic paper.

Some topics do not appear as individual chapters, although it might seem justifiable considering their importance. This is done, if a specific topic is covered by another Chapter in a broader context. For example, there are several serious concepts on random access memories (RAM), such as molecular electronic RAM, carbon nanotube based RAM, and phase-change RAM, which are not included in the corresponding Part IV. Molecular electronic is also considered for logic devices and, hence, the entire topic is covered in Chapter 20 in Part III. Carbon Nanotubes are extensively reported in Chapter 19. And phase-change RAM is discussed in Chapter 26, which is about Phase-change based re-writtable DVDs.

Using the Text

Prerequisites to understanding this book are undergraduate courses in sciences, particularly in physics and mathematics and an elementary introduction to electromagnetism, electronic materials, devices and circuits. Part I provides some additional background which is needed to understand successive chapters. It also intends to adapt the state of knowledge of readers with different background. Physics students may find it helpful to use Chapter 7 as an introduction to digital circuits, while students of Electrical Engineering and Information Technology will appreciate Chapter 3 to update their knowledge on Quantum Mechanics. Both may like the brief summaries on Organic Chemistry (Chapter 5) and on the molecular biology of nerve cells (Chapter 6).

There are different ways in which this book can be used. One possibility is to work through it as a classical textbook, chapter by chapter. Other readers might pick selected chapters. This is supported by the fact, that the chapters are relatively self-contained. In some cases it might be advisable to read the corresponding basics chapter first. A further possibility is to use the book as an introduction to a specific class of electronic materials, and to select an according subset of the chapter. A table which relates the various topics to the major material classes is provided in the Introduction to Part I.

Of course, no book of this size is without misprints, semantic errors, and ambiguities. This is specially true for a first edition. On behalf of all authors, the Editor apologizes for these deficiencies. We are eager to learn about what could be improved. Therefore, the reader is encouraged to send us comments and hints. Corrections as well as updated information of this textbook is displayed at our web site *http://www.emrl.de*. It is suggested to check this site from time to time.

Conventions

We tried to handle *symbols and constants* of a general nature consistently throughout the book. They are used according to the IUPAP convention. Few specific choices were made to take into account that this book is dealing with electronic materials, devices, and systems. Therefore, e. g., the symbol of energy (work) is W to make a clear distinction to the electric field E. According to common practice, physical and mathematical symbols and constants are written in italics, vectors are shown bold, in addition. In the Appendix, a list of symbols and constants is provided. In a few cases, different symbols or constants have been used in different chapters. This is taken into account in the list too.

Within the Chapters, *abbreviations* are defined at their first occurrence. In addition, an overall list of abbreviations is compiled in the Appendix. In few cases, there are identical abbreviations for very different meanings, steming from very different topics. This must be regarded as a tribute to the highly interdisiplinary character of the field.

Since this is a textbook and not a research review, only representative *references* are included at the end of each chapter. By no means, do they try to cover all original publications of the respective topic.

In order to assist the navigation within the book, the *Introductions to the Parts* of the book are printed on a grey background while the *Chapters* are printed on white.

Acknowledgements

This book owes its final shape and form to the hard work of many people. First of all, I would like to express my sincere appreciation to the authors, who provided wonderful contributions and have been willing to accept many requests for modifications on the route to increase the consistency throughout the book.

I am thankful to many colleagues who have contributed, sometimes unwittingly, to this textbook. Specifically, I am deeply indebted to Tobias Noll (RWTH Aachen) for a considerable amount of his precious time which he spent with me for exciting discussions and manifold suggestions, and for patiently explaining me essentials of those areas, which are distant to my own area of expertise, such as design and architecture of densely integrated digital circuits. Also, I am indebted to Heinrich Kurz (RWTH Aachen) for giving me insight into some of the excitement and challenges of nanotechnology in general and nanoelectronics in particular. With his keen mind, he provided a clear distinction between visions and fictions.

The following colleagues are especially thanked for highly stimulation discussions and for reading and constructively criticizing various parts of the book: George Bourianoff (Intel Corp.), Christoph Buchal (FZ Jülich),. Rainer Bruchhaus (Infineon Technologies), Ramón Campano (European Commission), Andreas K. Engel (University of Hamburg), Karl Goser (University of Dortmund), Gernot Güntherodt (RWTH Aachen), Wolfgang Hönlein (Infineon Technologies), James Hutchby (Semiconductor Research Cooperation), Benjamin Kaupp (FZ Jülich), Angus Kingon (North Carolina State University), Christoph Koch (Caltech), Marija Kosec (Josef-Stefan Institute, Slowenia), Hans Lüth (FZ Jülich), Thomas Mikolajik (Infineon Technologies), Jürgen Rickes (Agilent Technologies), Nava Setter (EPFL, Switzerland), Bernd Spangenberg (RWTH Aachen), Stephan Tiedke (aixACCT Systems), Werner Weber (Infineon Technologies), Stanley Williams (Hewlett Packard Research Laboratories), Victor Zhirnov (Semiconductor Research Cooperation), Karl Zilles (FZ Jülich and University of Düsseldorf).

In one way or another, all of my co-workers at the Research Center Jülich and the RWTH Aachen University contributed to the project. For the tedious work of checking the symbols and formulas within the chapters, as well as for comprehensive editorial and technical assistance in the compilation of the chapters, I would like to gratefully acknowledge: Dierk Bolten, Ulrich Böttger, Peter Ehrhart, Ulrich

Ellerkmann, Fotis Fitsilis, Andreas Gerber, Peter Gerber, Christiane Hofer, Susanne Hoffmann-Eifert, Silvia Karthäuser, Carsten Kügeler, Ralf Liedtke, René Meyer, Yacoub Mustafa, Christian Ohly, Rob Oligschlaeger, Andreas Roelofs, Matthias Schindler, Sam Schmitz (CAESAR,Bonn), Peter Schorn, Simon Stein, Stefan Tappe. I highly appreciate the selfless participation in the project of all of them.

To every project, there is a core team, for coordinating the immense work and endless details, for solving the difficult tasks, and for pushing the project forward with clear view on the target:

Thomas Pössinger who had the responsibility of preparing the overall and detailed layout, of assembling the different text formats, photos, and illustrations in our publishing system, and of supplying a complete copy-ready version of the entire book to the publisher. Despite all this, he kept the spirit of the team high with his unique recitations of Bavarian comedians. Ulrich Böttger kept an overview of all the endless details and solved many problems before they became obvious to others, while stimulating many people involved with his good humor. The heart and soul of this project has been Maria Garcia. With her enthusiasm and optimism together with her dedication (even in her night dreams) to cope with difficult and time-consuming tasks, she really drove the project.

And in addition to these three persons, there has been a dedicated team who worked days and (almost all) nights at the institute during the last phase before the very final deadline, kept alive by the delivery services of Hallo Pizza and of Restaurant Brunnenhof, Aachen. The team consisted of Dierk Bolten, Ulrich Ellerkmann, Martina Heins, Carsten Kügeler, Dagmar Leisten, René Meyer, and Rob Oligschläger.

Drawings and illustration which have not been provided by the authors or reproduced from the literature, were prepared with high artistic flair by Britta Meyer (Plastiktulpen, Kerpen), Dagmar Leisten, and Thomas Pössinger. Format conversions of the text have been carefully executed by Georg Ortmanns and Marianne Scholz. Furthermore, I am indebted to Caroline Zurhelle and Janet Carter-Sigglow for language-checking the entire text.

At my publishers, Wiley-VCH, I want to acknowledge Alexander Grossmann, the publishing director of this textbook, for his support and encouragement, and for the freedom he gave us in all aspects of design, and Ron Schulz, who carefully did much of the correction work.

I am deeply grateful to the Research Center Jülich (FZ Jülich) and specifically to the Institute for Solid State Research (IFF), who generously provided support for the organization of the Spring School in 2001, and for the realization of this textbook which helped in keeping the price of this book reasonably low making it affordable for students. In addition, the Research Center Jülich as well as the RWTH Aachen University have provided a stimulating intellectual environment that encourages interaction between basic sciences and application, an essential condition for editing this interdisciplinary textbook.

Jülich and Aachen
December 2002

Rainer Waser

Preface to the Second Edition

A new edition is needed much earlier than anticipated before, due to the high sales of this textbook. At this stage, we have focussed on correcting misprints. In addition, corrections and supplements to the scientific contents have been made where absolutely necessary. The editor again gratefully acknowledges the help by many colleagues in spotting errata and inconsistencies in the book.

Jülich and Aachen
January 2005

Rainer Waser

General Introduction

Contents

1 Properties of Information — 13

2 Mathematical Definition of Information — 16

3 Processing of Information — 18
3.1 Boolean Algebra and Switching Circuits — 18
3.2 Switching Algebra and Switching Circuits — 19
3.3 Multivalued Logic — 20
3.4 Irreversible and Reversible Logic — 20

4 Areas of Information Technology — 22

General Introduction

Firstly, we will briefly look at information theory and logic. We will take a bird's eye view and show the links of information theory to physics. By using this approach, we try to provide a basic background for the wide variety of concepts utilizing advanced electronic materials and novel devices, which are employed in or considered for nanoelectronics, in particular, and for information technology, in general. In the last Section, we will give a sketch of the areas of information technology in the light of the Parts and Chapters of this textbook.

For a more detailed introduction to information theory, the interested reader is referred to a great selection of excellent textbooks such as, for example, Feynman *Lectures on Computation* [1].

1 Properties of Information

A verbal message in a natural language, a printed page of a book, a table of results of physical measurements, a data file of a customer in the computer network of a bank, a mathematical expression, an image or movie, a track of music, the smell of a pheromone of an insect, and the DNA of a biological cell are a few examples of *information*. The source as well as the recipient of information typically are human beings, but may also be animals, computers, and other appropriate technical or natural objects. The general operations applicable to information are:

- Transmission
- Processing, manipulation
- Storage

Common to all types of information indicated above and their relation to the numerous types of sources/recipients, we can define three levels of information following the designation of the American philosophers and logicians Charles S. Pierce (1839 – 1914) and Charles W. Morris (1901 – 79):

Syntactic level

Information on the syntactic level is concerned with the formal relation between the elements of information, the rules of the corresponding language, the capacity of communication channels and the design of coding systems for information transmission, processing, and storage. At this level, we are *not* interested in the meaning of the information and its practical significance. Throughout this book information is predominantly treated on the syntactic level.

Semantic level

This level relates the information to its meaning. Information given in a natural language is based upon the convention within a group of people. Semantic units such as words or group of words are given a more or less precisely defined meaning. In the case of technical (e.g. computer) languages, the meaning of the semantic units is exactly defined by properties and applicable operations. On the other hand, the semantic information of music may be seen in the emotions caused in appropriate recipients. According to Shannon, semantics is not required for correct processing of the syntactics of information. Still it may be helpful, e.g. to increase the efficiency of data compression.

Pragmatic level

Here, information is related to the practical value or usefulness of it. It strongly depends on the context of the recipient, and may be of economical, political or psychological value. Often it is a function of time. Often delayed information is less valuable, while early information or a correct prediction may be highly valuable.

The three levels represent a hierarchy: We can operate (transmit, process, storage) information on the syntactic level without knowing its meaning. And we can handle information on the semantic level independent of its pragmatic value.

Global elements of information in a given language are symbolic expressions which carry the meaning and the practical value. Depending on the type of information, these symbolic expressions may be sentences composed of words, melodies composed of notes, pictures composed of colored areas. The smallest i.e. irreducible elements of a

General Introduction

Type of Information	Basic Set of Characters (alphabet)	Number of Basic Characters
Morse code	o – space	3
written English language	letters: a b ... z A B ... Z digits: 0 1 ... 9 punctuation: , . ! ? ...	111 printable ASCII characters
CJK-Ideograms (Chinese, Japanese, Korean)	examples: 金 鉄 家 道 ...	27.496 (Unicode v. 3.0)
decimal numbers	0 1 2 3 4 5 6 7 8 9	10
binary numbers	0 1 *alternatively* false true *other alternatives*	2
hexadecimal numbers	0 1 2 3 4 5 6 7 8 9 A B C D E F	16
genetic DNA code	base groups: A(denine), C(ytosine), G(uanine), T(hymine)	4
classical music	Frequencies: 12 notes per octave, 6 octaves Amplitudes: 5 levels Duration: 6 values	12 x 6 5 6
general sound	any frequency in the range 20 Hz to 20 kHz any amplitude from 0 dB to 120 dB	continuous
printed images	pixel (size depending on print technique) color	discrete or
smells	chemial compounds ○ volatile ○ excitable to olfactory nerve cells	$> 10^5$

Figure 1: Global elements of information in a given language are symbolic expressions which carry the meaning and the practical value. The smallest i.e. irreducible elements of a language are called tokens or character. The basic set of characters of a language constitutes the alphabet or code of this language. This table illustrates some prominent examples.

language are called tokens or **characters**. The *basic set B* of characters of a language constitutes the **alphabet** or **code** of this language. For instance, the Morse code consists of three discrete basic characters, i.e. dot, dash, and space. The English language exhibits a character set of 26 letters, in upper- and lower case, ten numerical digits and various punctuation marks. The Chinese written language uses symbols for words as the basic characters. The official Japanese standard basic set comprises 1945 characters, called Joyo-Kanjis. The binary code consists of only two characters, e. g. "0" and "1", or "false" and "true". Since any language requires at least two discrete basic characters in order to represent information (see Sec. 2), the Boolean code is the most elementary language. This makes it most suitable for carriers with bi-stable states, frequently used in digital electronics.

The examples given above, all show a **discrete** number of basic characters. On the other hand, classical music is a "language" composed of tunes of distinct frequencies, lengths, and amplitudes (including, of course, a rich variety of harmonics and modulations). In general sound, the frequencies and amplitudes are on a **continuous** scale. Another example are images which often are composed of pixels (picture elements) characterized by a color and brightness which may be on continuous scales. Figure 1 shows a table with some examples of types of information and their basic character set. In many cases, the characters of the basic set *B* are *ordered*, i. e. every character has a unique position in an ordered list and there is a character of least value as well as a character of highest value.

In order to be transmitted, processed or stored, information always requires a physical **carrier** and a **medium** for the carrier. The information is **coded** onto the carrier by structuring or patterning the carrier in space or time. These deliberate structures of a physical carrier are called **signals**. Signals may be spatial structures of ink on paper as a carrier, showing letters or figures. If light is used as a carrier on wave guides as the medium, information is coded by signals which are, for example, structures of light intensity (amplitude modulation) or the wavelength (frequency modulation). Biological neurons are the medium of electrical signals in order to transmit excitation within the nervous system of an organism. Obviously there are carriers and media which are more or less suitable to perform different operations on information. Electromagnetic waves are especially suitable for information transmission, however they are much less practical to *store* information (though the latter is not impossible, because it is conceivable to store information in the structure of modes in dielectric resonators). Magnetic materials, on the other hand, are very useful for information storage and less useful for information transmission. In the framework of this textbook, we will focus on electronic and electronically addressable carriers and media.

From what we learned about information so far we can deduce, that we can change the code alphabet (e.g. the syntactics) without affecting the meaning (e.g. the semantics). For example, the characters of the English alphabet can be translated into the hexadecimal code. Common translation tables are given e. g. by the 7-bit ASCII (<u>A</u>merican <u>S</u>tandard <u>C</u>ode for <u>I</u>nformation <u>I</u>nterchange) standard (see Figure 2). Another example is sound which is translated into binary code by suitable analog/digital (A/D) conversion. A further example is the chemical DNA code typically represented by the letters abbreviating the names of the base groups; it is the sequence of base groups in the DNA that provides the information for the synthesis of specific proteins. Translations can either be complete, as in the case of the ASCII table, or incomplete in the case of the conversion of any continuous coded (i. e. analog) signal such as sound by an A/D converter having limited resolution. In addition, the efficiency or redundancy of the coding (see Sec. 2) may change drastically during the translation.

Although information always requires a carrier, it is independent of the *type* of carrier. For example, binary-code information may be transferred from a magnetic tape to a semiconductor memory and vice versa without changing the information (neither the syntactics, nor the semantics). Of course, the type of carrier determines the possible ways of structuring it and, hence, determines the type of codes it is suitable for. Printed letters, for instance, may be carried on paper, or plastic foil and any type of display as media, but they need to be translated into another code before they can be transferred onto a magnetic tape.

	0	1	2	3	4	5	6	7
0	NUL	DLE	☐	0	@	P	`	p
1	SOH	DC1	!	1	A	Q	a	q
2	STX	DC2	"	2	B	R	b	r
3	ETX	DC3	#	3	C	S	c	s
4	EOT	DC4	$	4	D	T	d	t
5	ENQ	NAK	%	5	E	U	e	u
6	ACK	SYN	&	6	F	V	f	v
7	BEL	ETB	'	7	G	W	g	w
8	BS	CAN	(	8	H	X	h	x
9	HT	EM	)	9	I	Y	i	y
A	LF	SUB	*	:	J	Z	j	z
B	VT	ESC	+	;	K	[	k	{
C	FF	FS	,	<	L	\	l	|
D	CR	GS	-	=	M	]	m	}
E	SO	RE	.	>	N	^	n	~
F	SI	US	/	?	O	_	o	DEL

(columns = most significant hexadecimal character; rows = least significant hexadecimal character)

Figure 2: 7-bit ASCII translation between the characters of the English alphabet (and some control characters, in grey) shown in the matrix and the two-digit hexadecimal code shown at the columns and rows.

2 Mathematical Definition of Information

Now that we have described what information means, the question arises: can we measure it? Before we open a book, we are completely *uncertain* about the arrangement of letters on the first page. Reading the page, i.e. obtaining the information, removes the uncertainty which existed before. Obviously, it is possible to measure information by the probability p that a certain pattern occurs. The information is the greater, the lower the probability. The news "It rains in the Sahara desert" contains more information than the expression "It rains in London", simply because the latter is much more probable. Thus, one requirement for a measure of information is that it increases monotonically with decreasing probability. A second requirement is that it is additive, i.e. two pages of a book contain twice the information than one page. As proposed by Claude Shannon, a function satisfying both requirements for measuring information I is the logarithm of p

$$I = -k \ln p \tag{1}$$

where k represents a constant which will be defined below. If we have a (fixed) number m of printing positions per page and each position can be taken, for example, by one of the $N = 27$ upper case letters (26 letters and "space") of the English language, then the number of possible arrangements (or configurations) Ω is

$$\Omega = N^m \tag{2}$$

The probability of a character pattern to occur on the page is

$$p = \frac{1}{N^m} \tag{3}$$

if we assume for simplicity that the individual probability p_i of all letters to occur on a given position is equal. In reality, there are large differences of course, e. g. "E" is much more frequent than "X". The information capacity of our page of m character positions is then

$$I_m = k \ln N^m = k m \ln N \tag{4}$$

For example, for $m = 3000$ print positions on a page, $N^m = 27^{3000}$ which is approx. 10^{4294}. It should be noted that this huge number of possible patterns is significantly reduced for a real English text because subsequent letters usually must produce reasonable words and subsequent words must give reasonable sentences. In considering the different probabilities (i.e. natural occurencies) of letters in the English alphabet and of the letters in words, an *averaged information*, also called expectancy value H, is defined:

$$H = -k \sum_{i=1}^{N} p_i \ln p_i \tag{5}$$

instead of Eq. (1).

In order to determine k in Eq. (1), we need to establish a reference for measuring I. For this, we consider only one letter position instead of m positions and we change the code from the English language to the binary code. The probability of a "0" and a "1", e.g. the probability of obtaining front ("0") or back ("1") upon throwing a coin is $p = 0.5$.

By convention, this is the information

$$I = 1 \text{ bit} \tag{6}$$

we deduce $k = 1/\ln 2$ (bit) from Eq. (1).

Changing the logarithm to base 2, short ld, leads to

$$I = -\frac{\ln p}{\ln 2} = -\text{ld}\, p \tag{7}$$

and

$$H = -\sum_{i=1}^{N} p_i \,\text{ld}\, p_i \tag{8}$$

If we consider binary coded information, the information which we can represent in m positions (e.g. a memory, on a bus of signal lines) depends on the constraints on the selection. There are three predominantly different cases:

1. If only one position may take an opposite binary state as, for example, in demultiplexed signal lines (1-out-of-m decoding), then the number of possible configurations Ω is identical to the number of lines,

$$\Omega = m \tag{9}$$

2. If there is a (fixed) number n out of m positions which have an opposite state, the number of possible configurations Ω increases to

$$\Omega = \frac{m!}{n!(m-n)!} = \binom{m}{n} \tag{10}$$

3. If n is allowed to vary between 0 and m, Ω becomes

$$\Omega = \sum_{n=0}^{m} \binom{m}{n} = 2^m \tag{11}$$

which is the same result as Eq. (2), with $N = 2$ representing the binary code.

There is an important link between information theory and statistical thermodynamics. Case (2) can be realized by an ideal gas of n molecules confined to a volume V in which the gas molecules can occupy m different positions. According to Boltzmann, the entropy of a system is defined as

$$S = k_B \ln \Omega \tag{12}$$

where Ω is the number of possible configurations (or microscopic states, short: **microstates**) and k_B the Boltzmann constant ($k_B = 1.38 \cdot 10^{-23}$ J/K). Obviously, except for a constant factor, expressions for information and entropy are identical. This is because both are based on the possible (microscopic) configurations Ω of a (macroscopic) system.

In the case of different probabilities, we have introduced the average information by Eq. (5). This equation is identical to the entropy of mixtures of ideal gases. It has a maximum, if all probabilities are equal. Equally, the maximum information is given by Eq. (5) with $p_i = p_j$ for all i, j:

$$H_{\max} \equiv I = -k \ln p_i \tag{13}$$

What can be learnt from the relation between information and physical entropy of a system? If we consider again an ideal gas in a volume V and we now reduce the volume to $V' = V/2$ under *isothermal* conditions, we do not change the inner energy of the ideal gas. However, we *did* reduce the entropy by one half, because in the new volume V' only half as many possible positions are available for the gas atoms. According to the second law of thermodynamics this entropy reduction ΔS in the system is achieved by a dissipation of thermal energy ΔW_Q to the environment according to

$$\Delta W_Q = T \Delta S \tag{14}$$

A stored bit in any information system can be in one of two states. If both states, in principle, have equal probability, a minimum entropy associated with this bit is

$$S = k_B \ln \Omega = k_B \ln 2 \tag{15}$$

Reducing the number of bits, e. g. by processing it through a binary logic gate with fewer output lines than input lines (see Sec. 3) reduces the number Ω of possible microscopic states, by one for each erased bit. This leads to a energy dissipation of

$$\Delta W_Q = T \Delta S = k_B T \ln 2 - k_B T \ln 1 = k_B T \ln 2 \tag{16}$$

per bit. We will return to this aspect in conjunction with the energy requirements for computation [16] (see Introduction to Part III) and in the context of reversible computation (Sec. 3.4).

Let us consider the change of information density by exchanging the code. On one printing position on the page of English text in capital letters discussed above, a maximal information of

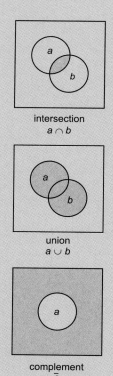

Figure 3: The algebra of sets as a Boolean algebra, illustrated by Venn diagrams.

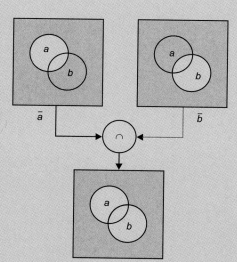

Figure 4: Venn diagram of a De Morgan's rule.

$$H_{max} = \text{ld}\, 27 = 4.76 \text{ bit} \tag{17}$$

can be stored. However, the probability of the letters in the English language is very different. This reduces the information capacity to approx. 4.13 bits. A further reduction in the information capacity per printing position occurs due to the facts that successive letters in a natural language do not occur independently and that successive words are interrelated, too. If all high level interdependencies are taken into account, an approximate value of $H = 1.7$ bit is found for an average collection of English texts. The relation

$$R = \frac{H_{max} - H}{H_{max}} \tag{18}$$

is called **redundancy**. For instance, for our example of the English language we calculate the redundancy to be approx. 65%. In natural languages, redundancy greatly contributes to the perceivability of the language in the daily life. In technical systems, redundancy is similarly used for *error detection* and *error correction* procedures.

More details about the aspects presented in this Section can be found, for example, in Refs. [2], [6], [9], [10] and [12].

3 Processing of Information

3.1 Boolean Algebra and Switching Circuits

After we have seen how we can measure information, we will briefly describe how information can be processed. Information processing is based on *proportional logic*, as founded by Aristotle. A system of logic introduces a basic set of symbols and basic rules for operations on these symbols. This group of basis symbols and operations represents an axiomatic system from which further formulas can be deduced and the huge world of logic relations and calculations is built. Within the system of logic, the symbols and operation remain *un*interpreted, corresponding to the processing of information on a syntactic level. George Boole (1815 – 1864) introduced an algebraic formulation of propositional logic [3].

A common representation of a Boolean algebra is based on a basic set of characters B and two operators, $\cap$ und $\cup$, for which the following axioms are set forth:

Axiom 1: Commutativity

For every a and b in B

$$a \cup b = b \cup a \tag{19}$$

$$a \cap b = b \cap a \tag{20}$$

Axiom 2: Distributivity

For every a, b, and c in B

$$a \cap (b \cup c) = (a \cap b) \cup (a \cap c) \tag{21}$$

$$a \cup (b \cap c) = (a \cup b) \cap (a \cup c) \tag{22}$$

Axiom 3: Existence of neutral elements

There exist unique elements "1" of highest value and "0" of lowest value in B such that for every a in B

$$a \cup 0 = a \tag{23}$$

$$a \cap 1 = a \tag{24}$$

Axiom 4: Existence of the complement

For every a in B there exists a unique element called $\bar{a}$ (complement of a) in B such that

$$a \cup \bar{a} = 1 \tag{25}$$

$$a \cap \bar{a} = 0 \tag{26}$$

The *algebra of sets* is a Boolean algebra in which "0" represents an empty set, "1" a complete set, ∪ is the union (disjunction), and ∩ is the intersection (conjunction). The intersection, union and complement of sets are illustrated as so-called Venn diagrams in Figure 3 (I. Venn, 1934 – 1923, English logician). If B is an *ordered* set, we can define

$$a \cup b := \max(a,b) \tag{27}$$

$$a \cap b := \min(a,b) \tag{28}$$

$$\bar{a} := 1 - a \tag{29}$$

where max (a,b) denotes the higher value out of a and b, while min (a,b) is the lower value out of a and b.

A range of rules can be deduced from the axioms of Boolean algebra. One example are *De Morgan's rules*:

$$\overline{a \cup b} = \bar{a} \cap \bar{b} \tag{30}$$

$$\overline{a \cap b} = \bar{a} \cup \bar{b} \tag{31}$$

Figure 4 shows the first of these rules in a Venn diagram.

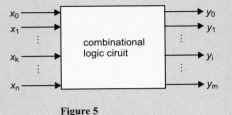

Figure 5

3.2 Switching Algebra and Switching Circuits

The axioms and deduced theorems of Boolean algebra presented above are given in general terms, without the number of characters of the set B being specified. Hence, the results obtained within this system of logic are valid for any Boolean algebra. Conventional digital computers today use a *two-valued logic*, i.e. $B = \{0,1\}$. The variables of this logic are called binary *switching variables*, the formalism is often referred to as *switching algebra*. Within this algebra, the operator "∪" is interpreted as OR operator (short "+"), the operator "∩" is interpreted as AND operator (short "•"), and the formation of the complement is called NOT operator (*inversion*). It can be shown by applying De Morgan's rules that a complete switching algebra can be built from the group of operators AND and NOT as well as from the group of operators OR and NOT. Some basic relationships of the switching algebra and the fundamentals of the conventional technical realization in voltage-controlled *switching elements* (short: gates) are described in Chap. 7. Concatenations of switching elements without memory elements are called **combinational logic circuits**. Typically, these circuits also do not contain feedback of signal lines. As a consequence, the state of the output signals is a unique Boolean function of the input signals (Figure 5). An output signal y_i can be described by a Boolean function of $x_0, x_1, ..., x_n$ as input signals

$$f_i(x_0, x_1, ..., x_n) = y_i \tag{32}$$

where $f_i()$ is any kind of Boolean expression. For n input signals, there can be 2^{2^n} Boolean functions. Figure 6 shows all 16 possible functions for two input variables.

Besides some trivial functions (see as f_0 and f_{15} in Figure 6) we find the AND operation as f_1 and the OR operation as f_7. Technically important are the function f_{14}, called NAND (= NOT(AND)), and the function f_8, called NOR (= NOT(OR)). Another important function is f_6 which represents an exclusive OR (short XOR or EXOR). An example of a more advanced combinational logic circuit is the *binary adder* (see Chap. 7). Every Boolean expression represents a Boolean function. Still, we do not know if the expression is the most simple representation of this function; i.e. the representation with the least number of logical gates, which will typically be the most economical one. For example, the function

$$f(x_0, x_1, x_2) = \bar{x}_0 \cdot x_1 \cdot x_2 + x_0 \cdot x_1 \cdot x_2 \tag{33}$$

x_0	0	0	1	1	realized by
x_1	0	1	0	1	
f_0	0	0	0	0	0
f_1	0	0	0	1	$x_0 \cdot x_1$
f_2	0	0	1	0	$x_0 \cdot \bar{x}_1$
f_3	0	0	1	1	x_0
f_4	0	1	0	0	$\bar{x}_0 \cdot x_1$
f_5	0	1	0	1	x_1
f_6	0	1	1	0	$x_0 \cdot \bar{x}_1 + \bar{x}_0 \cdot x_1$
f_7	0	1	1	1	$x_0 + x_1$
f_8	1	0	0	0	$\bar{x}_1 \cdot \bar{x}_1$
f_9	1	0	0	1	$x_0 \cdot x_1 + \bar{x}_0 \cdot \bar{x}_1$
f_{10}	1	0	1	0	$\bar{x}_1$
f_{11}	1	0	1	1	$\bar{x}_1 + x_0$
f_{12}	1	1	0	0	$\bar{x}_0$
f_{13}	1	1	0	1	$\bar{x}_0 + x_1$
f_{14}	1	1	1	0	$\bar{x}_0 + \bar{x}_1$
f_{15}	1	1	1	1	1

Figure 6

comprises four AND operations, one OR operation and one NOT operation. Applying the distributivity rule (axiom 2), one obtains

$$f(x_0, x_1, x_2) = (\bar{x}_0 + x_0) \cdot x_1 \cdot x_2 \tag{34}$$

The operation in brackets is always 1 according to axiom (4), and therefore the function reduces to

$$f(x_0, x_1, x_2) = x_1 \cdot x_2 \tag{35}$$

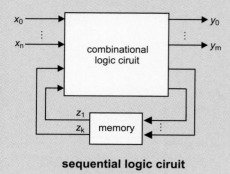

Figure 7

expressed by one AND operation. The switching theory offers a range for formal techniques which reduce every Boolean expression into their irreducible forms. For more details on this subject see, for example, Refs. [4], [5].

Switching circuits which contain memory elements are called **sequential logic circuits**, often referred to as **automata** (Figure 7). The memory is typically obtained in a feedback path, i.e. a backward concatenation, as realized, e.g. by flip-flops (Chap. 7). The output values of these circuits do not only depend on the input values but also on the inner state of the circuit, i. e. the switching history. An example of a simple sequential circuit is a *binary counter* which counts the number of signal changes it receives at a specific input. Since a counter needs to store the number of counts at any time, it needs memory.

3.3 Multivalued Logic

Besides switching circuits using binary Boolean logic, it is possible to design circuits in which there are more than two permitted logic states. If the basic set contains N characters, the corresponding system is called an N-valued logic or a multivalued logic [11][12].

The basic operations are the same as in the 2-valued logic, according to

$$\text{AND:} \quad x_0 \cdot x_1 = \min(x_0, x_1) \tag{36}$$

$$\text{OR:} \quad x_0 + x_1 = \max(x_0, x_1) \tag{37}$$

$$\text{NOT:} \quad \bar{x} = 1 - x \tag{38}$$

Figure 8 shows tables for these functions for a 4-valued logic. It should be noted, that Eqs. (36) to (38) represent only *one* possible choice of operations. Many others would be possible too.

In principle, we can continue to increase the number of levels until we arrive at a continuous character set for $N \to \infty$. As described in Sec. 1, there are various basic character sets which contain an infinite number of ordered characters, e.g. sets of amplitude-continuous and frequently-continuous signals. The logic operations according to Eqs. (36) to (38) can be applied to these logic systems as well. Figure 9 illustrates the basic operations for a 2-valued, 4-valued and ∞-valued logic.

The ∞-valued logic has been elaborated in the frame work of the so-called fuzzy logic. Fuzzy variables may take any value between 0.0 and 1.0 which can be interpreted as a certain grade (or probability) of falsehood or truthfulness. In the frame work of the set theory this means that variables are not restricted to be members of the set or non-members of the set. In fuzzy-logic, there is a membership function which maps each value of a variable into the range between 0 and 1. The function can be used to adopt the logic variable individually to its meaning (in the real world).

NOT		AND			OR		
x	$\bar{x}$	x_0	x_1	$x_0 \cdot x_1$	x_0	x_1	$x_0 + x_1$
0	1	0	0	0	0	0	0
1/3	2/3	0	1/3	0	0	1/3	1/3
2/3	1/3	0	2/3	0	0	2/3	2/3
1	0	0	1	0	0	1	1
		1/3	0	0	1/3	0	1/3
		1/3	1/3	1/3	1/3	1/3	1/3
		1/3	2/3	1/3	1/3	2/3	2/3
		1/3	1	1/3	1/3	1	1
		2/3	0	0	2/3	0	2/3
		2/3	1/3	1/3	2/3	1/3	2/3
		2/3	2/3	2/3	2/3	2/3	2/3
		2/3	1	2/3	2/3	1	1
		1	0	0	1	0	1
		1	1/3	1/3	1	1/3	1
		1	2/3	2/3	1	2/3	1
		1	1	1	1	1	1

Figure 8: Truth table of the basic operations for a 4-valued logic.

3.4 Irreversible and Reversible Logic

In all switching elements with more than one input discussed up to here, a loss of information occurs upon the information process. For example, at a conventional 2-valued AND gate it is not possible to reconstruct the information at the input by the information available at the output: a "0" of the output can be caused by three possible input signal configurations. For this reason, this kind of computing is called *irreversible*. As we have seen in the description of information (Sec. 2), any loss of information inherently leads to a (minimum) energy dissipation of $\Delta W_Q = T \Delta S = k_B T \ln 2$ per reduced bit in the case of an isothermal operation, because of the change in entropy ΔS of the system. Energy consumption of contemporary computers is still orders of magnitude larger than this inherent limit of $k_B T \ln 2$ per bit for irreversible computing. Transistor based gates dissipate approx. 10^{-14} J $\cong 3 \cdot 10^6 k_B T$ (data based on transistors in advanced VLSI circuits of the year 2000). Even the DNA copying mechanism in a human cell dissipates approx. 100 $k_B T$ of heat per bit copied, due to the chemical bonds that need to be broken and formed in the process [1]. As said, we will return to the issue of energy dissipation in the context of the limits of computation in the Introduction to Part III.

Still it is an interesting fundamental question, if information processing can be conducted in a *reversible* manner. Positive answers to this question were given by Bennett [14][15] and independently, by Fredkin. As mentioned above, reversible computation needs

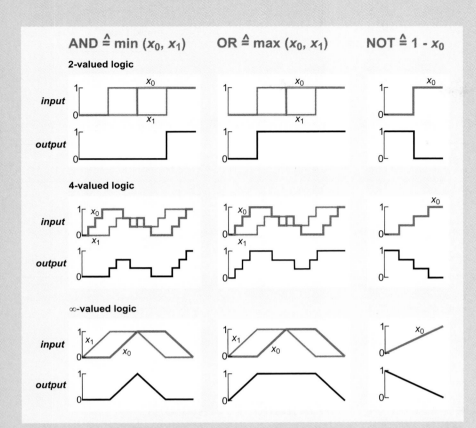

Figure 9: Illustration of the basic operations for a 2-valued, 4-valued and ∞-valued logic.

Figure 10: Matrix-type notation of the truth table for the NOT operation.

to be based on switching elements which do not loose information. Concerning the NOT gate which is a one-bit input/one-bit output gate, we have no problems, because one can always deduce the input signal from the output signal, and vice versa. Obviously, the conventional NOT gate is a reversible gate. This is easily seen if we write the truth table in the form of a matrix (Figure 10). The top row gives the possible input states and the side column lists the output states. The matrix is filled in a way that each row and column contains exactly a single "1", which is to be interpreted as a one-to-one mapping of the input to the output.

How about two-bit input/two-bit output gates? A possible realization is the so-called CONTROLLED NOT (short: C-NOT). Figure 11 shows an illustration and the matrix-type truth table of the C-NOT gate. The state of x_0 line controls the function of the ⊗ in the x_1 line. If $x_0 = 0$, then ⊗ reproduces an identity at the output, i.e. $y_1 = x_1$. If $x_0 = 1$, then ⊗ acts as a NOT function, i.e. $y_1 = \bar{x}_1$. In all cases, $y_0 = x_0$. Obviously, the C-NOT gate is a revesible gate. However, unlike the NAND gate in the irreversible Boolean algebra, it is *not* a universal gate in the sense, that one can build every conceivable Boolean function just from combinations of this gate. To realize this, we need to employ three-bit gates. Historically, two realizations have been proposed, the CONTROLLED-CONTROLLED-NOT gate (short: CC-NOT), also known as the Toffoli gate [17], or the Fredkin gate. They are illustrated in Figure 12 and Figure 13. The CC-NOT gate shows two lines which always remain unchanged: $y_0 = x_0$ and $y_1 = x_1$. The third line has a NOT function, which is only activated if both $x_0 = 1$ and $x_1 = 1$. The gate can realize an AND function by setting $x_2 = 0$, resulting in $y_2 = x_0 \cdot x_1$. As we have seen earlier, all conceivable Boolean functions can be realized from NOT and AND functions. Obviously, the CC-NOT gate, in fact, is a generic gate. The same is true for the Fredkin gate. Here, the first line controls whether the second and third lines are exchanged or not: If $x_0 = 1$, the output is $y_1 = x_2$ and $y_2 = x_1$.

In contrast to irreversible computing, reversible computing is not inherently accompanied by energy dissipation. We will return the thermodynamics of reversible computing in the Introduction to Part III, and, particularly, in the context of quantum computing in Chap. 18. Quantum computing inherently is reversible computing. In the context of nanoelectronics and molecular electronics circuits built from reversible gates are discussed. The idea is to guide the "waste" information away from the area of the circuit and to "destroy" it outside, in order to prevent heat dissipation within the nanoelectronic devices.

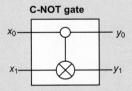

Figure 11: Symbol and matrix-type notation of the truth table for the C-NOT gate.

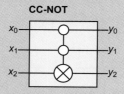
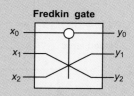

CC-NOT	output states $y_0y_1y_2$
input states $x_0x_1x_2$	000 001 010 011 100 101 110 111
000	1 0 0 0 0 0 0 0
001	0 1 0 0 0 0 0 0
010	0 0 1 0 0 0 0 0
011	0 0 0 1 0 0 0 0
100	0 0 0 0 1 0 0 0
101	0 0 0 0 0 1 0 0
110	0 0 0 0 0 0 0 1
111	0 0 0 0 0 0 1 0

Fredkin gate	output states $y_0y_1y_2$
input states $x_0x_1x_2$	000 001 010 011 100 101 110 111
000	1 0 0 0 0 0 0 0
001	0 1 0 0 0 0 0 0
010	0 0 1 0 0 0 0 0
011	0 0 0 1 0 0 0 0
100	0 0 0 0 1 0 0 0
101	0 0 0 0 0 0 1 0
110	0 0 0 0 1 0 0 0
111	0 0 0 0 0 0 0 1

Figure 12: Symbol and matrix-type notation of the truth table for the CC-NOT gate.

Figure 13: Symbol and matrix-type notation of the truth table for the Fredkin gate.

4 Areas of Information Technology

Although information can be described in a conceptional and mathematical manner, information in the *real* world is always bound to physical media. Operations on information – processing, manipulation, and storage of data – take place through processes which involve matter and energy. It is this relation between information and physics that builds the fundament for information technology and, hence, represents the starting point for this textbook.

Figure 14 illustrates the major areas of information technology and marks the parts of the textbook which deals with the areas. Raw information is provided by the world outside information technological systems, e. g. by other technical system, by humans, by nature. Sensors (Part VII) convert this raw information into signals appropriate for the specific information technological system. In the case of electronic systems, these sensors can be a keyboard, a microphone, a camera, a touch screen, an electron microscope, and many more. In biological systems these will be the senses which we use to obtain information to be processed in our brain or our central nervous system. Information transmission (Part VI) occurs at manifold stages, at various levels, and on very different distance scales, either within a system or between systems.

The core of the information processing can be divided into logic processes (Part III) and local, typically random access memory (Part VI). In the case of electronic systems, these are the dominant areas of the conventional microelectronics and the emerging nanoelectronic concepts. The definition of nanoelectronics varies throughout literature. Nanoelectronics in the broader sense is defined as electronics integrated on a lateral scale with a minimum feature size well below 100 nm. Although this is an arbitrary limit, decreasing the minimum feature size involves a number of novel issues, unknown in conventional microelectronics. Nanoelectronics in the more specific sense is regarded as electronics, in which quantum effects dominate due to the small feature sizes and are utilized for device functions. The quantizing may be due to quantum mechanical effects or the discreteness of charges. Both definitions are used to define the topics worked out in this textbook.

Information storage is performed in archival systems, called mass storage devices (Part V). In these systems, the storage principle is usually non-electronical. The result of information processing, information transmission, or information retrieval is fed back to the outside world by means of actuators in general, which may be displays, loudspeakers, motors, triggers of any action, etc. . In the context of this book, the focus is the operational principle of modern displays (Part VIII).

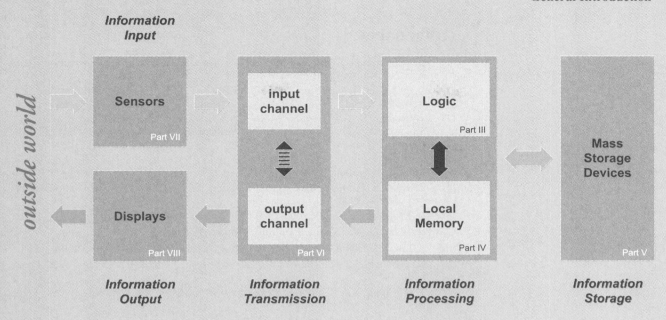

Figure 14

Obviously, Figure 14 shows a prototype structure of the areas of information technology. In reality, systems are typically much more complex. Often there is a hierarchy of sub-systems which by themselves exhibit structures similar to the one sketched in the figure. In other cases, the functional areas are split or nested or distributed. Information processing, which involves logic and local memory, is frequently embedded in sensors, communication systems, mass storage devices, and displays. Nevertheless, the areas discussed are comprehensive and, therefore, provide a reasonable structure for grouping the chapters of this textbook.

Acknowledgements

The editor gratefully acknowledges Tobias Noll (RWTH Aachen) for enjoyable and fruitful discussions, valuable suggestions, and a detailed review of this text. Thanks are due to Yacoub Mustafa (RWTH Aachen) for corrections.

References

[1] R. P. Feynman, *Feynman Lectures on Computation*, reprint with corrections, Perseus Publishing, Cambridge, MA, 1999.

[2] L. Brillouin, *Science and Information Theory*, 2nd Ed., Academic Press Inc., New York, 1962.

[3] G. Boole, *An Investigation of the Laws of Thought, on Which are Founded the Mathematical Theories of Logic and Probability*, 1849, Reprinted by Dover Publications, Inc., New York, 1954.

[4] H. Troy Nagle, Jr., B.D. Carroll, J.D. Irwin, *An Introduction to Computer Logic*, Prentice-Hall, Inc., New Jersey, 1975.

[5] Z. Kohavi, *Switching and Finite Automata Theory*, McGraw-Hill Computer Science Series, 1978.

[6] L.P. Hyvärinen, *Information Theory for System Engineers*, Springer Verlag, 1970.

[7] B. Sklar, *Digital Communications*, Prentice Hall, 1988.

[8] S. Lin, J. Costello, *Error Control Coding,* Prentice Hall, Englewood Cliffs, 1983.

[9] N. Gershenfeld, *The Physics of Information Technology*, Cambridge Univ. Press, 2000.

[10] T.M. Cover, J.A. Thomas, *Elements of Information Theory*, Wiley, New York, 1991.

[11] E. L. Post, Am. J. Math. **43**, 163 (1921).

[12] J. B. Rosser and A. R. Turquette, *Many Valued Logics*, North-Holland Publishing Company, Amsterdam, 1952.

[13] R. Balian, *From Microphysics to Macrophysics: Methods and Applications of Statistical Physics*, Springer-Verlag, New York, 1991.

[14] C. H. Bennett, IBM Journal of Research and Development **17**, 525 (1973).

[15] C. H. Bennett, Int. J. Theoretical Physics **21**, 905-40 (1982).

[16] R. Landauer, IBM Journal of Research and Development **5**, 183 (1961).

[17] T. Toffoli, *Reversible Computing*, Tech. Memo MIT/LCS/TM-151, MIT Lab. for Com. Sci. (1980).

Fundamentals

Contents of Part I

Introduction to Part I	27
1 Dielectrics	31
2 Ferroelectrics	59
3 Electronic Properties and Quantum Effects	79
4 Magnetoelectronics – Magnetism and Magnetotransport in Layered Structures	107
5 Organic Molecules – Electronic Structures, Properties, and Reactions	127
6 Neurons – The Molecular Basis of their Electrical Excitability	145
7 Circuit and System Design	165

Introduction to Part I

Contents

1 Interdisciplinarity 27
2 Prerequisites 27
3 Material Properties and Material Classes 29

1 Interdisciplinarity

Since the dawn of information technology and especially since the unrivaled rise of microelectronics, several disciplines have been involved in the design, fabrication, and application of the components and systems: solid state physics provides the fundamentals for the major operational principles, chemistry is involved in the manufacturing technologies, electrical engineering supplies the circuit design, simulation, and verification, and computer science assists the design of logical concepts of complex systems. Mathematics represents the backbone of all these disciplines.

Compared to former components, the novel and emerging devices – especially in the area of nanoelectronics – are characterized by

- an increasing number of different materials,
- an increasing density of integration, and
- an increasing functionality.

A typical contemporary task is, for example, the integration of very different classes of material (such as differently doped semiconductors, glasses, complex oxides, silicides, and metals) on a sub-100 nm scale while maintaining each of them in their optimal functionality and finding a process route which will not deteriorate other materials while optimising one of them.

This trend unequivocally contributes to a growing significance of interdisciplinarity. In this context interdisciplinarity is considered as the significant contribution of knowledge, methods, tools, and rules of different disciplines to the solution of a complex problem. Fruitful interdisciplinarity accepts the disciplines in their own rights and depths. It is based not only on an additive contribution from the disciplines, but by connecting their knowledge and methods, new ideas and concepts are created.

2 Prerequisites

The basic knowledge required for studying this book is acquired in undergraduate university courses including elementary physics, chemistry, electrical engineering, and material science. Those areas, in which more advanced knowledge is needed to understand the device chapters, are elaborated in Part I.

Here we list some keywords concerning the prerequisites which should have been gathered by the reader in the undergraduate courses mentioned, starting with physics. *Mechanics*, in particular the mechanical properties of matter, elasticity and plasticity, are involved in some fabrication processes and in the mechanical stress limits of many devices. In some cases, such as atomic force microscopy these properties are directly related to the function. *Thermodynamics* and the *kinetic theory of gases* teaches about the behavior of particles in the gas phase, for example, in vacuum-based deposition techniques, and some aspects of plasmas. The entropy and the second law of thermodynamics comes into play through fabrication processes on one hand, and through its link to information theory on the other. In many cases, the thermal properties of matter, the heat conductivity and the heat capacity, determine the thermal stress limits of the devices. In some cases, such as pyroelectric infrared detectors, they are of major importance for the device performance. *Electrostatics* is the basis of the treatise on the *dielectric properties* of matter (Chapter 1) and, hence, deeply involved in all field-effect devices, single-elec-

I Fundamentals

	Dielectric materials[1]	Ferroelectric materials	Magnetic materials	Semiconductor	Metals	Superconductors	Organic materials & polymers	Biomolecular materials
LOGIC DEVICES[2] *(Part III)*								
Silicon MOSFETs *(Chap. 13)*	●			●	●			
Ferroelectric Field Effect Transistors *(Chap. 14)*		●		●				
Resonant Tunneling Devices *(Chap. 15)*				●				
Single-Electron Devices *(Chap. 16)*	●			●	●			
Superconductor Digital Electronics *(Chap. 17)*						●		
Quantum Computing using Superconductors *(Chap. 18)*						●		
Carbon Nanotubes for Data Processing *(Chap. 19)*	●			●	●		●	
Molecular Electronics *(Chap. 20)*	●			●	●		●	
RANDOM ACCESS MEMORIES *(Part IV)*								
High-Permittivity Materials for DRAMs *(Chap. 21)*	●			●	●			
Ferroelectric RAMs *(Chap. 22)*	●	●		●	●			
Magnetoresistive RAMs *(Chap. 23)*			●	●	●			
MASS STORAGE DEVICES *(Part V)*								
Hard Discs *(Chap. 24)*			●					
Magneto-Optical Discs *(Chap. 25)*	●		●		●		●	
Rewritable DVDs[3] *(Chap. 26)*	●			●			●	
Holographic 3-D Data Storage *(Chap. 27)*	●	●		●			●	
AFM-Based Mass Storage Devices *(Chap. 28)*				●	●		●	
DATA TRANSMISSION AND INTERFACES *(Part VI)*								
Transmission on Chip and Board Level *(Chap. 29)*	●			●	●		●	
Photonic Networks *(Chap. 30)*	●	●		●				
Microwave Communication Systems *(Chap. 31)*	●	●				●		
Neuroelectronic Interfaces *(Chap. 32)*				●				●
SENSOR ARRAYS AND IMAGING SYSTEMS *(Part VII)*								
Optical 3-D Time-of-Flight Imaging System *(Chap. 33)*				●				
Pyroelectric IR Imaging Arrays *(Chap. 34)*		●		●				
Electronic Noses[4] *(Chap. 35)*	●			●	●		●	
2-D Tactile Sensors *(Chap. 36)*	●	●		●	●		●	
DISPLAYS *(Part VIII)*								
Liquid Crystal Displays *(Chap. 37)*	●	●			●		●	
Organic LED Displays *(Chap. 38)*	●				●		●	
Field-Emission and Plasma Displays[5] *(Chap. 39)*				●	●			
Electronic Paper *(Chap. 40)*	●						●	

tron devices, charge based memories, etc. Furthermore, fundamental electrostatics is required in the understanding of *ferroelectricity* (Chapter 2) which are exploited in devices such as non-volatile ferroelectric random access memories and ferroelectric field-effect transistors, as well as related to piezoelectric and pyroelectric effects. *Electromagnetism* and *optics*, including the basics of refraction, interference, and diffraction, are further elaborated in Chapter 1 and are the precondition to understanding several fabrication techniques and device concepts, such as optical lithography, several analytical methods, microwave and optical data transmission and data storage techniques, optical imaging, including the time-of-flight 3-D imaging, and all display concepts. A basic understanding of (1) how the *quantum theory* leads to the electronic structure of atoms, of (2) the types of *chemical bonds*, of (3) *crystal structures*, and of (4) *electronic band structure of solids*, leading to the formation of insulators, semiconductors, and metals are needed as a prerequisite to Chapter 3 and all device chapters which relate to the electronic properties and electronic transport phenomena in semiconductors and metals. Some processes and devices rely on the electronic excitation of atoms, either in the gase phase (to form a plasma) or in the solid state, e. g. to utilize fluorescence effects. The understanding of excitation effects is also provided by the electronic structure of atoms and solids. In addition, Chapter 3 gives an introduction to superconductivity. The *magnetic properties* of matter, including a basic knowledge about the difference between hard and soft magnets, demagnetising fields, form anisotropy and crystal anisotropy are required for Chapter 4 as well as for those chapters which deal with magnetic storage devices.

In addition to a basic knowledge about the electronic structure of the elements, chemical bonds, and solids, there are some further aspects of *chemistry*, which are assumed to be known. These include the energy balance and the kinetics of chemical reactions, the law of mass action, including acid / base equilibria and reduction / oxidation (redox) equilibria, the mass transport by diffusion, the transport of ions in combined field and concentration gradients, and the formation of electrochemical potentials. Chapter 5 complements this by a brief overview of the *organic chemistry* emphasizing the electronic properties and some basic synthesis principles of organic molecules and polymers. Chapter 6 introduces the signal processing of neurons, without any specific pre-knowledge in biology.

Finally, the textbook requires an elementary background knowledge of the operation principle of simple electronic devices such as diodes and transistors as well as the standard fabrication process steps in the contemporary semiconductor industry. A brief introduction to standard digital logic and the operation of processors is given in Chapter 7.

There are numerous excellent textbooks on the areas mentioned above. Some examples are Refs. [1] – [10]. To broaden the areas which are outlined in Part I, it is recommended to use advanced textbooks on solid state physics [11], [12], semiconductor theory and devices [16], physical chemistry [13], organic chemistry [14], neural biology [15], and information theory [17].

Figure 1: Material classes and their involvement in the device concepts according to the Chapter number. The strength of the involvement is qualitatively related to the relevance of the material property for the function of the device. Minor contributions by auxiliary or structural material properties are omitted.

Remarks:

1. These are mainly dielectric oxides. In some cases, dielectric and optical polymers are involved and listed here, as well as under organic materials and polymers.
2. There are additional concepts of logic devices not addressed in their own chapters. Some of them, including the materials involved, are compiled in the Introduction to Part III.
3. Phase-change materials are multinary transition element alloys which can solidify in crystalline or an amorphous phase, depending on the process conditions. They are listed here as semiconductors, although the semiconducting properties are not directly utilized.
4. Among many other materials, ion conductors are employed in gas sensors and electronic noses.
5. Fluorescence materials used in plasma and field-emission displays are usually non-conducting, inorganic materials which are employed because of their specific excitation characteristics.

3 Material Properties and Material Classes

Figure 1 illustrates the involvement of the different material classes in the device concepts detailed in this textbook. Obviously, this table can only give overview and can also not be free of a certain ambiguity. However, it is interesting to observe the differences in the breadth to which material classes supply functionality to devices of the information technology.

The properties of *dielectric* and *optical materials* are outlined in Chapter 1. These materials find themselves predominantly in data communication (components of photonic networks, microwave components), the holographic data storage, and the concept of improving the gate capacitance of Si-based field-effect transistors as well as of the cell capacitors in DRAMs by materials of increasing permittivity. In addition, these materials contribute to the displays, to several type of sensor arrays, and to most of the logic devices.

The properties of *Ferroelectric materials* are outlined in Chapter 2. These materials are utilized in non-volatile ferroelectric random access memories, ferroelectric field-effect transistors, and some specific LCD systems. In addition, their piezoelectric, pyroelectric, and electro-optical properties are employed in some touch sensors, infrared imaging arrays, and in specific components of optical and microwave data transmission systems.

The properties of *magnetic* and *magnetoelectronic* materials are outlined in Chapter 3. These materials are used in non-volatile magnetoresistive random access memories, as well as in mass storage devices such as hard-discs and magneto-optical discs.

The properties of *semiconductors* are outlined in Chapter 4. Semiconductors cover the broadest spectrum of functionalities including major types of logic devices, such as field-effect devices and quantum electronic devices based on resonant tunnelling effect, parts of the function in carbon-nanotube and molecular demonstrators, random access memories, and in the control logic of a multitude of other classes of devices. Fabricated as MEMS, they are engaged in scanning probe based techniques and some types of sensors.

The properties of *metals* are outlined in Chapter 4. Metals have a wide range of applications, too, as they are typically employed as conductors for the electrical signals. Only in a few cases, are metals involved in the immediate function of the device.

The properties of *superconductors* are outlined in Chapter 4, too. In the classification given in Figure 1, they provide the function for superconducting logic devices and are used to introduce the concept of quantum computing. In addition, they are utilized to reduce the losses in specific microwave systems.

The structures and properties of *organic molecules* and *polymers* are outlined in Chapter 5. These materials cover a broad range of applications. The area of logic devices, they enter by carbon nanotubes and through molecular electronics. In addition, they are used in optical discs, holographic storage systems, the AFM based mass storage concept, various sensor types, liquid crystal displays, organic LEDs, and some parts of electronic paper concepts. As biomolecules, they also provide the molecular mechanism for the activity of neurons (Chapter 6) which are listed as *biomolecular materials* and are investigated for the signal transfer across the biological neuron – semiconductor electronics interface.

References

[1] D. Haliday, R. Resnick, J. Walker, *Fundamentals of Physics*, John Wiley & Sons, Inc., 2000.

[2] H. D. Young, R. A. Freedman, T. R. Sandin, and A. L. Ford, *University Physics*, 10th ed., Addison-Wesley, 2000.

[3] L. Jones and P. Atkins, Chemistry – Molecules, Matter, and Change, 4th ed., W. H. Freeman & Co., 1999.

[4] T. L. Brown, H. E. LeMay Jr., B. E. Bursten, *Chemistry – The Central Science*, 9th ed., Prentice Hall, Englewood Cliffs, 2002.

[5] D. R. Askeland, *The Science and Engineering of Materials*, 3rd ed., PWS Publishing Co., 1999.

[6] S. O. Kasap, *Principles of Electronic Materials and Devices*, 2nd ed., McGraw Hill, 2001.

[7] D. K. Ferry and J. P. Bird, *Electronic Materials and Devices*, Academic Press, 2001.

[8] R. E. Hummel, *Electronic Properties of Materials*, 3rd ed., Springer, 2000.

[9] W. H. Hayt, Jr., *Engineering Electromagnetics*, 6th ed., McGraw-Hill, 2001.

[10] J. G. Brookshear, *Computer Science*, Addison-Wesley Publishing, 7th ed., 2002.

[11] C. Kittel, *Introduction to Solid State Physics*, 7th ed., John Wiley and Sons, 1996.

[12] N. W. Ashcroft and N. D. Mermin, *Solid StatePhysics*, Saunders College Publishing, 1976.

[13] P. W. Atkins, *Physical Chemistry*, 5th ed., Oxford University Press, 1994.

[14] K. P. C. Vollhardt and N. E. Schore, *Organic Chemistry: Structure and Function*, 3rd ed., W. H. Freeman & Co, 1998.

[15] E. R. Kandel, J. H. Schwartz, and T. M. Jessell, *Principles of Neural Science*, 4th ed., McGraw-Hill/Appleton & Lange, 2000.

[16] S. M. Sze, *Physics of Semiconductor Devices*, 2nd ed., John Wiley and Sons, 2001.

[17] R. Feynman, *Lectures on Computation*, Perseus Publishing, 2000.

Dielectrics

Susanne Hoffmann-Eifert, Department IFF, Research Center Jülich, Germany

Contents

1 Introduction	33
2 Polarisation of Condensed Matter	34
2.1 Electrostatic Equations with Dielectrics	34
2.2 Microscopic Approach and the Local Field	35
2.3 Mechanisms of Polarisation	35
3 Frequency Dependence of the Polarisation Mechanisms	36
3.1 The Complex Dielectric Function	36
3.2 The Frequency Dependence of the Dielectric Function	37
3.3 Resonance Phenomena	37
3.4 Relaxation Phenomena	39
3.4.1 Debye Relaxation in the Frequency Domain	39
3.4.2 Debye Relaxation in the Time Regime	40
3.4.3 Maxwell-Wagner Relaxation	40
3.4.4 Dielectrics with a Distribution of Relaxation Times	40
4 Polarisation Waves in Ionic Crystals	42
4.1 Acoustic and Optical Phonons	42
4.2 Polar Optical Phonons	44
4.3 Polaritons	44
4.3.1 Propagation of Electromagnetic Waves in Condensed Matter	44
4.3.2 Longitudinal and Transverse Polaritons	46
4.3.3 Phonon-Polaritons	48
4.3.4 The Lyddane-Sachs-Teller Relation	50
4.3.5 Softening of the Transverse Optical Phonon	50
4.3.6 Damped Phonon-Polaritons	51
4.4 Characteristic Oscillations in Perovskite-type Oxides	51
4.5 Temperature Dependence of the Permittivity in Titanates	51
4.6 Voltage Dependence of the Permittivity in Ionic Crystals	52
5 Optical Properties of Dielectrics	52
5.1 Plane Electromagnetic Waves	52
5.2 Resonator and Waveguide Modes	53
5.3 Absorption	53
5.4 Polarisation, Reflection, and Refraction	54
5.4.1 Reflection and Refraction	55
5.4.2 Birefringence	55
5.5 Electro-optical Effects	56
5.5.1 The Pockels and Kerr Effects	56
5.5.2 Photorefractive Materials	56
6 Closing Remarks	57

Dielectrics

1 Introduction

Dielectric and optical materials historically have had and continue to have a strong influence on the evolution of today's electrical engineering, electronics, and information technology. The material classes involved are typically crystalline and amorphous oxides as well as organic compounds and polymers. They are employed as bulk materials and, most often, as thin films.

The range of new applications of these materials in the field of information technology is extremely wide. New low permittivity dielectrics are being developed as insulators on advanced CMOS circuits in order to enhance the signal transfer rate across the chips (Chap. 29). High permittivity dielectrics are being investigated for the cell capacitors of future DRAM devices (Chap. 21). Very low losses and a specific temperature dependence of the dielectric properties are required for new microwave dielectrics for oscillator and filter applications (Chap. 31). A variety of gate dielectrics for field-effect transistors are being studied for applications in short-channel MOSFETs (Chap. 13), carbon nanotube based FETs (Chap. 19), ion-sensitive FETs for electronic noses (Chap. 35), as well as in FETs used to contact biological neurons (Chap. 32). Even thinner dielectrics are used in tunnelling barriers, for example, for MRAM devices (Chap. 15), in Coulomb blockage devices (Chap. 16), some molecular electronic test systems as well as in Josephson junctions for superconducting logic circuits (Chap. 17) and in quantum computing systems (Chap. 18). The fundamentals of charge transfer by tunneling are covered in Chaps. 3 and 15. Gradients of the refractive index are used in optical wave guides (Chap. 30). The electrically switchable anisotropy of the refractive index (i.e. the birefringence) is employed in LCDs (Chap. 37) as well as in optical switches and modulators for data communication (Chap. 30). Holographic memories (Chap. 27) are based on the photosensitivity (either photorefraction or photoabsorption) of an optical storage material which means a photo-induced change of the tensor of the complex dielectric function. Additional properties which arise in ferroelectrics and related materials and which are utilised in a further class of devices will be discussed in Chapter 2.

The development and the optimisation of the materials themselves and the related processing technologies for insulating materials as well as high dielectric constant and ferroelectric materials in the form of single crystals, ceramics, glasses, and polymers has made possible the technical use of electronic phenomena at ever higher electrical fields, smaller circuit dimensions and for an increasing number of functions. The spectrum of materials ranges from ionic crystals, like e.g. oxide ceramics, through glasses to polymers. The morphology of the integrated films varies depending on the lattice match to the substrate, the process temperature, and the material itself from epitaxial and polycrystalline to amorphous (see Chap. 8). The future evolution of information technology will depend crucially on the possibility, reproducibility, and perfection of the integration of these new oxide dielectrics with semiconductor components.

The understanding of the basic concepts of the interaction of the electromagnetic field with matter is one of the key issues in understanding the principles of the design and operation of the new electronic and photonic devices. Therefore the present chapter will discuss the basics of the solid state physics of dielectrics with respect to the properties in the low frequency range, the microwave regime, and the optical region. In particular, we will focus on the solid state properties caused by microscopic and macroscopic charge inhomogeneities generated by electrical fields.

The description of the dielectric and optical properties given here mainly focuses on the requirements of subsequent chapters, specifically those in Parts III to VI. For further details or a broader view of the topic, the reader is referred to comprehensive textbooks either on solid state physics in general (e.g. [1] – [7]), electronic materials (e.g. [8] – [10]), or specifically on dielectric and optical properties (e.g. [11] – [14] and [21] – [23]).

Before we enter into the specific physical description, we will first give the following definitions which will be used throughout this chapter:

Dielectrics are insulating materials that are used technically because of their property of polarisation to modify the dielectric function of the vacuum, e.g. to increase the capacity (i.e. the ability to store charge) of capacitors. They do not conduct electricity due to the very low density of *free* charge carriers. Here, the electrons are *bound* to microscopic regions within the material, i.e. the atoms, molecules, or clusters, in contrast to being freely movable in and out of a macroscopic system under consideration.

Polarisation is the separation of positive and negative charge barycentres of bound charges. If this separation is induced by an applied electric field, it is called *dielectric* polarisation. The characteristic property of the material that is a measure of the ability to be polarised is called the dielectric constant (or: permittivity) ε_r. There are several mechanisms for the dielectric polarisation determined by the polarisable unit (atom, cation-anion pair, orientable permanent dipole, etc.).

Conductors such as metals do conduct electric current because of the *free* mobility of the electrons within the lattice. *Free* here means the charges can enter and leave a system in contrast to bound charges (or: polarisation charges) which can only be displaced to a lesser or greater extent within the system. The characteristic measure is the specific electrical conductivity σ.

Optical properties are determined by free and bound charges. The limiting cases of conductors (free conducting charges) and dielectrics (bound polarisation charges) blend in the case of fields of high frequencies because here the charges are accelerated in only one direction during one half of a period of the alternating field and thus are unable to travel long distances. The characteristics are described by optical parameters like the refractive index *n*.

The chapter will start with a brief review of the macroscopic and microscopic description of dielectric polarisation. Subsequently, resonance and relaxation phenomena will be outlined, followed by a discussion of phonons and their interaction with electromagnetic waves. Finally, some basic electro-optical effects are sketched.

2 Polarisation of Condensed Matter

2.1 Electrostatic Equations with Dielectrics

According to the *Poisson equation*, each *free* charge acts as a source for the *dielectric displacement* **D**:

$$\text{div } \boldsymbol{D} = \rho_{\text{free}}, \tag{1}$$

where ρ_{free} denotes the density of free (conducting) charges. Based on this relation, the overall charge neutrality of matter in an external field is described by:

$$\boldsymbol{D} = \varepsilon_0 \boldsymbol{E} + \boldsymbol{P}. \tag{2}$$

The term $\varepsilon_0 \boldsymbol{E}$ describes the vacuum contribution to the displacement **D** caused by an externally applied electric field **E**, and **P** represents the electrical polarisation of the matter in the system. This relation is independent of the cause of the polarisation. The polarisation may exist spontaneously (*pyroelectric* polarisation, see Chap. 2 / Sec. 2.4), it may be generated by mechanical stress (*piezoelectric* polarisation, see Chap. 2 / Sec. 2.5), or it may be induced by an external electric field (*dielectric* polarisation).

Figure 1 illustrates the insertion of matter into a parallel plate capacitor. The matter is polarised to an extent described by **P**. If the insertion is performed at a *constant applied voltage*, i.e. **E** = const. (Figure 1b), additional free charges need to flow into the system to increase **D** according to Eq. (1). If the insertion is performed for *constant charges* on the plates, i.e. **D** = const. (Figure 1c), the electric field **E** and, hence, the voltage between the plates will decrease according to Eq. (2).

In the case of a *dielectric* polarisation, the polarisation of the matter is related to the electric field by

$$\boldsymbol{P} = \varepsilon_0 \chi_e \boldsymbol{E}, \tag{3}$$

which leads to

$$\boldsymbol{D} = \varepsilon_0 (1 + \chi_e) \boldsymbol{E} = \varepsilon_0 \varepsilon_r \boldsymbol{E} \tag{4}$$

with the material properties

χ_e : electrical susceptibility

ε_r : relative permittivity (or: dielectric constant).

If χ_e or ε_r themselves are field-dependent, e.g. become reduced for high electric fields, tunable dielectrics are achieved (see Sec. 4.6).

2.2 Microscopic Approach and the Local Field

We are now going to discuss what happens *inside* the dielectric material when it is put into an electric field – in other words, we want to find a correlation between the *macroscopic* polarisation P and the *microscopic* properties of the material.

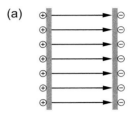

The macroscopic polarisation P is the vector sum of the individual dipole moments p of the material, like polarised atoms, molecules, ions, etc.:

$$P = N_a \cdot p \tag{5}$$

with N_a being the density of dipoles.

In the case of dielectric polarisation, the dipole moments are induced by the local electric field E_{loc} at the site of the particle:

$$P = N_a \cdot p = N_a \, \alpha \, E_{loc}, \tag{6}$$

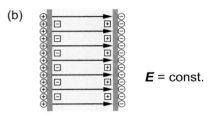

where α is the *polarisability* of an atomic dipole. In diluted matter like gases, where there is *no* interaction between the polarised particles, the local electric field is identical to the externally applied, global electric field E_a, i.e. $E_{loc} = E_a$.

In condensed matter, the density and therefore the electrostatic interaction between the microscopic dipoles is quite high. Hence, the local field E_{loc} at the position of a particular dipole is given by the superposition of the applied macroscopic field E_a and the field of all other dipoles:

$$E_{loc} = E_a + \sum E_{dipole}. \tag{7}$$

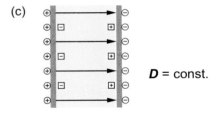

The local field E_{loc} can be calculated by the method of Clausius and Mossotti (see e.g. [6]). The calculation reveals a relation between the *atomic polarisability* α and the *macroscopic permittivity* ε_r. For example, for cubic crystal structures

$$\frac{N_a}{3\varepsilon_0}\alpha = \frac{\varepsilon_r - 1}{\varepsilon_r + 2} \tag{8}$$

is obtained. This is often referred to as the *Clausius-Mossotti equation*. It is important to remember that Eq. (8) is developed only for induced dipoles (ionic and electronic polarisation).

Figure 1: Empty parallel plate capacitor (a) and capacitor after inserting a dielectric material between the electrode plates at constant voltage (b) and at constant charge on the plates (c). Note: it is not necessary for the polarisation P to be induced by the electric field E.

2.3 Mechanisms of Polarisation

So far we have considered the correlation between the macroscopically measurable relative permittivity ε_r and the microscopic atomic polarisability α in an ensemble of induced dipoles. We now extend our scope to both induced and permanent dipoles. In the following, the different physical mechanisms of polarisation in the solid state will be sketched. Details are described in textbooks such as Ref. [3].

In general, one can distinguish between four different types of polarisation:

- *Electronic polarisation* describes the displacement of the negatively charged electron shell against the positively charged nucleus. Since all matter is built from atoms, this is true for dielectrics in general. The electronic polarisability α_{el} is approximately proportional to the volume of the electron shell. Thus, large atoms have a large electronic polarisability. Since the atomic radius is temperature-independent, generally the temperature dependence of α_{el} can be neglected.

- *Ionic polarisation* is observed in materials with ionic bonds (i.e. ionic crystals) and describes the mutual displacement of the positive and negative sublattices under the influence of an applied electric field. In general the temperature dependence of the ionic polarisability α_{ion} is weakly positive because of the thermal expansion of the lattice.
- *Orientation polarisation* describes the alignment of permanent dipoles. In many substances there are molecules – either regular constituents or impurities – which carry a (permanent) electric dipole moment. If these dipoles are mobile or, at least, able to re-orient themselves by rotation, they do contribute to the dielectric polarisation by the so-called orientation polarisation. At ambient temperatures, usually all dipole moments are mutually compensated because of the orientational disorder, i.e. the statistical distribution of their directions. An electric field, on the other hand, generates a preferred direction for the dipoles, while the thermal movement of the atoms perturbs the alignment. The average degree of orientation is a function of the applied field and the temperature. The solution is given by the so-called *Langevin function* [6]. For all technically applicable cases, the polarisation is far from saturation and is proportional to the applied field. In this case, the average polarisability originating from permanent dipole moments p is given by

$$\langle \alpha_{\text{or}} \rangle = \frac{p^2}{3 k_{\text{B}} T}, \tag{9}$$

where k_{B} denotes the Boltzmann constant and T the absolute temperature measured in Kelvin.

The strong temperature dependence is one of the main characteristics of the orientation polarisation.

- *Space charge polarisation* describes a polarisation effect in a dielectric material which shows spatial inhomogeneities of charge carrier densities. Space charge polarisation effects are not only of importance in semiconductor field-effect devices (see ref. [7]), they also occur in ceramics with electrically conducting grains and insulating grain boundaries (so-called *Maxwell-Wagner polarisation*, see ref. [9]) as well as in composite material systems in which metallic particles are isolated in polymer or glass matrices.

3 Frequency Dependence of the Polarisation Mechanisms

3.1 The Complex Dielectric Function

We now look at the behaviour of condensed matter in alternating electric fields. Moving charges cause a frequency-dependent phase shift between applied field and charge displacement. To express this mathematically, the relative dielectric permittivity is written as a complex function:

$$\underline{\varepsilon}_r = \varepsilon'_r + i \varepsilon''_r. \tag{10}$$

The real part ε'_r characterises the displacement of the charges, and the imaginary part ε''_r the dielectric losses. Analogously, the electrical susceptibility is also written complex:

$$\underline{\chi}_e = \chi'_e + i \chi''_e. \tag{11}$$

The loss tangent is defined as

$$\tan \delta := \frac{\varepsilon''_r}{\varepsilon'_r}. \tag{12}$$

In addition to the losses caused by dipole reorientation $(\tan\delta)_{\text{dipole}}$, the residual leakage current $(\tan\delta)_{\text{cond}}$ of the non-perfect insulator is added so that in general $\tan\delta$ and hence ε''_r, respectively, becomes the sum of both contributions:

$$\tan \delta = (\tan \delta)_{\text{dipole}} + (\tan \delta)_{\text{cond}}. \tag{13}$$

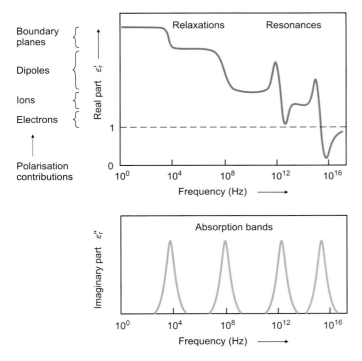

Figure 2: Frequency dependence of real (top) and imaginary (bottom) part of the dielectric function.

The term $(\tan\delta)_{cond}$ can be separated experimentally because of its distinct frequency dependence. For microwave ceramics (see Chap. 31) frequently a quality factor Q is quoted:

$$Q := \frac{1}{\tan\delta}. \tag{14}$$

3.2 The Frequency Dependence of the Dielectric Function

The total polarisation of a dielectric material results from the four contributions discussed above. Each contribution stems from a short-range movement of charges that respond to an electric field on different time scales and, hence, through a Fourier transform, in different frequency regimes. The dispersion of the real and imaginary part of the dielectric function is shown in Figure 2, covering the entire frequency spectrum.

The space charge polarisation is caused by a drift of mobile ions or electrons which are confined to outer or inner interfaces. Depending on the local conductivity, the space charge polarisation may occur over a wide frequency range from mHz up to MHz. The polarisation due to the orientation of electric dipoles takes place in the frequency regime from mHz in the case of the reorientation of polar ligands of polymers up to a few GHz in liquids such as water. It is often possible to distinguish between space charge and orientation polarisation because of the temperature dependence of $\langle \alpha_{or} \rangle$. In the infrared region (10^{12} - 10^{13} Hz), we find the resonance of the molecular vibrations and ionic lattices constituting the upper frequency limit of the ionic polarisation. The resonance of the electronic polarisation is around 10^{15} Hz. It can be investigated by optical methods.

As can be seen from Figure 2 the different polarisation mechanisms not only take place on different time scales but also exhibit a different frequency dependence. Depending on whether the oscillating masses experience a restoring force or not, we distinguish between resonance effects and relaxation effects, respectively. Resonance effects are observed for the ionic and electronic polarisation, while relaxation effects are found for orientation polarisation and space charge polarisation.

3.3 Resonance Phenomena

Let us first consider the resonance phenomena associated with electronic and ionic polarisation. In an electric field the charged species (either electrons and nuclei or positively and negatively charged ions) are displaced from their equilibrium position. This is described by the equation of motion:

$$m_i^* \frac{d^2 u}{dt^2} + m_i^* \gamma_i \frac{du}{dt} + m_i^* \omega_{0,i}^2 u = q_i E_{loc} \quad (15)$$

Where u is the displacement from the equilibrium position, m_i^* is the effective mass of the oscillating species, γ_i the damping, and $\omega_{0,i}$ is the resonance angular frequency of the undamped oscillation. The terms on the left of Eq. (15) describe the force of inertia, the friction force, and the restoring force due to electrostatic Coulomb interaction. The term on the right represents the driving force caused by the electric field.

If E_{loc} is a dc field and this is switched off at a given moment, the electric charges are pulled back into their original position driven by the Coulomb force. If the friction force is negligible, we arrive at the limiting case of *undamped resonance*, i.e. we observe an oscillation with the resonance frequency $\omega_{0,i}$. If, on the other hand, the driving force is significantly smaller than the friction force, we find a supercritical damped oscillation which is comparable to a *relaxation process* with a time constant $\tau_i = \gamma_i/\omega_{0,i}^2$ (see Sec. 3.4).

We now consider the complete scenario where the induced dipoles are excited to *damped oscillations* by the application of an alternating electric field of the angular frequency $\omega = 2\pi f$:

$$\underline{E}_{loc} = E_{loc,0} \cdot e^{i(\mathbf{kr}-\omega t)} \quad (16)$$

where $\mathbf{k}$ denotes the wave vector, and $\mathbf{r}$ the position.

With the ansatz

$$\underline{u} = \underline{u}_0 \cdot e^{i(\mathbf{kr}-\omega t)} \quad (17)$$

we obtain the frequency-dependent complex amplitude of the damped oscillation

$$\underline{u}_0 = \frac{(q_i/m_i^*) \cdot E_{loc,0}}{\omega_{0,i}^2 - \omega^2 + i\gamma_i \omega}, \quad (18)$$

and with the frequency-dependent dipole field

$$\underline{p}_i = q_i \underline{u}_0 \cdot e^{i(\mathbf{kr}-\omega t)} = \underline{\alpha}_i E_{loc,0} \cdot e^{i(\mathbf{kr}-\omega t)} \quad (19)$$

it then follows for the complex *dynamic* polarisability:

$$\underline{\alpha}_i(\omega) = \frac{q_i^2/m_i^*}{\omega_{0,i}^2 - \omega^2 + i\gamma_i \omega} = \alpha_i'(\omega) + i\alpha_i''(\omega), \quad (20)$$

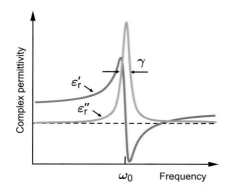

Figure 3: Spectral plot of $\varepsilon_r'(\omega)$ and $\varepsilon_r''(\omega)$ for a dipole oscillator.

with the real part

$$\alpha_i'(\omega) = \frac{q_i^2}{m_i^*} \frac{\omega_{0,i}^2 - \omega^2}{(\omega_{0,i}^2 - \omega^2)^2 + \gamma_i^2 \omega^2} \quad (21)$$

and the imaginary part

$$\alpha_i''(\omega) = \frac{q_i^2}{m_i^*} \frac{\gamma_i \omega}{(\omega_{0,i}^2 - \omega^2)^2 + \gamma_i^2 \omega^2}. \quad (22)$$

For $\omega = 0$ we obtain the *static* polarisability as

$$\alpha_{i,s} := \alpha_i'(0) = \frac{q_i^2}{m_i^* \omega_{0,i}^2} \quad \text{and} \quad \alpha_i''(0) = 0. \quad (23)$$

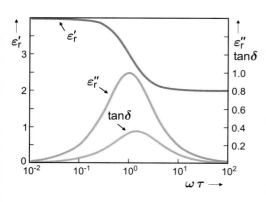

Figure 4: Frequency dependence of ε_r', ε_r'', and $\tan\delta$ for the case of a Debye-type relaxation process.

The frequency dependence discussed above holds for the electronic α_{el} as well as for the ionic polarisation α_{ion}. The resonance frequency of the electronic polarisation $\omega_{0,el}$ typically lies in the ultraviolet region at about 10^{15} Hz. The ionic polarisation is correlated to the lattice vibrations corresponding to the polar optical phonon branches. Therefore, the resonance frequency of the ionic polarisation $\omega_{0,ion}$ lies in the infrared regime between 10^{12} and 10^{13} Hz.

The frequency dependence of the complex dielectric function in the frequency range of $f > 10^{11}$ Hz follows directly from Eqs. (21) and (22) and the equation of Clausius and Mossotti (8). From this we obtain:

$$\underline{\varepsilon}_r(\omega) = \varepsilon'_r(\omega_{0+}) + \frac{\varepsilon'_r(\omega_{0-}) - \varepsilon'_r(\omega_{0+})}{1 - (\omega/\omega_0)^2 + i\gamma\omega/\omega_0^2} \quad (24)$$

where $\varepsilon'_r(\omega_{0+})$ and $\varepsilon'_r(\omega_{0-})$ denote values of the relative permittivity at frequencies significantly above and below, respectively, the resonance frequency ω_0 of the oscillator. The spectral dependence of the real and imaginary part of $\underline{\varepsilon}_r$ is shown in Figure 3.

Analogous to $\underline{\varepsilon}_r$, the *optical refractive index* $\underline{n}(\omega)$ is complex and frequency-dependent. For an insulator ($\sigma = 0$) with negligible magnetisation ($\mu_r = 0$) we obtain the refractive index from Maxwell's law of dispersion (see Sec. 4.3.1):

$$\underline{n}(\omega) = \sqrt{\underline{\varepsilon}_r(\omega)} \; . \quad (25)$$

Hence we can describe the optical properties of matter as electric properties under the influence of alternating fields of high frequency 10^{12} Hz $< f < 10^{18}$ Hz, as will be discussed in Sec. 5.

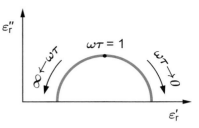

Figure 5: Cole-Cole diagram for Debye relaxation.

3.4 Relaxation Phenomena

3.4.1 Debye Relaxation in the Frequency Domain

Let us now deal with the relaxation phenomena which are observed in case of orientation polarisation as well as for space charge (especially Maxwell-Wagner) polarisation.

First we focus on *Debye relaxation*, which denotes a system with a single relaxation time τ. This is found, for example, for the orientation polarisation in a material with *one* type of permanent dipole which can be oriented by an external electric field. We immediately obtain the complex permittivity from the general solution, Eq. (24), if we omit the term due to the driving force:

$$\underline{\varepsilon}_r(\omega) = \varepsilon'_r(\omega_{0+}) + \frac{\varepsilon'_r(\omega_{0-}) - \varepsilon'_r(\omega_{0+})}{1 + i\gamma\omega/\omega_0^2} = \varepsilon'_r(\omega_{0+}) + \frac{\Delta\varepsilon'_r}{1 + i\omega\tau} \quad (26)$$

with

$$\Delta\varepsilon'_r := \varepsilon'_r(\omega_{0-}) - \varepsilon'_r(\omega_{0+}) \quad \text{and} \quad \tau := \gamma/\omega_0^2 \; .$$

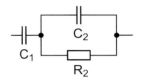

Figure 6: Equivalent circuit representing the Debye relaxation process.

The relaxation step $\Delta\varepsilon'_r$ is defined as the maximum value of the permittivity originating from a relaxation process.

For the case of orientation polarisation the magnitude of the relaxation step is temperature-dependent in the same way as the mean value of the polarisability:

$$\Delta\varepsilon'_{r,\text{or}} \propto \langle\alpha_{\text{or}}\rangle \propto T^{-1} \; . \quad (27)$$

From Eq. (26), the real and imaginary parts of the permittivity are calculated:

$$\varepsilon'_r(\omega) = \varepsilon'_r(\omega_{0+}) + \frac{\Delta\varepsilon'_r}{1 + \omega^2\tau^2} \; ; \quad \varepsilon''_r(\omega) = \frac{\omega\tau\Delta\varepsilon'_r}{1 + \omega^2\tau^2} \; . \quad (28)$$

The corresponding graphs are shown in Figure 4.

The relaxation time τ is often given by a thermally activated process:

$$\tau = \tau_0 \cdot e^{W_0/k_B T} \; , \quad (29)$$

which leads to an exponential shift of the relaxation frequency with temperature.

In case of a Debye-type relaxation process, the Kramers-Kronig integral relationship between the real and imaginary part of the permittivity simplifies to

$$\varepsilon'_r(\omega) = \varepsilon'_r(\omega_{0+}) + \frac{\varepsilon''_r(\omega)}{\omega\tau} \quad \text{and} \quad \varepsilon''_r(\omega) = \left(\varepsilon'_r(\omega) - \varepsilon'_r(\omega_{0+})\right)\omega\tau \; . \quad (30)$$

In the case of a pure Debye-type relaxation the plot of ε''_r vs. ε'_r, the so-called *Cole-Cole diagram*, results in a semi-circle (s. Figure 5).

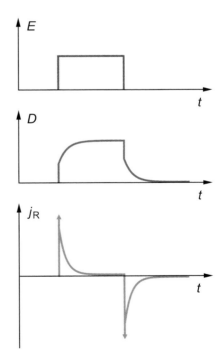

Figure 7: Response of dielectric displacement D and relaxation current density j_R to a step of the applied electric field E for a Debye relaxation. Due to the sudden change of D at $t = 0$ and $t = t_0$ originating from the induced polarisation we observe a peak of the relaxation current density.

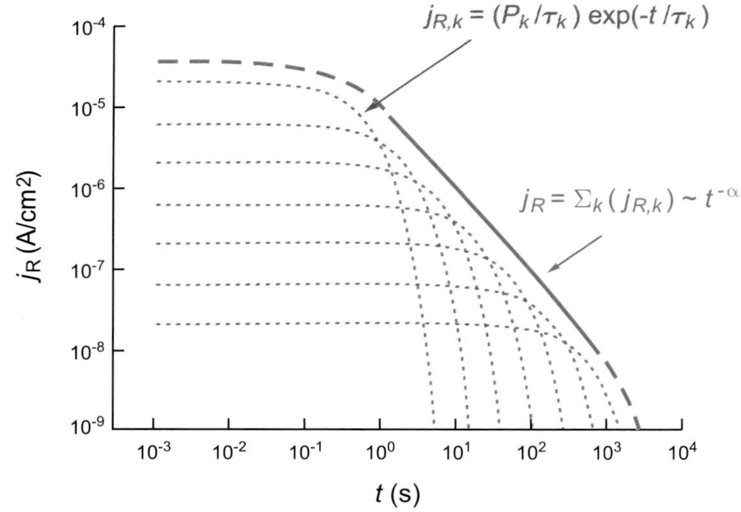

Figure 8: A non-exponential decay of relaxation currents can be described by a superposition of exponential decays having different time constants.

3.4.2 Debye Relaxation in the Time Regime

The Debye relaxation process, which is described by a dispersion of the permittivity as given in Eq. (26), can be represented by an equivalent circuit depicted in Figure 6.

The current response to a dc voltage step applied to this circuit can be easily calculated and gives the time dependence of the dielectric displacement and of the displacement current, which is shown in Figure 7.

For the dielectric displacement follows:

$$D = \varepsilon_0 \varepsilon_r'(\omega_{0+})E + \varepsilon_0 \Delta\varepsilon_r' E\left(1 - e^{-t/\tau}\right) \tag{31}$$

and for the displacement current density, which is also called the *relaxation current density*, we get

$$j_R = \dot{D} = \varepsilon_0 \Delta\varepsilon_r' \tau^{-1} E e^{-t/\tau}. \tag{32}$$

3.4.3 Maxwell-Wagner Relaxation

Maxwell-Wagner-type relaxation processes occur in inhomogeneous dielectrics with regions of different conductivity. This is the case for polycrystalline ceramics with slightly conducting grains and highly insulating grain boundary regions (see ref. [9]). The corresponding equivalent circuit of such a two-phase dielectric material can be transformed into one which is identical with the equivalent circuit that describes the Debye relaxation (see Figure 6).

3.4.4 Dielectrics with a Distribution of Relaxation Times

In a more general case of a heterogeneous dielectric we find a distribution of relaxation times τ_k. The resulting relaxation current of such a system can be interpreted as the sum of exponential decays according to Debye-type processes yielding a non-exponential curve as is shown in Figure 8. The corresponding equivalent circuit is given in Figure 9.

Figure 9: A network representing a capacitor filled with a lossy material. R_1 and C_1 represent a relaxation element with a single relaxation time, R_l a leakage resistance, C_∞ the high-frequency capacitance, and a series of R_k and C_k elements with different $R_k \cdot C_k = \tau_k$ products representing a distribution of relaxation times.

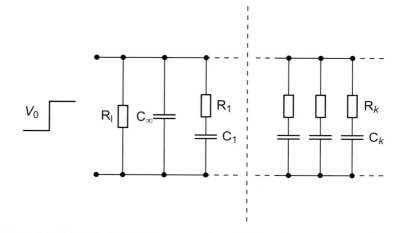

The voltage-step response of such a heterogeneous dielectric with a distribution of relaxation times is described by the *Curie-von Schweidler* behaviour:

$$j_R(t) = \beta \cdot t^{-\alpha} + \delta(t) \tag{33}$$

where j_R denotes the polarisation charging current of the dielectric material and the δ-function corresponds to the high-frequency polarisation processes (ionic and electronic polarisation).

The Curie-von Schweidler law describes a class of dielectrics with a power-law dependence of the dielectric relaxation current on time. It is typically observed for partially or completely disordered materials such as glasses and polymers [12]. It has been observed in titanate dielectric thin film capacitors as well [19]. As the relaxation current sets a limit for the application of high-permittivity titanate thin films in future DRAMs (see Chap. 21), we will discuss the time and frequency dependence of this relaxation in more detail.

In general, the exponent α in the Curie-von Schweidler dependence (Eq. (33)) is close to unity. As we will show in the following, the deviation from a value of one causes a dispersion in the frequency dependence of the susceptibility.

For time-invariant linear systems the frequency domain can be derived from the time domain by a *Fourier transform*. We obtain the frequency dependence of the electrical susceptibility of the system by a Fourier transform of the relaxation current response to a step function of the electric field:

$$\underline{\chi}_e(\omega) = \frac{1}{\varepsilon_0 E_0} \cdot \mathfrak{F}\{j_R(t)\} \tag{34}$$

with

$$\mathfrak{F}\{j_R(t)\} = \int_0^\infty j_R(t) \cdot e^{i\omega t} dt \tag{35}$$

and

$$E(t) = \begin{cases} 0 & \text{for } -\infty \leq t < 0 \\ E_0 & \text{for } 0 \leq t \leq \infty \end{cases} \tag{36}$$

Inserting Eq. (33) gives

$$\underline{\chi}_e(\omega) = \frac{1}{\varepsilon_0 E_0} \cdot \left(\beta \int_0^\infty t^{-\alpha} \cdot e^{i\omega t} dt\right). \tag{37}$$

Solving the integral yields:

$$\underline{\chi}_e(\omega) = \chi'_e(\omega) + i\chi''_e(\omega) = \frac{\beta \, \Gamma(1-\alpha)}{\varepsilon_0 E_0} \, \omega^{(\alpha-1)} \left(\sin\left(\frac{\alpha\pi}{2}\right) + i\cos\left(\frac{\alpha\pi}{2}\right)\right) + \chi_\infty \tag{38}$$

with the gamma function

$$\Gamma(x) = \int_0^\infty e^{-z} z^{x-1} dz. \tag{39}$$

From Eq. (38) we can derive the frequency dependence of the real part of the relative permittivity:

$$\varepsilon'_r(\omega) \approx \chi'_e(\omega) = \frac{\beta \, \Gamma(1-\alpha)}{\varepsilon_0 E_0} \cdot \omega^{(\alpha-1)} \cdot \sin\left(\frac{\alpha\pi}{2}\right) + \chi_\infty. \tag{40}$$

The $\omega^{(\alpha-1)}$ term leads to a dispersion of $\varepsilon'_r(\omega)$ which is increasingly pronounced as α deviates from the value of one.

The loss factor of a system which shows a Curie-von Schweidler behaviour is constant over the entire frequency range:

$$\tan\delta(\omega) = \left|\frac{\varepsilon''_r(\omega)}{\varepsilon'_r(\omega)}\right| \approx \left|\frac{\chi''_e(\omega)}{\chi'_e(\omega)}\right| = \cot\left(\frac{\alpha\pi}{2}\right). \tag{41}$$

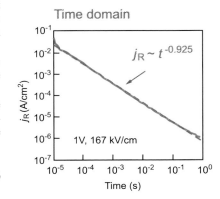

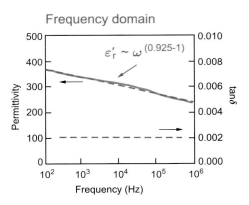

Figure 10: The dielectric response of $(Ba_{0.7}Sr_{0.3})TiO_3$ thin films in the time and frequency domain after [19].

As an example, Figure 10 shows the Curie-von Schweidler-type relaxation behaviour found in a thin film of barium strontium titanate prepared by MOCVD [19]. We can see that for $\alpha = 0.925$ the dispersion of $\varepsilon'_r(\omega)$ is in the range of 2 %, whereas the loss tangent is constant over five decades in frequency.

4 Polarisation Waves in Ionic Crystals

After having covered relaxation phenomena at low frequencies in some detail, we will now look at the properties of dielectrics at higher frequencies, in particular at microwave, infrared and optical frequencies.

The properties of dielectrics at frequencies in the microwave range between 1 GHz and 100 GHz are particularly dependent on the ionic polarisation, whereas the optical properties ($f \approx 10^{14}\,\text{Hz}$) are determined by electronic polarisation.

In Sec. 2.3 we described how electric fields displace the ionic sub-lattices and the electronic shell, thus creating induced polarisation. In this section we focus on the nature of these *lattice oscillations* and, after some basic concepts, we will discuss the characteristic oscillations of technically relevant perovskites such as strontium titanate or barium titanate.

4.1 Acoustic and Optical Phonons

Waves that can propagate in a crystal lattice are characterised by wave vectors $\mathbf{q}$, which can be reduced to the first Brillouin zone of the reciprocal lattice [17]. Therefore q has a value between zero and π/a (a: lattice constant).

In a 3-dimensional *primitive* lattice every $\mathbf{q}$-value fits three lattice vibrations with different frequencies, which belong to two transverse and one longitudinal branch of the *acoustic phonons* (see Figure 11). In the limiting case of long wavelengths the acoustic phonons represent the macroscopic sound waves in the crystal. The dispersion of these waves is described by the sound velocity v_s: $\omega = v_s \cdot q$. For low frequency sound the group velocity ($\text{d}\omega / \text{d}q$) is equal to the phase velocity (ω / q).

Acoustic waves in the lattice cannot be excited directly by an electromagnetic wave. The reason is that the sound velocity is much lower than the velocity of light ($v_s \ll c_0$), and thus for a given angular frequency ω one cannot find a sound wave with the same value of q as the periodicity of the electromagnetic wave. An indirect coupling of these waves is possible, however, as for example in piezoelectric crystals [10].

In *non-primitive* lattices the different atoms of the elementary cell can vibrate against each other, thus allowing frequencies $\omega \neq 0$ also in the case of $q = 0$. These are the characteristic vibrations, which are also called *optical phonons*. The opposite movement of neighbouring atoms generates strong electric dipoles as soon as the atoms have non-equal electronegativities, i.e. their chemical bonds show some polar character. This permits a strong coupling to electromagnetic waves (see Sec. 4.2).

In order to achieve further understanding of the acoustic and optical phonons we will discuss a simple model in more detail. For a fundamental discussion of the properties of phonons, we refer to the literature (e.g. [13] – [15]). In the limiting case of long wavelengths a lattice with two kinds of atoms can be simplified by a linear chain of periodically arranged atoms A and B (compare Figure 11). The atoms are connected to their neighbours by small springs representing the linear restoring force due to the chemical

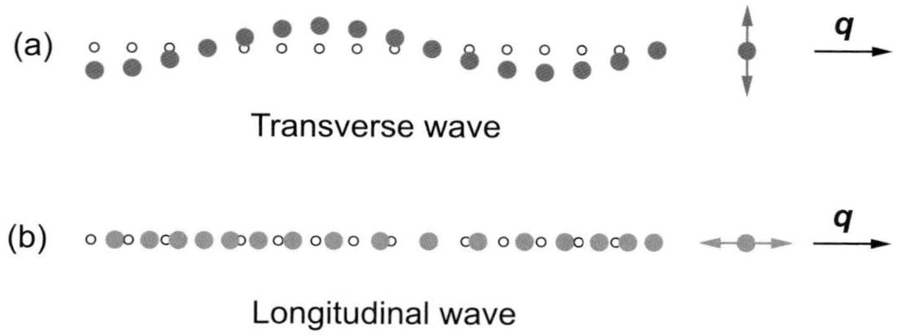

Figure 11: Different types of elastic waves in condensed matter:
(a) transverse wave and
(b) longitudinal wave.
The equilibrium positions of the atoms are marked by small circles.

bond within the lattice. In this model an oscillation of a single atom can spread over the whole crystal due to the connecting springs. This represents a system of many coupled *harmonic oscillators*.

The two types of atoms have masses of m_A and m_B, and at equilibrium they are separated by a distance $a/2$ along the x-axis. The crystal is periodic with a lattice constant of a. To calculate the deflection of the atoms we set up a system of coupled differential equations that can be solved with the ansatz of plane waves.

The solution yields a *dispersion* equation, i.e. a relation between the angular frequency ω and the wave number q with two different values ω_+ and ω_- belonging to every q, as:

$$\omega_\pm^2(q) = \frac{K}{\mu}\left[1 \pm \sqrt{1 - \frac{4\mu}{m_A + m_B} \cdot \sin^2\left(\frac{qa}{2}\right)}\right] \quad (42)$$

where K denotes the force constant of the connecting springs, and $\mu = (m_A m_B)/(m_A + m_B)$ the reduced mass of the vibrating system.

The two solutions $\omega_\pm(q)$ are plotted in Figure 12. They belong to different types of oscillations that show fundamentally different dispersion curves. The *optical branch* ($\omega_+(q)$) shows only a weak dependence of q and has its maximum frequency at $q = 0$. In contrast to this, at $q = 0$ the *acoustic branch* ($\omega_-(q)$) has the frequency $\omega_-(0) = 0$. The underlying oscillations are shown in Figure 13. Consecutive atoms are deflected in the same direction in the acoustic mode and in opposite directions in the case of optical oscillation. Figure 13 shows the similarity of the acoustic oscillation with a sound wave.

In general, a lattice consisting of N atoms in the elementary cell can vibrate with $3N$ characteristic oscillations in three-dimensional space. Three of the vibrations are simple displacements of the elementary cell as a whole corresponding to the acoustic phonons. The remaining $3(N-1)$ vibrations belong to the optical phonons.

Two types of phonons can be distinguished differing with respect to the direction of the deflection $\boldsymbol{u}$ relative to the propagation vector $\boldsymbol{q}$. These are
- the *longitudinal phonons* characterised by $\boldsymbol{u} \parallel \boldsymbol{q}$, and
- the *transverse phonons* described by $\boldsymbol{u} \perp \boldsymbol{q}$.

In the case of the *pure elastic forces* considered here, the longitudinal optical and the transverse optical oscillations are degenerated with one resonance frequency only:

$$\omega_{LO} = \omega_{TO} = \omega_+ \quad (43)$$

with the solution for ω_+ from Eq. (42).

This degeneracy is removed in most cases since the local field $\boldsymbol{E}_{loc}$ (see Sec. 2.2) caused by the adjoint atoms interacts with the polarisation. As an example Figure 14 shows the phonon dispersion curves of Ge, which exhibits a simple lattice with two atoms per elementary cell. The depicted curves correspond to a propagation direction parallel to the edge of the cubic elementary cell [100] and parallel to the cell diagonal [111], respectively.

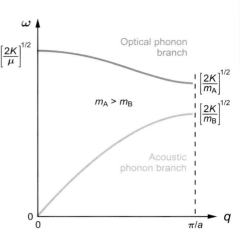

Figure 12: Optical and acoustic phonon branch of the dispersion relation of a linear chain consisting of two different types of atoms.

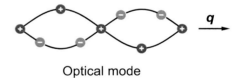

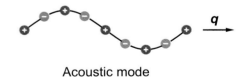

Figure 13: Transverse optical and transverse acoustic waves of a linear chain consisting of two different types of atoms, illustrated by the deflection of the atoms for two oscillations of the same wavelength.

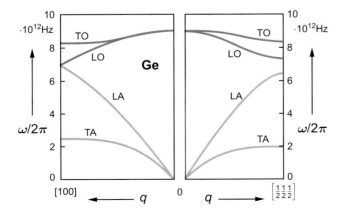

Figure 14: Phonon dispersion curves of germanium [16].

4.2 Polar Optical Phonons

We now deal with the optical oscillations in an ionic crystal with cubic structure. In the ionic crystal the different atoms in the elementary cell carry different charges. In a cubic structure, the charges compensate each other so that the elementary cell does not show a macroscopic polarisation. But, in the case of lattice vibrations, the opposite movement of vicinal atoms with complementary charge generates a strong electric dipole moment (see also Figure 13). This causes a strong interaction of the optical phonons with electromagnetic waves. The electric dipole moment also acts on the valence electrons, thus generating a retardation of the propagating wave within the system. The consideration of this retardation leads to the concept of *polaritons*. These are oscillations of the elastically vibrating crystal coupled to the electromagnetic field.

Also covalent systems can show a dipole moment due to distortion as soon as the atoms have non-equal electronegativities, i.e. their chemical bonds exhibit some polar character. In addition some elementary crystals with more than two atoms per elementary cell can show a polarisation due to a lattice distortion, as for example tellurium.

All optical lattice vibrations which coincide with an electric dipole moment are called *polar optical phonons*. These phonons are described quantitatively by an effective charge Q^*, which defines the dipole moment $\boldsymbol{p} = Q^* \cdot \boldsymbol{u}$.

In the case of base molecules with more than two atoms the vector $\boldsymbol{u}$ does not represent a simple deflection any more, but is a combination of deflections of the different atoms of the elementary cell which correspond to a characteristic vibration. The direction of the dipole moment $\boldsymbol{p}$ does not necessarily coincide with a simple deflection, therefore in the case of polar optical phonons we must distinguish their *lattice dynamical* polarisation (see Chap. 4.1) from their *electrical* polarisation. In general, again one can find two types of polarisation: the longitudinal polaritons with $\boldsymbol{p} \parallel \boldsymbol{q}$ and the transverse polaritons with $\boldsymbol{p} \perp \boldsymbol{q}$.

4.3 Polaritons

The properties of solid matter in the microwave and optical frequency regime are derived from the spectral structures of the material, as for example resonance lines or absorption and reflection edges, respectively. These properties control the suitability of material systems for applications in microwave devices (see Chap. 31) and optical components (see Chaps. 27 and 30).

In the frequency range beyond 1 GHz the materials have to be characterised by wave propagation, and the properties have to be derived from the interaction between the electromagnetic wave and the material. For these investigations it is necessary to derive from the *Maxwell equations* and the constitutive equations a wave equation which describes the coupling of consecutive regions in the solid. This self-consistent description of the interrelationship between microscopic excitation and the accompanying electromagnetic fields then leads to the concept of polaritons.

4.3.1 Propagation of Electromagnetic Waves in Condensed Matter

The interaction between polar matter and an electromagnetic field results in a retardation of the exciting signal. The time delay results in frequency-dependent material properties (see Sec. 3.2), while the spatial delay leads to wave-number-dependent ($k = 2\pi/\lambda$) properties which require a description by an action at a distance theory.

In the following we will discuss the simplest case where the response of the medium only depends on the fields at neighbouring sites. For this case we derive a wave equation in which the phase velocity c depends on the local material properties. Thus the frequency ω and the wave number k are not fully independent, but follow from the wave equation as $\omega / k = c$.

We start with Maxwell's equations in a medium with free electric charges and currents:

$$\begin{aligned} \nabla \times \boldsymbol{H} &= \dot{\boldsymbol{D}} + \boldsymbol{j} \\ \nabla \times \boldsymbol{E} &= -\dot{\boldsymbol{B}} \\ \nabla \cdot \boldsymbol{D} &= \rho_{\text{free}} \\ \nabla \cdot \boldsymbol{B} &= 0 \, , \end{aligned} \qquad (44)$$

where $\boldsymbol{D}$ denotes the electric displacement, $\boldsymbol{B}$ the magnetic flux density, $\boldsymbol{E}$ the electric field, and $\boldsymbol{H}$ the magnetic field, $\boldsymbol{j}$ the current density, and ρ_{free} the density of free charges; $\nabla\cdot$ and $\nabla\times$ are the divergence and the curl operations, and the 'dot' denotes the time derivative, e.g. $\dot{\boldsymbol{D}} \equiv \partial \boldsymbol{D}/\partial t$.

The relation between the field quantities and the flux densities, which depend on the material properties, is given by the *constitutive* equations:

$$\begin{aligned} \boldsymbol{D} &= \varepsilon_0 \boldsymbol{E} + \boldsymbol{P} \\ \boldsymbol{B} &= \mu_0 (\boldsymbol{H} + \boldsymbol{M}), \end{aligned} \quad (45)$$

where $\boldsymbol{P}$ is the polarisation density and $\boldsymbol{M}$ the magnetisation density. The constants ε_0 and μ_0 are the electric permittivity and the magnetic permeability of free space.

In order to derive the *wave equation* for the components of the fields $\boldsymbol{E}$ and $\boldsymbol{H}$ we have to first consider the properties of the medium itself which are reflected in the constitutive equations (45). For reasons of simplicity, we will restrict ourselves to non-magnetic media, which means $\boldsymbol{M} = 0$ and $\boldsymbol{B} = \mu_0 \boldsymbol{H}$.

The nature of a dielectric medium is exhibited in the relation between the polarisation density $\boldsymbol{P}$ and the electric field $\boldsymbol{E}$, where generally $\boldsymbol{P} = \boldsymbol{P}(r,t)$ and $\boldsymbol{E} = \boldsymbol{E}(r,t)$ are functions of position and time. We will first consider the simplest case of a source-free, dielectric medium, which is linear, nondispersive, homogeneous, and isotropic. Then, the relation is given by:

$$\boldsymbol{P} = \varepsilon_0 \chi'_e \boldsymbol{E} \quad (46)$$

where the electrical susceptibility χ'_e is a scalar constant and the vectors $\boldsymbol{P}$ and $\boldsymbol{E}$ are parallel and proportional at any position and time.

Equations (45) then simplify to:

$$\begin{aligned} \boldsymbol{D} &= \varepsilon_0 \varepsilon'_r \boldsymbol{E} \quad \text{with the dielectric permittivity} \quad \varepsilon'_r = 1 + \chi'_e \\ \boldsymbol{B} &= \mu_0 \boldsymbol{H}. \end{aligned} \quad (47)$$

Maxwell's equations (44) for this simple case reduce to

$$\begin{aligned} \nabla \times \boldsymbol{H} &= \varepsilon_0 \varepsilon'_r \dot{\boldsymbol{E}} \\ \nabla \times \boldsymbol{E} &= -\mu_0 \dot{\boldsymbol{H}} \\ \nabla \cdot \boldsymbol{E} &= 0 \\ \nabla \cdot \boldsymbol{H} &= 0. \end{aligned} \quad (48)$$

A necessary condition for $\boldsymbol{E}$ and $\boldsymbol{H}$ to satisfy Maxwell's equations is that each component satisfies the wave equation, which reads for $\boldsymbol{E}$:

$$\nabla^2 \boldsymbol{E} - \frac{1}{c^2}\frac{\partial^2 \boldsymbol{E}}{\partial t^2} = 0, \quad (49)$$

and analogously for $\boldsymbol{H}$.

In this case the phase velocity of the electromagnetic wave in the material is given by

$$c = \frac{1}{\sqrt{\varepsilon_0 \mu_0 \cdot \varepsilon'_r}} = \frac{c_0}{n} \quad (50)$$

with the speed of light in free space:

$$c_0 = \frac{1}{\sqrt{\varepsilon_0 \mu_0}}, \quad (51)$$

and the refractive index:

$$n = \sqrt{\varepsilon'_r}. \quad (52)$$

Based on the simple example presented in equations (46) to (52) we now discuss the cases for which one or more of the properties of linearity, nondispersiveness, homogeneity, and isotropy are not satisfied.

Generally, for an *inhomogeneous* dielectric medium (as for example a graded index medium) the proportionality equations (46) and (47) remain still valid, but the material properties are functions of the position, $\chi'_e = \chi'_e(r)$, $\varepsilon'_r = \varepsilon'_r(r)$, and $n = n(r)$. As long as $\varepsilon'_r(r)$ varies in space at a much slower rate than the field $E(r,t)$ the wave equation (49) remains applicable.

In *anisotropic* media, the relation between polarisation and electric field depends on the direction of the vector E, in particular P and E are not necessarily parallel. In the mathematical formulation, the dielectric properties of the medium are described by a matrix $\{\chi'_{e,ij}\}$ of 3×3 elements known as the *susceptibility tensor*. Correspondingly, ε'_r and n also change into tensors with elements $\{\varepsilon'_{r,ij}\}$ and $\{n_{ij}\}$. As waves with different polarisation directions travel at different velocities and undergo different phase shifts, the total polarisation vector is changed as the wave propagates through the material. Therefore, anisotropic materials provide useful components of optical devices (see Sec. 5.4.2).

In a *dispersive* medium, the relation between $P(t)$ and $E(t)$ is governed by a dynamic linear system described by an impulse-response function corresponding to a frequency-dependent susceptibility $\underline{\chi}_e = \underline{\chi}_e(\omega)$ (see Sec. 3.2). Consequently, the characteristic quantities $\underline{\varepsilon}_r$ and $\underline{n}$ become complex.

In a *non-linear* dielectric medium, the polarisation is some non-linear function of the electric field $P = F(E)$, as for example given by the most general Taylor expansion $P = a_1 E + a_2 E^2 + a_3 E^3$, where a_1, a_2, a_3 are constants. The wave equation (49) is not applicable to electromagnetic waves in non-linear media. Instead, Maxwell's equations have to be solved to derive a non-linear partial differential equation for these waves. For non-linear systems the principle of superposition is no longer applicable and optical waves interact with each other.

In general the complex refractive index $\underline{n}$ is divided into the real refractive index n and the absorption index κ:

$$\underline{n} = n + i\kappa. \tag{53}$$

For non-magnetic media, where

$$\underline{n} = \sqrt{\underline{\varepsilon}_r} \tag{54}$$

we find for the real and imaginary part of the relative permittivity:

$$\varepsilon'_r = n^2 - \kappa^2 \quad \text{and} \quad \varepsilon''_r = 2n\kappa. \tag{55}$$

These important equations enable the dielectric function to be determined from experimental investigations of the propagation behaviour of electromagnetic waves in the material.

4.3.2 Longitudinal and Transverse Polaritons

In the previous section we introduced Maxwell's equations and the constitutive equations adapted to the material under investigation which are needed to derive the wave equations in matter. The problem now is to find a self-consistent solution for the combination of field and constitutive equations. The material properties, as for example $\underline{\varepsilon}_r$, have to be determined from experiment and have to be explained by microscopic models. The allowed solutions for (ω, q) (frequency and wave number) lead to the possible microscopic excitations, which are accompanied by macroscopic electromagnetic fields. This type of excitation is called polariton.

In the following we discuss possible excitations for the case of *plane waves*:

$$E = E_0 \exp[i(qr - \omega t)], \tag{56}$$

where E represents a complex field quantity under study.

With this harmonic trial solution, we obtain for the wave equation (49) of a non-magnetic dielectric medium without free charges:

$$q \times (q \times E) = -\omega^2 \mu_0 (\varepsilon_0 E + P) - i\omega \mu_0 j \tag{57}$$

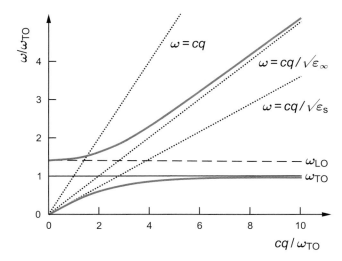

Figure 15: Dispersion curve for a phonon-polariton. The plotted part of the q-values (wave vector) is small in comparison to a reciprocal lattice vector. The lattice dispersion can therefore be neglected.

For the further discussion of equation (57) all field vectors are separated into a transverse (T) and a longitudinal (L) component with respect to the propagation direction $\boldsymbol{q}$. For the electric field (the same procedure can be performed for the magnetic field as well) this reads:

$$\boldsymbol{E} = \boldsymbol{E}_\mathrm{T} + \boldsymbol{E}_\mathrm{L} \quad \text{with} \quad \boldsymbol{q} \cdot \boldsymbol{E}_\mathrm{T} = 0 \quad \text{and} \quad \boldsymbol{q} \times \boldsymbol{E}_\mathrm{L} = 0. \tag{58}$$

With this separation, equation (57) falls into two independent equations for the longitudinal and the transverse components, respectively.

For the *longitudinal E-waves* we find:

$$\omega(\varepsilon_0 \boldsymbol{E}_\mathrm{L} + \boldsymbol{P}_\mathrm{L}) + i\,\boldsymbol{j}_\mathrm{L} = 0, \tag{59}$$

which means that the displacement current and the conduction current have to compensate each other.

To derive the necessary properties of a material in which longitudinal E-waves can be obtained, we replace $\boldsymbol{P}$ and $\boldsymbol{j}$ by the constitutive equation (46) and the *Ohm's law*

$$\boldsymbol{j} = \sigma \boldsymbol{E}, \tag{60}$$

where σ denotes the specific conductivity. For the isotropic case, where the electrical susceptibility $\underline{\chi}_\mathrm{e}$ and the conductivity σ are scalar quantities, the complex dielectric function $\underline{\varepsilon}_\mathrm{r}$ of the material has to have a zero point at the frequency ω_LO of the longitudinal E-wave:

$$\underline{\varepsilon}_\mathrm{r}(\omega_\mathrm{LO}) = 1 + \underline{\chi}_\mathrm{e}(\omega_\mathrm{LO}) + i\frac{\sigma}{\varepsilon_0 \omega_\mathrm{LO}} = 0. \tag{61}$$

The solution shows no restriction with respect to the wave number q. Therefore for every frequency which corresponds to a zero point in $\underline{\varepsilon}_\mathrm{r}$ any q-value is allowed. Due to this, in the ω-q-diagram the *longitudinal excitations* are horizontal lines (see Figure 15).

Furthermore, we see that due to the fact that polarisation and conduction currents compensate each other, the longitudinal E-wave is not accompanied by a magnetic field. Such longitudinal waves are *not* able to interact with the transverse light waves.

For the *transverse E-waves* we find from equation (57):

$$\underline{q}^2 \boldsymbol{E}_\mathrm{T} = \omega^2 \mu_0 (\varepsilon_0 \boldsymbol{E}_\mathrm{T} + \boldsymbol{P}_\mathrm{T}) + i\omega \mu_0 \boldsymbol{j}_\mathrm{T}. \tag{62}$$

The material properties which allow the propagation of *transverse E-waves* are discussed after introducing the complex refractive index (see Eq. (54))

$$\underline{n}^2 = \frac{\underline{q}^2}{q_0^2} = \frac{c_0^2}{\underline{c}^2}, \tag{63}$$

which relates the wave numbers q, q_0 or the phase velocities c, c_0 of a wave in matter and in vacuum, respectively. The velocity of light in vacuum is (see Eq. (51)) $c_0 = 1/\sqrt{\mu_0 \varepsilon_0}$, and the corresponding wave number is given by $q_0 = \omega/c_0$.

With this we find from equation (62)

$$\underline{n}^2 \boldsymbol{E}_T = \boldsymbol{E}_T + \frac{\boldsymbol{P}_T}{\varepsilon_0} + i\frac{\boldsymbol{j}_T}{\varepsilon_0 \omega}. \tag{64}$$

The combination of (64) with the constitutive equation (46) and Ohm's law (60) yields the *dispersion relation* for transverse *E*-waves for the isotropic case:

$$\underline{q}^2 = \frac{\omega^2}{c_0^2}\underline{n}^2 \quad \text{with} \quad \underline{n}^2 = \underline{\varepsilon}_r = 1 + \underline{\chi}_e + i\frac{\sigma}{\varepsilon_0 \omega}. \tag{65}$$

For frequencies $\omega = \omega_{LO}$, where the longitudinal *E*-waves can propagate through the material ($\underline{\varepsilon}_r = 0$), the transverse waves degenerate to pure vibrations because from equation (65) it follows $k = 0$ or $c \to \infty$, which means that all elements have the same phase.

The dispersion curve of a polariton ($q(\omega)$) in a dielectric medium depends on the dielectric properties of the material given by the relation $\boldsymbol{P} = \boldsymbol{F}(\boldsymbol{E})$. For a linear dielectric medium this is given by the complex frequency-dependent permittivity $\underline{\varepsilon}_r(\omega)$, or the susceptibilty $\underline{\chi}_e(\omega)$, respectively.

4.3.3 Phonon-Polaritons

We now discuss the propagation of electromagnetic waves in an ionic crystal in which polar optical phonons are observed [18]. Here we have to distinguish between the local electric field $\boldsymbol{E}_{loc}$ (see Eq. (7)) and the external applied field $\boldsymbol{E}_a$. For a cubic crystal structure we find from the Clausius-Mossotti equation (8):

$$\boldsymbol{E}_{loc} = \boldsymbol{E}_a + \frac{1}{3}\frac{\boldsymbol{P}}{\varepsilon_0}. \tag{66}$$

In equation (66) $\boldsymbol{P}$ denotes the total polarisation with contributions from both the distorted lattice and the dislocation of the valence electrons, which is proportional to the polarisability of the base molecule α (see Eq. (6)). In the following these contributions are called the *phononic* and the *electronic* polarisation, respectively.

According to this, the electrical susceptibility $\underline{\chi}_e$ defined by the medium equation (46) can be divided into two parts:

$$\underline{\chi}_e = \underline{\chi}_{VE} + \underline{\chi}_{PH}, \tag{67}$$

the first one originates from the valence electrons and the second from the lattice vibrations. For frequencies much lower than the electron resonance, as for example in the microwave or infrared region, the contribution of the valence electrons has a real and constant value, which is often abbreviated to $\underline{\chi}_{VE} \equiv \chi_\infty$.

Solving the set of differential equations assuming a classical *damped oscillator* dispersion model for the phononic contribution and considering the retardation yields for the total susceptibility:

$$\underline{\chi}_e = \frac{\left(-\omega^2 - i\omega\gamma_{TO} + \omega_0^2\right)3N_a\alpha + 3\omega_p^2}{\left(-\omega^2 - i\omega\gamma_{TO} + \omega_0^2\right)(3 - N_a\alpha) - \omega_p^2}, \tag{68}$$

where γ_{TO} denotes the damping frequency of the transverse optical phonons, ω_0 the resonance frequency of the optical phonons, ω_p the plasma frequency of the ions, N_a the concentration, and α the polarisability (see Eq. (6)).

In the following we rewrite equation (68) to get a simpler approximation. First, we look at the limit of very high frequencies ($\omega \to \infty$), for which we obtain from Eq. (68):

$$\chi_\infty = \frac{3N_a\alpha}{3 - N_a\alpha}, \tag{69}$$

which is exactly the contribution of the valence electrons to the susceptibility.

With the abbreviation

$$\underline{\chi}_0 := \frac{\omega_p^2}{\omega_0^2} \cdot \frac{\omega_0^2}{\omega_0^2 - \omega^2 - i\omega\gamma_{TO}} \qquad (70)$$

which equals the susceptibility of a classical damped oscillator without retardation, (68) becomes:

$$\underline{\chi}_e = \frac{3(N_a\alpha + \underline{\chi}_0)}{3 - (N_a\alpha + \underline{\chi}_0)}. \qquad (71)$$

Comparing this with equation (67), we obtain for the contribution of the lattice vibrations:

$$\underline{\chi}_{PH} = \frac{9\underline{\chi}_0}{(3 - N_a\alpha - \underline{\chi}_0)(3 - N_a\alpha)}. \qquad (72)$$

From equation (72) we see that the contribution of the phonons is not only given by the distortion of the lattice, represented by $\underline{\chi}_0$, but is considerably influenced by the displacement of the valence electrons, given by $(N_a\alpha)$.

With the characteristic quantities:

$$\begin{aligned}\omega_p^{*2} &:= \omega_p^2 \left(\frac{\chi_\infty + 3}{3}\right)^2, \\ \omega_{TO}^2 &:= \omega_0^2 - \frac{1}{3}\omega_p^2 \cdot \frac{\chi_\infty + 3}{3}\end{aligned} \qquad (73)$$

we now can rewrite equation (68) in the form of a classical damped oscillator dispersion relation:

$$\underline{\chi}_e = \chi_\infty + \frac{\omega_p^{*2}}{\omega_{TO}^2} \cdot \frac{\omega_{TO}^2}{\omega_{TO}^2 - \omega^2 - i\omega\gamma_{TO}}, \qquad (74)$$

where $\varepsilon_\infty = 1 + \chi_\infty = n^2$ is the high frequency optical permittivity, which takes into account the contributions of the vibrations of the valence electrons, and ω_{TO} and γ_{TO} denote the eigenfrequency and damping of the transverse phonon mode, respectively. The term $(\omega_p^{*2}/\omega_{TO}^2)$ equals the strength of the transverse phonon mode.

The derived formula (Eq. (74)) for the electrical susceptibility of an ionic crystal in which polar optical phonons can exist can now be inserted in the dispersion relation for the longitudinal and transverse phonon polaritons, respectively, given in equations (61) and (65). The solutions for an ionic lattice or, in other words, a system of harmonic oscillators, yields the dispersion relation $\omega(q)$ depicted in Figure 15 for the undamped oscillation ($\gamma_{TO} = 0$).

The frequency of the longitudinal polaritons is given by the zero point of the dielectric function (61). Insertion of $\underline{\chi}_e$ thus leads to

$$\omega_{LO}^2 = \omega_0^2 + \frac{2}{3}\omega_p^2 \cdot \frac{\chi_\infty + 3}{3\chi_\infty + 3}, \qquad (75)$$

and for the frequency of the transverse polariton (ω_{TO}) we have the result of equation (73).

The frequency of the longitudinal polaritons ω_{LO} is slightly larger than the resonance frequency ω_0 of the optical phonons at $q = 0$. This effect is due to the polarisation contributions of χ_∞.

In the gap between the resonance frequencies of the transverse and longitudinal polaritons, ω_{TO} and ω_{LO}, the complex refractive index (Eq.(54)) of the transverse polaritons turns into an imaginary number, which is equal to the existence of a stop band.

Fundamentals

4.3.4 The Lyddane-Sachs-Teller Relation

For the static limit ($\omega \to 0$) we obtain from equations (69), (70), and (71) the static susceptibility $\chi_s := \chi'_e(0)$, and with this the static limit of the dielectric function:

$$\varepsilon_s := \varepsilon'_r(0) = 1 + \chi_\infty + \frac{\omega_p^{*2}}{\omega_{TO}^2}. \tag{76}$$

Insertion of the characteristic quantities (73) and of the resonance frequency of the longitudinal polaritons (75) results in the *Lyddane-Sachs-Teller (LST)* relation

$$\frac{\varepsilon_s}{\varepsilon_\infty} = \frac{\omega_{LO}^2}{\omega_{TO}^2}. \tag{77}$$

In 1941, Lyddane, Sachs, and Teller developed this formula for the dependence of the phonon-polariton frequencies and the dielectric properties in the case of ionic polarisation. In this way, a correlation is given between the ratio of the square of the longitudinal and transverse optical mode frequencies at $q = 0$ and the ratio of the values of the real part of the relative dielectric permittivity at frequencies much lower (ε_s) and higher (ε_∞) than the resonance frequency of the ionic relaxation. Again, $\varepsilon_\infty = 1 + \chi_\infty = n^2$ is the high frequency optical permittivity, which takes into account the contributions of the vibrations of the valence electrons.

The LST relation implies that a large contribution of the polar optical phonons to the value of the dielectric function is directly correlated to a wide stop band for the propagation of transverse polaritons. This stop band is given by the gap between the eigenfrequencies ω_{LO} and ω_{TO} of the polariton (see Figure 15). The size of this stop band is also called the *LO-TO splitting*. Remember that in non-polar crystals phonon-polaritons do not exist. The long-wavelength optical phonons ($q = 0$) oscillate at the frequency ω_0, no LO-TO splitting is observed (see, for example, the dispersion curve of Ge in Figure 14).

The LST relation shows that a large value of the permittivity in ionic crystals is connected with a wide gap between the frequencies ω_{LO} and ω_{TO}, especially with a low value of the resonance frequency of the transverse optical phonon ω_{TO}.

Figure 16: Perovskite crystal structure of strontium titanate.

4.3.5 Softening of the Transverse Optical Phonon

As a further consequence of the retardation ($E_{loc} \neq E_a$) in systems where polar optical phonons exist, it follows that the polarisation fields act in different ways on the longitudinal and transverse modes of the vibrating system. Even for the simplest case of rigid ions ($\chi_\infty = 0$) we find from equations (73) and (75), respectively, a hardening of the longitudinal modes and a softening (*mode softening*) of the low frequency transverse mode:

$$\omega_{LO}^2 = \omega_0^2 + \frac{2}{3}\omega_p^2 \quad \text{and} \quad \omega_{TO}^2 = \omega_0^2 - \frac{1}{3}\omega_p^2. \tag{78}$$

In the case of the longitudinal optical mode the polarisation field enhances the mechanical restoring force. In contrast, the low-frequency transverse optical mode is characterised by a partial compensation of short-range lattice forces and long-range electrical fields; the mode becomes *soft*. Under certain temperature and pressure conditions the restoring forces for the transverse optical mode are very weak and phase transition is induced (see Chap. 2).

Figure 17:
(a) Oscillation of the linear O^{2-} - Ti^{4+} - O^{2-} chain against the remaining lattice at about 540 cm^{-1};
(b) Oscillation of the Ti^{4+} and O^{2-} ions against the Sr sub-lattice at about 180 cm^{-1};
(c) Oscillation of the oxygen octahedron against the sub-lattice constituted by the Sr^{2+} and Ti^{4+} ions. This oscillation takes place at about 87 cm^{-1} and is called the *soft mode*.

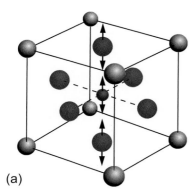
(a)

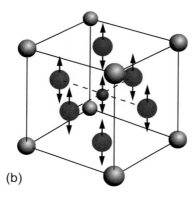
(b)

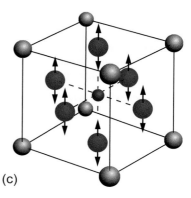
(c)

4.3.6 Damped Phonon-Polaritons

In the oscillator model we have taken damping into account by the parameter γ (see for example Eq. (70)). To a small degree, damping is caused by electromagnetic radiation of the vibrating dipoles. A larger part is, of course, given by a decay of the polariton mode in other lattice vibrations due to an anharmonicity of lattice forces. Thus, the susceptibility changes into a complex quantity, and so does the refractive index for the transverse polaritons (see Eq. (63)). The properties of the transverse polaritons are studied by measurements of the reflection and transmission of a sample in the infrared spectral region. The excitation of a sample with an electromagnetic wave generates polaritons in the crystal which propagate with a defined degree of damping depending on the crystal's properties. Thus, the wave vector q has to be a complex number. With the complex refractive index, this yields:

$$\underline{q} = q' + iq'' = (n + i\kappa) \cdot \frac{\omega}{c_0}. \tag{79}$$

4.4 Characteristic Oscillations in Perovskite-type Oxides

In the previous sections we described the lattice vibrations of an ionic crystal with cubic structure and two atoms per elementary cell. We have seen that the electric properties of these crystals in the infrared frequency range are derived from the dispersion of the optical phonon modes. In addition, we learned that the low-frequency tail of the imaginary part of the susceptibility caused by ionic polarisation is responsible for an inherent contribution to the dielectric losses in the frequency range between 1 GHz and 100 GHz, thus the ionic polarisation losses limit the quality factor of microwave dielectrics. This is important for the selection of dielectric materials for microwave applications as is explained further in Chap. 31.

In the following, some basic properties of the technically important alkaline earth titanates are shown which can be explained by the model of optical phonons. The alkaline earth titanates exhibit a perovskite crystal structure which is depicted in Figure 16 for the case of strontium titanate as an example. The temperature of the phase transition from the cubic to the distorted or tetragonal structure is approximately 105 K for $SrTiO_3$, and 396 K for $BaTiO_3$, respectively. The latter shows a distortion in the tetragonal lattice cell by a displacement of cations and anions which gives rise to the ferroelectricity of the material (see Chap. 2).

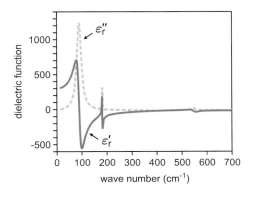

Figure 18: Dielectric function of $SrTiO_3$ in the infrared spectral region with real (outlined) and imaginary part (dotted line).

In this chapter, we focus on the dielectric properties, and thus we will restrict ourselves to the high-temperature cubic phase of the different titanates. In the cubic structures of $SrTiO_3$ and $BaTiO_3$ we can distinguish between three different infrared active modes. Here, infrared active means that the crystal exhibits a dipole moment induced by the displacement of the ions which can interact with the light wave. At the highest frequency this is an oscillation of the linear O^{2-} - Ti^{4+} - O^{2-} chain against the remaining sublattice (Figure 17a). As a second oscillation, the O^{2-} octahedron together with the Ti^{4+} ion move against the other ions as shown in Figure 17b. At the excitation with the lowest frequency, i.e. the *soft mode*, the O^{2-} octahedron oscillates against the Ti^{4+} and Sr^{2+} ions, respectively, so all negative ions are displaced against all positive ions (Figure 17c).

The frequency dependence of the real and imaginary parts of the complex dielectric function of single crystalline $SrTiO_3$ in the infrared region measured at room temperature is shown in Figure 18. The wave number is defined as the reciprocal of the wavelength.

The soft mode at 87 cm^{-1} exhibits the greatest strength of oscillation and the lowest damping compared with the other transverse optical phonons at 180 cm^{-1} and 540 cm^{-1}. The *softening* of the 87 cm^{-1} TO mode describes the fact that the mode's frequency ω_{TO} decreases with decreasing temperature when approaching the phase transition temperature coming from high temperatures. For $\omega_{TO} = 0$ the phase becomes unstable due to the vanishing restoring force.

4.5 Temperature Dependence of the Permittivity in Titanates

Taking $SrTiO_3$ as a representative example, the temperature dependence of the low-frequency dielectric constant $\varepsilon'_r(T)$ of this material in the cubic phase obeys the empirical *Curie-Weiss* law

$$\varepsilon'_r(T) \approx \chi'_e(T) = \frac{C}{T - \Theta}, \tag{80}$$

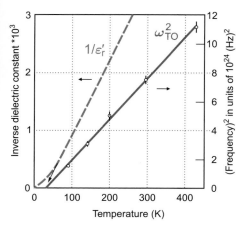

Figure 19: Temperature dependences of the reciprocal of the relative permittivity ($1/\varepsilon_r'$) (dotted line) and of the square of the frequency of the transverse optical oscillation (ω_{TO}^2) with $k = 0$ derived from neutron scattering experiments, (solid line) in SrTiO$_3$ [3].

where C is the Curie constant and Θ is the Curie temperature, which in general is smaller than the temperature T_C of the phase transition:

$$\Theta < T_C. \tag{81}$$

According to the Lyddane-Sachs-Teller relation (77), the Curie-Weiss law, i.e. the increase of the static dielectric constant with decreasing temperature (in the cubic phase), is caused by the decreasing frequency of the transverse optical phonon, which obeys a square root dependence as long as ω_{LO} can be assumed to be independent of the temperature:

$$\omega_{TO} \propto \sqrt{T - \Theta} \quad \leftrightarrow \quad \omega_{TO}^2 \propto (T - \Theta) \tag{82}$$

The correlation between the temperature dependences of the relative permittivity (Eq. (80)) and the frequency of the soft-phonon mode (Eq. (82)) has been confirmed for SrTiO$_3$, as is shown in Figure 19.

4.6 Voltage Dependence of the Permittivity in Ionic Crystals

In Sec. 4.1, we discussed an idealised ionic lattice built from harmonic oscillators. A harmonic potential means that the restoring force is a linear function of the displacement. In the case of a real ionic lattice, the local field generated by the neighbouring atoms leads to an *anharmonic* potential for each ion, and thus to a non-linear restoring force. The consequence of this is that the linear dependence between the dielectric displacement D and the electric field E (see Eq. (4)) no longer holds. Instead, we have to introduce a *field-dependent* permittivity $\varepsilon_r(E)$. The effect of this non-linearity becomes significant at high electric fields. Hence, the effect is more frequently observed in thin films than in bulk dielectrics because high electric fields are more easily reached in films at moderate voltages. In addition, for a given electric field the effect is more pronounced for higher permittivities. A high permittivity corresponds to a smaller restoring force between the ions of the lattice and a large atomic displacement at a given field. An example of the field dependent permittivity is shown in Figure 20. The non-linear electrical permittivity $\varepsilon_r(E)$ is exploited for voltage tunable microwave devices (see Chap. 31).

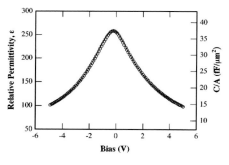

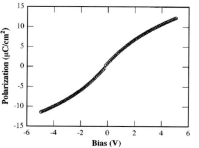

Figure 20: Non-linearity of capacitance and polarisation of a (Ba$_{0.7}$Sr$_{0.3}$)TiO$_3$ thin film prepared by MOCVD [20].

5 Optical Properties of Dielectrics

In the final section we now deal with some selected optical properties of dielectrics which are important for microwave (see Chap. 31) and photonic (see Chaps. 27 and 30) devices. For a deeper insight into the optical properties of matter, the reader is referred to the textbooks [22] – [24].

5.1 Plane Electromagnetic Waves

We first introduce the most important example of a monochromatic electromagnetic wave – the plane wave. Besides this, spherical waves and Gaussian beams are other common types (see [23]). The medium is assumed to be linear, homogeneous, isotropic, non-dispersive, and non-magnetic according to Sec. 4.3.1.

Consider now a monochromatic electromagnetic wave, whose electric and magnetic field components are plane waves of wavevector q as described by equation (56). Substituting the corresponding equations for E and H into Maxwell's equations (48), we obtain

$$\begin{aligned} \boldsymbol{q} \times \boldsymbol{H}_0 &= -\omega \varepsilon_0 \varepsilon_r' \boldsymbol{E}_0 \\ \boldsymbol{q} \times \boldsymbol{E}_0 &= \omega \mu_0 \boldsymbol{H}_0 \, . \end{aligned} \tag{83}$$

It follows from Eq. (83) that E, H, and q are mutually orthogonal. Since E and H lie in a plane normal to the direction of propagation q, the wave is called a *transverse electromagnetic* (TEM) wave. From a comparison of the two equations in (83) we find the equality $q = \omega/c = n\omega/c_0 = nq_0$, which is, in fact, the condition to satisfy the wave equation (49).

The ratio between the amplitudes of the electric and magnetic fields is therefore

$$\frac{E_0}{H_0} = \frac{\sqrt{\mu_0/\varepsilon_0}}{n} =: \eta, \qquad (84)$$

which is defined as the impedance of the medium.

The flow of the electromagnetic power which is governed by the complex *Poynting* vector

$$\underline{S} = \tfrac{1}{2}\, \underline{E} \times \underline{H}^* \qquad (85)$$

is parallel to the wave vector q ($\underline{H}^*$ denotes the complex conjugate of $\underline{H}$). The magnitude of the time-averaged Poynting vector equals the optical intensity of the TEM wave, which is therefore proportional to the squared absolute value of the complex envelope of the electric field:

$$I = \frac{|E_0|^2}{2\eta}. \qquad (86)$$

5.2 Resonator and Waveguide Modes

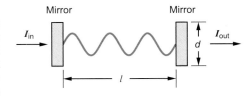

Figure 21: Standing wave inside an optical resonator.

In the microwave and optical devices described in chapters 27, 30 and 31, we have to deal with electromagnetic waves which are amplified in resonator systems or guided by means of wave guides. As the simplest example, we briefly discuss the case of a TEM plane wave inside a resonator consisting of two plane-parallel mirrors as is sketched in Figure 21. If the diameter d of the mirrors is large compared to their distance l, the resonator is identical to a *Fabry-Perot* interferometer (see [22]). For the frequencies $f_\xi = \xi \cdot c/2l$ ($\xi = 1,2,3,..$) standing waves are obtained between the mirrors. The frequency difference between successive standing waves $\Delta f = f_{\xi+1} - f_\xi = c/2l$ is also called the basic frequency since it is identical with the lowest eigenfrequency of the resonator ($\xi = 1$). The transmission of the interferometer for a plane, perpendicular incident beam of intensity I_{in} is a function of the frequency f:

$$\frac{I_{in}}{I_{out}} \propto \frac{1}{\sin^2(2\pi f \cdot l/c)}. \qquad (87)$$

For the resonance frequencies $f_\xi = \xi \cdot c/2l$ the transmission has a maximum value.

The standing waves of resonance frequency f_ξ are also called *longitudinal* or *axial cavity modes*, because the standing waves are set up along the cavity or z-axis, and the resonance frequency only depends on the corresponding number of nodes.

In contrast to the ideal Fabry-Perot interferometer, in practical cavities the assumption of infinitely large mirrors ($d \gg l$) is not valid. Moreover, we find just the opposite case, namely that the diameter of the front planes (or mirrors) is small compared to the distance ($d \ll l$). The limited dimensions of the mirrors cause a collimation of the ray inside the resonator thus resulting in diffraction of the wave. This diffraction leads to *transverse modes* which are sustained in the resonator. Since the fields are almost normal to the z-axis they are known as TEM_{mn} modes (transverse electric and magnetic). The m and n subscripts are the integer number of transverse nodal lines in the x- and y-directions across the emerging beam, which means that the beam is segmented in its cross section into one or more regions. The form of the different TEM_{mn} modes depends on the shape of mirror (plane or confocal) as well as on the form of its cross section (rectangular or circular). Examples of typical TEM_{mn} mode configurations are shown in Figure 22.

5.3 Absorption

The absorption of electromagnetic waves in a dielectric medium is strongly related to the dielectric properties described by the complex permittivity $\underline{\varepsilon}_r$ (see Eq. (10)). In absorbing media the wave equation (49) remains applicable, but a complex wave number is used to account for the losses:

$$\underline{q} = \omega \sqrt{\underline{\varepsilon}_r (\varepsilon_0 \mu_0)} = \sqrt{\varepsilon'_r + i\varepsilon''_r} \cdot q_0, \quad q_0 = \frac{\omega}{c_0}. \qquad (88)$$

Fundamentals

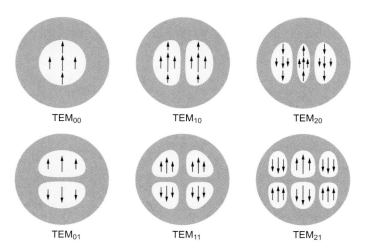

Figure 22: Mode configurations of a confocal optical resonator with rectangular symmetry (after [22]).

In terms of the complex refractive index (Eq. (53)), $\underline{q}$ reads:

$$\underline{q} = \underline{n} \cdot q_0 = (n + i\kappa) \cdot q_0 . \tag{89}$$

Looking at a monochromatic plane wave travelling in the -z-direction through an absorbing medium

$$\underline{E} = E_0 \exp\left[i(\underline{q}z - \omega t)\right] = E_0 \cdot \exp(-\kappa q_0 z)\exp\left[i(nq_0 z - \omega t)\right], \tag{90}$$

we find a phase velocity of $c = c_0/n$ and an exponentially decaying amplitude.

The intensity is defined by the square of the electric field (see Eq. (86)). For the intensity of the transmitted wave we derive from equation (90) the *Lambert-Beer's* law:

$$I(z) = I_0 \cdot e^{-az} , \tag{91}$$

where a is known as the absorption constant.

For weakly absorbing media, we find from inserting equations (88) and (90) into (91):

$$a(\omega) = \frac{\omega \, \varepsilon_r''(\omega)}{c_0} . \tag{92}$$

With Eq. (92) we have a particularly simple relation between the empirical absorption constant and the dielectric response function of a system.

5.4 Polarisation, Reflection, and Refraction

The polarisation of light is determined by the direction of the electric field vector $E(r,t)$. For transverse electromagnetic waves (TEM waves) the electric field vector lies in a plane perpendicular to the propagation direction (see Sec. 5.1). Generally, the wave is said to be *elliptically* polarised. When the ellipse degenerates into a straight line or a circle, the wave is called *linearly* or *circularly* polarised, respectively. Figure 23 shows, as an example, the planes of oscillation of the electrical and magnetic field vectors of linearly polarised light generated by reflection.

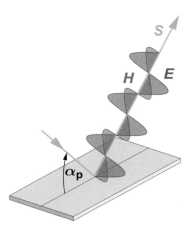

Figure 23: Planes of oscillation of the electrical and magnetic field vectors of linearly polarised light generated by reflection.

Polarisation plays an important role in the interaction of light with matter, because it is mainly the electrical field vector E which interacts with matter (Sec. 4.3). In the context of optical applications, some important examples are listed below:

- The polarisation of the incident beam determines the amount of light reflected at the boundary between two materials.
- Light scattering from matter is polarisation-sensitive.
- The refractive index of anisotropic media, and thus the phase velocity and the phase shift of the wave, depend on the polarisation.
- Optically active materials have the ability to rotate the plane of polarisation of polarised light.
- The absorption constants of certain materials are polarisation-dependent.

5.4.1 Reflection and Refraction

The polarisation-dependent reflection and refraction of light at a boundary between two dielectric media is of fundamental importance for the functionality of optical devices. The phenomenon of total reflection is the basic principle of light pipes.

Here we will examine the reflection and refraction of a monochromatic plane wave of arbitrary polarisation incident at a planar boundary between two dielectric media with refractive indices n_1 and n_2 as depicted in Figure 24. The media again are assumed to be linear, homogeneous, isotropic, nondispersive, and nonmagnetic.

Considering the angles of the different beams with respect to the axis of incidence, the following relations hold (see [22]):

- the angle of reflection equals the angle of incidence: $\theta_3 = \theta_1$;
- the angles of refraction and incidence satisfy *Snell's* law: $n_1 \sin\theta_1 = n_2 \sin\theta_2$.

To calculate the *reflection* and *transmission coefficients* one has to relate the amplitudes and the polarisations of the three waves taking into account the boundary conditions required by the electromagnetic theory (tangential components of **E** and **H** and normal components of **D** and **B** are continuous at the boundary, see also [6]). The calculations result in a dependence of the coefficients on the polarisation of the incident beam. Therefore, we have to distinguish between *transverse electric* (TE) or *s-polarisation*, for which the electric fields are orthogonal to the plane of incidence, and *transverse magnetic* (TM) or *p-polarisation*, where the electric fields are parallel to the plane of incidence.

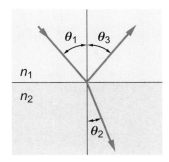

Figure 24: Incident, reflected and refracted beam at the boundary between two dielectric media.

The resulting expressions for the reflection and transmission coefficients, the *Fresnel* equations, read for *s*-polarisation:

$$r_s = \frac{n_1 \cos\theta_1 - n_2 \cos\theta_2}{n_1 \cos\theta_1 + n_2 \cos\theta_2} \quad \text{and} \quad t_s = 1 + r_s, \tag{93}$$

and for *p*-polarisation:

$$r_p = \frac{n_2 \cos\theta_1 - n_1 \cos\theta_2}{n_2 \cos\theta_1 + n_1 \cos\theta_2} \quad \text{and} \quad t_p = \frac{n_1}{n_2}(1 + r_p). \tag{94}$$

From (94) it follows that the reflection coefficient for a *p*-polarised beam vanishes at a certain angle $\theta_1 = \theta_B$, the so-called *Brewster angle*, which is given by

$$\theta_B = \tan^{-1}\left(\frac{n_1}{n_2}\right). \tag{95}$$

The property that *p*-polarised light is not reflected at the Brewster angle is used for the design of polarisers, so called Brewster windows.

The reflection and transmission coefficients r and t are ratios of the complex amplitudes. The power reflectance R and transmittance T are defined as the ratios of the power flow of the reflected and transmitted wave to that of the incident wave. For the power reflectance, we obtain:

$$R = |r|^2. \tag{96}$$

For the power transmittance the conservation of energy requires:

$$T = 1 - R. \tag{97}$$

For normal incidence the reflectance for both *s*- and *p*-polarisation is identical and given by the expression:

$$R_\perp = \left(\frac{n_1 - n_2}{n_1 + n_2}\right)^2. \tag{98}$$

5.4.2 Birefringence

We now deal with the case that medium 2 in Figure 24 is no longer isotropic, but *anisotropic* (see for comparison the discussion in Sec. 4.3.1). Thus we examine the refraction of a plane wave at the boundary between an isotropic medium, as for example air ($n_1 = 1$), and an anisotropic crystal like e.g. quartz or rutile (for further examples see [22]). Due to its *optical anisotropy*, which is a consequence of different electronic bonds in different directions, medium 2 supports two modes of distinctly different phase velocities. Therefore, each incident wave gives *two* refracted waves with two different directions and different polarisations (see Figure 25). The effect is called *birefringence*.

As a simple example, we take a uniaxial crystal and a plane of incidence parallel to the optical axis (for further discussions see [23]). The two refracted waves which satisfy the phase-matching condition at the boundary are:

- an ordinary wave of orthogonal (*s*-) polarisation at an angle $\theta = \theta_o$;
- an extraordinary wave of parallel (*p*-) polarisation at an angle $\theta = \theta_e$.

With respect to optical device applications, anisotropic crystal plates serve as polarising beam splitters, creating two laterally separated rays with orthogonal polarisations.

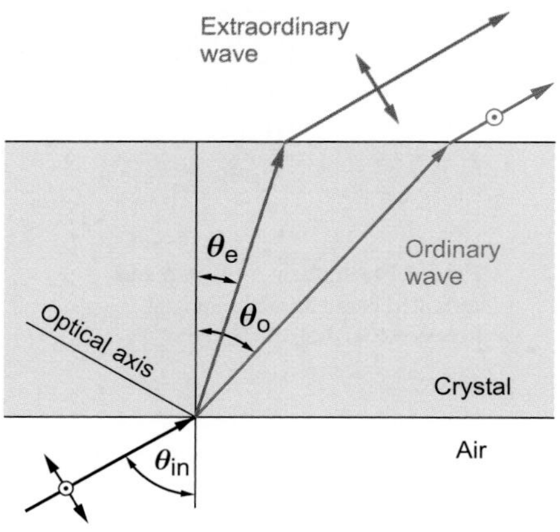

Figure 25: Birefringence through an anisotropic, uniaxial medium.

5.5 Electro-optical Effects

In the field of linear optics various effects are used for guiding and modulating electromagnetic waves in optical devices. In general, the light wave interacts with the electrons that are bound elastically in the material.

5.5.1 The Pockels and Kerr Effects

In certain materials like ionic crystals the positions of the ions and the shapes of the electronic orbitals are distorted when the material is subjected to an electric field (see Secs. 4.3 and 4.6). The *electro-optical effect* is the change in the refractive index (caused by a change in the dielectric properties) due to the application of a dc or low-frequency electric field.

The dependence of the refractive index on the applied electric field takes two forms:

- the refractive index changes in proportion to the applied field $n \propto E$, which is called the *linear* electro-optical or the *Pockels effect*. This effect is observed for ferroelectric materials with a preferential axis,
- the refractive index changes in proportion to the square of the applied field $n \propto E^2$, which is known as the *quadratic* electro-optical or the *Kerr effect*. This effect is observed for dielectrics (see Sec.4.6).

Especially the Pockels effect is used for many electro-optical devices as, for example, phase modulators, switches, spatial intensity modulators, and the Pockels readout optical modulator (PROM) [23].

5.5.2 Photorefractive Materials

Photorefractive materials combine photoconductive and electro-optical behaviour. They are thus able to detect and store spatial optical intensity distributions in the form of spatial patterns of altered refractive index. Photo-induced charges create a space-charge distribution that produces an internal electric field, which, in turn, alters the refractive index due to the electro-optical effect. Photorefractive devices therefore permit light to control light [23].

In photoconductors, free charge carriers are generated under illumination due to the absorption of photons, and the conductivity increases. In the dark, the electron-hole pairs recombine and the conductivity decreases.

When a photorefractive material is exposed to light, free charge carriers are generated by excitation from impurity energy levels to the conduction band, at a rate proportional to the optical power. Carriers diffuse into areas of low intensity where they are trapped by other ionized impurities. The result is an inhomogeneous space charge distribution generated by a light intensity pattern. The charge distribution creates an internal electric field that modulates the refractive index by the Pockels effect. The image may be accessed optically by monitoring the spatial index pattern with a probe optical wave. The pattern can be erased by illumination or heating of the photorefractive material. Important material candidates are barium titanate ($BaTiO_3$), bismuth silicon oxide, lithium niobate, and also gallium arsenide. The photorefractive effect is the basis for modern holographic storage devices. This is the topic of Chap. 27.

6 Closing Remarks

This chapter presented a short review of the basic concepts of the interactions between electromagnetic fields and dielectric matter. The entire spectrum from the dc response to the optical frequencies was covered. The dc response is important for capacitors and electronic devices, permitting the design of capacitors with increased or variable capacitance. The dc and the low-frequency ac response are dominated by displacing charged ions within the material. As a consequence, lattice distortions, vibrations and the phonon dynamics play an important role and sound waves are easily coupled to alternating electrical fields. At optical frequencies, only the electrons are able to follow the rapid changes of the electrical field and the ionic lattice provides the background of a periodic static charge distribution. Within a transparent polarisable material, it is the interplay between the asymmetric built-in lattice charge distribution and the electric field of the propagating electromagnetic wave which acts on the valence electrons of the crystal in a complex and very interesting way, leading to electro-optic effects, to acousto-optic devices, to nonlinear optics, to frequency conversion and even to holographic storage applications.

Acknowledgements

The author acknowledges Christoph Buchal (Research Center Jülich) for critical review and his helpful suggestions. Special thanks are due to Christian Ohly (Research Center Jülich) for his editorial and technical assistance in the compilation of this Chapter.

References

[1] N.W. Ashcroft, N.D. Mermin, D. Mermin, *Solid State Physics*, Holt, Rinehart and Winston, New York, 1976.
[2] H. Ibach, H. Lüth, *Solid-state physics: an introduction to principles of materials science*, Springer, Berlin, New York, 1996.
[3] C. Kittel, *Introduction to Solid State Physics*, Wiley, New York, 1996.
[4] O. Madelung, *Introduction to Solid State Theory*, Springer Series in Solid State Sciences, Vol. 2, Springer, Berlin, Heidelberg, New York, 1978.
[5] L. Bergmann, C. Schaefer, *Constituents of matter: atoms, molecules, nuclei and particles*, edited by Wilhelm Raith, W. de Gruyter, Berlin, New York, 1997.
[6] R.P. Feynman, *The Feynman Lectures on Physics 'Mainly Electromagnetism and Matter'*, Calif. Addison-Wesley, Redwood City, 1989.
[7] S.M. Sze, *Physics of Semiconductor Devices*, John Wiley & Sons, New York, 1981.
[8] A.J. Moulson and J.M. Herbert, *Electroceramics: materials, properties, applications*, Chapman and Hall, London, New York, 1990.
[9] R.C. Buchanan, *Ceramic materials for electronics: processing, properties, and applications*, M. Dekker, New York, 1991.
[10] B. Jaffe, W.R. Cook Jr., and H. Jaffe, *Piezoelectric ceramics*, Academic Press, London, New York, 1971.
[11] H. Fröhlich, *Theory of dielectrics: dielectric constant and dielectric loss*, Clarendon Press, Oxford, 1986.
[12] A.K. Jonscher, *Dielectric relaxation in solids*, Chelsea Dielectrics Press, London, 1983.
[13] M. Born and K. Huang, *Dynamical theory of crystal lattices*, Clarendon Press Oxford, 1988.
[14] R.K. Singh, S.P. Sanyal, *Phonons in condensed matter physics*, Wiley, New York, 1990.
[15] H. Bilz, W. Kress, *Phonon dispersion relations in insulators*, Springer, Berlin, New York, 1979.
[16] R. Geick, Phys. Rev. **138**, A 1495 (1965).

[17] L. Brillouin, *Wave Propagation in Periodic Structures*, Dover Publications, New York, 1953.

[18] R. Claus, L. Merten, J. Brandmüller, *Light Scattering by Phonon-Polaritons*, Springer Tracts in Modern Physics, Vol. 75, Springer, Berlin, Heidelberg, New York, 1975.

[19] S.K. Streiffer, C. Basceri, A.I. Kingon, S. Lipa, S. Bilodeau, R. Carl, and P.C. van Buskirk, 1995 MRS Fall Meeting, Mat. Res. Soc. Symp. Proc. **415**, 219 (1996).

[20] C. Basceri, S.K. Streiffer, A.I. Kingon, and R. Waser, J. Appl. Phys. **82**, 2497 (1997).

[21] L. Bergmann, C. Schaefer, *Optics of waves and particles*, edited by Heinz Niedrig, W. de Gruyter, Berlin, New York, 1999.

[22] E. Hecht, *Optics*, Addison-Wesley Pub. Co., Reading Mass., 1987.

[23] B.E.A. Saleh, M.C. Teich, *Fundamentals of photonics*, Wiley, New York, 1991.

[24] J. Feinberg, *Photorefractive nonlinear optics*, Physics Today (1988).

Ferroelectrics

Dieter Richter, Department IFF, Research Center Jülich, Germany

Susan Trolier-McKinstry, Department MATSE, Pennsylvania State University, USA

Contents

1 Introduction — 61

2 Spontaneous Polarization — 62
2.1 Symmetry Considerations — 62
2.2 Phenomenology of Ferroelectrics — 62
2.3 Antiferroelectricity — 64
2.4 Pyroelectricity — 64
2.5 Piezoelectricity — 64

3 Theory of the Ferroelectric Phase Transition — 65
3.1 Landau-Ginzburg Theory — 65
3.1.1 Paraelectric Phase — 66
3.1.2 Ferroelectric Phases – Second-Order Phase Transition — 66
3.1.3 Ferroelectric Phases: First Order Transition — 67
3.2 Soft Mode Approach of Displacive Phase Transition — 68

4 Ferroelectric Materials — 69

5 Ferroelectric Domains — 71
5.1 Origin of Domains — 71
5.2 Static Domain Configurations in Bulk Systems — 72
5.3 Static Domain Configurations in Thin Films — 73
5.4 Dielectric Small Signal Behavior — 74
5.5 Reversible and Irreversible Polarization Contributions — 75
5.6 Switching for Ferroelectric Domains — 76

6 Summary — 77

Ferroelectrics

1 Introduction

This chapter deals with dielectric crystals which show a **spontaneous electric polarization** and in which the direction of the polarization can be reoriented between crystallographically defined states by an external electric field. Materials which show these properties are called ferroelectrics. The topic of this chapter is comprehensively covered in textbooks such as [1] – [3].

Among the different ferroelectrics, oxides showing a perovskite structure or a related structure are of particular importance. Figure 1 displays the crystal structure of **barium titanate**, often regarded as the archetypical structure of a ferroelectric. In the cubic high temperature phase this material does not display any spontaneous polarization and, hence, the system is paraelectric. Upon cooling, a **phase transition** occurs during which the positive and negative metal ions displace with respect to each other, leading to a tetragonal deformation. Due to the asymmetry in this displacement (e.g. the titanium ion moves close to one of the six oxygen neighbors, yielding a relative displacement of the center of positive charge from the center of negative charge within the unit cell), a spontaneous polarization in the direction of the tetragonal axis appears. Additional distortions are also possible with this structure. Rhombohedral and orthorhombic distortions are widely reported, in which the central ion displaces along the <111> or the <101> cubic axes, respectively.

Ferroelectric materials have a lot of useful properties. High dielectric coefficients over a wide temperature and frequency range are used as dielectrics in integrated or in SMD (surface mounted device) capacitors, see Chapter 21. The large piezoelectric effect is applied in a variety of electromechanical sensors, actuators and transducers. Infrared sensors need a high pyroelectric coefficient which is available with this class of materials (Chapter 34). Tunable thermistor properties in semiconducting ferroelectrics are used in PTCR (positive temperature coefficient resistors). The significant non-linearities in electromechanical behavior, field tunable permittivities and refractive indices, and electrostrictive effects open up a broad field of further different applications. In addition, there is growing interest in ferroelectric materials for memory applications, where the direction of the spontaneous polarization is used to store information digitally, see Chapter 22.

The chapter is structured in the following topics:

- First we will deal with the **phenomenology** of ferroelectric materials. We will briefly relate piezo-, pyro- and ferroelectric properties with the symmetry class of crystals and then display key phenomenological properties of such materials.

- Thereafter, we will discuss the phase transitions of ferroelectrics in more detail. This we will do in terms of a mean field or **Landau-Ginzburg theory** which, as a consequence of the long-range Coulomb interaction, is able to describe such phase transitions very well.

- Then we will turn to a **microscopic description** of ferroelectric behavior. We will have a look on the optical phonons, and understand the displacive phase transitions as a polarization catastrophe.

- A short overview about the most relevant **ferroelectric materials** and their properties will be given.

- Finally, we will turn to the origin of **ferroelectric domains**. In particular, we will consider typical static domain configurations for single crystals, bulk ceramics, and thin films. In addition, the motion of domain walls and the basics of the ferroelectric switching, i. e. the polarization reversal, will be outlined.

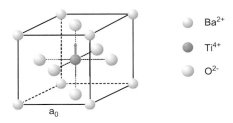

Figure 1: Unit cell of cubic $BaTiO_3$ (perovskite structure). The central Ti^{4+} ion is surrounded by six O^{2-} ions in octahedral configuration. The arrow schematically indicates one of the possible displacement of the central Ti^{4+} ion at the transition to the tetragonal ferroelectric structure that leads to a spontaneous polarization and, hence, to the ferroelectricity of tetragonal $BaTiO_3$ [4]. In reality all ions are displaced against each other.

2 Spontaneous Polarization

2.1 Symmetry Considerations

The ability of a crystal to exhibit spontaneous polarization is related to its symmetry. Of the 32 point groups, which describe all crystalline systems, 11 are centrosymmetric and contain an **inversion center**. In such structures, polar properties are not possible because any polar vector may be inverted by an existing symmetry transformation. Among the 21 point groups without an inversion center, all except the point group *432* can exhibit piezoelectricity. Thus an external strain leads to a change in electric polarization. The phenomenon of piezoelectricity was discovered by J. Curie in 1880. Piezoelectric properties will be briefly described in Sec. 2.5.

Among the 21 point groups without an inversion center, there are 10 polar groups with a unique polar axis. Such crystals may display spontaneous polarization parallel to the polar axis. In its tetragonal phase, barium titanate is such a material (see Figure 1). In the cubic phase on the other hand, the central titanium atom serves as an inversion center – then spontaneous polarization is not possible. Only with the occurrence of a tetragonal deformation, where the positively charged barium and titanium ions are displaced with respect to the negatively charged oxygen ions, is a polar axis formed in the direction of the tetragonal deformation, which marks the direction of the spontaneous polarization.

Following Maxwell's equations, spontaneous polarization is connected with surface charge density

$$P_S = \sigma \qquad (1)$$

where P_S is the spontaneous polarization and σ the surface charge density. Such surface charges in general are compensated by charged defects. A temperature change changes the spontaneous polarization and thus also the surface charges. This effect may be measured and is called the pyroelectric effect. Such a pyroelectric effect was first described by Theophrastus in the year 314 B.C. on tourmaline. Since this effect is the basis of pyroelectric infrared imaging arrays covered in Chap. 34, the basic relationships will be presented in Sec. 2.4.

2.2 Phenomenology of Ferroelectrics

If it is possible to reorient the spontaneous polarization of a material between crystallographically equivalent configurations by an external electric field, then in analogy to ferromagnetics one speaks about ferroelectrics. Thus, it is not the existence of spontaneous polarization alone, but its reorientability by an external field which defines a ferroelectric material. Figure 2 displays a characteristic **hysteresis loop** occurring during the reversal of the polarization in a ferroelectric.

In the case of an ideal single crystal, the polarization vs. field behavior, *P-E*, can be explained by a simple superposition of two contributions: firstly, the (non-ferroelectric) **dielectric ionic** and **electronic polarization** and, secondly, the **spontaneous polarization,** which is reoriented when the electric field E applied opposite to the polarization exceeds the coercive field E_C leading to the unidirectional jumps in the *P-E* curve.

In polydomain ferroelectric materials, especially in ceramics, initially there is a statistical distribution of domains (see Sec. 5.2) before it is polarized for the first time. Then the relation between the polarization P and the electric field E is characterized by the graph AB. Starting with a polarization $P = 0$, P increases with increasing field until it reaches saturation at point B. The saturation polarization P_S is obtained by extrapolating the graph BC to $E = 0$. If after saturation one now reduces the electric field again then at $E = 0$ a remanent polarization P_R is found. P_R relates to the domain structure in the material. In order to bring the polarization to zero, a negative electric field (the coercive field E_C) has to be applied. If the negative field is further increased, then the hysteresis loop is followed in the reverse sense. All the phenomenology of the hysteresis shows close analogy to the magnetic case.

As an illustration of the types of properties demonstrated by many ferroelectric materials, barium titanate will be considered. Figure 3 displays its temperature dependent relative permittivity ε_r. On cooling from high temperatures, the permittivity rises strongly and reaches values well above 10,000 at the **phase transition temperature** $T_C \sim 132°C$. If the inverse susceptibility $\chi^{-1} \cong (\varepsilon_r^{-1})$ is plotted as a function of temperature, a linear relation is obtained with χ^{-1} reaching zero below the **Curie temperature**, Θ.

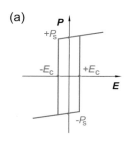

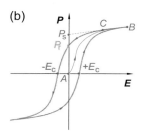

Figure 2: Hysteresis of the polarization P as a function of the field E for ferroelectric materials:
(a) Single domain single crystal recorded in the polar direction. The remanent polarization P_r and the spontaneous polarization P_s are identical. An electrical field amplitude $E > E_C$ is needed to reverse the polarization.
(b) Polycrystaline sample. The line A-B gives the initial polarization curve. Extrapolation of the line B-C towards zero electric field gives the saturation polarization P_s at $E = 0$. The hysteresis curve cuts the P axis at $E = 0$ giving the remanent polarization P_r. In order to reduce the polarization to zero, a coercive field E_c is necessary.

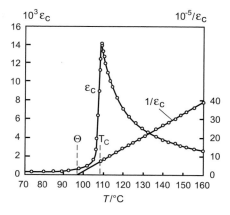

Figure 3: Dielectric permittivity ε_c for a barium titanate single crystal measured parallel to the c axis with an alternating field of a frequency $f = 1$ kHz. The transition temperature T_C depends on the purity of the sample. θ is the Curie-Weiss temperature.

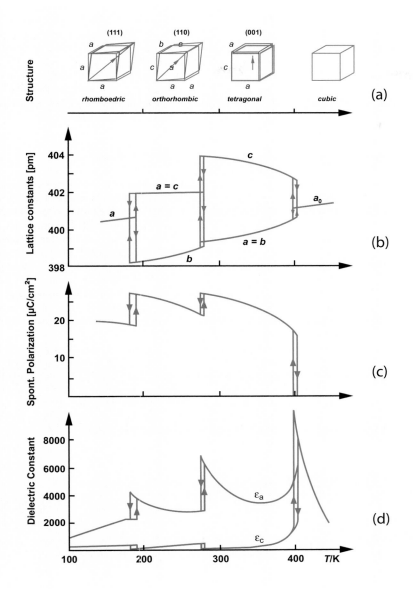

Figure 4: Various properties of barium titanate as a function of temperature. Anisotropic properties are shown with respect to the lattice direction.
(a) Structure,
(b) Lattice constants,
(c) Spontaneous polarization P_s
(d) Relative permittivity ε_r for barium titanate.

$$\chi \approx \varepsilon = \frac{C}{T - \Theta} \qquad (2)$$

The phase transition in barium titanate is first order in character, and as a result, there is a discontinuity in the spontaneous polarization, spontaneous strain, and many other properties, as becomes clear in Figure 4.

It is also clear from the figure that there are three phase transitions in barium titanate. In all cases, there is a small thermal hysteresis of the transition temperature, which depends on many parameters such as the rate of temperature change, mechanical stresses, crystal imperfections etc.. At the first transition upon cooling from high temperatures, the system transforms from a cubic to a **tetragonal structure** with a spontaneous polarization in the [001] direction as shown in Figure 4. The abrupt change of the spontaneous polarization of $\Delta P_S = 0.18$ C/m^2 at the cubic-to-tetragonal transition temperature T_C clearly demonstrates a first-order phase transition. The second phase transition transforms the tetragonal structure to an **orthorhombic structure**. In this case, the polarization direction is in the [101] direction of the prototype cubic cell. Finally, at –90°C a further phase transition takes place which deforms the orthorhombic to a **rhombohedral structure**. There, the polarization is in the [111] direction. From a crystal chemical perspective, this series of phase transitions can be viewed as a consequence of the Ti^{4+} ion being somewhat too small to occupy the interstice created by the

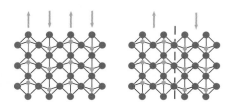

Figure 5: Sketch of perovskites structures. The white circles present the O^{2-} lattice, the black dots the sublattices with positive charges.
(a) antiferroelectric structure,
(b) two oppositely polarised ferroelectric domains.

Ba-O framework. As a result, the series of phase transformations takes place to reduce the Ti cavity size. Certainly, the radii of the ions involved impact the propensity for forming ferroelectric phases; thus both $PbTiO_3$ and $BaTiO_3$ have ferroelectric phases, while $CaTiO_3$ and $SrTiO_3$ do not [5].

We note that the optical properties of ferroelectric materials are characterized by **birefringence**. Barium titanate is isotropic only in the cubic phase. The tetragonal and the rhombohedral phases are uniaxially birefringent while the orthorhombic phase exhibits birefringent behavior with two axes. Figure 4d displays the temperature dependence of the permittivity in barium titanate over the full temperature range.

2.3 Antiferroelectricity

As in the case of magnetism, the electric dipole moments may orient themselves in a parallel or antiparallel fashion. Figure 5 displays schematically two different polarization patterns. In Figure 5a the positively and negatively charged ions are displaced alternately in the downward and upward direction. The associated dipoles create an **antiferroelectric order**. Functionally, a material is referred to as antiferroelectric if it can be field-forced to a ferroelectric state (i.e. the free energies of the ferroelectric and antiferroelectric states must be similar). In contrast to this case, Figure 5b displays the behavior of the displacements close to a domain wall of a ferroelectric phase. Whether a given structure forms ferroelectric or antiferroelectric order depends on the overall lattice forces and dipolar interactions. E.g. a calculation of the dipolar interactions in perovskites would suggest an antiferroelectric order [6]. The additional lattice forces then determine the actual structures.

Figure 6 displays the polarization dependence on the electric field in an antiferroelectric. First with a low electric field only a weak polarization is exhibited. Only if a critical field E_c which breaks the antiferroelectric order is surpassed, a major polarization is built up. Around this critical field hysteresis effects are observed in a similar way as they occur in ferroelectric materials around $E = 0$ (see Figure 2), although in this case the hysteresis is due to the field - forcing a phase transition from the antiferroelectric to a ferroelectric phase. An example of an antiferroelectric is $PbZrO_3$.

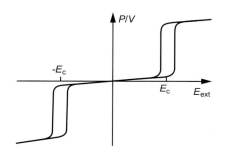

Figure 6: Antiferroelectric hysteresis loop. For $|E_{ext}| > E_C$ the system transforms into a ferroelectric state.

2.4 Pyroelectricity

The polarization charges of the surface of polar materials are usually screened by free charges causing residual currents (since no material has infinite resistivity) or charges captured from the ambient if the polar sample has been resting at a given temperature for some time.

However, since the spontaneous polarization, P_S, is temperature-dependent (for example, see Figure 4), any temperature change ΔT at a rate larger than the screening process will lead to uncompensated polarization charges

$$\Delta P = p_{py} \Delta T \tag{3}$$

where p_{py} denotes the **pyroelectric coefficient**.

These changes in the surface charge $\Delta Q = \Delta P \cdot A$ can be electrically detected as a current, I, in an external circuit if electrodes (of area A) are attached to the polar material (Figure 7).

Some infrared detector arrays operating at room temperature are based on integrated pyroelectric materials (see Chap. 34).

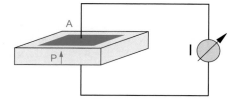

Figure 7: Slab of a pyroelectric crystal with the polarization vector and electrodes sketched. A temperature change will lead to a current I.

2.5 Piezoelectricity

All polar crystals show piezoelectricity, since any mechanical stress T will result in a strain S because of the elastic properties of the material. And the strain will affect the polarization since the polarization is caused by a displacement of the charge centers of the anions and cations. For small changes of the stress T, the relation

$$P = d \cdot T \tag{4}$$

is called the *direct* **piezoelectric effect**, where d denotes the piezoelectric coefficient. Because of the piezoelectric property of polar materials a *converse* effect is observed, if an external electrical field, E is applied, a strain

$$S = d \cdot E \tag{5}$$

is observed. The direct piezoelectric effect is employed for mechanical sensors, while the converse effect is used for mechanical actuators. In general, **P** and **E** are vectors, **S** and **T** are second-rank tensors, resulting in third-rank tensors for the **piezoelectric coefficients** **d**. The huge number of components of this tensor is significantly reduced for simple symmetries. For a material which has symmetry 8m, e.g. of poled ferroelectric ceramics, only three different d components remain:

- the **parallel component** d_{33} (Figure 8a) for a dielectric displacement (polarization), if a stress is applied in the same direction or for a strain, if the electric field is acting in the same direction,
- the **perpendicular component** d_{31} (Figure 8b) for a dielectric displacement (polarization), if a stress is applied in the perpendicular direction or, for a strain, if the electric field is acting in the perpendicular direction.
- the **shear component** d_{15} (Figure 8c) for a dielectric displacement (polarization), if a shear stress is applied or, for a shear strain, if the electric field is acting.

The matrix notation for the d coefficients is described in more detail in [7]. Both the parallel and the shear effect are employed in the piezo-response mode of the AFM technique (see Chap. 12).

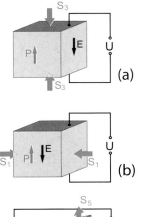
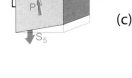

Figure 8: The three different configurations of the direct piezoelectric effect for a material with symmetry 8m.

3 Theory of the Ferroelectric Phase Transition

3.1 Landau-Ginzburg Theory

In this Section, we describe the thermodynamics of the ferroelectric phase transitions in terms of a Landau-Ginzburg theory. This theory is equivalent to a **mean field theory**, where the thermodynamic entity – here, the dipole – is considered in the mean field of all the others. Such a theory is a good approximation if the dipole interacts with many other dipoles. As a consequence of the long-range Coulomb interaction this condition is fulfilled and we can expect that the Landau-Ginzburg theory will be well suited to describe the ferroelectric phase transitions.

The Landau-Ginzburg theory introduces an **order parameter** P – in our case the polarization – which for a second order phase transition diminishes continuously to zero at the phase transition temperature T_C. Close to the phase transition, therefore, the **free energy** may be written as a functional of powers of the order parameter

$$F(P,T) = \int dV \left[\frac{1}{2} g_2 P^2 + \frac{1}{4} g_4 P^4 + \frac{1}{6} g_6 P^6 + \frac{1}{2} \delta (\nabla P)^2 - \frac{1}{2} PE \right]. \tag{6}$$

In this expansion the odd powers of P do not occur because of symmetry reasons. We shall see that by the inclusion of powers up to the sixth order we will also be able to describe first order phase transitions. The gradient term $(\nabla P)^2$ in Eq. (6) penalizes spatial inhomogeneities. It will become important when we consider ferroelectric domains. At present we will neglect this term and replace the functional of Eq. (6) by a polynomial.

$$F(P,T) = \frac{1}{2} g_2 P^2 + \frac{1}{4} g_4 P^4 + \frac{1}{6} g_6 P^6 - \frac{1}{2} PE. \tag{7}$$

The coefficient g_6 needs to be larger than zero because otherwise the free energy would approach minus infinity for large P. All coefficients depend on the temperature and in particular the coefficient g_2 may be approximated by

$$g_2 = C^{-1}(T - \Theta). \tag{8}$$

Eq. (8) is a result of a temperature expansion around Θ where the **Curie temperature** Θ is equal to or less than the **phase transition temperature** T_C. The Landau-Ginzburg theory leads to power law expressions for the thermodynamic quantities close to the phase transition. The exponents are called critical exponents and assume values which are independent of the system under consideration. Such critical exponents are characteristics of phase transitions in general. Depending on the universality class of the statistical mechanical ensemble they assume values independent of the particular system.

In the following we will consider the thermodynamic states with the conjugated field $E = 0$. Stable states are characterized by minima of the free energy

$$\frac{\partial F}{\partial P} = 0 = P\left(g_2 + g_4 P^2 + g_6 P^4\right) \quad (9)$$

$$\frac{\partial^2 F}{\partial P^2} = \chi^{-1} = g_2 + 3g_4 P^2 + 5g_6 P^4 > 0. \quad (10)$$

Eq. (9) and (10) are the necessary and sufficient conditions for a minimum of F. Eq. (9) is solved by $P = 0$, the condition of the paraelectric phase. Further solutions exists for $P_s > 0$. These are the ferroelectric phases.

3.1.1 Paraelectric Phase

In this case we have $P = 0$. Inserting Eq. (8) into Eq. (10) we immediately see that above T_C the coefficient g_2 needs to be larger than zero in order to obtain stable solutions. A comparison of Eq. (8) and (10) shows that g_2 is expressed by the susceptibility χ, for which a **Curie-Weiss law** is found. In the language of the critical exponents close to T_C the susceptibility follows a power law in $(T - \Theta)$. For the critical exponent γ, the Landau-Ginzburg theory gives $\gamma = 1$. In this derivation we have assumed that the temperature dependencies of g_4 and g_6 are comparatively small close to T_C:

$$\chi(T) = \frac{C}{T - \Theta} \propto (T - \Theta)^{-\gamma} \; ; \; \gamma = 1. \quad (11)$$

3.1.2 Ferroelectric Phases – Second-Order Phase Transition

We now consider the second-order phase transition to the ferroelectric state. In this case we have to take $g_4 > 0$ and to neglect the coefficient g_6. We have

$$\frac{\partial F}{\partial P} = P\left(C^{-1}(T - \Theta) + g_4 P^2\right) \quad (12)$$

with the solutions

$$P = 0 \quad \text{or} \quad P_s^2 = -\frac{(T - \Theta)}{g_4 C}. \quad (13)$$

For $T < \Theta$, a spontaneous polarization exists. The Curie temperature Θ is equal to the phase transition temperature T_C:

$$P_s = \left(\frac{1}{C g_4}\right)^{1/2} (T_C - T)^{1/2} \propto (T_C - T)^{\beta}. \quad (14)$$

The order parameter, namely the spontaneous polarization, depends with a square root law on the distance from the phase transition. In the language of the critical exponents the Landau-Ginzburg theory again predicts a power law now for the order parameter. The critical exponent β assumes the value ½.

Figure 9 schematically displays the free energy close to the second order phase transition for different temperatures as a function of the order parameter P_S. For $T > T_C$ a minimum is found for $P_S = 0$. At $T = T_C$, this minimum shifts continuously to final values of the polarization.

In [8] the authors report about the spontaneous polarization in **triglycine sulfate** (TGS) and show that the predicted square root law is well fulfilled up to 10 K below the phase transition.

Inserting Eq. (14) into Eq. (10) we obtain the temperature dependence of the susceptibility below the phase transition temperature:

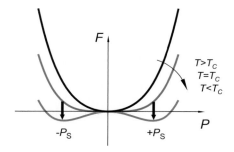

Figure 9: Free energy of a ferroelectric with a second-order phase transition. Below T_C, the minimum of the free energies continuously shifts towards finite values of P_s.

$$\chi^{-1}_{T<T_c} = 2\,\frac{(T_C - T)}{C}. \qquad (15)$$

Comparing Eq. (15) with Eq. (11) we realize that the prefactor of the susceptibility changes at a phase transition by a factor of two. Figure 10 presents the critical behavior of the inverse susceptibility in triglycine sulfate. Below and above the phase transition the experiments reveal an exponent $\gamma = 1$. At the same time at T_C the slope changes by a factor of two. It is also important to note here that the susceptibilities calculated in this manner are isothermal susceptibilities [1], while typical measurements of the susceptibility using a capacitance bridge under small alternating currents are adiabatic measurements. As a result, the predicted ratio in slopes is not always observed experimentally.

The contribution of the polarization to the **entropy** of the ferroelectric may be found by the derivative of the free energy, Eq. (9), with respect to the temperature at constant polarization.

$$S = -\left(\frac{\partial F}{\partial T}\right)_p = \begin{cases} 0 & T > T_C \\ \dfrac{T - T_C}{2C^2\, g_4} & T < T_C \end{cases}. \qquad (16)$$

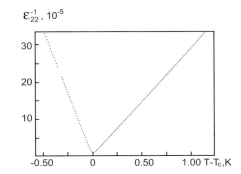

Figure 10: Inverse dielectric constant for TGS as a function of temperature close to the phase transition ($T_C = 49.42 \pm 0.05$C).

The polarization-induced contribution to the entropy decreases continuously towards zero if one approaches the phase transition from below – there is no latent heat for a second-order phase transition. The **specific heat** is obtained by the derivative of the entropy with respect to the temperature. This gives

$$\begin{aligned} C_p &= T\frac{\partial S}{\partial T} = T\,\frac{1}{2C^2\,g_4} & T < T_C \\ C_p &= 0 & T > T_C \end{aligned} \qquad (17)$$

at the phase transition the specific heat jumps as is shown experimentally for triglycine sulphate in [8] by

$$\Delta C_p = T_C/2\,C^2 g_4. \qquad (18)$$

3.1.3 Ferroelectric Phases: First Order Transition

We will now discuss first order transitions in ferroelectric systems. For that purpose in the free energy, Eq. (6), we have to choose $g_4 < 0$ and $g_6 > 0$. The stable states again will be defined by Eq. (9) with solutions $P = 0$ or

$$P_s^2 = \left\{|g_4| + \sqrt{g_4^2 - 4C^{-1}(T - \Theta)\,g_6}\right\}/2g_6. \qquad (19)$$

The positive sign and the bracket are required in order to obtain a stable solution. We now consider the temperature behavior of the free energy at the value of the spontaneous polarization:

(i) for $P = 0$ we have $F = 0$.
(ii) for P_s from Eq. (19) we obtain after insertion into Eq. (6)

$$F[P_s, T] = \frac{1}{24\,g_6^2}\left\{|g_4|\left(6g_2 g_6 - g_4^2\right) - \left(g_4^2 - 4g_2 g_6\right)^{3/2}\right\}. \qquad (20)$$

The free energy becomes zero for

$$g_2 = \frac{3 g_4^2}{16 g_6} = \frac{(T_C - \Theta)}{C}. \qquad (21)$$

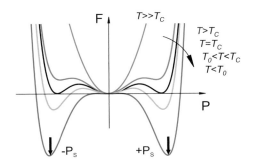

Figure 11: Schematic presentation of the free energy as a function of polarization for a ferroelectric with a first order phase transition as a function of temperature.

Figure 11 schematically displays the free energy as a function of polarization for some relevant temperatures. Above a temperature T_1 the free energy assumes a parabolic shape with a minimum corresponding to a stable paraelectric phase. During cooling, secondary minima at finite polarizations become visible. Their energy level at the beginning, however, is higher than that at $P = 0$. In this regime the paraelectric phase is stable and the ferroelectric phase metastable. Lowering the temperature further, at $T = T_C$ we reach the situation where all three minima of the free energy are at the same level. For the temperatures below T_C, F becomes negative and favors a finite spontaneous polarization. In the temperature regime between T_C and Θ the paraelectric phase coexists with the ferroelectric phase with the paraelectric phase being metastable. Somewhere during cooling through this regime, the first order phase transition to the ferroelectric state will occur with a corresponding jump of the spontaneous polarization from zero to a finite value.

We now consider the susceptibility. Following Eq. (10) for $T > T_C$ again a Curie-Weiss law is found with an apparent critical temperature Θ which does not coincide with the first order phase transition temperature T_C. The ratio of the susceptibilities below and above T_C now assumes a value of four. At T_C the susceptibility jumps by $9/16\, g_4^2/g_6$ Other than in Eq. (16) at T_C a jump of the **entropy** takes place.

$$\Delta S = \left(\frac{\partial F}{\partial T}\right)_- - \left(\frac{\partial F}{\partial T}\right)_+ = -\frac{3}{8}\frac{g_4}{g_6 C} \quad (22)$$

which is connected to the **latent heat** by $\Delta W = T_C\, \Delta S$. Subscript "–" denotes the $T < T_C$ regime, subscript "+" the $T > T_C$ regime. Figure 12 schematically displays the susceptibilities calculated in terms of the Landau-Ginzburg theory close to the phase transition. A comparison of Eq. (7) and the susceptibilities for barium titanate reveals a good qualitative agreement.

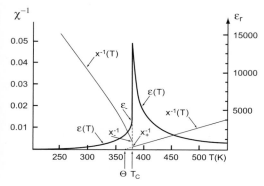

Figure 12: Schematic presentation of the prediction from the Landau-Ginzburg theory, for the susceptibility or the dielectric constant for a system with a first-order phase transition.

3.2 Soft Mode Approach of Displacive Phase Transition

The existence of the local electric field in ionic crystals leads to a splitting of the optical vibration modes, see Chapter 1. It was shown that the polarization fields act in different ways on the longitudinal and transverse modes of the vibrating system. The **longitudinal mode** frequency is shifted to higher frequencies while the **transverse mode** is softened. If we look at the zone center, we realize the large splitting of the longitudinal and transverse optic modes. In the case of the longitudinal optical mode the polarization field enhances the mechanical restoring force. The **softening** (*mode softening*) of the transverse modes originates from a partial compensation of the short-range lattice (elastic) forces on the one hand and the long-range electric fields on the other hand. This effect is strongest at the zone center. We will now look at the consequences of this for the transition into the ferroelectric phase.

If the compensation effect between the elastic and the electric forces is complete, then the transverse optic mode frequency of Chapter 1, Eq. (73) becomes zero, caused by a decrease of temperature,

$$\omega_{TO}^2 := \omega_0^2 - \frac{1}{3}\omega_p^2 \cdot \frac{\chi_\infty + 3}{3} \rightarrow 0 \quad (23)$$

and the **soft phonon** condenses out so that a phase transition to a state with spontaneous polarization takes place.

In order to make the mechanism clearer we should consider the optical modes again. In Figure 13 a and b the orientation of the polarization is with respect to the wavefronts of a LO and TO wave for $q \ll \pi/a$. The polarization is only homogeneous in planes whose thickness is small compared to $\lambda/2$. At the zone center ($q \rightarrow 0$) the region of homogeneous polarization becomes infinite, i.e. $\lambda \rightarrow \infty$, see Figure 13 c. In the case of the softening of the TO mode the transverse frequency becomes zero and no vibration exists anymore („frozen in").

Figure 13: Direction of the electric polarization with respect to the wave fronts of a TO wave as well as of an LO wave. The polarization is only homogeneous in planes the thickness of which is small compared to q.

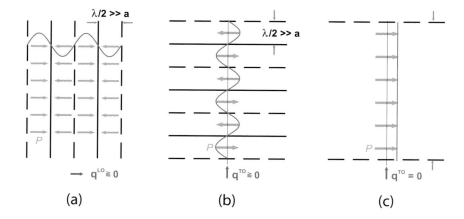

(a)　　　　　　(b)　　　　　　(c)

This soft mode phase transition can be studied by neutron scattering, where the phonon dispersion can be studied as a function of reciprocal lattice vector q. Figure 14 displays results for the phonon dispersion in SrTiO$_3$. For small reciprocal lattice vectors q the transverse optic phonon softens significantly at the zone center. Furthermore, such a softening is also observed for the transverse acoustic mode at the zone boundary in (111) direction.

Figure 15 displays ω_{TO}^2 at the zone center as a function of temperature. A linear relation is found suggesting that the temperature dependence of the optic mode frequency relates to the phase transition. In accordance with the **Lyddane-Sachs-Teller relation**, see Chapter 1, Sec.. 4.3.4, ω_T^2 relates directly to the dielectric constant and thereby to the susceptibility. From the extrapolation according to Eq. (11) a phase transition at $T_C = 50$ K would be expected. This phase transition, however, does not really take place. It is dominated by a competing displacive phase transition at the zone boundary. The softening of the acoustic zone boundary phonon can be read off from the dispersion relations in the (111) direction.

4 Ferroelectric Materials

Table 1 presents a number of typical ferroelectric crystals. According to Landoldt and Börnstein [9], [10], there exist about 600 ferro- and antiferroelectric materials. The crystals in Table 1 may be divided into three different groups. Group I comprises **hydrogen bonded systems** like KDP. In such systems ferroelectricity is created by a preferential occupation of the hydrogen sites within the hydrogen bond. The associated dipole moment creates the polarization. In the second class, the **ionic crystals** are found. Among them the perovskites, such as barium titanate are the most important group. Finally **narrow gap semiconductors**, like GeTe, may exhibit ferroelectric properties.

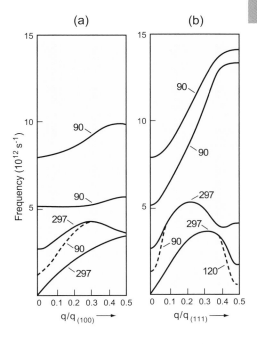

Figure 14: Phonon dispersion relation for strontium titanate.
(a) (100) direction
(b) (111) direction. Phonons soften not only at the zone center but for the (111) direction also at the zone boundary.

Material	Chemical formula	T_c [°C]
Barium titanate	BaTiO$_3$	120
Lead titanate	PbTiO$_3$	~ 490
Lead zirconate titanate, PZT (at the MPB)	Pb(Zr$_{0.52}$Ti$_{0.48}$)O$_3$	~ 370
Lithium niobate	LiNbO$_3$	1210
Strontium bismuth tantalate	SrBi$_2$Ta$_2$O$_9$	570
Yttrium manganate	YMnO$_3$	~ 640
Potassium dihydrogen phosphate (KDP)	KH$_2$PO$_4$	− 150
Ammonium fluoberyllate	(NH$_4$)$_2$BeF$_4$	− 98
Rochelle salt	KNaC$_4$H$_4$O$_6$ · 4H$_2$O	− 18; 24
Triglycine sulfate (TGS)	(NH$_2$CH$_2$COOH)$_3$ · H$_2$SO$_4$	49
Trisarcosine calcium chloride	(CH$_3$NHCH$_2$COOH)$_3$ · CaCl$_2$	− 146
Sodium nitrite	NaNO$_2$	164
Lead germanate	Pb$_5$Ge$_3$O$_{11}$	180
Germanium tellurium	GeTe	400

Table 1: Selection of ferroelectric materials.

Ferroelectric materials are widely used in applications such as capacitors (e.g. BaTiO$_3$, where the high dielectric constant is utilized), electromechanical transducers (e.g. Pb(Zr$_{1-x}$Ti$_x$)O$_3$, which is attractive for its high piezoelectric coefficients), pyroelectrics (e. g. modified PbTiO$_3$, (Sr,Ba)Nb$_2$O$_6$), and electrooptic components (e.g. LiNbO$_3$). In ferroelectric capacitors for memory applications, see also Chapter 22, Pb(Zr$_{1-x}$Ti$_x$)O$_3$ or bismuth layer structure ferroelectrics as SrBi$_2$Ta$_2$O$_9$ or Bi$_4$Ti$_3$O$_{12}$ are the most promising candidates.

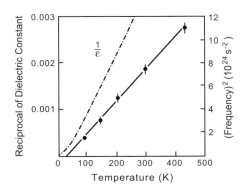

Figure 15: Frequency of the transverse optic zone center mode for strontium titanate as a function of temperature (solid line). Dashed-dotted line inverse dielectric constant.

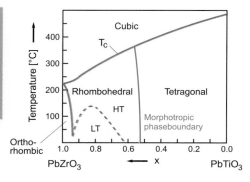

Figure 16: Phase diagram for PZT showing the morphotropic phase boundary between rhombohedral and tetragonal phases after [6].

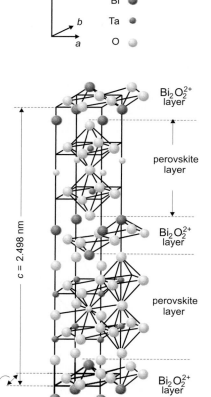

Figure 17: An example of a bismuth layer structure compound (Aurivillius phase). In these materials, perovskite slabs are separated by bismuth oxide layers.

There are a limited number of structures that are adopted by the majority of the commercially important ferroelectric oxides. In each of these structures, the ferroelectricity is tied to distortion of the coordination polyhedra of one or more of the cations in the structure. One example of this was already given in Sec. 1 for the perovskite structure. Cations that seem to be especially susceptible to forming such distorted polyhedra include Ti, Zr, Nb, and Ta. All of these ions lie near crossover points between the stability of different electronic orbitals, and so may be likely to form distorted coordination polyhedra [5]. Polarizable cations such as Pb and Bi are also common to many ferroelectric materials. In this case, it has been suggested that the lone pair electrons may play an important role in stabilizing ferroelectric structures. Thus the ferroelectric transition temperature and spontaneous distortion of $PbTiO_3$ is much larger than that of $BaTiO_3$.

One of the most important ferroelectric and piezoelectric materials is the **$PbTiO_3$-$PbZrO_3$** (PZT) solid solution. Over the entire solid solution range, PZT adopts distorted versions of the perovskite structure, as shown in Figure 16. At the so-called **morphotropic phase boundary** (MPB), the tetragonal phase and the rhombohedral phase are both observed. This leads to the presence of 14 polarization directions (6 along the <001> family from tetragonal phase, and 8 along <111> for the rhombohedral polytype which are equivalent, or nearly equivalent, in free energy. The result is that the morphotropic phase boundary corresponds to a highly polarizable state. Thus, it is reasonable that the dielectric and piezoelectric properties show strong maxima near this composition. Recently, it has been suggested that at low temperatures, a monoclinic phase is stabilized near the MPB [11]. It is unknown whether such a phase also contributes to the enhanced polarizability near the MPB. Whatever the source, the high **piezoelectric coefficients** (i.e. d_{33} values of ~ 250 to 400 pC/N, depending on the doping), coupled with a high transition temperature are the main reason that PZT ceramics are so widely used as piezoelectric sensors and actuators. The orthorhombic phase near the $PbZrO_3$ side of the PZT phase diagram corresponds to an antiferroelectric distortion of the perovskite structure, in which the polarization is cancelled on a unit cell level. As a result, the material has no spontaneous polarization at low applied electric fields, although the ferroelectric phase can be induced by large applied electric fields.

For ferroelectric memory applications utilizing PZT, most researchers to date have concentrated on $PbTiO_3$-rich compositions. This is due in part to the large spontaneous polarization available, and partly to the fact that the hysteresis loops of <111> oriented films are more rectangular for tetragonal than for rhombohedral compositions. The <111> orientation is promoted by the Pt bottom electrodes widely used in these applications. This typically results in remanent polarizations in excess of 30 $\mu C/cm^2$. As a result, low density memories based on PZT are now in commercial production, and there is a good prospect for scaling to high densities, given the large available spontaneous polarization.

An alternative structure that has also been widely investigated both for high temperature piezoelectric, as well as for ferroelectric memory applications is the **bismuth layer structure** family. As shown in Figure 17 for the case of **$SrBi_2Ta_2O_9$**, (SBT) the structure can be envisaged as layers of corner-connected octahedrally coordinated ions, separated by $Bi_2O_2^{2+}$ layers. The layers with the octahedrally coordinated cations look like slabs of the perovskite structure. As a function of stoichiometry, it is possible to stabilize bismuth layer structure compounds with perovskite blocks of different thicknesses. It has been shown [12] that when the perovskite block is an even number of octahedra thick, the symmetry imposes a restriction on the polarization direction, confining it to the $a - b$ plane. In contrast, when the perovskite block is an odd number of octahedra thick, it is possible to develop a component of the polarization along the c axis (nearly perpendicular to the layers). This has important implications for the use of bismuth layer structure compounds in ferroelectric memories; since there are comparatively few allowed directions for the spontaneous polarization, the remanent polarization is rather small for many film orientations as shown in Figure 18. As a result, memories based on SBT may be inherently more difficult to scale to high densities unless a method for producing well-oriented films is instituted.

It is also important to realize that **thin films** may differ in some substantial ways from bulk ceramics or single crystals of the same composition. One source of these differences is the substantial in-plane stresses that thin films are typically under, ranging from MPa to GPa [13]. Because many ferroelectric materials are also ferroelastic, imposed stresses can markedly affect the stability of the ferroelectric phase, as well as the ease with which polarization can be reoriented in some directions. Figure 19 shows an example of this, where the **phase diagram** for tetragonal $BaTiO_3$ is re-calculated for the case of a thin film without misfit dislocations [14], [15]. It is clear there that the phase diagram is considerably complicated by the presence of a dissimilar substrate.

5 Ferroelectric Domains

5.1 Origin of Domains

As in ferromagnetic materials, ferroelectrics form domain structures. A domain is a region where there is a uniform direction for the spontaneous polarization. In the case of tetragonal $BaTiO_3$, when the material undergoes ferroelectric transition, the Ti^{4+} ion displaces towards one of the neighboring oxygens. Because the Ti^{4+} is octahedrally coordinated, there were 6 possible directions for the Ti to move (for example: up, down, forward, backward, left, and right). Consequently, there are six possible domain states for tetragonal $BaTiO_3$. Similarly, in other systems, the allowed **domain states** are governed by the crystallography of the system and by the symmetry elements lost on transforming from the prototype state.

The boundaries between domains are referred to as **domain walls**. Domain walls in ferroelectrics are typically quite thin (~1 – 10 lattice parameters across) and so can be regarded as abrupt changes in the polarization direction. Domain walls are characterized by the angle between the polarization directions on either side of the wall. Thus a 180° domain wall demarks a boundary between antiparallel domains, while a 90° wall in tetragonal $BaTiO_3$ would be formed at the boundary between domains pointed "up" and "left", for example. The allowed angles for domain walls depend on the orientations of the spontaneous polarization allowed by symmetry. Thus, in rhombohedrally distorted perovskites, there are no 90° domain walls, but instead 71° and 109° walls. A more complete picture of the way the polarization changes as a domain wall is crossed is given by Cao and Cross [16].

Domain walls typically appear along specific crystallographic planes that correspond to conditions of mechanical compatibility [17]. That is, as you move along the domain wall, stable domain walls should not develop large strains. There are relatively few such orientations. For example, again in tetragonal $BaTiO_3$, the {101} family of domain walls corresponds to cutting the unit cell along a face diagonal, and then reassembling the crystal after rotating one piece ~90°. A second constraint on the angles between domains corresponds to the fact that in a highly insulating material, it is not energetically favorable to arrange domains in a head-to-head configuration [16]. Such a configuration can be stabilized, however, by the accumulation of a compensating charge at such a domain wall.

For a fully compensated **ferroelectric single crystal** (i.e. one which is electroded, or which has picked up sufficient surface charges from some other source to compensate the polarization) of a suitable orientation, a single domain state is the lowest free energy. However, in most other cases, domain formation will be driven by either the electrical or mechanical boundary conditions. For instance, at the surface of an uncompensated ferroelectric, the divergence of the polarization results in the appearance of a depolarizing electric field [3]. The energy associated with this can electrically drive domain forma-

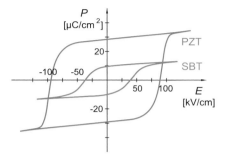

Figure 18: Hysteresis loops of ferroelectric thin films on Pt/Ti/Si wafer with a film thickness of 140 nm. The PZT (30/70) film is textured in (111) and was prepared by CSD. The SBT film is without any texture and was prepared by MOD. The figure indicates a trend. The exact values of P_r and E_c depend not on the ferroelectric material but also on the substrate, electrode, microstructure and texture (see also Chapter 22).

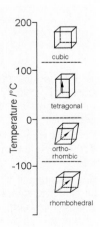 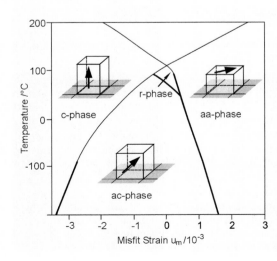

Figure 19: Phase diagram for a (001) $BaTiO_3$ epitaxial film as a function of misfit strain [14]. The notations refer to the polarization direction preferred by a single domain film. The r phase has components of the polarization in all three directions.

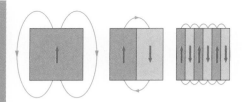

Figure 20: Domain formation driven by reducing the electrical depolarization energy.

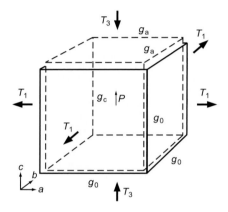

Figure 21: Spontaneous tetragonal deformation of a cubic grain. The cube edges correspond to the crystallographic axis.

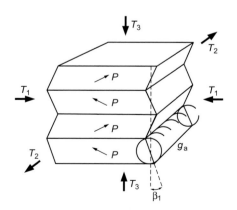

Figure 22: The spontaneous deformation is reduced by twinning. Serration appears at the grain boundary. Homogeneous stresses T as indicated can restore the gross cubic shape.

tion (see Figure 20). Similarly, in ferroelectrics which are ferroelastic as well, mechanical stresses can also affect the equilibrium domain formation as described in Section 5. It should be noted, however, that it is energetically costly to put a domain wall into a system, since the domain wall, like any other surface, has an associated surface free energy. As a result, the material will not continue to subdivide into smaller and smaller domains. Instead, a balance is reached between the energy required to create the wall and the energy gain from the reduction in energy. This energy balance will also depend on the grain size of the system [18], so the equilibrium domain size typically drops as the grain size is reduced.

It is interesting to consider first how domain configurations are determined in bulk ferroelectrics, and then to consider how these are modulated in thin films.

5.2 Static Domain Configurations in Bulk Systems

In general domain structures in equilibrium are formed such as to minimize the total energy in the crystal, including electric and mechanical strain fields.

$$W_{\text{tot}} = W_M + W_E + W_{DW} + W_S = \min \quad (24)$$

where W_M denotes the **elastic**, W_E the **electric**, W_{DW} the **domain wall** and W_S the **surface energy**.

Thus, the equilibrium domain structure of a ferroelectric should depend on both the electrical and mechanical boundary conditions imposed. As was described in the previous section, for uncompensated domains, electrostatic considerations can drive the formation of antiparallel domains in many ferroelectrics. In addition, it is also important to recall that ferroelectric materials typically develop a spontaneous strain (deformation) as a result of the appearance of the spontaneous polarization. For example, each barium titanate unit cell deforms from a cubic to a tetragonal shape on cooling below the Curie temperature. Figure 21 displays the tetragonal distortion of the barium titanate crystal at the transition.

If the material is completely clamped, then compressive stresses T_3 and tensile stresses T_1 would be necessary in order to keep the material in its original shape. These clamping stresses may be calculated on the basis of the strain tensor, describing the tetragonal distortion

$$S_0 = \begin{pmatrix} S_a & & \\ & S_a & \\ & & S_c \end{pmatrix}. \quad (25)$$

With $S_a = (a-a_0)/a_0$, $S_c = (c-a_0)/a_0$ and $a_0 = (a^2 c)^{1/3}$. The clamping stresses follow from Hooke's law

$$T_i = \sum c_{ij} S_j. \quad (26)$$

where c_{ij} are the elastic constants. The stresses $S_{1,2} = -S_a + S_x$ and $S_3 = -S_c + S_x$ have to be chosen such as to minimize the elastic energy:

$$W = \frac{c_{11}}{2}\left(S_1^2 + S_2^2\right) + c_{12} S_1 S_2 + c_{13}\left(S_1 S_3 + S_2 S_3\right) + \frac{c_{33}}{2} S_3^2. \quad (27)$$

With the elastic constants and the tetragonal distortions for barium titanate $W = 2.08 \times 10^6$ J/m^3 evolves. For the tensile and compressive stresses $T_1 = 190 f_{11}$ MPa and $T_3 = -380 f_{31}$ MPa are found. The coefficients f_{11} and f_{31} account for possible elastic depolarization effects, which come about if the environment also deforms.

Another possible way to reshape the tetragonally distorted cube is a shear in the (110) direction combined with a longitudinal deformation in the (100) direction. In general a grain can reduce its energy by twinning as shown in Figure 22.

Given the electrical and mechanical driving forces for forming domains, it is clear that there is an overall reduction in the free energy of many ferroelectric samples as domain formation proceeds. As mentioned above, however, there is also an energy cost associated with forming the domain wall. Thus, the system will typically drive to an equilibrium domain width. It can be shown that the thickness of the twin should be proportional to the square root of the size of the grain.

The situation becomes even more complicated when a **polycrystalline** ensemble is made. Now, each region of the material is not free to deform, but is constrained by the material around it. That is, any deformation will be counteracted by clamping effects of the environment and will cause high internal stresses. To minimize this, complicated domain patterns form.

Figure 23a displays a typical domain pattern for grains in a polycrystalline ceramic. The stripes which are visible under a polarizing microscope indicate the twin structures of the domains within the grain. Figure 23b displays the same grain under free surface conditions. Under these conditions, stresses may relax and a completely different domain structure evolves.

5.3 Static Domain Configurations in Thin Films

Because the equilibrium domain configuration depends explicitly on the mechanical and electrical boundary conditions for the ferroelectric, it is not surprising that domain configurations in thin films (and even the thermodynamic stability of the ferroelectric phase) can change as one moves from a bulk to a thin film sample [20], [21]. A good example of this is given in the work of Pertsev et al. where revised phase diagrams for $BaTiO_3$ and $PbTiO_3$ epitaxial films were derived (see Figure 19) as a function of the **misfit strain** between the substrate and the film [14]. Among the consequences were changes in the order of the phase transition (from 1st to 2nd order), stabilization of the rhombohedral phase to unusually high temperatures, and constraints on the allowed domains. Many of these same ideas are observed even in polycrystalline films.

Thus, PZT and $BaTiO_3$ films that are under considerable **tensile stress** on cooling through the phase transition temperature typically have the polarization tilted substantially into the film plane, while films under compressive stress show large out-of-plane polarizations. Many perovskite films deposited on Si substrates are under appreciable levels of in-plane tensile stress, given that the thermal expansion coefficients of the ferroelectric phase exceed those of Si, at least above the Curie temperature.

In Figure 24 possible domain patterns of different textures of tetragonal films of $PbZr_{0.52}Ti_{0.48}O_3$ are depicted. In the case of **compressive stress**, predominantly (001) orientation, 90° as well as 180° domains are expected. Such orientation could be realized by deposition of tetragonal PZT on magnesium oxide substrates [22]. Under the influence of an electric field the number of 180° domains is decreased. The resulting pattern predominantly consists in 90° domains. Tensile stress, predominately (100) orientation, is achieved by using a buffer layer of Yttrium-stabilized Zirconium and an oxide electrode of Lanthanum Strontium Cobaltate or by depositing on a (100)-$SrTiO_3$-substrate with a $SrRuO_3$ electrode [23]. The change of the domain structure by poling is similar to the (001) orientation but, the a-axis orientation is still preferred.

In standard systems for ferroelectric thin films, e.g. PZT with platinum electrodes on oxidized silicon wafers, the orientation of the crystallographic axes of PZT is in the (111) direction. Tuttle et al. have also shown that for many thin films the ferroelastic domain structure developed on cooling from the prototype phase is largely retained at lower temperatures, so that switching on non-180° domains is limited [24]. Poling should evoke a new multi-domain state as shown while the "head-to-tail" configuration is required. This may well limit the magnitude of the switchable polarization perpendicular to the film plane, and so is an important consideration in ferroelectric memory films.

The three-dimensional piezoresponse force microscope (PFM, see Chapter 12) enables the visualization of the domains in thin films and their polarization directions. In epitaxial PZT thin films which were grown on a (001) single crystalline $SrTiO_3$ substrate

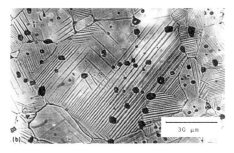

Figure 23: Representative barium titanate domain patterns of a grain.
(top) When the pattern is formed inside the ceramic body with three-dimensional clamping, (bottom) the same grain when the pattern is formed under free surface conditions [19].

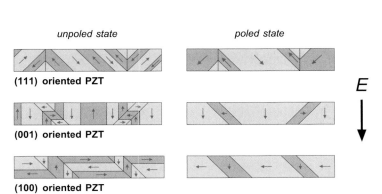

Figure 24: Domain structures of tetragonal PZT with different orientations (see text).

I Fundamentals

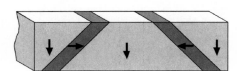

Figure 25: Out-of-plane and in-plane piezoresponse images of as-grown epitaxial PZT. The contrast corresponds to different orientations of polarizations in the domains.

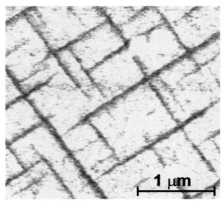

Out-of-plane signal   In-plane signal

coated with $La_{0.5}Sr_{0.5}CoO_3$ oxide layer a **self polarization** mechanism was found. Detailed investigation show that the out-of-plane polarization in c domains points preferentially towards the bottom electrode and a resulting remanent polarization exists without applying an external field [25]. The domain configuration is always of the "head-to-tail" type, see Figure 25.

The influence of **clamping** in thin ferroelectric films becomes clear when dense films are compared with single separated grains. Applying the PFM technique to lead titanate (PT) films it was found that the PT grains in dense films contain laminar 90° domain walls, whereas separated PT grains show more complicated structures of mainly 180° domain walls. For grains smaller than 20 nm, no piezoresponse was observed, which could be due to the transition from the ferroelectric to the superparaelectric phase that has no spontaneous polarization (Figure 26). When the thickness of a dense epitaxial thin film is reduced, the piezoelectric activity is observed down to 4 nm [27]. The difference can be explained if one takes into account that ferroelectricity is a collective phenomenon. The one-dimensional shrinking (as in dense films) leads to different stresses and to varying stabilities of the ferroelectric phase compared with the three-dimensional phase (as in single grains).

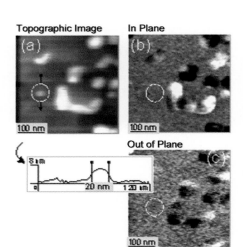

Figure 26: The topographic image
(a) shows eleven PTO grains of sizes from 100 nm down to 20 nm indicated by the circles. In the linescan over the grain denoted by an arrow, shown at the bottom, the size of the grains can be determined. In the PFM images
(b) in-plane and
(c) out-of-plane piezoresponse of the grain of the size of 20 nm is not visible leading to the assumption they do not have any permanent polarization [26].

5.4 Dielectric Small Signal Behavior

Under sufficiently small electric fields all dielectrics follow a linear relation, described by $D = \varepsilon E$. This dielectric **small signal response** is caused by reversible contributions of the electronic and ionic polarization processes (intrinsic), as shown in Chapter 1. In ferroelectric polycrystalline materials additional extrinsic mechanisms exist due to the **movement of domain walls** and the alignment of defects.

$$\varepsilon_{tot} = \varepsilon_{intrinsic} + \varepsilon_{extrinsic} \tag{28}$$

It has been shown that especially the electromechanically active non-180° domain walls, i. e. 90° domain walls in tetragonal structures, 71° and 109° domain walls in rhombohedral structures and so on, are responsible for a considerable contribution to the **dielectric coefficient** as well as to the piezoelectric and the elastic coefficients. The shift of the wall is clearly displayed in Figure 27 whereby favorably oriented domains with respect to the applied field grow at the expense of unfavorably oriented domains. This leads to a change of the electric and the elastic dipole moment (Δp and $\Delta \nu$) and, therefore, to contributions to the dielectric and piezoelectric response. In the coordinate system of a tetragonal domain twin the change is given by

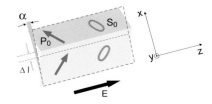

Figure 27: Shift of a 90° domain wall Δl changes the electrical and the elastical dipole moments causing a shear of the domain twin.

$$\Delta p = \sqrt{2}\, P_0\, A\, \Delta l \begin{pmatrix} 1 \\ 0 \\ 0 \end{pmatrix} \tag{29}$$

$$\Delta \nu = S_0\, A\, \Delta l \begin{pmatrix} 0 & 0 & 1 \\ 0 & 0 & 0 \\ 1 & 0 & 0 \end{pmatrix}. \tag{30}$$

A separation of the intrinsic polarization process from that of the non-180° domain wall is possible by dielectric high frequency measurements in the GHz range. There, these kinds of domain walls do not contribute to ε_{total}. The mechanism is caused by the fact that a vibrating non-180° wall acts as an emitter of elastic shear waves (see Figure 27) propagating at the shear wave velocity c_{Sh} through the crystallite. When the frequency of the applied electric field corresponds to c_{Sh} the vibration of the domain wall is suppressed and a strong dielectric relaxation is observed, as shown in Figure 28.

In ferroelectric thin films, the relaxation step is not found, see Figure 28. This is further evidence that non-180° domain wall processes in ferroelectric thin films are strongly limited, as mentioned in Sec. 5.3. Very recent results show that ferroelastic non-180° domain walls become able to move again, when the clamping effect is significantly reduced by patterning the ferroelectric film into discrete *islands* using a focused ion beam [28].

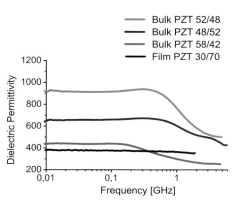

Figure 28: Dielectric high-frequency behavior of PZT bulk ceramic as well as thin films on silicon substrate.

5.5 Reversible and Irreversible Polarization Contributions

When the electric field is increased, the dielectric behavior of ferroelectric material changes from linear to non-linear and hysteretic. This is especially important when the ferroelectric becomes very thin as in the case of multilayer capacitors or integrated thin films capacitors. At a film thickness of 100 nm, operation voltages even below 1 V exceed the coercive voltage.

The origin of the ferroelectric hysteresis is the existence of irreversible polarization processes, i.e. the irreversible polarization reversal of a single ferroelectric lattice cell as explained by the Landau-Ginzburg theory, see Sec. 3.1. However, the exact interplay between this fundamental process, domain walls, defects and the overall appearance of the ferroelectric hysteresis is still not precisely known. In addition, the above mentioned reversible processes in ferroelectrics have to be taken into account.

The separation of the total polarization into reversible and irreversible contributions that has long been appreciated in the study of ferromagnetic materials [29] might facilitate the understanding of ferroelectric polarization mechanisms. These two contributions are of importance for the design of external circuits of FeRAMs. Especially, the irreversible processes are decisive, since the reversible processes cannot be used to store information.

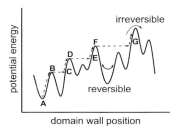

There are two major mechanisms possible for **irreversible processes**. First, lattice defects which interact with a domain wall and prevent it from returning into its initial position after removing the electric field that initiated the domain wall motion ("pinning") [30]. Second, the nucleation and growth of new domains which do not disappear after the field is removed again. In ferroelectric materials the matter is further complicated by defect dipoles and free charges that also contribute to the measured polarization and can also interact with domain walls [31].

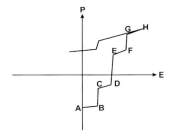

The motion of the domain wall under an external electric field takes place in a statistical potential generated by their interaction with the lattice, point defects, dislocations, and neighboring walls. **Reversible movement** of the wall is regarded as a small displacement around a local minimum. When the driven field is high enough, irreversible jumps above the potential barrier into a neighboring local minimum occur (see Figure 29). Based on these assumptions the measurement of the large signal ferroelectric hysteresis with additional measurements of the small signal capacitance at different bias voltages are interpreted in terms of reversible and irreversible parts of the polarization. As shown in Figure 30, the separation is done by substracting the reversible part from the total polarization, i. e. the integrated $C(V)$-curve [32]:

Figure 29: Motion of a domain wall in the statistical potential (top) and correlated hysteresis loop (bottom).

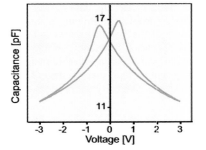 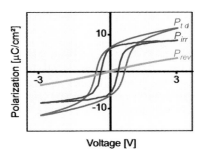

Figure 30: $C(V)$ - curve (left) and reversible and irreversible contribution to the polarization (right) of a ferroelectric SBT thin film.

$$P_{irr}(V) = P_{tot}(V) - \frac{1}{A}\int_0^V C(V')dV'. \tag{31}$$

Typical hysteresis loops are dynamically recorded at certain frequencies. If slow reversible polarization mechanisms also contribute to the total polarization P_{tot} the shape of the hysteresis loop, especially the coercive field, becomes frequency-dependent (see also Sec. 3.3). In order to overcome this influence, the measurement should be performed with the lowest frequency possible, i. e. quasi-statically [33].

5.6 Switching for Ferroelectric Domains

Since many applications of ferroelectric materials require either that the material be poled once, as in a piezoelectric device, or that it be switched repeatedly, as in a memory, it is interesting to consider how domain reversal proceeds. As described by Merz, switching takes place by processes of domain nucleation and domain wall motion. Apparent sideways motion of domain walls typically occurs by nucleating a step on the domain wall. This protrusion then rapidly grows along the length of the wall, effectively increasing the size of the more favorably oriented domain. This differs from the continuous sideways motion of domain walls observed in many ferromagnetic systems.

One way to monitor the **switching process** is to measure the current flow through the ferroelectric as a function of time. When an electric field of the same polarity is applied to a fully poled single crystal, there is an initial current flow that corresponds to charging the capacitor. If an opposite polarity field is now applied, in addition to the charging current, additional current flows as a consequence of domain reorientation. Subtracting these two signals enables the current flow associated with the switching process itself to be isolated. Alternatively, by appropriately adjusting the RC time constant of the circuit, it may be possible to separate the charging and switching components of the charge flow.

If measurements are made under conditions where the electric field is constant during the reversal process, then the maximum displacement current, i_{max}, and the switching time, t_s, are given by [34]:

$$i_{max}(E) = i_0 \exp\left(-\frac{\alpha}{E}\right) \tag{32}$$

$$t_S(E) = t_0 \exp\left(+\frac{\alpha}{E}\right) \tag{33}$$

where i_0 and t_0 are constants, E is the applied electric field, and α can be regarded as the activation field for switching, see Figure 31. At very high electric fields, the switching time is better described by:

$$t_S(E) = t_0^* E^{-n}. \tag{34}$$

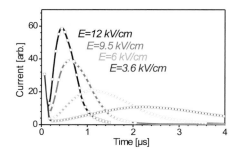

Figure 31: Polarization switching current of TGFB single crystal at different electrical fields.

where n ranges from 1 to 7 for a variety of materials [34]. As expected, the constants that describe the switching process (i.e. α and n) depend on temperature. This is sensible; since domain wall motion is thermally activated, poling and switching will be enhanced as the transition temperature is approached. It is also important to mention that ferroelectrics typically do not have well-defined switching fields, so that the amount of polarization switched depends both on the field amplitude, and the length of time the field is applied.

An alternative method of measuring switching is to monitor the current flow as sinusoidal or triangular waveforms are applied. Fundamentally, this is a measurement of the switching while the full hysteresis loop is traversed. When the field dependence of the switching is properly accounted for, it is possible to model the data by treating the domain reorientation process as a phase transformation problem, following Kolmogorov and Avrami. Ishibashi included the field dependence of the polarization switching and modeled the D-E hysteresis loops depending on the excitation frequency [35]. While the mathematics here are more complicated, the result can provide considerable insight into the switching kinetics. Ishibashi found a good agreement to experimental data of tri-glycine sulfate (TGS) single crystals, as shown in Figure 32, especially the pronounced **frequency dependence** of the coercive field E_C:

$$E_C \propto f^{-\beta}. \tag{35}$$

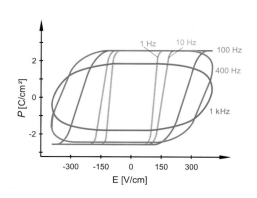

Figure 32: Hysteresis loops at different frequencies of TGS single crystal.

6 Summary

In this chapter we have learned that the ability of a crystal to exhibit spontaneous electrical polarization is related to the point group describing the structure and requires a symmetry group with a polar axis. Crystals without an inversion center, with one exception, exhibit piezoelectricity, where a mechanical strain induces electrical polarization. Crystals which carry spontaneous polarization are pyroelectric, i.e. temperature changes induce a change of the surface charge. Finally, ferroelectric materials are those where the application of an external field may reverse the direction of polarization. Thus, all ferroelectric materials are subsets of the polar crystal classes, and are both piezoelectric and pyroelectric.

Among the different ferroelectrics, the perovskites are the most commercially important. Multiple distortions of the cubic perovskite prototype are known, leading to the possibility of multiple phase transitions as a function of temperature, as in barium titanate, or multiple phases as the composition is changed, as in PZT. When morphotropic phase boundaries are present, peaks in the dielectric and piezoelectric properties are typically observed. Finally, as in the case of magnetism also antiferroelectric order is observed, where adjacent electric dipoles point in opposite directions.

The Landau-Ginzburg theory describes thermodynamic phase transitions in terms of an order parameter expansion of the free energy close to the phase transition temperature. The theory is a good approximation for systems with long-range interactions, as occurs for dipoles interacting via Coulomb forces. Some of the thermodynamic quantities are predicted to follow power law relations on the distance $(T - \Theta)$ from the phase transition. For second-order phase transitions, mean field critical exponents for the susceptibility $\gamma = 1$ and for the order parameter $\beta = ½$ are predicted. Also relations between the prefactors of the susceptibility below and above the transition temperature are calculated. For first-order phase transitions, above the phase transition temperature a Curie-Weiss law for the susceptibility is predicted, and the polarization is discontinuous at the phase transition. The theoretical predictions are in good qualitative agreement with experimental observations on ferroelectrics.

In ferroelectrics, optical lattice vibrations exist where the differently charged sublattices vibrate against each other. In particular for the transverse optical mode at the zone center, the restoring forces for these modes are such that electric Coulomb forces and elastic lattice forces work against each other. As a consequence the transverse optical zone center modes soften. If this compensation effect is complete and Coulomb and elastic forces cancel each other, then the corresponding phonon condenses out and a ferroelectric displacive phase transition takes place. The underlying lattice vibrations may be uniquely accessed by inelastic neutron spectroscopy, where the phonon dispersion relations can be directly measured.

Domains in ferroelectrics are formed in order to minimize the total energy. Thus, the static domain patterns depend on both the electrical and elastic boundary conditions for the system. For any material, the allowed domain variants are controlled by the crystallography of the ferroelectric phase. Domain walls mark the boundaries between domains, and are characterized by the angle between the polarization vectors on either side of the wall. Due to the large anisotropy of ferroelectric crystals, domain walls in general are thin, if we stay away from the phase transition.

The motion of domain walls is strongly influenced by pinning effects. A microscopic description of the motional process yields similarities to nucleation phenomena. The switching time for a domain of given size experiences low and high field regimes exhibiting distinctly different field dependencies.

Acknowledgements

The editor gratefully acknowledges Ulrich Böttger (RWTH Aachen) for his comprehensive editorial and technical assistance in the compilation of this chapter. Furthermore, he would like to thank Peter Schorn (RWTH Aachen) for checking the symbols and formulas in this chapter and Andreas Rüdiger (FZ Jülich) for corrections.

References

[1] F. Jona and G. Shirane, *Ferroelectric Crystals*, New York, Dover Publications, 1993.

[2] B. A. Strukov, A. P. Levanyuk, *Ferroelectric Pheonomena in Crystals*, Springer-Verlag Berlin, Heidelberg, New York, 1998.

[3] M. E. Lines and A. M. Glass, *Principles and Applications of Ferroelectrics and Related Materials*, Clarendon Press, Oxford, 1977.

[4] B. C. Fraser, H. Danner, R. Papinsky, Phys. Rev. 100, 745 (1955).

[5] R. E. Newnham, *Structure-Property Relations*, Springer-Verlag, New York, 1975.

[6] B. Jaffe, W. Cook, and H. Jaffe, *Piezoelectric Ceramics*, Academic Press, London, 1971.

[7] J. F. Nye, *Physical Properties of Crystals*, Clarendon Press, Oxford, 1979.

[8] S. Triebwasser, IBM Jour. Res. Dev. **2**, 617 (1958).

[9] K.-H. Hellwege and A. M. Hellwege, ed., *Complex Perovskite-type Oxides*, Landolt-Bornstein: Oxides, Vol. 16a, Springer-Verlag, 1981.

[10] E. Nakamura and T. Mitsui, ed., *Complex Perovskite-type Oxides*, Landolt-Bornstein: Oxides, Vol. 28, Springer-Verlag, 1990.

[11] B. Noheda, D. E. Cox, G. Shirane, R. Guo, B. Jones, and L. E. Cross, Phys. Rev. B **63**, 014103, (2001).

[12] R. E. Newnham, R. W. Wolfe, and J. F. Dorrian, Mat. Res. Bull. **6**, 1029 (1971).

[13] T. M. Shaw, S. Trolier-McKinstry, and P.C. McIntyre, Annu. Rev. Mater. Sci. **30**, 263 (2000).

[14] N. Pertsev, A. Zembilgotov, A. Tagantsev, Phys. Rev. Lett. **80**, 1988 (1998).

[15] N. Pertsev, A. Zembilgotov, S. Hoffmann, R. Waser, A. Tagantsev, J. Appl. Phys. **85**, 1698 (1997).

[16] W. Cao and L. E. Cross, Phys. Rev. B**44**, 5 (1991).

[17] J. Fousek and V. Janovec, Appl. Phys. Lett. **40**, 135 (1969).

[18] W. Cao and C. Randall, Solid State Communications **86**, 435 (1993).

[19] G. Arlt, J. of Materials Science **25**, 2655 (1990).

[20] G. A. Rossetti, L. E. Cross, K. Kushida, Appl. Phys. Lett. **59**, 2524 (1991).

[21] S. Streiffer et al., J. Appl. Phys. **83**, 2742 (1998).

[22] K. Nashimoto, D. K. Fork, and G. B. Anderson, Appl. Phys. Lett. **66**, 822 (1995).

[23] K. Nagashima, M. Aratani, and H. Funakubo, J. Appl. Phys. **89**, 4517 (2001).

[24] B. A. Tuttle et al., in *Science and Technology of Electroceramic Thin Films,* ed. O Auciello, R Waser, Kluwer Academic Publishers, 1995.

[25] A. Roelofs et al., Appl. Phys. Lett. **80**, 1 (2002).

[26] A. Roelofs, accepted to be published in Appl. Phys. Lett.

[27] T. Tybell, C. H. Ahn, and J. M. Triscone, Appl. Phys. Lett. **75**, 856 (1999).

[28] V. Nagarajan et al., Nature Materials, **2**, 43 (2003).

[29] S. Chikazumi and S.H. Charap, *Physics of Magnetism*, John Wiley & Sons, New York, 1964.

[30] T. J. Yang, V. Gopalan, P. J. Swart, and U. Mohideen, Phys. Rev. Lett. **82**, 4106 (1999).

[31] O. Boser and D.N. Beshers, Mat. Res. Soc. Symp. Proc.**82**, 441 (1987).

[32] O. Lohse, D. Bolten, M. Grossmann, R. Waser, W. Hartner, and G. Schindler, Mater. Res. Proc. **267**, (1998).

[33] D. Bolten, O. Lohse, M. Grossmann, and R. Waser, Ferroelectrics **221**, 251 (1999).

[34] L. E. Cross, in Encyclopedia of Chemical Technology 10, ed. Kirk Othmer, Wiley, 1980.

[35] S. Hashimoto, H. Orihara, and Y. Ishibashi, J. Phys. Soc. Jap. **64**, 1601(1994).

Electronic Properties and Quantum Effects

Hans Lüth, Department ISG, Research Center Jülich, Germany

Contents

1 Introduction	81
2 Electronic Properties of Crystals	81
2.1 Basics of Quantum Mechanics	81
2.2 Periodic Crystal Lattices and their Electronic Band Structure: Metals and Semiconductors	84
2.3 Fermi Statistics in Metals and Semiconductors	89
3 Dissipative Electronic Transport: The Electrical Resistance	92
4 Interfaces and Heterostructures	95
5 Low-Dimensional Structures	97
5.1 Confined Electronic States	97
5.2 Quantum Transport	100
6 Superconductivity	102
7 Conclusions	106

Electronic Properties and Quantum Effects

1 Introduction

Microelectronics has so far been essentially based on semiconductors, mainly on silicon (Si). Nevertheless, III-V semiconductors play a crucial role in optoelectronics and high-frequency devices. Superconductors are used for highly sensitive magnetic field sensors (SQUIDs) and may gain some importance as a niche technology for extremely high frequencies and in connection with quantum computing. For passive devices and passive parts of circuits such as wiring, dielectric interlayers etc., of course, other materials such as metals, oxidic layers and even polymers are attracting more and more attention. In all these fields, and irrespective of the material class, nanostructures are becoming more and more important, which is expressed in the term "nanoelectronics". It is the general assumption that nanoelectronics will follow – and already has in certain aspects – microelectronics, when further miniaturization and higher integration leads to ever smaller feature sizes of the devices on a chip. Even though device physics has mostly been able to avoid quantum mechanical aspects for the simulation of device performance so far, the understanding of semiconductor, metal and superconductor material properties was only possible on the basis of quantum mechanics [1].

This situation will change and has already changed with the advent of nanoelectronics, where the device dimensions have reached the length scale of the electron de Broglie wavelength in a semiconductor. Here, the wave nature of carriers cannot be neglected and in the near future device simulations will be confronted with a real quantum mechanical description rather than with (semi) classical models. This has, of course, already been true for some time for quantum devices, such as e.g. resonant tunneling devices and lasers and in the field of superconductivity, where quantum mechanical behavior extends into mesoscopic and macroscopic dimensions because of the large-scale coherence of the superconducting Bose condensate of Cooper pairs (Sec. 6). At the beginning of this book, therefore, a brief overview of the present quantum mechanical understanding of solids, in particular, their electronic properties, and quantum effects related to nanostructures is necessary.

The basis of it all is quantum mechanics, which describes the behavior of matter on the atomic scale.

2 Electronic Properties of Crystals

2.1 Basics of Quantum Mechanics

The atomistic understanding of solids, in particular semiconductors and superconductors, is only possible on the basis of quantum mechanics, which governs the dynamics of atoms, electrons and atomic nuclei that constitute a crystal. The underlying basic principle, in contrast to classical physics, is that matter has simultaneously a particle and wave-like character. First indications of this are already found in Einstein's quantum hypothesis that light consists of quanta having the energy $W = \hbar\omega$ with ω being the frequency of the electromagnetic radiation, Evidence came from numerous diffraction experiments with material particles (electrons, neutrons etc.), where interference patterns can only be explained by attributing to a moving particle such as an electron, a neutron etc. a wave with frequency ω and energy W

$$W = \hbar\omega \tag{1}$$

and a wavelength

$$\lambda = \frac{2\pi\hbar}{p} \tag{2}$$

the de Broglie wavelength ($p = mv$).

In this quantum mechanical description a free particle (moving in a spatially constant potential) can no longer be described by a classical trajectory $r(t)$ but rather by a plane wave

$$\psi(r,t) = c e^{i(k \cdot r - \omega t)} \tag{3}$$

with $W = mv^2/2 = \hbar\omega$ being the energy of the particle and $p = mv = \hbar k$ its momentum vector. Particle propagation is wave-like whereas the detection process reveals the particle character by transferring a well-defined quantum of energy to the detector at a particular site in space. The only reasonable interpretation of the wave function (3) in connection with the particle picture is in terms of a probability amplitude, where $|\psi(r,t)|^2 = \psi^*\psi$ is the probability (density) of detecting a particle in r at a time t. Since the wave function (3) has non-vanishing values everywhere in space, the propagation of a particle has to be described by a wave package, a superposition of many plane waves (in its simplest form a Gaussian package), i.e. in one dimension:

$$\psi(x,t) = \int dk \; e^{-\left(\frac{k-k_0}{2\Delta k}\right)^2} e^{i(kx-\omega t)} = \int dk \; a(k) \; e^{i(kx-\omega t)} \tag{4}$$

This wave package is centered in k-space around one center wave vector k_0 and in real space its maximum amplitude propagates with the group velocity

$$v_0 = \left.\frac{\partial \omega}{\partial k}\right|_{k_0} \tag{5}$$

This group velocity v_0 is identified with the classical velocity of the particle.

The description of particle propagation as the propagation of a wave package already includes the uncertainty principle of quantum mechanics. Assuming a Gaussian distribution $a(k)$ in (4) for the different plane waves contributing to the wave package $\psi(x,t)$ the spatial half width Δx of the peaked function $\psi(x,t)$ is inversely related to the half width Δk of the k-distribution by

$$\Delta x \; \Delta k \cong 1 \tag{6a}$$

Assuming the particle picture this is identical to

$$\Delta x \; \Delta p \cong \hbar \tag{6b}$$

which means that a well-defined space coordinate x of a particle implies a highly undefined momentum p. Location x and momentum p of an atomic particle, so-called complementary variables, cannot be simultaneously determined as sharp values. The product of their margins of error equals Planck's constant $\hbar = h/2\pi$.

This directly leads to the fundamental fact in quantum mechanics that given a general wave function $\psi(r,t)$ the important quantities, location r, momentum p and energy W, of a particle can no longer be simple sharp numbers as in classical mechanics, but rather they must result as statistical mean values from corresponding measurements of space, momentum, energy etc. The adequate mathematical description is that of observables which are attributed to the measured quantities x, p, W etc. From the wave function (3) of a freely moving particle and from the particle-wave relations (1) and (2) we easily find out that momentum $p = \hbar k$ and energy $W = \hbar\omega$ can be regained from the ψ-function by applying the operations $\hbar/i \cdot \partial/\partial x$ and $-\hbar/i \cdot \partial/\partial t$ on the wave function. The observables momentum and energy of a particle are therefore in general defined by the operators

$$\hat{p} = \frac{\hbar}{i}\frac{\partial}{\partial x} \tag{7}$$

$$\hat{H} = \frac{-\hbar}{i}\frac{\partial}{\partial t} \tag{8}$$

where $\hat{H}$ is called the Hamilton operator for the energy observable. In this description the space operator $\hat{x}$ is a simple multiplication by the coordinate x. Given a general wave function $\psi(x,t)$ in one dimension, the results of x, p and W-measurements are then statistical mean values so-called expectation values (they are expected from the outcome of the experiment).

In general a wave function is normalized, i.e.

$$\int dx\ \psi^*(x,t)\ \psi(x,t) = 1 \tag{9}$$

since the total probability of finding the particle somewhere (obtained by summing up over the entire volume under consideration) must be one (certainty). Thus the expectation values of the $\hat{x},\hat{p},\hat{H}$ operators are written as

$$\langle x \rangle = \int dx\ \psi^*\ x\ \psi = \langle \psi|\ x\ |\psi\rangle \tag{10a}$$

$$\langle p \rangle = \int dx\ \psi^*\ \frac{\hbar}{i}\frac{\partial}{\partial x}\ \psi = \langle \psi|\ \frac{\hbar}{i}\frac{\partial}{\partial x}\ |\psi\rangle, \tag{10b}$$

$$\langle W \rangle = \int dx\ \psi^*\ \frac{-\hbar}{i}\frac{\partial}{\partial t}\ \psi = \langle \psi|\ \frac{-\hbar}{i}\frac{\partial}{\partial t}\ |\psi\rangle \tag{10c}$$

It is useful to apply the so-called bra-ket Dirac notation where the integral is described by the bracket (bra-ket) symbol $\langle\ |\ \rangle$ and the bra wave function $\langle\ |$ is always the complex conjugate of the ket $|\ \rangle$.

The fundamental dynamic equation for the determination of the ψ-function (replacing $m\dot{v} = F$ in classical mechanics) is the so-called Schrödinger equation:

$$i\hbar\frac{\partial}{\partial t}\psi(\boldsymbol{r},t) = \left[-\frac{\hbar^2}{2m}\Delta + V(\boldsymbol{r})\right]\psi(\boldsymbol{r},t) \tag{11a}$$

This is a linear differential equation which has been proven so far to be the best non-relativistic equation for the description of one-particle dynamics in atomic and solid-state physics. It is the basis for the one-particle description of metals and semiconductors, which is treated in the following chapters.

With $\hat{H} = -(\hbar^2/2m)\Delta + V(\boldsymbol{r})$ as the Hamilton operator for the energy observable (11a) is written

$$i\hbar\frac{\partial}{\partial t}\ \psi(\boldsymbol{r},t) = \hat{H}\ \psi(\boldsymbol{r},t) \tag{11b}$$

Since for stationary problems $\hat{H}$ does not depend on time we can separate space and time with an ansatz

$$\psi(\boldsymbol{r},t) = f(t)\ \varphi(\boldsymbol{r}) \tag{12}$$

and find

$$\frac{1}{f(t)}i\hbar\frac{\partial}{\partial t}f(t) = \frac{1}{\varphi(\boldsymbol{r})}\hat{H}\varphi(\boldsymbol{r}) = W = const \tag{13}$$

where the left side is only dependent on time, while the right side only depends on space.

From (13) follows the solution

$$f(t) = \exp\left(-i\frac{W}{\hbar}t\right) \tag{14}$$

I Fundamentals

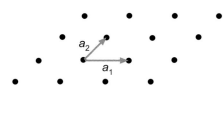

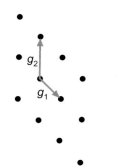

Figure 1: A plane oblique lattice (above) and its corresponding reciprocal lattice (below). The reciprocal lattice vectors g_1 and g_2 lie perpendicular to a_2 and a_1 in real space.

for the time dependence of all stationary solutions, while the spatial part of the wave function $\varphi(r)$ obeys an eigenvalue equation for the Hamilton operator

$$\hat{H}\varphi(r) = W\varphi(r) \tag{15}$$

All operators describing observables in quantum mechanics have real eigenvalues (in Eq. (15) the energy eigenvalues $W = W_i$) and a set of eigenfunctions $\varphi_i(r)$ which are obtained as the result of a measurement of that particular observable. When the wave function of a particle is a particular eigenfunction $\varphi_i(r)$ of an observable, measurement of that observable yields the single sharp eigenvalue (W_i in (15)) instead of the statistical mean values or expectation values (10), when the initial wave function is not an eigenfunction of the operator ascribed to the observable (particular type of measurement).

2.2 Periodic Crystal Lattices and their Electronic Band Structure: Metals and Semiconductors

A crystal, such as an ideal Si-crystal, has translational symmetry, its atoms are ordered along lines in the crystal such that the potential in which the electrons of the crystal move can be written as

$$V(r) = V(r + n_1 a_1 + n_2 a_2 + n_3 a_3) \tag{16}$$

with n_i as integer numbers and a_i the elemental translation vectors (spanning the elementary cell of the crystal). A crystal volume of 1 cm³ contains about 10^{23} electrons. A simple but very powerful approximation is the single-particle approximation, where the dynamics of only one single electron is considered in the periodic crystal potential (16). However, then the potential contains not only the Coulomb fields of the positive atomic cores but also in an indirect way the screening action of all other electrons except the one we are considering. With this drastic approximation we can use the one-electron Schrödinger equation (11a) or when we consider stationary states the time-independent equation (15), where $\hat{H}$ contains the periodic potential (16).

Since $V(r)$ and also $\hat{H}$ have crystal periodicity, one can evaluate all interesting quantities including $V(r)$ in a Fourier series

$$V(r) = \sum_G V_G \, e^{i\,G \cdot r} \tag{17}$$

rather than in a Fourier integral, where $G = h g_1 + k g_2 + l g_3$ are discrete points of the so-called reciprocal space (h, k, l integer) of wave vectors k, with g_i the spanning vectors of the reciprocal elementary cell.

For every Bravais lattice in real space one defines a reciprocal lattice (Figure 1) with primitive vectors g_i by

$$g_i \cdot a_j = 2\pi \, \delta_{ij}, \quad i,j \in (1,2,3) \tag{18a}$$

This is satisfied by

$$g_1 = 2\pi(a_2 \times a_3)/V_z, \quad g_2 = 2\pi(a_3 \times a_1)/V_z, \quad g_3 = 2\pi(a_1 \times a_2)/V_z \tag{18b}$$

where $V_z = a_1 \cdot (a_2 \times a_3)$ is the volume of the elementary cell in real space. For a cubic lattice the reciprocal lattice is again cubic with $g_i = 2\pi/a_i$. In general, vectors in real and reciprocal space (of wave vectors) are related to each other by

$$G_{hkl} \cdot (n_1 a_1 + n_2 a_2 + n_3 a_3) = 2\pi \, m, \quad m \text{ integer.} \tag{19}$$

For the solution of the Schrödinger equation (18) with periodic potential (16) one evaluates both wave function $\varphi(r)$ and potential $V(r)$ in plane waves $\exp(i\,k \cdot r)$. The periodic potential is then the Fourier series (17) and the whole problem turns out to be periodic in the reciprocal space of wave vectors k. The general time-independent wave function of the electron has the general form

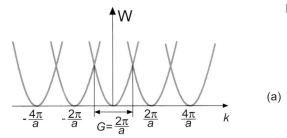
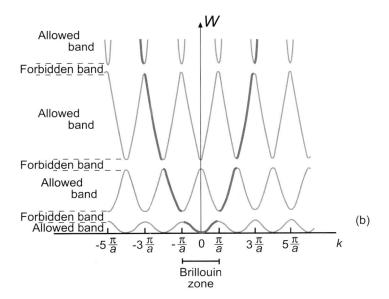

Figure 2:
(a) The parabolic energy curves of a free electron in one dimension, periodically continued in reciprocal space. The periodicity in real space is a. This $W(k)$ dependence corresponds to a periodic lattice with a vanishing potential ("empty" lattice).
(b) Energy dispersion curves $W(k)$ for a one-dimensional lattice (lattice constant a) in the extended zone scheme. As can be seen, the quasi-free electron approximation gives rise to forbidden and allowed energy regions due to the opening of band gaps. The parts of the bands corresponding to the free-electron parabola are indicated by the thick lines.

$$\varphi(\mathbf{r}) = u_\mathbf{k}(\mathbf{r})\, e^{i\,\mathbf{k}\cdot\mathbf{r}} \tag{20}$$

with $u_\mathbf{k}(\mathbf{r})$ being periodic in real space, i.e. a plane wave with amplitude modulation $u_\mathbf{k}(\mathbf{r})$; this is called a Bloch wave, and in the same way as for simple plane waves being solutions of (15) in a constant potential, they can be characterized by their wave vector $\mathbf{k}$. The translational symmetry of real space transfers into reciprocal space such that the energy eigenvalues of the electron $W(\mathbf{k})$ exhibit translational symmetry in reciprocal space:

$$W(\mathbf{k}) = W(\mathbf{k} + \mathbf{G}_{hkl}) \tag{21}$$

This symmetry property allows a qualitative insight into the electronic properties characteristic of electrons moving in a periodic crystal lattice.

We consider a one-dimensional crystal of macroscopic dimension (length L) with lattice constant a and assume at first that the periodic potential $V(x)$ is negligibly small, i.e. $V \cong const\ (=0)$. Then the energy eigenvalues $W(k)$ are those of a free electron, $W = \hbar k^2/(2m)$, a parabola in k-space. Now, in a gedanken experiment, the periodic potential is "switched on" gradually, such that the energetic effect remains negligible, but the symmetry property of translational symmetry starts to be effective, i.e. the energy eigenvalues now have the property (21)

$$W(k) = W(k + G_h) \tag{22}$$

In reciprocal space the energy parabola of the free electron has to be repeated on the k-axis, each time originating at multiples of the reciprocal lattice vector $G = 2\pi/a$ (Figure 2a). The periodicity volume in k-space centered around the $k = 0$ (Γ) point is called the 1st Brillouin zone. At the edges of the Brillouin zone, i.e. at $+G/2 = \pi/a$ and $-G/2 = -\pi/a$ two neighboring parabolas intersect and there is a degeneracy of the energy eigenvalues. The two corresponding solutions of the Schrödinger equation belonging to the two parabolas are equivalent. Since the Schrödinger equation is a linear

Fundamentals

Figure 3:
(a) Qualitative form of the potential energy $V(x)$ of an electron in a one-dimensional lattice. The positions of the ion cores are indicated by the points with separation a (lattice constant).
(b) Probability density $\rho_+ = \psi_+^* \psi_+$ for the standing wave produced by Bragg reflection at $k = \pm\pi/a$, the upper edge of band (1) in d).
(c) Probability density $\rho_- = \psi_-^* \psi_-$ for the standing wave at the lower edge of band (2) at $k = \pm\pi/a$ in (d).
Splitting of the energy parabola of the free electron (- - -) at the edges of the first Brillouin zone ($k = \pm\pi/a$ in the one-dimensional case). As a first approximation the gap W_g is given by twice the corresponding Fourier coefficient V_G of the potential. Periodic continuation over the whole of k-space gives rise to continuous bands (1) and (2), shown here only in the vicinity of the original energy parabola.

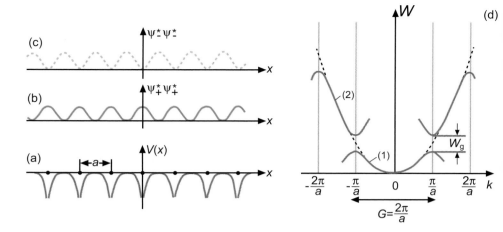

differential equation the most general solution at this k-value is a superposition of both particular solutions. For negligible potential variations the corresponding plane waves are

$$e^{iGx/2} \quad \text{and} \quad e^{i[(G/2)-G]x} = e^{-iGx/2} \tag{23}$$

The most general solutions at $G/2$ are therefore the two linear superpositions

$$\psi_+ \propto \left(e^{iGx/2} + e^{-iGx/2}\right) \propto \cos(\pi\frac{x}{a}), \tag{24a}$$

$$\psi_- \propto \left(e^{iGx/2} - e^{-iGx/2}\right) \propto \sin(\pi\frac{x}{a}) \tag{24b}$$

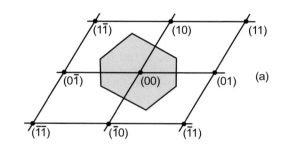

Figure 4:
(a) Construction of the first Brillouin zone for a plane oblique lattice. The zone boundary consists of perpendicular bisectors of the shortest reciprocal lattice vectors.
(b) The Brillouin zones of the face centered cubic, body-centered cubic and hexagonal lattices. Points of high symmetry are denoted by Γ, L, X etc. The surfaces enclosing the Brillouin zones are parts of the planes that perpendicularly bisect the smallest reciprocal lattice vectors. The polyhedra that are produced by these rules of construction can be drawn about every point of the reciprocal lattice. They then fill the entire reciprocal space. The cell produced by the equivalent construction in real space is known as the Wigner-Seitz cell. It can be used to describe the volume that one may assign to each point of the real crystal lattice.

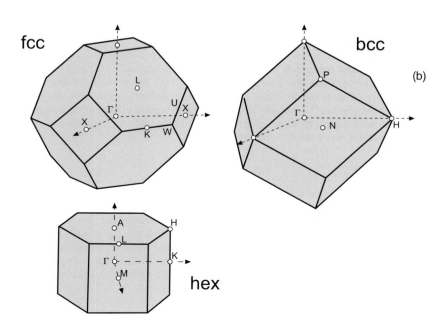

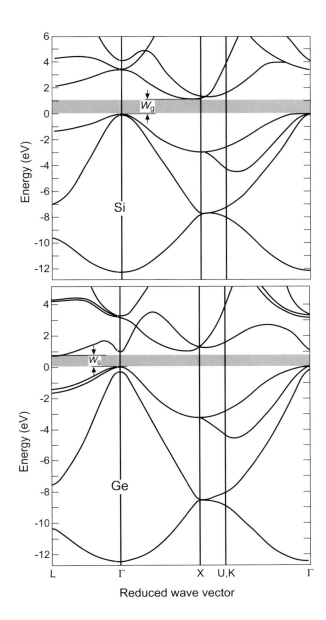

Figure 5: Calculated band structures of silicon and germanium. For germanium the spin-orbit splitting is also taken into account (After [2]). Both semiconductors are so-called indirect semiconductors, i.e. the maximum of the valence band and the minimum of the conduction band are at different positions in the Brillouin zone. The minimum of the conduction band of silicon lies along the $\Gamma X = [100]$ direction and that of germanium along the $\Gamma L = [111]$ direction. The calculation for Ge is relativistic. The spin-orbit splitting in the upper valence band near Γ is therefore resolved in contrast to Si. W_g is the width of the forbidden energy band.

According to Figure 3 the probability density $\psi_+^* \psi_+$ accumulates negative electronic charge at the location of the positive atomic core (minimum of $V(x)$) while $\psi_-^* \psi_-$ accumulates negative charge in between the positive atomic cores. In comparison with the state of a traveling wave $\exp(ikx)$, where the charge density is homogeneous all over the crystal, the ψ_- thus means an increase of the energy of the electron, whereas ψ_+ is related to a lower electronic energy. The originally degenerated states at $\pm G/2$ split off and a forbidden gap in the spectrum of one-electron states is opened at the Brillouin zone boundaries (Figure 2). As a consequence the electronic states in a periodic crystal, unlike the case of the parabola $W(k)$ of the free electron, now form allowed and forbidden bands on the energy scale. Within a band of allowed states $W(k)$ shows an oscillatory behavior as a function of the wave vector k of the Bloch states (20). Because of periodicity in k-space the $W(k)$ dependence can be restricted for practical reasons to the 1st Brillouin zone. Near the lower and upper energetic edges of the bands a parabolic description of the $W(k)$ curves is adequate. The origin of allowed electronic bands in a crystal can also be traced back to the interaction of neighboring atoms by overlap of the bonding atomic orbitals. Electronic bands can thus be characterized by the atomic orbitals from which they originate, e.g. sp-bands or d-bands etc.

The extension to three dimensions (3D) is straightforward, but much more complex $W(\mathbf{k})$ functions in the 3-D reciprocal $\mathbf{k}$-space are obtained. Even for the most common 3-D bcc and fcc lattices Brillouin zones as periodicity volumes in $\mathbf{k}$-space are complex polyhedra (Figure 4). $W(\mathbf{k})$ dependencies are then plotted as curves along symmetry

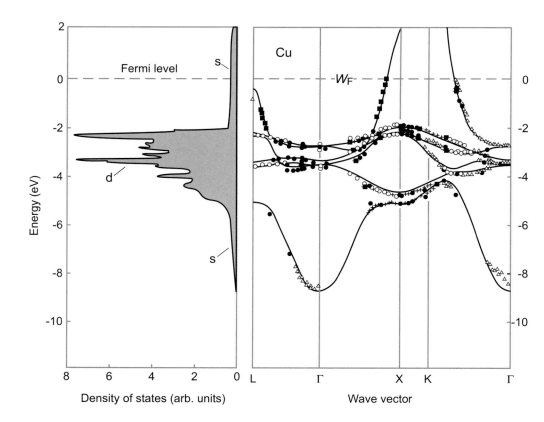

Figure 6: Band structure $W(k)$ for copper along directions of high crystal symmetry (right). The experimental data were measured by various authors and were presented collectively by Courths and Hüfner [3]. The full lines showing the calculated energy bands and the density of states are from [3]. The experimental data agree very well, not only among themselves, but also with the calculation. They correspond in density of states to s and d-bands (left).

directions, e.g. Γ–X, X–W etc., to give a rough overview of the Bloch state energies as a function of the ***k***-vector. As an example the calculated band structure of Si along symmetry lines L–Γ–X is shown in Figure 5 [2]. Shown in red in this figure is a totally forbidden band which is characteristic of semiconductors.

The important distinction between a semiconductor and a metal is made on the basis of the occupancy of electronic states $W(k)$. So far only the quantum mechanically possible electronic states have been calculated. Their occupancy with electrons is controlled by the Pauli principle, which is valid for all kinds of Fermions, i.e. elementary particles with spin 1/2. Electrons of course are Fermions, they obey the so-called Fermi statistics. According to the Pauli principle non-interacting Fermions can occupy a quantum mechanical state such as ψ_+ or ψ_- only once. Since in our consideration so far the two spin orientations of the electrons have not been taken into account, the electronic states described by the band structure $W(\boldsymbol{k})$ can be occupied by two electrons (opposite spins) at maximum. Up to a certain energy the states of the band structure $W(\boldsymbol{k})$ are thus filled, each with two electrons. When the highest occupied state is identical with the upper energetic edge of an electronic band which is separated from higher bands by an absolute gap, as for Si in Figure 5, then this material is a **semiconductor**. In order to excite electrons they have to overcome the gap in the band structure, since no continuum of empty states is available in the direct energetic neighborhood of the highest occupied states.

When the highest occupied energy level falls in the continuum of an allowed energy band, this band is partially occupied and partially empty. Electrons can be excited with infinitely small energy quanta and the material is a **metal**. As an example the band structure of Cu is shown in Figure 6 along some symmetry lines [3]. The highest occupied energetic state in a metal (at zero temperature T) is called the Fermi level W_F. For Cu the Fermi energy W_F crosses the energetically wide continuum of the $W(\boldsymbol{k})$ band which is derived from the atomic Cu s-states. Characteristic of transition metals are the energetically sharp, relatively flat electronic bands, which are derived from the atomic d-states. In Cu these bands are all occupied, i.e. the Fermi level W_F lies approximately 2 eV higher than the upper d-band edge. For other transition metals, such as Fe, Ni, Pt, the d-bands are partially empty and the Fermi energy W_F lies within the d-bands.

For many purposes it is not necessary to know the complete $W(\boldsymbol{k})$ dependence of the electronic band structure but rather the number of states dZ lying between W and $W+dW$, i.e. the density of states, is interesting (Figure 6, left). This is easily calculated by an integration in ***k***-space

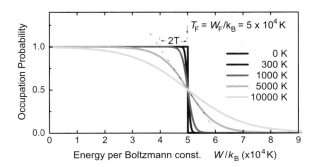

Figure 7: The Fermi distribution function at various temperatures. The Fermi temperature $T_F = W_F^Q/k_B$ has been taken as 5×10^4 K. The tangent at the point of inflection (- .. -) intersects the energy axis at $2kTk_B$ above W_F^Q at all temperatures.

$$dZ = \frac{V}{(2\pi)^3} \int_W^{W+dW} d^3k = D(W)\, dW \quad (25)$$

Here we have to take into account the fact that our crystal does not have infinite dimensions, but rather a macroscopic length of L or a volume $V = L^3$ with L of typically 10^8 Å = 1 cm. This means that the electronic waves, the Bloch states, can only have wave vectors k or wavelengths which fit as multiples into the length L. This leads to the restrictions $k_i = 2\pi n_i/L$ for the 3 coordinates ($i = 1,2,3$) with n_i as integers and discrete possible k_i values having distances $2\pi/L$ from each other. Since, however, the 1st Brillouin zone has a diameter of $G = 2\pi/a$, where a, the atomic distance, is about $10^{-8} L$, k can nevertheless be assumed to be quasi-continuous in k-space. But in k-space one has to attribute a volume of $(2\pi)^3/L^3 = (2\pi)^3/V$ to each quantum state; the inverse value shows up in (25). If the k-volume element d^3k is separated into an area element dS_W on the energy surface and a component $dk_\perp$ normal to this surface, i.e. $d^3k = dS_F\, dk_\perp$, then with $dW = |\text{grad}_k W| dk_\perp$ one has for the density of states (related to the crystal volume V)

$$D(W)\, dW = \frac{1}{(2\pi)^3} \left[\int_{W(k)=const} \frac{dS_W}{|\text{grad}_k W(k)|} \right] dW \quad (26)$$

This density of states $D(W)$ is high in regions of flat bands (e.g. d-bands in Figure 6) since $|\text{grad}_k W(k)|$ is small. Near maxima or minima of the band structure $W(k)$ has a parabolic k-dependence $W \propto k^2$, thus $|\text{grad}_k W|^{-1} \propto k^{-1}$, but the integration over the energy surface restores a k-dependence of $D(W)$. The density of states near a maximum or minimum thus behaves as $D(W) \propto k \propto \sqrt{W}$, similar to a free electron gas.

2.3 Fermi Statistics in Metals and Semiconductors

According to the Pauli principle each electronic state can only be occupied by one electron or by two if spin orientation is not taken into account. If we plot a probability function $0 \leq f(W) \leq 1$ for the occupancy of states with energy W for vanishing temperature T, it must be a step function, which switches from the value 1 (occupied) to 0 (non-occupied) at the Fermi energy W_F. Here W_F is given in a metal with continuously lying states by the maximum number of electrons which can occupy the states in the band. For increasing temperatures electrons deep below W_F cannot be excited, they find no empty states in their energetic neighborhood. Only electrons slightly below W_F can be excited to empty states slightly above W_F. For higher temperatures $f(W)$ therefore must exhibit a "smeared out" step near W_F (Figure 7). A detailed thermodynamic derivation of $f(W)$ based on the Pauli principle and the indistinguishability of electrons yields the mathematical expression.

$$f(W,T) = \frac{1}{e^{(W-W_F)/k_BT} + 1} \quad (27)$$

In order to get an impression of the shape of the Fermi function near W_F, one has to consider the typical energetic width of an electronic band. For Cu (Figure 6) W_F lies about 8 eV above the s-band minimum. In contrast, at room temperature k_BT amounts to

Fundamentals

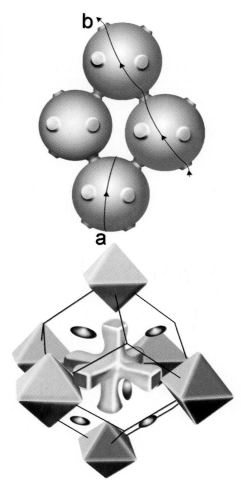

Figure 8:
(top) Fermi surface of copper (Cu), calculated on the basis of experimental data. The shape of the surface is shown in four neighboring Brillouin zones of the reciprocal k-space. In addition, two different carrier orbits are plotted: (a) a closed particle orbit; (b) an open orbit, which continues in the same general direction indefinitely in the repeated zone scheme (After [4]),
(bottom) proposed Fermi surface for tungsten (W). The six octahedron-shaped pockets at the zone corners correspond to holes. They are all equivalent; any one can be taken into any direction by a translation through a reciprocal lattice vector (After [5]).

(1/25) eV, i.e. even at room temperature $f(W)$ has a relatively sharp Fermi edge: The region over which $f(W)$ deviates significantly from the step function is of the order of $2\,k_B T$ to each side of W_F (Figure 7).

When in a metal the Fermi energy W_F cuts the band structure $W(\mathbf{k})$ in the lower part of the band, where $W(\mathbf{k})$ is essentially parabolic $W(\mathbf{k}) \cong \hbar^2 k^2/(2m^*)$ (m^*, the so-called effective mass, formally describes the reciprocal curvature of the energy parabola), the Fermi surface $W_F(\mathbf{k}) \cong \hbar^2 k_F^2/(2m^*)$ is essentially a sphere with radius k_F, the so-called Fermi wave vector. The Fermi surface is fully contained in the 1st Brillouin zone. This case is given for monovalent alkali metals. For noble metals such as Cu, Ag and Au the Fermi surface touches the boundary of the Brillouin zone. A purely parabolic energy dependence of $W_F(\mathbf{k})$ is no longer given and the Fermi surface becomes a deformed sphere with extensions into the neighboring Brillouin zones (Figure 8a) [4]. For transition metals, where W_F crosses the band structure in regions of complex d-band structures, the Fermi surfaces become very complicated ("monsters") shapes in k-space (Figure 8b) [5].

In a semiconductor (an insulator is a semiconductor with a large band gap $W_g > 3$ eV) there is a totally forbidden gap in the spectrum of allowed single electron energies. The lower band, called the valence band, is fully occupied at $T = 0$, whereas the allowed band above the forbidden gap, called the conduction band, is unoccupied at $T = 0$. Where is the Fermi energy W_F located in this case? The answer is simply given by considering a finite temperature, where thermally excited electrons originating from the valence band occupy states in the lower conduction band (Figure 9). The density of states both in the valence band $D_V(W)$ and in the conduction band $D_C(W)$ show a $\sqrt{W}$ dependence but are in general different. However, all the electrons missing in the valence band $D_V(W)[1-f(W)]$, holes in the valence band, are now found in the conduction band, where the corresponding electron density n is described by $D_C(W)f(W)$. Because of the symmetry of the Fermi function $f(W)$ around W_F the Fermi level W_F must therefore be located within the forbidden band, in the case of larger valence band density D_V in comparison to D_C, in the upper half of the gap $W_g = W_C - W_V$. W_C and W_V are the lower and upper edges of the conduction and valence bands, respectively. With m_n^* and m_p^* as the effective masses (inverse curvatures of bands) of electrons in the conduction band and holes in the valence band, respectively, the densities of states (Figure 9) are given as

$$D_C(W) = \frac{(2m_n^*)^{3/2}}{2\pi^2 \hbar^3}\sqrt{W - W_C},\ (W > W_C) \tag{28a}$$

$$D_V(W) = \frac{(2m_p^*)^{3/2}}{2\pi^2 \hbar^3}\sqrt{W_V - W},\ (W < W_V) \tag{28b}$$

Since the "width" of the Fermi edge ($\sim 2\,k_B T$) is, at normal temperatures, small compared to the gap width of a semiconductor (≥ 1 eV), the Fermi function can be approximated by Boltzmann occupation statistics within the bands ($W > W_C$ and $W < W_V$), i.e. for the conduction band

$$f(W) = \frac{1}{\exp[(W - W_F)/k_B T] + 1} \approx \exp\left(-\frac{W - W_F}{k_B T}\right) \ll 1 \tag{29}$$

for $W - W_F \gg k_B T$

This Boltzmann approximation for "non-degenerate" semiconductors simplifies the expression for the densities of electrons and holes in the conduction and valence band considerably. By means of (29) the density of free electrons in the conduction band

$$n = \int_{E_C}^{\infty} D_C(W)\,f(W,T)\,dW \tag{30}$$

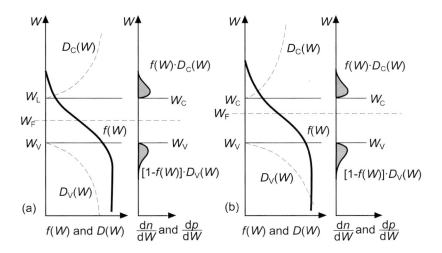

Figure 9:
(a) Fermi function $f(W)$, density of states $D(W)$ and electron (n) and hole (p) concentration, respectively, for equal densities D_c and D_v in conduction and valence band (schematically).
(b) Same plot, but now with unequal densities of states D_c and D_v in conduction and valence band. The densities of electrons in the conduction band must equal the density of holes in the valence band which shifts the Fermi energy W_F away from mid gap.

is obtained as

$$n = 2\left(\frac{2\pi m_n^* k_B T}{h^2}\right)^{3/2} \exp\left(-\frac{W_C - W_F}{k_B T}\right) = N_{\text{eff}}^C \exp\left(-\frac{W_C - W_F}{k_B T}\right) \quad (31a)$$

and analogously the density of holes in the valence band

$$p = 2\left(\frac{2\pi m_p^* k_B T}{h^2}\right)^{3/2} \exp\left(\frac{W_V - W_F}{k_B T}\right) = N_{\text{eff}}^V \exp\left(\frac{W_V - W_F}{k_B T}\right) \quad (31b)$$

The use of the so-called "effective densities of states" N_{eff}^C and N_{eff}^V in (31) allows the formal interpretation in which the whole conduction (valence) band can be characterized by a single energy level W_C (W_V) (i.e. the band edges) with the density of states N_{eff}^C (N_{eff}^V) (temperature-dependent!). The occupation densities n and p with electrons and holes of these bands are then simply determined by a Boltzmann factor (31).

Pure, so-called intrinsic semiconductors, with unintentional impurity concentrations below 10^{15} cm^{-3} are essentially insulators. For Si, e.g., the band gap energy of about 1.1 eV allows only tiny densities of thermally excited free electrons in the conduction band from states in the valence band. For device applications the controlled incorporation of electrically active impurities, so-called dopants, is necessary. In Si, e.g., dopant atoms such as As, P etc. with 5-fold valency can be built in into the lattice on Si sites and can give up their additional 5th valence electron into the conduction band, these dopants are called (electron) donors. Similarly built-in B, Ga or In atoms with 3-fold valency called acceptors are eager to accept electrons from the Si valence band. Empty states in the valence band are created which act as positive carriers, holes in the valence band. The excitation energies for electrons from donor states into the conduction band or from valence band states into acceptor levels are in the order of 30 meV, small in comparison to the forbidden band ($\approx$ 1 eV). Their energy levels W_D or W_A therefore have to be located close to the conduction band edge W_C or slightly above the valence band edge W_V (Figure 10b).

In the case of a doped semiconductor, e.g. n-doped with only donors, at moderate temperatures electrons in the conduction band first originate from excitation from the donor levels at W_D. All electrons in the conduction band originate from empty donor states. The Fermi level W_F at these temperatures must therefore lie between W_C and W_D; the density of free electrons in the conduction band increases exponentially with temperature and an activation energy $(W_C - W_D)/2$ (Figure 10). This regime is called the freeze-out range. At a certain temperature all donors are exhausted, in this so-called saturation regime the concentration of free electrons n saturates, the Fermi-level position shifts to energies below W_D until finally at elevated temperatures the generation of free electrons mainly from the valence band becomes important. In this intrinsic regime n increases again exponentially with temperature, but now with an activation energy $W_g/2$ determined by the width of the forbidden band W_g (Figure 10a).

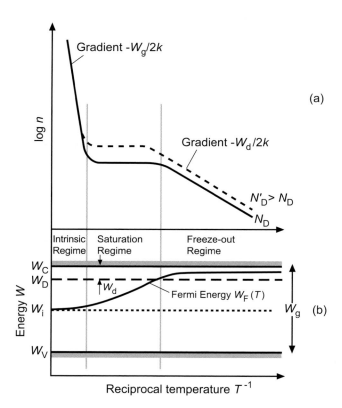

Figure 10:
(a) Qualitative temperature dependence of the concentration n of electrons in the conduction band of an n-type semiconductor for two different donor concentrations $N'_D > N_D$. The width of the forbidden band is W_g and W_d is the ionization energy of the donors.
(b) Qualitative temperature dependence of the Fermi energy $W_F(T)$ in the same semiconductor. W_C and W_V are the lower edge of the conduction band and the upper edge of the valence band, respectively, W_D is the position of the donor levels and W_i is the Fermi level of an intrinsic semiconductor (intrinsic neutrality level).

3 Dissipative Electronic Transport: The Electrical Resistance

Electrical conduction arises from electric-field-induced propagation of free electrons in a partially filled band of a metal or in semiconductors of free electrons in the conduction band and/or free holes (empty electronic states) in the valence band. In all these cases electrons and holes are described quantum mechanically by wave packages of Bloch waves (20) and the particle velocity $v(k)$ is the group velocity $(1/\hbar)\nabla_k W$ of the wave package. Given the electronic band structure $W(k)$ the electron takes up a certain amount of energy δW in the external electric field E in order to enhance its energy within the band by

$$\delta W = \nabla_k W(k)\delta k = \hbar\, v \cdot \delta k \tag{32a}$$

In this semiclassical description

$$\delta W = -e\, E \cdot v\, \delta t \tag{32b}$$

and

$$\hbar \dot{k} = -eE \tag{32c}$$

From (32) the simple semiclassical dynamic equation for crystal electrons follows as

$$\dot{v}_i = \frac{1}{\hbar}\frac{d}{dt}(\nabla_k\, W)_i = \frac{1}{\hbar}\sum_j \frac{\partial^2 W}{\partial k_i \partial k_j}\, \dot{k}_j \tag{33a}$$

$$\dot{v}_i = \frac{1}{\hbar^2}\sum_j \frac{\partial^2 W}{\partial k_i \partial k_j}\, (-eE_j) \tag{33b}$$

The inverse curvature of the energy bands can be interpreted as a so-called effective mass tensor

$$\left(\frac{1}{m^*}\right)_{ij} = \frac{1}{\hbar^2}\frac{\partial^2 W(\boldsymbol{k})}{\partial k_i \partial k_j} \qquad (34)$$

which replaces in a crystal lattice the free electron mass; all effects of the solid crystal interaction (chemical bonding) are included. In a simple cubic lattice the effective mass is

$$m^* = \frac{\hbar^2}{d^2 W/dk^2} \qquad (35)$$

These effective masses, in a semiconductor for electrons in the conduction band m_n^* and for holes in the valence band m_p^* describe, as essential band structure parameters, the curvatures of $W(\boldsymbol{k})$ near the band edges and thus also the densities of states (28) near the conduction and valence band edges, respectively.

Since Bloch waves (20) are stationary solutions of the Schrödinger equation in a crystal with perfect translational symmetry, such a state, once excited in an electrical field, would never decay. An infinite electrical conductance would result. A finite electrical resistance is due to scattering processes of the free electrons on perturbations of the ideal translational symmetry of the crystal lattice. Such perturbations are defects, e.g. impurity atoms, dislocations etc. and thermally excited lattice vibrations (phonons). Electron-electron scattering usually plays a negligible role; because of the Pauli principle only electrons within a shell of about $2 k_B T$ around W_F can contribute. When two electrons scatter from each other, one typically becomes faster and the other slower. But states for slower electrons are already occupied. The process is thus forbidden. Since defects can be assumed in a first approximation to be immobile, defect scattering is essentially elastic. Lattice vibrations are time-dependent oscillations of the lattice atoms. Kinetic energy can be exchanged between these vibrations and the carriers; electron scattering on lattice vibrations is thus inelastic.

The accelerating action of an external electric field E is in a stationary condition of DC current flow balanced by scattering processes such that the carriers assume an average, so-called, drift velocity v_D on top of their random, undirected thermal velocity, which is determined by W_F in a metal. In the simplest Drude-type description scattering is described phenomenologically by a friction term ($\propto v_D$), such that the semiclassical equation of motion for a carrier follows as:

$$m^* \dot{v}_D + \frac{m^*}{\tau} v_D = -eE_x \qquad (36)$$

For stationary conditions ($\dot{v}_D = 0$) the drift velocity v_D is obtained as

$$v_D = -\frac{e\tau}{m^*} E_x \qquad (37)$$

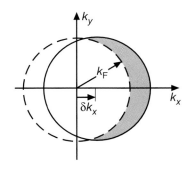

Figure 11: Schematic representation of the action of a constant electric field E_x on the distribution of quasi-free electrons in reciprocal k-space. The Fermi sphere of the equilibrium distribution (dashed line, centered around (0,0,0)) is shifted by $\delta k_x = -e\tau E_x/\hbar$ in the state of stationary current flow (full line).

where τ is the relaxation time which describes the decay time, after which the non-equilibrium carrier distribution relaxes into the equilibrium distribution after switching off the external field. For the calculation of the current density j Drude originally (in about 1900) assumed the participation of all free carriers in a metal. Quantum mechanics now comes into play here. According to the Pauli principle only a tiny amount of the whole free carrier density n can contribute to the current, namely carriers in the direct energetic neighborhood of the Fermi energy W_F. The situation is best described in reciprocal k-space, where in a simple metal the Fermi sphere is shifted on the average by a tiny k-displacement δk_x due to the superimposed drift velocity (Figure 11). From (37) and

$$m^* v_D = \hbar \, \delta k_x \qquad (38)$$

one obtains

$$\delta k_x = -\frac{e\tau(W_F)}{\hbar} E_x \qquad (39)$$

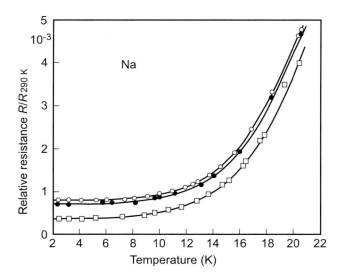

Figure 12: Electrical resistance of sodium compared to the value at 290K as a function of temperature. The data points (o, ●, □) were measured for three different samples with differing defect concentrations (After [6]).

Only electrons near W_F have to be taken into account [$\tau = \tau(W_F)$] and only the electrons on states in the shaded area (shell) of Figure 11 contribute to the current. The current density in x-direction therefore is obtained as

$$j_x = -\frac{2e}{8\pi^3} \int_{\text{Shell}} d^3k \; v_x(\boldsymbol{k}) \tag{40}$$

With the shell volume in k-space $\pi k_F^2 \delta k_x$ and by replacing v_x (on the Fermi surface only) by $v(W_F)$, the mean value on the whole Fermi sphere, one obtains

$$j_x = \frac{e^2 \tau(W_F)}{3\pi^2 \hbar} v(W_F) k_F^2 E_x = \frac{e^2 \tau(W_F) k_F^3}{3\pi^2 m^*} E_x \tag{41}$$

Since the volume of the Fermi sphere is given by the total electron density ($k_F^3 = 3\pi^2 n$) (41) yields the old Drude expression for the current density

$$j_x = \frac{e^2 \tau(W_F) n}{m^*} E_x = \sigma \; E_x \tag{42}$$

Even though only the electrons near W_F contribute rather than all the electrons. The total density n enters (42) via the integration over the Fermi sphere. Ohm's law expressed in (42) for a metal also holds formally for semiconductors, be it for free electrons in the conduction band, or for free holes in the valence band. The relaxation time τ as well as the effective mass m^* is then ascribed to electrons or holes near the band edges W_C and W_V, respectively. According to (41) electrons and holes are distributed according to Boltzmann statistics within the conduction and valence band and, in contrast to a metal, the whole densities of free electrons n and/or holes p near W_C and/or W_V contribute to the conductivity σ in (42).

In order to discuss the temperature dependence of σ(T) for metals and semiconductors, it is convenient to introduce the carrier mobility

$$\mu = \frac{e\tau}{m^*} \tag{43}$$

and to write (42) as

$$j_x = e\mu n E_x = \sigma E_x \tag{44}$$

For metals the electron density n, being given by the volume of the Fermi sphere, is essentially independent of temperature and the temperature effect on σ enters through temperature-dependent scattering processes described by μ(T) and τ(T), respectively.

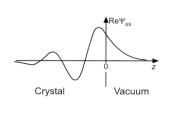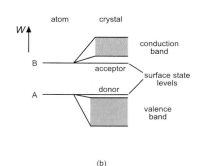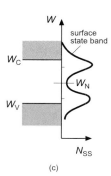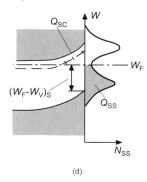

Elastic scattering on impurities gives rise to a temperature-independent so-called residual resistance ρ_R, while inelastic scattering on phonons is enhanced with increasing temperature, since the scattering cross section increases as a first approximation with the increasing vibrational amplitude of the atoms ($\propto k_B T$ as a first approximation). The superposition of both scattering mechanisms yields the characteristic temperature dependence of $\rho(T)$ for metals in Figure 12 [6].

In semiconductors both factors μ and n (or p for holes) in (44) depend on temperature. The carrier concentrations n and/or p, however, depend exponentially on temperature according to (33) and also for dopant-induced conductivity (only with smaller activation energy). The mobility $\mu(T)$, on the other hand, being determined by impurity and phonon scattering, depends only slightly on temperature, $\propto T^{3/2}$ for impurity scattering (dominant at low temperature) and $\propto T^{-3/2}$ for phonon scattering (dominant at high temperature). For semiconductors the conductivity $\sigma(T)$ therefore essentially reflects the exponential dependencies $n(T)$ (or $p(T)$ for holes) of the carrier densities (Figure 10a). Only slight modifications appear due to $\mu(T)$, essentially in the saturation regime of constant $n(T)$.

Figure 13:
(a) Real part Re(ψ_{SS}) of a surface state wave function localized at the surface ($z = 0$).
(b) Qualitative explanation of the origin of electronic surface states in the tight binding picture. Two atomic levels A and B form the bulk valence and conduction bands, respectively. Surface atoms have fewer bonding partners than bulk atoms and thus give rise to electronic levels that are closer to those of free atoms, i.e. surface state levels split off from the bulk bands. Depending on their origin these states have acceptor- or donor-like charging behavior.
(c) Due to their 2-D translational symmetry within the surface plane the acceptor- and donor-type surface states exhibit dispersion along $k_\parallel$ and form broader bands in the density of states N_{SS} with a neutrality level W_N in between.
(d) Because of charge neutrality near the surface the Fermi level W_F must cross the density of surface states N_{SS} close to the neutrality level $W_N \approx W_F$. For an n-type semiconductor (in the bulk: W_F close to the conduction band edge) a positively charged electron depletion layer builds up, the positive space charge Q_{SC} of which (due to ionized bulk donors) is compensated by negative charge in the surface states (W_F slightly above W_N).

4 Interfaces and Heterostructures

Surface effects have been neglected in all considerations of macroscopic, bulk-like solids with more than 10^{22} atoms per cm^3 volume. The electronic band structure $W(\mathbf{k})$ was calculated assuming an infinitely extended solid. This is certainly correct for macroscopic structures, when one envisages the fact that only 10^{15} atoms are surface atoms per cm^2 surface area. But surface and interface effects certainly have to be considered in more detail when in nanostructures with dimensions of several 10 nm the surface or interface in a heterostructure becomes more and more important in comparison with the bulk of the solid [7].

On a 2-D surface of an otherwise extended bulk solid the chemical bonds to neighboring atoms are broken. Due to the changed chemical surroundings the electronic structure of surfaces and interfaces between two different solids being e.g. epitaxially grown on each other is considerably changed with respect to the bulk interior. Surface and interface atoms generally assume crystallographic equilibrium positions different from those of the bulk material. Also the surface, or interface, periodicity in two dimensions (2D) might be changed with respect to the bulk. So-called reconstructions with 2-D superstructures occur at surfaces and interfaces. A great deal of work in surface studies by means of LEED (low electron energy diffraction), RHEED (reflection high energy electron diffraction) and STM (scanning tunneling microscopy) is devoted to the problem of surface reconstruction. Only a combination of several experimental techniques together with theoretical calculations leads to the identification of the atomic geometry, i.e. structure models, of a surface or interface. Reconstructions of a surface are usually described by notations such as (2×1) or (7×7). The numbers in brackets give the ratio of the lengths of the primitive translation vectors of the 2-D superstructure and those of the substrate (bulk) unit mesh.

Simultaneously with crystallographic surface modifications the electronic structure at surfaces and interfaces is different from that in the bulk. So-called electronic surface states or interface states are present, whose wave functions are locally restricted to the surface or interface. Their amplitude decays within several angstroms into the vacuum

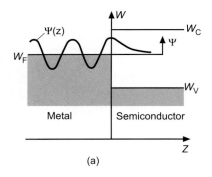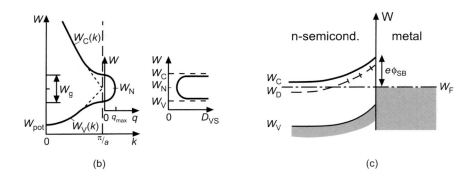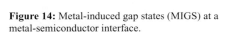

(a) (b) (c)

Figure 14: Metal-induced gap states (MIGS) at a metal-semiconductor interface.
(a) Origin of the MIGS: a metal Bloch state near W_F decays into the semiconductor, where its tails must be represented by an expansion of valence and conduction band wave functions.
(b) For the simple model of a one-dimensional chain of atoms [real band structure $W_C(k)$ and $W_V(k)$] the exponentially decaying surface states (Figure 13a) with imaginary wave vectors $\kappa = -iq$ fill the gap of the semiconductor between W_C and W_V. In the simplest case of symmetrical bands their density of states. N_{SS} is symmetrical with poles near W_C and W_V and with the neutrality level W_N at midgap.
(c) In a metal-semiconductor Schottky contact the Fermi level W_F must cross the density of MIGS near W_N and a depletion space charge layer with a Schottky barrier $e\phi_{SB}$ similarly builds up as in Figure 13d for surface states on the clean surface.

and/or into the bulk of the solid (Figure 13). The energetic distribution of these surface and interface states is different from those of the bulk states described by the bulk band structure $W(k)$. For a surface this is easily seen since for the topmost surface atoms the bonding partners on one side are totally missing, which means that their wave functions have less overlap with wave functions of neighboring atoms. The splitting and shift of the atomic energy levels due to interaction with neighboring atoms is thus smaller at the surface than in the bulk where much more overlap to neighbors is given (Figure 13b).

Every atomic orbital involved in chemical bonding and producing one of the bulk electronic bands also gives rise to one surface state level. Surface states, which are derived from the bulk band structure, so-called intrinsic surface states, have 2-D translational symmetry along the surface of an ideal crystal. Their wave functions ψ_{SS} therefore are of a Bloch-type character for coordinates $r_\parallel$ and wave vectors $k_\parallel$ within the surface plane

$$\psi_{SS}(r_\parallel, z) = u_{k_\parallel}(r_\parallel, z)\ \exp(i k_\parallel \cdot r_\parallel) \tag{45}$$

while their amplitude decays exponentially into the vacuum and into the bulk crystal. The energy eigenvalues of intrinsic surface states thus form a 2-D band structure $W_{SS}(k_\parallel)$ in the 2-D reciprocal space $k_\parallel$, which is attributed to the surface. The bands $W_{SS}(k_\parallel)$ usually show dispersion along $k_\parallel$ directions, which gives rise to broader surface state energy bands in a density-of-states plot $N_{SS}(W_{SS})$ as in Figure 13c. Since the surface states must be considered to be derived from bulk valence and conduction band states, they have different charging character depending on their bulk origin. States which are derived from the conduction band usually lie closer to the conduction band edge and are negatively charged when they are occupied by an electron; they are therefore called acceptor-like. States derived from the valence band (closer to the valence band edge) are donor-like, since they are neutral if occupied by an electron and positive if empty. Thus there exists a so-called neutrality energy W_N somewhere within the band of surface states. When the Fermi energy W_F crosses W_N, the surface as a whole is neutral. The location of W_F in the bulk of the crystal is determined by the doping level. For an *n*-type crystal as shown in Figure 13d, therefore, upward band bending near the surface must result in order to avoid strong charging effects near the surface. The amount of band bending is determined by the condition that the positive charge Q_{SC} residing in the emptied bulk donors of the depletion space charge layer is balanced by an equal negative charge Q_{SS} in the surface state band. W_F must therefore be located slightly above the neutrality level W_N. For common surface state densities N_{SS} in the order of 10^{15} cm^{-2}(eV)$^{-1}$ the deviation $(W_N - W_F)$, however is tiny, in the order of 10^{-2} to 10^{-3} eV, such that usually the Fermi level W_F can be assumed to be "pinned" very close to the neutrality level W_N at the surface. It does not change its energetic position with varying bulk doping or temperature. The pinning position $(W_F - W_V)_S$ at the surface is characteristic of the particular type of surface; $(W_F - W_V)_S$ is sometimes called surface potential.

The ideas developed here for a vacuum/solid interface can easily be transferred to Schottky barriers, i.e. metal/semiconductor interfaces and semiconductor heterostructures, where two semiconductors with more or less good lattice match such as AlAs/GaAs or InGaAs/InP are grown epitaxially on each other.

In an ideal metal/semiconductor junction (Schottky contact) the metal (Bloch) wave functions tail into the semiconductor in the energy range in which the conduction band of the metal overlaps the forbidden band of the semiconductor (Figure 14a). Since in the forbidden band no electronic states of the semiconductor exist, the tails of the metallic Bloch states have to be represented by a superposition of bulk semiconductor valence and conduction band states. Similar to surface states these "metal-induced gap states

(MIGS)" are locally restricted to the metal/semiconductor interface region (extension 5–10 Å to both sides) and they are composed of bulk valence and conduction band states (Figure 14b). They are thus donor-like in the lower part of the forbidden band and acceptor-like in the upper part. Their neutrality level W_N usually lies somewhere near midgap. The Fermi level W_F is usually pinned near W_N and both for *n*-tpye and *p*-type semiconductors electron and hole depletion layers result below the metal contact. These depletion layers constitute the Schottky barrier $e\phi_{SB}$, which has to be overcome by electrons or holes injected from the metal into the semiconductor (Figure 14c).

Similar rules as for Fermi-level pinning on a vacuum/solid interface or on a Schottky contact govern the electronic properties of semiconductor heterostructures. When two semiconductors with differing band gap are grown epitaxially on each other, there are energetic regions ΔW_C and ΔW_V (valence band offsets, Figure 15), where the continuum of conduction or valence band states of one semiconductor (with smaller band gap e.g. GaAs) touches the forbidden band of the large band gap semiconductor (e.g. AlAs). As in a Schottky contact the Bloch states of the low gap material tail into the forbidden band of the high gap semiconductor. Similar reasoning as for Schottky contacts leads to the conclusion that the band structures of both semiconductors match each other such that the neutrality levels W_N of both semiconductors nearly have to be matched at the interface (apart from slight deviations due to charge transfer within the chemical interface bonds). This rule determines the band offsets ΔW_V and ΔW_C as characteristic values for a special semiconductor combination.

The location of the Fermi level W_F with respect to the band edges is determined deep in the bulk of both semiconductors by the doping level. Furthermore, in thermal equilibrium W_F must be equal on both sides of the semiconductor heterojunction. In order to fulfill both requirements, material-specific well-defined band offsets ΔW_V, ΔW_C and fixed Fermi-level positions deep in the bulk of both semiconductors, particular band bendings with certain types of space charge layers, have to occur on both sides of the semiconductor heterointerface. For a special situation, *n*-doped high gap semiconductor on nearly intrinsic low-gap material, the interface electronic band scheme is shown in Figure 15c.

Due to the upward band bending within the wide gap semiconductor, bulk donor states are emptied within the space charge depletion layer, electrons originating from these donor levels are collected in a potential well at the interface within the narrow gap semiconductor. These electrons are squeezed together within a layer of thickness 1 to 2 nm, they form a quasi-two-dimensional electron gas (2 DEG) with free electron movement parallel to the interface. The narrow gap material is not intentionally doped; the electrons which form the 2 DEG originate from ionized donor atoms which are spatially separated from the 2 DEG carriers. Impurity scattering even at high dopant concentrations plays a negligible role. At low temperatures the electrons in the 2 DEG reach mobilities which are higher by orders of magnitude than in bulk semiconductors, e.g. up to 10×10^6 cm^2/Vs in AlGaAs/GaAs heterostructures at low temperature. These heterostructures, preferentially based on AlGaAs/GaAs, InGaAs/InP and AlGaN/GaN material systems are the basis for the fabrication of the fastest field-effect transistors (FETs), so called HEMTs (high electron mobility transistors). Their application in wireless telecommunications and radar systems is also promoted by the extremely low noise figures due to the two-dimensionality of the channel together with the lack of impurity scattering centers. Furthermore, the high mobility 2 DEGs are basic systems for studying quantum effects such as (fractional) quantum Hall effect and transport in low-dimensional structures.

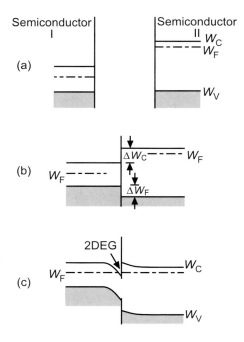

Figure 15: Formation of a semiconductor heterostructure shown qualitatively in its band-scheme representation.
(a) The two semiconductors: (I: moderately n-doped and II: highly n-doped) are separated in space.
(b) The two semiconductors are brought into contact, but are not in thermal equilibrium as the Fermi levels do not coincide.
(c) The two semiconductors are in contact with thermal equilibrium established; ideal interface without any interface states in the common forbidden band. The band bending is again determined by charge neutrality near the interface. In a simple model based on interface states in the energy range of the band offsets ΔW_C and ΔW_V (similar to the MIGS in Figure 14) the bulk neutrality levels of both semiconductors I and II have to coincide (approximately). This condition determines ΔW_C and ΔW_V as characteristic values for a particular interface. The band bending is additionally determined by the bulk doping in semiconductor I and II.

5 Low-Dimensional Structures

5.1 Confined Electronic States

The high mobility 2-DEG at a modulation-doped AlGaAs/GaAs heterostructure is one example of a low-dimensional semiconductor system with exciting new transport properties which are relevant for application. Meanwhile, there is a whole variety of low-dimensional systems with electron confinement such as 2-D quantum wells (films), 1-D quantum wires or 0-D quantum dots which exhibit interesting new physical properties. 0-D quantum dots, for example, might be of great interest for better electron confinement within the active layers of a semiconductor laser. Most of these low-dimensional structures are prepared by epitaxial layer growth in MBE (molecular

Fundamentals

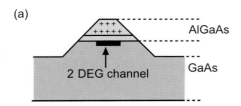
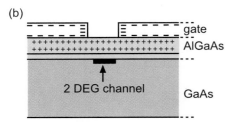
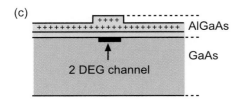

Figure 16: Schematic cross sectional views of three different ways to define narrow 2-DEG channels in an AlGaAs/GaAs heterostructure. The positively ionized donors and the negatively charged 2-DEG channel as well as the negative Schottky gate electrode (b) are indicated.
(a) Lithographically structured wire on a modulation doped (AlGaAs n-doped) AlGaAs/GaAs heterostructure.
(b) 2-DEG channel formed in the 2-DEG at an AlGaAs/GaAs heterostructure by the action of two evaporated metal gates. If biased negatively they repel the electrons in the 2-DEG below.
(c) A similar effect on the electron concentration in the 2-DEG is obtained by spatially varying ionized donor concentration (positive charges) in the upper AlGaAs layer.

beam epitaxy) or MOVPE (metal organic vapor phase epitaxy) and subsequent lateral structuring by electron beam lithography or lateral self-organization in particular growth modes (e.g. Stranski-Krastanov). Figure 16 shows some selected examples for the preparation of narrow 2-DEG channels (quantum wires) in an AlGaAs/GaAs heterostructure.

Confinement of carriers within these 0-D, 1-D, 2-D structures becomes important when the spatial extent of the confined dimension l is of the order of the Fermi wavelength $\lambda_F = 2\pi/k_F$ that depends on the electron density. For metals with high free electron concentrations in the 10^{22} cm^{-3} range λ_F is in the order of or below 1 nm, whereas in moderately doped semiconductors with carrier concentrations in the 10^{17} cm^{-3} range λ_F reaches values of about 50 nm. This length scale is accessible in modern nanostructuring techniques, and therefore semiconductor nanostructures are more easily used in order to study confinement effects.

In order to model confinement effects within nanoscopic length scale dimensions l we assume infinitely high potential barriers on both sides of l (thickness of quantum well or diameter of quantum dot) while in the other directions of macroscopic extent L we use periodic boundary conditions as in standard bulk solid state physics [see eq. (26)]. We then have standing wave solutions in the confined dimension(s). With the boundary condition that at $x = 0$ and $x = l$ the wave function must vanish we obtain in the case of 2-D macroscopic dimensions (quantum well or film):

$$\psi_{n,k_y,k_z} = \sqrt{\frac{2}{l}} \, \sin\left(\frac{n\pi}{l} x\right) \frac{1}{L} e^{i(k_y y + k_z z)} \qquad (46)$$

and energy eigenvalues

$$W_{n,k_y,k_z} = W_n + W_{k_y,k_z} \qquad (47a)$$

with

$$W_{k_y,k_z} = \frac{\hbar^2}{2m_\perp^*}(k_y^2 + k_z^2), \quad W_n = \frac{\hbar}{2m_\parallel^*} \frac{n^2\pi^2}{l^2} \qquad (47b)$$

$m_\perp^*$ and $m_\parallel^*$ are the effective masses of the carriers perpendicular and parallel to the quantum well (confinement sheet). While the spectrum of W_{k_y,k_z} is quasi-continuous because of the macroscopic L the spectrum of W_n is truly discrete because of l being mesoscopic or nanoscopic (typically 1 to 10 nm). For 1-D or 0-D confinement free motion is only possible along one (quantum wire) or zero dimension (quantum dot). We observe that the reduced dimensionality of the electron gas has a strong influence on the form of the density of states $D(W)$ (26). Both the dispersion relation $W(k) = \hbar^2 k^2/2m$ and the shape of the volume V_k enclosed by an energy surface in k-space (sphere in 3D, circle in 2-D, line in 1-D, dot in 0-D) enter the general formula for $D(W)$:

$$D(W) \, dW = 2\frac{1}{(2\pi)^\nu} \int_{V_k} d^\nu k \, \delta(W - W(\boldsymbol{k})) \qquad (48)$$

The front factor 2 takes into account the two possible spin orientations and ν counts the dimensions ($\nu = 0, 1, 2, 3$). In 3-D the well known $\sqrt{W}$ dependence of $D^{(3)}(W)$ is obtained by

$$D^{(3)}(W) \, dW = \frac{2}{(2\pi)^3} 4\pi k^2 dk = \frac{m}{\pi \hbar^3} \sqrt{2mW} \, dW \qquad (49)$$

while in 2-D the density of states becomes constant according to

$$D^{(2)}(W) \, dW = \frac{2}{(2\pi)^2} 2\pi k \, dk = \frac{m}{\pi \hbar^2} dW \qquad (50)$$

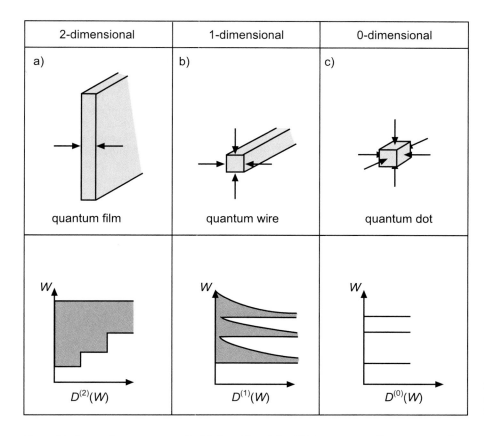

Figure 17: Confined structures in 2, 1 and 0 dimensions (top) and corresponding densities of states $D^{(i)}(W)$ (bottom).

Since this is true for every so-called 2-D subband W_n (47a) corresponding to z-quantisation perpendicular to the 2-D sheet, a step-like total density $D^{(2)}(W)$ results for a 2-DEG (Figure 18a). In one dimension for a quantum wire we obtain

$$D^{(1)}(W) \, dW = \frac{2}{2\pi} dk = \frac{m}{\pi \hbar} \frac{1}{\sqrt{2mW}} dW \tag{51}$$

i.e. a strongly peaked density for each subband (Figure 17b). This peaking is fully developed for a quantum dot in 0D, where sharp quantized energy levels result as in an artificial atom (Figure 17a).

These different shapes of the densities of states have dramatic effects on the electronic properties of confined states. While in 3-D structures energetic continua of electrons and holes can recombine during optical emission in lasers, e.g., in 1-D or 0-D structures electrons and holes are arranged in more or less sharp energy levels. The emitted photon energy is much sharper, i.e. centered around a mean value.

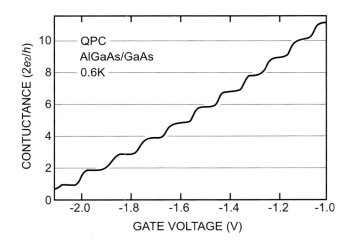

Figure 18: Quantized conductance of a quantum point contact (QPC) at 0.6K prepared at a AlGaAs/GaAs interface (2-DEG). The conductance was obtained from the measured resistance after subtraction of a constant series resistance of 400 Ω (After [8]).

5.2 Quantum Transport

Electronic transport in one dimension, e.g. in a quantum wire, is a particularly interesting phenomenon. Scattering effects can be neglected for sufficiently short wires, where the mean free path of electron transport λ_f is larger than the length of the wire. For a high mobility 2-DEG at an AlGaAs/GaAs heterointerface at low temperature λ_f amounts to several micrometers along which ballistic transport is possible. 1-D channels, i.e. quantum wires, might be prepared within 2-DEGs by several techniques (Figure 16). According to Figure 17b the density of states $D^{(1)}$ in such a system consists of several peak-like structures ($D^{(1)} \propto W^{-1/2}$) which are attributed to the several quantized subbands in which electronic transport can occur as in 1-D channels. We can quite easily calculate the resistance or conductivity of such a 1-D channel for ballistic transport : When we apply a certain voltage V to the channel the potential drop is given by

$$W_{F,l} - W_{F,r} = eV \tag{52}$$

where $W_{F,l}$ and $W_{F,r}$ are Fermi energies (i. e. the chemical potentials of the electron) on the left and right hand side of the quantum wire. The total current for one subband ν then follows as

$$I_\nu = \int_{W_{F,r}}^{W_{F,l}} eD_\nu^{(1)}(W)\ v_\nu(W)\ dW \tag{53}$$

with $v_\nu(W) = h^{-1}(dW_\nu/dk)$ as the electron group velocity in that channel. Since for one dimension the density of states can be written as $D_\nu^{(1)} = (2\pi)^{-1}(dW_\nu/dk)^{-1}$ for one spin direction one obtains:

$$I_\nu = \frac{e}{h}(W_{F,l} - W_{F,r}) = \frac{e^2}{h}V \tag{54}$$

The conductance of a ballistic 1-D channel is therefore quantized in so-called conductivity quanta e^2/h (corresponding to the resistivity quantum of $h/e^2 = 25.812807$ kΩ), which are observed in several experiments with 1-D conduction channels. For two-spin orientations (spin degeneracy) the 1-D conductance is, of course, quantized in units of $2e^2/h$. This conductance quantization has been observed experimentally (Figure 18) [8] in short 1-D channels prepared in an AlGaAs/GaAs 2-DEG by the action of metallic gate electrodes deposited on top of the layer structure (Figure 16b).

A general description of non-diffusive quantum-mechanical transport in nano- and mesoscopic structures has been developed by Landauer [9] and Büttiker [10]. In the idealized model two reservoirs 1 and 2 characterized by their chemical potentials $W_{F,1}$ and $W_{F,2}$ are connected through two ideal 1-D wires of length L. In these wires the electronic states are plane waves $\psi(x) = \exp(ikx)/\sqrt{L}$, which can have positive and negative k-vectors and two spin orientations. With the quantum-mechanical expression for the current density in one dimension

$$j = \frac{e\hbar}{2im}\left(\psi^*\nabla\psi - \psi\nabla\psi^*\right) \tag{55}$$

the corresponding current for one k-vector and one spin orientation is obtained as

$$I = \frac{e\hbar k}{mL} \tag{56}$$

Between the two wires an energetic barrier for the electrons is assumed, which is characterized by its quantum-mechanical reflection coefficient R and its transmission coefficient T, with $T + R = 1$ (Figure 19).

Due to the difference $W_{F,1} - W_{F,2}$ of the chemical potentials a current through the wires is induced. It results from electrons with energies $W_{F,1} \geq W \geq W_{F,2}$ and k-vectors in positive forward direction, which occupy the electronic states in the left wire. Part of the current is reflected at the barrier and the other part is transmitted. The reflected current is absorbed in reservoir 1 while the transmitted part is absorbed in reservoir 2. Only within these reservoirs does energy dissipation occur. The total current in the positive k-direction in the left wire is thus obtained by adding up all occupied states and using (56) as

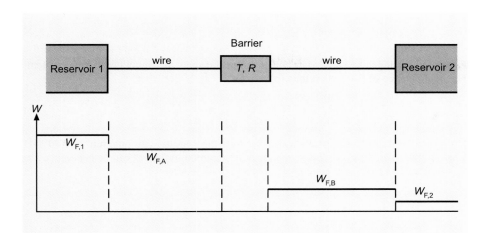

Figure 19: One-channel model for the derivation of ballistic quantum transport through a barrier between two reservoirs. The barrier is described by its transmittance T and its reflectance R. μ_1, μ_2, μ_A and μ_B are the chemical potentials in the different regions.

$$I_{\rightarrow} = \frac{e\hbar k}{mL} L D^{(1)}(W)(W_{F,1} - W_{F,2}) \qquad (57)$$

$D^{(1)}(W)$ is the 1-D density of states (per wire length L) according to (51). With $W = \hbar^2 k^2 / 2m$ this yields

$$I_{\rightarrow} = \frac{e}{\pi\hbar}(W_{F,1} - W_{F,2}) \qquad (58)$$

and for the transmitted net current in the right wire

$$I = \frac{e}{\pi\hbar} T (W_{F,1} - W_{F,2}) \qquad (59)$$

The reflected current I_R in the left wire is accordingly

$$I_R = \frac{e}{\pi\hbar} R (W_{F,1} - W_{F,2}) \qquad (60)$$

For the determination of the chemical potentials the total number of states in the wires, with positive and negative k-values, has to be taken into account, i.e. $2D^{(1)}(W)(W_{F,1}-W_{F,2})$. In the right wire the current, which is induced by $(W_{F,1}-W_{F,2})$, corresponds to a complete occupation of states between $W_{F,B}$ and $W_{F,2}$ (Figure 19), such that

$$TD^{(1)}(W)(W_{F,1} - W_{F,2}) = 2 D^{(1)}(W)(W_{F,1} - W_{F,2}) \qquad (61)$$

Within the left wire both the currents $I_{\rightarrow}$ and I_R have to be considered and the resulting occupation of states is assumed to correspond to the occupation of states between $W_{F,A}$ and $W_{F,2}$, such that

$$(1+R) D^{(1)}(W)(W_{F,1} - W_{F,2}) = 2 D^{(1)}(W)(W_{F,A} - W_{F,2}) \qquad (62)$$

With R+T = 1 the difference between (61) and (62) yields

$$W_{F,A} - W_{F,B} = R(W_{F,1} - W_{F,2}) \qquad (63)$$

With V as the voltage between both wires and $eV = W_{F,A} - W_{F,B}$ one obtains from (59) and (63) for the current through the wires

$$I = \frac{e^2}{\pi\hbar}\frac{T}{R}V = \frac{2e^2}{h}\frac{T}{R}V \qquad (64)$$

This is the analogue to Ohm's law for quantum transport through a nanoscopic system. The conductance of the system thus follows as

$$G = \frac{2e^2}{h}\frac{T}{R} \qquad (65)$$

This so-called Landauer formula again contains the conductivity quantum e^2/h of 1-D quantum transport. The Landauer formalism for quantum transport can be generalized to a network where several wires connect a barrier with reservoirs.

6 Superconductivity

Normal electronic transport under the action of an electric field is diffusive and connected with a finite electrical resistance. The major scattering processes are defect and phonon scattering as outlined in Sec. 3. In 1910 Kammerlingh Onnes detected the phenomenon of superconductivity on Hg. This unusual phenomenon subsequently appeared to be due to a phase transition in which a large number of materials lose their electrical resistance below a certain critical temperature T_c and simultaneously become ideal diamagnets (Meissner-Ochsenfeld effect). An external magnetic field is completely pushed out from a superconductor, the second characteristic property of this unusual state of matter. According to London and London a phenomenological description of these effects is possible by modifying the Drude-type transport Eq. (36) with the assumption of negligible scattering, i.e. vanishing resistance ($\tau \to \infty$, $\rho \to 0$). Thus the 1st London equation for the supercurrent j_s is obtained from (36) as

$$\frac{d}{dt}\bm{j_s} = \frac{n_s e^2}{m^*}\bm{E} \qquad (66)$$

with n_s as the density of superconducting electrons. Using Faraday's law of induction

$$\nabla \times \bm{E} = -\dot{\bm{B}} \qquad (67)$$

one obtains with (66):

$$\frac{d}{dt}\left(\frac{m^*}{n_s e^2}\nabla \times \bm{j} + \bm{B}\right) = 0 \qquad (68)$$

With the ad-hoc assumption of London that the integration constant of (68) is zero, the 2nd London equation follows as:

$$\nabla \times \bm{j_s} = -\frac{n_s e^2}{m^*}\bm{B}, \quad \lambda_L = \frac{m^*}{n_s e^2} \qquad (69)$$

From both London equations (66) and (69), together with Maxwell equations, one obtains:

$$\Delta \bm{B} = \frac{\mu_0}{\lambda_L}\bm{B}, \quad \Delta \bm{j_s} = \frac{\mu_0}{\lambda_L}\bm{j_s} \qquad (70)$$

These equations predict that supercurrents and magnetic fields in a superconductor can only exist in a layer of thickness $\Lambda = \sqrt{\mu_0/\lambda_L}$ at the surface. $\bm{B}$ and $\bm{j_s}$ decay exponentially into the interior of the superconductor ($\Lambda \sim 10^2 - 10^3$ Å for $T \ll T_c$). The superconducting current within the outermost layer of the superconductor is responsible for building up an internal magnetic moment which is opposite to that of the external field and thus compensates the external field in the interior. The interior of the superconductor is free of a magnetic field, the situation of an ideal diamagnet.

Only in 1959 were Bardeen, Cooper and Schrieffer (BCS) able to explain the effect atomistically. In the atomistic BCS theory of superconductivity the ground state of a Fermi gas of free electrons is unstable in the presence of a weak electron-phonon interaction (Fröhlich interaction). An electron traveling through the crystal lattice leaves behind a deformation trail in the lattice, which can be regarded as an accumulation of the positively charged ion cores, i.e. as a tiny compression of the lattice planes. This means that, temporarily, a region of enhanced positive charge is created behind the electron, and this exerts an attractive force on a second electron. This electron-electron coupling thus reduces the total energy of the system. Compared with the high electron velocity v_F ($10^7 - 10^8$ cm/s), the lattice follows very slowly (estimated by the Debye frequency ω_D). Thus the coupling of two electrons into a Cooper pair ($k\uparrow, -k\downarrow$) occurs over distances of more than 1000 Å, the order of magnitude of the extension of a Cooper pair in a BCS superconductor. This extension is also a lower limit for the coherence length ξ in a type I (BCS) superconductor. The interaction described is thus related to lattice-mediated electron-electron scattering. It is assumed to be active within an energy shell of thickness $2\hbar\omega_D$ symmetrically around the Fermi energy W_F. In the simplest approximation, the interaction energy is assumed to be a constant V_0 which reduces the total energy of the system. In calculating the ground state energy W_{BCS} of the ensemble of Cooper pairs one has to take into account that Cooper-pairs, because of their opposite electronic spins, are bosons (spin 0) and therefore assume one and the same many-particle ground state rather than a ladder of states as in a Fermi gas of free electrons in a metal. BCS calculate a ground state energy in which only that part of the total energy responsible for Cooper pairing is taken into account. All other energy contributions, e.g. those due to phonons and chemical bonds, are assumed to be identical in the normal and the superconducting phase. The new ground state energy of the superconductor, i.e. the energy of the many-Cooper-pair condensate W_{BCS}, is reached by phonon-mediated electron-electron scattering leading to Cooper pairing, which decreases the total energy of the whole system. On the other hand, Cooper pairing due to electron-electron scattering requires a certain occupation of single electron states above W_F and some empty states below W_F; even at $T = 0$ the otherwise sharp Fermi edge must be "washed out" in the occupation statistics of a superconductor. Nature risks a certain loss in kinetic energy by electrons being excited above W_F "in the hope" of winning even more energy by electron scattering with Cooper pairing. The total ground state energy W_{BCS} thus contains a first term

$$W_{BCS}^{(1)} = 2\sum_k v_k^2 W_k \qquad (71)$$

with $W_k = \hbar^2 k^2/2m - W_F$ as the surplus energy of an electron above W_F, which describes the energy loss due to electrons which are excited to states above W_F with probability v_k^2 in order to form Cooper pairs ($k\uparrow, -k\downarrow$). When these states above W_F are being filled up, it happens in pairs: if the state $k\uparrow$ becomes occupied, the state $-k\downarrow$ must also be occupied. A second term $W_{BCS}^{(2)}$ describes the energy gain (minus sign) by electron-electron scattering related to Cooper pairing. This term is the average potential energy calculated for the many-electron ground state $\phi(r_1, r_2 ... r_\nu ... r_N)$

$$W_{BCS}^{(2)} = -\int \phi^* \hat{V} \phi \, d^N r \qquad (72)$$

The ground state wave function can be expressed as a series

$$\phi = \sum_n c_n \psi_n(r_1, ... r_\nu ... r_N) \qquad (73)$$

of many body wave functions ψ_n which each describe one possible configuration for Cooper pairing. It is important that for scattering from ($k\uparrow, -k\downarrow$) into ($k'\uparrow, -k'\downarrow$) the states $k\uparrow, -k\downarrow$ must be occupied and the states $k'\uparrow, -k'\downarrow$ empty. One single function ψ_n is described by one set of possible ($k\uparrow, -k\downarrow$), ($k'\uparrow, -k'\downarrow$) configurations. The mutual scattering of two coupled electrons ($k\uparrow, -k\downarrow$) to another state ($k'\uparrow, -k'\downarrow$) can then be described as a transition from the state ψ_n to another state ψ_m and (72) is obtained as

$$W_{BCS}^{(2)} = -\int \phi^* \hat{V} \phi \; d^N r = -\sum_{m\,n} c_n^* c_m \int \psi_n^* \hat{V} \psi_m \; d^N r \qquad (74)$$

According to BCS the interaction matrix element is assumed to be constant ($= V_0$) within the energy range $2\hbar\omega_D$ and $W_{BCS}^{(2)}$ contains essentially the probability amplitudes c_m, c_n for the states ψ_m, ψ_n. Expressed by probabilities v_k^2 and u_k^2 for occupation and non-occupation of the Cooper-pair state ($k\uparrow$, $-k\downarrow$), respectively, the probability amplitudes c_m, c_n can be written as

$$a_n = \sqrt{v_k^2(1-v_{k'}^2)} = v_k u_{k'} \quad \text{with} \quad u_k^2 = 1 - v_k^2, \qquad (75)$$

$$a_m = v_{k'} u_k$$

and one obtains

$$W_{BCS}^{(2)} = -V_0 \sum_{k,k'} v_{k'} u_k v_k u_{k'} \qquad (76)$$

The total energy of the superconducting ground state

$$W_{BCS} = 2\sum_k v_k^2 W_k - V_0 \sum_{k,k'} v_{k'} u_k v_k u_{k'} \qquad (77)$$

contains the amplitudes u_k, v_k so far as free parameters. They will be determined by minimizing W_{BCS}. This requires that v_k^2 satisfies the equation $\partial W_{BCS}/\partial v_k^2 = 0$ with the additional condition

$$v_k^2 + u_k^2 = 1 \qquad (78)$$

Some calculation gives the result

$$v_k^2 = \tfrac{1}{2}\left(1 - \frac{W_k}{\sqrt{W_k^2 + \Delta^2}}\right) \qquad (79)$$

which indeed resembles a "washed-out" step function if one plots v_k^2 as a function of energy.

The energy of a single electron is obtained as

$$\varepsilon_k = \sqrt{W_k^2 + \Delta^2} \qquad (80a)$$

with

$$\Delta = V_0 \sum_k v_k u_k \qquad (80b)$$

From (80) we infer that the lowest possible energy of a free electron ε_k in a superconductor ($W_k = 0$) must be Δ (above the ground state). In a superconductor a gap energy Δ opens between the Cooper-pair ground state W_{BCS} and the excited one-electron states. The gap Δ is temperature-dependent. It has its maximum at $T = 0$ and vanishes at the critical temperature T_c, where superconductivity disappears ($v_k = 0$) (Figure 20). In the BCS approximation the gap energy Δ is calculated as

$$\Delta \cong 2\hbar\omega_D \exp\left(-\frac{1}{D(W_F)\;V_0}\right) \qquad (81)$$

where ω_D is the Debye frequency and $D(W_F)$ the electronic density of states at the Fermi level.

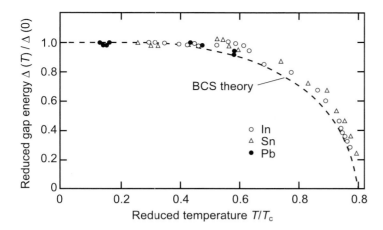

Figure 20: Temperature dependence of the superconductor gap energy $\Delta(T)$ relative to the value $\Delta(0)$ at $T = 0$K for In, Sn and Pb. Experimental data are compared with a calculated curve predicted by BCS theory (dashed line). (After [11]).

The gap energy $\Delta(0)$ at $T = 0$ is, of course, related to the specific critical temperature T_c, above which superconductivity disappears. Within the BCS approximation the simple relation

$$\Delta(0)/k_B T_c = 1.764 \qquad (82)$$

can be derived, which is approximately fulfilled for classical so-called type I superconductors such as Sn, In, Al, Zn etc. Typical critical temperatures for these materials are in the range of a few Kelvin, while their gaps $\Delta(0)$ amount to some meV.

The existence of the temperature dependent gap $\Delta(T)$ between the superconducting ground state W_{BCS} of boson-like Cooper-pairs and the first excited one-electron states is responsible for the negligible electrical resistance below T_c. Electrons which are paired in Cooper pairs in a superconductor cannot undergo inelastic scattering processes unless the electrons gain enough energy so that the superconductor gap Δ is overcome. Inelastic scattering is "switched off", it cannot give rise to electrical resistance. Since at least two electrons have to be excited in order to destroy a Cooper pair, the energy gain in the external field must be equal to 2Δ. Once electrons have gained enough energy in an external electric field so that Cooper pairs are broken, superconductivity breaks down when enough single electrons are excited above Δ to carry a normal current. This explains the existence of an upper critical current density

$$j_c \simeq \frac{e n_s \Delta}{\hbar k_F} \qquad (83)$$

beyond which superconductivity breaks down. n_s is the density of superconducting electrons and k_F the Fermi wave vector. Since a current-carrying wire is surrounded by an external magnetic field, there is also an upper critical magnetic field strength, H_c, which can be sustained by a supercurrent. External magnetic fields exceeding H_c destroy superconductivity.

How is the Meissner-Ochsenfeld effect, i.e. ideal diamagnetism in superconductors, explained? As already pointed out, the wave function of a Cooper pair has dimensions of about 10^3 to 10^4 Å, i.e. a volume of about 10^{-12} cm^3. Assuming 10^{19} to 10^{20} electrons being coupled in Cooper pairs, between 10^6 and 10^7 other Cooper pairs must have their center of mass within the volume of that one Cooper pair. Cooper pairs are not independent of each other, they are "anchored" to each other and highly correlated in their behavior; they form a coherent Bose condensate-like ground state, which is rigid in itself. Their quantum mechanical many-body-wave function extends macroscopically over the whole superconductor. As a rough approximation, where the internal coordinates of the Cooper-pairs are no longer considered (the Bose condensate behaves as a whole rigid entity) the ground state wave function is assumed as

$$\phi = \sqrt{n_s} \exp(i\varphi(r)) \qquad (84)$$

such that the probability density $\phi^*\phi$ equals the density of superconducting electrons n_s. The only coordinate dependence is then given in the phase $\varphi(r)$ and the supercurrent density j_s can be calculated by means of the quantum mechanical current density operator (55).

In the presence of an external magnetic field $\boldsymbol{B} = \boldsymbol{\nabla} \times \boldsymbol{A}$ the momentum operator $\hat{\boldsymbol{p}}$ has to be replaced by

$$\hat{\boldsymbol{p}} = \frac{\hbar}{i}\nabla + e\boldsymbol{A} \qquad (85)$$

and the current density is obtained as

$$\boldsymbol{j}_s = -\frac{e}{2m^*}n_s\left[2\hbar\nabla\varphi(\boldsymbol{r}) + 2e\boldsymbol{A}\right] \qquad (86)$$

Because of $\boldsymbol{\nabla} \times (\boldsymbol{\nabla}\varphi) = \boldsymbol{0}$ one obtains

$$\nabla \times \boldsymbol{j}_s = -\frac{n_s e^2}{m^*}\boldsymbol{B} \qquad (87)$$

With the assumption of a rigid Cooper pair condensate as the superconducting ground state (84) one easily derives the 2nd London equation (69), (87). This, on the other hand, directly leads to the property of ideal diamagnetism as an essential aspect of superconductivity.

7 Conclusions

The present short overview of basic concepts of electronic properties of solids and some quantum effects related to low-dimensional structures is far from being complete. It merely intends to be of some help for a deeper understanding of the following chapters of the book which are closer to present research topics.

Acknowledgements

The editor would like to thank Matthias Schindler (RWTH Aachen) for checking the clarity and consistency of this chapter. Furthermore, he would like to thank Ralf Liedtke (RWTH Aachen) for checking the symbols and formulas in this chapter.

References

[1] H. Ibach and H. Lüth, *Solid State Physics*, Springer, Berlin, Heidelberg, New York, 1996.
[2] J.R. Chelikowsky, M.L. Cohen, Phys. Rev. B**14**, 556 (1976).
[3] R. Courths, S. Hüfner, Phys. Rep. **112**, 55 (1984) and H. Eckhardt, L. Fritsche, J. Noffke: J. Phys. F**14**, 97 (1984).
[4] J.R. Klauder and J.E. Kunzler, *The Fermi-Surface*, Harrison and Webb, eds., Wiley, New York, 1960.
[5] D. Schoenberg, *The Physics of Metals-1, Electrons*, J.M. Ziman, ed., Cambridge, 1969.
[6] D.K.C. McDonald, K. Mendelssohn, Proc. R. Soc. Edinburgh, Sec. A**202**, 103 (1950).
[7] H. Lüth, *Solid Surfaces, Interfaces and Thin Films*, Springer, Berlin, Heidelberg, New York, 2001.
[8] B.J. van Wees, H. van Houten, C.W.J. Beenakker, J.W. Williamson, L.P. Kouwenhoven, D. van der Marel, C.T. Foxon, Phys. Rev. Lett. **60**, 848 (1988).
[9] R. Landauer, IBM J. Res. Dev. **1**, 223 (1957).
[10] M. Büttiker, Y. Imry, R. Landauer and S. Pinhas, Phys. Rev. B**31**, 6207 (1985).
[11] I. Giaever, K. Megerle, Phys. Rev. **122**, 1101 (1961).

Magnetoelectronics – Magnetism and Magnetotransport in Layered Structures

Daniel E. Bürgler and *Peter A. Grünberg*
Department IFF, Research Center Jülich, Germany

Contents

1 Introduction	109
2 Special Anisotropies at Surfaces and Interfaces	110
2.1 Interface (Surface) Anisotropy	111
2.2 Exchange Anisotropy	112
3 Interlayer Exchange Coupling (IEC)	114
3.1 Phenomenological Description	114
3.2 Microscopic Picture: Quantum Well States	116
4 Giant Magnetoresistance (GMR)	118
4.1 Phenomenological Description	118
4.2 Microscopic Picture: Spindependent Scattering	119
4.3 Means to Improve the GMR Effect	120
5 Tunnel Magnetoresistance (TMR)	121
5.1 Phenomenological Description	121
5.2 Microscopic Picture: Spindependent Tunnelling	121
5.3 Influence of Interfaces and Barrier Material	123
6 Current-Induced Magnetic Switching	123
6.1 Phenomenological Description	123
6.2 Microscopic Picture: Spin Accumulation	124
7 Summary	125

Magnetoelectronics – Magnetism and Magnetotransport in Layered Structures

1 Introduction

Electronic circuits and devices generally rely on the fact that electrons possess an electric charge which makes it possible to control the flow of electrons, i.e. electric currents, for example by electric fields. In the context of solid state physics, the term *Electronics* comprises all effects on electric transport that depend on the charge of the electron. The spin is a further fundamental property of electrons. The electronic spins correspond to magnetic moments, which give rise to the magnetism of solids. On the other hand, the spin provides a means to act with a magnetic field on the electrons. Thus, we define the term *Magnetoelectronics* to comprise all influences of the electron spin on electronic transport phenomena.

In ferromagnetic materials the motion of an electron may depend on its spin orientation with respect to the local magnetization. The effects that arise from this fact will be discussed in this Chapter. They are strongest or only appear, if the spin is conserved during the processes of interest. The characteristic length scale for spin conservation, the spin diffusion length, varies in the range from few nanometers (e.g. $Ni_{80}Fe_{20}$ alloy, also called Permalloy) up to several tens of nanometers (e.g. Co) for magnetic alloys and metals and exceeds 100 nm for nonmagnetic metals (e.g. Cu). For this reason thin films, multilayers, and also interfaces play a crucial role in magnetoeletronics. Granular systems, which provide another opportunity to introduce interfaces are not considered here.

The first experiment on thin magnetic films was performed by A. Kundt in 1884. He proved that there is a rotation of the polarization of light when it is transmitted through ferromagnetic metals like Fe, Co, or Ni [1]. In these experiments thin films were a means to overcome the problem of the strong absorption of light in metals rather than the main subject of research. Before, Faraday had seen a similar rotation of the polarization – nowadays called *Faraday rotation* – in a specimen of glass exposed to a magnetic field. For almost a century the investigation of this effect became the main driving force for research in thin magnetic films. Kundt established the proportionality between the angle of rotation and the magnetization component parallel to the light beam. This is called Kundt's law, the proportionality factor being Kundt's constant. While Kundt used electrochemical deposition for the preparation of his films, thermal evaporation became the favored deposition technique due to the tremendous improvements in vacuum techniques by 1950. It enabled research on a more reliable basis and allowed studying of many physical properties of thin magnetic films. As a result, in 1968 surface anisotropy (or more generally interface anisotropy, see Sect. 2) was experimentally observed for the first time [2], after it had been predicted by Néel already in 1954. In fact, interface anisotropy was the first intrinsic, magnetic interface property discovered.

In Sect. 2 we will first discuss interface and exchange anisotropy as examples for phenomena that solely appear at interfaces or in layered structures. The main focus will then be on the new effects, namely Interlayer Exchange Coupling (IEC, Sect. 3), Giant Magnetoresistance (GMR, Sect. 4), Tunnel Magnetoresistance (TMR, Sect. 5), and Current-Induced Magnetic Switching also termed Non-Equilibrium Exchange Interaction (NEXI, Sect. 6). They have been discovered only during the past 15 years and immediately attracted a lot of interest due to their high potential for applications in magnetoelectronics.

2 Special Anisotropies at Surfaces and Interfaces

Magnetic anisotropy describes the fact that in a magnetic solid certain directions of the magnetization – the so-called magnetic easy axes – are energetically more favorable than others. It can be expressed by a free energy function of the system that depends on the direction of the magnetization with respect to a coordinate system attached to the sample, e.g. the crystal axes. There are several sources of magnetic anisotropy. Most important in the present context are the shape anisotropy and the magnetocrystalline (or Néel-type) anisotropy.

Shape anisotropy arises from the demagnetising field H_{demag}, which is associated with a magnetized object. It is described by the demagnetising tensor N

$$H_{\text{demag}} = -NM, \tag{1}$$

where M is the magnetization of the sample. The demagnetising field tends to demagnetise the sample in order to lower the fringing field energy

$$W_{\text{demag}} = \frac{\mu_0}{2} V M N M. \tag{2}$$

V is the volume of the sample. Obviously, W_{demag} depends on the direction of M and the shape of the sample (via N). For rotational ellipsoids, N can be calculated and becomes diagonal. The components for the two orthogonal directions perpendicular to the rotation axis are equal and smaller (bigger) than the component parallel to the rotation axis for lens-shaped (cigar-shaped) ellipsoids. Here, we are interested in thin layers with a lateral extent much larger than the film thickness. They can be considered as the limiting case of an extremely flat lens-shaped ellipsoid, for which all components of N are zero except the one for the direction perpendicular to the layer (the rotation axis), which becomes unity. We can now derive the areal energy density (per unit area A) of the demagnetising field from Eq. (2)

$$\sigma_{\text{shape}} = \frac{W_{\text{demag}}}{A} = \frac{\mu_0}{2} d M^2 \cos^2(\vartheta). \tag{3}$$

The magnetization M includes the angle ϑ with the film normal, and d is the film thickness. σ_{shape} is minimal when the film is magnetized in-plane ($\vartheta = 90°$) and maximal for $\vartheta = 0°$. Thus, the film normal is a magnetic hard axis, and the shape anisotropy forces the magnetization to lie in the plane of the film.

Magnetocrystalline anisotropy arises from spin-orbit interaction, which links the spin direction – i.e. the magnetization direction – to the direction of the orbital momentum. In a simplistic picture the quantum number of the orbital momentum characterizes electronic orbitals (e.g. $d_{z^2}, d_{xy}, d_{x^2-y^2}$) of different symmetry. These orbitals are attached to the crystalline lattice in different orientations. Therefore, the spin-orbit interaction is dependent on the direction of the magnetization with respect to the crystal axes. Each disturbance of the crystalline lattice has an influence on the orbitals, in particular on their symmetry, which makes spin-orbit coupling and thus the magnetocrystalline anisotropy sensitive to the local order and symmetry of the lattice. Symmetry reduction due to strain, for instance, causes additional *magnetoelastic anisotropy*.

Phenomenologically, one describes this angular dependence of the free energy by an expansion into power series of the vector components of M. Taking into account the symmetry of the crystalline lattice and neglecting higher order terms one obtains for example for cubic systems (e.g. Fe, Ni)

$$\sigma_{\text{cryst}} = K_1 d \left[\sin^2(2\vartheta) + \sin^2(2\phi) \sin^4(\vartheta) \right] + K_2 d \ldots \tag{4}$$

K_1 and K_2 are phenomenological anisotropy constants. For cubic symmetry, the lowest order term is square and the next term is a power of 6, and thus a third order term. The second order term (i.e. power of 4) is forbidden by symmetry. If the symmetry is broken, the anisotropy constants change and initially forbidden terms may reappear, and the overall anisotropy behavior may change.

Typical magnetocrystalline anisotropy constants (K_1) are of the order of 10^5 J/m^3. According to Eq. (3) the corresponding constant of the shape anisotropy in the limit of a thin film is given by $\mu_0/2\, M^2$ which is of the order of 10^6 J/m^3. Thus, in the limit of thin films shape anisotropy prevails volume magnetocrystalline anisotropy.

2.1 Interface (Surface) Anisotropy

At a surface or an interface the symmetry of the crystal's volume is locally broken and thus gives rise to an additional magnetocrystalline anisotropy, the so-called interface (or surface) anisotropy. It can be predicted from bulk data on magnetocrystalline anisotropy and magnetostriction which both are modified by the presence of an interface [2]. This interface-induced anisotropy may prefer a magnetization direction in the sample plane or perpendicular to it. The latter is of particular interest, because it results in the opposite of what one would expect from the shape anisotropy.

Microscopically, interface and surface anisotropy can be understood by considering the fact that interface or surface atoms experience a different environment (e.g. a lower number of neighbours at a surface) than bulk atoms. This breaking of the symmetry modifies the atomic orbitals and thus also the spin-orbit interaction, which as a result shows an additional angular dependence. Since the spin-orbit interaction is part of the free energy of the system, we end up with a new type of anisotropy: the interface and surface anisotropy.

Extended numerical work of various theory groups established a relation between anisotropy and spin-orbit coupling and hence more generally with the electronic band structure. On this basis the strong perpendicular anisotropy in Co/Pd multilayers reported in 1985 could be explained theoretically. Since Ni has the same number of valence electrons as Pd, the calculations were extended to Co/Ni structures, and it was predicted that a strong interface anisotropy with an easy axis perpendicular to the sample plane should exist at the Co/Ni interface, too. Indeed, experiments on $(Co_1/Ni_2)_{20}$ layered structures with a total thickness of 120 Å showed strong perpendicular anisotropy which orients the magnetization spontaneously perpendicular to the sample plane. This result is displayed in Figure 1, where the magnetization saturates in small fields applied perpendicular to the sample plane ($\perp$) and at large fields when applied in the plane ($\parallel$).

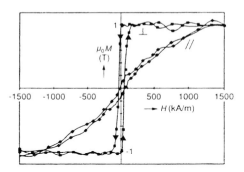

Figure 1: Remagnetisation curves of a $(Co_1/Ni_2)_{20}$ layered structure (the subscripts 1 and 2 in this notation denote the number of monolayers, and the subscript 20 the number of bilayers) that exhibits perpendicular anisotropy. From Ref. [2].

Phenomenologically, interface anisotropy is described by

$$\sigma_{\text{interface}} = K_S \cos^2(\vartheta) \quad (5)$$

Co/Pd	Co/Pt	Co/Ni	Co/Au	Ni/UHV	Ni/Cu	Fe/Ag	Fe/Au	Fe/UHV
-0.92	-1.15	-0.42	-1.28	0.48	0.22	-0.79	-0.54	-0.89

Table 1: Values for K_S in mJ/m² as defined by Eq. (5) for different interface configurations. The free surface is indicated by "UHV". From Ref. [2].

Note that the areal energy density $\sigma_{\text{interface}}$ does not scale with the film thickness d, because we are dealing with a purely interfacial effect. Table 1 displays values for K_S. For negative K_S the minimum of Eq. (5) is obtained for $\vartheta = 0°$ and $\vartheta = 180°$, and the normal to the sample plane is the favoured axis of the interface anisotropy. Therefore, a negative K_S competes with the shape anisotropy. The reorientation from out-of-plane to in-plane takes place at the thickness d_c where the total anisotropy $\sigma_{\text{interface}} + \sigma_{\text{shape}}$ becomes zero. Combining Eqs. (3) and (5) we find

$$d_c = -\frac{2K_S}{\mu_0 M^2}. \quad (6)$$

For Fe on Au we obtain a spontaneous orientation of the magnetization perpendicular to the sample plane for Fe thicknesses below $d_c \approx 2.9$ Å ≈ 2 monolayers (using $M = 1.714 \times 10^6$ A/m and K_S of the Fe/Au system from Table 1).

Apart from choosing the right materials one can increase the influence of the interfaces on the total anisotropy by increasing their density. This was brought to the extreme by alternating just one monolayer of Fe with one monolayer of Au or Pt in Ref. [3]. In-plane saturation fields were around 2 T in the case of $(Fe/Au)_{100}$ films and more than 6 T in the case of $(Fe/Pt)_{100}$. The relationship between electronic structure and interface (surface) anisotropy has recently been demonstrated for the Cu/Co system [4]. In a thin Cu film grown on Co, quantum well states form (see Sect. 3) and give rise to a thickness dependent electronic structure, e.g. thickness dependent density of states at the Fermi level. As a result $\sigma_{\text{interface}}$ of the Cu/Co interface oscillates and even changes sign as a function of the thickness of a Cu layer.

Figure 2:
(a) Hysteresis loop $M(H)$, and
(b) magnetoresistance $\Delta R/R_P$ of a $Fe_{20}Ni_{80}$ (6 nm) / Cu (2.2 nm) / $Fe_{20}Ni_{80}$ (4 nm) / FeMn (7 nm) GMR spin valve at room temperature. The FeMn layer is antiferromagnetic (AFM). The spin valve structure is schematically shown in the inset of (a). Orange pairs of arrows indicate the relative alignment of the magnetizations of the magnetic films. Note the large coercive field H_C of the pinned layer. From the exchange bias field H_E as indicated, $d_{FM} = 4$ nm, and $\mu_0 M_{FM} \approx 1$ T for permalloy, we obtain from Eq. (7) $\sigma_{EB} = 0.13$ mJ/m^2. After Ref. [6].

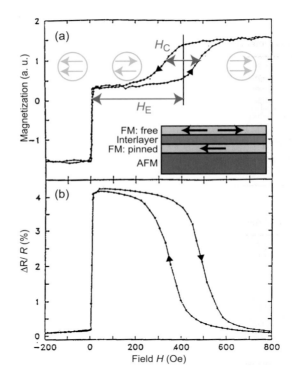

2.2 Exchange Anisotropy

Another type of anisotropy, which can also be classified as an interface anisotropy, is the so-called **exchange anisotropy** that also gives rise to **exchange bias** [5]. It was first seen in 1956 in fine Co particles covered by antiferromagnetic Co oxide, but was soon reproduced in structures of thin films, too. By means of this effect, it is possible to shift the hysteresis curves of ferromagnetic samples on the field axis. This effect is called exchange bias. An example is shown in Figure 2 for a so-called *spin valve structure* [6]. The *free* layer remagnetises in small fields, whereas the hysteresis curve of the *pinned* layer is shifted to positive fields by the **exchange bias field** H_E. The rather wide hysteresis loop (large coercivity H_C) of the pinned film is a further effect that always comes along with exchange anisotropy. The coercivity of the permalloy film that switches at about zero external field is so small that it cannot be resolved in Figure 2. The GMR effect of this spin valve structure shown in Figure 2 (b) will be discussed in Sect. 4.

We can use the interface areal energy density σ_{EB} due to the interaction of H_E with the magnetization M_{FM} of the ferromagnetic film for a description of the strength of the effect. The exchange bias field H_E then is given by

$$H_E = \frac{\sigma_{EB}}{\mu_0 M_{FM} d_{FM}} \qquad (7)$$

d_{FM} is the thickness of the ferromagnetic film adjacent to the antiferromagnet. H_E acts on the FM layer like an external magnetic field H. Thus, the FM layer experiences zero total field, when $H = -H_E$. This is the reason for the shifted hysteresis loop. Some representative values for σ_{EB} are given in Table 2 together with the Néel temperature T_N and the blocking tempearture T_B of the antiferromagnetic material. T_N is the temperature below which antiferromagnetic order among the magnetic moments exists, and T_B is the temperature below which the anisotropy (also existent for antiferromagnetic materials!) is sufficiently large to block the thermal motion of the magnetic moments with respect to the crystal axes. Exchange bias can only be observed at temperatures lower than T_N and T_B.

In a simplistic picture of exchange anisotropy one can assume an antiferromagnet consisting of atomic planes parallel to the surface. In each of them all moments are parallel to each other, but the direction alternates from one plane to the next (Figure 3 (a)). The exchange anisotropy and the shift of the hysteresis loop are due to the exchange interaction between the FM layer and the interface moments of the antiferromagnet which define a preferred direction.

If the alignment in the antiferromagnet is rigid (i.e. very high magnetocrystalline anisotropy in the antiferromagnet corresponding to a high blocking temperature T_B), the exchange bias effect can be estimated for a completely uncompensated surface (i.e. all

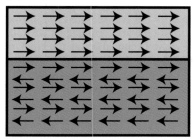
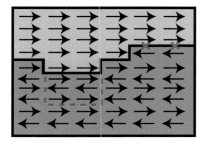
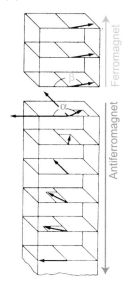

Figure 3:
(a) Exchange anisotropy at an ideal interface: All interface spins of the antiferromagnet are uncompensated and act in the same way on the ferromagnet. Here we assume ferromagnetic coupling across the interface.
(b) When the exchange coupling is stronger than the anisotropy, noncollinear (see angles α and β) configurations are likely: A domain wall in the antiferromagnet is formed when the ferromagnetic film is magnetized to the right. Only one spin sublattice of the antiferromagnet is shown. After Ref. [7].
(c) At real surfaces with roughness the nearest neighbour exchange couplings cannot all be fulfilled simultaneously. Therefore frustration (red crosses) and domain walls (dashed line) occur and the number of uncompensated spins at the interface is reduced, here from 6 in (a) to 2 in (c).

surface spins point in the same direction, see interface in Figure 3 (a)) to be comparable to direct nearest neighbor exchange between the ferromagnet and the antiferromagnet. However, this estimation of σ_{EB} is more than two orders of magnitude too large compared to the experiments. Furthermore, even completely compensated surfaces (the surface layer contains equal numbers of atoms from both spin sublattices, see vertical surface in Figure 3 (a)) show an exchange bias effect which cannot be explained in the simple situation sketched above.

Antiferromagnetic material	σ_{EB} (mJ/m^2)	Néel point T_N(°C)	Blocking temperature T_B(°C)
Fe$_{50}$Mn$_{50}$ (poly-ann)	0.05–0.47	217	150
Ni$_{50}$Mn$_{50}$ (poly-ann)	0.16–0.46	797	497
Pt$_{50}$Mn$_{50}$ (poly-ann)	≤ 0.32	207	127
Ir$_{18}$Mn$_{82}$	0.19	417	265
NiO	0.05–0.29	252	180
CoO	0.14–0.48	20	≤ 20

Table 2: Areal energy density σ_{EB} associated with the exchange bias effect due to various antiferromagnetic materials, as well as related Néel points and blocking temperatures (see text). *poly-ann* means polycrystalline after annealing. Mainly from Ref. [5].

Various models have been discussed in order to cope with these discrepancies. Here, we mention only the domain wall model and the random field model. The domain wall model was introduced by Mauri et al. [7] and attributes the apparent weakening of the interfacial exchange to the fact that the moments in the antiferromagnet are not completely rigid. The interface moments in the antiferromagnet are attached more strongly to those of the ferromagnet than to the antiferromagnetic neighborhood, and domain walls parallel to the sample plane can form in the antiferromagnet (Figure 3 (b)). Note that in this model, the reduction of the apparent ferromagnet/antiferromagnet interface exchange occurs even in the limit of an ideally smooth interface. The exchange bias effect is related to the domain wall energy in the antiferromagnet, rather than to the interfacial exchange coupling strength. This is different for the random field model invented by Malozemoff [8] which is based on interface roughness. In the case of an intrinsically uncompensated surface, roughness leads to the formation of terraces with opposite spin direction, and hence to a mesoscopic compensation (Figure 3 (c)). In the case of an intrinsically compensated surface, however, a small number of uncompensated spins appears due to the step edges. In the compensated and uncompensated case one can assume that planar domains form in the antiferromagnet when the ferromagnet is ordered in an external field and the system is cooled below the blocking temperature. In both models, the formation and motion of domain walls in the antiferromagnet upon field reversal give rise to the experimentally observed increased coercivity.

In general, the exchange bias effect is thought to be due to uncompensated spins at the surface of the antiferromagnet, but their number is much smaller than for an ideal uncompensated surface. A realistic description must include roughness and grain size effects as well as noncollinear spin configurations because any deviation from the ideal situation leads to conflicting interactions: direct exchange between neighbors in the ferromagnet and in the antiferromagnet as well as across the interface (with a certain material dependent sign). Frustrated spin configurations and domains are the result. Perpendicular effective interface coupling where the ferromagnetic moments are oriented perpendicular to the easy axis of the antiferromagnet is such a frustrated configuration. It has been observed in several systems [5] and demonstrates the complexity of the exchange bias effect in real samples.

With the recent invention of X-ray magnetic dichroism spectro-microscopy using synchrotron radiation a new method has become available which allows element selective magnetic domain observation on ferromagnets as well as on antiferromagnets. Using these techniques the domains of an antiferromagnet and the adjacent ferromagnet could be visualized and directly compared [9]. This new type of experiment promises more detailed insight into exchange anisotropy in the near future.

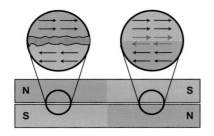

Figure 4: Two permanent magnets align antiparallel due to their fringing fields. In a real experiment (left inset) the two magnets are separated due oxides, contamination, and roughness at the surface. For ideal surfaces (right inset) the antiparallel alignment leads to conflicts at the interface as indicated by the red arrows.

3 Interlayer Exchange Coupling (IEC)

3.1 Phenomenological Description

What happens when two ferromagnets are brought in close proximity? One can try to address this question in a macroscopic experiment with two permanent magnets. They will arrange themselves in an antiparallel manner because the north pole of the first magnet will attract the south pole of the second and vice versa (Figure 4). This type of coupling is due to the fringing fields and thus of dipolar nature. The antiparallel alignment minimizes the fringing field energy. However, in this macroscopic experiment we do not really address the question of interest because an oxide layer, contamination, and roughness prohibit that the ferromagnets come in close proximity, i.e. at a separation of a few Ångstroems, where direct exchange interaction between spins plays a role (left inset of Figure 4). In an idealized experiment, as sketched in the right inset of Figure 4, where we neglect the disturbing effects such that the two ferromagnets can come in direct contact without forming an interface anymore, we arrive in a conflicting situation: neighboring spins of a ferromagnetic material do not align parallel! Therefore, interesting physics may be involved in the problem.

Thin ferromagnetic films separated by a structurally and chemically well defined spacer layer with a controlled thickness in the nanometer range allow to study ferromagnets in close proximity. As we will see below, this arrangement reveals a new type of magnetic interaction, which – as it is often the case on atomic length scales – is a quantum effect as it reflects the wave nature of electrons.

Ferromagnetic films can couple across nonmagnetic interlayers in various ways. When the lateral dimensions are sufficiently small, magnetostatic coupling aligning the magnetizations antiparallel can arise due to the fringing fields at the edge of the sample, similar to the macroscopic situation discussed above. Ferromagnetic interlayer coupling trying to align the magnetizations parallel, on the other hand, can occur as a result of local stray fields produced by interface roughness (Figure 5). This so-called *orange peel* – or *Néel type* – coupling is also of dipolar nature. It is probably present in many cases, but not our main interest here. Generally it is difficult to trace the origin of ferromagnetic interlayer coupling. There is always the extrinsic possibility that pinholes in and/or ferromagnetic bridges across the nonmagnetic spacer exist which give rise to direct exchange between the ferromagnetic films.

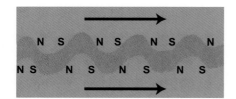

Figure 5: Illustration of ferromagnetic coupling between two magnetic layers due to interface roughness. Stray fields emerging from the protrusions are represented by local north (N) and south (S) poles. The shown parallel alignment of the two magnetic layers minimizes the stray field energy because different poles from the upper and lower film oppose each other. Other interface morphologies may induce antiferromagnetic or even 90°-coupling.

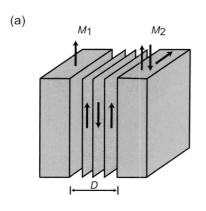
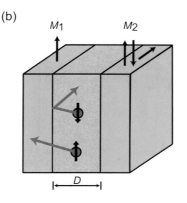

Figure 6: Illustration of two types of interlayer coupling depending on the nature of the interlayer.
(a) The interlayer is assumed to be antiferromagnetic with spin alignment as shown.
(b) A paramagnetic or diamagnetic material for the interlayer is assumed. The magnetization M_1 points upwards. Due to the coupling M_2 can show the alignments parallel and antiparallel (or even perpendicular) with respect to M_1.

An instrinsic coupling mechanism can be obtained, if we extend the idea of direct coupling at interfaces as described in the context of exchange bias in the previous section to a layered structure consisting of a thin antiferromagnetic layer with ferromagnetic material on both sides. This situation is displayed in Figure 6 (a). Obviously, the coupling of the ferromagnets across the antiferromagnet will be antiferromagnetic (ferromagnetic) for two (three) monolayers and any higher even (odd) number of monolayers. This oscillation of the coupling between antiferromagnetic and ferromagnetic with a period of two monolayers has indeed been observed for interlayers of antiferromagnetic Cr [10] and Mn [11] when the growth is sufficiently good. This mechanism is referred to as **proximity magnetism** [12].

However, for somewhat reduced growth quality, longer coupling periods of the order of 10 Å are observed which cannot be explained on the basis just described. This new aspect is even more distinct in the cases of coupling across metallic paramagnetic or diamagnetic interlayers, where a description based on static magnetic order in the interlayer is not possible. As we will see below, in these cases coupling is due to spindependent electron reflectivity at the interfaces as sketched in Figure 6 (b). In 1990 the oscillatory nature of this new type of coupling – the interlayer exchange coupling (IEC) – was recognized as a general phenomenon [10].

Experiments showed that IEC of ferromagnetic 3d metals across interlayers can phenomenologically be described by an areal energy density σ_{IEC} by

$$\sigma_{\text{IEC}} = -J_1 \cos(\vartheta) - J_2 \cos^2(\vartheta). \tag{8}$$

Here, ϑ is the angle between the magnetizations of the films on both sides of the spacer layer. The parameters J_1 and J_2 describe the type and the strength of the coupling. (J_1 and J_2 should not be confused with magnetic polarizations. We use the letter J for the coupling parameters in order to concur with literature.) If the term with J_1 dominates, then from the minima of Eq. (8) the coupling is ferromagnetic (antiferromagnetic) for positive (negative) J_1. If the term with J_2 dominates and is negative, we obtain 90°-coupling. The first term of Eq. (8) is often called bilinear coupling and the second biquadratic coupling. Biquadratic coupling is thought to be mainly due to interface roughness and will not be further considered here (see Ref. [10]).

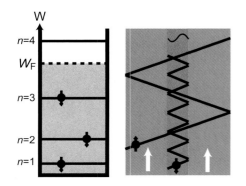
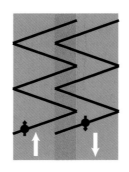

Figure 7: Illustration of spindependent reflectivity at the nonmagnetic/magnetic interfaces for the explanation of oscillatory coupling
(a) for parallel and
(b) for antiparallel alignment. The discrete energy levels of the quantum well states are also shown in (a).

I Fundamentals

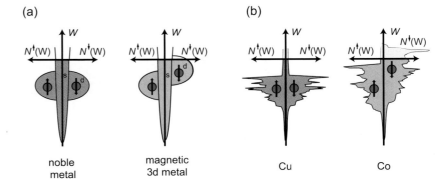

Figure 8:
(a) Schematic spinsplit density of states (DOS) for a noble metal and a 3d transition metal representing the spacer and magnetic layer, respectively. The relative positions of the bands for spin-up and spin-down electrons give rise to spindependent reflectivity at the interfaces.
(b) Realistic spinsplit DOS for Cu and Co showing the qualitative features sketched in (a).

3.2 Microscopic Picture: Quantum Well States

IEC is believed to be due to an indirect exchange interaction mediated by the conduction electrons of the spacer layer. We consider the itinerant nature of electrons in transition metal ferromagnets, which gives rise to a spinsplit band structure and spindependent reflectivities at the paramagnet/ferromagnet interfaces. The spindependent reflectivity is illustrated in Figure 7, where it is assumed that majority (minority) electrons, i.e. electrons with spin parallel (antiparallel) to the local magnetization, are weakly (strongly) reflected at the interfaces. The reason for this behavior is seen in Figure 8. For the spin-up (majority) electrons there is a good match of the states in the ferromagnet and the interlayer (here represented by a noble metal) as indicated by the same position of the bands on the energy scale. A good match of the states means that states with similar symmetry, k vector, and energy exist in both materials. Therefore, electrons in these states can more or less easily move from one material to the other. As it is always the case for an itinerant ferromagnet, the spin-down (minority) bands are shifted on the energy scale with respect to the spin-up (majority) bands due to the exchange interaction associated with the ferromagnetic order. Therefore, the good match with the bands in the interlayer is lost. In other words, minority electrons experience a higher potential step at the interface than majority electrons and are thus reflected with a higher probability. For the minority electrons this gives rise to quantum well states (QWS), i.e. there are spindependent interference effects like the formation of standing electron waves for certain interlayer thicknesses as indicated in the upper part of the interlayer in Figure 7 (a). But, QWS in the interlayer only form for parallel alignment of the magnetizations of the ferromagnetic layers (Figure 7 (a)) because only in this case the minority electrons are reflected on both sides of the spacer.

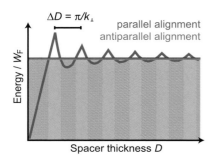

Figure 9: Spacer thickness dependence of the energy of the electronic system (normalized to the Fermi energy W_F) as a function of the spacer thickness D for the parallel and antiparallel alignment. Green (red) areas indicate ferromagnetic (antiferromagnetic) coupling for the corresponding spacer thickness ranges. After Ref. [13].

The description of QWS is similar to electrons in a one-dimensional potential well, except that for the time being we only consider states with a given momentum component perpendicular to the layers $k_\perp$. A justification for this restriction and a method to find the right $k_\perp$ will be given below. The condition to form a standing wave in a well (i.e. a QWS) of thickness D is

$$k_\perp^{(n)} = n\frac{\pi}{D} \quad ; \quad n=1,2,.... \quad (9)$$

These QWS correspond to discrete energy levels (left part of Figure 7a)

$$W_n = \frac{\hbar^2}{2m}\left(k_\perp^{(n)}\right)^2 = n^2\frac{\hbar^2\pi^2}{2mD^2} \quad ; \quad n=1,2,.... \quad (10)$$

Upon increasing the spacer thickness D, these levels move downwards on the energy scale. Each time when a level crosses the Fermi energy W_F, the corresponding QWS are populated, and the energy of the electronic system increases. When the QWS level moves further below the Fermi energy, the energy again decreases until the next QWS level approaches the Fermi energy. For the parallel alignment the energy oscillates as a function of spacer thickness as shown by the green curve in Figure 9. For the antiparallel alignment (Figure 7 (b)) the energy of the system does not show the oscillatory behaviour (red curve in Figure 9). In order to always take the configuration with the lowest energy the alignment switches between parallel and antiparallel, and hence the coupling oscillates. The oscillation period ΔD follows from Eq. (9) for $\Delta n = 1$

$$\Delta D = \frac{\pi}{k_\perp} \quad (11)$$

Due to the similarity of the arrangement in Figure 7 with an optical Fabry-Perot interferometer this model is sometimes called the Fabry-Perot model of oscillatory coupling. In the same way as the transmission of an optical Fabry-Perot for a given wavelength of the light oscillates as a function of the mirror distance, here the coupling oscillates as a function of the interlayer thickness.

For the whole discussion we have assumed that the spin direction of the electrons is conserved inside the spacer layer. Since there is scattering in the spacer layer as in any normal metal, this assumption only holds when the spacer layer is thinner than the characteristic length scale for spin conservation, the spin diffusion length. This condition implicates that IEC can only exist for spacer thicknesses of the order of a few nanometers at most (see Sect. 1).

A more detailed theoretical treatment of IEC is given in Ref. [14]. The basic result is that the oscillations period(s) of the interlayer coupling can be predicted for realistic electronic band structures by considering the Fermi surface of the spacer material. One then finds that oscillatory coupling is related to a so-called critical spanning vector Q in reciprocal space with the following properties: (i) Q points perpendicular to the interface, (ii) Q connects two sheets of the Fermi surface which are coplanar to each other, and (iii) Q is in the first Brillouin zone. The last condition follows from Bloch's theorem and reflects the atomic periodicity of the spacer material. The oscillation period is given by $2\pi/Q$. For real materials, several Q_i (i = 1,2,...) may exist, each of them corresponding to a different oscillation period $2\pi/Q_i$. In this case, the experimentally measured coupling *versus* thickness curve is the superposition of all these oscillations.

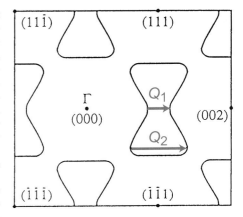

Figure 10: Cross section of the Fermi surface of Au with the critical spanning vectors Q_1 and Q_2 in [001] direction.

As an example we consider an Au spacer layer grown in [001] direction. For the Fermi surface of Au as shown in Figure 10, there are two critical spanning vectors (Q_1 and Q_2) in the [001] direction. The periods of the oscillatory coupling are given by $\Delta D_i = 2\pi/Q_i$ and thus are determined solely by the electronic properties of the interlayer material. Figure 11 (a) shows the result of an evaluation of remagnetization curves for a Fe/Au-wedge/Fe structure grown on an Ag-buffered GaAs(100) substrate. The coupling is strongly ferromagnetic ($J_1 + J_2 > 0$) for small spacer thickness D_{Au}, probably due to pinholes and magnetic bridges. For increasing D_{Au} the ferromagnetic coupling quickly decreases until there are oscillations around zero. Two periods of oscillation are superimposed with amplitudes that are attenuated as a function of the interlayer thickness. Measurements of the coupling strength in a Fe/Au-wedge/Fe trilayer grown on a Fe whisker, by observing the disappearance of antiferromagnetically coupled domains in a Kerr microscope, are shown in Figure 11 (b). Two periods of oscillatory coupling, 2.48 ML and 8.6 ML, were determined from the data in excellent agreement with the predictions based on Figure 10. Both samples used in Figure 11 (a) and (b) were grown very carefully. The stronger coupling for the Fe whisker sample is indicative of the better growth occurring on that almost ideal substrate.

The data of Figure 11 (b) were further analysed taking into account thickness fluctuations of the spacer to obtain *nonaveraged* values of the coupling strength for comparison with the theory. The coupling strengths for the short and long period oscillations were found to be 60% and 15% of the calculated values, respectively. These numbers represent the best agreement between experiment and theory achieved so far. Detailed studies of the influence of the interface morphology [16], [17] on the coupling reveal a strong dependence which is likely to be the reason for the at least one order of magnitude different coupling strengths in Figure 11 (a) and (b).

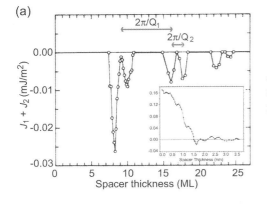
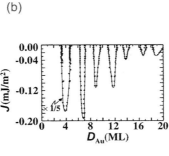

Figure 11: Coupling strength of Fe/Au-wedge/Fe(100) trilayers as a function of the Au interlayer thickness (a) on a Ag-buffered GaAs substrate [15] and (b) on an Fe(100) whisker [10]. The inset in (a) includes ranges where the coupling is ferromagnetic ($J_1 + J_2 > 0$).

I Fundamentals

Apart from the interface quality the strength of the coupling depends also on many details of the participating Fermi surfaces. For a compilation of observed values of J_1 and the associated coupling period(s) see Table 3. As was pointed out experimentally [18] and theoretically [19], favourable conditions for a high contrast in spindependent reflectivity and therefore strong coupling can occur when the material of the magnetic films and the interlayer are taken from the same – or from close – columns of the periodic table. This has to do with the associated equal – or similar – number of valence electrons. Indeed, a record value of -34 mJ/m^2 for the antiferromagnetic coupling has been found for the coupling of Co across Rh (Table 3). Apart from deviations from the most favourable band positions, the interference effects – and hence the strength of the IEC – will also be diminished by electron scattering due to structural imperfections or even an amorphous structure of the interlayer. The coupling across an amorphous interlayer has indeed been found to be very weak ($J_1 \approx -0.04$ mJ/m^2) [20].

Sample	Maximum strength $-J_1$ in mJ/m^2 (at spacer thickness in nm)	Periods in ML and (nm)
Co/Cu/Co (100)	0.4 (1.2)	2.6 (0.47); 8 (1.45)
Co/Cu/Co (110)	0.7 (0.85)	9.8 (1.25)
Co/Cu/Co (111)	1.1 (0.85)	5.5 (1.15)
Fe/Au/Fe (100)	0.85 (0.82)	2.5 (0.51); 8.6 (1.75)
Fe/Cr/Fe (100)	> 1.5 (1.3)	2.1 (0.3); 12 (1.73)
Fe/Mn/Fe (100)	0.14 (1.32)	2 (0.33)
Co/Ru(0001)	6 (0.6)	5.1 (1.1)
Co/Rh/Co (111)	34 (0.48)	2.7 (0.6)

Table 3: Selection of observed bilinear coupling strengths and periods collected from the literature [10].

4 Giant Magnetoresistance (GMR)

4.1 Phenomenological Description

The giant magnetoresistance effect describes the finding that in layered magnetic structures the resistivity depends on the relative alignment of the magnetizations of adjacent ferromagnetic layers [23]. The first experiments are displayed in Figure 12. At zero field adjacent Fe layers align antiparallel due to antiferromagnetic interlayer exchange coupling (Sect. 3) across the Cr spacer, whereas a large enough external magnetic field saturates the sample and forces the Fe layers into a parallel configuration. The transition from the antiparallel to the parallel alignment is accompanied by a drastic change of the resistivity. In the lower part of Figure 12 (b) the so-called **anisotropic magnetoresist-**

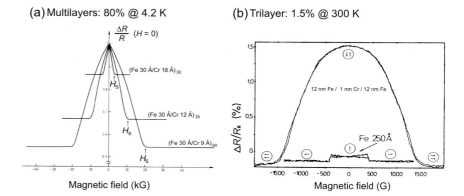

Figure 12: First observations of the GMR effect in (a) Fe/Cr multilayers [21] and (b) Fe/Cr/Fe trilayers [22].

ance (AMR) effect of a 250 Å Fe layer is shown for comparison. AMR describes the dependence of the electric resistivity on the angle between the current and the magnetization direction. AMR is a volume effect discovered in 1857 and applied in read heads since the 1970's. The much larger response of the layered structures is the reason why the new effect was dubbed **giant magnetoresistance** (GMR).

Apart from antiferromagnetic IEC the antiparallel alignment at small fields can also be achieved by hysteresis effects. In the latter case one film is magnetically pinned (e.g. by the exchange bias effect due to an antiferromagnet as discussed in Sect. 2) whereas the magnetization of the other is free to rotate when an external field is applied. Such arrangements are called **spin valves** and are relevant for applications. Figure 2 (b) shows the GMR signal of the spin valve displayed in the inset of part (a) in the same figure. The steep slope of resistance near zero field provides a sensitive signal to measure small magnetic fields.

If we denote by R_P the resistance for parallel alignment of adjacent ferromagnetic films and by R_{AP} the same for antiparallel alignment, then the strength of GMR effects is usually quoted in terms of

$$\frac{\Delta R}{R_P} = \frac{R_{AP} - R_P}{R_P} \qquad (12)$$

Mostly, the resistance is highest for antiparallel alignment yielding a positive $\Delta R/R_P$ corresponding to the so-called normal GMR effect. But there are also cases where the situation is reversed and $\Delta R/R_P$ becomes negative. This is called the *inverse GMR effect*.

The GMR effect has been investigated in two different geometries, namely the **CIP** (Current In Plane) and the **CPP** (Current Perpendicular Plane) geometry. The relative effect is stronger in the CPP geometry. However, due to the extremely unfavorable geometric conditions (lateral dimensions some orders of magnitude larger than the film thickness), the voltage drop perpendicular to the layers – CPP geometry – is very difficult to detect without special structuring. On the other hand, GMR in the CPP geometry can become sufficiently strong in suitably structured devices to be of interest for applications, e.g. GMR based MRAMs. Representative and record values for the GMR effect as defined by Eq. (12) both in the CIP and the CPP geometry have been compiled from the literature in Table 4.

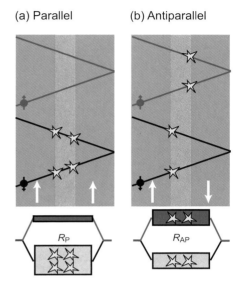

Figure 13: Simplistic picture of spindependent scattering for the explanation of the GMR effect. Only minority electrons are scattered as schematically indicated by the yellow stars. Majority electrons are not scattered and cause a short circuit effect, which appears for parallel alignment of the magnetizations (a) but not for antiparallel alignment (b). The substitutional circuit diagrams in the lower part for the total resistivities, R_P and R_{AP}, yield the relation $R_P < R_{AP}$ and hence a normal GMR effect. This picture holds for both the CIP and CPP geometry.

Sample	$\Delta R/R_P$ (%)	Temperature (K)
[Fe(4.5)/Cr(12)]$_{50}$	220 42	1.5 300
[Co(15)/Cu(9)]$_{30}$	78 48	4.2 300
[Co(8)/Cu(8.3)]$_{60}$	115 65	4.2 300
[Co(10)/Cu(10)]$_{100}$	80	300
Co$_{90}$Fe$_{10}$(40)/Cu(25)/Co$_{90}$Fe$_{10}$(8) ...	7	300
NiFe(100)/Cu(25)/Co(22)	4.6	300
[Co(15)/Cu(12)]$_n$ CPP	170	4.2
[Co(12)/Cu(11)]$_{180}$ CPP	55	300

Table 4: Representative values for GMR. Geometry is CIP unless specially marked with CPP (see text). Auxiliary layers not directly active in the GMR effect are mostly omitted. Numbers in brackets indicate the layer thickness in Å. Compiled from [24].

4.2 Microscopic Picture: Spindependent Scattering

The mechanism leading to GMR can be understood within Mott's two current model [23], [25] which assumes two independent current channels for spin-up and spin-down electrons. Due to their Fermi velocity the conduction electrons propagate with high speed but arbitrary direction through the layered structure. A current results from a much smaller drift velocity in the direction of the applied electric field. In Figure 13 paths between two reflections at outer surfaces are shown with scattering events in between. In order not to confuse the picture the changes in direction due to the scattering events are suppressed. Because of the dominance of the Fermi velocity, the schematic representation and the substitutional circuit diagrams in Figure 13 hold for both CIP and CPP geometry. The scattering processes are the cause of electric resistance. Only states near the Fermi energy contribute to the electric conductivity because they can reach empty final states just above the Fermi energy after a scattering event. In order to demonstrate how spin-dependent scattering leads to the GMR effect, we use in the following a simple – albeit unrealistic – consideration whose main argument is nevertheless valid in reality. In Figure 13 (a) it is assumed that only minority electrons (spin antiparallel to the local magnetization) are scattered at the magnetic/nonmagnetic interfaces. The origin for the spindependent behavior can again be found in the spinsplit DOS of 3d transition metals (Figure 8). They show different numbers of final states (density of states near the Fermi energy) for majority and minority electrons and, hence, different spindependent scattering probabilities. Thus, in our simplified picture, for parallel alignment of the magnetizations, majority electrons are not scattered at all, leading to a short circuit ($R = 0$) of the associated current. Therefore, the resistivity for the total current vanishes, too, as can be seen in the lower part of Figure 13 where the two spin channels are represented by two resistors in parallel connection. For antiparallel alignment of the magnetizations (Figure 13 (b)) there are scattering events for both types of electrons. Hence, the resistivity for the total current is finite. It is clear that even if the above strict condition is relaxed, the resistivity will be higher for antiparallel alignment as compared to the parallel one. For the whole discussion we have assumed that the spin direction of the electrons is conserved inside the spacer layer. Since there is scattering in the spacer layer as in any normal metal, this assumption only holds when the spacer layer is thinner than the characteristic length scale for spin conservation, the spin diffusion length. This condition implicates that GMR can only exists for spacer thicknesses of the order of a few nanometers at most (see Sect. 1). An animation explaining this simple picture of GMR and how GMR is used in read heads of high-density hard disks is available on the internet [26].

For proper material combinations – in particular in the case of low-resistivity interlayers like in Co/Cu/Co structures – it has alternatively been shown that CIP-GMR can also be explained on the basis of spindependent interface reflectivity [27]. For parallel alignment of the magnetizations there can be an electron channeling effect in the interlayer when QWS form due to interface reflectivity (see Sect. 3 and Figure 7). If the interlayer material has the lower resistivity, then there can be an overall low resistance. For antiparallel alignment of the magnetizations spin-down as well as spin-up electrons penetrate in both the ferromagnetic layers and the interlayer, and thus the *quasi shortcircuit effect* due to channeling in the interlayer disappears.

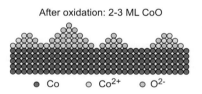

Figure 14: Sketch of surface smoothing of a metal (here: Co) due to preferential oxidation of spikes and bumps.

4.3 Means to Improve the GMR Effect

The larger GMR effect in *multilayers* as compared to trilayers apparent from Figure 12 and from the GMR values in Table 4 can be explained on the basis of an increased spindependent scattering probability when the electrons have to pass many interfaces instead of only two as in Figure 13. Therefore, it should be possible to increase the GMR ratio of trilayers by improving the specular reflectivity for the electrons at the outer surfaces. Under these conditions a trilayer should become equivalent to a multilayer. A method based on oxidation to smooth the outer surfaces and thus to increase the specular reflectivity has been shown to be possible by Egelhoff [28]. It exploits the fact that the oxidation of a rough transition metal surface preferentially removes bumps and spikes, which are converted into an insulating oxide. Hence, the surface of the conducting part of the material becomes smoother (Figure 14). As a result of the increased specular reflectivity, record GMR values ($\Delta R/R_p$) for trilayer spin valves of 19% have been achieved [28]. A spectacular increase of the GMR ratio due to the same effect was recently obtained by Veloso et al. [29] for the $Co_{90}Fe_{10}/Cu/Co_{90}Fe_{10}$ system. These authors do not obtain as high values as in Ref. [28], but the demonstration of the effect is more impressive because they start from a lower value of 6% and receive an increase to 12% by introducing so-called Nano Oxide Layers (NOLs). The NOLs are formed by interrupting the deposition of each $Co_{90}Fe_{10}$ layer and oxidizing its surface. The NOLs are about 15 Å thick, and it is likely that they are magnetic.

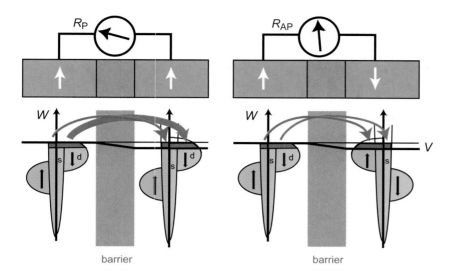

Figure 15: Assuming energy and spin conservation during the tunnelling process, the tunnelling current can be decomposed into spin-up (green arrows) and spin-down (red arrows) contributions. Their magnitudes (represented by the thickness of the arrows) is determined by the number of available initial and final states for each spin channel, here given by the simplified DOS of 3d metals. Hence, the total current and the resistance of the tunnel junction depend on the alignment of the magnetizations.

5 Tunnel Magnetoresistance (TMR)

5.1 Phenomenological Description

The basic configuration for **tunnel magnetoresistance** consists of two ferromagnetic electrodes – here in the form of thin films – separated by an insulating or semiconducting barrier as shown in the upper part of Figure 15. In most cases AlO_x barriers are used. If a voltage V (several tens to hundreds mV) is applied across the stack a small quantum-mechanical tunnelling current can flow across the barrier. This means that – unlike GMR – the TMR effect is always observed in CPP geometry. The magnitude of the tunnelling current is related to the overlap of the exponentially decaying wave functions inside the barrier. Therefore, the current exponentially decreases with the barrier thickness. Typical barrier thicknesses are of the order of 1 nm.

The tunnelling resistance is found to depend on the relative orientation of the magnetizations on both sides of the barrier [30]. Like in the case of GMR we denote by R_P the resistance for parallel magnetizations, and by R_{AP} the resistance for antiparallel alignment. The size of the TMR effect is determined in the same way as for GMR (compare Eq. (12))

$$\frac{\Delta R}{R_P} = \frac{R_{AP} - R_P}{R_P}. \qquad (13)$$

Figure 16 displays TMR curves which are representative for the current state of the art [31]. Record TMR values around 50% have been obtained at room temperature [30].

5.2 Microscopic Picture: Spindependent Tunnelling

The TMR effect can be understood on the basis of spinpolarized tunnelling. When the spin is conserved during tunnelling, a spin-up (spin-down) electron can only tunnel from an initial spin-up (spin-down) state to an unoccupied spin-up (spin-down) final state. As we will see, TMR arises from the imbalance between the number of spin-up and spin-down electrons that contribute to the tunnelling current. Therefore, we define the spin polarizations P_L and P_R of the left and right electrodes

$$P_{L,R} = \frac{N_{L,R}^\uparrow - N_{L,R}^\downarrow}{N_{L,R}^\uparrow + N_{L,R}^\downarrow} \qquad (14)$$

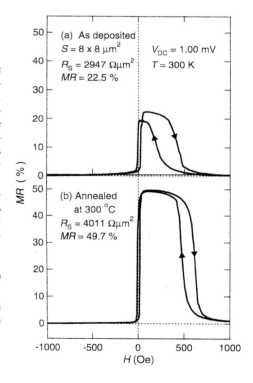

Figure 16: TMR curves measured at room temperature for films of $Co_{75}Fe_{25}$ (4 nm) across barriers of Al_2O_3 (0.8 nm). S = sample area, R_S = sheet resistance for parallel magnetization alignment, V_{DC} = bias voltage, $M_R = \Delta R/R_P$. The observed increase of TMR from the as deposited state (a) to the annealed state (b) is an often observed effect and is likely to be related to improvements of the interface configuration. The asymmetry with respect to $H = 0$ arises from exchange bias due to a 10 nm thick, antiferromagnetic IrMn layer. From Ref. [31].

Fundamentals

Figure 17:
TMR of Co/X/La$_{0.7}$Sr$_{0.3}$MnO$_3$ structures with
(a) X = SrTiO$_3$,
(b) X = Ce$_{0.69}$La$_{0.31}$O$_{1.845}$,
(c) X = Al$_2$O$_3$, and
(d) X = Al$_2$O$_3$/SrTiO$_3$.
The material dependent change from the inverse (a,b) to the normal (c,d) TMR effect indicates the influence of the Co/insulator interface. After Ref. [33].

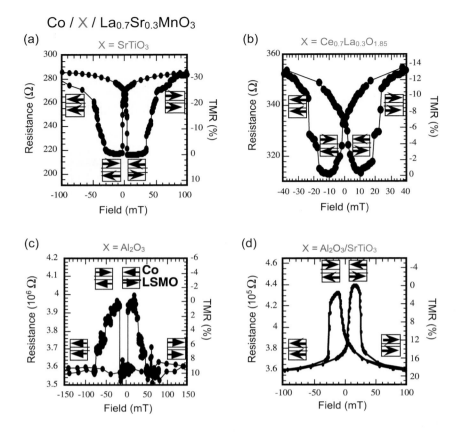

Here, $N^\uparrow_{L,R}$ and $N^\downarrow_{L,R}$ denote the number of states in an energy window at the Fermi level with a width given by the applied voltage V. Only states within this window can contribute to the tunnelling current. In Figure 15 the dark green and red coloured areas in the vicinity of the Fermi level correspond to these quantities. It is assumed that a positive voltage V is applied to the right electrode. The green and red arrows represent the spin-up and spin-down tunnelling currents, respectively, with their thickness indicating the magnitude of the currents. For instance, the spin-up current in the parallel configuration is proportional to the product $N^\uparrow_L N^\uparrow_R$ (green arrow in the left hand part). Obviously, the parallel alignment on the left hand side of Figure 15 gives rise to a larger total current and, thus, to the smaller tunnelling resistance. R_P and R_{AP} are inversely proportional to the total current (i.e. the sum of the spin-up and spin-down currents) and can be written as

$$R_P \propto \frac{V}{N^\uparrow_L N^\uparrow_R + N^\downarrow_L N^\downarrow_R} \quad ; \quad R_{AP} \propto \frac{V}{N^\uparrow_L N^\downarrow_R + N^\downarrow_L N^\uparrow_R} \tag{15}$$

After inserting these expressions in Eq. (13) and some rearranging, $\Delta R/R_P$ can be expressed by the polarizations defined in Eq. (14)

$$\frac{\Delta R}{R_P} = \frac{R_{AP} - R_P}{R_P} = \frac{2 P_L P_R}{1 - P_L P_R}. \tag{16}$$

Usually ΔR is positive, and therefore the TMR effect is called normal. For the inverse effect [32], [33], which has also been observed, ΔR is negative. In Figure 15 the effect turns out to be normal, but an inverse effect can result if the magnetic electrodes on both sides of the barrier were different in such a way that the P_L and P_R have opposite signs (see Eq. (16)). Examples are given below.

The TMR effect usually decreases as a function of bias voltage and temperature, the origin of both effects is so far not clear. Spinscattering in the interlayers as well as the excitation of spinwaves have been considered.

5.3 Influence of Interfaces and Barrier Material

There is a controversy on how to obtain the relevant values for P_L and P_R. Most likely, only the polarizations right at the interfaces, which of course may deviate from the bulk polarizations, are important. Additionally, the spindependent transition probability through the barrier as determined by the complex Fermi surface of the barrier material – which does not appear explicitly in Eq. (16) – has to be taken into account, too. The main effect is that the barrier material changes the decay length of the wave functions as compared to a vacuum barrier. This influence may change the relative contribution of different states to the tunnelling current, and thus the degree or even the sign of the polarization of the tunnelling current.

An impressive demonstration of the influence of the interlayer has been given by De Teresa et al. [33] and is shown in Figure 17. In these experiments the two ferromagnetic electrodes are always made of Co and $La_{0.7}Sr_{0.3}MnO_3$, but different barrier materials are used. For $SrTiO_3$ and $Ce_{0.69}La_{0.31}O_{1.845}$ barriers the TMR was found to be inverse (Figure 17 (a) and (b)), whereas it was normal for Al_2O_3 and $Al_2O_3/SrTiO_3$ barriers (Figure 17 (c) and (d)). These experiments clearly prove that TMR also depends on the barrier material. It is now believed that the spin polarizations P_L and P_R have to be related to interface states which play a major role for the chemical bonding at the interfaces.

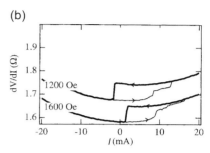

6 Current-Induced Magnetic Switching

6.1 Phenomenological Description

The occurrence of spindependent reflectivity as discussed before in the context of interlayer exchange coupling (Sect. 3) and GMR due to channelling in the interlayer (Sect. 4) enables a further effect, namely nonequilibrium exchange interaction (NEXI) due to spin accumulation which leads to current-induced magnetic switching [34]. This type of experiments was proposed and stimulated by a theory of Slonczewski [35].

An experimental arrangement for the observation of current-induced switching is displayed in Figure 18 (a). The sample consists of a column of layers from various materials stacked on top of each other as shown. A current can be fed in by leads I^- and I^+, and the voltage drop is measured at V^- and V^+. There is a thin Co layer, Co 1, with a thickness of 2.5 nm and a thick Co layer, Co 2, of 10 nm thickness. The Cu spacer in

Figure 18:
(a) Schematic pillar device with two thin Co layers (Co 1 and Co 2) separated by a 6 nm thick Cu layer.
(b) The dV/dI measurements as a function of the current through the column device yields the relative alignment of the magnetic layers via the GMR effect. At positive bias electrons flow from the thin Co 1 to the thick Co 2 layer. For large enough currents the Co 1 layer switches to antiparallel alignment as indicated by the higher resistance. For negative bias parallel alignment and a lower resistance is observed. An external field as indicated is applied to fix the magnetization direction of Co 2. After Ref. [34].

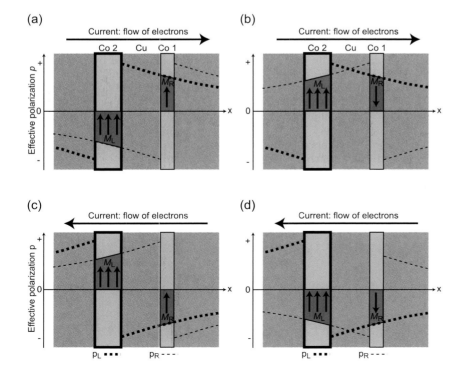

Figure 19: Explanation of NEXI in terms of effective polarizations $p_{L,R}$ due to spin accumulation interacting with the magnetic moments $M_{L,R}$ of the layers. For contributions with M parallel (antiparallel) to p the areal energy density σ_{NEXI} is increased (decreased) as represented by the red (green) color of the dark colored areas:
(a) is energetically stable, whereas the antiferromagnetic alignment in
(b) is unstable and will switch to the situation (a). For the opposite current direction the antiferromagnetic alignment in
(d) is the stable configuration.

between is 6 nm thick. The lateral diameter of the column is around 100 nm. This lateral restriction is required to obtain the necessary high current density (of the order of 10^8 A/cm^2) to establish a steady (constant current) nonequilibrium situation. As shown in Figure 18 (b), the relative orientation of the Co layers can be measured via the GMR effect of the Co 1/Cu/Co 2 trilayer. At negative bias electrons flow from the thick to the thin Co layer and stabilize the parallel magnetization alignment which yields a low dV/dI. At positive bias the parallel alignment is destabilized, Co 1 switches to the antiparallel alignment at a sufficiently large current, and dV/dI increases. Upon reducing the current (thick line in Figure 18 (b)), hysteretic behaviour is observed such that Co 1 switches back at a smaller current. The magnetization direction of the Co 2 layer is fixed by an external field.

6.2 Microscopic Picture: Spin Accumulation

An extended and comprehensible theoretical treatment has recently been given by Heide et al. [36]. These authors consider the energetics of the system in the nonequilibrium situation when a current is flowing and call the resulting effective coupling between the magnetic layers **nonequilibrium exchange interaction (NEXI)**. In the following we restrict ourselves to describe this model.

The effect can be understood on the basis of spindependent reflectivity as already discussed in the context of quantum well states in Figure 7 (a). When a current is passed through a ferromagnetic layer the spindependent reflectivity leads to a spin imbalance on both sides of the layer: For an upward magnetized layer spin-down electrons *accumulate* in front of the layer and spin-up electrons dominate behind it. This effect is called **spin accumulation** and produces on both sides a spin current with an effective polarization p, which decays exponentially with the distance from the ferromagnetic layer due to spinflip scattering. The characteristic length scale is the spin diffusion length. In Figure 19 spin accumulation is shown for the experimental situation introduced in Figure 18. The effective polarization of the left layer $\boldsymbol{p}_L$ (thick dashed curves) interacts with the magnetic moment of the right layer $\boldsymbol{M}_L$ and vice versa, and the total areal energy density is given by:

$$\sigma_{\text{NEXI}} \propto \boldsymbol{p}_R \boldsymbol{M}_L + \boldsymbol{p}_L \boldsymbol{M}_R \tag{17}$$

In Figure 19 the two contributions to the energy are given by the dark coloured areas. Red and green colours indicate energy cost and gain, respectively. The effective polarization of a layer $\boldsymbol{p}$ only depends on the spindependent reflectivity, which is an interface effect. Therefore, it is the same for both layers although they have different thicknesses – of course, $\boldsymbol{p}$ may be reversed depending on the relative orientation of $\boldsymbol{M}_L$ and $\boldsymbol{M}_R$. However, $\boldsymbol{M}_L$ and $\boldsymbol{M}_R$ depend on the thickness of the layers, and the energy contribution due to the magnetic moment of the thicker layer (the first term in Eq. (17)) dominates. For the current direction of Figure 19 (a) and (b) the total energy is higher for the parallel configuration of Figure 19 (a). Thus, NEXI acts as ferromagnetic coupling. When the current direction is reversed, both effective polarizations $\boldsymbol{p}_L$ and $\boldsymbol{p}_R$ in Eq. (17) reverse sign, and the antiparallel alignment becomes the energetically favoured situation which is reached by rotation of the thinner layer (Figure 19 (c) and (d)). For this situation NEXI acts as antiferromagnetic coupling. For identical magnetic layers the two terms in Eq. (17) always cancel, and NEXI is not acting. This is very reasonable as for the symmetric case *left* and *right* cannot be distinguished anymore.

Up to now only very few experiments concerning NEXI and current-induced switching have been reported, all dealing with the Co/Cu system. Future experiments employing different material combinations, geometries, and switching detection schemes are necessary to verify the theoretical model(s) and to find optimal implementations.

7 Summary

We have presented an overview of magnetic and magnetotransport phenomena in layered structures involving magnetic layers: exchange anisotropy, interlayer exchange coupling, GMR, TMR, and nonequilibrium exchange interaction. For each effect we first presented a phenomenological description and then developed a physical picture with emphasis on the basic mechanism. All effects only occur at interfaces or for thin enough films, which also require the presence of interfaces. Therefore, layering of magnetic and non-magnetic materials provides *new materials* with new properties. The discussed effects are nowadays summarized by the term *Magnetoelectronics* as they are used to build new electronic devices which – in addition to the charge of an electron – use its spin to control electronic transport properties.

Apart from featuring interesting basic research, magnetoelectronics has a high potential for applications mainly in data storage technology. First ideas for a *Magnetic Random Access Memory* (MRAM) based on permalloy films dates back to 1955. Since the 1970s thin magnetic films play an important role in hard disc technology as storage media as well as in the read and write heads. The new phenomena described above fit nicely into this tradition: The application of the GMR effect in read heads of hard disk drives was realized only 10 years after its discovery and allowed a significant increase of the storage density. Antiferromagnetically coupled (AFC) disk media using IEC to increase the thermal stability of the magnetic bits [37] led to a further increase of the storage density. Modern MRAMs employ the TMR effect and can therefore be realized as highly integrated solid state devices. They have the potential to replace semiconductor based memories (DRAMs) because of their nonvolatility, the lower energy consumption, and the higher scalability. GMR based MRAMs [38] which possibly will employ NEXI for the writing process are currently in development. Angle and position sensors based on GMR, e.g. for applications in the automotive industry, are available already since 1996. Further applications in magnetocouplers, strain sensors, reprogrammable logic devices, and biochips [39] are in development.

Acknowledgements

We thank M. Breidbach and T. Damm for proof reading this manuscript. The editor would like to thank Ulrich Ellerkmann and Carsten Kügeler (RWTH Aachen) for checking the clarity and consistency of this chapter. Furthermore, he would like to thank Simon Stein (FZ Aachen) for checking the symbols and formulas in this chapter.

References

[1] A. Kundt, Wied. Ann. **23**, 228 (1884).
[2] Reviews: U. Gradmann, in *Handbook of Magnetic Materials*, K. H. J. Buschow, (ed.), Elsevier, 1993; H. J. Elmers, Int. J. of Mod. Phys. **9**, 3115 (1995); articles by G. H. O. Daalderop et al. and by W.J.M. de Jonge et al. in *Ultrathin Magnetic Structures I*, J. A. C. Bland and B. Heinrich, (ed.), Springer, 1994.
[3] S. Mitani et al., J. Magn. Magn. Mater. **156**, 7 (1996).
[4] W. Weber et al., Phys. Rev. Lett. **78**, 3424 (1996).
[5] Reviews: J. Nogues and I. Schuller, J. Magn. Magn. Mater. **192**, 203 (1999); A. E. Berkowitz and K. Takano, J. Magn. Magn. Mater. **200**, 552 (1999).
[6] B. Dieny, J. Magn. Magn. Mater. **136**, 335 (1994).
[7] D. Mauri, H. C. Siegmann, P. S. Bagus, and E. Kay, J. Appl. Phys. **62**, 3047 (1987).
[8] A. P. Malozemoff, Phys. Rev. B **35**, 3679 (1987).
[9] F. Nolting et al., Nature **405**, 767 (2000).
[10] Reviews: D. E. Bürgler et al., in *Handbook of Magnetic Materials*, K. H. J. Buschow, (ed.), Elsevier, 2001; P. A. Grünberg and D. T. Pierce, in *Encyclopedia of Materials: Science and Technology*, Elsevier, 2001.

[11] S. S. Yan et al., Phys. Rev. B **59**, 11641 (1999).
[12] J. C. Slonczewski, J. Magn. Magn. Mater. **150**, 13 (1995).
[13] M. Stiles, private communication.
[14] J. Mathon et al., J. Magn. Magn. Mater. 121, 242 (1993); P. Bruno, Phys. Rev. B **52**, 411 (1995); M. Stiles, J. Magn. Magn. Mater. **200**, 322 (1999).
[15] Q. Leng et al., J. Magn. Magn. Mater. **126**, 367 (1993).
[16] D. T. Pierce, J. A. Stroscio, J. Unguris, and R. J. Celotta, Phys. Rev. B **49**, 14564 (1994).
[17] C. M. Schmidt et al., Phys. Rev. B **60**, 4158 (1999).
[18] S. S. P. Parkin, Phys. Rev. Lett. **67**, 3598 (1991).
[19] J. Mathon et al., J. Magn. Magn. Mater. **121**, 242 (1993).
[20] D. E. Bürgler et al., Phys. Rev. Lett. **80**, 4983 (1998).
[21] M. N. Baibich et al., Phys. Rev. Lett. **61**, 2472 (1988).
[22] G. Binasch, P. Grünberg, F. Saurenbach, and W. Zinn, Phys. Rev. B **39**, 4828 (1989).
[23] Reviews: P. M. Levy, Solid State Phys. **47**, 367 (1994); A. Fert et al., J. Magn. Magn. Mater. 140 – 144, 1 (1995); A. Barthélémy et al., in *Handbook of Magnetic Materials*, vol. 12 ed. by K. H. J. Buschow, Elsevier (1999); A. Barthélémy et al., in *Encyclopedia of Materials: Science and Technology*, Elsevier (2001); M. A. M. Gijs and G. E. W. Bauer, Adv. Phys. 46, 285 (1997).
[24] P. A. Grünberg, Sensors and Actuators A **91**, 153 (2001).
[25] J. Barnas et al., Phys. Rev. B **42**, 8110 (1990).
[26] http://www.fz-juelich.de/iff/staff/Buergler_D/D_Buergler_gmr.html.
[27] W. H. Butler et al., Phys. Rev. Lett. **76**, 3216 (1996).
[28] W. F. Egelhoff et al., J. Appl. Phys. **79**, 8603 (1996); ibid 82, 6142 (1997).
[29] A. Veloso et al., Appl. Phys. Lett. **77**, 1020 (2000).
[30] Review: J. S. Moodera and G. Mathon, J. Magn. Magn. Mater. **200**, 248 (1999).
[31] X. F. Han et al., Jpn. J. Appl. Phys. **39**, L439 (2000).
[32] M. Sharma, S. X. Wang, and J. H. Nickel, Phys. Rev. Lett. **82**, 616 (1999).
[33] J. M. De Teresa et al., Science **286**, 507 (1999).
[34] J. A. Katine et al., Phys. Rev. Lett. **84**, 3149 (2000).
[35] J. C. Slonczewski, J. Magn. Magn. Mater. **159**, L1 (1996).
[36] C. Heide, P. E. Zilberman, and R. J. Elliott, Phys. Rev. B **63**, 064424 (2001).
[37] E. E. Fullerton et al., Appl. Phys. Lett. **77**, 3806 (2000).
[38] J.-G. Zhu, Y. Zhang, and G. A. Prinz, J. Appl. Phys. **87**, 6668 (2000).
[39] M. M. Miller et al., J. Magn. Magn. Mater. **225**, 138 (2001).

Organic Molecules – Electronic Structures, Properties, and Reactions

Peter Atkins, Lincoln College, Oxford University, Great Britain
Rainer Waser, Department IFF, Research Center Jülich and Institute of Electronic Materials, RWTH Aachen University, Germany

Contents

1 Introduction 129

2 Hydrocarbons 129

3 Electronic Structure of π-Conjugated Systems 131
3.1 LCAO Theory 131
3.2 Hückel Approximations 133

4 Functional Groups and Structures of Molecules 135
4.1 Types of Functional Groups and Their Dipole Moments 135
4.2 Chiral Centers and Stereoisomers 138

5 Basic Principles of Chemical Synthesis 138
5.1 Substitution 140
5.2 Additions 141
5.3 Eliminations 142
5.4 Skeletal Rearrangements 143

6 Summary 143

Organic Molecules – Electronic Structures, Properties, and Reactions

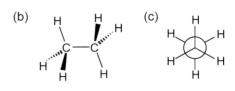

1 Introduction

Several chapters of this book describe the utilization of organic molecules for novel electronic devices: Carbon Nanotubes for Data Processing (Chap. 19), Molecular Electronics (Chap. 20), Neurobiological Interface (Chap. 32), Liquid Crystal Displays (Chap. 37), and Organic LEDs (Chap. 38). For this reason, here we give a very brief introduction to organic chemistry, focusing on the requirements of the devices chapters.

Section 2 summarizes the main classes of hydrocarbons and bond types, then Sec. 3 applies the methods of quantum mechanics (Chap. 3) to hydrocarbons. Finally, Sec. 4 introduces the functional groups and the major types of reaction mechanisms. Because space is limited, the reader should refer to the standard textbooks on organic chemistry and physical chemistry, for example, Refs. [1] – [6], in order to obtain a deeper understanding of the subject.

Figure 1: Representations of ethane, C_2H_6.
(a) Structural representation of the bonds.
(b) Sawhorse presentation.
(c) Stereo projection along the C-C axis.

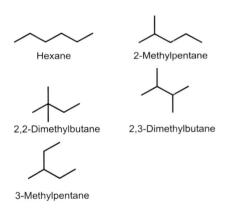

Figure 2: Structural isomers of hexane C_6H_{14}. According to the rules of the IUPAC (International Union of Pure and Applied Chemistry), the names of the isomers are based on the longest continuous carbon chain as the root, with branching groups named as substituents. The position in the root to which an alkyl group is attached is designated by number, starting at the end closest to the attachment. For this figure, the line notation of the organic chemistry is used. This notation represents C-C bonds by single lines. The presence of C-H bonds is inferred as needed to meet the valence requirements of carbon.

Figure 3: Structural representation of cyclohexane, C_6H_{12}.

2 Hydrocarbons

Saturated hydrocarbons are called **alkanes** and have molecular formula C_nH_{2n+2}. The chemical bonding in alkanes can be described as sp³ hybridization of the atomic orbitals (AO) at the carbon atoms, corresponding to a tetrahedral arrangement of the σ-bonds with the neighboring C and H atoms. As an example, Figure 1 shows the configuration of ethane, C_2H_6. Free rotation is possible around the C—C σ-bond with an activation energy of about 0.1 eV. For alkanes with $n > 3$, there are structural **isomers** (different arrangements of the same numbers of atoms) in the form of chains and branched molecules, as illustrated in Figure 2 for hexane, C_6H_{14}. Alkanes with a ring-type structure are called **cycloalkanes**. Figure 3 shows the structure of cyclohexane, for example.

Hydrocarbons with C=C double bonds are called **alkenes**. The sp² hybridization of the carbon AOs corresponds to a trigonal planar configuration of the σ-bonds with the neighboring C and H atoms. The remaining p-AOs at neighboring C atoms take part in π-bonding, for which the electron density is not concentrated along the axis connecting the two bonded atoms as in the σ-bonds, but is located above and below the axis (Figure 4) with its plane of maximum electron density normal to the plane of σ-bonds. The length of a C=C double bond is 134 pm whereas that of a C-C single bond is

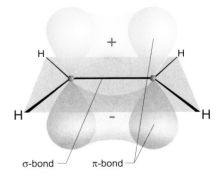

Figure 4: Illustration of the π-bond of ethene, C_2H_4, resulting from the co-planar p-AOs of the sp² hybridized C atoms. For clarity of the picture, the σ-bonds are not shown in their full shape.

129

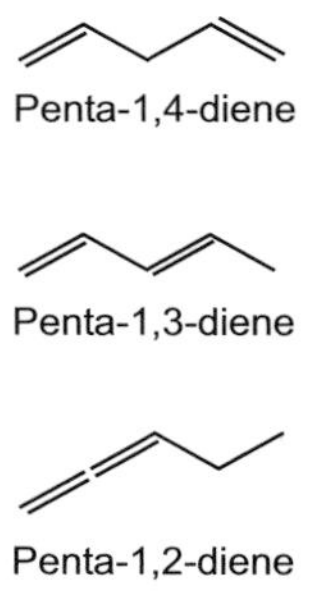

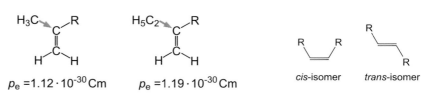

Figure 5: Dipole moment for alkenes in which the double bond is located asymmetrically within the molecule.

Figure 6: *Cis*- and *trans*-isomers of an alkene molecule. R indicates an alkyl substituent.

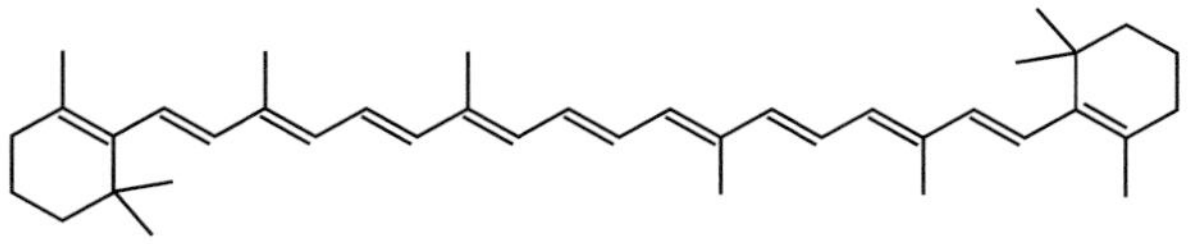

Figure 7: Types of polyenes illustrated by different pentadienes. Penta-1,4-diene shows isolated double bonds, penta-1,3-diene shows conjugated double bonds, and penta-1,2-diene shows cumulated double bonds. Since the C atom in the center of the cumulated double bond is sp-hybridized, this double bond is linear.

Figure 8: β-Carotene as an example for a polyene with long conjugated double bonding system.

154 pm. In contrast to single bonds, free rotation is not possible for double bonds since this would require the breaking of the π-bond. The physical properties of alkenes, such as melting and boiling temperatures and solubilities in various solvents, are similar to those of the alkanes. However, whereas alkanes are virtually non-polar, i.e. they have zero permanent electric dipole moment, p_e, asymmetric alkenes are slightly polar. This indicates that alkyl groups exhibit a weak electron donating effect (+*I* effect, see below), as illustrated in Figure 5. When different groups are attached on each side of a double bond, we distinguish between *cis*- and *trans*-isomers (Figure 6).

If there are more than one C=C bond in a molecule, the compounds are called **polyenes**. Depending on the different relative positions of the C=C bonds, the bonds are referred to as isolated, conjugated, and cumulated as shown in the example of pentadiene in Figure 7. As will be discussed later, conjugated C=C bonds play an important role in electron conduction because π-electrons are delocalized over the entire extent of conjugation. β-Carotene is an example of a naturally occurring polyene (Figure 8), and contributes to the colour of vegetation.

Cyclic polyenes with conjugation that spreads over an entire ring of atoms sequence constitute the class of **aromatic** hydrocarbons, or **arenes**. If they are planar (which permits overlap of adjacent p-AOs) and possess 4n+2 electrons with n = 1, 2, ... in the π-system (the Hückel criterion of aromaticity), they exhibit a high energetic stability. The π-electrons are delocalized over the entire ring and single and double bonds can no longer be distinguished: one consequence is the identity of bond lengths round the ring. The most prominent and important representative is cyclohexa-1,3,5-triene, which is universally called benzene (Figure 9). The electronic structure and the origin of the stability of aromatic rings with six π-electrons (i.e. n = 1 for the Hückel criterion) as in benzene will be discussed in Sec. 3.2. The pronounced stability and the delocalization of the π-electrons is maintained for fused rings, the so-called **polycyclic aromatic molecules**, (Figure 10), and for smaller or larger rings provided the number of π-electrons is kept at six (an electron sextet). As examples of the latter, Figure 11 shows a cyclopentadienyl anion and a cycloheptatrienyl cation. In addition, azulene can be regarded as a fused-ring of a cycloheptatriene and a cyclopentadiene. The Hückel stability criterion based on electron sextets can be extended to heterosubstituents in the carbon ring as shown in Figure 12 for some common hexagonal and pentagonal ring systems.

Hydrocarbons with C≡C triple bonds are called **alkynes**. The C atoms participating in the triple bond are sp-hybridized and the bond consists of one σ-bond and two π-bonds. The σ-bond connecting triply bonded C atoms is surrounded by a cylindrical cloud of π-electron density (Figure 13). The bond length is 120 pm, which is shorter than for single and double bonds. Because of the linearity of a triple bond with its adjacent single bonds, no *cis/trans* isomerism exists. Conjugated triple bonds show the same electron delocalization as conjugated double bonds. Because of this delocalization and their simple linear structure they are sometimes regarded as rigid rods and are of great interest for molecular electronic systems.

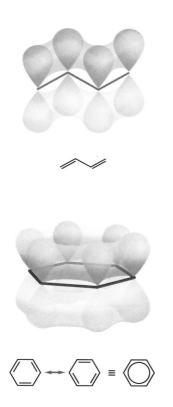

Figure 9: Structure and illustration of the electron density of the π-conjugated system of benzene. The two mesomeric Kekulé formulas according to the valence-bond theory as well as the frequently used Robinson ring symbol are shown.

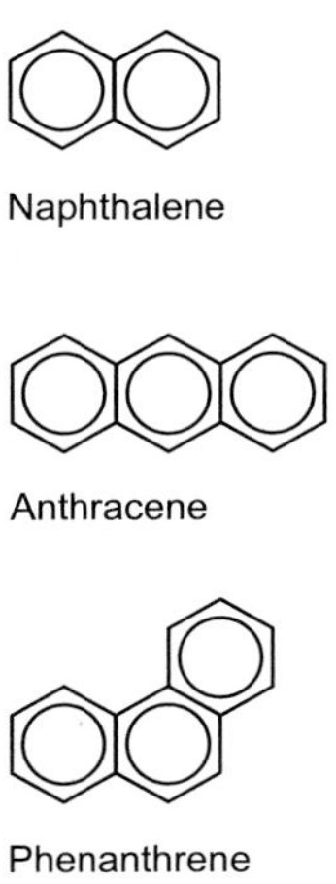

Naphthalene

Anthracene

Phenanthrene

Figure 10: Some examples of polycyclic aromatic molecules.

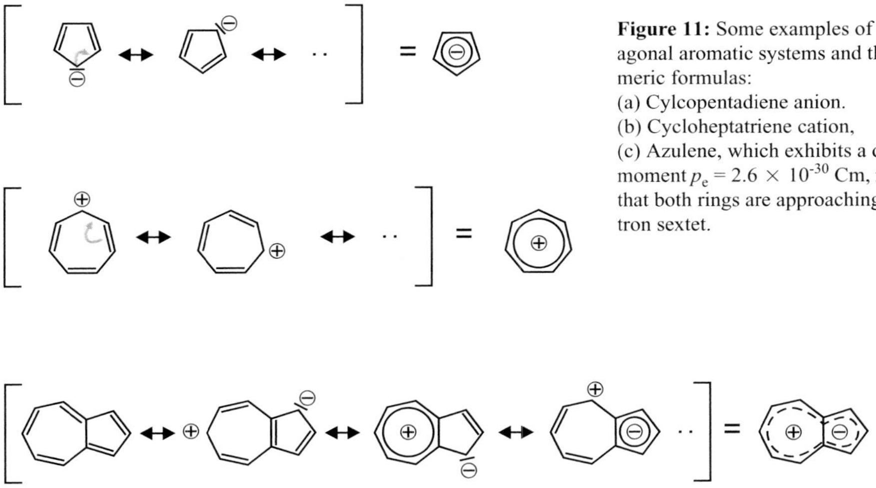

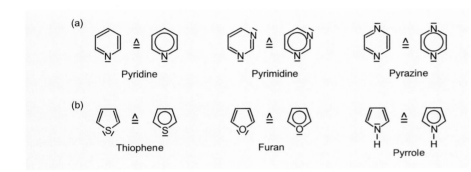

Figure 11: Some examples of non-hexagonal aromatic systems and their mesomeric formulas:
(a) Cylcopentadiene anion.
(b) Cycloheptatriene cation,
(c) Azulene, which exhibits a dipole moment $p_e = 2.6 \times 10^{-30}$ Cm, indicating that both rings are approaching an electron sextet.

(a) Pyridine Pyrimidine Pyrazine

(b) Thiophene Furan Pyrrole

Figure 12: Aromatic molecules based on heterocyclic systems.
(a) Hexagonal rings in which the free electron pair of the N atoms occupy sp² orbitals and do not participate in the π-conjugated electron system.
(b) Pentagonal rings in which one free electron pair participates in the π-conjugated system, i.e. the S, O, and N atoms contribute two electrons to the π-conjugation in contrast to the C atoms which contribute one electron each. In this manner, the stable electron sextet is reached in a pentagon. In all cases, only one of the mesomeric structures is shown.

3 Electronic Structure of π-Conjugated Systems

3.1 LCAO Theory

Electrons in molecules occupy **molecular orbitals** (MO) described by quantum mechanical wavefunctions, Ψ (see e.g. [6] – [9]). The probability density is given by Ψ^2 and for each Ψ there is an associated eigenenergy that, in principle, can be calculated from the Schrödinger equation. However, the complexity of this problem is so great that approximations are essential. Most frequently an approximation based on a **linear combination of atomic orbitals** (LCAO) is used, in which the MO is expressed as the linear combination

$$\Psi = \sum_{i=1}^{n} c_i \psi_i \qquad (1)$$

where ψ_i are the AO wave functions of the atoms involved in the molecule, c_i denote the contribution of the specific AOs, and n is the total number of AOs. The square of the coefficients c_i show (through their squares) to what extent the specific AOs are contributing to the MO.

The energy of the MOs is obtained by multiplying both sides of the Schrödinger equation by Ψ (more precisely, by the complex conjugate wavefunction Ψ^*) and integrating over the total volume

$$\Psi \hat{H} \Psi = \Psi W \Psi \qquad (2)$$

$$W = \frac{\int \Psi \hat{H} \Psi dV}{\int \Psi^2 dV} \qquad (3)$$

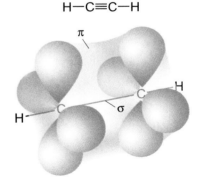

Figure 13: A three-dimensional view of the σ- and π-bonding constituting the C≡C triple bond in ethyne (also called acetylene).

Substitution of the LCAOs gives

$$W = \frac{\int \left(\sum_i c_i \psi_i\right) \hat{H} \left(\sum_i c_i \psi_i\right) dV}{\int \left(\sum_i c_i \psi_i\right)^2 dV} \quad (4)$$

Rearranging for the sums yields

$$W = \frac{\sum_j \sum_k c_j c_k \int \psi_j \hat{H} \psi_k dV}{\sum_j \sum_k c_j c_k \int \psi_j \psi_k dV} \quad (5)$$

in which $j, k = 1 \ldots n$.

Henceforth, we use the abbreviations

$$H_{ik} = \int \psi_j \hat{H} \psi_k dV \quad (6)$$

and

$$S_{jk} = \int \psi_j \psi_k dV \quad (7)$$

in which $H_{jk} = H_{kj}$ and $S_{jk} = S_{kj}$ for symmetry reasons. For $j = k$, the integrals H_{jk} are called **Coulomb integrals** and represent the effective energy of an electron in the field of one nucleus and the other electrons in the system, i.e. it is the energy of an AO. The integrals H_{jk} for $j \neq k$ are known as **resonance integrals** and express the energy of an electron moving under the influence of two different nuclei. Hence, they are a measure of the bonding energy. The S_{jk} are called **overlap integrals** because for $j \neq k$ they describe the overlap of the individual AOs. To make progress, we use the **variation principle**, which states that if an arbitrary wave function is used to calculate the energy, then the value calculated is never less than the true energy. Hence, we have to vary the coefficients c_i and seek for the energy minimum in order to obtain the best approximation for the true energy, i.e. we set $\partial W/\partial c_i = 0$ for all i and have to solve the resulting set of simultaneous equations. Rewriting Eq. (5)

$$W \sum_j \sum_k c_j c_k S_{jk} = \sum_j \sum_k c_j c_k H_{jk} \quad (8)$$

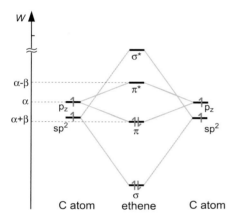

Figure 14: Sketch of the energy levels of the σ- and π-bonds of ethene generated by binding two sp² hybridized C atoms. The energy values α and β refer to the LCAO theory explained in the text.

and differentiating with respect to c_i gives a set of n equations:

$$W \sum_k c_k S_{ik} + W \sum_j c_j S_{ji} = \sum_k c_k H_{ik} + \sum_j c_j H_{ji} \quad (9)$$

with $i = 1 \ldots n$. Because $H_{jk} = H_{kj}$ and $S_{jk} = S_{kj}$, Eq. (9) can be written as

$$W \sum_k c_k S_{kj} = \sum_k c_k H_{jk} \quad (10)$$

or

$$\sum_k c_k \left(H_{kj} - W S_{kj}\right) = 0 \quad (11)$$

Equation (11) represents a set of linear homogeneous equations. The condition for non-trivial solutions of this set is that the so-called **secular determinant** vanishes:

$$\begin{vmatrix} H_{11} - WS_{11} & H_{12} - WS_{12} & \ldots & H_{1n} - WS_{1n} \\ H_{21} - WS_{21} & H_{22} - WS_{22} & \ldots & H_{2n} - WS_{2n} \\ \vdots & & & \vdots \\ H_{n1} - WS_{n1} & H_{n2} - WS_{n2} & \ldots & H_{nn} - WS_{nn} \end{vmatrix} = 0 \quad (12)$$

We will now apply the LCAO concept to the types of organic molecules introduced in Sec. 2. To calculate the electronic properties, which are relevant for applications in electronic devices, it is justifiable to consider only the π-MOs because the binding σ-MOs are localized and have a much lower energy (Figure 14). For this reason, they do not contribute in any charge transfer or optical excitation process. In the terminology of solid-state physics, this means that we only deal with electronic states that are not too far from the Fermi energy. Hence, in the following treatment we consider planar molecules for which the framework of σ-bonds is fixed. The LCAO approximation is used to calculate the energies and electron density distributions of the π-MOs from the $2p_z$-AOs of the sp^2-hybridized C atoms.

3.2 Hückel Approximations

In the 1930s, Hückel [10] introduced a set of drastic approximations that considerably simplify the calculation of the secular determinants. The procedure is now called the **Hückel MO method** (HMO method). We will use the HMO method because of its educational value and because it gives surprisingly reasonable physical results.

Assumption 1

The Coulomb integrals H_{ii} that represent the energy of an electron in a $2p_z$-AO are treated as identical for all sp^2-hybridized C atoms and denoted α:

$H_{ii} = α$ for all i

The value of α is given by the first ionisation energy of a C atom (approximately -10 eV).

Assumption 2

The resonance integrals H_{jk} depend on the length of the bond involved. The HMO method neglects this dependence and sets

$H_{jk} = β$ for adjacent, σ-bonded atoms, and
$H_{jk} = 0$ for non-adjacent atoms.

As we shall see from the calculation of the ethene, β represents the binding energy of an isolated π-bond and is approximately -0.77 eV [6]. The assumption that β is the same for all bonds implies that all bond lengths are equal.

Assumption 3

All overlap integrals are neglected:

$S_{ii} = 1$ (all atomic orbitals are normalized) and
$S_{jk} = 0$ for $j \neq k$.

This assumption looks quite severe; however, the effect on the MO energies and, especially, on the relative position of the energy levels is rather small.

The application of these assumptions to ethene leads to a secular determinant

$$\begin{vmatrix} α - W & β \\ β & α - W \end{vmatrix} = 0 \quad (13)$$

Division of both rows by β and substitution of $x = (α - W)/β$ gives

$$\begin{vmatrix} x & 1 \\ 1 & x \end{vmatrix} = 0 \quad (14)$$

from which we derive

$$x^2 - 1 = 0 \quad (15)$$

and, hence, $x = \pm 1$. It follows that

$$W = α \pm β \quad (16)$$

where $W = α + β$ is the energy of the bonding (energy lowering) π-MO and $W = α - β$ the energy of the antibonding (energy raising) π*-MO.

To calculate the coefficients c_i of the wavefunctions, we insert the energy values into the secular equations

$$\begin{aligned} c_1(α - W) + c_2 β &= 0 \\ c_1 β + c_2(α - W) &= 0 \end{aligned} \quad (17)$$

$W = α + β$ gives $c_1 = c_2$, and $W = α - β$ gives $c_1 = -c_2$. The normalization condition

$$\int \Psi^2 dV = 1 \quad (18)$$

together with Eq. (1) and the Hückel assumptions, Eq. (18), i.e.

$$\sum_j \sum_k c_j c_k S_{jk} = \sum c_i^2 = 1 \quad (19)$$

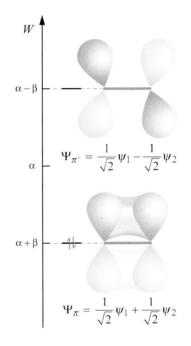

Figure 15: Result of the HMO calculation for ethene: energy levels and sketch of the wave functions of the π-MOs. The occupation of the energy levels is shown for the ground state.

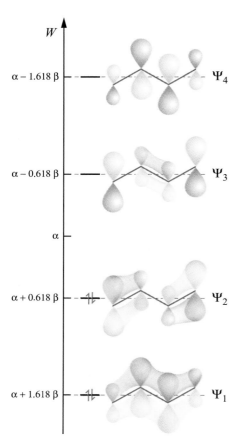

Figure 16: Result of the HMO calculation for buta-1,3-diene: energy levels and sketch of the wave functions of the π-MOs. The occupation of the energy levels is shown for the ground state.

leads to $|c_1| = |c_2| = 1/\sqrt{2}$ (Figure 15).

For buta-1,3-diene, the HMO method leads to the secular determinant

$$\begin{vmatrix} x & 1 & 0 & 0 \\ 1 & x & 1 & 0 \\ 0 & 1 & x & 1 \\ 0 & 0 & 1 & x \end{vmatrix} = 0 \qquad (20)$$

The fourth-order polynomial derived from Eq. (20) gives four eigenvalues, namely $x = \pm 0.618$ and ± 1.618. From these values, the coefficients c_i can be calculated. They are used to determine the phases and contributions of the four AOs to the four MOs (Figure 16). The electrons fill the bonding π-orbitals 1 and 2, and the antibonding π-orbitals remain empty. The total energy of the π-electrons is $4\alpha + 4.48\beta$. If this value is compared to the energy of two isolated π-bonds, $4\alpha + 4\beta$, we find that the energy due to delocalization is 0.48β.

In Figure 16, the **highest occupied orbital** (HOMO) Ψ_3 and the **lowest unoccupied orbital** (LUMO) Ψ_2 may be compared with the conduction band edge and the valence band edge, respectively, of a semiconductor. The band gap of the semiconductor corresponds to the **HOMO-LUMO gap** (HLG) in the MO picture of π-conjugated organic molecules. Accordingly, the energetically lowest optical absorption band is related to the HLG energy. Figure 17 shows the result of HMO calculations for conjugated π-systems of linear hydrocarbons. For odd numbers of C atoms, an MO occurs at the energy α, i.e. it is neither bonding nor antibonding. For neutral molecules, these **nonbonding MOs** are occupied by a single electron, and the molecule is a radical.

For benzene, the most important aromatic hydrocarbon, the secular determinant is given by

$$\begin{vmatrix} x & 1 & 0 & 0 & 0 & 1 \\ 1 & x & 1 & 0 & 0 & 0 \\ 0 & 1 & x & 1 & 0 & 0 \\ 0 & 0 & 1 & x & 1 & 0 \\ 0 & 0 & 0 & 1 & x & 1 \\ 1 & 0 & 0 & 0 & 1 & x \end{vmatrix} = 0 \qquad (21)$$

The MOs and their energies are illustrated in Figure 18. Note that Ψ_2 and Ψ_3 are degenerate, as are Ψ_4 and Ψ_5. The total energy of the π-electrons is $6\alpha + 8\beta$, which should be compared with the energy of the linear hexa-1,3,5-triene, $6\alpha + 6.98\beta$. The difference reveals that aromaticity yields an additional stabilization of approximately β compared

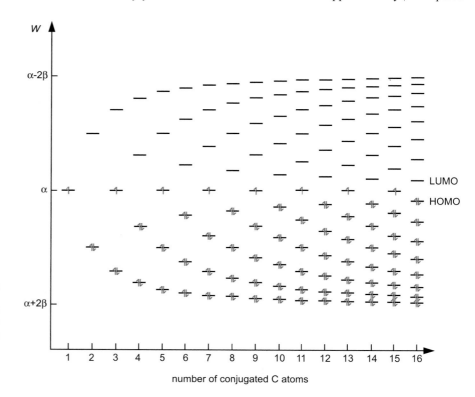

Figure 17: Energy levels and electron occupation for conjugated hydrocarbon with unbranched chain structures of one to 22 C atoms as calculated by the HMO theory. The system with one C atom represents a sp²-hybridized CH₃· radical.

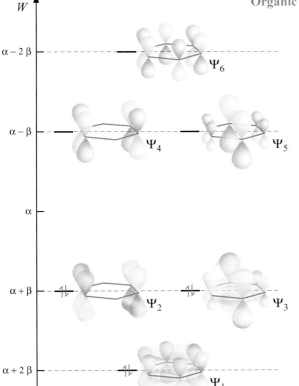

Figure 18: Result of the HMO calculation for benzene: energy levels and sketch of the wavefunctions of the π-MOs. The occupation of the energy levels is shown for the ground state.

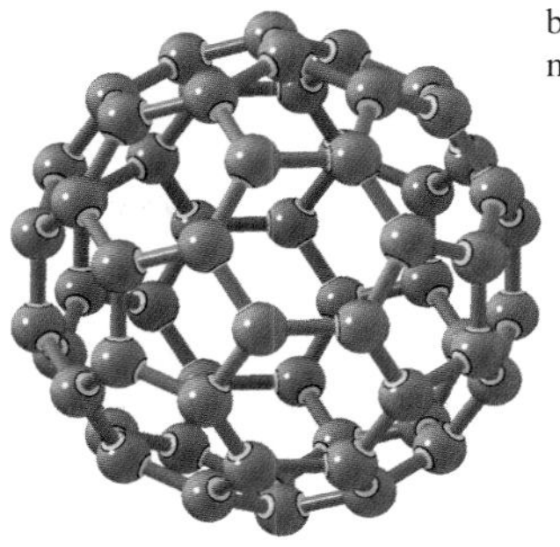

Figure 19: The buckminsterfullerene molecule C_{60}.

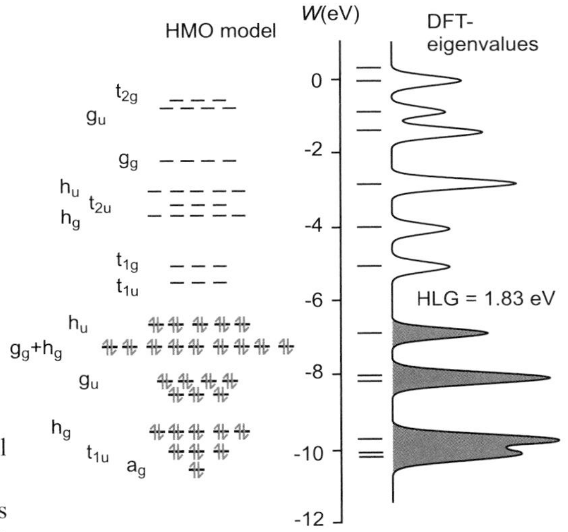

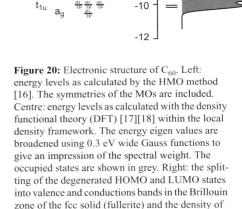

to the conjugated linear system of six C atoms. Tables of the MO energies and orbital coefficients for a large selection of hydrocarbons are collected in Ref. [11].

For increasing numbers of rings in polyaromatic compounds, the HLG decreases and finally vanishes for *graphene*, a two-dimensional infinite layer of fused benzene rings, the building unit for graphite. As a consequence, graphite possesses metallic conduction in the in-plane directions.

Another example of a conjugated carbon system is the C_{60} buckminsterfullerene molecule, the most prominent representative of the soccer-ball shaped **fullerenes** [12]. The C_{60} molecule has 12 regular pentagons and 20 hexagons (Figure 19). There are two types of bonds: one set is common to two hexagons and the other set is common to a hexagon and a pentagon. The lengths of these two kinds of bond, 139 pm and 145 pm, respectively, are intermediate between those for C-C single bonds and C=C double bonds (see Sec. 2). The C atoms can be regarded as approximately sp^2-hybridized. The σ-bonds make up the mechanically very stable cage and the π-electrons are almost completely delocalized over the molecule, with slightly enhanced amplitudes at hexagon adjacencies. Before C_{60} had been discovered experimentally, an HMO calculation had already identified the completely filled HOMO [13] and the energy eigenvalues [14]. Figure 20 shows the energy states calculated by the HMO method and an *ab initio* density functional theory approach. Further details are reported in Ref. [15]

Figure 20: Electronic structure of C_{60}. Left: energy levels as calculated by the HMO method [16]. The symmetries of the MOs are included. Centre: energy levels as calculated with the density functional theory (DFT) [17][18] within the local density framework. The energy eigen values are broadened using 0.3 eV wide Gauss functions to give an impression of the spectral weight. The occupied states are shown in grey. Right: the splitting of the degenerated HOMO and LUMO states into valence and conductions bands in the Brillouin zone of the fcc solid (fullerite) and the density of states $Z(W)$.

4 Functional Groups and Structures of Molecules

4.1 Types of Functional Groups and Their Dipole Moments

The replacement of carbon or hydrogen in hydrocarbons by other atoms (so-called "heteroatoms") leads to the formation of **functional groups**. Typically, these are groups that contain N-, O-, S-, or halogen atoms, although multiple bonds between carbon atoms are also sometimes included. The heteroatoms can be singly, doubly, or triply bonded to the

Fundamentals

Groups involving nitrogen

R—CH₂—NH₂ amine
R—CH=NH imine

R—C(NH)(NH₂) amidine
R—C≡N nitrile

R—N=N—R' azo compound

Groups involving oxygen

R—CH₂—OH alcohol

R—CHO aldehyde
R—COOH carboxylic acid

Groups involving halogenes

R—X   (X = F, Cl, Br, I)
halogenide

Groups involving different atoms

R—NO₂ nitro

R—CO—X carboxylic halogenide
R—SO₃H sulfonic acid

Charged groups

R—NR₃⁺ tetra-substituted ammonium cation
R—COO⁻ carboxylic acid anion

	Dipole moment (10^{-30} Cm)
H_2O	6.0
CH_3-OH	5.7
H_3C-Cl	6.3
H_5C_2-Cl	6.7
$(CH_3)_2CH$-Cl	6.9
$(CH_3)_3C$-Cl	7.2
Toluene	1.3
t-Butylbenzene	2.3

Table 2: Dipole moments of selected molecules

Table 1: Examples of functional groups.

H_3C-$\bar{N}H_2$ $\delta+$ $\delta-$

Figure 21: Partial charges at the amino methane molecule caused by the electron attraction of the nitrogen along the CN σ-bond. The arrow indicates the dipole moment.

neighboring C atoms. In the case of heteroatoms of the third period elements (e.g. S, P) empty d-orbitals may participate in the bonding. Table 1 lists a selection of functional groups.

Covalent C-C and the C-H bonds are nonpolar, whereas the covalent bonding between functional groups and the rest of the molecule usually exhibits a polar character on account of the different electronegativities of the atoms involved. The **electronegativity** of an atom determines to what extent electrons are attracted within a covalent bond and is determined by the electron affinity and the ionization energy of the atom [6]. Within a period of the periodic table the electronegativity increases with increasing order number, e.g. along the series C, N, O, F. For example, the C-N bond in amines is polarized so that the electron density is shifted towards the nitrogen, further enhancing the dipole moment due to the electron lone pair (Figure 21). This effect is strongly affected by the valence of the atoms involved and additional atoms within the functional group. For example, N in a nitro group, -NO_2, behaves differently from an N in an amino group, -NH_2.

Polarization effects that are caused by functional groups and act electrostatically along σ-bonds, are called **inductive effects** (*I* effects). If the functional group attracts electrons (σ-acceptor) and carries a negative partial charge, then the effect is called a **-*I* effect**; otherwise, it is denoted as a **+*I* effect** (σ-donor). For molecules with only one functional group, the strength of the *I* effect can be inferred from the dipole moment (Table 2). The relative strength of the *I* effect of various groups is shown in Figure 22. In the case of unsaturated and aromatic molecules, functional groups may also affect the electron density through another mechanism called the **mesomeric effect** (*M* effect, or resonance effect) and involves the π-electron system. The functional group may either attract charge density (-*M* effect, π-acceptors) from the π-system or donate partial charges (+*M* effect, π-donor) into the π-system from its own π- or non-bonding electrons (Figure 23). Because of the delocalization of π-electrons in conjugated or aromatic systems, the mesomeric effect is transferred to positions that may be quite distant from the functional group. In addition, the effect shows a modulation in the sense that the partial charge is especially pronounced at every second C atom within the conjugated system. The mesomeric structures caused by a nitro group at a benzene system and the resulting partial charges are illustrated in Figure 24.

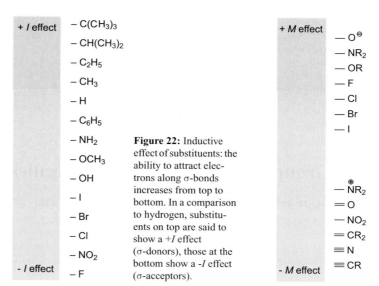

Figure 22: Inductive effect of substituents: the ability to attract electrons along σ-bonds increases from top to bottom. In a comparison to hydrogen, substituents on top are said to show a +*I* effect (σ-donors), those at the bottom show a -*I* effect (σ-acceptors).

Figure 23: Substituents showing mesomeric effects.

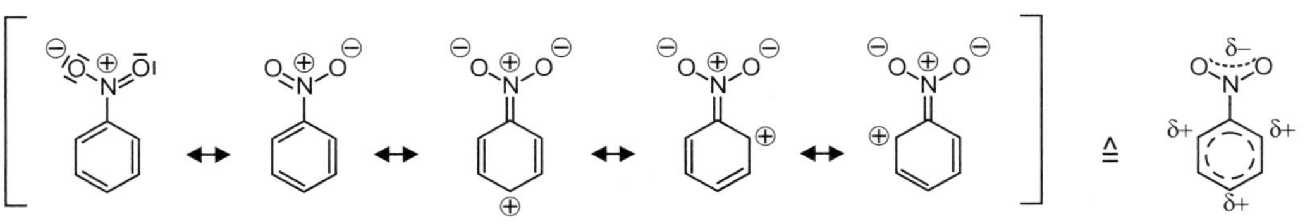

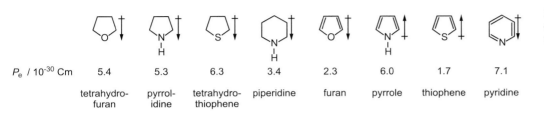

Figure 24: Mesomeric effect of substituents at aromatic compounds: the π-acceptor effect (–M effect) of the nitro group of nitrobenzene is illustrated by the mesomeric structures. The –M effect leads to positive partial charges at the o- and p-position of the phenyl ring but not at the m-position.

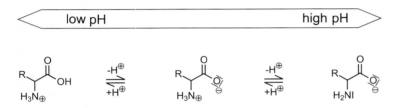

Figure 25: Some simple heterocyclic compounds and their dipole moments.

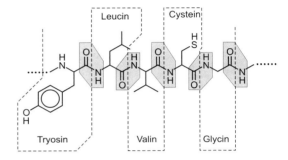

Figure 26: Illustration of the ambivalent acid-base behavior of α-amino acids: the protonation of the functional groups depends on the pH of the solution (and, more precisely, also on the concentration).

Figure 27: Fraction of the amino acid of insulin of beef.

Heteroatoms may be a part of cyclic systems as mentioned already in Sec. 2. The distribution of the electron density and, hence, the direction and strength of the dipole moment of heterocyclic molecules strongly depends on the type of ring, i.e. if it is aliphatic or aromatic (Figure 25). In aliphatic rings, there is a strong dipole moment directed towards the heteroatoms and its lone pairs. This is also true for aromatic rings, in which the lone pair does not participate in the π-system, as in pyridine. Here, the lone pair occupies an sp^2-AO of the N atom that is perpendicular to the π-MOs. If a lone pair of the heteroatom participates in the aromaticity (see Sec. 2), as in pyrole or furane, the dipole moment is much weaker (furane) or even reversed in sign (pyrole) compared to that of the aliphatic counterparts.

All except the simplest organic molecules have more than one functional group. Among the most important representatives of such multifunctional compounds are the **α-amino acids**, compounds of the form $RCH(NH_2)COOH$, which contain both the **amino** (-NH_2) and **carboxyl** (-COOH) functional groups (Figure 26). The amino group is a base (a proton acceptor) and the carboxyl group is an acid (a proton donor). Their importance extends far beyond the existence of basic and acidic characteristics in the same molecule, for the twenty naturally occurring amino acids are the building blocks of the **proteins**, the molecules that act as structural components of organisms (including cartilige, hair, and muscle), transport mechanisms (haemoglobin and the molecules that form ion channels in neuron membranes), and as regulators of biochemical processes (enzymes). A typical protein molecule consists of around a hundred or so linked amino acid molecules, which in this context are called **peptide residues**, in a definite sequence …NHCH(R)CONHCH(R′)CO… called the **primary structure** of the protein (Figure 27). The –CONH– group is called a **peptide link**. The polypeptide chain folds into a series of shapes, including helical and sheetlike regions, that is called the **secondary structure** (Figure 28). The secondary structure is due largely to the hydrogen bonds that form between the N and O atoms of the peptide link. Finally, these helical and sheetlike regions are themselves twisted into distorted shapes, and the whole molecule wrapped into, typically, a globular **tertiary structure** by other interactions, such as the formation of **disulfide links** (-S—S-). Synthetic analogues of the proteins include the **nylons**, which are condensation products formed from a dicarboxylic acid, HOOCCHR-COOH, and a diamine, $NH_2R′NH_2$, and have the much more monotonous form …OCCHRCONHR′NHCOCHR… .

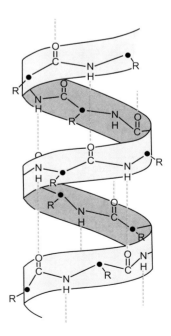

Figure 28: Helical shaped secondary structure of a protein, established by hydrogen bonds between NH and O groups.

I Fundamentals

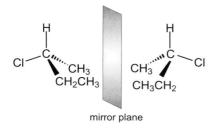

Figure 29: A chiral C atom. The C atom 2 of 2-chlorbutane is asymmetrical since it possesses four different substituents. The two stereoisomers are mirror images of each other, and hence are enantiomers.

4.2 Chiral Centers and Stereoisomers

The constitution of a molecule is defined by the type of bonds and the atoms that are present. Molecules of the same constitution may still be different as we have seen for the *cis*- and *trans*-isomers of alkenes (Sec. 2), and correspond to different **structural isomers**. Another type of isomerism, **stereoisomerism**, occurs if an sp³ hybridized C-atom has four different substituents (Figure 29). In this case, there are two possible isomers that are not superimposable (without breaking and reforming bonds) and the central C atom is classified as **chiral**. Molecules may possess more than one chiral C atom. For molecules of N chiral C atoms, up to 2^N stereoisomers may exist. Pairs of stereoisomers that are mirror images of each other are called **enantiomers**, other pairs are named **diastereomers**. Enantiomers show identical physical properties with the exception that they are **optically active**; that is, they rotate the plane of polarized light in opposite directions.

5 Basic Principles of Chemical Synthesis

A chemical reaction is controlled by thermodynamics, specifically the tendency of the system to attain lower Gibbs (free) energy, and by kinetics, the rate of the process as determined by the ability to overcome the intermediate energy barriers along the reaction path. Because organic molecules are covalently bonded, all chemical reaction entail *bond forming, bond breaking,* or both. Depending on what happens to the electrons, these basic processes can be classified into one of the following categories:

Polar reaction

If the bond forming or breaking does not separate the electron pair involved and, hence, involves ions as intermediates or – at least – in the transition state, the reaction is called a polar reaction:

$$A-B \rightarrow A^\oplus + {}^\ominus B \quad \text{heterolysis}$$
$$A^\oplus + {}^\ominus B \rightarrow A-B \quad \text{heterogenesis} \tag{22}$$

A positively charged C ion is called **carbocation**, a negative C ion is called a **carbanion**.

Radical reaction

If the reaction does involve the separation of the electrons of a bonding pair, then the reaction is classified as a radical reaction:

$$A-B \rightarrow A\cdot + B\cdot \quad \text{homolysis}$$
$$A\cdot + \cdot B \rightarrow A-B \quad \text{homogenesis} \tag{23}$$

Pericyclic reaction

Electron movements may occur in a concerted manner, without intermediate radicals, carbanions, or carbocations.

Most reactions used for the synthesis of organic compounds used in the context of this book, e.g. for liquid crystals displays, organic light-emitting or optically active systems, for studies on molecular electronics, and for molecular biology, are based on polar reactions. Polar reactions rely on electrostatic interactions, such as **nucleophiles**, species that seek centres of positive charge, interacting with at least partially positively charged regions of the target molecule and **electrophiles,** species that seek centres of negative charge, interacting with regions of high electron density. A nucleophile must have at least one lone pair of electrons that it can donate to the target molecule. Species with high nucleophilicity are often negatively charged or possess a high negative partial charge: they can be regarded as Lewis bases, electron-pair donors, that react with the target molecule behaving as a Lewis acid, an electron-pair acceptor. Electrophiles are either positively charged or have low-lying unoccupied orbitals.

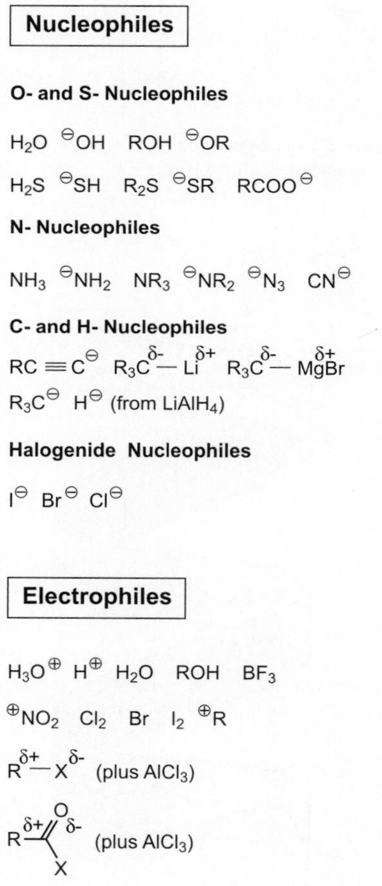

Figure 30 shows some common nucleophiles and electrophiles. Some reagents are ambivalent: for example, water can act as a nucleophile through its oxygen atom (which carries a negative partial charge), whereas the protons of water are electrophilic. The

Nucleophiles

O- and S- Nucleophiles
$H_2O \quad {}^\ominus OH \quad ROH \quad {}^\ominus OR$
$H_2S \quad {}^\ominus SH \quad R_2S \quad {}^\ominus SR \quad RCOO^\ominus$

N- Nucleophiles
$NH_3 \quad {}^\ominus NH_2 \quad NR_3 \quad {}^\ominus NR_2 \quad {}^\ominus N_3 \quad CN^\ominus$

C- and H- Nucleophiles
$RC\equiv C^\ominus \quad R_3 \overset{\delta-}{C}-\overset{\delta+}{Li} \quad R_3\overset{\delta-}{C}-\overset{\delta+}{MgBr}$
$R_3C^\ominus \quad H^\ominus$ (from $LiAlH_4$)

Halogenide Nucleophiles
$I^\ominus \quad Br^\ominus \quad Cl^\ominus$

Electrophiles

$H_3O^\oplus \quad H^\oplus \quad H_2O \quad ROH \quad BF_3$
$^\oplus NO_2 \quad Cl_2 \quad Br_2 \quad I_2 \quad {}^\oplus R$
$R\overset{\delta+}{-}X^{\delta-}$ (plus $AlCl_3$)
$R\overset{\delta+}{-}\overset{O^{\delta-}}{\underset{X}{\|}}$ (plus $AlCl_3$)

Figure 30: Examples of common nucleophiles and electrophiles.

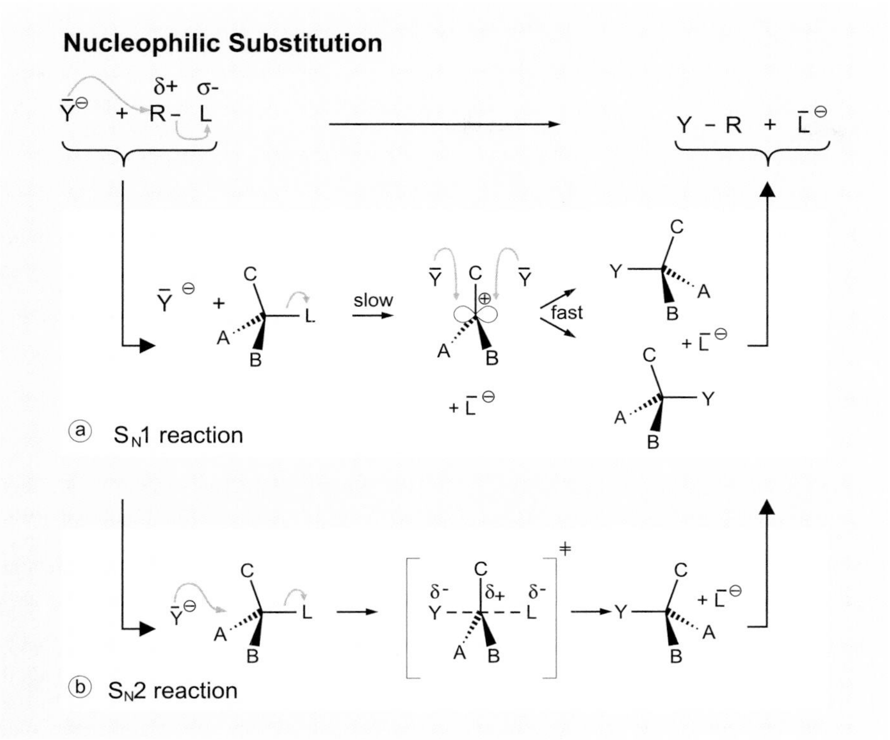

Organic Molecules – Electronic Structures, Properties, and Reactions

Nucleophilic Substitution

(a) S$_N$1 reaction

(b) S$_N$2 reaction

Figure 31: Scheme of a nucleophilic substitution at sp^3-hybridized C atoms. The mechanism is govered (a) by a two-step reaction in which the leaving group, L, departs with the electrons of the C-L bond in the first slow step followed by a fast attack by the substituting nucleophile, Y, or
(b) by a concerted reaction started through a backside attack of Y.

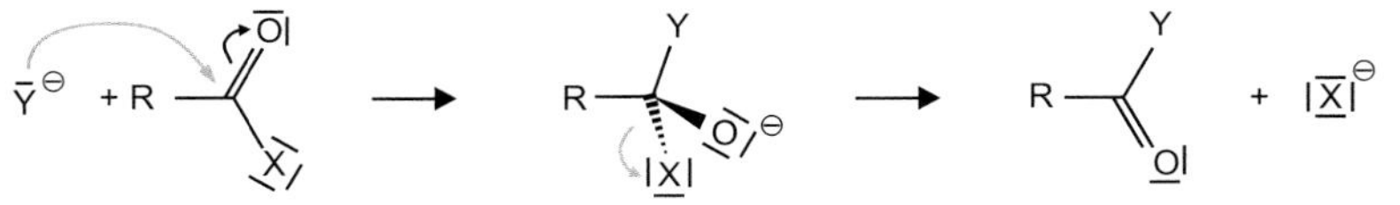

Figure 32: Nucleophilic substitution at the C atom of a carbonyl function.

strength of the nucleophilicity and electrophilicity is determined by the local electron density at the active region of the reagent, by geometrical aspects (how exposed this region is), and by the polarizability of the species.

The overall chemical reaction, which is used in a synthesis strategy, typically is classified according to the type of change it brings to the constitution of the target molecules:

- **Substitution,** in which one substituent of the target molecule is exchanged for another,
- **Addition,** in which a π-bond of a multiple bond is opened and two substituents are added,
- **Elimination,** in which two substituents on neighboring atoms are removed and an additional π-bond is formed,
- **Rearrangement** of the skeleton of the molecule.

We shall discuss the principal aspects of these reaction types below. Keep in mind, though, that because the field of organic synthesis is extremely broad and rich in variants and exclusions from general rules, we can give only some idea of the extent of this huge field.

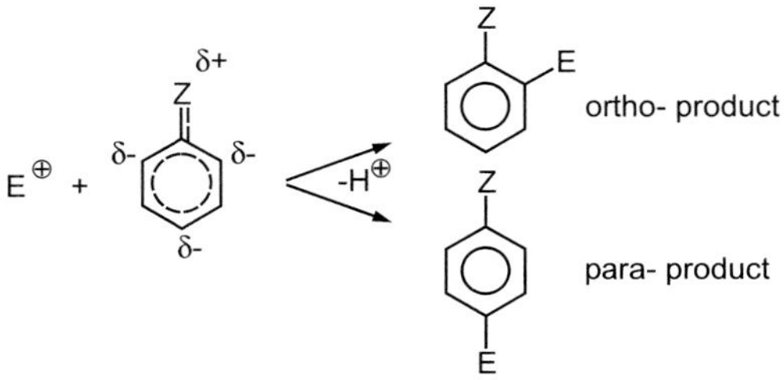

Figure 33: Electrophilic substitution at aromatic C atoms. The electrophile E$^+$ is replacing a proton. The electron donating (i. e. +M effect) substituent Z accelerates the reaction and directes it into the ortho- and the para-position.

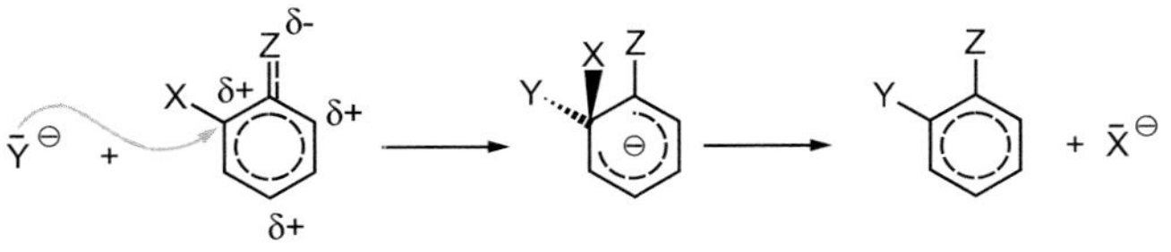

Figure 34: A strongly electron accepting (i. e. -M effect) substituent Z facilitates a nucleophilic substitution at aromatic C atoms.

139

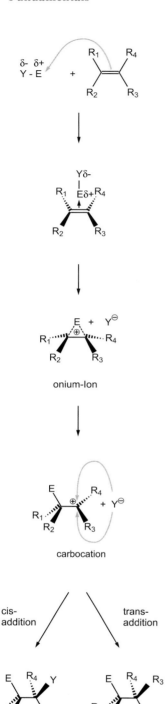

Figure 36: Electrophilic addition to a C=C double bond.

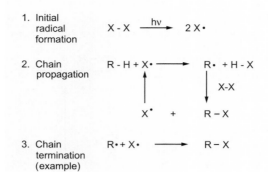

Figure 35: Chain reaction by a radical substitution.

5.1 Substitution

A **nucleophilic substitution** (S_N) at sp³-hybridized C atoms occurs if a nucleophile attacks a positively polarized C atom, i.e. a C atom with a substituent showing a -I effect (Figure 31). There are two limiting cases of the path of S_N reactions. In an **S_N1 reaction**, reaction takes place in two consecutive steps. In the first (slow) step, the leaving group L (which is usually highly reactive) is eliminated and a carbocation is formed that reacts in a second (fast) step with the nucleophile Y (Figure 31a). If the target C atom is chiral and a single enantiomer undergoes a reaction, the optical activity is lost because nucleophilic attack may occur at either face of the planar, sp²-hybridized carbocation and, hence, both possible stereoisomers are formed. The loss of optical activity is called **racemisation**. In an **S_N2 reaction**, a concerted attack of Y and departure of L leads to a positively charged transition state (Figure 31b). Because Y can only attack from one side, the backside of L, S_N2 reactions proceed with conservation of sterochemical identity at a chiral C target atom.

The rate of an S_N reaction as well as its type, is controlled not only by the nucleophilicity of Y but also by the ease with which the leaving group L is removed. Easily leaving groups are halogens –X (leaving as X⁻) as well as a protonated hydroxyl group -OH_2^+ (leaving as H_2O) and a protonated alcoxy group –ORH^+ (leaving as ROH). In addition, other factors such as the solvent, neighboring groups at the target molecule, play a significant role. Nucleophilic substitutions frequently occur also at carboxyl (sp²-hybridized) C atoms, as shown in Figure 32.

The π-system of an aromatic ring is a region of high electron density and, hence, attracts electrophiles. Because of the enhanced energetic stability of aromatic systems, electrophilic reactions will be substitutions at the aromatic ring in contrast to addition reactions, which often take place at nonaromatic multiple bonds (Sec. 2). Substituents of the aromatic system that exhibit a +I or a +M effect further increase the electron density in the aromatic π-system and, therefore, increase the rate of electrophilic substitution. In this case, the +M effect controls the position of the substitution since the electrophile tends to act where there is a negative partial charge. These centres are located in the ortho- and para-positions, as can be rationalized in a similar way to the opposite case, i.e. a -M substituent, demonstrated in Figure 24. Figure 33 illustrates the **electrophilic aromatic substitution**. If Z = Br, for example, a nitro group is easily introduced by $E^+ = NO_2^+$.

If an aromatic compound has substituents with strong -I or -M effects, the electron density of the π-system may be sufficiently reduced to allow nucleophilic aromatic substitution. Figure 34 shows one of the possible reaction paths. Again, the M effect encourages substitution at the ortho- and para-positions, provided that there are favourable leaving groups X (e.g. – Cl, – Br, – I leaving as anions, or – N_2^+, leaving as N_2).

Substitutions can also take place by a radical process. With few exceptions, radicals are chemically very aggressive and may react with relatively inactive molecules. Radicals are often formed in a thermolytic or UV photolytic process. The most common radical substitution at sp³-hybridized C atoms is the halogenation of alkanes (Figure 35). Because the intermediate radical R· generates new halogen radicals X·, it is a chain reaction that terminates only if two radicals react with each other.

5.2 Additions

Because multiple bonds are regions of enhanced electron density, they are likely to undergo electrophilic attack. If the electrophile can also act as a nucleophilic, an **electrophilic addition** to the multiple bond typically occurs. The mechanism involves the formation of a π-complex, followed by the dissociation of the attacking reagent (Figure 36). The site of the addition of the electrophile is controlled by *I* effects and *M* effects of the substituents neighbouring the multiple bond. The subsequent nucleophilic addition of Y⁻ may occur from both sides to the intermediately formed carbocation. In some cases, Y⁻ attacks the intermediate carbonium ion directly, which leads to *trans*-addition. Depending on the compounds and the reaction conditions, the intermediate carbocation itself may attack unreacted target molecules, which leads to a longer skeleton and new carbocation. This process can continue in the same manner and lead to polymers (Figure 37).

Nucleophilic additions are common at C=X bonds, with X = O, N, or S, on account of the strong positive polarization of the C=X double bonds (Figure 38). As in the case of nucleophilic substitutions at the carbon of carboxylic acid derivates, the first step is a nucleophilic attack. Depending on the nature of the leaving groups and on the presence of electrophiles, the subsequent course may be addition instead of a substitution. The reaction can be used to extend the skeleton of the target molecule by using a **Grignard reagent**, RMgBr, in which the alkyl group R is negatively polarized and, hence, acts as a nucleophile. Radicals may also attack multiple bonds. This is the first step to a **radical addition**, which often leads to polymerization.

Another class of addition reaction, a **pericyclic reaction**, proceeds through a concerted movement of all the bonds involved. During the entire course of the reaction, the symmetry of the molecular orbitals involved remains unchanged. The classic example is the Diels-Alder **cycloaddition** of ethene to buta-1,3-diene. As illustrated in Figure 39, the LUMO of the diene reacts with the HOMO of the ethene in a way that orbitals of the same sign interact with each other. The same is true for the HOMO of the diene and the LUMO of the ethene. This cycloaddition converts three π-bonds in the reactants into two σ-bonds and one π-bond in the product. The reaction is energetically favoured because the intermediate involves six electrons (and conforms to the Hückel criterion of aromaticity) in the transition state. The reaction is one out of a huge number of variants and is exhibited by suitable substituents that shift the energies of the MOs to achieve maximum interaction. It should be noted that the orbital symmetry is different for photolytically conducted cycloadditions because in photolytic reactions we have to consider excited states instead of ground states.

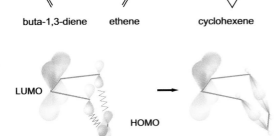

Figure 38: Nucleophilic addition at the C atom of a C=X double bond, e. g. a C=O bond. This is the also the first step of the nucleophilic substitution at a carbonyl C atom.

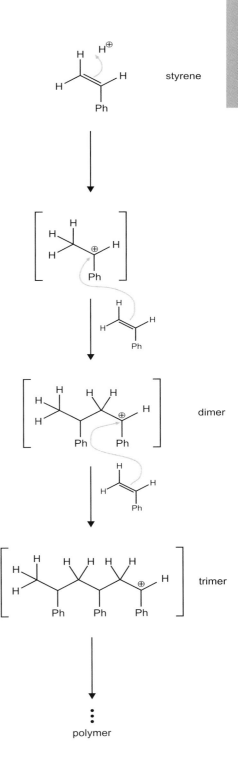

Figure 39: The Diels-Alder cycloaddition of ethene to buta-1,3-diene as an example of a concerted, pericyclic reaction. The aromatic six-electron transition state is characteristic for this class of reactions.

Figure 37: Carbocations generated by proton release of an alkene can themselves act as electrophiles to attack other alkene molecules. The examples shows the polymerisation of styrene.

Figure 40: Scheme of an elimination reaction. The mechanism is governed
(a) by a two-step reaction in which the leaving group, L, departs with the electrons of the C-L bond in the first step followed by the abstraction of a proton by the nucleophile, Y, or
(b) by a reaction in which an intermediate π complex is formed.

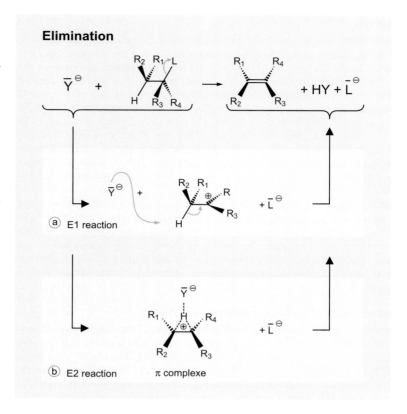

5.3 Eliminations

Elimination reactions are the reverse of addition reactions. Two substituents, A and B, of a molecule are eliminated and an additional bond is formed. In most cases, A and B are attached to neighbouring atoms and the newly formed bond is a π bond. For example, a C-C single bond is turned into a C=C double bond. Typically, one of the eliminated groups is a hydrogen atom and the other is a leaving group like those we have already encountered in S_N reactions. Also, the elimination is started by a nucleophile, as in an S_N reaction (Figure 40). Two reaction pathways are possible. In the case of sequential steps (type E1), the formation of the carbocation is slow. In the case of a concerted elimination (type E2), an intermediate π-complex is formed. Nucleophilic substitutions and eliminations are often in competition. The dominance of one over the other depends on many factors such as geometry of the target molecule, the type of nucleophile, and the dipole moment of the solvent molecules.

It should be noted that the presence of a nucleophile is not always required as, for example, in the case of the pyrolysis of esters:

$$\text{(24)}$$

The mechanism of this reaction is concerted. In contrast to the E2 reaction, however, it proceeds through a cyclic transition state:

$$\text{(25)}$$

As in the case of substitution and addition reactions, many elimination reactions are known and are widely used in synthesis strategies.

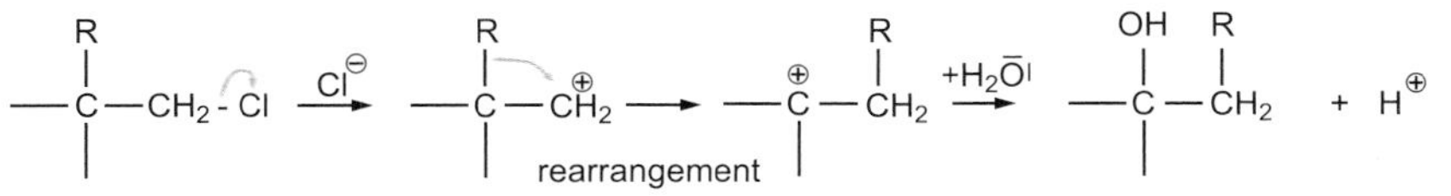

Figure 41: Example of a skeleton rearrangement through a carbocation intermediate stage.

5.4 Skeletal Rearrangements

The constitution of an organic molecule may change without any (net) attachment or detachment of a substituent. These **rearrangements** usually change the skeleton of the molecule, as in:

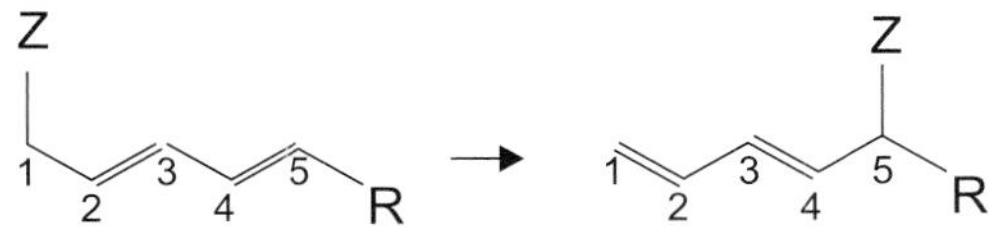

$$(26)$$

Here, Z is the rearranged or wandering group. Rearrangements can be classified as:
- **Anionotropic rearrangements**, in which the group Z is moving with the binding electron pair.
- **Cationotropic rearrangements**, in which the group Z is moving without the binding electron pair. If Z = H, this type of reaction is called a *prototropic rearrangement*.
- **Radical rearrangements**, in which the binding electron pair is broken homolytically. This type of reaction is rare.

In Eq. (27), the B atom may have only an electron sextet instead of a normal, more stable octet. To obtain this configuration, there is often a preceding reaction in which a detachment of a substituent from B occurs. In a similar manner, the A atom of the product molecule usually undergoes attachment of a nucleophile to obtain an electron octet again. An example is shown in Figure 41, which can be regarded as an S_N reaction with an intermediate anionotropic rearrangement of the alkyl group R.

Rearrangements are also possible through pericyclic reactions; these reactions are controlled by the symmetry of the molecular orbitals. Pericyclic arrangements are characterized by the movement of a single bond adjacent to one or two double bonds, as in the [1,5] rearrangement.

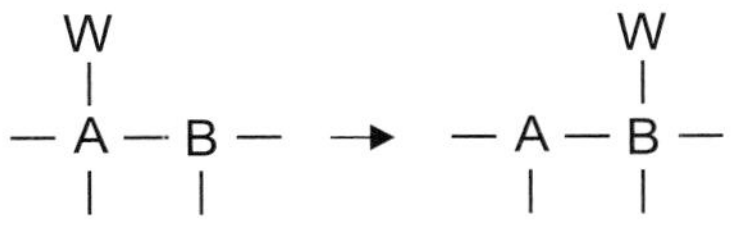

$$(27)$$

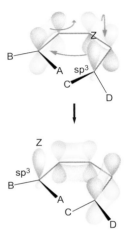

Figure 42: A pericyclic rearrangement takes place at one side of the π-system.

If the cyclic transition state is aromatic, the rearrangement takes place at one side of the π-system. This is the so-called **supraficial movement** (Figure 42).

6 Summary

Organic chemistry is a rich source of new materials despite its reliance on a single element, carbon. The ability of carbon to form such a variety of compounds depends on several factors. One is its centrality in the periodic table: it is neither particularly electronegative nor electropositive. Another factor is the readiness with which carbon can form multiple bonds not only to other elements but also to itself. Because carbon is neither electropositive nor electronegative, the electron density at it responds to the presence of other substitutents in the molecule. Finally, it is highly responsive to the identity of an attacking group, and may act as a source of electrons or a centre for attracting them. All these features can be rationalized, and they can all be employed in strategies of synthesis. This is the fundament for the large variety of organic compounds that can be employed in electronic devices.

Acknowledgements

The editor would like to thank Matthias Schindler (RWTH Aachen) for checking the symbols and formulas in this chapter.

References

[1] L. L. Jones and P. W. Atkins, *Chemistry – Molecules, Matter, and Change*, 4th ed., W. H. Freeman and Company, 1999.

[2] K. P. C. Vollhardt and N. E. Schore, *Organic Chemistry, Structure and Function*, 3rd ed., W. H. Freeman and Company, 1999; in German: *Organische Chemie*, 3rd ed., Wiley-VCH, 2000.

[3] M. A. Fox and J. K. Whitesell, *Organic Chemistry*, 2nd ed., Jones and Barlett Publishers, Boston – London, 1997.

[4] F. A. Carey, *Organic Chemistry*, 4th ed., McGraw-Hill, 2000.

[5] M. Loudon, *Organic Chemistry*, 3rd ed., Addison Wesley & Benjamin Cummings, 1995.

[6] P. W. Atkins, *Physical Chemistry*, 7th ed., Oxford University Press, Oxford, and W. H. Freeman & Co, New York, 2002.

[7] P. W. Atkins and R. S. Friedman, *Molecular Quantum Mechanics*, Oxford University Press, 1997.

[8] W. B. Smith, *Introduction to Theoretical Organic Chemistry and Molecular Modeling*, Wiley-VCH, 1995.

[9] N. Isaacs, *Physical Organic Chemistry*, 2nd ed., John Wiley & Sons, 1995.

[10] E. Hückel, Z. Physik **70**, 204 (1931) and **76**, 628 (1932).

[11] E. Heilbronner and H. Bock, *Das HMO-Modell und seine Anwendung*, Verlag Chemie, 1968.

[12] H. W. Kroto, J. R. Leath, S. C. O′Brien, R. F. Curl, and R. E. Snalley, Nature **318**, 162 (1985).

[13] D. A. Bochvar and E. G. Gal′pern, Dokl. Akad. Nauk SSR **209**, 610 (1973) [Proc. Acad. Sci. USSR **209**, 239 (1973)].

[14] R. A. Davidson, Theoret. Chim. Acta **58**, 193 (1981).

[15] M. S. Dresselhaus, G. Dresselhaus, and P. C. Eklund, *Science of Fullerenes,* Academic Press, New York, 1995.

[16] R. C. Haddon, L. E. Brus, and K. Raghavachari, Chem. Phys. Lett. **125**, 459 (1986).

[17] W. Kohn, Reviews of Modern Physics **71**, 1253 (1999).

[18] J. Hafner, Acta Mat. **48**, 71 (1999).

[19] J. H. Weaver and D. M. Poirier, in: Solid State Physics **48**, 1 (1994); W. E. Pickett, in: Solid State Physics **48**, 225 (1994), Academic Press.

Neurons – The Molecular Basis of their Electrical Excitability

U. Benjamin Kaupp and *Arnd Baumann*
Institute for Biological Information Processing
Research Center Jülich, Germany

Contents

1 Architecture and Basic Signaling Capabilities of a Neuron   147

2 Membrane Potential   150

3 What Determines the Resting Membrane Potential?   151

4 How is the Action Potential Generated?   154

5 Recording Electrical Signals from Neurons   155
   5.1 The Voltage-Clamp Technique   156
   5.2 The Patch-Clamp Technique   156

6 Signal Propagation along the Axon   158

7 How do Action Potentials Evoke Neurotransmitter Release?   159

8 Molecular Structure and Function of Ion Channels   160

9 Biochemical Aspects of Learning and Memory   163

Neurons – The Molecular Basis of their Electrical Excitability

1 Architecture and Basic Signaling Capabilities of a Neuron

Most biological cells are electrically excitable. The best studied examples are nerve cells, or *neurons*. In the following, we will explain the molecular basis of cellular excitability. This is a tutorial for physicists interested in information processing of the nervous system. There are several superb textbooks available that comprehensively cover this research area. The text and figures of this review are, in part, based on the text book "Principles of Neural Sciences" by Kandel et al. [1] (Chapters 2, 6-9) and the text books "Molecular Neurobiology" by Hall [2] and "Neuroscience" by Purves et al. [3]. The interested reader is also referred to other textbooks cited at the end of the text [4]-[7].

Two types of cells constitute the *central nervous system* (*CNS*) of all higher and lower eukaryotes. These are the glia cells and the neurons. A human brain typically contains 10^{11} neurons. Their main task is signalling. Neurons register information from the environment, integrate and evaluate this information, and decide whether electrical signals are transmitted to „downstream" targets. The typical architecture of a neuron is shown in Figure 1. As every cell in the body, neurons are surrounded by a thin sheet, or *plasma membrane*, that is formed by a phospholipid bilayer. The membrane is a diffusion barrier for hydrophilic substances; it profoundly contributes to the electrical properties of the neuron (see below). At rest, neurons maintain a difference in the electrical potential on either side of the plasma membrane. This is called the *resting membrane potential*. A typical value of the membrane potential at rest is –65 mV; but it can range between –40 mV and –80 mV. Excitable cells, such as neurons and muscle cells, differ from other cells in that their membrane potential can be altered quickly and transiently. The electrical impulse can be utilized as a signaling mechanism.

Independent of size, shape, and function, most neurons can be described by a model neuron that has four functional components: a local input (receptive) component; a trigger component (summation and integration of incoming signals); a long-range conducting (signaling) component; and an output (secretory) component. The four functionally defined components by and large correspond to four morphologically defined regions of the neuron: the dendrites, the cell body, the axon, and the presynaptic terminals. The *cell body* (or *soma*) is the metabolic center of the cell; it contains the nucleus, the organelle harboring the genetic information. The soma is also the site of protein synthesis and production of energy (*ATP*). The cell body gives rise to two kinds of cellular extensions or processes: several short *dendrites* (receptive component) and a single long *axon* (conducting component). The dendrites branch out in a tree-like fashion and are specialized to receive incoming signals from other neurons. Signals gathered by the dendrites are integrated and evaluated in other parts of the neuron. Axons are thin tubular structures with diameters ranging from 0.2 – 20 µm and a length of up to 3 m. The axon extends away from the cell body and is carrying electrical signals to other neurons. It is the main conducting part of the neuron (Figure 1). The electrical signals are called *action potentials*. Action potentials constitute the signals by which the brain receives, analyses, and conveys information. The signals are highly stereotyped throughout the nervous system, even though they are initiated by a great variety of events in the environment that impinge on our bodies. Action potentials are brief (~1 ms duration), invariant, and large (~100 mV in amplitude) electrical impulses. The action potential is *all-or-none*: While stimuli that do not reach a certain threshold value of the membrane potential produce no action potential, all stimuli above the threshold invariably generate the *same* signal, i.e. action potential. The action potential is propagated without decaying along the axon at a speed of up to 150 m/s. Action potentials are generated at a specialized trigger region,

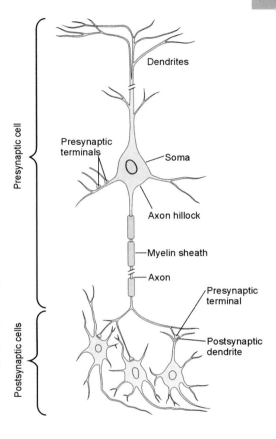

Figure 1: Blueprint of a neuron. A typical neuron contains branched cellular processes (dendrites) which are specialized to receive information from neighboring cells. The dendrites extend from the soma that contains the nucleus and the machinery for protein biosynthesis. Signals received in the dendrites are integrated and translated into electrical signals. The decision whether an action potential will be sent down the axon to the presynaptic terminal, is made at the axon hillock (adapted from [1]).

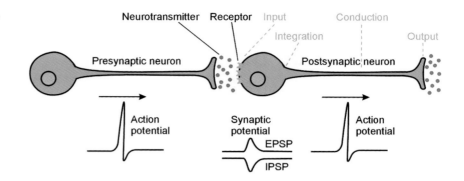

Figure 2: Communication between neurons. Neurons convey information by a combination of chemical and electrical signals. At chemical synapses, an action potential causes release of neurotransmitters onto the postsynaptic cell. The neurotransmitters bind to specific receptors and evoke electrical signals (synaptic potentials) in the postsynaptic cell. These synaptic potentials eventually trigger an action potential which travels down the axon to the nerve terminal and causes neurotransmitter release onto the next cell (adapted from [2]).

called the *axon hillock* (summing or integrative component), where the axon emerges from the cell body (Figure 1). From there, the action potential travels down the axon without failure, distortion or attenuation.

Near its end, the axon divides into fine branches that make contact to neighboring neurons. The point of contact between two communicating cells is called the *synapse*. The nerve cell transmitting a signal is called the *presynaptic cell*. The presynaptic cell transmits signals from the swollen ends of its axonal branches, called *presynaptic terminals* (output component). The cell receiving the information is called the *postsynaptic cell* (see Figure 1). The presynaptic cell does not actually touch or anatomically communicate with the postsynaptic cell because the two neurons are separated by a small space, called the *synaptic cleft*. Most presynaptic terminals end on the dendrites of the postsynaptic neuron; but the terminals may also target the cell body, or less frequently, the beginning or end of the axon of the receiving cell.

How is the action potential transmitted from the presynaptic to the postsynaptic neuron? Communication between neurons is achieved by a combination of electrical and chemical signals (Figure 2). When the action potential enters the presynaptic terminal, it stimulates the release of packets of small organic molecules referred to as chemical messengers or *neurotransmitters*. The transmitters are stored in tiny organelles (*synaptic vesicles*). The release of transmitters at the presynaptic terminal occurs at the active zone, a region that is directly apposed to the postsynaptic neuron (Figure 2). The release of transmitters serves as the output signal, translating the electrical signal into a chemical signal. The transmitter molecules traverse the synaptic cleft and bind to specific receptors in the membrane of the postsynaptic neuron. The binding of transmitter to receptors evokes a local electrical signal, called *synaptic potential*. The synaptic potentials – unlike the action potential – are not actively propagated; instead they passively spread to the trigger zone and eventually produce an action potential. The synaptic potentials are graded in size – depending on the strength of the stimulus. They can be either more positive than the resting potential (*depolarization*) or more negative (*hyperpolarization*). Because depolarization enhances a cell's ability to generate an action potential, it is *excitatory*. In contrast, a hyperpolarization makes a cell less likely to generate an action potential and, therefore, is *inhibitory*.

The molecular mechanisms underlying the resting potential, the action potential, and the various synaptic potentials will be described below in more detail.

At rest, no current flows from one part of the neuron to another; so the resting potential is the same throughout the cell. The neuron becomes excited – „fires" action potentials – by a *graded* local change in the membrane potential produced by the input components. In many neurons, the input component – the dendrites – are stimulated either by other neurons or by sensory cells, for example photoreceptors in the retina or taste cells on the tongue. However, some sensory cells possess a specialized *transduction* (or input) compartment that transforms a physical (stretch, light) or chemical („odorants") stimulus into a graded response, i.e. a brief change in membrane potential (Figure 3). For example, olfactory sensory neurons in the nose are endowed with thin filamentous structures, called cilia. The cilia are the sites where binding of odorants to receptor proteins (odorant receptors) elicits local ionic currents across the ciliary membrane, due to the opening of ion channels. The ionic current disturbs the resting potential, driving the membrane potential to a new level, called *receptor potential*. The amplitude and duration of the receptor potential depends on the strength of the stimulus, i.e. the concentration of the odorant. The larger and longer-lasting the stimulation by odorants, the larger and longer-lasting the receptor potential. Thus the receptor potential is the first representation of the stimulus to be coded by the nervous system. It is a purely *local* signal that spreads passively to the trigger region of the olfactory neuron, where –

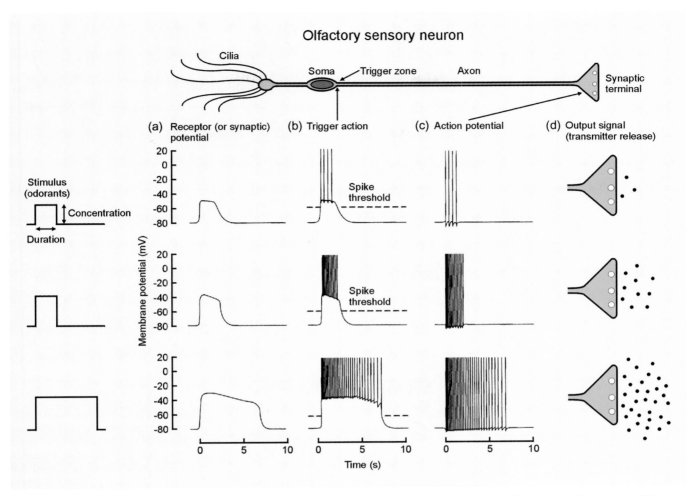

Figure 3: An olfactory neuron transforms a chemical stimulus into electrical activity in the cell. Sensory neurons in the olfactory epithelium which lines the nasal cavity consist of a cell body, a small dendrite, and a long axon that projects to the olfactory bulb in the brain. Approximately 10-15 thin cylindrical cilia protrude from the swollen end of the dendrite. The cilia harbour the molecular machinery to register odorants and transduce the stimulus into an electrical signal.

(a) The input signal is a change in the resting potential – called the receptor potential – elicited by stimulation of the cell with an odorant. The receptor potential is graded in amplitude and duration. The relation between stimulus strength (odorant concentration) and receptor potential amplitude can be very steep due to several stages of biochemical amplification inside the cilia.
(b) The trigger zone integrates the input signal and produces an action potential. The action potential is only generated if the receptor potential is greater than a certain threshold. Once the input signal surpasses this threshold, any further increase in amplitude of the input signal augments the frequency with which the action potentials are generated, not their amplitude or shape. The duration of the odorant stimulus determines the duration of the input signal and thereby the length of the active period, i.e. the number of action potentials fired.
(c) The action potentials are invariant in size and shape. Thus every action potential has the same amplitude, duration and waveform.
(d) When the action potential enters the synaptic terminal, a chemical messenger or neurotransmitter is released that represents the output signal of the cell. The released neurotransmitter molecules elicit a synaptic potential in the postsynaptic cell. The synaptic potential represents the input signal to that cell.

if it is large enough – it generates an action potential. The amplitude and duration of the input signal (= receptor potential) determines the number of action potentials. Thus the graded *amplitude-encoded* nature of the input signal is translated into a *frequency code* of invariant action potentials. In other words: the information in the signal is only represented by the frequency and number of spikes, not by their amplitude and shape.

Neurons come in different shapes and sizes. Bipolar cells in the retina possess a single dendrite that receives signals from photoreceptor cells. In contrast, Purkinje cells in the cerebellum which participate in the coordination of movements possess highly elaborated and wide branching dendritic trees (Figure 4). Also the lengths of axons vary considerably, ranging from just a few µm for some interneurons in the CNS to even meters for motor neurons passing signals from the spinal cord to muscle fibers in legs or toes. Independent of the size, shape, transmitter content, and physiological function (sensory neuron, motor neuron, interneuron), most neurons can be described by the generalized *model neuron* shown in Figure 2.

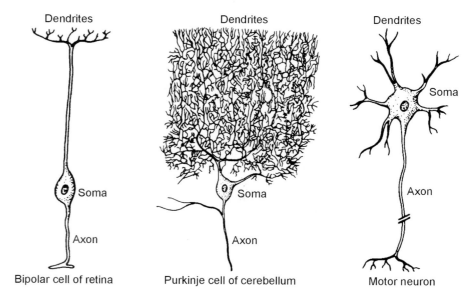

Figure 4: Examples of neurons from the vertebrate nervous system. The figure illustrates the diversity of sizes and shapes of individual neurons (adapted from [1]).

2 Membrane Potential

The cell membrane of a neuron consists of a double layer of (hydrophobic) lipid molecules. Due to the physico-chemical properties of these molecules, a separation of charges is established and maintained across the cell membrane. A thin layer of positively and negatively charged ions is spread over the inner and outer surfaces of the cell membrane. At rest, a neuron has an excess of positive charges on the outside of the membrane and an excess of negative charges on the inside. The charge separation gives rise to a difference in electrical potential, or voltage, across the cell membrane, called the *membrane potential* (V_m). It is defined as:

$$V_m = V_{in} - V_{out} \qquad (1)$$

where V_{in} is the potential on the inside of the cell and V_{out} the potential on the outside. The membrane potential of a cell at rest is called the *resting membrane potential* (V_r). Since, by convention, the potential outside the cell is defined as zero, the resting potential V_r is equal to V_{in}. In neurons V_r typically ranges from –60 mV to –70 mV. All electrical signaling in a neuron involves brief changes from the resting membrane potential. This is achieved by the transient opening and closing of ion channels leading to alterations in the flow of electrical current across the cell membrane. A schematic drawing of a cell membrane, ion channels, and hydrated cations is depicted in Figure 5.

The electric current that flows into and out of the cell is carried by positively charged (*cations*) and negatively charged (*anions*) ions. The direction of current flow is conventionally defined as the direction of *net* movement of *positive charge*. Thus, in an ionic solution cations move in the direction of the electric current, anions in the opposite direction. Whenever there is a net flow of cations or anions into or out of the cell, the charge separation across the membrane is disturbed, altering the polarization of the membrane. A reduction of charge separation, leading to a less negative membrane potential, is called *depolarization*. An increase in charge separation, leading to a more negative membrane potential, is called *hyperpolarization*. Changes in membrane potential that do not lead to the opening of ion channels, are called *electrotonic potentials* and are said to be passive responses of the membrane. Hyperpolarizing responses are almost always passive, as are small depolarizations. However, when depolarization of the neuron approaches a critical level, called the *threshold*, the cell responds actively with the opening of voltage-gated ion channels and an action potential is produced.

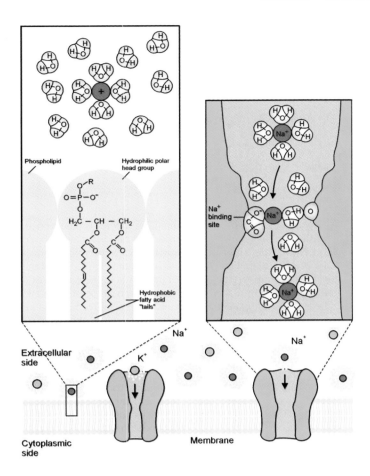

Figure 5: The ionic permeability properties of the membrane are determined by the interactions of ions with water, the membrane lipid bilayer, and ion channels. Ion channels are integral membrane proteins that span the lipid bilayer, providing a pathway for ions to cross the membrane. Phospholipids form bilayers that are the basis for all cellular membranes. Phospholipids have a hydrophilic head and a hydrophobic tail. The hydrophobic tails join to exclude water and ions, while the polar hydrophilic heads face the aqueous environment of the extracellular fluid and cytoplasm. *Left enlargement:* Ions in solution are surrounded by a cloud of water molecules. In the illustration, a positively charged ion attracts the electronegative oxygen atoms of the surrounding water molecules. This cloud is carried along by the ion as it diffuses through solution. It is energetically unfavorable, and therefore improbable, for the ion to leave this polar environment to enter the nonpolar environment of the lipid bilayer. The inset also shows the structure of a phospholipid. It is composed of a backbone of glycerol in which two of its –OH groups are linked by ester bonds to fatty acid molecules. The third –OH group of glycerol is linked to phosphoric acid. The phosphate group is further linked to one of a variety of small, polar head groups (R). *Bottom:* A model showing how the ion channels are able to select for either K^+ or Na^+ ions. Potassium channel (left): Although a Na^+ ion itself is smaller than a K^+ ion, its effective diameter in solution is larger because its local field strength is more intense, causing it to attract a larger cloud of water molecules. Thus, a channel can select for K^+ over Na^+ by excluding hydrated ions whose diameter is larger than the pore. Sodium channel (right): Sodium channels have a selectivity filter somewhere along the length of the channel, with a site that weakly binds Na^+ ions. According to the hypothesis developed by Bertil Hille and colleagues, a Na^+ ion binds transiently at an active site as it moves through the filter (*right enlargement*). At the binding site the positive charge of the ion is stabilized by a negatively charged amino acid residue on the channel wall and also by a water molecule that is attracted to a second polar amino acid residue on the other side of the channel wall. It is thought that a K^+ ion, because of its larger diameter, cannot be stabilized as effectively by the negative charge and therefore will be excluded from the filter (adapted from [1]).

3 What Determines the Resting Membrane Potential?

The membrane potential is determined by the ionic conductances in the cell membrane and by the distribution of at least four ion species across the membrane. There are three large families of proteins that actively participate in ion transport across the cell membrane: 1. energy-consuming ion pumps, 2. voltage controlled (= voltage-gated) ion channels, and 3. chemically controlled (= ligand-gated) ion channels. No single ion species is distributed equally on the two sides of a nerve cell membrane. While Na^+ and Cl^- ions are more concentrated outside the cell, K^+ ions and organic anions (A^-) are more concentrated in the cytoplasm. The organic anions are primarily amino acids, proteins, and nucleotides. Table 1 shows the distribution of these ions inside and outside of one particularly well-studied nerve cell, the giant axon of the squid. Although the absolute values of the ionic concentrations for vertebrate nerve cells are two- to threefold lower than those for the squid giant axon, the concentration *gradients* (the ratio of the external ion concentration to internal ion concentration) are about the same.

The unequal distribution of ions raises several important questions. How do ionic gradients contribute to the resting membrane potential? How are they maintained? What prevents the ionic gradients from dissipating by diffusion of ions across the membrane through passive (resting) channels? We shall answer these questions by considering two examples of membrane permeability: the resting membrane potential of a cell, which is permeable to only one species of ions, and the resting membrane potential of a cell, which is permeable to at least three.

Species of ion	Concentration in cytoplasm (mM)	Concentration in extracellular fluid (mM)	Equilibrium potential[1] (mV)
K^+	400	20	-75
Na^+	50	440	+55
Cl^-	52	560	-60
A^- (organic anions)	385	--	--

[1] The membrane potential at which there is no net flux of the ion species across the cell membrane.

Table 1: Distribution of the major ions across a neuronal membrane at rest: the giant axon of the squid.

The overall selectivity of a membrane for individual ion species is determined by the relative proportions of the various types of ion channels in the cell that are open. Consider the simple case of a cell that has a membrane which is exclusively permeable to K^+ ions. Because K^+ ions are present at high concentration inside the cell, K^+ ions tend to diffuse down their chemical concentration gradient from the inside to the outside of the cell. As a result, the outside potential of the membrane becomes more positive due to the slight excess of K^+, whereas the inside potential becomes more negative because of the deficit of K^+ and the resulting slight excess of organic anions. Due to the resulting electrical field, the diffusion of K^+ out of the cell is self-limiting. Thus, ions are subject to different forces driving them across the membrane: (1) a *chemical driving force* that depends on the concentration gradient of the ion across the membrane and (2) an *electrical driving force* that depends on the electrical potential difference across the membrane. Once K^+ diffusion has proceeded to a certain point, a potential is reached at which the electrical force driving K^+ into the cell exactly balances the chemical force driving K^+ ions out of the cell. That is, the outward current of K^+ (driven by its concentration gradient) is equal to the inward current of K^+ (driven by the electrical potential difference across the membrane). This potential is called the *potassium equilibrium potential*, V_K. In a cell that is permeable to K^+ ions only, V_K determines the resting membrane potential.

The equilibrium potential for any ion X can be calculated from an equation derived in 1888 from basic thermodynamic principles by the German physical chemist Walter Nernst:

$$V_X = \frac{k_B T}{z e} \ln \frac{[X]_o}{[X]_i} \qquad \text{Nernst Equation} \qquad (2)$$

where k_B is the Boltzmann constant, T the temperature, z the valence of the ion, e the elementary charge, and $[X]_o$ and $[X]_i$ are the concentrations of the ion outside and inside of the cell, respectively. (To be precise, chemical activities should be used rather than concentrations.)

Since $k_B T/e$ is 25 mV at 25 °C, and the factor for converting from natural to base-10 logarithms is 2.3, the Nernst equation can also be written as:

$$V_X = \frac{58 \text{ mV}}{z} \log \frac{[X]_o}{[X]_i} \qquad (3)$$

Using the K^+ concentrations of the squid giant axon given in Table 1:

$$V_K = \frac{58 \text{ mV}}{1} \log \frac{[20]}{[400]} = -75 \text{ mV} \qquad (4)$$

The Nernst equation can be used to calculate the equilibrium potential of any ion that is present on both sides of the cell and can permeate the cell membrane (see Table 1).

In our discussion, we have treated the generation of the resting potential by the diffusion of ions down their chemical gradients as a passive mechanism, one that does not require the expenditure of energy by the cell, for example through hydrolysis of *adenosine triphosphate* (*ATP*). Energy supplied by *ATP* hydrolysis, however, is required to set up the initial ionic concentration gradients and to maintain them. The ionic concentration gradients are produced by so-called ion pumps, or *ATPases*. These pumps are proteins in the cell membrane that couple active ion transport across membranes with the hydrolysis of *ATP* to *ADP*, even against steep chemical concentration gradients. Some pumps transport Na^+ ions from inside the cell to the outside and – at the same time – K^+ ions from the outside to the inside. These pumps are referred to as Na^+/K^+-*ATP*ases. Some pumps transport only Ca^{2+} ions or protons; they are called Ca^{2+}-*ATP*ases and H^+-*ATP*ases, respectively. The ionic gradients across cell membranes are determined by yet another class of transport proteins – ion exchangers. Ion exchangers work without *ATP* hydrolysis. These proteins couple the flux of one ion species down its chemical gradient to the flux of another ion species up its chemical gradient. An important exchanger molecule for neurons is the Na^+/Ca^{2+} exchanger. It is responsible for the extrusion of Ca^{2+} ions from the neuron after periods of activity, when the intracellular Ca^{2+} concentration is rising.

In contrast to most cells of the body, nerve cells at rest are permeable to Na^+ and Cl^- ions in addition to K^+ ions. How do the three ionic gradients interact to determine the cell's resting membrane potential?

To answer this question, it will be easiest to examine first the diffusion of K^+ and Na^+. Let us return to the simple example of a cell having only K^+ channels, with concentration gradients for K^+, Na^+, Cl^-, and A^- as shown in Table 1. Under these conditions the resting membrane potential, V_r, is determined solely by the K^+ concentration gradient and will be equal to V_K (–75 mV).

Now consider what happens if a few Na^+ channels are added to the membrane, making it slightly permeable to Na^+. Two forces drive Na^+ into the cell. First, Na^+ is more concentrated outside than inside and therefore it tends to flow into the cell down its chemical concentration gradient. Second, Na^+ is driven into the cell by the negative potential across the membrane. The influx of positive charge, i.e. Na^+, slightly depolarizes the cell from the K^+ equilibrium potential (–75 mV). The new membrane potential, however, does not come even close to the Na^+ equilibrium potential of +55 mV, because the open K^+ channels outnumber the Na^+ channels in the membrane.

As soon as the membrane potential begins to depolarize from the value of V_K (–75 mV), K^+ flux is no longer in equilibrium across the membrane. The reduction in the negative electrical force driving K^+ into the cell will result in an efflux of K^+ out of the cell, tending to counteract the Na^+ influx. The more the membrane potential is depolarized and moves away from the K^+ equilibrium potential, the greater is the electrochemical force driving K^+ out of the cell and consequently the greater is the K^+ efflux. Eventually, the membrane potential reaches a new resting potential at which the outward movement of K^+ just balances the inward movement of Na^+. This balance point (usually –60 mV) is far from the Na^+ equilibrium potential (+55 mV) and is only slightly more positive than the equilibrium potential for K^+ (–75 mV).

To understand how this balance point is determined, bear in mind that the magnitude of the flux of an ion across a cell membrane is the product of its *electro-chemical driving force* (the sum of the electrical and the chemical driving forces) and the conductance of the membrane to the ion:

ion flux = (electrical driving force + chemical driving force) x membrane conductance.

At rest, a cell has relatively few Na^+ channels open, so the conductance to Na^+ is quite low. Therefore, despite the large chemical and electrical forces driving Na^+ into the cell, the influx of Na^+ is small. Since there are many more K^+ channels open at rest, the membrane conductance of K^+ is relatively large. As a result, the small outward force acting on K^+ at the resting membrane potential is enough to produce a K^+ efflux equal to the Na^+ influx.

Although Na^+ and K^+ fluxes largely determine the value of the resting potential, V_r is neither equal to V_K nor V_{Na}. As a general rule, when the membrane potential V_m is determined by two or more species of ions, the influence of each species is determined not only by the concentrations of the ion inside and outside the cell but also by the ease with which the ion crosses the membrane. In terms of electrical current flow, the membrane's conductance provides a convenient measure of how readily the ion crosses the membrane. Another convenient measure is the permeability (P) of the membrane to an ion in units of velocity, cm/s. This measure is similar to that of a diffusion constant, which describes the rate of solute movement in solution. The dependence of membrane potential on ionic permeability and concentration is given quantitatively by the Goldman-Hodgkin-Katz (GHK) equation:

$$V_\mathrm{m} = \frac{k_\mathrm{B}T}{e}\ln\frac{P_\mathrm{K}[\mathrm{K}^+]_\mathrm{o} + P_\mathrm{Na}[\mathrm{Na}^+]_\mathrm{o} + P_\mathrm{Cl}[\mathrm{Cl}^-]_\mathrm{i}}{P_\mathrm{K}[\mathrm{K}^+]_\mathrm{i} + P_\mathrm{Na}[\mathrm{Na}^+]_\mathrm{i} + P_\mathrm{Cl}[\mathrm{Cl}^-]_\mathrm{o}} \qquad \textit{GHK Equation} \qquad (5)$$

This equation applies only when V_m is not changing. It states that the greater the concentration of a particular ion species and the greater its membrane permeability, the greater its role in determining the membrane potential. When the permeability to one ion species is exceptionally high, the GHK equation reduces to the Nernst equation for that ion. So, if $P_\mathrm{K} \gg P_\mathrm{Cl}$ and P_Na, the equation becomes:

$$V_\mathrm{m} \cong \frac{k_\mathrm{B}T}{e}\ln\frac{[\mathrm{K}^+]_\mathrm{o}}{[\mathrm{K}^+]_\mathrm{i}} \qquad (6)$$

For further reading, we recommend Benedek and Villars [4] and Hille [5].

4 How is the Action Potential Generated?

Every neuron has a unique region adjacent to its soma, the axon hillock. Here, the density of voltage-gated Na^+ and K^+ channels is very high. When V_m at the axon hillock is disturbed to slightly more positive values, the membrane permeability for Na^+ and K^+ ions, P_Na and P_K, changes due to the sequential opening of voltage-gated Na^+ and K^+ channels. When a certain threshold value is reached, a brief voltage pulse (~ 1-2 ms duration) is generated, the action potential. It travels along the axon of the neuron down to the synapse (see Figure 2). The timing and nature of the ionic currents involved in the generation of the action potential have been elucidated by Alan Hodgkin, Andrew Huxley, and Bernhard Katz. Figure 6 shows the typical waveform of an action potential. It is generated by a sequence of molecular events. The depolarization of the membrane causes the rapid opening of voltage-gated Na^+ channels. Due to the large electro-chemical driving force, an inward Na^+ current develops. This Na^+ current causes further depolarization of the cell. More Na^+ channels will open (i.e. g_Na becomes larger) and the Na^+ inward current increases further (Figure 6). This positive feedback cycle eventually drives the membrane potential to the peak of the action potential. If the membrane conductance would be solely determined by Na^+ ions ($g_\mathrm{Na} \gg g_\mathrm{K}$), then, theoretically V_m might become V_Na (+55 mV, see Table 1) at the peak of the action potential.

The depolarization of the membrane potential not only leads to the opening of voltage-gated Na^+ channels, but also causes them to close again, a process called *inactivation*. The inactivated state of the Na^+ channel is different from its closed state – although both states are non-conductive. The Na^+ channel must proceed from the inactivated to the closed state in order to be able to open again. When Na^+ channels become inactivated, no more Na^+ current can enter the cell. At about the time when Na^+ channels inactivate (~1 ms after Na^+ channel opening), the delayed opening of voltage-gated K^+ channels occurs. K^+ ions will leave the cell according to their electro-chemical driving

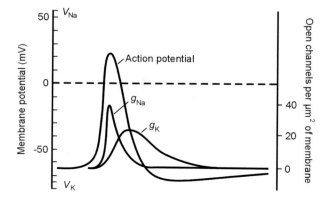

Figure 6: Shape of an action potential. The sequential opening of Na^+ and K^+ channels and the resulting conductance changes (g_Na, g_K) determine the shape of the action potential (adapted from [1]). The change in conductance is expressed as number of open Na^+ and K^+ channels per µm² of membrane on the ordinate at the right hand of the panel.

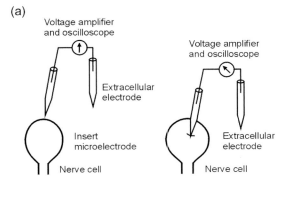
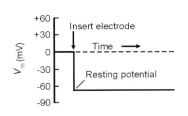
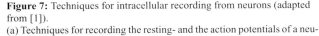
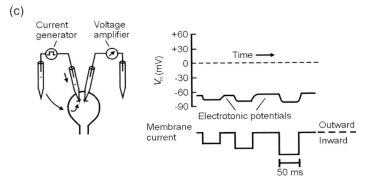

Figure 7: Techniques for intracellular recording from neurons (adapted from [1]).

(a) Techniques for recording the resting- and the action potentials of a neuron make use of glass micropipettes filled with a concentrated salt solution that serve as electrodes. These microelectrodes are placed on either side of the membrane. Wires inserted into the back ends of the pipettes are connected via an amplifier to an oscilloscope, which displays the amplitude of the membrane potential in volts. When both electrodes are outside the cell no electrical potential difference is recorded. But when one microelectrode is inserted into the cell the oscilloscope shows a steady voltage, the resting membrane potential.

(b) The membrane potential can be experimentally changed using a current generator connected to a second pair of electrodes – one intracellular and one extracellular. When the intracellular electrode is made positive with respect to the extracellular one, a pulse of positive current from the current generator will cause current to flow into the neuron from the intracellular electrode. This current returns to the extracellular electrode by flowing outward across the membrane. As a result, the inside of the membrane becomes more positve; the outside of the membrane becomes more negative. This decrease in the normal separation of charges is called depolarization. The small depolarizing current pulses evoke electrotonic potentials in the cell. The size of the change in potential is proportional to the size of the current pulses. Sufficiently large depolarizing current causes the opening of voltage-gated ion channels. The opening of these channels leads to the generation of an action potential.

(c) Reversing the direction of current flow – making the intracellular electrode negative with respect to the extracellular electrode – makes the membrane potential more negative. The increase in charge separation is called hyperpolarization. The responses of the cell to hyperpolarization are usually purely electrotonic and do not trigger an active response of the cell.

force. Because positively charged K$^+$ ions leave the cell and Na$^+$ ions cease to enter the cell, the membrane potential returns to its resting value (*repolarization*). Often, the action potential is followed by a transient hyperpolarizing afterpotential. This is caused by K$^+$ channels that stay open during the late phase of the action potential. The K$^+$ efflux from the cell is larger than during the resting state and V_m becomes slightly more negative than the resting membrane potential V_r.

When a neuron has just generated an action potential, the cell is unable to generate another one within a certain time span. This *refractory phase* lasts a few millisconds and is caused by the time it takes for the Na$^+$ channels to recover from inactivation.

5 Recording Electrical Signals from Neurons

Reliable techniques for recording the electrical potential across cell membranes were developed in the late 1940s. These techniques allow accurate recording of both the resting and the action potentials. The techniques make use of glass micropipettes, filled with a concentrated salt solution that serve as electrodes. These microelectrodes are placed on either side of the cell membrane. Wires inserted into the back ends of the pipettes are connected via an amplifier to an oscilloscope, which displays the amplitude of the membrane potential in volts. Because the tip diameter of a microelectrode is very small (< 1 μm), it can be inserted into a cell with little damage to the cell membrane.

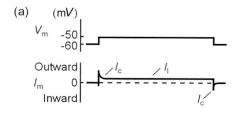
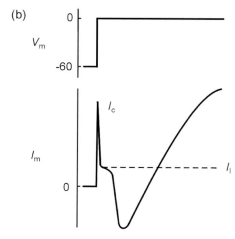
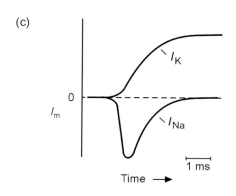

Figure 8: A voltage-clamp experiment demonstrates the sequential activation of two types of voltage-gated channels.
(a) A small depolarization is accompanied by capacitive and leakage currents (I_c and I_l, respectively).
(b) A larger depolarization results in larger capacitive and leakage currents, plus an inward current followed by an outward current. The inward current flows through Na^+ channels and the outward current through K^+ channels.
(c) Depolarizing the cell in the presence of tetrodotoxin (which blocks the Na^+ current) and again in the presence of tetraethylammonium (which blocks the K^+ current), reveals the pure K^+ and Na^+ currents (I_K and I_{Na}, respectively) after subtracting I_c and I_l (adapted from [1]).

When both electrodes are outside the cell, no difference in the electrical potential is recorded. But as soon as one microelectrode is inserted into the cell, the oscilloscope shows a steady voltage, the resting membrane potential. In most nerve cells at rest, the membrane potential is around –65 mV. The neuron can be impaled with a second electrode that allows the injection of positive or negative currents into the neuron. The changes in membrane potential that are produced by the current injection are shown in Figure 7 and explained in the figure legend.

5.1 The Voltage-Clamp Technique

Prior to the availability of the voltage-clamp technique, attempts to measure Na^+ and K^+ conductance as a function of membrane potential had been limited by the strong interdependence of the membrane potential and the gating of Na^+ and K^+ channels. The voltage-clamp technique was developed by Kenneth Cole in 1949 to stabilize the membrane potential of neurons for experimental purposes. It was used by Alan Hodgkin and Andrew Huxley in the early 1950s in a series of experiments that revealed the ionic mechanisms underlying the action potential.

The voltage clamp permits the experimenter to „clamp" the membrane potential at predetermined levels. The voltage-gated ion channels continue to open or close in response to changes in membrane potential, but the voltage clamp prevents the resultant changes in membrane current from influencing the membrane potential. This technique thus permits measurement of the effect of changes in membrane potential on the conductance of the membrane to individual ion species.

A typical voltage-clamp experiment starts with the membrane potential clamped at its resting value (Figure 8). If a 10 mV depolarizing potential step is commanded, an initial but very brief outward current instantaneously discharges the membrane capacitance by the amount required for a 10 mV depolarization (Figure 8a). This *capacitive current* (I_c) is followed by a smaller outward ionic current that persists for the duration of the pulse. At the end of the pulse, there is a brief inward capacitive current, and the total membrane current returns to zero. The steady ionic current that persists throughout the depolarization is the current that flows through the resting ion channels of the membrane and is called the *leakage current*, I_l. The total conductance of this population of channels is called the *leakage conductance* (g_l). These resting channels are responsible for generating the resting membrane potential. In a typical neuron most of the resting channels are permeable to K^+ ions; the remaining channels are permeable to Cl^- or Na^+ ions.

If a larger depolarizing step is commanded, the current record becomes more complicated (Figure 8b). The capacitive and leakage currents both increase in amplitude. In addition, shortly after the end of the capacitive current and the start of the leakage current, an inward current develops; it reaches a peak within a few milliseconds, declines, and gives way to an outward current. This outward current reaches a plateau that is maintained for the duration of the pulse.

A simple interpretation of these results is that the depolarizing voltage step sequentially turns on voltage-gated channels for two separate ions: one type of channel for inward current and another for outward current. Because these two oppositely directed currents partially overlap in time, the most difficult task in analyzing voltage-clamp experiments is to determine their separate time courses. This can be accomplished by toxins or blockers that specifically eliminate either the Na^+ current I_{Na} or the K^+ current I_K (Figure 8c).

5.2 The Patch-Clamp Technique

The patch-clamp technique is a refinement of voltage clamping and was developed in 1976 by Erwin Neher and Bert Sakmann to record current flow from single ion channels. A small glass micropipette is pressed against the membrane of a cell. The pipette is filled with a salt solution resembling that normally found in the extracellular fluid. A metal electrode in contact with the electrolyte in the micropipette connects the pipette to a special electrical circuit that measures the current flowing through channels in the membrane under the pipette tip. In 1980 Neher discovered that applying a small amount of suction to the patch pipette greatly increased the tightness of the contact between the pipette and the membrane. The result was a *seal* with extremely high resistance between the inside and the outside of the pipette. The seal lowered the electronic noise and extended the utility of the technique to the whole range of channels involved in electrical excitability. Since this discovery, the patch-clamp technique has been used to

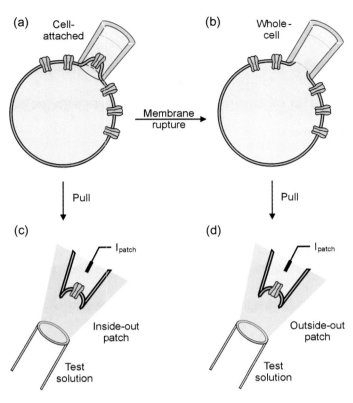

Figure 9: Patch-clamp configurations. The figure illustrates the experimental set-up of the cell-attached (a) and whole-cell (b) configuration and two versions of the excised-patch configuration, the inside-out (c) and the outside-out configuration (d).

study all major classes of voltage-gated, ligand-gated, and mechanically-gated ion channels. The patch-clamp technique has been employed in various configurations (Figure 9).

In the *cell-attached* mode, current is recorded that flows through channels in the patch of membrane that is electrically isolated by the tip of the glass electrode. In the *excised-patch* mode, a piece of membrane is excised from the cell membrane underneath the pipette. This recording mode comes in two configurations: the *inside-out* and the *outside-out* configuration. In the *inside-out* configuration, the intracellular side of the membrane patch is facing the bath solution – and vice versa. The two excised-patch configurations allow superfusion of the channel from both sides. In the *whole-cell* configuration, the membrane patch underneath the pipette tip is destroyed, thereby establishing an electrical contact between the cell interior and the pipette solution. In this configuration, the current flowing across the entire cell membrane is measured.

Figure 10 shows current fluctuations recorded from an excised membrane patch that contains a single ion channel gated by cyclic nucleotides. Cyclic nucleotides are an important class of chemical messengers *inside* cells that bind to certain types of ion channels and thereby activate these channels ([8], [9]). When the membrane patch is superfused with a cyclic nucleotide solution, the current rapidly and randomly fluctuates between two levels, which represent the closed and the open state, respectively, of the channel. When the cyclic nucleotide concentration is very high, i.e. all binding sites on the channel protein are occupied by the ligand, the channel spends most of it´s time in the open state. Thus the ligand determines the probability of the channel being open.

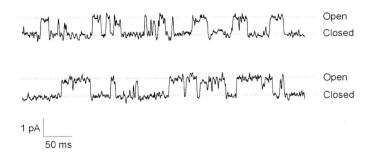

Figure 10: Patch-clamp record of the current flowing through a single cyclic nucleotide-gated ion channel ([8], [9]) that switches between closed and opened states.

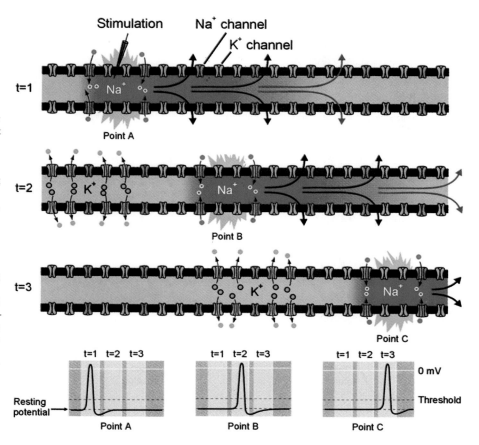

Figure 11: Mechanism of propagation of the action potential along the axon. Conduction requires both *active* and *passive* current flow. Active current flow occurs through ion channels across the membrane; passive or electrotonic flow of current occurs inside the axon. Depolarization at one point along an axon opens Na⁺ channels locally and produces an action potential in this region (point A) of the axon (time point (t) = 1). The resulting inward current flows passively along the inside of the axon, depolarizing the adjacent region (point B) of the axon. At a later time (t = 2), the depolarization of that adjacent region has opened Na⁺ channels and produced an action potential at this site. More inward current is produced that again spreads passively to the adjacent region (point C) farther along the axon. At a still later time (t=3), the action potential has propagated even farther. This cycle of events continues along the full length of the axon until the action potential reaches the synaptic terminal. As the action potential propagates, the membrane potential repolarizes due to the delayed opening of K⁺ channels and the inactivation of Na⁺ channels, leaving a „wake" of refractoriness behind the action potential that prevents its backward propagation (with permission from [3]).

6 Signal Propagation along the Axon

The mechanism by which the action potential is propagated without decrement is easy to grasp once one understands how an action potential is produced and how current passively flows along the axon. A depolarizing stimulus – either a synaptic potential, or a receptor potential, or current injection during an experiment (see Figure 7) – locally depolarizes the axon, thus opening Na⁺ channels in that region. The resulting Na⁺ inward current depolarizes the cell even further and opens still more Na⁺ channels, generating an action potential at that site. Because the excited segment of the axon adopts for a brief period of time a potential that is different from that of the rest of the cell, some of the local current produced by the action potential will flow passively down the axon and depolarize the membrane potential in the adjacent regions thereby opening new Na⁺ channels. The continuing cycle spreads until the action potential reaches the end of the axon (Figure 11).

The propagation of the action potential involves two forms of current flow: the passive flow of current along the axon and the active currents flowing through Na⁺ channels. The velocity of propagation is determined by both of these mechanisms. The conduction velocity increases with the square root of the diameter of an axon (because the internal resistance to passive current flow effectively decreases). This presumably explains why many invertebrates such as squids have evolved giant axons with diameters up to 1 mm. Another means to enhance the passive flow of electrical current is to insulate the axonal membrane, reducing the ability of current to leak out of the axon. Insulation increases the distance along the axon that a given local current can flow passively. This strategy has resulted in the evolution of *myelination*, a process by which specific glia cells wrap the axon in *myelin*, an insulating sheet consisting of multiple layers of flattened glia membrane. By acting as an electrical insulator, myelin greatly speeds up the conduction of the action potential. Conduction velocities of „naked" unmyelinated axons range from about 0.5 to 10 m/s, whereas myelinated axons can conduct at velocities up to 150 m/s. However, if the entire surface of an axon was insulated, there would

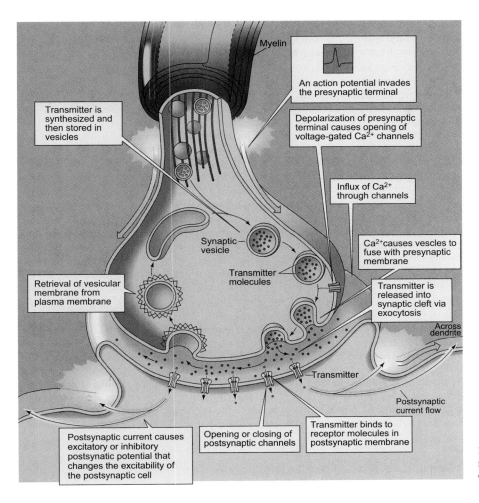

Figure 12: Sequence of events during signal transmission at a typical synapse (see text for more explanations (with permission from [3]).

be no place for current to flow in or out of the axon and action potentials could not be generated. The axon, therefore, is not completely insulated; the myelin sheet is interrupted at certain points along the axon, called *nodes of Ranvier*. At these sites, the density of Na^+ and K^+ channels is particularly high, whereas there are no or only few channels underneath the insulating myelin. The current generated by an action potential at a node of Ranvier flows passively within the myelinated segment until the next node is reached. Because the action potential is generated only at the nodes, this type of signal propagation is also called saltatory, meaning that the action potential jumps from node to node. The major reason for the larger conduction velocity is that the time-consuming process of the generation of the action potential is restricted to small segments along the axon.

7 How do Action Potentials Evoke Neurotransmitter Release?

How is the action potential transmitted from a neuron to a neighboring cell across the synaptic cleft? Neurons have developed a system of chemical signal transmission. The action potential triggers the release of neurotransmitters from the presynaptic terminal into the synaptic cleft (Figure 12). Neurotransmitters are either small organic molecules or larger peptides. Neurotransmitters can be grouped according to the synaptic potentials that they generate, i.e. inhibitory or excitatory. One group, by binding to their cognate receptors excites the postsynaptic cell, whereby – eventually – this cell may generate an action potential at the axon hillock. Excitatory transmitters are the *amino acid glutamate* and the *quaternary amine acetylcholine (ACh)*. Another group of transmitters causes inhibition of the postsynaptic cell. Binding to and activation of inhibitory receptors by the *amino acid glycine* and *γ-amino butyric acid* (*GABA*) impede the postsynaptic neuron to generate an action potential.

Neurotransmitter release is a complex process that involves several consecutive, highly regulated biochemical reactions. Neurotransmitters are stored in small organelles, the synaptic vesicles. When the action potential invades the presynaptic terminal, the vesicles fuse with the plasma membrane. The fusion process requires Ca^{2+} ions. The action potential causes opening of voltage-gated Ca^{2+} channels which mediate Ca^{2+} influx into the cell. These channels are located in close proximity to the fusion and release sites of the synaptic vesicles (*active zone*). Only a few vesicles will fuse and release their content per action potential. Within a couple of milliseconds after this process, called *exocytosis*, the vesicles are retrieved from the plasma membrane by a complex mechanism, called *endocytosis,* and refilled with neurotransmitter.

What happens to the released neurotransmitters? The neurotransmitters freely traverse the synaptic cleft and bind to receptors on the surface membrane of the postsynaptic cell. These receptors are ligand-gated ion channels that open upon binding of their cognate neurotransmitter. Activation of the ligand-gated ion channels leads to either depolarizing inward currents (*ACh-*, glutamate receptor) or hyperpolarizing outward currents (*GABA*, glycin receptor). The inward currents are carried by either Na^+ or Ca^{2+} or both; the outward currents are carried by Cl^- ions. The postsynaptic cell will respond with a small depolarization (*excitatory postsynaptic potential, EPSP*) or hyperpolarization (*inhibitory postsynaptic potential, IPSP*).

By means of it´s dendritic tree, a neuron can receive many – up to several thousand – input signals that are either excitatory (depolarizing) or inhibitory (hyperpolarizing). The receiving neuron sums and integrates these signals. The size and time course of the integrated response depends on the mixture of ion channels, their density and surface distribution as well as on the number and distribution of synaptic contacts on the dendrites, the axon, and the cell body. This large diversity endows each neuron with a specific computational power. In conclusion, the electrical properties of the activated receptors determine the electrical response of the postsynaptic cell.

8 Molecular Structure and Function of Ion Channels

The advent of molecular biology techniques, in conjunction with advances in electrophysiological recording techniques have uncovered a bewildering diversity of structures and mechanisms of activation and modulation of ion channels.

Broadly speaking, ion channels fall into two groups: voltage-gated channels that are activated by a change in membrane voltage and ligand-gated channels that become activated by binding of a small ligand to a receptor site on the channel polypeptide. Ligand-gated channels are either activated by binding of ligands on the external side of the cell (neurotransmitter receptors) or by *intracellular messengers*, such as Ca^{2+} ions, *adenosine 3',5'-cyclic monophosphate* (*cAMP*), *guanosine 3',5'-cyclic monophosphate* (*cGMP*), and *inositol-1,4,5-trisphosphate* (*IP₃*). Channels are set apart from each other by their ion selectivity (remember Na^+ or K^+ channels). Most voltage-gated channels are ion selective in that they sharply discriminate between cations and anions and also between different cations. However, selectivity among alkali ions or for monovalent versus divalent cations is not perfect. Relative ion permeabilities P_x/P_y (for example: x = Na^+, y = K^+) are a common measure of selectivity and range roughly between 1 : 10 and 1 : 1000. In addition, some ion channels even carry mixed currents, i.e. Na^+/Ca^{2+}- or Na^+/K^+ currents. We refer to this class of channels as *non-selective (cation) channels.*

The majority of K^+-, Na^+-, and Ca^{2+} channels genetically belong to the superfamily of voltage-gated channels. Curiously, ion channels that are gated by intracellular ligands, i.e. *cAMP, cGMP,* or *IP₃*, phylogenetically also belong to this large family of channel genes. Another family of genetically related genes encode ligand-gated ion channels which are activated by extracellular binding of ligands. Some well known members of this group are the *nicotinic acetylcholine receptors*, the *GABA receptors*, the *glutamate receptors*, the *glycine receptors*, and the *5-hydroxytryptamin (5-HT₃)-receptor.*

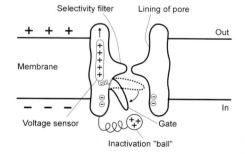

Figure 13: Functional domains of voltage-gated ion channels (see text for more explanations).

Functional channels are large heteromeric complexes that are usually composed of several distinct polypeptides (*subunits*). The subunits that build the complex may be identical or different. Functional channels may either contain four, five or six subunits. Each of these subunits contains certain domains that contribute to channel function (Figure 13). Most voltage-gated channel subunits embody positively charged segments that reside within the membrane and serve as *voltage sensor*. Upon depolarization, the voltage sensor traverses the membrane and causes a conformational change that gates open the pore of the channel. The protein structure that moves out of the pathway of the permeating ions is designated *channel gate*. A short hairpin-like structure lines the interior of the aqueous pore and interacts with the permeating ions. It determines which ions can pass through the channel, i.e. it serves as a *selectivity filter*. A *ball-and-chain structure* serves as an *inactivation gate* that, once open, moves into the mouth of the channel pore to block the ionic pathway.

Recently, the first structure of a K^+-selective channel from bacteria has been resolved on an atomic scale (Doyle et al. [10], Morais-Cabral et al. [11], Zhou et al. [12]). It shows that many structural features of channels that have been previously inferred from experiments using electrophysiological, genetic, or biochemical techniques were correct (Figure 14).

The large (molecular mass, M_w = 180 - 250 kD) channel-forming subunits of Na^+- and Ca^{2+} channels contain four internal homologous repeats, which fold into a pseudotetrameric structure (Figure 15). K^+ channel subunits are roughly 4-fold smaller (M_w = 60 - 100 kD) and assemble into functional tetramers (Figure 15). Each of the repeats of Na^+- and Ca^{2+} channels and K^+ channels harbor six hydrophobic segments that traverse the membrane. The structural and combinatorial variability of K^+ channels is enormous. Apart from the prototypical K^+ channel subunit with six membrane-spanning segments, several other subunits have been discovered that form K^+-selective pores. These subunits have either one or two transmembrane segments. Additional subunits have been discovered with either four or eight transmembrane segments and two pore-forming regions (Figure 15). The combinatorial diversity is dramatically enhanced as these subunits form K^+ selective pores by themselves or by co-assembly with other subunits.

Most voltage-gated ion channels are activated by depolarization of the cell membrane. There is, however, one interesting exception that deserves mention. A subclass of voltage-gated channels is activated by hyperpolarization rather than by depolarization. These *hyperpolarization-activated* channels share the general transmembrane architecture of K^+ channel subunits. They feature a positively charged voltage sensor motif that, however, operates differently than that of channels that are opened by depolarization. Moreover, these channels are also gated by the internal messenger *cAMP*. In this respect the hyperpolarization-activated channels are phylogenetic chimera that incorporate functional domains from both voltage-gated channels and ligand-gated channels [13].

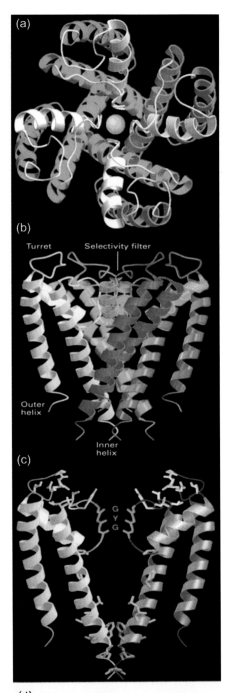

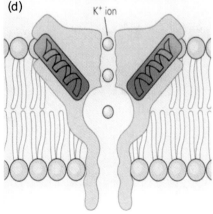

Figure 14: The X-ray crystal structure of a bacterial member of the inward rectifying K^+ channel family (with permission from Doyle et al. [10]).
(a) A view looking down at the channel from the outside of the membrane. Each of the four subunits contributes two long membrane spanning helices (in blue and red). The P-region is shown in white. It consists of a short α-helix (pore helix) and a loop that forms the selectivity filter of the channel. A K^+ ion is shown in the middle of the pore.
(b) A view of the channel in cross section in the plane of the membrane. The four subunits are shown, with each subunit in a different color. The membrane-spanning helices are arranged as an inverted teepee.
(c) Another view in the same orientation as in (b), showing only two of the four subunits. The selectivity filter (red region) is formed by three carbonyl oxygen atoms from the main chain backbone of three amino acid residues – glycine (G), tyrosine (Y), and glycine (G). Other residues important for binding of channel blocking toxins and drugs are labeled in white.
(d) A side-view of the channel illustrating three K^+ sites within the channel. The pore helices contribute a negative dipole that helps stabilize the K^+ ion in the water-filled inner chamber. The two outer K^+ ions are loosely bound to the selectivity filter formed by the P-region.

Fundamentals

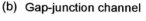

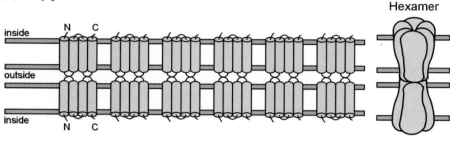

Figure 15: Families of ion channels.
(a) Several ligand-gated channels, including the nicotinic acetylcholine (Ach) receptor-channel, have five subunits, and each subunit consists of four transmembrane segments (S1-S4). Each cylinder represents a single transmembrane α-helix. A three-dimensional model of the channel is shown on the right.
(b) The gap-junction channel, found at electrical synapses, is formed from a pair of hemichannels in the pre- and postsynaptic membranes that join in the space between two cells. Each hemichannel is made of six subunits, each with four transmembrane segments. A three-dimensional model of the two hemichannels is illustrated on the right.
(c) The voltage-gated Na^+ channel is formed from a single (α) polypeptide chain that contains four homologous domains or repeats (motifs I-IV), each with six α-helical transmembrane segments (S1-S6) and one P-region thought to line the pore. The figure on the right shows the hypothetical model of the channel.
(d) Voltage-gated K^+ channels are composed of four polypeptide chains. Each subunit corresponds to one repeated domain of a voltage-gated Na^+ or Ca^{2+} channel, with six transmembrane segments and a P-region.
(e) Inward-rectifier K^+ channels are composed of four polypeptide subunits. Each subunit has only two transmembrane segments, connected by a P-region loop. A third family of K^+ channels has a characteristic subunit structure corresponding to two repeats of the inward-rectifier K^+ channel architecture, with two P-regions in tandem. The subunit composition of these channels is not known.

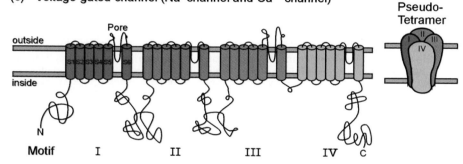

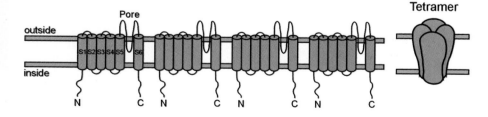

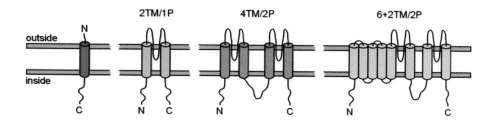

9 Biochemical Aspects of Learning and Memory

The ability of the nervous system to change and adapt is referred to in general terms as neural plasticity and in particular as *learning* and *memory*. Although the mechanisms of learning and other plastic changes are not completely understood in precise molecular terms, it is agreed that these phenomena are based on the regulation of *synaptic strength* or *efficacy* of extant synapses. These mechanisms differ from the plastic changes during the development of an organism, which are largely based on the wiring between cells – making new connections or removing existing ones. How do signaling events between neurons induce memory formation? What happens on a molecular level when we are learning or when we remember/recall a face, a place, or a situation? One can distinguish between two forms of memory: *short-term memory* (*STM*), lasting for minutes to hours and *long-term memory* (*LTM*), lasting for days to a lifetime. It is thought that short-term synaptic plasticity is underlying *STM*, whereas long-term synaptic plasticity is underlying *LTM*. Several different forms of synaptic plasticity have been described. The synaptic strength can be either enhanced or diminished; the changes can be of short duration or long-lasting.

Some changes in synaptic efficacy arise acutely as a result of previous synaptic activity. *Facilitation* is a short-term increase in synaptic efficacy when several action potentials enter the synaptic terminal in close succession. Each succeeding action potential releases more neurotransmitter; consequently, the amplitude of the synaptic potential increases progressively with each action potential. The facilitation is likely to be caused by an elevated level of the Ca^{2+} concentration in the synaptic terminal. Recall that the rate of fusion of synaptic vesicles and subsequent release of transmitter requires influx of Ca^{2+}. Each action potential incrementally increases the Ca^{2+} concentration because the clearance of Ca^{2+} from the cell is slow compared to the rapid Ca^{2+} influx occurring within a few milliseconds. Thus more vesicles fuse and more neurotransmitter is released. Another mechanism of short-term facilitation involves biochemical modification(s) of synaptic proteins. For example, phosphorylation (= adding a phosphate group) of ion channels profoundly alters their probability of being open. Upon repeated stimulation of a synaptic pathway, K^+ channels in the presynaptic terminal become phosphorylated, which locks the channel in the closed state. When an action potential enters the presynaptic terminal, repolarization of the cell is slower because fewer K^+ channels open, i.e. the duration of the action potential is longer and more Ca^{2+} ions enter the cell.

Another form of short-term plasticity is *synaptic depression*, following the repeated use of a synapse. Many action potentials in rapid succession release so much neurotransmitter that the vesicle pool becomes exhausted and the mechanisms for vesicle reuptake and recharging with neurotransmitter are overwhelmed. Thus excessive activity leads to progressive decline of synaptic strength.

Changes in synaptic plasticity that last for days, months, or even years rely on mechanisms referred to as *long-term potentiation* (*LTP*) or *long-term depression* (*LTD*). It is beyond the scope of this brief tutorial to describe the underlying cellular events in great detail. Suffice it to say that, while the mechanisms of short-term plasticity are located in the presynaptic terminal, the mechanisms of long-term plasticity seem to operate at the level of the postsynaptic cell. The induction of *LTP* also requires the persistent elevation of Ca^{2+} in the post-synapse that stimulates the synthesis of new proteins. Molecular studies have shown that phosphorylation of *transcription factors* in the cell nucleus leads to the activation of genes whose protein products are involved in *LTM* formation and consolidation. Some of these genes encode ion channels, transcription factors, or proteases. These proteases assist in keeping the enzymatic activity of kinases high. Thus more transcription factors will be phosphorylated and more genes are activated. In biochemical terms, this is a *positive feedback loop*.

In this short discourse, we have highlighted the molecular and physico-chemical aspects of neuronal cell function. Despite tremendous efforts in the past five decades, many questions remain unanswered. We trust that the forthcoming years will reward us with a wealth of new information on higher brain functions, of which learning and memory is only one facette.

Acknowledgements

The editor would like to thank Matthias Schindler (RWTH Aachen) for checking the symbols and formulas in this chapter.

References

[1] E.R. Kandel, J.H. Schwartz, T.M. Jessel, eds., *Principles of Neural Sciences*, 4th Edition, McGraw-Hill Companies, 2000.

[2] Z.W. Hall, *Molecular Neurobiology*, Sinauer Associates Inc., 1992.

[3] D. Purves, G.J. Augustine, D. Fitzpatrick, L.C. Katz, A.-S. LaMantia, J.O. McNamara (eds.), *Neuroscience*, Sinauer Associates Inc., 1997.

[4] G.B. Benedek, F.M.H. Villars, *Physics*, Vol. 3: Electricity and Magnetism. Addison-Wesley Publishing Company, 1979.

[5] B. Hille, *Ionic Channels of Excitable Membranes*, 3rd Edition, Sinauer Associates Inc., 2001.

[6] C. Koch, *Biophysics of Computation*, Oxford University Press, Oxford, 1999.

[7] F.M. Ashcroft, *Ion Channels and Disease*, Academic Press, San Diego, 2000.

[8] U.B. Kaupp, R. Seifert, *Cyclic nucleotide-gated ion channels*, Physiol. Reviews **82**, 769 (2002).

[9] R.S. Molday, U.B. Kaupp, *Handbook of Biological Physics*, Vol. 3: Vision (Part I), eds. D.G. Stavenga, W.J. de Grip, E.N. Pugh, Jr., Elsevier North-Holland, 2000.

[10] D.A. Doyle, J.M. Cabral, R.A. Pfuetzner, A. Kuo, J.M. Gulbis, S.L. Cohen, B.T. Chait, R. MacKinnon, Science **280**, 69 (1998).

[11] J.H. Morais-Cabral, Y. Zhou, R. MacKinnon, Nature **414**, 37 (2001).

[12] Y. Zhou, J.H. Morais-Cabral, A. Kaufman, R. MacKinnon, Nature **414**, 43 (2001).

[13] U.B. Kaupp, R. Seifert, Annu. Rev. Physiol. **63**, 235 (2001).

Circuit and System Design

Michael Dolle
Infineon Technologies, Munich, Germany

Contents

1 Introduction	167
2 MOSFET	167
3 CMOS Circuits	168
3.1 Inverter	168
3.2 Nand Gate	169
3.3 Nor Gate	169
3.4 Combined Gates	169
3.5 Tri-State Output	170
3.6 SRAM Cell	170
4 Digital Circuits	171
4.1 Flip-Flops	171
4.2 Multiplexer	172
4.3 Demultiplexer	173
4.4 Barrel Shifter	174
4.5 Adder	174
5 Logic Arrays	175
6 Circuit Simulation	175
6.1 Circuit Modeling	175
6.2 DC Analysis	176
6.3 Transient Analysis	176
6.4 AC Analysis	177
7 Microprocessor	177
7.1 Microprocessor Systems	177
7.2 Basic Principle	178
7.3 Pipelining	179
7.4 Instruction Set	180
7.5 Addressing Modes	180
7.6 Microcontroller Example	181
8 Digital Signal Processors	181
8.1 Classification of DSP Processors	182
8.2 Features of DSP Processors	182
8.3 DSP Example	184
9 Performance and Architectures	184

Circuit and System Design

1 Introduction

To complement the fundamentals part of this book for non-engineers (e. g. physicists, chemists, material scientists), this Chapter gives a brief and basic introduction into the characteristics of MOS transistors and the concept of standard digital logic circuits. It furthermore describes some fundamental digital components which represent the main building blocks for more complex systems. Finally, it outlines the basic principles of embedded microprocessors and digital signal processors which are used in many digital systems.

Since its first introduction in the 1960s the CMOS technology has evolved into the mainstream integrated circuit technology. Due to the complementary electrical characteristics of n-MOS and p-MOS transistors, CMOS technology is especially well suited for high-speed and low-power digital circuits. The scalability of MOS devices, i. e. the sustaining technology development towards smaller feature sizes, enables the implementation of more and more transistors and faster circuits on a single die, such as e. g. complete *systems-on-a-chip* (SoC) with hundreds of million of transistors.

This Chapter will *not* cover aspects of the fabrication technology, which is the topic of Part II of this book. As an extension to the standard digital logic introduced in this Chapter, a survey of more general logic concepts is presented in the Introduction to Part III. The physics behind selected types of alternative logic devices will be treated in the Chapters of Part III, while dynamic random access memories (DRAM) and selected non-volatile RAM types are covered by Part IV.

2 MOSFET

Figure 1 shows the device symbol and the basic operating voltages of a n-MOS enhancement transistor. The device has mainly three terminals: The *gate* terminal G is used to control the current flow between the *source* terminal S and the *drain* terminal D. The symmetry of the MOS transistor usually does not allow one to distinguish between the source and drain in a zero bias state; the roles of the two terminals are defined only after the terminal voltages are applied. The source is defined to be the terminal at the lower potential and is used as reference for the remaining bias arrangements. The drain-source voltage is denoted as V_{DS} and is defined to be positive. Similarly, V_{GS} describes the gate-source voltage. The drain current I_D is defined to be positive flowing into the drain.

For a gate-source voltage of $V_{GS} = 0$ the n-MOS enhancement transistor is switched off, and there is no current flow between source and drain ($I_D = 0$). This operating mode is called *cutoff* mode. If the gate-source voltage is increased to a value larger than a threshold voltage V_{th}, a conducting channel electrically connects drain and source. The n-MOS transistor is now in *active* mode. Depending on the relative value of the drain-source voltage V_{DS} with respect to $(V_{GS} - V_{th})$ two distinct current flow characteristics can be distinguished. For values of $V_{DS} < (V_{GS} - V_{th})$ the n-MOS transistor is said

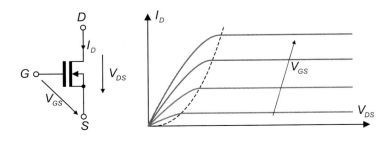

Figure 1: n-MOS transistor: Device symbol and bias arrangements, and current characteristic $I_D(V_{DS})$.

I Fundamentals

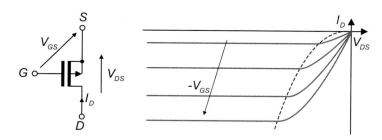

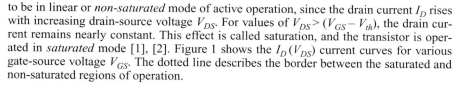

Figure 2: p-MOS transistor: Device symbol and bias arrangements, and current characteristic $I_D(V_{DS})$.

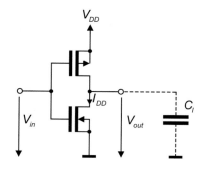

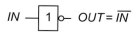

Figure 3: CMOS Inverter.

to be in linear or *non-saturated* mode of active operation, since the drain current I_D rises with increasing drain-source voltage V_{DS}. For values of $V_{DS} > (V_{GS} - V_{th})$, the drain current remains nearly constant. This effect is called saturation, and the transistor is operated in *saturated* mode [1], [2]. Figure 1 shows the $I_D(V_{DS})$ current curves for various gate-source voltage V_{GS}. The dotted line describes the border between the saturated and non-saturated regions of operation.

Enhancement mode n-MOS transistors have a positive threshold voltage $V_{th} > 0$. In contrast, *depletion* mode MOS transistors have a negative threshold voltage $V_{th} < 0$, that means they already conduct for a gate-source voltage of $V_{GS} = 0$. They are, for example, used as *load* devices in n-MOS designs [1].

The operational behavior of a p-MOS *enhancement* transistor represents the complement of the operation of the n-MOS device. Figure 2 depicts the voltage arrangements as well as the current characteristics for the p-MOS device. Compared to the n-MOS transistor, basically all voltage quantities are reversed.

The physical and technological background of MOSFETs as well as the scaling potential are described in Chap. 13.

3 CMOS Circuits

The complementary electrical characteristics of n-MOS and p-MOS transistors make them well suited for the design of logical gates used in many digital designs.

3.1 Inverter

The *inverter* represents the simplest CMOS circuit. Figure 3 shows the circuit diagram of the CMOS inverter consisting of a n-MOS and a p-MOS transistor. The input voltage is connected to the gates of both transistors. The two drain terminals are tight together and form the output of the inverter [1].

In steady-states, that means for a stable input voltage of either $V_{in} = 0$ or $V_{in} = V_{DD}$, only one of the transistors conducts while the other one is switched off. For $V_{in} = 0$ the p-MOS transistor conducts and the n-MOS transistor is switched off, hence the output is charged to the power supply voltage level V_{DD}. In case of $V_{in} = V_{DD}$ the n-MOS transistor conducts while the p-MOS transistor is switched off and the output voltage becomes zero. This characteristic has the great advantage that there is no current flow in static states (besides a very small leakage current) and represents the main reason for the low power dissipation of CMOS circuits.

During the switching process, that means for a change of the input voltage from a *low* voltage level to a *high* level, or from a *high* level to a *low* level, a crossover current flows through both transistors, since both transistors conduct simultaneously. Additionally, the load capacitance C_l has to be charged or discharged, respectively, resulting in an additional current component. Figure 4 depicts the voltage transfer curve $V_{out} = f(V_{in})$ and the corresponding current I_{DD}. The main characteristic is a relatively steep transfer region, where both transistors are conducting.

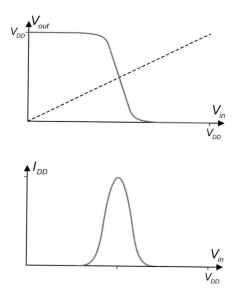

Figure 4: CMOS inverter: Voltage transfer curve $V_{out} = f(V_{in})$, and inverter current I_{DD}.

3.2 NAND Gate

Figure 5 shows the circuit diagram of a two-input NAND gate. With input variables A and B, the circuit gives the logical output of

$$Q = \overline{A \cdot B} \tag{1}$$

where "·" is used to denote the boolean operation AND, and the bar is used to denote the inverting.

The circuit uses full complementary structuring where each input is connected to both a n-MOS and a p-MOS transistor. Two n-MOS transistors are connected in series and two p-MOS transistors are connected in parallel to implement the boolean function. If both inputs A and B are high, then the two n-MOS transistors conduct current and discharge the output to ground level. However, if either A or B is low (or if both are low) then the path to ground is blocked. In this case at least one p-MOS device conducts to the power supply and the output is charged to a high level. The truth table in Figure 5 lists all possible logic combinations, where a "1" represents a high voltage level and a "0" represents a low level.

The two-input NAND gate shown in Figure 5 can easily be expanded to an n-input gate by adding additional transistors into the series n-MOS path, and by adding additional parallel p-MOS transistors, respectively.

3.3 NOR Gate

Figure 6 shows the circuit diagram of a two-input NOR gate. With input variables A and B, the circuit gives the logical output of

$$Q = \overline{A + B} \tag{2}$$

where "+" is used to denote the boolean operation OR.

Again, each input is connected to both a n-MOS and a p-MOS transistor. Compared to the NAND gate, there are now two p-MOS transistors connected in series and two n-MOS transistors are connected in parallel to implement the boolean function. If both inputs A and B are low, then the two p-MOS transistors conduct current and charge the output to power supply level. However, if either A or B is high (or if both are high) then the path to the power supply voltage is blocked. In this case at least one n-MOS device conducts and the output is discharged to a low level.

3.4 Combined Gates

NAND gates, NOR gates, and inverters are the basic elements of digital circuits. With these types of gates further logic operations can be constructed. For example, the combination of a NAND gate and an inverter results in an AND operation, and a NOR gate followed by an inverter gives the basic OR operation.

Moreover, according to De Morgan's law [3]

$$\overline{A \cdot B} = \overline{A} + \overline{B} \quad \text{and} \quad \overline{A + B} = \overline{A} \cdot \overline{B} \tag{3}$$

basic AND and OR operations can be realized with both types of gates. An AND function can be realized as follows:

$$A \cdot B = \overline{\overline{A \cdot B}} = \overline{\overline{A} + \overline{B}} \tag{4}$$

which means either through a NAND gate with a subsequent inverter or a NOR gate where the input signals are inverted. Accordingly, an OR function

$$A + B = \overline{\overline{A + B}} = \overline{\overline{A} \cdot \overline{B}} \tag{5}$$

can be realized with a NOR gate followed by an inverter or with a NAND gate where the input signals are inverted.

De Morgan's law also shows that all kind of logic operations can be constructed with a single type of gate, either NAND gates or NOR gates.

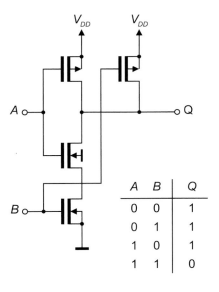

Figure 5: Two-input NAND gate: Circuit diagram and truth table.

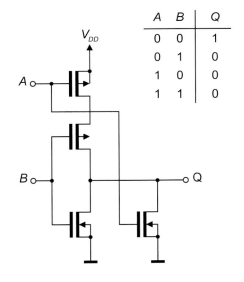

Figure 6: Two-input NOR gate: Circuit diagram and truth table.

Figure 7: AND-OR-Invert gate.

Logic operations are quite often expressed as sum of products like

$$Q = AB + CD \tag{6}$$

Applying De Morgan's law, this sum of two products can also be implemented by three NAND gates:

$$Q = AB + CD = \overline{\overline{AB} \cdot \overline{CD}} \tag{7}$$

An inverting form of a sum of two products, also known as AND-OR-Invert gate, can also be implemented as a single gate as shown in Figure 7. Eight transistors are required, thus saving four transistors compared to the three NAND gate implementation (or two transistors if an additional inverter is taken into account to achieve that same output polarity).

3.5 Tri-State Output

The logic gates described so far always drive their outputs to an active state, either to a high voltage level or to a low voltage level. This also implies, that the outputs of multiple gates must never be connected together. For some applications such as bus systems, however, it is more advantageous to connect outputs together. In order to make this possible, the gates must have the capability to switch their outputs into a *high-impedance* state.

Figure 8 shows the circuit diagram of a *Tri-State* buffer. An enable input *EN* determines whether the output of the gate is actively driving or whether it is switched into a high-impedance state. For *EN* = 0, both driver transistors are switched off regardless of the state of the input signal *A*, and hence, the output represents a high-impedance state *Z*. For *EN* = 1, the output stage is enabled, and the output reflects the value of the input signal, as shown in the truth table in Figure 8. Since the high-impedance state *Z* represents a third state besides the two driving states One and Zero, this output is named *tri-state* output. Although the outputs of multiple gates with tri-state capability can be connected together, it must be ensured that only one gate is enabled at a time.

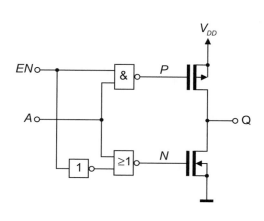

EN	A	P	N	Q
0	0	1	0	Z
0	1	1	0	Z
1	0	1	1	0
1	1	0	0	1

Figure 8: Tri-State Buffer.

3.6 SRAM Cell

Figure 9 shows the circuit diagram of a static random access memory (SRAM) cell. The key storage element consists of two cross-coupled inverters. This circuit has two stable states, designated as logic one and logic zero. Two pass transistors connect the memory cell to two bitlines, which are controlled by a wordline signal [6].

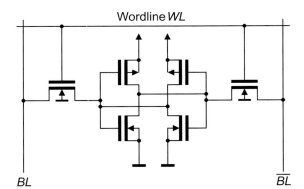

Figure 9: SRAM memory cell.

The wordline selects the memory cell for reading and writing data. A READ operation is performed by starting with both bitlines being pre-charged to a high level and selecting the word line. As a result, the data in the cell will pull one of the bit lines low. A sense amplifier connected to the bitlines will detect the differential signal, amplify it and feed it out to the output buffer. The READ operation is non-destructive, that means the logic state of the cell remains unchanged.

For a WRITE operation, the bitlines are activated with the bit data to be written. Note, that both bitlines must be activated with opposite values. Then the word line is activated. This forces the cell into the state represented by the bit lines, so that the new data is stored into the cross-coupled inverters.

A general description of RAM devices is given in the Introduction to Part IV.

4 Digital Circuits

The basic gates described above, like inverters, NAND gates, and NOR gates are the basic elements of digital circuits and are commonly used to construct more complex digital functions.

4.1 Flip-Flops

More complex logic systems also require *storing* logic states. The circuits used for this purpose are generally called *flip-flops* [3]. The simplest storage element, a RS-Flip-Flop, can be obtained by arranging two NOR gates as shown in Figure 10.

The two complementary outputs Q and $\overline{Q}$ are each fed back to an input of the other gate. The two inputs are S (Set) and R (Reset). For $S = 1$ and $R = 0$, the Q output will be set to one and the complementary $\overline{Q}$ output to zero. Accordingly, for $R = 1$ and $S = 0$, we obtain the reversed case, Q becomes zero and $\overline{Q}$ becomes one. In other words, a one signal at the S input *sets* the flip-flop and a one signal at the R input *resets* the flip-flop. For $R = S = 0$, the old output states are maintained, which represents the main principle of the RS-Flip-Flop as storage element. For $R = S = 1$ both outputs become zero. However, when both inputs return again to zero simultaneously, the output state is not defined any more, and therefore the input state $R = S = 1$ is usually not allowed.

According to (3), a RS-flip-flop can also be implemented with NAND gates instead of NOR gates as shown in Figure 11.

Synchronous digital systems require that a flip-flop reacts only at a certain point of time on a change of the input signals determined by a *clock* signal. Figure 12 shows the circuit diagram of a transparent D-flip-flop, also called *D-latch*. It mainly consists of a NAND based RS-flip-flop. The inverted set and reset inputs are each gated by a NAND gate which is controlled by the clock signal C. Therefore, the set and reset inputs can only become active if the clock signal is one. For $C = 0$ the actual output state of the RS-flip-flop is preserved. That means the flip-flop stores its output states for half a clock cycle, and is *transparent* for the other half of the clock cycle.

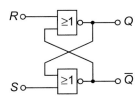

S	R	Q	$\overline{Q}$
0	0	Q_{-1}	$\overline{Q}_{-1}$
0	1	0	1
1	0	1	0
1	1	(0)	(0)

Figure 10: RS-Flip-Flop.

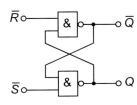

$\overline{S}$	$\overline{R}$	Q	$\overline{Q}$
0	0	(1)	(1)
0	1	1	0
1	0	0	1
1	1	Q_{-1}	$\overline{Q}_{-1}$

Figure 11: RS-Flip-Flop implemented with NAND gates.

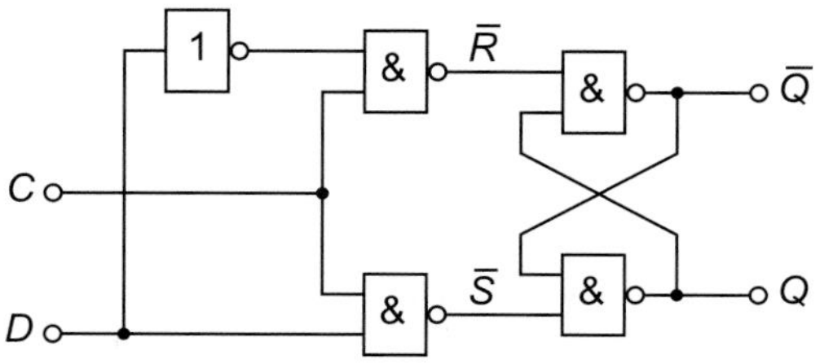

Figure 12: Transparent D-flip-flop (D-Latch).

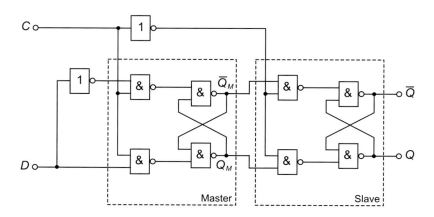

Figure 13: Edge-triggered D-flip-flop.

In order to store a data signal for a whole clock cycle, *single-edge triggered flip-flops* will be applied. They are implemented by connecting two transparent D-flip-flops in series as shown in Figure 13. The first stage is called the *master* flip-flop, and the second stage is called the *slave* flip-flop. The two stages are controlled by complementary clock signals. For $C = 1$, the master flip-flop is transparent and its output signal Q_M follows the input signal D, while the slave flip-flop stores the old state. When the clock changes from "1" to "0", the master stores the actual state of the input signal D. Since the change of the clock signal also causes the slave to become transparent, the output of the master flip-flop Q_M is now passed to the output of the slave flip-flop. In summary, the flip-flop only stores that state of the input signal D, which exists during the falling *edge* of the clock signal. For the remaining period of the clock signal the output state is preserved.

4.2 Multiplexer

Multiplexers are commonly used to select one out of n input signals D_i. A set of control signals S_i selects the desired input signal. For $n = 4$ the behavior can be described by the following boolean equation:

$$Q = S_1 D_1 + S_2 D_2 + S_3 D_3 + S_4 D_4 \tag{8}$$

If n can be represented by 2^a (here: $4 = 2^2$), then the a dedicated select signals S_i can be encoded into an address signal requiring only n lines, e.g. for $n = 4$:

$$Q = \overline{s_1}\,\overline{s_2} D_1 + s_1 \overline{s_2} D_2 + \overline{s_1} s_2 D_3 + s_1 s_2 D_4 \tag{9}$$

Figure 14 shows an implementation of a 4-to-1 multiplexer based on NAND gates.

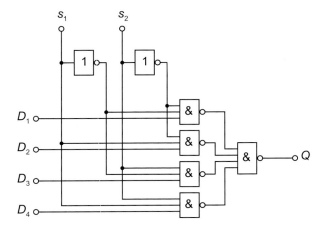

Figure 14: 4-to-1 Multiplexer.

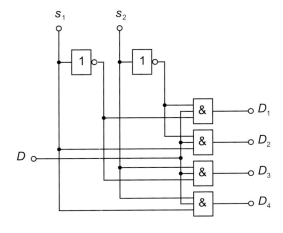

Figure 15: Demultiplexer.

The lines s_1 and s_2 represent the address lines for binary addresses between 00 and 11. This concept can be extended to any number a to realize n-to-1 multiplexers.

4.3 Demultiplexer

Demultiplexers provide the reversed function of multiplexers. A single input signal D will be distributed to one out of n outputs. Similarly to the multiplexer, a set of select signals addresses the desired output. Figure 15 shows the circuit diagram of a 1-to-4 demultiplexer.

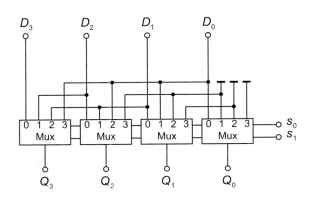

Figure 16: 4-bit Barrel Shifter (MUX denotes a 4-to-1 multiplexer).

4.4 Barrel Shifter

The multitude of applications requires operations where a bit pattern must be shifted by one or more positions. A circuit capable of shifting bit patterns within a single clock cycle is called a *barrel shifter* [3]. It basically consists of an arrangement of multiplexers as shown in Figure 16.

The control signals s_0 and s_1 determine the number of shift positions. For $s_0 s_1 = 01$, the input pattern will be shifted by one position, that means $Q_3 = D_2$, $Q_2 = D_1$, $Q_1 = D_0$, and $Q_0 = 0$. For $s_0 s_1 = 10$, the input pattern will be shifted by two positions, and for $s_0 s_1 = 11$ the pattern will be shifted by three position. While the higher bit positions will be lost, a fixed zero is inserted into the lower bit positions.

Applying multiplexers with n inputs, a shift operation of up to $n - 1$ positions can be achieved.

4.5 Adder

Another commonly required operation is the addition of binary digits. The addition of two one-digit binary numbers requires two input signals a_0 and b_0, and results in two output signals, one represents the sum s_0 of the two digits and the other one the carry c_1 into the next digit. The operation can be expressed as

$$s_0 = \overline{a_0} \cdot b_0 + a_0 \cdot \overline{b_0} = a_0 \oplus b_0 \tag{10}$$

and

$$c_1 = a_0 \cdot b_0 \tag{11}$$

Hence, the carry c_1 can be represented by an AND operation, and the sum s_0 by an exclusive OR (XOR) operation. Figure 17 shows the circuit diagram of the implementation of the *half-adder* [3].

For the addition of multi-digit binary numbers, the half-adder can only be applied for the least significant digit. For all other digits, three bits have to be added, since the carry from the last digit has to be considered. Expanding (10) and (11) taking into account the carry c_i from the last digit gives:

$$s_i = a_i \oplus b_i \oplus c_i \tag{12}$$

and

$$c_{i+1} = a_i \cdot b_i + a_i \cdot c_i + b_i \cdot c_i \tag{13}$$

Figure 18 shows the circuit diagram of a *full-adder* [3] implementing (12) and (13). It consists of two half-adders and an OR gate to generate the final carry output signal c_{i+1}.

In order to add multi-digit binary numbers a full-adder is needed for each digit, except for the least significant digit where a half-adder is sufficient. In simple implementations, the carry from a previous digit is used as input to the adder of the next digit. However, the main disadvantage of this implementation is its slow calculation time,

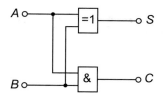

A	B	S	C
0	0	0	0
0	1	1	0
1	0	1	0
1	1	0	1

Figure 17: Half Adder.

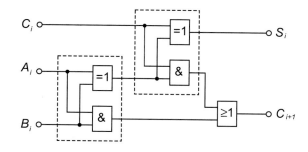

Figure 18: Full-Adder.

since all full-adders are effectively connected in series. For fast additions of multi-digit binary numbers, e.g. 32-bit number, a *carry-look-ahead* adder implementation is used. In this implementation, the carry bits for each digit are calculated independently.

5 Logic Arrays

Programmable logic devices (PLD) and *field programmable gate arrays* (FPGA) provide the elementary resources, like logic gates, flip-flops, and wires in a device that can be programmed [7]. This flexibility enables designers to make an application specific device replacing several standard parts, and is well suited for low to mid range volumes where it is not economically feasible to design custom devices. The available sub-micron silicon technology already enables programmable devices with hundreds of thousands of logic gates, thus, permitting designers to even program complex systems into a single gate array. This capacity together with the ease of re-programmability also enables another area of usage of gate arrays: The hardware verification of complex systems, such as microprocessors, before they will be implemented into standard or custom devices using expensive silicon manufacturing processes.

Programmable devices basically provide an array of user configurable macro cells with interconnect channels between the cells as schematically shown in Figure 19. Special I/O-blocks provide the connection to the pins of the package and can be programmed to act as input or output. The desired logical functions are established by connecting the inputs and outputs of the macro cells to the wiring channels, thus establishing direct connections between the outputs of one macro cell to inputs of other macro cells or I/O-blocks. The capability to configure each macro cell and its connection to the interconnect channels individually, provides a high degree of flexibility and allows to establish the desired functionality of the device. There are several mechanisms to program the required connections. Often PLD and FPGA devices use EEPROM or SRAM cells to establish the connections. This provides the capability to erase and to reprogram the device. In the case of SRAM cells, the required configuration must be loaded into the device each time after the power-supply voltage is supplied [7].

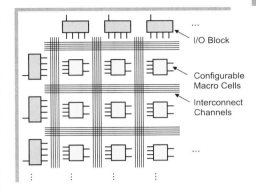

Figure 19: FPGA structure [8].

The structure and complexity of the macro cells and I/O-blocks, as well as the interconnect scheme varies from FPGA vendor to vendor. Figure 20 shows an example of a macro cell consisting of two six-input AND gates, four two-input AND gates, six two-to-one multiplexers and a D-flip-flop [8]. This structure allows implementing functions with up to 16 simultaneous inputs. Examples of functions which can be implemented are: One 16-input AND gate, two six-input AND gates plus two four-input AND gates, one five-input XOR gate, and numerous sum-of-products functions with up to 16 inputs or 16 product terms. More complex logical functions require the use of multiple macro cells. For this purpose the flip-flop can be by-passed, that means the output of the combinatorial circuit is directly fed into the routing channel and is used as an input to another macro cell. There are also PLD and FPGA implementations available using far more complex macrocells or logic elements with larger combinatorial functions, multiple flip-flops, and even small SRAM memories [9],[10].

6 Circuit Simulation

Since prototype manufacturing of semiconductor devices is an expensive process, economical reasons limit the number of test devices before mass production can start. Ideally, the first implementation of a new device should work without major problems. Circuit simulation programs like SPICE (**S**imulation **P**rogram with **I**ntegrated **C**ircuit **E**mphasis) are therefore commonly used to predict the behavior of complex analog and digital circuits before they will be implemented into a physical device. They allow to perform various types of analyses, like DC, AC, or transient analyses in order to test the response of a circuit to different inputs. Hence, the circuit designer can make any necessary changes and optimizations without modifying any hardware. The computer simulation allows to check the operability of the circuit in real life simulations even taking into account temperature variations or manufacturing process variations to validate the viability of the circuit. Since all tests, designs and modifications are made over a computer, the designer can save a lot of money.

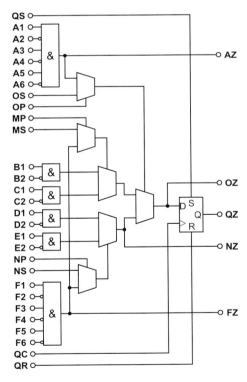

Figure 20: FPGA macro cell [8].

6.1 Circuit Modeling

The first step to perform a circuit simulation is the description of the circuit, either as a netlist or as a schematic diagram. For this purpose the simulation program provides a set of built-in models for passive components such as resistors, capacitors, and inductors,

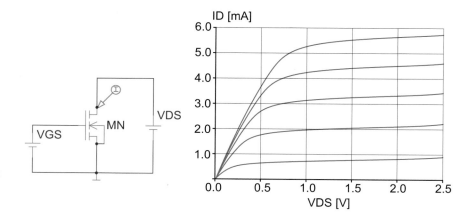

Figure 21: SPICE DC Sweep: Circuit diagram and simulated n-MOS current characteristic.

and active components such as diodes, bipolar transistors, and MOS transistors, etc. Independent current and voltage sources are used to model input signals to the circuit in order to simulate the response of the circuit.

The netlist or schematic diagram defines the required circuit components and its connections, and the respective model parameters describe the detailed behavior of the components. While passive components can be simply defined by their respective resistance or capacitance values, far more model parameters are required to model the more complex behavior of active components like MOS transistors. The SPICE circuit simulator provides a variety of built-in MOS transistor models with different levels of complexity and accuracy, from simple models implementing first order MOS equations up to complex BSIM models (**B**erkeley **S**hort-channel **I**GFET **M**odel) taking into account deep sub-micron effects and dependencies of important dimensional and processing parameters [12],[13].

6.2 DC Analysis

The DC analysis calculates the *steady-state* operating voltages and currents for all circuit nodes and branches, respectively. This is done by treating all capacitors as open circuits, all inductors as shorts, and using only DC values of voltage and current sources. Non-linear components, like diodes and transistors, are replaced by their DC resistance at the operating point. The simulation program then solves all current and voltage equations for the linearized circuit in order to find the steady-state node voltages and branch currents.

A DC *sweep* calculates a series of steady-state analyses where a certain circuit variable, like a source voltage or current, a model parameter, a component value, or the temperature is changed step by step. This type of analysis, for example, can be used to calculate transfer functions or transistor current characteristics as shown in Figure 21. The voltage value of an independent voltage source V_{DS} is used as parameter for the DC sweep. The value will be increased from zero to a maximum value step by step, and for each step a DC analysis is performed to calculate the appropriate current value I_D.

6.3 Transient Analysis

The transient analysis is used to simulate the timing behavior of a circuit and represents the most commonly used type of analysis in digital circuit design. Applying time-dependent voltage sources, such as step functions, to model the input signals to the circuit, allows to simulate the response of the circuit and to "measure" propagation delays of digital circuits.

An initial DC analysis computes the initial steady-state conditions of the circuit at time zero. The simulation program calculates all node voltages and branch currents of the circuit for a given point of time. The point of time will then be incremented by a

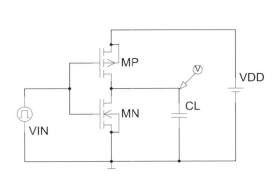
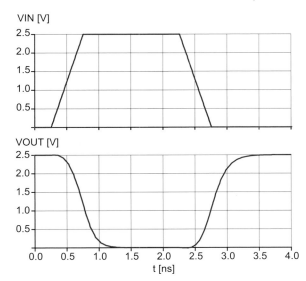

Figure 22: SPICE transient simulation: Circuit diagram and simulated voltage waveforms.

small timestep and all voltages and currents will be recalculated. Repeating the calculations over a specified period of simulation time results in a complete trace of node voltages and branch currents over time.

Figure 22 shows the circuit diagram of a CMOS inverter and the simulated waveforms of the input and output voltages.

6.4 AC Analysis

The AC analysis calculates the frequency response of a circuit. Non-linear devices are transformed to linear circuits about their bias point value. The circuit simulation program computes the partial derivatives at the bias point and uses these to perform a small-signal analysis. Since an AC analysis is a linear analysis, it only considers the gain and phase response of a circuit [13].

An AC sweep calculates a series of small signal analyses, where the frequency of an AC source will be swept from a start frequency to an end frequency. The resulting gain and phase response of the circuit, e.g. an amplifier, will then be plotted over the frequency.

7 Microprocessor

Microprocessors are used in a variety of *stand-alone* and of *embedded systems* applications. Embedded systems are single chip solutions which include one (ore more) processors, memories, interfaces, and other application specific hardware.

7.1 Microprocessor Systems

Microprocessors are the central processing units of modern computers. According to the bit width of their internal arithmetic-logical processing unit, they can be classified into 8-bit, 16-bit, 32-bit, and 64-bit microprocessors. A *microprocessor system* generally consists of the microprocessor, memories for storing instructions and data, and I/O devices or peripherals like parallel ports, serial ports, or analog-to-digital and digital-to-analog converters. If these functional blocks are realized on a single chip, the system is called a **microcontroller** or **embedded microcontroller**. The various subsystems like the processor itself, the memories, and the I/O devices, must have interfaces to each other to communicate. This is commonly done with a *bus system* [14]. A bus is a shared

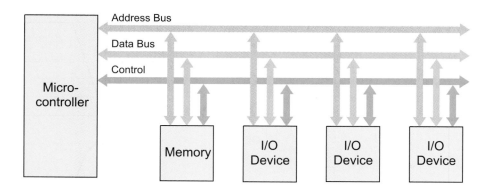

Figure 23: Microprocessor system.

communication link, which uses a set of wires to connect multiple subsystems. By defining a single connection scheme, new subsystems or peripherals can easily be added or moved from one place to another.

In general, a bus contains a set of data lines and a set of address and control lines as shown in Figure 23. The control lines are used to control the type of bus transaction and to determine what type of information is contained on the data lines at each point in the data transfer. They also indicate the direction of the bus transfer, whether it is a READ or a WRITE transaction. The address bus is used to select a certain device, and in particular to address a certain register or memory cell within that device.

At any given point of time, only one subsystem can be the *master* of the bus system, which is usually the microprocessor. The bus master controls the control signals and determines the type of bus transactions. Besides the shared control and data lines which connect all subsystems, there are a few dedicated lines between the bus master and each bus device, which are used by the master to select a particular device to participate on the current bus transaction.

7.2 Basic Principle

Figure 24 shows a simple block diagram and the main functional units of a microprocessor.

The processing starts by using the program counter (PC) to supply the instruction address to the instruction memory. After the instruction is fetched it is forwarded to the instruction decode and control unit for further processing. The decode unit examines certain bit fields in the instruction to identify the type of operation and the registers to be used. The instruction bits specifying the source and destination registers are directly used to address the register file. Depending on the type of instructions, the required source operands will be read from the internal register file, and forwarded to the arithmetic-logical unit (ALU). The ALU represents the key unit of the microprocessor, and determines the type of operation which can be performed within a single instruction. Commonly found sub-units of an ALU are a fast carry-look-ahead adder for performing additions and subtractions, a barrel-shifter for left-shift, right-shift, and rotate operations, and a logical unit for bit operations like AND, OR, XOR, etc. [14]. High-performance microprocessors also have a fast multiply-unit integrated into the ALU. The ALU can be used to compute a memory address for load/store operations, to compute an arithmetic result, or to compute a branch target address. If the instruction is an arithmetic-logical instruction, the result from the ALU is written into a destination register as specified in the instruction. If the operation is a load or store, the result from the ALU is used as an address to the data memory to either store a register value into the memory or to load a memory value into a register. After the instruction processing is finished, the next instruction cycle will be executed, that means the incremented program counter starts fetching the next instruction.

7 Circuit and System Design

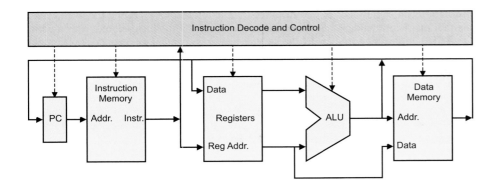

Figure 24: Simple block diagram of a processor showing the basic functional blocks.

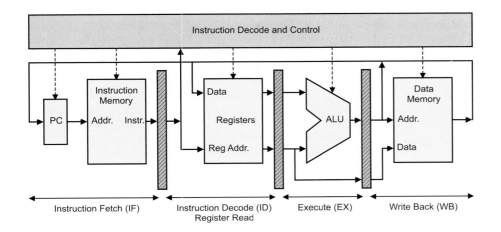

Figure 25: Microprocessor Block Diagram with Pipeline Register.

7.3 Pipelining

Pipelining is an implementation technique in which multiple instructions are overlapped in execution [14]. Figure 25 shows the block diagram from the previous section, where a pipeline register is placed in between the main functional units, thus creating a four stage pipeline (n. b. the program counter also represents a register). This means, that up to four instructions can be in execution during any single clock cycle.

According to their main tasks, the pipeline stages are named:

IF: Instruction Fetch
ID: Instruction Decode
EX: Instruction Execution
WB: Data Write Back

In the first cycle the instruction is fetched from the instruction memory and placed into the first pipeline register. During the second cycle, the instruction is decoded and the required operands will be read from the registers. During the same cycle, the next

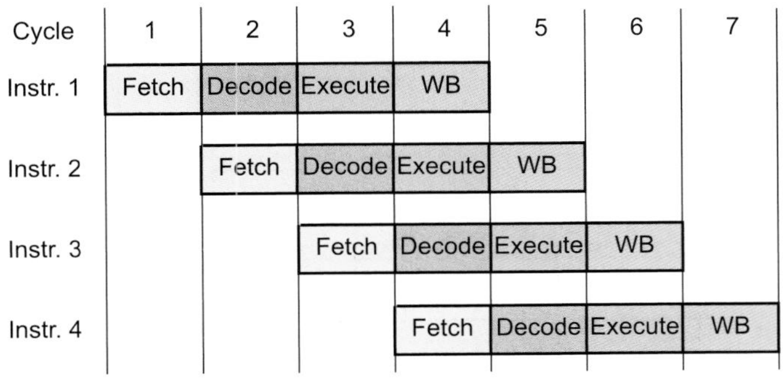

Figure 26: Pipelined Instruction Execution.

instruction can already be fetched from the instruction memory. During the third cycle, the first instruction enters the execute stage, meaning that the instruction is executed in the ALU. During the same cycle, the second instruction enters the decode stage, and a third instruction is fetched from the instruction memory, and so on.

Figure 26 illustrates the principle in more details. After a few cycles the pipeline is filled and up to four instructions are executed simultaneously. Hence, pipelining increases the rate at which instructions are started and completed. However, pipelining does not increase the time needed to complete an individual instruction: it still needs four clock cycles to complete the instruction. Pipelining improves the instruction *throughput* rather than the individual instruction *execution time*.

7.4 Instruction Set

The instruction set of the microprocessor determines the type of operations which can be executed within a single clock cycle, and depends on the available resources in the main execution unit. Nearly all 32-bit microprocessors are based on a RISC (**R**educed **I**nstruction **S**et **C**omputer) type load/store architecture [14]. That means all data processing instructions operate on internal register contents, they don't access a memory value directly. Dedicated load and store instructions are used to load memory values into a register or to store a register content into the memory, respectively.

The overall instruction set can be sub-divided into certain main classes of instructions: *load* and *store instructions* for accessing memory values, and *integer arithmetic instructions* for additions, subtractions, and multiplications of register values. The class of *logical instructions* includes multi-bit shift operations, bit-wise logic operations like AND, OR, XOR, etc, and also bit-field extract and insert instructions. *Address arithmetic instructions* are mainly required for the calculation of new memory addresses to be used by load/store instructions. It basically includes instructions for addition and subtraction, as well as instructions for address comparisons. Finally, *program flow control instructions* like unconditional and conditional branches, loop instructions, and instructions for calling sub-routines.

7.5 Addressing Modes

Addressing modes allow load and store instructions to efficiently access simple data elements within data structures such as records, randomly and sequentially accessed arrays, or stacks. Addressing modes supported by most microprocessors are: *Absolute addressing mode*, that means an absolute memory address is directly encoded into the instruction, *Register addressing mode*, that means the content of an address register is used as memory address, *Displacement addressing mode*, a constant is added to the content of

Figure 27: TriCore 32-bit Microcontroller showing the microprocessor block (center), the memory blocks (top and bottom), and the I/O block (right).

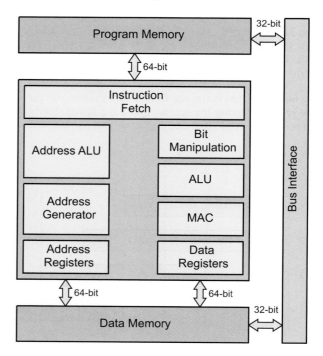

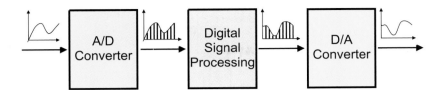

Figure 28: Digital Signal Processing System.

an address register to calculate the effective address, *Post-Increment* and *Pre-Increment addressing modes*, the content of the address register will be incremented after the memory access or before the memory access, respectively.

7.6 Microcontroller Example

Figure 27 shows the block diagram of the *TriCore* microcontroller architecture [15]. It implements a Havard architecture with separate address and data buses for program and data memories, that means instruction fetches can be handled in parallel with data accesses. The microprocessor architecture consists of two major pipelines of four stages. One pipeline is dedicated to load/store operations and the other pipeline to ALU operations. Since the pipelines operate in parallel, load/store instructions can be executed in parallel to integer operations.

There are two sets of register files: one is comprised of 16 address registers and the other is comprised of 16 data registers. Respective to the two parallel pipelines, there are two sets of execution units. The address ALU is part of the load/store pipeline, and is used for address calculations. The integer pipeline consists of an arithmetic-logical unit, a fast multiply-add unit, and a bit manipulation unit.

8 Digital Signal Processors

Digital Signal Processors (DSPs) are specialized processor architectures optimized for fast processing of signal processing algorithms [16]. Signals are represented digitally as sequences of *samples*. Often, these samples are obtained from physical signals through the use of analog-to-digital converters. After mathematical processing, digital signals may be converted back to physical signals via digital-to-analog converters as illustrated in Figure 28.

A key characteristic of DSP systems is its *sample rate*. Combined with the complexity of the algorithms, the sample rate determines the required speed of the DSP processor. Many DSP systems must meet extremely rigorous speed goals, since they operate on lengthy segments of real-world signals in real-time (e.g. speech processing in a digital cellular phone). Where other kind of systems may be required to meet performance goals *on average*, real-time DSP systems often must meet such goals in *every instance* (e.g. for every speech sample). In such systems, failures to maintain the necessary processing rates are considered a serious malfunction. Therefore, such systems are subject to *hard real-time constraints*.

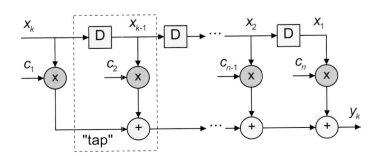

Figure 29: Finite Impulse Response (FIR) Filter.

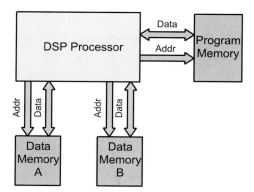

Figure 30: Modified Harvard Architecture with independent program and data memories.

8.1 Classification of DSP Processors

DSP processors can be generally classified into two classes: Fixed-Point DSPs and Floating-Point DSPs. Floating-Point DSPs primarily use floating-point arithmetics, where numbers are represented by the combination of *mantissa* and *exponent*. Fixed-Point processors represent the majority of DSP processors. In this case, numbers are represented either as integers or as fractions between −1,0 and +1,0. Most fixed-point DSP processors use a native data word size of 16 bits. There are also DSPs available with native data word sizes of 20 bits and 24 bits, which are often used for audio applications where a large dynamic range is required.

8.2 Features of DSP Processors

Digital signal processors have some special features which are designed to support fast execution of repetitive, numerically intensive tasks. To understand the need for special features, the example of a *finite impulse response* (FIR) filter will be used [17].

Figure 29 illustrates the principle of a basic FIR filter algorithm. The FIR filter is constructed from a series of *taps*. Each tap includes a multiplication and an accumulation operation. The blocks named D are *unit delay* operators. Their output is a copy of the input sample delayed by one sample period. At any given time, $n-1$ of the most recent input samples reside in the delay elements, where n is the number of taps in the filter. Each time a new sample arrives, the previously stored samples are shifted one place to the right. A new output sample is calculated by multiplying the newly arrived samples and each of the previously stored samples by the corresponding *coefficient* number. The results of each multiplication are added together:

$$y_k = x_k \cdot c_1 + x_{k-1} \cdot c_2 + \cdots + x_2 \cdot c_{n-1} + x_1 \cdot c_n \tag{14}$$

Fast Multiply Accumulate

As seen from the FIR example, the execution of multiply-accumulate operations (also called MAC) is one of the key requirements for DSP processors. In order to achieve this functionality, DSP processors include a hardware multiplier and an accumulator integrated into the ALU of the processor. Most DSP processors are able to execute a multiply-accumulate operation within a single clock cycle. That means, they are able to execute one tap of a FIR filter in one clock cycle. Additionally, to allow a series of multiply-accumulate operations without the possibility of arithmetic overflow, DSP processors provide accumulator registers with extra bits to accommodate growth of the accumulated results.

Multiple Memory Accesses

To achieve the required signal processing performance, DSP processor must be able to make multiple memory accesses within one clock cycle. Taking the FIR example, the processor must:

- Fetch the multiply-accumulate instruction
- Read the appropriate data value from the memory location
- Read the appropriate coefficient number
- Write the data value to the next location in memory (delay line shift)

Thus, the DSP processor must make three read accesses and one write access to memory in one clock cycle in order to compute an FIR filter at a rate of one tap per instruction cycle. This level of memory bandwidth is also needed for other important DSP algorithms besides the FIR filter.

In order to achieve multiple memory accesses, DSP processors provide multiple independent memory bus systems (*Harvard Architecture*). Figure 30 shows a modified Harvard Architecture which allows three independent memory accesses. One memory bank holds the program instructions and two further memory banks hold the data.

Addressing Modes

To enable arithmetic processing at maximum speed and to allow access of multiple operands in the same instruction cycle, DSP processors incorporate dedicated address generation units to calculate memory addresses. Address generation units typically support a selection of addressing modes tailored to DSP algorithms.

The most common addressing mode is the *register indirect addressing with post-increment*, which is used when commutation is performed on a sequence of data stored sequentially in memory. In this case, an internal address register holds the memory address, and with each memory access the address register will be automatically incremented. The address register then holds the memory address for the next access and no additional address operation is required.

Special addressing modes, like *circular* or *modulo* addressing are often supported to simplify the use of circular data buffers which are, for example, required for filter calculations. The circular addressing mode uses two address registers. One register holds the base address of the circular buffer and the second register holds the index into the buffer. The index register will be incremented with each memory access, and once it reaches the end of the buffer it will be set to zero again ("wrap-around").

Another special mode is *bit-reversed* addressing, which supports the implementation of fast Fourier transform (FFT) algorithms [17].

Execution Control

Since many DSP algorithms perform repetitive calculations, most DSP processors provide special support for efficient looping. For example, a special repeat instruction is provided that allows to implement a *for-next* loop without the requirement for additional instructions to update and test the loop counter or to jump back to the top of the loop.

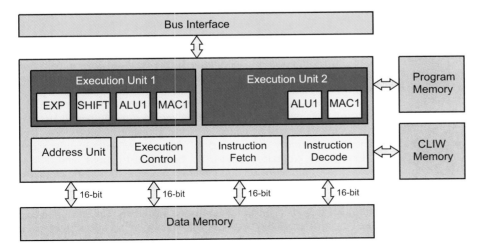

Figure 31: CARMEL DSP architecture.

8.3 DSP Example

Figure 31 shows the block diagram of the CARMEL DSP architecture [18]. This 16-bit fixed-point DSP architecture provides two execution units, and is therefore capable of executing two DSP instructions in parallel. Each execution unit consists of an ALU and a multiply-add unit (MAC). Additionally, one of the execution units provides a fast barrel-shifter and a unit for exponent calculations. An independent address calculation unit supports the required DSP addressing modes, like linear addressing, modulo, and bit-reversal addressing. The addressing unit gives access to up to four memory locations within a single cycle.

9 Performance and Architectures

Today's existing as well as emerging embedded system-on-chip (SoC) applications, such as GPRS and UMTS baseband chips for mobile phones, wireless LAN, automotive powertrain and infotainment applications, show an increasing demand for high-performance microcontroller and DSP capabilities.

The CMOS process technology is one of the key factors in enabling powerful and highly integrated solutions. The scalability of CMOS technology allows the introduction of a new process generation (identified by the length of the transistor gate) at a rate described by the empirical Moore's law. The challenges encountered for MOS transistors on the route of this development are described in Chap. 13.

Besides improvements in process technology further advancements in *system performance* are achievable through better microarchitectures. To compare the performance of architectures and systems an objective measure is required. A simple performance indicator such as MIPS (**M**illions **I**nstructions **P**er **S**econd) is not sufficient since it does not consider the efficiency of the architecture's instruction set. System performance, in general, can be measured in terms of the time it takes to complete a certain task, e.g. to perform a voice codec task in a mobile phone, and is not purely dependent on clock frequency f_{clk}. Such *benchmark programs*, as the SPEC benchmarks [20] or the EEMBC benchmarks [21] provide a set of frequently used algorithms and routines. The benchmark result, usually an absolute measure, can be easily used to compare the performance of different architectures and systems. Depending on the particular benchmark, the result values may be scaled or reciprocal, but in principal there are three main factors which influence the performance [19]:

$$Performance = (IPC \cdot f_{clk})/IC \tag{15}$$

IPC is the average number of instructions executed per cycle, and *IC* is the total number of instructions for the given task or benchmark and expresses the efficiency of the instruction set and also the efficiency of the compiler (see also: Introduction to Part III, Sec. 5.2). Hence, the system performance can be improved by increasing the frequency f_{clk}, by increasing the instructions per cycle *IPC*, and by decreasing the instruction count *IC*.

In general, higher clock frequencies can also be achieved by increasing the pipeline depths of the processor. The logical depth, i.e. the maximum number of gates in the critical path of a pipeline stage, is reduced by distributing the amount of work over several pipeline stages. However, it is not always possible to achieve a linear performance gain with longer pipelines. On one side, there is a certain overhead associated with each pipeline's stage due to the additional pipeline register and its clocking. On the other side, the average number of instructions per cycle *IPC* decreases due to growing pipeline stalls. Pipeline stalls or idle cycles are caused by dependencies, e.g. resource contentions, data or control dependencies, or memory delays. Control dependency stalls mainly occur when the control flow of the program changes, i.e. branch instructions, and a new instruction must be fetched from the instruction memory. The new instruction can only be fetched after the new branch target address has been calculated which occurs relatively late in the pipeline, especially for conditional branches. All instructions which are already in the pipeline after the branch instruction must be invalidated and execution can resume only after the new instruction has been fetched. Memory delays occur whenever a data value must be loaded from memory. Accessing memory can take between a few cycles and hundreds of cycles, and is one of the main bottlenecks in high-performance embedded system design, since memory access times increase much slower than processor frequencies. On-chip instruction and data caches are means to compensate for slow main memory access times and are already widely used in embedded microprocessor applications.

Another architectural measure to improve the system performance is to increase the *IPC* value by higher instruction level parallelism. *Superscalar* architectures, such as the *TriCore* microcontroller are able to execute two or more instructions in parallel: They fetch multiple instructions at a time, and depending on the available execution resources these instructions can then be fed into parallel execution pipelines. However, if there is a data dependency between the instructions, i.e. the result of one instruction is needed by the other one, parallel execution is not possible. Superscalar architectures schedule multiple instructions dynamically at run time into parallel execution units. In contrast, for *Very Long Instruction Word* architectures (VLIW) the parallelism will be determined at compile time, that means the compiler checks the data dependencies and assembles multiple instructions into a single long instruction word. The processor then fetches the long instruction word at once and feeds the sub-instructions into the respective execution pipelines. If it is not possible to fill all instruction slots of a very long instruction word due to data dependencies, the remaining slots will be filled with NOP (no operation) instructions and the respective execution pipeline remains unused.

Since superscalar and VLIW architectures consume more instructions at a time, a higher bandwidth to the instruction memory is required in order to supply the instructions to the processor.

One of the main challenges in embedded system design is to overcome the memory bandwidth limitations. On one side, the increasing application program code size demands for larger memories, and on the other side the on-chip memory space is limited due to cost reasons. The resulting demand for larger (slow) off-chip memories further increases the gap between processing speed and memory speed. It takes more and more cycles to load instructions from external memory into the cache after a cache miss, causing a large sequence of idle pipeline cycles. A possible solution to overcome these limitations is the use of *multi-threading* architectures. Instead of waiting a large sequence of clock cycles until the instruction is loaded from the external memory, the processor switches to an alternate thread or program task and continues execution. Once the missing instruction of the first thread is loaded into the instruction cache, the processor switches back and resumes execution of the original program. A multi-threading processor can therefore provide better performance, provided that the application program can be split into multiple independent tasks, and that the instructions of the alternate tasks are available in fast on-chip caches or memories.

Depending on the target application, each embedded system-on-chip solution looks different. The main challenge to achieve a cost efficient solution, lies in the proper system design, to choose the suitable microprocessor and DSP architectures, to find the right balance between programmable processors and dedicated hardware, and to design the right memory architecture.

More general aspects of the performance and the architecture of information processing systems will be described in the Introduction to Part III of this book.

Acknowledgements

The editor would like to thank Yacoub Mustafa (RWTH Aachen) for checking the symbols and formulas in this chapter.

References

[1] P. Uyemura, *Fundamentals of MOS Digital Integrated Circuits*, Addison Wesley, 1988.

[2] Y. Tsividis, *Operation and Modeling of the MOS Transistor*, McGraw-Hill, 1999.

[3] U. Tietze, C. Schenk, *Electronic Circuits, Design and Application*, Springer, 1991.

[4] P. Uyemura, *Circuit Design for VLSI*, Kluwer Academic Press, 1992.

[5] H. Veendrick, *Deep-Submicron CMOS ICs – From Basics to ASICs*, Kluwer Academic Publishers, 1999.

[6] A. Sharma, *Semiconductor Memories*, IEEE Press, 1997.

[7] J. Oldfield, *Field Programmable Gate Array*, John Wiley & Sons, 1995.

[8] QuickLogic, *pASIC 3 FPGA Family Data Sheet*, Nov 2000.

[9] Altera Corporation, *MAX 9000 Programmable Logic Device Family Data Sheet*, Nov 2001.

[10] Xilinx Inc., *Virtex Field Programmable Gate Arrays, Product Specification*, Nov 2001.

[11] C. Paul, *Fundamentals of electronic circuit analysis*, John Wiley & Sons, 2001.

[12] R. Heinemann, *PSPICE Elektroniksimulation*, Hanser 1998 (includes PSPICE Demo Software).

[13] Cadence Design Systems, Inc., *PSpice User's Guide*, Second Edition, May 2000.

[14] J. Hennessy and D. Patterson, *Computer Organization and Design*, Morgan Kaufmann, 1998.

[15] Infineon Technologies, *TriCore 32-Bit Single-Chip Microcontroller Architecture Manual*, 2000.

[16] P. Lapsey, J. Bier et. al., *DSP Processor Fundamentals*, IEEE Press, 1997.

[17] B. Porat, *A Course in Digital Signal Processing*, John Wiley & Sons, 1997.

[18] Infineon Technologies, *Carmel DSP Architecture Overview*, 2000.

[19] R. Ronen, et. al., *Coming Challenges in Microarchitecture and Architecture*, Proceedings of the IEEE, March 2001.

[20] Standard Performance Evaluation Corporation, www.spec.org.

[21] Embedded Microprocessor Benchmark Consortium, www.eembc.org.

Technology and Analysis

Contents of Part II

	Introduction to Part II	189
8	Film Deposition Methods	197
9	Lithography	221
10	Material Removing Techniques – Etching and Chemical Mechanical Polishing	247
11	Analysis by Diffraction and Fluorescence Methods	271
12	Scanning Probe Techniques	295

Introduction to Part II

Contents

1 Basic Concepts of Technology   189
2 CMOS Technology   191
3 Nanotechnological Approaches   192
4 Analysis Methods   195

1 Basic Concepts of Technology

Technology can be defined as the collection of techniques and methods including the related knowledge required to manufacture a class of products. In the context of this textbook, these products are devices and the hardware of systems for information processing, transmission, and storage. These methods can roughly be classified in the following way (Figure 1), one can distinguish between

- **additive methods** which include the deposition of material on a substrate to be processed, the assembly by techniques such as gluing, soldering or bonding, as well as the directed or self-controlled growth of desired features,
- **subtractive methods** which refer to the dedicated removal of material and include chemical and physical etching as well as mechanical milling and chipping techniques or radiation assisted ablation, and
- **modifying methods** in which material properties are changed, i. e. the concentration of carriers in semiconductors by doping (introduced by diffusion or ion implantation) or the micro- and nanostructure of the phase (single-crystalline, polycrystalline, amorphous) by radiation.

Of course, sometimes this classification may be somewhat arbitrary. For example, *doping* can be regarded as additive as well, since dopant atoms are added to a substrate. Nevertheless, the addition of material is minor here, and the modifying effect (on the conductance) typically is huge.

An additional, independent classification distinguishes between the nature of the effect which takes place in the technique under consideration. *Physical (mechanical, thermal)* effects can be utilized in the additive methods as for example, in the molecular

Additive Methods	Modifying Methods	Subtractive Methods
Thin Film Deposition • sputter deposition • molecular beam epitaxy • chemical vapor deposition • chemical solution deposition	Radiative Treatment • resist exposure • polymer hardening	Etching • wet-chemical etching • ion beam etching • reactive ion etching • focused ion beam etching
Printing Techniques • ink-jet printing • microcontact printing	Thermal Annealing • crystallization • diffusion • change of phase	Radiative and Thermal Treatment • laser ablation • spark erosion
Self-organized Growth • selective chemical reactions • biological growth of cells	Ion Beam Treatment • implantation • amorphization	Tool-Assisted Material Removal • chemical-mechanical polishing • chipping • drilling • milling • sand blasting
Assembly • wafer bonding • surface mount technology • wiring and bonding methods	Mechanical Modification • plastic forming and shaping • scanning probe manipulation	

Figure 1: Classification of technological methods and examples of different fabrication areas.

beam epitaxy for thin film depositon or in the wafer bonding. They are also found in subtractive methods, such as ion beam etching, milling, or sand blasting, and (physical) dissolution. And, there are modifying methods such as the crystallization of amorphous material by thermal annealing, or the amorphization by implantation using noble gas atoms. *Chemical* reactions are widely employed in technology too. Examples of additive methods are the chemical vapor deposition, reactive gluing techniques, as well as the controlled growth of biological neurons. Subtrative methods comprise all types of chemical etch techniques using etch gases or etching compounds in solution. *Radiation* of photons are typically used for subtractive techniques such as laser ablation and for modifying techniques such as the exposure of photo resist.

Especially for wafer-level technologies, when electromagnetic radiation or particle beams are utilized, a further criterion is the positional *selectivity*. For all three categories of Figure 1, the effect can be uniform, i. e. *non-selective* across the entire wafer (e. g. sputter deposition, thermal treatment, polishing) or it can be positional *selective* (e. g. ink-jet printing, electron beam writing of resist, focussed ion beam etching).

The fabrication route of any products in general comprises of many steps and a recurrent application of additive, subtractive, and modifying methods. Often, there is a choice of routes to arrive at the same product. The selection of the fabrication route usually is determined by economical criteria.

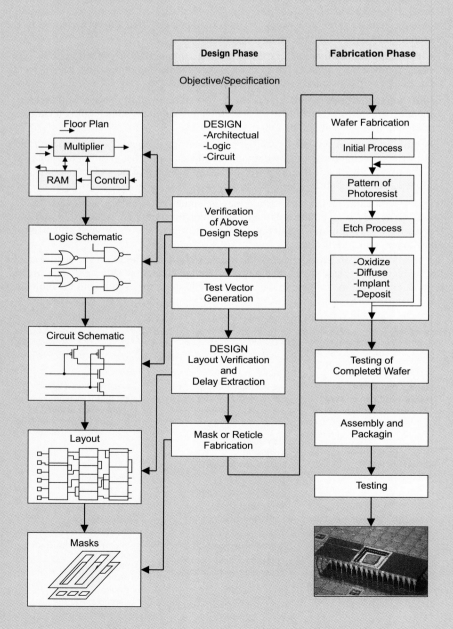

Figure 2: Steps required for the manufacture of integrated CMOS circuits (from Ref. [2] with modifications).

2 CMOS Technology

Figure 2 sketches the sequence of steps of the contemporary manufacturing of an integrated Si CMOS circuit. There are two phases, the **design phase** and the **fabrication phase**. As mentioned in the preface, design is a huge field of knowledge in itself which is not the focus of this textbook. We will only briefly describe the design steps using Figure 2. Detailed descriptions can be found, for example, in Refs. [1], [2]. The design starts at the system level by defining and interlinking functional blocks and sub-blocks. Subsequently, the sub-blocks are resolved into logic gates on the level of a logic schematic and – if necessary – further into transistor-level circuit schematics. Analog and/or digital simulation is used throughout the whole design phase to check and correct the design in order to catch errors as early as possible. Within the next design step, the circuit is initially layed out. The layout consists of sets of patterns and shows the precise place of all devices on the chip and the routing resolves all the precise paths of all interconnects. The initial layout is subjected to a set of design rule checks and is also used for parasitic extraction. Parasitic extraction allows the circuit designer to identify problems that would result from large parasitic capacitances or resistances. If problems are found, the design and/or layout has to be changed in order to correct them. At the same time, a test plan and test vectors are developed by test engineers who work closely together with the circuit designers. If possible, test vectors are generated directly from the design. These test vectors are later used to test the manufactured circuits. Once schematics and layout have passed all possible verification steps, the final layout is used to generate the mask patterns for the fabrication of the masks. The masks are required for the pattern-transfer steps to create the structures on the wafer surface. All design and simulation steps are computer-assisted.

The next phase is the *wafer fabrication* sequence. Generally silicon chips are prepared by an elaborate sequence of processing steps applied to a silicon wafer. After some initial process steps, the sequence comprises patterning by lithography and etching, and the application of modifying and/or additive processes (Figure 2, right). These last three steps are executed iteratively to fabricate first the transistors and, subsequently, the interconnect layers. To give an example, the transistor fabrication is briefly sketched in Figure 3. Prior to the situation shown in Figure 3 (left), the gate oxide (here: SiO_2) has been grown by an oxidation process, and a poly-Si layer as well as a photoresist layer have been deposited. By exposing the photo resist selectively through a mask, the pattern information of the gate area is transferred into the resist. After removing the exposed photoresist, the poly-Si and the gate oxide are removed by a plasma etching process (Figure 3, center). Subsequently, the ion implantation technique is used to form the source and drain area of the transistor. Here the implanted dopants change the electronic conductivity and these areas are now negatively doped silicon. Since the patterned polysilicon gate is used as an implantation mask this is called a *self-aligned* process (Figure 3, right). The shown processing steps are ensued by further processing steps to produce a working IC. Metallization layers and the layers of the interlevel dielectrics have to be processed by cycling through the deposition of metal, patterning the metal, applying the interlevel dielectrics, planarization by chemical-mechanical polishing, patterning the interlevel dielectrics for the vias. Finally, wafer probing is employed to run test procedures on the completely processed wafer. This is followed by a particulation into the dices, die placement on the carrier, bonding of the die, encapsulation, and a final test.

The major steps of the wafer fabrication as well as some potential alternatives are described in Chapters 8 to 11.

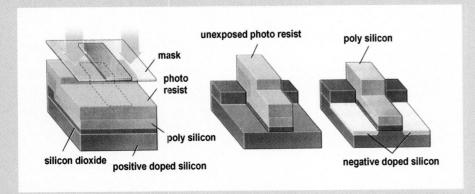

Figure 3: Illustration of the CMOS technology. Three essential steps in the fabrication of a MOSFET are shown: exposing photoresist with UV-light through a mask (left), after removing the exposed resist by a chemical solvent and ion etching to remove the unprotected poly-Si (center), after doping by ion implantation and removing the remaining resist (right).

The lateral resolution to which the smallest structures can be fabricated in a given technology generation is called the **minimum feature size** (short: F) and is determined by a given fraction of the wavelength of the light used in the photolithography. As a reference structure for the definition of F, often one half of the pitch of first metallization layer of DRAMs (*DRAM ½ half pitch*) is employed. Alternatively, the lithographically defined gate length of the smallest MOSFETs (*printed gate length*) is used. The latter is approx. 70% smaller than the DRAM ½ half pitch for sub-micron technology generations. In this textbook, we will use the DRAM ½ half pitch definition.

3 Nanotechnological Approaches

Conceptionally, there are two routes that lead to devices and circuits on the deep sub-100 nm scale. The first route is based on extending the conventional CMOS technology. It is called the **top-down approach** because manufacturing starts at the wafer level and patterning proceeds by lithography and etching, to obtain features on the μm and nm scale. The extension of the technological capabilities towards smaller minimum features sizes is predominantly facilitated by reducing the wavelength for photolithography, using extreme UV lithography (EUVL) or x-ray lithography (Chapter 9). An alternative top-down method is the imprint technique. Due to the required precision for all essential parts of the fabrication tools, the effort and the installation costs grow at an incredible pace on top-down routes.

Alternatively, the fabrication of nanosized structures can be started from individual atoms and molecules which are ordered physically or reacted chemically to obtain the desired features. This is called the **bottom-up** approach. It utilizes physical or chemical processes to arrive at nanosized structures which are grown from individual atoms or molecules on a regular order. This order extends usually only over short ranges, while it is difficult to achieve a long-range order as well because of disturbances in the formation process. To avoid these problems, **hybrid approaches** have been proposed, which employ a combination of a top-down approach for a coarse definition of the pattern and a bottom-up technique to realize short-range ordered nano-scale structures which *align* to the coarser, but in long-range order. In the framework of this textbook, we will summarize some of the major principles and give just a few examples, from a large number of reports published in recent years.

One principle, which facilitates the fabrication of nano-scale features using much coarser-scaled top-down technique, is based on flipping-up horizontal structures into the vertical direction. Since long, surface treatment and deposition techniques have been developed that make it possible to control the thickness of films and multilayers at the nm scale and even at the lattice spacing of crystals. If these structures are flipped-up into the vertical direction on a chip, differences in the etch rate of originally deposited material can be used to create laterally structured features on the same scale as the thickness of the original thin films. The most known variant of this *flip-up principle* is the so-called *spacer technique*. A hard-mask and/or functional material is deposited conformally as a thin film on coarse structures made out of a sacrificial material. This film thickness defines the final feature size. To realize a FET, for example, the Si nitride thin film has been deposited on photolithographically patterned Si dioxide (as sacrificial materials) on a poly-Si layer (Figure 4). The nitride film is carefully dry etched, leaving nitride spacers on the oxide block sidewalls. The sacrificial oxide is removed and the poly-Si is etched using the nitride spacers as hard masks. The remaining nitride / poly-Si ligament can serve fruther as a hard-mask for the dopant implantation. For details see, for example, [4] and [5]. The spacer technique can be employed to realize very different functions. For example, Cu interconnects of 40 nm width have been made, embedded in SiO_2 dielectrics to study the size-dependent resistivity of the metal [6]. Another variant of the flip-up principle has been reported by Heath [7]. A multilayer system of GaAs/AlGaAs was cut, precisely polished, and subjected to a selective etch process which resulted in a topological relief. This relief represents a stamp which can be used for pattern transfer. By evaporating a thin Pt layer onto the relief and printing it onto the Si surface, Pt nanowires down to 5 nm diameter with a spacing of 20 nm have been realized on Si substrates.

Bottom-up techniques are typically based on **self-organization** processes and, in particular, **self-assembly** processes. Self-assembly can be defined as a coordinated action of independent entities under local (i. e. not global, macroscopic) control of driving forces to produce larger, ordered structures or to achieve a desired group effect. Well known examples are found in biology, e. g. formation of cell membranes and the replica-

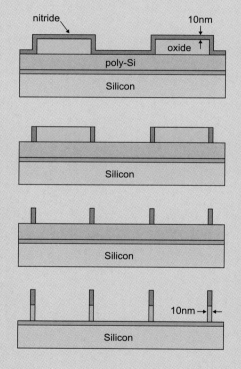

Figure 4: Spacer technique to realize the gate of a FET (from Ref. [4] with modifications).

Introduction

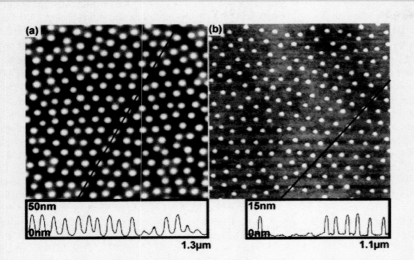

Figure 5: AFM topography image (top) and height profiles (bottom) of a monomicellar film cast from a solution of $HAuCl_4$ loaded diblock copolymer micelles on a glas substrate [8].
(a) as deposited micelles,
(b) film after oxygen plasma treatment, resulting in the bare Au particles on the glass substrate.

tion of cells by division and growth, and in chemistry, where appropriate molecules team up into supramolecular structures. The driving forces of self-organization processes are, in general, based on an interplay of thermodynamics and (chemical) kinetics, assisted by the structure of the system to be assembled or on which the assembly takes place, e. g. organic molecules, crystal structures, etc.

An example for a *chemically controlled self-assembly* process is the deposition of loaded diblock copolymer micelles [8]. One block of the copolymer, based on polystyrene is soluble in toluene, the other block, based on poly-(2-vinyl pyridine), is almost insoluble. Because of this solubility difference, spherical micelles of copolymer molecules are formed in toluene. These micelles can be loaded with compounds such as $HAuCl_4$. The loaded micelles have been deposited on planar substrates, where they form a densely packed, closed film. Oxygen plasma treatment pyrolyses the polymer and turns the Au compound into Au metal droplets. These nanosized Au dots remain at the positions of the original micelles and, hence, form a pattern which is very regular on short ranges (Figure 5). The distance of the dots can be controlled by employing copolymers of different chain length. In an hybrid approach, this method can be combined with conventional lithographical patterning, in order to obtain nanosized dots regularly arranged on the long-scale [9]. As shown in Figure 6, a resist is used as a template layer into which grooves and holes are fabricated by electron beam lithography. The width of the grooves and holes is adjusted to the size of the micelles. The deposition process of the micelles is controlled in a way, that only those micelles which are attracted into the grooves and holes by capillary forces finally remain on the surface. After the resist is

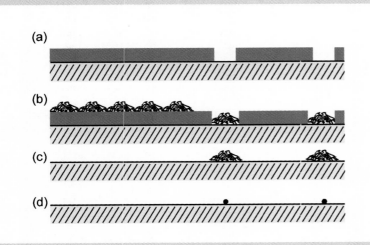

Figure 6: Schematic cross-section of the process by which nanometer sized gold dots are precisely located in periodic or artificial patterns.
(a) a template is created from an e-beam resist by writing a pattern in the resist and developing away the written areas which results in holes or grooves in the resist.
(b) Micelles are spincoated onto the template; those around the prestructures position themselves inside the grooves or holes due to capillary forces while those further away than the screening length remain on top of the unpatterned resist.
(c) The e-beam resist is dissolved with acetone (lift-off) and the micelles laying on the resist are removed while those in direct contact with the substrate surface remain.
(d) The substrate is exposed to a plasma which strips the micelles of the polymer and reduces the salt to a metal, gold in this case. This results in a substrate that is decorated with individual nanoclusters that are ordered according to the original pattern in the e-beam resist.

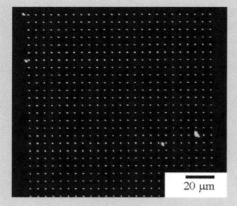

Figure 7: Optical dark field microscopy image of a square arrangement of Au-spots on Si. Every Au cluster is separated by 5 microns as a result and can be optically resolved [9].

Figure 8: Self-organized arrangement of Ge nanodots an a Si surface. The upper layer shows the AFM image (1.7 μm × 1.7 μm) and the lower layer exhibits the corresponding autocorrelation function. This function fades out towards the edges because of the fact, that the nanodot arrangement shows only a short-range order [13].

removed and the copolymer coat of the micelles is pyrolysed, a regular Au nanodot array is obtained. Due to the spherical structure of the micelles, the Au nanodots are found exactly in the center of the former holes and grooves.

A *physical self-organization* process is, for example, the formation of ordered nanoscale semiconductor dots by ion sputtering. For example, low-energy Ar^+ ion bombardment of an (100) GaSb surface resulted in the formation of nanodots which were arranged in a regular hexagonal lattice [10]. In a specific range of processing parameters, a negative surface tension effect occurs which supports the roughening of the surface. The final structure is built by a competition between this roughening instability and the smoothing of the surface due to surface diffusion. The size of the nanodots and their distance is primarily controlled by the ion flux. The hexagonal structure of nanodots obtained in this process exhibited a characteristic length over more then six periods of the dot lattice clearly demonstrating the self-organization potential of this method. Another method to produce self-organized semiconductor nanodots relies on the deposition instead of ion bombardment. A lattice mismatch between, e. g., Si and Ge is utilized as the driving force for the self-organization (e. g. [11]). At the initial state of epitaxial growth of Ge (or SiGe) on Si, a layer-by-layer growth takes place. Beyond a so-called critical thickness, the mechanism changes into a 3-D growth (so-called Stranski-Krastanov growth) to reduce the strain accommodated in the film. Because of the characteristic range of the elastic stress fields, the nanodots grow in a regular lattice on the surface. An example is shown in Figure 8 [13] and schematically in Figure 9 (left). In a hybride approach, this principle can be combined with lithographic methods. First, a mechanical strained epitaxial layer (material 2, Figure 9 middle) is grown on an unstrained substrate (material 1). The deposited layer is patterned by a standard lithographic process. If a second layer of material 1 is deposited the growth of the new layer is influenced by the strain field in the underlying layer (material 2). Like the Stranski-Krastanov growth as mentioned above, the difference in the mechanical strain and the lattice constants leads to different local layer thicknesses. If material 1 (e.g. Si) and 2 (e.g. SiGe) are deposited alternately, a series of self-aligned SiGe islands is produced (see Figure 9 right). In a final step, Ge is deposited which forms self-organized nanoislands. These islands form in contrast to a direct deposition on Si (Figure 9, left) a regular pattern, which is a necessary precondition for further processing steps. Such a self-aligned structure is a promising candidate to produce FETs consisting of nanoscaled dots, here called DOTFETs. The DOTFET device is based on a self-assembly of coherent, defect-free Ge dots on Si, which has been a major goal of strained layer epitaxy. Figure 10 gives a bright-field cross-section transmission electron microscopy (TEM) image of a 5-fold dot stack. The Ge dots cluster so closely that their individual strain fields can no longer be resolved. Instead, the five dots produce one common large lattice distortion. In the Si regions of this line, tensile strain is found, which reaches almost 1 % in the stack center. The compressive strain within the dots is reduced by partial elastic relaxation of the nanostructures. This relaxation generates the tensile strain in the sandwiched Si regions. An n-channel version of such a field-effect transistor under positive bias using dot arrangements is sketched in Figure 11. The DOTFET is designed in analogy to typical Si/SiGe MOSFETs, except that a thick and defect-rich buffer is substituted by a stack of defect-free Ge/Si islands, above which the n-channel is positioned. Vital is the flat geometry of the embedded dots to limit the strain changes and hence the scattering of the carriers flowing from source to drain. A fine tuning of Ge dot size relative gate length appears inevitable, but is certainly feasible.

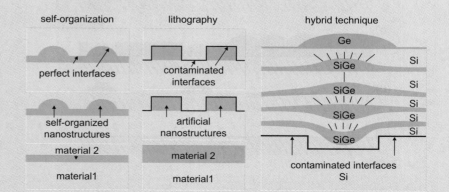

Figure 9: Sketch of the hybrid approach to Ge nanodots (redrawn from [13]). Since the SiGe is preferably deposited in the lithographically defined grooves and the subsequently deposited Si layer are set under tensile stress, the inhomogenity is vertically aligned and leads to the precise positioning of the Ge nanodots on top.

A further principle to produce ordered structures of functional material is based on crystal structures with large openings in the crystal lattice. These materials such as zeolithes or specific organic compounds can be used as templates for the controlled growth of a functional material. Figure 12 shows an example of this template technique. The fabrication of aligned Ag nanowires in the channels of a calix [4] hydroquinone (CHQ) crystal lattice is executed by an electro-/photochemical redox reaction in an aqueous solution [14].

The challenge for all these approaches is the controlled formation of a real interconnect system, in which functional elements are embedded in suitable positions. This interconnect system will be not perfect, especially if bottom-up techniques are involved which inherently show the influence of statistical fluctuations. The architecture of the electronic system fabricated along these routes and the software which is operated on the system need to incorporate an appropriate defect-tolerance.

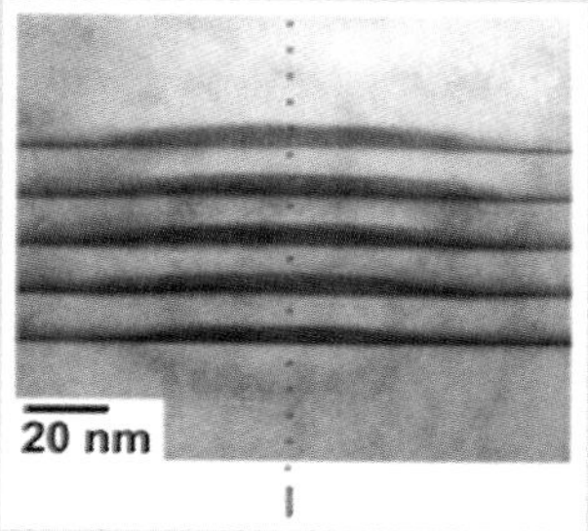

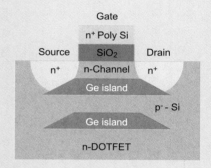

Figure 10: Bright field cross section TEM images of a 5-fold Ge/Si dot stack [12].

4 Analysis Methods

The substantial progress in information technology does not only depend on the advance in circuit manufacturing techniques but to a large extent on the capability of analysis and characterization methods, too. The fields, in which careful analysis of the chemical or physical state of a system is inevitable, comprise the circuit manufacturing processes themselves as well as the resulting properties of the electronic materials in the devices. In special cases analytical tools can even be used technologically in order to manipulate the samples in a predictable way, as for example focussed ion beam (FIB) or scanning probe microscopy (SPM) techniques.

The most relevant analysis techniques cover the fields of diffraction and fluorescence methods, described in Chapter 11, and scanning probe techniques which are the topic of Chapter 12.

Diffraction methods are generally used to analyse the structural properties of crystals, amorphous matrices or layer stacks. In order to resolve the structural characteristics of a sample, the wavelength of the probing beam has to be in the same order of magnitude as the length scale of the structures under investigation. The necessary variation of the wavelength is achieved by using different incident exciting probes such as photons (X-rays), electrons, ions or neutrons having different energies. X-ray diffraction (XRD), selected area electron diffraction (SAED), and neutron diffraction thus give information on the structural properties of materials. Transmission electron microscopy (TEM)

Figure 11: Schematic sketch of an n-channel DotFET [12]. The electron channel is created by the tensile strained Si above the embedded Ge dot.

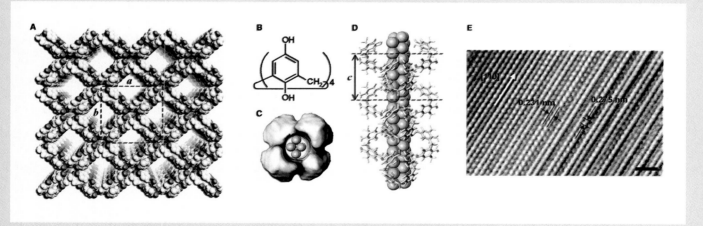

Figure 12: Organic nanotube templates and a silver nanowire inside the nanotube [14].
(A) A schematic view of CHQ nanotube arrays based on the X-ray analysis of the crystal are separated by 1.7 nm from the neighboring ones. Water molecules are not shown for clarity. O atoms are shown in red.

(B) CHQ monomer. (C and D) Top and side views, respectively, of a silver nanowire inside a CHQ nanotubetemplate. The solvent-accessible surface model in (C) and the stick model in (D) show schematic views of the CHQ nanotubes. The red color of the surface represents the negative electrostatic potential on oxygen atoms. The space-filled models (blue) represent the silver nanowires.

(E) HREM image of the aligned nanowires showing the coherent orientations of Ag atoms in different wires.

together with high resolution TEM (HRTEM) yield information about the sample's morphology, the structure at and near interfaces, the density of defects in the crystal lattice, etc.. Scanning electron microscopy (SEM) reveals the morphology of a sample, and together with energy or wavelength dispersive X-ray analysis of the diffracted beam (EDX or WDX) the stoichiometry of the material. Electron energy loss spectroscopy (EELS) and fluorescence (XRF) techniques are utilized to determine the stoichiometry of the sample. Surface analytical methods correspond to an analysis of near surface composition by secondary ion mass spectroscopy and Auger electron microscopy (SIMS and AES), and binding energy (XPS).

The most commonly used high-resolution surface analysis technique besides SEM is the scanning probe microscopy (SPM). Here, a sharp tip scans the surface of the sample. The interaction between surface and tip depends on the characteristic physical properties of the surface which should be analysed. Since the mid 1980s, a large family of SPM related techniques, based on various types of interactions between the tip and the sample, have been developed. Different SPM techniques such as atomic force microscopy (AFM), electrostatic force microscopy (EFM), magnetic force microscopy (MFM), scanning capacitance microscopy (SCM), near-field scanning optical microscopy (NSOM) and others were proved to be capable of measuring the local physical properties of materials with nanoscale resolution. Currently, SPM is established for nanoscale characterisation of materials by using mechanical, electrical, magnetic, optic and chemical interactions between the probing tip and the surface. In addition it has been demonstrated that the SPM approach allows manipulation of single atoms or molecules and the contacting of single grain electronic devices.

Acknowledgements

The editor gratefully acknowledges Susanne Hoffmann-Eifert (FZ Jülich) for writing Section 4 and Ralf Liedtke (RWTH Aachen) for writing Section 2 as well as assisting in the compilation. Furthermore, the editor would like to thank Carsten Kügeler (RWTH Aachen) for technical assistance and Dennis Bräuhaus (RWTH Aachen) for corrections.

References

[1] C. Y. Chang and S. M. Sze, *ULSI Technology*, McGraw-Hill Int. Editions, 1996.
[2] S. Wolf and R. N. Tauber, *Silicon Processing for the VLSI*, (Vol. 1), Lattice Press, 2001.
[3] J. F. Wakerly, *Digital Design*, 3rd edition, Prentice Hall, 2001.
[4] B. Doyle et al., Intel Technology Journal, **6**, no. 2, 42 (2002).
[5] Y.-K. Choi, T.-J. King, and C. Hu, IEEE Trans. Electron Devices, **49**, 436 (2002).
[6] W. Steinhögl, G. Schindler, G. Steinlesberger, and M. Engelhardt, Phys. Rev. B **66**, 075414 (2002).
[7] J. R. Heath, 3rd International Conference on Trends in Nanotechnology (TNT), Santiago de Compostela, Spain, Sep. 9-13, 2002.
[8] J. P. Spatz, S. Mössmer, C. Hartmann, M. Möller, T. Herzog, M. Krieger, H.-G. Boyen, P. Ziemann, and B. Kabius, Langmuir, **16**, 407 (2000).
[9] J. P. Spatz, V. Z.-H. Chan, S. Mössmer, and M. Möller, Advanced Materials (in press).
[10] S. Facsko, T. Dekorsy, C. Koerdt, C. Trappe, H. Kurz, A. Vogt, and H. L. Hartnagel, Science, **285**, 1551 (1999).
[11] K. L. Wang, J. L. Liu, and G. Jin, J. Crystal Growth, 1892, 237 (2002).
[12] O. G. Schmidt and K. Eberl, IEEE Trans. Electron Devices, **48**, 1175 (2001).
[13] O. G. Schmidt, Spektrum der Wissenschaft, no. 4, 8 (2002).
[14] B. H. Hong, S. C. Bae, C.-W. Lee, S. Jeong, K. S. Kim, Science, **294**, 348 (2001).

Film Deposition Methods

Peter Ehrhart, Department IFF, Research Center Jülich, Germany

Contents

1 Introduction	199
2 Fundamentals of Film Deposition	200
2.1 Gas Kinetics	200
2.2 Thermodynamics	201
2.3 Film Growth Modes	201
2.4 Strain Relaxation in Continuous Films	203
3 Physical Deposition Methods	203
3.1 Thermal Evaporation / Molecular Beam Epitaxy	203
3.1.1 Sources	204
3.1.2 Process Environment	204
3.1.3 In-situ Analytical Techniques	205
3.1.4 Example: Epitaxial Growth of Fe Films	205
3.2 Pulsed Laser Deposition	206
3.2.1 PLD Concept	206
3.2.2 Example: Optical Waveguides	206
3.3 Sputter Deposition	206
3.3.1 DC Sputtering	207
3.3.2 Sputtering Process	207
3.3.3 Magnetron Sputtering	207
3.3.4 RF Sputtering	207
3.3.5 Gas Pressure and Film Growth	208
3.3.6 Example: High Pressure Oxygen Sputtering	208
4 Chemical Deposition Methods	208
4.1 Chemical Vapour Deposition	208
4.1.1 Precursor Chemistry and Delivery	209
4.1.2 Reactor Design and Modelling	211
4.1.3 Growth Control	212
4.1.4 Conformal Deposition	213
4.1.5 Layer by Layer Growth and Ultrathin Films	213
4.2 Chemical Solution Deposition	214
4.3 Langmuir-Blodgett Films	215
5 Summary	216

Film Deposition Methods

1 Introduction

The deposition of thin-film functional layers on different substrates is an essential step in many fields of modern high technology, and applications range from large-area optical coatings on architectural structures to tribological layers, high-temperature superconductors, and finally to applications in micro- and nanoelectronics. Considering this broad spectrum of applications it is obvious that there cannot be one perfect deposition method which can be applied in all fields. In contrast, there is a wide spectrum of methods all on a high level of development and it is sometimes difficult to make the optimum choice. Even within the special field of information technology films of very different materials have to be considered: semiconductors, metals, especially magnetic layers, dielectric and ferroelectric oxides and organic layers. The deposition methods are dominated by depositions from the vapour phase and we will concentrate on the basic principles of these methods. A selection of vapour deposition methods is summarized in Table 1 with the generally accepted subdivision into physical and chemical methods. Physical methods may be characterized by a locally well-defined particle source and generally a free flight in vacuum to the substrate. For chemical methods, the so-called precursor molecules fill the reactor vessel as a vapour, dissociate at the hot substrate surface and release the atoms of interest. Some basic characteristics of the methods are summarized, however, the characterization is sometimes not very specific as the parameters depend strongly on the properties of the material of actual interest. Hence, there are many such comparisons in the literature showing some dependence on the personal bias of the author. A short introduction to depositions from solution is included in Sec. 4: i.e. chemical solution deposition (CSD) and the Langmuir-Blodgett method for monomolecular organic films. Of course, the list is not complete and we must refer to the literature for additional methods like electroplating (galvanic deposition) or thermal spray techniques.

	Physical Vapor Deposition			Chemical Vapor Deposition
	Evaporation / MBE	Sputtering	PLD	CVD / MOCVD
Mechanism of production of depositing species	Thermal energy	Momentum transfer	Thermal energy	Chemical reaction
Deposition rate	High, up to 750,000 Å/min	Low, except for pure metals	Moderate	Moderate Up to 2,500 Å/min
Deposition species	Atoms and ions	Atoms and ions	Atoms, ions and clusters	precursor molecules dissociate into atoms
Energy of deposited species	Low 0.1 to 0.5 eV	Can be high 1-100 eV	Low to high	Low; Can be high with plasma-aid
Throwing power a) Complex shaped object b) Into blind hole	Poor, line of sight Poor	Nonuniform thickness Poor	Poor Poor	Good Limited
Scalable to wafer size	up to large	up to large	limited	up to large

Table 1: Some Characteristics of Vapour Deposition Processes (modified after Bunshah [1]).

2 Fundamentals of Film Deposition

In this section, we introduce some fundamentals of film deposition which are valid for all methods. We start with the kinetics of gases as the residual gas pressure in the system determines the free path length of the deposited species and the possible incorporation of foreign atoms. Next, we discuss some basic thermodynamic data, like **vapour pressure** and **phase diagrams**, and their dependence on the residual gas pressure. Finally, we introduce the basic models for the nucleation and growth of thin films and the accommodation and release of lattice strain.

2.1 Gas Kinetics

The residual gas pressure in the system is one of the basic parameters to be controlled during film deposition as the residual gas atoms may collide with the depositing species or may hit the growing surfaces and may thus be incorporated in the film, Figure 1. For the simplest assumption that the gas atoms may be considered as not interacting masses with a Maxwell velocity distribution we obtain the mean free path length, λ, of the atoms or molecules

$$\lambda = \frac{1}{\sqrt{2}\pi N d^2} \qquad (1)$$

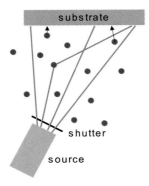

Figure 1: Schematics of a deposition system.

d = molecular diameter, N = concentration of the gas. With the law of the ideal gas: $N = p/k_B T$, k_B = Boltzmann constant, we obtain:

$$\lambda = \frac{k_B T}{\sqrt{2}\pi p d^2} \qquad (2)$$

For the example of air molecules we obtain a free path length which is of the order of a typical distance from source to substrate of about 20 cm at a pressure of $0.5 \cdot 10^{-3}$ mbar. This rather moderate vacuum level shows that the beam interaction is not a critical condition for the base vacuum. More critical is the number of residual gas atoms which hit the growing surface and limit the purity of the film if they are incorporated. This number can be expressed as

$$N_i = p_i \sqrt{\frac{1}{2\pi k_B m_i T}} \qquad (3)$$

m_i = atomic or molecular mass. Typical results are summarized in Tab. 2. Assuming a sticking coefficient of unity, the incorporation of residual gas atoms may be expressed in terms of monolayers and this growth rate may be rather high as compared to a typical growth rate of an epitaxial film i.e., one monolayer / s. Hence, for clean films **ultra-high vacuum** (UHV, better than 10^{-9} mbar) may be necessary.

A further aspect where the mean free path of the molecules becomes important is deposition into very small structures like small via or trenches, as they are becoming more and more important with increasing miniaturization of ICs, Figure 2. Comparing the dimensions of such submicron structures with the free path lengths of Table 2 we see that even for medium vacuum conditions the mean free path of the molecules becomes much larger than the structure size. Hence, for these conditions there is no gas collision within the holes, and for the distribution of the atoms reflection at the surfaces becomes important. For simulations of the deposition processes continuum gas dynamics must be supplemented by Monte Carlo methods.

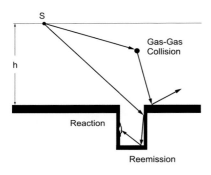

Figure 2: Deposition into a hole (after ref. 20). The example of an atom with starting position, S, with a height, h, above the substrate illustrates that either gas-gas collisions or reflection from the surface are necessary for a uniform deposition on bottom and sidewalls of a hole.

p, mbar	Mean free path, cm (between collisions)	Collisions / s (between molecules)	Molecules/(cm² s) (sticking surface)	Monolayer / s*
10^0	$6.8 \cdot 10^{-3}$	$6.7 \cdot 10^6$	$2.8 \cdot 10^{20}$	$3.3 \cdot 10^5$
10^{-3}	$6.8 \cdot 10^0$	$6.7 \cdot 10^3$	$2.8 \cdot 10^{17}$	$3.3 \cdot 10^2$
10^{-6}	$6.8 \cdot 10^3$	$6.7 \cdot 10^0$	$2.8 \cdot 10^{14}$	$3.3 \cdot 10^{-1}$
10^{-9}	$6.8 \cdot 10^6$	$6.7 \cdot 10^{-3}$	$2.8 \cdot 10^{11}$	$3.3 \cdot 10^{-4}$

* Assuming the condensation coefficient is unity

Table 2: Some facts about residual air at 25 °C in a typical vacuum used for film deposition (after Chopra [2]).

2.2 Thermodynamics

Phase-diagrams are the starting point for considering the deposition of a new material in order to see the stability range of the envisaged phase and the existence of concurrent phases. Standard phase diagrams are given at ambient pressure, however, changes with pressure must be considered for vacuum deposition methods. Figure 3 shows as a simple example the phase diagram of the completely intermixing binary system Si-Ge and the change from ambient pressure down to the UHV region [3]. At 1 mbar there is not much change compared to atmospheric pressure and we **observe** a wide range of stability of the mixed homogeneous crystalline phase, c, of Si-Ge (the decomposition of this homogeneous phase into two crystalline phases of different stoichiometry, c' and c'', at very low temperature is somewhat speculative). At higher temperatures the liquid, l, to solid (crystalline) phase transition is indicated and above 2000 K the liquid to vapour, v, transition is shown. With decreasing pressure there is a strong decrease in temperature of the l-v borderlines and even an overlap with the c-l lines. Finally, in the UHV region, 10^{-9} mbar, the liquid has disappeared and only direct sublimation, c→v, is left at temperatures around 1100 K. Hence, re-evaporation of the material under UHV conditions and high temperatures must be considered. In addition, a comparison with the deposition rates and gas pressures discussed along with Table 2 shows that the deposition of the films usually proceeds under high supersaturation, i. e. conditions far from thermodynamic equilibrium.

Vapour pressures and the related evaporation rates present another field of basic thermodynamic data, which are very useful for the control of the deposition of different compounds. Such data can be deduced from thermodynamic data tables (e.g., CODATA and JANAF) [4]. As an illustrative example we consider Pb-based perovskite oxides (such as $Pb(Zr,Ti)O_3$, PZT for short) which are the most important class of ferroelectrics for thin film applications. The deposition of these lead-based oxides is complicated by the fact that PbO is known as a very volatile oxide. Figure 4 shows the evaporation of PbO under atmospheric conditions and under UHV: under atmospheric pressure the volatile species is PbO and the vapour pressure of Pb is 3 orders of magnitude lower. In contrast, under UHV conditions the dominant species in the vapour phase is Pb.

2.3 Film Growth Modes

Nucleation and growth of a film proceeds from energetically favourable places on a substrate surface and even the cleanest polished surface shows some structure. Figure 5 shows schematically the structure of a well-polished single-crystal surface. The charac-

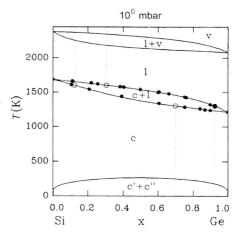

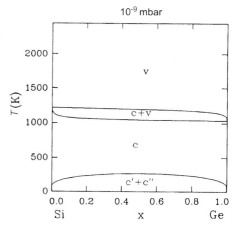

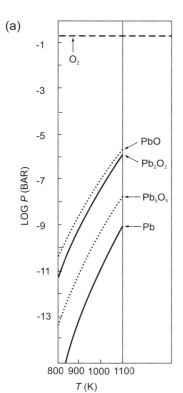

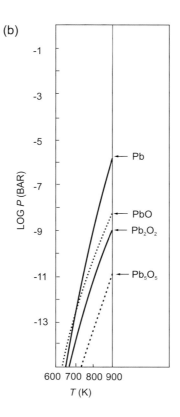

Figure 3: x–T phase diagram of the $Si_{(1-x)}$-Ge_x systems at 10^0 and 10^{-9} mbar [3].

Figure 4: Vaporization of PbO at different oxygen partial pressures, (a) 0.2 bar and (b) 10^{-15} bar [4].

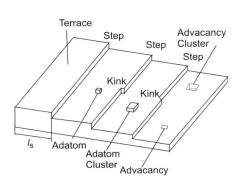

Figure 5: Schematic view of the elements of the surface morphology [3].

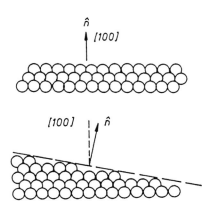

Figure 6: Change of the step distance, l_s, by cutting a surface at a small angle to a major crystallographic direction, i.e., forming a vicinal surface.

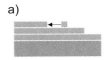
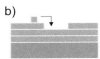

Figure 7: Growth modes of homoepitaxy: (a) step-propagation, (b) 2d-island growth, and (c) multi-layer growth.

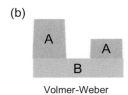
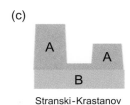

Figure 8: Growth modes of the hetero-epitaxy.

teristic features are the terraces of length, l_s, the steps and the kinks within the step line, which otherwise runs along well-defined crystallographic directions. If the surface diffusion is fast enough a randomly deposited adatom will diffuse to the energetically most favourable places like steps and especially kinks. If at lower temperatures the diffusion is slower, several mobile adatoms may encounter each other within a terrace and may form additional immobile adatom clusters within the terraces. Similarly, advacancies and their clusters might be formed at the end of the coverage of a terrace. By reducing the step distance and hence the diffusion length by vicinal surfaces, Figure 6, the step controlled growth may be extended to lower temperatures,.

The details of the growth modes for the simplest case of **homoepitaxy**, the growth of a film on a single-crystalline surface of the same material, is indicated in Figure 7. As discussed above, step propagation dominates at higher temperatures and/or small deposition rates and two-dimensional island growth will predominate if immobile clusters are formed by the encounters of mobile adatoms. This simple picture is, however, quite frequently modified: if the jump across the step is kinetically hindered multilayer growth will be observed. This enlarged activation energy for the jump across the step is called the Ehrlich-Schwoebel effect and can be understood in a simple model as the adatom is nearly dissociated from the surface in the saddle point of this jump.

If we want to grow an epitaxial film on a different substrate (so-called **heteroepitaxy**), two material parameters have to be considered in addition: the surface energy, γ, and the lattice parameter or lattice match of the two materials. For the case of good lattice match the difference in surface energy leads to two different growth modes as indicated in Figure 8a and b. As long as :

$$\gamma_{\text{layer}} + \gamma_{\text{substrate/layer}} \leq \gamma_{\text{substrate}} \tag{4}$$

we observe perfect wetting and pure layer by layer or *Frank-van-der-Merve growth*. For the opposite case, we observe island or *Volmer-Weber growth*. For this consideration the surface energies of the crystallographic orientations of actual interest must be applied, which are often not available in data reference tables. If there is a lattice mismatch between substrate and film, an additional growth mode may be observed as indicated in Figure 8c, *Stranski-Krastanov growth*. A first layer may grow matched to the substrate, which yields additional strain energy. With growing thickness this strain energy increases in proportion to the strained volume and an island formation may become more favourable in spite of the larger surface area.

The contributions of strain and surface energy can quite generally be described in a simple model and the resulting difference in energy between island growth and layer growth is given by Eq. (5) and illustrated in Figure 9.

$$\Delta W = W_{\text{surf}} + W_{\text{relax}} = const_1 \gamma d^2 - const_2 k \xi^2 d^3 \tag{5}$$

k = bulk modulus, ξ = strain

Considering films of the same volume content, the increased surface energy for the island growth Figure 8b, is proportional to the island area, d^2, whereas the energy released by relaxation of the lattice is proportional to the island volume, d^3. A relaxation mode which is characteristic of isolated islands is shown in Figure 10 for a case where the film material has a larger bulk lattice parameter than the substrate. The model predicts a critical value, d_{crit}, where the island growth is finally more favourable and a fast decrease of the energy for larger sizes. However, the limits of the model are reached in this region as the simple relaxation mode is obviously no longer valid for large sizes.

2.4 Strain Relaxation in Continuous Films

Along with film growth, the islands will overlap and a closed film will form, which can no longer relax by the mechanism discussed above. A possible mechanism for strain relaxation is the formation of **misfit dislocations** as schematically shown in Figure 11. As long as the film is rather thin, there is perfect epitaxy on the substrate, however, the unit cell is tetragonally distorted; as the in-plane lattice parameter is forced to smaller values, an expansion, according to Poisson's ratio, is observed in the direction perpendicular to the film. This tetragonal structure is manifested by the different in-plane and out-of-plane lattice parameters and by a tilt of the crystallographic angles, e.g., a deviation of the [110] direction from 45 °. This strain is relaxed by the formation of dislocations as indicated in Figure 11b, and the film returns, in principle, to the cubic structure, however, the interface is only semi-coherent.

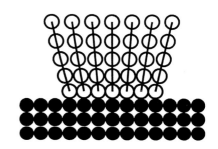

Figure 9: Energy contributions as a function of the island size [5].

3 Physical Deposition Methods

In this Section, we will give a short introduction to the basic principles of the different physical deposition methods and some comments on their advantages and drawbacks. Some special features of the different techniques will be demonstrated by examples which are selected to additionally demonstrate the wide field of thin-film applications in microelectronics and the wide spectrum of analytical tools for the characterization of thin films.

3.1 Thermal Evaporation / Molecular Beam Epitaxy

Molecular beam epitaxy (MBE) has evolved from simple thermal evaporation techniques by the application of UHV techniques to avoid disturbances by residual gases, and additionally includes many different sources. A schematic view of a system is shown in Figure 12 including several different sources which allow the controlled deposition of multi-element compounds. The main components and their use are summarized in Table 3: these are the different beam sources, the shutters, which are very important for controlling the growth of dopant profiles or multi-layers, the process environment and, very specific for MBE, the in-situ process control. Due to the UHV environment all UHV surface techniques might be applied, but only reflection high energy electron diffraction (RHEED) is included here as an example.

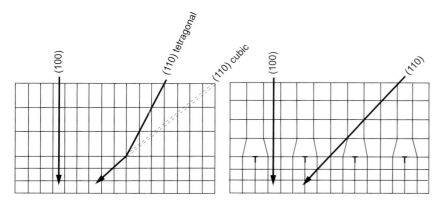

Figure 11: Strain relaxation by misfit dislocations for the example of two initially cubic crystals. As the film has a larger lattice constant than the substrate the forced matching at the interface yields a tetragonal distortion of the film. By misfit dislocations this strain can be relaxed and the film can re-approach its cubic structure.

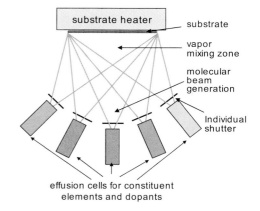

Figure 12: Schematic view of a MBE system for the growth of multi-element compound films.

II Technology and Analysis

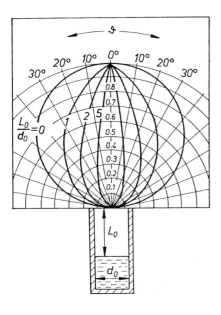

Figure 13: Schematics of a Knudsen cell and the distribution of the vapour beam intensity [7]. The distribution depends on the ratio L_0/d_0 and consequently on the filling level of the cell.

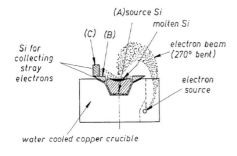

Figure 14: Schematics of a electron beam evaporator for Si evaporation [7]; B: Si guard ring, C: catcher for backscattered electrons.

Facilities	Components	Functions
Beam generators	Knudsen cells e^-- beam evaporators Gas or vapour cells	To provide stable, high-purity, atomic or molecular beams impinging onto substrate surface $\Rightarrow$ MOMBE
Beam interruptors	Fast-action shutters	To completely close or open line of sight between source and substrate. Action should be rapid (< 0.1 s) and should cause minimal thermal disruption of source
Process environment	Multichamber UHV system	To provide ultraclean growth environment, with residual gas species (e.g. O_2, CO, H_2O, CO_2) < 10^{-11} mbar
Beam and growth monitors	RHEED Beam monitoring ionization gauge mass spectrometer	To provide dynamic information on the surface structure on beam intensities and on compositional information

Table 3: Principle operative systems in MBE and their function (after Parker [6]).

3.1.1 Sources

The schematic of the classical MBE source, the Knudsen cell, is illustrated in Figure 13. The evaporation rate, N_e, is described by the Hertz-Knudsen (or Langmuir) equation:

$$N_e = \frac{p_e A_e}{\sqrt{2\pi m k_B T}} \qquad (6)$$

p_e is the equilibrium vapour pressure and A_e the area of the aperture [7]. Therefore, the source can be precisely controlled by a single parameter, the temperature. However, the technical details are very complex and involve more parameters than shown in Eq. (6).

Figure 14 shows the principle of an electron beam evaporator. The electron beam is magnetically deflected by 270° and is centred on the source material. In this way a melt of the source material is produced on a block of the same material which can be held in a water-cooled cold crucible in order to avoid contamination of the melt.

3.1.2 Process Environment

Larger MBE systems are composed of modular stainless steel building blocks and Figure 15 gives an example: the deposition chamber, the wafer or substrate preparation chamber and often an additional analytical chamber. All operative components are attached by flanges for service access. All materials used within the system need special consideration in terms of low gas desorption and high resistance to heating during out-gasing at elevated temperatures. Parts with high heat load are liquid-nitrogen-cooled ('cryo panel'). Pumping systems include turbomolecular pumps, especially in parts with larger gas load, cryopumps and ion getter pumps.

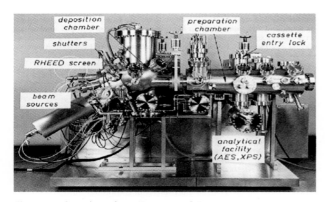

Figure 15: Overview of an MBE system [7].

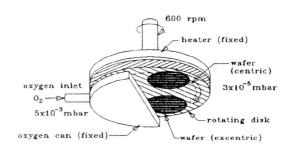

Figure 16: Control of the oxygen partial pressure by differential pumping [8].

Film Deposition Methods

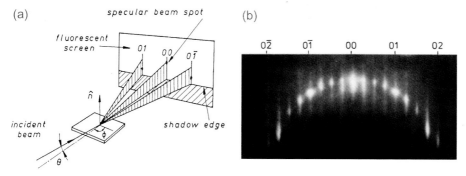

Figure 17:
(a) Schematics of the RHEED geometry; the elongated diffraction spots indicate the cut of the Ewald sphere with the 01, 00, and 0-1 reflections of the quasi-two-dimensional lattice,
(b) diffraction pattern of a reconstructed GaAs (001)-(2x4) surface [7].

The oxygen partial pressure for the deposition of some oxide films is a major problem for the UHV technique. However, several solutions have been used including differential pumping as indicated in Figure 16, i.e. by rotation of the substrate the film is temporarily exposed to oxygen whereas the sources and the vapour are under high vacuum conditions in order to avoid oxidation and deterioration of the sources [8].

3.1.3 In-situ Analytical Techniques

One of the decisive advantages of MBE is the possibility of implementing all UHV surface analytical techniques and controlling the growth process in situ. We will shortly introduce RHEED as a simple method which is most widely used. Figure 17a shows the schematic set-up. An electron beam of typically 10 keV hits the film under a flat angle and is reflected yielding a specular spot as well as diffracted spots. Figure 17b shows an example of the diffraction pattern of a GaAs surface.

The intensities of the diffraction spots yield detailed information on the surface structure, however, for monitoring the growth it is sufficient to monitor the intensity of the specular spot as shown in Figure 18. At the beginning of deposition there is a very flat substrate and a strong specular spot; with increasing coverage, θ, the intensity decreases and for coverage larger than $\theta = 0.5$ it increases again. Generally, the coverage of 1 is not exactly reached again as adatoms and vacancies accumulate during growth and yield a slightly larger roughness. Nevertheless, every individual layer can be controlled by the observed oscillations. In the example shown there are some slower oscillations after closing the shutter which indicate a re-evaporation under the given conditions as discussed in Section 2.2. RHEED shows, of course, only the changes and cannot distinguish between growth and re-evaporation

3.1.4 Example: Epitaxial Growth of Fe Films

Figure 19 gives an example of the application of MBE techniques for the deposition of metallic films as will be discussed in the lectures on magnetic layers. Fe-Cr layers are model systems for the investigation of magnetic exchange coupling and the figure shows a study of the growth of Fe layers. Perfect single crystalline substrates are available in the form of whiskers with step length in the order of 1 μm. The change of the film growth as a function of the deposition temperature is illustrated and the development of a process window for the deposition is indicated [9]. At room temperature we observe a fast decay of the amplitude of the oscillations, which indicates an increasing surface roughness along with the film growth, and this roughness is established by the scanning tunneling microscope, STM. There is a steady improvement with temperature and a perfect layer growth can be observed at 250 °C where the surface diffusion seems fast enough to allow perfect layer by layer growth.

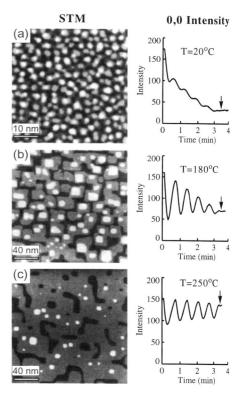

Figure 19: STM and RHEED results for the homo-epitaxial growth of Fe films on Fe(100) substrates. The growth was interrupted after 5 oscillations, as indicated by the arrow. The scale of the STM was changed between part a and b! The roughness of the films decreases strongly with temperature: rms (root mean square) amplitude 0.116 nm, 0.095 nm and finally at 250°C 0.06 nm [9].

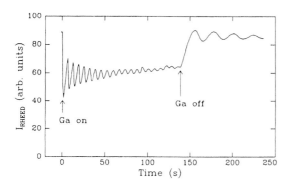

Figure 18: Growth and re-evaporation of GaAs as observed by RHEED [3].

II Technology and Analysis

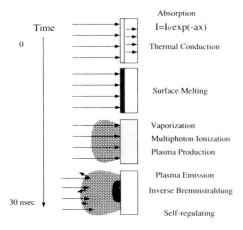

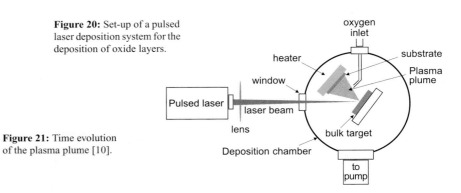

Figure 20: Set-up of a pulsed laser deposition system for the deposition of oxide layers.

Figure 21: Time evolution of the plasma plume [10].

3.2 Pulsed Laser Deposition

3.2.1 PLD Concept

The concept of a PLD experiment is basically simple and is shown in Figure 20. A short pulsed laser beam from an excimer laser is focused onto a target. The pulse energy of typically 1 J/pulse leads to the immediate formation of a plasma due to the high energy density of 3-5 J/cm^2 at the target surface. The plasma contains energetic neutral atoms, ions and molecules and reaches the substrate surface with a broad energy distribution of 0.1 to > 10 eV. Details of the laser ablation process [10] are summarized in Figure 21. A problem of the method is that, in addition to the energetic neutral atoms, ions and molecules, some small droplets may be deposited on the film ending up as so-called 'boulders'. Different methods have been developed to reduce these effects, e.g. time-of-flight selection of the deposits [11] as the heavier particles are slower, or the use of an off-axis geometry [12]. In off-axis PLD geometry the surface of the substrate is placed parallel to the expansion direction of the plasma. This geometry has the additional advantage that samples larger than the lateral extension of the plasma plume can be used, however, only if the target is rotated in order to obtain a homogeneous film.

Based on the extensive work on high-temperature superconductors the deposition of oxide films is well developed. Very satisfactory photon absorption within the oxide target is provided at UV wavelengths and characteristic deposition parameters are: wavelength of the laser beam of 248 nm or 193 nm and a repetition rate ≈ 50 Hz at a pulse length of 25 ns. For the growth of epitaxial oxide thin films sufficient ion mobility is needed. It is provided by a substrate temperature typically exceeding 750 °C, which is provided e.g. by an oxygen-resistant SiC heater. An ambient oxygen pressure of 0.3 to 1 mbar may be maintained within the chamber.

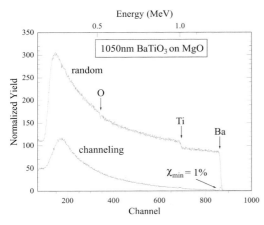

Figure 22: RBS/Channelling measurement of a BaTiO$_3$ thin film on MgO(100) [13].

3.2.2 Example: Optical Waveguides

Electro-optical thin films made of LiNbO$_3$, KNbO$_3$ and BaTiO$_3$ (BTO) with thickness of the order of microns were prepared using the pulsed laser deposition technique [13]. The deposition temperature and the oxygen pressure during the growth process are used to control the crystalline perfection, the orientation and the grain size of the film. As a typical example, a single crystalline thin BTO film on MgO(100) is shown in Figure 22. An RBS analysis is performed to characterize these BTO films. The RBS analysis provides information about the stoichiometry and the homogeneity. In connection with ion channelling measurements additional information is obtained about the crystal perfection of the investigated material. The RBS-channelling analysis shows the correct stoichiometry of the BTO film with a minimum yield of 1 %, which proves the single crystalline growth of BTO on MgO(100). Such a 1 μm thick BaTiO$_3$ film prepared on MgO(100) forms a waveguide showing optical losses as low as 3 dB/cm, which is an excellent result for such thin electro-optical BTO films.

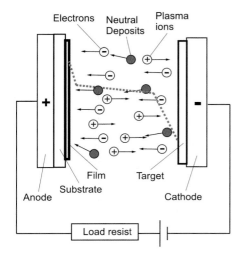

Figure 23: Schematics of a DC-sputter system; the red dotted line indicates the potential between anode and cathode.

3.3 Sputter Deposition

Sputtering of surface atoms has been known since 1852 when W.R. Grove observed this effect during his investigations of plasma discharges and it was soon recognized that this sputtering can be applied for the deposition of thin films. However, the large-scale technological application only developed during the last few decades (see e.g., [14], [15], [16] for review).

206

3.3.1 DC Sputtering

The simplest sputtering approach is so-called DC sputtering, which is schematically shown in Figure 23. In a vacuum chamber the target material, which is eroded, is at the cathode side (negative potential), and the substrate for the film is at the opposite anode side. The potential of several 100 volts between these plates leads to the ignition of a **plasma discharge** for typical pressures of 10^{-1}-10^{-3} mbar and the positively charged ions are accelerated to the target. These accelerated particles sputter off the deposits, which arrive at the substrate mostly as neutral atoms. The discharge is maintained as the accelerated electrons continuously ionize new ions by collisions with the sputter gas. The potential distribution between anode and cathode is indicated by the dotted red line: as the plasma is a good conductor there is no major potential drop in the plasma region, and due to the different mobility of electrons and ions the main voltage drop is observed at the cathode (darkroom). This potential distribution is advantageous as the acceleration of the sputtering gas ions proceeds directly in front of the target and not in a region far off, where the ions would undergo additional collisions and lose their energy on the long path to the target.

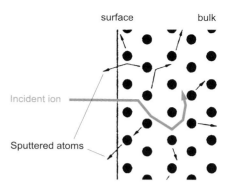

Figure 24: Principle of the sputtering of surface atoms by a collision sequence after the incident ion has hit the target surface [14].

3.3.2 Sputtering Process

The sputtering process is schematically illustrated in Figure 24. The ion, which has been accelerated within the darkroom with nearly the full voltage applied of 50 to 1000 eV, hits the surface atoms. The following **collision cascade** leads to a heating of the target and finally also to some back-reflected atoms which can leave the surface. The details of these collision cascades can be simulated very reliably by 'molecular dynamics' methods, and basically depend on the relative masses of projectile and target atom. Some results of the sputtering yield for incident Ar ions on different target materials are shown in Figure 25 [14]. The threshold energy for sputtering is much higher than the surface binding energy, W_b, of the atoms which is of the order of 4 to 8 eV. This difference can be directly understood as several collisions are necessary in order to obtain an atom in the backward direction. Hence, the threshold is observed at $4W_b$ to $8W_b$ corresponding to a threshold energy of 20 to 50 eV. A linear increase is observed for many conditions up to voltages of 1000 eV. At higher energies the ions penetrate too deeply into the target and the yield decreases again. As shown in Figure 25, different sputtering yields are observed for different target atoms and this difference in principle yields the deposition of a film of different stoichiometry. However, there is generally a good self-regulating mechanism: due to the very low penetration depth of the sputtering process the component eroded faster is denuded after a short initial time and finally in a quasi-equilibrium the difference in yield is compensated by the enrichment. Therefore, sputtering quite generally allows the deposition of films with the same stoichiometry as the target.

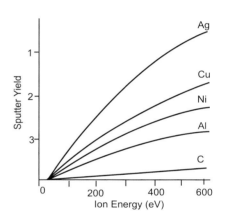

Figure 25: Dependence of the sputtering yield on the energy of the incident Ar ions and of the target properties [14].

3.3.3 Magnetron Sputtering

An ionization degree of less than 1% of the atoms is characteristic of a plasma and consequently a rather low sputter rate. To improve the ionization rate magnetic fields can be used which force the electron onto helical paths close to the cathode and yield a much higher ionization probability [10], [14]. This magnetron arrangement additionally allows a lower gas pressure, however, it has the disadvantage of more inhomogeneous target erosion than a simple planar geometry.

3.3.4 RF Sputtering

DC sputtering works very well as long as the target material shows some electrical conductivity. For insulating targets, however, a high-frequency plasma discharge, as shown in Figure 26, must be applied in order to avoid the accumulation of electric load. A typical frequency of 13.6 MHz is capacitively coupled to the target and there is only a small voltage decay across the electrode. As the electrons are much faster than the ions a negative potential at the electrodes as compared to the plasma potential evolves during each cycle. With a symmetrical arrangement of cathode and anode we would obtain similar re-sputtering rates and no film growth, however, non-symmetries, which yield some bias voltage, are introduced by the coupling of the RF and by differences in the geometry, i.e., different sizes of target and substrate, and especially by the generally applied grounding of the substrate and the deposition chamber. Nevertheless, deposition rates are much lower than for DC sputtering.

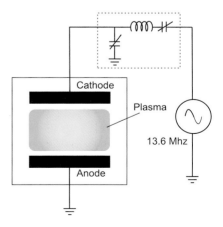

Figure 26: RF-plasma sputter system with RF matching network [14].

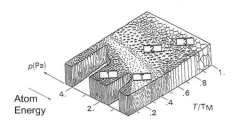

Figure 27: Structure zone diagram for sputter-deposited metals after Thornton [17].

3.3.5 Gas Pressure and Film Growth

Compared to the other physical deposition methods the partial pressure of the sputtering gas is an additional and important process parameter which must be considered and optimized. This pressure controls the free path length of the atoms and therefore their energy, angular distribution and finally also their incorporation in the film. For metals, so-called zone diagrams for the film growth have been developed [17], which show some systematic dependencies on the normalized temperature (the substrate temperature, T_s, normalized by the melting temperature, T_m) and the pressure, Figure 27.

Zone 1: $T_s < 0.2\ T_m$: at these low temperatures no bulk diffusion and only very limited surface diffusion is observed which would allow for crystallite rearrangement. Grain structure is composed of fibres whose size and orientation are determined by the initial random nucleation of the grains. Some shielding effects are visible depending on the angle of deposit incidence. This structure extends to higher temperatures, when the ion energy is reduced due to gas collisions. The size of the fibres increases with temperature mainly following the temperature dependence of the nucleation density.

Zone T_s: $(0.2 - 0.3\ T_m)$: In this transition zone, surface diffusion becomes effective and small crystals of energetically unfavourable orientation are eliminated, i.e. a competitive texture is observed.

Zone 2 and 3: $T_s > 0.5\ T_m$: Evolution of morphology and texture is determined by reconstruction, single crystalline columns are formed with increasing diameter, and, finally, texture is determined by the lowest free energy surface of the crystal. The influence of the Ar pressure decreases.

For oxide layers the gas pressure is even more important as negative oxygen ions are formed which can re-sputter from the film. As the re-sputtering yield also depends on the different elements the stoichiometry of the growing film is changed and these changes of the film are not compensated as was the case for the sputter yield from the target discussed above. This re-sputtering can be reduced by so-called high pressure oxygen sputtering [18] where the energy of the ions is reduced by the increased number of gas collisions connected to the shorter free path length; however, it must be considered that this high gas pressure also reduces the deposition rate.

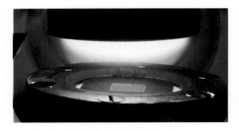

Figure 28: DC-sputter deposition in 'high pressure' oxygen.

3.3.6 Example: High Pressure Oxygen Sputtering

Figure 28 shows a high-pressure, a few mbar, oxygen sputtering system which is operated in DC mode for the given example of the deposition of electrically conducting oxide films [19]. Perfect control of the interface and epitaxial growth can be obtained by this 'low energy' sputtering process at rather slow growth rates. These perfect interfaces are demonstrated by the cross-sectional HRTEM view of the resulting multilayer structure, Figure 29, which is used as a junction barrier in high T_c devices [19].

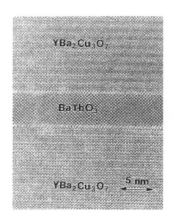

Figure 29: HRTEM micrograph of a sputter deposited trilayer system [19].

4 Chemical Deposition Methods

Chemical deposition generally includes **chemical solution deposition** (CSD) as well as **chemical vapour deposition** (CVD). In both cases, chemical precursors are employed which undergo chemical reactions for the formation of the film. We will place special emphasis on CVD as this method finally allows the deposition of ultrathin films and the conformal deposition on complex-shaped structures which are essential for ULSI. CSD includes sol-gel techniques and metal-organic decomposition MOD and typically uses spin-on techniques for the distribution of a solute film which is subsequently processed and crystallized. Finally, we give a short introduction to a very different method for deposition from solutions, the **Langmuir-Blodgett** (LB) technique. LB techniques allow the deposition of monomolecular organic films on different substrates making use of the hydrophilic/hydrophobic orientation of the molecules.

4.1 Chemical Vapour Deposition

The general principles of CVD are well established and a number of reviews and textbooks are available [20], [21], [22], which cover many generic issues common to any type of material. In CVD, film growth occurs through the chemical reaction of the component chemicals (i.e. precursors) which are transported to the vicinity of the substrate via the vapour phase. The film-forming chemical reactions typically utilize thermal

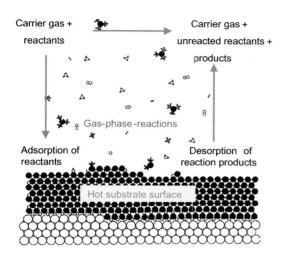

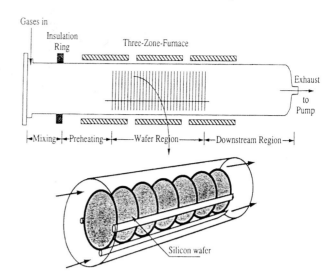

Figure 30: Schematics of the gas flow and the atomic scale chemical environment in the region of the growing film surface during a MOCVD process.

Figure 31: Schematic representation of a hot-wall multiple-wafer in-tube CVD reactor [20].

energy from a heated substrate as depicted schematically in Figure 30. Other more special methods, which cannot be discussed here, couple non-thermal energy sources such as RF or microwave power or light into the reaction process in order to reduce the thermal reaction temperature required. In order to complete the system, a delivery system for the precursors and, finally, an exhaust system must be added. The most straightforward type of CVD involves chemical precursor compounds that are sufficiently stable gases and such processes are standard processes in CMOS technology for the deposition of insulators and interlayer dielectrics like poly-Si, SiO_2, SiN_x and BPSG glasses. Figure 31 shows the schematics of a reactor for handling large batches of wafers simultaneously. Examples of the reactions involved are the thermal decomposition of silane, SiH_4, for the deposition of:

- Poly Si : $SiH_4 \leftrightarrows Si(s) + 2 H_2$ at 580-650 °C, and a pressure of $\approx$ 1 mbar;
- Si-Nitride: $3 SiH_4 + 4 NH_3 \leftrightarrows Si_3N_4 + 12 H_2$ at 700-900 °C, and atmospheric pressure;
- Si-Dioxide: $SiH_4 + O_2 \leftrightarrows SiO_2 + 2 H_2$ at 450 °C;

These SiO_2 films are usually under high stress and are not conformal. Therefore alternative routes using organic precursors have been developed e.g., the TEOS (tetra-ethyl-ortho-silane) process:

$$Si(C_2H_5O)_4 + 12 O_2 \leftrightarrows SiO_2 + 8 CO_2 + 10 H_2O \text{ at } 700 \text{ °C}.$$

Similarly, for the processing of many metals and especially the group-II metals, special precursors in the form of organometallic compounds had to be developed and a special subgroup of CVD techniques, **metal-organic**-CVD, MOCVD, has therefore evolved. Efficient, reproducible MOCVD processes hinge critically upon precursors with high and stable vapour pressures and the chemistry is therefore the decisive step in the development of MOCVD.

4.1.1 Precursor Chemistry and Delivery

Only a sufficiently high vapour pressure enables vapour-phase mixing of precursor components and transport of the reactants to the growing film and a vapour pressure of > 0.1 mbar at 100 °C is considered to be a lower limit. Adequate molecular stability of the precursor vapour is required to prevent premature reaction or decomposition of the precursor during vapour phase transport; these requirements characterize a process window between vaporization and decomposition of the precursor. Additional requirements

II Technology and Analysis

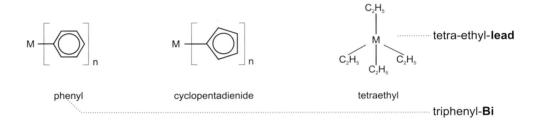

Figure 32: Selection of precursor molecules for oxide deposition [25].

are long-term stability of the precursors (e.g. low moisture sensitivity), complete decomposition (no contamination of the film e.g. by fluorine) and last but not least low toxicity and environmental regulatory requirements.

For complex oxides the precursor molecules typically contain the metal atoms, M, and organic groups, R. In most cases, such as in alkoxides, ketonates, and carboxylates, the metal atom is bound to the organic group through an oxygen atom, i.e. M-O-R. This is different from precursors used for the MOCVD of compound semiconductors where the precursors are conventional metal organic compounds, i. e. M-R systems. For heavy cations (especially group-II elements), a reasonable volatility is often reached only by organometallic precursors, e.g. by β-diketonates with several alkyl groups, such as in tetramethylheptadionates (thd) (and this type of organometallic precursor is the origin of an alternative acronym occasionally used: OMCVD). The two keto groups chelate the cation while the outer alkyl groups effectively shield any polar region of the molecule. Typically two (thd) ligands are reacted with one cation yielding Ba(thd)$_2$. Since the outer shell of this molecule consists entirely of alkyl groups there is only a weak van-der-Waals interaction between neighbouring molecules and, hence, a relatively low boiling temperature. For cations with a higher electronegativity (i.e. a lower tendency to form ionic bonds), alkyl compounds, such as Pb(C$_2$H$_5$)$_4$, and alkoxide compounds, such as Ti(OC$_3$H$_7$)$_4$, may give rise to a sufficient volatility [23].

The different types of precursors used for oxide deposition are summarized in Figure 32. The main applications for the present topics of IT are included in the figure: e.g., the diketonates for Ba and Sr, thd = tetramethylheptadionate, alkoxides for Ti (TIP = titaniumisopropoxide = Ti(O-i-Pr)$_4$) and Zr. Mixed precursors like Ti(O-i-Pr)$_2$(thd)$_2$ are also used to increase compatibility with other precursors. In addition, adducts or stabilizers are used in order to avoid reactions and oligomerization e.g., tetraglyme or pmdeta. Most of these precursors are liquid or solid at room temperature and reach a sufficient vapour pressure only at elevated temperature. Therefore, special delivery systems are necessary, i.e. bubblers or liquid source delivery systems.

A bubbler system as used for PZT deposition is shown in Figure 33. Three different liquid precursors are held at elevated temperature:

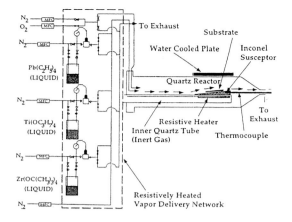

Figure 33: MOCVD research reactor with bubbler system and quartz tube horizontal flow reactor used for PZT deposition [25].

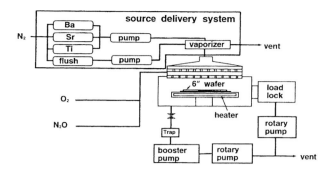

Figure 34: MOCVD production tool with liquid source delivery system showerhead reactor and load lock used for BST deposition [26].

$Pb(C_2H_5)_4$: bubbler held at 50 °C with a vapour pressure of 2 mbar
$Ti(OC_3H_7)_4$: at 95 °C and 5 mbar,
$Zr(OC_4H_9)_4$: at 65 °C and 1 mbar.

The vapour is transported by a carrier gas to the reactor where the mixing of the vapour occurs. The flow rate of the precursors, f_p, given in sccm (standard cm³ per minute) or mol/min is:

$$f_p = \frac{f_c p_p}{p_{tot} - p_c} \quad (7)$$

the index p refers to the precursor and c to carrier gas. The partial pressure of the precursors, p_p, can be calculated from the Clausius-Clapeyron equation:

$$\frac{d(\ln p)}{dT} = \frac{\Delta H}{k_B T} \quad (8)$$

where ΔH is the enthalpy of evaporation. For stable temperatures, the flow can be controlled by the flow of the carrier gas. The lines must be held at the bubbler temperatures in order to avoid condensation.

For group-II metals the precursors are solid up to higher temperatures and the vapour pressures are even lower e.g. $Ba(thd)_2$: 0.05 mbar at 200 °C and $Sr(thd)_2$: 0.20 mbar at 230 °C. As the precursors are not very stable at these temperatures a direct sublimation is not favourable and so-called liquid delivery systems have been developed. The precursors are dissolved in an appropriate solvent, e.g., butylacetate, and evaporated in close vicinity to the reactor. Such a system is shown in Figure 34. Here the liquids are mixed and finally flash-evaporated in a vaporizer on top of the reactor. Alternative evaporation systems use aerosol-assisted nebulizers or controlled single droplet injection systems [24].

4.1.2 Reactor Design and Modelling

The reactor must provide a controlled gas flow and heat distribution and for the precursors discussed here it should work at low pressure (10^{-1} to 10 mbar) in order to keep the number of gas collisions of the molecules very low. In addition, its construction materials must withstand oxygen and other aggressive chemicals even at elevated temperatures, and typical materials are e.g. quartz, SS and coated graphite. Figure 33 shows a small-size research reactor with quartz as the structural material and a simple gas flow system [25]. Figure 34 shows a reactor design for routine deposition of large wafers [26]. The evaporated precursors are homogeneously distributed by a so-called showerhead. The pumping system consists of a roots or booster pump and a rotary pump for the pumping of larger gas loads including the carrier gas. Finally, a load lock system is shown as is necessary for routine production. In order to increase the throughput, batch processing tools have to be developed also for oxide films and first results have been reported for an AIXTRON 2600G3 Planetary Reactor®, which can handle five 6-inch wafers simultaneously [27].

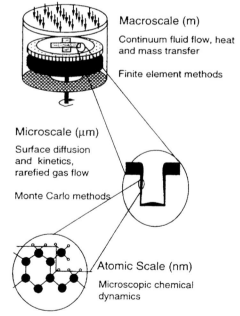

Figure 35: Schematic illustration of the coupling of micro- and macroscale phenomena in CVD processes [20].

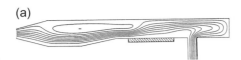

Figure 36:
(a) Calculated flow pattern in a horizontal reactor: the formation of transverse convection rolls is observed.
(b) Flow visualization with TiO_2 smoke demonstrating the existence of a transverse roll cell in the horizontal flow reactor and a dark space above the susceptor caused by thermophoretic migration of the smoke particles away from the hot susceptor [20].

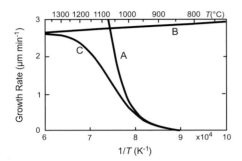

Figure 37: Schematic view of the control of the growth process: A: kinetic control, B: transport control, C: resulting behaviour; the actual temperature scale corresponds to the deposition of poly-Si from $SiCl_4$ [20].

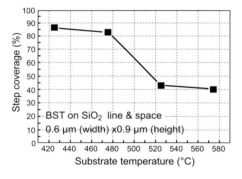

Figure 38: Dependence of the step coverage on the deposition temperature [26].

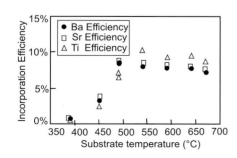

Figure 39: Dependence of the incorporation efficiency on the deposition temperature [29].

A complete understanding of the complex MOCVD process includes processes on very different length scales as indicated in Figure 35: Macroscale (turbulence, cold or hot spots), microscale (e.g., conformal deposition) and finally atomic-scale (chemical reactions, nucleation and growth, desorption). Considering the large number of design parameters, it is desirable to simulate the whole process in order to obtain an estimate of the reactor behaviour. Although complete modelling has to combine the simulation on all length scales, modelling can usually be performed separately on different scales and we restrict ourselves to some examples.

The deposition rate and the final composition of the film are primarily controlled by the spatial profiles of gas velocities and temperatures as well as of the partial pressure of the various precursors in the reactor. Collectively, these profiles are referred to as the reactor flow pattern. For CVD, a laminar gas flow pattern is generally required in order to obtain process reproducibility. The parameters which determine these profiles are either directly controllable (i.e. the total flow rates of gases the total reactor pressure, the flow rates of individual reactants, the temperature of the susceptor and occasionally the temperature of the reactor walls) or are fixed by the design of the system (i.e. the configuration of the gas injection manifold, the design of the exhaust flange, and the shape, size and materials of the deposition chamber and susceptor). For the modelling of this macroscale, the fluid flow in the reactor, the temperature gradient and the mass transfer finite element methods are widely used. By this modelling, disturbances of the laminar flow or regions of low temperatures, cold rolls, can be detected and corrected by changes of the reactor geometry. Results can be given in terms of dimensionless parameters like the Reynolds number, the Raleigh number and others, and can therefore be transferred to similar reactor types [20]. Figure 36 shows as an example the flow pattern in a horizontal flow reactor with the formation of a transverse convection cell and the experimental verification of such a cell by visualization through light scattering at TiO_2 smoke.

Although MOCVD is generally not a process under thermodynamic equilibrium conditions, the system's thermodynamic data can yield a lot of background information on phase stability assuming quasi-equilibrium phase distributions. As an example, Ref. [28] reports on the phase distribution under typical MOCVD deposition conditions for BST, and especially on the strong influence of the precursor mix on the formation of carbonate phases.

A typical example of calculations on the microscale is deposition into trenches or vias, as shown in Figure 2. It is obvious that the free path length of the gases (or the pressure) is an important parameter in this field and the optimum filling conditions for different geometrical details of the structure can be predicted, e.g., for a hole with a given aspect ratio, the ratio of the depth to the width.

4.1.3 Growth Control

Thermal energy, e.g. supplied by the heated substrate, is the primary source of energy fuelling film growth. Depending on the temperature, two different ranges for the limitation of the growth rate can be distinguished: at low temperatures the growth is usually limited by the reaction kinetics and at higher temperatures by the mass transport as shown schematically in Figure 37.

The reaction kinetics includes many different process steps ranging from precursor reactions, which may include many decomposition reaction steps, to the film growth kinetics. Hence, an effective activation energy, W_a, must generally be expected and the kinetically limited growth rate, j_k, may be written as:

$$j_k = const_1 N^\infty e^{-\frac{W_a}{k_B T}} \qquad (9)$$

N^∞ = precursor concentration at some distance from the substrate.

At medium and higher temperatures the exponential increase of the kinetically limited growth rate yields only a negligible limitation and the major limitations of the reaction are given by the transport of the precursors and reactants to the surface (i.e., by the diffusivity of the gases, D), and the so-called transport limited growth rate, j_t, is given by:

$$j_t = const_2 N^\infty \frac{\sqrt{D}}{T} \qquad (10)$$

where the diffusion constant of the gas is: $D = \lambda \langle u \rangle / 3$, and $\langle u \rangle$ is the average velocity of the gas atoms. In contrast to the diffusion in the solid state, the diffusivity in gases shows only a weak temperature dependence, and consequently also j_t (Figure 37). Within the simplest ideal gas approximation this yields $j_t \propto 1/T^{1/4}$, which is in reasonable agreement with more realistic calculations that also consider the reactor geometry and yield $j_t \propto 1/T^{1/6}$. The overall growth rate, j_g, may be obtained by adding the two reciprocal fluxes, which can be considered as two resistive elements in series:

$$\frac{1}{j_g} = \frac{1}{j_k} + \frac{1}{j_t} \qquad (11)$$

This means that the slower flux finally determines the growth rate. At even higher temperatures there is usually a decrease for different reasons: thermodynamics (exothermic reaction), precursor stability, pre-reactions etc. In spite of the complexity of the processes there are some general rules which may guide the selection of the appropriate temperature region (as the gas diffusivity depends on pressure, the limits depend on pressure, too).

Mass-transport–limited region: The deposition rate is insensitive to small temperature gradients and cold wall reactors are therefore often operated in this regime. However, the local deposition rate is very sensitive to the flow pattern, which may be complex for cold-wall reactors, and which is, as shown above, an indirect consequence of temperature gradients; hence uniformity problems may arise in this regime.

Kinetically limited region: The deposition rate is only weakly dependent on the flow homogeneity, therefore this regime is best suited for conformal deposition. However, the exponential temperature dependence necessitates a very high stability of the temperature as well as a good uniformity over large wafers.

Figure 40: MOCVD of $SrTiO_3$ thin films into a test hole of a SiO_2 layer with an aspect ratio of 1:6 and a width of 150 nm. The deposition was performed in a dome-type reactor at a wafer temperature of 420°C. The TEM cross section shows the very high conformality of the thin film [30].

4.1.4 Conformal Deposition

Conformal deposition is a prerequisite for applications in ULSI, e.g. for high-K dielectrics in DRAMs. The additional requirements for conformal deposition yield a strong restriction within the process parameter field for the optimization of the film properties. Conformality is expected in the regime of kinetically controlled growth, and as discussed above low temperatures and low pressures are favourable. As demonstrated by the example of (BaSr)TiO$_3$ (BST) depositions, Figure 38, the conformality is strongly improved at lower temperatures [25], however, the incorporation rate of the precursors decreases with lower temperatures [29] as indicated in Figure 39. In addition there are some differences for different elements which strongly affect the reproducibility of the stoichiometry. Hence, for these conditions there is only a very narrow process window to achieve conformality and reasonable deposition rates simultaneously. Nevertheless, very good conformality even in narrow holes with a high aspect ratio, i.e., ratio of width to depth, of 1:6 has been achieved [30] as demonstrated in Figure 40.

4.1.5 Layer by Layer Growth and Ultrathin Films

Similar to MBE, layer by layer growth is the final goal of the development of the technique and it has been achieved for metal-organic MBE (MOMBE) as a hybrid system. Although in situ control is limited with conventional MOVCD, optical methods and ion beam analyses in differentially pumped systems have been tested, and III-V compound multilayer systems for LED and laser diodes are deposited with high perfection. Additional approaches for the control of the layer growth include pulsed organometallic beam epitaxy (POMBE) and as a specific chemical approach, **atomic layer epitaxy** (ALE) and **atomic layer deposition** (ALD) which use the chemisorption of special polar molecules [32]. This process is illustrated in Figure 41 for the example of the deposition ZnS as used for phosphorus screens. In the first step the chemisorption of the precursor molecules, $ZnCl_2$, yields a saturated monolayer on the surface of the substrate. In the subsequent step the reactant, H_2S, reacts with this adsorbate and a monolayer of ZnS is deposited without external control of the reaction time. As there is no adsorption of the reactant on ZnS the next monolayer may be deposited similarly. Such a process yields atomic layers even without epitaxial growth. The deposition of ternary systems is of course much more difficult, nevertheless the development of a process for the deposition of BST has been reported [32].

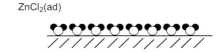
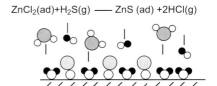
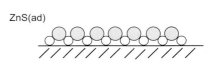

Figure 41: Principle of the ALD for the example of ZnS [31].

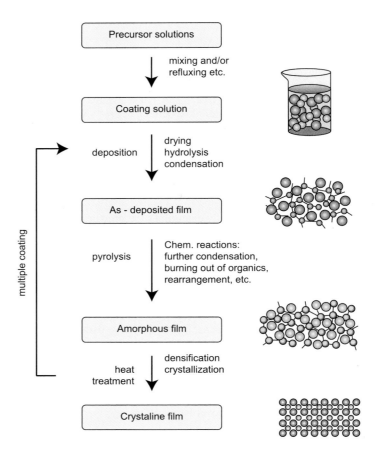

Figure 42: Flow chart of the chemical solution deposition technique. The bars describe the states during the CSD procedure, while the arrows indicate the treatment and the internal processes [23].

4.2 Chemical Solution Deposition

The chemical solution deposition (CSD) method comprises a range of deposition techniques and of chemical routes which have been reviewed recently [23], [33]. A generalized flow chart of the CSD of oxide thin films is shown in Figure 42. The process starts with the preparation of a suitable coating solution from precursors according to the designated film composition and the chemical route to be used. Besides mixing, preparation may include the addition of stabilizers, partial hydrolysis, refluxing, or else. The coating solution is then deposited onto substrates by:

- *spin-coating,* where typically a photoresist spinner is employed and which is suitable for semiconductor wafers,
- *dip coating,* which is often used in the optics industry for large or non-planar substrates, and
- *spray coating,* which is based on a misting of the coating solution and deposition of the mist exploiting gravitation or an electrostatic force.

The wet film may undergo drying, hydrolysis and condensation reactions depending on the chemical route. The as-deposited film possibly represents a chemical or physical network. Upon subsequent heat treatment, a further hydrolysis and condensation and/or a pyrolysis of organic ligands may take place, again depending on the chemical route. The resulting film consists of amorphous or nanocrystalline oxides and/or carbonates. Upon further heat treatment, any carbonate will decompose and the film will crystallize through a homogeneous or a heterogeneous nucleation. Typically, the desired final film thickness is built up by multiple coating and annealing.

Depending on the type and reactivity of the precursors, the chemistry shows a wide spectrum of reaction types. On the one hand, there are the pure sol-gel reactions i.e. alkoxide precursor systems, which undergo hydrolysis and condensation reactions. The formation of SiO_2 coatings starting from Si alkoxides is the classical example of this type of reaction. The condensation leads to a chemical gelation in which – under appropriate reaction conditions – no pyrolysis reaction of any organic ligand occurs. At the other extreme, there is **metal organic decomposition** (MOD), which typically starts

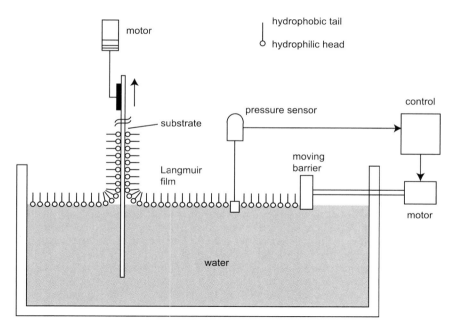

Figure 43: Trough for the controlled deposition of LB films; for details see text.

from carboxylates of the cations or, in special cases, from the nitrates. The carboxylates do not chemically react with water. Consequently, during heat treatment, first the solvents are evaporated, a process which is sometimes referred to as physical gelation. Upon further heating, the carboxylates pyrolytically decompose into amorphous or nanocrystalline oxides or carbonates. There is a wide spectrum of possible reaction routes between the pure sol-gel route and the MOD route. Depending on the type of alkoxides and a possible stabilization of the precursors, there may be a partial hydrolysis and condensation while some organic ligands remain in the gelated film and undergo pyrolysis upon further heat treatment. In the synthesis of multicomponent oxide films, often hybride routes are followed, i.e. there may be some precursors employed which tend to follow the sol-gel or partial sol-gel route while others undergo typical MOD reactions.

The microstructure formed during the CSD process strongly depends on the thermodynamics and kinetics of the solidstate reaction from the intermediate amorphous or nanocrystalline state after pyrolysis to the final equilibrium crystalline phase. This is controlled by the chemical composition of the film system (for details see e.g. ref. [23]).

Due to their low capital investment and simple processing, CSD techniques are widely used and are also applicable for micro- and nanoelectronics on a low level of integration, e.g. for present state FeRAMs.

4.3 Langmuir-Blodgett Films

Organic thin films can be deposited with all of the methods discussed above and especially CSD (spincoating) and CVD techniques are widely used for the deposition of low-K dielectrics in microelectronic integration technology [34]. With special consideration of the stability of the molecular reactions under high electronic excitation even sputter deposition [16] and PLD [12] have been considered. The aspects of heteroepitaxy of large molecules on inorganic substrates have been discussed in detail recently [35]. Organic light emitting diodes (OLED's) consisting of rather complex layer stacks have been deposited by MBE [36]. For organic molecules there is, however, an additional powerful deposition method which should be shortly introduced.

The Langmuir-Blodgett (LB) method is a classical method of surface chemistry for the deposition of molecular monolayers and multilayers and its development is closely connected with the oil on water problem. The organic molecules that are used in this type of deposition contain two types of functional groups. One end of the hydrocarbon chain is soluble in water (hydrophilic), e.g., acid or alcohol group, and the other end contains insoluble hydrocarbon groups (hydrophobic). As a result the molecules form a film on the surface of water (a Langmuir film) with the hydrophilic end at the water side.

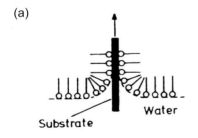
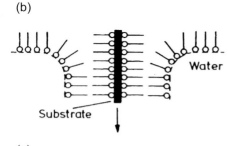
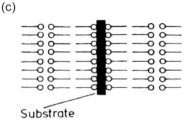

Figure 44: Y-type deposition sequence [39].

Pockels showed already in 1891 that such a film can be compressed to form a continuous monolayer on the liquid. The final step was achieved by Blodgett in 1935 when she demonstrated that the film adheres to a solid surface passing through the air-water interface.

Starting from this simple principle, remarkable technical progress has been achieved and starting in the mid-1970s fully automated troughs for building monomolecular layers and multilayers have been developed [37]-[39]. Figure 43 illustrates the basic principle: a substrate is removed from a bath which contains a monomolecular Langmuir film on a water surface. The Langmuir film is kept under a constant 2-D pressure by a moving barrier and a control unit. By this process, a monolayer of the organic molecules is transferred onto the substrate.

The details of the adsorption process and the underlying chemistry are complex and depend on many parameters, however, some basic processes can be specified. Firstly, we consider a hydrophilic substrate as shown in Figure 44: (a) The first monolayer is transferred as the substrate is raised through the subphase; if there is a strong hydrophilic interaction no layer adheres if the substrate is lowered through the compressed subphase and for weak interaction this result can only be obtained if the substrate is immersed before the floating layer is spread. Subsequently a monolayer is deposited on each traversal through the compressed layer, (b). The resulting stack is a head to head and tail to tail configuration, (c), and is referred to as Y type. Starting with a hydrophobic substrate the double layers are arranged analogously. Most of the substrates used are hydrophilic, e.g., transparent substrates like glass and quartz, metals like Al or Cr and their oxides, and also silicon wafers, which are most widely used. Care must be taken, however, in cleaning the surfaces, e.g. clean Si or oxidized Si are hydrophilic whereas a treatment with Cl (dimethyldichlorosilane) yields a hydrophobic Si surface.

So-called X- and Z-type films can be obtained if the films are deposited only during the immersion or the removal of the substrate from the liquid. In this case the interactions between adjacent monolayers are hydrophilic-hydrophobic and stable growth is most likely observed for weakly hydrophilic headgroups. Therefore these multilayers are less stable than the Y-type systems. Both X- and Z-type films are non-centrosymmetric and are therefore considered for applications in nonlinear optics. Depending on the chemistry and / or the deposition mode and speed there are also deviations from these ideal deposition modes. This different adsorption behaviour can be quantified by the ratio ϕ.

$$\phi = \frac{\theta_u}{\theta_d} \qquad (12)$$

θ_u and θ_d are the deposition (or transfer) ratios on the upward and downward passages as originally defined by Langmuir comparing the area occupied by the monolayer on the water and on the substrate. Thus $\phi = 1$ corresponds to the ideal Y-type deposition and X and Z-type are characterized by $\phi = 0$ and $\phi = \infty$, respectively.

Summarizing, LB methods allow the deposition of very differently structured films and multilayers in a highly controlled manner. Although the method is limited to organic films the importance for applications in the field of information technology is increasing rapidly. Applications range from films for photolithography with improved homogeneity, flat-panel displays, optical devices, especially nonlinear optical devices, and finally the wide field of future molecular electronics [39].

5 Summary

In this short overview the methods used for the deposition of thin films from the materials of interest for information technology have been introduced. Even this selection of the most basic properties of the techniques should give an impression of the wide range of possibilities which are available. We have given special emphasis to deposition methods for metallic/magnetic films as considered for MRAM application and perovskite oxide films for DRAM, FeRAM, FeFET, conducting oxide electrodes and optical waveguides. Nevertheless it should be kept in mind that these deposition methods have also been developed to the highest precision for semiconductor films and especially superlattices, e.g. Si/Ge alloys and III-V compounds. In addition, these methods can be applied for organic films, and especially MBE and CVD are being considered for large-scale production of organic light-emitting diodes. All of these deposition methods have reached very high technical standards and the deposition of high-quality films for most of the materials of interest can be achieved and has been achieved by several methods. Therefore, the optimum choice of method depends on details of the requirements and the differences between the requirements of research and production are often decisive. In addition to the properties listed in Table 1 we finally summarize some advantages and drawbacks of the methods.

MBE has been developed during the last 30 years to a high level of technical perfection and all components are commercially available. The method allows perfect layer by layer growth control and fairly independent control of the different deposit components and is therefore very flexible. Due to UHV conditions very clean films can be obtained and the method therefore is ideally suited for basic investigations of growth processes. For application as a production tool there are major drawbacks due to the generally rather slow layer by layer growth, the expensive UHV techniques and the large number of process parameters which have to be kept under control.

PLD is a well-developed thin film preparation method, and since the success of growing thin films from high temperature superconductors like YBaCuO the method has been shown to be especially suited for the deposition of oxides and other multicomponent materials. The PLD method offers several advantages compared to other deposition techniques, if the formation of multicomponent thin films is considered. The most important advantages of PLD are the stoichiometric transfer from a complex multi-elemental ceramic target material to the substrate without significant separation between the constituents and the rather high deposition rates. Therefore films can be optimized and deposited relatively fast. As the base vacuum is not critical for the process it may vary from UHV, e.g. for the additional option of in situ growth control, up to the oxygen partial pressures required for oxide growth. Disadvantages are the formation of droplets and boulders and the restriction to small substrate sizes, which can only be overcome by mechanical scanning of the substrates. For large-scale integration processes the low degree of conformal deposition is of concern.

Sputter deposition systems are well developed, commercially available and are standard techniques for metallization in CMOS production lines. Advantages are the large throughput, the large substrate sizes, good adhesion to the substrate due to the ion bombardment, often self-adjusting composition control, and less rigorous vacuum requirements than for thermal evaporation. The systems are very flexible and can be adjusted to many specific requirements. Additional modifications which could not be discussed here are e.g. bias sputtering in order to accelerate or decelerate the ions in a controlled manner, off-axis deposition in order to obtain more homogeneous deposition similar to PLD, and ion beam sputtering by the use of a controlled ion gun instead of the plasma as an ion source; as a further step of development multiple target systems, which are individually controlled, can be implemented. During sputtering of insulators there is a build-up of electrical load on the target which can be avoided by using an RF field; this RF field reduces the efficiency of the deposition and yields small deposition rates. Although only restricted conformality of the deposition can be achieved, this conformality is sufficient for many metallization requirements.

CVD techniques have been used for the deposition of thin films in many fields of modern technology and are standard processes in the CMOS technology for the deposition of insulators and interlayer dielectrics like SiO_2 and SiN_x and BPSG glasses. For the processing of many metals, however, special precursors in the form of organometallic compounds had to be developed in order to obtain a sufficient volatility and MOCVD emerged as a special subgroup of CVD techniques. MOCVD techniques are considered to be the primer deposition techniques in the field of high-K materials like $(Ba,Sr)TiO_3$ for advanced DRAM concepts as well as for ferroelectric materials like $Pb(Zr,Ti)O_3)$ for ferroelectric memories due to their excellent film uniformity, compositional control, high deposition rates, and amenability to large wafer-size scaling. Moreover, the need for a high degree of film thickness conformality over the complex device topographies common to ULSI-scale circuits makes MOCVD one of the most appealing film synthesis methods. Limitations of the methods arise from the availability of suitable precursors, the very complex process parameter field and the very limited in situ control of the processes.

In addition to these depositions from the vapour phase we have included two examples of methods for deposition from solutions. The *advantages* of CSD are excellent control of the film composition through the stoichiometry of the coating solution, a relatively low capital investment, and easy fabrication over large areas, up to multiple square metres for dip and spray coating techniques. The *disadvantages* are some specific obstacles to achieving epitaxial films, the lack of an opportunity to deposit atomic layer superstructures, and poor step coverage for narrow submicron 3D structures.

LB methods are well-developed deposition techniques in surface chemistry and although the method is limited to organic films the importance for applications in the field of information technology is increasing rapidly. Applications range from films for photolithography with improved homogeneity, flat-panel displays, optical devices, especially nonlinear optical devices, and finally the wide field of future molecular electronics.

Finally, it should be mentioned that the borderlines between the methods are not sharp and there is some overlap. As the necessary components are commercially available even hybrid methods can easily be achieved, e.g. MOMBE or a combination of thermal evaporation with simultaneous ion bombardment, ion-beam-assisted deposition, IBAD. Nevertheless, every new material and especially every new material combination presents a new challenge, and new ideas, like the recent applications of SURFACTANT's, and also very detailed optimization of the processes are necessary.

Acknowledgements

The editor would like to thank Sam Schmitz (CAESAR, Bonn) for checking the symbols and formulas in this chapter.

References

[1] R.F. Bunshah, *Handbook of Deposition Technologies for Films and Coatings*, Noyes Publ., Park Ridge, 1994.
[2] K.L. Chopra, *Thin Film Phenomena*, McGraw-Hill, NY, 1969.
[3] J.Y. Tsao, *Materials and Fundamentals of Molecular Beam Epitaxy*, Acad. Press, SanDiego, 1993.
[4] R.H. Lamoreaux, D.L. Hildenbrand, and L.Brewer, J. Phys. Chem. Ref. Data **16**, 419 (1987).
[5] B. Voigtländer, Experimente zum epitaktischen Wachstum, 28. Ferienkurs des FZ-Juelich, p.C3.1 (1997).
[6] E.H.C. Parker, (ed.), *The Technology and Physics of Molecular Beam Epitaxy*, Plenum Press, New York, 1985.
[7] M.A. Herman und H. Sitter, *Molecular Beam Epitaxy*, Springer Series in Materials Science, 2. Edition, 1996.
[8] P. Berberich, W. Assmann, W. Prusseit, B. Utz, H. Kinder, J. of Alloys&Compounds **195**, 271 1993.
[9] J.A. Stroscio, D.T. Pierce, and R.A. Dragoset, Phys. Rev. Lett. **70**, 3615 (1993).
[10] O. Auciello, A.I. Kingon, A.R. Krauss and D.J. Lichtenwalner, in *Multicomponent and Multilayerd Thin Films for Advanced Microtechnologies: Techniques, Fundamentals and Devices*, O. Auciello and J. Engemann, eds., Kluwer Acad.Press, 1993. p. 151 – 208
[11] D.B. Chrisey and G.K. Hubler, (eds.), *Pulsed Laser Deposition of thin Films,* John Wiley & Sons, NewYork, 1994.
[12] B. Holzapfel, B. Roas, L. Schultz, P. Bauer, G. Saemann-Ischenko, Appl. Phys. Lett. **61**, 3178 (1992).
[13] M. Siegert, J.G. Lisoni, C.H. Lei, A. Eckau, W. Zander, J. Schubert, and Ch. Buchal, Mat. Res. Soc. Proc. **597** 145 (2000).
[14] S.M. Rossnagel, in [10] p.1-20.
[15] K. Wasa and S. Hayakawa, *Handbook of Sputter Deposition Technology: Principles and Applications,* Noyes, Park Ridge, 1992.
[16] G.K. Wehner and G.S. Anderson, in *Handbook of Thin Film Technology*, L.I. Maissel and R. Glang, eds., McGraw-Hill, NY, 1970. p. 3.1 – 3.38
[17] J.A. Thornton, J. Vac. Sci. Technol. **11**, 666 (1974).
[18] U. Poppe, N. Klein, U. Dähne, H. Soltner, C.L. Jia, B. Kabius, K. Urban, A. Lubig, K. Schmidt, S. Hensen, S. Orbach, G. Müller, and H. Piel, J. Appl. Phys. **71**, 5572 (1992).

[19] U. Poppe, R. Hojczyk, C.L. Jia, M.I. Faley, W. Evers, F. Bobba, K. Urban, C. Horstmann, R. Dittmann, U. Breuer, H. Holzbrecher, IEEE Transaction on Appl. Supercond., **9**, 3452 (1999).

[20] M.L Hitchman. and K.F., Jensen, *Chemical Vapor Deposition, Principles and Applications,* Academic Press, 1993.

[21] H.O. Pierson, *Handbook of Chemical Vapor Deposition, Principles, Technology and Applications*, Noyes Publ. Park Ridge, NJ, 2nd ed. 1999.

[22] C.E Morosanu, *Thin Films by Chemical Vapor Deposition,* Elsevier, Amsterdam-Oxford-New York-Tokyo, 1990.

[23] R. Waser, T. Schneller, S. Hoffmann-Eifert, P. Ehrhart, Integrated Ferroelectrics **36**, 3 (2001).

[24] F. Felten, J.-P. Senateur, F. Weiss, R. Madar, and A. Abrutis, Journal de Physique, IV, C5-1079 (1995)

[25] C.M. Foster, in: *Thin Film Ferroelectric Materials and Devices,* R. Ramesh, ed., Kluwer Acad.Publ., Boston, 1997. p. 167 – 197

[26] C.S. Kang, H.-J. Cho, C.S. Hwang, B.T. Lee, K.-H. Lee, H. Horii, W. D. Kim, S. I. Lee , and M.Y. Lee., Jpn. J. Appl. Phys. **36**, 6946 (1997).

[27] P. Ehrhart, F. Fitsilis, S. Regnery, C.L. Jia, H.Z. Jin, R. Waser, F. Schienle, M. Schumacher, and H. Juergensen, Mat. Res. Soc. Proc. **655**, CC9.4.1 (2001).

[28] J.H. Han, H.-K. Ryu, C.-H. Chung, B.-G. Yu, and S.H. Moon, J. Electrochem. Soc. **142**, 3980 (1995).

[29] S.R. Summerfelt in [25], p.1-42.

[30] C.S. Hwang, J. Park, D.S. Hwang, and C.Y. Yoo, J. of the Electrochemical Soc. **148**, G636 (2001) and private communication.

[31] M. Leskelä and M. Ritala, J. de Physique **IV**, C5 –937 (1995).

[32] M. Vehkamäki, T. Hatanpää, T. Hänninen, M. Ritala, and M. Leskelä, Electrochem. and Solid-State Lett. **2**, 504 (1999).

[33] R.W. Schwartz, Chemistry of Materials **9**, 2325 (1997).

[34] T.M. Lu, S.P. Murarka, T.-S. Kuan, and C.H. Ting (eds.), *Low-Dielektric Constant Materials – Synthesis and Applications in Microelectronics,* Mat. Res. Soc. Symp. Proceed. **381** (1995).

[35] D.E. Hooks, T. Fritz, and M.D. Ward, Adv. Mater. **13,** 227 (2001).

[36] G. Gu, G. Parthasarathy, and S.R. Forrest, Appl. Phys. Lett. **74**, 305 (1999).

[37] G. Roberts, (ed.), *Langmuir-Blodgett Films,* Plenum Press New York, 1990.

[38] A. Ulman*, An Introduction to Ultrathin Organic Films, from Langmuir Blodgett to Self-Assembling,* Academic Press, Boston, 1991.

[39] A. Ulman, (ed.), *Organic Thin Films and Surfaces: Directions for the Nineties*; Thin Films, **20**, Acad Press, Boston, 1995.

Lithography

Shinji Okazaki, EUV Lithography Laboratory, ASET Atsugi Research Center, Japan

Jürgen Moers, Department ISG, Research Center Jülich, Germany

Contents

1 Survey   223

2 Optical Lithography   224
2.1 Illumination Methods and Resolution Limits   224
    Resolution Enhancement   226
2.2 Exposure Wavelength and Light Sources   228
2.3 Mask Materials and Optical System   230
2.3.1 Set-up of the Optical Path   230
2.3.2 Reflection optics for short wavelengths   232

3 Extreme Ultraviolet Lithography   232

4 X-Ray Lithography   234

5 Electron Beam Lithography   234
5.1 Electron Beam Direct Write   235
5.2 SCALPEL   236

6 Ion Beam Lithography   237
6.1 Focused Ion Beam   237
6.2 Ion Projection Lithography   238

7 Photoresist   239

8 Alignment of Several Mask Layers   241

9 Nanoimprint Lithography   242

10 Conclusions   244

Lithography

1 Survey

Greek λιθοσ 'a stone' and γραφειν 'to write', the art of printing from stone is one of the most important of the graphic arts ...[1]

This short explanation found in an encyclopaedia is followed by several pages describing the different methods of this art. This explanation is right, but does not cover the meaning which is the topic of this chapter. Here *lithography* addresses the key technology of semiconductor fabrication, the definition of lateral structures.

One of the fundamentals of our society is the storage and handling of information. The need for quicker processing of more and more data is the driving force for the development of microelectronic devices. In 2000 the critical dimensions of most important microelectronic devices, the Metal Oxide Semiconductor Field Effect Transistors (MOSFET), which are the heart of almost every integrated circuit, are about 180 nm; in 2001 this measure will be scaled down to 130 nm. The International Technology Roadmap for Semiconductors [2], which identifies the technological challenges and needs facing the semiconductor industry over the next 15 years, indicates a further shrinkage of this dimension down to 45 nm in the year 2013. This rapid development requires a huge amount of research and the advancement of new technologies.

A brief insight will be given here into the ongoing methods of lithography and the physical limitations. In Sec. 1 a survey is given, in Sec. 2 so-called optical lithography and its progress are discussed in detail. In Sec. 3 extreme ultra violet lithography and in Sec. 4 x-ray lithography are discussed. Another promising candidate for sub-100 nm patterning is lithography with electrons, which is described in Sec. 5, while lithography with ions is addressed in Sec. 6. Sec. 7 deals with the matter of resists and in Sec. 8 an insight in the issue of the alignment of several mask layers is given. As an example of non-lithographic patterning, "nano-imprint Lithography" is introduced in Sec. 9.

The term "lithography" describes the method with which a pattern is defined on a sample. A lithographic system consists of a radiation source, a resist-coated sample and an image control system that regulates which part of the sample is illuminated by the radiation and which is not (Figure 1a). The resist is changed by the illumination (Figure 1b). Depending on the type of resist, the exposed (positive tone process) or the unexposed resist (negative tone process) can be removed selectively by a developing process (Figure 1c). The pattern is now inscribed into the resist and can be transferred to the sample by a subsequent process step, e.g. an etching step. The name *resist* stems from this step, which is not actually part of the lithography process: The resist is resistant to the etching agent, so that the parts of the sample which are still covered by the resist are protected against etching.

Figure 2 shows a survey of the different types of lithography. They differ according to the type of radiation and the control system. However, the starting point of the process is the structure, which has to be transferred to the sample; it is normally given as a

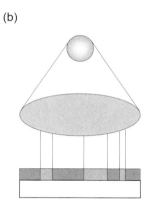

Figure 1: Schematic process flow of lithography with a positive tone resist:
(a) A lithographic system consists of a radiation source, an illumination control system and a resist-coated sample;
(b) Illumination Process: The resist is changed by the radiation,
(c) Development: The illuminated resist can be etched selectively to the unexposed resist.

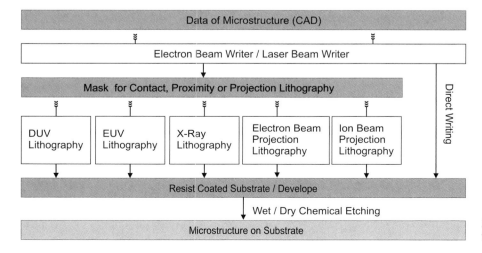

Figure 2: Survey of the different types of lithography.

CAD-file. The use of light as radiation yields so-called *optical lithography*. Depending on the wavelength, a distinction is made between ultraviolet (UV: 365 nm – 436 nm), deep UV (DUV: 157 nm – 250 nm), extreme UV (EUV: 11 nm – 14 nm) and x-ray (< 10 nm). The use of electrons or ions as radiation yields electron or ion lithography, respectively.

As the image control system either a *mask* is used, which yields *contact*, *proximity* or *projection lithography*, or the patterns are directly written into the resist by a focused beam (*laser*, *electron* or *ion beam lithography*).

With the first method, the mask consists of a carrier material, which is transparent for the radiation used, and an absorber layer, which is opaque. Into this opaque layer, the pattern is inscribed. The material depends on the radiation source and is addressed later. The radiation will only illuminate those parts of the sample where the corresponding part of the mask is transparent. Therefore only at those parts the resist is exposed and hence changed.

With *direct writing* a computer controls a focused beam of the used radiation. With deflecting units, the beam is scanned across the sample so that the pattern is written into the resist as if using a pen. Here every pattern has to be written after the other so that it takes a long time to finish a whole wafer. The development process is the same as for mask-based lithography. The relevancy of direct writing lies in its research purpose and in mask fabrication. There the mask itself is the resist-coated sample.

2 Optical Lithography

Optical lithography is the most important type of lithography. Originally the name referred to lithography using light with wavelength in the visible range. Nevertheless, gradually, the wavelength was driven down to 193 nm, which is used in semiconductor production nowadays, and even shorter wavelengths down to the sub-nm rage are under investigation.

The key issue of lithography is the resolution of the system, and hence the size of the smallest feature (minimum feature size: *MFS*) which can be defined on the sample. This *MFS* depends on the illumination method, the illumination wavelength λ, on the materials of the optical system and the resist used. In Sec. 2.1 the different illumination methods and their physical resolution limits are addressed, in Sec. 2.2 the wavelengths and the light sources are discussed, also for wavelengths below 15 nm, while lithography with these wavelengths is discussed in Sec. 3 and 4, and in Sec. 2.3 the materials and the forms of the optical system are dealt with.

2.1 Illumination Methods and Resolution Limits

Figure 3 shows a schematic view of the three different illumination methods *contact, proximity* and *projection lithography*. With all three, the light emitted by a light source passes a condenser optics so that a parallel beam is formed. With contact lithography, mask and sample are pressed together so that the mask is in close contact to the resist (Figure 3a). The resolution is limited by deflection and is expressed by the *MFS* which can be obtained. For contact lithography this is: $MFS = \sqrt{d \cdot \lambda}$, where d is the resist thickness and λ the wavelength. For a resist thickness of 1 µm and a wavelength of about 400 nm, this yields a minimum feature size of 600 nm. The major drawback of this method is that the quality of the mask suffers from contact to the resist, leading to failures in the structure. To avoid this problem, the second method was developed (Figure 3b). With *proximity lithography* there is a defined proximity gap g between sample and mask, so there is no deterioration of the mask. The drawback is the poorer resolution limit, which is proportional to $\sqrt{(d+g) \cdot \lambda}$. With same figures as above and a proximity gap of 10 µm, the *MFS* is 2 µm.

The method used today in industrial production is so-called *projection lithography* (Figure 3c). Here not the shadow of the mask is transferred to the sample as with the two other methods, but a picture of the mask is projected onto the sample. Therefore after passing the mask, the light is bundled by an optical system. The mask is not in contact with the sample, so there is no deterioration as in contact lithography, but the resolution is better than in proximity lithography. Furthermore it is possible to reduce the picture so the patterns on the mask are allowed to be bigger than the patterns on the sample. This is advantageous for mask fabrication: Errors are also reduced. If it is possible to obtain masks with an accuracy of 100 nm, then the error for a structure of 500 nm to be transferred onto a sample is 20 %, if it is transferred one by one. If the picture is reduced 4 times, then for a 500 nm feature on the sample, the feature on the mask has to be 2 µm; therefore the mask error is only 5 %. Because of the reduction, the wafer is not exposed in one exposure, but in several . This is done by so-called steppers, in which the wafer is adjusted under the mask by an x-y-table. The stepper moves the wafer from one exposure position to the next, while the mask is not moved.

In projection lithography the limiting factor to the MFS is diffraction. Consider a slit width b which is illuminated by a monochromatic plane wave. What will the intensity distribution look like on a screen at a distance l behind the slit? Therefore consider two Huygens waves, one from the lower rim of the slit, one from the middle. There will be an optical path difference between these two Huygens waves, depending on the angle of propagation Θ. The magnitude of the path difference (*PD*) is:

(a)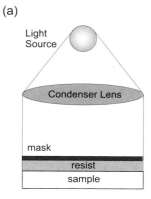

$$PD = \frac{b}{2}\sin(\Theta) \qquad (1)$$

The two Huygens waves will interfere destructively if the *PD* is an odd multiple of the half wavelength:

(b)

$$\frac{b}{2}\sin(\Theta_{\min}) = (2m+1) \cdot \frac{\lambda}{2} \qquad \text{with } m = 0, \pm 1, \pm 2, \ldots \qquad (2)$$

Under this condition, the Huygens waves from the lower part of the slit will interfere destructively with the ones from the upper part. At the angle $\Theta_{\min}$ there is a minimum of intensity.

The Huygens waves do interfere constructively resulting in a maximum of intensity when:

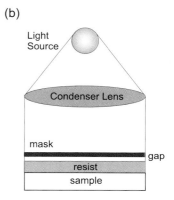

$$\frac{b}{2}\sin(\Theta_{\max}) = m\lambda \qquad \text{with } m = 0, \pm 1, \pm 2, \ldots \text{ holds.} \qquad (3)$$

In lithography the diffraction patterns of several structures are superimposed so the question leading to the MFS is the question of when two structures can be resolved. The first approach is given by the Rayleigh criterion [3]. When light coming from a point source passes an optical system a blurred diffraction pattern – the Airy disc – occurs. The Rayleigh criterion says that two ideal point sources (e.g. stars) can be resolved when the intensity maximum of the one Airy disc is in the first minimum of the other, so *MFS* is given as:

(c)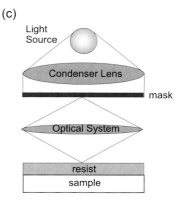

$$MFS = 0.61 \cdot \frac{\lambda}{NA} \qquad (4)$$

where *NA* is the numerical aperture of the optical system. Nevertheless the Rayleigh criterion is just a first approach to the *MFS* in microlithography. The mask patterns are not independent (i.e. incoherent) ideal point sources, on the contrary they have a finite width and the light is partially coherent. Nevertheless, the form of the criterion gives the right dependences. If the wavelength is decreased by 10 % or the *NA* is increased by 10 %, the *MFS* is improved by 10 %. Furthermore, it was derived only by properties of the optics although the photoresist also affects the MFS. Therefore more generally, the criterion is written as:

Figure 3: Lithography methods:
(a) contact,
(b) proximity and
(c) projection lithography.

$$MFS = k_1 \cdot \frac{\lambda}{NA} \qquad (5)$$

where k_1 is a constant (typically 0.5 – 0.9), which accounts for non-ideal behaviour of the equipment (e.g. lens errors) and the influences which do not come from the optics (resist, resist processing, shape of the imaged structures,...). Therefore k_1 is called the technology constant.

As a comparison, for a technology constant of 0.7 and a numerical aperture of 0.7, which are commonly used figures, the *MFS* is in the order of the wavelength λ. So it is better by about a factor of 0.66 than the *MFS* of contact printing.

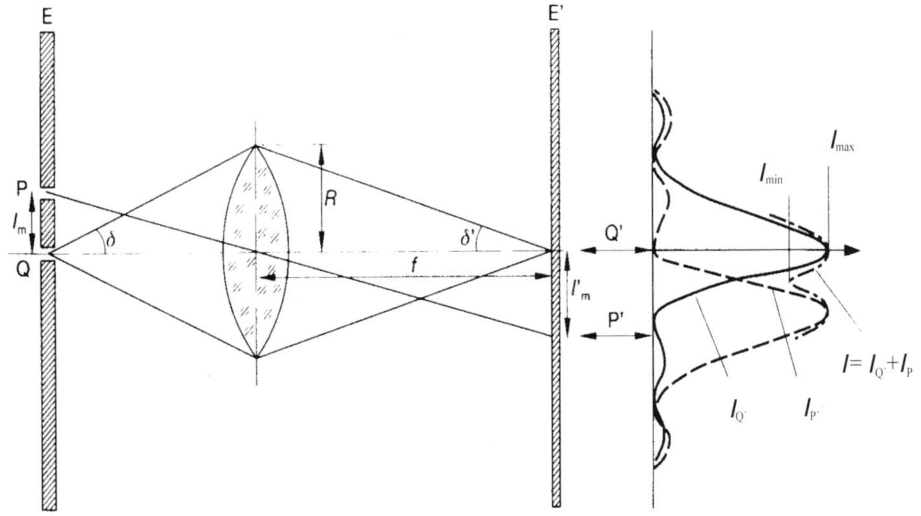

Figure 4: Intensity pattern of two features *P* and *Q* at projection lithography: The intensity distribution at the sample is broadened due to deflection [4].

Figure 4 clarifies the connection between mask, diffraction and intensity distribution in the image plane. Due to diffraction two sharp features, P and Q, on the mask give rise to an overall intensity distribution on the sample. To resolve these two features the intensity distribution has to have a minimum between the two main maximums. It is useful to define the so called modulation transfer function (*MTF*) as:

$$MTF = \frac{I_{\max} - I_{\min}}{I_{\max} + I_{\min}} \qquad (6)$$

The higher the value – the higher the difference between the maximum and minimum intensity – the better the contrast between exposed and unexposed areas, the better is the resolution of the equipment. It should be noted that the *MTF* is only derived by properties of the optical system. It is a measure of the capabilities of the lithographic tool in printing structures.

Resolution Enhancement

For a given tool and technology, the resolution is a given figure. There have been several attempts to improve this essential figure without any major changes to the tool (i.e. no other wavelength or *NA*). Figure 5 shows the possible places where changes can be made in the optical path to improve resolution. Two of these attempts are discussed in the next subsections, the phase shift techniques and off-axis illumination. In off-axis illumination, the effective light source is tailored, while in phase shifting techniques the wave edge of the illuminating light is tailored by the mask.

Phase Shifting Techniques

A huge improvement in resolution and/or in depth of focus can be obtained by improving the contrast by tailoring the phase differences of the wavefront. The phase difference is changed by varying the optical path length of the light passing through the vicinal structures, leading to constructive and destructive interference, which improves contrast (i.e. increase $I_{\max}$ or decrease $I_{\min}$). To understand the method the approach proposed by Levenson in 1982 [26] is discussed.

Consider a lines and spaces structure with pitch 2*p*. Figure 6 shows at the left hand side the amplitudes and intensities in the case of a conventional mask. At the mask itself, the normalized amplitudes are of rectangular shape (either +1 or 0) and give a proper image, but the light is diffracted into the dark regions and so the amplitude distribution is broadened as shown in Figure 6. The intensity of the light is the square of the sum of the amplitudes, so there is an intensity distribution with a significant $I_{\min}$ between the maximum intensities.

Now consider the case when the amplitudes of the light passing through the vicinal structures are out of phase by π (i.e. +1, 0 and –1) (Figure 6, right hand side). Again the light is diffracted into the dark areas, but now the light interferes destructively: There is a point where the sum of the amplitudes is zero, so the intensity is zero, too. These

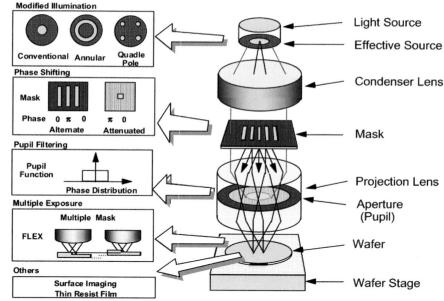

Figure 5: Survey of the resolution enhancement techniques.

so-called *Levenson* or *alternating phase shift masks* (PSM) can improve the resolution by 40 %. Unfortunately, this improvement is pattern-dependent; for a single structure there is no neighboring structure, so there is no light to interfere with. Even if there are structures which are not in a regular arrangement, there is no defined phase shift between these structures which could yield an improvement in the resolution of all structures.

The phase shift can be obtained by an additional transparent layer on the mask. If it has the refractive index n and thickness d, the phase shift is $\Phi = (n-1)2\pi d/\lambda$. So a shift of π is obtained, when the condition $d = \lambda/[2(n-1)]$ holds. On the other hand, it is also possible to recess the mask material so that the right optical path difference is obtained. But the etch depth can be controlled by the time only, and not, as in etching away an additional layer, by the thickness of the layer itself.

To deal with the drawbacks of alternating PSM, several other methods have been developed, which are described next. In rim-PSM, the whole mask is covered by a phase-shifter material and then with the resist. After development, the phase shifter is etched anisotropically and the masking layer is etched isotropically. By this a undercut under the phase shifter occurs at the rim of every structure. This also yields a resolution improvement, but not as much as with alternating PSM, although it is therefore not limited to certain structures.

Another way to engineer the optical path lengths is attenuated PSM. Here the opaque layer is replaced by a partially transparent (about 10 %) layer. The light passing these semi-opaque areas is not strong enough to expose the resist, by it can interfere with the light passing the transparent areas. So an improvement of resolution can be

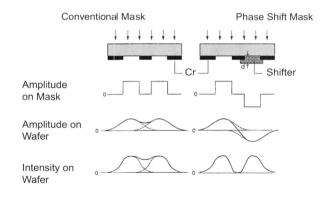

Figure 6: Comparison of the light amplitudes and intensities at the mask and on the wafer for a conventional and a phase shift mask. Note that the intensity on the wafer between the two features is zero for the phase shift mask [26].

Technology and Analysis

obtained. The advantage of attenuated PSM is the easier mask processing. There is no extra layer as in alternating or rim PSM. The technology to process the semi-transparent layer is in principle the same as with a *normal* opaque layer.

PSM techniques were introduced in 1982, but only from 1999 they have been used for industrial production. An example which illustrates the impact of PSM methods is the results published by INTEL on the International Electron Device Meeting (IEDM) in 2000. A 248 nm phase-shiftmask lithography tool was used to produce a MOSFET with a 30 nm gate length [36].

Off-Axis Illumination

To improve resolution without decreasing the wavelength or increasing *NA*, so-called off-axis illumination was applied. The method was already known as a contrast-enhancing technique for optical microscopes. With off-axis illumination, the light beam is directed from the mask towards the edge of the projection lens, and not, as in on-axis illumination, towards the center. In normal illumination with partially coherent light, there always is part of the light which is off-axis, but in the context here with off-axis illumination there is no on-axis component.

To understand the mode of operation of off-axis illumination, consider a line-and-spaces structure with pitch *p*. The incident light will be diffracted into a set of beams, of which only the undiffracted beam, the zero-order beam, travels in the direction of the incident light. The 1st order beam travels under the angle $|\theta_1| = \arcsin(\lambda/p)$. If *p* is too small, then $|\theta_{\pm 1}|$ is bigger than the acceptance angle α of the projection optics, then only the zero-order beam is projected to the sample (Figure 7a). But this does not carry any information of the pattern, and hence the pattern cannot be transferred onto the sample. At least the zero- and the 1st order beam have to be in the range of the aperture angle. If the incident light hits the mask under an angle $\Theta_0 < \alpha$ the undiffracted beam enters the projection lens at the edge, and the 1st order beam is still collected by the lens, and therefore a pattern transfer is still possible. The angle of incidence Θ_0 can be realized by inserting an aperture in the optical path between condenser and mask (Figure 7b).

Although the higher resolution is an advantage of off-axis illumination, the impact on the depth of focus (DOF) is of even greater value. In on-axis illumination, the beams of different deflection orders have to travel in different ways so they are phase-shifted to each other, which results in a lack of focus. In off-axis illumination, the zero order and 1st order beam reaches the projection lens at the same distance from the center, which means that their optical path length is the same. So the relative phase difference between these beams is zero, which increases the DOF dramatically.

Off-axis illumination is facilitated by an aperture (Figure 8) which is located in front of the condenser lens. It depends on the apertures shape which structures are improved. If there is an aperture as in Figure 8a, only the structures perpendicular to the arrangement of the apertures will be improved. The aperture shown in Figure 8b yields an improvement of structures which are adjusted to *good* angles – up/down or left/right direction. This is sufficient because in normal cases, the features are in a *good* arrangement. The aperture in Figure 8c even decreases this problem, but here the improvement in DOF is less.

When the resolution in principle has to be improved, then according to the Rayleigh criterion either the wavelength λ or the technology parameter k_1 have to be decreased, or the numerical aperture *NA* has to be increased.

Increasing *NA* means physically bigger lenses. Here the problem arises that it is difficult to produce huge lenses with the required quality; on the other hand the available materials also limit the physical size of the lenses. So there are still two possibilities of increasing the resolution: smaller λ and smaller k_1.

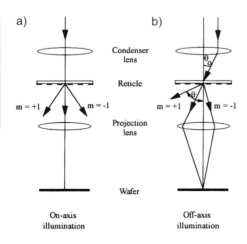

Figure 7:
(a) Optical path and deflection orders of on-axis and
(b) off-axis illumination. Note that with the same wavelength and structure size, the off-axis illumination allows the 1st order beam to pass the optical system [3]. A good description of off-axis illumination is also found in [6].

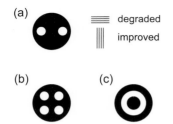

Figure 8: Apertures facilitating off-axis illumination.
(a) improvement of resolution perpendicular to the holes in the aperture,
(b) improvement in up-down and left-right direction, but not in diagonal direction and
(c) improvement in all directions [3].

2.2 Exposure Wavelength and Light Sources

Progress in optical lithography in the last few years was achieved by decreasing the exposure wavelength λ from 436 nm to 193 nm nowadays; research is in progress to push this boundary down to a few nm. In this section the different wavelengths, the methods of obtaining that light and the implications for the process are discussed. In Tab. 1 the wavelengths, the sources and the names of the wavelength ranges are given.

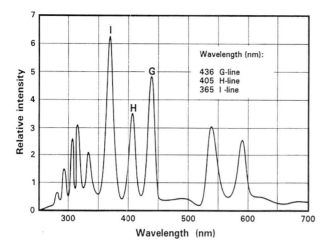

Figure 9: Spectrum of a high pressure Hg-lamp [46].

The first light used was the light emitted from Hg-arc lamps. It provides three lines, the G-line (436 nm), the H-line (405 nm) and the I-line (365 nm). A spectrum of a high-pressure Hg-lamp is shown in Figure 9. With typical k_1 and *NA* resolutions of ~400 nm were achieved. A further decrease of λ to 250 nm was obtained by a mixture of Hg and Xe, improving resolution to 300 nm, but the intensity at this wavelength is low.

To solve the intensity problem, at 250 nm a new light source occurs: the excimer laser. The word stems from *exited* and *dimer* and describes a molecule which only exists in an exited state. The gas mixture in an excimer laser is either KrF, ArF or F_2, resulting in the so-called deep UV (DUV) wavelengths of 248 nm, 193 nm and the vacuum UV (VUV) wavelength 157 nm, respectively. The excimer molecule consists of a noble gas and a halogen atom; in the ground state they cannot react, but if one or both are in an excited state, an exotic molecule can be formed. These dimer molecules decay into the ground state of both constituents with the emission of DUV light. The spontaneous decay time is long (i.e. nano- to microseconds), so inversion can be achieved by pumping the laser gas electrically.

In production DUV lithography with a resolution of 180 nm is used, while the boundary in research is being pushed down even further [36].

Wavelength [nm]	Source	Range
436	Hg arc lamp	G-line
405	Hg arc lamp	H-line
365	Hg arc lamp	I-line
248	Hg/Xe arc lamp; KrF excimer laser	Deep UV (DUV)
193	ArF excimer laser	DUV
157	F_2 laser	Vacuum UV (VUV)
~10	Laser-produced plasma sources	Extreme UV (EUV)
~1	X-ray tube; synchrotron	X-ray

Table 1: Illumination wavelengths, light sources and light ranges

Between 157 nm and ~13 nm is a huge gap where no usable wavelength exists. This is because all materials absorb light of that wavelength, so no masks, lenses and mirrors can be made. Nevertheless, when wavelengths in the range of 13 nm are used, it is possible to set up an optical path to do projection lithography. This range is called extreme ultraviolet (EUV). Shrinking the wavelength into the range of 1 nm leads to x-ray lithography. Here it is not possible to perform projection lithography because there is no material to set up an optical system (see Sec. 4).

II Technology and Analysis

The methods used to generate this light are the same for both ranges. All of them have to meet certain requirements such as being efficient enough at the desired wavelength and have low debris production (or feature a mechanism to avoid the contamination of tool and sample).

Firstly an x-ray tube can be used. In an x-ray tube, a metal anode is radiated with high energy electrons so that the characteristic x-ray radiation of the metal is emitted. The wavelength can be adjusted by the right choice of metal and electron energy. Unfortunately there are two drawbacks. On one hand, the intensity of these sources is very low leading to exposure times of several hours. This may match the requirements for research purposes, but surely not the ones of industrial production. On the other hand, the anode metal is sputtered and contaminates the exposure tool and the sample. This leads to unwanted loss of intensity, which decreases the lifetime of the tool and destroys the sample. In summary, x-ray tubes do not meet the requirements for EUV or x-ray lithography.

To a second group of methods the gas discharge sources belong. There are different types under investigation. While they differ in the concepts of plasma ignition, geometry of the source and discharge parameters, they all have in common that a plasma is generated by a fast discharge of electrical energy and emits thermal radiation in the EUV range. In common, gas discharge sources still suffer from debris. Nevertheless, in recent years some progress has been made with these issues [29], [30].

Another kind of sources is the laser-produced plasma sources (LPP). A pulsed laser focused on a target consisting of oxygen, fluorine, neon or xenon excites a plasma discharge in the desired wavelength region. The target is either solid (cryogenic), liquid or gaseous. The method is almost the same for all cases: The target is injected into a vacuum. Onto the jet a laser is focused, whereby the plasma is stimulated. The drawback is that the debris problem is not solved at all. For plasma excitation the gas has to have a certain pressure, therefore the distance between the plasma and the injector is too short so that the injector is sputtered. For liquid xenon targets [7] or a double gas target [8], which are under development, the debris problem is reduced. In a double gas target, the Xe-injector is surrounded by a second injector, from which a low-Z gas (He) is injected. The He encloses the Xe so that the distance between the injector and the plasma can be larger; therefore there is less debris and hence less contamination. Figure 10a shows a cross section of the injector, Figure 10b a space-integrated spectrum from a double-stream Xe/He target and Figure 10c the same for an ordinary Xe target.

The only light source which is luminous enough, at least in the 1 nm-wavelength range, and where no debris is produced is a synchrotron source. Electrons are accelerated onto a circular path and hence emit bremsstrahlung. But the synchrotron source is very expensive and therefore the cost of expanding lithography by one tool is very high. Also the shielding of the synchrotron is a drawback for these sources.

The question of which source will be used as an EUV or x-ray source is still not decided. While synchrotrons meet the optical requirements for wavelengths in the 1 nm-range, they are very expensive both with respect to purchasing costs and the cleanroom space they require. X-ray tubes are suitable only for research purposes, while LPP offers a sufficient light quality for development, but not for production. Discharge sources still suffer from debris, but they are cheaper to purchase cost of ownership. Good comparisons between the different sources are given in [31], [32].

2.3 Mask Materials and Optical System

2.3.1 Set-up of the Optical Path

It is possible to obtain light with wavelengths from <1 nm to 500 nm. But besides the light source supplying light of a certain wavelength, an optical system has to be set up to do lithography. Masks, mirrors and lenses are needed which are suitable for the used wavelength. This is expensive.

There are two concepts for setting up the optical system. On the one hand, it can be designed with refractive elements (e.g. lenses: refraction optics). In a refraction optics system, the light passes the mask and lenses. Therefore these components have to be made of transparent materials. But the transmission through the materials depends on the wavelength. While soda-lime glasses and borosilicate glasses were used in earlier IC fabrication, these glasses become too less transparent as the wavelength approaches 250 nm. Furthermore, these materials have a high thermal expansion coefficient ($\approx 93 \times 10^{-7}\,K^{-1}$), which leads to a significant expansion of the masks when it is heated by the absorbed light. Fused silica, used instead, is transparent enough for even 193 nm

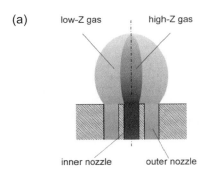
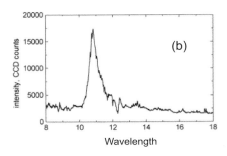
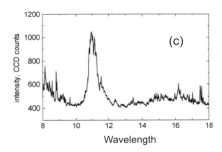

Figure 10:
(a) Double nozzle injector of a laser produced plasma source; through the inner nozzle the high-Z gas and through the outer nozzle the low-Z gas is injected; both gases form the target. Spectral distribution spectrum for the spectral image of
(b) a double stream Xe/He target and
(c) an ordinary Xe target [8].

(a)
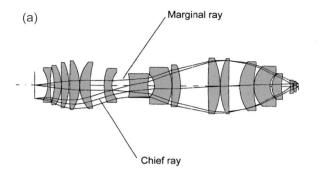

(c)
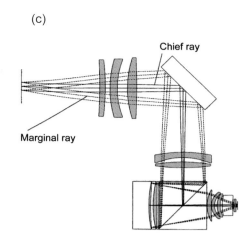

(b)
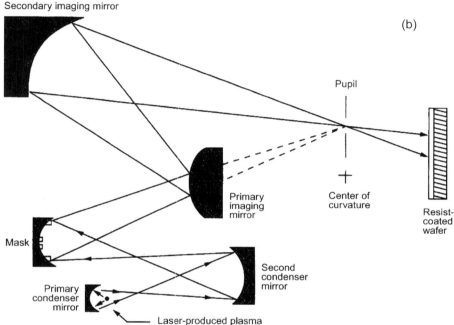

Figure 11: Setup of optical paths:
(a) refraction optics [3],
(b) reflection optics [37],
(c) catadioptric [41].

(Figure 12) and has a thermal expansion coefficient of about 1/20 of the other glasses. Nevertheless, even with the low coefficient of 5×10^{-7} K^{-1} a temperature change of 0.1 K results in a displacement error of 5 nm, which is more than 10 % of the allowed tolerances for 0.25 µm technology. Below the 193 nm boundary other materials have to be used. One candidate under investigation is CaF$_2$ and its alloys [9]. These materials also have to fit another boundary condition: The defect density of the material has to be as low as possible in the mask as well as in the lenses to avoid projection errors. Due to the high *NA* of the optics, this is a rather strong condition.

On the other hand, an optical path can be designed for the use of mirrors (reflection optics). With reflection optics, the light is bundled by a concave mirror, and even the mask is a mirror, whose reflectivity is spatially altered according to the desired structure. Regions with high reflectivity correspond to exposed regions on the wafer, regions with low reflectivity correspond to the unexposed regions. In catadioptric systems some parts of the optical path are made up of reflection and some of refraction optics elements. The advantage of reflection optics is the higher spectral bandwidth, so achromatic lens error can be avoided. On the other hand, such a system normally requires more than one optical axis, therefore it is difficult to align the elements.

It is also possible to combine both parts (catadioptric system). The design depends strongly on the wavelength in use (Figure 11).

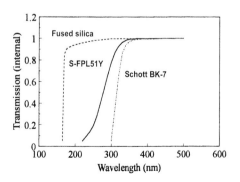

Figure 12: Transmission of different silica glasses. Below 170 nm normal quartz-based glasses become unsuitable for lithography [3], [42].

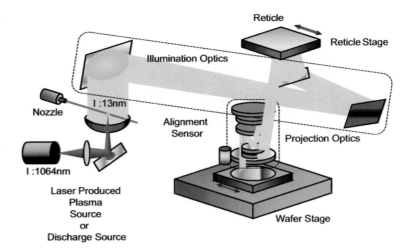

Figure 13: A schematic view of the EUV exposure system.

2.3.2 Reflection optics for short wavelengths

In the range from 10 to 15 nm (EUV) there is now material which is transparent enough to permit refraction optics, but there are no materials which can be used as a mirror. Therefore the mirrors are made up of a so-called *Bragg reflector*. Such a reflector is made of a layer stack, in which a high-refractive index material and a low-refractive index material are deposited onto each other in alternating order. The optical thickness of each layer is $\lambda/4$. At every interface some part of the incident beam will be transmitted and some part will be reflected. The transmitted beam travels to the next interface and again is transmitted or reflected. This second reflected beam will reach the first interface (again partly transmitted and reflected), and the now transmitted beam will interfere with the first reflected beam. The optical path difference is $\lambda/2$, but one of the reflected beams is reflected at an interface coming from a material with a higher refractive index and so there is a phase shift of π, i.e. $\lambda/2$. The whole path difference is therefore λ and the two beams interfere constructively. The optical path length is dependent on the angle of incidence, so the reflectivity also depends on that angle. This is the reason, why the optical path has to be designed for small angles of incidence as well.

3 Extreme Ultraviolet Lithography

EUV lithography is one of the promising candidates for next generation lithography (NGL). Efforts are being made in the USA (EUV-LLC/VNL) was formed in 1997 [33], in Japan there is an ASET program [34] and in Europe the EUCLIDES project has been implemented [35]. In EUV a wavelength of 13 – 14 nm is used. The resolution capability is determined by the Rayleigh criterion as in optical lithography. At such a short wavelength region, no refractive optics can be used, but only reflective mirror optics. Figure 13 shows the schematic view of the EUV exposure system.

As shown in the previous section, only a Bragg reflector (multilayer mirror) can be used in this wavelength region. For the 13 – 14 nm region, the mirrors are made of more than 40 repetitions of Mo/Si and show reflectivity of up to 70 %. Because of the low reflectivity (at least in comparison with optical mirrors) of these reflectors, the number

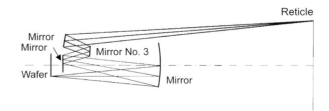

Figure 14: An example of the EUV 4 reflective mirror optical path.

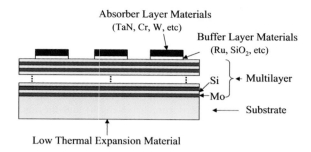

Figure 15: Schematic view of the multilayer mask.

of mirrors has to be kept small in the whole optical path. In the conventional projection optical lithography system, more than 25 lenses (namely more than 50 surfaces) are used for composing an aberration free projection optics. Almost all the surfaces of these lenses have a spherical shape. On the contrary, in the optics of EUV lithography system only 4 – 6 aspherical mirrors are used. The reason for the use of aspherical surfaces is to impose various aberration correction functions onto one mirror surface. Figure 14 shows an example of an EUV 4-mirror reflective mirror optical path. The biggest issue of the aspherical mirrors is the accuracy of the aspherical surface. Three major errors are minimized for the mirror: namely the figure errors, the mid-spatial frequency roughness and the surface roughness . Each error must be less than 0.20 nm, 0.15 nm, and 0.10 nm in rms respectively. The metrology of the figure error is very challenging, various visible and exposure metrology systems are being intensively developed.

In the case of reflection optics, the numerical aperture of the system cannot be so high as that of refraction optics because the incident light reflected by the mirror goes to the same direction as that from which the light ray comes. To avoid interference between the incident light and the reflected light, the usable solid angle is restricted. The maximum NA of the optics is 0.3 or smaller. However, even though such a small numerical aperture is used , as the exposure wavelength is very short, such as 13 – 14 nm, the resolution capability is expected to be very high. Currently it is thought that EUV lithography can be applied to 35 nm technologies and below.

The EUV mask must also be a reflective mirror mask. Figure 15 shows the schematic view of the mask structure. The very low thermal expansion material is used for the substrate. Mo-Si multilayer is deposited on the substrate and absorber patterns are made on the multilayer [12]. As an absorber layer, a material which shows very high EUV light absorption material should be adopted. Heavy metals are therefore good candidate materials. However, as the absorber patterns should be inspected by DUV light, DUV contrast should also be considered. Currently, Cr and TaN are thought to be good candidates. To delineate absorber materials, a buffer layer must be inserted between the absorber layer and multilayer to avoid damage to the multilayer during absorber patterning. For this purpose, SiO_2 and Ru are under investigation.

Defect reduction in the deposition process of the multilayer is one of the most critical issues in the multilayer mask process. The ion beam sputtering method is widely adopted for this purpose because the defect generation during the ion beam sputtering is lower than that of magnetron sputtering. However, the reflectivity of the multilayer deposited by the ion beam sputtering system is usually slightly lower than that deposited by the magnetron sputtering system. In the case of the multilayer of the optical mirror, as the reflectivity is more important than defect density the magnetron sputtering system is widely used.

Defect detection is a very important issue for the evaluation of the multilayer mask. A very small (2 – 3 nm in height) bump in the multilayer introduces a phase defect. A special defect inspection system is required for the inspection of the multilayer mask.

4 X-Ray Lithography

Decreasing the wavelength even further into the x-ray range yields so-called x-ray lithography. For these short wavelengths it is not possible to set up an optical path neither in reflection optics nor in refraction optics. On one hand, there is no material which is transparent enough to make lenses or masks from, and, on the other hand, it is not possible to make Bragg-reflectors. The individual layers in the layer stack have to have a thickness of $\lambda/4$, which corresponds to a layer thickness of ~0.3 nm. This is in the range of the thickness of one monolayer and is not achievable..

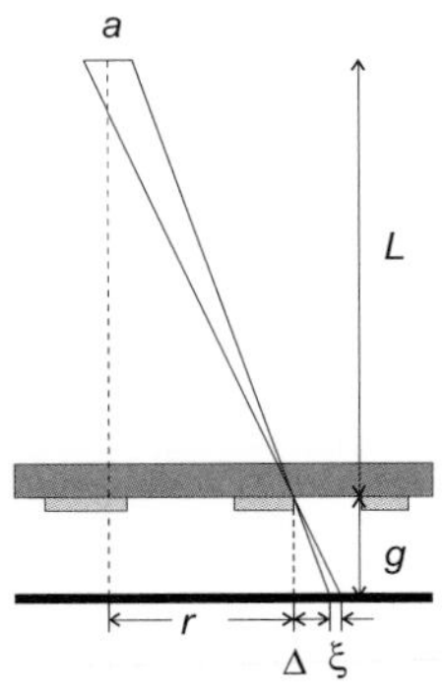

Projection x-ray-lithography is therefore not possible, but proximity x-ray lithography (PXL) is possible. The advantages are the high resolution limit ($\sim \sqrt{\lambda \cdot (g+d)}$), which is about 30 nm for 1 nm exposure wavelength) and the insensitivity to organic contamination. These contaminations (as all low atomic number materials) do not absorb the x-rays, and hence are not printed onto the sample.

But there are some limitations. Consider a source with diameter a of 1 mm at distance L of 1 m towards the mask and a proximity gap g of 10 µm. Then there is the so-called *penumbral blur* $\xi = a \cdot g / L \sim 10\,\text{nm}$, which limits the resolution (Figure 16). Furthermore, the pattern is not transferred correctly to the sample. Even if a point source is used, there is a displacement Δ of $\Delta = r \cdot g / L$, where r is the radial position on the sample (Figure 16). This error can be eliminated if it is taken into account when the mask pattern is generated.

Nevertheless, if synchrotron radiation is used, a high intense beam of parallel light is available so these errors do not occur. This parallel beam has another advantage: Due to the small deflection the exposure shows a high depth of focus of several µm, facilitating exposures of textured substrates or of thick resists (Figure 17).

The problem for PXL is the masks. Since there is no material which is as transparent to x-ray as quartz to DUV, the carrier layer has to be thin (1 – 2 µm). On the other hand, there is also no material which is as opaque to x-ray as chromium to DUV, so the masking layer has to be thick enough (300 – 500 nm). A carrier layer of 1 µm SiC only has a transparency of 57 %, while a masking layer of Au still lets 14 % of the light pass. The absorbed light will heat the mask so that it expands, which leads to another uncertainty in the pattern transfer. Furthermore, PXL is a non-reduction printing method, so the features on the mask are of the same size as on the sample. This makes the production of the masks very complicated when the target ist the sub-100 nm range.

The mask production sequence is as follows: On a silicon wafer, a thin membrane layer is deposited (e.g. SiC, Si_3N_4). Onto this layer, a chromium etch stop layer and the masking layer of 300 – 500 nm of a high-atomic number material is evaporated (e.g. Au, Ta). Then the mask is coated with an e-beam resist and exposed in an e-beam direct-write system. The resist is used to etch the masking layer with an etch stop on the chromium so the membrane is not hurt.

The commonly used DUV resists show good process aptitude.

Figure 16: Penumbral blur ξ and displacement error Δ for proximity x-ray lithography. L is the distance from source to mask, g is the proximity gap and a is the lateral diameter of the source [11].

5 Electron Beam Lithography

Another way to achieve sub-100 nm resolution is to change the type of radiation. In the foregoing sections only lithography methods were discussed which use light as the illuminating radiation. It is also possible to achieve the illumination with charged particles such as electrons. Electrons can be easily generated, either by thermionic or field effect emission, and focused to beams with a spot size of a few nanometers. This electron beam can be used to write the desired structure directly into the resist (Sec. 5.1) or using appropriate electron optics to perform electron projection lithography (Sec. 5.2). The chemical reactions in the resists are the same as in optical resists, only the reactive species has to be tuned to the electrons (nevertheless some optical resists can be used as electron beam resists, too).

5.1 Electron Beam Direct Write

In electron beam direct write electrons are formed to a beam and are accelerated to a determined position on the wafer surface, where the resist has to be exposed to form the pattern. An electron beam system consists of the electron source or electron gun, the electron-optical system (the electron column), a mechanical wafer stage and a controller system. A schematic view of an electron beam lithography tool is given in Figure 18.

The two types of electron guns which are commonly used are thermionic sources, on the one hand, and field emission sources, on the other hand. In thermionic sources the electrons are emitted by heating the source material, such as tungsten (W) or lantanum hexaboride (LaB_6). While LaB_6 offers a higher brightness (10^5(A/cm^2)/steradian)) and a longer lifetime (~1000 h) than W (10^4(A/cm^2)/steradian; ~100 h), W has the advantage that vacuum requirements are not as high as for LaB_6. Nevertheless, LaB_6 has become the standard source for thermionic e-beam sources.

In field emission sources the electrons are extracted from a sharp tip by a high electric field. Though these sources have a high brightness (10^7(A/cm^2)/steradian)), they are unstable and require a ultrahigh vacuum. Therefore they have not been widely adopted in electron beam lithography systems.

In the electron column the extracted electrons are formed to a beam with a definite diameter or shape. Therefore different electron-optical elements as focusing and defocusing lenses and apertures are employed. Further parts of the column are a beam blank to switch the beam on and off and a beam deflection system, with which the beam is positioned on the wafer.

Since the deflection system can only address a field of 400 – 800 μm (depending on spot size and tool), it is necessary to move the sample under the beam from one exposure field to the next by a mechanical wafer stage. The position of the stage is measured by an interferometer, so it is possible to adjust the beam with an accuracy of ~5 nm.

The whole system has to be under vacuum to enable the electron beam to be formed and has to be isolated from vibrations. Further requirements are low electromagnetic stray field, because this would hamper the positioning of the beam.

The pattern, which is given as a CAD file, is translated into movements of the electron beam/wafer stage by a computer. During an illumination, the tilt of the sample is measured continuously and the focus is adjusted. There are two exposure schemes: In the first one, the raster scan scheme, the deflection system and the wafer stage address every point of the sample, but the beam is switched on and off according to the structure. In the second scheme, the vector scan scheme, only the points which have to be illuminated are addressed. Hence the vector scan scheme is less time-consuming than the raster scan scheme.

The time needed for the illumination of a whole wafer depends on the pattern, but because the electron beam direct write is a serial method, it is time-consuming and not suitable for the industrial mass production of microelectronic circuits. Nevertheless, because the resolution is pushed to a few nanometers, it has a high impact on research activities and is the method of choice for defining the pattern on the masks used for optical lithography.

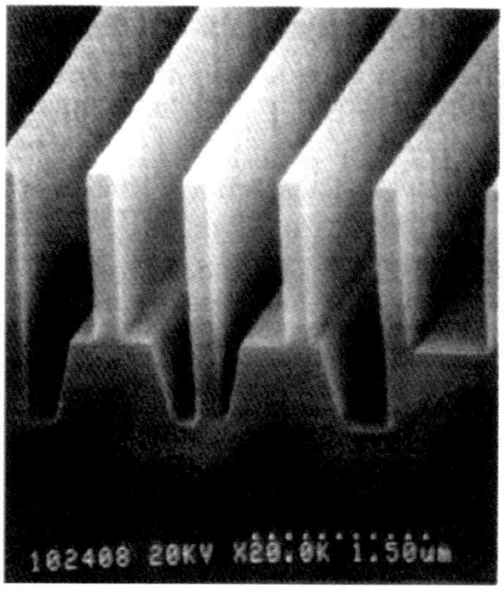

Figure 17: Resist structures on sample with high topography visualizing the high depth of focus of x-ray proximity lithography [10].

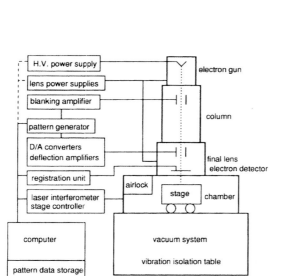

Figure 18: Schematic view of an electron beam lithography tool [16].

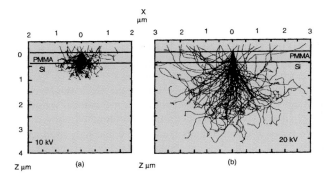

Figure 19: Monte Carlo simulations of the electron path in the resist and in silicon for 10 kV (left hand side) and 20 kV (right hand side) acceleration voltage. It is seen that there is a significant intensity of backscattered electrons in the vicinity of the written pattern [45].

The resolution is not limited by the deflection of the 1–100 keV electrons, but by the beam spot size (~5 nm achievable) and by the backscattering of electrons. Figure 19 shows Monte Carlo simulations of the electron paths, which have an energy of a) 10 keV and b) 20 keV. The electrons lose their energy slowly and a significant fraction of them are backscattered to the surface, where they expose the resist even at positions a few microns apart from the location of incidence. This so-called proximity effect leads to the fact that the effective exposure dose, with which the resist is exposed at one location, depends on the shape of the pattern in the vicinity and has to be taken into account when the pattern are developed (*proximity correction*).

5.2 SCALPEL

The drawback of electron beam direct write is the serial character of the method. In mass production, where throughput is concerned, exposure times of several hours are not acceptable. Though there are electron optics which could enable projection lithography analogously to optical projection lithography, this method suffers from the huge penetration depth of electrons. The masking layers have to be thick to stop a significant part of the electrons.

One method of circumventing this problem is the SCALPEL method (scattering with angular limitation in projection electron beam lithography). In SCALPEL a broad beam of electrons, 2 to 3 mm in diameter, is scanned across a mask consisting of a silicon-nitride membrane layer (~100 – 150 nm), on which a patterned scattering layer (25 to 50 nm of gold or tungsten) is situated (Figure 20a). The electrons, which only strike the membrane layer, will pass this layer mostly unscattered, while the electrons, which strike the scattering layer, will be distracted strongly from their path. The unscattered electrons are focused through an aperture and projected onto the wafer, while the scattered electrons will be blocked. So a high contrast image can be achieved.

As a projection lithography method, SCALPEL offers the advantage of image reduction thus making mask fabrication easier. The mask itself consists of silicon struts, between which the membrane layer is clamped (Figure 20b). The width of the membrane corresponds to the diameter of the electron beam, while it is a few cm in length. By means of the projection optics behind the aperture the electrons coming from two different membrane areas separated by a silicon strut can be stitched together at the wafer, so circuits of 2 cm times 3 cm can be exposed.

Figure 20:
(a) Electron path through a SCALPEL tool. A parallel beam of electrons passes through the mask; a scattering layer in which the pattern is inscribed scatters the electrons, so that they are not focused through an aperture by the electron optical system; only the unscattered electrons will pass the aperture. These electrons are projected onto the sample [16], [43].
(b) Top view of a mask and
(c) cross-sectional view of the mask. The masks are strips and separated by silicon struts. The masks are illuminated in series and the pictures of the masks are projected onto the adjacent sample.

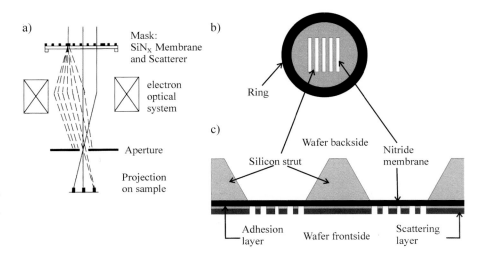

There are still other approaches to Electron Projection Lithography, from which the PREVAIL (Projection Reduction Exposure with Variable Axis Immersion Lenses) method has to be mentioned. In PREVAIL the optical axis of the electron system is shifted, so aberrations are reduced, enabling larger scan fields. Theory and experimental data show, that PREVAIL can be able to meet the requirements for the 70 nm and 50 nm technology nodes [13], [14].

SCALPEL is able to resolve structures in the sub-100 nm region. Like other e-beam lithography tools, the limiting fact is not the wavelength of the electrons, but the proximity effect due to the significant backscattering of electrons. To circumvent these problems, heavier particles, e.g. ions, should be used. Ions will dissipate their energy much quicker than electrons and are stopped within the very vicinity of the location of incidence. Therefore there is no proximity effect.

6 Ion Beam Lithography

As electrons, ions can be used as radiation for lithography, too. As with electrons, a focused ion beam (FIB) can be used as the *pen*, or a broad beam can be modulated by a mask and than projected onto the surface (ion projection lithography).

6.1 Focused Ion Beam

The setup of a focused ion beam (FIB) tool is similar to an electron beam lithography tool, but instead of an electron beam a focused ion beam is used either to expose a resist locally, as in electron beam lithography, or to modify the substrate directly. The heavy ions impinging on the surface will sputter the material or, depending on energy, will intermix the layers at the surface of the sample. By means of this so-called ion milling the properties of the material at the surface will be altered. Another possibility is the local deposition of an additional layer. The impinging ions can induce the decomposition of a gas. As in a Chemical Vapor Deposition (CVD) process, where the decomposition of the process gasses is induced globally by thermal activation (Low Pressure CVD) or by a plasma (Plasma Enhanced CVD), this local decomposition leads to a local deposition of the material.

Besides a certain impact on the structure definition in the research environment, the direct modification of the surface, the sputtering as well as the deposition, enables the method to be used in the most important application of FIB in industry, namely mask repair. Mask production is very expensive and due to some failure in the processing (e.g. dirt sticking on the mask or a mistake in the electron beam pattern generator) a mask can be faulty. Either some parts of the masking layer, which should have been removed, are still present, or some parts of the masking layer are removed in excess. These faults can be cured by FIB.

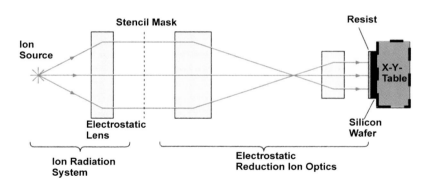

Figure 21: Schematic view of an ion projection lithography tool.

II Technology and Analysis

6.2 Ion Projection Lithography

Recently, ion projection lithography (IPL) has also been attracting increasing attention. Figure 21 shows a schematic view of an IPL tool. It consists of an ion source, electrostatic condenser optics, the mask carrier, projection optics and the wafer mount. The ions are emitted by an ion source and a parallel beam is formed by the condenser optics. This parallel beam of ions then illuminates the mask. Because of the low penetration depth of ions, the transparent part of the mask has to be made of nothing; even the thinnest membrane would stop the ions. The opaque parts consists out of a silicon wafer, which is protected against ion sputtering by a protection layer. This whole mask is called a stencil mask.

The mask technology for IPL is elaborate. The masks were fabricated from a silicon-on-insulator (SOI) wafer (a silicon wafer which has a buried oxide layer under a surface silicon layer). The patterns are inscribed on the top silicon. Then the bulk silicon and the oxide are etched. Only the patterned top silicon remains, serving as the absorber. Onto that layer the ion beam protective layer (carbon) is deposited, which hampers sputtering of the mask by the ions themselves.

Because of the stencil structure, it is not possible to define every structure in one run, e.g. it is not possible to write a circle: The middle will fall off. Therefore two masks with corresponding layout have to be used for one lithographic step. Furthermore the

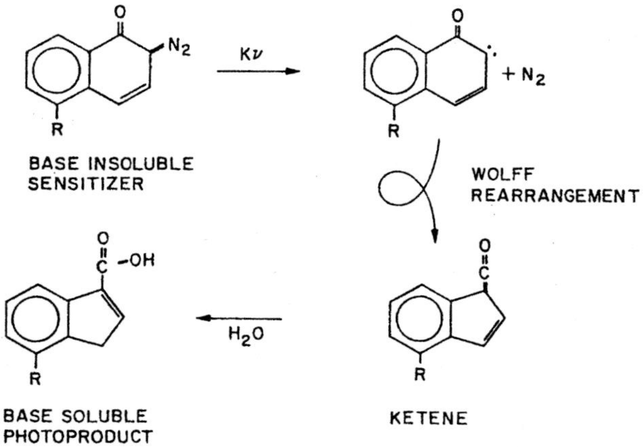

Figure 22: Exposure process of positive tone DNQ-resists [38], [39].

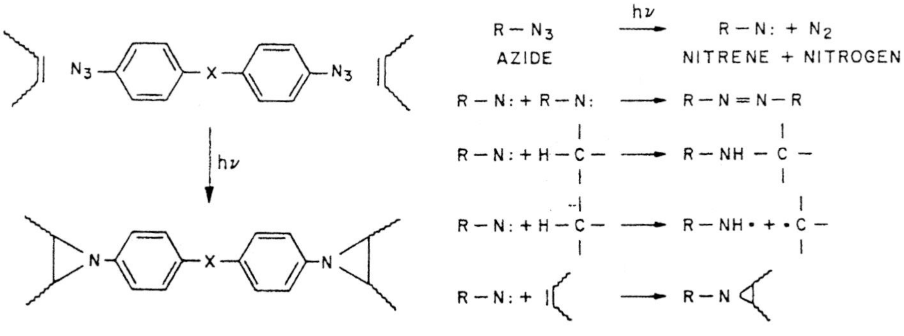

Figure 23: Reaction cycle of a negative tone resist during exposure [38], [39].

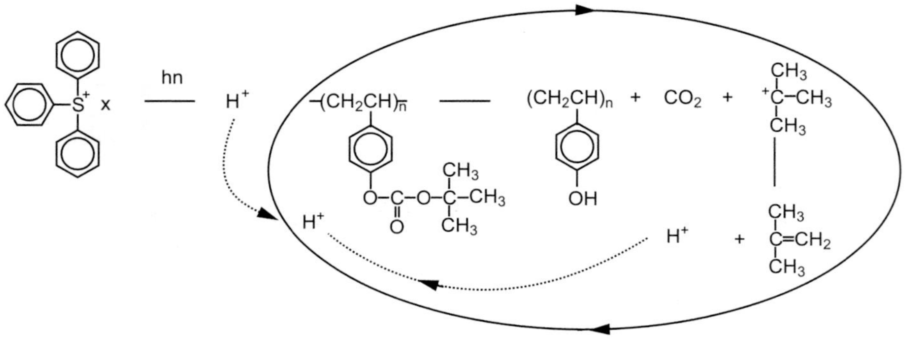

Figure 24: Chemical amplification cycle in a CAR [40].

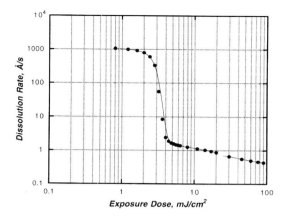

Figure 25: Contrast curve of the Shipley UVN30 DUV negative tone resist. The contrast curve shows the development rate of the resist versus the exposure dose. The steepness of the transition is high for a high contrast resist [44].

effect of gravity has to be taken into account. The distortion of the mask is dependent on the pattern itself, and so the splitting of the pattern into different masks is also a highly sophisticated problem. A good survey article is given in [17].

The IPL-tools are able to resolve 75 nm [27], but technical improvements can push this boundary down further. As an example of direct modification of the surface, the development of magnetic hard discs is given. Here a multilayer of Pt and Co is exposed by the projected ion beam. Using heavy ions, the layers will intermix and change their magnetic behavior, so the magnetic domains are formed [28].

7 Photoresist

Photoresists are also an integral part of lithography. The performance of the resist is the determining factor for the magnitude of the technology factor k_1. In general, photoresists are polymers which react when exposed to light. There are two different types of resists: With positive tone resists, the exposed areas of the resist will dissolve in the developer, with negative tone resists, the exposed areas will remain.

Positive tone resists consist of three components, a resin, which serves as a binder and establishes the mechanical properties, a photoactive compound (PAC), and a solvent to keep the resist liquid. The resin is not normally responsive to the exposure. The commonly used positive tone resist system for g- and i-line lithography is the novolac/diazonaphthoquinones (DNQ) system. The novolac is the resin material and dissolves in aqueous bases. The DNQ is the PAC, but when unexposed it acts as a dissolution inhibitor. Figure 22 shows the reaction cycle of the DNQ upon exposure. Upon exposure N_2 is split off the molecule. After a rearrangement, the molecule undergoes a reaction with the H_2O, which stems from the air. The reaction product now does not behave as a dissolution inhibitor, but as a dissolution enhancer. Therefore the exposed areas of the resist will dissolve about 100 times quicker than the unexposed areas.

Negative tone resists also consist of the three compounds: resin, photoactive compound and a solvent to keep the resist liquid. The resin consists of a cyclic synthetic rubber, which is not radiation-sensitive, but strongly soluble in the developer (non-polar organic solvents). The PAC is normally a bis-arylazide. Figure 23 shows the chemical structure of a rubber resin and a PAC. Upon exposure, the PAC dissociates into *nitrene* and N_2. These nitrene molecules are able to react with the rubber molecules, so a cross-linking between two rubber molecules can be established. Thus a three-dimesional cross-linked molecular network is formed, which is insoluble in the developer.

As device dimensions are scaled down further, the g-line steppers as well as the novolac/DNQ resists have been improved, so the features for 350 nm generation could be printed. But reaching the 250 nm generation, the illumination wavelength was shifted to 250 nm, too. However, at this wavelength novolac and DNQ do strongly absorb the light, therefore another class of resists had to be developed. Furthermore, the intensity of the mercury lamps at 250 nm was very low, so a high sensitivity of the resist was needed. The so called chemically amplified resists (CAR) use a chemical reaction to improve sensitivity and they are compatible with a 250 nm exposure wavelength. In a CAR, one of the compounds is a photo acid generator (PAG). Upon exposure an acid is released by PAG. During the post-exposure bake (PEB) – a heating of the sample after exposure – this acid reacts with the resin so that, firstly, the resin becomes soluble to a developer, and, secondly, a new acid is released. With this catalytic reaction it is possible to get 500 to 1000 reactions from one photogenerated molecule (Figure 24). Positive and negative tone CARs are available.

How do the resists affect the resolution? We have to consider the contrast of a resist: At what exposure dose is the resist exposed, and at what dose it is not yet exposed? As explained earlier, the light intensity at the sample is not of a rectangular shape, but is a diffraction pattern, so the resist at the sample is exposed according to that distribution. In Figure 25 a type of a contrast curve of a negative tone resist is given. In that curve, the etch rate of the resist in the developer is shown against the exposure dose. In an equivalent representation of the contrast curve, the normalized remaining resist thickness after a defined development process is plotted linearly against the logarithm of the exposure dose. For low exposure doses, the resist still behaves like an unexposed resist, for high doses it is fully activated. The transition between these two regions has to be steep to get a high contrast. To clarify this point refer to the intensity distribution of two patterns next to each other given in Fig. 4. The maximum intensity I_{max} has to be on the right side of the transition in Figure 25, the minimum intensity I_{min} on the left side.

The measure of the contrast is the contrast parameter γ. The higher γ the higher is the contrast, the closer I_{max} and I_{min} can be. It is determined as follows: The transition range of the contrast curve is fitted linearly the region where the resist is not etched by the developer (Figure 26a for negative tone resists, Figure 26b for positive tone resists). Where these two evens intersect, there is the dose D_u for positive and D_d for negative tone resists. The dose where the thickness of the resist becomes zero is D_d for positive and D_u for negative tone resists. The parameter γ is simply the negative slope of the contrast curve in the transition region on a logarithmic scale:

$$\gamma_\pm = -\frac{t_d - t_u}{\log_{10} D_d - \log_{10}(D_u)} = \frac{\pm 1}{\log_{10}\left(\frac{D_d}{D_u}\right)} \quad (7)$$

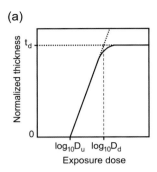

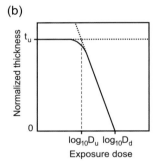

Figure 26: Schematic view of a contrast curve for (a) negative tone resist and (b) positive tone resist.

where t_d is the normalized thickness of the exposed resist, which means that t_d is 0 for positive tone resist and 1 for negative tone resist, and t_u is the normalized thickness for unexposed resist, so t_u is 1 for positive tone resist and 0 for negative tone resist; the sign stands for positive/negative tone resist, respectively. This characteristic can be expressed in terms of the so-called *critical resist modulation transfer function* $CMTF_{resist}$ which is defined as:

$$CMTF_{resist} = \frac{D_d - D_u}{D_d + D_u} = \frac{10^{\pm\frac{1}{\gamma}} - 1}{10^{\pm\frac{1}{\gamma}} + 1} \quad (8)$$

where the + stands for positive tone resist, and the – for negative tone resist. Now let us consider the exposure system with a given *MTF* with which a positive tone resist should be exposed. Choosing the exposure time s_e, with $I_{max} \cdot s_e = D_d$, $I_{min} \cdot s_e$ has to be less than D_u (otherwise I_{max} and I_{min} do not lie on the different sides of the slope of the contrast curve). Now :

$$MTF = \frac{I_{max} - I_{min}}{I_{max} + I_{min}} \cdot \frac{s_e}{s_e} = \frac{D_d - D_{min}}{D_d + D_{min}} \geq \frac{D_d - D_u}{D_d + D_u} = CMTF_{resist} \quad (9)$$

For a given *MTF* of the exposure system, the *CMTF* has to be smaller, and this means that the contrast of the resist has to be high enough.

Now let us consider the path of light in the resist. During exposure, the resist is not only traversed by the incoming beam, but due to reflection at the resist/sample interface, there are also reflected beams, giving rise to standing waves. The intensity variations emerging from these standing waves can be seen in the resist profile after development

(Figure 27a) and degrade the resolution. With CAR these ripples in the resist can be abolished by a thermal treatment after exposure (i.e. during a PEB). The reacted photoactive complexes diffuse, smoothening the side walls.

Furthermore, the exposure is not independent of the sample itself. Figure 27b shows simulated values of the linewidth transferred into the resist for two different samples – silicon wafers with oxide layers differing by the oxide thickness. It is seen that the linewidth depends strongly on the resist thickness and on the nature of the sample. This phenomenon stems from the reflections at the oxide/silicon interface and at the resist/oxide interface. To prevent these influences, anti-reflex coatings (ARC) are applied. It is possible to apply an ARC before the resist (bottom ARC) or afterwards (top ARC). With the BARC the reflectivity of the resist/substrate interface and with the TAR the reflectivity of the resist/air interface is minimized.

There are two possibilities to achieve this aim for BARC. First, the BARC could show a high absorption of the incident light. Unfortunately, if the absorption is high, the reflectivity of the BARC/resist interface will increase, degrading the effectiveness of the BARC. The second possibility is to match the BARC thickness and the refractive index, so that the optical path length of the BARC is $\lambda/4$. If the sample is exposed, the incident beam hits the resist/BARC interface, and is partially reflected and partially transmitted. The transmitted beam transverses the BARC, is reflected at the BARC/substrate interface and travels back to the BARC/resist interface. This beam and the first reflected beam are out of phase by $\pi/2$, so they interfere destructively, and the reflectivity at the BARC/resist interface is reduced. The TAR also has an optical path length of $\lambda/4$, so the interferences of the reflected beams are destructive.

BARC materials depend on the illumination wavelength and the samples. Spins on BARCs are primary absorption-type and consist of polymers. Oxinitrides deposited by PECVD can be used for index-matching type BARCs. The BARC layer has to be developed after lithography, requiring an additional process step. Figure 28a shows a resist structure without, Figure 28b with a BARC.

TAR can be spun on after spin coating. After exposure the TAR can be removed before development. Some TARs are designed to be developed in aqueous-based developer solution, which does not affect the resist. The handling of these TARs are much less complicated than the handling of BARC.

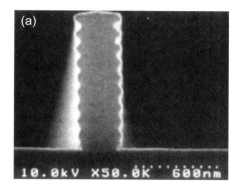

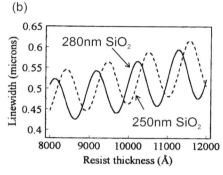

Figure 27:
(a) Impact of standing waves on the developed resist.
(b) Simulated linewidth as a function of resist thickness and substrate [3].

8 Alignment of Several Mask Layers

A modern microelectronic circuit needs several mask layers, which have to be properly aligned. For that purpose, every mask layer has so-called alignment marks: Special features on the mask whose positions are precisely known, and which are transferred to the sample by the subsequent etching or deposition step. The next mask layer also has an alignment mark at the corresponding position. Consider an exposure tool, in which the mask is loaded in a mask mounting fixture and the wafer on a movable wafer stage.

There are two systems of alignment in use. The off-axis alignment was developed first. The alignment marks on the sample were observed by a separate microscope using broadband non-actinic light as illumination to prevent the resist from being exposed. The wafer alignment marks were adjusted to marks etched into the microscope's objective. The mask was aligned independently to marks on the mask mounting fixture. This procedure would be enough for a single exposure if the mask and the wafer were separately aligned properly. But what is to be done if the wafer has to be exposed during several exposures as in a modern stepper? Admittedly, you know the structures to be transferred and you know the exposure positions on the wafer so it is possible to move the wafer stage to every exposure position, but the movement of the stage has to be very precise, and the long-term stability of the distance between alignment position and first exposure position, the so-called *base line* is difficult to achieve.

The second system is the through-the-lens alignment: Here, the image of the alignment mark on the sample is projected onto the corresponding mark on the mask and they are compared directly. One problem occurring with this system is the alignment illumination. A He-Ne-laser is used for this purpose so the resist will not be exposed, but the optics is not designed for that wavelength. Therefore the lens errors for that wavelength have to be corrected by additional lenses, which are brought into the optical path.

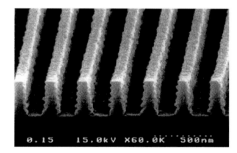

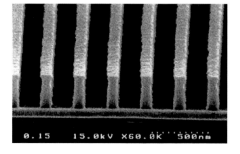

Figure 28:
(a) Resist structure without antireflex coating.
(b) Resist structure with ARC [15].

In contact and proximity lithography, the mask and wafer are aligned at every exposure. In modern projection lithography tools (e.g. a stepper), where the wafer is not exposed in one exposure, but in several exposures, attention has to paid for a accurate, quick and space efficient alignment. The wafer is moved by the wafer stage, while the mask is fixed. The position of the stage can be measured very precisely by three laser interferometers . Therefore some alignment strategies have been developed.

There are three degrees of freedom: x-shift, y-shift and rotation Θ. The first steppers used a technique called *two-point global exposure alignment*. In this strategy, one mark is used to adjust x- and y-shift, and the second mark is used to adjust Θ. Afterwards, the wafer is blindstepped through every exposure field. Here the movement of the stage has to be precise. To avoid this problem, a *zero-level alignment mark* can be used: this is a mark which is etched into the wafer before the process starts. Every mask-layer is aligned to that mark.

The next approach is the *site-by-site* alignment – that means performing the alignment at every exposure position. But this is time-consuming and, even a bigger drawback, there have to be alignment marks at every exposure position, which is a waste of space. Furthermore, the alignment marks on every site have to be small, so it is more difficult to detect them resulting in an increase in overlay error. So the *site-by-site* alignment strategy does not provide any advantage in comparison to a global mapping strategy as the 2-point-global-alignment strategy.

In the *enhanced global alignment strategy* 5 to 10 alignment marks are aligned and the stage position is measured several times. From the positions a least squares fit is computed. Based on these data, the positions of the exposure sites are corrected. These corrections are assumed to be stable for the time needed to expose the whole wafer.

The question of how to find the alignment marks and how to align them properly to each other is not discussed here, but the reader is referred to Ref. [47] and references therein for a survey of that topic.

9 Nanoimprint Lithography

There are several approaches for patterning structures without lithographic methods, e.g. a silicon surface can be modified by depassivation by the tunneling current in a UHV-STM (Ultra High Vacuum Scanning Tunneling Microscope [20], [21], or the surface can be modified by the movement of an Atomic Force Microscope (AFM)-tip . A certain interest has been focused on the nanoimprint lithography (NIL), which is described in more detail in this section.

With the NIL, a mold is processed by conventional technology, i.e. e-beam lithography and etching techniques, and is pressed onto a resist coated substrate. The structures in the mold are transferred into the resist and can be utilized after removing the mold. There are two different kinds of NIL, the hot embossing technique and a UV-based technique. A sketch of both techniques is given in Figure 29.

Hot Embossing Technique

Here the sample is heated above the glass transition temperature of the resist, which is a thermoplastic polymer. Above that temperature the polymer behaves as a vicous liquid and can flow under pressure. The mold itself can be made of different materials, usually a silicon wafer with a thick SiO_2 layer is used. This SiO_2 layer is patterned and structured by e-beam lithography and anisotropic reactive ion etching. The aspect ratio of the features are 3:1 to 6:1, and the mold size is several cm^2. As thermoplastic polymers either PMMA (a well known e-beam resist) or novolak resin-based resists are in use. PMMA has a small thermal expansion coefficient of $\sim 5 \times 10^{-5}\,K^{-1}$ and a small pressure shrinkage coefficient of $\sim 3.8 \times 10^{-7}\,psi^{-1}$. To ensure a proper removal of the mold, the resist is modified by release agents, which decrease the adhesion between mold and resist. Resist layers between 50 and 250 nm thickness are used. The imprint temperature and pressure are dependent on the resist. For PMMA the glass transition temperature is about 105°C, so the temperature at which the sample and the mold are heated is between 140 and 180°C. Then the mold is pressed onto the sample with pressures of about 40 – 130 bar. The temperature is then lowered below the glass transition temperature and the mold is removed. The features of the mold are now imprinted in the resist. The residual resist layer in these features is removed by anisotropic reactive ion etching.

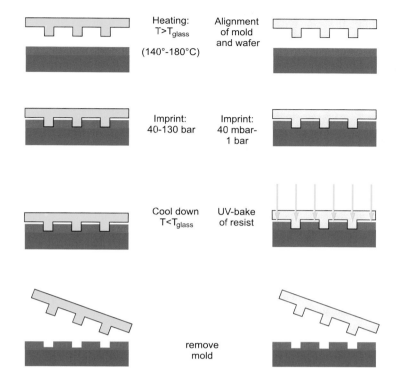

Figure 29: Nanoimprint lithography: hot embossing technique (left hand side) and UV nanoimprint (right hand side).

Afterwards, the structures can be transferred to the substrate either by direct etching or by metal deposition and lift-off. Structures down to a feature size of 10 nm for holes and 45 nm for mesas are imprinted with a high accuracy [22]–[24].

UV-based NIL

Heating and cooling of mold and sample is time-consuming. Therefore to achieve a somehow higher throughput, curing of the resist by UV irradiation is used. The thermoplastic resist is replaced by UV-curable monomers. The mold has to be fabricated of a UV-transparent material, e.g. quartz. The features are transferred to the mold by e-beam lithography and a Ti/PMMA resist stack. The patterned PMMA is used to transfer the features into the Ti, and the Ti is used to structure the quartz mold. The resists are acrylate- or epoxide-material systems, which can be modified with respect to low viscosity, UV curability, adhesion to the substrate and detachment from the mold. The low viscosity is essential for using low imprint pressures of 40 mbar – 1 bar. After pressing the mold on the sample, the sample is irradiated by UV-radiation through the mold and a baking, and hence a polymerization of the resist is initiated. This step lasts only about 90 seconds. After detaching the mold, the residual resist is removed by RIE and the further pattern transfer can be done. Again mold areas of several square centimeters can be imprinted in one run, and one imprint step takes about 10 minutes. The minimum feature size reported in the literature is 80 nm for dots. [25].

NIL offers the opportunity to define decananometer features in a rather *simple* manner, at least in comparison to the advanced lithography methods described above. The field size of ~2×2 cm^2 is comparable to a die which is illuminated by a stepper. On the other hand, this method is time-consuming (>10 min for one imprint) and up to now only structures on a plain surface have been investigated, while advanced lithography is able to define structures on textured substrates. Nevertheless, because of its technological simplicity, the NIL will be an alternative for research and small series production.

10 Conclusions

There are different approaches for achieving resolutions in the deep sub-100 nm region. EUV is being investigated by several projects (EUV-LLC in the USA, EUCLIDES in Europe and an ASET-program in Japan). But still the community is not sure about which method will be the most powerful. The SCALPEL method is being pushed forward by Lucent Technologies, and Infineon Technologies is involved with the development of IPL. All three methods are capable of resolving these fine dimensions, which have to be achieved within the next decade, but all three methods have their advantages and drawbacks. The EUV suffers from a very costly optics and mask technology, while for IPL and SCALPEL it is the small image area. The next five years will show which method will prevail.

For a deeper insight into the area of lithography there is a huge amount of good literature. Recommendable are [3], [10], [16], the Materials Research Society symposia, e.g. [18], and the annual conference Micro and Nano Engineering (MNE), from which e.g. [7] is taken. For theoretical purposes [19] gives a detailed description. For the whole environment of semiconductor technology refer to [11], for the future of semiconductors to [2].

Acknowledgements

The editor would like to thank Julio Rodriguez (FZ Jülich) for checking the symbols and formulas in this chapter.

References

[1] taken from *Chamber's Encyclopaedia*.
[2] International Technology Roadmap for Semiconductors (ITRS): http://public.itrs.net .
[3] H.J. Levinson and A. Arnold, in *Handbook of Microlithography, Micromachining and Microfabrication, Vol. 1,* SPIE, The International Society for Optical Engineering, Bellingham, WA, 1997.
[4] B. Hoppe, *Mikroelektronik,* Vogel, Würzburg, 1998.
[5] W. Waldo, in *Handbook of VLSI Microlithography,* Noyes Publications, Park Ridge, New Jersey, 1991.
[6] B. El-Kareh, *Fundamentals of Semiconductor Processing Technology,* Kluwer Academic Publishers, 1995.
[7] B.A.M. Hansson, L. Rymell, M. Berglund and H.M. Hertz, Microelectronic Engineering **53**, 667 (2000).
[8] H. Fiedorowicz et al., Optics Communications **184**, 161 (2000).
[9] E. Sarantopoulou, Z. Kollia and A.C. Cefalas, Microelectronic Engineering **53**, 105 (2000).
[10] F. Cerrina in *Handbook of Microlithography, Micromachining and Microfabrication, Vol. 1,* SPIE, The International Society for Optical Engineering, Bellingham, WA, 1997.
[11] C.Y. Chang and S.M. Sze, *ULSI Technology,* McGrawHill, 1996.
[12] D.G. Stearns, R.S. Rosen and P. Vernon, J. Vac. Sci. Technol. A**9**, 2662 (1991).
[13] H. C. Pfeiffer, Jap. J. Appl. Phys. **38**, 7022 (1999).
[14] K. Suzuki et al., Proc. SPIE **4343**, 80 (2001).
[15] S.H. Hwang, K.K. Lee and J.C. Jung, Polymer **41**, 6691 (2000).
[16] M.A. McCord, M.J. Rooks in *Handbook of Microlithography, Micromachining and Microfabrication, Vol. 1,* SPIE, The International Society for Optical Engineering, Bellingham, WA, 1997.
[17] R. Kaesmaier and H. Löschner, Microelectronic Engineering **53**, 37 (2000).

[18] MRS Symposium Proc. Vol. 584, *Materials Issues and Modeling for Device Nanofabrication*, MRS, Warrendale, PA, 2000.

[19] K.A. Valiev, *The Physics of Submicron Lithography*, Plenum Press, New York and London, 1992.

[20] J.W. Lyding, T.-C. Shen, J.S. Hubacek, J.R. Tucker, G.C. Abeln, Appl. Phys. Lett. **64**, 2010 (1994).

[21] J.W. Lyding et al., Appl. Surf. Sci. **130**, 221 (1998).

[22] S.Y. Chou, P.R. Krauss, P.J. Renstrom: Appl. Phys. Lett. **67**, 3114 (1995).

[23] S.Y. Chou, P.R. Krauss, P.J. Renstrom: J. Vac. Sci Technol. B **14**, 4129 (1996).

[24] S. Y. Chou, P.R. Krauss, W. Zhang, L. Guo, L. Zhuang: J. Vac. Sci. Technol B **15**, 2897 (1997).

[25] M. Bender, M. Otto, B. Hadam, B. Vratzov, B. Spangenberg, H. Kurz: Microelectronic Engineering **53**, 233 (2000).

[26] M. Levenson et al., IEEE Trans Electron Dev. ED-29, 1828 (1982).

[27] W.H. Bruenger, MRS Fall Meeting 2001. Symposium Y3.5.

[28] R. Berger et al., MRS Fall Meeting 2001, Symposium Y3.4.

[29] M.A. Klosner, W.T. Silvfast, Opt. Lett. **23**, 1609 (1998).

[30] K. Bergmann, O. Rosier, R. Lebert, W. Neff, R. Poprawe, Microelectronic Engineering **57**, 71 (2001).

[31] R. Lebert, et al., Microelectronic Engineering **57**, 87 (2001).

[32] V.Y. Banine, J.P.H. Benschop, H.G.C. Werij, Microelectronic Engineering **53**, 681 (2000).

[33] R.H. Stulen, Microelectronic Engineering **46**, 12 (1999).

[34] S. Okazaki, Proceedings SPIE **3676**, 238 (1999).

[35] J. Benschop, U. Dinger, D. Ockwell, SPIE **3997**, 34 (2000).

[36] R. Chau et al., International Electron Device Meeting, Technical Digest. IEDM (2000).

[37] A.M. Hawryluk, L.G. Seppola, J. Vac. Sci. Technol. B**6**, 2162 (1988).

[38] S. Wolf, R.N. Taubner, in *Silicon Processing for the VLSI Era*, Lattice Press, 2000.

[39] J. Bowden, in *Materials for Microlithography, Advances in Chemistry*, Series No. 266, American Chemical Society, Washinton D.C., 1984.

[40] R.D. Allen, W.E. Conley, R.R. Kunz, in *Handbook of Microlithography, Micromachining and Microfabrication,* Vol. 1, SPIE, The International Society for Optical Engineering, Bellingham, WA, 1997.

[41] US Patent ,4,953,960, *Optical reduction system*, David Williamson, inventor, filed July 1988 and granted Sept. 1990.

[42] Ohara i-line Glasses, Ohara cooperation, Kanagawa, Japan, Juni 1993; Schott Catalogue of Optical Glasses No. 10000 on Floppy Disc, Edition 10/92, Schott Glass Technology, Duryea, P.A.

[43] S.D. Berger et al., J. Vac. Sci. Technol. B**9**, 2996 (1991).

[44] Shipley DUVN30 product catalogue, Shipley Company, Marlborough, MA.

[45] D.F. Kayser, N.S. Viswanathan, J. Vac. Sci. Technol. **12**, 1305 (1975).

[46] HBO Mercury Short Arc Lamps for Microlithography *Technology and Application Guide* 2000/2001, OSRAM GmbH, Munich.

[47] A. Moel, E.E. Moon, R.D. Frankel, H.I. Smith, J Vac. Sci. Technol. B**11**, 2191 (1993).

Material Removing Techniques – Etching and Chemical Mechanical Polishing

Stefan Schneider, Unaxis Balzers AG, Balzers, Liechtenstein

Simon McClatchie, Lam Research, Fremont, USA

Contents

1 Introduction	249
2 Etch Techniques	249
2.1 General Aspects	249
2.2 Wet Etching	250
2.3 Dry Etching – Basic Methods	251
2.4 Plasma Generation	252
2.5 Reactor Types	253
2.6 Typical Dry Etch Applications	255
2.7 Plasma Diagnostic Methods	258
2.8 Challenges in Dry Etching	260
2.9 Future Trends	260
3 CMP – Chemical Mechanical Polishing	262
3.1 CMP Mechanisms	262
3.2 Dielectric CMP	263
3.3 Metal CMP	263
3.4 Commercial CMP Tool Configurations	264
3.5 Endpoint Detection	265
3.6 Pads	266
3.7 Slurry	266
3.8 Post CMP Cleaning	267
3.9 Some CMP Problems	267
3.10 Defects	267
3.11 Future Technology Trends / Challenges	267

Material Removing Techniques – Etching and Chemical Mechanical Polishing

1 Introduction

In this section we are going to discuss the two main techniques of material removal used in modern microelectronic device manufacturing: etching and polishing.

Two different etch mechanisms can be distinguished – physical and chemical etching. **Physical etching** or sputtering, as it is sometimes called, relies on the momentum transfer from particles hitting and eroding the surface, whereas in **chemical etch** reactions, products have to be formed that are either soluble in the etch solution or volatile at low pressures. The most prominent example of chemical etching is **wet etching**, i. e. etching in a solution. A major domain is bulk etching of silicon or compound semiconductors, e.g. in the micromachining of MEMS (microelectromechanical systems). The etch rates are typically quite high and a large number of recipes can be found to etch virtually every material. The crystal face selectivity of the etch rate can be used to define groves. Plasma-assisted etch techniques (**dry etching**) make use of partially ionised gases produced in low pressure (1 μbar – 10 mbar) discharges, where ions, electrons, and activated neutrals are produced from relatively inert molecular gases. Historically, dry etching started as a physical etch process, since only inert gases like argon could be used to erode the surface. Today, dry etching is a sometimes delicate combination of physical and chemical material removal. Inert gases have been replaced by e. g. fluorine-, chlorine-, and bromine-containing gases. Dry etching became increasingly important as the feature sizes shrunk below one micron and vertical feature profiles were required.

The role of **chemical mechanical polishing** (CMP) only emerged in the past decade. Due to the increasing integration density the device dimensions have shrunk to levels where photolithography comes closer to its physical limits. As a consequence, the depth of focus becomes smaller, and height differences on the wafer surface cannot be tolerated so easily any more. Furthermore, the increasing complexity of modern microchips requires at the same time an increasing number of metal layers to be built. The standard approach adopted by the industry to solve this task in a cost-effective way is to coat the transistor level with a dielectric passivation layer and polish this layer back to an even surface. Holes – so called vias – are subsequently etched to establish contacts. The interconnects are formed in a dual damascene process, where a dielectric material is deposited. The lines to establish the interconnects and the vias to form the contacts to the layer below are etched and filled with metal in a subsequent step. Finally, the metal is polished back to start the process all over again by depositing another dielectric layer (see Chap. 29 for details).

2 Etch Techniques

2.1 General Aspects

Removing material from a substrate by a chemical reaction or by ion bombardment is referred to as an **etch process**. To create a pattern in a thin film, specific regions on the wafer surface are protected by a **mask** and the material not covered by the mask is removed.

The rate of material removal by etching is known as the **etch rate** (ER), expressed in nm/min. For a high production throughput etch rates above 50 nm/min are desired. Depending on the etch process material removal occurs in both horizontal and vertical directions. The **anisotropy** of an etch process is given by

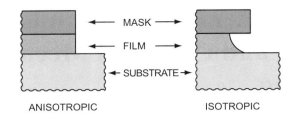

Figure 1: Comparison of the feature evolution in an isotropic/anisotropic etch process.

$$A = 1 - \frac{ER_L}{ER_V}, \qquad (1)$$

where ER_L and ER_V are the lateral and vertical etch rates, respectively.

Chemical etch processes are typically of an **isotropic** nature, as the horizontal and the vertical etch rates are equal and A becomes zero. An ideal etch process would transfer the mask to the underlying film with zero bias, creating vertical profiles.

If etch proceeds only in the lateral direction A becomes one. Most dry etch processes are of an **anisotropic** nature with values of $A < 0.1$. The effect of an isotropic etch rate is to create undercuts below the mask (see Figure 1).

Typically the mask material as well as the substrate are also attacked from the etch process. The ratio of the etch rates of different materials is called the **selectivity** of an etch process. The selectivity with respect to the mask as well as to the substrate is an important characteristic of an etch process. If the mask is etched too quickly, control over the feature size is lost, as the mask dimension shrinks during the process. To cope with film thickness and etch rate non-uniformities an additional **overetch** is often run after the main etch step to ensure the film is etched completely on the whole wafer. In those cases a high selectivity with respect to the substrate is needed to be able to stop the etch accurately at the desired level.

2.2 Wet Etching

Wet etching is a purely chemical process consisting of three different steps. The reactive species present in a solution have to move to the surface, a reaction yielding soluble etch products has to take place, and finally the etch products need to move away from the surface.

A distinct advantage of wet etch processes is that a large number of recipes exists to etch virtually every material (e.g. see Köhler [1]). They can typically be tailored to have a very high selectivity. As no ion bombardment takes place the damage to the substrate, induced by the etch process is rather low.

The etch solution's concentration needs to be controlled tightly on a microscopic scale to have a constant and uniform etch rate over large areas. Agitation and cycling of the solution or even continuous spraying is sometimes used in automated wet etch benches. High-purity chemicals and filtration need to be employed to avoid particle contamination of the surface. A drawback is the large consumption of chemicals compared to dry etching and the handling of toxic waste. Chemical etch processes are not regarded as practical if features sizes below one micron have to be patterned, as the etch is isotropic and undercuts develop easily.

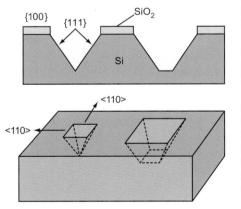

Figure 2: Orientation-dependent etch rate of Silicon for an ethylenediamine (EDA) wet etch process.

2.2.1 Crystal Face Selectivity

Orientation-dependent etching solutions have been developed which etch much faster in one crystallographic direction than in another. They are widely used in the field of micromachining of MEMS. The different density of atoms as well as the varying bonding strength is responsible for the anisotropy of the etch rate. The highest density of atoms for silicon is found in the <111> direction.

A 1:1 solution of KOH and water, for example, etches silicon in the <110> direction about 700 times faster than in the <111> direction. Adding propanol gives an additional selectivity to the <100> plane of 100:1. Ethylenediamine (EDA), which is both orientation-dependent and concentration-dependent (allowing the etch rate to be controlled in a wide range), additionally has a very high selectivity over SiO_2. Due to the crystal face selectivity, the etch virtually stops on <111> planes and the etch depth is determined by the mask opening and the etch time only (see Figure 2).

2.2.2 Important Materials and Etch Solutions

A common wet etch process is the removal of native oxide (**SiO₂ etching**) in a dilute solution of hydrofluoric acid (HF), as it offers an extremely high selectivity to silicon of about 1:100. The overall reaction for etching SiO_2 is

$$SiO_2 + 4\,HF \rightarrow SiF_4 + 2\,H_2O \qquad (2)$$

To maintain a constant HF concentration and in turn a constant etch rate, buffered solutions (BHF) are commonly used. Ammonium fluoride as a buffering agent maintains a constant HF concentration by the dissolution reaction

$$NH_4F \rightarrow NH_3 + HF \qquad (3)$$

More information on SiO_2 etching can be found in the CMP section.

The **photoresist mask** must be removed after each patterning step. Additionally, **polymer depositions**, generated, for example, during the etch process to obtain a desired steep etch slope or a high selectivity, may have to be cleaned away. The main objective here is to remove (to strip) the mask material and the polymer quickly without any damage to other films. Even though **dry ashing** in an oxygen plasma is an option, wet stripping is still used frequently.

Organic strippers break down the structure of resist material and remove any residue from the wafer surface. Typically the formula is designed for the specific application, taking the last process step into account as well as all the different materials present on the wafer surface (see Figure 3). As **inorganic strippers** (oxidising strippers) solutions of H_2SO_4 and oxidising agents like H_2O_2 are used. When heated to higher temperatures (>120 °C) even hardened resist can be removed residue-free.

To pattern **compound semiconductors**, like GaAs, etches to pattern thin layers that are selective to $Al_xGa_{1-x}As$ are required to make contacts. Solutions of hydrogen peroxide (H_2O_2) diluted with H_3PO_4 or H_2SO_4 to control the pH are used frequently. For more details refer to [1].

Defect-selective etching is used to emphasis (decorate) defects in the wafer. Invisible defects become observable and can be counted in a microscope. Useful recipes have been proposed by Schimmel [2].

2.3 Dry Etching – Basic Methods

Plasma-assisted etch techniques (dry etching) make use of partially ionised gases produced in low pressure (1 μbar –10 mbar) discharges, where ions, electrons, and activated neutrals are produced from relatively inert molecular gases [3].

Dry etching is performed in two main reactor configurations. In **plasma etching** the wafer is in direct contact with the plasma, exposed to the same pressure level. The chambers are typically compact and have low volume. Whereas in **ion beam etching** (IBE) the wafer is held at low pressure in a large chamber. A plasma is formed in a separate ion source and the ions are accelerated onto the wafer by applying appropriate voltages to a set of grids in front of the source.

Two distinctly different etch mechanisms are encountered in dry etching. In **sputter etching** ions of inert gases like argon are accelerated to the surface. Due to momentum transfer surface atoms are knocked off. The etch products are not volatile and the process is not very selective. **Chemical etching** is found if the surface material is spontaneously etched from the neutral or activated gas and forms volatile etch products. The process can be very selective. Prominent examples are silicon etching in a fluorine atmosphere or photoresist mask removal (so called **dry ashing**) in an oxygen ambient. The feature evolution for either process mechanism if shown in Figure 4.

In **reactive ion etching** (RIE) a combination of mechanisms is used. Instead of inert gases reactive gases containing fluorine, chlorine, bromine, or oxygen are employed. The role of the accelerated ions can be either to make a low-reactive surface more reactive due to damaging or to supply additional energy for etch products to desorb from the surface. As the gas phase species can react more readily and form volatile etch products, the etch rate increases. On the other hand, the process can be very selective.

Reactive gases can also be used in a totally different function such as only to increase the etch rate. Depending on the substrate material and the gas composition, etch regimes can be found where the etch products condensate on the sidewalls of the mask and the etched material forming protective layers. These processes, being typically very anisotropic as a lateral etch is inhibited, are designed to have a high selectivity.

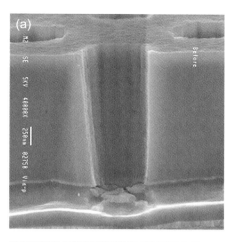

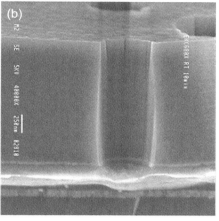

Figure 3: SEM micrograph showing the effect of a wet clean process to remove residues from a dry via etch step. Before clean (a), post clean (b) -courtesy of EKC Technology.

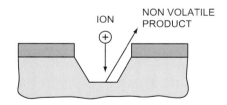

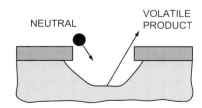

Figure 4: Dry etch regimes: sputtering and chemical etching.

II Technology and Analysis

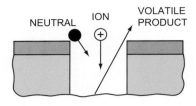

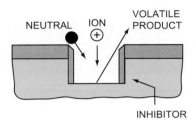

Figure 5: Dry etch regime: reactive ion etching.

Basically all the mentioned etch mechanisms can be used in either a plasma etcher or an ion beam tools. Over the time different acronyms have developed to denominate a specific combination. *Ion beam etching* (IBE) is performed in an ion beam tool using inert gases (like argon), whereas in **reactive** IBE (RIBE) the argon is replaced by reactive gases. The term **chemical-assisted** IBE (CAIBE) was historically used to describe a configuration where an ion beam etcher was used with inert gases in the source while reactive gases were introduced into the chamber by means of gas rings above the wafer. *Sputter etching* would take place in a plasma tool run with inert gases. And *reactive ion etching* (RIE) describes the combination of a plasma tool with the use of reactive gases (Figure 5). For further information the reader is referred to more specific textbooks covering this and many other issues [7] - [9]. The approach of using halogen gases together with oxygen for in situ sidewall passivation may also be used to generate nano-patterns in silicon. A thin, but etch resistant two-layer resist system in conjunction with a SF6/O2 process was, for example used by Spangenberg to pattern 30 nm features [10].

2.4 Plasma Generation

For commercial plasma etch reactors the **plasma density** (number of ions per cm^3) is in the range from 10^9 to some 10^{12} ions/cm^3. Typically only a very small fraction of the atoms and molecules in the phase is ionised (one charged particle per 1000 to 10^6 neutral species). A typical process pressure is in the range of about 10^{-3} - 0.1 mbar.

The negatively charged species are mostly electrons. As they are very light, they can obtain high energies since the energy transfer to other species is inefficient. This allows high-temperature-type reactions to form very reactive free radicals, while the neutral gas remains at a low temperature. Supplementary information may be found in Chapter 8, Section 3.

When a high voltage is applied between two electrodes in the plasma chamber, the gas molecules are accelerated and ionised by collisions forming a plasma. In principle, DC as well as AC voltages can be used to generate a plasma. As insulation substrates (like silicon oxide) would block DC currents, only AC or rf generators (as they operate in the radio frequency range) are used. Typically industrial frequencies such as 13.56 MHz or 400 kHz are used.

New particles are created in a plasma by means of **inelastic collision**, as energy and impact are transferred and dissociation and recombination may take place:

- Ion and electron formation
  $e^- + A_2 \rightarrow A_2^+ + 2\,e^-$
- Atom and radical formation
  $e^- + A_2 \rightarrow A\bullet + A\bullet + e^-$
- Heat and light formation
  $e^- + A_2 \rightarrow A_2^* + e^- \rightarrow A_2 + e^- + h\nu$

Figure 6: *I-V* characteristic of the plasma boundary giving rise to the bias voltage formation when applying a rf-signal.

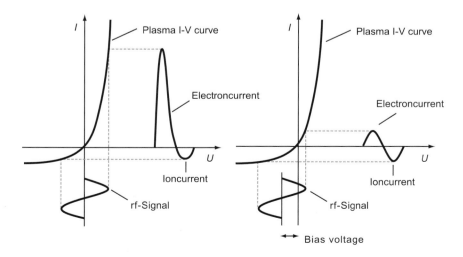

The main loss of species mechanisms are:

- Electron ion recombination
  $e^- + AB^+ \rightarrow A + B$
- Ion-ion recombination
  $AB^+ + C^- \rightarrow AB + C$
- Charge transfer
  $A^+ + B \rightarrow A + B^+$
- Attachment
  $e^- + AB \rightarrow A^- + B$
- Diffusion
  $e^- \rightarrow$ Wall (loss)
  $A \rightarrow$ Wall (loss/ reaction)

Due to the higher mobility of the electrons, the net flow of electrons toward an insulating surface is higher than that of the ions until the walls charge up negatively and repel excess electrons flowing to the wall ($J_i = J_e$). The bulk of the plasma becomes more positive as the walls (**plasma potential** V_p) and the *I-V* curve has a diode like characteristic.

If an rf-voltage is applied to a plasma an excess electron current flows to the walls to establish a DC potential difference (called **bias voltage** V_b) between the walls and the plasma bulk. The net current of electron and ions goes back to zero again (see Figure 6). This bias voltage accelerates the ions from the plasma onto the substrate.

2.5 Reactor Types

As the reactor types used in plasma etch became more sophisticated the more demanding the etch processes had to be to meet the challenges of an increasing integration density. In the early days (1970s) only batch-type barrel reactors where used. As the minimum feature sizes decreased and the wafer sizes increased, single-wafer reactors where introduced to gain more precision and maintain uniformity.

The next step (1980s) was to separate the plasma generation from the ion acceleration, giving a means of determining the plasma density independently of the ion energy. Further device integration meant that features sizes kept decreasing and reaching the order of the mean free path in the plasma (deep sub-micron region) making low-pressure operation a necessity. More efficient ways to couple the energy into the plasma were needed to sustain a stable plasma operation at low pressures. The most successful either use microwave radiation or couple the energy inductively into the plasma.

Ion beam etchers are a different class of plasma tools, where the plasma generation inside a source region is geometrically separated from the etched substrate.

2.5.1 Barrel Reactor

Barrel reactors (see Figure 7), once very common, are able to handle batches of wafers at one time. An rf generator is used to couple energy capacitively or inductively into a quartz tube (which looks like a barrel). Perforated shields are sometime employed to protect the wafers against ions to reduce damage. Only neutrals and radicals can reach the wafer and perform the etch.

Today this configuration is almost only used for non-critical applications as photoresist mask removal (plasma ashing) and surface cleaning due to the typically rather low plasma density of 10^{10} ions/cm^3. Oxygen and sometimes CF_4 or SF_6 are used as reactive gases. The etch is mainly chemical.

2.5.2 Plasma Reactors

The basic configuration is the diode-type reactor, with one rf-generator capacitively coupled to the reactor. The reactor can be run in two different modes depending where the energy is coupled to. If the substrate is grounded, the plasma develops only a very low potential, merely allowing for a chemical etch regime. Coupling the rf-power to the wafer adds a physical component to the etch, as the ion are accelerated from the bias potential. The reactor is characterised by a strong interaction of the pressure and the obtainable ion energy.

To gain an independent control of the plasma density and the ion energy, two separate energy sources are needed to feed power to the plasma. This configuration is also known as a triode reactor. Typically two different frequency are also used in the generators. As it becomes easier to ionise molecules at higher operation rf-frequencies, generators above 2 MHz are used to control the plasma density. A low frequency source (e.g. 400 kHz) coupled to the substrate is used to control the ion energy and accelerate the ions onto the wafer.

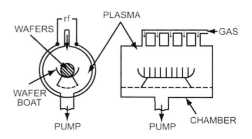

Figure 7: Barrel reactor configuration – The reactor is configured with only one power source to generate the plasma, thus the ion density and energy cannot be independently controlled.

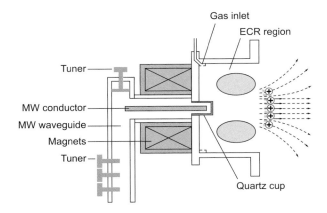

Figure 8: ECR tool configuration – Magnets force electrons onto circular orbits with the same frequency as the microwave radiation coupled into to the reactor. The resonant absorption generates a high-density plasma torus. However, the plasma diffuses down to the wafer level. A second power supply coupled to the wafer electrode may be used to accelerate ions onto the wafer.

The **electron cyclotron resonance** (ECR) source was the first modern source to operate at low (<10 mTorr) pressures. The energy from the microwave radiation (typically 2.45 GHz) is coupled into the plasma very efficiently by resonant absorption: Magnets force electrons onto circular orbits with the same orbital (gyration) frequency ω_g as the applied electric field.

$$\omega_g = \frac{e \cdot B}{m}, \qquad (4)$$

where e is the electronic charge, B the magnetic field and m the electron mass.

The schematic additionally shows an ECR source with the microwave guide and the solenoid coil producing the magnetic field (see Figure 8). Though the plasma is most intense in the so-called ECR zone, it extends to the whole chamber with ion densities of 10^{12} ions/cm^3. Sometime gas rings in the vicinity of the substrates are added. The ions are accelerated to the wafer by an additional rf-generator coupled to the substrate.

As the energy is coupled into the plasma without high voltages internal erosion is reduced allowing a clean operation. On the other hand, the microwave radiation introduces more complexity to the etch system.

To achieve uniform etching the distance from the source to the substrate has to be large to allow for the plasma density to homogenise. Therefore it becomes difficult to handle large substrates like 300 mm wafers.

An alternative is to couple the energy magnetically into the plasma, which led to the development of **inductively coupled plasma** (ICP) sources, first described by Thonemann [8]. A coil wrapped around a dielectric cylinder generates a magnetic field extending into the chamber. This alternating magnetic filed in the discharge region in turn generates an electric field which strikes a plasma by accelerating electrons. ICP sources are very clean sources, too, as no high voltages are used. Plasma densities are found around 10^{11} ions/cm^3. However, sometimes it becomes difficult to start the plasma at very low pressures due to the lack of high voltages.

A further development to reduce the chamber volume of the ICP source and to couple the energy more efficiently is the **transformer coupled plasma** (TCP$^©$) source (see Figure 9), from Lam Research. A flat coil above a dielectric window is used to sustain a high-density plasma directly above the wafer surface. This plasma acts as a secondary

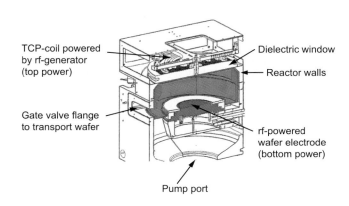

Figure 9: TCP reactor configuration – The power is coupled into the reactor by means of a planar coil on top of a dielectric window determining the plasma density. A second power supply coupled to the wafer electrode accelerates the ions onto the wafer. The compact chamber allows for high-density plasmas.

coil, as in a transformer. The plasma density is tuned by the rf-power fed to the coil, while the ion energy depends on the rf-power supplied by a separate generator to the substrate.

TCP-like source designs have the advantage of a good scalability to larger wafer sizes. They are known to produce high-density plasmas (10^{12} ions/cm^3) even at lower pressures.

2.5.3 Ion Beam Reactors

Historically ion beam reactors used to be equipped with Kaufman sources, generating the plasma by means of hot filaments. The presence of a hot filament allowed only for the tool to be used with inert gases for sputter etching or deposition applications. A set of grids in front of the source is used to determine the ion energy and the ion density by applying appropriate voltages.

To take care of the charge neutrality an additional electron gun is needed to compensate the positively charged ion beam. The wafer can be loaded onto a substrate holder to be tilted versus the beam while being rotated (see Figure 10). Employing ICP or ECR sources enables an operation with reactive gases.

RIBE tools are very useful for research applications, as the beam parameters are known.

Due to their limited throughput and complexity, they only found a few commercial niche applications like etching nonvolatile materials, e.g. magnetic read heads, or patterning mirror structures for compound semiconductors. A good introduction to ion beam sources is given by Wolf [9].

2.6 Typical Dry Etch Applications

2.6.1 Silicon Etch

Gate Etch

Gate-conductor etching of linewidths below 0.25 μm has introduced new challenges. The principal material is polysilicon or a compound gate stack of tungsten silicide (WSi$_x$) and polysilicon to lower overall conductor (wire) resistivity. The need to control the etched profile, etching selectivity, and overall **critical dimension** (CD) is the driving force behind these challenges. Using the chemical approach, HBr/Cl$_2$/HCl -based etching processes afford good selectivity to silicon oxide.

The most critical aspect of the gate-conductor etching process is control of the critical dimension (CD) – the gate length – that defines the final performance of the device. Gate linewidth variations for *nested lines* (tightly spaced) or *isolated lines* have to be strictly controlled. Furthermore, dual-doped (n and p type regions) polysilicon etching requires 1:1 selectivity. Fluorine-containing gas is used to minimise n/p profile differences (see Figure 11).

SEM cross-section micrograph of a gate structure with a metal (or silicide) gate. The etch has to be very selective to the oxide layer. The differently doped region (n and p type) often pose a challenge to etch uniformity.

Deep Trench

The deep-trench storage node capacitor used in some DRAM structures is another challenging etch application. Trench etching, typically carried out with a SiO$_2$ hard mask, requires a high etching rate even for ultrahigh-aspect-ratio (1:30) features and the control of trench taper angles to very tight limits.

The main factors controlling the taper angles are the O$_2$ partial pressure in a HBr/NF$_3$/Cl$_2$ etching process and the wafer surface temperature. The oxygen reacts with the exposed silicon to form a SiO$_x$ layer which is not spontaneously etched by the halogen chemistry in contrast to a pure silicon surface. Furthermore, silicon containing etch products (mainly SiBr$_y$, SiCl$_z$,) are incorporated in this sidewall passivation layer and help blocking a further etch attack. The passivation layer at the bottom of the feature is instantaneously destroyed due to the high arriving rate and energy of the impinging ions, allowing the etch to progress only in this direction.

In contrast to deep silicon etching (see below), the sidewall protecting film is generated *in situ* and one obtains a very tight control over the critical feature size. The etch rate is controlled by the plasma density and the ion energy.

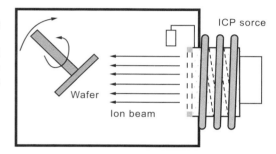

Figure 10: Reaction ion beam tool configuration – The wafer is clamped to a platen that can be rotated and tilted towards the ion beam, which is generated in a separate source. The power is mostly inductively coupled into the source to generate a plasma. Ion are extracted out of this plasma and accelerated by applying appropriate voltages to a set of grids in front of the source.

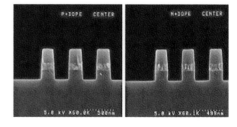

Figure 11: SEM micrograph of a gate etch for dual doped wafer – courtesey of Lam Research.

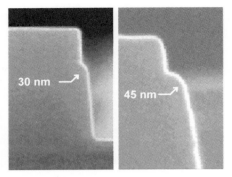

Figure 12: SEM micrograph of a STI structure – courtesy of Lam Research.

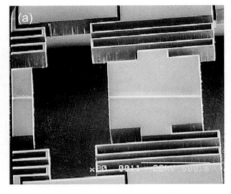

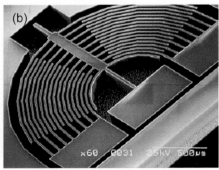

Figure 13: (a), (b), and (c) SEM micrograph of different MEMS structures – courtesy of Alcatel.

Shallow Trench Isolation (STI)

At high integration densities, the horizontal isolation of transistors becomes an important issue. The shallow trench isolation (STI) approach consumes less area on the wafer surface, while offering superior transistor properties to those of the formerly used local oxidation (LOCOS) isolation, due to reduced stress. Consequently, STI is the mainstream horizontal transistor isolation method today for all advanced logic or memory devices at 0.25 µm and below.

Shallow trenches are etched into the silicon wafer in close vicinity to the transistors using a SiO_2 hardmask. It is very important to achieve a rounding at the top and the bottom of the structures (see Figure 12), which are to be filled with a dielectric material in a subsequent step providing the isolation, to avoid stress affecting the transistor properties.

Deep Silicon Etching

In microelectromechanical systems (MEMS) sensors and actuators are integrated together with the logic in the same device. Bulk micromachining of silicon becomes necessary as trenches have to be etch several hundreds of micrometers deep with vertical sidewalls to form features like gyroscopes, suspended seismic masses or rotor blades with deep channels as shown in Figure 13a) to c).

The original work for deep and anisotropic silicon etching dates back to Robert Bosch GmbH in 1993 and is often referred to as the **Bosch Process**. Surface Technology Systems continued to develop and enhanced this method and also launched the **Advanced Silicon Etch** (ASE®) process. The process has tree alternating passivation etch steps (see also the schematical drawing in Figure 14):

- **Passivation step**: At the beginning of each cycle a C_4F_8 based plasma is used to conformally deposit a few monolayers of PTFE-type fluorocarbon polymer across all surfaces exposed to the plasma.
- **Etch step 1**: The plasma is then switched to SF_6 to create a plasma chemistry that isotropically etches silicon. Through the application of a DC bias to the wafer electrode, ions from the plasma bombard the surface of the wafer, removing the polymer. Increased ion energy in the vertical direction results in a much higher removal of fluorocarbon polymer from surfaces parallel to the wafer surface.
- **Etch step 2**: Following selective polymer removal, the silicon surfaces at the base of each trench are exposed to reactive fluorine-based species that isotropically etch the unprotected silicon. The remaining fluorocarbon polymer protects the vertical walls of the trench from etching.

By repeating the etch/passivation cycles and carefully controlling the etch time during each step, the degree of lateral etch is limited, allowing the trench to be etched vertically through the wafer. In contrast to deep trench etching, the etch rate is higher and only non toxic feed gases are used. But control over the feature size is not as tight.

2.6.2 Etching of Dielectrics

Dielectric etching applications, especially silicon dioxide and silicon nitride, typically rely on the competing influences of polymer deposition and reactive ion etching to achieve vertical profiles as well as etch-stopping on underlying layers. Vertical profiles are achieved by sidewall passivation, typically by introducing a carbon-containing fluorine species into the plasma (e.g. CF_4, CHF_3, C_4F_8).

Anisotropic dielectric etching is carried out in two ways. In the first way, a dielectric is used as a masking layer for patterning underlying materials for subsequent processing steps. Here, the integrity of the image transferred from the photoresist mask into the SiO_2 hardmask is most important. One example is the shallow trench isolation process, where a SiO_2 mask is used.

In the second, a dielectric material must be patterned without transferring the pattern to an underlying layer. For example, via holes in silicon-oxide based isolation layers must often be opened for electrical contacts to reach underlying features. A silicon nitride surface acts as an etch-stop layer to isolate underlying features. Here, selectivity is the biggest concern, and the etch process is chosen to provide sufficient polymerisation on the silicon nitride. Examples are self-aligned contacts (SAC) or dual damascene applications.

Traditional CHF$_3$/CF$_4$ gas combinations have been used to provide selectivity during oxide etching to silicon underlayers. Modern recipes contain a mixture of different gases. Noble gases like argon or helium are used to obtain a high etch rate or for dilution purposes. Gases like C$_4$F$_8$, and C$_2$F$_6$ control the polymerisation process. The ring molecule C$_4$F$_8$ tends to dissociate in CF$_2$ species, while the linear molecule C$_2$F$_6$ provides for a higher fraction of CF$_3$ species, which terminate polymer chains. Additions of hydrogen are used to reduce the free fluorine (reaction to HF) to obtain selectivity to silicon. Oxygen additions, on the other hand, can be used as a polymer scavenger gas (the reaction with the carbon to CO/CO$_2$ reduces the polymerisation tendency).

Self-aligned contact (SAC) schemes are used to allow for a larger miss alignment error of the lithography in printing the contacts to connect down to the doped silicon (source or drain region of the transistor). As the gate structures are protected by silicon nitride caps, the via holes can be etched down into the isolating silicon oxide inter level dielectric layer to the transistor level regardless of a slight mask miss alignment error.

The process is required to have a high selectivity at the bottom of the feature (drain source region), on the side walls, and the corners of the silicon nitride cap (see Figure 15). As the etch processes are very temperature-sensitive, care has to be taken in process control, since if the process drifts to a less selective regime, one can easily punch through the nitride cap.

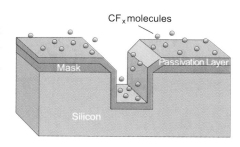
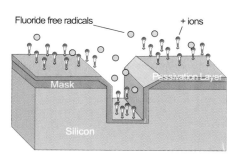
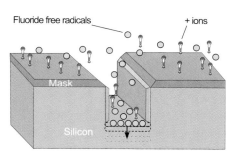

2.6.3 Etching of Metals

Aluminium Etching

Metals Al(0.5% Cu) are used to form interconnects (wiring) in the integrated circuits. The critical aspects of the processes are profile control during etching and corrosion prevention of the metal wires after etching. Fluorine based chemistries cannot be used to etch aluminium as AlF$_3$ is a non-volatile etch product. On the other hand Cl$_2$ etches pure aluminium at room temperature – even without a plasma being present – and forms Al$_2$Cl$_6$. Consequently chlorine-based chemistries are most common for aluminium etching despite being more hazardous than fluorine based chemistries. Although AlBr$_3$ is a volatile at room temperature, too, using bromine-based plasma chemistries for Aluminium etching gives no advantage.

The difficulty in etching aluminium arises from the native oxide (Al$_2$O$_3$) film on the surface (typical thickness ~3 nm). The reaction is thermodynamically "up hill". Thus, to break through the Al$_2$O$_3$, a reducing chemistry and an ion bombardment is needed. The most widely used metal etch source gases are BCl$_3$ and Cl$_2$.

Unfortunately the etched wafer surface is very reactive with oxygen and H$_2$O. To prevent corrosion from the water vapour of the atmosphere (which forms Levis acids with residual chlorine on the metal surface), a post-etch treatment is necessary. Typical approaches include in situ (i.e. before the wafer is exposed to the atmosphere) resist removing/plasma ashing and water rinse, or exposing the wafer surface to a fluorine plasma to replace the residual chlorine by fluorine.

The profile in metal etch is achieved by controlling the polymer deposition on the side walls by N$_2$ addition, preventing isotropic etching of the Al(Cu) by Cl radicals. However, it must remain thin enough so as not to introduce taper in the side wall profile.

Figure 14: ASE mechanism – courtesy of STS.

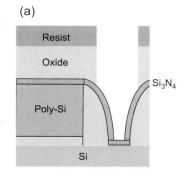

Noble Metal Etching

Noble metals like platinum or iridium are increasingly discussed as electrode materials in contact with complex oxides for future DRAM or FRAM applications. Their major drawback is that under conventional process conditions no volatile etch products exist. Even with reactive gases the etch remains mainly physical, rendering this process regime unsuitable for high-density integration as the control of the feature size is difficult and etch products redeposit easily.

One measure to cope with the low vapour pressures of the etch products is to operate the etch process at the lowest possible pressure. Unfortunately, plasma discharges become unstable below a pressure of roughly 1 µbar in most technical reactor configurations. The gas flow is usually increased as much as possible at the same time, to reduce the *residence time*, the mean time of a particle being in the chamber. Purging the reactor with a high gas flow reduces the redeposition probability of the etch products. However, the combination of a low pressure high gas flow process regime renders the pumping system for the chamber excessively expensive for commercial production.

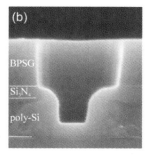

Figure 15: (a) Schematic and (b) cross-section SEM migrograph of a SAC structure.

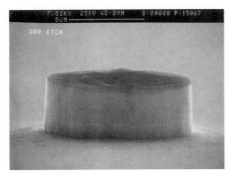

Figure 16: SEM micrograph of 5 μm deep GaAs/ AlGaAs ICP etch using a chlorine chemistry.

A more reliable and cost effective solution has been found by employing high-temperature etch processes, with wafer temperature of 200 °C and more in combination with temperature-stable hard masks. Chlorinated or fluorinated etch chemistries, for example, can produce volatile etch products under these conditions.

Oxygen-containing additives like oxygen, carbon monoxide, or carbon dioxide can be used to alter the plasma chemistry and increase the chemical etch rate of noble metals even more. In mixed Cl_2/O_2, Cl_2/CO, and Cl_2/CO_2 plasma discharges chlorine ions are produced more efficiently as in chlorine only processes due to the high dissociation and ionization potential difference of the gas additives compared to chlorine. In the presence of the oxygen containing gas additives the dissociation of chorine becomes the favourite electron impact reaction in the plasma. Furthermore, the gas additives can be used to adjust the selectivity to other layers.

For the Cl_2/CO plasma chemistry, an additional carbonyl reaction channel increases the platinum etch rate even more. At elevated temperatures of 300°C, for example, a 2-3 fold increase over the room temperature etch rate is observed [14].

2.6.4 Etching of Compound Semiconductors

Compound semiconductors are becoming increasingly important for optoelectronic applications. Elements like lasers, transistors, waveguides or distributed Bragg reflectors (DBR) are formed on the same chip.

An important requirement for etching this material class is step-free etching even of the frequently encountered heterostructures. To maintain the optical properties the side walls need to remain very smooth, a difficult task if etches several micrometer deep are required in the case of waveguides.

Figure 16 shows as an example a successful deep etch of a reflector made of a GaAs/AlGaAs hetrostructure [15].

The key to selecting the etch chemistry is the volatility of the etch products and avoiding a preferential etching of one element. Chlorine chemistries are used as well as CH_4/H_2 gas mixtures. Depending on the application, the etch rate may vary between a few nanometers and several micrometers per minute. Most processes are a combination of chemical and physical etch attack.

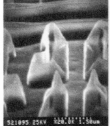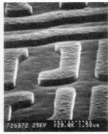

Figure 17: SEM micrograph showing the effect of a dry clean process to remove residues from a poly silicon etch step. Conventional oxygen strip with polymer remaining (left), residue-free process result (right). Courtesy of Mattson Technology.

2.6.5 Photoresist Ashing and Polymer Removal

The mask material has to be removed from the wafer surface after the etch is completed. Dry processes are the method of choice in the semiconductor industry so as to eliminate wet solvent steps and increase the process yield. Organic masks, like photoresist can easily be removed (ashed) in an oxygen plasma.

But as some etch processes generate – or even rely on – a polymer formation on the wafer surface, residues will remain after conventional oxygen strips. Figure 17 shows the cleaning of etch polymer residues from a poly-silicon etch step, which would remain on the wafer surface after a conventional oxygen ashing step. The wafer is treated with a CF_4 plasma using an Aspen Strip ICP system from Mattson Technology to remove the $Si_xCl_yBr_z$-rich polymer.

2.7 Plasma Diagnostic Methods

Monitoring the plasma is the best method to gain more knowledge about the process. While in academia the emphasis is more on methods that help to understand the process, industry likes to control established processes and catch a drift or some other abnormalities. Monitoring the plasma is often used to derive an **endpoint signal** from the measurements to stop the process at the desired etch depth. An extensive treatment of various plasma diagnostic methods is give by Manos and Dylla [3]. There are a variety of different methods that can be employed:

2.7.1 OES and Interferometry

Optical emission spectroscopy (OES) is a widely used technique in plasma processing for real-time monitoring. The emitted light from the plasma can be collected in an integral way for endpoint application. For process development and diagnostic application it may even be worth collecting the emission pointwise by scanning a fibre over the wafer to get a spatial profile.

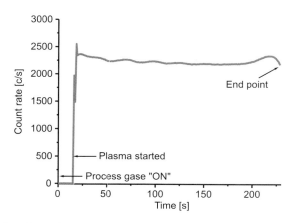

Figure 18: Endpoint trace.

In principle, all species present in the gas phase that can be excited by the plasma can be monitored, e.g. radicals, neutrals, ions. For endpoint detection one either looks at the emission from species coming from the film being etched, expecting a sharp drop off in the signal intensity when the film clears off (see Figure 18 after 225 s). Otherwise one looks for species from the film one likes to stop on, being prepared to switch the plasma off as soon as it's emission signal increases.

Quantitative species concentration can be measured if the observed species intensities can be compared to emissions from a small amount of an additional inert gas acting as an **actinometer**, usually argon, whose emission properties are known.

Monochromatic light from a laser or from a variable light source with a monochromator is directed onto the wafer in **interferometry**. The light reflects of the top surface of the thin film and the film-substrate interface, resulting in interference and modulation of the reflected intensity. The reflection is monitored while the film is etched. The film thickness d can be calculated, using the equation

$$d = \frac{\lambda}{2 \cdot n \cdot cos\theta}, \quad (5)$$

where θ is the angle from the normal incidence, n refractive index of the film, and λ the wavelength of the light. For further information on optical plasma diagnostic techniques, the reader is referred to [7].

2.7.2 Quadrupole Mass Spectroscopy

Quadrupole mass spectrometers are very powerful probes for analysing a plasma, as they allow for a direct sampling of the species in the plasma. Ions and their energy distribution can be measured. Together with an ioniser stage, neutrals and radials can be analysed.

The condition of the chamber and the purity of process gases can be easily controlled by routinely monitoring their status. Fragmentation and recombination reactions of the process gases under plasma conditions can be monitored to understand the reaction mechanisms. Detecting reaction products is a way of generating an endpoint signal.

2.7.3 Electrical Measurements

Direct measurement can be done at the generator, in-line (the match box), or at the plasma chamber (wafer electrode). Most commonly monitored are the generators power output, the reflected power, and the phase shift. At the wafer electrode the DC bias and the RF peak-to-peak voltage are monitored.

Plasma probes (*Langmuir* probes) can measure the electron density, electron temperature, and plasma potentials accurately and in a spatially resolved manner. As they are intrusive they can only be used for research purposes and not for production monitoring.

2.8 Challenges in Dry Etching

2.8.1 Feature Size Control

With an increasing wafer and a decreasing device size, the uniform feature size control over the whole wafer surface and from run to run becomes challenging. Attention has to be paid to different effects.

The etch rate of a material depends on the surface area of the etchable material present in an etch chamber (**macroscopic loading**). The etch rate changes depending on the number of wafers loaded into a barrel reactor. Even in a single-wafer reactors the etch rate for the same material will be a function of the area covered by the mask.

The term **microloading**, on the other hand, describes the local dependence of the etch rate on the local feature density. A large unmasked surface area exposed to the plasma consumes more etch species than a single trench, resulting in a local modulation of the of the plasma chemistry, and finally the etch rate. Also nested lines will show a different feature evolution than single lines (see Figure 19).

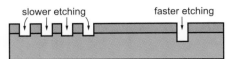

Figure 19: Microloading.

Aspect ratio dependent etch rate (ARDE) is also called **RIE-lag**. The etch rate – and in turn the etch depth – depend on the ratio of feature size and feature depth (see Figure 20). The smaller the feature becomes, the more difficult it is to bring enough species to the bottom of the feature being etched and remove the etch products. The limitation of the conductance for long small features is well known in vacuum science.

For most materials the etch rate is a function of the angle of the incident ion. As collisions of the ions are still common at the pressure levels in plasma tools, the ions do not all hit the surface perpendicularly. Depending on the angular ion distribution the vertical etch rate of the mask will increase and cause **facetes** to develop.

Redeposition of material can be another point of concern. If a process is run in a region to form an inhibitor protecting the sidewalls, sometimes even a slight process drift can be the cause of excessive **polymerisation**. This gives rise to a severe sloping of the sidewalls, if the etch does not stop at all.

For some materials the vapour pressure of the etch products is very low and at the same time their sticking coefficient is very high. This may cause the etch products to condense on any surface they encounter, in the worst case on the mask. The material will remain in place even after removing the mask material, forming free standing walls of residues (**fences**) on the wafer surface.

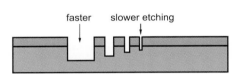

Figure 20: Aspect ratio dependent etch rate: Notice the different etch depths for different feature sizes.

2.8.2 Damage

The ions arrive at the wafer surface with an angular distribution and are reflected at the mask of the sidewall of an already etched feature. This will cause the ion intensity at the bottom of the feature to be higher than average. Depending on the dominant etch mechanism, the etch rate may increase in turn and **trenches** will develop. In severe cases the etch may even **punch through** a layer and extend to the one below.

Charging occurs in dielectric etch, when the charges from the impinging ions cannot flow to the substrate quickly enough and accumulate at the top of a feature. They deflect the trajectory of further ions coming down from the plasma and give rise to an inhomogeneous etch pattern at the foot of the feature, called **notching**. If the accumulated charge is high enough to develop large electric fields, they may even cause the gate dielectric material in the layers below to break down.

The incoming ions are stopped within a few atomic layers in the surface. Depending on their energy, they will create a **soft X-ray** emission. These x-rays may be the reason for a drift in the device characteristics, as e.g. additional electron-hole pairs will be generated to shift the threshold of a transistor.

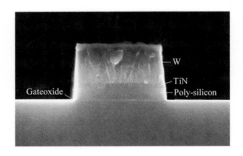

Figure 21: SEM micrograph cross-section of a dual metal gate structure etch.

2.9 Future Trends

2.9.1 Increasing Material Complexity

The complexity and the number of different material to be etched will increase in the future. For a device generation of 0.15 µm and beyond, for example, dual gates will be standard to replace poly-silicon as the only gate material. Metals like tungsten are introduced because of their low sheet resistance.

Polysilicon is still used in contact to the gate oxide to obtain a low threshold voltage. As both materials are isolated by a WN or TiN barrier three materials instead of one have to be etched, while simultaneously the feature size decreases with each device generation.

Figure 21 shows the result of a W/poly-silicon gate etch in an inductively coupled DPS tool from Applied Materials [16]. A combination of a NF_3/Cl_2-and HBr-based chemistry was used to etch the tungsten and the polysilicon, respectively.

2.9.2 Etching of Complex Oxides

The integration of complex oxides, used as high dielectric materials like $(Ba,Sr)TiO_3$ (BST) or ferroelectric materials like $Pb(Zr, Ti)O_3$ (PZT), into semiconductors is receiving more attention. Compared to the materials used up to now, this new class suffers from the common draw-back that their possible etch products have a very limited volatility, even if reactive gases are used.

To compensate the different vapour pressures of the etch products, mixtures of reactive gases are frequently used. The BST film shown in Figure 22 was etched with a Cl_2/HBr process chemistry in a TCP reactor [13]. Due to the reduced volatility of the etch products, the process has to be run at low pressures (1-5 mTorr). The process gives a steep sidewall and no surface contamination remains after the etch and mask removal.

Special attention has to be paid to avoid etching damage. Preferential etching is common, as the components have different volatilities, giving rise to surface damage, which has to be annealed in a subsequent step. Wafers etched under reactive process chemistries have to be specially treated, e.g. with rinse processes, before they can be exposed to the atmosphere again. Residual process gases like chlorine would otherwise react with the water vapour and corrode the surface.

From a production standpoint the etch processes are expensive, as the throughput is limited and the chambers have to be cleaned frequently after only etching a few hundred wafers – compared to several thousand for standard materials.

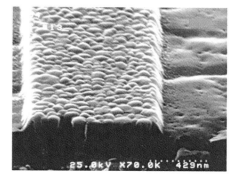

Figure 22: SEM micrograph cross-section of an etched BST line structure.

2.9.3 Structured Approach for Process Optimisation

Process optimisation sometimes involves six or more independent controllable variables, like pressure, gas composition, power delivered from the generators, and temperature. A structured approach based on statistical **design of experiment** (DoE) provides a powerful tool for process optimisation. It is more efficient than the traditional *change one variable at a time and keep all others constant*, as complete information over the parameter range is obtained and fewer experiments have to be run to obtain the same amount of knowledge.

Socalled **screening tests** are performed to identify the most important controllable variables (factors) by comparison with the experimental noise. **Response surface** experiments give the operator a detailed response variation of the process and can be used to predict the optimum process region. Ultimately, this data serves to construct a process model.

For more detailed information the reader is referred to the specific literature [11], [12].

2.9.4 Simulation

As even small improvements in the yield will result in major cost saving, the ability to simulate the process in advance has increasingly become more important. It will allow the process engineer to better design the process and avoid material loss due to mis-processing. Simulation has to be done on various length scales, ranging from a reactor-scale model down to a feature-scale model. Analytical as well as Monte-Carlo models are employed.

Reactor-scale models mainly analyse the plasma chemistry, the particle transport under the applied electrical field, and the wall interaction at low pressures. The plasma density and uniformity can be predicted (see Figure 23).

Feature-scale models like SPEEDIE from Stanford University are used for feature profile evolution [17]. The fluxes of particles from the plasma bulk to the wafer – such as the **ion energy distribution function** (IEDF) and the **ion angular distribution function** (IADF) – are needed as input data to calculate etch rates and wafer uniformity. Additionally, the **electron energy distribution function** (EEFD) needs to be known to calculate surface charging and profile artefacts like undercuts.

These models are very helpful to assess whether the equipment now in use is able to run next-generation processes with smaller feature sizes or to identify deficiencies.

II Technology and Analysis

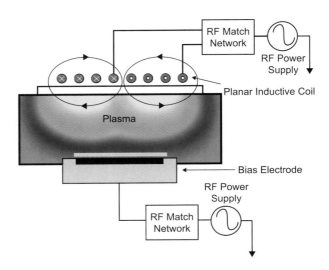

Figure 23: Model of a Lam TCP reactor. The chamber pressure and the process gases, the plasma density and uniformity can be modelled as a function of the TCP and bottom power.

3 CMP – Chemical Mechanical Polishing

Chemical mechanical planarization (CMP) is used to integrate more metal layers into a chip. Depositing an insulating layer will replicate, if not increase, the step height of the covered topology. Creating a wiring level on such an uneven surface becomes increasingly difficult the smaller the line width becomes and the more metal layers have to be integrated. The changing metal line cross section at the edges of the raised features will result in resistivity changes. Even more important now is that an uneven surface will cause problems with the small depth-of-focus window of modern lithography, already pushed to its limits.

Prior to CMP, techniques such as **thermal re-flow** (depositing a heavily doped silicon glass and performing a high temperature anneal that would smooth out step heights), **resist etchback** (depositing a mask with an inverse pattern to etch back the high spots) or **spin-on glass** (filling up the lower regions) were employed to provide local planarization. These techniques typically provided acceptable local planarization for interconnect systems with three or fewer levels of metal, but as the number of interconnect metal layers increased, they were no longer able to provide the necessary performance.

CMP has been employed for several years now to provide the global planarization required by advanced IC designs (see Figure 24) [16], [18].

The CMP process is achieved by mounting a wafer face down in a carrier assembly, which is held above a moving polishing pad. Depending on tool configuration the pad is mounted either on a rotating platen or a linear belt. The carrier itself and the wafer are rotated. A chemically active abrasive liquid (**slurry**) is introduced onto the pad, and the carrier assembly is then pressed down bringing the wafer into contact with the pad (& slurry). This chemical mechanical action results in the selective removal of material from the raised feature (high spots) on the wafer and the film being polished becomes planarized. The CMP process then continues until the desired amount of material has been removed.

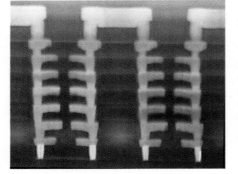

Figure 24: Use of CMP to manufacture TSMS's first 0.13 μm mixed-signal process with 8 levels of Cu/oxo-silicate-glass.

3.1 CMP Mechanisms

The CMP process, as its name implies, consists of two components: chemical etching and mechanical abrasion. The key to a successful CMP process is to achieve an appropriate balance between the chemical and mechanical components of the polishing process. This is not as simple as it sounds as the mechanisms of CMP are quite complex, and not that well understood. Mechanisms may differ by film and slurry type, and process results may be influenced by many input parameters. The removal rate for example is dependent on many variables including: *down force* (the force applied between the wafer and the pad), *pad* velocity, pad asperity, pad hardness, *slurry* type (chemistry, *pH*, particle type), slurry delivery rate, slurry distribution, *temperature*, configuration of polishing tool as well as film type. Clearly each of these variables must be precisely controlled in order to maintain a well-behaved polishing process.

The following sections provide a brief overview of some of the proposed mechanisms for dielectric and metal CMP:

3.2 Dielectric CMP

The mechanisms of CMP for silicon dioxide (SiO_2) are generally not well understood, although similarities do exist between the CMP of SiO_2 and the polishing of optical glass.

A relationship that models the optical glass polishing rate was proposed by *Preston*, and is known as Preston's law [20]. Preston's law states that the polishing rate, R_p (defined as the thickness removed per unit time) is proportional to the applied pressure (down force), P, and linear velocity, v, at which the pad moves relative to the wafer. This can be written as:

$$R_p = K_p \cdot P \cdot v, \qquad (6)$$

where K_p (Preston's coefficient) is constant for a given set of conditions. K_p is a function of the hardness and Young's modulus of the material being polished, as well as the particular slurry and polishing pad that is being used.

Various mechanisms have been proposed to explain optical glass polishing: Izumitani [21] proposed that optical glass polishing "proceeds by the formation of a hydrated layer by means of a chemical reaction between the glass surface and water and then the removal of this hydrated layer by the abrasive particles". This mechanism is consistent with the fact that the polishing rate is much higher in the presence of water. The hydrated layer that forms on the glass surface is usually produced by the ion exchange between hydrogen (or hydronium) ions and alkali (or alkaline earth) ions and generally has a much lower hardness and mechanical strength than the glass beneath it. (Optical glasses usually contain alkali or alkaline earth elements, which can exchange with the hydrogen ions in water to produce the hydrated surface layer) Plastic deformation of the hydrated layer results in mechanical removal (by a ploughing effect) when it is rubbed against harder abrasive particles in the slurry.

The polishing rate of SiO_2 is also greater in the presence of water, indicating that hydration plays an important role in the removal mechanism. However, SiO_2 does not contain any components that can exchange with hydrogen ions, and therefore hydration of the surface layer is more difficult. It has been proposed [22] that polishing instead proceeds by pressure-assisted water diffusion into the upper material layers, which allows the modified (hydrated) upper layers to be easily removed by abrasion. Abrasive particles, trapped in the pad pores, concentrate the applied pressure at points where the particles contact the SiO_2 surface causing plastic deformation of the SiO_2 layer. Frictional heating enhances this plastic deformation. The plastically deformed SiO_2 readily undergoes hydration, which generally reduces its hardness and mechanical strength. The removal of the resulting softer (mechanically weaker) hydrated surface layer then takes place by the ploughing action of the abrasive particles.

The degree of deformation is increased as the pressure applied to the pad (and particles) is increased, (also by frictional heating). The ploughing action is increased as the linear velocity of the pad is increased. The removal rate is thus dependent on both P and v as predicted by Preston's law. Dielectric CMP is typically used in the shallow trench isolation (STI) process flow (see Figure 25).

3.3 Metal CMP

The mechanisms of CMP for metals currently used in ULSI processing (W, Cu) are relatively simple in comparison to those of SiO_2. In general, the surface of the metal is first chemically oxidized, and then the metal oxide is mechanically removed from the surface by abrasion.

The CMP of many other metals is also possible, although care must be taken to choose the correct combination of chemical and abrasive content of the slurry. Metals that do not readily form oxides may be polished by more abrasive means. Abrasive particles suspended in the slurry are brought into contact with the surface being polished, causing dislocation and strain within the surface, and even the dislodgment of atoms or clusters of atoms from the surface. Chemicals in the slurry react with the dislodged materials (which have a larger surface area) or with the strained surface layers at accelerated rates and remove such material into the slurry. Chemicals in the slurry also play an important secondary role of passivating the surface being polished until mechanical abrasion removes the passivating layer allowing further etching to take place. This passivation mechanism helps in the reduction of topography on the surface.

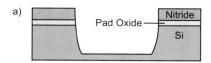
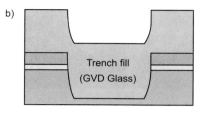
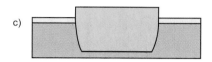

Figure 25: Simplified shallow trench isolation (STI) process flow
(a) etching of the trenches with a nitride mask,
(b) filling with CVD glass,
(c) planarization with subsequent stripping of the nitride mask.

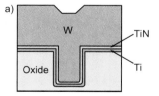

Figure 26: W plug formation using CMP:
(a) W deposition over Ti/TiN barrier layers,
(b) W and Ti/TiN CMP leaves recessed plugs,
(c) oxide CMP eliminates recess.

Figure 27: Modern commercial CMP tool: Reflexion© from Applied Materials.

3.3.1 Tungsten (W) CMP

Tungsten CMP has been developed to replace the tungsten etchback step previously used in the formation of W plugs (see Figure 26). The polishing of W proceeds by the oxidation of W to form WO_3, which is then removed by the mechanical abrasion of the slurry particles and pad. Typical chemicals used are ferric nitrate, potassium iodate and hydrogen peroxide.

A key issue in W CMP arises because the W films are deposited on top of a liner (typically TiN/Ti), which must also be removed during the same process as the W film. The polish rates of TiN/Ti however, are typically much lower than that of W. This is compounded by the necessity of having high selectivity to the underlying oxide in order to preserve the thickness of the oxide surrounding the plug. The liner film could be removed by simply overpolishing the W, but this would result in dishing of the W plug and the loss/damage of field oxide.

To overcome these issues, multi-step processes are often employed. For example the polishing could be accomplished in three steps. The first step removes the W layer stopping on the liner. The wafer is then moved to a second polish module where the liner is removed. Finally the wafer is moved again to a third polish module where an oxide-buff step is carried out to remove the oxide damage caused by the previous liner removal step.

3.3.2 Copper (Cu) CMP

Copper is replacing Aluminum as the conductor of choice in many high-performance ICs and is typically employed as the conductor in dual damascene process schemes. The damascene scheme typically uses Cu deposited on top of a Ta (or Ta/TaN) barrier metal (or liner). During the CMP process it is necessary to remove both the Cu and barrier metals whilst preserving the underlying dielectric thickness. This presents significant challenges since copper is a relatively soft metal that is readily oxidized with standard chemistry (e.g. hydrogen peroxide, hydrogen persulfate) and is easily removed with CMP, whilst Ta is a relatively hard metal that is much more difficult to remove by CMP.

The solution is again to use a multi-step process. A typical copper CMP process is carried out in three stages: The first stage removes the copper overburden from the wafer, stopping on the underlying diffusion barrier, typically Ta/TaN. Since tantalum's polishing properties differ from copper, a second stage using a different slurry is commonly used to remove the tantalum. Finally, the wafer is buffed to remove any damage to the oxide caused by the barrier removal step, and cleaned to passivate the surface and prevent corrosion.

3.4 Commercial CMP Tool Configurations

CMP tools have evolved considerably from the early machines that were used to polish bare silicon substrates. The earliest CMP processes were carried out with a stand-alone polisher, which was supported by a series of additional stand-alone systems that together provided the components of the CMP process. The modern CMP system provides fully integrated process capabilities with polishing, endpoint detection and cleaning included in a single system (see Figure 27 and Figure 29). The tools are typically based around a central robot handling system, which enables multiple polishing modules to be included on the same machine, thus allowing the efficient use of multi-step processing (see Figure 28).

For the **rotary** polisher the wafer is mounted face down into a carrier assembly, which is held above a moving polishing pad. The polishing pad is mounted on a rotating platen. The carrier itself and the wafer are also rotated. Slurry is introduced onto the pad, and the carrier assembly is then pressed down into contact with the pad (see Figure 30). Uniformity control is achieved by means of a relatively complex backside wafer carrier design that exerts differential forces on the backside of the wafer.

Orbital tools are similar to the rotary tool except that orbital motion is employed rather than rotary motion. **Linear** CMP systems are again similar to rotary systems except that the pad moves past the rotating wafer carrier in a linear motion rather than a rotary motion (see Figure 31). The potential benefit of a linear tool is that higher belt speeds can be employed (100 – 500 ft/min). A lower down-force (pressure) can also be employed for a given polish rate, due to the increased belt speed.

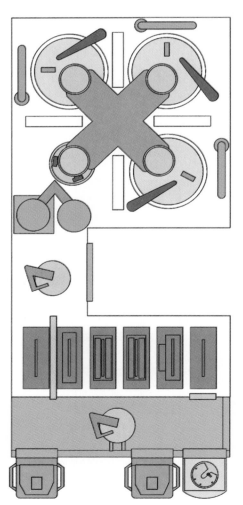

Figure 28: Schematic drawing of AMAT's Reflexion.

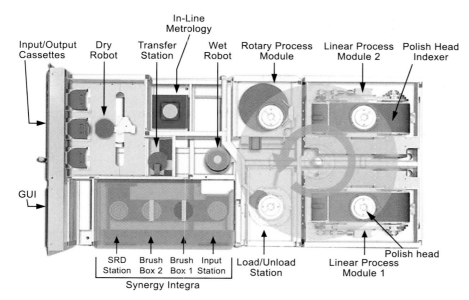

Figure 29: Modern commercial CMP tool: Teres© from Lam Research.

The linear system also allows the use of a radial zoned air bearing beneath the polishing belt, which controls uniformity via differential forces applied to the front-side of the wafer (see Figure 32). This architecture allows the wafer to be kept flat during polishing, which is a potential advantage for low-k applications.

3.5 Endpoint Detection

Endpoint detection is required to enable the CMP process to stop at the desired point in the process. Several methods are used to detect endpoint in CMP, the most useful of which use real-time in situ endpoint detection methods.

The laser interferometry technique uses a laser and detector mounted into the polishing table. The laser is aimed at the front side of the wafer through a window in the polishing pad, and the detector is used to sense the reflected light. The reflections are sampled and processed using sophisticated signal processing algorithms to remove unwanted noise and enable the endpoint to be detected.

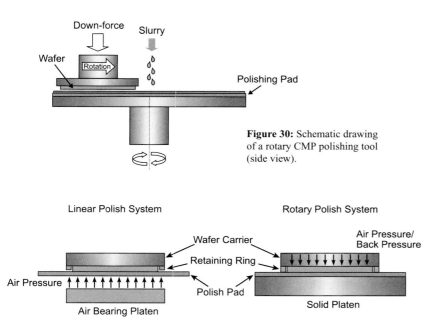

Figure 30: Schematic drawing of a rotary CMP polishing tool (side view).

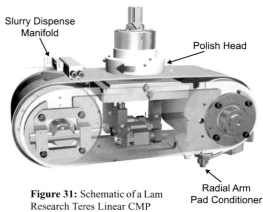

Figure 31: Schematic of a Lam Research Teres Linear CMP polishing tool (side view).

Figure 32: Comparison of the wafer holding and the force application in linear and rotary polish systems.

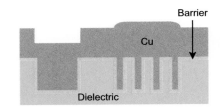

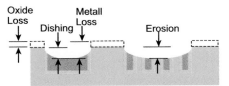

Figure 33:
(a) Before CMP: local topography after Cu deposition.
(b) After CMP: dishing erosion, and metal loss.

The spectrometer reflectivity technique uses a similar principle but enhances the technique by employing a multiple wavelength light source that is capable of monitoring film thickness for transparent films including very thin metals.

The motor current monitoring technique has also proven successful. For this technique the current supplied to the motor that drives the carrier rotation is monitored. Since the carrier is rotated at constant speed, to maintain constant removal rates, the motor current is free to vary to compensate for any load changes. Thus the motor current is sensitive to changes in friction at the wafer surface, such as when one material has been removed exposing another (CMP stop layer).

3.6 Pads

Polishing pads are typically made of cast and sliced polyurethane (with filler) or polyurethane-coated material. The surface of the pad may be smooth or shaped (grooved, perforated, porous etc.). Grooves in the pad surface aid the distribution of slurry across the pad and pores on the pad surface further aid slurry distribution. It is generally accepted that the hardness and porosity of the pad are key parameters in determining the final polish results. Harder and stiffer pads are known to provide better local planarization capability, whereas softer pads provide better uniformity of removal rate.

Stacked pads are often employed as a compromise, consisting of hard upper pad, providing planarization efficiency at the local scale, and soft pad, capable of following the shape of large substrate features and suppressing the substrate-related large-scale planarization.

One critical aspect of CMP is that the pad requires constant **conditioning** in order to maintain consistent removal rates [23]. The pad surface can become glazed during the polishing process reducing pad asperity and hindering slurry distribution. To overcome this effect pad conditioning is employed, whereby microscratches are intentionally formed in the pad surface during polishing. This conditioning is typically carried out by sweeping the surface of the pad with a "pad conditioner", the surface of which is impregnated with diamonds. Conditioning allows each wafer to be exposed to a pad with the same surface condition.

3.7 Slurry

A slurry is a stabilised suspension of small (<100 nm) abrasive particles in water with some dissolved chemicals. The most widely used abrasive materials are alumina or silica for metal-CMP, and silica or cerium oxide (ceria) for oxide-CMP. The abrasive particles are chosen to have roughly the same hardness as the material being polished. Dissolved chemicals are introduced to provide better wetting conditions, modify the polished surface, stabilise the suspension, and influence the friction coefficient and the solubility of the process by-products.

The chemical composition and the concentration of abrasive particles (usually quoted in wt %) are key parameters in determining the performance of the slurry. The size and size distribution of the particles is also important.

Slurry management is a key issue for the CMP process. The slurry must be mixed, stored and then dispensed to the CMP system in a reliable and repeatable way.

As already described a slurry consists of a suspension of small particles in water with some dissolved chemicals. Unfortunately the particles have the tendency to agglomerate into clusters or clumps or to settle out into the bottom of the storage container. For silica-based slurries the pH of the solution is carefully controlled to prevent this. At a pH above 7.5 the silica particles attain sufficient surface charge to generate electrostatic repulsion which maintains particle dispersion. For metal CMP slurries it is necessary to provide constant mechanical agitation to maintain the particle suspension.

Slurries have a finite shelf life ranging from one week to several months. After this time the agglomerate level becomes excessive and the slurry becomes unusable. The combination of particles and aggressive chemicals also presents a harsh environment to the components of the delivery system, requiring stringent design constraints to be employed.

Figure 34: An example of line corrosion in Cu CMP.

3.8 Post CMP Cleaning

CMP is an inherently dirty process introducing defects onto the wafer surface that can detrimentally affect wafer yield. Slurries tend to dry rapidly on the wafer once they are removed from the wet environment of the CMP tool, and, once dry, subsequent cleaning of the wafer becomes extremely difficult. The wafer is therefore kept wet throughout the process.

Directly following CMP the wafer is contaminated with a combination of residual slurry particles, CMP by-products, metal contaminants and chemicals, all of which can cause problems for subsequent processing. The wafer is therefore subjected to post CMP cleaning steps that remove these contaminants and return the wafer to a clean state acceptable for further processing. The cleaning technology of choice is double-sided brush scrubbing. This technique may sometimes be used in conjunction with megasonic cleaning, where sonic waves ($f = 0.7 - 1.0$ MHz) are employed to remove particles.

The cleaning typically starts with a de-ionized (DI) water rinse immediately following the CMP step. The water rinse removes the majority of slurry from the wafer surface and stops any further chemical reaction from taking place. The wafer is then moved into the brush scrubber where soft polyvinyl alcohol (PVA) brushes are brought into contact with both sides of the wafer. The brushing action, in conjunction with chemistry, facilitates the removal of unwanted material. Following scrubbing the wafer is moved into a spin rinse drying station to rinse and dry the wafer.

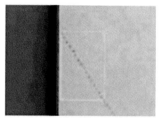
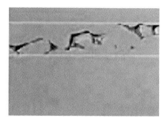

3.9 Some CMP Problems

Dishing is the term given to the thinning (or loss in thickness) of the material being polished. Conventional dishing occurs due to the flexing of the polishing pad into a trench, which then forms a concave surface which propagates with the polish front leaving the trench material "dished" at the end of the process (see Figure 33). Dishing is generally greater for larger features.

Erosion is the term given to the local loss of "supporting material" in the area surrounding the trench. Erosion is typically greater for small-pitch, high-density lines. **Oxide/nitride loss** is defined as the loss in thickness beyond the targeted thickness. **Metal loss** is defined as the sum of **dishing, erosion** and **oxide loss**. In copper CMP for example it is desirable to minimizes metal loss since the thinning of the copper lines beyond target thickness gives rise to increased line resistance and RC delay time, degrading device performance.

Figure 35: Examples of Cu scratches.

3.10 Defects

A whole host of defects can be induced or revealed by the CMP process [24]. CMP-induced defects include metal residues, corrosion, scratches (see Figure 34 and Figure 35), pinholes and microcracks. Other defects may be revealed by CMP, such as voids, seams, embedded particles, incomplete patterning (caused by errors in either lithography or etch steps), and random electroplate thickness variations (see Figure 36).

3.11 Future Technology Trends / Challenges

CMP challenges involve addressing incoming film non-uniformity and materials issues while minimising dishing, erosion and overall material loss of the underlying features. Current consumables and CMP hardware are struggling to keep up with the ever-increasing demands of the industry roadmaps, and considerable effort is being expended to improve their performance.

For copper CMP the material properties of copper and low-k dielectrics significantly impact the process. Since copper is softer than oxide and tungsten, it tends to scratch more easily and is more prone to dishing in open areas. Low-k materials exhibit inferior mechanical properties to SiO_2 in terms of hardness, modulus, and adhesion to underlying or overlying films. Since CMP can subject the wafer to significant shear forces during processing, the integrity of the low-k material and its interfaces to underlying or overlying films may be compromised, leading to possible cracking of the dielectric or delamination from adjacent films. As porosity is introduced to low-k materials to further reduce the dielectric constant; the mechanical properties of these materials are further compromised, leading to significant challenges for CMP. To address the issue of

Figure 36: Residual barrier pools created by prior level non-planarity – difficult to remove.

cracking and delamination of low-k films, and to improve planarization capability, lower down force capability and advanced edge control (3.0 – 0.5 PSI) that reduce the shear forces to which the wafers are subjected to are being explored. New low-abrasive or abrasive-free slurries are also being explored. These provide better microscratch performance and defect control, lower dishing and erosion, and better planarization, when compared to traditional abrasive-based slurries.

Low-k materials also present challenges for post CMP cleaning since their surfaces are typically hydrophobic and difficult to clean using conventional techniques.

For STI the trend is towards the adoption of direct STI processes (rather than reverse mask processes), which place stringent demands on the consumables in terms of selectivity and defect control.

On the hardware side the trend is towards the inclusion of in-situ metrology to enable film thickness and defects to be monitored. Pre and post CMP thickness measurements will be possible, allowing the potential for the CMP process to be fine tuned to each individual wafer.

Fixed abrasives (FA) have also received some attention, whereby the abrasives are built-into or permanently fixed at the polishing pads rather than being introduced within the slurry.

Finally, alternatives to CMP, such as electrochemical polishing [25] are under investigation.

Acknowledgements

Our sincere thanks are due to numerous colleagues at Alcatel, Applied Materials, EKC Technology, Lam Research, Mattson Technology, Oxford Instruments Plasma Technology, and Surface Technology Systems who helped us to write this chapter and contributed material. The editor gratefully acknowledges the presentation of Dr. B. Spangenberg (Institute of Semiconductor Technology, RWTH Aachen University) as a part of the 2001 IFF Spring Course and the very useful comments on the preparation of this chapter. Thanks are due to Carsten Kügeler and Stefan Tappe (RWTH Aachen) for corrections.

References

[1] M. Köhler, *Ätzverfahren für die Mikrotechnik*, Wiley-VCH, 1998.
[2] D. G. Schimmel, J. Electrochem. Soc. **126**, 479 (1979).
[3] D. M. Manos and D. L. Flamm (Ed.), *Plasma Etching*, Academic Press, 1989.
[4] S. A. Campbell, *The Science and Engineering of Microelectronic Fabrication*, Oxford University Press, 1996.
[5] H. Frey and G. Kienel, *Dünnschichttechnologie*, VDI-Verlag, 1987.
[6] G. E. McGuire, *Semiconductor Materials and Process Technology Handbook*, Noyes Publication, 1988.
[7] Gary S. Selwyn, *Optical Diagnostic Techniques for Plasma Processing*, AVS Monograph Series M-11, AVS Press, 1993.
[8] P. C. Thonemann, Progress in Nuclear Physics, 219 (1953).
[9] B. Wolf (Ed.), *Handbook of Ion Sources*, CRS Press, 1995.
[10] S. Hu, S. Altmeyer, A. Hamadi, B. Spangenberg, and H. Kurz, J. Vac. Sci. Technol. **B16(4)** (1998).
[11] S. Wolf and R. N. Tauber, *Silicon Processing for the VLSI Era*, Lattice Press, 1986.
[12] D. C. Montgomery, *Design and Analysis of Experiments*, John Wiles & Sons, 1997.
[13] St. Schneider, M. Kennard, and R. Waser, Mat. Res. Soc. Symp. Proc. **596**, Y9.6 (2000).
[14] St. Schneider, H. Kohlstedt, and R. Waser, Mat. Res. Soc. Symp. Proc. **688**, C5.6 (2002).
[15] Oxford Plasma Technology, Application Lab, personal communication.
[16] T. Lill and J. Holland, Applied Materials, personal communication.
[17] J.P. McVittie, S.A.Alibeik, D.S. Bang, J.S. Han, K. Hsiau, M.V. Joshi, P. Kapur, J. Li, B.P. Shieh, J. Zheng, K.C. Saraswat, *SPEEDIE*, Stanford University, 1998.
[18] Michael A. Fury, Solid State Technology **May**, 81 (1997).
[19] J.M.Steigerwald, S.P.Muraka, R.J.Gutmann, *Chemical Mechanical Planarization of Microelectronic Materials*, John Wiley & Sons, 1996.
[20] F.W.Preston, J. Soc. Glass Technology **11**, 214 (1972).
[21] T.Izumitani, S.Harada, J. Soc. Glass Technolology **12**, 131 (1971).
[22] Minoru Tomozawa, Solid State Technology **July**, (1997).
[23] Iqbal Ali, Solid State Technology **June**, (1997).
[24] Sumit Guha et. al., Solid State Technology **April**, (2001).
[25] D.H.Wang, P.H.Yih, S.S.Chiao, AVS 3rd International Conference on Microelectronics and Interfaces, February 11-14, 97 (2002).

Analysis by Diffraction and Fluorescence Methods

Oliver H. Seeck, Research Center Jülich, and DESY, Hamburg, Germany

Contents

1 Introduction — 273

2 X-ray Analysis Methods — 274
2.1 Introduction — 274
2.2 Principles of X-ray Scattering Experiments — 275
2.3 X-ray Scattering at Crystals — 276
2.4 An Example: Buried $CoSi_2$-Layers Implemented in a Si-Substrate — 278
2.5 X-ray Scattering at Small q — 280
2.6 An Example: Metallic Thin Films — 280
2.7 X-ray Fluorescence and Absorption Spectroscopy — 281

3 Electron Analysis Methods — 282
3.1 Introduction — 282
3.2 Electron Optics — 282
3.3 Transmission Electron Microscope (TEM) — 283
3.4 Scanning Electron Microscope (SEM) — 285
3.5 Electron Energy Loss Spectroscopy (EELS) — 287

4 Surface Sensitive Analytical Methods — 289
4.1 Introduction — 289
4.2 Secondary Ion Mass Spectrometry (SIMS) — 289
4.3 X-ray Photoelectron Spectroscopy (XPS) and Auger Electron Spectroscopy (AES) — 290
4.4 Low-Energy Electron Microscopy and Related Techniques — 292

5 Some Other Methods — 293

Analysis by Diffraction and Fluorescence Methods

1 Introduction

In the last few years, the use of advanced electronic materials has become crucial for the progress in the information technology. Knowledge about the properties of these materials and how to control the process of making them is of decisive importance for the whole industry. The resulting devices such as hard disks, micro machines, laser diodes or integrated circuits are not just simple one component systems, in most cases. Usually, they are made in several steps by stripping, deposition, or implementation of material from, onto, or into a substrate.

To become operative the devices must be prepared very carefully. Even very small mistakes during the manufacturing process lead to malfunctioning or even to complete uselessness of the device. To determine the success of the processing *near-surface sensitive analytical methods* must exist which are capable in reliably investigating the physical and chemical properties of the single components of the devices. Most important are lateral distances, layer thicknesses, and structures, surface and interface quality, crystallinity, and optical, electronic and magnetic properties.

To investigate the desired properties the adequate analytical method has to be chosen. First of all, the interaction of the probe (photons or particles) with the sample has to be considered. Usually, a sufficiently strong interaction is needed to get reliable results and the probe has to be chosen adequately. For instance, to investigate magnetic properties neutrons are suitable because there is a strong interaction of magnetic fields with the spin of a neutron. In contrast, choosing ions as a probe would not give any results as no magnetic interaction occurs.

Additionally, other aspects have to be considered. The following list shows one possible way to classify the methods. By no means is it complete but it visualizes the difficulties which arise when looking for the *proper* experiment.

- **Sensitivity**
  E.g. to specific chemical elements, magnetism, electronic or optical properties.
- **Resolution**
  E.g. spatial resolution, energy resolution for electronic or inelastic measurements.
- **Penetration Depth**
  From macroscopic scale (millimeter) to investigate the bulk to sub-nanometer (angstrom, 1 Å = 0.1 nm) scale to investigate the surface properties only.
- **Destructivity**
  YES such as *Transmission Electron Microscopy* (TEM) or *Secondary Ion Mass Spectrometry* (SIMS). Or *NO* such as most scattering methods, *X-ray Photoemission Spectroscopy* (XPS) or *Scanning Tunneling Microscopy* (STM). For some methods (e.g. X-ray scattering) the destructivity is not always clear but depends on the samples and other parameters.
- **Sensibility**
  E.g.: to impurities in Parts Per Million (ppm) or in percent range only.
- **Interpretation**
  The properties of the samples are directly visible (most microscopic methods) or the measurements have to be analyzed carefully with theory functions (fitting, refinements) to obtain the desired results (most scattering methods).

Another possibility for classification is the allocation of the method to the incident exciting probe and the emitted analyzed response. Most frequently, electrons, photons or ions are used to illuminate the sample and for detection. The information about the sample can mostly be derived form the change of the energy, the momentum, the polarization or the spin of the probed particles. Table 1 lists some analytic methods.

II Technology and Analysis

Table 1: Selected near surface sensitive investigation methods.
Abbreviations:
TEM: Transmission Electron Microscopy
SEM: Scanning Electron Microscopy
LEEM: Low Energy Electron Microscopy
EELS: Electron Energy Loss Spectroscopy
AES: Auger Electron Spectroscopy
IPES: Inverse Photo Electron Spectroscopy
EDX : Energy Dispersive X-ray Analysis
RBS: Rutherford Backscattering Spectrometry
SIMS: Secondary Ion Mass Spectrometry
XPS: X-ray Photoelectron Spectroscopy
PEEM: Photo Emission Electron Microscopy
XRD: X-Ray Diffraction
XRF: X-Ray Fluorescence
EXAFS: Extended X-ray Absorption Fine Structure

		Emitted and analyzed response		
		Electrons	Ions	Photons
Incident excitation probe	Electrons	TEM, SEM, LEEM, EELS, AES		IPES EDX
	Ions		RBS SIMS	
	Photons	XPS PEEM		XRD EXAFS XRF Ellipsometry

In this article the most important and common analytical methods are presented. As there are many of them, most descriptions will remain somehow qualitative. In this case the reader is encouraged to view the references for further information. The *X-Ray Diffraction* methods (XRD) and the microscopic methods using electrons (SEM, TEM) will be explained in more detail. Some attention is also given to other methods such as *Electron Energy Loss Spectroscopy* (EELS) and fluorescence methods (XRF) and, particularly to surface analytical methods (SIMS, XPS, AES, LEEM, PEEM). Very briefly mentioned will be other methods such as *Rutherford Backscattering Spectrometry* (RBS). Scanning probe methods (STM, AFM) are not topic of this article. They are discussed in Chapter 12.

2 X-ray Analysis Methods

2.1 Introduction

Nowadays, X-ray (or neutron) diffraction methods are widely used to investigate the structural properties of crystals, amorphous samples or layer systems. The power of these methods has its basis in the enormous amount of information which can be extracted from the data. Depending on the parameters which are chosen for the scattering experiment X-ray scattering can be bulk sensitive or surface sensitive only. It can easily be used with a resolution in the sub-angstrom range but it can also monitor distances on a micrometer scale. Some methods are extremely sensitive to impurities or deviations from the equilibrium state other methods are not. Finally, it can be used to distinguish between the chemical elements (at least roughly) or between charge or magnetic properties. However, the great disadvantage of X-ray diffraction methods is, that the data has to be analyzed in order to get the sample properties. Also, in many cases the result is not clear or unique, mostly due to the fact that some information gets lost during the scattering process (as described later on).

For an X-ray scattering experiment the sample is illuminated by electromagnetic waves. The wavelength of the photons λ is usually smaller than 10 nm (= 100 Å) which corresponds to an energy of $W_{h\nu}$ larger than 100 eV (via $W_{h\nu} = 12398.53$ eV/λ[Å]). The low energy photons from 100 eV to about 5 keV are called *soft X-rays*, high energy photons ($W_{h\nu} > 30$ keV) *hard X-rays*. A typical X-ray diffraction experiment is carried out at a photon energy of about $W_{h\nu} \approx 10$ keV which corresponds to a wavelength of about $\lambda \approx 1$ Å. To produce X-ray photons the effect is used that accelerated charged particles (such as electrons or positrons) radiate. For many general purposes with no special requirements on the primary beam, X-ray tubes are utilized (see Figure 1). Electrons with an energy around 50 keV are shot onto a target (called *anode*) usually made from copper or molybdenum. When the electrons hit the target they are drastically decelerated. The resulting radiation contains X-ray photons with an energy up to 50 keV with maximum flux at the characteristic lines of the anode material. They are located at e.g., $W_{h\nu} = 8.048$ keV for CuKα_1-radiation or $W_{h\nu} = 17.479$ keV for MoKα_1-radiation.

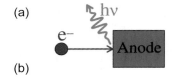
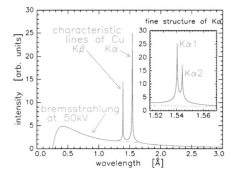

Figure 1: (a): Principle of an X-ray tube. Electrons with an energy W_m hit the anode (usually Cu or Mo). Due to the decelerating at the anode surface the electrons cause radiation with a maximum photon energy of $W_{h\nu} = W_m$. This bremsstrahlung spectrum is superposed by characteristic lines of the anode material (b). The intensity of the lines is much higher than the bremsstrahlung background and is used for the scattering experiments.

If extreme high intensity, polarized beam or almost energy independent flux is required an X-ray tube is not sufficient as a photon source. Instead, synchrotron radiation facilities can be used. They are available all around the world (e.g.: ESRF [Grenoble, France], HASYLAB [Hamburg, Germany], APS [Chicago, USA], SPRING8 [Nishi Harima, Japan]). In this case, an electron (or positron) beam is accelerated with specially designed magnetic devices (bending magnets, wigglers, undulators) [e.g. 1]. The radiation covers photon energies from the visible light range (some eV) up to some 100 keV depending on the synchrotron. A sketch how the X-rays are produced by a bending magnet is shown as an inset in Figure 2.

For relativistic electrons it is evident that the total radiated power $P_{BendMag}$ from a bending magnet depends on their energy W_e (so on the relativistic mass), the magnetic field B and the number of electrons (the current I) passing through the device. In good approximation one gets

$$P_{BendMag}[kW] = 26.5 W_e^3[GeV] B[T] I[A] \tag{1}$$

where the brackets denote the units of measure to be used. Thus, the total power strongly depends on the energy of the electrons. Typical values are $W_e = 4$ GeV, $B = 1$ T and $I = 100$ mA and it follows $P_{BendMag} \approx 100$ kW. This enormous power gives photons with an almost white spectrum down to a critical wavelength

$$\lambda_c[\text{Å}] = \frac{18.64}{B[T] W_e^2[GeV]} \tag{2}$$

For shorter wavelengths the intensity dramatically drops (see Figure 2). The quasi-white beam is usually adapted to the experimentators needs by optical devices such as focusing mirrors, apertures and monochromators. Afterwards, the primary beam has a well defined collimation and energy bandwidth.

The intensity of the synchrotron radiation can be increased easily by installing many bending magnets with flipping polarity one after the other. The electron beam wiggles through the device (which is therefore called wiggler) and the resulting radiation is the incoherent sum of the radiation produced at the sub-devices. In this way, the intensity can be amplified by a factor of 50 ... 100 compared to a single bending magnet. By choosing proper magnetic fields and distances between the single components a so-called undulator is gained. Radiation of particular wavelength adds up coherently and the resulting brightness is a function of N^2 where N is the number of sub-devices. Therefore, the peak intensity can be 10^4 times larger than observed from a single bending magnet [1].

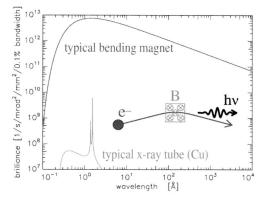

Figure 2: Comparison of the brilliance between a typical tube-source and a bending magnet. The 0.1 % bandwidth is according to the selected energy. The inset shows the principle of producing X-ray photons at a synchrotron radiation source. An electron beam is bent by a magnetic device. At the bending position X-ray photons are emitted.

2.2 Principles of X-ray Scattering Experiments

If an X-ray beam hits a sample the photons are scattered by charged particles in the illuminated volume [2]. These can be electrons or the nuclei of the atoms. However, the intensity of the scattered beam is proportional to $1/m^2$ in which m denotes the mass of the particles (*Thomson scattering*). Therefore, only the electron density $\rho(r)$ has to be considered. Figure 3 displays a sketch of an X-ray scattering experiment.

For a simple description of a scattering experiment some approximations have to be considered. First of all, elastic scattering is assumed. This means, that the energy (and wavelength, respectively) of the scattered photon does not change during the scattering process. In other terms: the mean value of the incident wave vector $|k_i|$ equals the mean value of the outgoing wave vector $|k_f|$, thus $k_{if} = |k_i| = |k_f| = 2\pi/\lambda$. In the case of photon energies around 10 keV this approximation is very well fulfilled. An energy loss or gain due to involving phonons (inelastic scattering) is usually less than 1 eV and this effect becomes negligible at 10 keV. Furthermore, multiple scattering processes will be neglected. This means, that a photon is scattered not more than once in the illuminated sample volume.

In simplest approximation, a sample is illuminated by a plane wave and each charged particle of the sample scatters the plane wave causing a spherical wave $\exp(ik \cdot r)$ with the wave vector k. Summing up all scatterers the scattered intensity is in the limit of this so-called *Born approximation* [2] simply given by

$$I(q) \propto \left| \int \rho(r) \exp(iq \cdot r) d^3 r \right|^2 \\ \propto \left| FT\{\rho(r)\} \right|^2 \tag{3}$$

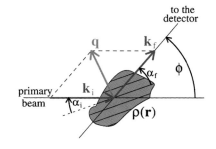

Figure 3: Sketch of a typical X-ray scattering experiment. The monochromatic beam (all photons have the same energy) hits the sample with the electron density $\rho(r)$. The incident wave vector is denoted by k_i. The intensity along the direction of the outgoing wave vector k_f is detected under an angle ϕ with respect to the incident beam. The vector q is called *wave vector transfer*. It determines the information which can be extracted from the scattered signal. The sample orientation is defined by additional angles (such as α_i or α_f in the sketch).

with the *wave vector transfer* q defined by

$$q = k_f - k_i \tag{4}$$

$$q = |q| = 2k_{if} \sin\frac{\phi}{2} = \frac{4\pi}{\lambda}\sin\frac{\phi}{2} \tag{5}$$

considering the scattering angle ϕ which is shown in Figure 3. The operator $FT\{\}$ in Eq. (3) denotes the 3-dimensional *Fourier transformation*.

If $I(q)$ is monitored for all scattering angles ϕ and all sample orientations (so for all possible q-values) the real space electron density can be deduced by performing a Fourier backtransformation. It has to be mentioned, that this backtransformation does usually *not* yield a unique result. Reason is the missing phase information due to the absolute square performed in Eq. (3). However, this lack of information can mostly be compensated be including pre-knowledge of the sample properties during the analyzing process.

2.3 X-ray Scattering at Crystals

Many solid samples, especially when they are used in semiconductor industry, have crystalline structure. This means that the electron density is periodical in three space directions. A simple example of a so-called *cesium chloride* (CsCl)-*lattice* is displayed in Figure 4. It shows a cubic lattice with a periodicity of a (the *lattice constant*) in each space direction x, y and z. The cell which is spanned by the vectors $a_1 = (a,0,0)$, $a_2 = (0,a,0)$, and $a_3 = (0,0,a)$ is called *unit cell*.

In the case of non-cubic lattices these vectors may have different lengths and/or directions. The only requirements on the set $a_{1,2,3}$ are, that non of them is 0 and that they are pairwise linearly independent from each other. Using the lattice vectors $a_{1,2,3}$, the electron density of a crystal can be expressed as

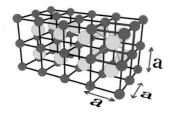

Figure 4: Sketch of a CsCl-lattice. The small spheres are the Cs$^+$-ions the large the Cl$^-$-ions. The electron density repeats in all three dimensions with a periodicity of a which is the lattice spacing (lattice constant). One cube with dimension a is called unit cell. In summary, it contains two ions: A Cs$^+$-ion in the center and 8 times 1/8 Cl$^-$-ions at the corners (which is equivalent to one Cl$^-$-ion at one corner). The arrangement of these two ions in the unit cell is called base.

$$\rho(r) = \sum_{n=(0,0,0)}^{N=(N_1,N_2,N_3)} \rho_{\text{cell}}(r + R_n) \quad \text{with} \quad R_n = n_1 a_1 + n_2 a_2 + n_3 a_3 \tag{6}$$

in which ρ_{cell} is the electron density of one unit cell and N_j is a measure of the size of the crystal given by the number of lattice vectors in each lattice direction. With this expression of the electron density Eq. (3) can be rewritten:

$$\begin{aligned} I(q) &\propto \left| \sum_n^N \int \rho_{\text{cell}}(r + R_n) \exp(iq \cdot r) \mathrm{d}^3 r \right|^2 \\ &\propto \left| \sum_n^N \int \rho_{\text{cell}}(r) \exp(iq \cdot r) \exp(-iq \cdot R_n) \mathrm{d}^3 r \right|^2 \\ &\propto \left| \left(\int \rho_{\text{cell}}(r) \exp(iq \cdot r) \mathrm{d}^3 r \right) \sum_n^N \exp(-iq \cdot R_n) \right|^2 \\ &\propto |F(q)|^2 \left| \sum_n^N \exp(-iq \cdot R_n) \right|^2 \end{aligned} \tag{7}$$

The function $F(q)$ is called *structure factor* and is the Fourier transformation of the electron density of the unit cell. Equation (7) can be simplified even more.

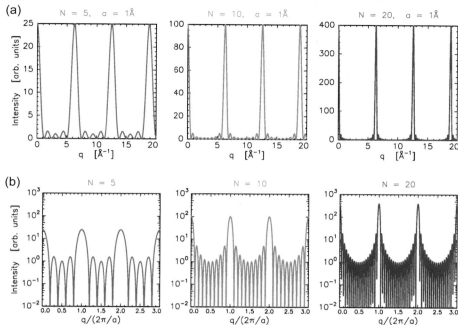

Figure 5:
(a) One-dimensional Laue functions calculated for different numbers of unit cells ($N = 5, 10, 20$ from left to right). The larger the value of N the more pronounced is the peaking character of the maxima which appear at $q = 2\pi/a$ with the lattice constant a. Also, the oscillations between the peaks seem to vanish for increasing numbers of N.
(b) Same calculation as shown in figure 5a, but in a log-scale and with normalized q. It can be seen, that the frequency of the Laue oscillations between the peaks increases with increasing N. Also, the intensity of the peaks increases.

$$I(\mathbf{q}) \propto |F(\mathbf{q})|^2 \left| \sum_{n_1 n_2 n_3}^{N_1 N_2 N_3} \exp\left(-i\mathbf{q}\cdot[n_1\mathbf{a}_1 + n_2\mathbf{a}_2 + n_3\mathbf{a}_3]\right) \right|^2$$

$$\propto |F(\mathbf{q})|^2 \prod_{j=1}^{3} \left| \frac{\sin\left(\frac{1}{2}N_j\mathbf{q}\cdot\mathbf{a}_j\right)}{\sin\left(\frac{1}{2}\mathbf{q}\cdot\mathbf{a}_j\right)} \exp\left(-i\left[\frac{N_j}{2}-1\right]\mathbf{q}\cdot\mathbf{a}_j\right) \right|^2 \quad (8)$$

$$\propto |F(\mathbf{q})|^2 \prod_{j=1}^{3} \frac{\sin^2\left(\frac{1}{2}N_j\mathbf{q}\cdot\mathbf{a}_j\right)}{\sin^2\left(\frac{1}{2}\mathbf{q}\cdot\mathbf{a}_j\right)}$$

Equation (8) shows that the scattered intensity breaks up into two functions: The above mentioned structure factor and the product expression which is called *Laue function*. One-dimensional Laue functions are depicted in Figure 5a on a linear scale and in Figure 5b on a logarithmic scale.

It can easily be seen that the on-dimensional Laue function exhibits pronounced maxima at the so-called *Bragg condition* $q_{\text{Bragg}} = 2\pi n/a$ with n being an integer number. In the general three-dimensional case this condition becomes

$$\mathbf{q}_{\text{Bragg}} = 2\pi h \frac{\mathbf{a}_2 \times \mathbf{a}_3}{\mathbf{a}_1 \cdot (\mathbf{a}_2 \times \mathbf{a}_3)} + 2\pi k \frac{\mathbf{a}_3 \times \mathbf{a}_1}{\mathbf{a}_1 \cdot (\mathbf{a}_2 \times \mathbf{a}_3)} + 2\pi l \frac{\mathbf{a}_1 \times \mathbf{a}_2}{\mathbf{a}_1 \cdot (\mathbf{a}_2 \times \mathbf{a}_3)} \quad (9)$$

with $h, k, l \in \mathbb{Z}$.

The integer numbers h, k, and l are called *Miller indices*, each term in Eq. (9) denotes a reciprocal lattice vector which spans the reciprocal lattice. For large crystals (so for very large numbers of N_1, N_2, and N_3) the Bragg peaks at $\mathbf{q}_{\text{Bragg}}$ become very intense and narrow delta peaks. If only very few unit cells are involved in the scattering process, the peaks show characteristic full widths at half maximum (see the linear plot in Figure 5a) and Laue interference at their tails (see the log-representation in Figure 5). Thus, from this information the size of the scattering particles can be deduced [2].

II Technology and Analysis

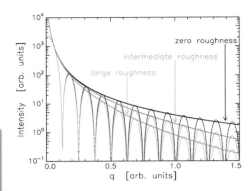

Figure 6: Calculated scattering of a single crystal film of 50 monolayers. The drop of the scattered intensity in the regime of the Laue oscillations depends on the surface roughness of the film. Zero roughness means a complete top monolayer. Intermediate roughness corresponds to a not fully occupied top monolayer. Large roughness means that the two topmost monolayers are not completely occupied. All units are arbitrary. The envelope lines are modified Lorentzians including surface roughness but excluding the Laue interferences.

An example of the one-dimensional case presented in Figure 5a,b would be a scattering experiment with the wave vector transfer q perpendicular to the surface of an epitaxially grown thin film of a simple cubic material with one atom in each unit cell. In this case, N corresponds to the number of the atomic layers and the Laue-oscillations are connected with the thickness of the film.

The quality of the film can also be deduced from the measurements: The perfectness of crystallization and the roughness at the surface determine the envelope of the intensity between the Bragg peaks. For perfectly smooth layers without any defects it follows a Lorentzian shape in good approximation. Surface roughness or defects in general lead to a characteristic drop of the scattered intensity for large wave vector transfers (see Figure 6).

The intensity of the Bragg peaks is modified by the absolute square of the structure factor (see Eq. (8)). It contains the information about the electron density $\rho_{cell}(r)$ of the unit cell whereas the Laue function contains the information about the size of the unit cell and the size of the scattering particles. In special cases, the structure factor can be zero at a Bragg condition which means that some Bragg reflections do not exist. As an example, this may be explained by using the lattice shown in Figure 4 but with the additional condition that both atoms (ions) are the same. This so-called *Body Centered Cubic (bcc) structure* appears e.g., in chromium, manganese or iron. For bcc-structured lattices the electron density of the unit cell $\rho_{cell,bcc}$ can be written using the electron density of an atom ρ_{atom} as

$$\rho_{cell,bcc}(r) = \rho_{atom}(r) + \rho_{atom}\left(r + \frac{1}{2}\begin{pmatrix} a \\ a \\ a \end{pmatrix}\right) \quad (10)$$

Therefore, the structure factor at the Bragg condition of a bcc-lattice becomes:

$$F_{bcc}(q_{Bragg}) = \left(\int \rho_{atom}(r)\exp(iq_{Bragg}\cdot r)d^3r\right)\left(1 + \exp\left[-\frac{iq_{Bragg}}{2}\cdot\begin{pmatrix} a \\ a \\ a \end{pmatrix}\right]\right) \quad (11)$$

Using Eq. (9) this simplifies to

$$F_{bcc}(q_{Bragg}) = \left(\int \rho_{atom}(r)\exp(iq_{Bragg}\cdot r)d^3r\right)\left(1 + \exp[-i\pi(h+k+l)]\right) \quad (12)$$

It can be seen from Eq. (12) that the structure factor vanishes for all odd $h+k+l$. Therefore, for example in bcc-lattices the reflection with $h = 1$, $k = 0$, and $l = 0$ is forbidden whereas the $h = 1$, $k = 1$, and $l = 0$ reflection is allowed. These so-called *selection rules* (or also *extinction rules*) are different for each type of crystal lattice and are usually more complicate than in the presented case of a bcc-lattice.

2.4 An Example: Buried CoSi$_2$-Layers Implemented in a Si-Substrate

In this section an example is presented how the measurement of Bragg reflections can be used to determine the quality and the properties of a thin buried metallic layer in a single crystal silicon substrate. A *buried layer* is not prepared on top of the surface of the substrate but deposited inside with a depth D which may by several 10 nm large. The thickness of the layer d is usually in the range of some 10 nm or even less. These layers can be used as highly conducting paths which connect semiconductor circuits in Si-chips.

As a metal CoSi$_2$ can be chosen. It is a very good conducting material. The crystalline structure almost matches the structure of silicon: Both are cubic, the lattice constant of silicon is a_{Si} = 5.431 Å that one of CoSi$_2$ is a_{CoSi2} = 5.365 Å. However, this small mismatch may lead to different kinds of crystallinity of the CoSi$_2$-layers depending on the thickness of the layer. E.g., directly at the silicon surface it is expected that the lattice constant of the CoSi$_2$ *in-plane* equals the silicon lattice constant. To preserve the volume of the unit cell the CoSi$_2$ lattice constant *perpendicular* to the interface has to shrink. Thus, a distorted so-called *pseudomorphic* lattice appears which is no longer cubic.

Of course, this distorted layer exhibits a larger free energy than a completely relaxed crystal. Therefore, if the film becomes very thick the lattice will finally flip to its characteristic lattice structure. However, this relaxation process includes lattice faults which are also not favorable in terms of minimum free energy. In summary, both effects

(a)
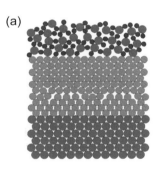

(b)
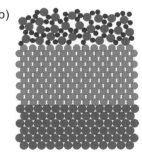

Figure 7: General example of a buried layer (green, small circles) on a substrate (red, large circles) and covered by an amorphous oxide layer (top layer, molecules). As the covering layer is amorphous no lattice strain can be observed at the oxid-'buried layer' interface.
(a) Relaxed layer growth. Directly at the substrate interface the lattice of the layer is distorted. After a few atomic distances it has relaxed by forming lattice faults.
(b) Pseudomorphic layer growth. The lattice of the layer remains distorted (in this case non-cubic) over its whole volume. No lattice faults appear.

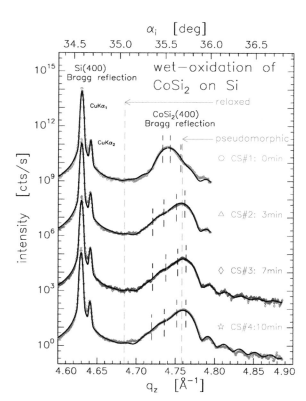

Figure 8: Measurements of the lattice constants of buried $CoSi_2$-layers in a silicon matrix depending on the time of the applied wet-oxidation process (given in minutes). The thicknesses of the top oxide layer are (from top to bottom): CS#1: 60 Å, CS#2: 700 Å, CS#3: 1050 Å and CS#2: 1400 Å. The data have been taken at a laboratory source with copper anode. The pronounced peak at $q_z = 4.626$ Å^{-1} is the Bragg reflection of the Si-substrate with the Miller indices $h = 4$, $k = 0$ and $l = 0$. The small feature on the right tail is the same peak caused by a different wavelength which was not completely suppressed for this experiment (unfortunately). However, it is negligible as it is three orders of magnitude smaller than the main signal. At $q_z \approx 4.75$ Å^{-1} the Bragg reflection of the $CoSi_2$-layer appears. The film is quite thin which means that N_1 is small [see Eq. (8)]. Therefore, the Laue oscillation at the tails are visible, especially at large q_z (compare with Figure 5b). It turns out that four different Laue functions with different lattice spacing (marked by the dashed lines) have to be used to completely explain the data. All $CoSi_2$-layers do not exhibit relaxed growth. Instead, they are in a not very well defined pseudomorphic state. The results do not depend on the time of oxidation (which is equivalent to the depth of deposition). Significantly different is only the unoxidized sample CS#1.

(lattice distortion and relaxation) have to be balanced and usually for each system a critical film thickness exist below which the layer is pseudomorphic. A sketch of relaxed and pseudomorphic growth is displayed in Figure 7 a, b.

In the following an example of buried $CoSi_2$-layers in a single crystal Si-matrix is presented. They have been prepared by depositing a 30 nm thick $CoSi_2$-film on top of a polished Si-substrate via molecular beam allotaxy and subsequent wet-oxidation of the sample [3], [4]. The oxidation process leads to a growth of an amorphous silicon oxide top layer and causes the $CoSi_2$-layer to sink into the substrate. Depending on the time of oxidation the $CoSi_2$-film is finally deposited in a depth of up to $D = 200$ nm inside the sample.

Figure 8 depicts some measurements of the scattered intensity versus the wave vector transfer q_z perpendicular to the sample surface [5]. The data, taken for different depths D, show two pronounced features: A sharp peak at $q_z = 4.626$ Å^{-1} (with a small peak at its right tail) and a broad maximum around $q_z \approx 4.75$ Å^{-1} with oscillations at the right tail.

The first one is caused by the (400)-reflection of the Si-substrate (the small peak is a contamination by the fine structure of the used X-ray spectrum, see also Figure 1b). No Laue oscillations are visible which is due to the large number of illuminated Si-atoms. The penetration depth of the X-rays into the substrate is several micrometers which corresponds to several 1000 monolayers, so $N_j > 10^3$ in Eq. (8). In this case the Laue oscillations become too narrow to be resolved by the scattering experiment.

The broad maximum is the scattering from the (400)-reflection of the $CoSi_2$-layer. As mentioned above the layer thickness is around 30 nm = 300 Å which corresponds to 50..60 lattice constants of $CoSi_2$. Therefore, Laue-oscillations are visible, in this case only on the right side of the reflection. On the left side the signal is swamped by the intensity of the Si-reflection.

None of the $CoSi_2$-reflections match the position of the completely relaxed lattice at $q_z = 4.685$ Å^{-1}. Instead, the layers show pseudomorphic growth with the reflection at $q_z = 4.759$ Å^{-1}. Therefore, the lattice of the $CoSi_2$ is no longer cubic but tetragonal with the x- and y-axis matching the lattice constant of silicon and the z-component of $a_z = 4 \cdot 2\pi/(4.759$ Å$^{-1}) = 5.278$ Å. The tetragonal splitting cannot be seen in the data because the wave vector transfer is always pointing along a_z and there is no sensitivity for in-plane properties of the sample.

It is also evident that the layers are not perfectly ordered. Basically, four different Laue functions at slightly different positions are necessary to fit the data which means domains of slightly different lattice constants in z-direction. However, the quality of the film does not suffer from increasing oxidation time and it seems to be possible to reproducible bury $CoSi_2$-layers with sufficient quality into Si-substrates.

2.5 X-ray Scattering at Small q

Sometimes, only the mesoscopic properties of a layer system such as film thicknesses or interface roughnesses are of interest rather than the exact crystallinity. E.g., in the above mentioned example there is no way to determine the deposition depth D because the cap layer is formed by amorphous silicon oxide which does not exhibit Bragg reflections. Furthermore, there seems to be no direct method to directly determine the exact thickness d of the $CoSi_2$-layer. With Eq. (8) it can only be estimated from the averaged lattice constant of the film and the number N_l of the Laue fits.

However, there is one *Bragg reflection* which always exist: The (000)-reflection (all Miller indices are zero) which is identical to the through-going primary beam. Of course, this *reflection* does not contain much information (except may be about the absorption of the sample). But the tails are similarly modified by the sample properties as it is for the *real* Bragg reflections. To measure in the regime of the (000)-reflection, the scattering angle ϕ has to be very small, typically smaller than 5°, and the mean value of the corresponding wave vector transfer remains also small. The scattering experiment is not sensitive to the particular crystal structure but only to the averaged electron density and even amorphous layers become visible. The principles of X-ray scattering at small q is qualitatively explained in Figure 9.

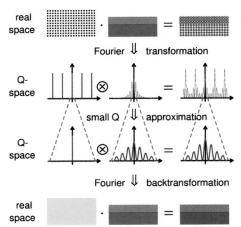

On the right hand side of the first line a monolayer system of two single crystal materials is shown in real space. It can mathematically be expressed by a product of an infinite lattice function (left hand side) and a homogeneous density function (center). The resulting density has to be Fourier transformed to obtain the scattered intensity (see Eq. (3)). For this, the convolution theorem of Fourier transformations can be applied. It states, that the Fourier transformation of a product equals the convolution of the Fourier transformed of the single components:

$$FT\{g_1 \cdot g_2\} = FT\{g_1\} \otimes FT\{g_2\} =: f_1(q) \otimes f_2(q) =: \int f_1(Q) f_2(Q+q) \mathrm{d}^3 Q \qquad (13)$$

The Fourier transformation of an infinite lattice (Eq. (3)) leads to the well known Bragg reflections which are delta-functions (see Figure 9, second line, left) the Fourier transformation of the density function shown in the example gives an oscillating function (see second row, center column). The period depends on the film thickness and is *not* due to Laue interference. The convolution of both functions is displayed in the right column. The third row only shows the small q range as a magnification. In this case, all Bragg reflections are neglected except for the 000-reflection which is the primary unscattered beam. A Fourier backtransformation of this cut data yields the fourth row. Due to the limited range of q the information about the lattice structure has vanished (the Fourier backtransformation of a single delta-peak is just a constant). Only the information of the density function itself is left.

A special case of a small q (or small angle, respectively) X-ray scattering experiment is the reflectivity from a surfaces or a layer system. For this, the incident and the exit angle with respect to the sample surface have to be equal (see Figure 10).

A reflectivity measurement has a wave vector transfer perpendicular to the sample surface. Therefore, $q = (0,0,q_z)$ holds with $q_z = (4\pi/\lambda)\sin\alpha_i$. This means that no explicit in-plane information can be extracted from a reflectivity measurement due to the vanishing q_x- and q_y-components. But the reflectivity measurements are extremely sensitive to the electron density profile perpendicular to the surface (the z-direction), averaged in the x-, and y-directions. Using the assumption of a *homogeneous* electron density (the crystal structure is neglected) and the existence of a surface Eq. (3) can be rewritten. In Born approximation the reflected intensity as a function of q_z is given by

Figure 9: Principles of small angle scattering. The first row shows the product of an infinite lattice (left) with a homogeneous monolayer density function (center). The result is shown on the right hand side. After performing the Fourier transformation it turns out that due to the density function all Bragg peaks are modified with an oscillating shape due to the monolayer thickness (see second row). The third row depicts a magnification of the scattered intensity around the origin of q. All other Bragg reflection are neglected. The Fourier backtransformation of the limited q-range is shown in the last row. The information about the crystalline structure has completely vanished.

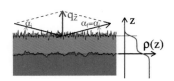

Figure 10: Sketch of a monolayer system with rough interfaces. The incident and the exit wave vectors have the same angle with respect to the sample surface (reflectivity experiment). The in-plane averaged electron density of the sample is shown on the right. Due to the interface roughnesses the shape is not a step function but smeared.

$$I(q_z) \propto \frac{1}{q_z^4} \left| \int \frac{\mathrm{d}\rho(z)}{\mathrm{d}z} \exp(i q_z z) \mathrm{d}z \right|^2 \qquad (14)$$

which means that the reflectivity is actually sensitive to *changes* in the density profile [6]. Therefore, every interface between two components with different electron density will be visible in the data.

2.6 An Example: Metallic Thin Films

To visualize the sensitivity of X-ray reflectivity a measurement of a metallic thin film system is chosen. The sample is prepared by depositing first a 10 nm thick iron layer on a silver substrate then a 1 nm chromium layer and finally a 10 nm thick iron cap layer. However, the top part of the cap layer was oxidized so that the final layer system was

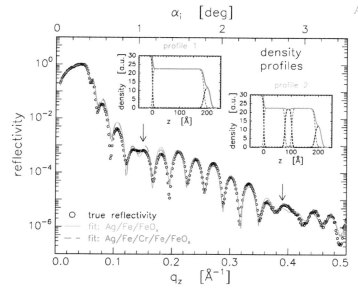

Figure 11: X-ray reflectivity of an Ag/Fe/Cr/Fe/FeO$_x$-layer system. The symbols denote the data, the solid line a fit with the assumption that the Cr-layer is *not* visible (see inset profile 1) and the dashed line a fit with the true profile (see inset profile 2). The difficulties to distinguish between both profiles is obvious. The reason is the low contrast of the Fe/Cr interface (see Eq. (14)). However, a careful analysis (see with arrows marked areas) finally yields the parameters of the Cr-layer.

Ag/Fe/Cr/Fe/FeO$_x$. The values of the layer thicknesses are only estimates from the processing parameters. The reflectivity measurement should give the real numbers and even the interface roughnesses.

The difficulty in investigating this layer system are the Fe/Cr interfaces which have been of particular interest: The electron density of both elements differs only about 7 % whereas the contrast between Ag and Fe is about 25 % and between Fe and FeO$_x$ almost 100 % (and so for the FeO$_x$-air surface). Therefore, the signal is dominated by these interfaces and the Cr-layer is hardly visible (see also Eq. (14)). However, a careful data analysis definitely shows the existence of the Fe/Cr interfaces (see Figure 11) and yields the exact density profile $\rho(z)$ with all layer thicknesses and interface roughnesses.

2.7 X-ray Fluorescence and Absorption Spectroscopy

From the structure factor (Eqs. (7) and (11)) it can be seen, that the intensity of the Bragg reflections depends on the Fouriertransformation of the electron density of the single atoms, the so-called form factor. However, in an exact calculation the form factor is a more complicated expression: It contains dispersion and absorption corrections which are due to the special electronic properties of the atoms. These corrections are very element specific and depend strongly on the wavelength of the used X-rays. They have their origin in the electron band structure. For Eqs. (7) and (11) an ideal electron gas is assumed. In reality, the electrons occupy bands and/or shells with more or less well defined energy, so electrons cannot occupy arbitrary energy levels. This energy quantization causes deviations of the absorption and the emission of photons (the fluorescence) from the expected behavior.

Figure 12 explains this behavior using copper as an example. The figure displays the simplified shell/band structure of copper. Electrons can only occupy the levels (and of course every level above the 0 eV-line, which means that they are completely free and no longer bounded to the Cu-atom). Therefore, no photon with an energy smaller than e.g. 8979 eV can remove electrons from the K-shell, except for the very unlikely possibility that some outer shell is free and the photon energy exactly matches the energy difference between both levels. On the other hand, when scanning the photon energy and observing the absorption of the material so-called absorption edges can be seen. For Cu the K-absorption edge is located at 8979 eV. Photons with smaller energy can only be absorbed by L- and M-shell electrons. Photons with higher energy can also be absorbed by the K-shell electrons, which means a change in absorption at 8979 eV.

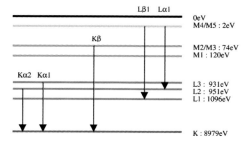

Figure 12: Simplified band structure of copper. The arrows symbolize the allowed transitions of an electron falling from an outer level into another free inner level The excess of energy is emitted as a photon. All photons form a spectrum with the so-called fluorescence lines.

To perform an X-ray absorption experiment, the sample is illuminated by a monochromatic X-ray beam where the energy of the photons is the scan parameter. The through going beam is monitored and normalized to the incident flux. Absorption edges can be seen as sudden drops of the passing intensity. As mentioned above the position of the edges are element specific (e.g.: Cr-K : 5989 eV, Mn-K : 6539 eV, Fe-K : 7112 eV, Co-K : 7709 eV, Ni-K : 8333 eV, Cu-K : 8979 eV). In this way, for unknown compositions of the samples most elements can be identified if the concentration (and thus the change in absorption at the edge) is sufficient.

When performing a careful energy scan around an edge the high energy side usually exhibits oscillations in the absorption (see Figure 13a, taken at the wiggler beamline W1.1 at HASYLAB/DESY). This so-called Extended X-ray Absorption Fine Structure (EXAFS) is caused by secondary processes of backscattered low energy photoelectrons. The analysis of an EXAFS spectrum gives information about the distances and coordination numbers of the atoms which surround the absorbing atom [7].

Absorbing a photon means removing an electron from a shell. This hole will be filled by another electron from an outer shell whereby several transitions are not allowed. Special selection rules have to be obeyed. E.g. in the above given example the transition L1→K is not allowed. The excess of energy $W_{initial} - W_{final}$ is emitted as a photon carrying exactly this energy difference. As the result a line spectrum with the so-called fluorescence lines is observed. Examples, also taken at the wiggler beamline W1.1 at HASYLAB/DESY are shown in Figure 13b.

For a fluorescence experiment the sample is illuminated with X-rays of an energy well above the absorption edges which should be excited. E.g. if the fluorescence lines of Cu for the transitions...→K are to be determined, the energy of the incident photons has to be larger than 8979 eV. An energy resolving detector monitors the intensity which is not scattered in forward direction. In the example in Figure 13b the energy resolution of the chosen detector was only 100 eV which explains the relatively large width of the peaks. However, the $K\alpha,\beta$-lines of Co and Zn can easily be distinguished. With higher resolution devices such as Lithium drifted Silicon (SiLi) detectors an element analysis can be made without too much effort from any sample of unknown composition.

In contrast to EXAFS the fluorescence method is also capable of surface sensitive measurements. For this the sample surface is illuminated under extreme grazing incidence. By changing the incident angle (typical values are less than 0.3°) the penetration depth of the X-rays can be tuned from about 10 nm up to several micrometer and a depth profiling of element concentrations inside the sample is possible [8].

(a)
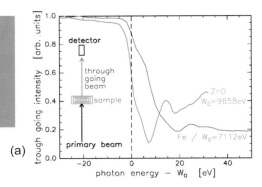

(b)
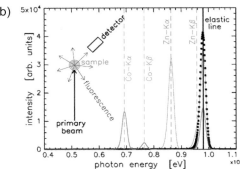

Figure 13:
(a) X-ray absorption measurements. The experimental setup is shown as an inset. The energy scale is normalized to the absorption edges W_0 of the materials.
(b) Examples of X-ray fluorescence measurements. The inset shows the experimental setup. The symbols denote data taken with a polymer sample which does not show fluorescence in the displayed energy range. The Zn-Kβ line can be seen as a shoulder close to the elastic line.

3 Electron Analysis Methods

3.1 Introduction

X-ray or neutron scattering methods are powerful tools to investigate condensed matter as mentioned above. However, the interpretation of the data is at no time easy and does not give a direct picture of the sample. Moreover, due to the lack of the phase information the result is often not unique and additional knowledge about the sample has to be considered. Analytic methods using electrons as a probe can be performed as diffraction or spectroscopy experiments where the direction of the scattered electrons, the energy change (*Electron Energy Loss Spectroscopy*, EELS) or even the spin change (*Spin Polarized EELS*, SPEELS) is analyzed. Additionally, they open up the possibility of getting information independently by real space imaging of the samples. This can be done by electron optical methods similar to an optical microscope or by scanning a very well focused electron beam over the sample area. Of course, the imaging and the diffraction/spectroscopy can be combined. In this section the most important methods, the *Transmission Electron Microscopy* (TEM), the *Scanning Electron Microscopy* (SEM), and EELS with its varieties are discussed. They require electron optics which will qualitatively be explained first. Some attention is also put on the sample preparation which has to be done accurately in order to get reasonable data.

3.2 Electron Optics

Electron microscopy requires a source to illuminate the sample with an electron beam. The electrons can be produced by heat emission of a filament (see Figure 14), by field emission or a combination of both. The source is put inside a Wehnelt cylinder with an aperture at the bottom which acts as a shielding. The biasing between the filament and the Wehnelt shielding can be controlled to optimize the focal spot [9].

The key parameters of the electron source are the virtual source size, the brightness and the energy spread of the emitted electrons. The source size (and shape) determines the resolution, the brightness is equivalent to the intensity of the beam and a large energy spread causes chromatic aberrations at the lenses.

Basically four different sources with different properties can be used. They are listed in Table 2. The tungsten filament produces the electrons simply by heat emission from the tip of a hairpin shaped wire. This source is actually *antiquated*. It is most inefficient and has the worst source parameters however, the system requirements such as vacuum conditions are quite relaxed. Still in use are guns where the electrons are produced by a heated tip of a LaB_6 single crystal. The tip radius is about 5 µm. Tips of single crystal material, usually tungsten, are also used with the field emitters or the Schottky emitter. In this cases the tip radius is extremely small (< 0.5 µm) and the work function for the electrons to leave the crystal is reduced by an external electrical field (cold emission) or a ZrO coating (Schottky emitter). The best beam properties are supplied from a cold emission gun, however the vacuum requirements are very high. The Schottky emitter is a good compromise in efficiency and operating expenses.

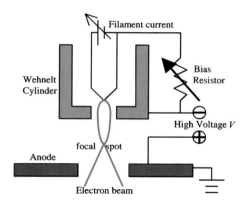

Source type	brightness [A/cm²/sr]	source size [µm]	energy spread [eV]	Operating vacuum [mbar]	operating temperature [K]	lifetime [h]
tungsten filament	10^5	25	2..3	10^{-6}	2700	100
LaB_6 tip	10^6	10	2..3	10^{-8}	1700	500
Schottky emitter	10^8	0.02	1.0	10^{-9}	1800	> 1000
cold field emitter	10^9	0.005	0.2	10^{-10}	300	> 1000

Table 2: Properties of electron sources commonly used in electron beam tools.

Figure 14: Sketch of a self-biased electron gun [9]. The electron beam is produced by heat emission. Due to the negatively charged Wehnelt cylinder the electron beam is focused to a focal spot if the bias resistor is set to a suitable value. The high voltage V is needed to accelerate the electrons to the desired energy. In the case of a field emission gun the filament is replaced by very tiny tip with a radius of curvature of less than 1 µm. The Wehnelt cylinder is then called suppresser, the anode is called extractor.

It is well known, that the resolution of an optical microscope is determined by the wavelength of the used light. A similar consideration also holds for an electron microscope where the wavelength is defined by the *deBroglie wavelength* of the electrons λ_{dB} given by

$$\lambda_{dB} = \frac{h}{\sqrt{2mW}} = \frac{h}{\sqrt{2meV}} \quad (15)$$

Eq. (15) is a function of the Planck constant h, the electron mass m and the kinetic energy $W = eV$ of the electrons given by the electron charge e and the voltage V which was used to accelerate them. Therefore, the resolution of an electron microscope strongly depends on the values for V which are typically in the range of 100 kV, thus $\lambda_{dB} \approx 0.037$ Å = 3.7 pm. However, the actual resolving power of the instruments is not that good and depends on the apparatus but it can still be in the angstrom level.

Similarly to optical microscopes lenses and apertures are used to define the beam and to magnify the image. The lenses work electromagnetically: The electrons are deflected by specially designed magnets. The focal length of the lenses are tunable via the current which is used to drive the magnets. The lenses can be classified as condensers, objectives and projectors. An additional device which is used in an SEM is a deflecting coil which can move the beam on the sample surface. Figure 15a-c depict overviews of the beam paths of a TEM and a SEM [9]. Some details of the difference between the imaging mode and the diffraction mode when using a TEM are given in the next section. Subsequently, the SEM is explained.

3.3 Transmission Electron Microscope (TEM)

At a TEM instrument, the electron beam is defined by a condenser lens system (in Figure 15 only one lens is shown but up to three may be used). The beam hits the sample which has to be transparent for electrons. After passing the sample the electrons are collimated by an objective aperture and an objective lens (see Figure 15). The number of diffracted spots which contribute to the final image can be restricted here. The purpose of the objective is comparable to light microscope setups. The sector lens can be used to magnify the image or the diffraction pattern. Therefore, two different modes are possible using a TEM, the *microscopic* imaging mode (Figure 15a) and the *analyzing* diffraction mode (Figure 15b and reference [9], [10]). Finally, the projector lens images the electrons on a screen or a two dimensional camera.

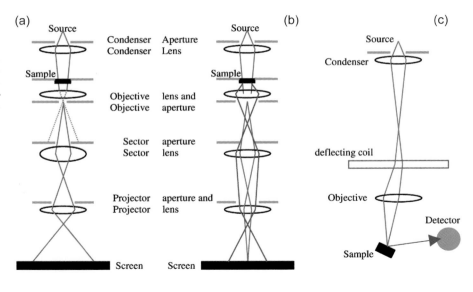

Figure 15:
(a) Schematic beam paths of a TEM in an imaging mode.
(b) Schematic beam paths of a TEM in a diffraction mode to see crystallinity. The blue lines correspond to the diffracted Bragg peaks of the lattice. The red lines correspond to the red lines of (a). As the magnification in diffraction mode is very small, they do not form a microscopic picture on the screen but just a spot in the center.
(c) Schematic beam paths of an SEM.

Figure 16: Schematic procedure to prepare a planar sample for TEM. After preparing the sample onto a substrate the latter may be etched. The final thickness of some 100 nm can be achieved by ion milling. The up pointing arrows denote the direction of the TEM beam.

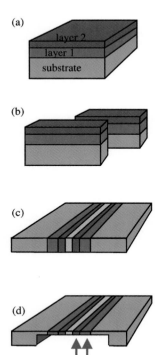

Figure 17: Cross section TEM sample of a bilayer sample. After preparation (a) the sample is cut (b) and glued together with the surfaces in contact ((c) gray part in the center: glue). The whole sample is polished down to about 50 µm thickness and finally ion milled to the desired thickness (d). The arrows denote the direction of the TEM beam.

As electrons are highly absorbed by any kind of material the specimens have to be extremely thin. For low resolution experiments a thickness of 100 nm is sufficient. To achieve atomic resolution the sample has to be as thin as 10 nm or even less. This means that the samples have to be specially prepared for the measurements (TEM is a highly destructive method). During the preparation one has to be extremely careful that the thinning process does not lead to artifacts in the sample which may mislead the investigator when interpreting the image [10].

Comparably easy to prepare is a planar sample where in-plane information is of interest. E.g. a thin layer of the sample can be deposited on a material which can be etched subsequently. An additional procedure, called *ion milling,* can be applied to reduce the sample thickness if desired. It is basically a kind of sputtering process where ions (e.g. argon) are shot onto the sample and *sand blasting* it (see Figure 16).

It is also possible to make a cross-section TEM sample [11]. In this case, not the in-plane structure but the structure perpendicular to the sample surface is of interest. This can be achieved by slicing the sample (e.g. a multilayer system) into two pieces of a millimeter thickness, gluing them together (with the surfaces in contact) and finally polishing and ion milling them (see Figure 17).

As mentioned above, the TEM can be used in two different modes: The imaging mode with large magnification, where the instrument works similar to an optical microscope (see Figure 15a) and the diffraction mode with small magnification where the atomic structure of the sample leads to Bragg peaks or diffraction rings similar to X-ray scattering (see the outer lines in Figure 15b).

The diffraction mode (which can also be called *Small Angle Electron Diffraction*, SAED) can easily be understood when considering that the electrons act like photons with the deBroglie wavelength λ_{dB}. Therefore, for a crystal Bragg reflections are expected regarding Eq. (8). The intensity of the reflections is given by the structure factor. It is determined by the interaction potential of the sample atoms with the probe. As X-rays are different from electrons, also the interaction potential is different. Therefore, the electron structure factor differs from the X-ray structure factor which has to be taken into account when interpreting the data. If the sample is not crystalline no Bragg reflection exist. However, calculations yield that circular intensity maxima around the through going primary electron beam appear. These rings are connected with averaged neighbor-neighbor distances of the molecules in the sample or with long ranged electron density modulations. Typical TEM images and TEM diffraction patterns are depicted in Figure 18.

TEM-images with resolutions much worse than atomic distances can be interpreted directly similar to light microscopy. With *High Resolution* TEM (HRTEM) the magnification is larger so that atomic distances can be resolved. However, the images are no longer direct pictures of the sample (even if they look like those) [10].

The reason is, that the electron beam passes through quite a bit of material (much more than one atomic layer) so that a projection of the specimen to a 2-dimensional image is expected. Furthermore, on an atomic length scale quantum mechanic effects have to be taken into account for the electrons. This means using a coherent superposition of the electron wave functions to obtain the final wave function which will appear on the screen. The final wave function does not only depend on the sample but also very strongly on the beam properties and the optics. Therefore, to reasonably interpret a HRTEM-image a lot of experience and many model calculations are necessary. Figure 19 depicts an example how a small change of the focus creates very different pictures of exactly the same sample.

3.4 Scanning Electron Microscope (SEM)

As shown in Figure 15c the SEM does not work in transmission but in reflection mode: An electron beam (usually with an energy of 1 keV – 20 keV) is focused on the sample surface to a size of approximately 1 nm – 10 nm. Spatial information can be achieved by scanning the focus on the sample. This is done by a scanning (or deflecting) coil which is put in the optics of the microscope. When the electron beam hits the sample several processes occur. The most important are mentioned in the following list [9]. A sketch of all processes is also depicted in Figure 20.

1. **Elastic Backscattering**
   Many electrons are backscattered from the sample without changing their original W_0 energy.
2. **Inelastic Backscattering**
   Some electrons are backscattered with a different energy. An energy loss of less than 50 eV is usually due to plasmon excitation. These excitations are very specific to the matter which is illuminated because they strongly depend on the electronic band structure on the crystal structure on interfaces or defects.
3. **Characteristic X-rays**
   Electrons can ionize atoms by removing a shell electron. The vacancy is immediately filled by another electron from an outer shell. To conserve the energy a photon can be emitted with an energy equal to the difference in energy of both shells: A characteristic line spectrum is emitted from the surface which is identical to the fluorescence line spectrum mentioned in Section 2.7. By an *Energy Dispersive X-ray analysis* (EDX) an element specific surface analysis can be performed.
4. **Auger Electrons**
   Instead of emitting a characteristic X-ray photon the energy excess mentioned in point 3 can be transferred to another shell electron. This so-called *Auger electron* leaves the atom which is now 2 times charged. The Auger electrons are highly element specific. The energies range from 100 eV to 1 keV.
5. **True Secondary Electrons**
   All excited electrons which undergo multiple scattering events on their way to the surface of the sample. They have a broad energy distribution with a maximum below 50 eV. They are usually used for the spatial imaging.

Therefore, by using different detectors, which are sensitive to electrons of different energies or even to photons, SEM-images are specific to special properties of the specimen.

As the SEM does not work like an optical microscope but by scanning the surface of the sample the resolution of the instrument is determined differently. Of course, the focus of the incident beam on the sample surface is of importance. The smaller the focus the better the theoretical achievable resolution. However, it is not necessarily true that all electrons and photons in the list above are emitted only from the illuminated area. In fact, the primary electrons penetrate quite a bit of the sample and exhibit a comparatively large halo of about $R_{pe} \approx 1\mu m$ diameter around the focal spot (see Figure 20). From this halo all characteristic X-rays are emitted therefore, the X-ray image will never have better resolution than 1μm. A similar argument applies also for the backscattered electrons. However, because of the compared to the primary electrons smaller energy the reach R_{be} of the backscattered electrons is around half times smaller than R_{pe}. The sec-

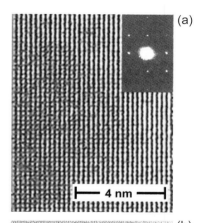

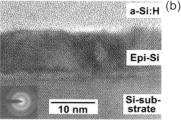

Figure 18:
(a) High resolution TEM image of a gold single crystal where the lattice planes can be resolved [9]. The inset shows the diffraction pattern (compare Figure 15b) of the sample. The Bragg reflections due to the crystalline structure of the sample surround the through going electron beam.
(b) Cross sectional TEM picture (compare Figure 15a) of an amorphous silicon/crystalline silicon junction [12]. The diffraction pattern (compare Figure 15b) only shows the small angle regime therefore the Bragg peaks of the substrate cannot be seen. Clearly, the typical ring pattern of the amorphous top layer is visible.

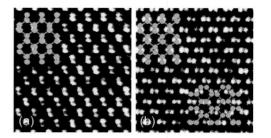

Figure 19:
a) HRTEM image of a silicon sample looking in (110)-direction. Each white spot means a column of Si-atoms one after the other. The small distance between the paired columns is 0.14 nm. Figure 19 is a direct picture of the crystal. This example is taken from the homepage of FEI Company, USA.
b) Same sample with same orientation as in Figure 19a but with different focus of the electron beam. This is an indirect picture of the sample because it has to be corrected for the TEM-beam parameters. There is no way to directly match the crystal properties. Without correction one would get a wrong interpretation.

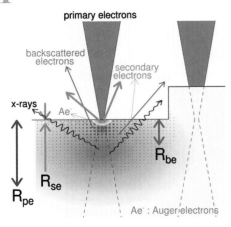

Figure 20: Sketch of the processes appearing at SEM imaging. A focused beam of primary electrons hits the sample. The situation on the right hand side shows that the size of the illuminated surface area can dramatically change if the sample has hills or valleys. This effect changes resolution during a scan. On the left hand side the halo of the primary electrons with the reach R_{pe} is shown. The characteristic X-rays are emitted from the whole volume of this halo. The primary electrons are backscattered with smaller energy. Therefore, the *backscattering halo* (marked by red squares) has a smaller reach $R_{be} < R_{pe}$. For the secondary electrons R_{se} is very small and their halo (small, green) is restricted very close to the illuminated surface. Some secondary electrons may also be produced by backscattered electrons which smears the *secondary halo*. Auger electrons are only emitted directly from the surface.

Figure 21: The different types of contrast. Compare the different numbers of small arrows (which represent the reflected electrons) of each picture with the four arrows in the *undisturbed* very left picture: Each case will yield different intensity in the detector.

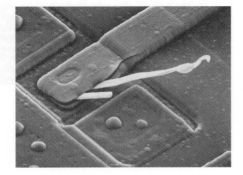

Figure 22: 3-dimensional SEM image of a broken semiconductor device. The junction between the wiring and the device itself is perturbed. Clearly visible are edge-, shadow-, decline-, and material-contrast. The picture was taken at the Brockhouse Institute for Material Research, McMaster University, Canada.

ondary electrons have only a very limited reach R_{se} in the material. Thus, most of them are only emitted very close to the focal spot. Some of them are produced at the surface by backscattered electrons which may reduces the resolution (see again Figure 20). Finally, Auger electrons are only emitted directly from the surface.

Another effect which can limit the resolution is determined by the focusing of the incident beam. In order to best compensate for various lens aberrations a short distance between the last optical element and the sample, a strong final lens and relatively wide open apertures are required. This also means that the opening angle (the divergence) of the primary beam is large. Therefore, the illuminated surface area can dramatically change if it is not placed in focus plane which may occur when scanning 3-dimensional objects. In this case, the resolution depends on the sample area which is scanned (see also Figure 20).

One of the most important parameter of an image is the contrast between two areas [10]. In an SEM picture this value is a function of the number of electrons which are collected by the detector when the primary spot illuminates the sample. The most important contrast is the *edge contrast*. If an edge is present in the focal spot much more secondary electrons are emitted compared to a simple smooth surface. The *shadow contrast* is due to the fact that the detector is directed: If the illuminated area is not pointing towards the detector less electrons contribute to the signal and the spot is dark. The emittance of the secondary electrons is also a function of the angle of the incident beam with respect to the sample surfaces. This causes a *decline contrast*. The *material contrast* depends on the examination of the inelastic backscattering or Auger electrons. Finally, when very low energy electrons are detected even a *potential contrast* is seen: Differently charged parts of the sample (e.g. a charged capacitor on a semiconductor device) have different electrical fields which may favor or prevent the secondary electrons to be detected. The different contrasts are schematically shown in Figure 21.

As electrons are used to illuminate the sample and also electrons are detected major difficulties arise when the sample is becoming charged. For metal specimens this can simply be prevented by grounding. For semiconductors or insulators the energy of the primary electrons have to be adapted: If the energy is too low very few secondary electrons are emitted and finally the sample is negatively charged which heavily affects the imaging.

The same argument holds, if the primary electron energy becomes too large. In an

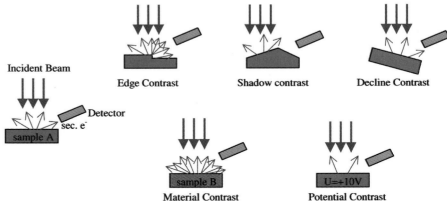

intermediate region the surface is slightly positively charged because the amount of incoming electrons and secondary electrons is almost equal. The slight positive charge is usually of minor importance for the imaging process. In some cases, if only the spatial information is of interest the samples can be coated with a thin gold layer to prevent charging effect.

An SEM image of a broken semiconductor or shown in Figure 22. It shows the various contrasts such as edge-, shadow-, decline-, and material- contrast.

Analysis by Diffraction and Fluorescence Methods

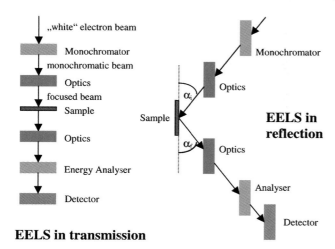

Figure 23: Schematic drawings of typical electron energy loss experiments. The left picture shows the *transmission case* where the setup is quite similar to a TEM-device. It is bulk sensitive and can be used for investigation of intermediate and high energy losses. The *white* electron beam denotes electrons with a broad energy distribution. The sketch also displays EELS in reflection mode where small energy losses can be examined. This is a surface sensitive method. The incident angle α_i and the exit angle α_f can be different.

3.5 Electron Energy Loss Spectroscopy (EELS)

As mentioned in the previous sections, when illuminating a sample with an electron beam many of the electrons interact with the sample. They can change their travel direction by an elastic scattering process without changing the energy. But they can also lose kinetic energy during the scattering (inelastic scattering). Basically, a classification of three different ranges of energy loss can be introduced [10]:

Energy losses can be caused by vibrational processes of the crystal lattice which are called *phonons*: The incoming electron hits an atom which is displaced from its mean position and creates a sound wave within the solid. Because of the crystalline structure of the sample the phonons are quantized. The energy to excite the phonon is in the range of some 10 ... 100 meV and is taken from the kinetic energy of the incident electron. Therefore, after the scattering process the electron carries its original energy minus the phonon energy.

In the range of 1 ... 50 eV so-called *plasmons* can be created. They are collective excitations of electrons in the electron gas and may be localized at the sample surface or at interfaces. The energy of the plasmons basically depends on the electron density of the solid. Also, single electron excitations which are related to the joint density of states between the conduction and valence bands can be seen in this energy range.

Energy losses of more than 100 eV are due to excitation of various inner atomic shells to the conduction band. Their fine structure is characteristic for the elements and they can therefore be used for element spezific analysis.

An EELS experiment can be performed in different ways depending on the information of the sample which is desired. For all a monochromatic electron beam is required. Thus, the incident electrons have a well defined energy W_0 with a very small mean deviation ΔW_0. The value of ΔW_0 determines the energy resolution of the experiment. Depending on the absolute value of W_0 the outgoing electrons can be detected in transmission or in reflection where the transmission experiments are sensitive to the bulk and the reflection experiments are sensitive to the vibrational surface properties. The scattered electrons are monitored after passing an energy analyser which sorts them with respect to their final energy. Optional, the direction of the outgoing electrons can be determined. If the spot size of the primary beam is kept at very small dimensions, a scanning picture of the sample is feasible. Figure 23 shows the principal experimental setups.

As seen in Figure 23 and already explained the monochromator for the incident electron beam and the energy analyser for the outgoing electrons are of eminent importance for the efficiency of the EELS-device. Especially, in reflection mode when the vibrational structure of adsorbates or the clean surface is of interest energy losses of about 100 meV occur and an energy resolution of better than 10 meV (*High Resolution EELS, HREELS*) has to be achieved.

To separate electrons with different energy static electric or magnetic fields can be used. E.g. in a homogeneous magnetic field $\boldsymbol{B}$ a moving electron experiences a force $\boldsymbol{F} = e[\boldsymbol{v} \times \boldsymbol{B}]$ perpendicular to $\boldsymbol{B}$ and the moving direction (Lorentzian Force).

This results in a circular movement of the electrons with the radius

$$R = \frac{mv_0}{eB} = \frac{\sqrt{2mW_0}}{eB} \tag{16}$$

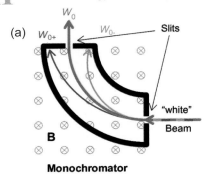

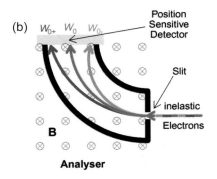

Figure 24: Sketch of a monochromator (a) and analyser setup (b) for an electron beam with a broad energy distribution. At the monochromator the slits select one energy. The energy resolution for the analyser is given by the pixel width of the position sensitive detector.

in the case of non-relativistic speed v_0. Therefore, the larger W_0 the large is R. For monochromating a couple of slits can be used to pick out a special electron energy. Energy analysis can be achieved by a position sensitive detector (see Figure 24).

The standard EELS in transmission is based on a TEM-apparatus and has therefore a very high resolution in space. It is frequently used for microchemical investigations of metals, alloys, semiconductors or ceramics. The limit of detection is very small and can be a few atoms in the illuminated volume. To distinguish between the different kinds of atoms the energy loss larger than 100 eV of the through going electrons is analyzed. The energy loss is not arbitrary as mentioned above. The incoming electrons lose exactly that amount of energy which is necessary to lift core electrons from inner shells to the conduction band (comparable schemes of those processes are given later on in the section about XPS and AES). Only at the surface, the excited electrons can also be removed completely from the sample. Depending on the shells the energy loss spectra around these ionization edges have a fine structure. The position of the ionization edges and their fine structure is typical for each element. Even the chemical bounding of the target atoms can be seen because the bounding changes the characteristics of the edges in a typical way.

A serious problem are multiple scattering processes where the incoming electrons interact more than once with the sample electrons. The whole energy loss is than a result of single events which cannot be distinguished any more. However, if the samples are very thin multiple scattering becomes unlikely and in the best case from an energy loss spectrum the exact composition and chemical state of the sample can be extracted. When performing a surface sensitive HREELS experiment in reflection basically two different types of scattering occur:

- **Dipole Scattering**
  A long-range effect mediated by the Coulomb field. The incoming electron interacts with a vibrating dipole at the surfaces and can deposit energy in a vibrational mode. The dipole selection rules apply which means that specularily reflected electrons (incident angle α_i equals exit angle α_f, see Figure 23) can only excite dipoles which vibrate perpendicular to the surface. However, strict calculations show that the selection rules break down for very large scattering angles so that parallel vibrating dipoles can be excited, too.

- **Impact Scattering**
  A short range scattering process where the electron hits ion cores. The scattering is more isotropic.

Figure 25: HREELS spectrum of a Ni (111) and Pt (111) surface with a (4 × 2)-layer of adsorbed CO. The separated small peaks at the zero-loss position are scaled down by a factor of 270. They are elastically scattered electrons. The width determines the energy resolution. On the Ni surface the vibration spectrum only indicates a single CO species on a site of high symmetry. In contrast, on the Pt surface two different sites must be occupied with different vibrational modes. The sketches of the occupied sites are drawn for each case. For Ni the possible vibrations of the CO-molecule are shown as molecular sketches. Only perpendicular modes appear in the spectrum due to the dipole selection rules, therefore the excitations of 78 meV, 24 meV, 6 meV and 66 meV cannot be seen in the spectra,. For Pt another six modes exist for the CO-molecules located on the top sites.

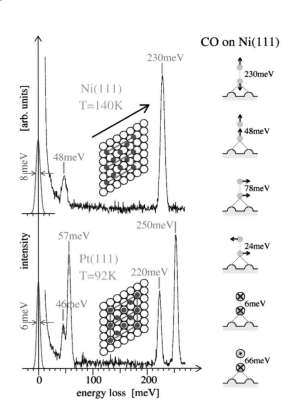

HREELS is especially sensitive to surface vibrations of adsorbed molecules as shown in Figure 25. It displays energy loss measurements of a Ni(111) and Pt(111) surface respectively, both covered with half a monolayer of CO ordered as a (4 × 2) overlayer. First of all, the spectra show a strong zero-loss peak (energy-loss equals zero). Secondly, it turns out that less peaks appear than expected from the possible vibrations of the CO-molecules on the substrates. This is due to the dipole selection rules which favor perpendicular exciting.

Most important is the difference in the number of peaks for Ni and Pt though the CO-layer has the same (4 × 2)-structure in both cases: In the case of Ni all occupied positions are identical in symmetry and exhibit therefore the same vibrational modes. However, on Pt half of the CO-molecules occupy the top sites the other half the twofold bridges of the surface.

The above mentioned methods are sensitive to the electronic or vibrational structure of the samples. But it is also possible to investigate the magnetic properties: Every incident electron carries a spin (which is somehow comparable to the polarization of a electromagnetic wave). The direction of this spin can flip during the scattering process due to spin exchange with a sample electron. In magnetic materials the spins of electrons in special bands/shells are all oriented. E.g. if they all point in the same direction the material is called ferromagnetic. In this cases, a spin-flip process of the energy-loss electrons can reflect the magnetic properties of the sample.

Usually, for EELS the spin directions of the primary electrons are arbitrary, but for the experiment which is called *Spin-Polarized Electron Energy Loss Spectroscopy* (SPEELS), the sample is illuminated with an incident electron beam of well defined polarization. The outgoing electrons are not only analyzed with respect to their energy loss but also with their polarization direction (spin-flipped or not). With this information the exact magnetic structure of the sample can be determined.

4 Surface Sensitive Analytical Methods

4.1 Introduction

The analytical methods mentioned so far penetrate the samples at least some 10 nanometers. This is obvious for X-ray scattering where even at glancing angles the penetration depth Λ is never smaller than 10 nm. At larger angles it can be on the order of several micrometers. For TEM the beam is passing through a sample of at least 10 nm thickness and even for SEM most secondary electrons are emitted from a depth of several 10 nm, depending on the energy of the incident electrons. However, in the latter case it turns out that the escape depth of the Auger electrons is on the order of 1 nm – 3 nm which means quite good surface sensitivity if only Auger electrons would be monitored.

In the following three different types of analytical methods are explained which are particularly surface sensitive. The first one is the *Secondary Ion Mass Spectrometry* (SIMS) which is a destructive method to investigate the composition of surfaces. Afterwards, the *X-ray Photoelectron Spectroscopy* (XPS) and the *Auger Electron Spectroscopy* (AES) are illustrated qualitatively. These methods are not used to investigate the atomic or mesoscopic structure of the sample but to study the electronic properties such as electronic band structures or binding energies.

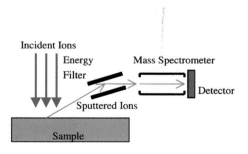

Figure 26: Sketch of a SIMS apparatus. The incident ions hit the sample and cause ions to escape from the sample. Their energy and mass is analyzed before they hit the detector.

4.2 Secondary Ion Mass Spectrometry (SIMS)

Using SIMS means to erode the surface of the sample by an ion beam [11], [13]. The incident ions (mostly used are Cs^+, O^{2+}, Ar^+ and Ga^+) are accelerated to energies between 1 keV and 30 keV. In this energy range the penetration depth of the ions into the sample usually remains smaller than 10 nm which means that they sputter the surface rather than to be implanted into the material (as used in ion implantation technology). The primary ions kick out particles which can be neutral, positively or negatively charged and typically carry energies of up to some hundred eV. The charged particles from the specimen are passing through an energy filter and are analyzed using a mass spectrometer (see Figure 26). From the concentration and the kind of the eroded ions conclusion can be made concerning the composition of the investigated material. In some case the sensitivity of this method is very high. E.g., it is possible to detect boron impurities in a silicon matrix which are in the parts per billion (ppb) range.

II Technology and Analysis

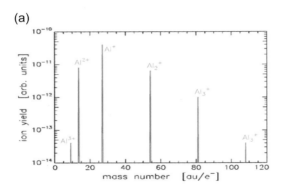
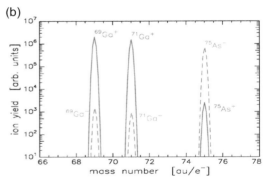

Figure 27:
(a) Secondary ion cluster spectrum from pure aluminum. Predominant is Al$^+$ but other particles also appear [11].
(b) Positive (solid lines) and negative (dashed lines) ion yield from GaAs. The sensitivities are extremely different [11].

SIMS can be used in a scanning mode which means that the incident ion beam is scanned on the sample surface (similarly to the SEM) to achieve spatial in-plane resolution. Also, from the time of illumination of each spot and the beam parameters the number of atomic layers can be estimated which are removed by the eroding process. Therefore, SIMS is also a depth sensitive method.

The major problem is how to interpret the ion yield. E.g. sputtering a simple single-element surface does not yield just one single sort of outgoing ion. In the case of aluminum a wide range of differently charged particles with different masses can be detected (see Figure 27a). Also, the sensitivity to single components can be extremely different in a multi component system (see Figure 27b).

Furthermore, the incident ions may not only sputter the surface. Some may also be implanted and others may drive target atoms into the bulk of the sample which leads to a mixing and a change of the original distribution of elements. This has to be taken into account in the case of a depth profiling.

Finally, any covering with oxygen does not only generate the respective oxygen ions but also strongly affects the yield of the other components (this is of course true for the existence of any other electropositive or electronegative ions at the target surface). In the case of a thin oxide layer the yield of the sputtered ions is usually drastically increased. This can be understood by considering the electronic surface band structure. An uncovered metal surface allows neutralization of positive ions leaving the surface by transferring electrons from the conduction band. These neutral particles are lost for the SIMS analysis because they cannot be detected. In contrast, a thin oxide layer causes a change the band structure at the surface via forming a large forbidden band gap. Thus, only very few electrons are available for neutralization and the amount of ions remains large compared to the clean surface.

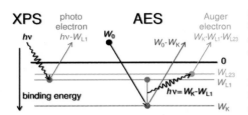

Figure 28: Schemes of the processes which occur for XPS (left hand side) and AES (right hand side) experiments. XPS requires a photon of known energy $h\nu$. The removed photoelectron carries the energy $W_{h\nu} - W_b$. For the Auger process in this example a K-electron is removed by the incident electron. The vacancy is filled by an L1-electron which transfers its energy via a photon to an L23-electron to kick it out of the atom. The energy of the Auger electron is $W_K - W_{L1} - W_{L23}$.

4.3 X-ray Photoelectron Spectroscopy (XPS) and Auger Electron Spectroscopy (AES)

As mentioned in the introduction of this section XPS and AES are usually not used to get spatial structural information about the specimens (Micro- and Nano-AES methods exist, but they are not discussed here). Instead, the electronic properties such as band structures or binding energies are studied. Using XPS means to illuminate the sample with a monochromatized X-ray beam [14]. The photons, which have a very well defined energy, can be absorbed by the material and cause ejection of electrons, the so-called *photoelectrons*. The kinetic energy of these photoelectrons is related to the binding energy in the atoms and the original energy of the photon.

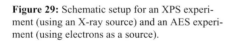
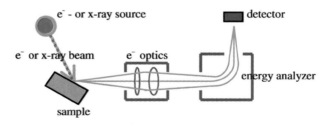

Figure 29: Schematic setup for an XPS experiment (using an X-ray source) and an AES experiment (using electrons as a source).

The AES method requires an electron gun to illuminate the specimen. As described in section about the SEM, the primary electrons cause the removal of core electrons (inner shell) of the atoms. These holes can be filled by electrons from lower levels whereby the excess of energy is transferred to another electron the so-called *Auger electron* which is freed from the sample and can be detected. A scheme of the excited electrons in the band structure for the photoelectrons and Auger electrons is displayed in Figure 28.

The experimental setups of XPS and AES are quite similar (see Figure 29) and actually only differ with respect to the source. As the XPS process also removes inner shell electrons Auger electrons can be seen in the XPS spectra, too. However, an electron source is more efficient to produce Auger electrons.

Both, the photoelectrons and the Auger electrons are element specific and both methods are highly surface sensitive. The investigation depth is determined by the path length at which an electron in the sample can travel without being scattered or absorbed (the so-called *inelastic mean free path*). This length depends on the material but for metals typically exhibits a minimum at some 10 eV for the electron energy (see Figure 30). For photo- or Auger electrons the length is a few nanometers.

An XPS or Auger spectrum is taken as the number of detected electrons versus the electron energy. Distinct peaks appear in the spectrum which are material specific. For XPS the position of the energy of these peaks is related to the binding energy of the electrons: Removing an electron with the binding energy W_b from the specimen by a photon of energy $W_{h\nu}$ means that the final photoelectron has a kinetic energy of $W_k = W_{h\nu} - W_b$. Thus, the structure of the spectrum can be associated with core level electrons and even with electrons from the valence and conduction bands which are freed by the photoeffect. As the binding energy of the electrons are related with the density of states of the electrons XPS yields a direct picture of the density of states. An example for a palladium sample illuminated with MgKα-radiation ($W_{h\nu}$ = 1254 eV) is shown in Figure 31.

An Auger spectrum looks similar to an XPS spectrum. However, the lines appear at different positions due to the different electron emission mechanism. An Auger electron carries the energy of an electron which has fallen from an outer shell to an inner one *minus* the energy which is necessary to remove the Auger electron from its own shell. E.g.: if a K-shell electron was kicked out by the incident beam an L_1-electron may fill this hole. The energy $W_K - W_{L1}$ is transferred to the Auger electron which may sit in the $L_{2,3}$-shell. To free it from the atom the energy W^*_{L23} is needed (the asterisk is to remember, that the atom is already excited due to the hole at L_1). Therefore, the final energy of this particular Auger electron would be $W_{KL1L23} = W_K - W_{L1} - W^*_{L23}$. The problem is now to determine the proper values for the asterisked energy to obtain the band or shell energies. However, Auger spectra from the elements are very characteristic and can be used to be very element specific (see Figure 32).

Finally, AES and XPS can be used in semiconductor or comparable industry to investigate the exact composition of the materials and even some electronic properties such as band widths or band gap distances.

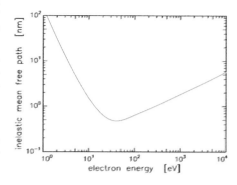

Figure 30: Typical inelastic mean free paths of electrons in a metal. The increase at small energies is due to the fact, that the energy of the electrons is not sufficient to excite any other electron, phonon, plasmon and so on. Therefore, no inelastic scattering is allowed.

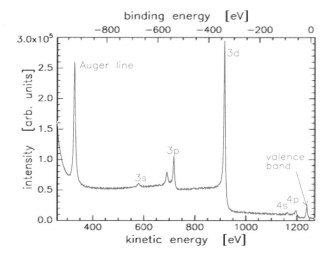

Figure 31: Simulation of an XPS spectrum of a palladium sample using MgKα radiation ($W_{h\nu}$ = 1254 eV). At very low binding energies $W_b \approx -10$ eV the valence band shows up as a peak the weak peaks close to it (at $W_b = -54$ eV and $W_b = -88$ eV) are the 4p and 4s states, respectively.
At $W_b = -335$ eV the most intense peak is due to the 3s level of palladium, the following double peak at $W_b = -534$ eV and $W_b = -561$ eV is associated to the 3p shell. Finally, at $W_b = -673$ eV the 3s level is visible. All other lines are due to Auger electrons.

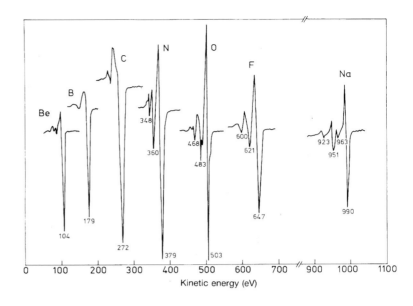

Figure 32: Auger spectra for light elements. The main peaks arise from the $KL_{23}L_{23}$ transition [13].

4.4 Low-Energy Electron Microscopy and Related Techniques

If a sample is biased with a negative electrical potential (i. e. as a cathode), it may emit electrons which can be used to image the sample surface by an appropriate electron optical system. Because of the mean free path of electrons in solids is in the range from 0.3 to 3 nm for low electron energies in the range of 1 to 100 eV, the information is very surface sensitive. The excitation of the electrons can be by illumination of the surface with low-energy electrons (*Low-Energy Electron Microscopy*, LEEM), by illumination with photons (*Photoemission Electron Microscopy*, PEEM), by high temperature (thermionic emission), etc. The lateral resolution can be as low as 5 nm.

A LEEM system is schematically sketched in Figure 33 [15], [16]. The main difference to a TEM (Figure 15a, b) is given by the fact, that the electrons from the source travel coaxially in the opposite direction within the objective lens. For this reason, a magnetostatic beam separator is introduced which enables the electron illumination and image beams to fold back upon themselves after specular reflection at the surface. Due to this configuration, the sample can be considered as an integral part of the objective lens system whose aberrations determine the limit of the achievable lateral resolution. The source and the sample are on a very similar electrostatic potential which permits the energy of the incident electrons to be controlled in the required range. The remaining part of the optical system is at a potential of approx. 15 kV. At these energies, electrons are much less sensitive to magnetic stray fields and it is easier to build up an electron optical system. This means the electrons emitted by the source are accelerated by the condensor system and decelerated short before the sample surface by the objective system. There are several contrast mechanisms, similar to TEM and SEM. In the case of crystalline samples, the incident low-energy electron beam is diffracted due to the crystal structure of the surface (*Low-Energy Electron Diffraction*, LEED). Selected diffraction spots may be used for generating the imagine. Those pictures usually show dark and bright areas. Only those regions on the sample which contribute to the diffraction spot are bright, the rest is dark. Therefore, these pictures are called *dark field images*. In addition, the specular reflected beam can be used which gives the so-called *bright field image* (comparable to the usual microscopy). Furthermore, inelastically scattered or secondary electrons may be used for the analysis.

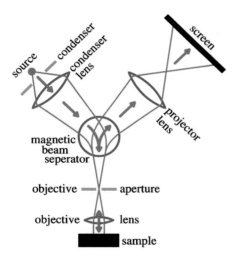

Figure 33: Basic configuration of a low-energy electron microscope (LEEM), emphasizing the key elements of a LEEM that are necessary for surface imaging.

A PEEM system is schematically sketched in Figure 34. Similar to LEEM the heart of the instrument is the objective lens system with the sample integrated into it. UV or soft X-ray photons are used for the illumination. The excitation mechanism is the same as described for XPS (Sec. 4.3). Electron emission occurs when the photon energy exceeds the threshold given by the illumination wavelength and the local work function. The photoemission process can be made very specific to the surface chemistry if the illumination wavelength is tuned into excitation resonances (i. e. chosen to match the excitation band) of the material under study. In addition, the surface magnetization can be imaged if polarized light is used (see Chap. 23).

A more detailed overview of the LEEM based techniques is provided by Ref. [17].

5 Some Other Methods

To investigate the properties of materials many other methods exist. They cannot all be explained in more detail but few of them will shortly be mentioned in this section, how they work principally and what kind of information can be extracted.

A very important method is the *Rutherford Backscattering Spectrometry* (RBS). From the experimental point of view it is similar to EELS in reflection: A sample is bombarded by a beam of high energy particles (such as 2 MeV He^{2+}-ions). Most of the them are implanted but a small fraction undergo a direct collision of near surface located atoms of the sample. The backscattered particle is measured with respect to its energy and direction. It turns out, that the ratio between the energy of the scattered and the incident particle is a quite simple function of the ratio between the mass of the incident particle and the target particle. Therefore, RBS can be used for surface near elemental analysis. As the cross section of the scattering process is proportional to the square of the atomic number of the target atoms, the method is very sensitive to heavy elements and less useful to investigate light elements such as oxygen [18].

To determine film thicknesses, dielectric properties or densities of layer systems *Ellipsometry* can be used. For this the sample surface is illuminated under different angles of incidence with linearly polarized light of different wavelength. The specularly reflected beam is elliptically polarized with the shape and orientation of the ellipse depending on the sample properties (mainly the complex dielectric function and film thicknesses) [19].

Similar results compared to EELS methods can be obtained by *Micro-Raman Spectroscopy*. A sample is illuminated with a high brilliant photon source (usually a laser beam). Some light is scattered inelastically by changing the wavelength (the so-called Raman-shift). The energy was transferred to vibrational or rotational modes of the illuminated molecules. As a big advantage, for Raman Spectroscopy no special treatment or preparation of the sample is necessary unlike for EELS methods. The method is most useful to investigate materials with pronounced chemical bonds such as organic materials. Pure metals are not Raman scatterers. The spatial resolution depends on the focal spot size which can be about 1μm.

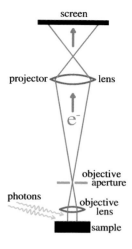

Figure 34: Basic configuration of a photoemission electron microscope (PEEM).

Acknowledgements

The editor would like to thank Susanne Hoffmann-Eifert, Peter Ehrhart, and Christian Ohly for a careful review of this chapter. Furthermore, he acknowledges Marlies Giesen (FZ Jülich) for reviewing the section on the PEEM and LEEM technique.

References

[1] K. Wille, *The Physics of Particle Accelerators*, Oxford University Press, Oxford, 2000.

[2] B. E. Warren, *X-Ray Diffraction*, Dover Publications Inc., New York, 1990.

[3] S. Mantl, M. Dolle, S. Mesters, P. F. P. Fichtner, H.L. Bay, Appl. Phys. Lett. **67**, 3459 (1995).

[4] S. Mantl, J. Phys. D, **31**, 1 (1998).

[5] I. D. Kaendler, O. H. Seeck, J.-P. Schlomka, M. Tolan, W. Press, J. Stettner, L. Kappius, C. Dieker, S. Mantl, J. Appl. Phys. **87**, 133 (2000).

[6] J. Als-Nielsen, *Topics in Current Physics. Structure and Dynamics of Surfaces*, Springer, Berlin, 1986.

[7] S. J. Gurman, J. Synch. Rad. **2**, 56 (1995).

[8] K. H. A. Janssens, F. C. V. Adams, A. Rindby, *Microscopic X-Ray Flourescence Analysis*, John Wiley & Sons, New York, 2000.

[9] T. G. Rochow, P. A. Tucker, *Introduction to Microscopy by Means of Light, Electrons, X-Rays, or Acoustics*, Plenum Press, New York, 1994.

[10] S. Amelinckx, D. van Dyck, J. van Landvyt, G. van Tendeloo, *Electron Microscopy: Principles and Fundamentals*, John Wiley & Sons, New York, 1997.

[11] L. C. Feldmann, J. W. Mayer, *Fundamentals of Surface and Thin Film Analysis*, Elsevier Science Publishing, New York, 1986.

[12] A. Elshabini-Riad, F. D. Barlow III, *Thin Film Technology Handbook*, McGraw-Hill, 1998.

[13] J. C. Rivière, *Surface Analytical Techniques*, Clarendon Press, Oxford, 1990.

[14] S. P. Wolsky, A. W. Czanderna, *Methods of Surface Analysis*, Elsevier Science Publishing, New York, 1989.

[15] E. Bauer, Ultramicroscopy **17**, 51 (1985).

[16] R. M. Tromp, M. Mankos, M. C. Reuter, A. W. Ellis, M. Copel, Surf. Rev. Lett. **5**, 1189 (1998).

[17] L. H. Veneklasen, *Rev. Sci. Instrum.* **63**, 5513 (1992).

[18] McIntyre, Leavitt, and Weller, *Handbook of Modern Ion Beam Materials Analysis*, Materials Research Society, Pittsburg, 1995.

[19] H. G. Tompkins, *A User's Guide to Ellipsometry*, Academic Press, Boston, 1993.

Scanning Probe Techniques

Philipp Ebert and *Kristof Szot*, Department IFF, Research Center Jülich, Germany

Andreas Roelofs, Seagate Technology, Pittsburgh, USA

Contents

1 Introduction — 297

2 The Scanning Tunneling Microscope — 298
2.1 Theoretical Fundamentals of the Scanning Tunneling Microscope — 298
2.2 Operating Modes of the Scanning Tunneling Microscope — 300
2.3 Experimental Realisation of a Scanning Tunneling Microscope — 300
2.4 Applications of the Scanning Tunnelling Microscope — 301

3 The Scanning Force Microscope — 303
3.1 Theoretical Principles of the Scanning Force Microscope — 303
3.2 The Operation Principle of a Scanning Force Microscope — 304
3.3 Applications of the Scanning Force Microscope — 306
3.3.1 Material-Specific Effects — 306
3.3.2 Magnetic Scanning Force Microscopy (MFM) — 307
3.3.3 Electrostatic Scanning Force Microscopy (EFM) — 308
3.3.4 Piezoresponse Force Microscopy (PFM) — 309

4 Imaging of soft organic or biological Samples — 311

5 Manipulation of Atoms and Molecules — 312
5.1 Lateral Manipulation — 313
5.2 Vertical Manipulation — 315
5.3 Effects induced by the Tunnel Current — 315
5.4 Complex Chemical reactions with the STM — 316

Scanning Probe Techniques

1 Introduction

Scanning probe microscopy (SPM) techniques have developed into important tools for surface physics and the characterisation of surface structures in recent years. The different types of scanning probe microscopes provide the possibility of investigating periodic and non-periodic electrical, topographic, optical, magnetic, and many other types of surface properties with atomic resolution. Depending on the mode of operation on which the scanning probe microscopes are based, they can be used for conducting as well as non-conducting materials. In contrast to scattering experiments, scanning probe microscopy techniques have the advantage to provide images of a variety of surface structures in real space. Therefore, they are also frequently used as a complement to scattering experiments.

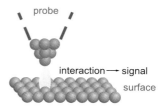

Figure 1: Schematic of the principle of a scanning probe microscope.

A scanning probe microscope works according to a simple principle (Figure 1): A probe is scanned over the surface of interest at a small distance, where an interaction between the probe and the surface is present. This interaction can be of various nature (electrical, magnetical, mechanical, etc.) and provides the measured signal (tunnel current, force, etc). Depending on the quality and type of probe, the measurement signal can be observed to reproducibly vary at atomic distances during scanning of the surface. If these scanning processes are put together line by line, an "*image of the surface*" with the highest resolution is obtained. Such an image shows the spatial variation of the measured parameter, e.g. of the tunnel current.

The development of scanning probe microscopes started in 1981 with the invention of the scanning tunnelling microscope, which still provides the highest resolution of all scanning probe microscopes that have been developed since that time. It is mostly used for semiconductors, metals, and superconductors since it constitutes a probe for electrical surface properties and thus requires electrically conducting surfaces. It measures the *tunnel current* between a metallic, extremely fine tip and the surface (hence scanning *tunnelling* microscope, STM [1]). Among the different scanning probe microscopes, in particular, the scanning tunnelling microscope allows imaging of surfaces with atomic resolution *as a standard feature*. Figure 2 shows as example a scanning tunnelling microscope image of the InP(110) surface. Each bright local peak represents an atom on the surface. Black holes indicate missing atoms, i.e. vacancies. Thus, one can recognize atomic rows and individual point defects. The microscope most frequently used is in fact not the scanning tunnelling microscope, but rather the scanning force microscope (SFM, also known as atomic force microscope, AFM), which probes the force between a tip mounted on a special spring and the surface. The force involved can arise from van der Waals, electrostatic, magnetic, or repulsive atomic interactions. Since the initial development of these two basic scanning probe microscopes, a number of additional ones were invented, that basically only differ by probing different sample properties, such as temperature distributions (scanning thermal microscope), acoustic properties (scanning acoustical microscope), etc.. Of all those maybe the scanning near-field optical microscope (SNOM) gained a larger importance, because it utilises the special focusing of the optical near field at a pointed light guide to probe optical surface properties.

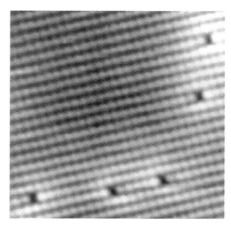

Figure 2: Scanning tunnelling microscope image of several phosphorus vacancies on a *n*-doped InP(110) surface.

This presentation will focus on scanning tunnelling and scanning force microscopy. To begin with, scanning tunnelling microscopy will be dealt with since all the other microscopes are based on this principle. Following this, the most widely used scanning force microscopy will be presented. The measuring possibilities of each scanning probe microscope technique will be illustrated by examples after giving an overview of the operating principle and the theoretical interpretation of the measurement results. It should also be mentioned that SXM is generally used as the acronym for scanning probe microscopes, where "S" stands for scanning, "M" for microscope and "X" alternatively for the type of probe, i.e., for example, T for tunnelling and F for force. Finally, we present the application of scanning probe microscope in the manipulation of individual atoms as one of the bottom-up approaches in building novel nanostructures.

II Technology and Analysis

2 The Scanning Tunneling Microscope

2.1 Theoretical Fundamentals of the Scanning Tunneling Microscope

How does a measuring instrument function that allows us to see *single atoms?* In the case of a *scanning tunnelling microscope* a fine metallic tip is used as probe (called tunnelling tip) (see Figure 3). This tip is approached toward the surface until a current flows when a voltage is applied between the tip and the sample surface. This happens at distances in the order of 1 nm. The current is called tunnel current since it is based on the quantum-mechanical *tunnel* effect. After a tunnelling contact is established, the tip is moved over the surface by a piezoelectric *scanning* unit, whose mechanical extension can be controlled by applying appropriate voltages. The scanning unit is typically capable of scanning an area of a few nm up to several μm. This allows us to obtain a *microscopic* image of the spatial variation of the tunnel current. Hence the name *scanning tunnelling microscope*.

- A metallic tip is moved as probe towards a conducting surface up to a distance of about 1 nm
- With an applied voltage a current flows due to the **tunnel** effect (tunnel current)
- The spatial variation of the tunnel current is measured by **scanning** over the sample surface
- A **microscopic** image of the surface is produced

At this stage we have to ask what kind of atomic-scale structures can be made visible by the scanning tunnelling microscope utilising the tunnel effect? These structures must by nature correspond to electrical states from or into which the electrons can tunnel. In the tunnelling process, the electrons must tunnel through the vacuum barrier between tunnelling tip and sample, which represents a potential barrier. The tunnel effect allows a particle (here an electron) to tunnel through this potential barrier even though the electron's energy is lower than the barrier height. The probability of such a process decreases exponentially with the geometrical distance between the tip and the sample and with increasing barrier height. An experimental apparatus making use of the tunnel effect must therefore minimise the potential barrier to be tunnelled through. This is realised in the scanning tunnelling microscope configuration by moving the tip very close (about 1 nm) to the surface. The electrons can then pass between the surface and the tip. The direction of the tunnel current is fixed by applying a voltage between sample and tip.

In order to explain and interpret the images of the surface states obtained in this way, efforts to develop a theory were made very soon after the invention of the scanning tunnelling microscope. One of the possible theoretical approaches is based on Bardeen's idea of applying a transfer Hamiltonian operator to the tunnelling process [2]. This had the advantage of adequately describing the many-particle nature of the tunnel junction. In the model, a weak overlap of the wave functions of the surface states of the two electrodes (tunnelling tip and sample surface) is assumed to allow a perturbation calculation. On this basis, Tersoff and Hamann developed a simple theory of scanning tunnelling microscopy [3], [4]. Hence follows the tunnel current:

$$I \sim V \cdot \rho_{\text{tip}}(W_\text{F}) \cdot \rho_{\text{sample}}(r_0, W_\text{F}) \qquad (1)$$

The tunnelling tip is assumed to be a metallic s-orbital as shown schematically in Figure 4. In addition, it is assumed that low voltages V (i.e., much smaller than the work function) are applied. $\rho_{\text{tip}}(W_\text{F})$ is the density of states of the tip and $\rho_{\text{sample}}(r_0, W_\text{F})$ is that of the sample surface at the centre r_0 of the tip orbital and at the Fermi energy W_F. Eq. (1) shows that at low voltage *the scanning tunnelling microscope thus images the electronic density of states at the sample surface near the Fermi energy*. However, this result also means that the scanning tunnelling microscope images do not directly show the atoms, but rather the electronic states bound to the atoms. As can be seen in Eq. (1), the tips density of states enters in the measurement in the same way as the density of states of the sample. It is therefore desirable to know the exact electronic state of the tip, but unfortunately, in practice, every tip is different and the details remain unknown.

- Weak overlap of the wave functions of the surface states of the two electrodes (tunnelling tip and sample surface)
- Tunnelling tip approximated as an s-orbital
- Low voltages ($V \ll$ work function)

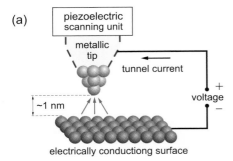

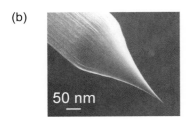

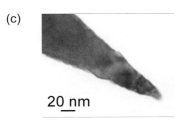

Figure 3:
(a) Schematic representation of the scanning tunnelling microscope: the tunnel current is used as measurement signal.
(b) and (c) show a scanning and transmission electron microscope image, respectively, of a typical metallic tip used for a scanning tunnelling microscope.

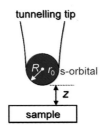

Figure 4: Schematic representation of the tunnelling geometry in the Tersoff-Hamann model.

- The tunnel current is proportional to the local density of states of the sample
- The scanning tunnelling microscope images the electronic local density of states of the sample near the Fermi energy.

In a first approximation the density of surface states decreases exponentially into the vacuum with the effective inverse decay length k_{eff}

$$k_{\text{eff}} = \sqrt{\frac{2m_e B}{\hbar^2} + |\mathbf{k}_\parallel|^2} \qquad (2)$$

m_e is the electron mass and $\mathbf{k}_\parallel$ is the wave vector parallel to the surface of the tunnelling electrons. B is the barrier height, which is approximately a function of the applied voltage V and the work functions Φ_{sample} and Φ_{tip} of the sample and tip [5], respectively:

$$B = \frac{\Phi_{\text{tip}} + \Phi_{\text{sample}}}{2} - \frac{|eV|}{2} \qquad (3)$$

The tunnel current thus decreases exponentially with the tip-sample distance z:

$$I \sim \exp[-2k_{\text{eff}} z] \qquad (4)$$

The exponential current-voltage dependence is quite essential for the high measurement accuracy of a scanning tunnelling microscope, since even small changes in distance may cause a large change in the tunnel current. Thus the tip just needs one microtip, which is only about 0.1 nm closer to the surface than the next one, and still all current flows over only the closest microtip. Thus even apparently wide tips can yield atomic resolution via one microtip.

The description of the tunnel current by Eq. (1) however, has an important restriction: it strictly speaking only applies to low voltages V. In particular for the investigation of semiconductor surfaces voltages of the order of 2 to 3 V are required due to the existence of a band gap. Thus the theory must be extended. The simplest extension yields:

$$I \sim \int_{W_{F,\text{tip}}}^{W_{F,\text{tip}}+eV} \rho_{\text{tip}}(W) \rho_{\text{sample}}(W+eV)\, T(W,V)\, dW \qquad (5)$$

$T(W,V)$ is a transmission coefficient which depends on the energy of the electrons and the applied voltage. The tunnel current is composed of the product of the density of states of the tip and sample at all the different electron energies that are allowed to participate in the tunnelling process (Figure 5). For example, an image measured at −2 V applied to the sample, consequently shows all occupied sample states with an energy between the Fermi energy and 2 eV below the Fermi energy. Tunnelling at a positive voltages analogously provides a measurement of the empty surface states in an energy interval determined again by the voltage.

In order to illustrate this effect more clearly, in the following the InP(110) surface will be presented. On InP(110) surfaces two electrical states exist near the surface: an occupied state below the valence band edge and an empty state above the conduction band edge (Figure 6). All the other states are located geometrically deeper in the crystal or energetically deeper in the bands. They thus contribute little to the tunnel current.

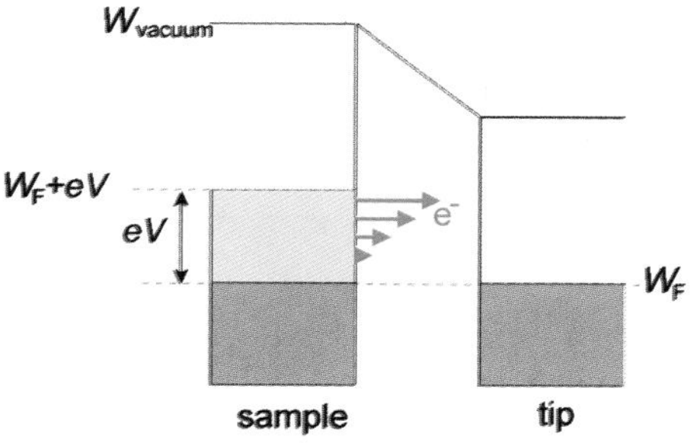

Figure 5: At high voltages not only the states near the Fermi energy W_F contribute to the current but all states whose energy ranges between W_F and W_F+eV.

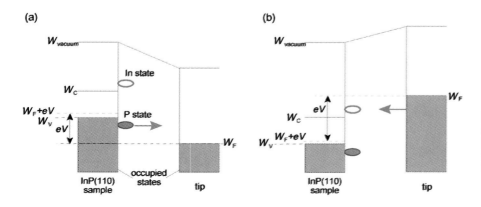

Figure 6: Schematic of the tunnelling process at (a) negative and (b) positive voltages applied to the InP(110) surface.

II Technology and Analysis

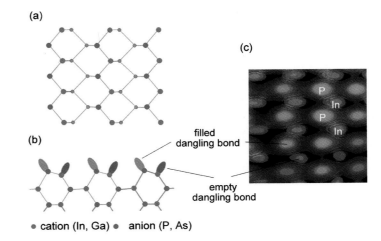

Figure 7:
(a) Schematic top view and
(b) side view of the (110) surfaces of III-V compound semiconductors.
(c) Superposition of two scanning tunnelling microscope images measured at positive (red) and negative (green) voltage. The density of state maxima correspond to the surface states at the In and P atoms, respectively.

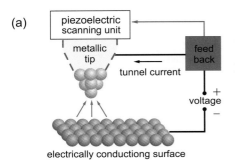

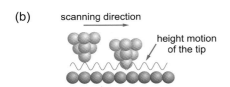

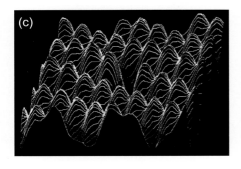

Figure 8:
(a) and (b) Schematic of a scanning tunnelling microscope with feedback loop.
(c) Tunnel image produced by recording individual scan lines of the tip-sample separation.

In the special case of the InP(110) surface, the occupied surface state is spatially located above the P atoms, whereas the empty state is bound to the In atoms (Figure 7a,b). The P and In atoms are alternately arranged in zigzag rows. At negative sample voltages, the scanning tunnelling microscope probes the occupied states located at the P sublattice, whose electrons tunnel into the empty states of the tunnelling tip (Figure 6a). Conversely, only the empty surface states at the In sublattice are probed at positive voltages applied to the sample (Figure 6b) [6] – [8]. If the voltage polarity is changed every scan line, i.e. the occupied and the empty states are probed each alternating scan line, the two resulting images can be superimposed and the zigzag rows of alternating "In" and "P" atoms become visible (Figure 7c).

Apart from the spatial distribution of the density of states, its energy dependence is also of interest, and it should be possible to determine this dependence from current-voltage characteristics using Eq. (5) In order to do so, however, information is required about the transmission coefficient, which turns out to be a great obstacle even if approximations [9] are used. Therefore, in most cases, an experimentally viable approach is used, in which the density of states is approximated as follows [10], [11]:

$$\rho_{\text{sample}}(eV) \approx (dI/dV)/(I/V) \qquad (6)$$

It is thus possible to experimentally measure the density of states as a function of the energy relative to the Fermi level.

2.2 Operating Modes of the Scanning Tunneling Microscope

Up to now, the theoretical background of a scanning tunnelling microscope has been presented, but nothing has been said about the experimental operation of a scanning tunnelling microscope. The simplest way to obtain a scanning tunnelling microscope image is to directly measure the variation of the tunnel current as a function of the scanning position while keeping the distance between tip and sample surface constant. A so-called current image is then obtained. Instead of directly recording the lateral variation of the current, however, the usual procedure is to keep the tunnel current constant while scanning over the surface. This is done by changing the distance between tip and surface using a feedback loop (Figure 8). In order to get an image, the voltage required at the piezoelectric crystal to adjust the distance is recorded. One obtains a so-called constant-current STM image.

2.3 Experimental Realisation of a Scanning Tunneling Microscope

A large variety of different types of scanning tunnelling microscopes were developed in recent years, in order to build the best suited systems for special research projects. Of course, it is not possible to discuss all the designs here and we refer to a selection of references [12], [13]. In the following, a slightly modified design developed at the Research Center Jülich will be discussed. The general operating principle is the same as for other scanning probe microscopes. In particular, the surface is always scanned with the aid of piezoelectric adjusting elements.

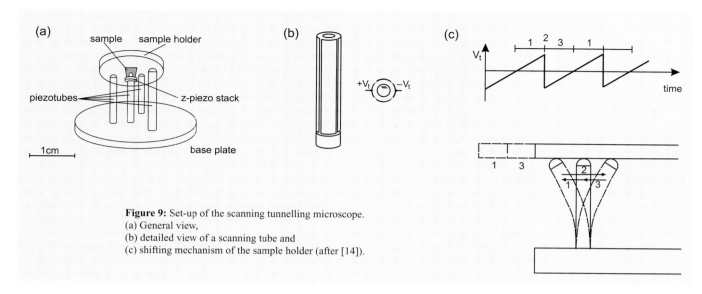

Figure 9: Set-up of the scanning tunnelling microscope.
(a) General view,
(b) detailed view of a scanning tube and
(c) shifting mechanism of the sample holder (after [14]).

In order to obtain atomically resolved images of surfaces using the scanning tunnelling microscope, a horizontal resolution limit below one lattice constant is needed. High demands must therefore be made on the mechanical stability of the scanning tunnelling microscope, such that no uncontrolled movements take place between the tunnelling tip and the sample surface during the measurement. How critical the mechanical design of a microscope is, may be recognised by the fact that the tunnelling tip must be positioned relative to the sample surface with a precision one order of magnitude better than the measuring accuracy required, i.e. horizontally within approx. 10 pm and vertically within 1 pm. The desired mechanical properties are achieved, for example, with a microscope that consists of radially polarized piezotubes arranged to form an equilateral triangle (Figure 9). A fourth tube, the scanning tube, is glued in the centre of the triangle. One of the tubes is mounted on a mobile base plate, whereas the other two and the scanning tube are mounted on a common fixed base plate. The three outer tubes carry the sample holder. A small z-piezo, which holds the tunnelling tip, is mounted on the scanning tube for decoupling the z-motion from the x- and y-scanning motions. The inner metallization of the piezotubes is electrically connected via the base plates to ground. The piezotubes are externally provided with four additional metallizations to which the scanning voltages are applied. Due to the radial polarisation of the piezoelectric material, the tubes can be bent, elongated or shortened. A coarse approach is achieved by raising and lowering the outer piezotube, which is mounted on the mobile base plate.

The microscope with coarse approach rests on several damping rings in a ultrahigh vacuum chamber, which is positioned on a compressed-air-damped table of approx. 1 t weight. The measurements are performed at a pressure of less than 1×10^{-8} Pa in the vacuum chamber to ensure that the surfaces remain clean. Figure 10 shows such a scanning tunnelling microscope viewed through one of the window flanges of the vacuum chamber.

The preparation of the tunnelling tips is one most crucial part is operating a scanning tunnelling microscope, because the tunnelling tip as probe is directly affecting the quality of the measurement results. One possibility of preparing tunnelling tips is the electrochemical etching of polycrystalline tungsten wire with NaOH. The tips produced in this manner have a radius of curvature of only 5 nm as shown in Figure 3.

Figure 10: View of a scanning tunnelling microscope used at Research Center Jülich through one of the window flanges of the vacuum chamber.

2.4 Applications of the Scanning Tunnelling Microscope

The scanning tunnelling microscope covers a wide field of applications wherever information about the surface structure is required in real space. The applications are so widely distributed over many research fields, ranging from biology to crystallography, that it is essentially impossible to provide a full overview of the possibilities to apply the scanning tunnelling microscope. For a overview of we refer to Refs. [15] – [17]. Here we present only examples, which illustrate the potential of a scanning tunnelling microscope.

Due to its high spatial resolution, the scanning tunnelling microscope is an ideal instrument for the examination of lattice defects. In particular, the properties of point defects such as vacancies can be measured, which has not yet been possible with other methods on the atomic scale. As an example, P vacancies on InP(110) and GaP(110) surfaces will be described. Both surfaces have the same geometrical structure, as shown in Figure 6 and Figure 7 in the preceding section for the InP(110) surface. In order to produce a vacancy, an atom must be removed from the surface. Therefore three bonds are broken and so-called dangling bonds, i.e. unsaturated bonds, are left. These unsaturated bonds do not represent the energetically most favourable configuration and they reconstruct forming three defect energy levels. Defect energy levels are electron states located in a vacancy. In the case of high defect concentrations, the defect energy levels change the electrical properties of whole crystals, which is utilized e.g. in doping semiconductor crystals with impurities. As shown in Figure 7, a scanning tunnelling microscope only images either the occupied or the empty states. Since the occupied states correspond to the positions of the P atoms in the surface, a missing occupied surface state is the signature of a *P* vacancy. This missing occupied surface state can be seen in Figure 11 for all four vacancies. The four vacancies are all P vacancies which, however, are observed on different crystal surfaces. The upper images show a positively (a) and a negatively (b) charged P vacancy on *p*- and *n*-doped GaP(110) surfaces, respectively. The bottom images show the corresponding positively (c) and negatively (d) charged P vacancies on the *p*- and *n*-doped InP(110) surfaces [18], respectively.

What causes the different charges of the vacancies on crystal surfaces (e.g. InP(110))? In order to understand this, it is necessary to draw the defect energy levels of the P vacancies. It has already been mentioned that a P vacancy has three energy levels located in the valence band, the band gap and the conduction band, respectively (Figure 12). Only the energy level in the band gap is of interest here.

If the material or surface is *p*-doped, the Fermi level is at the valence band edge. In this case, the energy level in the valence band is occupied by two electrons, whereas the energy level in the band gap in empty. On the other hand, if the sample is *n*-doped, the Fermi level is at the conduction band edge and the defect energy level in the band gap is occupied by two electrons. An uncharged P vacancy has three electrons in the three defect energy levels. Thus, the P vacancy on *p*-doped surfaces is positively charged (a negatively charged electron is missing in comparison to the uncharged electron configuration) and negatively charged on *n*-doped surfaces. With the changes in the electron occupancy of the defect levels the vacancy also changes its geometric structure. This can be seen in Figure 11.

If the material or surface is *p*-doped, the Fermi level is at the valence band edge. In this case, the energy level in the valence band is occupied by two electrons, whereas the energy level in the band gap in empty. On the other hand, if the sample is *n*-doped, the Fermi level is at the conduction band edge and the defect energy level in the band gap is occupied by two electrons. An uncharged P vacancy has three electrons in the three defect energy levels. Thus, the P vacancy on *p*-doped surfaces is positively charged (a negatively charged electron is missing in comparison to the uncharged electron configuration) and negatively charged on *n*-doped surfaces. With the changes in the electron occupancy of the defect levels the vacancy also changes its geometric structure. This can be seen in Figure 11.

Figure 11: Phosphorus vacancies on
(a) *p*- and
(b) *n*-doped GaP(110) and
(c) *p*- and
(d) *n*-doped InP(110). On the left, two positively charged vacancies are shown, whereas on the right they are negatively charged.

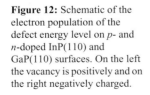
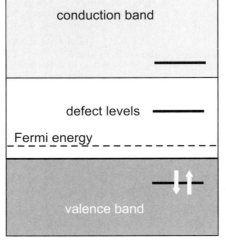
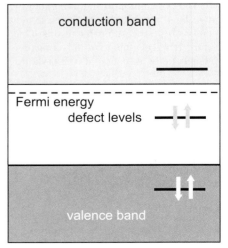

Figure 12: Schematic of the electron population of the defect energy level on *p*- and *n*-doped InP(110) and GaP(110) surfaces. On the left the vacancy is positively and on the right negatively charged.

This example thus shows that it is possible to investigate the electronic properties and geometrical symmetries of individual lattice defects with the aid of a scanning tunnelling microscope, although the defects may only consist of one missing lattice atom in the surface.

Further more Figure 13 is an example for STM reaching atomic resolution in liquids. The adsorption of 2,2' Bipyridin on the Au(111)/aqueous and Au(100)/aqueous electrolyte interface is shown [19], [20].

3 The Scanning Force Microscope

The design and development of the scanning force microscope (SFM) (or frequently called atomic force microscope AFM) is very closely connected with that of scanning tunnelling microscopy. The central component of these microscopes is basically the same. It is a fine tip positioned at a characteristic small distance from the sample. The height of the tip above the sample is again adjusted by piezoelectric elements. The images are taken by scanning the sample relative to the probing tip and measuring the deflection of the cantilever as a function of lateral position.

Initially, the height deflection was measured by a scanning tunnelling microscope piggybacked on the spring. However, this method is very demanding. Nowadays, optical techniques are preferred for measuring the deflection. A rich variety of forces can be sensed by scanning force microscopy. In the non-contact mode (of distances greater than 1 nm between the tip and the sample surface), van der Waals, electrostatic, magnetic or capillary forces produce images, whereas in the contact mode, ionic repulsion forces take the leading role. Because its operation does not require a current between the sample surface and the tip, the SFM can move into potential regions inaccessible to the Scanning Tunnelling Microscope (STM). For example image fragile samples which would be damaged irreparably by the STM tunnelling current. Insulators, organic materials, biological macromolecules, polymers, ceramics and glasses are some of the many materials which can be imaged in different environments, such as in liquids, under vacuum, and at low temperatures.

In the non-contact mode one can obtain a surface analysis with a true atomic resolution. However, in this case the sample has to be prepared under UHV conditions. Recently, it has been shown that in the tapping mode (a modified non-contact technique) under ambient conditions it is possible to observe, similar as in STM investigations, single vacancies or their agglomeration (Figure 14) [22]. Additionally, a non contact mode has the further advantage over the contact mode that the surface of very soft and rough materials is not influenced by frictional and adhesive forces as during scanning in the contact mode, i.e. the surface is not "scratched".

3.1 Theoretical Principles of the Scanning Force Microscope

As already mentioned above, van der Waals forces lead to an attractive interaction between the tip on the spring and the sample surface. Figure 15 shows schematically the van der Waals potential between two atoms. The potential can be described in a simpler classical picture as the interaction potential between the time dependent dipole moments of the two atoms. Although the centres of gravity of the electronic charge density and the charge of nucleus are exactly overlapping on a time average, the separation of the centres of gravity is spatially fluctuating in every moment. This produces statistical fluctuations of the atoms' dipole moments. The dipole moment of an atom can again induce a dipole moment in the neighbouring atom and the induced dipole moment acts back on the first atom. This creates a dipole-dipole interaction on basis of the fluctuating dipole moments. This interaction decreases with d^{-6} in the case of small distances d (Lenard-Jones potential). At larger distances, the interaction potential decreases more rapidly (d^{-7}). This arises from the fact that the interaction between dipole moments occurs through the exchange of virtual photons. If the transit time of the virtual photon between atoms 1 and 2 is longer than the typical fluctuation time of the instantaneous dipole moment, the virtual photon weakens the interaction. This range of the van der Waals interaction is therefore called retarded, whereas that at short distances is unretarded.

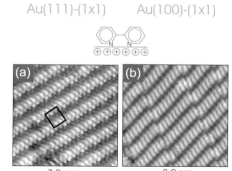

Figure 13: Unfiltered high resolution images of the 2,2' Bipyridin on the Au(111)/aqueous in 0.5 M Na_2SO_4 and Au(100)/aqueous in 0.05 M H_2SO_4, (tunnel current = 2 nA) [19], [20].

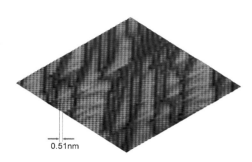

Figure 14: Scanning force microscope image with atomic resolution obtained in the tapping mode under ambient conditions of $SrO-SrTiO_3$ [22].

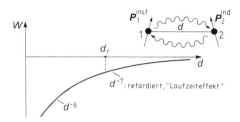

Figure 15: The van der Waals potential W between two atoms. d_r is the critical distance above which the transit time effects weaken the interaction [23].

The scanning force microscope is not based on the interaction of individual atoms only. Both the sample and the tip are large in comparison to the distance. In order to obtain their interaction, all forces between the atoms of both bodies need to be integrated. The result of this is known for simple bodies and geometries. In all cases, the summation leads to a weaker decrease of the interaction. A single atom at distance d relative to a half-space leads to an interaction potential of

$$W = -\frac{C\pi\rho}{6} \cdot \frac{1}{d^3} \quad (7)$$

where C is the interaction constant of the van der Waals potential and ρ the density of the solid. C is basically determined by the electronic polarizabilities of the atoms in the half-space and of the single atom. If one has two spheres with radii R_1 and R_2 at distance d (distance between sphere surfaces) one obtains an interaction potential of

$$W = -\frac{AR_1R_2}{6(R_1+R_2)} \cdot \frac{1}{d} \quad (8)$$

where A is the so-called Hamaker constant. It is materials specific and essentially contains the densities of the two bodies and the interaction constant C of the van der Waals potential. If a sphere with radius R has a distance d from a half-space, an interaction potential of

$$W = -\frac{AR}{6} \cdot \frac{1}{d} \quad (9)$$

is obtained from Eq. (8). This case describes the geometry in a scanning force microscope best and is most widely used. The distance dependence of the van der Waals potential thus obtained is used analogously to the distance dependence of the tunnel current in a scanning tunnelling microscope to achieve a high resolution of the scanning force microscope. However, since the distance dependence is much weaker, the sensitivity of the scanning force microscope is lower.

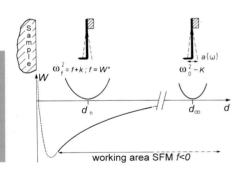

Figure 16: Schematic representation of the effect of the van der Waals interaction potential on the vibration frequency of the spring with tip. As the tip approaches the surface, the resonance frequency of the leaf spring is shifted. (from [23]).

The dynamic operation method of a scanning force microscope has proved to be particularly useful. In this method the nominal force constant of the van der Waals potential, i.e. the second derivative of the potential, is exploited. This can be measured by using a vibrating tip (Figure 16). If a tip vibrates at distance d, which is outside the interaction range of the van der Waals potential, then the vibration frequency and the amplitude are only determined by the spring constant k of the spring. This corresponds to a harmonic potential. When the tip comes into the interaction range of the van der Waals potential, the harmonic potential and the interaction potential are superimposed thus changing the vibration frequency and the amplitude of the spring.

This is described by modifying the spring constant k of the spring by an additional contribution f of the van der Waals potential. As a consequence, the vibration frequency is shifted to lower frequencies as shown in Figure 17. ω_0 is the resonance frequency without interaction and $\Delta\omega$ the frequency shift to lower values. If an excitation frequency of the tip of $\omega_m > \omega_0$ is selected and kept constant, the amplitude of the vibration decreases as the tip approaches the sample, since the interaction becomes increasingly stronger. Thus, the vibration amplitude also becomes a measure for the distance of the tip from the sample surface. If a spring with low damping Q^{-1} is selected, the resonance curve is steep and the ratio of the amplitude change for a given frequency shift becomes large.

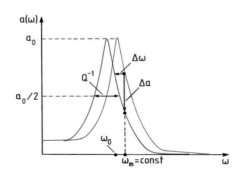

Figure 17: Resonance curves of the tip without and with interaction with a van der Waals potential. The interaction leads to a shift $\Delta\omega$ of the resonance frequency with the consequence that the tip excited with the frequency ω_m has a vibration amplitude $a(\omega)$ attenuated by Δa [23].

In practice, small amplitudes (approx. 1 nm) in comparison to distance d are used to ensure the linearity of the amplitude signal. With a given measurement accuracy of 1 %, however, this means that the assembly must measure deflection changes of 0.01 nm, which is achieved most simply by a laser interferometer or optical lever method.

3.2 The Operation Principle of a Scanning Force Microscope

The main electronic components of the SFM are the same as for the STM, only the topography of the scanned surface is reconstructed by analysing the deflection of the tip at the end of a spring. Today, the interferometrical and optical lever method dominates commercial SFM systems. The most common method for detecting the deflection of cantilever is by measuring the position of a reflected laser-beam on a photosensitive detector. The principle of this optical lever method is presented in Figure 18a. Without cantilever displacement both quadrants of the photodiode (A and B) have the same irradiation $P_A = P_B = P/2$ (P represents the total light intensity). The change of the irradiated area in the quadrants A and B is a linear function of the displacement

$$\delta \propto \Delta d = 2\sin(\Theta) \cdot S_2 = 2\Theta \cdot S_2 = 3S_2 \cdot \delta/L \qquad (10)$$

For small angles $\sin(\Theta) \approx \Theta$ and Θ may be evaluated from the relation $\Theta = 3\delta/2L$ (Figure 18b). For P_A and P_B one would get approximately $P_A = P/2 \cdot (d + \Delta d)/2$ and $P_B = P/2 \cdot (d - \Delta d)/2$. Using the simple difference between P_A and P_B would lead to

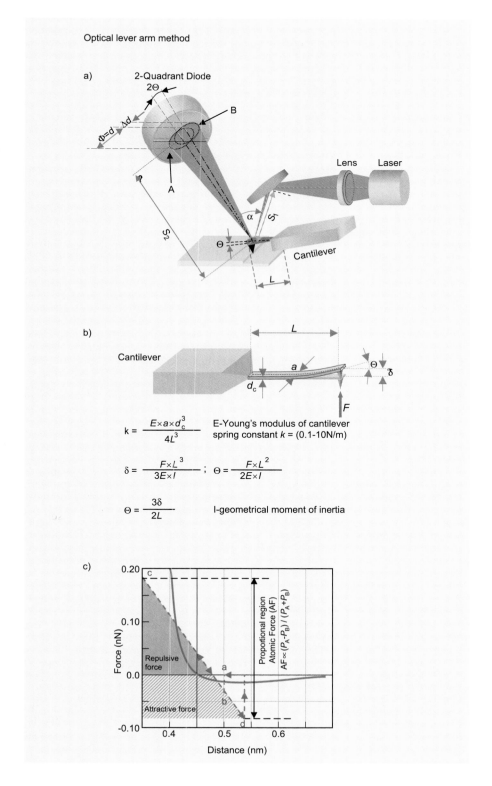

Figure 18: The amplification of the cantilever motion through the optical lever arm method.
(a) Optical laser path in the standard AFM set-up.
(b) Cantilever beam in bending.
(c) Cantilever force as a function of the distance tip – sample distance.

$\Delta P = P \cdot 3S\delta/(Ld)$ but in this case one cannot distinguish between the displacement δ of the cantilever and the variation in the laser power P. Hence the normalised difference is used, which is only dependent on δ:

$$\frac{P_A - P_B}{P_A + P_B} = \delta \cdot \frac{3S_2}{Ld} \qquad (11)$$

The "lever amplification" $\Delta d/\delta = 3S_2/L$ is about a factor of one thousand. On the basis of this kind of technique one is able to detect changes in the postion of a cantilever of the order of 0.01 nm.

For large distances between the tip and the sample the bending of the cantilever by attractive forces is negligible. After the cantilever is brought closer to the surface of the sample (point "a" Figure 18c) the van der Waals forces induce a strong deflection of the cantilever and, simultaneously, the cantilever is moving towards the surface. This increases the forces on the cantilever, which is a kind of positive feedback and brings the cantilever to a direct contact with the sample surface (point "b"). However, when the cantilever is brought even closer in contact to the sample, it actually begins to bend in the opposite direction as a result of a repulsive interaction ("b-c"). In the range ("b-c") the position of the laser beam on both quadrants, which is proportional to the force, is a linear function of distance. On reversal this characteristic shows a hysteresis. This means that the cantilever loses contact with the surface at a distance (point "d") which is much larger than the distance on approaching the surface (point "a").

Up to now, the actual probe, i.e. the tip of the leaf spring, has not been discussed in detail. Its preparation is particularly demanding since the tip and the sensitive spring should be one piece. Moreover, the cantilever should be as small as possible. Nowadays, such scanning tips are commercially available (in contrast to the tunnelling tips, which you should prepare yourself). Figure 19 shows such a spring with tip (cantilever) made of Si. The characteristic parameters of a cantilever have been presented in Figure 18b. The spring constant $k = Ead_C^3/4L^3 \sim 0.1$ N/m – 10 N/m of the cantilever enables topographical analysis with atomic resolution.

For the realisation of a scanning force microscope, the force measurement must be supplemented by a feedback control, in analogy to the scanning tunnelling microscope. The controller keeps the amplitude of the vibration of the cantilever (the tip), and thus also the distance, constant. During scanning the feedback controller retracts the sample with the scanner of a piezoelectric ceramic or shifts towards the cantilever until the vibration amplitude has reached the setpoint value again. The principle of height regulation is exactly the same as for the scanning tunnelling microscope. *The scanning force micrographs thus show areas of constant effective force constant.* If the surface is chemically homogeneous and if only van der Waals forces act on the tip, the SFM image shows the *topography of the surface*.

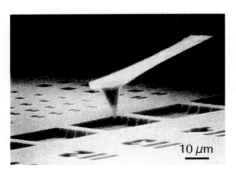

Figure 19: Scanning electron micrograph of a cantilever made of Si. [24].

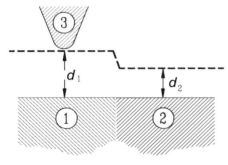

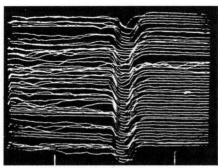

Figure 20:
(a) Sketch concerning material dependence and (b) isogradient lines of a photo resist test structure of 90 nm thickness on Si. (from [23].

3.3 Applications of the Scanning Force Microscope

To a large degree the scanning force microscope (SFM) is applied to determine routinely topographic information of surface structures. The advantage of the SFM lies, however, in the wide applicability by changing the interacting force probed. One of the most prominent examples is the use of magnetic interaction in the SFM. In the following, we present several examples of applications of the SFM, which surely cannot cover the full width of applications (for that we refer to Ref. [15]), but provide a first insight into (i) the effect of dissimilar materials on the topographic image, (ii) the use of magnetic forces to probe magnetic structures, and (iii) the electrostatic SFM.

3.3.1 Material-Specific Effects

If the topography of a surface is scanned using the non-contact SFM, a purely topographic image is only obtained if the surface consists everywhere of the same material. This restriction is due to the fact that the material-specific Hamaker constant A enters into the interaction potential (Eq. (9)). Lets assume for simplicity a sample consisting of two different materials (labelled 1 and 2) and a perfectly flat surface. This sample has no topographic structure, but still one can observe a contrast in the scanning force microscope images due to the change of the sample's material. This can be understood as follows. The height of the tip above the sample is fixed in the non-contact mode by the condition of a constant force constant (see Figure 16 and Figure 17) $f_{1,tip}(d_1) = f_{2,tip}(d_2)$ with $f_{i,tip} \sim A_{i,j}/d_i$. Thus one obtains with $A_{i,j} \approx (A_{i,i} \times A_{j,j})^{0.5}$ the condition:

$$\frac{d_1}{d_2} = \left(\frac{A_{1,1}}{A_{2,2}}\right)^{1/6} \tag{12}$$

Eq. (12) shows that a material-dependent distance change occurs. An example of the effect of different materials on the scanning force microscope image is shown in Figure 20. A Si wafer has been partially covered with a photoresist layer. The nominal thickness of the photoresist layer is 90 nm. Photoresist and Si have a Hamaker constant of 5×10^{-20} and 30×10^{-20} J, respectively [25]. Hence, for a working distance of 10 nm the apparent step has a height of about 3 % of the working distance. The material-dependent effect is thus relatively small, but needs to be taken into account in case of strongly inhomogeneous and otherwise flat surface. For rough surface structures the effect of different materials can, however, be neglected, as the topography changes are much larger than any possible influence of different materials.

3.3.2 Magnetic Scanning Force Microscopy (MFM)

If a magnetic tip is used in the scanning force microscope, magnetic structures can be imaged. Magnetic scanning force microscopy is of interest, in particular, for the investigation of magnetic storage media. In the most general case, the magnetic force between sample and tip is

$$F_{\text{mag}} = -\nabla \int_{\text{tip}} M_{\text{tip}} \cdot H_{\text{sample}} \, dV \tag{13}$$

or

$$F_{\text{mag}} = (m_{\text{tip}} \nabla) B_{\text{sample}} \tag{14}$$

where H_{sample} and B_{sample} are the magnetic stray field and the magnetic induction of the sample, respectively. M_{tip} and m_{tip} are the magnetisation and the magnetic moment of the tip, respectively. Since in most cases the exact magnetic structure of the tip is not known, a model tip magnetization must be assumed. In the simplest case, the tip is a spherically structured magnetic single domain with the magnetisation M_{tip}. Of particular interest are the stray fields of magnetic storage media which consist of different domains. Since the important aspect in force microscopy is not the force but the force gradient, a pronounced variation of the signal is found near the domain walls, but not inside a domain. This situation is sketched in Figure 21. The parameter of the two curves shown (solid and broken lines) is the ratio of the working distance d and the radius R of the magnetic domain of the tip.

Figure 22a shows an experimentally measured picture of four different oriented magnetic domains. Images b and c show the fine structure of a 180° domain. Alternating bright and dark contrasts can be seen. These contrast changes show that the domain wall consists of segments with different wall orientation. This example illustrates that magnetic SFM is well suited for imaging magnetic structures that are commonly used in today's storage media.

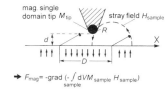
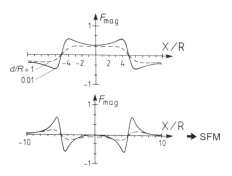

Figure 21: Principle of magnetic scanning force microscopy. On the left, the tip-sample configuration is shown and on the right the force and nominal force constant as a function of distance for this configuration. Two domain walls exist at position X/R = +5 and – 5 (after [23]).

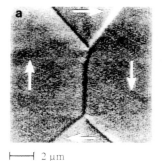

Figure 22: Magnetic SFM image of magnetic domains.
(a) shows four domains of a Landau-Lifshitz structure in which the domain walls are the dark and bright lines. (b) and (c) show the fine structure of a 180° domain wall. The domain wall consists of segments with different wall orientation. Arrows denote the domain orientation. (after [26]).

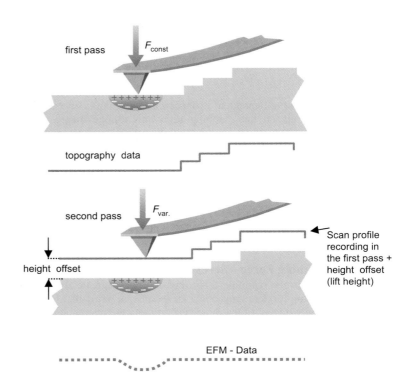

Figure 23: Schematic view of the two-pass SFM technique.

3.3.3 Electrostatic Scanning Force Microscopy (EFM)

The scanning force microscope can also be used for electrostatic or Coulomb images. In this case, however, the tip must be insulated. A modern version of electric force microscopes uses a complicated lock-in and phase loops electronics for the determination of surface charge or surface potential. The essential modification with respect to the old apparatus is the so called two-pass technique (LiftMode). In this method each line must be scanned twice. On the basis of two line scans, in which the first represents the topography of the surface in a contact mode and the second is taken at a fixed distance relative to the surface (Figure 23), one can precisely reconstruct the distribution of the charge or the potential on the surface without topographical error. In Figure 24 an example case is given, which highlights the possibilities of this modern tool for microelectronics. Another example showing an image of a broken carbon nanotube is represented in Figure 25.

Figure 23 shows the effect of the charge when the tip comes into the region of the local surface charge. An attractive Coulomb interaction exists between these charges and the charge induced in the tip. The electrostatic force as a function of distance is given by

$$F_{\text{el}} = -\pi\varepsilon_0 V^2 \frac{r^2}{d^2} \tag{15}$$

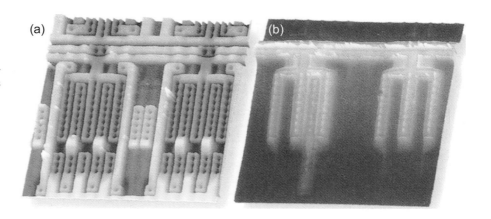

Figure 24: Topography of an IC with passivation layer on (a). Electric Force Microscopy image shows defect transistors in saturation (bright regions) (b) (after [27]).

Imaging break in a SWNT

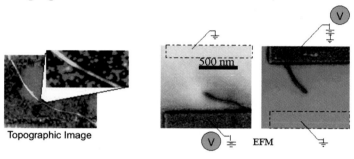

Topographic Image   EFM

Figure 25: Single wall nanotube (SWNT). The topographic image shows the SWNT between the electrodes. In the EFM image a break in the nanotube leading to a voltage drop is visible [28].

with $d/R \gg 1$. This force is plotted in Figure 26 in comparison to the van der Waals force as a function of distance. It can be seen that the electrostatic forces have a much larger extension range than the van der Waals forces and are also stronger.

3.3.4 Piezoresponse Force Microscopy (PFM)

The interest in ferroelectric materials is strongly increasing due to their potential use in Random Access Memories. In order to get better insight into the dielectric, electric, and mechanical properties of these ferroelectric thin films, their domain structure needs to be studied on the nanometer scale. Scanning probe microscopy provides three different methods to obtain information about the domain configuration. Simple topographic imaging is limited to visualize ferroelastic domains indirectly via the wedge-shaped surface caused by the mechanical stress relief. Secondly the different polarization charge induced on the surface can be measured using conventional electric force microscopy (EFM). A more direct observation is applied in the piezoresponse force microscopy (PFM), where the local inverse piezoelectric effect is monitored, providing a method to image the three dimensional distribution of the ferroelectric polarization.

The experimental setup consists of a Scanning Force Microscope SFM equipped with a four-quadrant photodetector using a conductive cantilever, allowing to determine the perpendicular and the lateral deflections from the cantilever (Figure 27). Electronic feedback is established by keeping the average force on the cantilever constant (contact mode). This facilitates accurate tracking of the sample surface topography while recording the domain distribution.

The contrast in PFM results from the interaction between the electrically biased SFM tip (ac-voltage $V = V_0 \cdot \sin(\omega t)$) and the polarization within the ferroelectric sample measured by piezoelectric probing. Cantilever oscillation may stem from two different origins: the inverse piezoelectric effect and Maxwell stress. The second origin of tip oscillation is the electrostatic interaction between the cantilever and the polar sample: the voltage leads to an oscillating charge $q = C \cdot V_0 \sin(\omega t)$ on the tip, where C denotes the capacitance between the tip and the bottom electrode. This charge q may now interact with the stray field E at the sample surface arising from any polarization within the sample. Since q varies with ω, an oscillation force on the cantilever exits, which is given by $F = q \cdot E = CVE\sin(\omega t)$.

Both Maxwell stress and an inverse piezoelectric effect may be present simultaneously. Nevertheless, it is possible to determine the major contribution, since the two effects are phase-shifted by 180°, at least for the out-of-plane areas [29]. Hence, the contrast depends on the relative orientation between the externally applied electric field E and the internal polarization P in the following manner:

$$E \cdot P = |E| \cdot |P| \cos(\theta) \tag{16}$$

where θ specifies the angle between E and P. Since E is applied between the tip and counter electrode (i.e. simplified perpendicular to the sample surface) any surface areas with a polarization vector parallel to the sample surface (i.e. in-plane polarization) have $\theta = 90°$ which results in no coupling normal to the sample surface, as deduced from Eq. (16). Nevertheless, a perpendicular arrangement of E and P in an in-plane polarized domain gives rise to a lateral shear stress exerted on the ferroelectric crystal (Figure 27 II). Since the conducting tip stays in contact with the sample surface we get

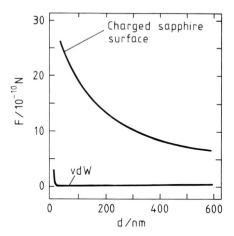

Figure 26: Comparison of the distance dependence of the electrical and van der Waals forces between a tip and a local electrical surface charge of $2 \cdot 10^{-16}$ C (after [23]).

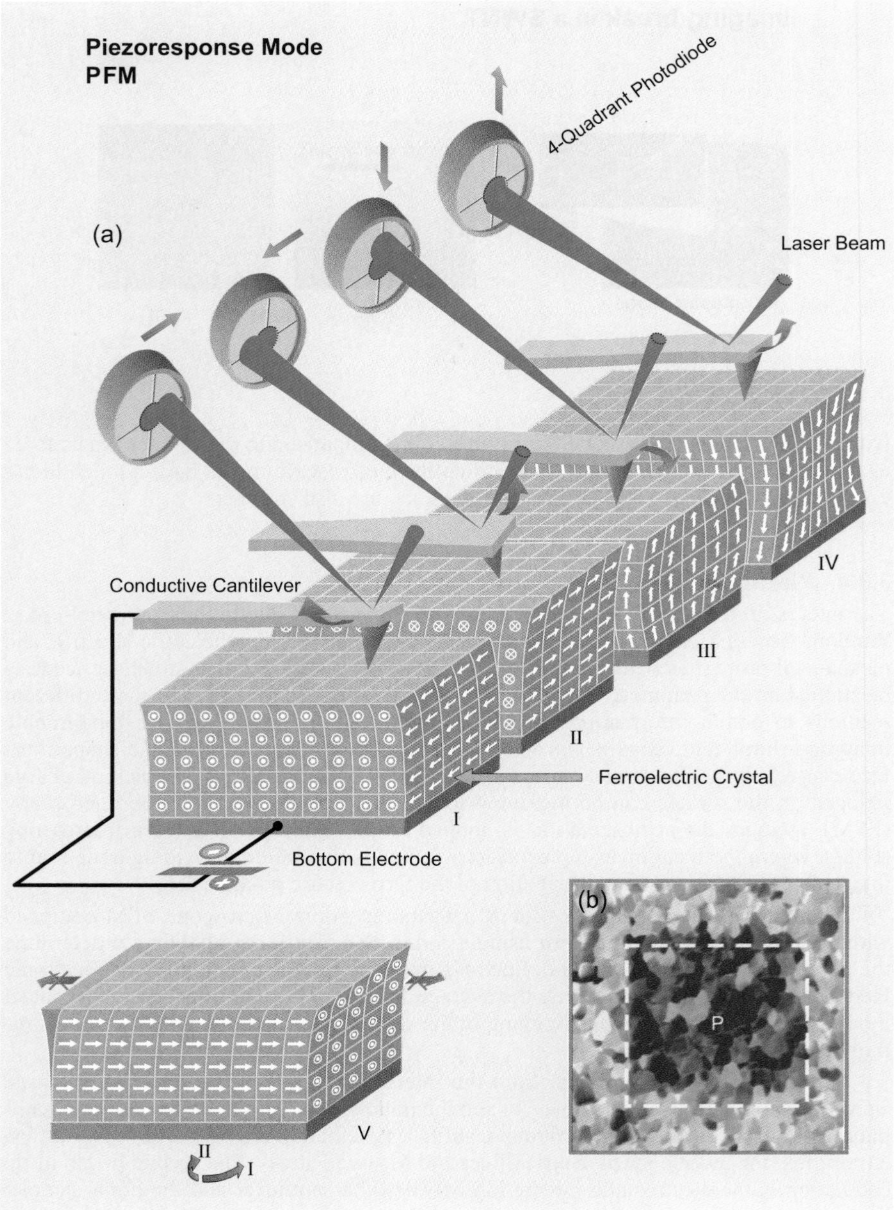

Figure 27: The different kind of cantilever bending as an effect of the interaction between tip bias and different oriented vectors of spontaneous polarisation in ferroelectric crystals are presented.
(b) An example of a piezoresponse image showing a polarised region (P) in PbTiO$_3$ thin films [30].

a torsional bending of the cantilever along the x-axes (the z-axis is always perpendicular to the sample surface). Using two lock-in amplifiers operating at the oscillation frequency ω, the amplitudes of the thickness and lateral oscillations can be evaluated by demodulating the feedback error signal and the frictional force signal, respectively. The output of the two lock-in amplifiers, referred to as polarization signals P_x and P_z, represent the direction of the polarization vector of the domain below the cantilever.

By simultaneously recording the bending along x- and z-axes we obtain a map of the effective polarization elements P_x and P_z. However, due to the cantilever asymmetry, y-axis polarized domains can only be recorded by physically rotating the sample by 90°. Taking all the information together allows to determine all three components of the polarization vector P_x, P_y, and P_z. Figure 27b shows the P_z component of the polarization in a PbTiO$_3$ thin film after polarizing the squared region (P).

4 Imaging of soft organic or biological Samples

Dealing with organic, biological or in general soft material can lead to anomalies due to the compressibility of the structures [31]. The normal forces of the tip acting on the sample easily deform the fragile molecules. Furthermore in many cases the molecule-substrate adhesion is weak, which can lead to artifacts due to the lateral dragging of material. For this reasons the interaction forces between the tip and the sample have to be rather low.

Presently the most frequently used method for imaging soft materials is the tapping mode or the intermittent contact mode [32], [33], although normal forces in air are in both cases tens of nN [34]. Hence, the SFM image quality in air is often not satisfactory and the biological systems show lower heights than expected. This frequently results in a weak contrast of the molecular systems on the substrate. For instance, the height of DNA molecules typically appears lower than its theoretical value of 2 nm [35], and different groups give different values ranging from 0.28 nm to 1.7 nm [31], [36]. Better resolution is obtained by working in liquid allowing to work with loading forces in the pN range [37]. Nevertheless, in some cases it is desirable to measure in air, such as when the molecules would be washed away by the liquid due to the weak adhesion, as well as for several technological applications [38].

To solve this problem one can operate the scanning force microscope in the regime of attractive forces by using a special feedback circuit (called Q-control) as proposed by the group of Fuchs [39], [40]. To prove the operation mode a self-assembled oligomeric nanostructure consisting of bisbiotinylated DNA fragments connected by the protein streptavidin (STV) was used. The structure of DNA molecules can be considered as rods with a typical diameter of 2 nm, while STV molecules are almost spherical particles with a typical diameter of about 5 nm [39]. For comparison images in the attractive as well as in the repulsive mode were monitored.

Figure 28 presents a SFM image first taken in the attractive force regime (Figure 28a), then the SFM was switched to the repulsive force regime and the same area was scanned again (Figure 28b). Working in the repulsive mode gives rise to a weaker contrast as a result of the reduced height of the structures. In this mode the heights of the DNA and the STV molecules were 0.5 nm and 1.7 nm respectively, whereas the heights in the attractive mode were 1.1 nm for the DNA and 3.3 nm for the STV molecules. The height values obtained in the attractive regime are still smaller than the diameter of these molecules in liquid solution. Possible reasons for this discrepancy can be related to sample-substrate adhesion effects as well as tip-sample energy dissipation or sample preparation effects.

An improved method for studying self-assembled monolayers of alkanethiols has been developed by Bumm et al. [41]. Here a home-built STM employing two modulation frequencies is described, which is used to record images in constant-current mode and at high tunnelling gap impedances ($\sim 10^{12}\ \Omega$). This ensures a large tip-sample separation and therefore a minimal contact between the probe tip and the monolayer.

This special STM technique has been used for example to study the phase separation within a binary self-assembled monolayer (SAM) driven by an amid-containing alkanethiol [42]. It could be shown that the introduction of hydrogen-bonding functionality induces the formation of single-component domains. In Figure 29 well ordered, phase-separated domains of both 3-mercapto-N-nonylpropionamide and n-decanethiolate species on gold terraces are resolved. The amide-containing molecules correspond to the topographically higher regions in the image (roughly 0.15 nm higher). This high difference could be clearly distinguished from monoatomic gold steps, which are typically 0.24 nm in height.

In addition single organic molecules can also function as nano-switch as shown in [43]. Donhauser et al. employed phenylene ethynylene molecules in a dodecanethiolate monolayer matrix which exhibit a bistable conductance state. The more conductive state is denoted as *ON*- the less conductive one as *OFF*-state. The different states originate from an internal conformational twist of the molecules. When individual or bundle molecules are observed over time, they frequently change their conductance state and switch from *ON* to *OFF* and back (Figure 30). The switching in conductance is accompanied by a change in the apparent height by STM topography of approx. 0.3 nm. Because topographic STM images represent a convolution of the electronic and topographic structure of the sample surface, the apparent height change can thus be due to a change in the conductance of the molecule, a change in the physical height of the molecule or both. A time lapse series of images acquired over several hours recorded the long-time behavior of the molecules (Figure 30). In Figure 30a a large scan area is shown from which the data in Figure 30b and c were extracted. Furthermore individual molecules

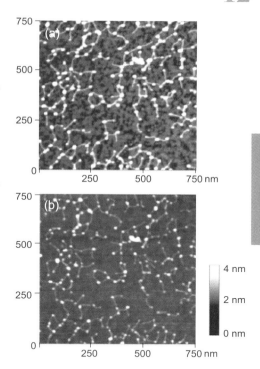

Figure 28: SFM images of bisbiotinylated DNA-STV complexes. (a) scanned in the attractive regime with an active feedback module (Q-control), (b) taken in the repulsive regime [39].

Figure 29: STM image of a $30 \times 30\ nm^2$ area of a phase-separated self-assembled monolayer (SAM) formed by coadsorption from an equimolar solution of 1ATC9 and *n*-decanethiol (1mM in total thiol). The image was recorded at a sample bias of +1.0 V and a tunneling current of 1.0 pA. Topographically higher regions correspond to the brighter areas. [42]

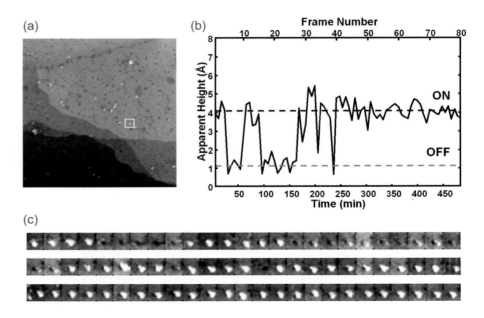

Figure 30: (a) A 150 nm by 150 nm topographic STM image acquired at a sample bias of −1.4 V and a current of 0.2 pA. The vertical scale is ~3 nm. Several molecules are inserted at the gold substrate step edges and the characteristic dodecanethiol SAM defect sites. The area in the small square is extracted from a sequential series of images to assemble the data in (b) and (c). (b) Height versus time for the molecule in the extracted area. Height is determined as described in (24). (c) Extracted frames of the selected molecule. The time interval between frames is ~6 min [43].

can be addressed by the STM tip and by applying an electric field they can be switched from the *ON*- to the *OFF*-state. After switching the *OFF*-state was stable for more than one hour and then switched back to its initial state.

5 Manipulation of Atoms and Molecules

Scanning probe microscopes do not only have the ability to image individual atoms. The interaction needed for imaging the surface can also be used to manipulate individual adatoms, molecules, or the surface structure itself on the atomic scale. Indeed a large number of works concentrated on the manipulation of individual atoms and in the following novel nanostructures were built. Here we briefly show the work by Eigler and coworkers [44], [45] as examples, followed by a more subtle tip-induced manipulation of atoms, and the tip-induced migration of defects by tip-induced excitement of defects [46], [47]. More recently the group of Rieder could even perform full chemical reactions with single molecules [48]. Three different manipulation modes can by distinguished: the lateral and the vertical manipulation as well as the tunnel current induced changes. The combination of all three modes enables to achieve tip controlled chemical reactions.

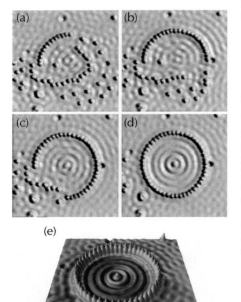

Figure 31: (a) to (d) show the process of building a quantum corral consisting of 48 Fe atoms positioned on a Cu(111) surface. The resulting structure and the standing waves induced by the quantum confinement of surface electrons in the structure is visible in the three-dimensional view (e) of the quantum corral [44].

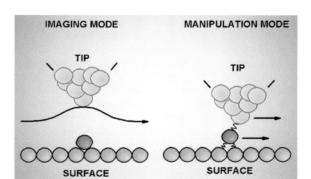

Figure 32: In the STM imaging mode the tunnel current is kept constant and the cantilever is raised. For manipulation the tip is lowered above an atom dragging it to the desired position. By lifting, the cantilever loses interaction with the atom [44].

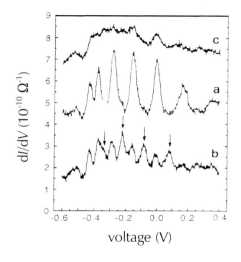

Figure 33: Scanning tunnelling spectra obtained for different positions in the quantum corral of Figure 16. The spectra were obtained
(a) in the circle centre,
(b) 0.9 nm from the circle centre, and
(c) outside the circle. A clear quantization of states can be seen inside of the circle. [44].

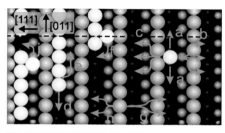

Figure 34: Top and side views of a sphere model of the Cu (211) substrate. The Cu atoms are represented as spheres with the deepest atoms shaded darkest. The ideal surface consists of (111) terraces and (100) steps. The small letters indicate the different manipulation procedures.

5.1 Lateral Manipulation

In the lateral manipulation mode a particle on the sample surface is moved along the surface to the desired location without losing contact to the surface. The motion can be obtained either by dragging or pushing. Figure 31a to d shows the build up of a quantum corral by manipulating individual Fe atoms on a Cu(111) surface at 4 K [44]. The final structure of 15 nm diameter consists of 48 Fe atoms. At this stage one may ask how such a fine manipulation can be achieved. The procedure is as follows: Fe atoms are evaporated onto a Cu(111) surface cooled to 4 K. The surface containing statistically distributed Fe atoms is then examined by STM. Normally no atoms are displaced, but if the distance between tunnelling tip and a Fe atom is reduced, the tunnelling tip exercises an attractive force on the Fe atom and the Fe atom can be dragged by the tunnelling tip to the desired location on the surface [44], [45]. Once the desired location is reached, the tip is retracted. Increasing the distance between the tip and the sample reduces the tip-Fe atom interactions and, hence, the Fe atom remains at its new position (Figure 32). Figure 31e shows that in this way a whole circle of iron atoms can be built up. The artificially build nanostructure shown in Figure 31 confines the electrons of the two-dimensional surface electron gas on Cu(111). Therefore, as soon as the circle is complete, the electrons are scattered in the circle and form standing electron waves due to quantum mechanics. Figure 31e thus illustrates the wave nature of the electrons.

In a quantum well, due to the quantization of the electron states, not only standing electron waves but also discrete energy values of the electrons are expected. Consequently, increased electron densities should occur at specific energies. As already described, the density of states can be estimated from experimentally measured current-voltage characteristics by calculating $(dI/dV)/(I/V)$. For metals, however, I/V is generally constant and the sample density of states is therefore proportional to dI/dV. The variation of dI/dV as a function of voltage reflects the density of states variation as a function of energy. Figure 33 shows the density of states thus obtained for three different surface positions. At the centre of the circle, as expected, peaked energy levels occur (curve a), whereas outside the circle no structure in the density of states is measurable (curve c). If the density of states is measured at a distance of 0.9 nm from the circle centre, even more energy levels occur as shown by the arrows in curve b.

This example of spectroscopic measurements and of the spatial distribution of the electron waves in a potential well provides a particularly illustrative picture of quantum mechanics. The construction of different quantum structures by an atom-by-atom manipulation approach using scanning probe microscopes nowadays allows a new look into the quantum world and a direct spatial measurement of the electron waves.

Instead of evaporating foreign atoms onto a copper surface it is also possible to reconstruct the substrate surface itself, which is more difficult due to the higher coordination number and binding energy of the atoms located in the surface or in steps [49]. The experiments shown here are carried out on Cu(211) substrates at 30 – 40 K. In Figure 34 a sphere model of the copper surface is shown, whereby the atoms are shaded darker the deeper they lie. Lateral manipulation of single Cu atoms parallel and perpendicular to step edges is presented in Figure 35 [49]. A measure for the minimum force necessary to move a copper atom is the tunnel resistance which displays the distance

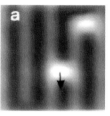
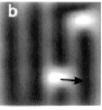

Figure 35: STM image showing the lateral manipulation of a single Cu atom (a) parallel and (b) perpendicular to a step. The motion is indicated by an arrow. In the upper part of the image a single copper atom serves as marker. The processes involved correspond to the motion shown in Figure 34a and b [49].

Figure 37: The molecular structure of hexa-*tert*-butyl decacyclene (HB-DC). (a) the top and (b) is showing the side view. The C atoms are blue and the H atoms are white [50].

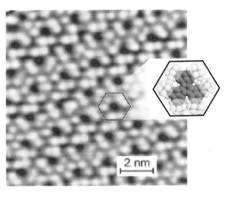

Figure 38: STM image of an Cu(100) surface after exposure to a full monolayer coverage of HB-DC molecules at room temperature. Image area is 11.4 nm by 11.4 nm [50].

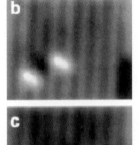

Figure 36:
(a) Single copper atom serving as marker. (b) – (c) Two copper atoms are dug out one by one. The current needed for manipulation is 130 nA and for imaging 1.35 nA. The sample bias voltage is 12 mV [49].

between tip and sample. The tunnel resistance used for motion along a step edge is approx. 700 kΩ and ~500 kΩ for moving them over a step edge. Figure 36a – c demonstrate that it is even possible to "dig out" single copper atoms from higher coordinated sites. The single Cu atom (Figure 36a) is used as a marker. Figure 36b, c show the drag out of single Cu atoms leading to corresponding vacancies in the initial site of the atoms.

Furthermore instead of moving single atoms, the lateral manipulation technique is also capable to move entire molecules. Gimzweski et al. deposited hexa-*tert*-butyl decacyclene (HB-DC) molecules onto a Cu(100) surface [50]. The decacyclene core of the HB-DC is equipped with six bulky *t*-butyl-legs (Figure 37). At monolayer coverage, the molecules are immobile, forming a two dimensional van der Waals crystal (Figure 38). Separated HB-DC molecules on a Cu(100) surface are extremely mobile, making it impossible to get STM images with atomic resolution.

For this reason a coverage of just less than one monolayer is chosen and STM images resemble those of the immobilized 2-D lattice at full monolayer coverage. However, there are some random voids. In this layer the molecules can be at sites with different symmetry with respect to the surrounding molecules (Figure 39). Molecules at sites of lower symmetry rotate at speeds higher than the scan rate used for imaging and therefore appear as torus (Figure 40a). The molecules at the higher symmetry sites are observed as six-lobed images, proving that they are immobile (Figure 40b). Gimzweski uses the lateral manipulation to drag a rotating HB-DC molecule from a low symmetry site into a higher symmetry site and the six lobes of the immobilized molecule can again be clearly observed.

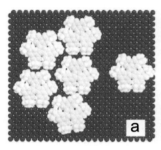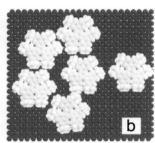

Figure 39: Model of a molecular mechanics simulation.
(a) The HB-DC molecule in a higher and a (b) lower symmetry site acting as a molecular rotator.

5.2 Vertical Manipulation

In the vertical manipulation process, the adparticles are transferred from the sample surface to the tip apex and vice versa [51], [52]. The first experiments on vertical STM manipulation were carried out by Eigler "picking up" Xe atoms [51]. The group of Rieder showed that transferring a Xe atom to the tip apex leads to markedly improved resolution [53]. The single Xe atom obviously "sharpens" the tip.

In Figure 41(a) [54], [55] a schematic presentation of the *pick up* process of a CO molecule from Cu(111) is shown. It is well known that CO molecules stand upright on a Cu(111) surface [56] with the carbon atom bonding to the copper atoms. Due to occasional contact between the tip and the surface some copper atoms are transferred to the tip apex. During the transfer of the CO molecule to the tip, the molecule must consequently rotate. A reliable procedure for transferring the CO to the tip and back to the surface requires ramping of the tunneling voltage and the simultaneous decrease of the tip-surface distance. Figure 41b and c show that scanning with a CO molecule on the tip apex leads to a clear chemical contrast. Figure 41b is scanned with a clean metal tip and all adsorbents appear as depressions. After the transfer of the CO molecule to the tip apex (indicated with an white arrow) and rescanning the area, Figure 41c shows that all CO molecules changed their appearance to protrusions. Only the oxygen atom in the upper left part of Figure 41b and c retains its appearance.

In [57] it has been described how to combine the potential of single atom manipulation of STM and single atom sensitivity of an atom probe mass spectrum to realize an ultimate technique for surface science. The System used by Shimizu et al. consists of an STM, an atom probe, load lock chambers and a mechanism to transfer tip and sample. The tip can be transferred reversibly between the STM and the atom probe stages. To investigate the pick-up of Si atoms during manipulation, a clean Si surface is approached with a clean tungsten tip applying a bias of + 2 V and 0.3 nA at the sample. After manipulation the tip is transferred from the STM to the atom probe. The atom probe analysis shows the formation of two different layers on top of the tip apex. The topmost layer is WSi_2 and the next layer is W_5Si_3 and finally the clean tungsten surface appeared. Using this combination of an STM and an atom probe proves that the tips during manipulation do not only adsorb atoms but furthermore, depending on the conditions, alloys can be formed.

5.3 Effects induced by the Tunnel Current

It is also possible to excite atoms by tip-sample interactions. Figure 42 shows a set of consecutive STM images acquired with 8 s time interval. The images show that the defects change their lattice positions [58], [59]. The tip can excite defects by several physical mechanisms. The case shown here is based on a field-induced migration due to the strong electrostatic field penetrating into the semiconductor. Defects can, however, also be excited by tunnelling of minority carriers into defect states followed by a charge

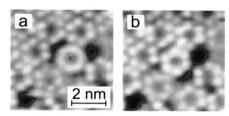

Figure 40: STM images of a Cu(100) surface after exposure to a coverage just below one complete monolayer of HB-DC molecules at room temperature. In (a) the molecule is imaged as a torus and is in a location where it is not in phase with the overall 2D molecular overlayer. The molecule is rotating. (b) The same molecule is translated by 0.26 nm and imaged as a six-lobed structure in registry with the surrounding molecules. Image area is 5.75 nm by 5.75 nm.

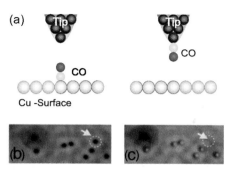

Figure 41:
(a) A sketch of the picking up procedure of CO molecules on Cu(111) surfaces. Notice that the CO molecule stands upright with the carbon atom attached to the surface and has to switch its orientation when being transferred to the tip.
(b), (c) STM images showing the pick up of a CO molecule. Notice the chemical contrast after the pick up [54].

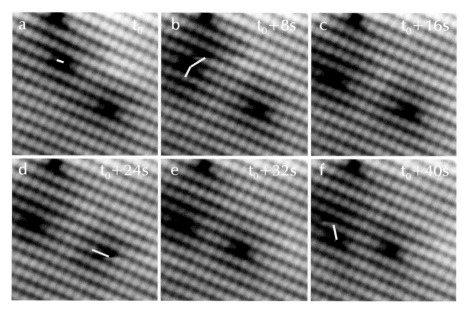

Figure 42: Migration of three phosphorus vacancies on the GaP(110) surface. The changes in the lattice positions of the vacancies is induced by the tip of the scanning tunnelling microscope. In this particular case the jumps are field-induced.

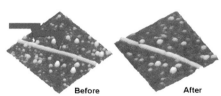

Figure 43: Cutting of a carbon nanotube with an AFM [61].

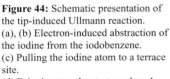
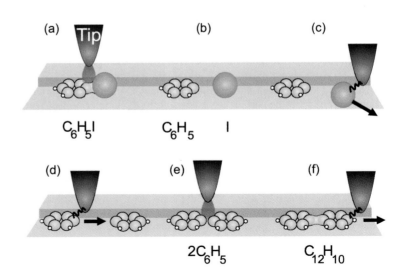

Figure 44: Schematic presentation of the tip-induced Ullmann reaction.
(a), (b) Electron-induced abstraction of the iodine from the iodobenzene.
(c) Pulling the iodine atom to a terrace site.
(d) Bringing together to two phenyl molecules by lateral manipulation and (e) electron-induced chemical association to biphenyl.
(f) Pulling the synthesized molecule by its front end to prove the association.

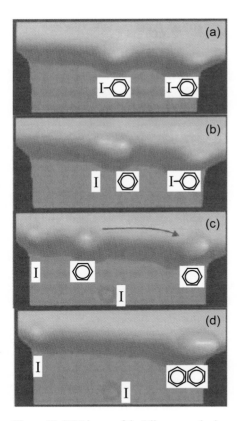

Figure 45: STM image of the Ullmann synthesis induced by the tip.
(a) two iodobenzene molecules are absorbed at a Cu(111) step edge. Introducing a voltage pulse through the tip abstracts the iodine from the phenyl molecules
(b) (the left molecule).
(c) by lateral manipulation the molecules are further separated and
(d) the phenyl molecules are moved together to prepare for their association. (scan area 7×3 nm²) [48]

carrier recombination with electron-phonon coupling [47]. There are surely even further mechanisms which may possibly excite atoms on the surface. Which of those will take place depends on the measurement conditions.

Finally, in Figure 43 the cutting of a carbon nanotube with an AFM is presented [60]. Earlier experiments controlling the length of carbon nanotubes were carried out using a STM [61]. At first the 600 nm² area is scanned, then the AFM cantilever is lowered at the positions marked in the left image and for cutting a voltage pulse of –6 V is applied. The image on the right hand side shows the carbon nanotube after cutting.

5.4 Complex Chemical reactions with the STM

In 1904, Ullmann et al. heated iodobenzene with copper powder as catalyst and discovered the formation of biphenyl with high purity [62]. This aromatic ring coupling mechanism is now nearly 100 years old and known as the Ullmann reaction. Combining the presented STM manipulation methods, namely the moving of adsorbents and the influence of increased tunnel current, it is possible to control this complex chemical reactions at low temperatures step by step. S. H. Hla et al. presented the synthesis of to one biphenyl molecule out of two iodobenzene on a copper surface at 20 K [48].

The synthesis consist of three different steps. First two iodobenzene molecules (C_6H_5I) have to be dissociated into phenyl (C_6H_5) and iodine (Figure 44a and b). Secondly the two phenyl rings have to be located one to another (Figure 44d and e) and finally in the third step, through tunnelling electrons the two phenyl rings are associated to biphenyl (Figure 44e).

To abstract the iodine from the iodobenzene the STM tip is positioned right above the molecule at a fixed height and the sample voltage is switched to 1.5 V for several seconds. The energy transfer from a single electron causes the breaking of the C-I bond Figure 45a – c, [62]. As the bond energies of the C-H and C-C bonds are two and three times higher than the C-I bond, it is not possible to break them with a single electron process at this voltage. After preparing two phenyl reactants and moving away the

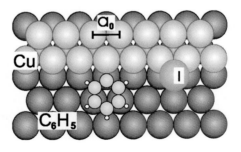

Figure 46: A sphere model of the chemical constituents illustrates the adsorption sites of phenyl and iodine as shown in Figure 45(b). On the left site the phenyl, the iodine and on the right site the iodobenzene are absorbed at a (100)-type step edge [48].

iodine, the left phenyl in Figure 45c is brought close to the other one by lateral manipulation using the tip adsorbate forces in Figure 45d. Though the two phenyls are close together they do not join at 20 K. The two phenyls can easily be separated again by lateral manipulation. Both phenyls are still bond to the Cu step edge via their σ_{C-Cu} bonds. Figure 46 shows a model where the phenyl is lying with its ring on the terrace while one of its C atoms is pointing towards the step edge and σ-bonding to a Cu atom. The final reaction step to associate the two phenyls to biphenyl is done by positioning the tip right above the centre of the phenyl couple and increasing the current drastically. The successful chemical association can be proved by pulling the synthesized molecule by its front end with the STM tip [62].

Acknowledgements

The editor would like to thank Silvia Karthäuser (FZ Jülich) for checking the clarity and consistency of this chapter. Thanks are due to Frank Peter (FZ Jülich) for corrections.

References

[1] G. Binning and H. Rohrer, Spektrum der Wissenschaft, Sonderdruck (1986).
[2] J. Bardeen, Phys Rev. Lett. **6,** 57 (1961).
[3] J. Tersoff and D.R. Hamann, Phys. Rev. **B 31**,805 (1985).
[4] J. Tersoff and D.R. Hamann, Phys. Rev. Lett. **50**, 1998 (1983).
[5] J.A. Stroscio, R.M. Feenstra, D.M. Newns, and A.P. Fein, J.Vac. Sci. Technol. A **6**, 499 (1988).
[6] R.M. Feenstra, J.A. Stroscio, J. Tersoff, and A.P. Fein, Phys. Rev. Lett. **58**, 1192 (1987).
[7] Ph. Ebert, B. Engels, P. Richard, K. Schroeder, S. Blügel, C. Domke, M. Heinrich, and K. Urban, Phys. Rev. Lett. **77**, 2997 (1996).
[8] Ph. Ebert, G. Cox, U. Poppe, and K. Urban, Surfac. Sci. **271**, 587 (1992).
[9] A. Selloni, P. Carnevali, E. Tosatti, and D.C. Chen, Phys. Rev. B **31**, 2602 (1985), idem **34**, 7406 (1986).
[10] J.A. Stroscio, R.M. Feenstra, and A.P. Fein, Phys. Rev. Lett. **57**, 2579 (1986).
[11] C.J. Chen, J. Vac. Sci. Technol. A **6**, 319 (1988).
[12] K. Besocke, Surf. Sci. **181**, 145 (1987).
[13] Y. Kuk and P.J. Silverman, Rev. Sci. Instrum. **60**, 165 (1989).
[14] G. Cox, *Untersuchung von Grenzflächen und Gitterbaufehlern in GaAs mit Hilfe der Rastertunnelmikroskopie,* Dissertation RWTH Aachen als Forschungszentrum Jülich GmbH Bericht 2382 (1990).
[15] H.-J. Güntherodt and R. Wiesendanger, *Scanning Tunneling Microscopy*, Vol. 1 and R. Wiesendanger and H.-J. Güntherodt, *Scanning Tunneling Microscopy*, Vols. **2** and **3**, Springer Series in Surface Science **20**, **28**, and **29**, R. Gomer, ed., Springer, Berlin, 1992, 1993.
[16] S. Chiang (Ed.), Special Issue of Chemical Reviews **97**, June 1997.
[17] Ph. Ebert, Current Opinion in Solid State and Materials Science **5**, 211 (2001).
[18] Ph. Ebert, K. Urban, and M.G. Lagally, Phys. Rev. Lett. **72**, 840 (1994).
[19] Th. Dretschkow, D. Lampner, Th. Wandlowski, J. Electroanal. Chem. 458 (1998).
[20] Th. Dretschkow, D. Lampner, Th. Wandlowski, J. Electroanal. Chem. 467 (1999).
[21] K. Szot, W. Speier, Phys. Rev. B **60**. 5909 (1999).
[22] K. Szot, W. Speier, U.Breuer, R.Meyer, J.Szade, R.Waser, Surf. Sci. **460**, 112 (2000).
[23] R.-H. Robrock, *Rasterkraftmikroskopie.* IFF-Bulletin **37** (1990).
[24] O. Wolter, T. Bayer, and J. Grescher, J. Vac. Sci. Technol. B **9**, 1353 (1991).
[25] J.N. Israelachvili, *Intermolecular and Surface Forces.* Academic Press, 1985.
[26] M. Schneider, *Untersuchung mikromagnetischer Strukturen in einkristallinen Eisenschichten mit einem kombinierten Kerr-Kraftmikroskop.* Dissertation Univ. Köln, Forschungszentrum Jülich GmbH Bericht 3059 (1995).
[27] F.M. Serry, K. Kjoller, R.J. Thornton, R.J. Tench, and D. Cook, Digital Instruments, Veeco Metrology Group AN27 (1999).

[28] A. Bachtold, M. S. Fuhrer, S. Plyasunov, M. Forero, Erik H. Anderson, A. Zettl, and Paul L. McEuen, Phys. Rev. Lett. **84**, 6082 (2000).

[29] L.M. Eng, H.-J. Güntherod, G.A. Schneider, U. Köpke, and J. Muòoz Saldaòa, Appl. Phys. Lett. **74**, 233 (1999), M. Abplanalp, L. M. Eng, P. Günter, Appl. Phys. A **66**, 231 (1998).

[30] A. Roelofs, U. Böttger, R. Waser, F. Schlaphof, S. Troisch, and L.M. Eng, Appl. Phys. Lett. **77**, 3444 (2000).

[31] D. Etienne, A. Fourcade, J.-C. Poulin, A. Barbin, D. Coulaud, E. Le Cam, E. Paris, Microsc. Microanal. Microstruct. **3**, 457 (1992).

[32] D. Anselmetti, M. Dreier, R. Lüthi, T. Richmond, E. Meyer, J. Frommer, H.-J. Güntherodt, J. Vac. Sci. Technol. B **12**, 1500 (1994).

[33] J. Li, C. Bai, C. Wang, C. Zhu, Z. Lin, Q. Li, and E. Cao, Nucleic Acids Res. **26**, 4785 (1998).

[34] F. Zenhausern, M. Adrian, B. Tenheggelerbordier, L.M. Eng, and P. Descouts, Scanning **14**, 212 (1992).

[35] L. Stryer. *Biochemistry* (Freeman, New York 1995).

[36] H.G. Hansma, R. Golan, W. Hsieh, C.P. Lollo, P. Mullen-Ley, D. Kwoh, Nucleic Acids Res. **26**, 2481 (1998).

[37] H.G. Hansma, J. Vesenka, C. Siegerist, G. Kelderman, H. Morrett, R.L. Sinsheimer, V. Elings, C. Bustamante, P.K. Hansma, Science **256**, 1180 (1992).

[38] M.A. Reed, C. Zhou, C.J. Muller, T.P. Burgin, J.M. Tour, Science **278**, 252 (1997).

[39] B. Pignataro, L.Chi, S.Gao, B. Anczykowski, C. Niemeyer, M.Adler, H. Fuchs, Appl. Phys. A **74**, 447 (2002).

[40] B. Anczykowski, J.P. Cleveland, D. Krüger, V. Elings, H. Fuchs, Appl. Phys. A **66**, 885 (1998).

[41] L. A. Bumm, J. J. Arnold, L. F. Charles, T. D. Dunbar, D. L. Allara, and P. S. Weiss, J. Am. Chem. Soc. **121**, 8017 (1999).

[42] Rachel K. Smith, Scott M. Reed, Penelope A. Lewis, Jason D. Monnell, Robert S. Clegg, Kevin F. Kelly, Lloyd A. Bumm, James E. Hutchison, and Paul S. Weiss, J. Phys. Chem. B **105**, 1119 (2001).

[43] Z. L. Donhauser, B. A. Mantooth, K. F. Kelly, L. A. Bumm, J. D. Monnell, J. J. Stapleton, D. W. Price Jr., A. M. Rawlett, D. L. Allara, J. M. Tour, P. S. Weiss, Science **292**, 2303 (2001).

[44] M.F. Crommie, C.P. Lutz, and D.M. Eigler, Science **262**, 218 (1993).

[45] D.M. Eigler and E.K. Schweizer, Nature **344**, 524 (1990).

[46] Ph. Ebert, M.G. Lagally, and K. Urban, Phys. Rev. Lett. **70**, 1437 (1993).

[47] G. Lengel, J. Harper, and M. Weimer, Phys. Rev. Lett. **76**, 4725 (1996).

[48] S. W. Hla, L. Bartels, G. Meyer, K.-H. Rieder, Phys. Rev. Lett. **85**, 2777 (2000).

[49] G. Meyer, L. Bartels, S. Zöphel, E. Henze, and K. H. Rieder, Phys. Rev. Lett. **78**, 1512 (1997).

[50] J. K. Gimzewski, C. Joachim, R. R. Schlittler, V. Langlais, H. Tang, I. Johannsen, Science **281**, 531 (1998).

[51] D.M. Eigler, C.P. Lutz, and W.E. Rudge, Nature (London) **352**, 600 (1991).

[52] Y. Hasegawa and Ph. Avouris, Phys. Rev. Lett. **71**, 1071 (1993).

[53] B. Neu , G. Meyer, and K. H. Rieder, Mod. Phys. Lett. B **9**, 963 (1995).

[54] G. Meyer, et al., Jpn. J. Appl. Phys. **40**, 4409 (2001).

[55] L. Bartels, G. Meyer, and K. H. Rieder, Appl. Phys. Lett. **71**, 213 (1997).

[56] S. Ishi, Y. Ohno, and B. Viswanathan, Surf. Sci. **162**, 349 (1985).

[57] T. Shimizu, J. T. Kim, H. Tokumoto, Ultramicroscopy, **73**, 157 (1998).

[58] Ph. Ebert, Surf. Sci. Rep. **33**, 121 (1999).

[59] Ph. Ebert and K. Urban, Ultramicroscopy **49**, 344 (1993).

[60] L. C. Venema, J. W. G. Wildöer, H. L. J. Temminck Tuinstra, C. Dekker, A. G. Rinzler, and R. E. Smalley, Appl. Phys. Lett. **71**, 2629 (1997).

[61] J.Y. Park, Y. Yaish, M. Brink, S. Rosenblatt, and P. L. McEuen, Appl. Phys. Lett. **71**, 2629 (2002).

[62] F. Ullmann, G. M. Meyer, O. Loewenthal, and O. Gilli, Justus Liebig's Annalen der Chemie **331**, 38 (1904).

Logic Devices

Contents of Part III

Introduction to Part III	321
13 Silicon MOSFETs – Novel Materials and Alternative Concepts	357
14 Ferroelectric Field Effect Transistors	385
15 Quantum Transport Devices Based on Resonant Tunneling	405
16 Single-Electron Devices for Logic Applications	423
17 Superconductor Digital Electronics	443
18 Quantum Computing Using Superconductors	459
19 Carbon Nanotubes for Data Processing	471
20 Molecular Electronics	499

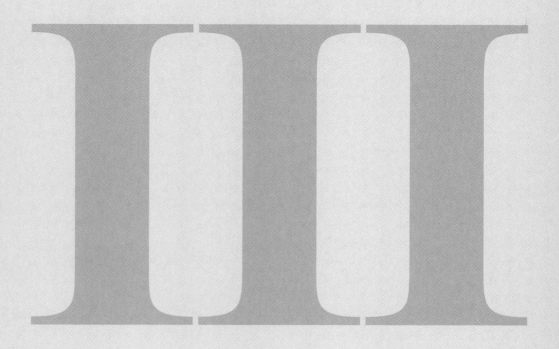

Introduction to Part III

Contents

1 Fundamentals of Logic Devices — 322
1.1 Requirements for Logic Devices — 322
1.2 Dynamic Properties of Logic Gates — 325
1.3 Threshold Gates — 325

2 Physical Limits to Computation — 326

3 Concepts of Logic Devices — 330
3.1 Classifications — 330
3.2 Two-Terminal Devices — 330
3.3 Field Effect Devices — 331
3.4 Coulomb Blockade Devices — 332
3.5 Spintronics — 333
3.6 Quantum Cellular Automata — 335
3.7 Quantum Computing — 337
3.8 DNA Computer — 338

4 Architectures — 338
4.1 Flexibility of Systems for Information Processing — 338
4.2 Parallel Processing and Granularity — 341
4.3 Teramac – A Case Study — 342

5 Performance of Information Processing Systems — 343
5.1 Basic Binary Operations — 343
5.2 Measures of Performance — 345
5.3 Processing Capability of Biological Neurons — 346
5.4 Performance Estimation for the Human Brain — 348

6 Ultimate Computation — 351
6.1 Power Dissipation Limit — 351
6.2 Dissipation in Reversible Computation — 352
6.3 The Ultimate Computer — 353

In the General Introduction, we briefly looked at information processing and logic in an abstract manner, i.e. without considering the physical realization. Chap. 7 gave a short survey of the two-valued digital logic, realized by conventional CMOS technology – from individual gates to microprocessors. Here, we will broaden our viewpoint again. Firstly, we will discuss the fundamental requirements for logic devices. These are requirements which are valid for CMOS devices, as well as for biological neurons, superconducting logic devices, and any other conceivable physical implementation of logic. Subsequently, we will review the physical limitations of computing and we will give an overview of physical implementation concepts. Here, we will also introduce those concepts which are – for various reasons – not covered by individual chapters in this textbook. We will then briefly report on some major aspects of architecture and we will present general estimations of the performance of information processing systems. Finally, we will try to sketch the ultimate computer.

III Logic Devices

1 Fundamentals of Logic Devices

1.1 Requirements for Logic Devices

In order to perform information processing in the real, physical world, the logical states must be mapped onto physical properties. This mapping may take place

- on the *amplitude* of physical properties, as e.g. the voltage levels in the CMOS circuits, or
- on the *time* evolution, i.e. the timing of pulses of a physical property. This is employed, for example, in the RSFQ logic (Chap. 17), and, in some aspects, in biological neurons (Chap. 6).

The physical properties mentioned here can be selected without much restrictions, as long as the devices fabricated fulfill the requirements below. The mapping of logical states will always be on *intervals* of physical properties (for example, the voltage) in order to build up interference resistance, i.e. immunity against noise and signal scatter (Figure 1).

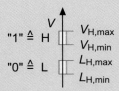

Figure 1: As an example, the voltage level representation of 2-valued logical states is shown, as realized in conventional digital CMOS circuits. Since the high values (H) correspond to the "1" and the low values (L) to the "0", this is an example of positive logic. In negative logic, the relationships are mutually exchanged.

There is a range of indispensable requirements for logic devices which make it possible to assemble *sequential switching circuits* (including processors of any kind) of, in principle, unlimited size. Some of the requirements may be relaxed, if the size of the circuit, i.e. the number of gates, is kept within certain limits. In order to demonstrate the general applicability of these requirements, we will discuss them using the Si-based CMOS technology on the one hand and biological neurons on the other hand as examples.

Requirement #1: Non-linear characteristics

Logic gates need to show a *non*-linear transfer characteristic (Figure 2). This is required to maintain a sufficient signal-to-noise ratio even in unlimited long chains of gates. The output signal intervals are smaller than the input signal intervals, i.e. the difference between "0" and "1" is increased, because the output signal intervals must fit into the input signal intervals also in the presence of noise and signal scatter. To achieve this, the gate will amplify the signal (i.e. $\nu > 1$) in the center region. The noise is reduced since it occurs in the region of the characteristics with $\nu < 1$, leading to a compression of the intervals.

In principle, it is possible to create logic processing systems also from linear amplifiers [1]. These systems are called *analog computers* and can be used to solve e.g. differential equations up to a certain size. However, since every amplifying stage inherently adds to the noise, the signal-to-noise ratio decreases in long concatenated chains of stages, if no non-linear element is build in.

The non-linearity of digital CMOS gates stems from the output characteristics of the MOSFETs in combination with the external circuit in these gates (Figure 3). In the case of neurons, one important non-linearity is obtained by the threshold function which guarantees that only signal voltages exceeding the threshold voltage trigger action potential pulses which are then transferred along the axon. This *threshold gate* operation will be described in some more detail in Sec. 1.3. There are many additional non-linearities in biological neurons, as for example the transfer function of signals across the synapses, which play an important controle in neuronal information processing.

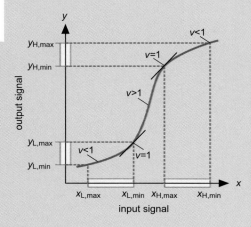

Figure 2: Transfer characteristic of a logic gate showing the non-linear characteristic. Due to the center region of $\nu > 1$, and signal level region of $\nu < 1$, the allowed input signal window (between $x_{L,min}$ and $x_{L,max}$ as well as $x_{H,min}$ and $x_{H,max}$) are mapped onto smaller windows on the output signal axis. This improves the signal-to-noise ratio. Remark: for educational reasons, this simple unity gate is shown here, although it is not performing a logic operation. It may only used as a buffer. An inversion of the characteristics leads to a NOT gate.

Requirement #2: Power amplification

In order to maintain the signal level during logic processes over long concatenated chains of gates, power amplification is necessary. This amplification balances signal losses which are unavoidable during processing. It is not sufficient to have only signal amplification (e.g. voltage amplification), since every logic process is a switching process, and switching inherently involves energy. In the case of digital CMOS circuits, the output of a logic gate must not only fulfil the *voltage* amplification set by requirement #1, but has to drive *currents* to charge and discharge the line capacitance and input capacitances of the subsequent gates. In addition, it needs to drive at least two inputs to facilitate the branching of signals in complex circuits. The number of inputs one output can drive is called FAN-OUT.

Reversible computing means that the energy required for a switching process is not dissipated but recovered to facilitate a subsequent switching process. In Sec. 6.2, we will see that there is an inherent energy dissipation even for reversible computation at a finite rate. Hence, in principle, the requirement of power amplification is valid for reversible computing, too.

In biological neurons, the power amplification is performed by the voltage-triggered (voltage-gated) ion channels along the axon which utilize the electrochemical potential difference between the inside and the outside of the cell membrane based on the concen-

Figure 3: Sketch of the fundamental requirements of logic devices and their realization in CMOS logic (left) and biological neurons (right). In many aspects, the illustrations are simplified to emphasize the key points. In some cases, identity gates are used as an example of CMOS logic. The CMOS-based AND and OR function are shown only to illustrate the principle. In real CMOS circuits, NAND and NOR gates (without resistors) are used.

Introduction III

| *Example* CMOS Logic | *Example* Biological Neuron |

Non-Linearity

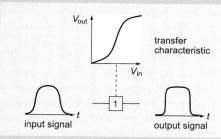 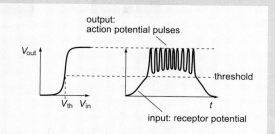

Power Amplification

  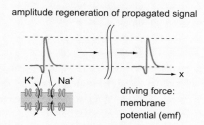

Concatenability

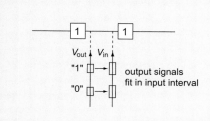 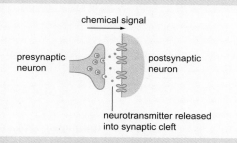

Feedback Prevention

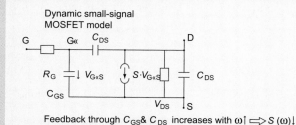 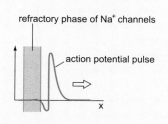

Basic Logic Functions

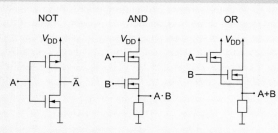 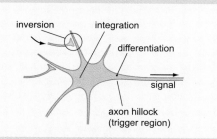

tration difference of the Na$^+$ and K$^+$ ions (Chap. 6). The operation of these ion channels ensures a constant amplitude of the action potential pulse while it is travelling along the axon which may be more than 1 m long. The concentration difference is restored by the Na$^+$/K$^+$ pump channels which consumes chemical energy (supplied as ATP).

Requirement #3: Concatenability

This requirement simply means that the input signal and the output signal must be compatible, i.e. they are based on the same physical property and value range. For instance, if output signals were optical and the input signals were electrical, a logic circuit could not be built. A signal converter would be required, the addition of which would fulfil the requirement. Similarly, the signal levels must fit. If, for instance, a gate adds a significant offset to the signal, the output signal intervals will not meet the allowed input signal intervals (see Figure 2) and, again a converter, would be needed.

In biological neurons, the concatenability is given by the fact that the neurotransmitter molecules released (output) by the synaptic terminal into the synaptic cleft find highly specific receptors (input) at the ligand-gated ion channels of the postsynaptic neuron (Chap.6).

Requirement #4: Feedback prevention

In order to execute a calculation, we need a directed flow of information. If a signal at a given position in a logic circuit ran forwards and backwards with equal probability, no information process would ever be completed. Especially in circuits made for reversible computing, the direction of the information flow is not determined by the circuit (see General Introduction, Sec. 3.4) and, hence, must be determined by the gates. Obviously, an adequate feedback prevention must be built into the logic gates.

In CMOS gates, feedback prevention is very effectively performed by the MOSFET. A certain feedback is given due to the gate-drain capacitance, C_{GD}, and the gate-source capacitance, C_{GS}. Obviously, the feedback prevention decreases with increasing signal rate and vanishes above the transit frequency of the MOSFET.

In the axons of biological neurons, backward propagation of action potential pulses is excluded due to the refractory periode of the voltage gated Na$^+$ channel (Chap. 6). It should be noted, however, that also antidromic spike propagation from the axon hillock into the soma and up, into the dendritic tree occurs. This seems to be very important for handshake and learning [2]. Still, in addition, the neurotransmitter transfer in the cleft only works in forward direction, contributing to an effective feedback prevention.

It must be noted that, of course, the feedback prevention only concerns the *flow of information in the device*. Often there is *control* signal feedback within the device, which is needed for the correct operation. And on the *circuit* level, feedback loops are frequently employed in logic – in CMOS as well as in biological neural systems.

Requirement #5: Complete set of Boolean operators

As mentioned in the General Introduction, a basic set of Boolean operators is needed to realize a complete Boolean algebra. A generic set consists, at least, of a unifying operator (disjunction, OR) and an inversion (NOT) or it consists of a intersecting operator (conjunction, AND) and an inversion (NOT). In both cases, the operators can be combined into one, such as the NOR or the NAND.

The realization of Boolean function in CMOS technology has been explained in Chap. 7.

Biological neurons act as threshold gates (see Sec. 1.3). The unifying operation is performed by the integration of the receptor potentials. The trigger region of the axon hillock shows an intersecting (threshold) operation. The inversion function is localized in the inhibitory synapses.

As already stated, the complete set of fundamental requirements #1 to #5 have to be fulfilled to build logic circuits of unlimited size. For circuits of limited size, some requirements can be relaxed. Most importantly, power amplification is not required if the circuit is limited to a size which still ensures a sufficient signal-to noise ratio at the output of the longest chain of gates in the circuit. This is important for the array based logic in which logic gates are configured through dedicated connections made at the nodes of a e.g. cross-bar array (Sec. 4). Obviously, such an array may be passive, i.e. it may incorporate no amplifying element. Still it is possible to build small logic circuits, if power amplification is implemented at the inputs and outputs of the passive arrays. Using actively amplifying input/output devices, small circuits can again be combined to circuits of unlimited size.

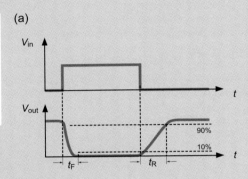

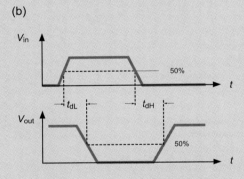

Figure 4:
(a) Response V_{out} of a CMOS NOT gate upon an ideally rectangular input signal, V_{in}. t_F: fall time, t_R: rise time.
(b) Definition of the propagation delay times t_{dL} and t_{dH} of the gate. The average delay time is given by $t_{pd} = (t_{pdL} + t_{pdH})/2$.

1.2 Dynamic Properties of Logic Gates

The time dependence of signals at real logic gates is sketched in Figure 4 using a CMOS inverter (NOT gate) as an example. If an ideal V_{in} signal is applied to the input (Figure 4a), the output signal shows a fall time, t_F. The trailing edge of the V_{in} signal leads to a rise time t_R of the output signal.

In digital circuits, the signals are usually sketched as shown in Figure 4b, showing propagation delay times for the H(igh) and L(ow) state transitions, t_{dH} and t_{dL}.

1.3 Threshold Gates

Here, we will give a short introduction to the operation principle of threshold gates. The intention is twofold. Firstly, it will be demonstrated how the seemingly different logic worlds of Si-based CMOS gates and biological neurons can be linked. Secondly, threshold gates are the basis for the so-called **neuromorphic logic**. Although these devices do not play an important role in advanced microelectronics, they are interesting from a fundamental point of view and there are ideas in nanoelectronics research which are based on neuromorphic logic.

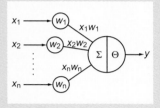

The concept of threshold gates was inspired by the understanding of the operation of biological neurons in the 1960s (see e.g. [3],[4]). A linear threshold gate is a logic device that has n two-valued inputs, $x_1, x_2, ..., x_n$, and a single two-valued output y.

The Boolean function of a threshold gate

$$y = f(x_1, x_2, ..., x_n) \quad (1)$$

Figure 5: Threshold gate with binary inputs $x_1, x_2, ..., x_n$, the weights $w_1, w_2, ..., w_n$, the weighted sum χ and the threshold value Θ.

is determined by the **weights** $w_1, w_2, ..., w_n$ which are associated with the inputs and the **threshold value** Θ at the output (Figure 5). The Boolean function is given by

$$y = \text{sign}(\chi - \Theta) = \begin{cases} 1 & \text{if } \chi \geq \Theta \\ 0 & \text{if } \chi < \Theta \end{cases} \quad (2)$$

where χ is the weighted sum

$$\chi = \sum_{k=1}^{n} w_k \cdot x_k, \quad x_k = \{0,1\} \quad (3)$$

The threshold value Θ and the weights w_k may be any real, finite, positive or negative number. In any reasonable implementation, a weight will have a finite resolution, i.e. a finite number, N, of possible values between the minimum and the maximum values the weight may assume.

As an alternative to two-valued inputs and internal weights, a threshold logic may have multiple-valued inputs which simply means that the weights are no internal parameters of the gate but part of the input signals. However, the representation of threshold gates according to Eq. (2),(3), using two-valued inputs is more robust against disturbing signals which may interfere with the logic circuits.

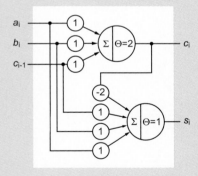

The basic advantage of threshold logic compared to conventional digital logic is the inherent parallel processing due to the internal multiple-valued computation of the weighted sum. Hence, circuits constructed of threshold gates usually consist of fewer components than the corresponding circuits implemented with conventional Boolean gates.

Every Boolean function can be realized by threshold gates. An AND function for n inputs, for example, is realized by a single threshold gate using $w_1 = ... = w_n = 1$ and a threshold value in the interval $n-1 < \Theta < n$. To realize an OR function using the same weights, the threshold value would have to be in the range $0 < \Theta \leq 1$. A two-input NAND gate is realized by $w_1 = w_2 = -1$ and $-1 < \Theta < -2$. More complex Boolean functions require networks of threshold gates. A full adder with two operands is shown in Figure 6. If compared to its conventional binary logic counterpart (Chap. 7, Sec. 4), the threshold gate adder uses less gates. In addition, threshold gates with more inputs can be used to realize larger adders [3],[4]. However, the requirements on the precision of setting the threshold value increases with the number inputs. As a consequence, threshold gates are less robust against circuit parameter variations which are inevitable in any technological realization. This is the major drawback of the threshold logic and the reason why it did not prevail in the standard CMOS technology. Still, there are concepts such as resonant tunneling devices (Chap. 15) which are suitable to implement a threshold logic [5]. In addition, the basics of threshold logic is needed if one wants to estimate the processing capabilities of biological neural networks.

Figure 6: Full adder realized by threshold logic.

III Logic Devices

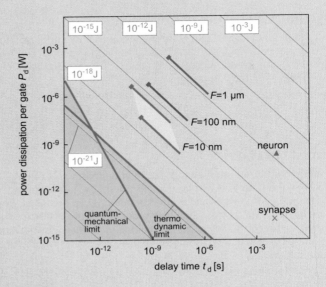

Figure 7: Average dissipated power per gate P_d versus transition delay time t_d. The red area is inaccessible due to fundamental limits of thermodynamics (boundary: $W_{TD,min}$) at $T=300$ K and quantum mechanics (boundary: $W_{QM,min}$). The device limits for CMOS gates of the 1000-nm, the 100-nm, and the projected 10-nm technology generations are illustrated. Furthermore, estimated values for biological neurons in the human brain and synapses of these neurons are shown.

2 Physical Limits to Computation

There are three *fundamental* limits to the performance of logic devices. These limits are derived from physical principles of

- **thermodynamics**
- **quantum mechanisms**
- **electromagnetism**

In addition, there is a hierarchy of limits given by the *materials* employed, the *device* type used, the *circuit* concept, and the *system* configuration. Details of these topics are described by Meindl in Ref. [6], [7] and references cited therein.

The major parameters limiting the **performance** of logic devices are: the **time** and the **energy** required for a logic operation. Typically, the performance of devices as well as limits for their performance are illustrated in a diagram which shows the average power dissipation per logic gate, P_d versus the average delay time, t_d, for the switching of the gate [6]. The average energy dissipated during a logic operation, W_d, is then given by

$$W_d = P_d \cdot t_d \tag{4}$$

as long as the static power can be neglected (see (8)). In Figure 7, lines of equal energy are plotted.

In the General Introduction, we have shown that the fundamental limit imposed by thermodynamics is the (minimum) energy required for a binary switching transition at a given operating temperature

$$W_{TD,min} = k_B T \ln 2 \approx 3 \cdot 10^{-21} \text{ J/bOp} \tag{5}$$

where bOp stands for a binary operation. This is also the mimimum energy to move a bit of information through a communication channel [8], described by Shannon's classical theorem for channel capacity [11].

In the terminology of statistical physics, Eq. (5) means that a bit at the output of a logic gate has to go somewhere. If the information goes into observable degrees of freedom of the system, i.e. an adjacent logic gate, then it has not been erased but merely moved. But if it goes into unobservable degrees of freedom, such as the random microscopic motion of molecules, it results in an increase of entropy of (at least) $k_B \ln(2)$ [12]. The first case represents reversible computing, the second irreversible computing. In other words, for irreversible computing, Eq. (5) is equal to the minimum energy, dissipated for every bit lost during the operation [13]. The limit imposed by Eq. (5) is shown in Figure 7 for $T = 300$ K.

The Heisenberg uncertainty principle of quantum mechanics imposes a second fundamental limit. This principle states that the energy of a state with a life time Δt can only be determined with a precision of ΔW, given by

$$W_{\text{QM,min}} \equiv \Delta W \geq h/\Delta t \tag{6}$$

where h is Planck's constant. Precisely stated this means that a quantum state with spread in energy ΔW takes at least a time $\Delta t = h/\Delta W$ to evolve into an orthogonal (and hence distinguishable) state [14]. It has been shown that Eq. (6) holds not only for the energy spread, but also for the average energy [15]. For very short switching times $t_d < 10^{-12}$ s, Eq. (6) determines the fundamental limit of the minimum required power, as shown in Figure 7 for $T = 300$ K. The fundamental limit of the minimum spatial distance of wells for electrons as the physical realization of bits is described in [16],[17].

We will again use Si-based CMOS logic and biological neurons to find typical areas in this diagram.

Figure 8 shows a CMOS inverter as a representative example of a digital logic gate. The load capacitance C_L

$$C_L = C_{\text{out}} + C_{\text{con}} + C_{\text{in}} \tag{7}$$

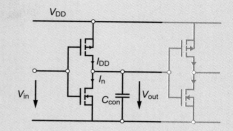

Figure 8: CMOS inverter gate for illustrating the dynamic power dissipation, the load capacitance C_L comprises the output capacitance of the gate, the interconnect capacitance, and the input capacitance of the subsequent gate (shown in grey).

comprises the output capacitance of the gate, C_{out}, the capacitance of the interconnect to the subsequent logic gate, C_{con}, and the input capacitance of this subsequent gate, C_{in}. The input capacitance of a MOSFET is primarily determined by the gate capacitance of the device (Chap. 13). In the case of short interconnects, *none* of the three contributions to C_L in Eq. (7) can be neglected. The power dissipation P_d of a gate can be separated into three components

$$P_d = P_{\text{dyn, CL}} + P_{\text{dyn, SC}} + P_{\text{stat}} \tag{8}$$

Figure 9 illustrates the dynamic components during a switching cycle.

The major contribution is the dynamic power dissipation due to the load capacitance:

$$P_{\text{dyn, CL}} = \sigma f \, C_L \, V_{\text{DD}}^2 \tag{9}$$

where σ is a prefactor in the range from approximately 0.25 to 1.5, stating the average number of charge and discharge events per clock cycle, determined by the circuit design and the processed data. f is the clock frequency. The maximum clock frequency is given by

$$f_{\text{max}} = \beta / t_d \tag{10}$$

where β typically is in the range from 0.1 to 0.3. *Lower-power, dedicated (hard-wired) logic* uses the smallest transistors which are available in a given technology generation. The interconnect capacitances C_{con} are very small, since the subsequent logic gates are located in the immediate vicinity. As a consequence, very low $P_{\text{dyn,CL}}$ values can be obtained. In *programmable logic*, typically employed in general purpose microprocessors, the logic signal typically have to be propagated over much larger distances on the chip, for example between processing and temporary storage areas. For this reason, the C_L values are larger and the transistors are much larger to drive the required currents for high speed processing. The dynamic power $P_{\text{dyn,SC}}$ of CMOS circuits dissipated during the transition of a logic signal results from the fact that the semiconductor channels of the complementary FETs are both partially conducting during the transient time. Typically, $P_{\text{dyn,SC}}$ is in the 10% range of $P_{\text{dyn,CL}}$.

The static power dissipation P_{stat} is caused by the off-currents of the MOSFETs, the pn-junctions, and the gate leakage current. In former generations (min. feature size $F > 250$ nm; reference: DRAM ½ pitch), P_{stat} has been in the 1% range with respect to P_{dyn}. Due to the decreasing operating voltages, the margin for the threshold voltage V_T shrinks and, hence, the off-currents are rising. In addition, the trend to thinner gate oxides leads to significantly enhanced leakage (tunneling) currents through the gates (Chap. 13). For this reason, P_{stat} plays an increasingly important role with decreasing feature sizes F.

For the cases discussed, the constant energy lines $W_d = P_d \cdot t_d$ are shown in Figure 7. There are minimum delay times t_d which are settled by the device and the circuit. The transition of the device can be estimated from the channel length and the carrier velocity v

$$t_{d,\text{FET}} = L_{\text{ch}}/v \tag{11}$$

The carrier velocity is given by the carrier mobility and the applied electric field and reaches a saturation velocity of $v_s \approx 10^7$ cm/s for, both, electrons and holes, in Si at electric field of approx. $E > 3 \cdot 10^4$ V/cm.

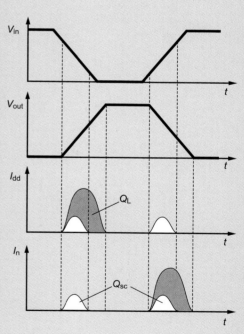

Figure 9: Switching cycle for a CMOS inverter (Figure 8). The I_{dd}-t and the I_n-t curves show the changes during the transitions. Q_{SC} is the charge due to the transient conduction of both FETs, Q_L is the charge transferred onto C_L during the rising edge of V_{out} and further to ground during the falling edge of V_{out}.

III Logic Devices

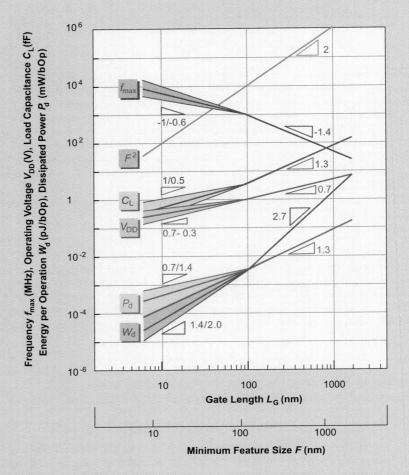

Figure 10: Various properties of low-power CMOS logic circuits versus feature size F (DRAM 1/2 pitch) and printed gate length L_g. Shown are the operation voltage V_{DD}, the max. operating frequency f_{max}, the load capacitance C_L, the dissipated energy per gate W_d, and the dissipated power per gate P_d at frequency f_{max}. The basic data (V_{DD}, C_L and f_{max}) are taken from [19] for $L_G = 1000$ nm and 100 nm, and from [25] for the prospected $L_G = 10$ nm scenarium. The extrapolations to small feature sizes do not include proposed ultralower power (slow) logic concepts based on a subthreshold logic [20].

This results in $t_{d,FET}$ values of approx. 1 ps for channel lengths of 100 nm. For FET test devices with $L_{ch} < 20$ nm, $t_{d,FET}$ values in the sub-ps regime have been measured [18]. The over-all gate delay time t_d in the critical path of a physically optimized, low power circuit is primarily determined by the saturation current I_{sat} through the channel and by the average C_L per logic gate within a circuit, such as a full-adder. For the 100-nm generation, for example, t_d is approx. 100 ps [19]. These minimum times are included on the W_d lines in Figure 7 for the 1000-nm and the 100-nm, and a projected 10-nm generation. If circuits are optimized for speed, transistors are made considerably wider. This increases the saturation current and decreases t_d correspondingly. For the 100-nm generation, gate delay times < 30 ps can be achieved by this strategy.

Figure 7 also includes estimated data for the average biological neuron in the human brain, as well as per average synapse of such neuron. The rational for these numbers is given in Sections 5.3 and 5.4.

Figure 10 illustrates the evolution of the power dissipation and the operation frequency for the physical optimized low-power, CMOS logic according to the relationship discussed above. In technology generations with features size $F > 100$ nm, the total dissipated power is dominated by the *dynamic* power needed to charge and discharge the load capacitance C_L. In first approximation, the scaling has been $C_L \propto F$ and $V_{DD} \propto F$ which resulted in $W_d \propto F^3$, according to Eq. (7) and (9). The more realistic evolution can be fitted by $W_d \propto F^{2.7}$ [19]. On the route from $F = 100$ nm to $F = 10$ nm, this capacitance will not decrease at the same pace as before, because of the requirements on the cross section aspect ratios of the interconnect lines. At very low F, C_L may even rise again because of a possible increase of C_{con}.

The over-all delay time t_d is dominated by charging of C_L by the saturation current through the channel, $t_d \approx C_L \cdot V_{DD}/I_{sat}$. In first approximation, the scaling in the 1-μm to 100-nm regime has been approximately $t_d \propto F^{1.4}$ corresponding to approximately $f \propto F^{-1.4}$. The scaling of t_d in the sub-100-nm regime is expected to be less pronounced.

Standard microprocessors and digital signal processors (DSP) operate in a synchronous mode, i.e. they have a clock signal which is distributed in a tree-like fashion into every tiny part of the chip. Due to resistance of the interconnects for distributing the clock signal and the buffer amplifiers at the branching positions of the tree, the clock signal network consumes a high power in high-performance processors operated at GHz frequencies. For example, approx. 30% of the total power consumption of the Pentium® IV/2 GHz is solely used for the clock signal distribution. In order to save considerable fractions of this power, the buffer amplifiers can be used to activate the clock signal only for those parts of the chip which are used at this moment. An alternative approach is the asynchronous operation of digital logic chips. No clock signal is used and an interlinking of the individual processes is facilitated by, for example, hand-shaking protocols [21].

There is another fundamental aspect, which limits the information processing rate especially over large distances on a chip. This is due to the *electromagnetic* character of the signal and the *finite speed of light* [22]. The delay time τ of a signal travelling via an interconnect of length L is expressed by

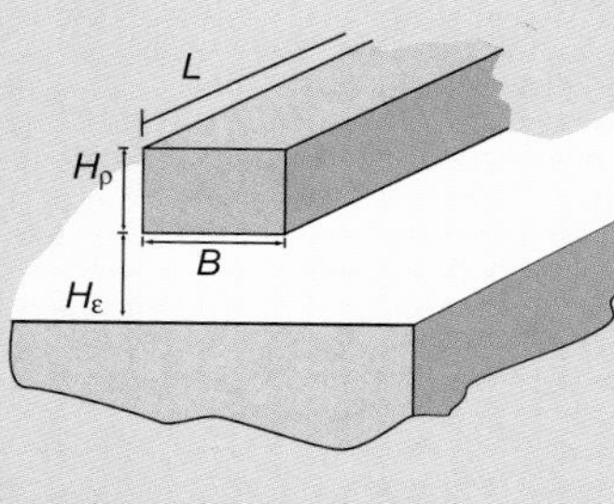

$$\tau = \frac{L}{c_0}\sqrt{\varepsilon_r} \qquad (12)$$

Figure 11: Sketch of the geometries assumed for the model interconnect.

where ε_r is the relative permittivity of the dielectric surrounding the interconnect and c_0 the speed of light. Since the delay time is also determined by the resistance R of the interconnect and its capacitance C, another limit is given by the material and geometry of the interconnect line.

The latency of a single global interconnect is expressed by the distributed RC time as

$$\tau = \alpha RC = \alpha r c L^2 \qquad (13)$$

where r and c are the distributed resistance per unit length and distributed ground capacitance per unit length, respectively (Figure 11):

$$r \approx R\frac{B}{L} = \frac{\rho}{H_p} \qquad (14)$$

$$c \approx C\frac{1}{BL} = \frac{\varepsilon}{H_\varepsilon} \qquad (15)$$

(ρ/H_p) denotes the conductor sheet resistance in Ω and $(\varepsilon/H_\varepsilon)$ is the sheet capacitance in F/cm². In Eq. (13), α is a factor in the order of 0.5 which accounts for the distributed nature of the RC network [23]. Because of Eq. (13), L^{-2} is typically plotted versus the delay time τ revealing the diagonal as the locus of constant distributed RC product. The electromagnetic speed limit, Eq. (12), is shown in Figure 12 for $\varepsilon_r = 1$ (vacuum) and a dielectric with a permittivity $\varepsilon_r = 2$. The RC limit, Eq. (13), is plotted for an interconnect thickness H_ρ and dielectric thickness H_ε of $H \equiv H_\rho = H_\varepsilon = 30$, 100 nm, and 300 nm. Obviously, interconnects with $L \leq 100$ µm are not affected by these limits, but more global interconnects are. This issue will be discussed in Chap. 29.

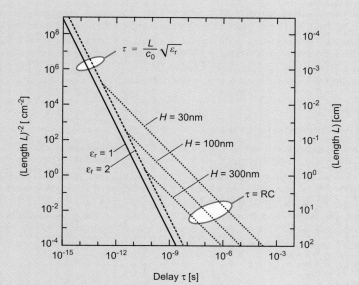

Figure 12: Reciprocal interconnect length squared, L^{-2}, versus interconnect delay time τ, assuming a copper-polymer technology ($\varepsilon_r = 2$, $\rho = 1.7 \cdot 10^{-6}$ Ωcm) (after [6] with modifications).

3 Concepts of Logic Devices

As a framework to Part III, this Section will give a brief overview of the physical principles which may be employed to realize logic devices. Since we cannot expand every conceivable principle into a full chapter due to space limitations in this textbook, a short introduction into *those* subjects will be given here, which are not described in detail later in Part III. Due to the diversity of concepts, a pragmatic classification will be used in this Section although the criteria are not completely independent of each other. A more detailed overview is compiled in Refs. [5], [24] – [28], [61].

3.1 Classifications

As mentioned above, logic devices operate on logical states which are represented by physical properties modulating a carrier and are used as input and output signals. In conceivable logic devices, these physical properties are typically either the *transport* of, e. g., charge or the static *configuration* of, e. g., the electrostatic potential, the spin orientation, the phase of a wave function. Figure 13 lists some possible combinations. For example, CMOS logic is based on field-effect transistors which use an electrostatic potential as the input signal and the source-drain current as the output signal. In biological neurons, the transport of very specific chemical compounds (neurotransmitters) can be considered as the input signal, leading to an appropriate change of the membrane potential which is finally conducted along the axon as action potential spikes. In Quantum Cellular Automata (QCA, see Sec. 3.6), the configuration of electrical charges, spin orientations, or phases of a superconducting current are the input and output signals. The geometrical configuration of mechanical latches can be employed as input and output states in the macroscopic world and, potentially, also in the microscopic world.

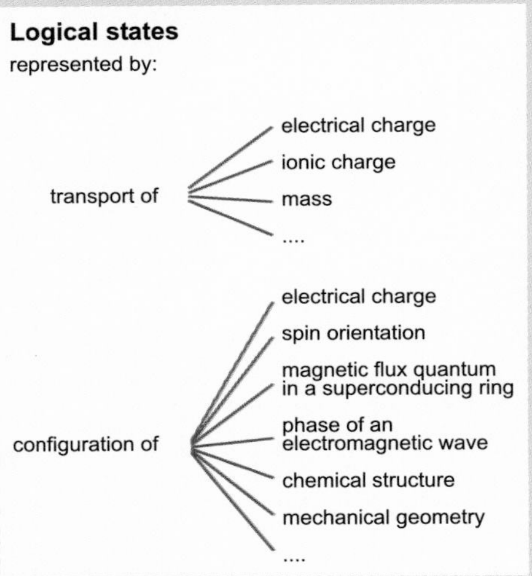

The physical properties representing logical states must arise from a non-linear behavior of the carrier in order to generate the discrete logical states of the digital logic. The physical origin of the non-linearity or discretisation does not matter. It may be

- the classical non-linearity of the function (e.g. the I-V characteristic of a FET in combination with its external circuit),
- the discreteness of the electrical charge (as in single electron devices, see Chap. 16), and
- the quantum-mechanical discreteness of energy states in microscopic systems (as in resonant tunneling devices, see Chap. 15).

In nanoelectronics, obviously the relevance of the latter two cases is much more pronounced.

Another classification refers to the number of terminals:

- **two-terminal devices** in which the input signal to modify (write) the output state and the reading of the output signal (as well as the energy supply) use the same terminals; examples are: switches and diodes.
- **three-terminal devices** (in general: multi-terminal devices) in which the input signal uses a separate terminal than the output signal; example: transistors, biological neurons.

Logic devices are used to build logic gates. The latter must comply with the set of requirements listed and described in Sec. 1.1. In some cases, a single logic device represents already a logic gate (e. g. a biological neuron), in other cases a minimum number of logic devices need to be connected to build a gate. For example, the realization of a NAND in CMOS technology requires four MOSFETs (Chap. 7). Also different devices (e. g. tunneling diodes and transistors) can be combined to build a logic gate.

3.2 Two-Terminal Devices

The conventional logic devices in CMOS technology and in biology can be considered as three- (or multi-) terminal devices. In digital logic, the MOSFET is an electrical switch (between the source and the drain) which is controlled by the electrical potential at the third terminal, the gate. The most simple biological neuron would have only one input synapse (at the dendritic tree) and one output synapse (at the end of the axon). At first glance, this looks like a two-terminal device. However, the ambient electrolyte connected to the interior of the neuron by numerous ion channels in the membrane must be

Logical states represented by:

transport of
- electrical charge
- ionic charge
- mass
- ....

configuration of
- electrical charge
- spin orientation
- magnetic flux quantum in a superconducing ring
- phase of an electromagnetic wave
- chemical structure
- mechanical geometry
- ....

Figure 13: Examples of input and output signals of logic devices.

regarded as a third, although distributed, terminal. The electrolyte serves as a potential reference (and power supply, see Chap. 6). Obviously, in both cases the input signal and the output signal are decoupled which makes it easy to accomplish feedback prevention.

On the other hand, real two-terminal devices are attractive for building densely integrated, nanoelectronic circuits because the lower number of terminals immediately reduces the huge interconnect problem significantly.

Reconfigurable molecular switches and resonant tunneling diodes are examples of two-terminal logic devices. The challenge is that the logic state of the device must be set through the same terminals as it is read. This can be solved by using different signal values for the two processes, e. g. large (positive and negative) voltage amplitudes to set the state of the switch and small amplitudes to read the resistance of the switch without changing it (see Chap. 20). The problem of the feedback prevention can be solved by introducing diodes (which are also two-terminal devices). Resonant tunneling diodes (RTD) show a regime of a negative differential resistance (NDR). This can be employed for power amplification which is needed for signal regeneration in large circuits. In addition, combinations of RTDs facilitate the implementation of a generic set of logic functions. Thus, in principle it is possible to construct logic circuits solely from two-terminal devices (RTDs and diodes) [29]. However, several clock signals as well as modulated voltage supplies are needed to drive the circuit. With the exception of this special case, circuits made from two-terminal devices suffer from the absence of power amplification. This must be provided in the periphery of the circuit. The maximum size of the circuit will be determined by its RC times and the operating frequency f.

3.3 Field Effect Devices

The basic principle of field effect devices is the charging of a gate electrode which creates an electric field in the channel between the source and drain. Depending on the polarity of the gate potential and the characteristics of the channel, this field leads to an enhancement or a suppression of the conduction.

The by far most important device in digital logic is the **Si-based Metal-Oxide-Semiconductor Field Effect Transistor** (MOSFET). The challenges on the route to further reduced sizes and the potential of new materials and devices concepts are described in Chapter 13.

The introduction of ferroelectrics as a gate oxide (**Ferroelectric FET**) gives the chance to conserve the charge on the gate electrode if the supply voltage is switched off (Chap. 14).

Carbon Nanotubes can also be employed as channels of a field effect device. A gate electrode made from any conductor (metal, aqueous electrolyte, etc.) attached to the tube wall can be used to control the current flow. These **Carbon Nanotube FETs** are explained in Chapter 19.

Organic semiconductors are used as a thin-film channel material for **organic FETs** (OFETs, also called **organic thin-film transistors**, OTFTs) [30] – [34]. Because the carrier mobilities are at least two orders of magnitude smaller for thin-film organic and polymeric conductors compared to good inorganic semiconductors, OFETs are no alternative to the very high switching speed devices and high integration density circuits based on monocrystalline Si, Ge or compound semiconductors. Still, they represent the basic devices of a low cost – low performance polymer electronics. They are candidates for applications, which – besides low cost – require large area coverage, flexible substrates (such as polymer foils), low weight, and low processing temperatures. For example, this includes the backplanes of all kinds of active matrix displays (see Part VIII) in which inorganic TFTs based on hydrogenated amorphous silicon (a-Si:H) may be substituted. Because of the low-temperature processing, OFETs can extend this application to transparent polymer substrates. Furthermore, smart identification tags on goods, large area sensor arrays, and other applications have been proposed.

Organic semiconductors and conductors are based on π-conjugated oligomers and polymers. Common p-type conducting molecules are pentacene as well as thiophene compounds. Tetracyanoquinodimethane (TCNQ) is an example of a n-type conductor. More details and examples are compiled in Ref. [30]. Intramolecular charge transport along the conjugated π-orbitals is relatively easy and high local conductivities are obtained (see Chaps. 6, 19, 20). However, the charge transport *between* the molecules is much more difficult. Reports in the literature about mobilities in organic semiconductor thin films cover many orders of magnitude. The highest reported values are slightly above 1 cm^2/Vs, for both p- and n-type conductors. These can be used to fabricate complementary logic circuits [35]. The preparation conditions are of crucial importance.

III Logic Devices

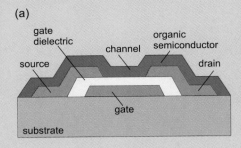

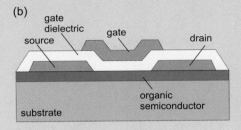

Figure 14: OFET configurations.
(a) bottom gate device,
(b) top-gate device.

Specific impurities must be avoided with great care and a crystallinity with a low density of structural defects must be established in order to obtain high mobility values. It was proposed to describe the conduction mechanism as a superposition of a thermally activated hopping and thermally activated tunneling of polarons, which occur between localized states that are disordered both in space and in energy [36], [32]. The two mechanisms show opposite signs of the temperature dependence which may explain the small over-all temperature dependence of the mobility of high-quality films. Figure 14 schematically shows the cross-section of two device configurations. In general, attention must be paid to the quality of the surface onto which the organic film is deposited and on the avoidance of thermal and chemical stress after the deposition. As an example of the bottom-gate configuration, Ni gate electrodes were deposited on borosilicate glass and on plastic substrates, followed by SiO_2 gate dielectrics, and – low work function – Pd source/drain electrodes [31]. Ion-beam sputtering is used for the deposition because of the low process temperatures and low surface roughnesses. Finally, a pentacene film of approx. 40 nm thickness was deposited by thermal evaporation. Typical test structures are shown in Figure 15. Alternative deposition methods are a carrier gas assisted sublimation or the deposition from solutions by stamping or printing. While the latter is highly attractive because it could eliminate lithography from the device fabrication, the quality of the films obtained from solution is inferior to those made by vacuum techniques. The electrical characteristics of OFETs are determined by operation voltages, which are relatively high compared to standard MOS devices. The gate voltages and the source-drain voltages are in the order of 10 to 30 V, the I_{ON}/I_{OFF} ratio is typically 10^5 to 10^6, and the max. currents are in the range of several milliamps for a device with a footprint area of 10^{-4} cm^2. The dynamic response times facilitate circuits which run at clock frequencies in the range of a few kHz, perhaps approaching 1 MHz. This is sufficient for the designated applications areas mentioned above.

3.4 Coulomb Blockade Devices

The voltage at and the charge on a macroscopic capacitor are linearly and continuously related according to

$$V = \frac{Q}{C} \tag{16}$$

The energy of this capacitor is given by

$$W = \frac{Q^2}{2C} \tag{17}$$

For nanometer sized capacitors, these relations change to highly non-linear step-like functions because of the discreteness of the electronic charge, $Q = n\,e$, where n is the number of electrons and e is the unit charge:

$$V_n = \frac{ne}{C} \tag{18}$$

and

$$W_n = \frac{n^2 e^2}{2C} \tag{19}$$

To observe these non-linearities, the energy steps $\Delta W = W_{n+1} - W_n$ need to be significantly larger than the thermal energy $k_B T$. An energy step

$$\Delta W = \frac{e^2}{2C}(2n+1) \tag{20}$$

is related to the electrode area A of a plate capacitor with a given dielectric thickness d

$$C = \varepsilon_r \varepsilon_0 \frac{d}{A} \tag{21}$$

by

$$\Delta W = \frac{e^2 d}{2\varepsilon_r \varepsilon_0 A}(2n+1) \tag{22}$$

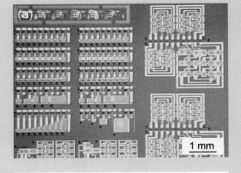

Figure 15: Photographs of OFET test devices.
(a) transistors with different gate widths and lengths. The gate contact is in the center of the squares. Scale: the horizontal edge represents approx. 2 mm;
(b) test devices on a flexible polymer substrate. (Courtesy of Siemens AG, Erlangen, Germany).

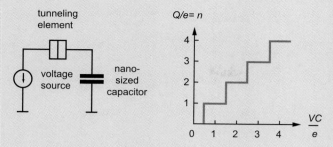

Figure 16:
(a) Circuit diagram of a voltage source connected to a nanosized capacitor by a tunnelling element.
(b) Normalized charging characteristics of the capacitor.

For the first electron to charge the capacitor (from $n = 0$) and the requirement $\Delta W \gg kT$, this leads to electrode areas

$$A \ll \frac{e^2 d}{2\varepsilon_r \varepsilon_0 kT} \qquad (23)$$

At $T = 300$ K and a SiO_2 dielectric of 3 nm thickness, the right term reveals an area of approx. $2.6 \cdot 10^{-16}$ m^2, e. g. a square of 16 nm edge size. Consider such a tiny capacitor connected to a voltage source a tunneling device (Figure 16). If the capacitor is discharged in the beginning and the voltage is raised from 0 V, the tunneling of the first electron onto the capacitor only occurs if this process leads to a decrease of the free energy of the system. This is elaborated in the first section of Chapter 16.

Because of the small size of the nanosized capacitors, the energy levels are discrete as determined by the quantum mechanics. This effect superimposes on the Coulomb blockade effect and is described in Chapter 16, too.

Concepts for *nanoelectronic devices* have been developed which exploit the non-linear characteristic of nanosized capacitors for logic devices and memories [37]. The major concepts investigated are the following [24]:

- **Single electron transistors** (SET) are three-terminal devices, which consist of a nanosized island (nanodot) made from metal or a semiconductor, connected to source and drain electrodes by tunneling junctions. The quantization of the charges on the island and the voltage at a third electrode (gate) in the vicinity controls the transfer characteristic of the SET. Details are described in Chapter 16.
- **Nanowire memories** are two-terminal devices, in which ultrathin poly-Si wires are arranged in a cross structure separated by an oxide. The naturally formed nanosized crystallites obviously can be processed to show a Coulomb blockade behavior [38].
- Non-volatile **nanodot memories** are three-terminal devices, which are constructed similar to flash memory elements. The floating gate of the conventional flash transistor is split into separate nanosized islands. See also: Introduction to Part IV and references cited therein.

The scaling behavior of MOSFETs and SETs is very different. Due to the reduction of the operating voltage and of the minimum feature size F, which determines the channel length and the gate capacitance, the required energy which is dissipated during the switching process decreases strongly along the reduction of F (see Sec. 6). In the same manner, the number of electrons stored on the gate capacitor reduces. Conceptually, if this finally arrives at one electron on the gate electrode, the MOSFET turns into a single electron device. On the other hand, if the size of island of a SET shrinks, the required energy to transfer an electron onto the island increases according to Eq. (22) and (23). This is schematically illustrated in Figure 17, where the capacitor area is set $A = F^2$ for simplicity. Also shown in this figure is a line for $10\, k_B T \ln(2)$ for $T = 300$ K, which can be regarded as a limit for a reasonably reliable operation of logic devices (see Sec. 6).

The processing speed of SET is relatively low because the impedance levels involved are quite high (approx. $10^5\, \Omega$) and their product with the interconnect capacitance determines the delay time [10].

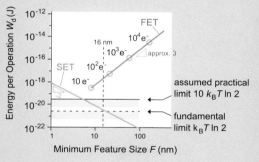

Figure 17: Dissipated switching energy W_d versus minimum feature size F for MOSFETs and SETs. The area below the horizontal boundary at $10 \cdot W_{TD,min}$ is not accessible for reliable information processing at $T = 300$ K. Note: the line for FETs is calculated from the *device* data estimated in the ITRS 2001. Hence, the energy per switching operation is much lower than the energy in Figure 10, which refers to an average logic gate in a *circuit*. The number of electrons for FETs are the estimated excess electrons in the inversion channel.

3.5 Spintronics

Besides their electric charge, electrons show another fundamental property which can be utilized in information technology and nanoelectronics: their spin. The discovery of a very pronounced effect of the spin orientation on electronic transport properties in thin metallic ferromagnetic/non-ferromagnetic/ferromagnetic multilayers (Giant Magneto-Resistance effect, GMR) led to highly sensitive, fast, and very small magnetic field sensors, which are used, for example, in the read heads of hard disk drives (Chap. 24). The

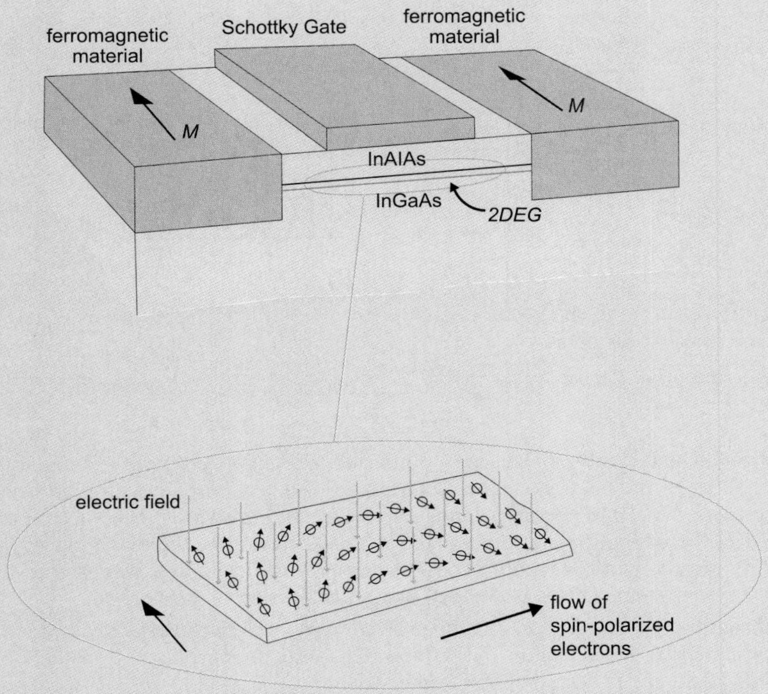

III Logic Devices

Figure 18: Hypothetical spin FET. Source and drain are oriented ferromagnets acting as polarizer and analyser of spin-polarized electrons. An electric field rotates the spin polarization direction of the electrons travelling in the channel with relativistics speeds.

spin dependence of the tunneling current through ultrathin insulating films (Tunnel Magneto-Resistance, TMR) is utilized, for example, in magnetic random access memories (Chap. 23). The fundamentals of these magnetoelectronic effects, established by the interrelation between spin polarization and electronic transport are introduced by Grünberg and Bürgler in Chapter 4.

Since the beginning of the 1990s, there have been several proposals to engage magnetoelectronic effects also in active logic devices such as transistors, giving rise to the name *spintronics*. For example, Datta and Das [39] proposed a theoretical concept which is based on the field effect on a spin-polarized current (Figure 18). The source and drain are supposed to be ferromagnets with, e. g., identical magnetization direction while the channel is made by a heterojunction of an appropriate compound semiconductor, establishing a channel region with a highly mobile 2-D electron gas (2-DEG). The source is supposed to inject a spin-polarized current into the channel. Without applied voltage at the gate, the spin polarization would remain unchanged and the interface resistance towards the drain contact would be low. The electrons travel at high velocity (approx. 1% of the speed of light) through the 2-DEG channel. If an electric field in the channel region is created by applying a voltage to the gate, the electrons experience an magnetic field because of their relativistic speeds. This magnetic field can rotate their spin direction. Consequently, at the channel/drain interface, the electron spin is not aligned anymore with the spin orientation in the drain which would lead to an increase in the scattering probability and, hence, an increase of the resistance. Such a spin FET device could potentially have some advantages over conventional FETs. For instance, the energy required to flip the spin polarization may be less than the energy needed so far to deplete (or accumulate) the charges in the channel (of contemporary FETs – not at the physical limit). There are also other spin transistors which have been proposed. However, at the publication time of this book, a spin transistor with power amplification has not been reported yet.

Spin transistor concepts are based on three effects: (1) electrons injected into the active region of the transistor need to show a high degree of spin polarization, (2) there must be a control signal (electric, magnetic, optical) which makes it possible to tune the spin polarization, (3) the spin polarization must sustain the traveling time and distance in the active region. A long coherence time is also interesting for the idea to use the electron spin as a quantum bit (short: qubit) for quantum computation (see Sec. 3.7 and Chap. 18). The electron spin precesses in a vertically applied magnetic field and the spin vector can be rotated into the horizontal axis. A horizontally polarized electron spin can be regarded as a coherent superposition of the spin-up and the spin-down state and, hence, represents a qubit. It has been demonstrated that precessing electrons in com-

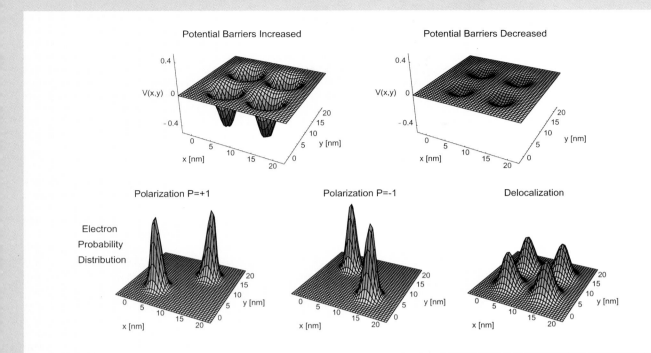

pound semiconductors such as GaAs maintain their coherence for several 100 ns and can be dragged more than 100 μm by an applied electric field (see e.g. [40]). This is sufficient to fulfill the requirements of spin transistors as well as to be promising for quantum computers. Furthermore, it has been shown that the coherence can be maintained across heterojunctions (e. g. GaAs/ZnSe junctions) and within GaN. Generation of highly spin-polarized currents by injection into non-ferromagnetic semiconductors remain challenging. Despite different approaches which include, for example, ohmic injection from ferromagnetic metal into semiconductors (e. g. InAs), tunnel injection through Schottky barriers, and ballistic injection, the generation of an efficient electron injection with a high degree of spin polarization at room temperature turned out to be quite difficult. The best value reported at the publication date of this textbook have been a spin polarization of 2% [41] and of 13% [42] at room temperature for electrons injected from Fe into GaAs. On the other hand, a plethora of new magnetoelectronic effects and materials are being discovered which involve optical manipulation of the spin coherence, electro- and photoluminescence effects, and promising new ferromagnetic semiconductors such as Mn doped III-V semiconductors, as potential seeds for future spintronic devices. Details can be found in review articles [43], [44] and references cited therein.

Figure 19: Upper row: potential barriers for the electrons in a QCA cell made from four quantum dots. If the system is clocked, the potential barriers may be decreased in order to enable the transition of the polarization state. Lower row: electron probability distribution of polarization state "0" (left), "1" (center), and a delocalized state, e. g. by loading with four electrons or as time-average of an isolated QCA (right), shown by Goser and Pacha [47].

3.6 Quantum Cellular Automata

In all logic devices discussed so far, information transfer in information processing system was based on the *flow* of particles, specifically electronic charges (or on the spin states of moving electrons, Sec. 3.5). An alternative is to implement logic states by means of discrete stationary states of a microscopic system and to employ the interaction via fields to move information through the system and to set the logic states. The basic idea of Quantum Cellular Automata (QCA) is an elementary cell of two stable states, representing "0" and "1", which can be toggled by fields emerging from neighboring cells. Since there is no (static) flow of charges or particles and single atomistic entities (electrons, electron spin, small molecules) solely change their position in a potential well, an ideal QCA circuit would, in principle, operate near the thermodynamic limit of information processing. In addition, the interconnect requirements on the nanometer scale are relaxed.

Consider an electrostatically operating cell [45] made from four quantum dots arranged in a square (Figure 19). The dots are close enough that electron tunneling between the dots may take place. If such a cell is loaded with two electrons, the electrons will occupy diagonally opposite dots because of their Coulomb repulsion. There are two diagonal states with opposite polarizations representing the two logic states (Figure 20).

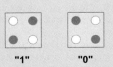

Figure 20: The two logic states of a QCA cell.

III Logic Devices

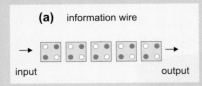
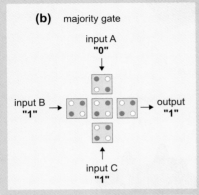
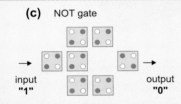

Figure 21: QCA gates.
(a) a linear row of QCA cells for transferring a logic state,
(b) a majority gate,
(c) a NOT gate. The signal is branched into two identical one first (upper and lower branch), before it is affecting the output cell, since the inversion needs to be across a corner. The interaction across two corners makes the gate more reliable.

In an isolated cell without external fields, the two states have identical electrostatic energy and, hence, are occupied with equal probability. If there are other polarized cells in the immediate vicinity, the energetically favorite state is determined by fields originating from neighboring cells. A linear series of cells acts as a "wire" which can be used to transfer a logic state. If the state is toggled at one side, all subsequent cells toggle as in a row of dominoes (Figure 21a) because adjacent cells try to be in the same polarization state to minimize the electrostatic energy. This principle can also be employed to create logic gates. Figure 21b shows a majority voting gate, as an example. The state of the output cell is determined by the majority of states at the three input cells. By fixing one of the inputs cell in the state "0" or "1", one obtains an AND or an OR gate, respectively. The gate can also be regarded as a three-input threshold gate, in which the weights are all one and the threshold value is $\Theta \geq 2$. Figure 21c shows a NOT gate.

Obviously, QCA circuits fulfill some but not all requirements of logic devices. Firstly, there is no power amplification, hence, the circuits may not have infinite size. Secondly, the feedback is not prevented, i. e. circuits can be operated equally in both direction. In order to compensate for these aspects, there should be actively driving input and output circuits as in the case of the passive cross bar arrays (Sec. 4.1).

Some simple electrostatic QCA gates have been realized, based on metal-insulator tunnel junctions, which operate at $T = 70$ mK [48], and on Si quantum dots embedded in a SiO_2 matrix [49], which operate at a temerpature of some hundred mK. The limitations on the operating temperature are imposed by the requirement that the electrostatic dipole interaction energy must be significantly larger than the thermal energy $k_B T$. For a room temperature operation, the quantum dots need to be smaller than 5 nm and the edge of a cell must be < 25 nm. The relative requirement on the fabrication precision is extremely high (a fraction of 1 nm). To circumvent this difficulty, fine tuning is possible by leads positioned in the vicinity of each cell and by applying adjustment voltages. Because this concept is hardly practical, clocking signals could be used which lower the tunneling barriers between the dots within the cells at defined clock periods) [47].

In the framework of the general QCA concept, different interacting fields between the cells and, hence, different kinds of dominos have been prosposed:

- electrostatic domino cells, as described above,
- magnetostatic domino cells, as described in Ref. [50], and
- mechanical domino cells, as described in Ref. [51] for an example on the nanometer scale.

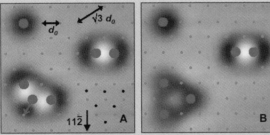

Figure 22: STM images (1.9 nm × 1.9 nm) of CO molecules on a Cu(111) surface [51]. The gray-scale images represent the curvate of the tip height, so local peaks appear light and local valleys appear dark. Solid red circles indicate locations of CO molecules. Blue dots indicate surface-layer Cu atoms, and black dots indicate second-layer Cu atoms. d_0 is the Cu-Cu distance, 0.255 nm.
(A) An isolated CO molecule (top left), a dimer (right), and a trimer in the chevron configuration (bottom left). The arrow indicates how the central CO in the center of the chevron will hop spontaneously, typically within a few minutes at 5 K.
(B) The same area after the CO molecule has hopped. Spontaneous hopping is prevented in domino cells stablizing each others in rows or logic gates.

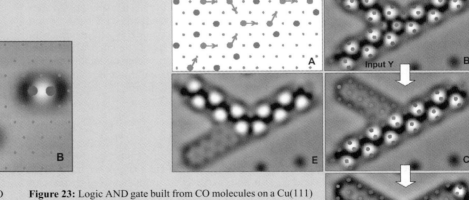

Figure 23: Logic AND gate built from CO molecules on a Cu(111) surface [51].
(A) Model of the AND gate. Large blue dots indicate CO molecules that hop during device operation, and green dots indicate positions after hopping. Both input cascades (left) have to be triggered to set up a chevron in the center, which then triggers the output cascade.
(B to D) Sequence of STM images (5.1 nm × 3.4 nm) showing the operation of the AND gate.
(B) Initial configuration.
(C) Result after input X was triggered manually by moving the top left CO molecule with the STM tip.
(D) When input Y was triggered, the cascade propagated all the way to the output.
(E) The result of manually resetting all CO molecules to the initial configuration of (B), and then manually triggering input Y. Subsequently triggering input X yielded configuration (D).

The concept of mechanical dominos is based on the idea that the toppling of a row of standing dominos can be used to perform mechanical computation. A row of standing dominos can be tipped and that causes the sequential toppling of all dominos, transferring one bit of information from the beginning to the end of the row. The toppled and untoppled states of a domino represents "0" and "1", respectively. Appropriate configurations can be used to realize logic gates. The energy for the computation is stored in the standing dominos. Before every calculation, all dominos must be reset to their standing position. The dominos reported in Ref. [51] have been CO molecules on a Cu(111) surface. A cell is realized by three CO molecules which are in a chevron configuration, representing the standing up domino (Figure 22A), and in a relaxed position, in which one of the CO has hopped to a energetically favorable position (Figure 22B). The starting configuration is achieved by STM manipulation (Chap. 12). Upon triggering the inputs of a logic gate configuration, the output value is obtained after few seconds (Figure 23). The experiments were conducted at 5 K. The hopping motion of the CO molecules during the toppling can be described as a quantum tunneling. A three-input sorter has been realized to demonstrate on how small an area logic circuits can be fabricated (Figure 24). The area occupied is $2.6 \cdot 10^5$ times smaller than for a realization as an physically optimized CMOS circuit using the 130-nm technology. One computational run of this sorter represents approx. 100 bOp. The CO domino version dissipates approx. $2 \cdot 10^{-19}$ J per computational cycle (at 5 K), while the CMOS counterpart needs $2 \cdot 10^{-14}$ J (at 300 K).

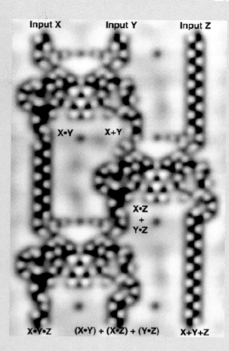

Figure 24: STM image (12 nm × 17 nm) of a three-input sorter in the initial setup [51]. The symbol + denotes logic OR, and • denotes logic AND. Images with one or more inputs triggered are not shown.

3.7 Quantum Computing

Conceptionally, QCA circuits can be extended into **Quantum Computers** (QC), by replacing the cells by so-called **qubits**. To describe the difference between bits and qubits, consider a set of two as the simplest case. Two bits can represent $2^2 = 4$ different states, i. e. they can adopt any of the states 00, 01, 10, or 11. In contrast, the two qubits are represented by individual wavefunctions which superimpose to result in a wavefunction of the set. Through this superposition, the two qubits represent all possible logic states *simultaneously*. Any logic operation on the qubits will act on all possible states simultaneously, too. A classical computer starts with a certain set of binary values as in the input variables and acts on them according to the algorithm to obtain a solution. The solution for another set of input values requires to run the computer with the same algorithm again, to obtain the next solution. A quantum computer processes all possible states *at once*, and hence, delivers the solutions for all possible combinations of input values simultaneously. This results in an ideal parallel processing of the information. In other words, a system of N qubits represents 2^N possible states as in a classical computer. In contrast to the later, the 2^N different states are superimposed and the quantum computer applies the algorithm to all these states simultaneously.

This suggests the following comparison: classical computers relate to quantum computers as ordinary photographs relate to holographs. Photos code the visual information by the *amplitude* (intensity) of the backscattered light. By taking a conventional picture, a large part of the information about the object, the *phase* relation between the waves emerging from the points of the object, gets lost. Using coherent light to illuminate the object and a coherent reference beam, the phase relations are turned into interference pattern which can be stored as an amplitude pattern again. Through this trick, the phase information is conserved. The full information, phase and amplitude, can be released again by illuminating the holographic picture by a coherent reference light again. As in a hologram, the quantum computer conserves the full information of the qubit wavefunctions during the logical operations. As a consequence, the potential solutions for the complete set of possible input values is calculated concurrently. Because of its principle, quantum computers would be especially suitable for problems which require to use the entire set of input values in order to find the solution (so-called *non-deterministic polynomial* problems, short: NP problems), such as the factoring of integers, the traveling salesman problem, and the cryptographic encoding. For problems with larger numbers of input variables quantum computer may deliver solutions which are unsolvable for classical computers, simply because the computation time would be astronomically large.

Different approaches have been taken to the implementation of quantum computers. These are as follows [25]:

- Bulk resonance quantum implementations including Nuclear Magnetic Resonance (NMR), linear optics, and cavity quantum electrodynamics,
- Atomic quantum implementations including trapped ions and optical lattices,
- Solid state quantum implementations including semiconductors, spin-polarized systems, and superconductors.

Chapter 18 gives a short general introduction into quantum computing and describes the approach to implement quantum computers on the base of superconductors.

3.8 DNA Computer

Another approach for computing is to encode information on a DNA molecule by a sequence of DNA bases (of the four distinct bases A, T, G, and C; see General Introduction). There are various well known biochemical methods for manipulating DNA; these can be used to execute parallel computation by modification of the information encoded by the DNA strands. This is known as DNA computation or more generally as biomolecular computation. Due to the small size of the molecule, a huge amount of information (approx. 10^{18} DNA molecules per cm^3 of solution) can be processed in parallel, providing a degree of massive parallelism far exceeding traditional computing technologies.

DNA computation was first experimentally demonstrated by Adleman [52] – [54] to solve non-deterministic polynomial problems (see Sec. 3.7). The idea of Adleman is to code all possible input values of a problem on different (single-stranded) DNA molecules. The problem is (complementary) coded on another set of DNA molecules. Now the sets are subjected to a standard enzyme catalyzed recombinant reaction in a test tube or on a chip on which one set of single-stranded DNA molecules has been immobilized. This reaction takes few ten minutes. During this time all potential solutions to the problem are synthesized as double-stranded DNA molecules, representing a massive parallel computation. The DNA representing the correct solution to the problem will be among a huge amount of incorrect ones. By either selectively accrete or capture the correct DNA molecules or enzymatically destroy the incorrect ones, both in a iterative sequence of steps, the correct DNA can be isolated and analyzed to read-out the final solution to the problem.

The use of DNA computation to solve non-deterministic polynomial problems is good for demonstration of small scale DNA computations, but it is not likely to be useful in practice for a number of reasons. This particular application of DNA computation exhibits a low flexibility, i.e. it cannot be used to realize an Universal Turing Machine (see Sec. 4.1), since any new problem and the whole set of possible input variables needs to be freshly coded by synthesizing a large number of different DNA molecules. Even if this task can be automized and miniaturized to a high degree, the approach may turn out to be not very efficient for very large problems. Also, the number of distinct DNA strands grows exponentially with the size of the problem statement, and the molecular-scale data storage is eventually swamped by the numbers of DNA strands required for large problems. For the type of problems discussed above, the likely upper limit is approximately 70 to 80 Boolean variables [55].

However there are other applications of DNA computation techniques that appear not to suffer from these scalability difficulties and potentially may provide large and quite unique benefits which conventional electronic computers can not provide. These include the use of DNA biotechnology to store extremely large databases of biological information and material and to execute associative searches on these databases [55].

4 Architectures

The architectures of information processing systems is an area which is comparably huge as the area of electronic materials and devices for micro- and nanoelectronics. In the context of this book, only a brief look at some relevant concepts and trends can be given. Some more details on architectures in the nanoelectronics era are provided, for example, by the Refs. [24], [25], [27], [59], [61].

4.1 Flexibility of Systems for Information Processing

There are numerous ways to assemble systems for the information processing from logic devices. In a first classification one can distinguish between:

- **free-programmable systems** in which an instruction flow (software) fed into the system controls the sequence of operations performed by the system,
- **reconfigurable systems** in which the hardwire configuration can be changed by corresponding instructions fed into the system, and
- **hardwired systems** in which the internal hardware structure is mainly fixed during the assembly of the system and cannot by changed by software or by reconfiguration.

This classification correlates to the **flexibility** (or *universality*) of the system. To estimate the flexibility, one considers the entire set of computable tasks, i.e. tasks which can in principle be solved by information processing systems. The flexibility then reflects how easy it is on the average to re-arrange the system in a way that it is able to solve any other task. Figure 25 illustrates the flexibility for some typical examples of systems [19]. The *general purpose processor* represents a highly flexible free-programmable system. By using appropriate operating systems and application programs it can be easily tuned to solve, in principle, every conceivable computational task. This is a result of the von Neumann principle according to which the structure of the computer is completely independent of the structure of the task (Chap. 7). The price for the high flexibility is a relatively low performance, since there is no optimization for a specific class of tasks, and even more a high power dissipation, for reasons given in Sec. 2. Still, the specific configuration should be regarded. If a physically optimized floating-point unit is integrated onto the processor chip, a very high performance in floating-point operations is obtained (Sec. 5.2).

Digital Signal Processors (DSP) are also relatively flexible but they feature some optimization for certain classes of tasks, i. e. signal compression and decompression, data coding and decoding, vector and matrix arithmetics, Fast Fourier Transformation, etc. Their application areas are speech and image processing, communication systems, multimedia, etc.. There is an additional aspect, which leads to a relatively low performance and high power dissipation of programmable processor systems. In these systems, the total number of binary operations (bOp) is much higher than the actual number of binary operations required to perform the data processing task. This is because in addition to the logic operations on the data, there is a *control and glue logic* which is needed to run the system without contribution to the direct processing of the data flow. Although this is true for every logic system, especially in programmable processors, this control and glue logic is much larger than the *arithmetic logic*.

Field Programmable Devices (FPD) or, more specifically, *Field Programmable Gate Arrays* (FPGA) are typical representatives of reconfigurable systems. They are typically built from large numbers of simple programmable logic units. An example of the latter is shown in Figure 26. It essentially consists of cross bar arrays which are used as look up tables (LUT). All crossing points are programmable switches. The input signals x_0, x_1, and x_2 are fed into the system in their inverted and non-inverted form. At the output of the AND array, the signals are fed into an OR array. Lines without connected switches must be pulled to "1" in the AND array and to "0" in the OR array. Depending on the type of FPD, the AND array or the OR array are programmable or both. By pro-

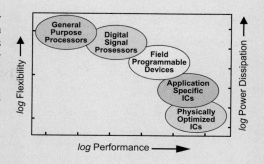

Figure 25: Types of realization of information processing systems sketched in a flexibility – performance – power dissipation diagram (adapted from [19]). The *flexibility* indicates the effort (measured, for example, in time or costs) required to prepare the system to solve new (arbitrarily selected) task. This axis spans approximately three orders of magnitude from changing an application program in the case of general purpose processors to designing and fabricating a completely new physically optimized integrated circuit. The abscissa illustrates the computational *performance* (for a comparable number of logic gates). Physically optimized integrated circuits can be approx. five orders of magnitude faster than general purpose processes performing the same dedicated task. The right ordinate illustrates the power dissipation for the same task. As described in Sec. 2, the *power dissipation* of physically optimized ICs can be up to five orders of magnitude lower than for general purpose processors.

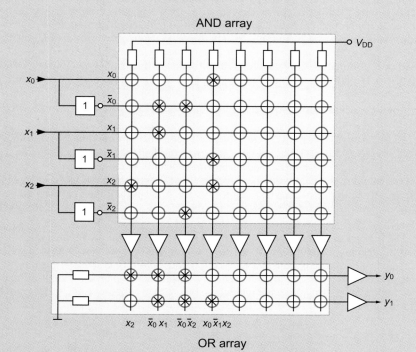

Figure 26: Example of a basic cell of a Field Programmable Device. In this variant, called Programmable Logic Array (PLA), the AND array as well as the OR array are configurable. There are others types, in which either the AND or the OR array are fixed to represent 1-of-n encoders or decoders. In the example shown here, the configuration represents Boolean functions according to:

$y_0 = x_2 + \bar{x}_0 x_1$

$y_1 = \bar{x}_0 x_1 + \bar{x}_0 \bar{x}_2 + x_0 \bar{x}_1 x_2$

Because the AND and OR arrays are configurable here, the right half of the arrays are unused and could be omitted. However, if the same Boolean functions were realized with either AND array or the OR array as 1-out-of-n decoders, the full arrays are required.

III Logic Devices

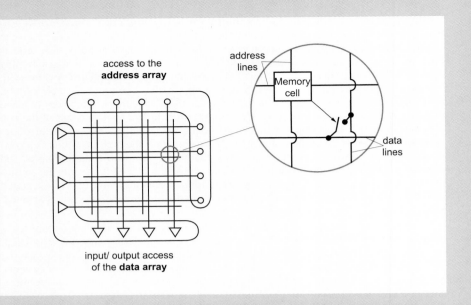

Figure 27: IIllustration of the cross bar array of an Programmable Logic Device. The double array structure consists of an address array for storing the configuration information in the memory elements and for controlling the switches of the actual data array. In standard CMOS technology, the memory elements and the switch control consist of six transistors (adapted from Ref. [57]).

gramming the arrays (Figure 27), a FPD unit can represent every conceivable Boolean function (see General Introduction). The data arrays of the FPD are completely passive, and – in the ideal case – do dissipate only little energy. ´Ideal´ refers to lines and closed switches without ohmic resistance and open switches with infinite resistance. The data arrays fulfill some requirements of logic (Sec. 1.1), however power amplification, feedback prevention, and the complete set of Boolean operators are missing. In the case of real arrays, there must be power amplifying gates at some stages. Often they are located in the periphery of the data array. These gates in the periphery are wired unequivocally as inputs and outputs and, hence, provide a clear data flow direction (feedback prevention). Since the data arrays cannot establish a NOT operation, appropriate inverters are located in the periphery too.

A contemporary CMOS-based FPGA may consist of many thousands or millions of simple programmable logic units, typically based on static random access memory (SRAM) LUTs, as well as fixed logic gates and flip-flops [56]. Usually the connections between these primitive units are programmable as well. Fabricating logic systems from FGPAs is highly attractive for nanoelectronics. This originates from the fact, that one could save the second cross bar array for addressing and programming the switches, if the programming can be performed through the same array as the array for the storing of the information. Molecular switches are conceivable, for example, which are opened and closed (i. e. configured) at relatively high voltages and operated at much lower voltages (Chap. 20). In addition, it is possible to built in defect tolerance into FPGA based logic systems (Sec. 4.3).

Another example of typical reconfigurable systems are *neural networks*, including the biological ones.

The next area in Figure 25 are (so-called semi-custom) *application-specific integrated circuits* (ASICs). Here, we need to design and fabricate integrated circuits if we want to solve a specific task. Since standard blocks as well as automatic synthesis, placement, and routing are used, the design process is faster than the full-custom design of physically optimized ICs. The flexibility is much lower than for field programmable devices, since a change of the task typically means a re-design and re-fabricating of the chip.

The highest performance and the lowest power dissipation is available through the design of integrated circuits which are *physically optimized* on the gate level. This kind of design is the most time-consuming, therefore the flexibility of these ICs is ranked the lowest. As mentioned in Sec. 2, however, the performance of physically optimized, dedicated logic can be many orders of magnitude higher than for processors, programmed to perform the same task. In addition, the power consumption is orders of magnitude lower, too.

There is a considerable overlap between the different types of logic alternatives shown in Figure 25, because of the large number of variants. Still, only processors and field programmable devices represent so-called Turing machines. As defined in 1936 by the British mathematician and logician Alan M. Turing, these are machines which are

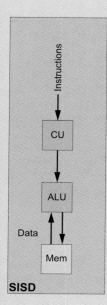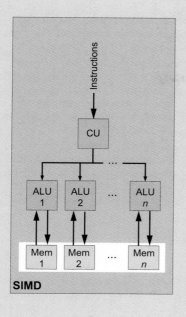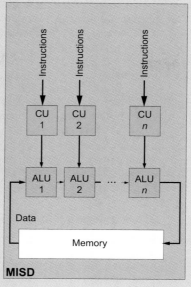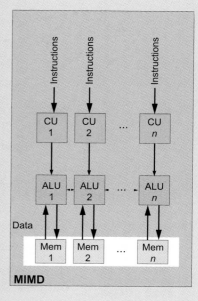

Figure 28: Classification of single and parallel processing systems after Flynn. Abbreviations: SISD Single Instruction – Single Data, SIMD Single Instruction – Multiple Data, MISD Multiple Instruction – Single Data, MIMD Multiple Instruction – Multiple Data; CU: Control Unit, ALU Arithmetic-Logic Unit, Mem Memory.

able to solve every computable problem, without changing the hardware, i. e. just by means of an appropriate program and the required data fed into the machine, on the condition that (in principle, indefinitely large) memory is available. The **Universal Turing Machine** (UTM) is made from a finite state machine and the (infinite) memory which can be sequentially read by the machine in arbitrary direction.

4.2 Parallel Processing and Granularity

If the type of a (single) logic system according to Figure 25 and the task to solve are defined, we can increase the speed, i. e. the computing performance (measured in bOp/s) by decreasing the delay times t_d of the elementary logic operations. For typical processors, this is equivalent to increasing the clock frequency $f_{max} = \beta/t_d$. As discussed in Sec. 2, there are limits to a decrease of t_d set by the given technology and, finally, by physics. In addition, there is a limit set by the maximum power dissipation density which is related to the maximum temperature a circuit may encounter and the thermal conductivity of the materials involved (Sec. 6).

As a consequence, if one wants to increase the computing performance of a system, the only choice is to employ more than one processing unit. This concept is called parallel processing and is used in mainframe computers and supercomputers for a long time. A very coarse classification of computers according to their data and instruction flows for single and parallel processing goes back to Flynn [60]:

- SISD: Single Instruction – Single Data
- SIMD: Single Instruction – Multiple Data
- MISD: Multiple Instruction – Single Data
- MIMD: Multiple Instruction – Multiple Data

This classification is not detailed enough to categorize the large variety of today's concepts. Still, it helps for an initial approach. Figure 28 illustrates the basic differences. The simple von-Neumann type single processor computers are SISD systems. A control unit (CU) uses a single instruction flow to control the arthimetic logic unit (ALU) which processes a single data stream.

Modern processors which employ pipelining can be regarded as a kind of MISD system, since a single data stream is processed simultaneously by several instructions (Chap. 7). It should be noted, however, that single processors today are usually classified as SISD. Systems which have been formerly called systolic arrays and which are used to carry out matrix multiplications, solving specific sets of differential equations, and similar tasks can also be considered as members of the MISD category.

SIMD systems are designed to execute an instruction simultaneously on several operands. For this reason, SIMD systems show one CU but several ALUs. Often, digital signal processors for high-performance video or multi-media applications, as well as VLIW processors (see Chap. 7) feature a SIMD structure. Another example are finger

III Logic Devices

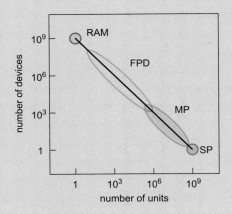

Figure 29: Illustration of the granularity of different computing systems in a comparison in which the number of logic devices times the number of units of the system is 10^9 (after [58] with modifications).
SP = single processor,
MP = multiprocessor system,
FPD = field-programmable device system.
A RAM device is included for comparison.

print sensors, where the instructions are transferred simultaneous to the 2-D arranged processing elements. The biological neural network of the retina of the eye can be regarded as a 2-D SIMD structure in a very similar fashion.

In MIMD systems, the individual processors have their own CU, ALU, and local memory. In addition, there is often a global memory and network of high bandwidth for information exchange between the individual units. Examples for MIMD structure are multiprocessor systems and parallel computers, including modern supercomputers. Due to the high flexibility of these systems, it is expected that these types of systems will gain a further increasing importance in the future.

Besides the structure, the degree of distribution of the total computational power of a system on parallel processing individual units is called *granularity*. Figure 29 sketches systems of different granularity at approximately the same total number of logic gates (diagonal in diagram Figure 29).

4.3 Teramac – A Case Study

With a growing number of devices per chip, two trends are obvious. Firstly, the required design effort strongly increases for processors as well as for optimized circuits. This problem in the semiconductor industry is known as the *design gap*. Only systems made from regular repetitive structures such as field programmable device based system show this problem to a much lower degree, because the actual structure is defined in the configuration phase *after* the system has been completed, before it is started to solve specific task. Secondly, the probability of defects statistically grows with the number of components if the fabrication process is conducted with the same degree of perfection. The defect probability for transistors in the current CMOS fabrication processes range from 10^{-7} to 10^{-9} [62] and it appears certain that this cannot be significantly improved. Thus, as the number of gates is growing towards a billion per chip and eventually beyond, defects will be hardly avoidable. In addition, new chemical fabrication techniques may provide relatively cheap but occasionally defective structures. Even when the defect probability is kept in the part per million range, these techniques will lead to some thousands of defective devices on a billion gate chip.

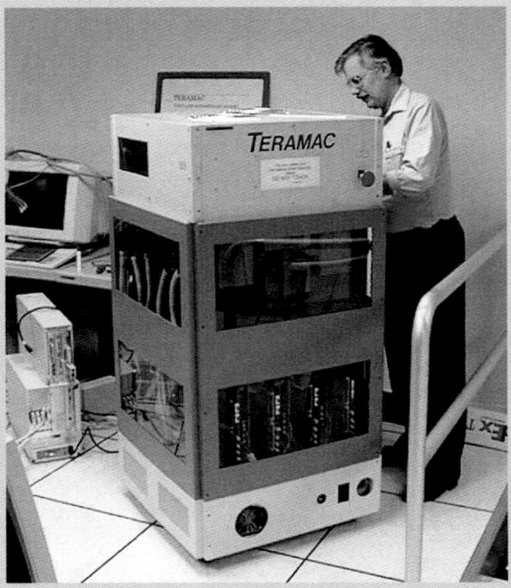

The so-called Teramac project was started in the mid-1990s at the Hewlett-Packard research laboratories to build a reconfigurable and defect-tolerant computer and to study its properties (Figure 30) [57]. The name *Teramac* originates from the fact that approx. 10^6 logic gates are incorporated. In combination with the (relatively slow) system clock of 10^6 Hz, this gives a formal performance of 10^{12} (1 "Tera") bOp. "mac" stands for *multiple architecture computer*. The system is build from a hierarchy of eight printed circuit boards (PCB), which carry four multichip modules (MCM) each, which in turn carry 27 FPGA chips each. These FPGAs contain 256 64-bit look-up tables (LUT) each. Of the 27 FPGAs per MCM, eight are used for representing the logic and 19 for the interconnect configuration. This relation indicates the importance of configurable signal and communication routing. A total of four Mb ($8 \times 4 \times 8 \times 256 = 65536$ 64-bit LUTs) of configuration memory is employed in Teramac.

The interconnect structure of the Teramac followed the so-called *fat-tree* concept (Figure 31). This highly redundant tree gives a high communication bandwidth which supports a high processing speed and it guarantees that there are always several routes from one junction to any other junction. Hence, signal paths can always be routed around any defective LUT and interconnect, providing the required defect-tolerance. Although approx. 70% of the FPGAs of the Teramac are used for communication purposes and, hence, have to be attributed as control and glue logic, the fraction of logic for the actual data processing is much higher than in general purpose processors and DSPs.

Figure 30: The Teramac system with Philip Kuekes, its principal architect [63].

Approximately 3 % of all FPGAs have been defective. For the configuration process, Teramac is connected to a workstation which either runs a test program or it loads a test program onto Teramac for a self-test procedure. The latter requires that certain (few) system components run reliably. The test program locates and cataloges the defective components. Following this preparation routine, a specific compiler is used to configure Teramac to perform a dedicated computational task. During this process, the catalog is used to route around the defects. After the configuration is completed, Teramac performs most tasks considerably faster than a top-end workstation at that time (late 1990s) and is perfectly reliable.

Despite the advantages (defect tolerance and processing speed), FPGA based systems such as Teramac will not economically substitute general processors or DSPs because they need almost one more order of magnitude transistors to provide an equivalent processing power. This is mainly due to the fact that the cross bar data array needs

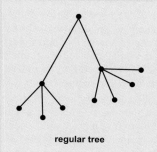
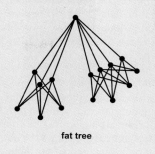

Figure 31: Tree-type interconnect structures of distributed computing systems.

an address array to control the switches and because the arrays must always be completely built although only a certain fraction of the switches is actually used. This situation might change if the switches could be programmed through the same cross bar array as they are read out and a cost-effective production of these passive arrays is found, where the switches are transistor-less, two-terminal devices, based, for example, on molecular electronic units (see Chap. 20).

5 Performance of Information Processing Systems

This section will deal with the number of logic operations per unit of time information processing systems are able to perform. As we shall see, determining the processing performance is straightforward only for simple automata. With increasing size and complexity of processing systems the number of possible permutations is growing so fast, that only limiting cases can be considered and estimations can be made. At the end of this Section, we will extend these estimations to biological neurons and the brain – emphasizing all the care which is indispensable in such attempts.

5.1 Basic Binary Operations

For given operations such addition and multiplication with operands of a fixed size, the required number of *two-input basic binary gates* can be determined. Conventional AND, OR, EXOR, NAND, and NOR are examples of basic binary gates. A basic binary operation (short bOp) is defined here as an approximation of the logic operation of a basic binary gate, deliberately neglecting the differences between the gates and the differences in their technological realization. To describe the rate, we introduce the number of bOp per second (bOp/s), which is independent of the sequential or parallel conduction of the operations.

As shown in Chap. 7, a half adder consists of an EXOR gate and an AND gate, i. e. 2 bOp. A full adder requires two half-adders and an OR gate, i.e. 2·2 bOp + 1 bOp = 5 bOp. The binary addition of two 16-bit operands, for example, requires a chain of 16 full adders (Figure 32), i.e. 16 × 5 bOp = 80 bOp. We are neglecting here, that the first in the chain could be a half adder. The time for addition is determined by the gate delays in a full adder for propagating the carry bit, c_i, and by the length of the chain. There are more time-efficient addition concepts which process all carry bits at once. However, these need more gates [64]. Such a trade-off is often encountered in the design of logic systems: improved speed can be traded for additional gates.

Binary multiplication can be composed of an array of additions, in which the multiplicand is shifted by one bit for each step [64]. For the multiplication of two numbers which are n bit long, n^2 gated full-adders are required. For example, a multiplier of two 16-bit numbers would contain 16 × 16 × 5 bOp = 1280 bOp. Binary floating point numbers are processed similarly. The binary exponents are added adequately. A standard 64-bit floating point operation (with a 52-bit mantissa and a 12-bit exponent) addition would require approx. 300 bOp, a multiplication approx. 16500 bOp. There is a range of more time-efficient adders and multipliers known, which are more difficult to implement, but are based on similar bOp values. (Remark: According to our definition of bOp, it is not the number of gates which count but only the number of actual binary operations).

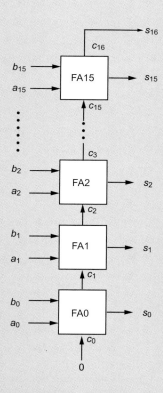

Figure 32: A 16-bit adder built from sixteen full-adders (FA). a_i and b_i (with $i = 0...15$) are the bits of the operands, s_i (with $i = 0...16$) denote the sum. c_i (with $i = 0...16$) are the carry bits, where $c_0 = 0$ and $c_{16} = s_{16}$.

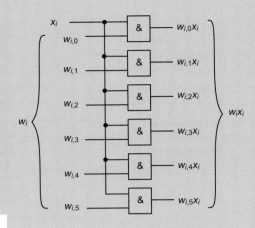

Figure 33: Binary logic representation of a weighted input of a threshold gate for a 6-bit resolution of the weight w_i.

The time for the operation necessarily depends on the type of circuit. In a combinational logic circuit, the processing time is given by the longest sequence k of subsequent gate propagation delays, Δt, as mentioned above for carry propagation of the adder. The binary operation rate would be formally calculated by dividing the bOp by $\Delta t \cdot k$. For example, if a 16×16 bit multiplier discussed above required 10 μs, the binary operation rate would be 1280 bOp / 10 μs = $1.28 \cdot 10^8$ bOp/s. Sequential logic circuits are usually synchronized by a clock. In this case, the time for an operation would be the clock period t_{clk} times the number of clock cycles k the operation requires, i.e. $t_{clk} \cdot k$.

In order to calculate the equivalent number of basic binary logic gates of a *threshold gate* with n inputs, and a resolution of N values which the weights may assume, we will use a formal equivalent circuit. This equivalent circuit is solely set up for our calculation purpose, it has no real counterpart! First, let us consider *one* input. Figure 33 shows the binary logic representation of an input using $N = 64$ as an example of the resolution. Note, we are using a power of 2 as the value N for the simplicity of the calculation. (here, $N = 2^p$ with $p = 6$). The value of w_i,

$$w_i = w_{i,0} \cdot 2^0 + w_{i,1} \cdot 2^1 + w_{i,2} \cdot 2^2 + w_{i,3} \cdot 2^3 + w_{i,4} \cdot 2^4 + w_{i,5} \cdot 2^5 = \sum_{r=0}^{r=p-1} w_{i,r} \cdot 2^r \quad (24)$$

is binary coded on the six digital lines representing the weighted input $w_i x_i$. If input $x_i = 0$, all lines of the weighted input are 0. If $x_i = 1$, the set of lines [0:5] shows the binary value w_i. According to Figure 33, the implementation of a weighted input requires p AND gates (here: $p = 6$).

Let us consider a threshold gate with $n = 4$ inputs. Then, in the subsequent step, the weighted inputs need to be added (Figure 34). This is performed in two steps, where 6-bit values are added in the first and 7-bit values in the second step, yielding a 8-bit result. In the final step, this result is compared to an 8-bit threshold value Θ. How many bOp do we need to represent the threshold gate? These are the contributions:

6	AND gates per weighted input =	6 bOp
4	input	× 4
		24 bOp
2	6-bit adders (= 6 × 5 bOp = 30 bOp)	+ 2 × 30 bOp
1	7-bit adders (= 7 × 5 bOp = 35 bOp)	+ 1 × 35 bOp
2	4-bit comparator	+ 130 bOp
		249 bOp

The equivalent number of bOp for the 8-bit comparator has been obtained from commercial 4-bit comparator (standard logic integrated circuit, type 7485) comprising approx. 65 internal bOp. Two 4-bit comparator are cascaded into a 8-bit comparator. According to this estimation, our threshold gate with 4 inputs and a 6-bit resolution of the weights can be compared to approx. 250 bOp. We will come back to this estimation in the context of discussing the processing performance of biological neural networks.

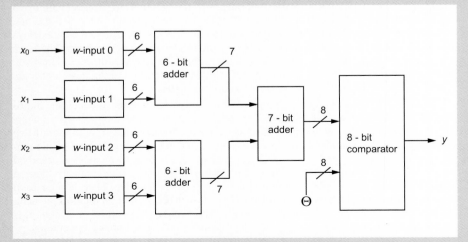

Figure 34: Binary logic representation of a complete threshold gate with four inputs. The blocks w-input stand for the weighted inputs according to Figure 33. The slashes across single lines denote set of lines (buses). The number of lines in a bus is given by the number at the corresponding slash. The threshold value Θ is fed into the circuit as a 8-bit value.

5.2 Measures of Performance

In contrast to the dedicated i.e. hardwired operations discussed up to now, processors work according to program instructions which are loaded and decoded before operations on data take place. An outline of the working principles of processors is given in Chap. 7. During the last decades different measures for the performance of processors and entire computer systems have been introduced. The most frequently used conventions will be briefly described. More details can be found, for example, in Ref. [65].

Processors run on a given clock frequency f_{clk}, the inverse of the cycle time $t_{clk} = 1/f_{clk}$. One measure of performance is

$$\text{performance} = \frac{f_{clk}}{CPI} = \frac{1}{t_{clk} \cdot CPI} \qquad (25)$$

where *CPI* is the average *number of cycles per instruction*. Sometimes, the average *number of instructions per cycle*, *IPC*, is used, defined as the reciprocal value of *CPI*. To be realistic, the averaging has to take into account the average depth of filling an instruction pipeline and parallel execution lines (see Chap. 7).

The unit of performance according to Eq. (25) is MIPS, or **native MIPS** (Million Instructions Per Second). The definition of native MIPS does not take into account the differences in performance per instructions. This difference becomes especially obvious in a comparison of CISC and RISC processors. One complex instruction on one processor may require three instructions on another to perform the same task. For this reason, the *relative performance* has been introduced which uses, by definition, the VAX 11/780 (first marketed in 1978) as a reference. The normalization factor

$$n_{VAX} = \frac{\text{average number of VAX instructions}}{\text{average number of instructions on system M}} \qquad (26)$$

shows $n_{VAX} > 1$ for a machine M which has a more powerful instruction set than the VAX 11/780. The *relative* performance is then given by

$$\text{performance} = \frac{f_{clk}}{CPI} \cdot n_{VAX} \qquad (27)$$

and measured in **relative MIPS**.

Despite this improvement, the usefulness of using MIPS is very limited since the performance depends strongly on the operating system, the compiler, and the entire system configuration – not only on the processor. Especially, MIPS does not take into account if a program is heavily using floating-point number operations or not.

To account for applications with a high load of floating-point operations, the unit **FLOPS** (floating-point operations per second) has been introduced. Because this convention relates to complete operations instead of instructions (which can be different fractions of an operation, depending on the architecture, etc.), it tends to be more meaningful for a comparison between machines. Since the execution time depends on the *type* of operation (a division may take approx. four times longer than a multiplication), the FLOPS are usually normalized with respect to the type of operation. Still, there are considerable differences in the implementation of floating-point units in different systems. Furthermore, commercial FLOPS statements of system manufacturers are always peak performances based on simple, optimized programs. In real applications, these performances will hardly be reached. The convention is only useful for a rough comparison between machines using the same program provided that differences in the compilers (etc.) have been analyzed and are considered. Often, FLOPS are used to state and compare the peak performance of multiple-processor super-computers. Figure 35 shows the development of FLOPS performances for selected processors and super-computers [66] – [70].

In order to arrive at a comparison of machines, which is more meaningful for real application program, **benchmark** tests are used. A benchmark is an application program, which is run on different machines to measure the execution time. Since there are extremely different application programs, typically benchmark *suites* are employed. These are collections of benchmarks which try weighted mixtures to represent the different requirements of an average user of a general purpose computer. Due to shifts of users customs and the general development of the computer systems area, benchmark suites are revised approx. every four years. Figure 36 shows the SPEC CPU2000 benchmark suite and gives a flavor of the type of application programs considered [70].

In all performance tests, despite intentions to judge processor performances, it is not only the processor capability which counts. Rather, it is the entire computer system, especially the size and access rate of memory, as well as the width of the system buses.

III Logic Devices

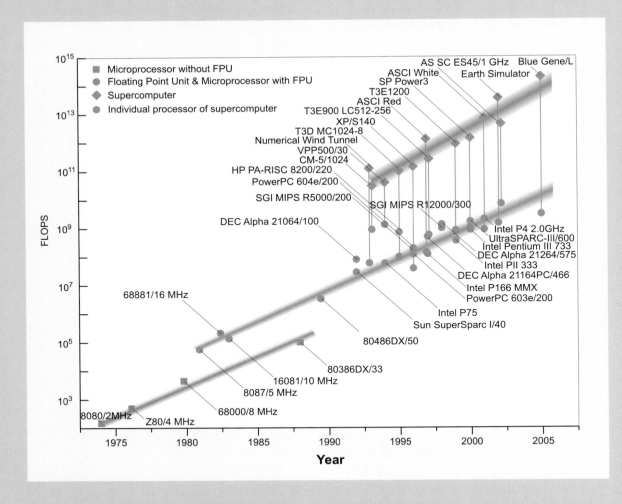

Figure 35: Development of FLOPS performance. (■) Microprocessors without floating-point unit, (●) microprocessor with external or integrated floating-point units (FPU), (◆) Super-computers with the number of processors involved, (●) super-computer performances divided by the number of processors involved.

5.3 Processing Capability of Biological Neurons

For a comparison of architectures, we will briefly discuss information processing capability of biological neurons. Obviously we cannot go into any detail of this topic which has been intensely studied for approx. hundred years – and is still not completely understood. The interested reader is referred to excellent textbooks such as [72], [73].

We use the threshold-gate as a starting point (Sec. 1.3). Compared to real biological neurons (Chap. 6), the threshold gate model is highly simplified mainly because

- its output is just binary
- it contains no temporal information (the threshold gate is a combinational circuit, e.g. time is no variable)
- it considers no noise and stochastic process.

In the following, we will complement the threshold gate model and introduce the fundamentals of more realistic models of biological neurons. The most obvious adaptation required concerns the nature of the output signal. Without further improvement of the resolution, a "1" state at the output of a biological neuron is represented by a train of action potential spikes, which are sent down the axon at maximum firing rate. This rate is determined by the spike length and the (minimum) refractory period between the spikes. Maximum fixing rates, f_{max}, are typically $\approx$ 500 Hz. The "0" state corresponds to an output signal without any spikes. Of course, there is no additional information in the *amplitude* and the *shape* of the action potential spike function, since all spikes of a given neuron look alike and are solely determined by the generation mechanism based on interacting ion channels. For this reason, the real spike is substituted by Dirac delta functions

$$\delta_i (t - t_i) \tag{28}$$

where t_i is the instant of generation of spike i, in all models which aim at the information processing of neurons.

In a first model refinement, the simple binary threshold-function h is replaced by an activation function g (Figure 36) which comprises the threshold for starting the firing as well as the maximum firing rate and, in addition, a strongly non-linear (but not simply step-like) function in which the firing rate increases if the membrane potential V_m exceeds the threshold voltage V_{th}. In the rising part of $g(V_m)$, before f_{max} is reached, the temporal distance between spikes encodes additional information.

This statement does not say, if the information is encoded in the timing of the discrete pulses or in a continuous firing rate. In fact, in the literature there are two groups of more detailed models, one are the **firing rate code models** and the other are the **spike code models**. There is a large amount of experimental data obtained in neural response experiments to sensory stimulation of animals which supports the former view and suggests that the average firing rate $<f(t)>$ of a neuron constitutes the major variable in the sensory periphery (see e.g. Refs. [74] – [77]). In this definition, the rate is estimated by averaging over a window ΔT that is large enough to obtain statistical information

$$\langle f(t) \rangle = \frac{1}{\Delta T} \int_{t}^{t+\Delta T} \sum \delta_i(t' - t_i) dt' \qquad (29)$$

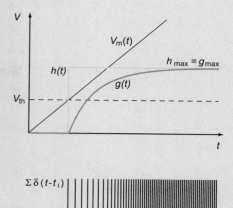

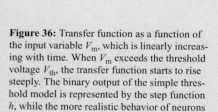

Figure 36: Transfer function as a function of the input variable V_m, which is linearly increasing with time. When V_m exceeds the threshold voltage V_{th}, the transfer function starts to rise steeply. The binary output of the simple threshold model is represented by the step function h, while the more realistic behavior of neurons is described by g. At high input values, g reaches a saturation value which is identical to that of the h function, i. e. $h_{max} = g_{max}$. Adapted from [72], [73] with modifications.

where the sum represents the train of Dirac delta functions. Experimentally, $<f(t)>$ is obtained by averaging, in addition, over a large number of identical trials. In conventional rate models, ΔT has been set to, for example, 100 ms or 1 s.

On the other hand, the fast reaction, of a fly, for example, is in the range of 30 to 40 ms which includes e.g. visual detection of an approaching large object (e.g. a hand …), analysis, and reaction (abrupt change of the flight direction). This is not compatible with long averaging times [81]. A number of similar experiments with different animals indicated that instantaneous firing rates averaged over periods of less than 10 ms appear to be relevant for neural information processing. These observations lead to models based on the "instantaneous" firing rate, for which ΔT is set to a value < 10 ms. For these small times, however, stating a "rate" is very problematic and should be replaced by a more complex strategy.

Still, in any firing code (instantaneous or long-term averaged), the timing of individual spikes in the spike train is not taken in account – it is only the rate which is considered to carry information. There is experimental evidence, however, that some additional information is contained in the correlation of the timing in a train of consecutive spikes (see survey in [82]).

Fast reaction times, correlation effects, and other findings lead to spike code models (see e. g. Refs. [78] – [81]). The most prominent representatives are the **integrate-and-fire models**, which will be used as the basis of the following discussion. Figure 37 summarizes the relevant aspects.

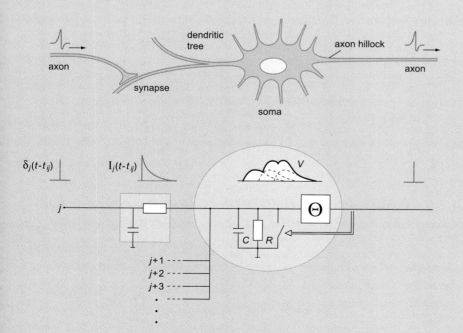

Figure 37: Sketch of the signal path through a real neuron (top) and a corresponding integrate-and-fire model of a neuron. The model is a combination from Refs. [72], [73] with modifications. A delta-type input signal arrives from the presynaptic neuron. The passive signal transfer through the dendrite is modelled by a RC element. This is a great simplification, of course. A more realistic model would be a distribution of RC elements which show a non-linear characteristic, in addition. The input currents from different dendrites charge the leaky capacitor C. If the voltage across C reaches the threshold voltage, a delta-type spike is generated and send into the axon on the right. The effect of an individual synaptic input on a postsynaptic neuron is in the range of 1 % to 5 % of the firing threshold, with important exceptions for specific neuron types [91].

An incoming presynaptic action potential spike leads to the release of neurotransmitter from vesicles. This takes place at release sites in the membrane of the presynaptic terminal to which the vesicles fuse to release the transmitter. The probability p of vesicle release at a given site is between 0 (no release) and 1 (release). Typically, $p \approx 0.1$ is an realistic assumption. Since the probability of vesicle release at one site in the membrane is independent of the release at other sites, the overall amount of neurotransmitter emitted into the synaptic cleft is given by the product $m = n \cdot p$. The number n of release sites varies strongly between the type of neuron. It is in the order of 1000 for motoneurons, and, hence, $m \approx 1000 \cdot 0.1 \approx 100$ reveals a high reliability in the muscle activation, e.g. in a frog [83]. In contrast, synaptic terminals of central neurons, e.g. in cortical and hippocampal regions of the brain of vertebrates show much smaller numbers: $n \approx 1 \ldots 10$. Thus, at most **one** vesicle is released, making this synapses quite unreliable, binary connections [85].

The neurotransmitter release leads to a receptor potential (EPSP or IPSP, see Chap. 6) which can be modeled by a current $I_j(t - t_{ij})$ determined by passive RC elements as shown in Figure 37. In addition, there are active conductances. The distance of the synapse to the axon hillock (among other characteristics) determines the time of arrival and the weight of the input signal. It is worthwhile noting, that the weight is determined by the membrane potential and the synaptic Ca^{2+} ionic conductances. Taking the noise into account (see below), the resolution of a synaptic weight is between 1 bit and up to approx. 6 bit, adapting at very different time scales, from milliseconds to seconds and longer [72]. The currents I_j arriving from a number of different presynaptic excitations (and inhibitions) in the dendritic tree are charging a capacitor C, which is leaky due to the resistor R. If the voltage V exceeds the threshold V_{th}, the transfer element Θ releases a delta signal output and resets (discharges) the capacitor. The gain function g (see Figure 36) is built into the characteristics of Θ.

Except in the axon, the information processing of the neuron is performed by continuous, analog signals (currents and voltages) [84]. Since the information content of analog signals is limited by noise, we will briefly mention the sources of **noise** and **stochastic processes** in neurons. One significant source is the *spontaneous* (i.e. not spike-triggered) *vesicle release* at the presynaptic junction. The spontaneous rate can be as high a few tens per second [86]. All ion channels are noisy and, hence, noise is introduced while the input signal is traveling down the dendritic tree. At the trigger point, the threshold voltage will show fluctuations and, thus, add noise in the firing process and the transfer function g. Furthermore, most neurons are spontaneously active, spiking at random intervals in the absence of input. Different neuron types show different spontaneous rates, ranging from few spikes per second to approx. 50 s^{-1} [91]. It should be mentioned, that noise does not only have seemingly disadvantageous effects such as limiting the resolution but, on the other hand, may assist (large) neural networks to escape from local minima (of calculation) during processing.

In neurophysiological experiments with living animals, a highly stochastic or random response is found if the signal of an (arbitrarily selected) cortical neuron is recorded upon repeatedly applied identical simuli [72]. Two opposing interpretations are discussed in the literature. One view regards the stochastic nature of the response signals as further support for the firing rate models, because averaging is required, to extract information from the responses. Another view considers the possibility that cortical neurons act as coincidence detectors, processing additional information by the correlation of signals between different neurons that can not be found by observing single neurons. Today, it is not yet clear to what extent these possible correlations affect the information processing in neural systems. Studies which analyze the influence of noise and the synaptic unreliability on information transfer [87],[85] result in an information content in the range from < 0.1 bits/spike [85] to approx. 3 bits/spike [81],[88]. The lower value is due to the interrelationship of spikes in a train, the higher value because of the information encoded in the temporal distance.

5.4 Performance Estimation for the Human Brain

The human brain is an organ which weights about 1.4 kg and can be subdivided into three major parts (Figure 38):
- forebrain (cortex, basal ganglia, hippocampus, thalamus) which consists of the two hemispheres (left and right) and accounts for approx. 80% of the brain's weight,
- the midbrain, and
- the hindbrain (pons, cerebellum, medulla).

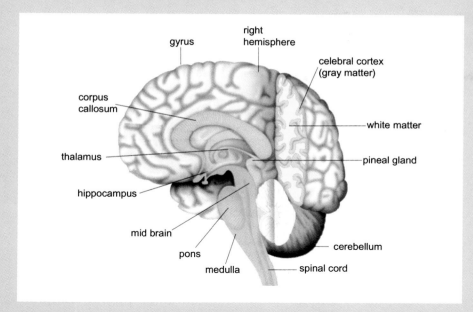

Figure 38: Parts of the human brain (from Ref. [90], with modifications).

In the following, we will list few facts which may be of interest for the discussion of the information processing capability of the cerebrum. For further details see e.g. Refs. [89],[91].

The cortex consists of an inner part, the white matter, and a highly convoluted outer part of grey matter. This grey matter is made of a high density of non-myelinated neurons (interneurons and others), while the white matter mainly consists of myelinated axons of projection neurons interconnecting the left and right hemisphere and other parts of the brain. The major bridge between the two hemispheres is called *corpus callosum* and contains (at least) $2 \cdot 10^8$ axons which pass through a cross section of approx 6 cm^2 in the center, indicating an average axon cross section of 3 µm^2. The total surface area of the cerebral cortex is approx. 1800 ± 200 cm^2 showing a thickness of approx. 2 to 3 mm. The number of neurons in the complete brain is estimated to be approx. 10^{11}, of which approx $2 \cdot 10^{10}$ are located in the **celebral cortex**. From this data we calculate a neuronal density of $4 \cdot 10^7$ neurons/cm^3. This corresponds to an average volume of $2.5 \cdot 10^4$ µm^3 per neuron, and an average linear distance of approx. 30 µm between neurons. Every neuron has between some thousands and a few ten thousands synpases. If we assume an average value of 10^4 synapses per neuron, we obtain an average volume of 2.5 µm^3 per synapse. Due to the high number of synapses per neutron and the distribution of the connections, each neuron is connected to every other neuron in the cortex via not more than four neurons, on the average. Compared to CMOS technology, which essentially realizes the transistor functions in an area, the cortex arrives at its enormous numbers of functional units at a relatively large distance between the units, because of the 3-D packing in cortical layer.

In a burst of action potentials, the firing rate is in the range of several 100 s^{-1}. On the average, approx. 1 % of the neurons are active at any one time [95], somewhat depending on state of the organism (sleep, awakeness, awareness, action). In order to obtain (purely *formal*!) numbers of the information processing capabilities of the cortex, one needs to multiply these rates by the approx. $2 \cdot 10^{10} \cdot 10^4 = 2 \cdot 10^{14}$ synapses and the number of bits transferred per spike (see Sec. 5.3). The formal result of the calculation must be corrected due to the *redundancies* in the processes and for what could be called the *control and glue logic* (see Sec. 4.1) within the cortical neural network, which could easily lead to a decrease by some orders of magnitude. The actual value is not known yet. Correlation effects, on the other hand, can lead to an increase of the processing efficiency. In any case, these numbers have to be regarded as upper limits for the formal processing rate. It is not at all accounting, for example, for the semantics of the processed information and its redundancy.

A direct comparison of the processing capabilities of a cortical neural network and a supercomputer is not very reasonable because of the very different architectures. To summarize some aspects of this difference and its implication: Supercomputers are based on a very coarse-grained parallel architecture using high-speed CMOS processors and the traditional separation of computing and memory function in the grains. They are optimized to reach extremely high FLOPS values, which are desirable for simulation

III Logic Devices

tasks, for example. In contrast, the very fine-grained neural network architecture of the cortex is built from stochastic elements, shows an integration of computing and memory functions on the microscopic scale, exhibits a constant reprogramming of the entire system, and has many characteristics of analog computing, using a menu of linear and non-linear operations [72]. This optimizes cortical neural networks for fast recognition and interpretation of complex pattern, non-linear associations, and all the features which we attribute to consciousness. On the other hand, the formally huge processing capability and memory can not be accessed randomly (as everybody knows from daily experience). In addition, it is known from detailed studies that only a tiny fraction of information can be transferred reliably from our ultra-short term and short-term memories into our long-term memory. The latter cannot be filled at rates larger than 0.2 bit/s to 2 bits/s, independent of the type of information (text, sound, images, etc.) [92]. This means that we are not able to store more than approx. 20 to 200 MB of information in our long term memory during an average life of 75 years, even if we were learning constantly at highest speed during our awake periods.

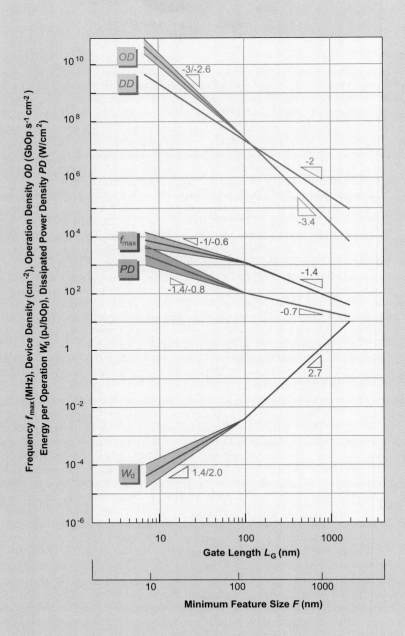

Figure 39: Operation density OD, device density DD, and density of the dissipated power PD for physically optimized, low-power CMOS circuits versus minimum feature size F as derived from the trends of the frequency f and energy per operation W_d shown in Figure 10. Note: For $F \lesssim 100$ nm, these are formal trends as no limit for PD is assumed (see text).

6 Ultimate Computation

6.1 Power Dissipation Limit

In Section 2, we considered the thermodynamic limit of irreversible computation and the trends of CMOS technology with respect to the minimum feature size F. However, we only discussed the features of individual circuits, neglecting their density on a chip. The device density DD is inversely proportional to F^2, where the proportionality factor X_{aa} is the average area which a logic gate occupies on the chip (measured in units of F^2):

$$DD = \frac{1}{X_{aa} F^2} \tag{30}$$

X_{aa} typically is in the order of 300 (e.g. for a NAND gate). From DD and f_{max}, one calculates the performance of a chip which will be called here the operation density

$$OD = DD \cdot f_{max} \tag{31}$$

measured in bOp/(s·cm^2). The density of dissipated power PD is given by

$$PD = DD \cdot P_d \tag{32}$$

Figure 39 uses the scaling of the energy per operation W_d, the frequency f_{max}, and the dissipated power per gate P_d as sketched in Figure 10, to reveal the scaling of OD and PD.

In the case of highly integrated circuits on large die sizes (chip sizes), the removal of the dissipated heat establishes a serious problem. A power density PD of approx. 100 W/cm^2 can be regarded as a reasonable limit, given by the thermal conductivity of materials and the geometry for setting up temperature gradients. This PD value is ten times above a common kitchen heating plate at full power. To realize larger PD values for Si-based chips, very sophisticated (and, hence, expensive) cooling arrangements are required, to be integrated into the chip package.

We will consider physically-optimized low-power CMOS circuits again and resume the discussion of Section 2. The entire development was driven by the technological reduction of the minimum feature size F, as sketched in Figure 40a (and Figure 41, regime I). This determined (1) the saturation current of the FETs I_{sat} through the device dimensions and material properties, (2) in combination with system considerations, it determined the operation voltage V_{DD}, and (3) it determines the average load capacitance C_L of a circuit, composed of the output capacitance of a gate, C_{out}, the capacitance of the interconnect to the subsequent logic gate, C_{con}, and the input capacitance of this subsequent gate, C_{in} (given by the gate capacitance of the MOSFETs). The circuit delay time t_d and, hence, the maximum operation frequency of the circuit f_{max} (see Eq. (10)) are settled by C_L and the ratio V_{DD}/I_{sat}. The energy per binary operation of a device W_d is calculated from C_L and V_{DD}, according to the combination of Eqs. (9) and (11). The operation frequency f and the energy W_d give the power P_d dissipated by one logic gate of the circuit. The device density DD, the operation density OD, and the power dissipation density PD are obtained through Eq. (30) to (32). As all other parameters of the logic circuit discussed here, PD is derived from the feature size F in this regime (F limited regime).

We have shown that PD increased with decreasing feature size in the range from $F = 1000$ nm to 100 nm. Now, as soon as PD reaches the ultimate limit for heat dissipation, 100 W/cm^2, it is supposed to grow no more. This changes the entire strategy of circuit optimization. Now, besides F also PD determines the other circuit parameters (Figure 40b). In this regime, which we call the PD controlled regime, the maximum operation density OD is *not* determined anymore by the device density and the frequency, but by the power density PD and energy per operation W_d:

$$OD = PD / W_d \tag{33}$$

The frequency turns out to be *dependent* variable now, determined by OD and DD:

$$f_{max} = OD / DD \tag{34}$$

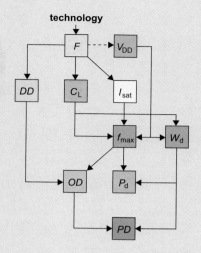
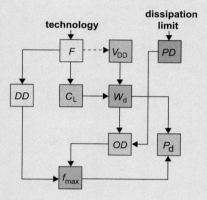

Figure 40: Interdependencies of the circuit parameters
(a) in the feature size controlled regime and
(b) in the power dissipation controlled regime.

III Logic Devices

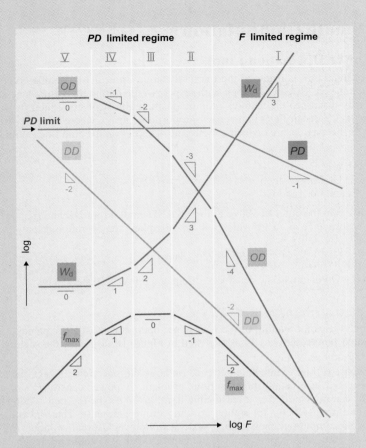

Figure 41: Illustration of the circuit parameters versus the feature size F. Regime I is the F controlled regime, regimes II to V are controlled by the limit of dissipated power density PD. In all regimes, a scaling of the device density DD according to the available area is assumed (i.e. $DD \propto F^2$).
Regime I: All characteristics are directly or indirectly determined by the min. feature size F. The scaling characteristics is a coarse approximation to the evolution of the CMOS technology in the $F=100$ regime (see Figure 40).
Regime II: Since PD is limited, the frequency f_{max} can only grow as much as allowed by the scaling of W_d and of DD.
Regime III: If the scaling of W_d is weaker (e.g. $\propto F^2$), then f_{max} may not scale at all anymore.
Regime IV: If the scaling of W_d is further reduced (e.g. $\propto F^1$), then f_{max} must drop with decreasing F, to meet a constant PD.
Regime V: if W_d shows no scaling, f must scale $\propto F^2$ to compensate for $DD \propto F^2$.
On the route from regime I to V, scaling of the operation density OD chances dramatically.

This makes sense, because when PD is fixed and DD is determined by the feature size, the operation frequency must be set to a value which guarantees, that PD is not exceeded. There are other remarkable effects (Figure 41): If the switching energy W_d shows a dependency on F which is less than F^2, then the operation frequency needs to be reduced with decreasing F (regimes IV and V in Figure 41). This is the opposite trend to what was observed in the range from $F = 1000$ nm to 100 nm for the conventional CMOS technology. If W_d is not dependent on F, the operation density OD is constant too (regime V). Since the device density may still grow with F^2 upon decreasing F, the frequenc must be reduced strongly (i.e. $\propto F^2$) to keep the dissipated power density constant. It must be emphasized, that this description shows the basic interrelationships, and *not* the projected evolution of CMOS technology or any other technology. Nevertheless, *any* technology needs to operate within the described parameters.

6.2 Dissipation in Reversible Computation

If we have stored a bit using an electron in an (ideal) double-well electrostatic potential (Figure 42), then ΔW will be required to switch the state, i.e. to take the electron from the left well to the right well. In Sec. 3 of the General Introduction, we have seen that ΔW must be at least $kT\ln 2$ to prevent thermal energy from destroying the stored information by letting the electron change the well arbitrarily. Now we will take a closer look at the switching process. To start a controlled switching, ΔW has to be fed into the system to move the electron from the left into the saddle point. This energy ΔW will, however, be gained back when the electron moves down into the right valley. The energy is then, in principle, available to perform a switch in any adjacent double-well. In that manner, ΔW remains available for subsequent switching processes. Alternatively, ΔW can be used to switch the system back into its original state. The forward-switching process has no preference over the backward-switching process. This is the nature of reversible computing. However, if there is no preference in the direction of computing the net speed of the computational process is zero, i.e. a calculation would never be finalized.

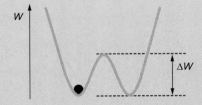

Figure 42: Double well potential to illustrate the two possible states of an electron and the energy barrier in between which must be overcome for a switch operation.

This is the same as for any other process in thermodynamics: a reversible conduction of the process requires an infinitely slow process rate. In order to have a finite process rate of computation in forward direction, the forward rate f must be higher than the backward rate b, i. e. $f > b$.

As has been explained in details by Feynman [93], the (minimum) energy cost for a computation at rate $r = f - b$ is

$$kT \ln \frac{f-b}{(f+b)/2} \qquad (35)$$

For large forward rates, $f \gg b$, Eq. (35) approaches the limit for irreversible computation, $kT \ln 2$. Obviously, there is a continuous transition between the ideal reversible computation without energy dissipation, but no net computation rate, and the ideal irreversible computation with a (minimum) energy dissipation of $kT \ln 2$ per bit and maximum forward computation rate.

It should be noted that, also for reversible computation, error correction during processing leads to the loss of information at the expense of increasing the reliability. This loss of information is accompanied by the dissipation of energy.

6.3 The Ultimate Computer

According to the brief explanations given in this introduction (and the more detailed descriptions in the literature, e. g. [61]), it appears that the ultimate information processing systems for general purpose exhibit the following features:

- homogeneous arrays, which are relatively fine grained
- parallelism at different hierarchical levels
- emphasis on local interconnects
- universal non-volatile memory
- defect and fault tolerance
- in addition: the systems should be small and light, cheap, fast, robust, work at room temperature.

Using some aspects of the physical limits to computation, architecture, and performance, we will sketch some specifications of an ultimate computer. These thoughts are inspired by an article of Lloyd [94]. In contrast to him, we will exclude nuclear conversion of matter into energy and stay at temperatures which are conceivable for mobile systems. In addition, we will restrict this consideration on irreversible computation, since for reversible computers there is a strong dependence on several features such as the degree of forward computation.

Let us assume that it might be possible once to realize a logic device which operates close to the thermodynamic limit. We may take $W_d = 10\, k_B T \ln 2 \cong 3 \cdot 10^{-20}$ J as a value for operation at room temperature, which guarantees a suitably reliable operation. For circuits at the power dissipation limit, this results in an operation density OD of approx. $3 \cdot 10^{21}$ bOp/(s·cm^2). Compared to a physically optimized low-power logic circuit based on a 100-nm CMOS technology, which may show $W_d \cong 3 \cdot 10^{-15}$ J (see Sec. 2) this is an improvement of five orders of magnitude. As indicated in Figure 25, general purpose processors require five more orders of magnitude larger energy to perform the same computational task. This additional energy requirement is partially due to larger transistors which are used to optimize the circuit for high speeds and to charge extended interconnect lines, common in processor architectures. This partly accounts for approx. one order of magnitude of the additional energy requirement. The other part of approx. four orders of magnitude is needed for the large control-and-glue logic of general purpose processors. This latter part cannot be recovered from progress in circuit technology. Potentially, the field programmable logic concept may present a compromise between a high operation density and a high flexibility, due to its reconfigurability.

Acknowledgements

First of all, the editor is most grateful to Tobias Noll (eecs, RWTH Aachen) for numerous suggestions and for so many hours of stimulating discussion. In this context, the editor would like to thank Holger Blume (eecs, RWTH Aachen) and Michael Gansen (eecs, RWTH Aachen) for simulations on the performance of CMOS circuits of different technologies and for a detailed comparison of various architectures. The following colleagues have been so kind as to look carefully through specific sections of this introduction: Hagen Klauk (Siemens AG) at the section on OFETs, Gernot Güntherodt (RWTH Aachen) at the section on spintronics, Michael Indlekofer (FZ Jülich) at the section on quantum cellular automata, Thomas Schäpers (FZ Jülich) at the section on quantum computing, and John Reif (Duke University) at the section on DNA computers. The editor is indebted to Christof Koch (Caltech) for a detailed review of the sections on neurons and the brain and for valuable suggestions, as well as to Andreas Engel (University of Hamburg) and Karl Zilles (FZ Jülich and University of Düsseldorf) for stimulating discussions. On the overall topic of this introduction, the editor had enjoyable and fruitful discussions with George Bourianoff (Intel Corp.), Mike Forshaw (University College London), Karl Goser (University of Dortmund), James Hutchby (Semiconductor Research Cooperation), Ramón Campano (European Commission), Stanley Williams (Hewlett Packard Laboratories) and Victor Zhirnov (Semiconductor Research Cooperation) which are highly acknowledged. The editor gratefully acknowledges Fotis Fitsilis (FZ Jülich), Yakoub Mustafa (RWTH Aachen), Jürgen Rickes (Agilent), Stephan Tiedke (aixACCT Systems) and René Meyer (RWTH Aachen) for comprehensive editorial and technical assistance in the compilation of this chapter.

References

[1] J. N. Warfield, *Introduction to Electronic Analog Computers*, Englewood Cliffs: Prentice Hall, 1959.

[2] C. Koch and I. Segev, *Nature Neuroscience* **3**, 1171 (2000).

[3] P. M. Lewis and C. L. Coates, *Threshold Logic*, John Wiley & Sons, New York, 1967.

[4] M. L. Dertouzos, *Threshold Logic: A Synthesis Approach*, The MIT Press, Cambridge Mass., 1965.

[5] C. Pacha, K. Goser, A. Brennemann, and W. Prost, Proc. ESSCIRC´98, 24th Eurpoean Solid-State Circuits conference, The Hague, September 1998.

[6] J.D. Meindl, Proc. of the IEEE **83**, 619 (1995).

[7] J. D. Meindl, Q. Chen, and J. A. Davis, Science **293**, 2044 (2001).

[8] J. Davis, J.D. Meindl, IEEE Trans. Electron Devices **47**, 2068 (2000).

[9] R.W. Keyes, Proc. of the IEEE **89**, 227 (2001).

[10] P. Hadley and J. E. Mooij, http://qt1.tn.tudelft.nl/publi/2000/quantumdev/qdevices.html .

[11] C. Shannon, Bell Syst. Tech. J. **27**, 379 (1948).

[12] S. Lloyd, Phys. Rev. A**39**, 5378 (1989).

[13] R. Landauer, IBM J. Res. Dev. **5**, 183 (1961).

[14] Y. Aharonov, D. Bohm, Phys. Rev. **122**, 1649 (1961).

[15] N. Margolus, L.B. Levithin, Physica D**120**, 188 (1998).

[16] R.K. Cavin, V.V. Zhirnov, J.A. Hutchby, and G.I. Bourianoff, Proceed. IEEE, scheduled for March 2003.

[17] R.K. Cavin, V.V. Zhirnov, J.A. Hutchby, and G.I. Bourianoff, submitted to Nanotechnology.

[18] B. Doyle, R. Arghavani, D. Barrge, S. Datta, M. Doczy, J. Kavalieros, A. Murthy, R. Chau, Intel Technology Journal **6**, 42 (2002).

[19] H. Blume, H. Feldkämper, and T.G. Noll, J. VLSI Signal Processing (in press).

[20] E.J. Nowak, IBM J. Res. & Dev. **46**, No. 2/3, 169 (2002).

[21] I. E. Sutherland and J. Ebergen, Scientific American, **287**, no. 2, 46 (2002).

[22] J.A. Davis, R. Venkatesan, A. Kaloyeros et al., Proc. of the IEEE **89**, 305 (2001).

[23] H. B. Bakoglu, *Circuits, Interconnections, and Packaging for VLSI*, Addison-Wesley Publ. Co., 1990.

[24] R. Compano (ed.), *Technology Roadmap for Nanoelectronics 2000*, European Commission IST Programme – Future and Emerging Technologies, Belgium, 2001.

[25] The International Technology Roadmap for Semiconductors (ITRS) 2001, Semiconductor Industry Association; http://public.itrs.net/ .

[26] J. A. Hutchby, G. I. Bourianoff, V. V. Zhirnov, and J. E. Brewer, IEEE Circuits and Devices Magazine, **18**, 28 (2002).

[27] R. Ronen, A. Mendelson, K. Lai, S.-L. Lu, F. Pollack, and J. P. Shen, Proc. IEEE, **89**, 325 (2001).

[28] K. L. Wang, J. Nanosci. Nanotechn., **2**, 235 (2002).

[29] S. C. Goldstein, Proc. IEEE Computer Society Workshop on VLSI 2001 (WVLSI '01).

[30] C. D. Dimitrakopoulos and P. R. L. Malenfant, Advanced Materials **14**, 99 (2002).

[31] H. Klauk and T. N. Jackson, Solid-State Technology, **43**, 63 (2000).

[32] E. Cantatore, Proc. SAFE/IEEE Workshop 2000, 27 (2000).

[33] K. Kudo, M. Iizuka, S. Kuniyoshi, and K. Tanaka, Thin Solid Films **393**, 362 (2001).

[34] H. Klauk, M. Halik, U. Zschieschang, G. Schmid, W. Radlik, R. Brederlow, S. Briole, C. Pacha, R. Thewes, and W. Weber, Technical Digest of the International Electron Devices Meeting, 2002.

[35] B. K. Crone, A. Dodabalapur, R. Sarpeshkar, R. W. Filas, Y. Y. Lin, Z. Bao, J. H. O'Neill, W. Li, and H. E. Katz, J. Appl. Phys., **89**, 5125 (2001).

[36] E. A. Silinsh and V. Cgapek, *Organic Molecular Crystals*, AIP, New York 1994.

[37] K. K. Likharev, IBM Journal of Research and Development, **32**, 144 (1988).

[38] K. Yano, T. Ishii, T. Hasimoto, T. Kobayashi, F. Murai, and K. Seki, Appl. Phys. Lett. **67**, 828 (1995).

[39] S. Datta and B. Das, Appl. Phys. Lett. **56**, 665 (1990).

[40] B. Beschoten, E. Johnston-Halperin, D. K. Young, M. Poggio, J. E. Grimaldi, S. Keller, S. P. DenBaars, U. K. Mishra, E. L. Hu, and D. D. Awschalom, Phys. Rev. B **63**, R121202 (2001).

[41] H. J. Zhu, M. Ramsteiner, H. Kostial, M. Wassermeier, H. P. Schönherr, and K. H. Ploog, Phys. Rev. Lett. **87**, 016601 (2001).

[42] A.T. Hanbicki, B. T. Jonker, G. Itskos, G. Kioseoglou, and A. Petrou, APL **80**, 1240 (2002).

[43] S. A. Wolf, D. D. Awschalom, R. A. Buhrman, J. M. Dauhton, S. von Molnár, M. L. Roukes, A. Y. Chtchelkanova, and D. M. Treger, Science **294**, 1488 (2001).

[44] D. D. Awschalom, M. E. Flatté, and N. Samarth, Scientific American **286**, 52 (2002).

[45] P. D. Tougaw, C. S. Lent, and W. Porod, J. Appl. Phys. **74**, 3558 (1993).

[46] V. I. Varshavsky, Proc. IEEE Comput. Soc. Press, **vii+147**, 134 (1996).

[47] K. Goser and C. Pacha, Proc. ESSCIRC'98, 24th European Solid-State Circuits Conference, The Hague, September 1998.

[48] I. Amlani, A. O. Orlov, G. Toth, G. H. Bernstein, C. S. Lent, G. L. Snider, Science **284**, 289 (1999).

[49] C. Single, R. Augke, F. E: Prins, D. A. Wharam, D. P. Kern, Semicond. Sci. and Technol. **14**, 1165 (1999).

[50] R. P. Cowburn and M. E. Welland, Science **287**, 1466 (2000).

[51] A. J. Heinrich, C. P. Lutz, J. A. Gupta, and D. M. Eigler, Science **298**, 1381 (2002).

[52] L. M. Adleman, Science **266**, 1021 (1994).

[53] L. M. Adleman, Scientific American **279**, 34 (1998).

[54] R. S. Braich, N. Chelyapov, C. Johnson, P. W. K. Rothemund, L. Adleman, Science **296**, 499 (2002).

[55] J. H. Reif, Science **296**, 478 (2002).

[56] V. George and J. M. Rabaey, *Low-Energy FPGAs – Architecture and Design*, Kluwer Academic Publishers, 2001.

[57] J. R. Heath, P. J. Kuekes, G. S. Snider, and R. S. Williams, Science **280**, 1716 (1998).

[58] K.F. Goser, *Von der Mikroelektronik zur Nanoelektronik* (in Ger.), Lecture Notes, Universität Dortmund, 2001.

[59] K. F. Goser, C. Pacha, A. Kanstein, and M. L. Rossmann, Proc. IEEE **85**, 558 (1997).

[60] M. J. Flynn, Proc. IEEE **54**, 1901 (1966).

[61] P. Beckett and A. Jennings, Proc. 7th Asia-Pacific Computer Systems Architecture Conference (ACSAC´2002), Melbourne, Australia.

[62] M. R. B. Forshaw, K. Nikolic, and A. Sadek, 3rd Annual report, Autonomous Nanoelectronic Systems With Extended Replication and Signalling, Tech. Rep., University College London, Image Processing group, London, U. K.; http://ipga.phys.ucl.ac.uk/research/answers/reports/3rd_year_UCL.pdf .

[63] http://www.hpl.hp.com/personal/Bruce_Culbertson/tm003.jpg

[64] J. F. Wakerly, *Digital Design*, 3rd ed., Prentice Hall, Upper Saddle River, 2000.

[65] J. L. Hennessy and D. A. Patterson, *Computer Architecture – A Quantitative Approach*, 3rd edition, Morgan Kaufmann Publishers, San Francisco, 2003.

[66] R. Vogt and R. Waser, Elektronik **20/83**, 85 (1983).

[67] http://thproxy.jinr.ru/file-archive/doc/specmarks/specmark.html

[68] http://performance.netlib.org/performance/html/PDStop.html

[69] www.top500.org

[70] http://www.spec.org

[71] W. S. McCulloch and W. Pitts, Bull. Math. Biophys. **5**, 115 (1943).

[72] C. Koch, *Biophysics of Computation*, Oxford University Press, 1999.

[73] W. Maas and C. M. Bishop (eds.), *Pulsed Neural Networks*, MIT Press, Cambridge Mass., 1998.

[74] E. D. Adrian, *The Basis of Sensation: The Action of the Sense Organs*, W. W. Norton, New York, 1928.

[75] J. P. Lettvin, H. R. Maturana, W. S. McCulloch, and W. H. Pitts, Proc. Inst. Rad. Eng. **47**, 1940 (1959).

[76] H. B. Barlow, Perception **1**, 371 (1972).

[77] W. T. Newsome, K. H. Britten, and J. A. Movshon, Nature **341**, 52 (1989).

[78] D. H. Perkel and T. H. Bullock, Neurosci. Res. Prog. Sum. **3**, 405 (1968).

[79] W. Bialek, F. Rieke, R. R. D. van Steveninck, and D. Warland, Science **252**, 1854 (1991).

[80] C. M. Gray and W. Singer, Proc. Natl. Acad. Sci. USA, **86**, 1698 (1989).

[81] F. Rieke, D. Warland, R. R. D. Stevenick, and W. Bialek, *Spikes: Exploring the Neural Code*, MIT Press, Cambridge Mass., 1996.

[82] B. J. Richmond and L. M. Optican, in: *Neural Networks for Perception*, H. Wechsel (ed.), 104-119, Academic Press, Boston, 1992.

[83] B. Katz, *The Release of Neurotransmitter Substances*, Liverpool University Press, 1969.

[84] More detailed studies show, that spikes propagate back from the axonal hillock into the dendritic tree, which carry a handshake-like signal and are probably critical for telling a synapse that "its" neuron just fired. Thus, this feedback signal internal to the neuron might be critical for synaptic learning [2].

[85] A. Manwani and C. Koch, Neural Computation **13**, 1-33 (2000).

[86] E. G. Strunsky, M. D. Borisover, E. E. Nikolsky, and F. Vyskocil, Neurochem. Res. **26**, 891 (2001).

[87] A. Zador, J. Neurophysiol. **79**, 1219 (1998).

[88] R. Wessel, C. Koch, and Gabbiani, J. Neurophysiol. **75**, 2280 (1996).

[89] K. Zilles, in: *The Human Nervous System*, George Paxinos and Jürgen Mai (eds.), Academic Press, San Diego, 2002 (in press).

[90] *Das visuelle Lexikon (*in Ger.), Gerstenberg Verlag, 1998.

[91] P. S. Churchland and Sejnowski, *The Computational Brain,* MIT Press, 1992.

[92] G. Raisbeck, *Information Theory*, MIT Press 1963.

[93] R. P. Feynman, *Feynman Lectures on Computation*, reprint with corrections, Perseus Publishing, Cambridge, MA, 1999.

[94] S. Lloyd, Nature **406**, 1047 (2000).

[95] S. B. Laughlin and T. J. Sejnowski, Science **301**, 1870 (2003); S. B. Laughlin, R. R. de Ruyter van Steveninck, and J. C. Anderson, Nature Neuroscience **1**, 36 (1998).

Silicon MOSFETs – Novel Materials and Alternative Concepts

Edward W. A. Young
International Sematech, Philips Semiconductors, p/a IMEC, Leuven, Belgium

Siegfried Mantl, *Department ISG, Research Center Jülich, Germany*

Peter B. Griffin, *SNF, Stanford University, USA*

Contents

1	**Introduction**	359
2	**Fundamentals of MOSFET Devices**	360
2.1	MOS Capacitor	360
2.2	MOSFET	362
3	**Scaling Rules**	364
4	**Silicon-Dioxide Based Gate Dielectrics**	367
4.1	High-K Materials for CMOS	368
4.2	Dielectric Properties	368
4.3	Thermodynamics	369
4.4	Electronic Properties	369
4.5	Microstructural Stability	370
4.6	High-K Deposition Tools and Chemistry	371
4.7	Process Compatibility	372
4.8	Examples of High-K Gate Stacks	373
5	**Metal Gates**	374
5.1	Polysilicon vs. Metal Gates	374
5.2	Metal Gate Materials Selection	375
6	**Junctions and Contacts**	375
6.1	Shallow Junctions	375
6.2	Junction Contacts	377
7	**Advanced MOSFETs Concepts**	378
8	**Summary**	381

Silicon MOSFETs – Novel Materials and Alternative Concepts

1 Introduction

The first transistor was built at Bell Laboratories by John Bardeen and Walter Brattain in Shockley's group [1], [2]. Figure 1 shows a picture of the original point contact transistor [2] in comparison with a transmission electron microscopy (TEM) cross-sectional micrograph of a research metal oxide semiconductor field effect transistor (MOSFET) made by INTEL in 2002 [3]. The point contact transistor consisted of a piece of n-type Ge several mm in size on a metal plate which served as the gate contact and two spring loaded top contacts, called emitter and collector. These contacts were made by gold foils attached to the two sides of the triangle. Biasing the back contact produced an inversion layer in the n-Ge on the top side and a hole current was observed between the emitter and negatively biased collector. The discovery of the transistor effect is described in Ref. [2]. In contrast, the research MOSFET of Figure 1 has a gate length of only 10 nm. Source, drain, and gate are fabricated on single crystalline silicon. The gate is made of polysilicon with a width of only 10 nm and a height of 50 nm. The central region of the transistor, consisting of doped silicon, the thin silicon dioxide layer and the silicided polysilicon contact, forms a MOS capacitor, which plays a key role in the performance of the transistor.

Many new materials have already been introduced in integrated circuits and many more will be needed for the next CMOS generations as illustrated in Figure 2. For example, Al has been replaced by Cu for the multilevel interconnects, which are nowadays embedded in dielectrics with low permittivity (called low-K materials), such as porous oxides. Various silicides have been introduced such as source, gate and drain contacts to lower the device resistance. $TiSi_2$, has mostly been replaced by $CoSi_2$, which maintains excellent electrical conductivity even at wire widths below 100 nm. Due to the continuous down-scaling of the device dimensions the thickness of the silicon dioxide gate dielectric is approaching the limit where direct tunnelling causes excessive gate leakage. Alternative gate dielectrics with higher permittivity (called high-K materials) will replace silicon dioxide and the polysilicon gates will be replaced by metal gates. For DRAM capacitors, high-K materials such as $(Ba, Sr)TiO_3$ (short: BST) are under consideration as well (Figure 2). In addition, strained silicon will be used for high performance devices. Parallel to the investigation of new materials alternative transistor concepts are under development.

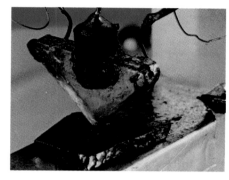

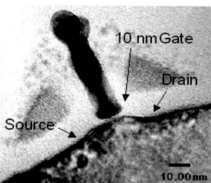

Figure 1: Photograph of the original point contact transistor in comparison with a TEM cross-section micrograph of a 10 nm gatelength research MOSFET [2], [3].

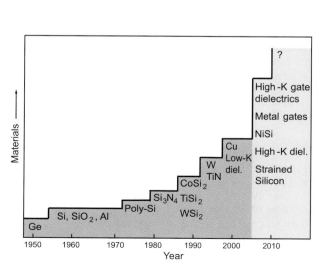

Figure 2: Implementation of new materials in CMOS processes [4].

III Logic Devices

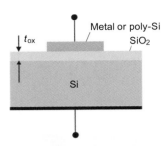

Figure 3: Schematic structure of a MOS capacitor.

As an introduction to this rapidly changing field, the following topics will be addressed:
- Fundamentals of MOSFET devices
- Scaling rules
- Silicon dioxide based gate dielectrics
- High-K materials for CMOS
- Metal gates
- Junctions and contacts
- Advanced MOSFETs concepts

2 Fundamentals of MOSFET Devices

2.1 MOS Capacitor

Figure 3 shows the structure of a MOS capacitor with the three components, the metal or polysilicon contact, the silicon dioxide with a thickness t_{ox} and the silicon. The corresponding band diagram is shown in Figure 4. Due to the 9 eV bandgap of the silicon dioxide and the large band offsets relative to the silicon, the potential barrier between the conduction band of the silicon and the silicon dioxide is large (≈ 3.1 eV). This barrier crucially controls possible charge transport through the dielectric layer in the presence of an applied voltage, and thus, determines the reliability of the dielectric-semiconductor interface. Frequently poly-Si is used as a contact material instead of a metal. For p-type poly-Si the work function is $\Phi_s = \chi + W_g/2q + \Psi_B \approx 4$ eV, where χ denotes the electron affinity, W_g the band gap energy, Ψ_B the difference between the Fermi potential W_F and the intrinsic potential W_i.

The energy band diagram of an ideal MOS capacitor with a p-type semiconductor is shown in Figure 5 ($q\Phi_{ms}$ is assumed to be zero, see Figure 4). When a negative gate potential $V_G < 0$ is applied the Fermi level of the metal increases and an electric field is created in the SiO$_2$, indicated by the slope of the conduction band of the SiO$_2$, and in the silicon. Because of the low carrier concentration the Si bands bend upwards at the SiO$_2$ interface, leading to an **accumulation** of excess holes. In order to conserve charge, an equivalent number of electrons is accumulated at the metal side of the MOS capacitor.

When a positive potential is applied at the gate contact, its Fermi level moves down leading to band bending in the silicon in the downward direction. As a consequence, the hole concentration near the interface decreases. This status is called the **depletion condition**. Charge neutrality requires the induction of an equivalent amount of positive charge at the metal-oxide interface Q_M as negative charge in the semiconductor Q_S, explicitly,

$$Q = -Q_M \quad \text{with} \quad Q_S = Q_d \tag{1}$$

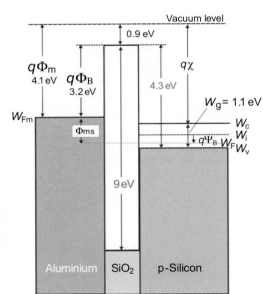

Figure 4: Energy-band diagram of the three components of a real MOS capacitor, consisting of an Al contact, silicon dioxide and p-type silicon. $q\Phi_m$ denotes the work function of the metal, $q\Phi_{ms}$ the workfunction difference of Al versus p-Si, χ the electron affinity of the silicon, W_g the band energy, W_c the conduction band, W_v the valence band of silicon, $q\Psi_B$ the difference between the intrinsic Fermi level W_i and the Fermi level W_F [5].

where Q_d originates from the ionized donor states. A further increase of the positive gate potential, enhances band bending such that at a certain gate potential the intrinsic Fermi level crosses the Fermi level as shown in Figure 5c. Energetically, it becomes now favourable for electrons to populate the newly created surface channel. The surface behaves like an n-type semiconductor where the doping was created by inverting the original p-type silicon with an applied field. This condition is called **weak inversion** and the corresponding onset gate voltage the threshold voltage V_T. The negative charge at the semiconductor interface Q_S consists of inversion charge Q_i (electrons) and ionized acceptors Q_d (Figure 5d)

$$Q = Q_i + Q_d \tag{2}$$

As indicated in Figure 5c, three regions develop within the semiconductor: a shallow inversion region, a depletion region with a maximum depth w_d and deeper in the substrate a neutral region. A further increase of the potential yields to **strong inversion** when the concentration of the electrons exceeds the hole concentration in the substrate ($Q_i > Q_d$). Then, the gate voltage V_G can be expressed by

$$V_G = V_{ox} + \psi_S = -\frac{Q_S}{C_{ox}} + \psi_S \tag{3}$$

where C_{ox} is the oxide capacitance per unit area and ψ_S is the surface potential, reflected by the band bending in Figure 5c. The surface potential and the total induced charge at the interface can be calculated by solving Poisson's equation with appropriate boundary conditions (see e.g. [6], [7]). Under extreme accumulation and inversion conditions, when V_G and V_{ox} are significantly larger than ψ_S, then Q_S can be approximated by

$$Q = -C_{ox} V_G, \quad \text{with} \quad C_{ox} = \frac{\varepsilon_{ox}}{t_{ox}} \quad (4)$$

since ψ_S is always less than W_g. Eq. (4) implies that the total induced charge at the interface increases with the gate capacitance (per unit area) C_{ox}. ε_{ox} denotes the permittivity and t_{ox} the thickness of the oxide layer.

The total capacitance of the MOS-capacitor C is a series combination of the oxide capacitance C_{ox} and the semiconductor capacitance C_S. Figure 6 shows a capacitance-voltage (C-V) curve for an ideal MOS capacitor at low and high frequencies, as well as under deep depletion conditions. Whereas C_{ox} is basically independent of the gate voltage, the semiconductor capacitance changes, due to the different charge states discussed above. At zero voltage the flat band capacitance C_{FB} is given by

$$C_{FB} = \frac{1}{\frac{t_{ox}}{\varepsilon_{ox}} + \frac{L_D}{\varepsilon_S}} \quad (5)$$

where L_D is the Debye length and ε_S the silicon permittivity. (For a real capacitor a voltage must be applied to flatten the bands, because $\Phi_{ms} \neq 0$ (see Figure 4)). At negative voltages an accumulation charge builds up with a capacitance $C_S = -dQ_S/d\psi_S$ (Figure 5c). Since ψ_S is limited to 0.1 to 0.3 V in accumulation the total capacitance rapidly reaches its saturation value C_{ox}. A small positive voltage produces a depletion layer which acts as a dielectric with a width w_d in series with the oxide. Thus, the total capacitance C is given by

$$C = \frac{1}{\frac{t_{ox}}{\varepsilon_{ox}} + \frac{w_D}{\varepsilon_S}} \quad (6)$$

decreases rapidly to a minimum C_{min}. When the gate voltage reaches the threshold voltage $V_T = 2\psi_S$ an inversion layer starts to form and C increases again. Analogous to Eq. (2), the semiconductor capacitance C_S can be broken up into a depletion charge capaci-

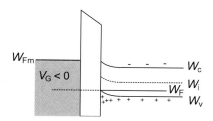

(a) Accumulation

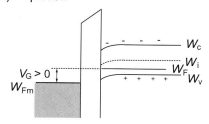

(b) Depletion

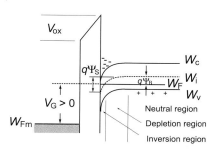

(c) Inversion

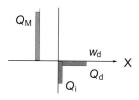

(d) Charge distribution

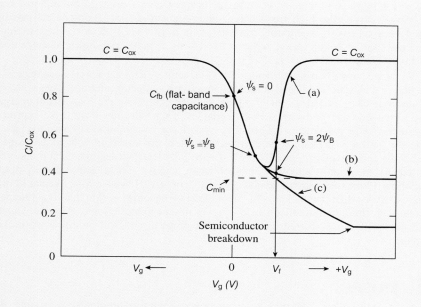

Figure 6: C-V curve of an ideal MOS capacitor under
(a) low frequency,
(b) high frequency and
(c) deep-depletion conditions [6].

Figure 5: Energy-band diagrams of an ideal MOS capacitor with a p-type semiconductor at various gate voltages, showing
(a) accumulation,
(b) depletion and
(c) inversion conditions and
(d) charge distribution under inversion conditions (very schematically).

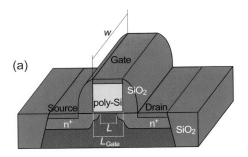

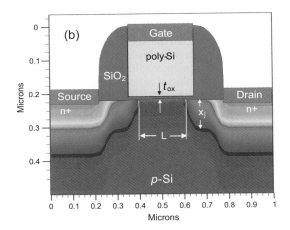

Figure 7:
(a) MOSFET device structure with the terminals, source, gate and drain. L_{Gate} denotes the (printed) gate length, L the channel length or physical gate length and w the gate width. The transistors are isolated with SiO_2 trenches on each side. The gate contact is isolated from source and drain with SiO_2 spacers on each side of the poly-silicon gate contact.
(b) Net doping profiles on a micrometer length and depth scale for a transistor with a gate length $L \cong 0.2$ μm and a gate oxide thickness t_{ox} as calculated with a device simulator (Silvaco). The colours reflect the net dopant concentrations in the Si, ranging from $\approx 10^{17}$ B cm^{-3} in the p-Si to $\approx 10^{20}$ As cm^{-3} near the source/drain silicide contacts. The depth of the n$^+$/p-junction at the extensions, indicated with x_j, is much shallower than the junction depth below the source and drain contacts (at the blue/dark red boundary).

tance C_d and an inversion layer capacitance C_i. C_d and C_i are parallel capacitances in series with C_{ox}, and thus an increase of C_i increases the total capacitance as shown in Figure 6 (curve a). In contrast to the accumulation condition, under the inversion condition the surface potential ψ_S may increase to about 1.0 V. Consequently, the concomitant inversion capacitance C_i can become much larger than the depletion capacitance C_d. Under strong inversion w_d reaches its maximum when the semiconductor is effectively shielded from further penetration of the electric field by the inversion layer. C reaches its maximum value C_{ox}.

If the capacitance measurement is performed at higher frequencies (> 100 Hz), curve (b) in Figure 6 is obtained because the inversion charge arising from minority carriers cannot respond to high frequencies, unless the surface inversion channel is connected to a reservoir of minority carriers as in a MOSFET device. Thus, at high frequencies the inversion charge remains fixed at its dc value and the capacitance does not show an increase at larger V_G.

So far we have discussed only ideal MOS-structures. Real capacitors have undesirable charges within the oxide and at the dielectric/semiconductor interface. These may be mobile ionic charges, like K$^+$ or Na$^+$ ions, trapped charges in the SiO_2, fixed charges close to the interface and interface-state charges. Their densities have to be kept at a minimum. C-V measurements are sensitive to such defects, and thus are used to characterize the dielectric layers. Oxide charges will affect the threshold voltage and consequently the performance of the MOSFET. The Si/SiO_2 interface has excellent properties, making silicon the most important semiconductor material. The interface density of the state of the art thermally grown oxides is 2×10^{10} cm^{-2}/eV. However, fundamental limitations will arise when the thickness of the oxide layer becomes so thin that direct tunnelling through the ultrathin silicon oxide causes unacceptable leakage. Alternative gate dielectrics with higher permittivities solve this problem and will be discussed in this chapter.

2.2 MOSFET

Figure 7a shows a basic MOSFET device structure with its terminals, the source, the gate and the drain. L_{Gate} denotes the gate length, actually the dimension of the gate contact (printed gate length), L the channel length or the physical gate length, and w the gate width. (For various definitions of the term channel length we to refer to [6]). Modern transistors are isolated with SiO_2 trenches, called shallow trench isolation. For an n-

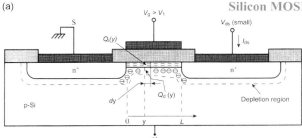
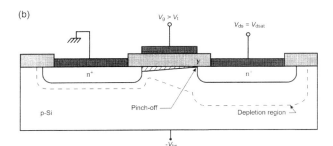
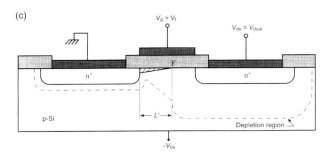

Figure 8: MOSFET operation at a gate voltage $V_g > V_t$ with increasing drain voltage V_{ds}.
(a) At low drain voltages the transistor is in the linear range. Q_i and Q_d are the inversion and the depletion charges, respectively.
(b) shows the pinch-off condition with the pinch-off point Y.
(c) saturation regime where the effective channel length is reduced to L' [6].

channel MOSFET the substrate is p-type and the source and drain regions are n$^+$-doped as shown in Figure 7b as obtained with a commercial device simulator (SILVACO) for a transistor with a gate length $L \approx 0.2$ µm. The dopants were introduced by ion implantation of B, As and P ions and activated by thermal annealing at temperatures around 1000°C. The drain and source regions are highly doped to form the n$^+$/p-junctions with the p-type substrate. The As concentration reaches 10^{20} cm^{-3} close to the silicide contacts. At greater depths the concentration falls off slowly (over ≈ 100 nm) due to ion range straggling and thermal diffusion for the given simulation parameters. The junction depth of the drain extension (indicated by x_j in Figure 7b) is significantly shallower than the junction depth below the source and drain contacts. For even smaller devices these highly doped regions have to be shallower and the concentration gradients very steep, as will be shown below in Table 2.

For the operation of the transistor a gate and a drain voltage are applied. A sufficiently large positive gate potential V_G induces a conducting inversion layer between the source and drain contacts as discussed for the MOS capacitor. When an additional drain voltage is applied, a current flows from source to drain along the dielectric/silicon interface. Figure 8 illustrates the operation of the MOSFET at various gate and drain voltages. At low drain voltages (Figure 8a) the drain current increases linearly as shown in the I-V curves of Figure 9. The channel acts as a resistor. Q_i and Q_d are the inversion and the depletion charge in Figure 8a. Since the drain-substrate n$^+$-p-diode is under reverse bias, the depletion region increases below the n$^+$-drain contact and extends under the gate region with increasing drain voltage (Figure 8a - c). Thus the quasi Fermi level of the drain is qV_D lower than the quasi-Fermi level in the p-type substrate so that inversion can no longer occur near the drain although the bands at the surface are bent (see Figure 5c). As a consequence the inversion charge at the drain side approaches zero. This condition is called **pinch-off** and the corresponding drain voltage the saturation drain voltage (Figure 8b). Since the channel resistance is increased, the drain current saturates (saturation region). The pinch-off point, determined by V_{Dsat}, moves towards the source contact with increasing drain voltage. The carriers now drift down the conducting channel and are injected into the surface depletion region at the pinch-off point near the drain (Figure 8c).

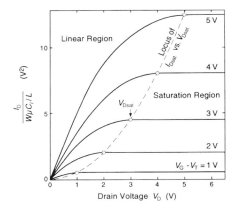

Figure 9: Idealized I-V curves of a MOSFET. The dashed line indicates the locus of I_{Dsat} vs. V_{dsat} [6].

For long channel devices (many microns) the drain current I_D can be approximated by the relations (see e.g. [6])

$$I_D \cong \mu_{\text{eff}} C_{\text{ox}} \frac{w}{L}(V_G - V_T)V_D \tag{7}$$

for the linear region and

$$I_D \cong \mu_{\text{eff}} C_{\text{ox}} \frac{w}{2L}(V_G - V_T)^2 \tag{8}$$

for the saturation region. μ_{eff} denotes the effective carrier mobility in the inversion channel and w the gate width of the transistor. μ_{eff} is generally much smaller than the bulk semiconductor mobility because of additional carrier scattering at the dielectric/semiconductor interface. Therefore, the interface quality crucially effects the channel mobilities.

An important parameter characterizing the small signal response of the transistor is the transconductance g per unit width defined as

$$g = \frac{1}{w}\frac{\partial I_D}{\partial V_G}\bigg|_{V_D = \text{const}} = 2\mu_{\text{eff}}\frac{C_{\text{ox}}}{L}(V_G - V_T) \tag{9}$$

for the saturation region. Eq. (8) and Eq. (9) show that both, the saturation current and the transconductance, scale with C_{ox}/L. For fast signal response a large transconductance is required, which can be achieved by decreasing the physical gate length L and increasing the oxide capacitance per area. However, as the lateral dimensions of the transistor shrink, the total gate oxide capacitance also decreases. We will show in the following section that this has to be taken into account when the transistor dimensions are scaled down.

3 Scaling Rules

Scaling of MOSFETs to smaller dimensions is required to increase package density and speed, as well as to a reduced power consumption. One of the major changes when the gate length of the MOSFET is reduced to submicron dimensions is that the electric field distribution in the channel region is transformed from one dimensional to two dimensional. In the long channel device the potential contours are nearly parallel to the oxide/silicon interface. The carriers experience a largely one-dimensional field. For short channel devices the drain voltage causes a two dimensional potential distribution, as illustrated in Figure 10, showing simulated constant potential contours for a short-channel device obtained with a simulation program (ATLAS, Silvaco). Therefore, numerous measures have to be taken to maintain the transistor function for very small dimensions. Small transistors tend to exhibit undesired effects, including a lack of saturation, gate oxide degradation due to hot electrons, threshold voltage shifts, gate-induced drain leakage and drain-induced barrier lowering (DIBL). The latter effect is caused by lowering of the potential barrier at the source-channel side with increasing drain voltage in short channel devices. Various scaling models have been developed to overcome these problems.

The simplest scaling concept of MOSFET transistors is **constant-field scaling**. The idea is to scale the device dimensions in horizontal *and* vertical directions, as well as the voltages by the same factor α, while proportionally increasing the substrate doping concentration, to keep the electric field pattern unchanged. The doping concentration has to be increased to decrease the depletion width (this follows from Poisson's equation). Constant field scaling implies that the *gate oxide thickness has to be decreased by* α to maintain the oxide field, while decreasing the gate voltage. This is necessary to maintain gate control of the channel and ensure good short channel behaviour. The thinner the gate oxide the smaller the two-dimensional effects become. Results of constant field scaling on dimensions and circuit parameters are given in Table 1. With respect to device performance, we see that the circuit delay time decreases by a factor of α and the power dissipation per circuit even by α^2, which is important to reduce the generation of heat. For advanced scaling models refer to [6], [8]).

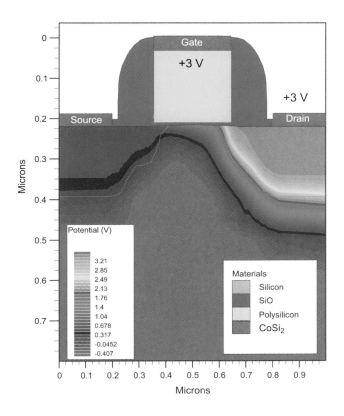

Figure 10: Simulated potential distribution of a 0.2 μm MOSFET with $V_G = 3$ V and $V_D = 3$ V. Near the drain region the potential lines are strongly affected by the drain voltage. The thin solid line indicates the n+/p-junctions.

	MOSFET device and Circuit Parameters	Multiplicative Factor ($\alpha > 1$)
Scaling assumptions	Device dimensions (t_{ox}, L, w, x_j)	$1/\alpha$
	Doping concentration (N_a, N_d)	α
	Voltage (V_D)	$1/\alpha$
Derived scaling Behavior of device Parameters	Electric field (E)	1
	Depletion-layer width (w_d)	$1/\alpha$
	Capacitance ($C = \varepsilon A/t_{ox}$)	$1/\alpha$
	Inversion-layer charge density (Q_i)	1
	Carrier velocity	1
	Current, drift (I)	$1/\alpha$
Derived scaling Behaviour of device Parameters	Circuit delay time ($\tau \sim CV_D/I_D$)	$1/\alpha$
	Power dissipation per circuit ($P \sim V_D I_D$)	$1/\alpha^2$
	Power-delay product per circuit ($P\tau$)	$1/\alpha^3$
	Circuit density ($\propto 1/A$)	α^2
	Power density (P/A)	1

Table 1: Device and circuit parameters resulting from constant field scaling [6].

Scaling rules are used to predict future device dimensions and performance parameters. There is a long tradition of collecting all these data in the International Technology Roadmap for Semiconductors (ITRS). Table 2 shows selected parameters from the ITRS 2001 [9].

III Logic Devices

DRAM ½ PITCH (nm)	130	100	80	65	45	32	22
MPU / ASIC ½ PITCH (nm)	150	107	80	65	50	35	25
MPU PRINTED GATE LENGTH (nm)	90	65	45	35	25	18	13
Physical gate length high-performance (HP) (nm)	65	45	32	25	18	13	9
Equivalent physical oxide thickness for high-performance t_{eq} (EOT) (nm)	1.3-1.6	1.1-1.6	0.8-1.3	0.6-1.1	0.5-0.8	0.4-0.6	0.4-0.5
Gate depletion and inversion layer quantum effects electrical thickness adjustment factor (nm)	0.8	0.8	0.8	0.5	0.5	0.5	0.5
t_{ox} electrical equivalent (nm)	2.3	2.0	1.9	1.4	1.2	1.0	0.9
Nominal power supply voltage (V_{dd}) (V)	1.2	1.0	0.9	0.7	0.6	0.5	0.4
Nominal high-performance NMOS sub-threshold leakage current, $I_{sd,leak}$ (at 25°C) (μA/μm)	0.01	0.07	0.3	1	3	7	10
Nominal high-performance NMOS saturation drive current I_D (at V_D, at 25°C) (μA/μm)	900	900	900	900	1200	1500	1500
Drain extension x_j (nm)	27-45	19-31	13-22	10-17	7-12	5-9	4-6
Maximum drain extension sheet resistance (PMOS)(Ω/sq)	400.0	550.0	770.0	760.0	830.0	940.0	1210.0
Contact (nm)	48-95	33-66	24-47	18-37	13-26	10-19	7-13
Silicide thickness (nm)	35.8	24.8	17.6	13.8	9.9	7.2	5.0
Contact silicide sheet $R_{contact}$ (Ω/sq)	4.2	6.1	8.5	10.9	15.2	21.0	30.3
Contact maximum resistivity (Ω/sq²)	4.10E-07	2.70E-07	1.80E-07	1.10E-07	6.40E-08	3.80E-08	2.40E-08
Parasitic source/drain resistance (R_{SD}) (Ω-μm)	190	180	180	140	110	90	80
Parasitic source/drain resistance (R_{SD}) percent of ideal channel resistance (V_D/I_D)	16 %	17 %	19 %	20 %	25 %	30 %	35 %
High-performance NMOS device delay time τ (ps)	1.6	1.1	0.83	0.68	0.39	0.22	0.15
Energy per ($w/L = 3$) device switching transition (fJ)	0.347	0.137	0.065	0.032	0.015	0.007	0.002
Static power dissipation per ($w/L = 3$) device (W/device)	5.6E-09	1.0E-08	2.6E-08	5.3E-08	9.7E-08	1.4E-07	1.1E-07

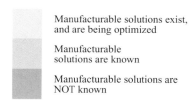

Manufacturable solutions exist, and are being optimized

Manufacturable solutions are known

Manufacturable solutions are NOT known

Table 2: Requirements for high performance logic (MPU, ASIC) and for memory (DRAM) applications taken from the International Technology Roadmap for Semiconductors 2001 [9]. MPU is the Micro Processor Unit and ASIC is a Application Specific Integrated Circuit. The pitch size denotes the closest distance of metal lines in the first metal layer and EOT the equivalent oxide thickness

In order to increase the IC package density, the CMOS transistor is shrinking in all dimensions. According to Tab. 2 the half pitch size also known as the "technology node" is shrinking as well as the gate length, the dielectric thickness and the junction depth. The operating voltage is being lowered continuously whereas the drain current is increased to allow an increase of the clock frequency. The issues of junction and contact fabrication will be discussed in Sect. 6.

The scaling requires an ever-increasing specific capacitance in the channel (C_{tot}/A). In the past this has been done by scaling down the thermally grown silicon oxide thickness. The use of ever-thinner silicon dioxide gates is limited by the exploding gate leakage as shown in Figure 11 with decreasing minimum feature size. Although the use of high-K dielectrics is anticipated, it is an industry standard to convert high-K dielectric thickness into SiO$_2$ *Equivalent Oxide Thickness* (EOT). This means that the high-K material can have a greater thickness t_x. It is converted into the equivalent oxide thickness t_{eq} by

$$t_{eq} = \frac{\varepsilon_{SiO_2}}{\varepsilon_{film}} t_x \qquad (10)$$

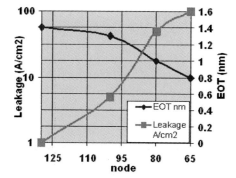

Figure 11: Gate leakage current and equivalent oxide thickness (EOT) versus technology node given in nm.

Here, ε_{film} represents the permittivity of the high-K material. Figure 11 shows the equivalent oxide thickness versus technology node.

4 Silicon-Dioxide Based Gate Dielectrics

Silicon dioxide has been used as a gate dielectric since 1957 [10]. For today's mainstream MOSFETs gate oxide thicknesses of $\cong$ 1.5 - 2 nm are used. These ultrathin dielectric layers are grown in a standard thermal diffusion furnace using pure oxygen or preferably a mixture of oxidizing gases at temperatures between 900°C and 1000°C. Before oxidation the silicon surface must be cleaned carefully with high purity chemicals. The SiO_2 layer grows controllably in a layer by layer mode, where the growth rate can be described by the Deal-Grove reaction model [11]. Figure 12 shows cross-sectional high resolution transmission electron microscopy (HRTEM) micrographs of MOS structures with a 2.7 and 2.4 nm thick SiO_2 layer respectively. Atomic steps of $\approx$ 0.2 nm on the silicon surface and on the oxide are typical for these interfaces. The upper image shows nicely the grain structure of the polycrystalline silicon (poly-Si) gate contact.

Thinning down the oxide thickness raises severe technological problems:

- Dielectric thickness variation
- Penetration of impurities, in particular boron, from the highly doped polysilicon gate
- Reliability and lifetime problems for devices made with ultrathin oxides
- Gate leakage current increases exponentially with decreasing thickness.

The problem of gate leakage is demonstrated in Figure 13 which shows the measured and simulated leakage currents versus the gate voltage for various oxide thicknesses ranging from 3.6 to 1.0 nm. The leakage current at a gate bias of 1 V changes from 1×10^{-12} A/cm^2 at 3.5 nm to 10 A/cm^{-2} at 1.5 nm, which is 13 orders of magnitude. The practical SiO_2 thickness limit in a MOSFET device is reached when the gate leakage becomes equal to the off-state source to drain sub-threshold leakage current. Beyond this limit direct electron tunneling from the p-Si through the thin oxide into the n$^+$-polysilicon becomes the dominant leakage mechanism. Tunneling of holes is less critical since the tunneling probability for holes is much smaller than for electrons.

For *low power applications* in the field of portable equipment (mobile phones, palm-tops etc) gate leakage in serious excess of the off current is unacceptable. The gate leakage current reduces battery life in stand-by mode. (Note CMOS inverters circuits are selected for it's low power consumption in stand-by mode. The on-current only flows during switching). Here, gate leakage currents above approx. 1 Acm^{-2} are not acceptable. For *high performance applications* (such as desktop computers) stand–by power consumption is less of an issue. However, power dissipation becomes critical at about 10 A/cm^2.

Numerous attempts have been made to improve the oxide quality. A successful approach used in advanced CMOS production is the introduction of nitrogen into SiO_2 which leads to the formation of SiO_xN_y (oxynitrides). Nitrogen slows down the thermal growth rate and improves the interface uniformity. Furthermore, it reduces boron penetration from the highly doped poly-silicon gate stack and makes the oxide less sensitive

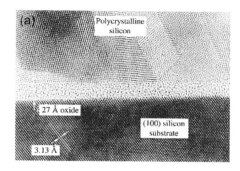

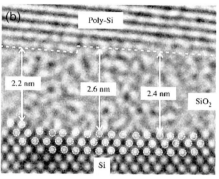

Figure 12: Cross-sectional HRTEM of a MOS structure with
(a) 2.7 and
(b) 2.4 nm thick SiO_2 layers embedded between single crystalline silicon and polysilicon [10].

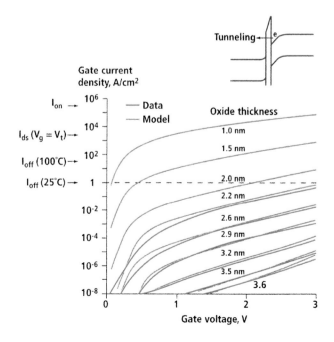

Figure 13: Measured (red curves) and simulated (blue lines) gate leakage current versus gate voltage for various oxide thicknesses. The leakage current increases exponentially as the oxide thickness is scaled down. The inset illustrates direct tunneling through the oxide [8].

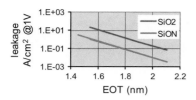

Figure 14: Comparison of gate leakage current of SiO$_2$ and SiON dielectrics as a function of thickness.

to hot electrons which may create defects in the dielectric and cause breakthrough. Boron incorporation into the oxide causes an undesirable shift of the threshold voltage. The nitrogen incorporation has to be carried out in a very controlled way, since a sufficient nitrogen content, preferably at the polysilicon side, is needed to stop boron penetration into the oxide but too much nitrogen at the silicon interface degrades carrier mobility and transistor transconductance. Films grown in nitrous oxide, N$_2$O, or nitric oxide, NO, exhibit superior quality. Also post-annealing of the oxide layer in nitrous oxide or nitric oxide may improve the quality. For details we refer the reader to [12] and references therein. State of the art nitrided silicon oxide reduces the gate leakage by one to two orders of magnitude compared to standard SiO$_2$ as shown in Figure 14. This will allow to extend the use of oxynitride to be extended to somewhat less than 1 nm EOT for high-performance MOSFETs [9] and thus the need for the integration of high-K materials is delayed by one to two chip generations.

4.1 High-K Materials for CMOS

High-K materials can solve the leakage problem of silicon-dioxide-based dielectrics since their thickness can be made significantly larger. The equivalent oxide thickness t_{eq} is given by Eq. (10). However, it should be noted that the leakage current is limited by the direct tunnel current in nearly perfect materials only. There are several other leakage mechanisms associated with material defects that can contribute to leakage far in excess of the fundamental tunnel current leakage limit. Furthermore, the tunnel current is low only if the energy band offsets of the dielectric relative to silicon are sufficiently high as discussed in Sect. 4.4.

Table 3 lists a significant number of potential high-K materials and the respective dielectric permittivities. New dielectrics have to fulfil a number of requirements in order to be compatible with the established silicon technology and to meet the specific properties required for ultra-short channel MOSFET devices. J.-P. Maria and A.I. Kingon have discussed these criteria in detail in their review article [13]. Some important aspects such as

- dielectric properties,
- thermodynamic stability,
- electronic properties,
- microstructural stability
- high-K deposition tools and chemistry and
- process compatibility

will be discussed in the following. In addition, examples of possible gate stacks will be presented We will see that only very few of the oxides listed in Tab. 3 meet the stringent requirements.

4.2 Dielectric Properties

The permittivity of the new dielectric should be considerably larger than that of SiO$_2$ (ε_{SiO2} = 3.9) and the thickness of the dielectric layer should be equivalent to a silicon dioxide thickness of 1 nm (or even less), which corresponds to an areal capacitance of 32 pF/μm^2.

The dielectric properties are closely related to the crystal structure and, hence, the microstructure of the oxides which will be discussed in Sect. 4.5. Amorphous oxides always show relatively low permittivities, while some of epitaxially grown oxides show really high permittivities. However, these high values are not required. Because of the superior low interface state densities of the Si/SiO$_2$ interface, the first monolayer of the gate dielectrics probably needs to be SiO$_2$ even if high-K oxides are employed. Since the capacitances of this ultrathin SiO$_2$ layer and the high-K layer are in series, there is a limit to the extent to which an increase of the permittivity of the high-K material has a significant effect on the capacitance on the total gate dielectrics. This aspect is shown in a simulation in Figure 15.

Binary oxides	
Al$_2$O$_3$	9-11.5
BaO	31-37
CeO$_2$	18-26
HfO$_2$	20-22
La$_2$O$_3$	25-30
Ta$_2$O$_5$	25-45
TiO$_2$	80-95
Y$_2$O$_3$	11-14
ZrO$_2$	22-25
Multicomponent oxide alloys	
HfO$_2$·SiO$_2$	10-13
La$_2$O$_3$·SiO$_2$	16-20
Y$_2$O$_3$·SiO$_2$	10-11
ZrO$_2$·SiO$_2$	10-13
HfO$_2$·ZrO$_2$	20-25
HfO$_2$·Al$_2$O$_3$	14-17
Multicomponent stoichiometric oxides	
LaAlO$_3$	25
SrZrO$_3$	25
SrTiO$_3$	< 250
(Ba,Sr)TiO$_3$	< 400

Table 3: Permittivity of selected high-K dielectrics at room temperature taken from [13], [15] and references cited therein. Note: BaO is highly hygroscopic. The values of the multicomponent oxide alloys refer to approx. 50 mol-% mixtures. The data for multicomponent stoichiometric oxides are for epitaxially grown films. The effective permittivities for SrTiO$_3$ and (Ba, Sr)TiO$_3$ are stated for thin films (< 30 nm). For further decreasing thicknesses, these values will drop significantly as discussed in Chap. 21.

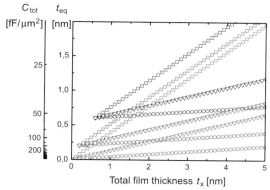

Figure 15: Calculated equivalent oxide thickness (EOT) t_{eq} and capacitance density plotted as a function of total dielectric thickness for a high-K dielectric film and a SiO$_2$ layer. Color code: thickness of the SiO$_2$ layer 0.6 nm (blue), 0.2 nm (green), and 0.0 nm (red). Symbol code: permittivity ε_r of the high-K 10 (squares), 30 (triangles), 100 (circles).

4.3 Thermodynamics

The high-K dielectric is in direct contact with silicon and the gate contact material (e.g. poly-Si or a metal). This sandwich is subjected to severe temperature treatments (up to $\approx 1000°C$). The gate stack can only maintain its properties if no chemical reactions take place. The most likely reactions are the growth of silicon dioxide, the formation of metal oxides and the formation of silicides. If the silicon oxide is more stable than the oxide of the metal M the following reaction may occur (unbalanced)

$$\text{Si} + 2\text{MO} \rightarrow \text{SiO}_2 + 2\text{M} \tag{11}$$

A further example, is the formation of a silicide phase accompanied with the simultaneous growth of SiO$_2$ following the reaction

$$(x/2+y)\,\text{Si} + x\,\text{MO} \rightarrow x/2\,\text{SiO}_2 + \text{M}_x\text{Si}_y \tag{12}$$

Table 4 lists the formation enthalpies ΔG (oxide/mole of oxygen) for a number of high-K oxides and gives the stability of the oxide with silicon.

Obviously, the formation of interfacial metallic layers should be avoided. An oxide should therefore be discarded as potential high-K candidate to be used in direct contact with silicon if one of the reactions has a negative free enthalpy of formation at process temperatures. It cannot be used in direct contact with neither silicon substrate nor a poly-Si-gate. ZrO$_2$ and HfO$_2$ are stable in direct contact with silicon, but Ta$_2$O$_5$ and TiO$_2$ are not. Any use of Ta$_2$O$_5$ and TiO$_2$ will have to rely on diffusion barriers such as SiN$_x$ between the (Si) gate and the substrate or the dielectric. The demands on the integrity of the ultrathin diffusion barrier will be extremely severe to withstand the subsequent high-temperature processing.

	ΔG kcal/mol per oxygen at 1000 K	Stable interface with Si
SiO$_2$	170	+
Ta$_2$O$_5$	155	-
TiO$_2$	180	- (TiSi$_2$)
Al$_2$O$_3$	215	+
La$_2$O$_3$	215	+
BaO	220	+
ZrO$_2$	250	+
HfO$_2$	250	+
Y$_2$O$_3$	245	+

Table 4: Gibbs free enthalpy ΔG associated with oxide formation per oxygen atom and interface stability (+/-) with silicon of various oxides [15].

4.4 Electronic Properties

In the process of selecting the appropriate high-K candidate material, the band gap energy and the conduction and the valence band-offsets have to be taken into account. In order to achieve acceptable low leakage these offsets should be at least 1 eV for a gate voltage ≤ 1 V. This means the band-gap of the material should be > 3.1 eV since a 1 eV barrier at the valence and the conduction band is needed apart from the silicon band-gap energy of 1.1 eV. Figure 16 shows optical band gap energies as a function of dielectric permittivity for selected dielectrics [15]. In spite of the large band gap energies of many oxides, the band alignment with silicon is often highly asymmetric as shown by calculations by Robertson in Figure 17, [16]. These results clearly indicate that compounds such as BaZrO$_3$, BaTiO$_3$, and Ta$_2$O$_5$ have a fairly large band-gap but lack sufficient barrier for the conduction band offset.

Experimental results for Al$_2$O$_3$ and ZrO$_2$ are shown in Figure 18. The band offsets were measured by means of photoelectron spectroscopy [17]. These results are in fair agreement with the theoretical predictions. Although Al$_2$O$_3$ has many favourable properties its use is questionably because of the rather small permittivity of 8-11 (Tab. 3).

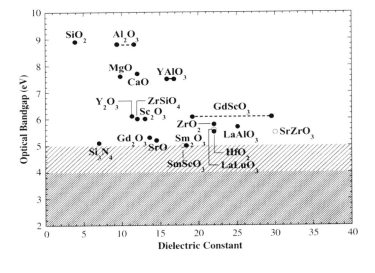

Figure 16: Optical band gap energies versus dielectric constant [15].

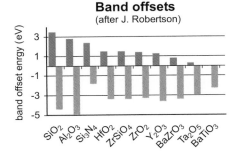

Figure 17: Conduction band (red bars) and valence band offsets (blue bars) with silicon calculated for high-K materials [16].

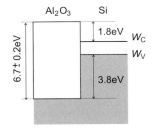

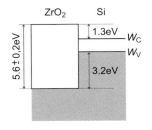

Figure 18: Band alignment of Al_2O_3 and ZrO_2 with silicon measured with photoelectron spectroscopy [18].

4.5 Microstructural Stability

The microstructure requirements of high-K materials are still the subject of debate. In principle, there are three possible microstructures:

- (perfectly) epitaxial,
- polycrystalline, or
- amorphous.

The structure should be stable throughout the full processing. Meeting the high temperature stability requirements is the most important issue. The preferred structure of the high-K dielectric is either perfectly epitaxial or amorphous. Polycrystalline layers are considered problematic because grain boundaries may give rise to detrimental diffusion reactions, enhanced leakage, and high density of interface states. Besides, device operation characteristics such as noise and matching will be adversely affected in the case of a polycrystalline nature of the gate dielectric.

Epitaxial high-K oxides offer a high potential when grown directly on Si, i. e. without a SiO_2 interface layer which reduces the total capacitance (Figure 15). One (tough) selection criteria for epitaxial dielectrics is that the interfacial lattice mismatch relative to silicon should be small in order to allow the growth of pseudomorphic layers. Otherwise, the formation of misfit and threading dislocations or excessive stress may deteriorate the carrier mobility and reliability of the MOSFET. Considering an acceptable lattice mismatch, a thermal stability against Si, a sufficiently high permittivity and an appropriate band alignment, $SrZrO_3$, $CaZrO_3$ and $LaAlO_3$ may be potential candidates (Tab. 3). An example of expitaxial growth of $SrTiO_3$ on Si is given in Sect. 4.8. Even if the growth problems are solved, it is not yet clear whether a sufficiently low interface state density at the (SiO_2-free!) Si/high-K oxide interface can be achieved.

Because of these huge challenges for epitaxial oxides, research and development focuses mainly on *amorphous* materials. However, binary oxides are not favourable for the amorphous phase. The high melting point of the oxides and the relatively high symmetry of their crystal lattices act as a driving force for crystallisation at relatively low temperatures. For this reason compounds with low melting point are preferred. Many mixed oxide systems show a low melting point in combination with high temperature stability of the amorphous phase. Mixed oxides of either hafnia or zirconia with silica or alumina have attracted a lot of attention, because they can be formed directly on silicon as will be shown in Sect. 4.8. Nevertheless, microstructural stability is an issue for these aluminates and silicates. The silicates show a tendency for phase segregation at higher temperatures. In comparison, the aluminate phases seem to be more stable. The difference in the driving force for this phase segregation can be found in the respective phase diagrams (Figure 19). The stable liquid immiscibility zone in the ZrO_2/SiO_2 and HfO_2/SiO_2 systems extends into the amorphous phase and phase segregation is unavoidable. The ZrO_2/Al_2O_3 and HfO_2/Al_2O_3 systems do not show this immiscibility zone and show no tendency of segregation. Further examples will be discussed in 4.8. The X-ray diffraction (XRD) data of Figure 20 show the thermal stability of Zr aluminate [18]. The amorphous phase is stable up to 900°C. Above 900°C the film becomes polycrystalline as indicated by the XRD signals for the higher temperatures.

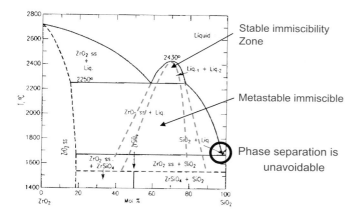
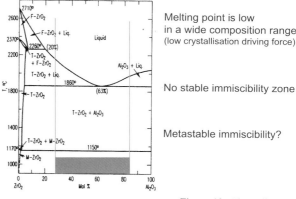

Figure 19: Phase diagrams for ZrO_2/SiO_2 and ZrO_2/Al_2O_3 systems [18].

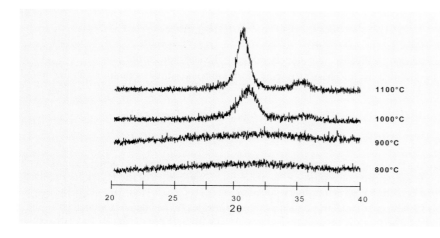

Figure 20: XRD of thick mixed oxide samples after consecutive anneals during 30 min showing the crystallization above 900°C.

Though many materials issues are still open, mainstream high-K research concentrates on a limited set of materials. Favorite materials are $Hf(Zr)O_2$, $Hf(Zr)$ silicates and $Hf(Zr)$ aluminates. All of which have specific advantages and disadvantages. The permittivity of pure alumina might be too low. The polycrystalline nature of pure hafnia and zirconia could be detrimental. Real stable silicates are silicon-rich and consequently show relatively low K values. Aluminates show electrical instability. However, interface preparation, deposition technology and post deposition anneals can have a large impact on the final performance of the gate material.

4.6 High-K Deposition Tools and Chemistry

Whichever material is selected as a high-K dielectric for CMOS, an adequate deposition technology must be made available to the semiconductor manufacturing industry. Dielectric deposition should of course be reliable and safe and should meet the uniformity requirements for gate materials of 2 % on 200 and 300 mm wafers, with scaling options to 450 mm sizes. Processing should be fast (typically < 5 minutes).

In principle, four deposition technologies (see Chap. 8) can be considered:
- Evaporation, including molecular beam epitaxy (MBE)
- Sputter deposition
- Chemical vapor deposition (CVD) and
- Atomic layer deposition (ALD).

Molecular beam epitaxy is generally considered expensive, slow and requires enormous maintainance. Therefore, it is rarely used in semiconductor industry. Sputter deposition has too many damage hazards and will not easily meet the uniformity target. CVD based technologies, such as metal organic chemical vapor (MOCVD) deposition and atomic layer deposition (ALD) are attractive for the deposition of ultrathin layers of high-K oxides.

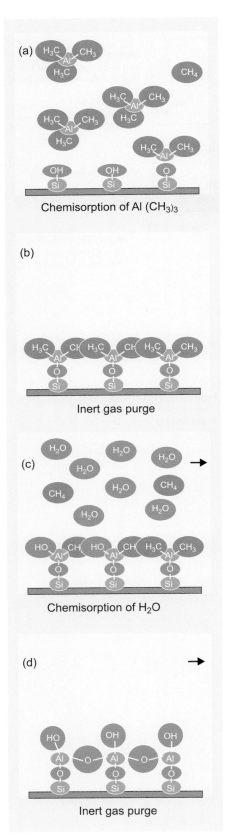

Figure 21: The ALD process [19].

Presently commercially available are:

MOCVD	$Hf(Zr)O_2$
	$Hf(Zr)$silicates
ALD	Al_2O_3
	$Hf(Zr)$aluminates

MOCVD is performed at temperatures of typically 500°C. For zirconia, metal organics such as TDEAZ (tetradiethylaminozirconium) [21] and oxygen are used as a precursor.

ALD is a special CVD application where the precursors are not mixed prior to introduction in the reactor. Instead, alternating pulses of reactants are introduced in the reactor. In a first pulse, a (sub) monolayer of the first reactant is absorbed on the wafer surface, the reactor is subsequently purged, then a second reactant is introduced to react with the adsorbed layer. In the ideal case, monolayers can be grown repeatedly in a controlled manner. For ZrO_2 deposition typically $ZrCl_4$ and water are used at a deposition temperature of about 300°C. The two-step reaction can be expressed in a simplified manner by:

$$ZrCl_4 + 2OH(surface) \rightarrow ZrCl_2O_2(surface) + 2HCl$$
$$ZrCl_2O_2(surface) + 2H_2O \rightarrow ZrO_2(OH)_2(surface) + 2HCl \quad (13)$$

where "surface" indicates a covalent bonding to the substrate surface.

For alumina from TMA (trimethylaluminium) a similar reaction can be noted (Figure 21)

$$Al(CH_3)_3 + OH(surface) \rightarrow Al(CH_3)_2O(surface) + CH_4$$
$$Al(CH_3)_2O(surface) + H_2O \rightarrow AlO_2(surface) + 2CH_4 \quad (14)$$

Current ALD and MOCVD processes are operated at relatively low temperatures. As a result, precursor decomposition can be incomplete and purity of the layers (particularly C and Cl contaminates) is a primary concern.

So far new gate dielectrics, such as the oxynitrides, have been introduced by replacing a gate furnace or the process on a given furnace. In the case of high-K the gate oxidation furnace or gate rapid thermal process (RTP) tool has to be replaced by a high-K deposition tool. In addition, the thin (high-K) dielectrics cleaning might be time critical because the thin dielectrics are likely to be sensitive to air exposure. A cluster tool with process chambers for *cleaning, deposition, anneal*, and *deposition* of poly-Si or metal gate seems to be required here.

4.7 Process Compatibility

High-K (Gate) etch

For most high-K materials removal in a (dry) etch process is of prime concern. Unlike Si, the high-K metals (oxides) do not have volatile fluorides. However, for some of the metals, Hf, Zr, and Al in particular, chlorides and bromides do show some volatility. Etch chemistry should be Cl/Br-based rather than F-based. For HfO_2 and ZrO_2 in particular, dry etch recipes have been developed. However, selectivity of the etch process is a severe problem. Remarkably though, successful use of CF_4 has been reported [20]. Yttrium and lanthanum oxides might be a lot more difficult to etch.

System on a Chip

System On a Chip (SOC) is an increasing trend in semiconductor manufacturing processing (Figure 22). This means that two or even three gate oxide processes are employed on a single chip to combine high speed logic (EOT 1 nm), low power logic (EOT 2 nm) and input/output (I/O) logic (EOT ≥ 3.5 nm) applications on a single device.

The two (or three) oxide thicknesses are generated by two (or three) subsequent oxidation steps. First the thick oxide and then the thin oxide are grown. The thick oxide will increase slightly in thickness during the subsequent gate oxidation process. Implementing a high-K gate dielectric in SOC processing adds a new set of requirements to high-K process compatibility as illustrated in Figure 23.

	Cl [°C]	Br [°C]	F [°C]
SiO_2	57	154	−86
Al_2O_3	200*	255	>1000
ZrO_2	317	323	970
HfO_2	331	360	912
Y_2O_3	>1000	>1000	>1000
La_2O_3	>1000	>1000	>1000

Table 5: High-K etch chemistry showing the boiling points of metal halides for some binary metal oxides. Compounds with low boiling points become volatile during etch and are readily removed [21].

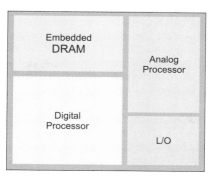

Figure 22: System on a Chip (SOC).

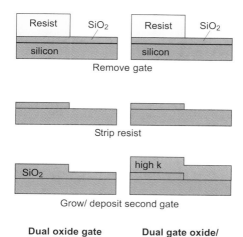

Figure 23: SoC process with dual oxide gate (left) and with high-K materials (right). The dielectric film (red) is thinned by etching at selected areas by the use of patterned photoresist.

4.8 Examples of High-K Gate Stacks

In the following we present examples of different high-K gate stacks: HfO_2 with a thin interlayer on Si, a hafnium silicate directly deposited on silicon and epitaxial $SrTiO_3$ on Si(100).

Figure 24a shows a XTEM image of a vertical replacement gate transistor with a gate length of 50 nm with HfO_2 as the gate dielectric. Unlike the standard transistor process, in the replacement gate process source and drain are manufactured before the gate dielectric is deposited. The advantage is that high-temperature anneals (1000°C) of the gate dielectric can be avoided. Here, the high-K dielectric was formed by first growing an ultrathin underlayer (UL) on the vertical silicon channel followed by the deposition of HfO_2 by atomic layer deposition (ALD) [22]. The high-resolution XTEM micrograph of Figure 24b shows the highly conformal HfO_2 layer at the top corner of the gate. The underlayer has a thickness of 0.7 nm and the HfO_2 of 4.5 nm. The high-resolution image also reveals the polycrystalline nature of hafnia. Nevertheless, good transistor performance was achieved [22]. Typical C-V measurements for p- and n-MOS capacitors with a similar gate stack, HfO_2 with a thickness of 3.8 nm and an interlayer of 0.6 nm, are shown in Figure 25 [23]. Here, the film was DC reactively sputtered from a Hf target at temperatures of 550 °C. A minimum equivalent oxide thickness $t_{eq} = 1.2$ nm for this layer was deduced from the C-V measurements.

Silicates, such as $HfSi_xO_y$ and $ZrSi_xO_y$, can be formed *directly on silicon* as shown in Figure 26. $HfSi_xO_y$ was sputter deposited directly on Si heated to 500°C during growth [24]. The cross-section HRTEM image shows a 3 nm thick smooth $Hf_7Si_{29}O_{64}$ layer with abrupt interfaces (a) after deposition and (b) after annealing at 1050°C for

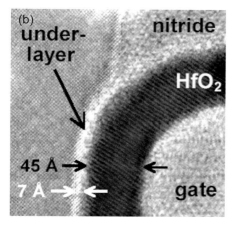

Figure 24:
(a) Bright field XTEM image of a vertical replacement gate transistor with conformal HfO_2 as gate dielectric. PSG denotes a phosphorsilicateglass which is used for isolation. Source, drain and the vertical silicon channel are indicated.
(b) High resolution XTEM micrograph of the underlayer (UL) and the HfO_2 layer at the upper corner of the gate. The lattice planes of a HfO_2 grain are clearly visible indicating its polycrystalline microstructure [22].

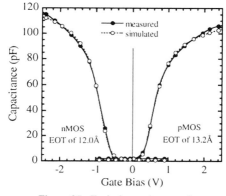

Figure 25: Typical capacitance-voltage curves for n- and p-MOS capacitors with 3.8 nm HfO_2 and a 0.6 nm interlayer on Si. The corresponding equivalent oxide thicknesses (EOT) are indicated [23].

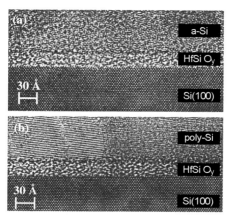

Figure 26: Cross-section HRTEM micrograph of a 5 nm thick $Hf_7Si_{29}O_{64}$ layer (a) after deposition and (b) after RTA at 1050°C for 20 s in N_2 [24].

III Logic Devices

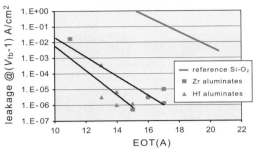

Figure 27: Leakage current of Hf and Al aluminates and reference SiO$_2$ [23].

20 s in N$_2$. The amorphous Si film on top of the silicate crystallized during the anneal but the film remained amorphous and the interfaces sharp. HfSi$_x$O$_y$ films with a thickness of 5 nm exhibited an equivalent oxide thickness t_{eq} = 1.8 nm, breakdown fields of 10 MV/cm and interface state densities of 1 - 5 × 10^{11} cm^{-2} eV^{-1}. Promising properties were obtained for Hf concentrations of 3 - 8 at %. Figure 27 provides an example of leakage current measurements of Hf and Zr aluminates in comparison to SiO$_2$. As expected the leakage current is many orders of magnitude lower than the leakage current of SiO$_2$ with comparable EOT [23].

An interesting example of a single crystalline perovskite on Si is SrTiO$_3$ commensurately grown on silicon [25]. Figure 28 shows a cross-sectional image of a SrTiO$_3$/Si(100) heterostructure using Z-contrast transmission electron microscopy which produces a strong material contrast at the interface. The inset at the left of Figure 28 shows the atomic positions in a model calculation. The structure was grown by molecular beam epitaxy on Si(100). First the alkaline metal, here Sr, was deposited at a substrate temperature of 850°C formings a commensurate 1-2 monolayer thick Sr silicide layer. Then, the substrate temperature was reduced to 200°C, oxygen was introduced into the growth chamber and Sr and Ti were co-deposited. A MOS capacitor with a 15 nm thick SrTiO$_3$ film exhibited a capacitance of 40 fF/μm^2 at a voltage of –3 V. The interface trap density shows a peak at 0.11 eV above the valence band with values up to 6×10^{11} cm^{-2}. Despite these promsing results, it is unlikely that epitaxial SrTiO$_3$ will be used. The conduction band offset to silicon is much too small (Figure 17) and also the growth technique is too complicated to be introduced in a production line.

5 Metal Gates

5.1 Polysilicon vs. Metal Gates

In the standard CMOS process, heavy doped poly-Si is applied as a gate contact material in the transistor MOS capacitor. The main advantage of poly-Si is that the work function can be adjusted by doping for p- and n-channel devices (see Sect. 2.1). In addition, the process is well known and provides high yield. The disadvantage of poly-Si use is the formation of a depletion layer within the poly-Si and its high resistivity. The resistivity limits the drive current and the depletion requires the use of thinner gate dielectrics. For high current transistor operation, a high charge in the channel in inversion is the key parameter. Thus, a high capacitor equivalent thickness (CET) is required. The depletion in the semi metallic poly-Si adds "dielectric thickness" to the true dielectric thickness, the EOT. Typically a 0.4 nm "thickness" is added to the CET value in inversion (see Tab. 2) Figure 29 shows C-V curves for a poly-Si gate stack with different doping concentrations of the poly-Si. Under inversion conditions the effect of poly-Si depletion becomes very pronounced for the lower doped poly-Si contacts. Because of the high carrier density in metals the depletion effect is negligible. Hence, the use of metal gates relax the requirements of the high-K dielectrics. For a given high-K material, the gate dielectrics can be made thicker and the gate leakage can be reduced.

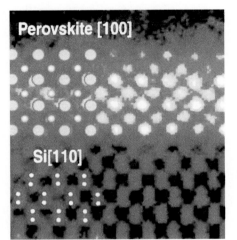

Figure 28: Z-contrast image of epitaxial SrTiO$_3$ on Si(100). The inset in the left side shows a model of the perovskite/silicon projection [25].

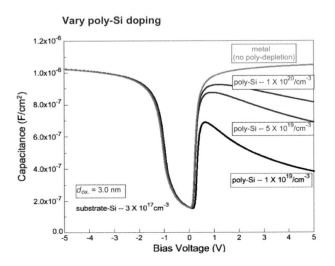

Figure 29: Poly depletion effect in inversion [26].

5.2 Metal Gate Materials Selection

A similar set of requirements as for high-K materials can be listed for metal gates. Importants aspects are

- thermodynamic stability,
- electronic properties, and
- process compatability.

It is important to note that different metals are needed for the gates of PMOS and NMOS transistors, in order to obtain a suitable band alignment (see Figure 4). The work function of pure metals is well known [27]. Theory and experimental data are in fair agreement. Figure 30 gives a summary of metal work function data for PMOS. Taking a 4.9 eV to 5.3 eV range value for PMOS, metals such as Be, Co, Au, Re, Ni, Te, Ru, and Os (possibly extended to Ir) can be selected. Figure 31 gives NMOS candidate work function data. In the selected range 3.9 to 4.4 eV metals Ta, Nb, Cd, Zn, Pb, Ga, Sn, Bi, Al, V, Ag, Mn and Cu can be chosen for NMOS.

As in the case of high-K materials selection, the metal gate is subjected to high-temperature processing. Annealing conditions required to activate source and drain implants and hydrogen passivation anneal should be considered. For NMOS, the only metals with a suitable work function and a sufficiently high melting point ($\gg 1000°C$) are Ti, Ta and Nb. As for hydrogen anneal compatibility, Ti might store the hydrogen and expand due to the hydrogen content. On the basis of similar arguments, the high-melting metals Co, Re, Ni and Ru, may be considered for PMOS gate contacts. The conducting oxides of Ru and Pt are unlikely to survive hydrogen passivation, since they are thermodynamically unstable in hydrogen.

High-melting metallic alloys are also potential candidates. Other candidates are silicides and nitrides. Silicides have the advantage of poly-Si compatibility, chemical and thermal stability. Silicidation has an effect on the work function. Ni- and Ta-silicides are also good candidates for PMOS and NMOS gates, respectively [28]. Generally, nitrides show larger work functions than the corresponding pure metals [29]. Nitrogen itself has a work function as high as 6.89 eV, Ti of 3.6 eV and TiN is reported to have 4.6 eV, somewhat above mid gap energy. Transition metal nitrides, such as ZrN and HfN might be options for NMOS. Maybe even LaN could be considered if the nitride can be made conductive. For PMOS, MoN and CrN might be explored.

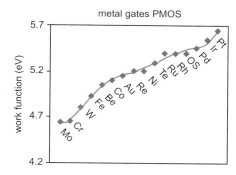

Figure 30: Work functions of selected metals for PMOS metal gates [27].

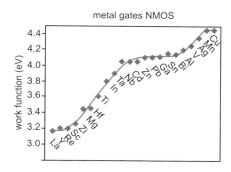

Figure 31: Work functions of selected metals for NMOS metal gates [27].

SOI and mid-gap metal gates

Integration of dual metal gates in standard CMOS processing is very difficult. Many process integration issues remain to be solved. On the other hand, fully depleted silicon on insulator (SOI) MOSFETs (see Sect. 7), one of the alternative transistor structures, requires just one metal gate, a mid-gap gate electrode. TiN is a good mid-gap metal gate candidate. It is stable, high-melting and compatible with poly-Si. Both, the performance advantages of SOI for high frequency applications and the technological problems encountered with a dual metal gate approach, make SOI devices highly attractive.

6 Junctions and Contacts

6.1 Shallow Junctions

The series resistivity of a MOSFET associated with the contacts, the shallow junctions and the channel are illustrated in Figure 32. Ideally, the current drive in a MOSFET is limited by the intrinsic channel resistance. All other resistances in Figure 32 degrade the intrinsic device capability. Normal design procedures require these other resistances to total less than 10 % of the channel resistance. Simple first-order MOS device physics gives the channel resistance as

$$R_{\text{chan}} = \left[\frac{w}{L}\mu\frac{\varepsilon_{\text{ox}}}{t_{\text{ox}}}(V_G - V_T)\right]^{-1} \qquad (15)$$

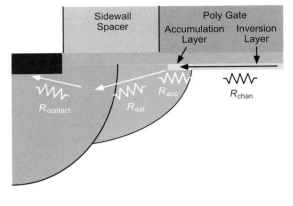

Figure 32: Schematic cross section of the source/channel boundary and part of the channel region. Arrows indicate the current flow path and the resistors illustrate the various regions that can affect the current drive capability of the device.

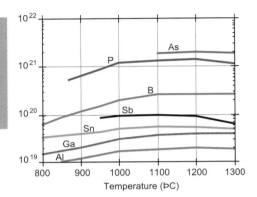

Figure 33: Solid solubility curves for various dopants in silicon. Values are the equilibrium solubilities [cm^{-3}] at each temperature and may not be achieved in device doped regions [30].

As device geometries are scaled down, ideal scaling suggests that w, L, t_{ox}, V_G and V_T all decrease at the same rate (Tab. 1). In this scenario then, R_{chan} would remain constant as the technology is scaled. Therefore, if this were done, the various parasitic resistances in Figure 32 would simply need to remain relatively constant in value from generation to generation. However, this becomes difficult as doping concentrations become limited by their solubility in silicon. In addition, ideal scaling has not been followed in the past because higher performance has been a specific objective of scaling the technology and the parasitic resistances have been required to decrease in order to keep R_{chan} as the dominant resistive component in the device.

The extension (or tip) region links the contacts in the deep source/drain to the channel while presenting a shallow effective source/drain junction to the channel. There are two issues that dominate the constraints placed on the source/drain extension regions: parasitic resistance and short channel effects. Short channel effects are a result of the drain electric field extending through the channel region and therefore modulating the channel potential near the source (see Figure 10). A shallow junction minimizes short channel effects as it makes it more difficult for the drain field to affect the source region. However, a shallow junction has a higher resistance than a deeper junction with the same doping, so it is important to maximize the doping in the extension to minimize the parasitic series resistance. Eventually, higher doping in the extension becomes impossible when the solubility of the dopant in silicon is reached and there is a trade-off between short-channel effects and series resistance.

When current leaves the channel inversion layer, it first flows through an accumulation region until it reaches a part of the extension where the doping exceeds the accumulation layer carrier density (Figure 32). It then gradually spreads and flows through the length of the extension, usually defined by a sidewall spacer. The accumulation and spreading resistance depend directly on the extension region lateral doping profile and are the dominant component of the extension resistance. The steeper the lateral profile is, the shorter the accumulation region will be and the lower the spreading resistance. Thus, the entries in Table 2 for abruptness indicate progressively sharper profiles as the technology progresses. There is an optimal lateral abruptness for a particular extension junction depth because of the complex interplay between series resistance and short channel effects. Two competing factors are at work. If the junction gradient is too grad-

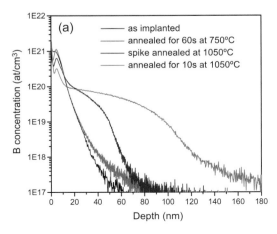
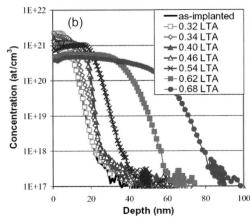

Figure 34:
(a) Boron depth profiles after implantation of 1 keV B$^+$ ions with an ion dose of 1×10^{15} cm^{-2} and rapid thermal processing as measured with SIMS.
(b) Boron depth profiles after preamorphisation and 2 keV B$^+$ implantation laser ($3 \cdot 10^{15}$ cm^{-2}) annealing with different laser power [J cm^{-2}]. The black solid line shows the as-implanted profile [32].

ual, the series resistance is too high. (Also, counter-doping of the channel by the junction gradient causes the threshold voltage at short channel lengths to decrease). Very abrupt junctions lower the series resistance, but present an effectively deeper junction to the channel edge that worsens short channel effects. The improvement in drive current by reduced series resistance is countered by the degradation in leakage currents due to the deteriorating short channel effects.

Practical profiles achievable with today's manufacturing techniques of ion implantation followed by a rapid thermal anneal to activate the dopants are typically limited in the lateral abruptness that they can achieve. They have an achievable sheet resistance that is a factor of two to three worse than an ideal box-shaped profile because implanted profiles are smeared out by ion range straggling and by diffusion of the dopant atoms during the activation anneal. In spite of this, it is the lateral abruptness that contributes to most of the resistance in the extension region. The key to forming shallow junctions is achieving maximum electrical activation with minimum diffusion. The maximum concentration of a dopant that can be dissolved in silicon under equilibrium conditions without forming a separate phase is termed the solid solubility [30]. The maximum concentrations of many of the elements used as dopants are shown in Figure 33. Though the solid solubility is the thermodynamic maximum concentration that can be accommodated in a solid without a separate phase forming, kinetics effects typically limit the electrically active dopant concentrations that can be achieved during normal anneals. By this we mean that small, electrically inactive clusters limit the electrical solubility to values considerably lower than the maximum solid solubility shown in Figure 33. It is the electrically active concentration that is most important to device designers.

Figure 34a shows Boron depth profiles after implantation of 1 keV B^+ ions with a dose of 1×10^{15} cm^{-2} into Si and different annealing procedures. The concentration depth profiles were measured with secondary ion mass spectroscopy (SIMS). The profiles smear out rapidly with increasing annealing temperature. When only a fast heating pulse, a so called spike anneal is used, fairly shallow profiles can be achieved. Nevertheless, Figure 34a implies that the ITRS requirements will not be achievable for the technology node of 100 nm. This is why red entries appear in Table 2 for the contact depth approx. at the 80 nm generation.

There may be opportunities in this area for new conceptual approaches. One possibility is the use of metastably doped silicon, that is, the incorporation of doping concentrations in excess of the thermodynamic solubility limits. It is well established that such doping levels can be achieved by laser annealing [31]. In these processes, the silicon is locally melted and dopants can be "frozen in" at electrically active concentrations above 10^{21} cm^{-3} during the very rapid cooling that occurs after the laser pulsings. Figure 34b shows B depth profiles after implantation and laser annealing [32]. The sample was implanted in two steps. First 10 keV Ge+ ions with a dose of 2×10^{15} cm^{-2} were implanted to produce a 20 nm thick amorphous Si layer, followed by a 1 keV B implantation with a dose of 3×10^{15} cm^{-2}. For the laser annealed 20 ns pulses with a wavelength of 532 nm were used. The amorphous surface layer melts at a lower temperature than crystalline silicon. This allows the melt depth to be controlled by adjusting the laser power. The laser power was varied between 0.2 and 0.68 Jcm^{-2}. As as result the depth profiles shown in Figure 34b are much shallower and steeper than after RTP processing (Figure 34a). However, such doping concentrations are metastable, and any subsequent heat cycles provide a huge driving force for precipitation and deactivation of the dopants [33].

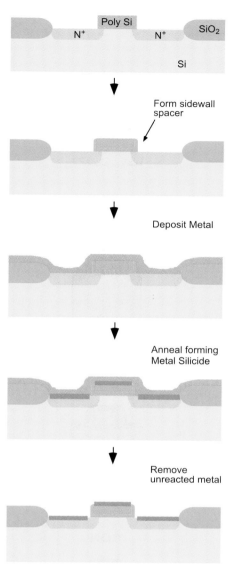

Figure 35: Process flow for the formation of self-aligned silicide contacts.

6.2 Junction Contacts

The other major resistance component in a device is the contact resistance. Contacts in today's device structures are normally made by self-aligned silicides contacting heavily doped silicon as illustrated in Figure 35. The self-alignment is achieved by depositing a metal, e.g. Co, which reacts during thermal processing with the underlying silicon to form the desired silicide phase (CoSi$_2$) but remains unreacted on all areas of the chip protected with silicon dioxide. The pure metal on the SiO$_2$ can be selectively removed by wet etching. This process provides an ohmic contact completely covering the area of the source/drain diffusion and, therefore, minimizes the contact resistance.

The effective contact resistance depends on the current flow pattern or, in other words, on the effective area of the contact. When the current flows into the entire length of the contact, the contact resistance is approximately given by

$$R_{\text{contact}} \approx \frac{\rho_c}{w_{\text{si}} L_c} \quad (16)$$

where ρ_c is the specific contact resistivity of the silicide/semiconductor contact (Ω cm^{-2}), w_{si} the contact width and L_c the contact length. Thus the contact resistance varies inversely with the contact area. The ideal silicide should have very low contact resistance, a low electrical resistivity, be thermally and structurally stable and should consume only little silicon during its formation. The latter may cause a problem when the junctions become ultrashallow and high doped silicon is consumed. The junction resistance may increase. For a tunneling contact, the specific contact resistivity ρ_c depends on the semiconductor-metal barrier height and the silicon doping [33].

$$\rho_c = \rho_{c0} \exp\left(\frac{2\sqrt{\varepsilon_S} m^*}{\hbar} \frac{\phi_B}{\sqrt{C_S}} \right) \quad (17)$$

where ρ_{c0} is contact resistivity for an infinitely high active surface doping concentration, C_S the actual active surface doping concentration and Φ_B the barrier height between the silicon and metal or silicide.

Required values in Table 2 are derived from a desire to limit the total parasitic resistance in the device structure to no more than 10 % of the channel resistance. If the area of the silicide/semiconductor contact is taken to be a square defined by the minimum feature size, then the implied contact resistance from the numbers in Table 2 is approximately 1000 Ω, independent of generation technology. With reducing feature sizes, this constant resistance is achieved by requiring the contact resistivity to scale directly with the contact area.

Based on Eq. (17) two key parameters in reducing ρ_c are the silicon doping and the barrier height Φ_B. If we take the barrier height as one half of the silicon bandgap (0.55 eV) and assume that the maximum electrically active dopant concentration in the silicon is 2×10^{20} cm^{-3}, then ρ_c is limited to about 1×10^{-7} Ωcm^{-2}. Thus, the required values beyond the 100 nm node are not achievable by the contacting schemes currently employed in CMOS technology. New approaches will be needed. The obvious areas to focus on are the barrier height and the dopant solubility. As mentioned above laser melting can be employed to increase C_s in excess of 10^{21} cm^{-3}. A further promising technique is the formation of elevated source and drain contacts by selective deposition of a high doped Si or Si-Ge layer by CVD as will be discussed in Sect. 7. The deposition of a Si-Ge alloy layer has two beneficial effects: the barrier height is lowered due to the reduced band gap energy of the Si-Ge alloy, and the solid solubility for dopants is higher than for pure Si [34].

7 Advanced MOSFETs Concepts

Research MOSFETs with gate lengths of 15 nm [36] or even 10 nm have been fabricated as shown in Figure 1 [3]. However, besides scaling the transistor to such extreme dimensions substantial improvements by the use of new materials and new transistor concepts are required. At small dimensions undesired short channel effects tend to become dominant, due to the distortion of the channel potential by the drain voltage as indicated in Figure 10. There are a number of alternative transistor concepts to reduce short channel effects and improve performance. Very promising is the use of SOI substrates instead of bulk Si wafers. SOI wafers are commercially produced either by wafer bonding and etch back (BESOI) or by ion implantation. The latter process, SIMOX (separation by implantation of oxygen) process involves implantation of a high dose of oxygen into a heated silicon substrate and subsequent annealing at very high temperature, typically ≈ 1250°C for 8 h [37]. Generally, SOI transistors have smaller parasitic capacitances, smaller source/drain leakage, are more immune to soft errors caused by alpha particles and allow higher speed and lower power consumption than bulk transistors. The buried, perfectly insulating SiO$_2$ layer with a typical thickness of several 100 nm eliminates several leakage paths. However, the SOI substrate is more expensive and the poor thermal conductivity of the buried silicon dioxide may create a heat problem. Nevertheless, many different transistor concepts are presently under investigation as shown in Figure 36 in

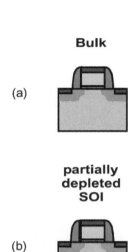
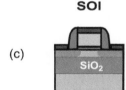
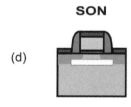

Figure 36: Conventional bulk MOSFET (a) compared with SOI (b, c) and SON-Transistor concepts (d).

comparison with a standard bulk transistor. Figure 36b shows a transistor fabricated on a SOI substrate where the silicon region under the gate is partially depleted. Under gate bias, the potential of this silicon region may float and cause undesirable threshold shifts. This problem is eliminated when the silicon layer on the silicon dioxide is made so thin that the region becomes fully depleted. Figure 36c shows a fully depleted transistor with an additional improvement, elevated source and drain contacts, which is an efficient measure to reduce the source and drain resistance. An example is given in Figure 37, showing a cross-section TEM micrograph of a 50 nm gate length transistor with a thin silicon layer of 30 nm [38]. The source/drain contacts are raised in order to lower the series resistance. This is made by selective deposition of silicon on the contact areas before the formation of the self-aligned silicide contacts. A comparison of transistors with and without raised contacts reveals substantial improvements. The I_{on} drive current could be increased by 20 to 30 % for n- and p-channel devices, respectively. This results in a larger I_{on}/I_{off} ratio (order of 10^6) and steeper subthreshold slopes.

Another concept is *silicon on nothing* (SON) where a region under the gate is removed and possibly refilled with a dielectric material (Figure 36d) [39]. A buried Si-Ge alloy layer is selectively underetched in a mesa structure to realize the SON structure. The advantage of this is that no expensive SOI wafers are necessary and the buried insulating layer is manufactured only at selected areas. This approach appears promising for SoC. Excellent performance can be achieved when the Si-channel layer is kept very thin (10 - 20 nm) [40].

Alternatively, **ultra-thin body** (UTB) transistors can be made on SOI. The structure is identical to the fully depleted SOI transistor shown in Figure 36c, except for the very small thickness of the Si body which will be less than 10 nm. This concept allows the fabrication of very densely packed, high performance transistors with record frequencies. Scaling the linear dimension of a transistor by 30 % reduces the area by 50 %, thus doubling the number of transistors per unit area becomes feasible. When the gate length is reduced to about 20 nm, frequencies in the THz range seem achievable [41].

In addition to the described concepts transistors with **double gates** (DG) are investigated, where two symmetric gates are arranged in a planar or in a vertical configuration as illustrated in Figure 38 [42], [43]. A thin silicon ridge serves as the heart of the transistor with symmetric gates on both sides. The direction of the current flow from the source to drain can be horizontal (Figure 38a, b) or vertical (Figure 38c). In the FinFET type source and drain are beside the gate region and the current flows along the silicon ridge (Figure 38b). The name stems from the silicon fin which forms the basic building block of the transistor. The advantage of the vertical DG device (Figure 38c) is that the physical gate length can be defined by layer growth or by ion implantation and diffusion and not by lithography. However, all DG transistors are difficult to fabricate, in particular, the planar DG transistor, where a single crystalline silicon channel layer has to be grown on top of a buried gate stack. The two gates have to be perfectly aligned, which is very difficult task in view of the extremely small dimensions (1-3 × 10 nm). The vertical DG transistors are somewhat easier to fabricate, nevertheless the technological problems, such as the etching of the nanostructures and the growth or deposition of a high-quality gate dielectric on the vertical side walls, is a major challenge (see e.g. Figure 24)

Both the DG and the UTB devices rely on the thickness of the silicon channel to obtain optimum gate control on the channel, suppression of short channel effects and minimization of leakage. Figure 39 shows the basic structure of these transistors, the corresponding band diagrams under inversion conditions and the charge distribution near the oxide layer. In a double gate configuration two symmetric gates produce two inversion layers on both sides of the silicon layer. Excellent transistor control is expected without channel dopants. This implies an interesting effect. Due to the lack of

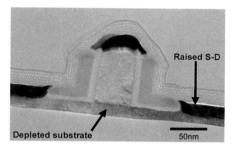

Figure 37: TEM cross-section of a 50 nm transistor on a thin silicon body with raised source/drain contacts [38].

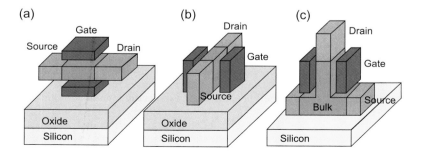

Figure 38: Various types of parallel double-gate MOSFETs [42].

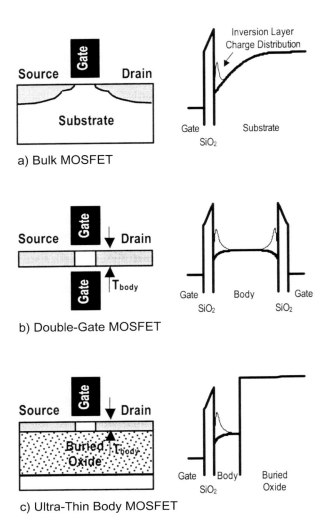

a) Bulk MOSFET

b) Double-Gate MOSFET

c) Ultra-Thin Body MOSFET

Figure 39: Standard bulk, double-gate and ultra-thin body MOSFETs and corresponding band diagrams and charge distributions under inversion conditions. T_{body} denotes the thickness of the silicon layer, named body [45].

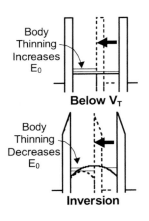

Figure 40: Energy band diagrams and energy levels for DG and UTB transistors as a function of silicon body thickness for two gate voltages, below and above the threshold voltage [45].

dopants no depletion charge ($Q_{depl} = 0$) exists. This yields a reduction of the effective vertical electric field, via $E_{eff} = (Q_{depl} + \eta\, Q_{inv})/\varepsilon_{Si}$ [44]. As a result, a higher carrier mobility in the channel is expected than in a doped channel, since the carrier mobility is no longer reduced by impurity scattering. The vertical electric field can be further lowered by thinning the silicon layer. At 1 V gate bias the effective electric field can be lowered by about 50 % for a 25 nm thick DG or a 10 nm UTB transistor as compared to a bulk device [44]. The smaller vertical field also reduces interface scattering for ultrathin gate dielectrics, and most important, reduces direct-tunneling through the gate dielectric. As a consequence, gate leakage is reduced and the requirements for the gate dielectric are less stringent.

At very small dimensions, quantum effects in terms of energy quantization, come into play in DG and UTB transistors. As illustrated in Figure 40 the energy ground state is determined by quantization when the thickness of the silicon body layer is sufficiently small. Below the threshold voltage the carriers are confined in a square well. Under inversion conditions, a triangular well forms with discrete energy levels. As the body is thinned, the confining electric field and depth of the potential well decrease and thus the eigenstate energies in the well are reduced as indicated in the lower part of Figure 40. This effect lowers the threshold voltage of the transistor. The effect of the reduction of the vertical field strength in ultrathin structures is more pronounced in the DG-case, because at very small body thicknesses merging of the two inversion charge distributions decreases the depth of the potential well. This effect promises further improved carrier mobility in the channel, since scattering at the silicon-dielectric interface should become less critical. However, for extreme scaling of the body thickness (≈ 5 nm) this advantage may vanish, since the potential well, responsible for carrier confinement, becomes shallower, causing carrier scattering at the interface and thus gate leakage to increase.

A further possibility of improving device performance without scaling the transistors to smaller dimensions, involves the use of strained silicon and strained $Si_{1-x}Ge_x$ layers. Strained silicon has considerably larger carrier mobilities than bulk silicon and thus faster devices can be made [46]. Thermodynamically the Si-Ge system is fully miscible. Ge has a 4.2 % larger lattice constant at room temperature than Si. This implies that $Si_{1-x}Ge_x$ layers grown on a Si substrate are strained. The elastic strain of the layer depends nearly linearly on the Ge concentration x. On the other hand, if Si is epitaxially grown on a strain relaxed $Si_{1-x}Ge_x$ layer, e.g. $x = 0.3$, the silicon grows tetragonally distorted as long as the thickness is below its critical thickness. The in-plane lattice parameter is maintained and the vertical lattice parameter adjusts elastically. The major problem is the production of these strained systems on the substrate with low defect densities. Wafer bonding [47] and He implantation are promising techniques for forming thin strain-relaxed $Si_{1-x}Ge_x$ layers [48]. Both, MOSFETs with strained Si channels [46] and modulation doped FETs using strained Si quantum wells as the active channel exhibit superior performance [49]. Leading chip manufacturers have decided to adopt strained silicon for high performance CMOS microprocessors [46], [50].

8 Summary

Research and development of high-performance MOSFETs with gate length < 100 nm focuses on three parallel issues: the scaling of the dimensions, the implementation of new materials and the investigation of new transistor concepts. In this chapter we addressed these issues with particular emphasis on the materials developments. The large number of new materials to be implemented in the next chip generations is a major challenge for material science. The materials have nanoscale dimensions and have to be investigated under extreme conditions. The most prominent case is the replacement of the silicon oxynitride gate dielectric by high-K materials combined with metal gates. The challenges for high-K materials have been described. High-K oxides are certainly capable of solving the leakage current problem envisaged for oxynitrides with thicknesses of less than 1.5 nm. However, many materials issues, e.g. interface state density, thickness uniformity, thermal stability and effect on the carrier mobility in the MOSFET channel remain subjects of research and development. Remarkable is also the rapid development of strained silicon as a new material for high speed devices. We also addressed the problems related to the scaling of the source and drain junctions and the silicided contacts. In addition, we presented an introduction into new transistor concepts.

Further issues, such as gate architecture, mobility requirements, noise and matching and many other process/performance-related topics can only partially be addressed now. Only truly scaled devices can really show all the hidden traps and drawbacks of the new materials. Here, research can anticipate a lot of possible problems.

Acknowledgements

The editor would like to thank Andreas Gerber (FZ Jülich) for checking the symbols and formulas in this chapter.

References

[1] J. Bardeen, W. H. Brattain, Phys. Rev. **74**, 230 (1948).

[2] I.M. Ross, Bell Labs Technical Journal, Autumn, 3 (1977).

[3] B. Doyle, R. Arghavani, D. Barrage, S. Datta, M. Doczy, J. Kavalieros, A. Murthy, R. Chau, Intel Technology Journal **6**, 42 (2002).

[4] conceptually taken from L. Risch, personal communication.

[5] T. Hori, *Gate Dielectrics and MOS ULSIs: Principles, Technologies and Applications*, ed. I. P. Kaminow, W. Engl and T. Sugano, Springer Verlag, Heidelberg, 1997.

[6] Y. Taur, and T. H. Ning, *Fundamentals of Modern VLSI Devices*, Cambridge University Press, 1998.

[7] S. M. Sze, *Physics of Semiconductor Devices*, John Wiley & Sons, 1981.

[8] S.-H. Lo, D. A. Buchanan, and Y. Taur, IBM J. Res. Develop. **43**, 327 (1999).

[9] Int. Technology Roadmap for Semiconductors, http://www.public.itrs.net.

[10] D. A. Buchanan, IBM J. Res. Develop. **43**, 245 (1999).

[11] B. E. Deal and A. S. Grove, J. Appl. Phys. **36**, 3770 (1965).

[12] S. H. Lo, D. A. Buchanan, Y. Taur and W. Wang, IEEE Electron Device Letter, 209 (1997).

[13] A. I. Kingon, J.-P. Maria and S. K. Streiffer, Nature **406**, 1032 (2000).

[14] T.B. Reed, *Free energy of Formation of binary compounds*, MIT, 1971.

[15] D.G. Schlom and J. H. Haeni, Mat. Res. Bull. **27**, 198 (2002).

[16] J. Robertson, J. Vac. Sci. Technology B**18**, 1785 (2000).

[17] E. Nohira, W. Tsai, W. Besling, E. Young, J. Petry, T. Conard, W. Vandervorst, S. De Gendt, M. Heyns, J. Maes, M. Tuominen, Journal of Non-Crystalline Solids **303**, 83 (2002).

[18] C. Zhao, O. Richard, E. Young, H. Bender, G. Roebben, S. De Gendt, M. Houssa, R. Carter, W. Tsai, O. Van, M. Heyns, Journal of Non-Crystalline Solids **303**, 144–149 (2002).

[19] ASM international, private communication.

[20] A. Agarwal, M. Freiler, P. Lysaght, L. Perrymore, R. Bergmann, C. Sparks, B. Bowers, J. Barnett, D. Riley, Y. Kim, B. Nguyen, G. Bersuker, J. E. Lim, S. Lin, J. Chen, R. W. Murto, H. R. Huff, E. Shero, Materials Research Society Symposium Proceedings Vol.670, page K2.1.1 (2000).

[21] *Handbook Chemistry and Physics*, 76 ed., CRC Press, New York.

[22] Sang-Hyun Oh1, J.M. Hergenrother, T. Nigam, D. Monroe, F.P. Klemens, A. Kornblit, W.M. Mansfield, M.R. Baker, D.L. Barr, F.H. Baumann, K.J. Bolan, T. Boone, N.A. Ciampa, R.A. Cirelli, D.J. Eaglesham, E.J. Ferry, A.T. Fiory, J. Frackoviak, J.P. Garno, H.J. Gossmann, J.L. Grazul, M.L. Green, S.J. Hillenius, R.W. Johnson, R.C. Keller, C.A. King, R.N. Kleiman, J.T-C. Lee, J.F. Miner, M.D. Morris, C.S. Rafferty, J.M. Rosamilia, K. Short, T.W. Sorsch, A.G. Timko, G.R. Weber, G.D. Wilk, and J.D. Plummer, Proc. of the IEEE Int. Electron Device Meeting (IEDM) (2000).

[23] Laegu Kang, Katsunori Onishi, Yongjoo Jeon, Byoung Hun Lee, Changseok Kang, Wen-Jie Qi, Renee Nieh, Sundar Gopalan, Rino Choi, and Jack C. Lee, Proc. of the IEEE Int. Electron Device Meeting (IEDM), (2000).

[24] G. D. Wilk, R. M. Wallace, and J. M. Anthony, J. of Appl. Phys. **87**, 484 (2000).

[25] R. A. McKee, F. J. Walker, and M. F. Chisholm, Phys. Rev. Lett. **81**, 3014 (1998).

[26] J.R. Hauser, K Ahmed, CP449 Charactization and Metrology for ULSI Techn., Intern. Conf., 235 (1998).

[27] F.R. de Boer, R. Boon, W. C. M. Mattens, A.R. Miedema, *Cohesion in Metals*, North Holland, 1988.

[28] You-Seok Suh, G. P. Heuss, V. Misra, Appl. Phys. Lett. **80**, 1403 (2002).

[29] M. Quin, V. M. C. Poon, S. C. H. Ho, J. Electrochem. Soc. **148**, G271 (2001).

[30] F. A. Trumbore, Bell Syst. Tech. J. **39**, 205 (1960).

[31] P. M. Rousseau, P. B. Griffin, W. T. Fang, and J. D. Plummer, J. Appl. Phys. **84**, 3593 (1998).

[32] R. Murto, K.S. Jones, M. Rendon and S. Talvar, Proc. of the Internat. Conf. on Ion Implantation Technology, Alpbach, Austria, IEEE 00EX432, 2000.

[33] K. S. Jones, E. Kuryliw, R. Murto, M. Rendon and S. Talwar, Proc. of the Internat. Conf. on Ion Implantation Technology, Alpbach, Austria, IEEE 00EX432, 2000.

[34] C. M. Osburn and K. R. Bellur, Thin Solid Films **332**, 428 (1998).

[35] M.C. Ozturk, N. Pesovic, I. Kan, J. Liu, H. Mo and S. Gannavaram, Extended Abstracts of the Second Intern. Workshop on Junction Technology (IEEE Cat. No. 01EX541C), Tokyo, Japan, 77 (2001).

[36] F. Bœuf, T. Skotnicki, S. Monfray, C. Julien, D. Dutartre, J. Martins, P. Mazoyer, R. Palla, B.Tavel, P. Ribot, E. Sondergard, and M. Sanquer., Proc. of the IEDM, 2001.

[37] J.-P. Colinge, *Silicon on Insulator: Materials to VLSI*, Kluwer Academic Publisher, Boston, 1991.

[38] R. Chau, J. Kavalieros, B. Doyle, A. Murthy, N. Paulsen, D. Lionberger, D. Barlage, R. Arghavani, B. Roberds and M. Doczy, Proc. of the IEDM, 2001.

[39] T. Skotnicki, Proc. ULIS Conf. Munich, 2002.

[40] S. Monfray, T.Skotnicki, Y. Morand, S. Descombes, M.Paoli, P.Ribot, A.Talbot, D.Dutartre, F. Leverd, Y.Lefriec, R.Pantel, M.Haond, D.Renaud, M-E. Nier, C. Vizioz, D. Louis, N. Buffet, Proc. of the IEEE Int. Electron Device Meeting (IEDM) (2001).

[41] www.intel.com/research/silicon.

[42] L. Risch, Infineon, private communication.

[43] H.-S. Philip Wong, Proc. of the 31st European Solid State Circuits Conference (ESSDERC 2001), Nürnberg, Germany, 2001.

[44] A.G. Sabnis et al., Proc. of the IEDM, 16 (1979).

[45] L. Chang, K. J. Yang, Y.-C. Yeo, Y-K. Choi, T-J. King, and C. Hu, Proc. of the IEDM, 2001.

[46] K. Rim, R. Anderson, D. Boyd, F. Cardone, K. Chan, H. Chen, J. Chu, K. Jenkins, T. Kanarsky, S. Koester, B. H. Lee, K. Lee, V. Mazzeo, A. Mocuta, D. Mocuta, P. Mooney, P. Oldiges, J. Ott, P. Ronsheim, R. Roy, A. Steegen, M. Yang, H. Zhu, M. Ieong, H-S.P. Wong, Proc. of ULIS Conference, Munich (2002).

[47] L. Huang, et al., Proceed. VLSI Symp. 57 (2001).

[48] B. Holländer, St. Lenk, S. Mantl, H. Trinkaus, D. Kirch, M. Luysberg, T. Hackbarth, H.-J. Herzog, and P. F. P. Fichtner, Nucl. Instr and Meth. in Phys. Res. **B175-177**, 357 (2001).

[49] H.-J. Herzog, T. Hackbarth, U. Seiler, U. König, M. Luysberg, B. Holländer and S. Mantl, IEEE Elect. Dev. Lett. **23**, 485 (2002).

[50] S. Thompson, N. Anand, M. Armstrong, C. Auth, B. Arcot, M. Alavi, P. Bai, J. Bielefeld, R. Bigwood, J. Brandenburg, M. Buehler, S. Cea, V. Chikarmane, C. Choi, R. Frankovic, T. Ghani, G. Glass, W. Han, T. Hoffmann, M. Hussein, P. Jacob, A. Jain, C. Jan, S. Joshi, C. Kenyon, J. Klaus, S. Klopcic, J. Luce, Z. Ma, B. Mcintyre, K. Mistry, A. Murthy, P. Nguyen, H. Pearson, T. Sandford, R. Schweinfurth, R. Shaheed, S. Sivakumar, M. Taylor, B. Tufts, C. Wallace, P. Wang, C. Weber, M. Bohr, Proc. of the Int. Electron Device Meeting (IEDM) (2002).

Ferroelectric Field Effect Transistors

Hermann Kohlstedt, Department IFF, Research Center Jülich, Germany

Hiroshi Ishiwara, Frontier Collaborative Research Center, Tokyo Institute of Technology, Japan

Contents

1 Introduction	387
2 Principles of Ferroelectric Field Effect Transistors	387
2.1 Design Structures for Ferroelectric Field Effect Transistor and Material Aspects	388
2.2 Ferroelectric Directly on Silicon	389
2.3 Buffer Layer between Ferroelectric Material and Silicon	390
2.4 Metal-Ferroelectric-Metal Floating Gate Structures	391
2.5 Metal-Ferroelectric on a Conductive Oxide	392
3 Electrical Characterization of FeFETs	392
3.1 MFIS Structures	393
3.2 MFMIS Structures	394
3.3 Optimization of a FeFET	396
4 Cell Designs and Device Modeling for FeFETs	398
4.1 Device Simulation of FeFETs	398
4.2 A 1T-2C Cell Design for FeFETs	399
4.3 A FeFET Memory Concept with Disturbance Free Operation Scheme	399
5 Neural Network Circuits with FeFETs	400
6 Summary and Outlook	401

Ferroelectric Field-Effect Transistors

1 Introduction

The bi-stable hysteresis of a ferroelectric material offers the possibility to develop electrically switchable, non-volatile data storage devices. The typically implemented ferroelectric data storage element is a capacitor consisting of a thin ferroelectric film in between two conductive electrodes. As described in Chap. 22, the polarization direction is set by a (positive or negative) voltage pulse to the ferroelectric capacitor defining logical "0" or "1". For readout, another voltage pulse is applied and the stored bit configuration determines whether or not the polarization switched direction. The ferroelectric capacitor suffers from the fact that data are destroyed during readout and bit reprogramming is required after each read cycle, i.e. the data read process is destructive. The result in commonly used FeRAM (Ferroelectric Random Access Memory) cells is a large accumulation of read/write cycles due to the destructive readout scheme which may reduce the lifetime of FeRAMs because of fatigue of the ferroelectric material. In addition, during read-processing data storage is volatile and data could be erased if power supply is lost during read. To overcome the destructive readout scheme of FeRAMs considerable efforts have focused recently on the development of a so-called ferroelectric-gate field-effect transistor (FeFET). This device offers a significant advantage in comparison to the ferroelectric capacitor. The data readout in an FeFET is non-destructive and needs no reprogramming. Further advantages are reduced power consumption and better scaling properties as in the case of usual FeRAM cells. In addition, the FeFET is obviously comprised of not only a memory function but also the logical (switching and amplifying) function of transistors. Ultimately, it can be considered as a logical device which memorizes its state in a non-volatile fashion. The principle of a ferroelectric-gate field-effect transistor is based on a conventional Si MOSFET (Metal Oxide Semiconductor Ferroelectric-Gate Field Effect Transistor) whose gate dielectric is a ferroelectric material. Therefore, we advise the reader, who is not familiar with the background of MOSFETs and ferroelectric capacitors, first to study Chap. 2 and Chap. 13 of this book. The concept of a FeFET is a logical consequence of those topics.

This chapter is organized as follows: Sec. 2 presents the concept of the ferroelectric field-effect transistor. Discussion will focus on the implementation of various buffer layers sandwiched between the semiconductor channel and the ferroelectric film to avoid interdiffusion and charge injection will be discussed. In Sec. 3 the most important electrical parameters of FeFETs are discussed. In Sec. 4, a review of the current status of FeFET device simulation tools is given including the most relevant cell design structures. Sec. 5 describes the attempts to use FeFETs for artificial neural network applications.

2 Principles of Ferroelectric Field Effect Transistors

As described in Chap. 22 ferroelectric memories are based either on the (MOS) 1 transistor - 1 capacitor (1T1C FeRAM cell) approach. The 1T1C memory (and varieties) has a cell-structure similar to that of a dynamic random access memory (DRAM). The transistor is separated by a thick interlayer dielectric from the ferroelectric capacitor. In other words, the capacitor and the transistor are two individual electronic devices and are typically separated by some 100 nm within a cell. Fabrication issues of the 1T1C cell are mainly concerned with the reliability of the ferroelectric film and the transistor. The reason for that is, that both devices need rather different deposition methods, etch-

III Logic Devices

ing conditions and post annealing treatments which may influence the device performance. From the point of view of integration, the idea to implement the ferroelectric capacitor directly as gate dielectric, is rather straightforward.

This is shown in Figure 1a–c. Figure 1a represents the layout of a 1T1C DRAM cell. For a 1T1C FeRAM cell, the dielectric (of the DRAM capacitor) is replaced by a ferroelectric material, e.g. $PbZr_XTi_{1-X}O_3$ or $SrBi_2Ta_2O_9$, as shown in Figure 1b. The ferroelectric field-effect transistor consists of a field-effect transistor whose gate dielectric is a ferroelectric material (Figure 1c). This means that the ferroelectric layer is in direct contact with the semiconductor channel. In comparison with the 1T1C FeRAM cell, the FeFET is a single element device.

The FeFET layout is shown schematically in Figure 2. Before discussing the interfaces in FeFETs in detail (in Sec. 2.1 – 2.5), the principle operation of a FeFET is described.

In Figure 3(a) and (b) the charge motion in a ferroelectric FET during one cycle of operation is shown [1], [2].

For an applied positive voltage pulse ($V_{GS} > V_c$) the polarisation vector $\boldsymbol{P}$ is directed towards the Si-channel (Figure 3a). Here, the coercive voltage V_c is defined as E_c/d_{Fe}. E_c is the coercive field of the ferroelectric material and d_F is its thickness. The negative charge within the ferroelectric material at the interface to the p-type semiconductor channel results in an inversion and an accumulation of electrons (in the semiconductor) near the interface. The source-drain channel is in the on-state because the channel resistance is low. This state is stable, (for $V_{GS} = 0$ V) as long as the remanent polarization $\boldsymbol{P}_r$ stays sufficiently large, i.e. the retention loss is small enough. The stable polarisation of the ferroelectric layer is responsible for non-volatile operation. Vice versa, if a negative voltage pulse is applied ($V_{GS} < -V_c$), $\boldsymbol{P}_r$ is directed opposite and electrons are depleted within the semiconductor channel. The channel conductance is high and the FeFET is in the off-state. This situation is schematically shown in Figure 3b. The non-destructive readout results from the fact that only the source-drain channel conductance has to be determined by a peripheral sense amplifier. In this respect, FeFETs are basic elements of resistance based RAMs (see: Introduction to Part IV). Thereby, the status of the ferroelectric polarization is undisturbed, while in a 1T1C FeRAM cell readout process, a polarisation reversal and reprogramming is necessary.

In short terms, interesting memory functions of ferroelectric field-effect transistors (FeFETs) [3] – [6] are:

- the non-volatile data storage,
- the non-destructive readout, and
- the compact cell design, with the potential of a high integration density.

2.1 Design Structures for Ferroelectric Field Effect Transistor and Material Aspects

The straightforward concept of a FeFET design, as sketched in Figure 2, is based on a stack composed a metal gate electrode, a ferroelectric, and the semiconductor (short: MFS). This design results in a range of requirements, which leads to some extreme challenges for the integration process. [6] – [9]. In order to develop highly competitive ferroelectric gate type transistors, it is necessary that the interface between the silicon and the ferroelectric material fulfills at least the following issues:
(a) the lattice mismatch between Si and the perovskite should be as small as possible, because misfit dislocations lead to uncontrolled interface states,
(b) chemical reactions and intermixing should be suppressed since this would lead to additional phases at the Si-ferroelectric interface,
(c) the number of interface states (traps) should be less than 10^{12} eV^{-1} cm^{-2} minimum,
(d) the formation of low-ε dielectric layers (e.g. SiO_2) during deposition should be avoided and
(e) the ferroelectric must form a pinhole free layer.

Once all these factors have been taken into account, only a very few perovskite oxides are suitable for growth directly on silicon. In the past it turned out, that the perfect growth of perovskites directly on silicon is a task with many obstacles (Sec. 2.2). Henceforth, the properties of the Si-perovskite interface are essential for the functioning and the characteristics of the ferroelectric field-effect transistor. Indeed, the ferroelectric-silicon interface is the most crucial part in fabricating a ferroelectric field-effect transistor.

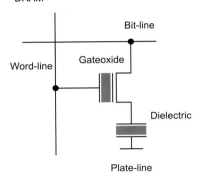

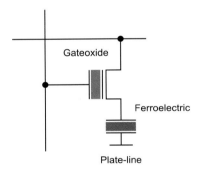

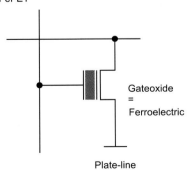

Figure 1: A conventional 1T1C DRAM cell (a), a typical 1T1C FeRAM cell (b), and a ferroelectrtic FeFET (c).

To by pass the problems of the perovskite Si interface, alternative gate stack layouts and various buffer layer configurations have been developed. A considerable number of experiments on different combinations of ferroelectric gate stacks which include buffer layers have been performed. Typically, the gate stacks of ferroelectric field-effect transistors are categorized in the following way:

MFS- metal (gate electrode) – ferroelectric – semiconductor
MFIS- metal – ferroelectric – insulator (buffer) – semiconductor
MFMIS- metal – ferroelectric– metal – insulator – semiconductor
MF-ABO$_3$- ferroelectric on a conductive oxide (no silicon transistor)

In Figure 4a – d, cross-sectional views are shown schematically for the most important FeFET device structures. In the following section some of the developments with respect to material issues and layout designs are summarised.

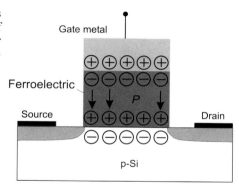

Figure 2: Schematic cross-section of a FeFET. The gate dielectric is a ferroelectric material. As an example a polarization state including surfaces charge layers is shown.

2.2 Ferroelectric Directly on Silicon

The principle operation of a FeFET was successfully achieved by Moll and Tarui [10]. The first attempt to fabricate a silicon based ferroelectric field-effect transistor was reported by Wu in 1974 [11]. The layer sequence is shown in Figure 4(a). Bismuth titanate (Bi$_4$Ti$_3$O$_{12}$) was deposited on highly doped silicon (100) by rf-sputtering. The device suffered from charge injection effects. This effect was identified because the change of the conductance of the semiconductor channel after polarization reversal (of the ferroelectric layer) was opposite to the expected direction. This result was explained by injection of holes and electrons from the semiconductor into the ferroelectric bismuth titanate layer.

To avoid the problem of the SiO$_2$ formation at the interface, also oxygen-free ferroelectrics have been studied. Sinharoy et al. [12] reported in 1991 on the deposition of BaMgF$_4$ on Si using sublimation under UHV (ultra high vacuum, p < 10^{-10} mbar) conditions. BaMgF$_4$ has an orthorhombic crystal structure and the spontaneous polarization vector is parallel to the (100) axis. It has been possible to demonstrate FeFET devices based on BaMgF$_4$ [13]. Nevertheless, problems such as a retention of only several hours and lacking of a definite coercive field during hysteresis programming remained.

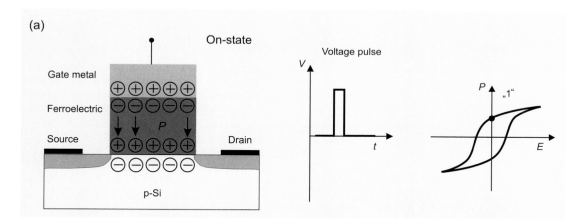

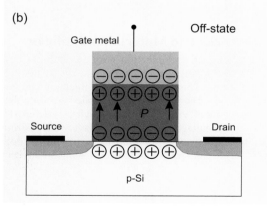

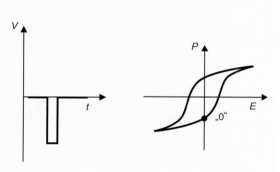

Figure 3: Charge diagram for the on- (a) and off-state (b) of a FeFET after applying a positive and subsequently a negative voltage pulse at the gate. The right figures show the corresponding polarization state [1], [2].

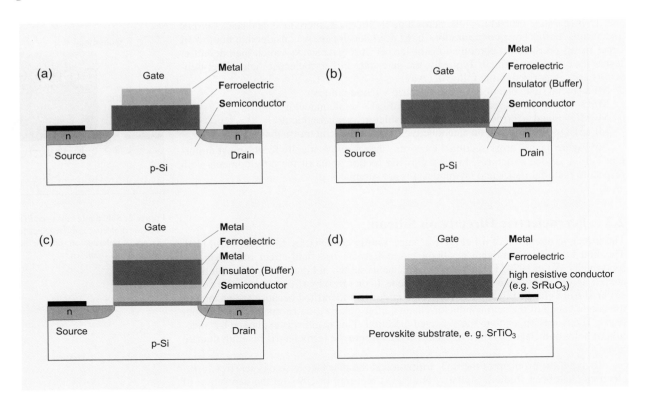

Figure 4: Variuos concepts for ferroelectric field effect transistors (a) MFS, (b) MFIS, (c) MFMIS, (d) MF-ABO$_3$ are shown.

Subsequently, a range of other deposition technologies and ferroelectric oxides have been tested. For example, LiNbO$_3$ has been tried due to its high spontaneous polarization of 71 µC/cm^2, sufficiently good epitaxial growth on Si (111) and because it is chemically relatively inactive [14].

The necessarily high deposition temperature in the order range of 500 °C to 850 °C, in combination with a high oxygen pressure (e.g. 10^{-1} mbar up to a few mbar), causes intermixing of Si with the perovskites and the formation of low-k dielectrics, typically SiO$_2$, at the interface with high trap densities. The TEM image in Figure 5 shows an unintentionally grown reaction layer [51]. The intermixing from Si in the perovskite leads to the degradation of the ferroelectric properties and causes unwanted properties as imprint, fatigue and retention (Chap. 22). The very high density of surface states, such which are called traps (mostly oxygen vacancies) [9], results in large injected charge densities, which preclude ferroelectric hysteresis in the electrical response of the device.

There has been a successful approach strategy, using the molecular beam epitaxy (MBE) deposition technique, for growing perfect grown perovskite-Si interfaces [8]. SrTiO$_3$ (which is a high-k dielectric but not ferroelectric) was perfectly grown on Si [7] using a processing trick to circumvent the formation of the thermodynamically more stable SiO$_2$. In addition, Motorola Labs. reported the growth of BaTiO$_3$ on Si with interface state densities of about 2×10^{11}eV^{-1} cm^{-2} using the same technique [37]. More details about the growth of high-k materials on Si are given in Chap. 13 in the discussion of novel gate oxides for MOSFETs.

2.3 Buffer Layer between Ferroelectric Material and Silicon

In 1975 Sugibuchi et al. [15], noticed that the charge injection problem in Bi$_4$Ti$_3$O$_{12}$ deposited on thermally oxidized silicon was suppressed. Nevertheless, the switching voltage was 15 V and to large for applications. This effect of charge injection can be minimized by employing an engineered buffer or barrier layer sandwiched between the silicon and perovskite layer. The cross-sectional view of such a device structure is shown in Figure 4b. This structure has the advantage that the buffer layer reduces the problem of intermixing between Si and the components of the ferroelectric material due to the buffer layer. In addition the buffer layer can reduce the lattice mismatch between the ferroelectric layer and Si. A disadvantage is that the gate oxide is comprised of two capacitors in series, the buffer layer and the ferroelectric layer. The effect of the buffer layer on the potential difference V_F across the ferroelectric layer can be calculated [16].

Using the Maxwell-equations the electric field at the interface between the insulating buffer layer B and the ferroelectric layer Fe is:

$$(\mathbf{D}_B - \mathbf{D}_{Fe}) \cdot \mathbf{n} = 0 \quad (1)$$

The electric field vector and the displacement vector $\mathbf{D}$ are perpendicular to the surface and parallel to the normal vector $\mathbf{n}$ ($\mathbf{D} \parallel \mathbf{n}$). Consequently,

$$\varepsilon_B E_B = \varepsilon_{Fe} E_{Fe} \quad (2)$$

Therefore, the voltage drop V_{Fe} across the ferroelectric layer is:

$$V_{Fe} = \frac{V_{gs}}{\frac{\varepsilon_{Fe}}{\varepsilon_B}\frac{d_B}{d_{Fe}} + 1} \quad (3)$$

Here, $V_{Fe} = E_{Fe} d_{Fe}$, where d_{Fe} is the ferroelectric layer thickness.

This means that in principle a buffer layer weakens the electric field across the ferroelectric. To minimize this effect, the buffer layer should be thin with a high dielectric constant ε_B. In addition, the total polarization P is:

$$P = P_r \frac{d_{Fe}}{d} \quad (4)$$

Therefore, the buffer layer causes a reduction of the permanent polarization P_r at zero voltage ($V_{GS} = 0V$) depending on the buffer layer thickness d_B ($d = d_{Fe} + d_B$).

PbZr$_x$Ti$_{1-x}$O$_3$ (PZT) and SrBi$_2$Ta$_2$O$_9$ (SBT) are preferred materials for FeRAM applications. Therefore these ferroelectrics have been used as ferroelectric gate oxides. For temperatures above 400°C Pb and Si easily interdiffuse [17]. Several kinds of buffer layers are currently under investigation as interdiffusion barriers, including Y$_2$O$_3$ [18], CeO$_2$ [19] and SrTiO$_3$ [20].

SrBi$_2$Ta$_2$O$_9$ directly deposited on Si showed a transition layer (e.g. SiO$_2$) with a low dielectric constant not suitable for FeFET applications [21]. This low-ε layer may originate from the reduction of Bi oxides by Si atoms. It was shown that a 2–3 nm thin Si$_3$N$_4$ buffer layer was effective in suppressing the formation of an interlayer [22]. CeO$_2$ as a buffer layer was not effective in improving electrical characteristics. Further studies were focused on SBT on SiN$_x$SiO$_2$/Si and SBT on SiON (silicon oxynitrid) [19].

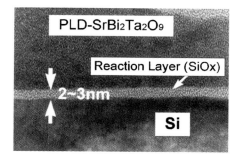

Figure 5: Cross-sectional TEM image of SBT/SiO2/Si structure [51]. An unintentionally grown SiO$_x$ reaction layer is clearly visible.

2.4 Metal-Ferroelectric-Metal Floating Gate Structures

To overcome the interface reaction problems in MFS and MFSI structures, Nakamura [23] proposed a gate layer sandwich of the form MFMIS (metal-ferroelectric-metal-insulator-semiconductor), as schematically shown in Figure 4c. In the MFMIS structure the MIS diode is perfectly separated from the ferroelectric MFM capacitance through a metal as interdiffusion barrier. Therefore, MIS diodes with SiO$_2$ or Si$_x$N$_y$ as an insulator with excellent interface properties can be used. In this case, the trap density at the insulator-semiconductor interface is as small as 10^{12} eV^{-1}cm^{-2}. On the other hand the MFMIS acts as a voltage divider. The gate voltage is divided according to the capacitance ratio of the MIS and MFM structures. Using Eq. (3), it can be seen that the capacitance of the MIS diode should be large enough to allow the polarization reversal of the MFM.

Nakao et al. [24] published data on MFMIS-FETs using PbZr$_{1-x}$Ti$_x$O$_3$ (F) and SiO$_2$ (I) as an insulator. Although the Si-SiO$_2$ interface properties were excellent, the device suffered from the relatively large voltage necessary to switch the ferroelectric capacitor, due to the low-ε SiO$_2$ buffer layer. More advanced material combinations have been used. Tokumitsu [25], [26] reported on Pt/SrBi$_2$Ta$_2$O$_9$/Pt/SrTa$_2$O$_6$/SiON/Si (MFMIS) structures. SrTa$_2$O$_6$ played the role of a high-ε dielectric buffer layer ($\varepsilon = 110$) and the SiON acted as a diffusion barrier against oxygen and, therefore, prevented SiO$_2$ formation at the Si interface. In addition, it is known that SiON on Si shows low trap densities. Those devices show rather promising electrical data and will be described in Sec. 4 in more detail.

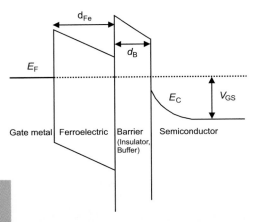

Figure 6: Simplified band-diagram of a MFIS structure.

2.5 Metal-Ferroelectric on a Conductive Oxide

In this approach, the semiconductor source-drain channel is completely replaced by a conductive oxide such as for example, $SrRuO_3$, or even a high-temperature superconductor such as $YBa_2Cu_3O_7$ (Figure 4d). Those materials have in some cases similar growth conditions as ferroelectric materials. The lattice mismatch is small and epitaxial growth is achieved. The aim is to modulate the conductivity of the conductive oxides by the polarization of the ferroelectric gate layer to get non-volatile function.

Although these devices were fabricated and tested successfully, applications are not yet considered seriously. The available crystalline substrates as for example $SrTiO_3$ are to expensive. Nevertheless, from the point of view of material compatibility and fundamental research, these kinds of field effect devices are a playground for basic science. A number of groups have successfully fabricated all-oxide FeFETs. Details can be found in [27] – [35].

3 Electrical Characterization of FeFETs

In Figure 6, the potential and energy-band diagram for an MFS structures (with an n-type Si substrate) is shown. Here we like to briefly discuss the essential difference between a MFIS and a conventional MOSFET without a ferroelectric layer. For a MFIS-type gate structure, the drain current I_D in the linear regime is given by:

$$I_D = -(W/L)\mu_h \left[P^* + C(V_{GS} - V_T)\right] V_{SD} \qquad (5)$$

with

$$P^* = PC_B/(C_B + C_{Fe}). \qquad (6)$$

Here L, W and μ_h are the gate (channel) length, gate (channel) width and effective hole mobility, respectively. The remanent polarization P is that of the ferroelectric layer. C is the gate capacitance per unit area which consists of the series connection of the capacitance per unit area of the ferroelectric layer C_F and of the buffer layer C_B. V_{GS} and V_{SD} denote the voltages for the gate-source and source-drain, respectively. V_T stands for the built-in threshold voltage and it is associated with effects other than the ferroelectric polarisation. The drain conductance and transconductance are given by:

$$g_D = -(W/L)\mu_h \left[P^* + C(V_{GS} - V_T)\right] \qquad (7)$$

and

$$g_m = -(W/L)\mu_h C V_{SD}. \qquad (8)$$

This means that the source-drain current vs. V_{SD} characteristic shows two separate curves as a result of the two polarization states $+/- P$ of the ferroelectric gate layer.

In addition, it is interesting to have a look at the voltage drop across the buffer layer with respect to the maximum polarisation of the ferroelectric film. The sheet carrier densities could be as low as $2 \times 10^{-12}/cm^2$ at the Insulator (I)-semiconductor(S) interface if the insulator is a high quality oxide buffer such as SiO_2 or Si_xN_y. This value corresponds to a charge density of $P_{cd} = 0.32\ \mu C/cm^2$. This means that the remanent polarization value of a ferroelectric layer should not necessarily to be large, if it is effectively used to induce inversion at the semiconductor surface. Typical remanent polarizations for PZT and SBT are 40 $\mu C/cm^2$ and 10 $\mu C/cm^2$, respectively, and considerably larger than P_{cd}. Moreover, large remanent polarisations cause large charge densities and are in conflict with the maximum charge density of the insulator I. For SiO_2 ($\varepsilon = 3.9$) the critical field for breakdown is 10 MV/cm by charge area-density of 3.5 $\mu C/cm^2$. If a PZT film with 50 $\mu C/cm^2$ is used, the voltage drop for a 10 nm SiO_2 would be 145 V! The consequence would be an electric field of 145 MV/cm, more than one order of magnitude larger than the critical field of 10 MV/cm. In this configuration only the sub loops (without tracing the complete P-E hysteresis) of the PZT a with maximum polarization of 3.5 $\mu C/cm^2$ are reasonable.

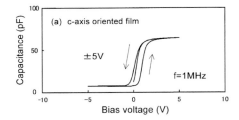

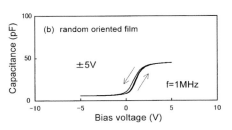

Figure 7: C-V characteristics of Pt/BLT/Si_3N_4/n-type Si diodes.
(a) c-axis-oriented film
(b) randomly oriented film [38].

In Sec. 2 various material aspects of FeFETs have been discussed. The material and interface properties of FeFETs, as in any other electronic devices, are strongly correlated with the electrical performance. In this section some of the most important electrical features of FeFETs will be described as representative examples. In this context the following electrical dependencies are essential for FeFETs:

the *C-V* (capacitance-voltage) characteristics,

the source-drain current I_{ds} vs. gate voltage V_g,

the memory window and threshold voltages,

the I_{on}/I_{off} ratio of the drain-source current, and

the retention time.

However, if a ferroelectric film is directly deposited on a Si surface, it is generally difficult to form a good interface between them, since constituent elements in both materials easily interdiffuse, and a transition layer is formed at the interface.

In order to improve the interface properties, an engineered buffer layer, which is composed of either a dielectric material (MFIS structure) or a stacked structure of conductive and dielectric materials (MFMIS structure) is often inserted between a ferroelectric film and Si substrate. Such gate layers with rather sophisticated and advanced layer sequences will be discussed in the following paragraphs.

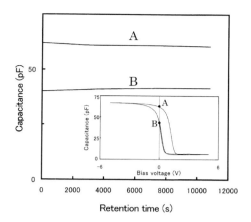

Figure 8: Variation of the zero bias capacitance of a Pt/100 nm-$Bi_{3.25}La_{0.75}Ti_3 = 12/3$ nm-Si_3N_4 Si MFIS diode with time at 300 K. The inset shows the C-V curve [38].

3.1 MFIS Structures

In MFIS structures, it is desirable from the viewpoint of the data retention characteristics to choose a ferroelectric film with a small remanent polarization and to use it in the saturated polarization condition. For this application, Kijima et al. [38] used $(Bi,La)_4Ti_3O_{12}$ (BLT) as an important candidate, because it is known to be fatigue-free [39] and its remanent polarization P_r and coercive field E_c are as small as 4 µC/cm² and 4 kV/cm along the c-axis. However, the P_r and E_c values are as large as 50 µC/cm² and 50 kV/cm along the a-axis. Thus, it is important in the use of this material to control the crystallite orientation of the film. Kijima et al. succeeded in achieving the following promising electrical results for a FeFET based gate structure consisting of a **Pt/75 nm-100 nm $(Bi,La)_4Ti_3O_{12}$/3 nm Si_3N_4/Si** stack.

Capacitance vs. voltage characteristics for the Pt/BLT (100 nm)/Si_3N_4 (3 nm)/Si structures are shown in Figure 7a, b. The memory window width for the c-axis-oriented (Figure 7a) film is about 1.2 V for a voltage sweep of +/–5 V. When the BLT film has a random orientation (Figure 7b), only a small minor loop can be used because the remanent polarization of the film is very large. Kijima et al. [38] concluded from these results that a c-axis-oriented BLT film is very suitable for MFIS-FET applications. The retention properties of the MFIS structure (100-nm-thick BLT film crystallized on p-type Si at 800°C) were examined at room temperature. After applying a writing bias voltage of +5 V or -5 V to the sample, the time dependence of the capacitance was measured at zero bias voltage. The result is shown in Figure 8, which shows that the capacitance change is negligible after 3 h. As can be observed from the inset of the figure, the measurement at zero bias voltage is not necessarily advantageous from the viewpoint of the retention time, because the *C-V* characteristics are shifted to the positive voltage direction and the point B in the figure is located in the depletion state, where the depolarization field becomes larger because of the existence of the depletion layer capacitance.

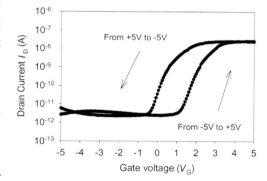

Figure 9: The memory window from a FeFET with PGO as ferroelectric [40].

$Pb_3Ge_5O_{11}$ (PGO) is known to be a ferroelectric, with a small remanent polarization P_r of the order of 10 µC/cm² and, therefore, an interesting candidate for MFIS gate structures. Li et al. [40] deposited PGO by the MOCVD (metal organic chemical vapour deposition) technique on **(Zr,Hf)O_2 buffer layer**. The memory window was determined to be 1.6 volt in a complete transistor structure. Figure 9 shows the drain current vs. gate voltage. In addition the retention time for this MFIS structure was about 60 min (Figure 10). As will be described in Sec. 3.2, PGO has been also used by the same authors for MFMIS gate structure.

Choi et al. [41] reported on the advantages of Al_2O_3 as a buffer layer. In conjunction with **$Sr_{0.85}Bi_{2.4}Ta_2O_9$ (SBT)** as the ferroelectric, interesting data have been obtained. The remanent polarization was 10.2 µC/cm² and the coercive field E_c was 37.5 kV/cm. By using 400 nm and 10 nm SBT and **Al_2O_3** films, a memory window of 2.2 V was achieved. The leakage current was about 7.6×10^{-9} A/cm² at 6 V and was two orders of magnitude lower than the Pt/SBT (400 nm)/Si stack without Al_2O_3. More details concerning the leakage behaviour are shown in Figure 11. The Al_2O_3 played an important role as diffusion barrier between SBT and Si.

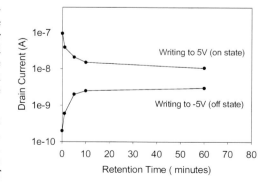

Figure 10: Retention time measurement [40].

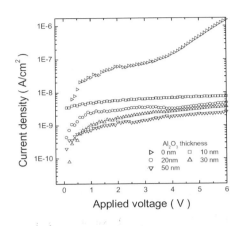

Figure 11: The leakage current density of the Pt/SBT/Al$_2$O$_3$/Si structures. Thickness variation of Al$_2$O$_3$ [41].

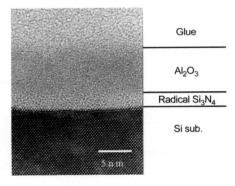

Figure 12: Cross-sectional TEM image of Al$_2$O$_3$/radical-Si$_3$N$_4$/Si structure [42].

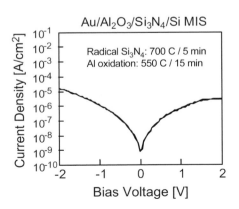

Figure 13: Current-voltage charactreristic with a Al$_2$O$_3$/radical-Si$_3$N$_4$ stacked insulator [42].

To avoid a possible oxidation at the Al$_2$O$_3$ interface Fujisaki and Ishiwara [42] introduced a **Si$_3$N$_4$ intermediate layer** between Al$_2$O$_3$ and Si. The Si$_3$N$_4$ film was prepared by the direct nitridation of a Si-substrate with a atomic nitrogen radical source. XPS spectra and TEM (Figure 12) images were carried and showed that the Si$_3$N$_4$/Si interface did not oxidize during post oxidation of the Al$_2$O$_3$ layer at 700°C. The current-voltage characteristic is shown in Figure 13.

Another way to suppress the formation of a low-k SiO$_2$ interfacial layer is the Y pre-metal layer method. By employing a two-step process starting with an ultra-thin Y metal deposition post-oxidation procedure of the Y layer to form Y$_2$O$_3$, Lim et al.[43] showed interface densities as low as 8.72×10^{10} cm^{-2}eV^{-1}. The approach seems to be another successful procedure to develop FeFETs.

3.2 MFMIS Structures

Now the C-V characteristics of a MFMIS structure is discussed. The considered gate stack is compromised by the aforementioned MIS-diode with a **Ti/Pt-Rh(M)/PbZr$_x$-Ti$_{1-x}$O$_3$(F)/Pt-RhTi(M)** sandwich on top [36]. Pt-Rh alloy acts as an intermixing barrier between Pb and Si. The advantage of the device is the high quality **SiO$_2$/Si interface** from the conventional MIS technology in combination with the inherent non-volatile characteristic of the MFM structure. The complete cross-sectional view is schematically shown in Figure 14.

The C-V curve (Figure 15) displays a pronounced difference in comparison to that of the MIS diode (Chap. 13). The threshold voltages of the capacitance change are determined depending on the tracing direction of the gate voltage. The two values of capacitance (max and min) are settled by the direction of the previously applied voltage.

The hysteretic character of the C-V curve is a result of the hysteresis in the polarization (P vs. E) curve of the ferroelectric layer. The memory window is defined by the voltage difference of the two threshold voltages, i.e. 2.4 V for the FeFET device in Figure 11. A large memory window (voltage) is desirable for the stable and reliable memory operation of a FeFET.

In Figure 16 another important feature of a FeFET is shown: the **source-drain current** I_d vs. gate voltage V_{GS} dependency.

The curve was measured on an **Ir/IrO$_2$/PbZr$_x$Ti$_{1-x}$O$_3$/Ir/IrO$_2$/poly-Si/SiO$_2$/Si (MFMIS) gate stack** [36]. The I_{on} and I_{off} currents were measured at -15V and $+15$V, respectively. The on-state to off-state ratio is 10^6 and is used for the non-volatile binary data storage. Whether the source-drain channel is in an on-state (binary "1") or in an off-state (binary "0") is detected by peripheral on-chip sense amplifiers. The memory window is visible (as in C-V curves) and determined to be 3.3 V. The drain-source voltage was fixed at -2V. The drain current on-off ratio vs. time is used to measure the **retention characteristic** of a FeFET. Figure 17 compares results of some examples based on retention of different FeFET concepts [26].

Drain currents and on/off ratios for **Pt(M)/SrBi$_2$Ta$_2$O$_9$(F)/Pt(M)/SrTa$_2$O$_6$(I$_1$)/SiON(I$_2$)/Si gate structures** are indicated by dotted lines and dash-dotted line (upper two curves). The ratio SM/SF is the area ratio of the MFM structure to the MIS structure. More explanation are given in the text below.

In this specific comparison the retention is clearly improved from MFS-FET (worse case) to MFIS-FET and finally, the MFMIS gate structure shows the best retention. The improvement is a result of better interface structure (low-trap density). Still the knowledge of interface structures is insufficient to draw any final and general conclusions on different FeFET concepts. In addition, it is desirable to extend the retention time to cover also the real non-volatile device applications.

Finally, we discuss the quantitative optimisation processes (which is a tricky business) for a FeFET on a device structure with the best performance developed by Tokumitsu [25].

In order to optimise the MFMIS stacked gate structure, the buffer layer (insulator I) capacitance must be as large as possible, i.e. a thin layer with a high dielectric constant (this is not the case for SiO$_2$) is preferable. On the other hand, the ferroelectric layer should have a remanent polarization which is not too large, as discussed above. The ε should be relatively small (in relation to that of I) and the thickness should be large to maintain a sufficient voltage drop across the ferroelectric layer. Fortunately the area A of the MFM part and the MIS part does not necessarily have to be the same. Figure 18 shows a MIMFS structure in which the MFM has a smaller area than the MIS. Using this structure, a MFMIS-FeFET was optimized and fabricated by Tokumitsu [25].

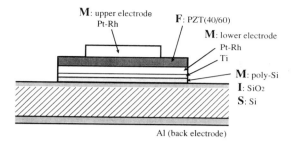

Figure 14: Schematic cross-section of a Ti/Pt-Rh(**M**)/PbZr$_x$Ti$_{1-x}$O$_3$(**F**)/Pt-RhTi(**M**) MFMIS sample [36].

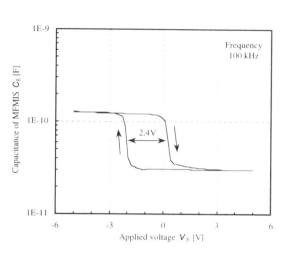

Figure 15: Metal-Insulator-Semiconductor C-V measurement. The structure was: Ti/Pt-Rh(**M**)/PbZr$_x$Ti$_{1-x}$O$_3$(**F**)/Pt-RhTi(**M**). The sweep frequency was 100 kHz with an amplitude of +/- 5 V [36].

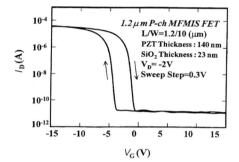

Figure 16: I_d-V_{gs} characteristics of 1.2 µm p-channel MFMIS using an Ir/IrO2/PbZr$_x$Ti$_{1-x}$O$_3$/Ir/IrO2/poly-Si/SiO$_2$/Si gate stack.[23].

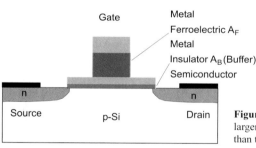

Figure 18: FeFET with a larger area A_B of the MIS than that of the MFM A_F.

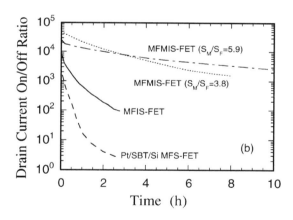

The gate stack was: **Pt(M)/SrBi$_2$Ta$_2$O$_9$(F)/Pt(M)/SrTa$_2$O$_6$(I$_1$)/SiON(I$_2$)/Si**. The reasons for using this rather complicated layer sequence are the following:

SiON (I$_2$) forms a good interface to Si with a low trap density. In addition it is a diffusion barrier for oxygen and prevents SiO$_2$ formation on the Si surface. SrTa$_2$O$_6$ (I$_1$) plays the role of a high-ε dielectric (ε = 110) insulator. SBT was used due to its low P_r. Pt is a common electrode material for ferroelectric capacitors. The MFM area was approximately 6 times smaller than that of the MI$_1$I$_2$S structure. In this way the equivalent remanent P_r is decreased further and so is the capacitance of the MFM. The whole P-E hysteresis was used for polarization reversal of the SBT even at low voltages. In addition the breakdown of the I$_1$-I$_2$ insulator double layer was not reached. Figure 19 shows retention measurements on this device.

The retention time was as long as 160 h. Currently, this is the best value for a FeFET. On the other hand a gate oxide structure with different areas leads to a somewhat more complicated fabrication and limited scaling.

Figure 17: Drain-source current on/off ratio as a function of data retention time for MFS, MFIS and MFMIS-FETs. Drain currents and on/off ratios for Pt(**M**)/SrBi$_2$Ta$_2$O$_9$(**F**)/Pt(**M**)/SrTa$_2$O$_6$(**I$_1$**)/SiON(**I$_2$**)/Si gate structures are indicated by dotted lines and dash-dotted line (upper two curves). The ratio S_M/S_F is the area ratio of the MFM structure to the MIS structure [25]. More explanation are given in the text below.

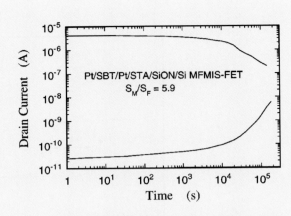

Figure 19: Retention data characteristics of a Pt(**M**)/SrBi$_2$Ta$_2$O$_9$(**F**)/Pt(**M**)/SrTa$_2$O$_6$(**I$_1$**)/SiON(**I$_2$**)/Si Gate stack with a ratio of the MIS area to the MFM capacitor area of $A_B/A_F = 5.9$ [25].

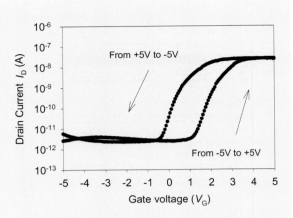

Figure 20: Memory window measurement from a MFMIS structure with PGO as ferroelectric layer [40].

The PGO ferroelectric has been integrated into MFMIS gate structures (**Pt/Pb$_3$Ge$_5$O$_{11}$/Ir/(Zr,Hf)O$_2$**) by Li et al. [40] from Sharp Laboratories/America. A typical drain-current vs. gate voltage curve is shown in Figure 20. A memory window of 1.8 V was measured. It was observed a retention time of more than 23 h min by writing voltage of +/− 4V from the on-state to the off-state. This curve is shown in Figure 21.

3.3 Optimization of a FeFET

In this section the optimisation of a FeFET is discussed and summarized. The short retention time in the MFIS and MFMIS structures originates from the fact that a dielectric capacitor is equivalently connected in series with the ferroelectric capacitor, as explained in the following. When the power supply is turned off and the gate terminal of the FET is grounded, the top and bottom electrodes of the two capacitors are short-circuited. At the same time, electric charges $\pm Q$ appear on the electrodes of the both capacitors due to the remnant polarization of the ferroelectric film and due to the charge neutrality condition at a node between the two capacitors (the FI interface in the MFIS structure or the floating gate M in the MFMIS structure). The Q-V relation for the buffer layer capacitor is $Q = CV$, and thus, the relation in the ferroelectric capacitor becomes $Q = -CV$ under the short-circuited condition, where C is capacitance of the buffer layer. That is, the direction of the electric field in the ferroelectric film is opposite to that of the polarization. This field is known as a depolarization field and it reduces the data retention time greatly, particularly when C is small.

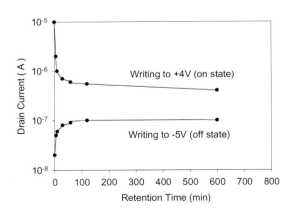

Figure 21: Retention time [40].

Another point to be considered is the leakage current of both the ferroelectric film and the buffer layer. If the charge neutrality at a node between the two capacitors is destroyed by the leakage current, electric charges on the electrodes of the buffer layer capacitor disappear, which means that carriers on the semiconductor surface disappear and the stored data can not be readout by drain current of the FET, even if the polarization of the ferroelectric film is retained.

- **Necessary conditions for buffer layer and ferroelectric materials.** In order to minimize the depolarization field in the ferroelectric film, the buffer layer capacitance C must be as large as possible. This condition means that a thin buffer layer with a high dielectric constant is preferable. Another important point is to reduce the leakage current of both a ferroelectric film and a buffer layer, as discussed above. That is, the thinnest limit of the buffer layer thickness is determined by the leakage current. Concerning the ferroelectric material, it is preferable to have a square-shaped P-V hysteresis loop, as well as a small leakage current. In a ferroelectric film with a square-shaped hysteresis loop it is expected that the polarization is not reversed even if a small depolarization field is generated. Concerning the polarization state of the ferroelectric film, the use of the fully polarized state seems to be more important for obtaining a long retention time unlike the case of the partially polarized state where the P-V hysteresis curve draws a minor loop.

- **Optimization of device structure.** Here improvement of the data retention characteristics is further discussed from a device structure viewpoint [44] – [45]. It is well known that dielectric constants of ferroelectric films are generally much higher than those of dielectric buffer layers. Thus, if both the ferroelectric and buffer layer capacitors are formed with the same size, most of the external voltage is applied to the buffer layer and only a little is to the ferroelectric film. In order to solve this dielectric constant mismatch problem, it is necessary to make the ferroelectric capacitor area smaller and also to make the film thicker. However, if a too thick ferroelectric film is used, the operation voltage of the FET becomes too high. Thus, there is a limit on the film thickness to be used.

The other problem is the mismatch of induced charge. The remnant polarization values of PZT and SBT are about 40 $\mu C/cm^2$ and 10 $\mu C/cm^2$, respectively, and they can induce the same density of positive and negative charges at the electrodes of a capacitor. These values are generally much larger than the maximum charge density induced by a dielectric film. For example, the maximum induced charge density of SiO_2 is about 3.5 $\mu C/cm^2$ for an electric field of 10 MV/cm and the film breaks down for the higher electric field.

Thus, if a ferroelectric capacitor with a large remnant polarization is connected in series with a SiO_2 capacitor with the same area, and if a sufficiently high voltage is applied across the both capacitors, the SiO_2 film breaks down before the saturation polarization of the ferroelectric film is reached. This situation is illustrated in Figure 22a for a combination of SBT and SiO_2. As shown in the figure, only a small hysteresis loop of SBT can be used under the condition that the SiO_2 buffer layer does not break down. It should be noted that this condition is independent on the film thickness of both capacitors.

In order to solve the mismatch problem of induced charge, it is necessary to form an MFMIS structure and to optimize the area ratio between the ferroelectric capacitor and the buffer layer capacitor. That is, in order to use the saturation polarization of the ferroelectric film effectively, it is important to make the ferroelectric capacitor area small. If the area of an SBT capacitor is reduced to 1/5 of an SiO_2 capacitor, the vertical scale of the P-V (polarization vs. voltage) characteristic equivalently becomes 1/5 as shown in Figure 22b, and the saturation polarization curve can be drawn in the region where the polarization value does not exceed the maximum induced charge density of SiO_2 ($\pm$ 3.5 $\mu C/cm^2$) [46]. These results suggest that combination of planar capacitors (MFM parts) and three-dimensional FETs (MIS parts) is important for integration of MFMIS-FETs with the optimized area ratio in high density. It is interesting to note that this design concept is opposite to that of the DRAM.

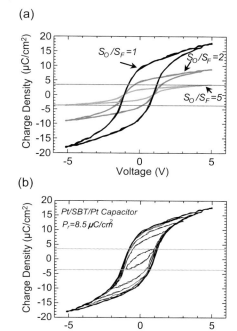

Figure 22:
(a) Induced charge density mismatch between SBT and SiO_2, and
(b) variation of the hysteresis loop by optimization of capacitor area.

4 Cell Designs and Device Modeling for FeFETs

The success of modern CMOS-VLSI circuits is based on thin film technology and simulation tools. In general, simulation tools are mathematical descriptions of devices and circuits. In case of a good knowledge of the device physics and electrical characteristics simulation, tools are able to describe device performance, including the behavior of the device within a circuit. In microelectronic engineering simulation tools are essential for the development, of new generations of individual devices and circuits by saving time and money because they help to reduce the number of real (and expensive) experiments to a minimum. Well known simulations tools in CMOS circuit design are Spice, Cadence etc. [47]

Despite an appropriate simulator, the cell structure is important for memory application of FeFETs. A skilful design of the cell may help to avoid specific problems of a device, such as the short retention time in the case of a FeFET. In the following two sections a short review about simulation tools for FeFETs and an example for a new cell structure, the 1T-2C cell, are given.

4.1 Device Simulation of FeFETs

There have been several attempts to develop simulation tools for the FeFET. Adequate simulator tools are essential for future FeFETs circuit designs. Only in this case will the FeFET will be considered as an industrially relevant device with a serious potential in the memory market.

The implementation of the ferroelectric $P(V)$ hysteresis loop into a CMOS device simulator is not an easy task. The reason for this is the still incomplete physical understanding of the switching characteristics of ferroelectric capacitors. Simply put, the properties of a ferroelectric hysteresis (like a magnetic hysteresis) depend on the history of the material. Therefore, only a deep understanding of the interplay of electric dipoles and domain walls is necessary to describe all features of a $P(V)$ hysteresis.

A straightforward approach was developed by Miller et al. [48] to describe the ferroelectric capacitor. They assumed a tanh relation for the $P(V)$ hysteresis. Nonsaturated loops were simulated by an arbitrary scaling factor as a consequence of a lack of solid state theory. More physical approaches were performed by Evans and Bullington [49] by modelling the $P(V)$ hysteresis using a one-dimensional distribution of domain wall coercive voltages. Recently Bartic et al. [50] applied the Preisach model for $P(V)$ hysteresis loops. Although the Preisach model was developed for magnetic particle ensembles, the mathematical framework is useful for describing the interaction of electric dipoles in case of ferroelectric materials. With some limits in accuracy the Preisach model was successfully combined with the Spectre® circuit simulator.

In order to model the ferroelectric transistor we start with modelling of a ferroelectric layer.

This is done following the Preisach approach. The polarization at a time **t** is given by

$$P(t) = -P_{sat} + \sum_i 2P_{sat} \cdot A_i \cdot a_i + P_{lin}(t), \quad a_i = \{-1, 1\} \quad (9)$$

where A is

$$A = \frac{\left(\arctan\left(\frac{E_{min} - E_c}{W}\right) + \frac{\pi}{2}\right) \cdot \left(\arctan\left(\frac{E_{min} - E_c}{W}\right) + \frac{\pi}{2}\right)}{\pi^2} \quad (10)$$

and

$$W = \frac{E_c}{\ln\left(\frac{1 + \frac{P_r}{P_s}}{1 - \frac{P_r}{P_s}}\right)} \quad (11)$$

$$P_{lin}(t) = \varepsilon \cdot E(t) \quad (12)$$

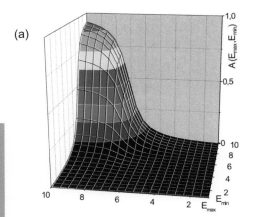

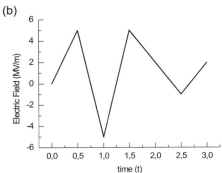

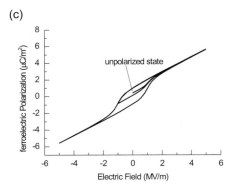

Figure 23:
(a) 3-dimensional surface plot based on the Preisach model.
(b) The time evaluation of the switching voltage to get a hysteresis loop.
(c) Simulated hysteresis loop.

The plot of this function is a 3-dimensional surface (Figure 23a). The value of the function gives the percentage of the dipoles polarized in the direction of the applied field and varies from 0, for the unpolarized state to 1, for the saturated state. The time evaluation of the switching voltage is shown in Figure 23b. The two parameters are the electric field minimum and maximum. By varying the field (voltage) we are moving on the surface on a certain path, i.e. following lines of constant maximum E_{max}, if we are moving away from a maximum, or following lines with an angle of 45°, if we surpass a maximum.

This way, the material history is taken into account and closed sub-loops are guaranteed as shown in Figure 23c. The model parameters are: remnant polarization P_r, saturation polarization P_s, coercive field E_c, layer thickness, dielectric constant.

The ferroelectric layer model can be implemented into a MOSFET model (BSIM3), resulting in the Ferroelectric Transistor Model (FerroFET). This new device combines the switching function of the transistor with the non-volatility of the ferroelectric capacitor. It can find applications in compact non-volatile memories. One other big advantage is the non-destructive readout that makes a refresh obsolete.

Nonetheless a reliable simulation for describing the performance of a FeFET is not yet available especially if important failure mechanisms such as fatigue, imprint and retention of the ferroelectric gate material should be included.

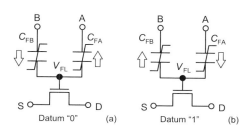

Figure 24: Schematic "read" operation for the 1T-2C cell [4].

4.2 A 1T-2C Cell Design for FeFETs

So far, two types of FeRAM cells have been proposed; one is a capacitor-type or 1T1C-type cell which is composed of a FET and a ferroelectric capacitor, and the other is a FET-type or 1T-type cell which is composed of a single ferroelectric-gate FET. The latter cell has the an advantage that stored data can be read out non-destructively. However, a significant problem with 1T-type cells is that the data retention time is too short. In order to solve this problem, a 1T2C-type cell [4], [5] has been proposed, in which two ferroelectric capacitors with the same area are connected to the gate electrode of a MOSFET, and the basic operation of the cell has been demonstrated. The cell structure and the polarization directions of the ferroelectric capacitors are shown in Figure 24. In order to write a data value in this cell, a positive or negative pulse is applied between the terminal A and B, so that the two ferroelectric films are polarized oppositely with respect to the gate electrode of MOSFET. In this case, the electric charges induced at the electrodes of both capacitors are cancelled each other and no charge is induced at the gate electrode of the FET, which means that no depolarisation field is generated in the ferroelectric film. For the readout operation, voltage pulses are applied to the terminal B, keeping the terminal A open. In this operation, when the stored value is "0", no polarization reversal occurs in the ferroelectric film and a little drain current flows through the MOSFET. On the contrary, when the stored value is "1", the polarization is reversed and a large drain current flows. In order to integrate these cells, an array structure shown in Figure 25 is proposed, in which each Si stripe formed on an insulating film corresponds to parallel connection of MOSFETs, and the capacitors CFA and CFB are formed between the second metal electrode and the floating gate electrode and between the first metal electrode and the floating gate electrode, respectively. In this structure, the film thickness of CFA and CFB is different, but the area is geometrically adjusted to be the same. Thus, in order to write a data value in this array, it is necessary to apply a voltage between the first and the second metal electrodes which are placed perpendicularly to each other. That is, no cell selection FET is used in this array. An effective writing method in such an array is the $V/3$ rule and the compensation operation is also known to be effective for minimizing the "data disturb" effect. Voltage pulses applied to the periphery circuit are shown in (Figure 26 a, b).

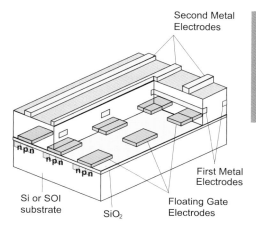

Figure 25: Integration of the 1T-2C-type on a SOI structure [4].

4.3 A FeFET Memory Concept with Disturbance Free Operation Scheme

A novel AND-type ferroelectric field effect transistor memory concept has been presented recently [52]. This cell structure is similar to semiconductor gate flash memories and best suited for non-volatile memory array circuits. The memory array uses global source lines and each is connected to its own sense amplifier. The AND-cell structure is shown in Figure 27. The memory cell concept is based on a p-type substrate and a 0.5 μm technology with a W/L ratio of 1. For the circuit simulation, the SBT layer thickness was 180 nm. The source and drain of the cell are connected to parallel bit lines

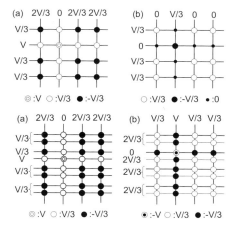

Figure 26: "Write" operation to a selected cell (a) and the compensation operation to minimize the "disturb" effect [5].

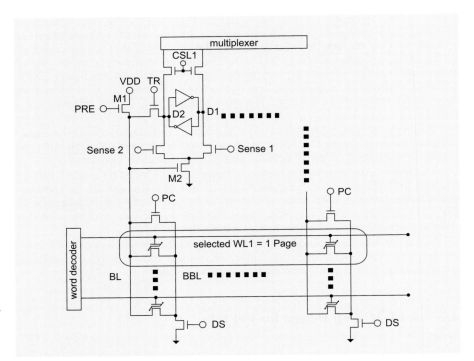

Figure 27: Memory architecture of a disturbance free operation scheme [52].

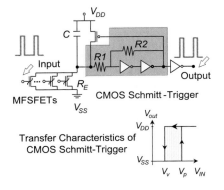

Figure 28: An adaptive-learning neuron circuit composed of ferroelectric-gate FET`s and a CMOS Schmitt-Trigger oscillator [4].

and a block is built by 128 parallel cells. The bits selected by one word line represent a 512 byte. Key components of the sense amplifier are a latch and a sense transistor (Figure 1).

The read operation is divided into two parts:

The first part is the page-read mode. Here the page data are transferred simultaneously to the sense latches. During the second part, there is a fast serial read out, whereby the data-registers are copied to the output buffers.

The write operation of a single FeFET is sensitive to disturbance problems because the ferroelectrics have no ideal threshold voltage. Any gate voltage will cause a change in polarization. Those problems can be avoided by the concept in which the FeFET works in the depletion regime. In case that the unselected cells are in the depletion regime, the change of the polarization state of those cells is minimized and therefore the disturbance problems are also reduced. More details about this concept can found in ref. [52] – [53].

5 Neural Network Circuits with FeFETs

A ferroelectric-gate FET is also useful as an analog memory [4] – [6]. It can typically be used as an electrically modifiable synapse device for storing the weight value in an artificial neural network. In this application, it is also possible to give an adaptive-learning function to the FET, where the term "adaptive-learning" means such a function that electrical or optical properties of a device are changed partly or totally after the device has processed a certain number of the usual signals. That is, when the pulse width of input signals is sufficiently narrow, polarization of the ferroelectric film is not completely reversed by application of a single pulse, but it is gradually changed by application of many pulses. It has been shown that the adaptive-learning neuron circuit using ferroelectric-gate FET's is particularly useful in such analog-digital-merged systems as the pulse width modulation (PWM) or the pulse frequency modulation (PFM) system, in which the input pulse height is kept constant but pulse width or pulse frequency is varied. In this case, synaptic connection among neurons is represented by an array of ferroelectric-gate FET's and the layout of synapse devices is essentially the same as that in Figure 1. Since the packing density of synapses is fairly high in Figure 1, this layout is considered to be particularly important in a large-scale neural network. Recently, the PFM-type

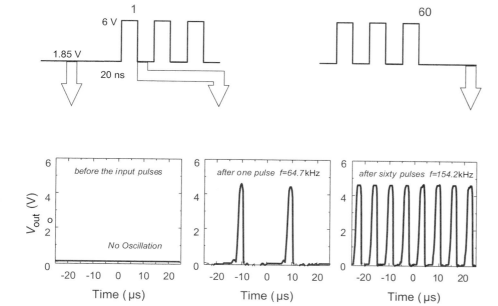

Figure 29: Output waveforms of the neuron circuit [4].

neuron circuit was actually fabricated and the adaptive-learning function was demonstrated. A diagram of the circuit is shown in Figure 28, in which ferroelectric-gate FETs are used as synapse devices. The output pulse interval of this circuit is mainly determined by the time constant of the capacitor C and the channel resistance RE of ferroelectric-gate FET's, and the charging and discharging operation of the capacitor C is controlled by the hysteretic transfer characteristic of a CMOS Schmitt-trigger circuit. A test circuit with a single synapse was fabricated on an SOI structure using a 5 μm design rule. In the fabrication process, an 150-nm-thick SBT ($SrBi_2Ta_2O_9$) film was directly deposited on Si and it was removed using a selective enchants composed of HCl: NH_4F, except for the channel region of the FET. In order to demonstrate the adaptive-learning function of the circuit, pulse input signals were applied to the gate terminal of ferroelectric-gate FET's. Their width and height were 20 ns and 6 V, respectively, and the power supply voltage was 5 V. During this measurement, a constant DC voltage of 1.85 V was also applied to the gate terminal. Under this condition, the neuron circuit did not show the oscillatory behavior before the first pulse was applied to the gate, as shown in the left graph of Figure 29. The center and right figures show typical output waveforms after a single pulse and sixty pulses were applied to the gate. It can be seen from these figures that the circuit starts to oscillate by application of a single pulse and the oscillation frequency increases as the number of input pulses increases. This behavior can be regarded as the adaptive-learning function of the circuit.

6 Summary and Outlook

The ferroelectric field effect transistor (FeFET) is an interesting alternative for non-volatile memory applications. The FeFET is in competition to conventional FeRAMs (Chap. 22), semiconductor non-volatile Flash memories and magnetic random access memories (MRAMs) (Chap. 23).

The advantages of a FeFET are the non-volatile data storage, the non-destructive readout, and the compact cell design, with high integration capability. Nonetheless, many interface and material aspects of the FeFET have been addressed in the past. Although considerable efforts were made, a commercial product using FeFETs is not available yet. The device performance suffer still from interface states (traps), complex fabrication issues and short retention times in the order of hours. Currently it is not possible to give a clear answer about the future of the FeFET. Newly developed materials for ferroelectric gate oxides and advanced layer sequences may help to overcome the existing problems.

Acknowledgements

The editor would like to thank Andreas Gerber (FZ Jülich) for checking the symbols and formulas in this chapter. Thanks are due to Adrian Petraru (FZ Jülich) for corrections.

References

[1] M. Ullmann, H. Goebel, H. Hoenigschmid, T. Haneder, G. E. Dietz, SISPAD Kyoto Japan, 1999.
[2] S.L. Miller and P.J. McWhorter, J. Appl. Phys. **72**, 5999 (1992).
[3] Y. Tarui, T. Hirai, K. Teramoto, H. Koike, K. Nagashima, Appl. Surf. Sci. **113/114**, 656 (1997).
[4] H. Ishiwara, Mat. Res. Soc. Symp. Proc. **596**, 427 (2000).
[5] H. Ishiwara, in: *Handbook of Thin Film Devices*, ed. Maurice H. Francombe, Academic Press, 2000.
[6] H. Ishiwara, Future Electron Devices (FED) **11**, 27 (2000).
[7] R.A. McKee, F.J. Walker, and M.F. Chisholm, Phys. Rev. Lett. **81**, 3014 (1998).
[8] H. Mori and H. Ishiwara, Jpn. J. Appl. Phys. **30**, L1415 (1991).
[9] K. Ito and H. Tsuchiya, Solid-State Electron. **20**, 529 (1977).
[10] J. L. Moll and Y. Tarui, Trans. Electron. Devices **ED-10**, 338 (1963).
[11] S. Y. Wu, IEEE Trans. Electron Devices **ED-21**, 499 (1974).
[12] S. Sinharoy et al., J. Vac. Sci. Technol. A **9**, 409 (1991).
[13] D. R. Lampe et al., Ferroelectrics **133**, 61 (1992).
[14] T. A. Rost, H. Lin, and T.A. Rabson, Appl. Phys. Lett. **59**, 3654 (1991).
[15] K. Suguchi, Y. Kurogi, and N. Endo, Jpn. J. Appl. Phys. **46**, 2877 (1975).
[16] R. Aidam, *Untersuchung des epitaktischen Wachstums dünner $Pb(Zr_{0.52}Ti_{0.48})O_3$-Schichten und ihre Anwendung in ferroelektrischen supraleitenden Feldeffekttransistoren*, Wissenschaftliche Berichte, KFK Karlsruhe, FZKA 6226, 1999.
[17] Y. Shichi, S. Tanimoto, T. Goto, K. Kuroiwa, and Y. Tarui, Jpn. J. Appl. Phys. **33**, 5172 (1994).
[18] B.-E. Park, S. Shioriki, E. Tokumitsu, and H. Ishiwara, Jpn. J. Appl. Phys. **37**, 5145 (1998).
[19] T. Hirai, K. Teramoto, K. Hagashima, H. Koike, and Y. Tarui, Jpn. J. Appl. Phys. **34**, 4163 (1995).
[20] E. Tokumitsu et al. IEEE Electron. Dev. **18**, 160 (1997).
[21] M. Noda, H. Sugiyama, and M, Okuyama, Jpn. J. Appl. Phys. **38**, 5432 (1999).
[22] H. Sugiyama, T. Nakaiso, Y. Adachi, M. Noda, and M. Okuyama, Jpn. J. Appl. Phys. **39**, 2131 (2000).
[23] Nakamura Y. Nakao, A. Kamisawa, and H. Takasu, Integrated Ferroelctrics **9**, 179 (1995).
[24] Y.Nakao, T. Nakamura, A. Kamisawa, and H. Taksu, Integrated Ferroelectrics **6**, 23 (1995).
[25] E. Tokumitsu, G. Fuji, and H.Ishiwara, Appl. Phys. Lett. **75**, 575 (1999); S.-M. Yoon, E. Tokumitsu, and H. Hiroshi, IEEE Electr. Devices **20**, 526 (1999); S.-M. Yoon, E. Tokumitsu, and H. Hiroshi, IEEE Electr. Devices **47**, 1630 (2000).
[26] E. Tokumitsu, G. Fuji, and H. Ishiwara, Jpn. J. Appl. Phys. **39**, 2125 (2000).
[27] Y. Watanabe, Appl. Phys. Lett. **66**, 1772 (1995).
[28] C. H. Ahn, R. H. Hammond, T. H. Geballe, and M. R. Beasley, Appl. Phys. Lett. **70**, 206 (1997).
[29] J. Mannhart, Superc. Sci. Techn. **9**, 49 (1996).
[30] S. Mathews, R. Ramesh, T. Venkatesan, and J. Benedetto, Science **276**, 238 (1997).
[31] M. W. J. Prins, K.-O. Grosse-Holz, G. Müller, J. F. M. Cillessen, J. B.Giesbers, R. P. Weening, R. M. Wolf, Appl. Phys. Lett. **68**, 3650 (1996).
[32] K.-O. Grosse-Holz, J. F. M. Cillessen, R. Waser, Appl. Surf. Sci., **96**, 784 (1996).
[33] W. Wu, K.-H. Wong, C. L. Mark, C. L. Choy, and Y. H. Zhang, J. Appl. Phys. **88**, 2068 (2000).

[34] S. Gariglio, C. H. Ahn, D. Matthey, and J.-M. Triscone, Phys. Rev. Lett. **88**, 067002 (2002).

[35] C. H. Ahn, S. Garigilio, P. Paruch, T. Zybell, L. Antognazza, and J.-M. Triscone, Science **284**, 1152 (1999).

[36] T. Kawasaki, Y. Akiyama, S. Fujita, and S. Satoh, IEICE Trans. Electron, **E81-C** (4), 584 (1998).

[37] Z. Yu et al., J. Vac. Sci. Technol. B **18**, 2139 (2000).

[38] Kijima, Jpn. Journal of Applied Phys. **40**, 2977 (2001).

[39] B H. Park, B. S. Kang, S. D. Bu, T. W. Noh, J. Leeand W. Jo, Nature **401**, 682 (1999).

[40] T. Li, S. T. Hu, B. Ulrich, and D. Evans, Mat. Res. Soc. Symp. Proc. **688**, C4.3. (2002).

[41] J.-H. Choi, J.-W. Kim, and T.-S. Oh, Mat. Res. Soc. Symp. Proc. **688**, F4.9. (2002).

[42] Y. Fujisaki and H. Ishiwara, Mat. Res. Soc. Symp. Proc. **688**, C11.4. (2002).

[43] D.-G. Lim, Bum-Sik Jang, S.I. Moon, D.-M. Jang, J. Heo, and J. Yi, Mat. Res. Soc. Symp. Proc. **688**, F7.7. (2002).

[44] T.Kijima and H.Matsunaga; Jpn. J. Appl. Phys. **38**, 2281 (1999).

[45] Fujimori, Y., Nakamura, T. and Kamisawa.A., Jpn. J. Appl. Phys. **38**, 2285 (1999).

[46] Ishiwara, H.; Mater. Res. Soc. Sympo. Proc. **596**, 427 (2000).

[47] Spice : MicroSim PSpice with Circuit Analysis, 2nd Ed., F. Monssen, Prentice Hall, Upper Saddle River, NJ, 1998. or Cadence: http://www.cadence.com/

[48] S. L. Miller, R. D. Nasby, J. R. Schwank, M. S. Rodgers, and P. V. Dressendorf, J. Appl. Phys. **68**, 6463 (1990).

[49] J. T. Evans and J. A. Bullington, IEEE 7th Intern. Symp. Applicat. of Ferroelectrics, New York, 1991.

[50] A. T. Bartic, D. J. Wouters, and H. E. Maes, J. T. Rickes, and R. M. Waser, J. Appl. Phys. **89**, 3420 (2001).

[51] T. Yamaguchi, M. Koyama, A. Takashima, and S. Takagi, Jpn. J. Appl. Phys. **39**, 2058 (2000).

[52] M. Ullmann, T. Haneder, W. Hoenline, and H. Goebel, Integrated Ferroelectrics, **40**, 23 (2001).

[53] M. Ullmann, H. Goebel, H. Hoenigschmid, T. Haneder, Integrated Ferroelectrics **34**, 37 (2001).

Quantum Transport Devices Based on Resonant Tunneling

Koichi Maezawa, School of Engineering, Nagoya University, Japan

Arno Förster, Department ISG, Research Center Jülich, Germany

Contents

1 Introduction	407
2 Electron Tunneling	407
2.1 Transfer Matrix Method	407
2.1.1 Tunneling Through a Single Barrier	409
2.1.2 Tunneling Through a Double Barrier Structure	409
3 Resonant Tunneling Diodes	410
3.1 Resonance Properties	410
3.2 Current-Voltage Characteristics	411
3.3 Interface and Growth Temperature	412
4 Resonant Tunneling Devices	413
4.1 Operation Speed of RTDs	413
4.2 Applications of RTDs	414
4.2.1 Resonant Tunneling Transistors	414
4.2.2 Concept of the Monostable-Bistable Transition Logic Elements (MOBILEs)	415
4.2.3 Integration Technology for MOBILEs	416
4.2.4 Examples of MOBILE Circuits	417
4.2.5 Extensions of the MOBILE Concept	420
5 Summary and Outlook	420

Quantum Transport Devices Based on Resonant Tunneling

1 Introduction

The complexity of microelectronic circuits is increasing in a dramatic way and simultaneously the dimension of each single device decreases rapidly. The decrease in the minimum feature size is expected to lead to the "quantum limit" within the next fifteen years. At present this limit seems to be the most fundamental restriction which cannot be exceeded. This does not necessarily stop the progress in microelectronics but it will focus the attention of research on new ways in which new fundamental ideas and solutions are needed. Even if one cannot remain under a certain limit of device dimensions it would be an advantage to work with higher operation frequencies. This is important in especially computer science and information technology. In this field new computer architectures and alternatives for the well established silicon based CMOS technology are necessary. At the moment most of the semiconductor products are CMOS based (93 % - 95 %). In the non-CMOS semiconductor field the huge hope for future innovative circuit concepts lies in the nm dimension in which quantum size effects are dominant. Although there are a lot of different promising new ideas for alternative electronics nobody can see at present a clear replacement of the classical microelectronics. Today the research on quantum size devices has already been started and the classical microelectronics goes towards the quantum size effect based nanoelectronics.

The actual quantum mechanical devices can be divided into two categories: one which is far away from the technical applications and one which is in active competition with conventional microelectronics. As a prominent device from the latter category we present in this article some general aspects of Resonant Tunneling Diodes (RTDs) with focusing on digital applications.

2 Electron Tunneling

2.1 Transfer Matrix Method

As a result of the wave character of electrons, the quantum mechanical transport phenomena become relevant in structures which dimensions are in the order of the electron wavelength. This is typically in the nanometer (nm) regime. One of the most exciting quantum mechanical phenomenon in semiconductor physics is the tunneling process of electrons through potential barriers. As known, electrons can penetrate into and traverse a potential barrier with a finite transmission probability independent of temperature. One can understand this behavior on the basis of quantum mechanics in which the wave character of the electron is important. In a classical picture the electrons can overcome a potential barrier only thermionically. Therefore a barrier would block all electrons with energies below the potential height Φ.

In solid state physics quantum mechanical calculations are based on the envelope function description of electron states in which the rapid changing electron potential is approximated with an envelope potential. The envelope function description is based on the effective mass approximation of the band structure and leads to the effective mass Schrödinger equation:

$$\left(-\frac{\hbar^2}{2} \frac{d}{dz} \frac{1}{m^*(z)} \frac{d}{dz} + \Phi(z) \right) \Psi(z) = W_z \Psi(z) \tag{1}$$

where $\Psi(z)$ is the electron wave function, W_z is the electron energy in z direction, m^* is the electron effective mass and $\Phi(z)$ is the potential energy at the conduction band minimum. In quantum mechanics the Schrödinger equation has the same importance for the electron statistics in small dimensions as the wave equation for optical experiments.

III Logic Devices

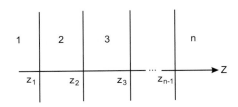

Figure 1: Sequence of n different layers; in each layer the effective mass m_i^* and the potential Φ_i was assumed to be constant.

Occupation probabilities can be predicted in the form of the absolute square of the wave function, $|\Psi(z)|^2$. In this sense in quantum physics all statements like "a single electron" can be interpreted only as occupation probabilities. Most of the quantum physical phenomena are no longer illustrative in a classical picture.

In order to establish a quantitative description of the quantum mechanical properties, a sequence of n different layers will be considered, with different potential energies Φ_i and electron effective masses m_i^* in each layer i. As a result of the solution of the Schrödinger equation the wave function $\Psi(z)$ can be written as a superposition of propagating waves in z and $-z$ direction, with amplitudes A_i and B_i, respectively.

In a system in which the solution of the effective mass Schrödinger equation is known in each region, the transmission probability can be evaluated with the transfer matrix method.

Suppose $\Psi_i(z)$ satisfies Eq. (1) in section 1 of Figure 1.

$$\Psi_i(z) = A_i e^{ik_i z} + B_i e^{-ik_i z} := A_i \Psi_{i+} + B_i \Psi_{i-} \tag{2}$$

where k_i is the electron wavevector in region i. The Schrödinger equation has to be solved with the boundary conditions at $z = z_i$.

The continuity of the wave function:

$$\Psi_i(z_i) = \Psi_{i+1}(z_i); \tag{3}$$

and of its derivative:

$$\frac{1}{m_i^*} \frac{d}{dz} \Psi_i(z_i) = \frac{1}{m_{i+1}^*} \frac{d}{dz} \Psi_{i+1}(z_i) \tag{4}$$

This connection rules express the conservation of the probability density and current and can be written in a matrix form:

$$TM_{1(z=z_1)} \begin{pmatrix} A_1 \\ B_1 \end{pmatrix} = TM_{2(z=z_1)} \begin{pmatrix} A_2 \\ B_2 \end{pmatrix} \tag{5}$$

$$TM_{2(z=z_2)} \begin{pmatrix} A_2 \\ B_2 \end{pmatrix} = TM_{3(z=z_2)} \begin{pmatrix} A_3 \\ B_3 \end{pmatrix} \ldots \tag{6}$$

$$\ldots TM_{n-1(z=z_{n-1})} \begin{pmatrix} A_{n-1} \\ B_{n-1} \end{pmatrix} = TM_{n(z=z_{n-1})} \begin{pmatrix} A_n \\ B_n \end{pmatrix} \tag{7}$$

with the matrix definition:

$$TM_i := \begin{bmatrix} \Psi_{i+} & \Psi_{i-} \\ \frac{1}{m_i} \Psi'_{i+} & \frac{1}{m_i} \Psi'_{i-} \end{bmatrix} \tag{8}$$

where Ψ_{i+}, Ψ_{i-} are defined in Eq. (2) and Ψ'_{i+}, Ψ'_{i-} are the derivatives in z, respectively.

A multiplication of Eq. (5) with $TM^{-1}_{2(z=z_1)}$ leads to the amplitudes A_2 and B_2 as a function of the amplitudes A_1 and B_1. Inserting the result in Eq. (6) and carrying out successively this procedure up to Eq. (7) finally the amplitudes of the propagating waves in z and $-z$ direction in the last layer are correlated via the transfer matrix TM with the amplitudes of the propagating waves in the first layer.

$$\begin{pmatrix} A_n \\ B_n \end{pmatrix} = TM \begin{pmatrix} A_1 \\ B_1 \end{pmatrix} \tag{9}$$

with the transfer matrix as a product of n matrices

$$TM = TM^{-1}_{n(z=z_{n-1})} \ldots TM_{2(z=z_2)} TM^{-1}_{2(z=z_1)} TM_{1(z=z_1)} \tag{10}$$

(a)
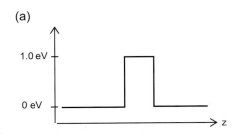

(b)
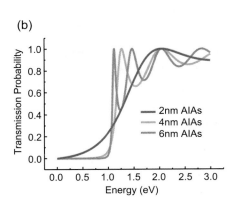

Figure 2: Schematic band diagram of a single AlAs barrier (a) and the corresponding tunneling transmission probability for different barrier thicknesses (b).

Eq. (10) yields the calculation rule for the transfer matrix where TM^{-1}_i denotes the inverse of the matrix TM_i.

The electron *transmission probability* T_c can be written as the ratio of the outgoing to the incoming quantum mechanical probability current

$$T_c = \frac{k_n}{k_1} \frac{m_1^*}{m_n^*} \frac{|A_n|^2}{|A_1|^2} \qquad (11)$$

Since we do not expect reflections behind the last layer ($B_n = 0$) the amplitude A_n can be written as follows:

$$A_n = \frac{\det TM}{TM_{22}} A_1, \qquad (12)$$

where TM_{22} is the corresponding transfer matrix element.

The single determinants of the transfer matrix TM in Eq. (10) are unless a factor linear independent Wronski determinants of two independent differential equations and hence independent of the z coordinate so that e. g.:

$$\det TM_{2(z=z_2)} TM_{2(z=z_1)}^{-1} = 1 \qquad (13)$$

This leads to the determinate of TM:

$$\det TM = \frac{k_1 m_n}{k_n m_1} \qquad (14)$$

Finally the transmission coefficient T_c can be written as follows:

$$T_c = \frac{k_1}{k_n} \frac{m_n^*}{m_1^*} \frac{1}{|TM_{22}|^2} \qquad (15)$$

where k_1 and k_n are the electron wavevectors of the incoming and outgoing waves before and behind the tunneling layer sequence, m_n^* and m_1^* denote the effective masses on the right and the left side of the layer sequence, respectively. This method allows the calculation of the coherent transmission probability T_c for arbitrary layer sequences like single tunneling barriers, double barrier or multi quantum well structures. The conduction band discontinuity between two different materials like GaAs and AlAs defines the potential energy of the electrons in the barriers. The effective masses are assumed to be constant in each layer.

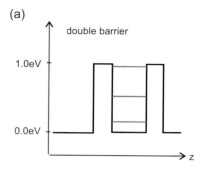

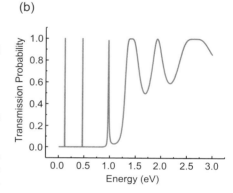

Figure 3: Schematic band diagram of a double barrier structure of AlAs embedded in GaAs (a) and the corresponding tunneling transmission probability (b).

2.1.1 Tunneling Through a Single Barrier

We consider the tunneling probability through a single potential barrier Figure 2. The experimental equivalent is an AlAs barrier embedded in GaAs. The electron transmission probability as a function of the electron energy was calculated according to Eq. (15) for three different thicknesses of the barriers. First we observe a finite transmission probability for electrons far below the potential height of 1.0 eV. This effect is known as the tunneling effect. The electron wave function in front of the barrier leaks out through the barrier and leads to a finite transmission. The smaller the barrier thickness, the higher is the tunneling probability of the electrons with energies below the potential energy of the barrier. In a classical picture the electrons could not penetrate the barrier. In addition we see a modulation of the transmission probability for electrons at energies above the 1.0 eV barrier height. In this region interference effects of transmitted and reflected electron waves appears, which demonstrate the wave character of the electrons.

2.1.2 Tunneling Through a Double Barrier Structure

To see the difference between the tunneling effect through a single barrier and the resonant tunneling effect, we discuss the case of a double barrier structure (see Figure 3). We consider two 4 nm thick AlAs barriers separated by a 5 nm GaAs well. In contrast to the transmission through a single barrier now electrons with very low energies can cross the double barrier structure with a transmission probability of 1. Three additional very sharp maxima appear below 1 eV in Figure 3b: they could be interpreted as quasi-bound states with a very narrow energetic bandwidth, through which electrons can tunnel like

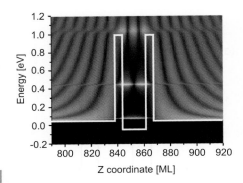

Figure 4: Gray-scale plot of the local density of states in a double barrier structure.

through open channels in the barrier. This is at first astonishing and not compatible with a sequential tunneling picture. In a sequential transport picture we would expect that the transmission probability through two barriers is very much smaller than through one barrier because the transmission through the first barrier is already much below 1. A completely new quantum mechanical system has been developed which can not be described by the behaviour of each single system. This may also be a drawback for quantum devices in general. Quantum mechanical devices can therefore not be placed extremely close to each other without changing the characteristics of the single device.

The wave character of electrons in a quantum device is impressively visible in the gray-scale plot of the electron *local density of states (LDOS)* in a double barrier structure shown in Figure 4. In this graph the absolute square of the wavefunction is plotted as a function of z-coordinate and energy. The distribution of the electrons is well visualized and the quantisation inside the potential well is evident. For such calculations the Schrödinger Eq. (1) has to be solved numerically to obtain the energy states as well as the wavefunctions.

3 Resonant Tunneling Diodes

3.1 Resonance Properties

The experimental realization of the double barrier heterostructure of Figure 3a with barrier thicknesses of only a few monolayers (MLs) is called a resonant tunneling diode (RTD). The transmission probability directly yields the energy and the half width of the quasi-bound states.

Figure 5 shows the behaviour of the resonances as a function of the supply voltage. A resonance can be considered as a "channel" which opens an electron flux while crossing the Fermi level. Therefore the current density first increases and than decreases. This effect causes a negative differential resistance in the *I - V* characteristics. The current density is strongly affected by the resonance half width Γ_0 of the resonance peak. The shape of the calculated resonances according to Eq. (15) is over a range of several orders of magnitude in very good agreement with a Lorentzian lineshape.

The half width of the resonance is exponentially dependent on the barrier thickness d_b since it is directly related to the degree by which the wave function "leaks out" of the well:

$$\Gamma_0 = c_1 e^{-2\alpha d_b} \tag{16}$$

where $\alpha = \sqrt{2 m_b^* (\Phi_b - W_r)}/\hbar$ is the barrier attenuation coefficient. The function c_1 is dependent on the effective electron mass in the well m_w^* and in the barrier m_b^*, the well width d_w, the potential barrier Φ_b and the resonance energy of the quasi-bound state W_r.

In the "thick barrier limit" the function c_1 can be written as (see Brown [1]):

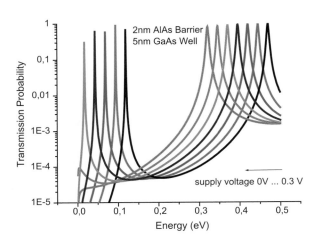

Figure 5: Transmission probability of a double barrier structure at different supply voltages.

$$c_1 = \frac{16\hbar k \gamma^2}{m_w^*(1+\gamma^2)\left[(1+\gamma^2)d_w + (1+\gamma^2 m_b^*/m_w^*)2\alpha^{-1}\right]} \quad (17)$$

with $k = \sqrt{2m_w^* W_r}/\hbar$ and $\gamma = (m_w^* \alpha)/(m_b^* k)$.

3.2 Current-Voltage Characteristics

The current density at a certain supply voltage can be calculated using the transmission probability together with the corresponding actual electron occupation densities. For the example of a double barrier structure the highly doped supply layers left and right of the double barrier can be described as free electron gases. The current density is obtained as the difference between the current density flux from the left to the right side of the double barrier and that one in the opposite direction. After Tsu *et al.* [2] this can be written as:

$$j = \frac{2|e|}{(2\pi)^3}\int_0^\infty dk_z \int_0^\infty dk_\perp \left(f_l(W) - f_r(W+eV)\right) T_c(W_z, V) \frac{1}{\hbar}\frac{\partial W_z}{\partial k_z} \quad (18)$$

where the coherent transmission probability $T_c(W_z, V)$ is a function of the supply voltage V and energy in z direction W_z, f_l and f_r are the Fermi distributions, left and right of the double barrier, and $k_\parallel$ and k_z denote the parallel- and z-component of the momentum, respectively. The integration of Eq. (18) leads to the current density expression containing the supply function:

$$j(V) = \frac{4\pi|e|m^* k_B T}{h^3}\int_0^\infty dW_z T_c(W_z, V)\ln\left|\frac{1+\exp\left(\frac{W_F - W_z}{k_B T}\right)}{1+\exp\left(\frac{W_F - W_z - eV}{k_B T}\right)}\right| \quad (19)$$

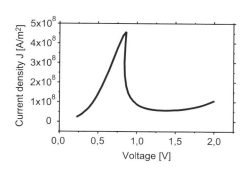

Figure 6: Typical calculation of the current-voltage characteristics of a resonant tunneling diode.

For an accurate calculation of the transmission probability the real potential profile across the device is required. The potential Φ includes the device energy band offset of the heterojunctions, the voltage drop across the structure and the contributions from the doping and mobile charges. By coupling the effective mass Schrödinger equation with the Poisson equation the potential Φ is obtained in a self-consistent manner.

Calculations of current voltage characteristics lead to a deeper insight into the physics of resonant tunneling diodes and are necessary in the device designing. A typical calculation is shown in Figure 6. The main characteristic feature is the existence of a negative differential resistance region which is the base of most of the RTD-applications. From the application point of view important parameters are: the peak current density, the valley current density, the peak to the valley current density ratio (PVR) and the peak voltage.

The peak current density decreases exponentially with the barrier thickness as the halfwidth of the resonance Eq. (16).

While the absolute peak-current densities resulting from simulations are in good agreement with experimental data, the calculated valley current densities are one or more orders of magnitude lower than the experimental ones. For AlAs/GaAs or AlAs/InGaAs diode structures on GaAs the experimental PVRs at room temperature are in the order of 6. The predicted PVR values from simulations are more than one order of magnitude higher (see Figure 7b and [10]). The reason for this discrepancy is the neglect of scattering effects in the calculation. Scattering effects broaden the resonance in the transmission probability while simultaneously damping it. The peak current density is nearly not sensitive to scattering effects but the valley current and the PVR are very strongly influenced.

An appropriate scattering model is based on the Breit-Wigner generalization of the Lorentzian form of the resonant transmission probability. Within this formalism resonant tunneling in one dimension is studied by Stone *et al.* [3] who derived the total transmission probability in the presence of inelastic scattering for a symmetric structure as:

(a)
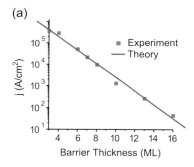

(b)
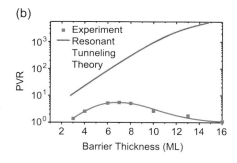

Figure 7: Comparison between theory according to Eq.(19) and experiment for a GaAs resonant tunneling diode
(a) peak current density,
(b) PVR.

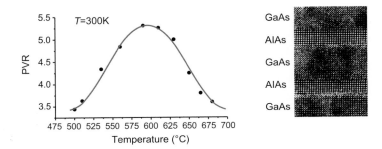

Figure 8: PVR of an AlAs/GaAs double barrier structure as a function of growth temperature (left). The optimum growth temperature of about 600 °C corresponds to the best quality of the interface in the HRTEM picture (right).

$$T_{\text{tot}} = \frac{\frac{1}{4}\Gamma_0\Gamma}{(W - W_r)^2 + \frac{1}{4}\Gamma^2} \tag{20}$$

where Γ_0 is the half width of the resonance in the coherent transmission probability and $\Gamma = \Gamma_0 + \Gamma_i$ is the total resonance half width, Γ_i representing the contribution to the broadening due to the inelastic scattering. Büttiker [4] has interpreted this total transmission probability as a sum of a coherent and sequential transmission probabilities

$$T_{\text{tot}} = T_c + T_i \tag{21}$$

In this picture of scattering the fraction of carriers penetrating the structure coherently is $T_c / T_{\text{tot}} = \Gamma_o / \Gamma$ and the fraction of carriers traversing the structure sequentially is $T_i / T_{\text{tot}} = \Gamma_i / \Gamma$. From these results one can infer that the smaller the elastic width Γ_0, the smaller is the amount of scattering needed to make the sequential tunneling current dominant. This means that in tunneling diodes with thick barriers (sharp resonances) in spite of a small scattering probability, considerable sequential tunneling contributions will be observed. Furthermore, Eq. (20) can be interpreted as a folding of the coherent transmission probability (Eq. (20)) with $\Gamma_i = 0$ with a normalized Lorentzian of half width Γ_i. In current density calculations this mechanism conserves the peak current density but affects the valley current very strongly resulting in lower PVR values. In this kind of treatment of I-V curves the effect of scattering is used as a fitting parameter to determine the resonance broadening at room temperature. For a typical RTD with 6 ML AlAs barriers and a 5 nm GaAs quantum well a resonance halfwidth of about 8 meV at room temperature was found (see [5]).

From the theoretical point of view this treatment of scattering is not satisfactory. Therefore a more complex approach is needed. In an enhanced calculation non-equilibrium Green-function theory is the base of the calculations in which self-consistent charging, incoherent and inelastic scattering, and the band structure is considered. Lake et al. [6] have developed a complex simulation package in which most of the relevant effects are taken into account. A real-space tight binding formulation provides an accurate synthesis of heterostructures on an atomic scale. It implies the consideration of inter-valley and inter-band transitions and gives a sophisticated description of electrons in the gap-region ("band-wrapping"). This approach was the base for the quantum device simulation package NEMO (NanoElectronic MOdeling) that simulates a wide variety of quantum devices, including RTDs, HEMTs, HBTs, superlattices, and Esaki diodes.

3.3 Interface and Growth Temperature

Most of the resonant tunneling diodes reported in the literature are based on AlAs(AlGaAs)/GaAs or on pseudomorphic InGaAs/GaAs(InP) systems and were grown by MBE. While for AlAs(AlGaAs)/GaAs the PVR does not exceed values higher than 7 at room temperature, Broekaert et al. [7] and Smet et al. [8] reported a PVR of 30 and 50, respectively, in $In_{0.53}Ga_{0.47}As/AlAs/InAs$ RTDs. A small InAs layer in the quantum well was used in these RTDs to decrease the resonance energy. Therefore the authors were able to reduce the well width leading to less scattering in the device. Although the reason for the much higher PVR of these diodes compared with those ones in the AlAs(AlGaAs)/GaAs system is not very clear one can also think of less interface scattering in this kind of diodes. Because of the high current densities needed in applications of RTDs, the barriers have to be thin (several monolayers) and the interface has to be very sharp. It has been shown that AlAs/GaAs interfaces do not exhibit chemically abrupt, completely flat interfaces – even under optimum growth conditions [9]. The transition between GaAs and AlAs occurs within 1 and 4 monolayers. The AlAs barrier can therefore be described by a barrier thickness distribution of a gaussian lineshape with a main barrier thickness and a variation of the barrier thickness depending on the roughness of the interface. This assumption leads to a mean transmission probability [10]. Even in the case of an interface roughness of 1 monolayer for the normal interface (GaAs/AlAs) and two monolayers for the inverted interface (AlAs/GaAs), which is a typical value of a very sharp GaAs/AlAs interface, the effect in the transmission probability is very strong. This assumption leads to a strong decrease in the PVR and explains the discrepancy between the resonant tunneling theory and experimental values of Figure 7. It is shown that this assumption is in agreement with experimental results up to barrier thicknesses of 6 monolayers. For larger barriers other scattering effects are relevant. In the AlAs/GaAs system e. g. the Γ - X scattering effect in AlAs barriers causes a strong decrease of the PVR [11].

A study of the performance of resonant tunneling diodes as a function of substrate temperature has shown that an optimum growth temperature for AlAs/GaAs diodes can be found. The PVR, serving as a figure of merit for the quality of the diode, increases with substrate temperature up to the temperature range between 580 °C and 600 °C. Above that temperature the PVR decreases rapidly. Using HRTEM one can show that the highest PVR coincides in fact with the sharpest interface between GaAs and AlAs (see Figure 8).

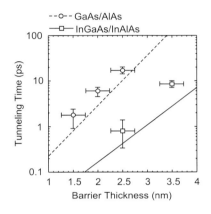

Figure 9: Tunneling time obtained from resonant tunneling bipolar transistors. The open circles and the open squares show those for the GaAs/AlAs RTDs, and the InGaAs/InAlAs RTDs, respectively.

4 Resonant Tunneling Devices

4.1 Operation Speed of RTDs

From the application point of view one of the most attractive features of RTDs is their potential for extremely high operation speed. RTDs with 712 GHz oscillation [12], response in the THz range [13] and 1.5 ps switching times [14] have been reported. In this section, we will briefly discuss the operation speed limit of the RTDs, and also we will compare it with Esaki diodes, which show similar I - V characteristics to RTDs. The discussion about the operation speed of RTDs is not so simple, and still controversial. First, it is important to differentiate two response times: the so-called "*tunneling time*" and the "*RC-time*". The former is the time it takes electrons to tunnel through the RTD structure, and is related to quantum mechanics. The latter is the time required to charge the capacitance of the RTD, and is related to circuit theory. Let us first consider the tunneling time [15]. Suppose that the electric field in the RTD structure changes from the non-resonant to the resonant state at a certain time t_0. The amplitude of the wave function in the quantum well changes to its steady state value in response to this change. The tunneling time is the time required for this change. This time is in the order of the resonant-state lifetime, t_{life}, or the escape time, which is the time it takes an electron in the quantum well to escape from it. From simple theory, this time is determined by the energy level width Γ_0 as

$$t_{\text{life}} = \hbar / \Gamma_0 \tag{22}$$

The energy level width Γ_0 is determined as the half-width of the transmission probability function through the resonant state, which was described in the previous section. Roughly speaking, Γ_0 exponentially decreases with increasing barrier thickness and height. This means a shorter tunneling time can be obtained with thinner and lower barriers, though there is a trade-off against the peak-to-valley current ratio. This time determines the fundamental speed limit for ideal RTDs, and it can be shorter than 0.1 ps.

However, various non-idealities in the real RTDs affect the tunneling time. These non-idealities include interface roughness, and inelastic scatterings. Several theoretical and experimental studies have been devoted to clarifying the tunneling time of RTDs. A time-resolved photoluminescent measurement using ultra-short pulses [16] has revealed that the escape time from a 2-dimensional well agrees reasonably well with Eq. (22). On the other hand, the tunneling time was directly estimated from the high-frequency characteristics of quantum-well-base bipolar transistors [17]. The devices used in this experiment were heterojunction bipolar transistors with a base layer consisting of a double barrier structure. Figure 9 shows the estimated tunneling time together with the theoreti-

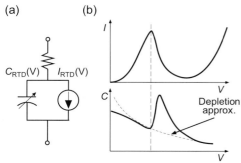

Figure 10: (a) The equivalent circuit model of a resonant tunneling diode.
(b) A schematic diagram of the capacitance-voltage curve with the current-voltage curve as a reference.

cally calculated lifetime. As shown in the figure, the tunneling time is in reasonable agreement with Eq. (22). The resonant lifetime described in Eq. (22) is a useful guideline for designing high-speed RTDs, though further studies to clarify the tunneling time in real systems are necessary.

Next, we will discuss the operation speed limited by RC-time. In most applications, the operation speed of RTDs is limited not by the intrinsic tunneling time but by the charging time of RTD capacitance. RTDs are well described by the equivalent circuit in Figure 10 (a). This circuit consists of a voltage-dependent current source $I_{RTD}(V)$, a voltage-dependent capacitor $C_{RTD}(V)$, and a series resistor R_s. Here, the parallel combination of $I_{RTD}(V)$ and $C_{RTD}(V)$ represents an intrinsic RTD, and the R_s is the sum of series resistances such as the contact resistances. An investigation of the capacitance $C_{RTD}(V)$ is extremely important in determining the maximum operation speed of RTD circuit. A schematic diagram of the capacitance-voltage curve is shown in Figure 10 (b) with the current-voltage curve as a reference. The capacitance of the RTD was extracted from the results of microwave S-parameter measurements [18]. In short, there are two main points. First, the capacitance is roughly equal to that calculated from the undoped spacer layer and the depletion layer of the device, except for the voltage near the peak. Second, there is an anomalous peak structure in the negative differential resistance (NDR) region, as shown in Figure 2(b). This peak is due to resonant electrons accumulated in the well. This must be taken into account to discuss operating speed precisely. Note that $I_{RTD}(V)$ and $C_{RTD}(V)$ do not depend on the frequency when the frequency is sufficiently smaller than the intrinsic limit determined by the lifetime discussed above. In addition to these times, the transit time across the collector depletion layer affects RTD response when large spacer layers are used to decrease its capacitance.

It is worth discussing here about the advantages of RTDs compared to *Esaki diode*s or tunnel diodes. The Esaki diodes consist of a heavily-doped pn-junction and show the *I - V* curves similar to RTDs. Several applications now studied for RTDs had been once studied for Esaki diodes. The one of most important advantages of the RTDs is the ability to obtain a high peak current density with a relatively low capacitance. For example, an extremely high current density of 6.8×10^5 A/cm^2 is obtained with a capacitance of about 1.5×10^{-7} F/cm^2 [14]. These values indicate that the speed index, which is defined as the capacitance per unit area divided by the peak current density C/J_p, is as small as 0.22 ps/V. (The speed index corresponds to the speed of voltage variation, when the RTD capacitance is charged by its peak current). This is much smaller than those of Esaki diodes, which are larger than 10 ps/V. This is possible because the current density of RTDs can be increased by changing the barrier and the well thicknesses, and this can be achieved without decreasing depletion layer thickness. On the other hand, one must increase impurity density to decrease tunnel barrier (= depletion layer) thickness in order to increase current density in Esaki diodes. Consequently, the maximum operation speed of RTDs can be much higher than that of Esaki diodes. Furthermore, RTDs can avoid degradation observed in Esaki diodes due to impurity diffusion at the heavily-doped pn-junction.

4.2 Applications of RTDs

Several possible RTD applications that exploit the negative differential resistance (NDR) are now being developed. Though most of them are similar to those once proposed for Esaki diodes, progress in related device technology and the ultra high-speed potential of RTDs open up new possibilities. Here, we will focus on the digital applications, in particular monostable-bistable transition logic element (MOBILE), though the microwave/millimeter wave analog applications are also promising [19].

4.2.1 Resonant Tunneling Transistors

An RTD has only two terminals, which restricts the use of resonant tunneling phenomena. Adding a control terminal to RTDs extends their usability to a variety of applications. The most straightforward way to do this is to merge RTD with conventional transistors to make a composite device. This approach has been used to build resonant tunneling bipolar transistors (RTBTs) [20], resonant tunneling hot electron transistors (RHET) [21] and gated RTDs [22], [23], [24]. The RTBTs have a resonant tunneling structure at the emitter/base junction region or in the base. An RHET is similar to RTBT and has a resonant tunneling structure at the emitter of the hot electron transistor. Consequently, these devices have negative transconductance in emitter-grounded transistor characteristics. These non-linear input-output characteristics can be applied to several

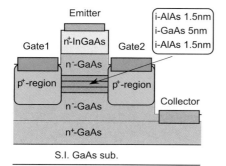

Figure 11: A resonant tunneling transistor having pn-junction gates around the emitter. This was used for the demonstration of the MOBILE.

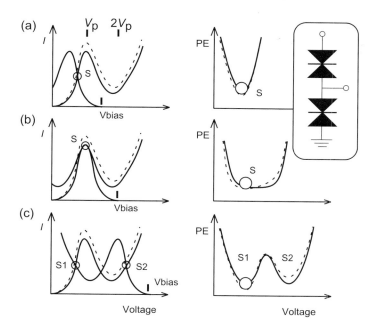

Figure 12: Concept of the MOBILE operation. Load curves (left) and corresponding potential diagrams (right) are shown together with the basic MOBILE circuit.
(a) $V_{bias} < 2V_p$,
(b) $V_{bias} \sim 2V_p$,
(c) $V_{bias} > 2V_p$.

circuits, such as an XOR logic gate with only one transistor. On the other hand, gated RTDs have Schottky or junction gates around the emitter to control RTD area, and show an NDR with controlled peak current. These devices are used for a functional logic gate called a MOBILE [22], on which we will focus in the next section. Figure 11 shows an example of the gated RTD which was used for MOBILEs. The emitter area, and thus the emitter current, can be modulated by the pn-junction gates surrounding the emitter. The other and more practical way to add the control terminal is to connect RTDs with conventional transistors to make parallel or series circuits. Integration of RTDs with conventional transistors is necessary in order to do this, but this has a significant advantage beyond merely adding a control terminal. That is, one can use both of RTD circuits and conventional transistor circuits according to their respective merits and demerits. Integration of RTDs with HEMTs or bipolar transistors has been proposed, and several circuits using such devices have been also reported [25], [26], [27].

4.2.2 Concept of the Monostable-Bistable Transition Logic Elements (MOBILEs)

In this section we will focus on the ultrahigh-speed functional logic gate, Monostable-Bistable Transition Logic Element (MOBILE), which exploits the NDR of the RTDs [22], [28]. There are three aspects to the operating principle: 1) to employ the monostable-to-bistable transition of a circuit consisting of two NDR devices connected serially, 2) to drive this circuit by oscillating the bias voltage to produce the transition, and 3) the NDR device(s) having the third terminal to modulate their peak currents. Figure 12 shows the load curves and the corresponding potential energy diagrams for the circuit. As shown in Figure 12 (a), the number of stable points is one when the bias voltage is smaller than twice the peak voltage ($2V_p$). This stable point splits into two branches, S_1 and S_2 (Figure 12 (c)), when the bias voltage increases through $2V_p$. A small difference in the peak current between the two NDR devices determines the circuit's state after the transition. For example, a larger peak current in the driver device results in the stable point S_1 (dotted lines). The difference in the peak current can be extremely small for switching, since at the transition point (Figure 12 (b)) the system is sensitive to the difference. With the oscillatory varying of the bias voltage, the circuit forms a logic gate. This oscillatory bias voltage works as a clock. The phase of the oscillating bias voltage for the next stage must be delayed from the first one to propagate signals properly.

At least one of the NDR devices must have a mechanism to modulate its peak current. Resonant tunneling transistors with junction gates as described in the previous section can be used for such an NDR device. Also, the parallel circuit of an RTD and a FET can be used for this NDR device. This mode of operation has several advantages. First, MOBILEs have a significant advantage in that a threshold logic operation for the

III Logic Devices

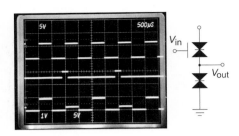

Figure 13: Circuit configuration and the operation wave form of the simple MOBILE inverter. Traces are the clock, input, and output voltages, from top to bottom, respectively.

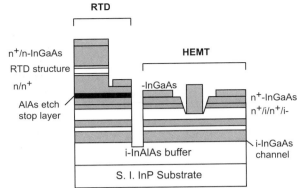

Figure 14: Schematic cross-sectional view of the RTD/HEMT integration structure.

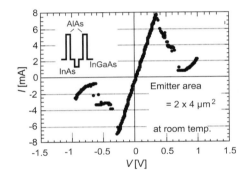

Figure 15: The current-voltage characteristics of the fabricated RTD with the schematic of the RTD structure.

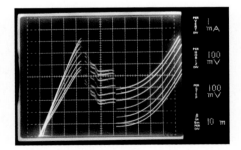

Figure 16: The current-voltage characteristics of the fabricated RTD/HEMT parallel circuit. The maximum gate voltage is 0.6 V with a 0.1 V step.

weighted sum of input signals is possible for a multiple input device. This is because the MOBILE uses the monostable to bistable transition and it does not output intermediate value. This operation has a wide range of applications, as will be shown later. In addition to these, edge-trigger and latching function are unique characteristics of the MOBILE, and can be applied for several circuits.

Figure 13 shows the circuit configuration of the simple MOBILE inverter and its operation waveform. The traces are Clock (bias voltage), Input, and Output, from top to bottom respectively. It is clearly shown in the figure that the switching occurs at the rising edge of the clock, and the output remains unchanged while the clock is high.

4.2.3 Integration Technology for MOBILEs

MOBILE needs NDR devices having control terminals to modulate their peak current. First, the gated RTDs with pn-junction gates were used for MOBILEs [22], [28]. The gated RTD is a simple and compact device, easily fabricated and suitable for demonstration of the basic operation. However, it has some disadvantages from practical point of view. First, the operation speed is limited by the parasitic capacitance under the gates. In this configuration, the pn-junction under the gate is irrelevant to the current modulation and the large capacitance related to this region degrades the speed of the MOBILE. Second, it is difficult to optimize the layer structure for both resonant tunneling and current modulation simultaneously. Third, the peak-to-valley current ratio (PVR) for this device was smaller than that for the simple RTD due to the lateral potential variation in the well. And finally, it is difficult to fabricate this type of resonant tunneling transistor using InGaAs/InAlAs material system lattice-matched to InP substrate, though a superior performance is expected for such RTDs due to the small effective electron mass. This is because the InGaAs has a small bandgap (~ 0.7 eV) and the leakage current for the pn-junction is much larger than that in GaAs. To overcome these problems the RTDs and HEMTs were integrated on the InP substrate. In the integrated device, an RTD is connected in parallel with a HEMT. The total source-to-drain current (I_{DS}) is equal to the sum of the currents passing through the RTD (I_{RTD}) and the HEMT (I_{HEMT}). Since the gate voltage can modulate I_{HEMT}, I_{DS} is also modulated by V_G. As a result, the peak current of the integrated device is effectively modulated by V_G.

Here, we will briefly describe the RTD/HEMT integration technology on an InP substrate [27]. The InGaAs/InAlAs material system lattice-matched to InP substrate is known as a promising system for ultrahigh speed electron devices. An ultrahigh-cutoff frequency of higher than 400 GHz has been reported for HEMTs fabricated on this material system [29]. In addition to this, a small effective electron mass and a large barrier height obtainable in this system permit us to fabricate high performance RTDs. A high peak current density higher than 1×10^5 A/cm^2, which is essential for high speed operation, and a high PVR larger than 10 can be obtained simultaneously in this system. These properties make the InGaAs/InAlAs system very promising for the MOBILE.

Figure 14 shows a schematic cross-sectional view of the RTD/HEMT integrated devices. First, epitaxial layers were grown by MBE on a semi-insulating InP substrate. The structure consists of the HEMT and the RTD, grown sequentially from the bottom to the top. The HEMT structure is a non-alloyed ohmic contact HEMT [30] grown at 510 °C, which consists of a 200 nm i-InAlAs buffer, a 15 nm i-InGaAs channel, a 2 nm i-InAlAs spacer, a 4 nm n$^+$-InAlAs (1×10^{19} cm^{-3}) carrier supply layer, a 20 nm i-InAlAs Schottky enhancement barrier, and then an n$^+$-InAlAs(15 nm)/InGaAs(20 nm) (1×10^{19} cm^{-3}) double-layer cap which enables the formation of the non-alloyed ohmic

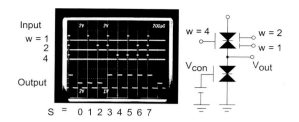

Figure 17: The circuit configuration and the operation of three-input MOBILE. The traces are the input 1 ($w = 1$), the input 2 ($w = 2$), the input 3 ($w = 4$), and the output, from top to bottom, respectively. The weighted sum (S) is shown at the bottom of the figure as a reference. The threshold value is selected by the control voltage, V_{con}, and it is chosen to be 2.5 in this figure.

contact for both the source and the drain. On top of the HEMT, a 2-nm i-AlAs layer was grown to serve as a selective etch stopper, followed by a 20 nm n$^+$-InGaAs interfacing layer between the HEMT and the RTD. The RTD consists of a 150 nm n-InGaAs (1×10^{18} cm^{-3}) cathode layer, an i-InGaAs spacer (1.5 nm), a double barrier quantum well (DBQW) with a In$_{0.53}$Ga$_{0.47}$As (1.2 nm)/InAs (2.8 nm)/In$_{0.53}$Ga$_{0.47}$As (1.2 nm) strained well sandwiched by AlAs (1.4 nm) barriers, an i-InGaAs (1.5 nm) spacer, an n-InGaAs anode layer (100 nm, 1×10^{18} cm^{-3}), and finally an n$^+$-InGaAs (50 nm, 2×10^{19} cm^{-3}) non-alloyed ohmic contact layer.

In this RTD structure, the key point is to use the strained AlAs barriers and the strained InAs subwell. The AlAs barrier has larger barrier height than that of InAlAs. This reduces the leakage valley current and increases the PVR. The InAs subwell reduces the ground level energy while less affects the first excited level. This makes the peak voltage lower and the PVR larger. Figure 15 shows the current-voltage characteristics of the fabricated RTD with the conduction band diagram of the RTD structure. A high current density of 1×10^5 A/cm^2 is obtained due to the small effective mass. And the high peak-to-valley current ratio of about 9 as well as the low peak voltage of 0.3 V can be obtained. It should be noted that there is no high resistive region at small voltages. This enables us to make good use of the bias voltage for logic swing. For the device fabrication, an RTD mesa was first defined by non-selective wet etching (sulfuric-acid-based) down to the n-InGaAs cathode layer below the RTD structure. Then a citric-acid-based selective etchant was used to etch to the AlAs etch stopper, which was followed by a slight non-selective etching (sulfuric-acid-based etchant was used again) to remove the etch stopper and expose the HEMT cap layer. The selective etch stopper is critical for obtaining a smooth HEMT surface, which leads to improved uniformities in threshold voltages and transconductances of the HEMTs. The device mesas were then defined by photolithography and wet etching. Using the same photoresist etching mask, the mesa-sidewall of the HEMT was selectively recessed using the citric-acid-based etchant to reduce the gate leakage current. Ti/Pt/Au were used as source and drain non-alloyed ohmic contact metals. After gate recess etching, Pt/Ti/Pt/Au were deposited as the gate electrode.

Figure 16 shows the typical I-V characteristics of the integrated device (parallel circuit of an RTD and a HEMT). The RTD area was 6 µm^2, and the gate length and width of the HEMT were 0.7 and 10 µm, respectively. The peak current is effectively modulated by the gate voltage.

4.2.4 Examples of MOBILE Circuits

In this section we will show some examples of MOBILE's circuits. First example is the weighted sum threshold logic function [22]. Figure 17 shows the circuit configuration and the operation of three-input MOBILE [7]. The pn-junction RT transistors were used in this experiment, the top view of which was shown in Figure 18. The gate widths are 12, 6, and 3 µm, for a weight ratio of 4:2:1. The total current change of the upper transistor is proportional to the weighted sum of input signals. Consequently, this logic gate performs a threshold logic function for the weighted sum of input signals as expressed by,

$$S = \sum_i w_i x_i \qquad (23)$$

$$O = u(S - L_{th}) \qquad (24)$$

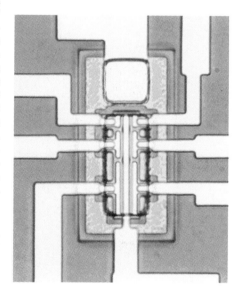

Figure 18: Top view microphotograph of the three three input RT transistor. The weights of the inputs are 1, 2, 4 (Two gates having a same gate width are connected in the experiment).

III Logic Devices

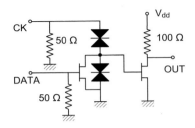

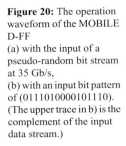

Figure 19: The configuration of the test circuit for high-speed operation. This circuit works as a D-FF.

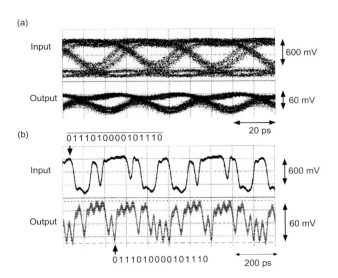

Figure 20: The operation waveform of the MOBILE D-FF (a) with the input of a pseudo-random bit stream at 35 Gb/s, (b) with an input bit pattern of (0111010000101110). (The upper trace in b) is the complement of the input data stream.)

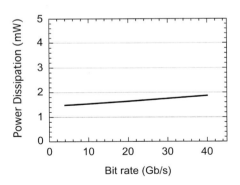

Figure 21: The calculated power dissipation of the core circuit shown in Figure 19.

Here, the w_i and x_i (= 0, 1) are the weight and the input signal of the i-th gate, and S, O and L_{th} are the weighted sum, output, and threshold value, respectively. The step function $u(x)$ equals 0 when $x < 0$, and 1 when $x > 0$. The result shown in Figure 17 is an example of operation when the threshold value $L_{th} = 2.5$. The L_{th} can be varied from -0.5 to 7.5 by changing the control voltage. This indicates that all of the input patterns ($2^3 = 8$) can be distinguished with this logic gate. Another advantage of MOBILEs is that both positive and negative weights are possible. That is, inputs to the upper RT transistor have positive weights, and those to the lower one have negative weights [31]. This type of operation is useful for simplifying circuits and also useful for fabricating artificial neural networks.

Next, we will show the demonstration of the high-speed operation [32]. Figure 19 shows the configuration of the test circuit. It consists of a MOBILE core and direct coupled FET logic-type (DCFL-type) output buffer. Although this is a simple circuit to test the high-speed operations of a MOBILE, it does have a practical use. The MOBILE core works as a delayed flip-flop (D-FF) with a return-to-zero (RZ) mode complement output. Therefore this circuit can be regarded as a "return-to-one" (RO) mode D-FF. The output buffer is indispensable in minimizing the influence of the measuring system on MOBILE operation, because the measuring system is 50 Ω-terminated. The areas of the RTDs were 6 and 7.4 μm^2 for lower and upper devices, and the gate length and the width of the HEMT were 0.7 and 10 μm, respectively. A data stream and a clock pulse up to 35 Gb/s (GHz) were fed into the circuit to test high-speed operation. Figure 20 (a) shows the results of operation with the input of a pseudo-random bit stream at 35 Gb/s. Clear RO mode eye-patterns were obtained at this bit rate. (The eye-pattern is a superposition of the output for various input signals, and the size of the clear hole in the pattern is a measure of the noise margin). The relatively small output amplitude of about 50 mV is due to the design of the output buffer circuit. The inner logic swing is estimated to be around 0.6 V. Figure 20 (b) shows the result at 35 Gb/s with an input bit pattern of (0111010000101110). (The upper trace is the complement of the input data stream). This figure confirms proper operation. This operation speed is close to the cutoff frequency of the HEMT used in the circuit, and extremely high-speed operation is expected with a short gate-length HEMT.

As this circuit uses a clock pulse as an oscillating bias voltage for the MOBILE, measuring power dissipation is not straightforward. The power dissipation of the MOBILE was therefore estimated from a simple equivalent circuit calculation. In this calculation the RTDs were modeled as a parallel circuit of voltage dependent capacitance and a voltage dependent current source. Schulman's model [33] was used for the current source, and it was fitted to the experimental I-V curve. The capacitance of the RTD was modeled with a simple approximation that all collector voltage drops in the collector layers. (Here, we ignored the peak structure in the C-V curve for simplicity). This was about 4 fF/μm^2 around the peak voltage. Figure 21 shows the calculated power dissipation of the core circuit, which consists of a MOBILE and the input capacitance of the next stage, as a function of the operation frequency. In this calculation, the clock amplitude was set to be 0.7 V_{p-p} with a dc bias of 0.55 V. We think this is close to the experimental condition, though the exact value could not be determined due to the impedance mismatching. The power dissipation is less than 2 mW/gate. This is

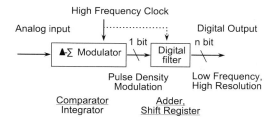

Figure 22: A block diagram of a $\Delta\Sigma$ ADC. It consists of a $\Delta\Sigma$ modulator and a digital filter.

extremely small compared to a few tens of mW for a high-speed source coupled FET logic (SCFL) gate [34]. Moreover, this difference extends several times if one compares it to the core circuit of an SCFL D-FF. This small power dissipation results from the small valley current as well as the small supply voltage. In fact, the large current corresponding to the peak current of the RTD flows only at the rising and falling edge of the clock, and the small current corresponding to the valley flows while the clock is high. Furthermore, this figure shows that the power dissipation has only limited dependence on the bit rate. This is because the capacitance of the RTDs is so small that, in this frequency range, the dynamic current is much smaller than the static current.

Finally, we will describe more specific application of the MOBILE. Among several circuits, the *analog-to-digital converter (ADC)* is one of the most promising applications of MOBILEs. A high-speed and high-resolution ADC is a key component for software defined radio (SDR) for future microwave communication, and considerable efforts are being made to realize such ADCs [35], [36], [37]. Using MOBILEs the circuit complexity and the operating frequency are expected to be much improved. Here, we describe a novel high-speed *delta-sigma ($\Delta\Sigma$) converter* using MOBILEs [38].

Figure 22 shows a block diagram of a $\Delta\Sigma$ ADC. It consists of a $\Delta\Sigma$ modulator and a digital filter. The former converts the input analog signal into the pulse density at a frequency much higher than the Nyquist rate, and reduces the quantization noise at low frequencies at the expense of its increase in high frequencies. The latter cuts the high-frequency component and then down-converts the pulse density signal into the high-resolution digital output at the Nyquist rate. This type of ADC has a significant advantage; higher resolution can be easily obtained by increasing the sampling rate. Furthermore, it does not require a high-accuracy analog component to achieve high resolution. MOBILE's unique characteristics make it possible to fabricate high performance $\Delta\Sigma$ modulator. Moreover, the digital filter, which consists of shift registers and adders, can be fabricated with MOBILEs in a very compact form because of the latching function and the weighted-sum threshold-logic function of the MOBILE.

The circuit configuration of the $\Delta\Sigma$ modulator using a MOBILE is shown in Figure 23. This is an extremely simple circuit consisting of a capacitor, an MOBILE, and three field-effect transistors (FETs). The capacitor C and the input FET Tr1 function as an integrator, and the MOBILE functions as a 1-bit comparator.

Here, we will briefly describe the operating principle of the proposed $\Delta\Sigma$ modulator. The current I_1 flows depending on the input voltage V_{in}, so that the charge stored at the capacitor gradually decreases. When the capacitor voltage, V_C, decreases to the threshold voltage of the MOBILE, it switches to the "HIGH" state and outputs a pulse. This output pulse charges the capacitor C through Tr2 (I_2), and it inhibits output of the pulse (negative feedback loop). This process repeats according to the clock, and the period of the process depends on the discharge rate of the capacitor C through Tr1. Consequently, the input voltage V_{in} is converted to the output pulse density.

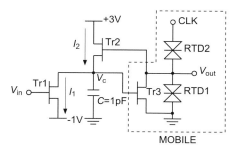

Figure 23: The circuit configuration of the $\Delta\Sigma$ modulator circuit.

The circuit complexity is reduced considerably due to the unique features of a MOBILE. First, the variation of the capacitor voltage V_C can be designed to be small so that the source voltage of the feedback transistor Tr2 is almost constant. This ensures a constant feedback when V_C is smaller than the threshold voltage of the MOBILE. This is possible because the MOBILE can switch according to an extremely small difference in peak currents of two RTDs. Second, the logic swing of the output of the MOBILE is constant independent of the input. This also ensures that the feedback is constant when it takes place. The advantage of this circuit is that it requires no extra clock cycle in the feedback loop, thus this circuit can be operated at an extremely high sampling rate.

We studied the operation of this circuit by numerical simulation using Simulation Program with Integrated Circuit Emphasis (SPICE). The values of the circuit parameters were chosen as follows, C = 10 pF, the gate lengths of Tr1, Tr2, and Tr3 were 0.7 μm. The clock frequency of MOBILE, that is, the sampling frequency f_S, was 10 GHz. The results are shown in Figure 24. This figure shows the output waveform in response to a ramp input. This figure clearly demonstrates that the input voltage is actually converted

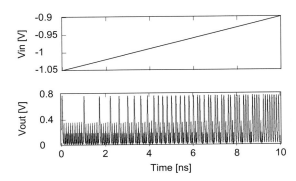

Figure 24: The simulated output waveform of the $\Delta\Sigma$ modulator in response to a ramp input.

to the pulse density. Important figures of the $\Delta\Sigma$ modulator, such as noise power density, dynamic range, were also calculated from the simulation results, and showed a good agreement with those for ideal $\Delta\Sigma$ modulator.

4.2.5 Extensions of the MOBILE Concept

In this section, we have described the concept and applications of the MOBILE, which exploits the unique properties of RTDs. Recently, various extensions of the MOBILE concept have been proposed. A multiple valued logic device, called MML (Monostable Multistable Logic), is one of such devices, which extends the MOBILE concept to the circuits consisting of three or more RTDs connected serially. This device can handle multiple valued logic. This idea was successfully applied to the flash-type AD converter using multiple valued quantizer [39]. Then, by replacing the HEMT to photodiode, optical input-electrical output MOBILEs were proposed for high speed optical communications. Akeyoshi and his coworkers used ultrahigh speed photodiode, called UTC-PD, to make optoelectronic MOBILE [40]. They reported the operating bit rates up to 80 Gb/s. Most recently, German groups proposed to replace the HEMT for input to RTHEMT, the RTD/HEMT serial combination [41]. This proposal has practical significance in future medium/large scale integration of the MOBILE because this improves the robustness against device parameter scattering.

5 Summary and Outlook

At the end of this chapter, we would like to emphasize the importance of reducing the complexity of circuits, and hence the importance of functional devices in the future. Ultrahigh-speed operation of circuits needs to minimize the wiring delay. Readers may think it is a problem only for large circuits as Si-VLSIs. (From this viewpoint, functional devices using quantum effect, such as RTDs, are very promising also for Si technology. Recently, Si-based NDR devices have been attracting much attention). However, the wiring delay has been already a severe problem even for small circuit blocks. Figure 25 shows the results of delay time analysis for small circuit blocks, clocked inverter and D-latch, consisting of InGaAs HEMTs, when operating at 60 GHz [42]. As you can see, the wiring delay time cannot be ignored; it is indeed large, about 50 % of the intrinsic delay time. So, reducing the circuit complexity and size has been already an important problem, and in future, its importance should increase further. One of the most important features of quantum effect devices is their functions resulting from unique I-V characteristics. Consequently, functional devices based on quantum effects including RTDs should play an important role in such ultrahigh-speed ICs.

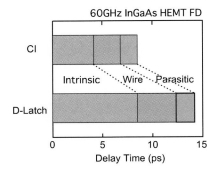

Figure 25: The results of delay time analysis for clocked inverter (CI) and D-latch operating at 60 GHz.

Acknowledgements

The editor would like to thank Christian Ohly (FZ Jülich) for checking the symbols and formulas in this chapter.

References

[1] E.R. Brown, in: *Hot Carriers in Semiconductor Nanostructures, Physics and Applications*, Jagdeep Shah, ed., Academic Press, 1992.
[2] R. Tsu and L. Esaki, Appl. Phys. Lett. **22**, 568 (1973).
[3] A. Douglas Stone and P.A. Lee, Phys. Rev. Lett. **54**, 1196 (1985).
[4] M. Büttiker, IBM J. Res. Develop. **32**, 63 (1988).
[5] Arno Föster, in: *Festkörperprobleme/ Advances in Solid State Physics*, Vol. 33, ed. R. Helbig, Vieweg, Brauschweig/Wiesbaden, 1994.
[6] Roger Lake, Gerhard Klimeck, R. Chris Bowen and Dejan Jovanovic, J. Appl. Phys. **81**, 7845 (1997).
[7] Tom P.E. Broekaert, Wai Lee and Clifton G. Fonstad, Appl. Phys. Lett. **53**, 1545 (1988).
[8] [Jurgen H. Smet, Tom P.E. Broekaert and Clifton G. Fonstad, J. Appl. Phys. **71**, 2745 (1992).
[9] D. Bimberg, F. Heinrichsdorff, R. K. Bauer, D. Gerthsen D. Stenkamp, D.E. Mars and J.N. Miller, J. Vac. Sci. Technol. B **10**, 1793 (1992).
[10] A. Förster, Proceedings of the Tenth International Workshop on the Physics of Semiconductor Devices, Allied Publisher LTD., New Delhi, 2000.
[11] H. Moroc, J. Chen, U.K. Reddy and T. Henderson, Appl. Phys. Lett. **49**, 70 (1986).
[12] E. R. Brown, J. R. Söderström, C. D. Parker, L. J. Mahoney, K. M. Molvar, and T. C. McGill, Appl. Phys. Lett. **58**, 2291 (1991).
[13] T. C. L. G. Sollner, W. D. Goodhue, P. E. Tannenwald, C. D. Parkar, D. D. Peck, Appl. Phys. Lett. **43**, 588 (1983).
[14] N. Shimizu, T. Nagatsuma, T. Waho, M. Shinagawa, M. Yaita, and M. Yamamoto, Electron. Lett. **31**, 1695 (1995).
[15] D. K. Ferry, F. Capasso (ed.), *Physics of Quantum Electron Devices*, Springer-Verlag, Berlin, 1990.
[16] M. Tsuchiya, T. Matsusue, and H. Sakaki, Phys. Rev. **59**, 2356 (1987).
[17] T. Waho, S. Koch, and T. Mizutani, Superlattices and Microstructures **16**, 205 (1994).
[18] N. Shimizu, T. Waho, and T. Ishibashi, Jpn. J. Appl. Phys. **36**, L330 (1997).
[19] K. Maezawa, MWE'99 Microwave Workshop and Exhibition, Yokohama, Japan, 1999, Microwave Workshop Digest, 143 (1999).
[20] F. Capasso, S. Sen, F. Beltram, and A. Y. Cho, F. Capasso (ed.), *Physics of Quantum Electron Devices*, Springer-Verlag, Berlin, 1990.
[21] N. Yokoyama, S. Muto, H. Ohnishi, K. Imamura, T. Mori, and T. Inata, F. Capasso (ed.), *Physics of Quantum Electron Devices*, Springer-Verlag, Berlin, 1990.
[22] K. Maezawa, T. Akeyoshi, and T. Mizutani, IEEE Trans. Electron Devices **41**, 148 (1994).
[23] P. H. Beton, M. W. Dellow, P. C. Main, L. Eaves, and M. Henini, Phys. Rev. B. **49**, 2264 (1994).
[24] Jürgen Stock, Jörg Malindretos, Klaus Michael Indlekofer, Michael Pöttgens, Arno Förster and Hans Lüth, IEEE Transaction on Electron Dev. **48**, 1028 (2001)
[25] A. C. Seabaugh, E. A. Beame III, A. H. Taddiken, J. N. Randall, Y.-C. Kao, IEEE Electron Device Lett. **14**, 472 (1993).
[26] J. Shen, S. Tehrani, H. Goronkin, G. Kramer, R. Tsui, IEEE Electron Device Lett. **17**, 94 (1996).
[27] K. J. Chen, K. Maezawa, T. Waho, M. Yamamoto, IEICE Trans. Electron. **E79-C**, 1515 (1996).
[28] K. Maezawa, and T. Mizuani, Jpn. J. Appl. Phys. **32**, L42 (1994).
[29] Y. Yamashita, A. Endoh, K. Shinohara, M. Higashiwaki, K. Hikosaka, T. Mimura, and S. Hiyamizu, IEEE Electron Device Lett. **22**, 507 (2001).
[30] K. J. Chen, K. Maezawa, K. Arai, M. Yamamoto, and T. Enoki, Electron. Lett. **31**, 925 (1995).
[31] T. Akeyoshi, K. Maezawa, T. Mizutani, Jpn. J. Appl. Phys. **33**, 794 (1994).
[32] K. Maezawa, H. Matsuzaki, M. Yamamoto, and T. Otsuji, IEEE Electron Device Lett. **19**, 80 (1998).
[33] J. N. Schulman, H. J. De Los Santos, and D. H. Chow, IEEE Electron Device Lett. **17**, 220 (1996).

[34] T. Otsuji, E. Sano, Y. Imai, and T. Enoki, Tech. Dig. 18th GaAs IC Symp., Florida, 14 (1996).

[35] T. Itoh, T. Waho, K. Maezawa, and M. Yamamoto, IEICE Trans. Inf. & Syst., **E82-D**, 949 (1999).

[36] T. P. E. Broekaert, B. Brar, J. P. A. van der Wagt, A. C. Seabaugh, F. J. Morris, T. S. Moise, E. A. Beam, III and G. A. Frazier, IEEE J. Solid-State Circuits **33**, 1342 (1998).

[37] T. P. E. Broekaert, B. Brar, F. Morris, A. C. Seabaugh and G. A. Frazier, *IEEE Proc. Ninth Great Lakes Symp. VLSI,* IEEE Press, Michigan, 1999.

[38] Y. Yokoyama, Y. Ohno, S. Kishimoto, K. Maezawa, T. Mizutani, Jpn. J. Appl. Phys. **40**, L1005 (2001).

[39] T. Ito, T. Waho, J. Osaka, H. Yokoyama, and M. Yamamoto, IEEE MTT-S Int. Microwave Symp. 197 (1998).

[40] K. Sano, K. Murata, T. Otsuji, T. Akeyoshi, N. Shimizu, and E. Sano, IEEE J. Solid State Circuits **36**, 281 (2001).

[41] C. Pacha, U. Auer, C. Burwick, P. Glösekötter, A. Brennemann, W. Prost, F.-J. Tegude, K. Goser, IEEE Trans. VLSI Systems **8**, 558 (2001).

[42] Y. Umeda, K. Osafune, T. Enoki, H. Yokoyama, Y. Ishii, and Y. Imamura, IEEE Trans. Microwave Theory and Tech. **46**, 1209 (1998).

Single-Electron Devices for Logic Applications

Ken Uchida
Advanced LSI Technology Laboratory, Toshiba Corporation, Japan

Contents

1 Introduction — 425

2 Single-Electron Devices — 425
2.1 Single-Electron Box — 426
2.2 Single-Electron Transistor — 428
2.3 Other Single-Electron Devices — 431
2.4 Fabrication of Single-Electron Devices — 431

3 Application of Single-Electron Devices to Logic Circuits — 433
3.1 Introduction — 433
3.2 Analytical Model of Single-Electron Transistor for Circuit Simulation — 433
3.3 Logic Circuits with Single-Electron Transistors — 436

4 Future Directions — 439

Single-Electron Devices for Logic Applications

1 Introduction

Scaling down of electronic device sizes has been the fundamental strategy for improving the performance of ultra-large-scale integrated circuits (ULSIs). Metal-oxide-semiconductor field-effect transistors (MOSFETs) have been the most prevalent electron devices for ULSI applications, and thus the scaling down of the sizes of MOSFETs [1][2] has been the basis of the development of the semiconductor industry for the last 30 years.

However, in the early years of the 21st century, the scaling of CMOSFETs is entering the deep sub-50 nm regime [3]. In this deep-nanoscaled regime, fundamental limits of CMOSFETs and technological challenges with regard to the scaling of CMOSFETs are encountered [4]. On the other hand, quantum-mechanical effects are expected to be effective in these small structured devices. Therefore, in order to extend the prodigious progress of LSI performance, it is essential to introduce a new device having an operation principle that is effective in smaller dimensions and which may utilize the quantum-mechanical effects, and thus provide a new functionality beyond that attainable with CMOSFETs.

Single-electron devices [5][6] are promising as new nanoscaled devices because single-electron devices retain their scalability even on an atomic scale and, moreover, they can control the motion of even a single electron. Therefore, if the single-electron devices are used as ULSI elements, the ULSI will have the attributes of extremely high integration and extremely low power consumption. In this respect, scalability means that the performance of electronic devices increases with a decrease of the device dimensions. Power consumption is roughly proportional to the electron number transferred from voltage source to the ground in logic operations. Therefore, the utilization of single-electron devices in ULSIs is expected to reduce the power consumption of ULSIs.

In this chapter, firstly, the operation and operation principle of single-electron devices is briefly explained. Since the operations of single-electron devices have been described in detail in many excellent books and monographs, for example by Devoret and Grabert [6], Averin [7], and Likharev [8], [9], the topics are restricted to those essential to an understanding of the operation of simple single-electron devices. Then, the advantages and disadvantages of single-electron devices over conventional MOSFETs are discussed. Next, the analytical device model of a single-electron transistor, which is a typical functional single-electron device, for circuit simulation is derived and the methodology of designing logic circuits with single-electron transistors is discussed.

2 Single-Electron Devices

Single-electron devices are devices that can control the motion of even a single electron and consist of quantum dots having tunnel junctions.

In this section, the operation and operation principle of single-electron devices are explained. Firstly, the concept of single-electron phenomena and Coulomb blockade effects are explained by referring to a single-electron box, the simplest single-electron device.

2.1 Single-Electron Box

Structure of Single-Electron Box

The smallest set of the functional single-electron device is composed of a quantum dot connected with two electrodes. One electrode is connected with the quantum dot through a tunneling junction. The other electrode, called the gate electrode, is coupled with the quantum dot via insulator through which electron cannot pass by quantum tunneling (Figure 1). Therefore, electrons are injected/ejected into/from the quantum dot through the tunneling junction.

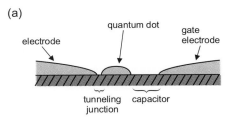

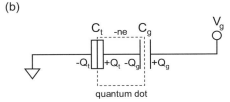

Figure 1:
(a) Schematic structure of single-electron box. The single-electron box consists of a quantum dot, an electrode connected to the dot through a tunneling junction, and an electrode coupled to the dot through an ideal, infinite-resistance, capacitor.
(b) Equivalent circuit of single-electron box.

Basic Operation of Single-Electron Box

As the size of the quantum dot decreases, the charging energy W_c of a single excess charge on the dot increases. If the quantum-dot size is sufficiently small and the charging energy W_c is much greater than thermal energy $k_B T$, no electron tunnels to and from the quantum dot. Thus, the electron number in the dot takes a fixed value, say zero, when both the electrodes are grounded. The charging effect, which blocks the injection/ejection of a single charge into/from a quantum dot, is called Coulomb blockade effect. Therefore, the condition for observing Coulomb blockade effects is expressed as,

$$W_c = \frac{e^2}{2C} \gg k_B T \;, \tag{1}$$

where C is the capacitance of the quantum dot and T is the temperature of the system.

However, it should be noted that by applying a positive bias to the gate electrode we could attract an electron to the quantum dot. The increase of the gate voltage attracts an electron more strongly to the quantum dot. When the gate bias exceeds a certain value an electron finally enters the quantum dot and the electron number of the dot becomes one. Further increase of the gate voltage makes it possible to make the electron number two. Thus, in the single-electron box, the electron number of the quantum dot is controlled, one by one, by utilizing the gate electrode (Figure 2).

Conditions for Observing Single-Electron Tunneling Phenomena

In order to observe single-electron tunneling phenomena, or Coulomb blockade effects, there are two necessary conditions. One condition is, as described above, that the charging energy of a single excess electron on a quantum dot is much greater than the thermal energy (Eq. (1)). The other condition is that the tunneling resistance R_t of the tunneling junction must be larger than resistance quantum h/e^2. This condition is required to suppress the quantum fluctuations in the electron number, n, of the dot so that they are sufficiently small for the charge to be well localized on the quantum dot. The condition is obtained by keeping uncertainty principle $\Delta W \Delta t > h$ while letting ΔW be the charging energy of the quantum dot, $\sim e^2/C$, and Δt be the lifetime of the charging, $R_t C$. Then, the uncertainty principle reduces to

$$\Delta W \cdot \Delta t \sim \frac{e^2}{C} \cdot R_t C = e^2 R_t > h \;. \tag{2}$$

As a result, one obtains the condition for the tunneling resistance R_t in order to observe the Coulomb blockade effects

$$R_t \gg \frac{h}{e^2} = 25.8 \text{ k}\Omega. \tag{3}$$

Bias Conditions for Coulomb Blockade Effects

The voltage range, which keeps the electron number at n in the dot, is extracted by considering the free energy of the system. The free energy of the system having n electrons in the island $F(n)$ is expressed as

$$F(n) = W_c(n) - A(n) \;, \tag{4}$$

where $W_c(n)$ is the charging energy and $A(n)$ is the work done by the voltage source connected to the gate electrode in order to make the electron number be from zero to n.

It is important to note that when tunneling phenomena do not occur the tunneling junction behaves as a normal capacitor and that the polarization charge on the capacitors does not have to be associated with a discrete number of electrons, n. This polarization charge is essentially due to a rearrangement of the electron gas with respect to the positive background of ions. Therefore, the polarization charge takes a continuous range of value, although the number of electrons in the quantum dot takes a discrete number of electrons, n. The polarization charges on the tunneling junction and gate capacitor are obtained from the following relationship.

$$Q_t - Q_g = -ne ,$$
$$\frac{Q_t}{C_t} + \frac{Q_g}{C_g} = V_g , \quad (5)$$

where Q_t and Q_g are the polarization charge on the tunneling junction and the gate capacitor, respectively. By using Q_t and Q_g, the charging energy $W_c(n)$ of the quantum dot is expressed as,

$$W_c(n) = \frac{Q_t^2}{2C_t} + \frac{Q_g^2}{2C_g} , \quad (6)$$

which reduces to

$$W_c(n) = \frac{e^2 n^2}{2C_\Sigma} + \frac{1}{2}\frac{C_t C_g V_g^2}{C_\Sigma} , \quad (7)$$

where $C_\Sigma = C_t + C_g$. In addition, the work, $A(n)$, done by the gate voltage source in order to make electron number of the quantum dot be from zero to n is expressed as,

$$A(n) = \int I(t) \cdot V_g dt = Q_g V_g = en\frac{C_g}{C_\Sigma}V_g + \frac{C_t C_g V_g^2}{C_\Sigma} . \quad (8)$$

In order to maintain the electron number in the quantum dot, the following conditions are required.

$$F(n) < F(n \pm 1) \quad (9)$$

From Eqs.(7) to (9), the voltage range, within which Coulomb blockade effects are in effect and the electron number of the dot takes a fixed value of n, can be obtained as follows.

$$\left(n - \frac{1}{2}\right)\frac{e}{C_g} < V_g < \left(n + \frac{1}{2}\right)\frac{e}{C_g} \quad (10)$$

This condition is also expressed with critical charge Q_c as follows.

$$|Q_t| < Q_c , \quad (11)$$

where Q_c is expressed as,

$$Q_c = \frac{e}{2}\left(1 + \frac{C_g}{C_t}\right)^{-1} . \quad (12)$$

Free energy change $\Delta F(n, n+1)$ that accompanies a transition of the electron number from n to $n+1$ is also simply expressed with critical charges Q_c as,

$$\Delta F(n, n+1) = F(n+1) - F(n) = \frac{e}{C_t}(Q_t - Q_c) . \quad (13)$$

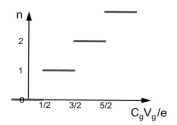

Figure 2: Electron number versus gate voltage characteristics of single-electron box. The number of electron in the quantum dot increases one by one as the gate voltage increases.

Application of Derived Formulas for Single-Electron Box to Other Single-Electron Devices: Thévenin's Theorem

The single-electron box is important not only because it is the simplest functional single-electron device and its operation is easy to understand, but also because any single-electron system consisting of tunneling junctions, capacitors, and voltage sources can be reduced to a simpler form for each individual tunneling junction. Using Thévenin's theorem, the circuit to which the tunneling junction is coupled is reduced to an equivalent capacitor C_e in series with a voltage source V_e [13] and thus the circuit is equivalently converted to the single-electron box having gate capacitance of C_e and the gate voltage of V_e. As a result, the Coulomb blockade condition for each tunneling junction can be obtained from Eqs. (10) and (12). In the next subsection, we apply Thévenin's theorem to single-electron transistors and extract the conditions for maintaining and destroying the Coulomb blockade effects.

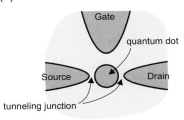
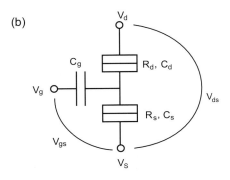

2.2 Single-Electron Transistor

Although a single-electron box can control the number of electrons in the quantum dot, it does not have the properties of a switching device. Since switching devices are essential elements of ULSIs, single-electron switching devices are required to utilize single-electron devices in logic circuits.

Structure of Single-Electron Transistors

Single-electron transistors (SETs) are three-terminal switching devices, which can transfer electrons form source to drain one by one. The schematic structure of SETs is shown in Figure 3. As shown in the figure, the structure of SETs is almost the same as that of MOSFETs. However, SETs have tunneling junctions in place of pn-junctions of the MOSFETs and a quantum dot in place of the channel region of the MOSFETs.

Operation of Single-Electron Transistors

The operation of single-electron transistors can be described by using Thévenin's theorem and applying derived Eqs. (10) - (12) for a single-electron box.

By using the Thévenin's theorem, the circuit connected to the tunneling junction of the source is transformed to the circuit shown in Figure 4a. From this equivalent circuit and Eq. (10), the condition to maintain the electron number at n in the dot is expressed as

$$\left(n-\frac{1}{2}\right)\frac{e}{C_g+C_d} < \frac{C_gV_g+C_dV_d}{C_g+C_d} < \left(n+\frac{1}{2}\right)\frac{e}{C_g+C_d}, \quad (14)$$

which reduces to

$$\frac{1}{C_d}\left(ne-\frac{e}{2}-C_gV_g\right) < V_d < \frac{1}{C_d}\left(ne+\frac{e}{2}-C_gV_g\right). \quad (15)$$

In the same manner, the circuit connected to the tunneling junction of the drain is transformed to the circuit shown in Figure 4b and the condition to maintain the electron number at n in the dot is expressed as

$$\frac{1}{C_s+C_g}\left(-ne+\frac{e}{2}+C_gV_g\right) > V_d > \frac{1}{C_s+C_g}\left(-ne-\frac{e}{2}+C_gV_g\right) \quad (16)$$

Figure 5a shows the relationship between the drain voltage V_d and the gate voltage V_g, which satisfies the conditions expressed by Eqs. (15) and (16). The gray areas shown in Figure 5a are Coulomb blockade regions, where the Coulomb blockade is effective and the electron number in the dot takes a fixed value indicated in the areas.

On the other hand, in other regions, the quantum dot can take at least two electron numbers. In the green regions shown in Figure 5a the quantum dot can take two electron numbers. For example, in the green region indicated by A, the electron number in the dot is zero or one. More precisely, the electron number of one is preferable for the tunneling junction of the source and the electron number of zero is preferable for the tunneling junction of the drain. Therefore, when a finite positive source-to-drain voltage V_{ds}, indicated by dashed line in Figure 5a, is applied between the source and drain electrodes and the gate voltage is $e/2C_g$, an electron transport process described below is observable. The initial electron number of the dot is assumed to be zero. For the tunneling junction of the source, the electron number of one is preferable so that an electron tunnels from the source to the dot and the electron number in the dot becomes one. However, for the tunneling junction of the drain, the electron number of zero is preferable so that an electron tunnels from the dot to the drain and the electron number in the dot becomes zero. As a result, an electron tunnels from the source to the drain, and source-to-drain current is observable at these bias conditions.

In the same manner, at the gate voltage of $ne/C_g + e/2C_g$, the source-to-drain current I_{ds} is observed, and thus oscillating I_{ds} versus V_g characteristics shown in Figure 5b is observed in single-electron transistors. The oscillating I_{ds} - V_g characteristics are called Coulomb oscillations.

Figure 3:
(a) Schematic structure of single-electron transistor.
(b) Equivalent circuit of single-electron transistor.

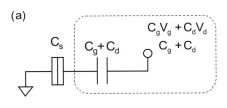
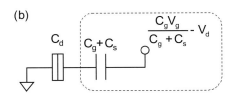

Figure 4: Application of Thévenin's theorem.
(a) Surrounded by the dashed line is the equivalent circuit connected to the tunneling junction of the source.
(b) Surrounded by the dashed line is the equivalent circuit connected to the tunneling junction of the drain.

The I_{ds} versus V_{ds} characteristics are also obtained in the same manner. Typical I_{ds}-V_{ds} characteristics are shown in Figure 5c in the case of gate voltage of zero and $e/2C_g$. The conductance suppression observed around $V_{ds} \sim 0$ when $V_g = 0$ is called Coulomb blockade characteristics.

Calculation of Current Through Quantum Dot

The source-to-drain current of single-electron transistors can be calculated by using the tunneling rate of an electron through the tunneling junction. The rate $\Gamma(n, n+1)$ of an electron tunneling through a tunneling junction, which accompanies a transition of the electron number in the dot from n to $n+1$, is given by [6][7]

$$\Gamma(n,n+1) = \frac{1}{e^2 R_t} \frac{\Delta F(n,n+1)}{1-\exp\left[-\Delta F(n,n+1)/k_B T\right]}, \quad (17)$$

where $\Delta F(n, n+1)$ is the free energy change that accompanies the tunneling and R_t is the tunneling resistance of the junction. In the following, the tunneling rate through the junction of the source is denoted by $\Gamma_s(n, n+1)$ and the tunneling rate through the junction of the drain is denoted by $\Gamma_d(n, n+1)$.

The probability p_n of finding n electrons in the dot may change by leaving this state or by coming into this state from the states $n-1$ or $n+1$ [6].

$$\frac{dp_n}{dt} = \Gamma_{tot}(n+1,n)p_{n+1} + \Gamma_{tot}(n-1,n)p_{n-1} - \left[\Gamma_{tot}(n,n+1) + \Gamma_{tot}(n,n-1)\right]p_n \quad (18)$$

where

$$\Gamma_{tot}(n,n+1) = \Gamma_s(n,n+1) + \Gamma_d(n,n+1). \quad (19)$$

There exists normalization condition

$$\sum_{n=-\infty}^{+\infty} p_n = 1. \quad (20)$$

The current I of single-electron transistor is obtained from

$$I = e \sum p_n \left[\Gamma_s(n,n+1) - \Gamma_d(n,n+1)\right]. \quad (21)$$

The calculations of Eqs. (18) - (21) are awkward. Therefore, the characteristics of single-electron devices are usually calculated with numerical simulators [11], [12].

Semiconductor Quantum Dot

It should be noted that when the quantum-dot size is comparable with the de Broglie wavelength of the electrons in quantum dots (this situation frequently occurs in the case of semiconductor nanoscaled quantum dots), the energy quantization becomes comparable with the charging energy. In this case, the energy difference due to the addition of a single electron to the dot is given not by the charging energy W_c but by the electron addition energy W_a, which is given by the following formula.

$$W_a = W_c + \Delta W \quad (22)$$

Here ΔW is the quantum energy level difference due to the addition of a single electron to the dot.

As a result, periodicity of Coulomb oscillations is modified as [10],

$$\Delta V_g = \frac{e}{C_g} + \frac{\Delta W}{e}. \quad (23)$$

Thus, in quantum dots holding just a few electron, the electron addition energy W_a can no longer parameterized with W_c, and the Coulomb oscillations are significantly modified by electron-electron interactions and quantum confinement effects. Therefore, in this case, quantum dots are regarded as *artificial atoms* [14].

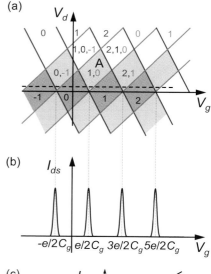
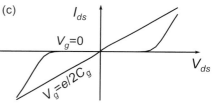

Figure 5:
(a) Relationship between the drain voltage V_d and the gate voltage V_g, satisfying the conditions expressed by Eqs. (15) and (16). The diamond-shaped structure along the x-axis is called Coulomb diamond.
(b) source-to-drain current I_{ds} versus gate voltage V_g characteristics of single-electron transistors.
(c) I_{ds} versus V_{ds} characteristics of single-electron transistors.

III Logic Devices

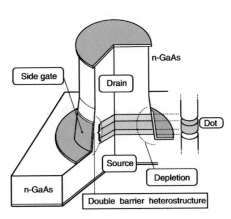

Figure 6: Schematic of a quantum dot in a vertical device (After Tarucha *et al.* [17], © 2001 IOP Publishing Co.).

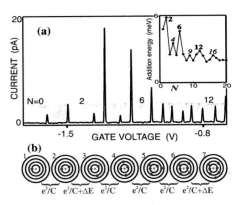

Figure 7: Current flowing through a two dimensional circular quantum dot on varying the gate voltage.
(a) The first peak marks the voltage where the first electron enters the dot, and the number of electrons, N, increases by one at each subsequent peak. The distance between adjacent peaks corresponds to the addition energies (see inset).
(b) The addition of electrons to circular orbits is shown schematically. The first shell can hold two electrons whereas the second shell can contain up to four electrons. It therefore costs extra energy to add the third and seventh electrons (After Kouwenhoven *et al.* [16], © 2001 IOP Publishing Co.).

By utilizing Eq. (23), the energy spectrum of a quantum dot, or an *artificial atom*, can be studied. Tarucha *et al.* have fabricated vertical single-electron transistors (SETs) having circular-disk-shaped dots with double heterostructure barriers and surrounding side gate, shown in Figure 6, and observed, in the transport measurements, atom-like properties such as "magic numbers" and "Hund's first rule" [15], [16], [17].

In the vertical SET, the quantum dot is located in the center of the pillar. The diameter of the dot is a few hundred nanometers and its thickness is about 10 nm. The dot is sandwiched with two non-conducting heterostructure barrier layers, which separate it from conducting material above and below. A negative voltage applied to the side gate around the pillar squeezed the effective diameter of the dot. Consequently, the number of electrons is reduced, one by one, until the dot is completely empty.

If the lateral confinement has the form of harmonic potential, the eigen-energy $W_{n,l}$ is expressed with radial quantum number n_r (=0, 1, 2, …) and the angular momentum quantum number l (= 0, ±1, ±2, …):

$$W_{n_r,l} = \left(2n_r + |l| + 1\right)\hbar\omega_0 , \quad (24)$$

where $\hbar\omega_0$ is the lateral confinement energy [16], [17]. Here, the Zeeman effect is neglected. Therefore, it should be noted that each state is spin degenerate.

Figure 7a shows the I_d-V_g characteristics, or Coulomb oscillation characteristics, of the vertical SET. The distance between the consecutive peaks is proportional to W_a, which is the energy difference between the transition point of (N to $N + 1$) and ($N + 1$ to $N + 2$) electrons and is equal to the difference of the ionization energy and the electron affinity [18]. The addition energies, W_a's, extracted form the I_d-V_g characteristics are summarized in the inset of Figure 7a. It should be noted that W_a is not constant, and larger energy is necessary to add an electron to the dot with 2, 6, and 12 electrons. The numbers in this sequence can be regarded as "magic numbers" for a two-dimensional harmonic potential dot [15].

The reason is explained as follows [15], [16], [17]. Figure 7b shows the two-dimensional orbits allowed in the dot. The orbit with the smallest radius corresponds to the lowest energy state ($W_{0,0}$), which has zero angular momentum and can have two electrons with opposite spin. The addition of the second electron thus only costs the charging energy, e^2/C. Extra energy ΔW is necessary to add the third electron, because the electron must go into the next energy state ($W_{0,-1}$, $W_{0,1}$), which has an angular momentum ±1 and can have four electrons. Therefore, extra energy is again necessary to add the seventh electron. The numbers in the above sequence can be thus regarded as *magic numbers* for a two-dimensional harmonic dot.

In addition, for the filling of electrons in the same orbit, parallel spins are favored by "Hund's first rule". This leads to another series of magic numbers of N = 4, 9, 16, … corresponding to the half filling of the second, third, fourth orbits, respectively [17].

Thus, the atomic-like features are successfully observed in the vertical SETs having circular disk quantum dots.

Since the above discussions concerning artificial atoms are based on ref. [16], [17], the interested reader is advised to refer to the original monographs [16], [17] and related articles, such as [15], [19], [20].

Advantages and Disadvantages of Single-Electron Transistors Compared with MOSFETs

As for the advantages, the SETs have the properties of low power consumption and good scalability, as describe in Sect. 1. These are the reasons why SETs are promising as future LSI elements.

On the other hand, as for the disadvantages, 1) operations of SET circuits are generally limited to low temperature. This is a negative aspect of the good scalability. In order to operate SET circuits at room temperature, the size of the quantum dot must be much smaller than 10 nm. With the present technology, fabricating a structure smaller than 10 nm is difficult. In addition, 2) SETs have the disadvantage of high output impedance, due to the high resistance of tunnel junctions, which must be much higher than 25.8 kΩ (Eq. (3)). Finally, 3) source-to-drain voltage of SETs should be smaller than gate voltage swing in order to use SETs as gate controlled switching device, because the potential of the dot is easily affected by the source-to-drain voltage. The effect of source-to-drain voltage on the switching characteristics of SETs will be quantitatively evaluated in Sect. 3.2.

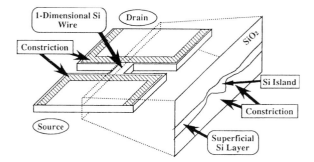

Figure 8: Schematic diagram of the 1-D wire and 2-D regions and its cross section after thermal oxidation (After Takahashi et al. [26], © 1996 IEEE).

2.3 Other Single-Electron Devices

In single-electron transistors, the one-by-one transfer of electrons can be achieved. However, the timing of the single-electron tunneling cannot be controlled in single-electron transistors. A single-electron turnstile [21] and a single-electron pump [22] are single-electron devices, which can control the timing of single-electron tunneling. The operations of these devices are not described here, because of space constraints regarding this chapter. The interested reader is advised to refer to other books and articles, such as [6].

2.4 Fabrication of Single-Electron Devices

There have been a number of reports on the fabrication of single-electron devices. Since single-electron phenomena can be observed in any conductive substances, single-electron devices are fabricated in a variety of materials such as aluminum [24], heterostructures [25], and silicon. However, in order to utilize single-electron devices as elemental devices of LSIs, the realization of single-electron devices made in silicon is essential. This can be achieved if fabrication techniques of nanometer-scaled silicon quantum dots are established.

Regarding silicon quantum dot formation, many approaches have been reported [26], [27], [28], [29], [30], [31] and they are generally categorized into two groups: patterning the silicon quantum dots by fine-lithography techniques and the growth of silicon quantum dots by deposition processes.

Using the former approaches, it is possible to accurately define the structures and positions of quantum dots. For example, Takahashi et al. proposed a novel silicon-quantum-dot fabrication process named pattern-dependent oxidation (PADOX) [26]. When a 1-D Si nano-wire, which has wide 2-D Si layers at its ends and is fabricated in silicon-on-insulator (SOI) wafer, is subjected to oxidation process, the oxidation process not only reduces the width and height of the 1-D Si wire, but also constricts the Si wire

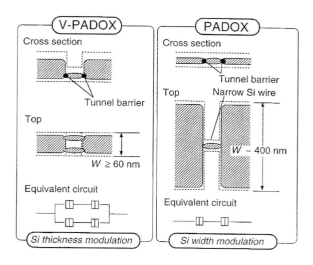

Figure 9: Si patterns and the corresponding circuits for V-PADOX and PADOX. In the cross-sectional and top views, broken lines represents preoxidation Si patterns and hatched regions represent islands and leads after oxidation (After Ono et al. [27], © 2000 IEEE).

III Logic Devices

at its ends. Since oxygen atoms penetrate not only from the surface oxide layer but also from the backside (the interface of SOI and buried oxide) through the pattern side, oxidation occurs more in the neighborhood of the pattern edges of the 2-D Si layers, as shown in Figure 8. Ono *et al.* have developed the vertical version of PADOX (V-PADOX) [27], shown in Figure 9. In PADOX, laterally broad 2-D regions are essential for tunnel-barrier formation. On the other hand, in V-PADOX, vertically broad, namely thick, 2-D regions are utilized for the tunnel-barrier formation. The advantage of the V-PADOX is that the V-PADOX makes it possible to form two tiny islands in a small area by utilizing not a lithographic process but the oxidation process, which induces the accumulation of stress in small structures. Thus, by utilizing V-PADOX, two SETs can be fabricated in an extremely small area, as shown in Figure 10.

The latter approaches are favorable from the viewpoints of throughput and fabricated quantum-dot sizes. In fact, Yano *et al.* successfully fabricated the room-temperature operating silicon single-electron memory by using the formation process of thin poly-silicon film, in which an array of 10-nm grains is naturally formed [28]. Tiwari *et al.* reported single-electron memory having Si nanocrystal storage [29].

Uchida *et al.* proposed another approach, where slight etching of an ultrathin SOI film with an alkaline-based solution is utilized [30], [32]. The proposed device structure is schematically illustrated in Figure 11a. As shown in the figure, the device structure is almost the same as that of conventional SOI-MOSFETs, but the SOI film has two key features: 1) its surface is intentionally undulated in nanoscaled dimensions as shown in Figure 11b by utilizing an alkaline-based solution; 2) the channel SOI thick-ness is thinned to a few nanometers. The nanoscaled undulation in the ultrathin film results in the formation of nanoscaled potential fluctuations due to the difference of quantum confinement effects from one part to another. Consequently, both the narrow electron channel through potential valleys and small potential pockets, storing memory information, are formed in the film as shown in Figure 11c. Since potential fluctuation still exists in the narrow channel, the channel effectively splits into several quantum dots. The quantum dots included in the channel are the origin of the SET operation. Thus, the device works as a single-electron transistor with nonvolatile memory function. Since the fabrication process of this SET is compatible with that of CMOSFETs. SET/CMOS hybrid circuit is fabricated on a chip, as shown in Figure 12, and its operation is successfully demonstrated even at room temperature [32].

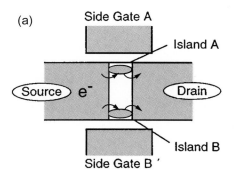

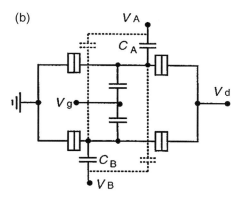

Figure 10: Fabrication of two SETs.
(a) Top view of the structure.
(b) Equivalent circuit (After Ono *et al.* [27], © 2000 IEEE).

Figure 11: Device structure and operation principle of single-electron transistor with nonvolatile memory function.

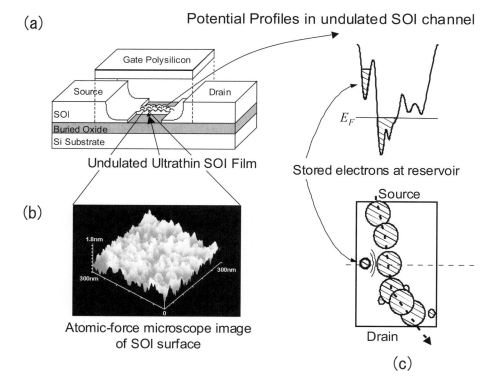

3 Application of Single-Electron Devices to Logic Circuits

3.1 Introduction

As described in Sect. 1, the implementation of single-electron devices is considered to contribute to increase of the packing density and to decrease of the power consumption of ULSIs. Therefore, many attempts [6], [23], [33], [34], [35], [36] have been made to develop logic circuits consisting of single-electron devices.

So far, there have been generally two methods of implementing logic operations in the circuits of single-electron devices: 1) by representing a bit by a single electron and using single-electron devices to transfer electrons one by one [23], [33] or 2) by representing a bit by more than one electron and using single-electron devices to switch the current on/off [34], [35], [36]. The former method is more attractive from the power consumption standpoint. However, in this case even one erroneous electron caused by noise or thermal agitation will completely alter the operation results. Therefore, from the viewpoint of operation stability, the latter method is preferable. In addition, from the viewpoint of operation, logic circuits realized by the former method resemble those consisting of charge-coupled devices (CCD), whereas logic circuits realized by the latter method resemble those consisting of MOSFETs. Since the present logic circuits consist of MOSFETs, the latter method is preferable. In fact, utilizing the design scheme of logic circuits of MOSFETs means that we can also utilize the computer-aided-design (CAD) tools of logic circuits of MOSFETs.

In order to design logic circuits correctly, the development of SET device model, which can be implemented into a circuit simulator, is essential. Therefore, we firstly derive an analytical device model of SETs [37] in Sect. 3.2. Then, the design scheme of logic circuits, in which SETs are used to switch the current on/off, is discussed [38] and the operation of designed logic circuits is confirmed by circuit simulations.

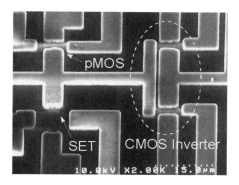

Figure 12: Micrograph of SET/CMOS hybrid circuit.

3.2 Analytical Model of Single-Electron Transistor for Circuit Simulation

The analytical SET models proposed so far [39], [40] are derived under the condition of an extremely small source-to-drain voltage V_{ds} ($V_{ds} \sim 0$), and thus, are insufficient for the analysis of realistic logic circuits where the bias condition that V_{ds} is comparable to the gate voltage swing should be taken into account.

In this section, we derive a compact, physically based, analytical SET model describing SET characteristics accurately over a wide source-to-drain voltage range of $|V_{ds}| < e/C_\Sigma$. This SET model is sufficiently accurate even at a relatively high temperature of $0.1 e^2 / 2 C_\Sigma k_B$. Because of its simplicity and accuracy over the wide V_{ds} and temperature ranges, this model is considered to be suitable for the design and analysis of realistic SET circuits.

The notations illustrated in Figure 3b are used throughout this section.

Assumptions

Our assumptions for deriving the analytical SET model are as follows. Firstly, we assume that all the source and drain terminals of SETs are connected to the capacitors whose capacitance is much larger than the total capacitance of the SET island or biased by constant-voltage sources. This assures that the SET characteristics are affected by neighboring circuit components only through the node voltages of SET terminals [41], [42]. Secondly, the source and drain resistances are assumed to be the same: $R_s = R_d = R_t$. Although this assumption puts a small restriction on the SET structures, it not only greatly reduces the calculation task but also renders the final formula simple and concise. Because only a few circuits utilizing a resistance mismatch have been proposed so far, it is considered that this restriction will not affect the usefulness of this model. Thirdly, at each given gate voltage, the two most probable numbers of the electrons in the SET island are taken into account. For example, when the gate voltage is 0 (e/C_g) and the drain voltage is positive, we consider that the number of electrons in the SET island is 0 or −1 (1 or 0).

Besides these assumptions, the tunneling resistance is supposed to be much larger than the quantum of resistance, $h/e^2 \sim 25.8$ kΩ, in order to suppress tunneling by quantum fluctuation, as described in Sect. 2.1.

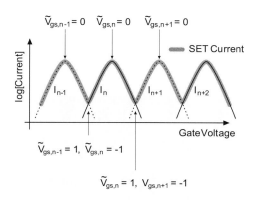

Figure 13: Schematic of model current $I_{n-1}, I_n, ..., I_{n+2}$ (solid and dotted lines) and SET current (thick gray line). I_n defines one period of Coulomb oscillations. The summation of I_n over a desired gate voltage range gives the Coulomb oscillations.

Derivation of the Model

We derive the analytical model of a double-junction SET on the basis of the "orthodox" theory and the steady-state master equation method. Based on the assumptions described above, the steady-state master equations can be accurately calculated, and thus, the $I-V$ characteristics of SET having n or $n+1$ electrons in its island is given by

$$I_n = \frac{e}{2R_\Sigma C_\Sigma} \frac{\left(\tilde{V}_{gs,n}^2 - \tilde{V}_{ds}^2\right)\sinh\left(\tilde{V}_{ds}/\tilde{T}\right)}{\tilde{V}_{gs,n}\sinh\left(\tilde{V}_{gs,n}/\tilde{T}\right) - \tilde{V}_{ds}\sinh\left(\tilde{V}_{ds}/\tilde{T}\right)} , \quad (25)$$

where

$$\tilde{V}_{gs,n} = \frac{2C_g V_{gs}}{e} - \frac{(C_g + C_s - C_d)\cdot V_{ds}}{e} - 1 - 2n , \quad (26)$$

$$\tilde{V}_{ds} = \frac{C_\Sigma V_{ds}}{e} , \quad (27)$$

$$\tilde{T} = \frac{2k_B T C_\Sigma}{e^2} \quad (28)$$

and $R_\Sigma = R_t + R_t$ ($R_t = R_s = R_d$). It should be noted that the dependence of I_n on $\tilde{V}_{gs,n}$ takes a symmetrical hump shape having a maximum at $\tilde{V}_{gs,n} = 0$. It is, therefore, reasonable to consider that I_n corresponds to one period of Coulomb oscillations. Consequently, by incorporating $\tilde{V}_{gs,n} = 0$ into Eq. (26), we obtain the gate voltage giving the peak of Coulomb oscillations as

$$V_{gs} = \frac{e}{2C_g} + \frac{ne}{C_g} + \frac{(C_g + C_s - C_d)\cdot V_{ds}}{2C_g} \quad (29)$$

Considering the period of Coulomb oscillations, e/C_g, we obtain the gate voltage range, for which the model current I_n holds, as

$$\frac{ne}{C_g} + \frac{(C_g + C_s - C_d)\cdot V_{ds}}{2C_g} < V_{gs} < \frac{(n+1)\cdot e}{C_g} + \frac{(C_g + C_s - C_d)\cdot V_{ds}}{2C_g} \quad (30)$$

which is equivalent to $-1 < \tilde{V}_{gs,n} < 1$. The summation of I_n over a desired gate voltage range gives the Coulomb oscillations as shown in Figure 13.

By utilizing the model, the I_d-V_{gs} characteristics of SET having the parameters of $C_s = C_d = 1$ aF, $C_g = 3$ aF, and $R_t = 10$ MΩ are calculated as shown in Figure 14. The results numerically calculated using the simulator CAMSET are also illustrated. There are two factors undermining our two-electron-number approximation; temperature and the drain voltage. However, the results obtained using the analytical model perfectly reproduce the numerically calculated ones even in cases of a relatively high drain voltage of $\tilde{V}_{ds} \sim 0.625$, namely $V_{ds} = 20$ mV, and a relatively high temperature of $T \sim 0.108$, namely $T = 20$ K. Here, it should be emphasized that the peak and the valley currents of Coulomb oscillations, as well as the phase shift of the V_{gs} dependence induced by drain voltage, are perfectly represented by the model, despite its simplicity.

In addition, it should be noted that this model is easily applicable to multigated and double-junction single-electron devices such as CMOS-like SETs [34] by properly rewriting Eq. (26).

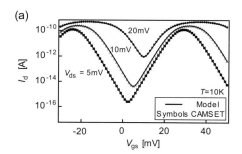

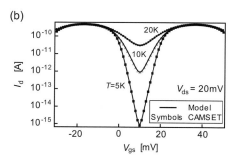

Figure 14: I_d-V_{gs} characteristics of a single-electron transistor, calculated using the SET model $I_{-1}+I_0$ (lines) and the reference simulator CAMSET (symbols) for various source-to-drain voltages (a) and temperatures (b). Here, $C_g = 3$ aF, $C_s = C_d = 1$ aF, $R_t = 10$ MΩ.

Circuit Simulations with the Model

The present SET model can be easily implemented in conventional circuit simulators, such as the simulation program with integrated circuit emphasis (SPICE), as a small circuit comprising voltage-controlled current-sources and capacitors. This implementation allows us to simulate circuits including SETs using a conventional circuit simulator alone. However, it should be noted that the circuits should satisfy the condition that all the source and drain terminals of SETs are connected to large capacitors whose capacitance is much larger than the total capacitance of the quantum dot of the SET.

As a typical example of SET circuits, the SET inverter [36] shown in Figure 15a is simulated using SILVACO *SmartSpice* [43], in which our model is implemented as a circuit of current-sources and capacitors. As shown in Figure 15b, the results obtained using the present model coincide well with those using CAMSET. Therefore, it is confirmed that our analytical model enables us to simulate SET circuits using a conventional circuit simulator alone.

Simple Formulas Describing Typical SET Characteristics

By utilizing the analytical SET model, useful formulas, which depict some typical aspects of SET characteristics, can be obtained. In this section, we describe the derivation of the gate/drain voltage swing needed to change the current in the Coulomb blockade region by one decade. In addition, we derive the formula depicting the shift of the gate voltage dependence of SET current due to the drain voltage. When this drain-induced effect is considerable, SETs could not be used as gate-controlled switching devices.

First, we derive the gate voltage swing needed to change the current in the Coulomb blockade region by one decade. It should be noted that as $|\tilde{V}_{gs,n}|$ approaches one, the Coulomb blockade works more effectively, and thus, an inequality

$$\left|\tilde{V}_{gs,n}\right| \gg \left|\tilde{V}_{ds}\right|, \tilde{T} \qquad (31)$$

holds in most cases in the Coulomb blockade region. Under this condition, $dV_g/d\log I_n$, which gives the gate voltage swing S needed to change the current by one decade, is reduced to

$$S \approx \ln 10 \cdot \frac{k_B T}{e} \cdot \frac{C_\Sigma}{C_g} \quad \text{for} \quad \left|\tilde{V}_{gs,n}\right| \gg \left|\tilde{V}_{ds}\right|, \tilde{T} \qquad (32)$$

This formula is the same as that obtained by Fujishima *et al.* [44] based on some physical considerations. In the same manner, the drain voltage swing S_D needed to change the *valley* current of Coulomb oscillations by one decade is obtained as

$$S_D \approx \ln 10 \cdot \frac{2k_B T}{e} \quad \text{for} \quad \left|\tilde{V}_{gs,n}\right| \gg \left|\tilde{V}_{ds}\right| \gg \tilde{T} \qquad (33)$$

These formulas are useful for designing SETs and SET circuits, as the subthreshold slope is in the case of MOSFETs. From Eqs. (32) and (33), we can obtain the information that the ratio of C_g/C_Σ should be greater than 0.5 in order to use SETs as gate-controlled switching devices; otherwise, the current would be controlled more by the source-to-drain voltage.

Next, we determine the formula describing the shift of the gate voltage dependence due to the drain voltage. The gate voltage dependence of SET current shifts horizontally with an increase in V_d as shown in Figure 3. This effect is highly problematic, because it greatly changes the switching states of SETs whether the output of a logic gate is at a "high" level or at a "low" level.

From Eq. (29), we can determine the shift of the gate voltage dependence ΔV_{gs} due to V_{ds} as

$$\Delta V_{gs} = \frac{(C_g + C_s - C_d) \cdot V_{ds}}{2C_g} \qquad (34)$$

Thus, in the case of a completely symmetric SET, where $C_s = C_d$, the shift of the gate voltage dependence reaches one-half of the drain voltage $V_{ds}/2$. It is, therefore, suggested that the value for the drain voltage should be much smaller than that for the gate voltage swing in order to use SETs as gate-controlled switching devices; otherwise, the switching state of SETs could not be controlled well by the gate voltage alone.

In this section, we have presented a compact, accurate, analytical SET model suitable for the design and analysis of SET circuits. The SET characteristics, as well as the SET inverter characteristics, are successfully calculated using this model. Moreover, the characteristics of the hybrid circuit of SETs and MOSFETs are also simulated well using the model implemented in a conventional circuit simulator. By utilizing the model, it is clarified that the drain-voltage induced shift of the gate voltage dependence of SET current reaches one-half of the drain voltage in the case of a completely symmetric SET.

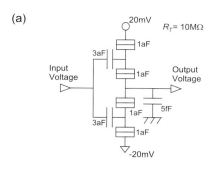

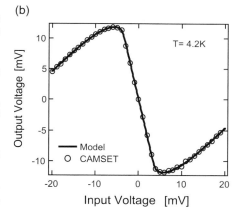

Figure 15:
(a) Schematic of a single-electron-transistor inverter.
(b) Characteristics of the SET inverter, calculated using the SET model (line) and the simulator CAMSET (symbols) at $T = 4.2$ K.

3.3 Logic Circuits with Single-Electron Transistors

The purpose of this section is to discuss a scheme for designing logic circuits in which single-electron transistors (SETs), typical examples of single-electron devices, are used to switch the current on/off. Firstly, the bias conditions that SETs must satisfy in order to be applicable as switching elements are discussed. The design scheme for SET logic circuits is then introduced. An example circuit, a four-way exclusive OR, and the confirmation of its operation by a computer simulator are shown. Finally, the advantages of the proposed SET logic circuits for hybridization with conventional circuits are briefly discussed.

Bias Conditions for Single Electron Transistors

In this section, we consider the bias conditions for turning SETs on/off. It is assumed that the tunneling resistances of source and drain junctions are the same.

In order to construct logic circuits with switching elements, both pull-up devices and pull-down devices are necessary. The pull-up devices are used for charging up load capacitors. The pull-down devices are used for discharging load capacitors. First, we discuss the bias conditions required to turn pull-up SETs on, and then those to turn pull-down SETs on. Finally, we briefly discuss the conditions required to turn both the pull-up and pull-down SETs off.

Conditions Required to Turn SETs On

Figure 16 shows a simple circuit in which a SET is used as a pull-up device. The supply voltage, V_{dd}, is applied to the SET. The SET is connected to the load capacitor. The other end of the load capacitor is connected to the ground. As shown in Figure 16, V_{ds} and V_{gs} are expressed as follows.

$$V_{ds} = V_{dd} - V_{OUT}$$

$$V_{gs} = V_g^{ON,up} - V_{OUT},$$

where $V_g^{ON,up}$ is the gate voltage of the pull-up SET to turn it on, and V_{OUT} is the output voltage of the node at which the SET and the load capacitor are connected. These equations can be reduced to

$$V_{gs} = V_g^{ON,up} - V_{dd} + V_{ds} \tag{35}$$

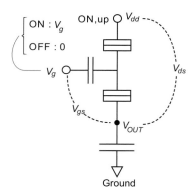

Figure 16: Circuit diagram of a single-electron transistor as a pull-up device. The output node is charge to V_{dd} via the single-electron transistor.

In order to completely charge up the load capacitor via the pull-up SET, it is necessary that the SETs are turned on even at a source-drain voltage of around zero. As shown in Figure 1b, this requirement means that the source-gate voltage V_{gs} is required to satisfy the following condition.

$$V_{gs} = \frac{e}{C_g}\left(\frac{1}{2} + n\right) \quad \text{at} \quad V_{ds} = 0 \tag{36}$$

where n is the number of electrons in the SET island. By substituting Eq. (36) into Eq. (35), $V_g^{ON,up}$ is calculated as

$$V_g^{ON,up} = V_{dd} + \frac{e}{C_g}\left(\frac{1}{2} + n\right) \tag{37}$$

Thus, the relationship between V_{gs} and V_{ds} is also calculated as

$$V_{gs} = V_{ds} + \frac{e}{C_g}\left(\frac{1}{2} + n\right) \tag{38}$$

Hence, we obtain the bias conditions required to turn the pull-up SETs on.

Next, we consider the conditions required to turn the pull-down SETs on. Figure 17 shows the case in which a SET is used as a pull-down device. As shown in Figure 17, the source voltage of the SET is fixed to the ground. By using the same procedure, the gate voltage of the pull-down SET to turn it on, i.e., $V_g^{ON,down}$, is calculated as

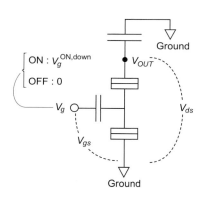

Figure 17: Circuit diagram of a single-electron transistor as a pull-down device. The output node is discharged to the ground via the single-electron transistor.

$$V_g^{ON,down} = \frac{e}{C_g}\left(\frac{1}{2} + n\right) \tag{39}$$

From Eqs. (37) and (39), it is deduced that when the gate voltage applied to turn them on is $V_g^{ON,down}$ or $V_g^{ON,up}$, SETs are used only as pull-down or pull-up devices, respectively. This is explained in detail below. We assume that SETs are biased by V_{dd} and $V_g^{ON,up}$ as shown in Figure 18 by an open circle. When the SETs are used as pull-up devices, the relationship between V_{gs} and V_{ds} is expressed by Eq. (38) and plotted as the dashed line in Figure 18. As shown in this figure, the SET is always in the on-state and can completely charge up the load capacitor. On the other hand, when the SETs are used as pull-down devices, the relationship between V_{gs} and V_{ds} is plotted as the dotted line in Figure 18. V_{ds} decreases with the decrease of V_{OUT} and finally the SET enters the off-state at the filled circle in Figure 18. Therefore, it cannot fully discharge the load capacitor.

Conditions Required to Turn SETs Off

Next, we consider the bias conditions required to turn the SETs off. In order to turn the SETs off at $V_{gs} = 0$, the V_{ds} of SETs is required to be

$$-\frac{e}{2(C_g + C_s)} < V_{ds} < \frac{e}{2(C_g + C_s)}. \tag{40}$$

If the polarities of V_{ds} and V_{gs} are the same, this condition means

$$\left|V_g^{ON,up}\right|, \left|V_g^{ON,down}\right| > |V_{ds}| \tag{41}$$

V_{ds} ranges from ground to V_{dd}. Therefore, Eq. (41) suggests that the supply voltage V_{dd} should be lower than $V_g^{ON,down}$ and $V_g^{ON,up}$.

In this section, we have shown that SETs can be used only as pull-down or pull-up devices when the gate voltage applied to turn them on is $V_g^{ON,down}$ or $V_g^{ON,up}$, respectively. We have also suggested that the supply voltage should be lower than the gate voltage swing of SETs.

Furthermore, we have to point out that a lower supply voltage to SETs than $V_g^{ON,down}$ and $V_g^{ON,up}$ has the following advantages: 1) The on/off ratio of Coulomb oscillations becomes higher with decreasing V_{dd}. 2) At finite temperature, SETs initially biased by sufficiently low V_{dd} can fully discharge load capacitors, even if the gate voltage is $V_g^{ON,up}$ [38].

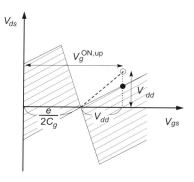

Figure 18: Relationship between V_{gs} and V_{ds} when the V_g is $V_g^{ON,up}$. The open circle shows the bias conditions for SETs. The dashed line shows the relationship between V_{gs} and V_{ds} when SETs are used as pull-up devices. The dotted line shows the relationship when the SETs are used as pull-down devices.

Design Scheme

In this section, we introduce a design scheme for SET logic circuits. First, logic trees should consist of pull-down SETs only. The SET logic tree implements logic operations. For example, two cascaded switches operate the logic function of AND (NAND) and two parallel SET switches operate the logic function of OR (NOR). By using complementary input, SET trees can implement all the logic operations. Furthermore, the output voltage swing varies from ground to supply voltage, regardless of the number of cascaded SETs, because the logic tree contains no pull-up SETs.

Second, the supply voltage to SET logic trees should be lower than the gate voltage swing of SETs. This is a requirement in order that the SETs can switch the current off, as described above.

Third, one pull-down device controlled by the clock is connected between the ground and the SET logic trees. One pull-up device controlled by the inverse clock is connected between the power supply and the SET logic tree. This guarantees the cyclic operation of the circuit. The pull-up device may be a p-type metal-oxide-semiconductor field-effect-transistor (pMOSFET)* or a SET.

Finally, the output voltages of logic trees are amplified to the same voltage as the gate voltage swing of SETs with conventional electronic devices, such as MOSFETs [45] and bipolar transistors, in order to drive the next gates. In any event, these conventional devices are mandatory to compensate for the poor driving capability of SETs [45].

Figure 19 shows a schematic of the SET logic circuits based on the above scheme. The circuit consists of the SET logic tree, a lower supply voltage than the gate voltage swing of SETs, the pull-down switch, the pull-up switch, the clock, and the CMOS amplifier biased by the same supply voltage as the gate voltage swing of SETs. The circuit operation is as follows.

* It should be noted that, when the pull-up device is a pMOSFET, the pull-up device is controlled by the clock, not by the inverse clock.

III Logic Devices

1. Clock: L (Precharge period)

The pull-up device between the load capacitor and the power supply is turned on and the pull-down device between the logic tree and the ground is turned off. Thus, the load capacitor is charged to lower supply voltage, regardless of the inputs of SETs.

2. Clock: H (Evaluation period)

The pull-up device is turned off and the pull-down device is turned on. Then, the input voltages are applied to the gates of the SETs, and the logic operation is performed depending on whether the node is discharged or not. Finally, the output voltage is amplified with the CMOS amplifier to the same voltage as the input voltage swing of the SETs so that it can drive the next gate.

Note that the circuit operation described above is quite similar to that of the CMOS dynamic logic, such as the P-E logic and CMOS domino logic [46]. In other words, our design scheme of SET logic resembles that of the CMOS dynamic logic, except for the following. First, logic trees consist of SETs. Second, the trees are biased by a lower supply voltage than the input voltage swing of the switching elements. Finally, a voltage amplifier is mandatory. Therefore, the methodology of the CMOS dynamic logic is useful for designing the SET logic.

Example Circuit: Four-Way Exclusive OR

Figure 20 shows an example of the proposed circuit, a four-way exclusive OR. This circuit is similar to the CMOS dynamic logic family, namely, the clocked cascode voltage switch logic [47]. The gate capacitance is designed so that SETs are turned completely on and off by the output voltage from the CMOS inverter. Therefore, the gate capacitance C_g of SETs is given by

$$C_g = \frac{e}{2V_{CMOS}} \qquad (42)$$

where V_{CMOS} is the supply voltage to the CMOS inverters. The total capacitance of the SET island is determined based on the same capacitance as the self-capacitance of a 1 nm diameter silicon sphere surrounded by silicon dioxide. The floating capacitance C_f at the nodes is large enough to suppress the Coulomb blockade effects there. Room temperature, i.e., 293 K, is also assumed. The ratio of thermal energy to charging energy is approximately 0.07. Figure 21 shows the simulation result for the circuit.

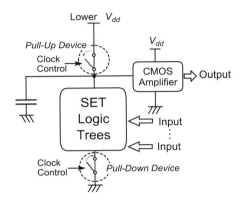

Figure 19: Schematic of SET logic circuits described in this section. The circuit consists of the SET logic tree, a lower supply voltage than the gate voltage swing of SETs, the pull-down switch, the pull-up switch, the clock, and the CMOS amplifier biased by the same voltage as the gate voltage swing of SETs.

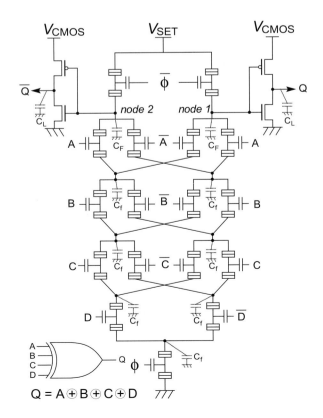

Figure 20: Example of SET logic circuit, four-way exclusive OR circuit. The complementary imputs and outputs are used.

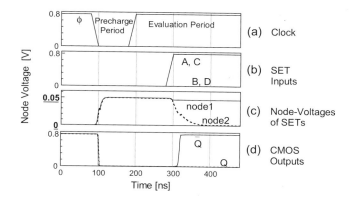

Figure 21: The simulated timing chart of the circuit shown in Figure 20. Here,
C_g = 0.1 aF,
$C_s = C_d$ = 0.06 aF,
R_t = 500 kΩ,
C_L = 10 fF,
C_F = 1 fF,
C_f = 50 aF,
V_{SET} = 50 mV, and
T = 293 K.

During the precharge period, the clock is lowered to "low" level (80 - 100 ns), and the nodes 1 and 2 are precharged. During the next evaluation period, the clock is pulled up to the "high" level (180 - 200 ns) and the input voltages are transferred to the gates of the SETs (280 - 300 ns). Thus, one of the two nodes is discharged and the other node remains at the "high" logic-level. These node-voltages are amplified by CMOS inverters to the power-supply voltage level of CMOS as shown in Figure 21d. We consider that the operation speed will be considerably improved by optimizing the capacitance of load capacitors connected to the outputs of the logic tree, reducing the tunneling resistance, and refining CMOS amplifier circuits.

4 Future Directions

At the beginning of the research of single-electron devices, whether single-electron devices can operate at room temperature or not was of great concern. However, recent reports on room-temperature operating single-electron devices seem to promise that single-electron devices with high on/off current ratio at room tempreture will be fabricated in a well-controlled manner in the near future. Thus, now our concerns are how to utilize single-electron devices and what benefits can be obtained by utilizing single-electron devices.

In the latter part of this chapter, one of the methods of replacing MOSFETs in logic LSIs with single-electron devices is discussed. By utilizing such a scheme, the power consumption is expected to be reduced and the packing density of devices in LSIs is expected to be increased. However, it should be noted that the functionality of LSIs, where single-electron devices are implemented according to the above sheme, is basically the same as the functionality of LSIs which do not have single-electron devices.

As a result, it is now important to develop an operation scheme, under which the functionality of SET circuits can surpass that of conventional CMOS circuits. A programmable single-electron transistor logic [32] and a mutiple-value logic [48] are attractive examples of the functionality improvement by the utilization of single-electron transistors.

Acknowledgements

The author would like to thank Prof. A. Toriumi of The University of Tokyo, Drs. A. Kurobe, S. Takagi, J. Koga, R. Ohba, and K. Matsuzawa of Toshiba Corporation, and Dr. A. Ohata for valuable discussions. This work was partly performed under the management of FED as a part of METI R&D of Industrial Science and Technology Frontier Program (Quantum Functional Devices Project) supported by NEDO. The Editor is indebted to Dr. B. Spangenberg (Institute of Semiconductor Technology, RWTH Aachen University) for his review of this chapter and his valuable comments. The editor would like to thank Christian Ohly (FZ Jülich) for checking the symbols and formulas in this chapter.

References

[1] R. H. Dennard, F. H. Gaensslen, H. Yu, V. L. Rideout, E. Bassous, and A. R. Leblanc, IEEE J. Solid-State Circuits **9**, 256 (1974).

[2] G. Baccarani, M. R. Wordeman, and R. H. Dennard, IEEE Trans. Electron Devices **31**, 452 (1984).

[3] R. Chau, J. Kavalieros, B. Roberds, R. Schenker, D. Lionberger, D. Barlage, B. Doyle, R. Arghavani, A. Mrthy, and G. Dewey, Technical Digest of Int. Electron Devices Meet., 45 (2000).

[4] International Technology Roadmap for Semiconductors (ITRS) 2001 Edition: http://public.itrs.net/Files/2001ITRS/Home.htm

[5] K. K. Likharev, IEEE Trans. Magn. **23**, 1142 (1987).

[6] H. Grabert and M. H. Devoret, *Single Charge Tunneling*, Plenum, New York, 1992.

[7] D. V. Averin and K. K. Likharev, in *Mesoscopic Phenomena in Solids*, B. L. Altshuler, P. A. Lee, R. A. Webb, eds., Elsevier, Amsterdam, 1991.

[8] K. K. Likharev, IBM J. Res. Develop. **32**, 144 (1988).

[9] K. K. Likharev, IEEE Proc. **87**, 606 (1999).

[10] C. W. J. Beenaker, Phys. Rev. **B44**, 1646 (1991).

[11] C. Wasshuber, H. Kosina, and S. Selberherr, IEEE Trans. Computer Aided Design **16**, 937 (1997).

[12] M. Kirihara, K. Nakazato and M. Wagner, Jpn. J. Appl. Phys. **38**, 2028 (1999).

[13] L. J. Geerlings and J. E. Mooij, in *Granular Nanoelectronics*, D. K. Ferry, J. R. Barker, and C. Jacoboni, eds., Plenum Press, New York, 1991.

[14] M. A. Kastner, Phys. Today **46**, 24 (1993); R. C. Ashrooi, Nature **379**, 413 (1996).

[15] S. Tarucha, D. G. Austing, T. Honda, R. J. Van der Hage, and L. P. Kouwenhoven, Phys. Rev. Lett. **77**, 3613 (1996).

[16] L. P. Kouwenhoven, D. G. Austing, and S. Tarucha, Rep. Prog. Phys. **64**, 701 (2001).

[17] S. Tarucha, D. G. Austing, S. Sasaki, L. P. Kouwenhoven, S. M. Reimann, M. Koskinen, and M. Manninen, in *Physics and Applications of Semiconductor Quantum Structure*, T. Yao, ed., IOP Publishing Co., Bristol, 2001.

[18] G. J. Iafrate, K. Hess, J. B. Krieger, and M. Macucci, Phys. Rev. B **52**, 10737 (1995).

[19] S. Tarucha, D. G. Austing, Y. Tokura, W. G. van der Wiel, and L. P. Kouwenhoven, Phys. Rev. Lett. **84**, 2485 (2000).

[20] S. Sasaki, S. De Franceschi, J. M. Elzerman, W. G. van der Wiel, M. Eto, S. Tarucha, L. P. Kouwenhoven, Nature **405**, 764 (2000).

[21] L. J. Geerlings, V. F. Anderegg, P. A. M. Holweg, and J. E. Mooij, Phys. Rev. Lett. **64**, 2691 (1990).

[22] H. Pothier, P. Lafarge, P. F. Orfila, C. Urbina, D. Esteve, and M. H. Devoret, Physica B **169**, 573 (1991).

[23] K. Nakazato, and J. D. White, Technical Digest of Int. Electron Devices Meet. 487 (1992).

[24] T. A. Fulton and D. J. Dolan, Phys. Rev. Lett. **59**, 109 (1987).

[25] U. Meirav, M. A. Kastner, and S. J. Wind, Phys. Rev. Lett. **65**, 771 (1990).

[26] Y. Takahashi, H. Namatsu, K. Kurihara, K. Iwadate, M. Nagase, and K. Murase, IEEE Trans. Electron Devices **43**, 1213 (1996).

[27] Y. Ono, Y. Takahashi, K. Yamazaki, M. Naruse, H. Namatsu, K. Kurihara, and K. Murase, IEEE Trans. Electron Devices **47**, 147 (2000).

[28] K. Yano, T. Ishii, T. Hashimoto, T. Kobayashi, F. Murai, and K. Seki, Technical Digest of Int. Electron Devices Meet. (1993); K. Yano, T. Ishii, T. Hashimoto, T. Kobayashi, F. Murai, and K. Seki, IEEE Trans. Electron Devices **41**, 1628 (1994).

[29] S. Tiwari, F. Rana, K. Chan, H. Hanafi, W. Chan, and D. Buchanan, Technical Digest of Int. Electron Devices Meet. (1995).

[30] K. Uchida, J. Koga, R. Ohba, S. Takagi, and A. Toriumi, Digest of 57th Device Research Conference, p. 138 (1999); K. Uchida, J. Koga, R. Ohba, S. Takagi, and A. Toriumi, J. Appl. Phys. **90**, 3551 (2001).

[31] A. Tilke, R. H. Blick, H. Lorenz, and J. P. Kotthaus, J. Appl. Phys. **89**, 8159 (2001).

[32] K. Uchida, J. Koga, R. Ohba, and A. Toriumi, Technical Digset of Int. Electron Devices Meet. (2000) 863; K. Uchida, J. Koga, R. Ohba, and A. Toriumi, ISSCC Dig. Tech Pap. 206 (2002).

[33] N. Asahi, M. Akazawa, and Y. Amemiya: IEEE Trans. Electron Devices **44**, 1109 (1997).

[34] J. R. Tucker, J. Appl. Phys. **72**, 4399 (1992).

[35] A. N. Korotkov, R. H. Chen, and K. K. Likharev, J. Appl. Phys. **78**, 2520 (1995).

[36] R. H. Chen, A. N. Korotkov, and K. K. Likharev, Appl. Phys. Lett. **68**, 1954 (1996).

[37] K. Uchida, K. Matsuzawa, J. Koga, R. Ohba, S. Takagi, and A. Toriumi, Jpn. J. Appl. Phys. **39**, 2321 (2000).

[38] K. Uchida, K. Matsuzawa, and A. Toriumi, Jpn. J. Appl. Phys. **38**, 4027 (1999).

[39] L. I. Glazman, and R. I. Shekhter, J. Phys. Condens. Matter **1**, 5811 (1989).

[40] C. W. J. Beenakker, Phys. Rev. B **44**, 1646 (1991).

[41] S. Amakawa, H. Majima, H. Fukui, M. Fujishima, and K. Hoh, IEICE Trans. Electron. **E81**-C, 21 (1998).

[42] Y. S. Yu, S. W. Hwang, and D. Ahn, IEEE Trans. Electron Devices **46**, 1667 (1999).

[43] SILVACO International, Santa Clara, CA 94054, USA.

[44] M. Fujishima, H. Fukui, S. Amakawa and K. Hoh, IEICE Trans. Electron. **E80**-C, 881 (1997).

[45] A. Ohata, A. Toriumi, and K. Uchida, Jpn. J. Appl. Phys. **36**, 1686 (1997).

[46] R. H. Krambeck, C. M. Lee and H. S. Law, IEEE J. Solid State Circuits **17**, 614 (1982).

[47] L. G. Heller, W. R. Griffin, J. W. Davis, and N. G. Thoma, ISSCC Dig. Tech. Pap. 16 (1984).

[48] H. Inokawa, A. Fujiwara, and Y. Takahashi, Technical Digest of Int. Electron Devices Meet. 147 (2001).

Superconductor Digital Electronics

Michael Siegel

Institute of Micro- and Nanoelectronic Systems, University of Karlsruhe, Germany

Contents

1 Introduction	445
2 Josephson Junctions	445
2.1 Josephson Effects: Basics	445
2.2 RSJ Model	446
2.3 Junctions Parameters	447
3 Voltage-State Logic	448
3.1 Switching Characteristics	448
3.2 Logic Gates	448
3.3 Memory Elements	448
4 Single-Flux-Quantum Logic	449
4.1 SFQ Basics	449
4.2 SFQ Logic Gates	450
4.2.1 Josephson Transmission Line	450
4.2.2 D flip-flop	451
4.2.3 Clocking of SFQ Circuits	451
5 Superconductor Integrated Circuit Technology	452
5.1 Low-Temperature Superconductor Technology	452
5.2 High-Temperature Superconductor Technology	453
6 Present Status of RSFQ Logic	455
6.1 Opportunities for Immediate Applications	455
6.1.1 Analog-to-Digital Converters	455
6.1.2 Digital-to-Analog Converters	455
6.2 Prospects for Future Applications	456
6.2.1 Ultra-fast Network Switches	456
6.2.2 Software-Defined Radio	456
6.3 HTS Implementation of RSFQ Circuits	456
7 Summary	457

Superconductor Digital Electronics

1 Introduction

The International Technology Roadmap for Semiconductors (ITRS) [1] predicts that the density of semiconductor integrated circuits will increase three more orders of magnitude during the next decade. However, the forecast for clock frequency predicts a much lower pace. The problem of limited speed in very large integrated circuits (VLSI) is not determined by the intrinsic switching properties of future CMOS devices which will be below 1 ps (see Chap. 13). In fact, a dominant part of the much longer, (sub)-1-ns-scale clock cycle in modern integrated circuits is spent on recharging the interconnect capacitance (C) by the on-currents (I) of logic gate transistors. The relative contribution of the gate capacitance to C is almost negligible [1], [2], [3]. The most effective way to speed up the clock frequency is to increase the output current of the driver by increasing the transconductance. However, this will immediately increase the power consumption of the CMOS circuits. Because of the dynamic power consumption it is convenient to use its average density (power per unit area)

$$P_0 = C_0 V^2 f_C \qquad (1)$$

where V is the logic swing (typically close to the power supply voltage V_{DD}), f_C is the clock frequency, and C_0 is the effective total interconnect capacitance per unit area. P_0 of a modern 1 GHz microprocessors is above 50 W on a 1-cm^2-scale chip area. Removal of such enormous power from a chip without its overheating presents a very serious technical challenge. The ITRS foresees only a slight increase of this value in the next 15 years. The only way to increase the speed of semiconductor devices and keeping power within the limits shown above, is to decrease the supply voltage. Another way to increase speed is to use novel semiconductor devices like resonant tunnelling transistors showing switching speed above 100 GHz [4]. However, the heat removal problem has to be solved also for these devices because they have voltage swing comparable to CMOS devices.

The saturation of clock frequency can be solved using *superconductor integrated circuits* which can be operated at frequencies above 100 GHz with a very low power consumption. Here, we give a short introduction to the basics of superconductor digital circuits. The potential of superconductor digital circuits for applications will be discussed.

2 Josephson Junctions

2.1 Josephson Effects: Basics

The most important device of superconductor integrated circuits is the Josephson junction [5], [6]. For simplicity, we will discuss the properties of a Josephson junction consisting of two superconductor electrodes which are separated by a thin (≈ 1 nm) oxide layer (see Figure 1). Josephson junctions reveal unusual non-linear dynamics because of the macroscopic quantum properties of Cooper pairs in superconductors [5]. For more details see Chapter 3. Cooper pairs have an integer spin and follow the Bose statistics, hence, may be described with a single wave function:

$$\psi(r, t) = |\psi| \exp\{i\varphi(r, t)\} \qquad (2)$$

with the amplitude $|\psi|$ and phase φ. Using the Schrödinger equation:

$$i\hbar \partial \psi / \partial t = H\psi \qquad (3)$$

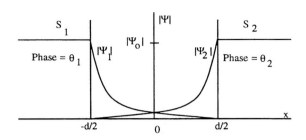

Figure 1: Schematic view of a tunnel junction consisting of two superconductor electrodes S_1 and S_2 separated by an oxide barrier layer with a thickness d. Complex wave functions from each superconductor electrode decay exponentially in the barrier region.

and the single wave function from Eq. (2) with a constant amplitude of ψ and a Cooper pair energy of $W = 2e\mu + \text{const.}$ (where μ is the electrochemical potential) we get the following equation for the phase:

$$\partial \varphi(\mathbf{r}, t)/\partial t = -(2e/\hbar)\mu(\mathbf{r}, t). \tag{4}$$

Subtracting Eqs. (4) for two arbitrary fixed points in both superconductor electrodes, we obtain the fundamental relation between phase $\phi = \varphi_1 - \varphi_2$ and the voltage $V = \mu_1 - \mu_2$ between both electrodes:

$$d\phi/dt = (2e/\hbar)V(t) = 2\pi V(t)/\Phi_0 \tag{5}$$

Here, $\Phi_0 = h/2e$ is the magnetic flux quantum. The phase difference $\phi(\mathbf{r}, t)$ determines the spatial and time dynamics of a Josephson junction. In particular, for $V = 0$ the Cooper pair current (supercurrent) is determined by the relation [5], [6]

$$I_S = I_C \sin \phi. \tag{6}$$

Here I_C is the maximum supercurrent (the critical current of the Josephson junction). Eq. (6) describes the so-called **dc Josephson effect**.

Eq. (6) continues to hold for $V > 0$, but now ϕ depends on the voltage across the junction according to Eq. (5). Integration of Eq. (5) yields in an alternating current:

$$I_S = I_C \sin(\omega_J t) \tag{7}$$

which oscillates at the Josephson Frequency:

$$f_J = \omega_J/2\pi = 2eV/h = V/\Phi_0. \tag{8}$$

This ac current is referred to as the **ac Josephson effect**. Essentially, a Josephson junction acts like a superconducting short for $V = 0$ but like an ideal voltage-controlled oscillator for $V \neq 0$. Eq. (5) follows directly from the fact that if there is a dc voltage across the junction, this raises the energy level of Cooper pairs on one side relative to the other by $\Delta W = -2eV$ (see Figure 2). Further Eq. (8) is an expression of the basic quantum relation that $\Delta W = hf$, where f is the frequency of a photon coupling of two energy levels in a quantum system.

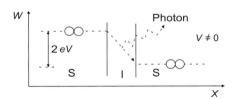

Figure 2: Diagram of energy levels of Cooper pairs on opposite sides of a Josephson junction with voltage V across junction. The Josephson oscillation occurs at a frequency f_J such that $hf_J = 2 eV$.

2.2 RSJ Model

The dynamical behaviour can be understood in an equivalent junction model proposed by Stewart and McCumber [7]. Figure 3 shows the equivalent circuit of a Josephson junction. The dynamical equation of the junction can be written as a sum of four components of current through the junction:

$$I(t) = I_C \sin \phi + C\, dV/dt + V/R + I_f(t) \tag{9}$$

Where C is the capacitance of the junction and R is its normal resistance. The fluctuation current is given by I_f. In the following we will neglect the term I_f. The system of equations (6) and (9) gives an implicit relation between the current and voltage in a Josephson junction. Moreover, due to the fundamental Eqs. (5) and (6) Eq. (9) can be rewritten and we get a second-order non-linear equation for the phase difference. A detailed analysis [6] shows that these devices allow generation of picosecond waveforms and switching on a ps-time scale. Josephson junctions may recover weak input pulses, restoring their waveform to a nominal value.

Figure 4 shows the current-voltage characteristic (IVC) of a Josephson junction. In the case of a current biasing scheme depending of the damping properties, the Josephson junction may exhibit a non-hysteretic (Figure 4b) or a hysteretic IVC (Figure 4a). Usually, the damping properties are described with the Stewart-McCumber parameter β_C given as:

$$\beta_C = 2eR^2CI_C/\hbar. \qquad (10)$$

For $\beta_C < 0.8$ the IVC exhibit no hysteresis and for $\beta_C > 0.8$ the IVC shows a clear hysteresis. In the stationary state Josephson junction does not dissipate energy, because at $\phi(t) = $ const the voltage V across the junction is $V = 0$. The power is finite only during the transient:

$$\Delta W = \int IV dt = (\Phi_0/2\pi) \int I d\phi \sim (\Phi_0/2\pi) I_C 2\pi = I_C \Phi_0 \qquad (11)$$

per each switching event. A single switching event requires an energy of about $5 \cdot 10^{-19}$ Ws.

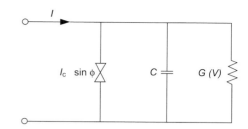

Figure 3: Equivalent circuit of a Josephson junction containing the Josephson junction, a capacitance C and a resistor with $R = G^{-1}(V)$. The current I is biased by a current source.

2.3 Junctions Parameters

Here, we will estimate some typical parameters of Josephson junctions. For a typical Nb/Al$_2$O$_3$/Nb tunnel junction with a barrier thickness of about 2 nm and a dielectric constant $\varepsilon_r \cong 10$ we obtain a capacitance per unit area $\varepsilon_0\varepsilon_r/d \cong 0.05$ pF/µm². Thus, for a typical junction area of 10 µm², we obtain $C \cong 0.5$ pF. With a typical current density of 10^3 A/cm² and the same size the critical current is $I_C \cong 100$ µA. Using the relation $I_CR_N = \pi\Delta/2e$ ($\Delta_{Nb} = 1.28$ meV) we can estimate the normal resistance $R_N \cong 20$ Ω. The characteristic time constant of this junction will be $RC \cong 10$ ps. Using Eq. (10) we calculate a damping parameter $\beta_C \cong 60$. This large value of β_C leads to a strong hysteresis in the current-voltage characteristics. For some applications like single-flux logic or **S**uperconducting **Qu**antum **I**nterference **D**evices (**SQUID**) junctions with non-hysteretic IVC are required. This can be achieved by using an additional shunt resistor R_S which reduces β_C. For $\beta_C < 0.8$ the shunt resistor should be $R_S \leq 6$ Ω leading to $I_CR_N \cong 200$ µV. Thus, a so-called shunted junction will have a characteristic time constant $RC \leq 1$ ps. A reduction of the switching time can be achieved by decreasing the capacitance of the junction. A reduction of the area of junction will require an increase of the current density to keep the critical current constant. Maximum values of $I_CR_N \cong 1,5$ mV for low-temperature superconductor (LTS) based on non-hysteretic junctions have been reached with a current density of $2 \cdot 10^5$ A/cm². The switching time for this device was 0.17 ps leading to a 770 GHz operation frequency of a T-Flip-Flop [9]. High-temperature superconductor (HTS) junctions have shown $I_CR_N \cong 5$ mV at $T = 4.2$ K and $I_CR_N \cong 3$ mV at $T = 20$ K. These high I_CR_N products which lead to very short switching times determine the interest in developing HTS Josephson junctions for very fast applications.

The energy dissipation of a junction with $I_C \cong 100$ µA can be calculated using Eq. (11) to be $\Delta W = I_C \cdot \Phi_0 \cong 2 \cdot 10^{-19}$ J. In comparison, the lowest energy dissipation in semiconductor devices is about $\Delta E = 10^{-16}$ J. Thus, superconductor devices may be at least a factor of 10 faster and lower in power consumption. This leads to a factor of 100 lower energy dissipation in superconductor circuits.

(a)
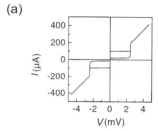

(b)
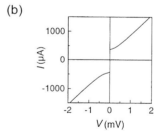

Figure 4: Current-voltage characteristic (IVC) of a Josephson junction in the current biased regime. Curve (a) represents a hysteretic IVC of a Nb/Al$_2$O$_3$/Nb junction with a small damping and curve,
(b) shows a non-hysteretic IVC of YBa$_2$Cu$_3$O$_7$ bicrystal junction with a small damping [8].

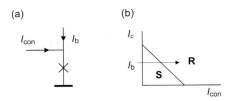

Figure 5: Switching curves of Josephson junctions.
(a) Current injection scheme for switching of a Josephson junction with current I_{con}.
(b) Threshold curve for schematics (a).

3 Voltage-State Logic

3.1 Switching Characteristics

As discussed in the previous section, the switching characteristics of Josephson junctions are different depending on their damping properties. Although a damped junction with $\beta_C < 0.8$ switches much faster, the early development of superconductor logic devices was performed on under-damped (hysteretic) junctions. The belief in the 1970s, when IBM initiated the development of a *Josephson computer*, was that hysteretic devices provide a more robust basis for digital logic [10].

A Josephson junction switches from the superconducting to the resistive state if the current through the junction exceeds the critical current. This switching can be realised either by controlling the current magnetically or directly. Figure 5 shows the current-injection scheme and the threshold characteristic. If a control current I_{con} is applied to the device, the operating point may cross the threshold into the resistive state, as indicated in Figure 5b. The difference between hysteretic and non-hysteretic junction is that a non-hysteretic junction will return to the zero-voltage state, whereas a hysteretic junction will remain in the voltage state after turning off the bias current. To switch a hysteretic junction back into the zero state (reset) the bias current has to be reduced to zero.

3.2 Logic Gates

We could have two or more control currents to produce additive effects in depressing the critical current. In particular, Figure 6 represents a NAND gate with two input lines. Here, the two control currents I_X and I_Y can exceed the critical current I_{C0} only together, because there is no bias current. The switching of the junction feeds the input currents to the output into the load R.

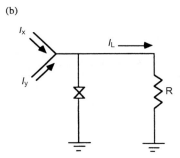

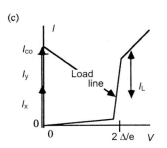

3.3 Memory Elements

Digital systems require memory elements which can store and retrieve binary data on the same time scale as the logical operation, e.g. for cache memory. The circulating current in a SQUID loop can be used as a natural basis and magnetic flux quantization may provide an encoding scheme: "0" or "1" for zero or one magnetic flux quantum in the loop. In addition a "READ" gate can be easily made by magnetic coupling of a read-out SQUID to the "WRITE" or store SQUID. Figure 7 shows a simple scheme of Josephson memory. Assume that the junction is in the zero voltage state and there is no circulating current in the loop. If we apply a bias current I_b the current will be split in the two parts of the SQUID loop. If now we apply additionally a writing current I_w which should depress the critical current of the junction, the junction switches to the voltage state which leads to a redistribution of the current in the loop. During this switching process one or more magnetic flux quanta have entered the loop and the junction returns into the zero voltage state. After removing the bias current I_b the magnetic flux is conserved and the loop contains the logical information "1". The circulating current can be used as a control current for a read-out SQUID. Such a memory cell acts as a *N*on-*D*estructive *R*ead-*O*ut (NDRO) cell during the read-out of data. Figure 7b represents the principal schematics of a memory array [11]. The disadvantage of the NDR memory is that it requires a large area for the write SQUID and for the readout SQUID. A memory cell with an inductance L is designed to store a magnetic flux quantum $\Phi_0 \approx LI_C/2$. An alternative memory cell which is much more compact uses a two-junction SQUID loop performing a *D*estructive *R*ead-*O*ut (DRO) [11].

The problem of latching logic is that the reversible switching from "1" to "0" is rather long, of the order of a few nanoseconds to avoid errors. Recently, some circuits have been tested at a few GHz [12].

Figure 6: NAND gate in latching logic.
(a) Standard schematics of the NAND gate.
(b) Equivalent scheme of the NAND gate.
(c) IVC of the junction with Load line.

4 Single-Flux-Quantum Logic

4.1 SFQ Basics

An alternative approach to use superconductors for digital logic is based on their property to quantise magnetic flux in a superconducting loop in multiples of the flux quantum Φ_0. Using Faraday's induction law:

$$d\Phi/dt = V \tag{12}$$

and Eq. (5) and integrating over time yields in a relation between magnetic flux and the phase difference of the Josephson junction:

$$\phi = 2\pi\Phi/\Phi_0 \tag{13}$$

Now, if we close the ends of the superconducting loop we have to require that the wave functions of electrons (in the identical points) have to coincide, beside of a phase difference of 2π. Immediately, Eq. (13) gives the well-known equation for magnetic flux quantization in superconductors:

$$\Phi = n\Phi_0, \qquad n = 0, \pm 1, \pm 2, \ldots \tag{14}$$

Let us consider the simplest, single-flux-quantum circuit (SFQ) with one Josephson junction, see Figure 8. Using Eqs. (9) and (13) we can derive the formula for the total magnetic flux in the loop:

$$\Phi = \Phi_{ex} - LI \tag{15}$$

and phase difference:

$$\phi + l \sin \phi = \phi_{ex} \tag{16}$$

with the following definitions:

$$l \equiv 2\pi L I_C/\Phi_0, \qquad \phi_{ex} \equiv 2\pi\Phi_{ex}/\Phi_0. \tag{17}$$

The external magnetic flux Φ_{ex} is created by an external current I_{ex} according to $\Phi_{ex} = M I_{ex}$, where M is the mutual inductance.

If the SQUID parameter $l > 1$, the phase difference ϕ, and hence the total magnetic flux Φ and the supercurrent may have several stable stationary states, $\phi_n \approx 2\pi n$, i.e. $\Phi_n \approx n\Phi_0$, for the same external field ϕ_{ex} (see Figure 9). The Josephson junction limits the number of possible flux quanta (or stable flux states) in the loop. In typical SFQ circuits $l \approx 2\pi$, ($LI_C \approx \Phi_0$), and one can operate with two flux states.

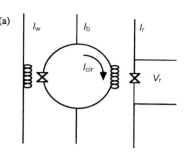

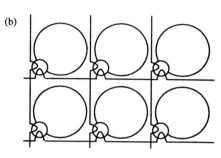

Figure 7:
(a) Non-destructive Read-Out (NDRO) memory cell, based on the presence (logical "1") or absence (logical "0") of a magnetic flux quantum in a SQUID loop. The writing current I_w is magnetically coupled to the storage SQUID loop inducing a circulating current I_{cir} which is magnetically coupled to the readout junction or SQUID.
(b) Array of a NDRO memory array using AND gates for the WRITE lines, permitting a row and column addressing of memory cells.

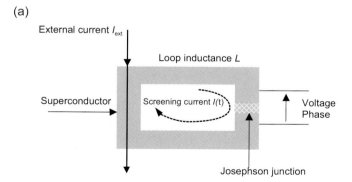

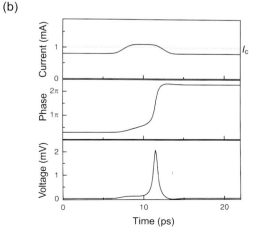

Figure 8: (a) Scheme of the simplest SFQ circuit ("SQUID"). (b) Dynamics of its switching in the moment when the externally applied flux induced by the slowly changing current I_{ex}, reaches its threshold. Inductive parameter $l = 2\pi$; $\beta_c = 1$. For typical junctions, the FWHM switching time is about $4\tau_0 \approx 1$ ps and the pulse amplitude is about 300 µV.

III Logic Devices

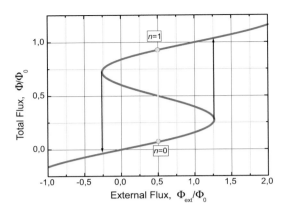

Figure 9: RF SQUID: total magnetic flux Φ as a function of applied flux Φ_{ex} as given by Eq. (15) for the LI_C product typical for "*quantizing*" loops in RSFQ circuits ($l = 2\pi$, i.e. $LI_C = \Phi_0 > \Phi_0/2\pi$). Arrows indicate flux-state switching induced by a slow change of the external field – for dynamics, see Figure 8b. Solid points show two stable states at the equilibrating value $\Phi_{ex} = \Phi_0/2$.

By fixing the dc flux bias at $\Phi_0/2$ (dashed line in Figure 9) these two states may have equal energy and stability. The switching between the two states may be achieved by changing the external flux Φ_{ex} via changing I_{ex}. The time of the leap (switching time) is determined by Eqs. (5) and (9) and of order of a few units of:

$$\tau = \max[\tau_0, RC], \qquad \tau_0 \equiv \hbar/2eI_C R = \Phi_0/2\pi I_C R. \tag{18}$$

In order to have no latching effects (non-hysteretic) the damping parameter of the junction should be $\beta_C \leq 0.8$.

The most convenient way to transfer digital information is the dynamic form, the so-called Rapid-Single-Flux-Quantum Logic (RSFQ). Indeed, according to Faraday's law, see Eq. (12), during the switching between two flux states a short voltage pulse is formed across the junction (Figure 8b). Since the magnetic flux is quantized, so the pulse area will be:

$$\int V(t)dt \approx \Phi_0 \approx 2\text{ mV·ps}. \tag{19}$$

4.2 SFQ Logic Gates

4.2.1 Josephson Transmission Line

The main use of a JTL in real designs is interconnecting more complex cells over short distances while insulating the cells from each other. In dynamic single-flux-quantum circuits SFQ pulses are passed to another part of the circuit via a Josephson Transmission Line (JTL) [13]. A JTL consists of a parallel array of identical junctions, each biased close to its critical current I_C and coupled by a superconducting line with an inductance L, see Figure 10. Let us start from the moment when junction J_1 switches. For the loop comprising that junction, inductance L_2+L_3 and junction J_2, this event is equivalent to the insertion of additional external flux $\Delta\Phi \approx \Phi_0$. This increase causes an immediate increase of the current in the loop by $\Delta I \approx \Phi_0/(L_{J1} + L_2 + L_3 + L_{J2})$, where L_J are effective inductances of the Josephson junctions. In the JTL, the segment inductances $L_2 = L_3 = \ldots = L$ are made small: $LI_C < \Phi_0$ (non-quantizing loops). Thus the new value of

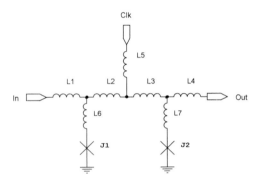

Figure 10: Josephson transmission line: schematic of a 2-stage fragment. Nominal parameters: $I_{C1} = I_{C2} = \ldots = I_C = 250\ \mu\text{A}$, $I_1 = I_2 = \ldots = I_{DC} = 175\ \mu\text{A}$, $L_2 = L_3 = \ldots = L = 4.0$ pH ($LI_C \approx 0.5\ \Phi_0$). For details and units see [35].

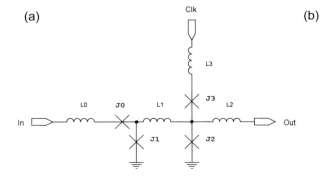
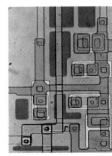

Figure 11:
(a) Equivalent circuit of a RSFQ D flip-flop [35].
(b) Layout of a RSFQ D flip-flop fabricated in a 3 μm technology with a clock frequency of 20 GHz [35].

current through J_2, which is the sum of ΔI and the dc current $I_{DC} \approx 0.7\, I_C$, exceeds I_C. As a result, this junction switches just like its predecessor, with a delay $\tau_D \approx 4\tau_0$. Now $\phi_2 \approx \pi/3 + 2\pi$, i.e. the difference between ϕ_1 and ϕ_2 is small again, while the difference between ϕ_2 and ϕ_3 is close to 2π. For the loop J_1-L_2-L_3-J_2 this leads to a *reduction* of the effective external flux by Φ_0, with all junction currents below I_C, and the cell becomes dormant, but for the next loop (J_2-L_5-J_3) it means an *increase* of flux by Φ_0, and a start of a similar switching process in J_3. Notice that the amplitude of the SFQ pulse is the same on each junction, despite the dissipation of energy $\Delta W \sim I_C\Phi_0$ during each switching event. (The necessary energy is picked up from the dc power supply providing dc currents $I_1, I_2,\dots$). This recovery/amplification of the SFQ pulse in the JTL (and all other RSFQ devices) is due to the fundamental quantization of flux and hence of the SFQ pulse area – see Eq. (19). However, this quantization doesn't influence the current scale of the SFQ pulse (and hence its impedance and energy scales) which may be regulated by the choice of I_C (at fixed LI_C product and I_{DC}/I_C ratio).

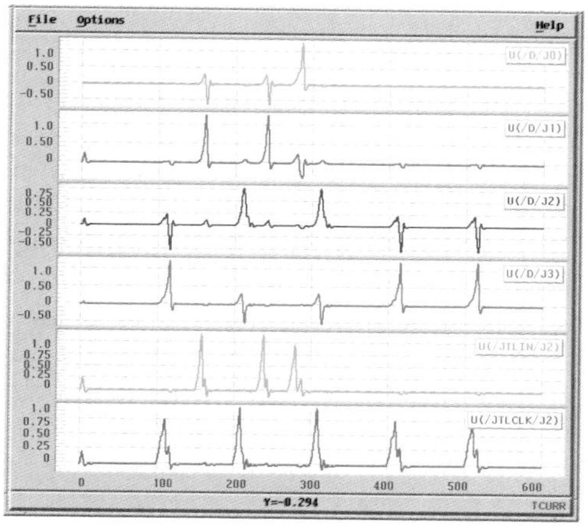

4.2.2 D flip-flop

Figure 11 (a) shows the simplest RSFQ latch, the D flip-flop [14], built around a quantizing loop J_2-L_1-J_3 which may be in either of two equilibrated flux states, "0" or "1" ", i.e. with and without a magnetic flux quantum inside. In state "0", an SFQ voltage pulse applied to the input "in" enters the SQUID through junction J1 and is stored inside. This switching leads to an insertion of one flux quantum into the loop J_1-L_1-J_2 because of its inductance ($L_1 I_C \approx \Phi_0$). Hence, in this state the persistent current ΔI circulates in the quantizing loop clockwise; in J_2 it subtracts from the initial dc bias current I_1 making this junction almost unbiased ($\phi_2 \ll \pi/2$). Thus, junction J_2 remains in the superconducting state. If another SFQ pulse is applied to the input "in", it flips junction J_0 and the latch remains in state "1". A new incoming SFQ pulse would lead to a switching of J_1 which will not change the flux stored in the loop. If, instead, we apply an SFQ pulse to the "clk" input when the latch is in state "1", junction J2 would flip, releasing the stored flux quantum and thus clearing the quantizing loop. In state "0" junction J3 is closer to its threshold value than J2. Thus an SFQ pulse at input "clk" flips J3, so that the latch remains in state "0". For a clocked operation in a larger design, when all inputs arrive from previous stages (as in a shift register) junction J0 is not necessary. The waveforms in Figure 12 show voltages across all 4 junctions of the latch as well as the input junctions /JTLIN/J2 and /JTLCLK/J2 of a connected Josephson transmission line.

Figure 12: Voltages across all 4 junctions of the latch. Voltages /JTLIN/J2 sand /JTLCLK/J2 correspond to junctions of connected Josephson transmission lines at input and clock lines, respectively, which are required in a testing or real circuit environment [36].

4.2.3 Clocking of SFQ Circuits

The signal protocol of RSFQ logic differs significantly from standard combination semiconductor logic. Two factors are responsible for that:

- "return-to-zero" nature of SFQ pulses, and
- natural internal memory function of quantizing loops.

Standard RSFQ circuits are based on the protocol shown in Figure 13.

In this convention, a signal in a data line is treated as binary "1" if it carries an SFQ pulse within the given clock period – see signal D_1. On the contrary, the absence of the pulse during this time interval (see signal D_2) is understood as binary "0". The Fan-out of a RSFQ cell is 1. This leads to additional Josephson transmission lines acting as drivers and special pulse splitters for the following cells.

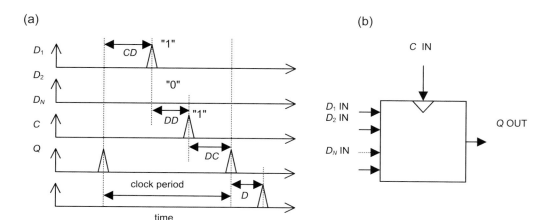

Figure 13: The standard RSFQ protocol and (b) a typical clocked gate. Timing parameters shown in (a) are discussed in detail in the text below. The delay times in the circuit are shown in τ_{ij}, where i and j are the different input and output lines.

5 Superconductor Integrated Circuit Technology

5.1 Low-Temperature Superconductor Technology

In recent years, an integrated circuit (IC) technology has been developed based on Nb films and Nb-trilayer Josephson junctions [15]. Niobium has a critical temperature of 9.2 K which allows operation of circuits at liquid helium temperatures ($T = 4.2$ K). Superconductor IC technology is ideally suited for ultra large scale integration of more than 10^7 devices per chip. The circuit density can be increased without losing speed because several superconducting ground planes are fabricated as an integral part of the circuit to form transmission lines for sub-ps pulses. The ultimate density of circuits is only limited by fabrication tools, processes and defect density. TRW Inc. has demonstrated a high current density fabrication process with 4 kA/cm². They have fabricated functional circuits with 3000 junctions per chip and clock speeds of 200 GHz. This operation speed does not require sub-μm lithography. For higher operation temperatures at $T \approx 10$ K, NbN is used as superconductor electrodes. The circuit fabrication technology is quite similar to the Nb process.

Typically, the whole process technology includes more than 10 layers. This includes the Nb trilayer for preparation of Josephson junctions and Nb-wiring and resistor layers. The wiring layers are separated by SiO_2 insulator layers. The standard pitch size is 3 μm and minimum junction and contact sizes are 1.75 μm and 1.0 μm. Figure 14 shows a cross section of a typical LTS circuit. The Josephson junction consists of a Nb-base and a Nb-counter electrode separated by a thin (≈ 1 nm) Al_2O_3 barrier layer. The formation of the Al-oxide barrier layer is the crucial process. The process involves depositing an ultra thin layer of metallic Al (≈ 6 nm) on top of a fresh sputtered Nb film without breaking the vacuum between both deposition processes. Then a controlled amount of oxygen is introduced into the system to oxidise the Al film forming a thin Al_2O_3 barrier layer. Finally, a Nb-counter electrode is sputtered on top of the barrier layer. This, so-called tri-layer, is patterned to form the Josephson junction using standard Reactive-Ion-Etching (RIE) processes. The entire circuit includes resistor layers made of Mo and Pd and Nb wiring layers. Details see Table 1.

	nominal value	inter wafer tolerance	on wafer tolerance	on chip tolerance
Josephson current density	1 kA/cm²	± 30%	± 10%	≤ ± 5%
Sheet resistance of Cr/Pt/Cr layer	1.0 Ω	± 7%	± 5%	≤ ± 2%
L_o:trilayer – groundplane	0.62 pH	± 10%	± 5%	≤ ± 2%
L_o:wiring – trilayer (over groundplane)	0.74 pH	± 10%	± 5%	≤ ± 2%
L_o:wiring – groundplane	1.03 pH	± 10%	± 6%	≤ ± 2%
I_c	200 μA	± 30%	± 10%	≤ ± 5%
$I_c R_N$	250 μV	± 30%	± 10%	≤ ± 5%

Table 1: Tolerances for SFQ circuits on the basis of externally shunted 3 μm linewidths Nb/Al-oxide technology (mainstream) for operation at 4.2 K.

The capability of simulating and designing superconducting circuits is a key prerequisite for the development of integrated circuits. A different situation is given in the LTS technology where large-scale integrated circuits have already become feasible. Processes with defined parameters, narrow tolerances, and established design rules allow for integration levels at which simulation as well as circuit design and layout development on a component level are no longer efficient. Instead, single functional modules are developed and assembled within a cell library. The single cells are not too complex and can therefore be designed and optimized. They are described in terms of structural, behavioural and geometrical data at different levels of abstraction and can be processed within contemporary design automation software. These systems support a hierarchical design style and are therefore capable of performing component-level, mixed-mode- and logic analysis. Tremendous performance gains in the simulation have been demonstrated this way. It has already been proven to be feasible to design complex digital circuits on a topological level using a top-down design methodology. Instead of establishing the electri-

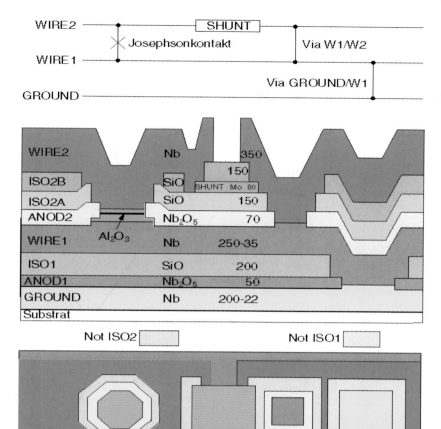

Figure 14: Schematic cross section and top view of a shunted and grounded Josephson junction for the IPHT Jena process. In the cross section the vertical scale is increased [34]. The thin film thickness in the cross section is given in nm. The full process requires 12 photo mask levels [34].

cal connections between the gates by a schematic entry, the design originates here from a behavioral description. This contains the desired specifications and is given e.g. as VHDL which has a clear, standardized representation and is easy to assemble. Subsequently, the synthesis program generates a gate netlist according to the cells available in the library. The synthesis can be controlled by means of specifying constraints. Commonly, the synthesis result can be optimal with respect to maximum switching speed or smallest chip area.

As an example, Figure 15 shows the RSFQ design of a superconductive shift register together with other functional circuits for implementation in Nb/AlO$_x$/Nb technology. The final design is entirely composed of sub-cells which are ordered hierarchically. A cell library of basic circuits was established in the framework of the FLUXONICS network. This was done in collaboration between TU Ilmenau and IPHT Jena as LTS foundry [34].

5.2 High-Temperature Superconductor Technology

The High-Temperature-Superconductor (HTS) Technology, mainly based on the YBa$_2$Cu$_3$O$_{7-x}$ (YBCO) material, is still in its infancy. HTS Josephson junctions offer a 10 times higher circuit speed than LTS circuits (due to the 10 times larger I_cR_n product of HTS) and have the additional advantage of being of the *self-shunted* type, eliminating the need for chip area-wasting external shunts. Additionally, HTS circuits may be operated at higher temperature at about 20 K. This reduces the cost for cooling of superconductor electronic systems significantly. At present, the main problem with HTS technology is a more than twice too large spread in the HTS Josephson junction parameters, which limits its use to circuits with low complexity (of 10-100 JJ's). In spite of these drawbacks HTS digital electronics is expected to have important niche applications in mobile equipment and airborne or satellite payloads, where low weight, size and cooling power consumption (as compared to those for LTS) is a decisive issue. In large-scale integrated RSFQ circuits, however, HTS faces fundamental as well as tech-

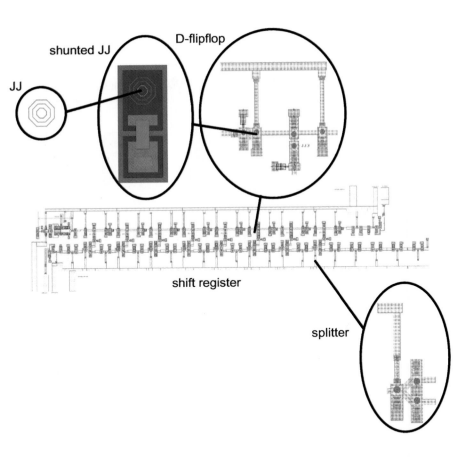

Figure 15: Design of a superconductive RSFQ shift register together with other functional circuits for implementation in Nb/AlO$_x$/Nb technology [34].

nological problems, as the advantage of operating at higher temperature is diminished by the increased thermal noise. The increased thermal noise can be combated by increasing the critical current of the junctions, but this, in turn, puts severe constraints on loop inductances and, consequently, on the circuit design. In spite of impressive improvement in HTS materials growth and Josephson junction technology in last couple of years, considerable progress is still required in order to reach an acceptable level of integration of, say, 10.000 junctions per RSFQ chip. Within the next 5-10 years it is therefore a safe prediction that all RSFQ circuits with a complexity exceeding 1000 junctions will use LTS, Nb-based technology.

There is no certainty that the preferred HTS junction technology has yet been invented. With the limited evidence available, standard deviations of critical current around 10-15 % are possible for all different technologies. At least a factor of 2 improvement is needed for medium complexity circuits to give economic yields. Further improvements in reproducibility could be translated into higher complexity or higher operating temperature.

At present, the most promising HTS technology is based on ramp-edge Josephson junctions. Ramp-edge junctions with YBCO electrodes and a heteroepitaxial barrier layer or an ion-beam modified interface barrier are fabricated from c-axis films. The schematics of the cross section of a ramp-edge junction based technology is shown in Figure 16. The first step is the deposition of a YBCO-insulator double layer having a very high homogeneity and smoothness. Next, a shallow ramp-edge is prepared with an angle less than 45°, to prevent the formation of grain boundaries. The ramp-edge is made by ion beam etching. A sufficiently shallow ramp-edge can be fabricated by using a shallow-edge resist mask made by re-flowing of photo resist at about 120°C. An important task is the pre-cleaning of the ramp-edge area prior to the deposition of the barrier and top YBCO layers. Recently, a new approach of forming a barrier layer has been developed using a special *plasma* and ion beam treatment to modify the YBCO interface [16]. After the last deposition, the multilayer is patterned into microbridges to form Josephson junctions by photolithography and ion beam etching.

The so-called interface-engineered, junctions can be fabricated in a wide range of J_C from 10^2 to 10^5 A/cm^2 and $I_C R_N$ products from 50 to 500 µV at 77 K. The 1σ-spreads of I_C of 5 % have been achieved and first simple RSFQ circuits were demonstrated [17], [18].

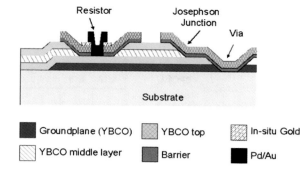

Figure 16: Cross section of a HTS circuit consisting of a ramp-edge Josephson junction, a via and a resistor.

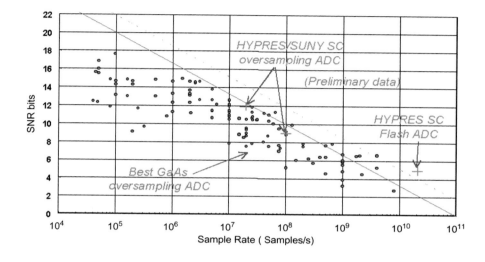

Figure 17: Performance of semiconductor and superconductor ADC. (•) plot of semiconductor data from [19], (+) show data from different groups for superconducting ADC [20].

6 Present Status of RSFQ Logic

6.1 Opportunities for Immediate Applications

6.1.1 Analog-to-Digital Converters

Very fast and accurate analog-to-digital converter (ADC) are mainly required for high-speed digital communication. Examples are image coding, clock jitter reduction and frequency synthesis. Here, ultra-fast ADCs reach the limit of operation speed of present semiconductor technology.

The performance of semiconductor Analog-to-Digital Converters (ADC) is improving at a rate of only ~ 1.5 bit per 6 years. Figure 17 represents the actual data of semiconductor ADC from [19] and includes new data on superconductor ADC. A number of different RSFQ ADC approaches have been designed, fabricated and tested. Among them are high resolution and fast flash-type architectures. A fully implemented high-resolution ADC is shown in Figure 18.

6.1.2 Digital-to-Analog Converters

There are very good prospects for the extension of the recent results to develop Digital-to-Analog Converters (DAC) with high accuracy. For example, in the next few years a multi-chip 20-bit DAC with a settling time better than 1 ms, an absolute accuracy better than 0.001 ppm, and a maximum output voltage of 1 V may be available. These converters may be used for AC voltage calibrators in metrological systems. The RSFQ integrated circuit solution is cheaper than a complex picosecond pulse sequence synthesiser. More details can be found in Refs. [20], [21], [22], [23], [24].

Figure 18: High-resolution RSFQ ADC chip (15-bit, 20 GHz clock, 200 MS/s output). It is fabricated using standard HYPRES's Nb process with 3 um min JJ size. About 6,000 JJs are used in this ADC. The chip is fully operational up to 19.6 GHz clock. The circuit contains ~6000 Josephson junctions, the power consumption is below 1 mW [37].

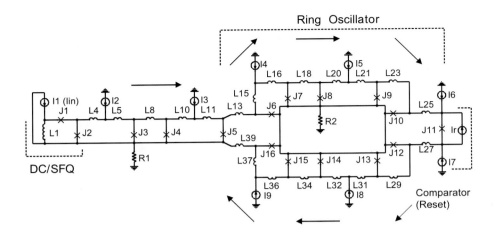

Figure 19: Equivalent circuit of a ring oscillator fabricated using HTS Josephson junctions [28]. The circuit was operated in the temperature range between 30 and 40 K. At 39 K and a frequency of 3 GHz a BER of 10^{-11} was determined [32].

6.2 Prospects for Future Applications

6.2.1 Ultra-fast Network Switches

The transmission capability of an optical fiber can be more than 1 Tbit/s. Although photons are a good match for sending signals, they make it difficult to switch electronic devices. Therefore it is difficult to make full optical switches. The optical I/O capability of RSFQ makes it possible to develop a full optical-electronic network switch. A 2-node SFQ crossbar network switch has been demonstrated with both data streams over 16 Gbit/s or a 16x16 design has been demonstrated at 3 GHz [25]. One analysis has shown that a 128 x 128 channel self-routing Batch-Banyan switching core implemented in a 0.8 μm RSFQ technology could provide a throughput close to 100 Gbit per channel and dissipate 10 mW of power on a 1-cm^2 chip [26].

6.2.2 Software-Defined Radio

The Software-Defined Radio (SDR) concept has attracted attention because it simplifies the receiver significantly, and allows fast reconfiguration of the hardware by software. The SDR concept relies on the digitalisation of waveforms as close to the antenna as possible, with a subsequent hardware and software digital-signal processing. For semiconductor ADC, one or two stages of down-conversion are still required to limit the bandwidth according to the performance of ADC. Thus a single input channel is mixed down to the lower frequency base band and handled digitally. Using fast wide band RSFQ ADC discussed in 6.1.1, however, a full band or several bands can be handled at once. This allows to skip one or all down-mixers. The ability to feed the digitised signals directly to a RFSQ digital pre-filter might enable programmable systems [27].

6.3 HTS Implementation of RSFQ Circuits

All basic cells which are necessary for designing the main functional blocks for RSFQ circuits have been implemented in HTS technology based on ramp-edge Josephson junctions [17], [29], [30]. Simple basic circuits as shift registers containing 50 junctions have been designed, fabricated and successfully tested [31]. The higher operation temperature leads to higher values of the critical current to reduce thermally-induced errors. In comparison to LTS SFQ circuits the critical current should be increased ~ 5 times. This leads to smaller inductances making the layout more difficult. Additionally, the increase of the London penetration depth with temperature close to the critical temperature limits the operation temperatures below 50 K. Thus, an important issue is the maximal possible operation temperature. Both, static and dynamic bit-error rates (BER) have been studied [32]. From static BER follows that operation in the temperature range between 30 K and 40 K is possible. The dynamic BER was studied using a SFQ ring oscillator. The circuit diagram of the ring oscillator is shown in Figure 19. The ring oscillator was operated up to frequencies of 10 GHz. The lowest dynamic BER of $\approx 10^{-11}$ was measured at 39 K. This value was limited by the stability of the temperature with time. These results show that HTS circuit can be operated with sufficiently small BER in the temperature range of 20...30 K. Recently, theoretical analysis of the BER in HTS circuits revealed that BER less than 10^{-20} can be obtained in the same temperature range [33]. Nevertheless, the issue of the influence of the higher thermal noise on proper circuit design has not been solved so far. Especially the reduction of operation margins with higher noise has to be studied in more detail.

The main challenge for HTS technology is to reduce the spread of critical current and normal resistance of Josephson junctions. Reducing the spread below a few percent will allow one to design circuits with more than ~ 1000 junctions.

7 Summary

One hundred years after the discovery of superconductivity in 1911 the prospect of practical applications of superconductor electronics are promising – at least for dedicated niche markets. The demand for 100-GHz speed performance generated by semiconductors may be fulfilled by superconductor integrated circuits. RSFQ technology is a key technology for these applications. RSFQ data converters are the fastest and most sensitive which have ever been demonstrated. Based on all functional RF and digital blocks as amplifiers, clocking, logic and memory this versatile technology may merge in System-on-Chip solutions. Moreover, as new technologies, like quantum computation, optical and biological systems mature, RSFQ may be an interesting approach with speed and integration capabilities to close the gap between classical and future electronics. The cryogenics nature of RSFQ systems will not be as much a technological issue, as a psychological issue for the user – it represents a real paradigm shift in the definition of an electronic system.

Acknowledgements

The editor would like to thank Matthias Schindler (RWTH Aachen) for checking the clarity and consistency of this chapter. The editor would like to thank Ralf Liedtke (RWTH Aachen) for checking the symbols and formulas in this chapter.

References

[1] *International Technology Roadmap for Semiconductors,* http://public.itrs.net/.
[2] G. A. Sai-Halasz, Proc. of IEEE **83**, 20 (1995).
[3] C. Wann, F. Assaderaghi, and Y. Taur, IEEE El. Dev. Lett. **18**, 625 (1997)
[4] S. L. Rommel et al., Appl. Phys. Lett. **73**, 2191 (1998).
[5] T. Van Duzer and C. W. Turner, *Principles of Superconducting Circuits*, Elsevier, New York, 1981.
[6] K. K. Likharev, *Dynamics of Josephson Junctions and Circuits*, Gordon and Breach, New York, 1986.
[7] W.C. Stewart, Appl. Phys. Lett. **12**, 277 (1968); D.E. McCumber, Appl. Phys. Lett. **12**, 3113 (1968).
[8] W. Chen et al., IEEE Trans. on Appl. Supercond. **9**, 3212 (1999).
[9] B. Ruck, *Entwicklung von Multilagenbauelementen für HTSL-RSFQ-Schaltungen und erste Messungen zur Fehlerrate*, Dissertation, UGH Wuppertal, 1997
[10] W. Anacker, ed., Special issue of IBM J. Res. Devel. **24**, 105 (1980).
[11] Y. Wada, Proc. IEEE **77**, 1194 (1989).

[12] M. Jeffery, W. Perold, and T. Van Duzer, Appl. Phys. Lett. **69,** 2746 (1996).
[13] K. K. Likharev, K.K., O. A. Mukhanov, and V. K. Semenov, IEEE Trans. on Magn. **23,** 759. (1987).
[14] S. V. Polonsky et al., IEEE Trans. on Appl. Supercond. **3,** 2566 (1993).
[15] L. Abelson, Q. P. Herr, G. L. Kerber, M. Leung, and S. Tighe, IEEE Trans. on Appl. Supercond. **9,** 3202 (1999).
[16] B. H. Moeckly and K. Char, Appl. Phys. Lett. **71,** 2526 (1997).
[17] T. Satoh, M. Hidaka, and S. Tahara, IEEE Trans. Appl. Supercond. **9,** 3141 (1999).
[18] J.-K. Heinsohn, R.H. Hadfield, and R. Dittmann, Physica C **326-327,** 157 (1999).
[19] R. Walden, IEEE J. Select. Areas Comm. **17,** 539 (1999).
[20] O. Mukhanov et al., Applied Superconductivity Conference (2000), to be published in IEEE Trans. on Appl. Supercond. **11,** No. 2 (2001).
[21] H. Sasaki et al., IEEE Trans. on Appl. Supercond. **9,** 3561 (1999).
[22] V.K. Semenov, Yu. A. Polyakov, and E. Wikborg, Applied Superconductivity Conference (2000); to be published in IEEE Trans. on Appl. Supercond. **11,** No. 2 (2001).
[23] A. Kidiyarova-Shevchenko, and D. Zinoviev, IEEE Trans. on Appl. Supercond. **5,** 2820 (1995).
[24] P. D. Dresselhaus, E. J. Dean, A. H. Worsham, J. X. Przybysz, and S. V. Polonsky, IEEE Trans. on Appl. Supercond. **9,** 3585 (1999).
[25] R. Sandell, J. Spargo, M. Leung, IEEE Trans. on Appl. Supercond. **9,** 2985 (1999).
[26] D. Zinovev, K. Likharev, IEEE Trans. on Appl. Supercond. **7,** 3155 (1997).
[27] E. Wikborg, V. Semenov, K. Kikharev, IEEE Trans. on Appl. Supercond. **9,** 3615 (1999).
[28] A. V. Rylyakov, D. F. Schneider, and Yu. A. Polyakov, IEEE Trans. on Appl. Supercond. **9,** 3623 (1999).
[29] B. D. Hunt, M. G. Forrester, J. Talvacchio, and R. M. Young, IEEE Trans. Appl. Supercond. **9,** 3362 (1999).
[30] B. Oelze et al., Appl. Phys. Lett. **68,** 2732 (1996).
[31] B. Oelze, B. Ruck, E. Sodtke, A.F. Kirichenko, M.Yu. Kuprianov, W. Prusseit, Appl. Phys. Lett. **70,** 658 (1997).
[32] B. Ruck et al., IEEE Trans. on Appl. Supercond. **9,** 3850 (1999).
[33] Th. Ortlepp, H. Toepfer, F.H. Uhlmann, Applied Superconductivity Conference (2000); to be published in IEEE Trans. on Appl. Supercond. **11,** No. 2 (2001).
[34] SCENET Roadmap for Superconductor Electronics,
[35] see http://orchidea.maspec.bo.cnr.it/working_groups/RSFQ_roadmap_final.pdf
[36] http://pavel.physics.sunysb.edu/RSFQ/
[37] O.A. Mukhanov et al., Supercond. Sci. Technol. **14,** 1065 (2001).

Quantum Computing Using Superconductors

Alexey Ustinov, Institute of Physics, University Erlangen-Nuremberg, Germany

Contents

1 The Principle of Quantum Computing   461
2 Computing with Qubits   462
3 Qubits: How to Realize them   463
4 Why Superconductors?   463
5 Charge Qubits   464
6 Flux Qubits   466
7 Other Qubits   467
8 Decoherence Mechanisms   468
9 Outlook   468

Quantum Computing Using Superconductors

1 The Principle of Quantum Computing

Quantum Computing (QC) has become a very hot topic in the past few years. It is exciting for many scientists from various areas, i.e. theoretical and experimental physics, computer science and mathematics. What is QC? Although the concept of information underlying all modern computer technology is essentially classical, physicists know that nature obeys the laws of quantum mechanics. The idea of QC has been developed theoretically over several decades to elucidate fundamental questions concerning the capabilities and limitations of machines in which information is treated quantum mechanically. In contrast to classical computing which we are all familiar with, QC deals with quantum information processing. In quantum computers the ones and zeros of classical digital computers are replaced by the quantum state of a two-level system. In short, QC is based on the controlled time evolution of quantum mechanical systems.

Classical computers operate with *bits*; quantum computers operate with quantum bits that have been named *qubits*. Unlike their classical counterparts, which have states of only 0 or 1, qubits can be in a complex linear superposition of both states until they are finally read out. For example, the states of a spin 1/2 particle can be used for quantum computation. For a qubit, the two values of the classical bit (0 and 1) are replaced by the ground state ($|0\rangle$) and the first excited ($|1\rangle$) state of a quantum two-level system.

Figure 1 illustrates the difference between a classical bit and a quantum bit. A classical two-state system can be prepared and stored in either of the states 0 or 1. This system is characterized by two stable states, e.g. as a particle placed in a double-well potential. In quantum mechanics, a particle is described by a quantum-mechanical wave function and it can tunnel under the barrier, which separates two wells. As a consequence, a particle can be in two or more states at the same time – a so-called superposition of states. A quantum system characterized by the double-well potential has the two lowest energy states $|0\rangle$ and $|1\rangle$. The wave function for the ground state $|0\rangle$ is symmetric, for the excited state $|1\rangle$ it is antisymmetric. Quantum theory predicts that a system prepared in a superposition state should follow coherent oscillations between the two wells. Once a measurement is performed, the probability of finding the particle in the specific well (left or right) oscillates periodically with time. The frequency ω of these coherent oscillations is proportional to the quantum tunneling rate between the wells. This leads to splitting of the lowest energy level by a so-called coherence gap $\Delta = \hbar\omega$. Many elementary books on quantum mechanics treat the physics of two-level systems that is essential for the understanding of QC.

While one classical bit of information is stored as either 0 or 1, a qubit can be in a weighted superposition of both states. For example, $a|0\rangle + b|1\rangle$, where a and b are complex numbers that vary with time t, and $|a|^2 + |b|^2 = 1$. Thus, not only 0 and 1, but all the states $|\psi(t)\rangle = a(t)|0\rangle + b(t)|1\rangle$ can be used to encode information in a qubit. This fact provides massive parallelism of QC due to *superposition of states*. When measured with a readout operator, the qubit appears to collapse to state $|0\rangle$ with probability $|a|^2$, and to state $|1\rangle$ with probability $|b|^2$. The state of two qubits can be written as a four-dimensional vector $|\psi\rangle = a|00\rangle + b|01\rangle + c|10\rangle + d|11\rangle$, where $|a|^2 + |b|^2 + |c|^2 + |d|^2 = 1$. The probability of measuring the amplitude of each state is given by the magnitude of its squared coefficient. In general, the state of n qubits is specified by $(2^{n+1} - 1)$ real numbers – an exponentially large amount of information – relative to the number of physical particles required. Most of these states are *entangled* – to create them requires some kind of interaction between the qubits, and the qubits cannot be treated entirely independently from one another. An entangled state cannot be written simply as a product of the states of individual qubits.

(a)
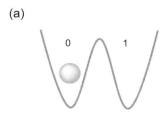

(b)
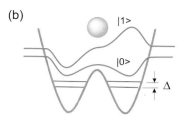

Figure 1:
(a) A classical computer manipulates with bits, which may take the values 0 or 1.
(b) A Quantum computer manipulates with quantum-mechanical two-level systems called qubits. The two quantum states are noted as $|0\rangle$ and $|1\rangle$.

2 Computing with Qubits

The great interest in QC is related to the fact that some problems, which are practically intractable with classical algorithms, can be solved much faster with QC. Factorization of large numbers, for which a quantum algorithm was proposed by P. Shor [1], is probably the best-known example in this respect. Shor showed that quantum computers could factor large numbers into prime factors in a polynomial number of steps, compared to an exponential number of steps on classical computers. What it means in practice can be illustrated by an example: Using a modern workstation cluster, a factorization of a number N with $L = 400$ digits will require 10^{10} years, which is larger than the age of the Universe. But a single hypothetical quantum computer should be able to do this job in less than 3 years! Shor's factoring algorithm works by using a quantum computer to quickly determine the period of the function $F(x) = a^x \bmod N$ (that means the remainder of a^x divided by N), where a is a randomly chosen small number with no factors in common with N. From this period, the techniques developed in the number theory can be used to factor N with high probability. The two main components of the algorithm, modular exponentiation (computation of $a^x \bmod N$) and the inverse quantum Fourier transform take only $\sim L^3$ operations.

Prime factorization is an essential part of modern public key cryptographic protocols, paramount to privacy and security in the electronic world. As quantum computers can, at least in theory, factor numbers in exponentially fewer steps than classical computers, they can be used to crack any modern cryptographic protocol. Another problem that can be treated very efficiently by QC is sorting. Quantum computers should be able to search databases in $\sim \sqrt{N}$ queries rather than $\sim N$ on an ordinary machine [2].

Let us briefly discuss the basic computational operations with a spin system of qubits as an example [3]. Manipulations of spin systems have been widely studied and nowadays nuclear magnetic resonance (NMR) physicists can prepare the spin system in any state and let it evolve to any other state. Controlled evolution between the two states $|0\rangle$ and $|1\rangle$ is obtained by applying resonant microwaves to the system but state control can also be achieved with a fast dc pulse of high amplitude. By choosing the appropriate pulse widths, the NOT operation (spin flip) can be established as

$$|0\rangle \rightarrow |1\rangle; \quad |1\rangle \rightarrow |0\rangle \tag{1}$$

or the Hadamard transformation (preparation of a superposition) as

$$|0\rangle \rightarrow (|0\rangle + |1\rangle)/\sqrt{2}; \quad |1\rangle \rightarrow (|0\rangle - |1\rangle)/\sqrt{2}. \tag{2}$$

These unitary *single bit* operations alone do not make a quantum computer yet. Together with single-bit operations, it is of fundamental importance to perform *two-bit* quantum operations; i.e., to control the unitary evolution of entangled states. Thus, a universal quantum computer needs both one and two-qubit gates. An example for a universal two-qubit gate is the controlled-NOT operation:

$$|00\rangle \rightarrow |00\rangle; \quad |01\rangle \rightarrow |01\rangle; \quad |10\rangle \rightarrow |11\rangle; \quad |11\rangle \rightarrow |10\rangle. \tag{3}$$

It has been shown that the single-bit operations and the controlled-NOT operation are sufficient to implement arbitrary algorithms on a quantum computer. *Quantum computers can be viewed as programmable quantum interferometers.* Initially prepared in a superposition of all the possible input states using the Hadamard gate (2), the computation evolves in parallel along all its possible paths, which interfere constructively towards the desired output state. This intrinsic parallelism in the evolution of quantum systems allows for an exponentially more efficient way of performing computations.

Without going into any detail due to the space limitation of this review, it is worth noting that the above mentioned Shor's algorithm uses two registers of $n = 2\lceil \log_2 N \rceil$ and $m = \lceil \log_2 N \rceil$ qubits. The algorithm is realized by five major computation steps, namely: (1) Initialization of both registers by preparing their initial state; (2) Applying a Hadamard transformation to the first n qubits; (3) Multiplying the second register by $a^x \bmod N$ for some random $a < N$ which has no common factors with N; (4) Performing the inverse quantum Fourier transformation (based on two-qubit controlled-phase rotation operator) on the first register; (5) Measuring the qubits in the first register. For detailed reviews devoted to algorithms of QC, we refer to [4] – [6].

3 Qubits: How to Realize them

It is common to adopt the spin 1/2 particle language for describing quantum algorithms (see Figure 2). Quantum theory predicts that if such a system is strongly coupled to the environment, it remains localized in one state and therefore behaves classically. Thus it is very important to have the quantum system decoupled from the rest of the world. Weak coupling to the environment damps the coherent oscillations between the states discussed in Sec. 1. The damping rate vanishes as the coupling to the environment goes to zero. The inverse of the damping rate is often called *decoherence time* τ_{dec}. This time is essentially the quantum memory of the system – after a long enough time $t > \tau_{dec}$ the system "forgets" its initial quantum state and is no longer coherent with it. In the ideal case, we would want to have $\tau_{dec} \to \infty$ for using a quantum system as qubit.

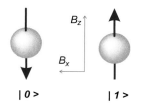

Figure 2: An example of a two-level system is a particle with the spin 1/2. The two basis quantum states $|0\rangle$ and $|1\rangle$ correspond to the spin orientations down and up with respect to the quantization axis B_z of the external magnetic field.

There are at least five important criteria that must be satisfied by possible hardware for a quantum computer [7]. To do QC, one needs:

1. Identifiable qubits and the ability to scale them up in number. This means that being able to build up only a few qubits is not sufficient for any useful quantum computation. For practical QC one would require very many (ideally, any desired number) of qubits in some controlled and reliable way.

2. Ability to prepare the initial state of whole system. All the qubits first have to be prepared in a certain state (like, e.g. $|0\rangle$ or $|1\rangle$) and only after that quantum computation can be started.

3. Low decoherence — the key issue, which rules out many of possible candidate systems for the quantum hardware. For quantum-coherent oscillations to occur, it is required that $t_{dec}\Delta/h \gg 1$. An approximate benchmark for sufficiently low decoherence is a fidelity loss of less than 10^{-4} per elementary quantum gate operation.

4. Quantum gates. The universal set of gates is needed in order to control the system Hamiltonian. After preparing a certain state, we have to be able to switch the interaction between them on and off in order to make qubits act together and do useful computation.

5. Perform a measurement. The final requirement for QC is the ability to perform quantum measurements on the qubits to obtain the result of the computation. Such readout transfers the information to the external world, i.e. to classical computers, in order to make the information useful.

Any candidates for quantum computing hardware should be assessed against this "DiVincenzo checklist" [7].

A number of two-level systems have been examined over the last few years as candidates for qubits and quantum computing. These include ions in an electromagnetic trap [8], atoms in beams interacting with cavities [9], electronic [10] and spin [11] states in quantum dots, nuclear spins in molecules [12], [13] or in solids [14], charge states of nanometer-scale superconductors [15], [16], flux states of superconducting circuits [17] – [19], quantum Hall systems [20], electrons on superfluid helium [21], and nanometer-scale magnetic particles [22]. Though all these systems fulfill some points of the checklist, some open questions remain. There is currently no clear favorite for quantum computing, analogous to the transistor for silicon-based classical computing. In addition to further work on existing systems, new candidates for quantum computing hardware should be explored.

Maintaining the coherence of a quantum device throughout the calculation is the major challenge for practical quantum computation. The device should be maximally decoupled from the environment in order to avoid decoherence and thus the loss of the quantum information.

4 Why Superconductors?

The advantage of microscopic quantum systems (atoms, spins, photons, etc.) is that they can be easily isolated from the environment, which reduces decoherence. The disadvantage is that the integration of many qubits into a more complex circuit in order to build a practical computer is a formidable task. From that point of view, macroscopic quantum systems offer much more flexibility to design a quantum computer using standard inte-

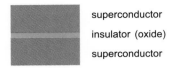

Figure 3: A Josephson tunnel junction is a structure formed by two superconductors separated by a very thin (2 - 3 nm) layer of a dielectric. Typically, Nb or Al is used as the superconducting material and Al_2O_3 as the dielectric.

grated circuit technology. Already proposed macroscopic qubits are based on nano-structured electronic circuits, which may consist of either quantum dots or superconducting Josephson junctions.

The large number of degrees of freedom associated with a solid-state device makes it more difficult to maintain the coherence. As yet, this problem has been met by either resorting to well-isolated spins (on quantum dots [11] or through deliberate doping of semiconductors [14]) or by making use of the quasi-particle spectrum in superconductors that is characterized by an energy gap.

All proposed superconducting quantum circuits are based on superconducting structures containing Josephson junctions. A Josephson junction is a structure consisting of two superconducting electrodes separated by a thin dielectric tunnel barrier (see Figure 3).

There are two possibilities for constructing a superconducting qubit. They differ by the principle of coding the quantum information. The first approach is based on very small Josephson junctions, which are operated by maintaining coherence between individual states of electron Cooper pairs. This type of qubit is called *charge qubit*. The charge states of a small superconducting island (a so-called electron box) are used as the basis states of this qubit. The second, alternative approach relies on the macroscopic quantum coherence between magnetic flux states in relatively large Josephson junction circuits. The latter qubit is known as the magnetic *flux (phase) qubit*. In fact, the flux qubit is based on a special realization of a superconducting quantum interference device (SQUID). An up-to-date review devoted to the implementation of quantum computation by means of superconducting nanocircuits has been recently published by Makhlin, Schön and Shnirman [23].

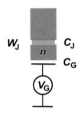

Figure 4: A Josephson charge qubit, in its simplest design, is formed by a superconducting electron box [23]. The box is separated from the superconducting reservoir by a Josephson tunnel junction.

5 Charge Qubits

These devices combine the coherence of Cooper pair tunneling with the control mechanisms developed for single-charge systems and Coulomb-blockade phenomena. The qubit is realized as a small (few 100 nm in dimensions) superconducting island attached to a larger superconducting electrode. The charge on the island, separated from a superconducting reservoir by a low-capacitance Josephson junction, is used in the qubit as the quantum degree of freedom. The basics states $|0\rangle$ and $|1\rangle$ differ by the number of superconducting Cooper pair charges on the island. The charge on the island can be controlled externally be a gate voltage. In the description of the charge qubits, I follow the guideline of review [23].

Quantum-coherent tunneling of Cooper pairs is, to some extent, similar to single-electron tunneling between very small conducting islands. These islands must be small enough so that the charging energy of a Cooper pair moving between the superconducting islands dominates all other characteristic energies in the system.

The simplest Josephson junction qubit is shown in Figure 4. It consists of a small superconducting island ("box") with n excess Cooper pair charges relative to some neutral reference state. The island is connected to a superconducting reservoir by a tunnel junction with capacitance C_J and Josephson coupling energy W_J. A control gate voltage V_G is applied to the system via a gate capacitor C_G. Suitable values of the junction capacitance, which can be fabricated routinely by present-day technologies, are in the range of femtofarad.

At low temperatures (in the mK range), the only charge carriers that tunnel through the junction are superconducting Cooper pairs. The system is described by the Hamiltonian:

$$H = 4W_C(n - n_G)^2 + W_J \cos\varphi \qquad (4)$$

Here φ is the phase of the superconducting order parameter of the island. The variable φ is the quantum mechanical conjugate of the number of excess Cooper pair charges n on the island:

$$n = -i\hbar \partial/\partial(\hbar\varphi) \qquad (5)$$

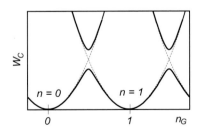

Figure 5: The plot shows the charging energy of the superconducting island as a function of the normalized gate charge nG for different numbers of extra Cooper pairs n on the island (dashed lines). Near degeneracy points, the weaker Josephson coupling mixes the charge states and modifies the energy of the eigenstates (solid lines) and the system reduces to a two-state quantum system [24].

Equation (5) is linked to the fundamental quantum-mechanical uncertainty relation for a Josephson junction between the superconducting grain and reservoir, which writes as $\Delta n \cdot \Delta \varphi \geq 1$. Thus, the superconducting phase difference φ between the island and reservoir cannot be determined simultaneously with the number of electron pairs n on the island. It is analogous to a condition that holds, e.g., for an optical pulse in a fiber – the number of photons in the pulse cannot be fixed simultaneously with the phase of the pulse.

In the charge qubit, the charge on the island acts as a control parameter. The gate charge is normalized by the charge of a Cooper pair, $n_G = C_G V_G /(2e)$, it accounts for the effect of the gate voltage V_G. For the charge qubit, the charging energy $W_C = e^2/(2(C_J+C_G))$ is much larger than the Josephson coupling energy W_J. A convenient basis is formed by the charge states, parameterized by the number of Cooper pairs n on the island. In this basis the Hamiltonian (4) can be written

$$H = \sum_n \left\{ 4W_C (n-n_G)^2 |n\rangle\langle n| + \tfrac{1}{2} W_J \left(|n\rangle\langle n+1| + |n+1\rangle\langle n| \right) \right\} \qquad (6)$$

For most values of n_G the energy levels are dominated by the charging part of the Hamiltonian. However, when n_G is approximately half-integer and the charging energies of two adjacent states $n = 0$ and $n = 1$ are close to each other, the Josephson tunneling mixes them strongly, see Figure 5.

The two states of the charge qubit differ by one Cooper pair charge on the superconducting island. In the voltage range near a degeneracy point only the two states with $n = 0$ and $n = 1$, play a role, while all other charge states having much higher energy can be ignored. In this case, the superconducting charge box behaves as a two-level (two-state) quantum system. In spin-1/2 notation its Hamiltonian can be written as

$$H = -\tfrac{1}{2} B_z \hat{\sigma}_z - \tfrac{1}{2} B_x \hat{\sigma}_x \qquad (7)$$

The charge states $n = 0$ and $n = 1$, shown in Figure 6, correspond to the spin basis states $|\downarrow\rangle$ and $|\uparrow\rangle$ illustrated in Figure 2. The charging energy, which is controlled by the gate voltage V_G, corresponds in spin notation to the z-component of the magnetic field

$$B_z \equiv 4W_C (1 - 2n_G) \qquad (8)$$

In its turn, the Josephson energy plays the role of the x-component of the magnetic field

$$B_x \equiv W_J \qquad (9)$$

The manipulations of charge qubits can be accomplished by switching the gate voltages [15] that play the role of B_z and modify the induced charge $2en_G$. The Josephson coupling energy W_J that corresponds to B_x can be controlled by replacing the single junction by two junctions enclosed in a superconducting loop (SQUID) [24], as shown in Figure 7. In this modified circuit, a current supplied through a superconducting control line that is inductively coupled to the SQUID induces a magnetic flux Φ_x, which changes the critical current and thus the Josephson coupling energy W_J of the device.

In addition to the manipulation of the qubit, its final quantum state has to be read out. For a Josephson charge qubit, this can be accomplished by coupling it to a single-electron transistor (SET). As long as the transport voltage is turned off, the transistor has only a weak influence on the qubit. When the voltage is switched on, the dissipative current through the SET destroys the phase coherence of the qubit within a short time.

Experimentally, the coherent tunneling of Cooper pairs and the related properties of quantum mechanical superpositions of charge states has been demonstrated in spectacular experiments by Nakamura et al. [16]. These authors observed in the time domain the quantum coherent oscillations of a Josephson charge qubit prepared in a superposition of eigenstates. The layout of their qubit circuit is shown in Figure 8. It includes a small superconducting grain (a Cooper pair "box") attached to a superconducting reservoir by two Josephson junctions as shown above schematically in Figure 7.

Using a dc gate, the Josephson charge qubit ("box") is prepared in the ground state far from the degeneracy point. In this regime, the ground state is close to the charge state, say, $|0\rangle$. Then the gate voltage is changed for a short time (less than one nanosecond) to a different value using the pulse gate. If it is switched to the degeneracy point, the initial state, a pure charge state, is an equal-amplitude superposition of the ground state $|0\rangle$ and the excited state $|1\rangle$, as is illustrated in Figure 5. These two eigenstates have different energies; hence, in time they acquire different phase factors.

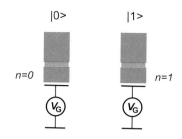

Figure 6: The basis states $|0\rangle$ and $|1\rangle$ of the superconducting charge qubit. They differ by the number of excess Cooper pairs n on the small superconducting island.

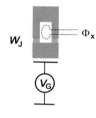

Figure 7: A charge qubit with tunable effective Josephson coupling. A flux-threaded SQUID replaces the single Josephson junction. A current carrying loop coupled to the SQUID controls the magnetic flux.

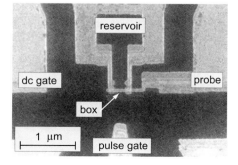

Figure 8: Micrograph of a Cooper-pair box with a magnetic flux-controlled Josephson junction and a probe junction (Nakamura et al. [16]).

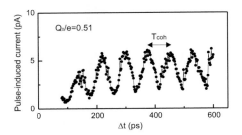

Figure 9: The coherent (Rabi) oscillations in the Josephson charge qubit observed in the experiments of Nakamura et al. [16].

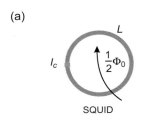

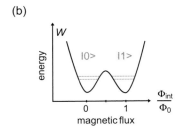

Figure 10:
(a) Sketch of MQC experiment with a SQUID, which is a superconducting ring containing a Josephson junction.
(b) SQUID energy as a function of the internal flux Φ_{int}. The external flux is equal to $\Phi_0/2$. Horizontal levels indicate mixed energy states.

The final state of the qubit in the experiment by Nakamura et al. [16] was measured by detecting a tunneling current through an additional probe-junction. Ideally, zero tunneling current implies that the system ended up in the $|0\rangle$ state, whereas maximum current is expected when the final state corresponds to the excited one state. In the experiment, the tunneling current shows an oscillating behavior as a function of pulse length, as shown in Figure 9. These data demonstrate the coherent time evolution of a quantum state in the charge qubit.

6 Flux Qubits

Since superconductivity is a macroscopically coherent phenomenon, macroscopic quantum states in superconductors offer a challenging option for quantum computing. There have already been experiments that demonstrated macroscopic quantum tunneling (MQT) of the superconducting phase in current-biased Josephson junctions and superconducting quantum interference devices (SQUIDs). Moreover, it has been found that the tunneling rate agrees well with the value predicted by the Caldeira-Leggett theory with a phenomenological treatment of the dissipation. Since MQT involves only a single potential well from which the tunneling of the system takes place, there is no issue of coherence between different quantum states attached to it.

A quantum superposition of magnetic flux states in a SQUID is called macroscopic quantum coherence (MQC). It is called macroscopic because the currents are built of billions of electrons coherently circulating within the superconducting ring. Figure 10 illustrates its main idea. If the magnetic flux bias applied to a SQUID is equal to $\Phi_0/2$ (where $\Phi_0 = \pi h/e = 2.07 \times 10^{-15}$ Wb is a magnetic flux quantum, h is Plank's constant, e is the electron charge), its potential energy has two symmetric minima. The flux in the SQUID loop can tunnel between the two minima. This implies that the degenerate ground state energy of the SQUID is split by the energy difference ΔW related to the tunneling matrix element, and the two states are mixed energy states. Therefore, if the coherence of this mixture can be maintained long enough, the magnetic flux will oscillate back and forth between the two states at the frequency $\Delta W/(2\pi h)$. Since the observation of MQT in Josephson structures in the 80's, there has been great interest in detecting MQC in SQUIDs. However, experiments were not successful and many of them were discontinued after the advent of high-temperature superconductivity in 1986.

A qubit can also be realized with superconducting nano-circuits in the limit $W_J \gg W_C$, which is opposite to charge qubits. The magnetic flux qubits are larger than the charge qubits, which makes them easier to fabricate and test. The flux qubit dynamics is governed by the superconducting phase difference across the junction rather than by the charge. The flux qubit consists of a SQUID as a macroscopic quantum coherent system.

The Hamiltonian of a single-junction SQUID (which is also called rf-SQUID) reads

$$H = -W_J \cos\left(\frac{2\pi\Phi}{\Phi_0}\right) + \frac{(\Phi - \Phi_x)^2}{2L} + \frac{Q^2}{2C} \quad (10)$$

Here, L is the self-inductance of the superconducting loop, and Φ is the magnetic flux in the loop. The externally applied flux is denoted by Φ_x. In the limit in which the self-inductance is large, the two first terms in the Hamiltonian form a double-well potential near $\Phi = \Phi_0/2$. The charge Q is a canonically conjugated variable to the phase difference across the junction $\varphi = 2\pi\Phi/\Phi_0$, see Eq. (5). The Hamiltonian (10) can be reduced to that of a two-state system. By controlling the applied magnetic field, all elementary operations can be performed.

Flux qubits seem more robust then charge qubits, they can be inductively coupled relatively easily. In the proposal by Mooij et al. [18], a qubit is formed by 3 junctions as shown in Figure 11. Flux qubits can be coupled by means of flux transformers, which provide inductive coupling between them.

The quantum mechanical properties of SQUIDs have been thoroughly investigated in the past, but only recently last year was the quantum superposition of different magnetic flux states evidenced experimentally [19] by the SUNY group at Stony Brook. One state corresponds to a persistent current in the loop flowing clockwise whereas the other corresponds to the current flowing anticlockwise. The major experimental result of the SUNY group is presented and briefly explained in Figure 12.

Nearly simultaneously with the SUNY team, the Delft group observed the quantum superposition of macroscopic persistent current states in their 3-junction SQUID [25]. Both experiments used a spectroscopic technique to detect the energy level splitting (more precisely, the level anti-crossing) due to the tunnel coupling between the two macroscopically distinct circulating current states of the circuit.

Coherent quantum oscillations in the time domain have not yet been detected in SQUID systems. To probe the time evolution, pulsed microwaves instead of continuous ones have to be applied. Observation of such oscillations would imply the demonstration of MQC, awaited since the 80's. The determination of decoherence time is the major remaining task to evaluate the feasibility of this type of flux qubits for practical quantum computing.

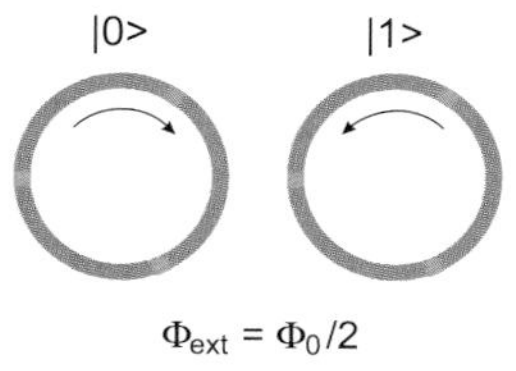

Figure 11: The basis states $|0\rangle$ and $|1\rangle$ of the superconducting flux (persistent current) 3-junction qubit [18]. They differ by the direction of the persistent current in the superconducting loop containing the junctions.

7 Other Qubits

Recently our group suggested using the *macroscopic quantum states of Josephson vortices* as a flux qubit for quantum computation [26]. Our original idea was to use the two distinct states of a fluxon trapped in a magnetic-field-controlled double-well potential inside a narrow long junction to design a qubit, as illustrated in Figure 13. Theory predicts that a fluxon in a double-well potential behaves as a quantum-coherent two-state system.

The physical principles of the fluxon qubit and the persistent current qubit are similar. It is possible by variation of the external field and the junction shape to form an arbitrarily shaped potential for a magnetic fluxon in the long Josephson junction. The amplitude of this potential can be easily varied by tuning the magnetic field. The superposition of two macroscopically distinct quantum states of the fluxon as quantum particle can be expected at low temperatures.

In the quantum regime, the coupling between the two states depends exponentially on the size of the energy barrier separating them. The energy barrier can be tuned in a wide range by changing the magnetic field applied to the junction. At low fields, the vortex tunnels through the barrier, and thus coupling between the two states appears. At high fields, however, tunneling is essentially suppressed and the vortex remains localized in one of the states. Thus, by applying a sufficiently large field the system can be switched into the classical regime in which the quantum states of the vortex correspond to their classical counterparts.

Recently we experimentally demonstrated a protocol for the preparation and read-out of the vortex qubit states in the classical regime [27]. We were able to manipulate the vortex states by varying the magnetic field amplitude and its direction, and by applying a bias current to the junction.

Other proposals using multi-junction loops for designing better flux qubits are under development. In particular, the use of π-junctions, which have an unconventional current-phase relation, is being considered for the design of qubits [28]. The hope here is that the combination of conventional and π-junctions in a single circuit may make it possible to design so-called "quiet" qubits, which do not require any external magnetic field for their operation. Thus, "quiet" qubits may be easier to decouple from the environment. However, a reliable technology for π-junctions still does not exist. The long-discussed approach to realizing a π-junction by making use of a copper-oxide d-wave superconductor is still very hard to realize in practice. The most promising approach in this respect seems to be so-called SFS (superconductor – ferromagnet – superconductor) junctions that are made with magnetic impurities in the Josephson channel [29].

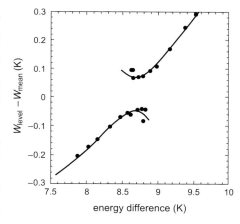

Figure 12: Experimental results of Friedman et al. [19]. The plot shows the energy of two spectroscopically measured levels, relative to their mean energy, as a function of the energy difference between the bottoms of the wells. At the midpoint of the figure, the measured tunnel splitting D between the two states in this "anticrossing" is about 0.1 K. The calculated energy levels are indicated by the lines.

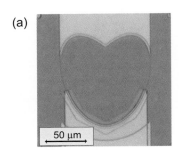

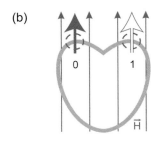

Figure 13:
(a) Photograph of a 300 nm wide heart-shaped long Josephson junction.
(b) The two degenerate vortex states in a heart-shaped junction are formed by applying an in-plane magnetic field. Arrows indicate the vortex locations that correspond to the states $|0\rangle$ and $|1\rangle$.

8 Decoherence Mechanisms

For performing quantum computing, it is very important that qubits are protected from the environment, i.e., from any source that could cause decoherence. This is a very difficult task because at the same time the evolution of the qubits also has to be controlled, which inevitably means that the qubit has to be coupled to control systems in the environment. Single atoms, spins and photons can be decoupled from the outside world. However, the large-scale integration that is needed to make a quantum computer useful seems to be impossible for these microscopic systems. Qubits made using solid-state devices (quantum dots or superconducting circuits), may offer the great advantage of scalability.

In their experiment with the superconducting charge qubit, Nakamura et al. [16] estimate the decoherence time to be about 2 ns. It may be speculated that the probe junction directly coupled to their circuit and the $1/f$ noise (presumably due the motion of background charges) are the main source of decoherence. In their absence (which so far has been difficult to accomplish), the main dephasing mechanism is thought to be spontaneous photon emission to the electromagnetic environment. Decoherence times of the order of 1 μs should then be possible for charge qubits.

The decoherence time for flux qubit has not been measured yet. In general, here estimates are more optimistic than for charge qubits. Decoherence times as large as milliseconds have been estimated. The 3-junction geometry has the advantage that it can be made much smaller than rf-SQUID with appropriate self-inductance L, so that it will be less sensitive to noise introduced by inductive coupling to the environment. Nevertheless, in all designs the measuring equipment coupled to qubits is expected to act destructively on quantum coherence.

9 Outlook

Superconducting tunnel junction circuits can be manipulated in a quantum coherent fashion in a suitable parameter range. Currently, they seem to be very promising for quantum state engineering and as hardware for future quantum computers. We have discussed their modes of operation in two basic regimes, dominated by the charge and the magnetic flux. There are several important constraints to overcome (mainly dephasing effects due to various decoherence sources) before a first useful QC circuit can be made. Nonetheless, nano-electronic devices have several important advantages as compared to other physical realizations of qubits; this gives us hope for the future.

If a quantum computer is ever to be made, it would require both an input and an output interface to interact with the external world. It is worth mentioning that such interface hardware does already exist for flux qubits. It can be designed using the rapid single-flux quantum (RSFQ) logic implemented in classical superconducting electronics. Indeed, the classic-computer RSFQ interface can be used for the preparation of initial states and for the read-out circuitry of the magnetic-flux carrying states. RSFQ is a well-developed technique that will be the natural choice for communicating between classic and quantum parts of superconducting quantum computer. Thus, all control and data exchange with classically operated electronics can be provided by high-speed on-chip RSFQ circuitry (see Bocko et al. [17]), and the external communication between RSFQ and room temperature semiconductor electronics can be realized by using optical fiber channels combined with MSM (metal-semiconductor-metal) switches and laser-emitting diodes.

In this brief review, I discussed the status of experiments to the beginning of 2002. Experimental observation of the macroscopic quantum coherent oscillations in a flux qubit, which is expected in the near future, should open the way for practical QC based on existing superconducting electronics technology [30].

Acknowledgements

I am grateful for discussions on this topic to G. Blatter, C. Bruder, M. J. Feldman, M. Fistul, J. R. Friedman, D. Geshkenbein, A. Kemp, Y. Makhlin, J. E. Mooij, Y. Nakamura, G. Schön, C. H. van der Wal, and A. Wallraff. The editor would like to thank Christian Ohly (FZ Jülich) for checking the symbols and formulas in this chapter.

References

[1] P.W. Shor, Proc. of the 35th Annual Symposium on the Foundations of Computer Science, ed. S. Goldwasser IEEE Computer Society Press, Los Alamitos, CA, 1994.
[2] L.K. Grover, in Proceedings 28th Annual ACM Symposium on the Theory of Computing, 1996.
[3] R. Fazio and H. van der Zant, Physics Reports **355**, 235 (2001).
[4] S. Lloyd, Science **261**, 1589 (1993).
[5] C.H. Bennett, Phys. Today **48**, 24 (1995).
[6] C.H. Bennett and D. P. DiVincenzo, Nature **404**, 247 (2000).
[7] D.P. DiVincenzo, Topics in Quantum Computers, in *Mesoscopic Electron Transport*, ed. L. Kowenhoven, G. Schön, and L. Sohn, NATO ASI Series E, Kluwer Ac. Publ., Dordrecht, 1997.
[8] J. I. Cirac and P. Zoller, Nature, **404**, 579 (2000).
[9] S. Haroche, M. Brune, and J. M. Raimond, Phil. Trans. R. Soc. Lond. A **355**, 2367 (1997).
[10] A. Ekert and R. Jozsa, Rev. Mod. Phys. **68**, 733 (1996).
[11] D. Loss and D. P. DiVicenzo, Phys. Rev. A **57**, 120 (1998).
[12] N.A. Gershenfeld and I. L. Chuang, Science **275**, 350 (1997).
[13] D. Cory, A. Fahmy, and T. Havel, Proc. Nat. Acad. Sci. **94**, 1634 (1997).
[14] B.E. Kane, Nature **393**, 133 (1998).
[15] A. Shnirman, G. Schön and Z. Hermon, Phys. Rev. Lett. **79**, 2371 (1997).
[16] Y. Nakamura, Yu. A. Pashkin, and J. S. Tsai, Nature **398**, 786 (1999).
[17] M.F. Bocko, A.M. Herr, and M.J. Feldman, IEEE Trans. Appl. Superconductivity **7**, 3638 (1997).
[18] J.E. Mooji, T. P. Orlando, L. Levitov, L. Tian, C. H. van der Wal, and S. Lloyd, Science **285**, 1036 (1999).
[19] J.R. Friedman, V. Patel, W. Chen, S.K. Tolpygo, and J. E. Lukens, Nature **406**, 43 (2000).
[20] V. Privman, I. D. Vagner, and G. Kventsel, Phys. Letters A **239**, 141 (1998).
[21] P.M. Platzman and M.I. Dykman, Science **284**, 1967 (1999).
[22] J. Tejada, E. M. Chudnovsky, E. del Barco, J.M. Hernandez, and T.P. Spiller, Nanotechnology **12**, 181 (2001).
[23] Y. Makhlin, G. Schön and A. Shnirman, Rev. Mod. Phys. **73**, 357 (2001).
[24] Y. Makhlin, G. Schön, and A. Shnirman, Nature **386**, 305 (1999).
[25] C.H. van der Wal, A.C.J. der Haar, F.K. Wilhelm, R.N. Schouten, C.J.P. M. Harmans, T.P. Orlando, S. Lloyd, and J. E. Mooij, Science **290**, 773 (2000).
[26] A. Wallraff, Y. Koval, M. Levitchev, M. V. Fistul, and A. V. Ustinov, J. Low Temp. Phys. **118**, 543 (2000).
[27] A. Kemp, A. Wallraff, and A. V. Ustinov, Physica C **368**, 324 (2002).
[28] L. B. Ioffe, V. B. Geshkenbein, M. V. Feigel'man, A. L. Fauchere, and G. Blatter, Nature **398**, 679 (1999).
[29] V.V. Ryazanov, V.A. Oboznov, A.Yu. Rusanov, A.V. Veretennikov, A.A. Golubov, and J. Aarts, Phys. Rev. Lett. **86**, 2427 (2001).
[30] The year 2002 brought a breakthrough in experimental studies of superconducting qubits. Large decoherence times have been measured by three groups using the charge-flux [D. Vion et al., Science 296, 886 (2002)] and phase qbits [Y. Yu et al., Science 296, 889 (2002)]; J. M. Martinis et al., Phys. Rev. Lett. 89, 117901 (2002)].

Carbon Nanotubes for Data Processing

Joerg Appenzeller, T. J. Watson Research Center, IBM Research Division, USA

Ernesto Joselevich, Weizmann Institute of Science, Israel

Wolfgang Hönlein, Corporate Research, Infineon Technologies, Germany

Contents

1 Introduction   473

2 Electronic Properties   474
2.1 Geometrical Structure   474
2.2 Electronic Structure of Graphene   474
2.3 Electronic Structure of Carbon Nanotubes   474
2.4 Transport Properties   476
2.5 Contacts   479

3 Synthesis of Carbon Nanotubes   480
3.1 Synthetic Methods   480
3.2 Growth Mechanisms   482
3.3 Processing and Functionalization   483
3.4 Assembly of Nanotube Arrays and Nanocircuitry   484

4 Carbon Nanotube Interconnects   485
4.1 Nanotubes in Vias   486
4.2 Maximum Current Density and Reliability   486
4.3 Signal Propagation in Nanotubes   486

5 Carbon Nanotubes Field Effect Transistors (CNTFETs)   487
5.1 Comparison to MOSFETs   487
5.2 Tailoring of Nanotubes   488
5.3 Back-gate CNTFETs   488
5.4 Complementary Carbon Nanotube Devices   490
5.5 Isolated Back-Gate Devices   490
5.6 Isolated Top Gate Devices   491
5.7 Comparison of Si-MOSFETs with up-scaled CNT-MOSFETs   491
5.8 Carbon Nanotube Circuits   492

6 Nanotubes for Memory Applications   493
6.1 CNT-SRAMs   493
6.2 Other Memory Concepts   493

7 Prospects of an All-CNT Nanoelectronics   494

Carbon Nanotubes for Data Processing

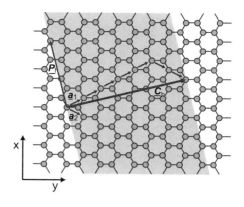

Figure 1: Arrangement of the C atoms in graphene with a roll-up stripe (grey) and the resulting nanotube. For a seamless tube, the circumference vector C_h must be a linear combination of the unit vectors a_1 and a_2 of the 2-D graphene lattice.

1 Introduction

Carbon Nanotubes (CNTs) have been discovered by Sumio Iijma of the NEC Tsukuba laboratory in a HRTEM study of carbon filaments [1]. A CNT can be thought of as a stripe cut from a single graphite plane (so-called **graphene**) and rolled up to a hollow seamless cylinder (Figure 1). As in graphite, the C atoms form a hexagonal network. They are trigonally coordinated because of their sp^2 hybridization. While in graphene, there is a pure sp^2 hybridization, some small contributions of sp^3 are mixed in, due to the curvature of the network in the case of CNTs.

With diameters between approx. 1 and 10 nm, the CNT cylinders can be tens of micrometers long. The ends may be open or capped with half a fullerene molecule in case of highly symmetrical nanotubes. Besides single-wall nanotubes (SWNT), there are multi-wall nanotubes (MWNT) which consist of numerous cylinders tightly stuck into another (Figure 2). In addition, ropes of CNTs are frequently encountered. These ropes are self-assembled bundles of (typically single-wall) nanotubes, in which the tubes line-up parallel to each other (Figure 3).

For electronic applications on the basis of carbon nanotubes or any other novel material the question to address is: "How can CNTs be employed in microelectronic and nanoelectronic devices?" The existing silicon based technology is extremely successful in producing the basic elements of integrated circuits, transistors and interconnects, at ever smaller dimensions in a highly parallel and cost effective process. Thus, a new technology, that wants to compete with a mature and reliable mainstream approach has to offer significant benefits. On the other hand, the constant scaling of length dimensions,

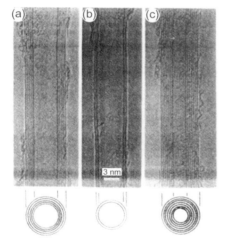
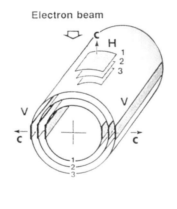

Figure 2: TEM image and imaging scheme of multi-wall nanotubes with various inner and outer diameters, d_i and d_o, and numbers of cylindrical shells N reported by Iijma [1]:
(a) $N = 5$, $d_o = 6.7$ nm,
(b) $N = 2$, $d_o = 5.5$ nm,
(c) $N = 7$, $d_i = 2.3$ nm, $d_o = 6.5$ nm.

which is the driving force behind the continuously improving silicon technology, will unavoidably lead to molecular and even atomic dimensions, where the toolkit developed for processing in the µm-range will eventually cease to be the optimum choice. The small size (diameter) of nanotubes in combination with their transport properties are very attractive in this context. CNT technology offers a new approach that may turn out to be more suitable for devices with nanometer-scale dimensions.

This Chapter will mainly focus on SWNTs although MWNT will be described too. It will describe the structure, the electronic properties, the synthesis, and first device concepts which have been published in recent years. We will restrict the discussion in this Chapter to nanoelectronics device applications although there are other interesting applications of nanotubes such as field emission displays, light sources, actuators, sensors and batteries. For a more comprehensive and more general coverage of CNT topic see, for example: Refs. [2], [3].

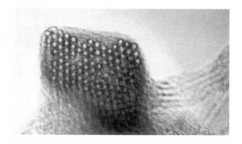

Figure 3: TEM cross section of a robe of SWNTs illustrating a hexagonal (i. e. densely packed) arrangement of the aligned tubes [4], [5].

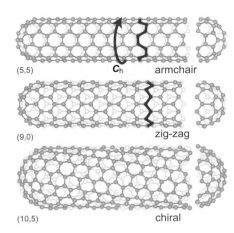

Figure 4: Examples of CNTs with different circumference vectors C_h [5].

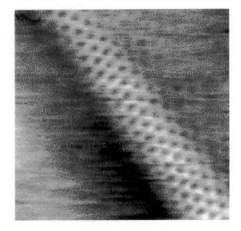

Figure 5: STM image at 77 K of a SWNT at the surface of a rope [6].

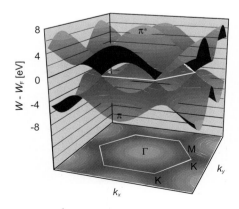

Figure 6: 3-D illustration of the dispersion relation of graphene.

2 Electronic Properties

2.1 Geometrical Structure

The structure of CNTs is described by the circumference vector or **chiral vector** C_h, which represents the full circumference of the tube. It is defined by

$$C_h = na_1 + ma_2 \tag{1}$$

where a_1 and a_2 are the unit vectors in the hexagonal lattice, and n and m are integers (Figure 1). C_h also defines the propagation vector P_h representing the periodicity of the tube parallel to the tube axis. Furthermore, it settles the so-called chiral angle which is the angle between C_h and a_1. If either n or m are zero, the chiral angle is 0° and the structure is called *zig-zag*. If $n = m$, the chiral angle is 30° and the structure is called *armchair* (Figure 4). All other nanotubes show chiral angles between 0° and 30°. They are known as *chiral* nanotubes because they produce a mirror image of their structure upon an exchange of n and m.

Experimentally, the diameter of nanotubes is frequently determined by TEM, STM or AFM. The chiral structure can be determined by STM (Figure 5).

2.2 Electronic Structure of Graphene

For the discussion of the electronic structure of CNTs, we start again with graphene. As an extension of the description of fused benzene (Chap. 5), in graphene, a bonding π-band and an anti-bonding π*-band is formed from the overlap between 2p$_z$-AOs of adjacent atoms. P. R. Wallace [7] derived an expression for the 2-D energy states, W_{2D}, of the π-electrons in the graphene plane as a function of the wave vectors k_x and k_y (see also [8]):

$$W_{2D}(k_x,k_y) = \pm\gamma_0 \left[1 + 4\cos\left(\frac{\sqrt{3}k_x a}{2}\right)\cos\left(\frac{k_y a}{2}\right) + 4\cos^2\left(\frac{k_y a}{2}\right)\right]^{1/2} \tag{2}$$

where γ_0 denotes the nearest-neighbour overlap (or: transfer) integral and $a = 0.246$ nm is the in-plane lattice constant. The two different signs in Eq. (2) represent the π- and π*-band. The calculations show that the π- and π*-band just touch each other at the corners of the 2-D Brillouin zone (Figure 6). In the vicinity of the Γ point, the dispersion relation is parabolically shaped, while towards the corners (K points) it shows a linear $W(k)$ dependence. At $T = 0$ K, the π-band is completely filled with electrons and the π*-band is empty. Because the bands only touch at the K points, integration over the Fermi surface (which is a line for a two-dimensional system) results in a vanishing density of states. On the other hand no energy gap exists in the graphene dispersion relation. This means we are dealing with the unusual situation of a gapless semiconductor. (The real graphite yet is a metal since the bands overlap by approx. 40 meV due to the interaction of the graphene planes.)

2.3 Electronic Structure of Carbon Nanotubes

For the description of the band structure of graphene, it has been assumed that the graphene plane is infinite in two dimensions. For CNTs, we have a structure which is macroscopic along the tube axis, but the circumference is in atomic dimensions. Hence, while the density of allowed quantum mechanical states in axial direction will be high, the number of states in the circumferential direction will be very limited. More precisely, the roll-up by the chiral vector C_h leads to periodic boundary conditions in the circumferential direction. Quantum mechanically, these boundary conditions define allowed modes (1-D states) along the tube axis according to:

$$C_h \cdot k = 2\pi j \quad \text{with} \quad |j| = 0, 1, 2, ... \tag{3}$$

In the case of arm-chair tubes, the periodic boundary condition yield allowed values for the wave vector in circumferential direction according to:

$$k_{y,j} = \frac{j}{q_y}\frac{2\pi}{\sqrt{3}a} \tag{4}$$

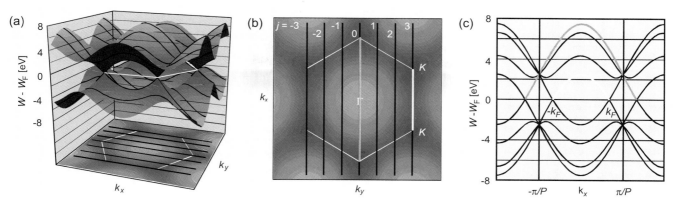

Figure 7: Dispersion relation of a (3,3) CNT.
(a) 3-D illustration of the dispersion relation for graphene including the allowed states for the (3,3) CNT. The periodic boundary conditions along the circumference of the tube result in a discrete set of allowed k_y values.
(b) Projection of the allowed states onto the first Brillouin zone of graphene. Obviously, the K points are allowed states for CNTs of this chirality.
(c) 2-D illustration of the dispersion $W(k_x)$. The states at the Fermi level indicate the metallic behaviour of this tube. The periodicity volume in the k-space is given by the interval from $-\pi/P$ to $+\pi/P$.

Figure 8: Dispersion relation of a (4,2) CNT.
(a) 3-D illustration of the dispersion relation for graphene including the allowed states for the (4,2) CNT. The periodic boundary conditions along the circumference of the tube result in a discrete set of allowed k values.
(b) Projection of the allowed states onto the first Brillouin zone of graphene. Obviously, the K points are no allowed states for CNTs of this chirality.
(c) 2-D illustration of the dispersion $W(k)$. the conduction band and the valence band are separated by a bandgap.

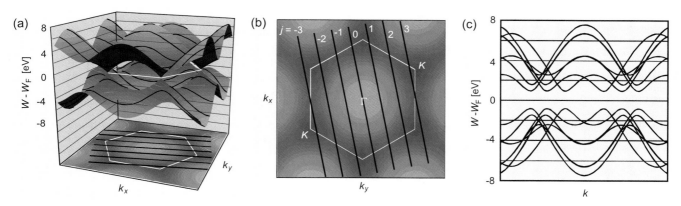

where $q_y = n = m$. For the armchair geometry, the tube axis is identical to the x-direction and the circumference represents the y-direction. As an example of an armchair tube, Figure 7 shows the dispersion relation, the projection of the allowed 1-D states onto the first Brillouin zone of graphene, as well as the $W(k_x)$ relation for a (3,3) tube. Due to the periodic boundary conditions, i. e. by inserting Eq. (4) into Eq. (2), the allowed states condense into lines (black lines in Figure 7a). Here, there are $q_y = 3$ lines on either side of the center of the Brillouin zone and an additional line going through the center. In case of a (3,3) tube the allowed states include the K points. Since the system is now one-dimensional in an electronic sense, different from the case of graphene, the integration over the Fermi surface (which is the sum over the Fermi points) yields a finite density of states at the Fermi energy. The (3,3) tube, and armchair tubes in general, show a *metallic* behavior.

As an example of a chiral tube, Figure 8 shows the dispersion relation, the projection of the allowed 1-D states onto the first Brillouin zone of graphene, as well as the $W(k_x)$ relation for a (4,2) tube. We will illustrate why the electronic properties of this (4,2) tube is very different from the (3,3) tube despite their very similar diameters. Again, due to the periodic boundary conditions, the allowed states condense into lines (black lines in Figure 8a). In contrast to the (3,3) tube, the C_h vector is not parallel to the y-direction and, hence, leads to a mixed quantization of k_x and k_y. The propagation of an electron along the tube axis is described by a combination of of k_x- and k_y-components. For this reason, the general letter k is used in Figure 8c, representing the momentum of the electron in the direction of propagation. The band structure of (4,2) tubes is deter-

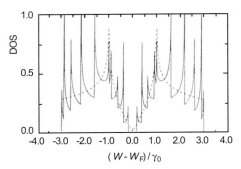

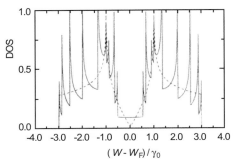

Figure 9: Calculated density of states (DOS) per unit cell of graphene as a function of energy, W, normalized by the overlap integral γ_0 [9].
(a) Result for a (10,0) tube, showing a bandgap around W_F.
(b) Result for a (9,0) tube, showing no bandgap. For comparison, the dashed lines represent the DOS for graphene.

mined by the fact that there are no modes which include the K points of the Brillouin zone of graphene (Figure 8b). The Fermi level is not dependent on the C_h vector, W_F is now in a bandgap, i. e. this type of tube is a *semiconductor*. The bandgap is of the order of a few eV (Figure 8c). In general, the bandgap decreases with increasing diameter of the tube.

In general, the semiconducting or metallic behavior of CNTs is controlled by the C_h vector and, hence, by the relation of n and m. Metallic behavior occurs for

$$n - m = 3q \tag{5}$$

where q is an integer. As a consequence, one-third of all CNTs types are metallic for a statistic distribution of chiralities including all armchair types, since $q = 0$ for them.

The periodic boundary conditions for zig-zag tubes, $(n,0)$ tubes and $(0,m)$ tubes, results in allowed wave vectors according to

$$k_{x,j} = \frac{j}{q_x} \frac{2\pi}{a} \tag{6}$$

The condition for metallic tubes, Eq. (5), is fulfilled for one-third of the tubes, i. e. if n or m are multiples of three. Figure 9 illustrates the density of state (DOS) for two zig-zag type CNTs [9], a (10,0) tube showing a bandgap and, hence, semiconducting behavior (Figure 9a), and a (9,0) tube showing no bandgap and, hence, metallic behavior (Figure 9b).

The discussion so far has been restricted to isolated SWNTs. Theoretical and experimental studies have shown that the intertube coupling within MWNTs and ropes of SWNTs [10], [11] have a relatively small effect on the band structure of a tube [12]. As a consequence, semiconducting and metallic tubes retain their character if they are a part of MWNTs or ropes. By statistical probability, most of the MWNTs and ropes show an overall metallic behavior, because one single metallic tube is sufficient to short-circuit all semiconducting tubes.

2.4 Transport Properties

For the discussion of the unperturbed transport of electrons in metallic SWNTs, we consider again the situation near the Fermi level shown by the 1-D dispersion spectrum, Figure 7c. The dispersion relation around the Fermi energy is linear. Furthermore, the energetical seperation, ΔW_{mod} between the modes at $\pm k_F$ is of the order of electron volts. It is this large energetical spacing between the 1-D subbands which prevents interband scattering to a large extend even at room-temperature. The transport is constrained to a single 1-D mode. Since there are subbands with positive and negative slope at both, $+k_F$ as well as for $-k_F$, one expects a Landauer conductance

$$G = 2 \cdot 2 \cdot \frac{e^2}{h} = \frac{4e^2}{h} \tag{7}$$

for an ideal, scattering-free i. e. **ballistic transport** of a metallic CNT. The degeneracy due to the spin is considered by a factor 2.

According to different authors (e.g. [50]) it is expected that ballistic transport properties are maintained in carbon nanotubes over distances of several micrometers. For transport on a larger scale scattering has to be taken into account.

We will first consider the elastic **scattering at impurities**. For 3-D metals, the probability of a scattering event, i. e. the scattering rate, is described by the classical Rutherford scattering theory:

$$\tau_{imp}^{-1} \propto v_F \cdot \frac{1}{v_F^4} = v_F^{-3} \tag{8}$$

where v_F denotes the Fermi velocity. In particular, τ_{imp}^{-1} is independent of temperature. This result relies on the fact that the velocity of electrons in an interval of approx. $4 k_B T$ around the Fermi energy is given by v_F in a very good approximation. This is in general a valid assumption for 3-D metals where W_F is large compared to $k_B T$. Eq. (8) holds exactly, if the dispersion relation is linear within the energy interval of approx. $4 k_B T$. This condition is fulfilled for CNTs up to room temperature and above. Figure 10 is a magnification of Figure 7c and shows the segment of the dispersion relation in the vicinity of the Fermi positions at $\pm k_F$. The arrows indicate the Fermi velocities of the elec-

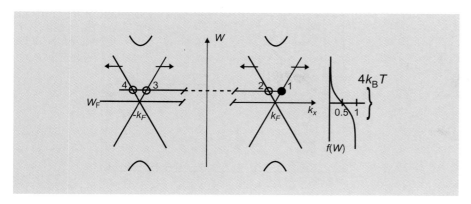

Figure 10: Illustration of impurity scattering processes on a metallic CNT. Dispersion relation in the vicinity of $-k_F$ and $+k_F$ as well as the Fermi-Dirac occupancy relation are shown. Position 1 (black circle) indicates an initial state of an electron, the positions 2 to 4 are potential final states after scattering events.

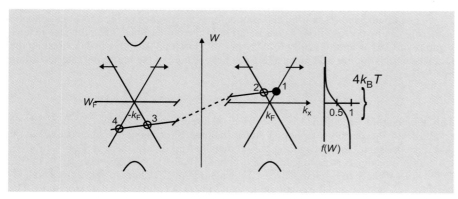

Figure 11: Illustration of an acoustic phonon scattering event on a metallic CNT. Dispersion relation in the vicinity of $+k_F$ as well as the Fermi-Dirac occupancy relation are shown. Position 1 (black circle) indicates an initial state of an electron, position 2 is the final state after the scattering event.

trons. Scattering at an impurity is an elastic scattering process. Consequently, an electron can only be scattered in another state at the *same* energy of the nanotube dispersion relation. The allowed final states for an electron in an initial state at position 1 are indicated as 2 to 4 in Figure 10. The change in electron momentum is accounted for by the impurities. Because of the one-dimensionality of the system small-angle scattering as present in 2-D or 3-D metals does not exist in carbon nanotubes. While scattering from 1 to 3 represents a forward scattering event without effect on the sample conductance, scattering from 1 to 2 or 4 is a backscattering event and will increase the CNT resistance. The two-terminal resistance of a CNT segment of length L will be

$$R_{\text{imp}} = \frac{h}{4e^2} \cdot \frac{L}{\lambda_{\text{imp}}} \qquad (9)$$

where λ_{imp} is the elastic mean free path which is roughly speaking the average distance between impurity centers. R_{imp} will be temperature independent in a good approximation.

For the discussion of the **phonon scattering** of electrons on CNTs, we have to distinguish scattering by optical and acoustical phonons. Starting point for describing the latter is the linear dispersion of the *acoustical phonons* with $W_{\text{ph}} = \hbar c_{\text{ph}} k_{\text{ph}}$ [13]. Since the velocity of sound, c_{ph}, is approx. 10^4 m/s, i.e. about two orders of magnitude smaller than the Fermi velocity of the electrons ($v_F \cong 10^6$ m/s), the scattering of an electron by an acoustic phonon results in a rather small electron energy change. Figure 11 illustrates a possible final state 2 for an electron scattered by an acoustic phonon from an initial state 1. The two states are connected through a line with a slope (c_{ph}) much smaller than the one of the electron dispersion relation reflecting the aforementioned difference in velocities. Scattering from the crossed dispersion region around $+k_F$ to $-k_F$ is suppressed because there are no empty, allowed states available around $-k_F$ even at room-temperature. This can be understood as follows: Scattering from $+k_F$ to $-k_F$ requires a Δk of approximately $2 k_F$ which has to be delivered by the phonon. This means $k_{\text{ph}} = \Delta k = 2 k_F$. The corresponding phonon energy is $W_{\text{ph}} = \hbar c_{\text{ph}} k_{\text{ph}} \cong 100$ meV. This value is much higher than $k_B T$ even at room-temperature and thus scattering from around $+k_F$ to $-k_F$ is suppressed. On the other hand, scattering from 1 to 2 remains possible even at lowest temperatures. This is true because the required energy transfer W_{ph} is always smaller than the electrical excess energy of the electron in position 1, since $c_{\text{ph}} \ll v_F$.

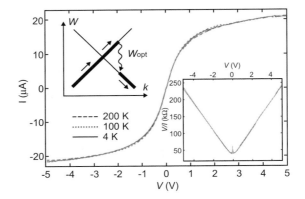

Figure 12: High-field *I-V* characteristics for metallic CNTs at different temperatures. The right inset plots $V/I = R$ vs. V. The left inset shows a section of the dispersion relation and illustrates the phonon emission [14].

A quantitative expression for the temperature dependence of the acoustical phonon scattering can be derived from a consideration of the dimensions. Since the Debye temperature of CNT is approx. 2000 K [14], the situation for T < 300 K is described by the Grüneisen relation. For 3-D metals, this is a dependence of the scattering rate according to

$$\tau_{ph}^{-1} \propto T^5 \tag{10}$$

This temperature dependence is composed of three factors. A first $\propto T^2$ dependence results from the density of phonon states in three dimensions. A second $\propto T^2$ term describes the small-angle scattering in 3-D metals. And finally, there is factor $\propto T$ from the energy transfer between electron and phonon. For 1-D systems such as CNTs, the first two terms have to be changed to $\propto T^0$, because the 1-D phonon DOS shows no temperature dependence and the small-angle scattering is not allowed in 1-D structures. The final $\propto T$ term remains and, hence, the resistance contribution due to acoustical phonon is given by

$$R_{ph} = \frac{h}{4e^2} \cdot \frac{L}{\lambda_{ph}} \quad \text{where} \quad \lambda_{ph} = v_F \cdot \tau_{ph} \quad \text{and} \quad \tau_{ph} \propto T^{-1} \tag{11}$$

For small electric excess energies of the electrons (approx. eV < 100 meV), scattering by *optical phonons* can be neglected since there are no unoccupied states at energies

$$W_{final} = W_{initial} - W_{opt} \tag{12}$$

where W_{opt} denotes the energy of optical phonons [14]. For large dc biases, i. e. high fields, electrons on the CNT are taking up energies well beyond values of $k_B T$, they become *hot*. Interestingly, hot in this case does not mean that the electron velocity has increased. Because of the special situation of a linear dispersion relation, the velocity of the carriers remains constant while their energy increases. This changes the situation completely. In the case of hot electrons, scattering by *optical phonons* can be the main contribution to the overall resistance of CNTs. Figure 12 shows the high-field *I-V* characteristics of metallic CNTs at different temperatures. The curves overlap almost completely, proving a temperature independent behavior. For small voltages, approx. V < 0.2 V, the *I-V* characteristics exhibits a linear behavior. For larger voltages, the *I-V* curve is strongly non-linear. For voltages approx. > 5 V, the current exceeds 20 µA which corresponds to a current density of more than 10^9 A/cm^2. Furthermore, it seems that a saturation current I_0 is approached at large bias. The resistance shows a constant value R_0 at small bias and increases linearly for $V > 0.2$ V (see right inset in Figure 12), i.e. it can be expressed by:

$$R = R_0 + V/I_0 \tag{13}$$

This behaviour can be explained by the inset picture in the left of Figure 12. Once an electron has gained enough energy to emit an optical phonon, it is immediately back-scattered [15]. A steady state is approached in which the electrons moving in forward direction have an energy W_{opt} higher than the backward moving ones. This leads to a saturation current

$$I_0 = \frac{4e^2}{h} \cdot \frac{W_{opt}}{e} \tag{14}$$

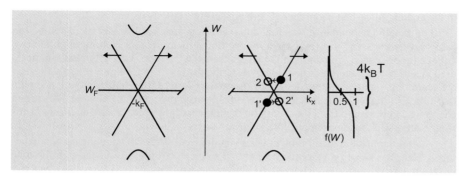

Figure 13: Illustration of an electron-electron scattering event on a metallic CNT. Dispersion relation in the vicinity of $+k_F$ as well as the Fermi-Dirac occupancy relation are shown. Positions 1 and 1′ (black circles) indicate the initial state of the two electrons, positions 2 and 2′ are the final states after the scattering event.

Here, the first term on the right represents the Landauer conductance as before and the second the voltage corresponding to W_{opt}. Using a value of approx. 160 meV [14], Eq. (14) leads to a saturation current of approx. 25 mA, in very good agreement with the experimental results. Accordingly, the mean free path l_{opt} for backscattering phonons is just the distance an electron needs to accumulate the threshold energy:

$$\lambda_{opt} = \frac{W_{opt}}{e} \cdot \frac{L}{V} \quad (15)$$

L denotes the electrode spacing and V the applied voltage, i.e. V/L is the electric field along the CNT. If this is combined with a field-independent scattering term (i. e. from impurities) with a mean free path λ_{imp}, an overall expression for the voltage dependent resistance can be obtained:

$$R = \frac{h}{4e^2} \cdot L \left(\frac{1}{\lambda_{imp}} + \frac{1}{\lambda_{opt}} \right) = \frac{hL}{4e^2 \lambda_{imp}} + \frac{h}{4eW_{opt}} \cdot V \quad (16)$$

The right equation is obtained by inserting Eq. (15) into the expression for λ_{opt}. Eq. (16) fully explains the empirical result, Eq. (13).

An additional mechanism that leads to a resistance contribution is the **electron-electron scattering**; a mechanism which may get especially pronounced in 1-D conductors. Typically, electron-electron scattering does not result in any measurable change in resistance for a normal conductor. This is true because energy and momentum conservation can only be fulfilled if as many electrons are backscattered as get scattered in the forward direction. The situation is drastically changed for a carbon nanotube. Figure 13 shows the dispersion relation in the vicinity of k_F. Because of the mode crossing, two electrons in the positions 1 and 1′, which contribute to the forward transport, may get scattered in positions 2 and 2′, respectively. The total energy and momentum is conserved during this process. The scattered electrons in 2 and 2′ now occupy states in the negative current direction. Hence, this process leads to a resistance increase. With raising temperature, the number of allowed initial and final states increases and so does the scattering probability. A detailed analysis [16] reveals a linear dependence of the electron-electron scattering rate on the temperature:

$$R_{e\text{-}e} = \frac{h}{4e^2} \cdot \frac{L}{\lambda_{e\text{-}e}} \quad \text{with} \quad \tau_{e\text{-}e} \propto T^{-1} \quad (17)$$

where $\lambda_{e\text{-}e}$ is the mean free path between scattering events and $\tau_{e\text{-}e}$ is the inverse scattering rate. The scattering rate also rises with increasing electric excess energies, eV, of the electrons because the number of allowed states increases.

2.5 Contacts

In order to perform any kind of electrical measurement on a carbon nanotube it is essential to create some kind of contact between the tube and the outside world. One way of doing this is to localize the position of a tube, or a rope of tubes by means of a scanning electron microscope (SEM) and then design the desired contact geometry by electron beam lithography, metal deposition and lift-off. A typical example of such a structure is shown in Figure 14.

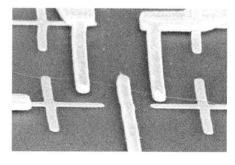

Figure 14: SEM image of a rope of single-wall nanotubes contacted by several gold fingers attached to the top of the tube.

As with any other system, a highly transmissive, minimum invasive contact is important to study the tube properties. But exactly this type of contact is hard to achieve in the context of carbon nanotubes. There are several reasons why this is the case.

For metallic nanotubes it was found that contacts attached to a carbon nanotube can cause severe backscattering. E.g. in a four-terminal measurement configuration the two voltage probes can significantly influence transport in the tube. This may result in measured electrical characteristics that do not describe the intrinsic tube properties but reflect the scattering situation in the vicinity of the contacts. Only if particular attention is paid to buffer the impact of the metal contacts, the intrinsic properties of the metallic carbon nanotube can be studied and the theory on the impact of different scattering mechanisms from Section 2.4 becomes applicable [17].

Another contact related problem occurs in the case of semiconducting carbon nanotubes. Here the situation is more complicated since electron transport through the tube has to be enabled prior to characterizing the contact quality. This is typically done in a *transistor-like* three-terminal configuration where a gate electrode is used to accumulate charge on the semiconducting tube thus allowing current transport through the same. When a metal/nanotube/metal system is investigated in this way, one would in general expect a Schottky barrier at the metal/nanotube interface to have a significant impact on the current flow. For a *regular type* 3-D metal / 3-D semiconductor interface (without a significant number of interface states) one could expect the barrier height to vary with the work function difference between the two materials. Following this argument, a number of groups have contacted semiconducting nanotubes with high work function metals as Ni or Au assuming that barrier free hole injection into the valence band of the nanotube would become possible [18], [19]. In fact, their results seem to support their assumption. Output as well as transfer characteristics strongly resemble those of conventional, barrier-free MOSFETs (see Sections below). On the other hand, recent studies on the impact of annealing versus doping of tubes [20], experiments on ambipolar nanotube devices [21], and detailed studies of transistor characteristics as a function of temperature and dielectric film thickness [22] support the existence of Schottky barriers and their importance for the current transport in carbon nanotube transistors. The controversy about the impact of contact barriers in carbon nanotube transistors is not yet settled. Further work will finally reveal the true nature of carbon nanotube/metal contacts.

3 Synthesis of Carbon Nanotubes

Single-wall carbon nanotubes (SWNTs) can be produced by several methods. Mass production and processing are still serious obstacles toward many proposed applications in industry, and the fabrication of ordered nanotube arrays with full control over their structural and electronic properties remains a major challenge toward applications in nanoelectronics. Consequently, both the synthesis and the assembly of carbon nanotubes continue to be hot research topics and new procedures are continuously being reported. Here, we focus on a few of the most common or promising methods of synthesis, processing and assembly of SWNTs, with a special emphasis on those that are more relevant for nanoelectronics.

3.1 Synthetic Methods

Electric arc discharge

The first carbon nanotubes, discovered by Iijima in 1991 [1] were multi-wall carbon nanotubes (MWNTs), as described in the beginning of the chapter. These nanotubes were produced by a an arc-discharge method similar to the one that was earlier used to synthesize the famous football-shaped C_{60} molecule [23] The method consists of applying a voltage between two graphite electrodes held close together in a chamber filled with an inert gas. The electrical discharge that takes places between the electrodes heats up the region to thousands of degrees, leading to the evaporation of carbon. The carbon vapor crystallizes on the end of the negative electrode, forming MWNTs with diameters ranging between 4 and 30 nm. The introduction of small amounts of transition metals such as Fe, Co, and Ni lead to the formation of single-wall carbon nanotubes (SWNT) [24], [25]. As an example for an efficient SWNT production method, the arc-discharge apparatus of Journet et al. [26] is schematically depicted in Figure 15a. Highest yields

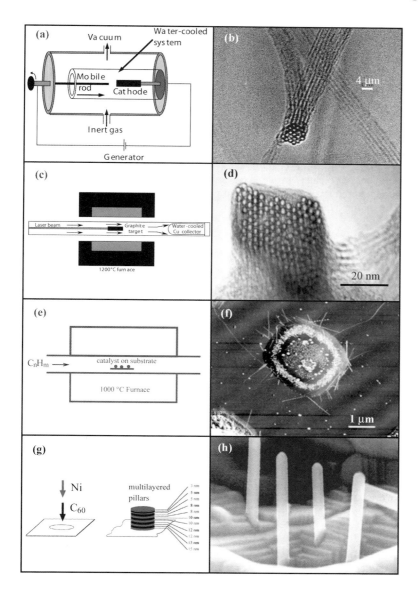

Figure 15: Synthetic methods for the production of single-wall carbon nanotubes:
(a) – (b) Electric arc discharge, and TEM image of the product;
(c) – (d) Laser vaporization, and TEM of the product;
(e) – (f) Chemical vapor deposition; and SEM image of SWNTs grown from lithographic catalyst islands on a Si wafer.
(g) – (h) Fullerene recrystallization, and TEM of the product (the rods have a diameter of ~40 nm).

(70 – 90 %) of SWNTs were obtained by filling the positive graphite electrode with 1 % Y and 4.2 % Ni, and gradually screwing it toward the negative electrode during the reaction, in order to keep a constant voltage drop of 30 V with a current of 100 A. The SWNTs in Figure 15b show an average diameter of 1.4 nm, and form crystalline ropes very similar to the ones obtained by the laser vaporization method.

Laser vaporization

A high-yield (70 %), large-scale production of SWNTs is also achieved by the laser-vaporization method which has been first reported in 1996 by the group of Smalley [27]. In this method, represented in Figure 15c, a target of graphite containing small amounts of Co and Ni powder is placed in the middle of a tube furnace at 1200 °C under a flow of argon, and hit by a series of laser pulses. With every laser pulse, a plume of carbon and metal vapors emanates from the surface of the target, and nanotubes start to grow in the gas phase. The nanotubes continue to grow while flying downstream along the tube, until they exit the furnace and are collected on a cold finger as a spongy black deposit. Each fiber of this material consists of a rope of 100 to 500 parallel SWNTs, close-packed as a two-dimensional triangular lattice, as shown in Figure 15d. This pulsed-laser ablation synthesis afforded SWNTs in large amounts, which made many physical and chemical studies possible for the first time, and it is still the production method preferred by several companies that commercialize carbon nanotubes.

Chemical vapor deposition

Both the electric arc discharge and laser vaporization methods produce ropes of SWNTs. The first production of individual SWNTs was reported in 1998 by he group of Hongjie Dai [28] and was based on a CVD technique (Figure 15e). SWNTs were grown *in situ* on silicon wafers having lithographically patterned catalytic islands of alumina (Al_2O_3) powders containg Fe and Mo catalytic nanoparticles. These substrates were placed in a tube furnace at 1000 °C under a flow of methane. The hydrocarbon worked as a carbon precursor, which decomposed on the catalyst, and the carbon crystallized in the form of individual SWNTs emerging from the catalyst islands. This simple CVD procedure opened up the possibility of producing prototype *nanotube chips* by growing individual SWNTs *in situ* on specific locations on a flat substrate (example: Figure 15f, see also Figure 18) and, hence, may prove very useful for *in situ* poduction of nanotube assemblies and nanocircuitry. It was also found that other gases could be used as carbon precursors, including carbon monoxide (CO), ethylene (C_2H_4) and benzene (C_6H_6) [29].

Fullerene recrystallization

SWNTs produced by the above-mentioned methods consist of random mixtures of nanotubes with different diameters and chiralities, that can be either metallic or semiconducting. This structural and electronic variety is a serious obstacle toward the application of SWNTs in nanoelectronics. A new synthesis method, reported in 2001 [30] has been the production of homogeneous single crystals of SWNTs by alternated layers of Ni and C_{60} evaporated on Mo or Si substrates through a shadow mask, forming an array of C_{60}/Ni multilayer pillar. However, detailed studies have shown that this initial idea proved to be not correct and different phases to be formed instead of SWNTs [31].

3.2 Growth Mechanisms

The mechanism of SWNT growth is still a debated issue. Since SWNTs were first observed, several growth mechanisms have been proposed, although none of them has yet received definitive experimental or theoretical support.

Catalytic vs. vapor-liquid-solid model

Initially, the fact that transition or rare metals, or mixtures of them, were always required for the formation of SWNTs, led to the proposition of catalytic mechanisms involving the coordination of metal atoms to the dangling bond of the growing nanotubes. Such was, for example, the *scooter mechanism* [32], where transition metal atoms bridging between two carbon atoms would go *scooting* around the edge of the nanotube, as it is growing. A rather different mechanism was based on the vapor-liquid-solid (VLS) macroscopic model, which was known from the 1960s to explain the growth of Si whiskers. In the VLS model, growth occurs by precipitation from a super-saturated catalytic liquid droplet located at the top of the filament. The atomic catalyst models were earlier supported because it was understood that if the VLS mechanism were correct, a catalyst particle should be left at the end of each nanotube, and this had not been observed. More recently, however, both the controlled growth of SWNTs from individual nanoparticles [33] by the CVD method, and meticulous observation of SWNT ropes under the TEM [34], allowed the unequivocal observation of metal clusters at the ends of SWNTs. Since then, a very plausible mechanism is actually a combination of the catalystic and the VLS models, referred to as the *root-growth* mechanism.

Root-growth mechanism

The root-growth mechanism for the growth of an individual SWNT from a metal nanoparticle on a substrate by the CVD method is schematically represented in Figure 16. First the hydrocarbon decomposes on the metal nanoparticle into hydrogen and carbon, which dissolves in the metal (Figure 16a). When the carbon becomes super-saturated in the nanoparticle, it starts to precipitate in the form of a graphitic sheet. Since the edges of the graphitic sheet are unstable, the emergence of *pentagon defects*, leading to the for-

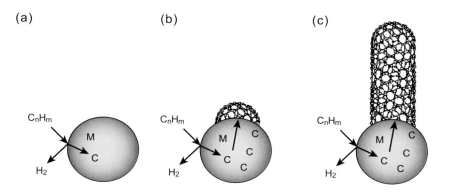

Figure 16: Root-growth mechanism for the formation of a single-wall carbon nanotube from a metal nanoparticle, by chemical vapor deposition:
(a) decomposition of the hydrocarbon on the nanoparticle and solubilization of the carbon therein.
(b) nucleation by formation of a fullerene cap.
(c) elongation of the SWNT by incorporation of further carbon into the metal-carbon bonds at the growing end.

mation of a curved fullerene cap (Figure 16b), becomes energetically favored, as it allows the dangling bonds of this cap to be stabilized by coordination with the metal. As in the catalytic models, the interaction between the partially filled 3d orbitals of the transition metal and the empty π^* orbitals of the carbon, may play a crucial role in stabilizing the dangling bonds of the fullerene cap. After the cap is formed, two things can happen. In one case, as shown in Figure 16a, more carbon atoms can insert into the metal-carbon bonds, leading to the elongation of the fullerene, and the growth of a SWNT. Otherwise (not shown), the fullerene cap can keep growing around the nanoparticle, eventually engulfing it, and preventing any further growth. The competition between these two pathways should determine the yield of SWNT growth. An analogous mechanism has been proposed for the growth of SWNT ropes from larger metal nanoparticles from carbon vapors, produced either by arc discharge or laser vaporization.

3.3 Processing and Functionalization

SWNTs obtained by both arc discharge and laser vaporization, are usually decorated with a significant fraction (10 – 30 %) of nanoscale impurities, including amorphous carbon, bucky onions, spheroidal fullerenes and residues of the metal catalyst. A convenient way of *purifying* these as-made SWNTs consists of first refluxing the material in nitric acid (HNO_3), and then suspending the nanotubes in a basic acqueous solution of a non-ionic surfactant followed by cross-flow filtration [35]. The resulting suspension is passed through a Teflon filter, and a black mat or *bucky paper* is pealed off the filter. This bucky paper consists of a dense tangle of clean SWNT ropes. These ropes usually have a length of several microns.

SWNTs are extremely insoluble in any solvent. In addition, the strong Van der Waals forces keep nanotubes aggregated in ropes and tangles. SWNT ropes can be suspended in aqueous solutions with appropriate surfactants, as mentioned above, but the ropes will usually not break apart into well-separated SWNTs. *Suspensions* of SWNTs in organic solvents were first obtained with thionyl chloride and octadecylamine [36]. Natural polymers extracted from the acacia tree have been used successfully to stabilize individual SWNTs in aqueous solutions [39].

Nanotubes are quite inert to covalent functionalization. *Covalent functionalization* of the nanotube sidewalls was only achieved by reaction with very reactive species, such as radicals, carbenes and nitrenes [37]. Otherwise, the non-covalent sidewall functionalization was obtained by adsorption of polycyclic aromatic hydrocarbon groups, which served for anchoring proteins to the nanotubes [38]. At the open end, nanotubes are somewhat more reactive. The terminal C atoms can be oxidized to carboxylic groups which can be used for typical chemical synthesis strategies.

3.4 Assembly of Nanotube Arrays and Nanocircuitry

For nanotubes to be used in nanoelectronics, it is the ability to assemble and integrate them in nanocircuitry, rather than mass production, what constitutes the most critical issue. Another important aspect is the control of diameter and chirality, or at least a selection between metallic and semiconducting nanotubes, which will play different roles in nanocircuitry.

Controlled deposition from solution

SWNT arrays lying on a surface have been produced by selective deposition on functionalized nanolithographic templates [40]. In spite of initial success, the extension of this *wet* approach proved to be rather difficult due to the tendency of SWNTs to aggregate due to van der Waals interactions. It is just hard to find anything that nanotubes would like to stick to, better than to each other. Nonetheless, if SWNT ropes are good enough for a particular use, microfluidics combined with electric fields has produced nice crossbar arrays of SWNT ropes [41] as shown in Figure 17a. With the recent discovery of new ways of suspending individual SWNTs, as mention above, the assembly of nanotube arrays by *wet methods* may now become more feasible.

Controlled growth of suspended networks

After the discovery that individual SWNTs could be grown *in situ* on silicon wafers by the CVD method, **controlled growth** became an attractive alternative to **controlled deposition**. The *in situ* approach has the advantage that it avoids nanotube aggregation. Dai et al. found that when SWNTs were grown from catalytic islands deposited on the top of microfabricated pillars, nanotubes stretched from one pillar to the next one, and so forth, forming amazing suspended networks, as shown in Figure 17b, [42]. Presumably, when a nanotube is growing from the top of a pillar, it "waves around" in every direction, but when it touches the top of another pillar, it gets pinned to it. Then, the same nanotube can keep growing and jumping from pillar to pillar for more than 100 µm. This approach can be used to build different nanotube architectures. The directionality of suspended SWNTs could also be enhanced by applying an electric field [43].

Lattice-directed growth

An interesting approach to the directional growth of supported SWNTs, was developed by Liu et al. [44], who found that when SWNTs were grown by CVD on etched Si wafers, the nanotubes preferred to grow parallel to the lattice directions of the crystalline surface. Thus, when SWNTs were grown on Si(100), the nanotubes were lying with angles of 90° and 180° between each other as shown in Figure 17c. On the other hand, when SWNTs were grown on Si(111), the nanotubes were lying with angles of 60° and 120° between each other. This directionality could be explained by the specific interactions of the SWNTs with the aligned rows of Si atoms of the wafer. This procedure may, in principle, lead to supported nanotube crossbars. However, it is not clear yet if surface-growth allows nanotubes to cross each other as they are both interacting with the surface.

Vectorial Growth

An approach for the production of supported nanotube arrays, reported by Joselevich et al. [45] is the concept of vectorial growth, where the growth of SWNTs lying on a surface is geometrically defined as a vector, having a particular position, direction and length. Ideally, one would also like to have control over the diameter and the chirality of the nanotubes, which determine whether they are metallic or semiconducting. SWNTs were produced *in situ* by CVD on Si wafers under the action of a local electric field parallel to the surface. The origin of the growth vector is defined by the position of patterned catalyst nanoparticles, while the direction of vectorial growth is defined by the electric field that is created by a pair of lithographic microelectrodes. The length is determined by the reaction time, and the diameter can be controlled by the catalyst nanoparticle size. When the nanotubes are longer than a critical size of the order of a micron, most nanotubes are well aligned with the electric field (Figure 17d). However, when the nanotubes are shorter than the critical length, only the metallic nanotubes are well aligned along the field, while the semiconducting ones are found in completely random orientations. This finding is consistent with theoretical calculations which showed

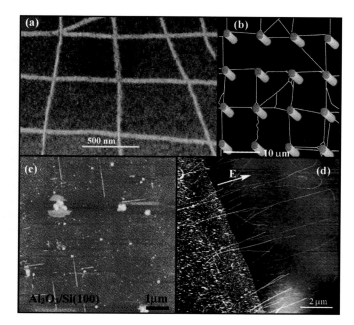

Figure 17: Methods for the assembly of carbon nanotube architectures:
(a) Controlled deposition from a polymer-stabilized organic solution in a microfluidic system with applied electric fields.
(b) Controlled growth of suspended structures from pillars.
(c) Lattice-oriented growth on Si(100).
(d) Vectorial growth from patterned catalyst nanoparticles under an electric field.

that when nanotubes are shorter than the critical length, only the metallic ones attain an induced dipole that is large enough to overcome the thermal fluctuations. Hence, sequential steps of vectorial growth could be used to produce complex and robust nanocircuitry. Moreover, this method may provide a means of selective synthesis or separation of metallic and semiconducting nanotubes, which is very important toward the application of carbon nanotubes in nanoelectronics.

4 Carbon Nanotube Interconnects

The lateral scaling of dimensions in silicon technology affects the transistors as well as their interconnects. In modern technology generations, chip wiring is produced using the damascene technique, where grooves are etched into a dielectric layer and subsequently filled with copper. Chemical mechanical polishing is used to remove the excess copper above the grooves. The result resembles inlay work or the patterned structure on a damascene sword which lends its name to the technique (see Chap. 29).

Scaling the width of the lines increases the resistance, not only because of the reduced cross section, but also due to increased scattering from the surface and the grain boundaries [46]. This problem can not be addressed by material innovation because the only metal with better bulk conductivity than copper is silver with only a 10 % improvement. If, however, wires could be made without intrinsic defects and with perfect surfaces, additional scattering might be avoided.

Carbon nanotubes may fulfill this requirements to a large extent. They offer unparalleled translational symmetry in one dimension with an intrinsically perfect surface. For metallic nanotubes the electron density is high and the conduction is easy along the tube axis due to the π-electron system. Moreover, it has been shown [50] that electron transport along the tubes is ballistic within the electron-phonon scattering length, which is of the order of micrometers at room temperature [17], [48] (Sec. 2.4). The absence of scattering also allows for much higher current densities than in metals.

Carbon nanotubes can be grown in different ways as described in Sec. 3. Most suitable for microelectronic applications is the catalyst mediated CVD growth [49]. This method can be used to grow CNTs at predefined locations if the catalyst is patterned by lithographic methods (Figure 18).

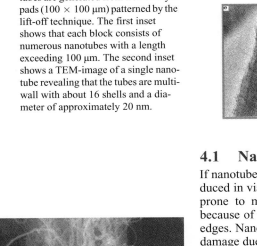

Figure 18: Nanotube blocks created by catalyst mediated growth. The tubes are generated from iron catalyst pads (100 × 100 µm) patterned by the lift-off technique. The first inset shows that each block consists of numerous nanotubes with a length exceeding 100 µm. The second inset shows a TEM-image of a single nanotube revealing that the tubes are multi-wall with about 16 shells and a diameter of approximately 20 nm.

4.1 Nanotubes in Vias

If nanotubes can be grown on a metal substrate in a similar way, they could also be produced in vias, which are the interconnects between two metallic layers. Vias are always prone to material deterioration such as void formation and subsequent breakdown because of the high current densities in small holes and current crowding effects at the edges. Nanotubes, that serve as interconnects in vias, would be much less susceptible to damage due to high current densities and permit vias to be made with smaller diameter.

Figure 19 (top) shows a SEM cross section through a via. The bottom contact is tantalum covered with the catalyst. It should be noted that the morphology of the catalyst plays a dominant role in nanotube growth and the deposition process must not change the catalyst's structure. It can be seen that the nanotube growth is selective and homogenous in the via. Figure 19 (bottom) a tungsten top contact has been formed by a focussed ion beam deposition.

The electrical characterisation of the nanotube via is shown in Figure 20. The current-voltage characteristic is ohmic indicating good contact between the nanotubes and the metal layers. Obviously the majority of the nanotubes contacted are metallic or semiconducting with high conductivity. An estimation of the number of tubes involved yields an average resistance per tube of 10 kΩ which is close to the optimum resistance of one shell and one spin degenerate conduction level [50]. Most probably more than one shell per tube contributes to the conductance, giving rise to additional conductive channels.

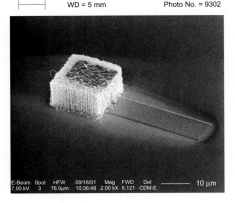

Figure 19: Homogenous selective growth of nanotubes in a narrow via. The bottom layer metal is tantalum covered by SiO$_2$ dielectric (top). The top contact is formed from a FIB-deposited tungsten layer (bottom).

4.2 Maximum Current Density and Reliability

Ballistic transport in a carbon nanotube implies that no scattering occurs within a characteristic scattering length. The power dissipation will be restricted to the contact region if the wire length does not exceed this scattering length. The nanotubes themselves should exhibit a much higher maximum current density than a polycrystalline metal. This has been confirmed in many publications where current densities up to 10^{10} A/cm^2 have been reported [55].

To date the reliability of multi-wall nanotubes has not been investigated systematically. However, first attempts, reported in ref. [56], have been made to monitor the maximum current as a function of time at elevated temperatures in air. Both tubes investigated carried current densities of 5×10^9 A/cm^2 and 2×10^{10} A/cm^2 for more than 300 h. For comparison, copper interconnects fail at current densities of ~10^7 A/cm^2. This result shows that carbon nanotubes have intrinsically much higher strength against deterioration effects like electromigration than metals.

4.3 Signal Propagation in Nanotubes

Signal propagation in conventional ohmic wires is determined by the propagation velocity of an electromagnetic wave in a dielectric and by the signal rise-time influenced by the resistance, the capacitance and the inductance of the wire. The exact treatment of signal propagation in nanotubes is a sophisticated task. However, for a first estimation of the differences between a classical ohmic wire and a nanotube we consider the nanotube as a wire with a length-independent resistance and a capacity that is modified by the electrochemical capacity to account for the intrusion of the electric field into the nanotube [57]. If we approximate the capacity for both ohmic and nanotube wires in the coaxial cylinder configuration, and neglect inductance and interaction with drive and load transistors, we can deduce the interconnect delay as function of the wire length [47]. Figure 21 shows the delay of a set of ohmic wires (black lines) with different cross sections A. The wires with cross sections smaller than $(100\,\text{nm})^2$ are corrected for additional surface scattering effects [46]. The nanotube case is represented by the two blue lines for quantum wires with two ($M = 2$) and ten ($M = 10$) occupied energy levels. The $M = 10$ case also represents a multi-wall nanotube with, e.g., five shells and two occupied energy levels in each shell. It can be easily seen that a multi-wall nanotube might show less delay than an ohmic wire of $(10\,\text{nm})^2$ cross section for wire lengths $> 300\,\text{nm}$. Thus, beneficial applications in nanoelectronics would be metallic multi-wall nanotubes with large diameters (large number of occupied levels).

5 Carbon Nanotubes Field Effect Transistors (CNTFETs)

5.1 Comparison to MOSFETs

The success of modern silicon technology is owed, to a large extent, to MOSFETs and the CMOS concept in which circuits draw current only during the switching action and are thus the indispensable ingredients for low power complex circuits like processors and controllers. The inversion channel of MOSFETs can be considered as a 2-D conduction system. Electron motion in the inversion channel is not *apriori* restricted in planes but due to high gate fields the motion perpendicular to the gate plane is quantized giving rise to subband formation. The conductive channel can be considered as a 2-D electron gas. At room temperature many subband levels contribute to the current transport enabling a high driving capability and switching speed.

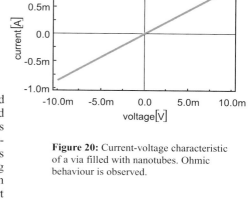

Figure 20: Current-voltage characteristic of a via filled with nanotubes. Ohmic behaviour is observed.

Semiconducting carbon nanotubes can be operated in a gate electrode configuration in a similar way to silicon MOSFETs. The electronic properties of CNTs have been described in Sec. 2. The excess electrons are delocalized and form a highly conductive π-electron system. Unlike in silicon MOSFETs the electron system of a nanotube is 1-D. Placing a field electrode next to the nanotube one can influence its conductivity by the accumulation or depletion of electrons provided that the electron density is not to high (the tube is semiconducting). This configuration is called CNTFET, in analogy to the silicon field effect transistor. As the electrical characteristics of carbon nanotubes strongly depend on their chirality, diameter and doping, the characteristics of CNTFETs can be controlled by choosing the appropriate morphology of the CNT. Single-wall semiconducting tubes are best suited for CNTFETs because their electron system is not bypassed by inner shells.

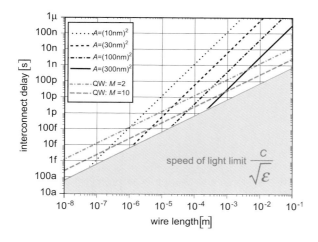

Figure 21: Signal delay in ohmic and ballistic wires as a function of wire length. The black lines represent the delays of ohmic wires with different cross sections A. Surface scattering has been taken into account. For wire lengths smaller than the intersections of black and red lines, the delay is governed by the speed of light and the relative permitivity of the dielectric. The nanotube wires display a completely different behavior to the ohmic wires, as shown by the blue lines. Due to the length-independent resistance, nanotubes exhibit better delay values for longer wires. M represents the number of conductive channels, that can be formed either by the number of occupied energy levels or by the number of shells of a multi-wall nanotube.

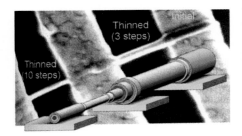

Figure 22: Controlled burn-off of the individual shells of a multi-wall nanotube between three adjacent contacts. In the first section three shells have been removed whereas ten shells have been stripped in the second section. The SEM picture and inset with schematic drawing are taken from [62].

5.2 Tailoring of Nanotubes

The production of single-wall nanotubes and their deposition is still a matter of arduous work and is by no means compatible with the requirements of a controlled and reproducible parallel production which is characteristic of integrated circuit (IC) technology processes. For IC technology an appropriate number of nanotubes with predefined characteristics must be placed at desired locations in a reproducible way. One step towards this goal is to use custom designed multi-wall nanotubes instead of the single-wall species. Since the diameter as well as the chirality of the shells determines the energy gap and the conduction type of each individual shell, it should be possible to choose the desired characteristics by contacting the appropriate shell. This was realized by a group at IBM who managed to successively burn-off the outer shells of a MWNT located on contacts [62]. The outer shell turned out to be the dominant conductive channel. Figure 22 shows a SEM picture of a multi-wall carbon nanotube placed on top of three contact strips. By applying a controlled current to the tube between two adjacent contacts it is possible to burn-off individual shells within this section by oxidizing them in air, as depicted in the sketch. After manipulating the carbon nanotube in the above described manner it was possible to investigate the electrical conductivity with a back-gate electrode.

Figure 23 shows the conductivity of the tube as a function of the back-gate voltage for 13 different shells that have been successively removed. The three outermost shells were burned-off between the right electrode pair whereas further thinning was done between the left pair of electrodes. The conductivity variation with the gate voltage is quite different for each shell, reflecting the different morphology of each layer. The outermost shell responds to the gate voltage, indicating its semiconducting nature whereas the next shell is metallic with no response. Note that the conductivity decreases as the tube diameter diminishes and that the innermost shells clearly show semiconducting behavior (The conductivity in the bottom diagram is displayed on a logarithmic scale). It can even be deduced that the energy gap widens as the tube diameter decreases, as expected from theory (see Sec. 2).

Figure 24 displays the current-voltage characteristics for high source-drain voltages. It can be seen that for each shell the current saturates and that each saturation value contributes equally in the high current case. For the innermost shells, and at smaller source-drain biases, an exponential behavior can be deduced which originates from tunneling contacts between outer and inner shells in the contact region.

This experiment provides a deep insight into the transport mechanisms in multi-wall nanotubes, at least a qualitative one. Moreover, to solve the problem of arranging the predefined types of nanotubes at the desired locations, it might also be feasible to deposit a multi-wall species and design the required characteristics by the controlled burn-off of individual shells. Thus, certain sections of the same tube can be defined to act as metallic interconnects, whereas others can perform as semiconducting FET devices. It can easily be imagined that circuits with higher complexity might be assembled exploiting controlled breakdown, at least for demonstrators.

It was also shown in [62] that a bundle of single-wall tubes with arbitrarily mixed conduction types can be separated from the metallic species by applying a back gate voltage to drive the semiconducting ones into depletion, while burning-off the metallic ones by an appropriate source-drain voltage.

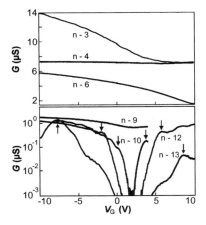

Figure 23: Conductivity of individual shells of a multi-wall nanotube (taken from [62]). The shells are removed by controlled current induced burn-off. The outermost shells show high conductivity with shell n-4 being metallic (top). As the diameter of the shells decreases, the conductivity at constant bias voltage V_G decreases likewise. The innermost shells show a pronounced gate voltage dependence and a widening of the energy gap, in accordance with theoretical expectations (bottom).

5.3 Back-gate CNTFETs

Having at hand the deposition techniques described in the previous sections for single-wall nanotubes and multi-wall nanotubes as well as the possibility to manipulate individual shells of multi-wall tubes we can now begin to construct carbon nanotube FETs. The simplest arrangement is to place a nanotube on top of a silicon wafer covered with a silicon dioxide dielectric layer. After contacting both ends of the nanotube with an appropriate electrode we can apply a gate voltage at the silicon bulk acting as an overall gate electrode. Tans et al. [58] were the first to investigate a carbon nanotube device which was modulated by a planar gate electrode and operated at room temperature. Slightly later two other groups presented similar results [28],[59]. In all three cases, the CNTs turned out to be p-type conductors.

The arrangement used by Tans et al. [58], which is shown in Figure 25, consists of a single-wall nanotube located on top of two platinum strips which serve as source and drain electrodes. The tube and the contacts are separated by a 300 nm thick SiO_2 layer grown thermally on top of a silicon wafer that acts as a back-gate electrode. The single-

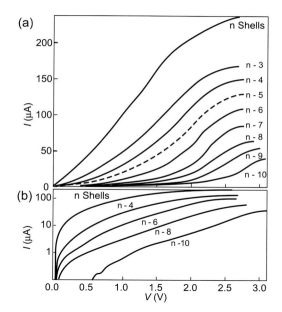

Figure 24: Conduction of individual shells of a multi-wall nanotube for the high bias case. The current is driven into saturation and the saturation value of each shell is ~20 mA. In the semi-log plot (bottom) exponential current behavior is observed for the innermost shells, most probably resulting from tunneling contacts between the shells in the contact region.

wall nanotubes were deposited from solution and lie on the contacts by chance. The authors managed to measure about 20 nanotubes, some of them showing metallic behavior with no dependence on the gate electrode voltage and linear current-voltage characteristics. Those showing gate electrode responses were identified as semiconducting and were used for further characterization.

Figure 26 shows the current-voltage characteristics of a single tube for a range of back gate voltages. It can be seen that for positive gate voltages no current flows for small source-drain voltages. By increasing the source-drain voltage a nonlinear current-voltage dependence can be observed for both polarities indicating an energy gap-like behavior with no energy states available for positive gate voltages and small biases. Decreasing the gate voltage to negative values opens the conductive channel even for small source-drain voltages. The insert in Figure 26 reveals that the conductance can be modulated over 6 orders of magnitude for small source-drain voltages, a behavior which is comparable to a silicon MOSFET. The saturation value of the resistance is similar to the resistance of the metallic nanotubes measured in the same experiment (~1 MΩ). The major part of the resistance is attributed to the contact resistance as one would expect a value close to the quantum resistance of 6.5 kΩ for a spin degenerate system with two occupied energy levels [50].

The presentation of the CNTFET by Dekker's group in 1998 set a landmark in the development of nanoelectronic devices since, for the first time, the function of a device with nanometer dimensions was demonstrated at room-temperature and which exhibited similar or even better characteristics than those expected for silicon-transistors with equivalent dimensions.

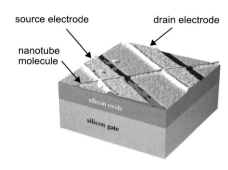

Figure 25: A carbon nanotube field effect transistor (CNTFET). The nanotube (red) is located on top of two platinum contacts (yellow). The back-gate-stack (blue) is formed by a silicon dioxide dielectric on top of a silicon wafer (colored AFM-image taken from [58]).

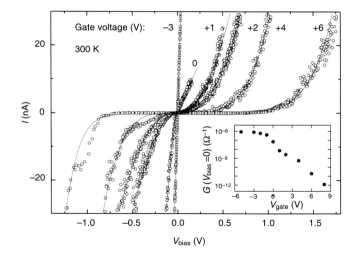

Figure 26: The current-voltage characteristics of a semiconducting single-wall carbon nanotube for different gate voltages (see Figure 25). For large positive gate voltages the conductance of the tube is very small for source-drain biases less than approximately 1 V. Changing the gate voltage to negative values increases the conductivity steadily until saturation is reached at approximately −3 V (see insert). The maximum conductivity is comparable to the values found for metallic tubes measured in the same experiment (taken from [58]).

III Logic Devices

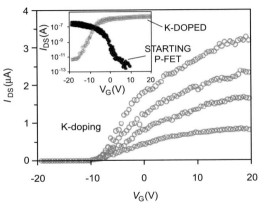

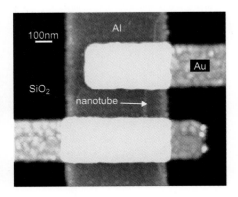

Figure 27: Inversion of the conductivity type of a semiconducting carbon nanotube. As displayed by the insert the starting device was a p-type CNT-FET. After exposure to potassium vapor, the conductivity type was reversed with current flowing for positive gate voltages (after [64]).

5.4 Complementary Carbon Nanotube Devices

As mentioned, the CNTFETs presented up to now were p-type only! As in silicon technology, however, it is highly desirable to have both p- and n-type CNTFETs available. Only complementary arrangements, for which one transistor always is in the off-state, show sufficiently low power consumption for large scale integration circuitry.

For carbon nanotubes the fabrication of p-n junctions within one nanotube has been achieved by covering one part with a resist and exposing the uncovered part to potassium vapor [51]. The first appearance of n-type behavior of nanotube ropes under positive back-gate bias was reported by Bockrath et al [52]. They also used potassium vapor in a doping vessel to diffuse donor atoms into the former p-type nanotubes. Following the concept of doping, Collins et al. [53] and Bradley et al. [54] interpreted the earlier results on p-type devices as unintentional oxygen doping of carbon nanotubes. However, it was not perfectly clear at that point, why only either electron *or* hole transport was observable and no ambipolar characteristics could be obtained despite the small band gap in case of nanotubes of only 0.6 eV. Martel et al were the first to succeed in producing CNTFETs with ambiploar conduction character depending on the polarity of the gate voltage. Their devices had Ti/TiC contacts providing sufficiently low barrier heights for both hole and electron transport. In addition, a thin SiO_2 passivation layer was used for stable operation in air [21].

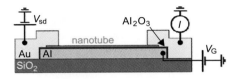

Figure 28: Isolated back-gate CNTFET. The gate is realized by an Al-strip on top of SiO_2. The gate oxide is the naturally grown thin Al_2O_3 layer on top of the Al-strip. Source and drain contacts are made by Au deposition ([65]).

Derycke et al. [64] used potassium doping to reverse the conduction type of a CNT-FET. Figure 27 (insert) shows the conversion of an originally p-type nanotube FET to n-type by K-doping and the resulting I_{DS} versus V_G characteristics. Potassium acts as an electron donor shifting the Fermi-level to the conduction band. As a result, the concentration of electrons in the conduction band increases for positive gate voltages. The IBM group was also able to prove that pure n-type nanotube behavior can be obtained by annealing devices in vacuum. The results were discussed as a consequence of the presence of Schottky barriers and their sensitivity to ambient gases such as oxygen. Nowadays understanding is that a combination of bulk doping and contact effects is likely to be responsible for the transport properties observed in nanotube devices.

5.5 Isolated Back-Gate Devices

The CNT-devices discussed so far were controlled by a common back-gate stack which consists of a thermal oxide layer on top of a doped silicon wafer. Thus, the gate electrode extends over the whole silicon wafer and covers also the large contact pad areas. Due to the risk of defects resulting in high leakage currents or circuit shorts the gate oxide thickness must be kept in the 100 nm range. An effective gate control can therefore only be achieved with high gate voltages and the gain of most devices is smaller than unity. However, the Delft University group succeeded in constructing an isolated back-gate device by exploiting thin, naturally grown Al_2O_3 on top of a patterned aluminum electrode as gate oxide with a thickness of a few nanometers [65]. The basic device is displayed in Figure 28. Source and drain electrodes are patterned by electron-beam lithography and subsequent metal deposition and lift-off on top of a 1 nm diameter single-wall nanotube pre-selected and positioned by AFM.

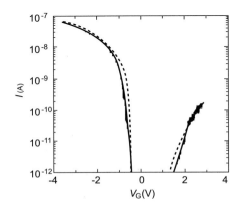

Figure 29: Source-drain current of a single-wall nanotube on Al_2O_3 gate dielectric for a source-drain voltage of 5 mV. p- and n-channels can be opened by altering the gate voltage (after [65]).

In Figure 29 the drain current for a fixed source-drain voltage is shown as a function of the gate voltage. Obviously the Fermi-level is shifted by the gate voltage from the valence band (accumulation) over the gap (depletion) into the conduction band (inversion). The conductance type can be varied electrostatically by the gate voltage from p-type to n-type. The minimum resistance can be deduced at high negative gate voltages to be 80 kΩ, indicating good contacts.

In Figure 30 the I/V_{sd} characteristics of a CNTFET are shown for different gate voltages. Saturation behavior similar to a silicon-MOSFET can be clearly seen. The gain of this device exceeds 10, making it suitable for driving other devices. The transconductance is approx. 0.3 µS and the on/off ratio covers more than 5 orders of magnitude. The maximum current is of the order of 100 nA, equivalent to a current density of approx. 10^7 A/cm² with an on-resistance of 26 MΩ.

5.6 Isolated Top Gate Devices

A further step towards compatibility with microelectronics was achieved by Wind et al, who presented an optimized CNTFET with Ti/TiC source-drain contacts and a thin (15 – 20 nm) gate-oxide which was deposited on top of the nanotube [66]. Due to the small thickness of the gate oxide and the top gate arrangement excellent gate control is achieved. Figure 31 shows the schematic cross section of the device and the drain current versus source-drain voltage characteristics. Note that the device can be operated with a gate voltage swing of only 1 V. N-type devices can be fabricated by annealing a p-type tube in inert atmosphere prior to gate oxide deposition. A thin passivation layer capping the device enabled stable operation in air. A record value of 3.25 µS was measured for the transconductance.

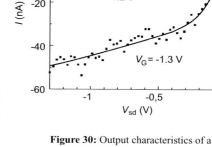

Figure 30: Output characteristics of a p-CNTFET. The drain current shows a saturation-like behavior at large source-drain voltages. The gain is above 10 (after [65]).

5.7 Comparison of Si-MOSFETs with up-scaled CNT-MOSFETs

Since their first introduction in 1998 the evolution of CNTFETs has yielded a steady improvement of the electrical characteristics, as has been described in the previous sections. A comparison with state of the art Si-MOSFETs has first been attempted by Martel et al [68]. Their data, as deduced from a back-gate CNTFET with Ti/TiC source-drain contacts, already shows competitiveness to state-of-the-art Si-MOSFETs. The top-gate transistor presented by Wind et al. [66] with carefully designed contacts and gate-oxide, shows unprecedented values for transconductance and maximum current drive (Table 1).

	p-type CNFET	Ref. 59	Ref. 60
Gate length (nm)	260	15	50
Gate oxide thickness (nm)	15	1.4	1.5
V_t (V)	–0.5	~ –0.1	~ –0.2
I_{ON} (µA/µm) ($V_{ds} = V_{gs} - V_t \approx$ -1 V)	2100	265	650
I_{OFF} (nA/µm)	150	< 500	9
Subthreshold slope (mV/dec)	130	~ 100	70
Transconductance (µS/µm)	2321	975	650

Table 1: Comparison of most important transistor data for a p-type CNTFET and two advanced Si-MOSFETs (reproduced from [66]).

The data for the CNTFET were deduced from a single nanotube and were virtually upgraded by parallel operation of an appropriate number of tubes to form a two-dimensional arrangement with dimensions comparable to two most advanced MOSFET devices. It has to be kept in mind that a grid of parallel nanotubes has not been realized up to now and that the gate field would be weakened by screening of densely packed tubes. There is also a difference in channel length between both devices. Assuming ballistic transport CNTFETs should be scalable in length. However, it is still unclear whether the location of the gate relative to the contacts has any influence on the transconductance.

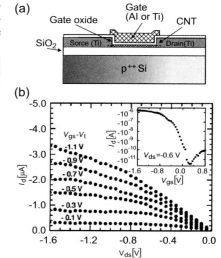

Figure 31:
(a) Schematic cross section of a top gate CNTFET.
(b) Output characteristics of a p-type device with Ti gate and a gate oxide thickness of 15 nm. Gate voltage ranges from –0.1 V to –1.1 V above the threshold voltage of –0.5 V. The insert shows the gate controlled drain current at a fixed source-drain voltage of –0.6 V [66].

III Logic Devices

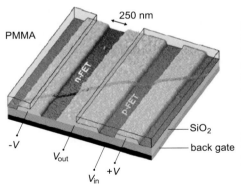

Figure 32: Intramolecular complementary CNTFET gate made by an n-type and p-type CNTFET in series, operated by a common back-gate. Complementary transistors are produced from one single nanotube by doping the section which is not covered by the PMMA resist with potassium (after [64]).

It can be seen from Table 1 that, even for transconductance to be half of the values deduced, the CNTFET still outperforms the MOSFETs. Further improvement can also be expected by reduction of the gate-oxide thickness and lowering of the contact resistances at source and drain.

Let us now consider the theoretical limits of a CNTFET. Performing a best case approximation and taking into account that the conductance is independent of the length (ballistic transport), the value for a spin-degenerate one level system is 153 µS per tube (!). Assuming equal lines and spaces and a tube diameter of 1.4 nm the maximum conductance would be more than 50000 µS/µm. This is unrealistically high because contact resistances and gate coupling were not considered. However, improvement of the contact resistance alone will further enhance the nanotube transconductance considerably.

Guo et al. have performed a thorough theoretical treatment of a CNTFET [67]. They showed that a coaxial gate would enhance the transconductance by a factor of 7 as compared to the plane gate electrode used for all devices up to now. They have also found that the maximum transconductance of their n-type device is 63 µS a value considerably higher than the best value reported so far for a p-type device (3.25 µS [66]).

This comparison shows, that there is lot of room for improvement of current nanotube device technology, which is by no means mature.

5.8 Carbon Nanotube Circuits

With the availability of devices with sufficient gain, on-off ratio and driving current separately addressable by a gate voltage, one can realize simple circuits – the first step for integration of CNTFETs. Derycke et al. succeeded in selectively doping part of one single carbon nanotube placed on top of three contacts in a back-gate arrangement by covering a section of the tube with PMMA and exposing the other to potassium vapor or annealing it in vacuum [64]. Figure 32 shows this *intramolecular* gate. The entire nanotube represents a n-type and a p-type FET in series, controlled by a common gate. By applying a gate voltage, either one of the two transistors is in the off-state with no current flowing in both logic high and low states. In Figure 33 the proper operation of the *CMOS-type* gate shown in Figure 32 is demonstrated. It can be seen that the output voltage changes within a narrow input voltage interval centered at 0 V. The straight line marks a voltage gain of 1. The blue line is an average of five consecutive measurements indicating that this device shows a gain higher than unity. This is a prerequisite for driving other gates.

Using the CNTFETs presented in the previous section and combining them with external resistors the Delft University group [65] realized an inverter, as shown in Figure 34 (left). When the gate voltage is switched from 0 V (logical 0) to –1.5 V the output voltage changes from –1.5 V to 0 V. In a similar way, by connecting two CNTFETs in parallel to a common resistor, they realized a NOR gate with the typical input-output scheme as displayed in Figure 34 (right). Due to the high resistances and capacitances involved, switching is slow compared to the typical values of today's microelectronics. However, it should be kept in mind that the switching element itself has lateral dimensions of only 1 nm and the on-resistance has to be compared with the lowest resistance that a quantum system can reach (Landauer resistance $R = h/2e^2 \approx 13$ kΩ for spin degeneracy and one occupied level) [50]. As can be deduced from Figure 34 (left), the voltage gain of this device is approximately 3.

By combining an odd number of inverter elements and feeding the output back to the input it was possible to realize a ring oscillator, as shown in Figure 35. As indicated by the output oscillations the ring oscillator operates at a frequency of 5 Hz. The low frequency is obviously due to the parasitic capacitances and the high series resistances of the nanotubes which can at least partly be reduced by a complete integration of all components on the substrate.

The realization of simple logic circuits represents a giant step towards the integration of carbon nanotubes. This proof of their operation further enhances the thrust driving CNT technology forward. However, it should be noted, that the circuits realized up to now were build out of p-type transistors and the only complementary gate was realized by Derycke et al. [64].

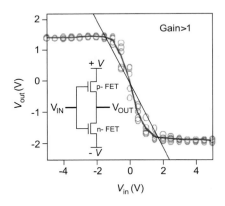

Figure 33: Input-output characteristic of a complementary CNTFET gate. The output voltage switches from logical 1 to 0 as the input voltage changes from negative to positive values. Red circles represent the data of five measurements on the same device. The blue line is the average of this measurements indicating a voltage gain > 1 (straight line).

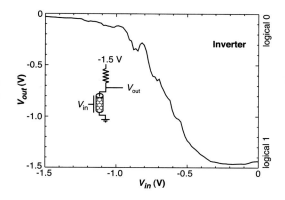
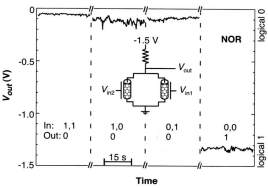

Figure 34: Left: The input-output characteristics of an inverter stage realized by a CNTFET and an external resistor.
Right: Two CNTFETs in parallel working on a common resistor. Input/output characteristics showing NOR Function (taken from [65]).

6 Nanotubes for Memory Applications

Memory devices for information storage play a dominant role in microelectronics applications. The most important characteristics of a state-of-the-art memory concept are high storage density, fast and random data access, low power consumption, low price per bit, easy integration into the mainstream IC technology, and preferably non-volatility of data after power-off. For the Si technology, this has been achieved through devices such as DRAM, SRAM, EPROM etc. which are reviewed in Introduction to Part IV. In the following sections, we will address some of the first attempts to realize memory devices with carbon nanotubes.

6.1 CNT-SRAMs

It has been shown in Section 5.8 that basic circuits can be built with nanotubes. Once having realized a NOR gate, it is possible, by adding another resistor and cross-coupling the outputs to the inputs, to build up a simple SRAM unit-cell, as shown by the insert in Figure 36. Static memory function was demonstrated by writing a logical 1 or a logical 0 to the input, disconnect the input voltage and tracing whether the output retained its logical state.

6.2 Other Memory Concepts

In principle, every information storage concept that exploits switching devices can also be realized with elements based on CNTFETs, as described in the previous section. However, it is challenging to look for new memory concepts which incorporate the specific characteristics of this macromolecular device in a more beneficial way. Such an attempt has been made by the Harvard University group of Lieber [69]. Their basic arrangement is a crossbar array of carbon nanotubes with one set of nanotube wires separated from the other by a small distance provided by non-conducting supporting blocks (Figure 37).

It can easily be seen from Figure 38 that, for an optimum initial wire separation (~ 2 nm), the upper wires have two stable positions, one in their minimum elastic energy positions without contact to the lower cross-point wires and the other one with the wires held in contact with the lower wires due to the van der Waals force (left dip). If the contact state is reached, the electrical resistance of the cross-point changes by orders of magnitude. This can be detected by a sensing matrix and be exploited for bit storage. Once in the contact state, the wires can only be driven apart by charging them transiently with the same potential. Thus, we have a nonvolatile memory device that retains its memory information after power turn-off. The minimum cell size depends on the elastic deformation potential of the nanotubes including their suspension. The authors estimate the minimum cell size to be a square of 5 nm length, enabling a packing density of 10^{12} elements/cm^2. The inherent switching time was estimated to be in the 100 GHz range. The bending of the nanotube does not exceed the elastic limit and reversible switching operation of an experimental crossbar device was sustained over several days.

Another, and much more speculative, nano-mechanical memory principle is proposed in [70] where a charged bucky ball incorporated into a short carbon nanotube is driven into two stable sites at both ends of a very short tube by an electrical field applied

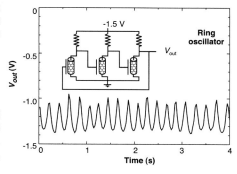

Figure 35: Combining an odd number of CNTFET NOR gates and feeding the output back to the input yields a ring oscillator.
Oscillation of the output voltage is demonstrated. However, due to the high capacitances and resistors involved the frequency is very low (taken from [65]).

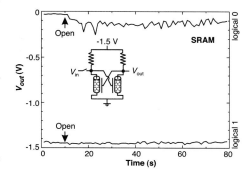

Figure 36: A simple SRAM cell made by two cross-coupled CNTFETs with external resistors. Writing either 1 or 0 to the input and disconnecting the input shows that the output remains in the stored state (taken from [65]).

III Logic Devices

Figure 37: Crossbar arrangement of rows and columns of nanotubes separated by supporting blocks. Application of an appropriate voltage between the desired column and row bends the top tube into contact with the bottom nanotube. The resistance drop is used for bit storage. Van der Waals forces maintain the contact even if the voltage is removed (after [69]). Separation of the CNTs is achieved by application of a voltage pulse of equal polarity at the crossing point.

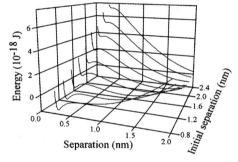

Figure 38: Total energy of the crossbar CNT memory cell system as a function of the initial tube separation (after [69]). For a distance of approx. 2 nm two minima can be realized representing the two stable configurations contact (left minimum) and non-contact (right minimum).

along the tube (Figure 39). A nonvolatile memory element can be constructed with this *bucky shuttle* device as memory node between an array of crossed lines. So far there is no report of a functional device based on this idea.

Following the principle of EEPROM memory, the authors of [71] propose to build a nanotube memory by combining semiconducting and metallic nanotube species in a crossbar configuration with a dielectric layer capable of storing charge in between. Each crossing point represents a transistor with gates formed from sections of the metallic tubes on top and the channel represented by the underlying section of the semiconducting tubes (Figure 40). By applying a high enough voltage at a cross-point, charges are induced into the $SiO_2/Si_3N_4/SiO_2$ (ONO) dielectric and are trapped in the nitride, modifying the threshold voltage of the transistor. The threshold voltage shift is used to represent the two logic states. As all transistors are connected in series, NAND operation is favorable. First realization of this principle using an ordinary gate and a one layer SiO_2 dielectric was achieved by Fuhrer et al [72]. They succeeded in reversibly injecting electrons from a nanotube into the gate dielectric. This extra charge shifted the threshold voltage of a CNTFET, which could be exploited for memory operation.

Information storage in nanotube devices undoubtedly seems to be attractive because the storage node can be scaled down to molecular dimensions and the inherent switching times are estimated to be extremely fast. However, the storage elements have to be connected to standard microelectronic circuits to detect the stored information and the timing cycles also have to comply with this external logic. It should also be noted that, though placement of the nanotubes in regular crossbar arrangements might be easier than in random networks, a reproducible method of deposition compatible with microelectronics technology requirements still has to be demonstrated.

7 Prospects of an All-CNT Nanoelectronics

In the preceding sections we have reviewed some of the most outstanding achievements that have been made in the last couple of years in the use of carbon nanotubes for devices which are the essentials of today's microelectronics: interconnects, transistors and even simple logic circuits and memories. We have also shown that these devices can compete with silicon devices or perform even better if we compare them at the appropriate length scale. It is therefore possible to conceive of a scenario in which all necessary elements of a functional microelectronics technology are made out of carbon nanotubes. For the moment we do not care that the most successful processing principle in silicon technology, i.e., the parallel processing of a great number of devices at the same time, has not yet been demonstrated for carbon nanotubes.

By assuming that it is possible to deposit the nanotubes with the desired characteristics on top of an arbitrary flat insulating substrate, one can imagine the integration scheme which is shown in Figure 41. Transistors as well as interconnects are made from the same multi-wall nanotubes with some sections being semiconducting and others being metallic. The semiconducting sections may be formed by controlled shell removal, as described in Section 5.2 until the desired characteristics are realized. Doping tube sections through a mask [63] would deliver the prerequisites for complementary devices. Even gate electrodes may be formed by the tips of nanotubes or by short sections of tubes lying parallel and close to the semiconducting ones.

Gate oxides and passivation layers may be deposited by conventional SiO_2/Si_3N_4-CVD, because nanotubes can withstand temperatures of more than 1000°C in inert atmosphere and ~400°C in oxygen. Chemical mechanical polishing would flat-

Figure 39: A bucky ball incorporated in a short closed carbon nanotube (right). The Van der Waals forces favor two stable positions at both ends of the tube (left) which can be exploited for bit storage. The bucky ball is positively charged by the incorporation of a potassium ion (after [70]).

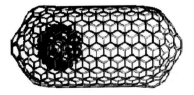
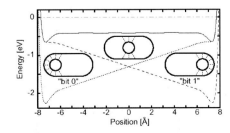

ten the surface allowing the stacking of more layers. Interconnects between the layers may also be constructed from nanotubes by exploiting the growth methods in vias described in Section 4.1. Thus, it would be possible to stack the desired number of layers by application of the processes already developed for semiconductor interconnect technology without the need for additional expensive processes for bonding prefabricated silicon chips on top of each other [73].

The ultimate step forward would be that, instead of exploiting the defect-free single crystal silicon surface for transistors, the nanotubes themselves provide the translational symmetry and quasi-crystalline perfection at the places we like to have it, and not just restricted to the silicon surface. Even wires would benefit from the crystalline structure, since surface and grain-boundary scattering is absent and electron-phonon interaction is low due to one-dimensional effects. The current carrying capacity is, therefore, orders of magnitudes higher than in conventional metallic wires. The resistance of the wires would be independent of the length, provided that the large electron-phonon scattering length is not surpassed. The thermal conductivity of nanotubes is also higher so the nanotube network itself can act as an efficient heat transportation system.

However, there is still a long way to go before this scenario can be realized, because a controlled parallel deposition of carbon nanotubes on a flat surface has not yet been demonstrated. On the other hand, it is well known that nanotubes can be considered as macromolecular structures and the deposition of certain predefined tubes might also be achieved using assembly methods derived from chemistry. Nanotubes can be functionalized with certain reactive groups [74]. If these groups are attached, e.g., to the ends of a semiconducting tube, it might be feasible to react them with appropriate molecules attached to specific sites on the substrate or to other nanotubes. After reaction the tubes may be coupled to other tubes or to predefined sites. Definition of the reaction site may be performed with lithographic methods. It can easily be imagined that, with this method, devices like CNTFETs or even circuits with interconnects might be constructed.

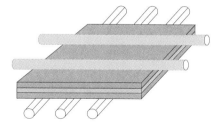

Figure 40: Crossbar arrangement of semiconducting (bottom) and metallic (top) nanotubes separated by an ONO dielectric. Each crossing point represents a CNTFET. Application of a sufficient voltage at a node drives carriers into the ONO dielectric where they are trapped in the nitride causing a threshold voltage shift in the transistor. This is used as bit storage principle.

Acknowledgements

The authors would like to thank Sam Schmitz (Caesar, Bonn) for checking the symbols and formulas in this chapter. One of the authors (Wolfgang Hoenlein) would like to thank his group at Infineon for their contributions and valuable discussions.

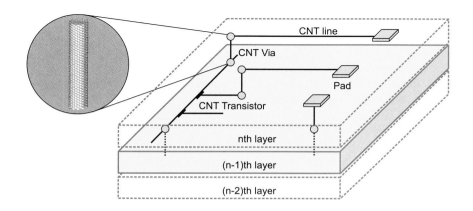

Figure 41: Concept for an all-CNT-based microelectronics technology. All active elements and interconnects are made out of nanotubes. Due to the quasi-crystalline nature of the nanotubes high quality devices are possible without the need for a single crystal silicon substrate. By the application of well known back-end-of-line techniques cost-effective stacking of a real 3-dimensional structure is feasible.

References

[1] S. Iijima, Nature **354**, 56 (1991).
[2] F. Harris, *Carbon Nanotubes and Related Structures*, Cambridge University Press, 1999.
[3] M. S. Dresselhaus, G. Dresselhaus, and Ph. Avouris, *Carbon Nanotubes*, Springer-Verlag, Berlin, Heidelberg, New York, 2001.
[4] A. Thess et al., Science **273**, 483 (1996).
[5] Picture is taken from the Image Gallery of the Center of Nanoscale Science and Technology at Rice University (http://cnst.rice.edu/pics.html).
[6] T. W. Odom, J. L. Huang, P. Kim, C. M. Lieber, Nature **391**, 62 (1998).
[7] P. R. Wallace, Phys. Rev. **71**, 622 (1947).
[8] R. Saito and H. Kataura, *Carbon Nanotubes*, M. S. Dresselhaus, G. Dresselhaus, and Ph. Avouris, eds., Springer-Verlag, Berlin, Heidelberg, New York, 2001.
[9] R. Saito, M. Fujita, G. Dresselhaus, and M.S. Dresselhaus, Appl. Phys. Lett. **60**, 2204 (1992).
[10] A.A. Maarouf, C.L. Kane, and E.J. Mele, Phys. Rev. B **61**, 11156 (2000).
[11] H. Stahl, J. Appenzeller, R. Martel, Ph. Avouris, and B. Lengeler, Phys. Rev. Lett. **85**, 5186 (2000).
[12] L. Forró and C. Schönberger, in: *Carbon Nanotubes*, eds. M. S. Dresselhaus, G. Dresselhaus, and Ph. Avouris, Springer-Verlag, Berlin, Heidelberg, New York, 2001.
[13] R. A. Jishi, M. S. Dresselhaus, and G. Dresselhaus, Phys. Rev. B **48**, 11385 (1993).
[14] Z. Yao, C. L. Kane, and C. Dekker, Phys. Rev. Lett. **84**, 2941 (2000).
[15] Z. Yao, C. Dekker, and Ph. Avouris, in: *Carbon Nanotubes,* eds. M. S. Dresselhaus, G. Dresselhaus, and Ph. Avouris, Springer-Verlag, Berlin, Heidelberg, New York, 2001.
[16] L. Balents and P. A. Fischer, Phys. Rev. B **55**, R11973 (1997).
[17] J. Appenzeller, R. Martel, P. Avouris, H. Stahl, and B. Lengeler, Appl. Phys. Lett. **78**, 3313 (2001).
[18] Ch. Zhou, J. Kong, and H. Dai, Appl. Phys. Lett. **76**, 1597 (2000).
[19] S. J. Tans and C. Dekker, Nature **404**, 834 (2000).
[20] V. Derycke, R. Martel, J. Appenzeller, and Ph. Avouris, Appl. Phys. Lett. **80**, 2773 (2002).
[21] R. Martel, V. Derycke, C. Lavoie, J. Appenzeller, K. Chan, J. Tersoff, and Ph. Avouris, Phys. Rev. Lett **87**, 256805, (2001).
[22] J. Appenzeller, J. Knoch, V. Derycke, R. Martel, S. Wind, and Ph. Avouris, Phys. Rev. Lett. **89**, 126801 (2002).
[23] M.S. Dresselhaus, G. Dresselhaus and P.C. Eklund, *Science of Fullerenes and Carbon Nanotubes,* Academic Press, New York, 1996.
[24] S. Iijima, T. Ichihashi, Nature **363**, 603 (1993).
[25] D. S. Bethune, C.H. Kiang, M. S. de Vries, G. Gorman, R. Savoy, J. Vazquez & R. Beyers, Nature **363**, 605 (1993).
[26] C. Journet, W. K. Maser, P. Bernier, A. Loiseau, M. Lamy de la Chapelle, S. Lefrant, P. Deniard, R. Leed & J. E. Fischer, Nature **388**, 756 (1997).
[27] A. Thess, R. Lee, P. Nikolaev, H. Dai, P. Petit, J. Robert, C. Xu, Y. H. Lee, S G. Kim, A. G. Rinzler, D. T. Colbert, G. E. Scuseria, D. Tomanek, J. E. Fischer, R. E. Smalley, Science **273**, 483 (1996).
[28] J. Kong, H. T. Soh, A. M. Cassell, C. F. Quate, H. J. Dai, Nature **395**, 878 (1998).
[29] H. J. Dai, Topics Appl. Phys. **80**, 29 (2001).
[30] R. R. Schlittler, J. W. Seo, J. K. Gimzewski, C. Durkan, M. S. M. Saifullah & M. E. Welland, Science **292**, 1136 (2001).
[31] Matthew F. Chisholm et al., Science **300**, 1236b (2003).
[32] J.-C. Charlier & S. Iijima, Topics Appl. Phys. **80**, 55 (2001).
[33] C. L. Cheung, A. Kurtz, H. Park & C. M. Lieber, J. Phys. Chem. B **106**, 2429 (2002).
[34] J. Gavillet, A. Loiseau, C. Journet, F. Willaime, F. Ducastelle & J.-C. Charlier, Phys. Rev. Lett. **87**, 275504 (2001).
[35] J. Liu, A. G. Rinzler, H. J. Dai, J. H. Hafner, R. K. Bradley, P. J. Boul, A. Lu, T. Iverson, K. Shelimov, C. B. Huffman, F. Rodriguez-Macias, Y.S. Shon, T. R. Lee, D. T. Colbart & R. E. Smalley, Science **280**, 1253 (1998).

[36] J. Chen, M. A. Hamon, H. Hu, Y.S. Chen, A. M. Rao, P. C. Eklund & R. C. Haddon, Science **282**, 95 (1998).

[37] M. Holzinger, O. Vostrowsky, A. Hirsch, F. Hennrich, M. Kappes, R. Weiss & F. Jellen, Angew. Chem. Int. Ed. **40**, 4002 (2001).

[38] M. Shim, NWS. Kam, R. J. Chen, Y. M. Li, H. J. Dai, Nano Lett. **2**, 285 (2002).

[39] R. Bandyopadhyaya, E. Nativ-Roth, O. Regev, R. Yerushalmi-Rozen, Nano Lett. **2**, 25 (2002).

[40] J. Liu, J. Casavant, M. Cox, D. A. Walters, P. Boul, W. Lu, A. J. Rimberg, K. A. Smith, D. T. Colbert, R. E. Smalley, Chem. Phys. Lett. **303**, 125 (1999).

[41] M. R. Diehl, S. N. Yaliraki, R. A. Beckman, M. Barahona, J. R. Heath, Angew. Chem. Int. Ed. **41**, 353 (2002).

[42] N. R. Franklin, H. Dai, Adv. Mater. **12**, 890 (2000).

[43] Y. Zhang, A. Chang, J. Cao, Q. Wang, W. Kim, Y. Li, N. Morris, E. Yenilmez, J. Kong, H. J. Dai, Appl. Phys. Lett. **79**, 3155 (2001).

[44] M. Su, Y. Li, B. Maynor, A. Buldum, J. P. Lu, J. Liu, J. Phys. Chem. B **104**, 6505 (2000).

[45] E. Joselevich and C. M. Lieber, NanoLetters **2**, 1137 (2002).

[46] W. Steinhögl, M.Engelhardt, G. Schindler and G. Steinlesberger, Phys. Rev. B **66**, 075414 (2002).

[47] W. Hönlein, Jpn. J. Appl. Phys. **41**, 4370 (2002).

[48] C.L. Kane, E.J. Mele, R.S. Lee, J.E. Fischer, P. Petit, H.Dai, A. Thess, R.E. Smalley, A.R.M. Verschueren, S.J. Tans and C. Dekker, Europhysics Letters **41**, 683 (1998).

[49] F. Kreupl, A.P. Graham, E. Unger, M. Liebau, Z. Gabric and W. Hönlein, in: *Electronic Properties of Molecular Nanostructures*, Proc. of the XV International Winterschool/Euroconference, Kirchberg, Tirol, Austria, 2001.

[50] S. Frank, P. Poncharal, Z.L. Wang and W.A. de Heer, Science **280**, 1744 (1998).

[51] C. Zhou, J. Kong, E. Yenilmez, and H. Dai, Science **290**, 1552 (2000).

[52] M. Bockrath, J. Hone, A. Zettl, P.L. McEuen, A.G. Rinzler, and R.E. Smalley, Phys. Rev. B **61**, R 10606 (2000).

[53] P.G. Collins, K.Bradley, M. Ishigami and A. Zettl, Science **287**, 1801 (2000).

[54] K. Bradley, S.-H. Jhi, P.G. Collins, J. Hone, M.L Cohen, S.G. Louie, A. Zettl, Phys. Rev. Lett **85**, 4361 (2000).

[55] Z. Yao, C. L. Kane and C. Dekker, Phys. Rev. Lett. **84**, 2941 (2000).

[56] B. Q. Wei, R. Vajtai and P.M. Ajayan, Appl. Phys. Lett. **79** 1172 (2001).

[57] M. Büttiker, H. Thomas and A. Pretre, Physics Letters A **180**, 364 (1993).

[58] S. J. Tans, A. R. M. Verschueren, and C. Dekker, Nature **393**, 49 (1998).

[59] R. Martel, T. Schmidt, H.R. Shea, T. Hertel and P. Avouris, Appl. Phys. Lett. **73**, 2447 (1989).

[60] B. Yu, H. Wang, A. Joshi, Q. Xiang, E. Ibok and M.-R. Lin, IEDM Techn. Digest, 937 (2001).

[61] R. Chau, J. Kavalieros, B. Doyle, A. Merthy, N. Paulsen, D. Lionberger, D. Barlage, R. Arghavani, B. Roberds and M. Dosczy, IEDM Techn. Digest, 621 (2001).

[62] P. G. Collins, M.S. Arnold and P. Avouris, Science **292**, 706 (2001).

[63] C. Zhou, J. Kong, E. Yenilmez and H. Dai, Science **290**, 1552 (2000).

[64] V. Derycke, R. Martel, J. Appenzeller, and Ph. Avouris, Nano Letters **9**, 453 (2001).

[65] A. Bachtold, P. Hadley, T. Nakanishi and C. Dekker, Science **294**, 1317 (2001).

[66] S.J. Wind, J. Appenzeller, R. Martel, V. Derycke and P. Avouris, Appl. Phys. Lett. **80**, 3817 (2002).

[67] J. Guo, M. Lundstrom and S. Datta, Appl. Phys. Lett. **80**, 3192 (2002).

[68] R. Martel, H.P. Wong, K. Chan and Ph. Avouris, Proc. IEDM (2001).

[69] Th. Rueckes, K. Kim, E. Joselevich, G. Y. Tseng, C. Cheung and Ch. M. Lieber, Science **289**, 94 (2000).

[70] Y. Kwon, D. Tomanek and S. Iijima, Phys. Rev. Lett. **82**, 1470 (1999).

[71] R. J. Luyken and F. Hofmann, in: *Electronic Properties of Molecular Nanostructures*, Proc. of the XV International Winterschool / Euroconference, Kirchberg, Tirol, Austria, 2001.

[72] M.S. Fuhrer, B.M. Kim, T. Dürkop, and T. Brintlinger, to be published in Nanoletters (2002).

[73] M. Engelhardt, Proc. 3rd International AVS Conference on Microelectronics and Interfaces, ICMC (2002).

[74] R.J. Chen, Y. Zhang, D. Wang and H. Dai, J. Am. Chem. Soc. **123**, 3838 (2001).

Molecular Electronics

Marcel Mayor, Heiko B. Weber
Institute for Nanotechnology, Forschungszentrum Karlsruhe GmbH, Germany

Rainer Waser
Department IFF, Research Center Jülich and Institute of Electronic Materials, RWTH Aachen, Germany

Contents

1 Introduction	501
2 Electrodes and Contacts	503
3 Functions	504
3.1 Molecular Wires, Insulators, and Interconnects	504
3.2 Diodes	505
3.3 Switches and Storage Elements	506
3.4 Three-Terminal Devices	508
4 Molecular Electronic Devices - First Test Systems	509
4.1 Scanning Probe Methods	509
4.2 Monomolecular Film Devices	510
4.3 Nanopore Concept	512
4.4 Mechanically Controlled Break Junctions	512
4.5 Electromigration Technique	514
5 Simulation and Circuit Design	515
5.1 Theoretical Aspects	515
5.2 Design Rules for Molecular Nanocircuits	517
6 Fabrication	518
6.1 Chemical Synthesis	518
6.2 Integration Processes	520
7 Summary and Outlook	520

Molecular Electronics

1 Introduction

The evolution of microelectronics into nanoelectronics and the requirements for future developments of logic circuits have been discussed in the Introduction to Part III. As we have seen there, the ongoing feature size reduction of Si-based technology will run into severe physical and economic limitations in the long term. Several potential alternatives to supplement or to replace this technology have been described in Chapters 13 to 19, each of which is based on a typical class of materials. The subject of this chapter is the ultimate miniaturization of logic circuits by utilizing single molecules which would act as electronic switches and storage elements [1]. Molecules in this context are considered as small, typically organic molecules. These molecules are several orders of magnitude smaller than present feature sizes. They may be tailor-made by chemical synthesis and their physical properties are tunable by their structures (see Chapter 5). In addition, they have the potential to organize themselves on surfaces to regular 2-D patterns as well as to well-defined 3-D supramolecular objects. In fact, they appear to be the ideal building blocks for designing future high-density electronic devices. Therefore, in visionary concepts it has already been predicted that complete systems for information processing may be built from basic functional units consisting of molecules acting as logic devices. This idea of using molecules to perform electronic functions was born in 1974. Aviram and Ratner suggested that a molecule with a donor-spacer-acceptor structure would exhibit a diode-like *I-V* characteristic similar to a traditional semiconductor pn-diode (in which the *spacer* is given by the depletion space charge region at the pn interface) [2]. At this time, any realization of a molecular device was completely unfeasible and their suggestion was solely based on theoretical assumptions. However, it marks the origin of the progressively developing research area entitled today *Molecular-Scale Electronics* or *Molecular Electronics*. In the literature the latter term is used for two very different approaches, which must be distinguished to prevent possible confusion. The main differences are sketched in Figure 1.

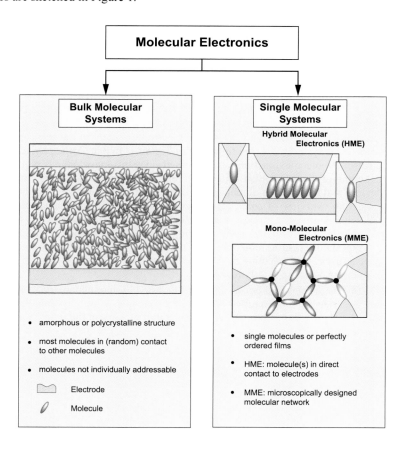

Figure 1: Sketch of the definition of conventional bulk molecular systems and single molecular systems. Only the latter is treated in this Chapter.

Bulk Molecular Systems for electronic devices are based on organic compounds with specific dielectric or electronic conduction properties. The organic compounds consist of small molecules, oligomers, or polymers and have found application in devices such as liquid crystal displays (Chap. 37), organic light-emitting diode displays (Chap. 38), and soft plastic transistors (Introduction to Part III). The volume of the organic compounds in these devices shows an amorphous or polycrystalline, often textured structure. The characteristic dimensions are much larger than the sizes of the molecules (as sketched in Figure 1, left side). Consequently, most of the molecules are in (arbitrary) contact to other molecules, instead of being directly contacted by external electrodes. At the electrode interface, a huge ensemble of molecules is contacted by a typically inorganic, electronically conducting phase.

Single Molecular Systems, in contrast, aim at individual contact to single molecules or small arrays of identical and perfectly ordered molecules. This approach differs radically from the bulk molecular system approach, as it tries to utilize the physical properties of single molecules for nanosized electronic devices. Due to the extreme difficulties in contacting and manipulating single molecules, research work during the first period according to the suggestions of Aviram and Ratner was focused on theoretical models and on studies of the electronic properties of molecular structures in solution or in bulk material. In recent decades, the invention and development of scanning probe techniques and many advances in micro- and nanotechnology have allowed the manipulation and operation on the level of small numbers or even a single molecule. The single-molecular systems approach (as sketched in Figure 1, right side) may be subdivided into hybrid molecular electronic (HME) devices and mono-molecular electronic (MME) devices [5]. HME systems are characterized by organic molecules directly contacted by inorganic electrodes. In the case of three-terminal devices, the gate electrode are of an inorganic nature, too. The HME approach is used to study and to manipulate individual molecules in test devices (see Section 4) and may provide the first platform for the integration of single molecular systems into contemporary high-density circuits. The vision of the MME approach reaches much further. It is based on the idea that all major functions of logic circuits can be integrated into molecules which are individually connected to each other. MME systems would need electrode contacts only for data exchange with the outside world and for an energy supply.

It should be mentioned that the use of small *organic* molecules shows an inherent advantage over small *inorganic* clusters, i. e. nanosized clusters of metal or semiconductor. Any object of few nm in size or smaller shows discrete (quantum) energy levels instead of energy bands which characterize the solid-state on a large scale. These energy levels will be crucial for the electronic properties of the nanosized devices made from these objects. Inorganic clusters made by self-assembly techniques or cutting-edge lithography and patterning will differ slightly in the number of atoms they consist of, and this scatter will be reflected in a scatter of their quantum energy levels. For future nanoelectronic devices, this scatter may be too large to be tolerated. On the other hand, organic molecules of a given compound are absolutely identical. Hence, there is no scatter in the quantum energy levels and all nanoelectronic functions based on a given type of molecule would show exactly the same characteristics.

The present Chapter deals with Molecular Electronics solely in the sense of single (organic) molecular systems, i. e. the right side of Figure 1. In Sec. 2, the physical properties of electrode-molecule junctions are discussed since this knowledge is required for the understanding of the subsequent text. In the following Section 3, the basic functions of logic devices and their potential realization by organic molecules are described. Section 4 presents an overview of some experimental approaches to molecular electronic devices. The challenges of a future circuit design for molecular electronic systems as well as the contribution of microscopic modeling and theory are covered in Section 5, while the synthesis of specific molecules and potential routes into suitable integration technologies is discussed in Section 6. At the end (Sec. 7), a summary is given and the challenges and prospects are briefly reviewed .

Several review articles and books may be suggested as supplementary literature to this chapter. A series of books entitled *Molecular Electronics* edited by Ratner, Aviram and Mujica [3], [4] provides a comprehensive collection of work in the field. In addition, the review articles [5], [6] are recommended.

2 Electrodes and Contacts

A basic requirement for molecular electronics is the connection of the molecule to the outside world. If we want to drive a current through individual molecules, we need an electrode pair with nanometer-sized spacing to contact them. If we are interested in molecular films, again we need contact electrodes which must be controlled on the scale of atomic length scales. The usual way is to use metallic or semiconducting electrodes, which yield hybrid (HME) devices. In the future, the replacement of metallic electrodes by molecular wires might be investigated (MME).

It turns out that the nature of the electrodes is also of importance for the electronic properties of the device. A molecule which is closely connected to an electrode has very different properties from a molecule in solution. Two main groups of links between molecules and solids can be distinguished: **covalent bonds** and **van der Waals interaction**.

A frequently used and up to now probably the best investigated covalent link is the bond between a thiol (sulfur) group on the molecule and a Au substrate. Au is favorable due to its proper and non-oxidizing surface. The thiol endgroup is one of the rather rare functionalities which form covalent bonds with the noble metal gold. Further requirements are good stability of the bond at room temperature, which must be, however, loose enough to allow for self-assembly (i.e. to rearrange continously until finally a completely ordered monolayer of molecules is formed). Other combinations like Se-Au or S-Ag have already been investigated as members of this family. Molecules with hydroxy groups are used on SiO_2 and TiO_2 substrates. Such couplings are of particular interest due to the use of these materials in traditional microelectronics and hence may form a bridge between the fields. However, they lack the advantage of subsequent self-organization of the molecules, due to the large stability of the covalent bond formed. Covalent bonds lead to a mechanically fairly stable and resistant connection between the molecule and the substrate.

Van der Waals interaction acts in particular between Langmuir-Blodget (LB) films of organic lipophiles and planar substrates. This technique results in very well organized films with the advantage of substrate diversity: the only requirement is a planar surface with appropriate wetting properties, i.e. a designated lipophilic or hydrophilic characteristic (depending on the type of molecules and the desired orientation). While for some types of molecules LB films might be suitable, in other cases they suffer from poor long-term stability due to the weakness of the Van-der-Waals interactions. A successful example will be discussed in Section 4.2: a LB-film sandwiched between two electrodes displays the characteristics of a (reconfigurable) switch. In other cases, electronic components with molecular building blocks require the stability of covalent linkages rather than the weak van der Waals bound interfaces.

These different contacts types also correspond to different electron transport mechanisms. Imagine a molecule with a *conductive* inner part (this can be realized with extended, *delocalized* π-electron systems, see Chapter 5) connected to a gold electrode. If the distance to the metallic surface is very short (of the order of bond lengths), the inner conducting orbitals and the outer metallic electronic states overlap to a certain extent. This yields a hybridisation of the inner and the outer extended wave functions and hence a common delocalized electronic wave function which extends over the whole junction. The junction then can be imagined as a waveguide for the electrons which are transmitted in a similar way to light passing through an optical fiber. This case is, for example, realized when thiol endgroups are attached to benzene rings: the π-orbitals of the benzene and the conduction band of the metal overlap at the sulfur atom. It should be noted that the sulfur is nevertheless an imperfect transmitter and acts as a bottleneck for the extended wave function. The influence of the bond on the molecule is complicated and not fully understood. In the theory Section 5.1 some of the challenges of this question will be addressed.

If the distance chosen is large or badly-conducting molecular units are in between, the wave function inside and outside do not overlap and can with good approximation be treated independently. This case corresponds to LB films. Electron transport can then better be imagined as particles tunneling from one electrode onto the molecule and, after a short dwell time, tunneling to the opposite electrode. In this case, the resistance per molecule is expected to be higher.

3 Functions

The visionary concept of *Molecular Electronics* in terms of HME and MME is more than the usage of molecular structures in electronic circuits. The perspective that the electronic properties of a device may be adjusted by design of the chemical structure is the real beauty of the approach. Early investigations have already demonstrated the strong interdependence of chemical structure and electronic transport properties of integrated molecules. Further, a more molecular junction should also be tunable by other stimuli e.g. voltage, light, or magnetic fields. Thereby a whole set of functions can be embedded in a circuit by appropriate choice of the molecule. This section gives a first impression of the functions and properties that are approachable by molecular structures. These functions based on electron transport properties through molecular structures have already been investigated without integrating molecules into electronic circuits, but in solution or in the solid materials. In the last decades numerous compounds consisting of a functional unit bridging two redox-active centers have been synthesized and investigated spectroscopically and electrochemically. Most of the chemical structural knowledge and working principles of the functional units presented in this chapter has been gathered by such experiments in solution and only a few experiments have shown the validity of the same concepts for individual contacted molecules.

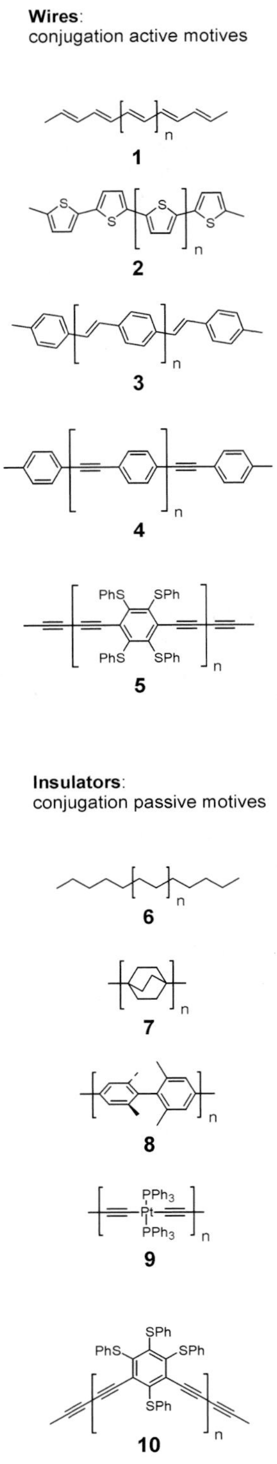

3.1 Molecular Wires, Insulators, and Interconnects

The most basic electronic function is a wire, a one-dimensional object that allows the transport of electric charge [7]. Transferred to a molecular scale, of particular interest are rod-like structures that transport electrons from one end to the other [8]. Electron transport is expected to take part through the frontier orbitals of a molecule, as these should be closest to the Fermi levels of the electrodes (see Sec. 5.1). In general, with increasing size of the π-system the energy difference between the frontier orbitals decreases and hence the energy difference to the Fermi level of the electrode. Promising candidates as molecular wires are therefore large delocalized π-systems. Structural motives that allowed the design of chains with delocalized π-systems were studied extensively in solution [9]. The simplest example of such a chain is polyene **1** (for the following molecules, cf. Figure 2) consisting of an alternating sequence of single and double bonds leading to a π-system over the whole length [10]. Many other examples consisting of aromatic building blocks like polybenzene [11], polythiophene **2** [12], polypyrrole and combinations of aromatic building blocks with conjugated double or triple bonds like polyphenylenevinylene **3** [13] or polyphenyleneethynylene **4** [14] were extensively studied. The enormous developments in nanoscale manipulations down to a single atom level, made it possible to investigate the electric current through selected π-systems, on surfaces or between electrodes, as will be discussed in detail in section 4 of this chapter. An experiment that illustrates the concept nicely is the incorporation of protruding rigid π-systems as molecular rods out of a self-assembled monolayer (SAM) of an thioalkane in Ref. [15] (Sec. 4.3). This first conductance investigations of single molecules demonstrated substantial differences in electron transport properties between the rod and the thioalkane SAM emphasizing the concept of delocalized π-systems as molecular wires.

Conductance is not always the property which has to be optimized. The opposite property, a rather insulating molecular structure, is of similar importance for particular applications. The very first HME-device considered, the rectifier by Aviram and Ratner, is based on a donor and an acceptor π-system linked together by a spacer [2]. They suggested a rigid adamantyl cage **7** as a non-conjugating linker between both π systems which was expected to behave as an insulating molecular unit. The choice of the spacer will be very important, as it has to be sufficiently insulating to preserve the energy differences between the π systems, but still allows to some extent electron transport. In this particular case, the authors suggested electron transport by tunneling through the insulating structure. Structural motives not consisting of delocalized π-systems are in general poor conductors but reasonably good insulators. However, on a molecular scale, the rigidity of the structure is of similar importance to prevent short circuiting of the separated units through space. Rigid molecular structures restricted to non-conjugating systems are rather limited. The above-mentioned adamantyl structure meets the conditions but is synthetically quite demanding. Alkanes **6** are known for their insulating properties, but lack the required rigidity. π-Systems meet the rigidity conditions but have already discussed as good conductors. However, the delocalisation of the π-system depends strongly on the torsion angles between the subunits. Two neighboring subunits with perpendicular π-systems reduce their electronic communication and, hence, the

Figure 2: Molecular structural motives described in the text. Motives **1-5** are delocalized conjugated systems with the potential to act as molecular wires. Motives **6-10** are barely conjugated - even so in the case of **7-10** still rigid - systems, rather acting as insulating spacers.

connecting single bond between them becomes a rigid and insulating connection on a molecular scale, as is the case for the tetramethylsubstituted biphenyl building block **8**. The transparency for electrons through a benzene core also depends on the relative positions of the linkages. While *ortho-* and *para-*connections are conjugation-active linkers, the *meta-*position is conjugation-passive [16]. This has been shown, for example, in the comparison of *para*-diacetylene connected thiophenyl-substituted benzenes **5** and the corresponding *meta-*connected building block **10** [17]. Another approach to meet the required rigidity and electronic passivity is the use of metalorganic complexes as linkers. The potential of this approach has already been demonstrated on a single- molecule level by the investigation of a *trans*-acetylene platinum(II) molecule **9**, which turned out to display the characteristics of a single molecule insulator, as will be discussed below (Section 4.2) in detail [18]. An additional motivation for using metal centers as connectors is the adaptability of the electronic transparency by the choice of the linking metal center, raising hopes of a rich future construction kit consisting of tailor-made linkages based on metal ions. An overview of molecular building blocks for rigid rods is given by Schwab, Levin and Michl [19].

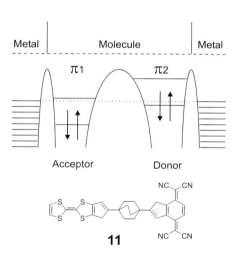

3.2 Diodes

As already mentioned several times in this chapter, the field of *Molecular Electronics* is closely related to the function of a diode, as was the first hypothetic function described of the field. In 1974 Aviram and Ratner described their visionary HME device as a molecule **11** between a pair of electrodes, performing the function of a rectifier. The working principle they suggested is based on the difference in energy of the frontier orbitals of two separated donor- and acceptor- π-systems, as shown schematically in Figure 3. To transfer electrons from the cathode to the acceptor and subsequently from the donor to the anode should be feasible at a smaller applied voltage than to transport electrons in the opposite direction. As discussed in the section 3.1 above, the donor- and acceptor π-system of the molecule is separated by a spacer, that preserves the energy differences of the frontier orbitals between both π−systems but allows to some extent electronic transport, as an electric conduction through the device shows a preferential direction.

The realization of this concept however, turned out to be difficult. The suggested molecule has a defined working direction and therefore its orientation between the electrodes is essential. This may be neglected with a test system based on a single molecule approach, as the orientation of a single molecule in an HME device may be adjusted by the wiring of the electrodes. However, to contact single molecules is still a challenge, as we will see in section 4. Therefore, first efforts were focused on films consisting of a single molecular layer. In these films it is crucial, that all molecules are lined up in the same direction, as a random orientation would equalize the directional effects from the individual molecules. Packing effects in a Langmuir-Blodget (LB) film results in perfectly oriented molecules. With this technique, LB-films consisting of donor-spacer-acceptor molecules **12** (Figure 4) were deposited on metallic surfaces as electrodes and subsequently covered by a second metallic top electrode, as will be discussed in more detail in section 4.3 [20]. Such devices displayed rectifying properties, but lacked the required stability due to the weak van-der-Waals interaction between the electrode and the molecules. However, it was the first attempt to verify the principle of an electronic device with its functionality based on a molecular structure. But was the difference in voltage threshold really due to the molecular structure? These devices consist of a tailor-made molecular structure consisting of donor-spacer-acceptor-alkyl chain. The alkyl chain is required for the LB-film formation. While in the final device the donor is deposited directly onto the electrode surface, the acceptor is separated from the top electrode by a layer of alkyl chains. The observed differences could be explained as a consequence of the different contacts of the LB-film to the two electrodes as well as by an effect of the separated donor and acceptor in the molecular structure. Moreover the two electrodes consisted of different metals and only very recently were findings on such a device published, that was made of two Au electrodes [21]. The nature of the contact realisation is at least of equal importance as the molecular structure in between [22]. In the case of sandwiched LB-films between two electrodes, the contacts to both electrodes are Van-der-Waals interactions, but to rather different molecular substructures. Is the interaction with the electrode on either sides of a different nature, an even stronger dependence on the current direction through the device is expected.

Another interesting device was proposed by Reed and Tour [23]. The rod like molecule **13** (cf. Figure 4) with a thiol function at one end was immobilized on a small Au surface in a Si_3N_4 pore – which yielded a self-assembled molecular film (SAM). This

Figure 3: The first approach to molecular electronics [2]. Molecules with a donor and an acceptor group, separated by an insulating spacer, are predicted to behave as diodes. The upper panel displays the electron energies in the system, when no bias voltage is applied: In the metallic electrodes outside, electrons are filled up to the Fermi energy. The π-systems of the donor and acceptor units are confined in two potential wells. If a positive voltage is applied, the potential of the left lead is slightly increased and the potential of the right lead is correspondingly lowered: current can flow from the left to LUMO1, then to HOMO2 and further to the right electrode, going towards lower energies at each step. If the opposite voltage is applied, conduction take place only at much higher voltages. This is the behavior of a diode with the favourable current direction from the acceptor to the donor.

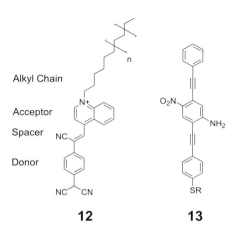

Figure 4: The LB-Film molecule **12** and the rod-like molecule **13**.

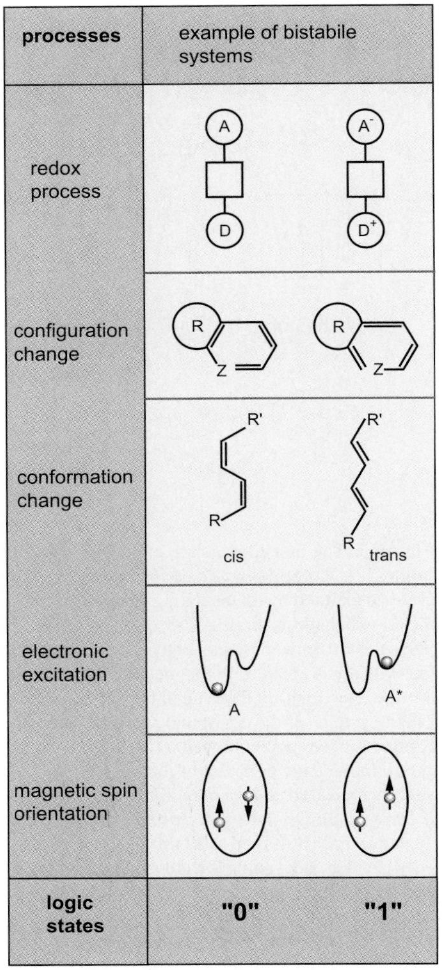

III Logic Devices

processes	example of bistabile systems	
redox process	A—□—D	A⁺—□—D⁺
configuration change	R-ring-Z (cis)	R-ring-Z (trans)
conformation change	R'/R cis	R'/R trans
electronic excitation	A (double well, ground)	A* (double well, excited)
magnetic spin orientation	↑↓	↑↑
logic states	"0"	"1"

Figure 5: Illustration of bistabile molecular structures.
(a) redox states, where A denotes an acceptor group and D is a donor group of the molecule;
(b) configurations obtained by a re-arrangement where Z is the wandering group;
(c) example of cis and trans conformations;
(d) ground state A and excited state A* of a molecule;
(d) parallel and anti-parallel spin states within a molecule.

SAM was subsequently covered with a Au electrode. The molecular rods were covalently bonded through the S-group to the Au bottom electrode while the linkage to the Au top electrode was undefined. In addition to the diode characteristic, the device displayed a **negative differential resistance** (NDR) in the electric current. A NDR is a decrease in current caused by an increase in voltage above a certain threshold. The observed effect was temperature-dependent, reversible and reproducible. The substituents on the molecular rods turned out to be crucial for the effect. Even though the device demonstrates the potential of the concept, the nature of the effect is not fully understood yet and is the topic of current investigations.

3.3 Switches and Storage Elements

There are classes of molecules, which are stable in two different states (so-called metastable). As a consequence physical properties like the conductance will differ for these two states. If this difference in the conductance is sufficiently high, these molecules may be used as *molecular storage elements* if good control of the addressing and switching of the molecules is realized. Figure 5 illustrates some possibilities of such **bistable molecular switches**. (a) A reduction-oxidation (redox) process may change between neutral and ionized acceptor/donor groups of a given molecule. (b) A configuration change may take place through a reversible re-arrangement reaction. (c) A conformation change may be controlled in such a manner that both conformations are sufficiently stable at the operation temperature of the designated storage element. (d) Electronically excited states may be used in the same way, if the lifetime of state A* is made sufficiently high. (e) The idea may also be applied to spin magnetic moments, as has been shown for Mn acetate [24]. In all cases of Figure 5, the bistable switches are characterized by double well potentials concerning their energy and exhibit a *hysteresis* when driven between the two states. In many cases, more than one of the above mentioned mechanisms must cooperate to achieve the hysteretic properties. Below, for example, we will report on a molecule which undergoes a conformational change due to a redox reaction. To control (meta-) stable physical properties of molecular structures by external stimuli has fascinated researchers for a long time and several structural motives as molecular switches have been published in the last decades. The book *Molecular Switches* edited by Feringa is recommended [25] as an overview for the interested reader.

In general, molecular switches may be classified by

- the stimulus that *triggers* the switch and
- by the *property* or function that is switched.

Frequently light or chemical parameters such as pH, for example, are used as triggers and the switched property is often a structural feature. Of particular interest in view of molecular electronic applications are systems, in which the two states of the switch display different current transport properties.

One of the best-studied examples in chemistry is the light-triggered switch by Masahiro Irie **14** [26], which is shown in Figure 6. Two methyl-thiophene units are linked with a hexafluorocyclopentene bridge in the open form of the switch. Irradiation with light in the wavelength range 200 – 380 nm leads to the closed conjugated form on the right side which is more favorable for charge transfer. Irradiation with light in the range 450 – 720 nm opens the switch back again. The potential of the switch has been studied extensively by synthetic systems possessing redox and/or photo-active units located at the thiophenes which allowed the investigation of electron delocalization over the switching structure (see Figure 7). Even though this system has been carefully studied and provides promising properties like excellent addressability and conversion into the desired switching state, the main drawback of the system is the use of light as switching trigger. For electronic applications switches that are addressable by voltage or charges are of much greater interest, as this would allow the integration of such a switching device into a nanoelectronic circuit.

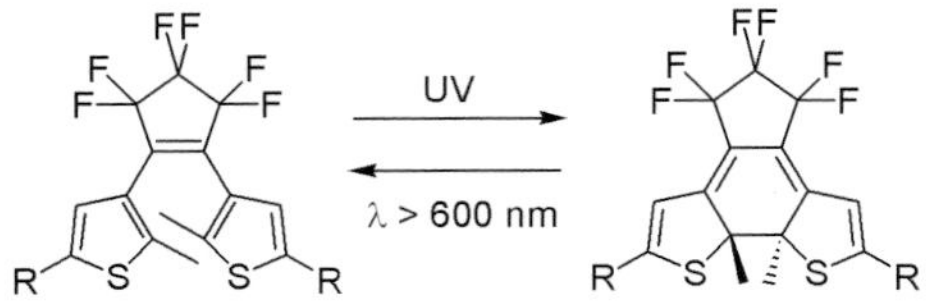

So far, only few switches fulfill these boundary conditions. An interesting approach comes from supramolecular chemistry. Rotaxanes and catenanes have been synthesized to switch as a function of an applied potential between two different states. An overview is given in the review article [27]. In particular, the catenane **15** shown in Figure 8 [28] has already been used to build up electronic memory devices [29], as we will see in Section 4.2. A catenane consists of two interlocked rings. In this particular case of two different rings, one is functionalized with two viologene units and the other with a dioxynaphtalene and a tetrathiafulvalene (TTF) unit. The interaction of the different redox

Figure 6: The light-triggered switch by Masahiro Irie [26].

14 open form ⇌ (UV / λ > 600 nm) 14 closed form

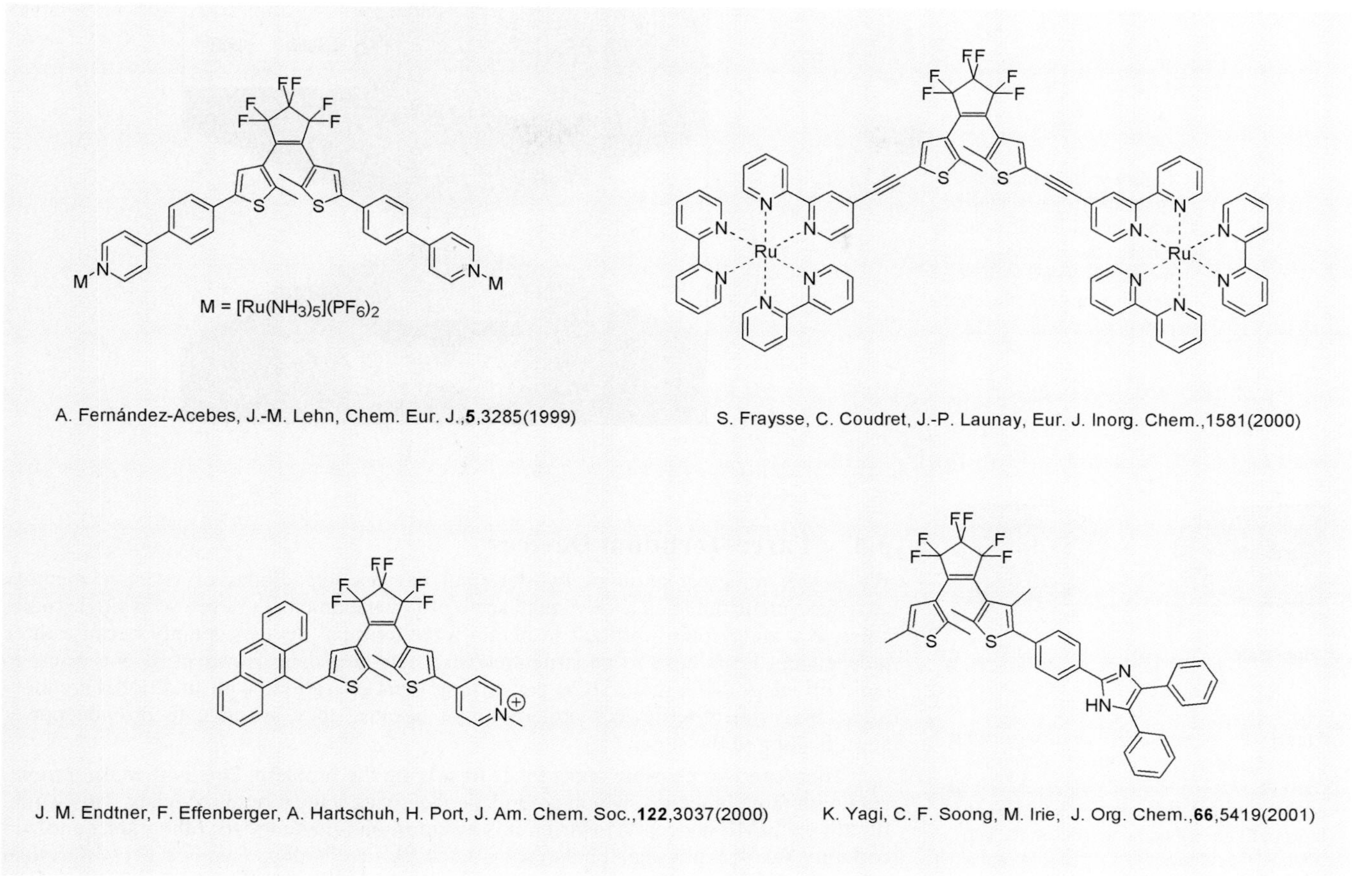

M = [Ru(NH3)5](PF6)2
A. Fernández-Acebes, J.-M. Lehn, Chem. Eur. J.,5,3285(1999)
S. Fraysse, C. Coudret, J.-P. Launay, Eur. J. Inorg. Chem.,1581(2000)
J. M. Endtner, F. Effenberger, A. Hartschuh, H. Port, J. Am. Chem. Soc.,122,3037(2000)
K. Yagi, C. F. Soong, M. Irie, J. Org. Chem.,66,5419(2001)

states of these units make it possible to rotate one ring within the other. Therefore this molecule is bistable at zero voltage: it may exist in two configurations. By applying a positive potential pulse to the solution it can be set into one configuration (dioxynaphtalene unit in the viologene ring) while applying a negative potential pulse results in the other configuration (TTF unit in the viologene ring). The switching mechanism is explained in details in Figure 8. The states of the molecule can be regarded as memory bits which can be written and erased.

Figure 7: Examples of redox and/or photo-active units connected to the switching unit discussed in Fig. 6. The extent of interaction between the peripheral units depends on the state of the switch.

(a)
15
'Co-conformer' [A0]
'Co-conformer' [B+]
(b)
[A+]
[B+]
[B0]
[A0]
[AB#]
Net Oxidizing
Net Reducing
0

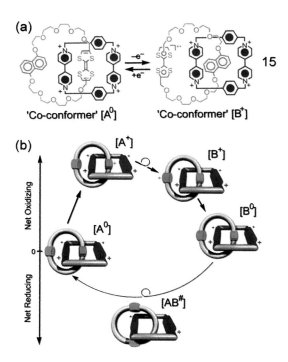

Figure 8: Hysteretic redox-triggered rearrangement of the catenane **15**. At zero bias the molecule is in the ground-state [A^0] with the TTF unit between the two doubly charged viologenes. Biasing the system into the net oxidizing direction by +2V leads to an ionization of the TTF unit into state [A^+]. Because of electrostatic repulsion, the ring will mechanically rotate into state [B^+]. Reducing the bias back to zero will remove the ionization and, hence, the ground-state [B^0] with the dioxynaphthalene between the two doubly charged viologenes is established. A bias pulse into the net reducing direction with an amplitude of –1.5 V reduces the two viologenes into the monocharged state [$AB^{\#}$] which causes a preference for the TTF unit between the viologenes. Resetting the bias to zero results again in the state [A^0] with the two positively charged viologenes surrounding the neutral TTF unit. (Reproduced from Collier et al. [29] with permission; copyright © 2000 by the American Association for the Advancement of Science).

III Logic Devices

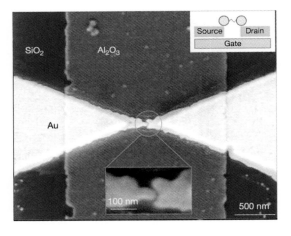

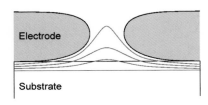

Figure 9: SEM picture of a Au electrode pair on top of an Al gate electrode, which is covered by an Al$_2$O$_3$ insulator. Such a setup was used as a molecular field effect transistor [48].

Figure 10: Scheme of an electrode gap with a buried gate electrode formed by the substrate (covered by an insulating layer). When a gate voltage is applied, the equipotential lines (red) are not homogeneous in the gap region. If the molecule is far way from the surface, high gate voltages are needed to affect the potential at the molecule's position. If the distance from the gate electrode is not well controlled, this will result in strongly varying switching voltages.

3.4 Three-Terminal Devices

While the implementation of two-terminal devices like diodes or resistive memory switches can be made on a very small scale of few nanometers or even with single molecules, it is much more difficult to obtain three-terminal devices, simply because three independent leads have to be structured on a few-nm scale. However, if one wants to build up integrated circuits (ICs) performing logic operations with traditional architectures, transistor-type devices are important because they are able to provide power amplification to the signal.

There are two possible approaches to solving the problem. One is to make a molecule with three branches, which are independently contacted by three leads. This would fit into the MME concept. However, it is an enormous challenge for lithography and will certainly involve physical phenomena much more complex than for a two-terminal molecular device. So far, no meaningful concept has been developed for this case. An HME-type approach is to place the third lead relatively far away (for example, buried in the substrate, see Figure 9), not in contact with the molecule, but able to modify the electrostatic potential inside the molecule by field effects. This has already been demonstrated [46], [48] in low-temperature experiments. The conductivity of single organometallic molecules was tuned by electrostatic gate electrodes and the properties of a *Single Electron Transistor* (cf. Chapter 16) were observed (Figure 22). The junction was switched from the Coulomb blockade regime with a strong suppression of the current to a conducting mode (which has still a high resistance: $R \sim 100$ MΩ). The gate voltage dependence demonstrates that indeed a single molecule could be selected in a few cases. A key issue is to make this third gate electrode sufficiently close to the molecule to enable operation at small voltages. One of the emerging problems which can be seen in Figure 10 is the reproducibility of the transistor, as the exact distance of the molecule with respect to the gate will affect the voltage necessary to switch the molecule's conductance. If, however, the molecular junction comprises of a large number of molecules, the tuning by electrostatic gating will be complicated by screening effects. The polarizability of the external molecules will weaken the field which acts on the inner molecules and the gate effect will be smeared out. This problem might be overcome by molecules which react to external stimuli with changes in the conformation, which might allow a cooperative switch of all molecules as soon as the external field reaches a certain threshold. This would involve nucleation statistics as an uncertainty.

A completely different approach was reported in Ref. [42]. C$_{60}$ molecules were deposited on a metallic substrate and investigated with an STM. After imaging the surface and detecting where C$_{60}$ could be found, the tip was used to simultaneously squeeze the C$_{60}$ and measure the conductance (Figure 11). Though only two terminals (tip and surface) were present, the mechanical force on the tip added a third parameter. It turned out that the conductance could be tuned by two orders of magnitude per nano-newton. Whether this setup is able to provide power amplification to a signal depends on the development of an appropriate nano-electromechanical system used to apply the force.

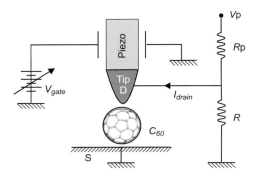

Figure 11: A nanomechanical three-terminal device [42]. The C$_{60}$ molecule is contacted by the metallic substrate on one side, and by the STM tip from the opposite side. The steering parameter which replaces the third electrode is the force which acts on the STM tip. By pushing on the C$_{60}$ molecule, the conductance varies by two orders of magnitude per nano-newton.

4 Molecular Electronic Devices - First Test Systems

Parallel integration of single molecular structures as functional units in electronic circuits as the long-term aim of molecular electronics is still unattainable. However, first sophisticated experimental setups allow controlled assemblies or even single molecules to be contacted and their properties investigated in an applied electric current. This section is devoted to these first test systems.

4.1 Scanning Probe Methods

One of the main triggers for the enormous boom in nanoscience was the development of the scanning probe methods. Properties such as shape, size, diffusion, conductivity etc. of individual molecules on surfaces could be achieved for the first time [34]. Scanning tunneling microscopes (STM), atomic force microscopes (AFM) and similar instruments provide an enormous variety of experimental applications (Chapter 12).

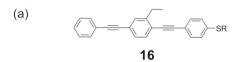

If we wish to study the conduction along a rod-like single molecule on a surface, while the surface acts as one electrode and the STM tip as the opposite electrode, we have to prepare the molecules such that they stand up right on the surface. However, isolated molecules often prefer to lie flat on surfaces due to Van-der-Waals interaction. To force configurations perpendicular to the surface, tripodal attachments to the molecules has been proposed such as tetrahedral-shaped molecules in which three of the four *legs* are terminated by S-groups as 'alligator clips' [35]. Another approach makes use of a carpet of upright standing *insulating* alkanethiols to embed the *conducting* molecules under investigation. A stabilization of the position and the orientation of the sample molecules at angles between 60° and 90° to the surface was thus achieved. It should be noted that most of these experiments are in the tunnel regime, where the tip-molecule contact is not a chemical bond.

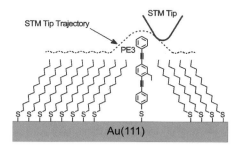

As an example, the molecular rod **16** (cf. Figure 12) is a fairly good electronic conductor, due to its conjugated π-system. To investigate its conduction properties it was embedded in a SAM matrix of electrically insulating molecules, such as dodecylmercaptan ($C_{12}H_{26}S$) as a long-chain thioalkane [30], [36]. The film preparation started with the preparation of the pure alkane SAM layer. As observed by STM, the SAM layer is organized in domains, which resemble 2-D crystals. Within one domain, the molecules are perfectly ordered. Subsequently, the SAM layer is treated in a diluted solution of **16**, which leads to a partial exchange at the Au surface. STM studies show that the exchange takes place at domain boundaries and triple points while **16** is not observed within the domains. Since STM is based on the tunneling current, which is determined by both, the thickness of the SAM layer (i.e. the length of the molecules) and the conductance of the molecules, additional physical information is needed to separate the contribution of the two properties. The combination of conventional DC-voltage STM with a microwave AC-STM (alternating current-STM) technique allowed at the same tip the two properties to be distinguished [36]. Based on this method, the topography of the SAM film was separated from the conductance mapping. In many spots (but not always), the topography maxima attributed to molecules **16**, which are approx. 0.7 nm longer than the dodecane molecules, coincide with conductance maxima.

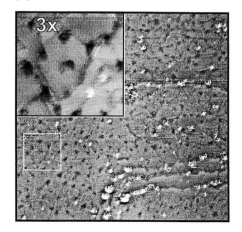

Scanning probe methods are not limited to imaging and electrical measurements. They may also be used as a working tool in the nanometer range, as is illustrated by the following example. To circumvent the arbitrary placement of **16** in the SAM matrix, the SPM tip was used to pattern the SAM layer as sketched in Figure 13 [37]. After SAM formation on gold, the specimen was inserted into a dilute solution of **16** and NH_3 in an STM liquid cell with the exclusion of oxygen. The SAM layer can be locally removed at the tip position by applying a short voltage pulse to the substrate (Figure 13 b) in an area of 10 nm diameter. After patterning a few sites, the film was monitored by STM imaging and a filling of the pits by **16** was observed. Each pit hosts approx. 400 molecules **16**.

Figure 12:
(a): Protruding molecular rod **16** out of a SAM layer of dodecylmercaptan.
(b) STM picture of the sample displaying protruding rods **16** as brighter spots. Correlated maxima in the conductivity and the topology indicate the increased conductance through the molecular rods **16**. [36]

Another approach to direct the placement of molecules by scanning probe methods is the manipulation by mechanical forces in nanometer dimensions by AFM (Atomic Force Microscopy). Single atoms, clusters and molecules can be moved and manipulated on the sample surface by this technique [38], [39], [40]. Electrochemical deposition makes it possible to create designed and defined structures in the few nanometer regimes using an AFM-tip as the local writing electrode [41]. Such structures may soon reach molecular length scales.

Altogether, scanning probe techniques are amazingly powerful tools to investigate substrates coated with molecules. They can be used for manipulating, imaging and measuring nano-objects with the same tip. While their contribution to the comprehension of the correlation between molecular structure and physical property is outstanding, the potential of the system for upscaling to parallel integration is still currently under investigation (Chapter 28).

III Logic Devices

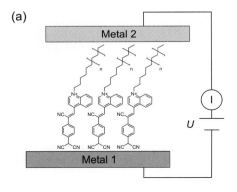

Figure 13:
Left: Illustration of the patterning of the SAM layer and the placement of electronically conducting **16** molecules in the alkanethiol matrix:
(a) normal STM imaging of the SAM surface with a tip bias V_b;
(b) SAM removal by applying a pulse V_p to the substrate;
(c) filling of the pit by **16** molecules from the solution (adapted from Ref. [37]).
Right: (top) the image of a dodecanethiol SAM surface after consecutive patterning, three pits show two peaks indicating adsorbed **16** and one pit without adsorption; (bottom) image taken few minutes later showing adsorption onto the third remaining pit.

4.2 Monomolecular Film Devices

While future devices based on single molecules are still highly challenging, devices based on molecular films are much more promising for the near term. The required films of molecules can be made available by numerous different techniques like self-assembly, vapor deposition techniques or LB films. The film may be sandwiched between two metallic leads (a bottom electrode and a subsequently evaporated top electrode). This allows a patterning of the leads by standard lithography techniques. The outcome is a 2-D architecture which can easily be combined with conventional circuits. The device can in principle be scaled arbitrarily to a few-nanometer lateral sizes. Having a large number of molecules, the I-V properties of the individual molecule are averaged out which enhances the reproducibility. A major problem with molecular film devices is defects. When the second electrode is evaporated on top of the molecular film or due to diffusion processes during the device's lifetime, metal atoms may penetrate the film and short-circuit the device. There is a need for elimination of this problem by appropriate engineering of the films. Experiments have revealed that with decreasing area covered by an electrode the probability of a defect is reduced. Several demonstration devices based on molecular films displaying a variety of physical properties like rectifying, negative differential conductance and switchable memories have been proposed. This section is only a short overview based on some prominent examples.

One of the very first examples of a device based on molecular films was the realization of the rectifying principle predicted in the seventies and demonstrated in the 1990s [49], [50], as has already been mentioned in Section 3.2. A LB film of molecules **12** consisting of an electron rich π-system as donor D linked to an electron-poor π-system as acceptor A decorated with a long alkyl tail has been sandwiched with a second top electrode (Figure 14, a). A torsion angle between the A and D π-systems reduces the overlap of the two π-systems and allows a zwitterionic ground state, T-D$^+$-π-A$^-$, where T is the hexadecyl 'tail' to support the LB film formation. It exhibits a pronounced different conductivity depending on the polarity of the voltage applied to it in an LB film as shown experimentally (Figure 14, b) [49], [50], [51] and studied theoretically [52]. A negative bias at the acceptor side triggers an intramolecular electron-transfer from the A$^-$ side to the D$^+$ side of the zwitterion as the initial step of the conduction process. This is energetically more favorable than electron injection from the metal contact into the D$^+$ side of the molecule under reverse bias

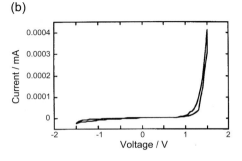

Figure 14:
(a) Schematic presentation of a rectifying device based on an LB-film of the donor-acceptor molecule **12**.
(b) The I-V curve of the sandwiched LB-monolayer displaying rectifying character.

as the initial step. The device impressively demonstrated the potential of molecular films as electronic components, but, lacked sufficient stability: The applied electric fields between the electrodes caused a reorientation of the molecules thus destroying the device.

An approach towards real integrated devices is based on the deposition of supramolecular systems like catenanes and rotaxanes on a substrate with prefabricated nanowires as bottom electrodes. For example, the electrochemically active components on the catenane rings make it possible to switch between two configurations by voltage pulses, as discussed in Section 3.3. As both configurations (which may serve as the two states of a single bit of information) differ in their tunnel currents the state of the switch can subsequently be read out with the tunnel current at low voltage (see Figure 15). This approach aims at a cross-bar random access memory with hysteretic, non-volatile molecular switches.

As explained in the Introduction to Part III, cross-bar arrays can be impressively utilized as RAM devices as well as field-programmable gate array (FPGA) based wired-logic devices. Since the cross-bar itself can not provide gain, it has to be supplemented by active devices outside or underneath the cross-bar system. Based on their cooperation on the molecular electronic variant of the Teramac concept [62], Heath [63] and Williams [64] presented different demonstrators of this concept. The fabrication starts by the deposition of either poly-Si [63] or Pt bottom electrodes [64] as parallel wires using e-beam lithography or nano-imprint techniques. Thereafter, a monomolecular film is deposited by the LB technique followed by a Ti layer as a top contact and to protect the molecular film from the subsequent integration steps. The final top electrodes (Al, Au or else) again are deposited by e-beam lithography or nano-imprint techniques. Finally the protective Ti layer is removed by etching between the top electrodes to avoid short-circuits.

In Figure 16 a section of a cross-bar array made by these technique is shown. Electrochemical switchable supramolecular systems like rotaxanes [63] and catenanes [29] have been deposited between these electrodes as LB-films. In both cases, the configuration of the supermolecules is switched by applied voltage pulses. Similar as in solution (see Section 3.3), the relative position of an interlocked ring as part of the supermolecules is assumed to be the different states of the switch. It was found that the rotaxanes employed in this concept need a relatively large footprint to show well-defined switching voltages. Obviously, the rings can not move individually if the footprint is too small. Voltages in the range from approx. 1 V to 2 V are required in the *write* operation to set the molecule into the high-resistive and low-resistive state. The *read* operation is executed at much lower voltages which do not change the resistive state of the switch and, in addition, contribute to a low power consumption of the device. The R_{on}/R_{off} ratio of individual nodes ranges from approx. 3 to 10 for the Si/molecule/Ti system. The integration of pn diode junction in the Si bottom electrode wires led to a considerable suppression of cross-talk. In addition to a small memory matrix, the demultiplexer and the multiplexer for the row and column lines have been realized by wired-AND and wired-OR in order to demonstrate the successful operation of a complete system (see Figure 29 in Sec. 6.2). Furthermore, using an EXOR gate (Figure 17) in combination with the (simpler) AND gate indicated that a half-adder function can be configured from a cross-bar circuit

Between Pt bottom electrodes and Ti/Al top electrodes the series of as LB film deposited molecules were supplemented by long alkyl chain carboxylic acid. These devices displayed very promising switching properties with R_{on}/R_{off} ratio of individual nodes up to 10^4. Although in the latter case, the R_{on} value as well as the R_{off} value shows a variation of up to one order of magnitude, still a R_{on}/R_{off} ratio of 10^2 remains in the worst case. However, the nature of this switching behavior is not clarified yet. As it works with rather simple molecules as well, the switching obviously is not due to a configuration change of the supermolecule in this case and the role of the molecules in this setup is not understood at the publishing date of this book.

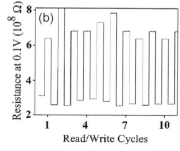

Figure 15: Switching of the catenane [56] (Figure 8). a) Molecular signature of the device upon stepwise variation of the writing voltage and read out at -0.2 eV. b) Repetitive switching of the device at room temperature. Switched with voltage pulses of -2 V and +2 V respectively and read out at +0.1 V.

Figure 16: An SEM image of a semiconductor crossbar structure utilizing molecular switch tunnel junctions. Each junction has an area of 0.007 μm² and contains about 5000 molecules. The inter-wire separation distance was kept large so as to simplify the task of contacting this circuit to external devices for testing [63].

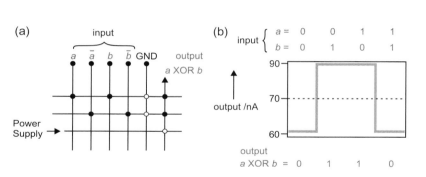

Figure 17:
(a) The wiring diagram and
(b) truth table of an EXOR gate. In the wiring diagram, ● represents closed switches and ○ represent open switches. The junctions that are not dotted, which would be diodes in an ideal molecular-switch crossbar circuit, are not connected. $\bar{a}$ and $\bar{b}$ are the complements of a and b. An AND (which is a simpler structure) and an EXOR function combine to yield a half-adder, with the EXOR representing the sum of two 1-bit numbers and the AND representing the carry. In the truth table, the green trace is the output signal recorded corresponding to four different input combinations [63].

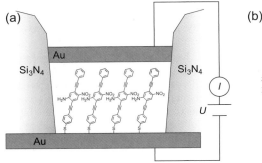
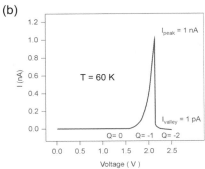

Figure 18:
(a) Schematic presentation of the NDR-device. SAM of **13** between two Au electrodes laterally limited by Si_3N_4 walls.
(b) The *I-V* characteristics of the device displaying an negative differential conductance (NDR). The Q values are suggested to refer to charge (in electrons) set by the bias voltage according to the theoretical studies [55].

4.3 Nanopore Concept

Another interesting device architecture is the use of tiny holes in a silicium membrane as the well-defined surrounding of self-assembled monolayers. The processing is based on standard microtechniques [53] and starts with the formation of a suspended Si_3N_4 membrane. A 50 nm low-stress Si_3N_4 is deposited by chemical vapour deposition on both sides of a double-side polished Si-wafer, followed by an opening of the nitride coating on the back surface in an area of 400 µm × 400 µm. An anisotropic etch (KOH solution at 85°C) was used to remove the Si wafer and to obtain a Si_3N_4 membrane, 40 µm × 40 µm in size. 100 nm SiO_2 was grown on the Si sidewalls to improve the electrical insulation. By means of e-beam lithography and RIE, a single hole of approx. 40 nm in nominal size was introduced into the membrane. The RIE rates were adapted to achieve a bowl shaped hole into the membrane, which led to a further reduced diameter of the pore on the bottom of the membrane. A Au top electrode was placed on the membrane by evaporation to fill the pore with Au. The specimen was immersed in a solution containing **13** leading to a SAM on the Au surface. After careful (low-temperature) application of the bottom Ti/Au or Au electrode, the specimens were electrically characterized. It was found that an amino and nitro substitution in the middle aryl ring leads to a strong NDR effect at 60 K with an I_{peak}/I_{valley}-ratio of approximately 1000 [23], exceeding the corresponding values of semiconductor tunneling diodes (Figure 18, b). Molecules without the amino group exhibit a NDR effect at room temperature, with a peak/valley ratio of 1.5 [54]. The *I-V* curve is fully reversible. The nature of this strong effect is still under debate.

4.4 Mechanically Controlled Break Junctions

Scanning Probe methods can in principle be used to make chemical contacts to both sides of the molecule. The advantage is that images and contacts can be made with the same instrument. However, the contacts are not symmetric, because the tip and the planar surface differ in both shape and (often) material. Another disadvantage is a lack of drift stability as soon as the distance-control feedback-loop is switched off. Complementary techniques to make stable test electrode pairs are required. The use of high-resolution lithography and shadow mask techniques allows the fabrication of metallic structures with a width in the order of 10 to 20 nm. These structures are still about one order of magnitude too large for single molecules, but can serve as a starting point for new methods. The goal of further processing is to reduce the size of the lithographically fabricated structures in a controlled manner. Metallic wires with predetermined breaking points are fabricated and subsequently treated to open the hyphenation points to tiny voids. Two examples of this type are the mechanically controlled break junction technique and the electromigration technique.

First attempts to immobilize molecules between a pair of electrodes of a mechanically controlled break junction (MCB) are reported by the groups of Tour and Reed [31]. A notched gold wire was mechanically broken while it was exposed to a 1,4-dithiobenzene solution. The gold surfaces are soon covered with strongly adhering self-assembled monolayers (SAM) of the dithiol compound. After the solvent evaporated, the tips were slowly moved together until the onset of conductance was achieved. Non-linear *I-V* curves were measured repeatedly. Unfortunately, up to now no microscopy technique is able to observe the molecules in the junction. Hence, the only information available is the conductance data. In this case, the authors suggested that the junction is formed by only one molecule bridging the gap. However, a final proof of this hypothesis was lacking.

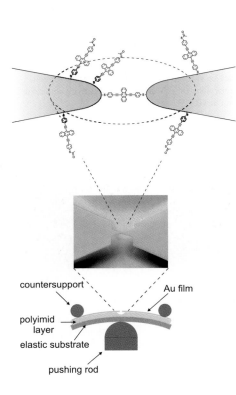

Figure 19: Scheme of a mechanically controlled break junction. On top of a bendable substrate a Au film is patterned with a freely suspended bridge in the center (see SEM micrograph in the middle panel). By bending the substrate in a three point support (lower panel: the pushing rod is driven by a motor), the Au bridge can be broken into two electrodes. By bending the substrate back and forth, the gap between the electrodes can be tuned with a distance resolution much better than Angstroms. This setup can be used to match the electrode gap precisely to the length of a molecule and finally to contact a molecule from two sides via well defined chemical S-Au bonds.

The use of high-resolution e-beam lithography provides access to metallic structures such as wires with much better defined predefined breaking-points than a notched gold wire. These high-quality MCBs were originally used and designed to observe single-atom junctions between metals at low temperatures [43]. Bourgoin and co-workers first reported the immobilization of molecules in such advanced MCBs [32]. They immobilized molecular rod-like structures consisting of three thiophene units functionalized at the ends with thiols. In comparison to a rather simple theoretical model, the number of investigated molecules is assumed to be very small or even a single one. However, there are no convincing experimental arguments with regard to the number of molecules.

To resolve this ambiguity, a systematic comparison of measurements with different molecules was very useful. In this experiment, the electrode pair was prepared in the following way (similar to the previously mentioned experiment): A flexible substrate was covered with an insulating polyimide layer. A gold structure consisting of two large areas connected with a thin gold bridge has been deposited lithographically on top. Reactive ion etching removes the polyimide around the gold structure, underetches the gold bridge and leaves it freely suspended. The final structure on the substrate is shown in Figure 19. The substrate is fixed in a three point set-up that makes it possible to mechanically bend the substrate (Figure 19). In a vacuum chamber the substrate is carefully bent while the electric resistance between the two large gold areas is monitored. In this way, the thin gold bridge is elongated until it finally breaks, which is observed as an immediate increase of the resistance. Release of the bending tension allows to approach the two broken ends of the gold structure. The extremely flat architecture results in a distance resolution of the two electrodes better than a tenth of an Ångstrom. The set-up turned out to be ideal to immobilize molecular rods with the length in the order of a few nanometers [33].

Two molecular rigid rods **17** and **18** consisting of *para* acetylene-connected aromatic cores as delocalized π-systems, with thiol functions at each end and a length of 2 nm have been synthesized as shown in section 6.1 [44]. Both molecules were very similar concerning their length, functional endgroups, and polarizability. The main difference of both structures, however, is their spatial symmetry. Because of the symmetry plane in the center of rod **17** perpendicular to its axis (cf. Figure 20) the conductance properties are expected to be independent of the current direction and hence the differential conductance $dI/dV(V)$ should be symmetric with respect to positive/negative voltages. This symmetry is absent for rod **18** due to the acetylamine and the nitro substituents at the central aromatic core. Thus, asymmetries in $dI/dV(V)$ are expected. The conductance data observed in the experiment reflected indeed the symmetry properties of the molecules (see Figure 21). This lead to the following important conclusions: First, it was proven that the junction was formed of the sample molecules (and not, for example, of undesired adsorbates). Further, the asymmetric shape of $dI/dV(V)$ which appeared randomly in approximately mirrored shape, indicated that only one molecule was contacted: an ensemble of many molecules would have resulted in a symmetric $dI/dV(V)$ because of averaging over randomly oriented molecules. Together with further considerations, this showed that indeed single molecules were observed. A very important property of single-molecule junctions was also reported in these experiments: all stable junctions displayed conductance curves which had a couple of common features, but differed nonetheless substantially. This is not surprising: as the molecule is chemically connected to both electrodes, the limitations of the molecule are not clear: the organic molecule, but also the randomly arranged metal atoms in its vicinity form the molecular wave function and determine the electronic properties of the junction. Because of small random variations in the atomistic electrode arrangements, sample-to-sample variations are observed, which are a typical and expected feature of a single-molecule experiment.

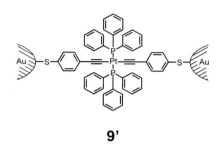

Figure 20: Rigid molecular rods **17'**, **18'** and **9'** immobilized between two gold electrodes with terminal sulfur groups.

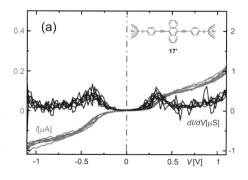
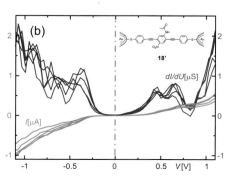
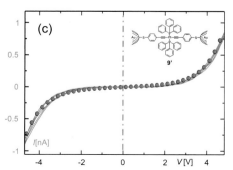

Figure 21: *I-V* curves (red) and differential conductance dI/dV (blue) for a) Au-**17'**-Au and b) Au-**18'**-Au. c) *I-V* curve (red) for Au-**9'**-Au and simulated current with a barrier height of 2.5 eV (blue circles) are shown.

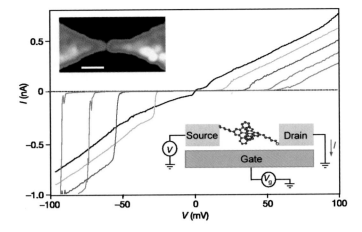

Figure 22: A single molecule transistor with an organometallic molecule [46]. The electrode pair with ~ 2 nm spacing has been manufactured by the electromigration technique (see SEM picture in the inset). The Si substrate served as gate electrode, separated by a SiO_2 insulating layer. If the charge of the central Co ion is fixed to either Co^{2+} or Co^{3+}, no current can flow at small bias voltage due to Coulomb blockade (colored lines). If the gate is tuned such that Co^{2+} or Co^{3+} have the same energy, the ion can be continuously charged or uncharged: current can flow even at small bias voltages (black line).

These findings, displaying the potential of the MCB technique to investigate single molecules, encouraged further research to control the electronic properties by careful design of the molecules structure. The first example is the adjustment of the resistance caused by the molecular structure in the MCB device. Both molecular rods discussed above are delocalized π-systems of very similar structural features like connections of the subunits and length. The similar overall resistance (~1MΩ) of the device with both structures is not surprising. If the conjugation of the molecular rod is interrupted, an increase of the overall resistance is expected. As *trans*-platinum(II) ethynyl complexes are known as conjugation-passive but rigid spacers, this motive was chosen to design the molecular rod **9** [18]. Its stiffness allows the design of rod-like structures and the pure σ-character of the Pt-C(sp) bonds is expected to divide the molecules' π-system. A single molecule junction formed with this rod-like molecule in a MCB displayed an increase of the overall resistance by several orders of magnitude (~5GΩ) compared with the conjugated structures **17** and **18**. The entire *I-V* curve displays the properties of an insulator and was best fitted with a simple model of a rectangular barrier (height 2.5eV). Remarkable is that both, the insulating effect and the experimental set-up, survives up to a bias voltage of 5 V, displaying the stability and therewith the interesting application potential of such molecular units as electronic components. In future HME and MME devices, such *trans*-platinum(II) complexes may serve as insulating linkers and as well defined tunnel junctions on a single molecule level.

4.5 Electromigration Technique

Another experimental set-up to observe transport features through single nanoscaled object is the electromigration technique. It benefits as well from the weakness of a hyphenation point in a fabricated metallic wire. Here, the hyphenation point is opened to a tiny void by application of a moderate electric current. The current flow causes electromigration of metal atoms and the metal wire breaks up at the bottleneck. The electrodes thus resulting have a distance of about 1 to 3 nm, which can easily be estimated by the resistance. In the experiment, several electrode pairs are fabricated and the most appropriate ones are selected. As these gaps can be made on a chip with a conducting substrate (covered with an insulator) which can serve as additional gate electrode, this set-up is suitable to make a nanoscale field effect transistor. Indeed, transistor properties have been described with colloidal cadmium selenide nanocrystals [45], [47], fullerene (C_{60}), and inorganic complexes (see Section 3.4 and Figure 22). The advantage of this method is that the electrode gap is close to the surface and the electrodes are stable. This allows for systematic studies like gate voltage dependence, temperature dependence or magnetic field dependences. In contrast to MCB experiments, the distance can not be manipulated further after the junction has formed. Hence, both methods give complementary opportunities for experiments.

The investigations described so far enable the conductance properties of single molecules to be investigated. The wealth of phenomena observed as well as the accuracy of the investigations are constantly increasing. One main focus of these activities was dedicated to the understanding of how different molecules generate different conductance properties. Thus a picture evolves which allows to design the electronic properties of a device by tailor-made molecules. The applicability of these experimental concepts as interface in a future hybrid device remains disputable.

Figure 23: Quantum chemistry can only be used in a finite system with a fixed electron number. To compute not only the organic molecule, but also the interaction with the electrode, one considers a *super-molecule*, including parts of the metallic electrode.

5 Simulation and Circuit Design

5.1 Theoretical Aspects

To understand electron transport through molecules a better theoretical description of the processes is required.

If the molecules are attached to the surface only by weak van der Waals forces, the molecule and the electrode can be treated independently in good approximation. In this case of weak coupling, electron transport is manifested by tunneling processes: an electron from one electrode is tunneling on the molecule, remains there for a while and continues to tunnel to the opposite electrode. A possible description of the current can be given by rate equations. Electrons are charged particles and Coulomb blockade comes into play (see Chapter 16), which is able to suppress the current at small bias voltages. The description is then similar to the description of a *quantum dot* with a large level spacing (of the order of 1eV). In such a set-up equivalent effects like in weakly coupled quantum dots can be found, such as single-electron tunneling [46].

The description becomes much more challenging when the molecule is better coupled to the electrodes. Then the molecular orbitals hybridize with the metallic states in the leads, which yields a broadening of the energy levels and a higher conductance. This occurs, for example, in thiol-ended molecules chemically connected to two metal electrodes. The electron transport adopts a more wave-like character which can be handled according to the following procedure.

The first task is to compute the wave function of the junction in equilibrium (i.e. when no current is flowing). This is already very difficult because traditional quantum chemistry has been developed to compute the wave function of an isolated molecule with fixed integer electron number. In our case, both requirements are not fulfilled: the molecule is chemically connected to the semi-infinite metallic leads and electrons are continuously exchanged between them. Thus, we have rather an open system with no fixed electron number and no borders. The usual workaround is that one separates the system into three separate parts: the two leads and the molecular junction itself. But where should the borderline between the lead and the molecule be drawn? The best approach is to involve the first metal atoms into the quantum chemical computation, forming a super-molecule, which describes the metal-molecule coupling on the well-controlled quantum chemistry level (Figure 23). At this point, some assumptions have to be made about the atomic form of the molecule-metal bonds as well as the shape of the electrodes (plane surfaces or sharp tips). It should be noted, that this choice strongly affects the results [44], [57]. Then, the coupling of the super-molecule to the leads can be done between metal and metal, which is less critical. The more metal atoms are involved in the super-molecule, the more accurate are the results with respect to the interaction between the molecule and the electrode surface.

We have so far only described how to calculate the equilibrium wave function at zero voltage. If we are interested in the conductance, some methods developed for the physics of nanostructures can be applied. In particular, the Landauer-Büttiker picture has successfully been used to describe the electronic conductance (cf. Chapter 3). Essentially, electrons are considered as plane waves, coming e.g. from the left side and which may be transmitted through the molecular junction to the right side with a certain proba-

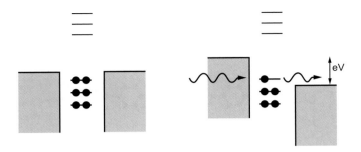

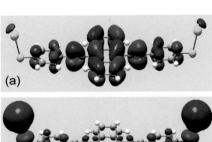

Figure 24: Simplified picture of the conductance process: At low voltages, no molecular orbital contributes to the current (left). As soon as the first molecular orbital comes in the energy window spanned by the bias voltage, conductance is possible (right).

bility or alternatively be reflected. The system is thus treated as a scattering object. The scattering matrix *t* can be computed when the wave function is known. The **transmission function** T (or transparency) results from

$$\text{Tr}(t^\dagger t) \qquad (1)$$

where t are the scattering matrix elements which describe the scattering from the states on one electrode to those of the opposite electrode.

Now the current I can be calculated by means of a Landauer-type formula [58], [59], [60]

$$I(V) = \frac{2e}{h}\int T(W,V)f(W+eV/2) - f(W-eV/2)dW \qquad (2)$$

where f is the Fermi-Dirac distribution function, W the energy, and V the applied voltage.

Some results of the theory:

- The potential difference V of the electrons in the left and the right electrode spans an energy window. The basic mechanism is shown in Figure 24, where the levels are for simplicity drawn as sharp levels. At low voltage V, the conductance is poor because no molecular level can carry electrons. When V is further increased, the first molecular orbital will enter into the energy window: the molecule starts to conduct. Hence, one would expect a step-like increase of the current as a function of the voltage. Due to the hybridisation of the molecule and the metallic states in the leads, however, all these molecular orbitals are strongly broadened in energy, which yields a rounding of the current steps and might also contribute to conductivity even at lowest voltages.

- The thiol-bonds, which are the anchor group between the molecule and the metal surface, act as tunnel barriers and suppress the amplitude of the current. When a finite voltage is applied, a large fraction of the voltage drops along these barriers. Therefore, the chemical environment of the tunnel barriers is very important for the transport properties: the conductance varies strongly, if the molecule-metal-bond is arranged differently.

- The contribution of the current through different molecular orbitals may strongly vary. This is illustrated by calculations shown in Figure 25. In this example, the contribution of the HOMO or LUMO is in certain examples not dominating because the charge of the HOMO is accumulated in the center of the molecule, while that of the LUMO is depleted in the center. Other MOs, although being far away in energy may carry much more current due to a favourable regular spatial structure.

- This procedure can now be refined in several aspects. For example, the influence of the applied voltage on the electronic structure and even the atomic structure of the molecule can be considered by recomputing the (super-)molecule at each voltage in the corresponding electric field. Furthermore, the current flowing at finite voltage may cause some stationary redistribution of the charge. This requires a so-called non-equilibrium transport theory, which has to be adapted to the special situation of a molecular junction.

Figure 25: Spatial distribution of the charge density of three different molecular orbitals (MO's), calculated with DFT for the example of the same molecule as in Figure 1. In this calculation, only one gold atom was considered to model the influence of the electrodes [67]. The HOMO (a) as well as the LUMO, (b) concentrate in the center / at the ends of this supermolecule and contribute little few to the current. The MO (c) although being ~1 eV below the HOMO, contributes considerably to the current due to its homogeneous charge distribution along the junction.

Consequently, it makes no sense to assign a conductance to a molecule. The conductance can only be defined for the compound electrode-molecule-electrode system. A lot of interesting questions have not yet been answered by theory. One important example is the heat arising in the molecule when a current is flowing and inelastic scatter processes become important. Where does the heat occur? How does the molecule carry the heat, i.e. the vibrational energy away? Of course, too much heat will inevitably destroy the organic molecule and therefore, this is a crucial problem for all technical applications.

5.2 Design Rules for Molecular Nanocircuits

In general, circuit design and analysis is based on the application of Kirchhoff's laws: (1) The charge conservation implies that the sum of all currents I_k into a circuit node must be zero

$$\sum_k I_k = 0 \qquad (3)$$

(2) The fact that the electrostatic potential is path-independent (in the absence of time-dependent magnetic fields) implies that the sum of all voltages in a mesh must be zero

$$\sum_i V_i = 0 \qquad (4)$$

In conventional solid state circuits, the impedances of the components in a circuit are independent of each other because the coherence lengths for electron transport in metals and semiconductors at room temperature are much smaller than the overall device dimensions. Therefore, the components are treated as lumped elements and the circuit design and analysis exploits the principle of superposition of the impedances of the individual components. In conventional electronics, the applicability of this **superposition principle** is restricted to frequencies below which the wavelength of the electromagnetic fields becomes comparable to device dimensions. At microwave frequencies this condition is not valid any more and, hence, microwave circuit design and analysis treat the components as distributed elements which make it necessary to take into account the geometrical layout and a simultaneous modeling of the entire circuit using the electromagnetic field theory.

Similarly, components can not be treated independently when molecules are connected in MME devices parallelly or serially. If, for example, two molecular branches are connected in parallel, the wave function of one branch hybridizes with the other branch and modifies thus the properties of both branches (Figure 26). This is fully consistent with the statement in the former section that it makes no sense to define the conductance only of a molecule, but rather of the compound junction. The following points may explain some mechanisms, as to why the *conductance* of one molecular subunit depends strongly on the other molecular subunits around and is thus badly defined:

1. effects on the density of states and the geometry of molecular orbitals (MOs),
2. Coulomb blockade effects due to the discrete nature of the electric charge,
3. interference effects.

Category 1. Adding a branch in a given molecule modifies the electronic structure of the entire molecule, i.e. the geometry and the energy levels of the MOs will be changed. We already pointed out in the last section (cf Figure 25) that the conductance is very sensitive to small details of the spatial electronic distribution.

The influence on the energy levels of the MOs upon changes of the structure are well known from traditional molecule spectroscopy. A very simple example of this fact is presented in Chap. 5 / Figure 17 in which the MO energy diagram is shown for polyenes of different chain length. The addition of each chain element (ethenyl group) increases the density of states and decreases the HOMO-LUMO gap.

Category 2. The Coulomb blockade effect of nano-sized capacitors and their impact on the charge transport through nanostructures are described in Chap. 16. In molecules, due to their smallness, these effects are particularly important. Hence, molecular subunits which are in close proximity, have very strong mutal capacitive interaction. This affects not only the static wave function, but in particular electronic transport: the *conductance* of a molecular subunit depends strongly on whether the next subunit is temporarily charged or not.

Category 3. If a wave-like current is flowing through different parallel branches of a molecule, the two partial waves may interfere constructively or destructively, yielding to a modification of the conductance. This can be compared with an optical analogue: the Fabry-Perot interferometer.

Hence, the understanding of MME devices is so far very poor. The design rules are not evident. Detailed theoretical studies which attempts to simulate electronic functions with MME devices for certain examples can be found in Ref. [65] – [66]. Presumably, one has to think in a modular way. Once the electronic function of a molecular unit is understood, it has to be connected to another unit by a well defined node which suppresses the mutual interaction. The search for such decoupling building blocks is therefore very important.

If HME devices are connected together and the anorganic part is sufficiently large such that phase coherence and charging of the electrode can be neglected (i.e. it serves as a *reservoir*), the interaction between the modules can be neglected and the conventional circuit rules apply.

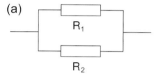

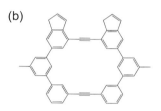

Figure 26:
(a) If two resistors are in parallel connection, the overall conductance G results as $G = G_1 + G_2$.
(b) Sketch of a MME device with two different branches. The molecular unit in the upper branch is strongly interacting with the lower branch. Thus, no G_1 or G_2 can be reasonably assigned to each branch.

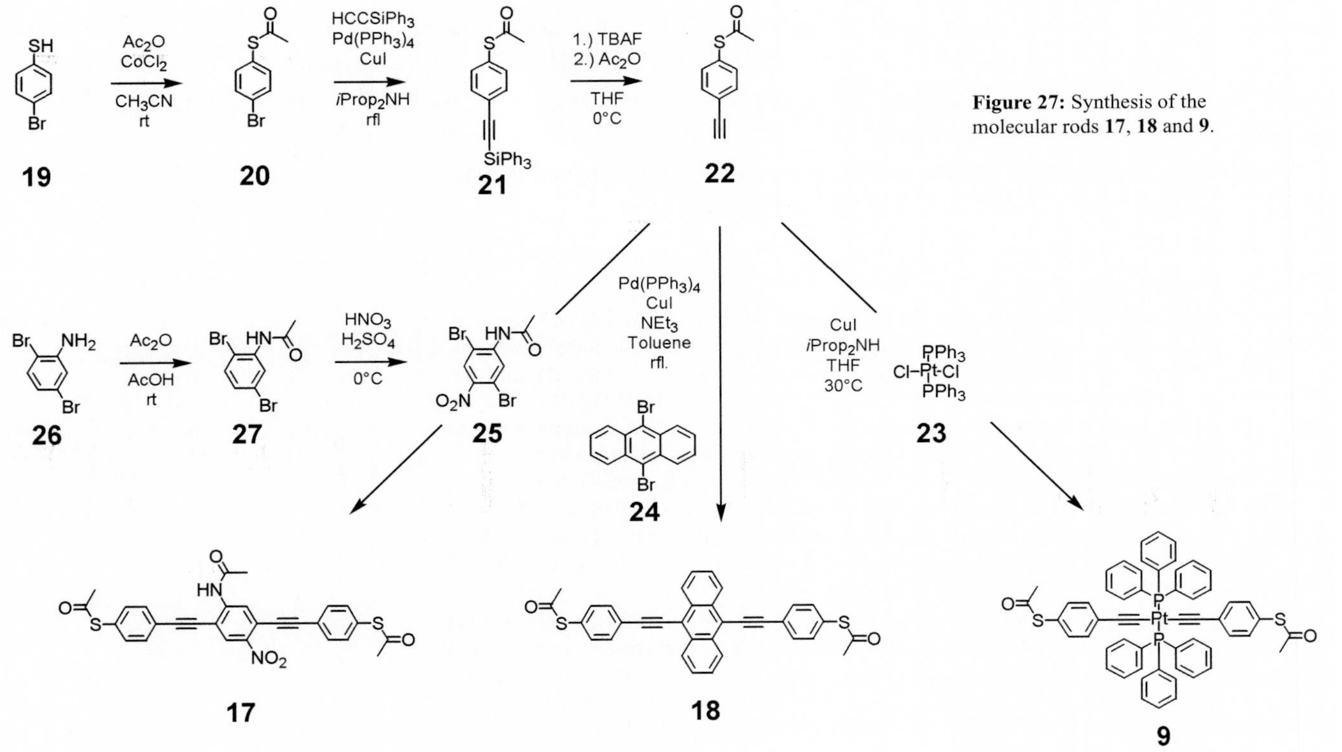

Figure 27: Synthesis of the molecular rods **17**, **18** and **9**.

6 Fabrication

The fabrication of molecular devices depends strongly on the individual type of device and a general description is hardly possible. However, both crucial steps of device fabrication, the synthesis of the molecular components and their integration will be discussed here using specific examples.

6.1 Chemical Synthesis

The routes towards the molecules employed in the electronic functions described above are based on elementary and advanced synthetic strategies and methods of organic, inorganic, metalorganic and supramolecular chemistry. A successful synthesis is often carried out over many steps. After each step, the reaction products need to be separated and analyzed in order to verify and characterize the molecular structure. The separation of the desired compound from unwanted by-products is frequently performed through selected distillation, crystallization, and chromatographic techniques. The analyses involves e. g. NMR spectroscopy, mass spectroscopy, elemental analysis, X-ray analysis as well as optical spectroscopy. In this section, synthetic aspects will be discussed with the help of four molecules, which are part of the devices already discussed .

In Section 4.4. the electric currents through three molecular rods were discussed extensively. Here we focus on the chemical synthesis that made these rods available. Every synthesis starts with the definition of the target compound. In this particular case, the synthetic targets were designed molecules, only made to be investigated in a MCB. Prior to their synthesis was a designing and development step in which the chemist tries to optimize a multitude of different aspects. Due to the limitations of this section only two key points are mentioned. The target structure must meet numerous physical requirements (stiff, rod like structure, anchor groups, symmetry options, conductivity, solubility etc.). And at least as important, the target structure must be achievable by chemical synthesis (in a reasonable effort by a reasonable number of synthetic steps, starting materials available, etc.). Often the careful balance of many aspects leads to the final choice of synthetic targets.

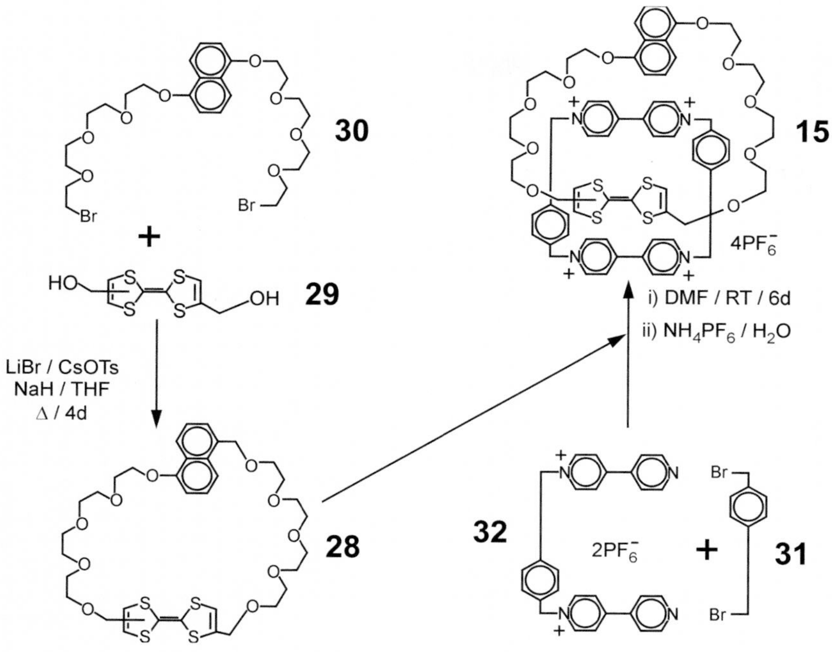

Figure 28: Synthesis of the catenane **15** [56].

The synthesis of the organic rods **17** and **18** and of the metalorganic rod **9** is displayed in Figure 27. It is striking that all three rods have a common intermediate, the phenylacetylene **22** with an acetyl protected thiol function in *para*-position. Often synthetic pathways are chosen, so that intermediates may lead to several target structures with interesting properties. The synthesis of the intermediate **22** starts with the acetyl protection of commercial *para*-bromothiophenol **19** to yield in its acetyl-protected form **20**. In a palladium(0) and copper(I) catalyzed *Sonogashira* acetylene coupling with triphenylsilylacetylene, the bromine on the aromatic core was substituted to yield in the desired *para*-thiophenylacetylene in its twice-protected form **21**. Treatment of **21** with fluorine anions deprotected both, the silyl protection group of the acetylene and the acetyl protection group on the sulfur. As the latter was required for the further steps, the crude reaction mixture was subsequently treated with acetylanhydride to reprotect the thiol to give the desired intermediate **22**. In a copper(I) catalyzed reaction with *trans*-bis(triphenylphosphine)dichloro platinum the organometallic target compound, the rod-like platinum complex **9** was synthesized. The symmetric organic rod **18** was again synthesized in a *Sonogashira* coupling as **21** above, with similar catalysts but with variations in solvents, bases and temperature. The challenge of organic synthesis is often not to find the right synthetic pathway, but to find the right conditions for a desired product. The acetylene functions of **22** replaced the bromines of commercial 9,10-dibromoanthracene **24** to yield in the symmetric organic rod **18**. To synthesize the asymmetric rod **17** the required dibromo-benzene **25** with an acetyl-protected amino group and a nitro group had to be synthesized in two steps. Commercial 2,5-dibromoaniline **26** was first acetyl-protected with acetylanhydride in acetic acid to its acetyl protected form **27**. To introduce the nitro group, **27** was treated with a 1/1 mixture of nitric acid and sulfuric acid at 0°C to yield the desired intermediate **25**. Similar coupling conditions as for the symmetric rod applied to **22** and **25** gave the asymmetric rod **17** as the desired target compound. All three molecular rods were synthesized with acetyl-protected terminal thiol functions, as these protection group turned out to be ideal for their immobilization between the Au electrodes of a mechanically controlled break junction (MCB). This protection group is removed spontaneously on Au surfaces but has the advantage of reducing the kinetic of SAM formation on the Au surface [61]. All target compounds and intermediates during the course of the synthesis were fully analyzed by classical chemical methods, like ^{1}H- and ^{13}C-NMR spectroscopy, mass spectroscopy and elemental analysis.

The second synthetic example demonstrates how supramolecular structures arise. Supramolecular chemistry describes the interaction between molecules beyond covalent bonds, which is a rather recent discipline in chemistry. It evolved in the 1970s of the last century and for their pioneering work in the field, Cram, Pedersen, and Lehn received the Nobel prize for Chemistry in 1987. A typical example of a supramolecular structure is Stoddard's catenane, discussed in the sections 3.3 and 4.2. Both ring molecules of the catenane are interlocked and not fixed by a covalent bond. But how does one interlock two molecules? From the very beginning, the two rings are carefully designed by Stoddart to interlock each other. The synthesis of the catenane is displayed in Figure 28. First the cycle **28** containing a redox active tetrathiofulvalene (TTF) unit is closed. Therefore a building block consisting of a TTF with two benzylic alcohols **29** reacts with a polyether with two terminal alkylbromines **30** as leaving groups. In a nucleophilic substitution reaction the deprotonated alcohol functions of **29** replaced the bromines of **30**. This reaction was carried out in sufficiently dilute conditions to favor the formation of the cycle **28** rather than chain-like polymers. The formation of the second ring was based on a nucleophilic substitution reaction too. The bromines of 1,4-di(brommethyl)benzene **31** were replaced by the terminal nitrogen atoms of the 1,4-di(bipyridinium)benzene $(PF_6)_2$ salt **32** to yield the second ring. The trick necessary to get the catenane, is to make this second ring closure reaction in the presence of the first ring **28** under appropriate reaction conditions. Due to supramolecular interactions, the precursor **32** is preorganized in the ring **28** and hence, substantial amounts of the second ringclosure reaction resulted in the desired catenane **15**. Such supramolecular structures are purified and analyzed with the same established chemical methods as mentioned above. The structure of catenene **15,** for example could even be analyzed by X-ray diffraction.

6.2 Integration Processes

The integration of the basic logic and memory function within the MME concept is, indeed, performed by chemical synthesis. All synthesized molecules would be able to perform the designated functions. The remaining task is to interconnect them among each other and to the outside world. This task is similar to the tasks required for fabricating HME devices. The steps of this fabrication are similar to the standard micro- and nanoelectronic processes (see Part II). For circuit fabrication deep in the (few) nanometer regime, bottom-up techniques based on self-assembly and self-organization steps will gain importance. The additional challenge is to integrate a – typically monomolecular – *organic* film or *organic* molecules into the entire device. This fact imposes several severe limitations on the selection of the process steps and process parameters because of the low temperature budget of the organic molecules, their sensitivity to oxygen at elevated temperatures, and their mechanical softness. The process strategy requires all high-temperature process steps to be performed prior to the deposition of the organic film. An example of a fabrication strategy which, in principle, is suitable for mass-production has been sketched in Section 4.2. A sequence of pictures of the cross-bar Pt/molecule/Ti system is shown in Figure 29 for a wide range of magnification factors. As mentioned above, an interpretation of the observed characteristic is not yet possible.

7 Summary and Outlook

This chapter presents the potential of carefully designed molecular structures to act as specific electronic components. They may be the core of very small devices, which may be fabricated by means of self assembly.

In the past, important preparatory work came from the chemistry of molecules in solution. Oligomers as molecular conductors and molecular switches have been investigated spectroscopically and by electrochemistry in solution. Some correlations between molecular structure and physical properties were elaborated.

In more recent years, the immobilisation of molecules on metallic electrodes allowed experiments, where a steady-state current flows through the molecules. Molecular films, sandwiched in between metallic electrodes, were shown to act as diodes, configurable switches (memories), transistors and negative differential resistance devices. Single molecule junctions have been investigated with STM techniques on surfaces and by means of the mechanically controlled break junction technique. It was demonstrated that the electronic properties of the device depend strongly on the choice of appropriate molecules, which opens up an infinite space of possible variations and design possibilities. The properties of the junction, however, also strongly depend on the electrode material, its shape and the chemical bonds between the molecule and the leads. The challenge is to understand these complex systems and to convert this increasing knowledge into useful devices. To reach this goal, also theoretical advances in these interesting systems are required.

Reproducibility and stability still remains a challenge in molecular electronics. The molecule itself is often pretty stable and might not be the reason for short lifetimes. It is rather the metallic electrodes that provide instability because the movement of few metal atoms may strongly affect the conductance properties of the whole device. While single molecule devices are a very interesting tool for basic research, they do not seem to be suitable for technological applications. The lifetime, in particular when a voltage is applied and a current is flowing, is insufficient. A molecular film device, representing a larger ensemble, might be more suitable and might be integrated soon into existing circuitry. Nevertheless, a good reproducibility requires the absence of defects and high stability of the electrodes and the molecular film. A great effort is necessary to better engineer the electrodes.

The key problem of molecular electronics, however, remains the interface to the molecules and thereby their integration and addressability. To overcome this problem great efforts are still required. The implementation of molecular electronics into traditional silicon architectures requires a 100% yield on a chip. This might not be achievable when self-assembly is used. New, fault-tolerant architectures are probably needed. A promising approach to cope with huge but imperfect computer systems is the Teramac project of Agilent Technologies [62], as outlined in Section 4.2.

In conclusion, while the ongoing miniaturisation in electronic circuits continues, the use of molecules in electronic circuits seems to be likely. Despite the many challenges to be met, the concept of molecular electronics shows an outstanding potential for the long-term future of the information technology because of the ultimate density of logic and memory functions, their low fabrication costs, their low power consumption, and the huge number of opportunities different by the plethora of possible molecules.

Acknowledgements

The authors are grateful for discussions with Ari Aviram, Detlef Beckmann, Jean-Philippe Bourgoin, Luisa De Cola, Michelle Di Leo, Mark Elbing, Ferdinand Evers, Harald Fuchs, Carsten von Hänisch, Jim Heath, Matthias Hettler, Christian Joachim, Jean-Pierre Launay, Jean-Marie Lehn, Hilbert von Löhneysen, Mark Ratner, Mark Reed, Joachim Reichert, Mario Ruben, Günter Schmid, Christian Schönenberger, Ulrich Simon, Christophe Stroh, Daniel Vanmaekelbergh, Florian Weigend, Wolfgang Wenzel, René Williams, Stanley Williams, Olaf Wollersheim, Sophia Yaliraki. Thanks are due to Stephan Kronholz, Björn Lüssem, Lars Müller-Meskamp (FZ Jülich) for corrections.

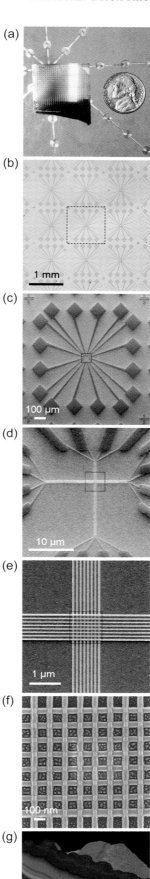

Figure 29: Cross-bar array of reconfigurable switch junctions fabricated at Hewlett-Packard laboratories. An outline of the fabrication process is given in Sec. 4.2. The photos show an increasing magnification from (a) to (g). (a), (b) optical micrographs, (c) – (f) SEM micrographs, (g) AFM micrograph. Courtesy of Yong Chen at Hewlett-Packard Laboratories, Palo Alto.

References

[1] J. M. Tour, Acc. Chem. Res. 33, 791 (2000).
[2] A. Aviram and M. R. Ratner, Chem. Phys. Lett. **29**, 277 (1974).
[3] A. Aviram and M. R. Ratner, Ann. N. Y. Acad. of Sci., **852** (1998).
[4] A. Aviram, M. R. Ratner and V. Mujica, Ann. N. Y. Acad. of Sci., **960** (2002).
[5] C. Joachim, J. K. Gimzewski, A. Aviram, Nature **408**, 541 (2000).
[6] M. A. Reed and J. M. Tour, Scientific American **282** (2000) 86.
[7] W. B. Davis, W. A. Svec, M. A. Ratner, M. R. Wasielewski, Nature **396**, 60 (1998).
[8] J.M. Tour, Chem. Rev. **96**, 573 (1996).
[9] K. Müllen and G. Wegner, *Electronic Materials: The Oligomer Approach,* Wiley-VCH, 1998.
[10] R. Kuhn, Angew. Chem. **50**, 703 (1937).
[11] W. Kern, M. Seibel, H.-O. Wirth, Makromol. Chem. **29**, 164 (1959).
[12] W. Steinkopf, R. Leistmann, K.-H. Hofmann, Lieb. Ann. Chem. **546**, 180 (1941).
[13] G. Drefahl, G. Plötner, Chem. Ber. **94**, 907 (1961).
[14] G. Drefahl, G. Plötner, Chem. Ber. **91**, 1274 (1958).
[15] Z. J. Donhauser, B. A. Mantooth, K. F. Kelly, L. A. Bumm, J. D. Monnell, J. J. Stapleton, D. W. Price Jr, A. M. Rawlett, D. L. Allara, J. M. Tour, P. S. Weiss, Science **292**, 2303 (2001).
[16] M. Uno, P.H. Dixneuf, Angew. Chem., **110**, 1822 (1998); Angew. Chem. Int. Ed. Engl. **37**, 1714 (1998); H. Fink, N.J. Long, A.J. Martin, G. Opromolla, A.J.P. White, D. J. Williams, P. Zanello, Organometallics **16**, 2646 (1997).
[17] M. Mayor, J.-M. Lehn, K.M. Fromm, D. Fenske, Angew. Chem., **109**, 2468 (1997); Angew. Chem. Int. Ed. Engl., **36**, 2370 (1997); M. Mayor, J.-M. Lehn, J. Am. Chem. Soc. **121**, 11231 (1999).
[18] M. Mayor, C. von Hänisch, H. B. Weber, J. Reichert, D. Beckmann, Angew. Chem. Int. Ed. **41**, 1183 (2002).
[19] P. F. H. Schwab, M. D. Levin, J. Michl, Chem. Rev. **99**, 1863 (1999).
[20] A. S. Martin, J. R. Sambles, G. J. Ashwell, Phys. Rev. Lett. **70**, 218 (1993); C. M. Fischer, M. Burghard, S. Roth, K. von Klitzing, Appl. Phys. Lett. **66**, 3331 (1995); R. M. Metzger, Acc. Chem. Res. **32**, 950 (1999).
[21] T. Xu, I. R. Peterson, M. V. Lakshmikantham, R. M. Metzger, Angew. Chem. Int. Ed. **40**, 1749 (2001).
[22] D. R. Stewart et al., Nano Letters (2004), **4**, 133–136
[23] J. Chen, M. A. Reed, A. M. Rawlett, J. M. Tour, Science **286**, 1550 (1999).
[24] R. Sessoli, D. Gatteschi, A. Caneschi, and M. Novak, Nature **365**, 141 (1993).
[25] B. L. Feringa, *Molecular Switches*, Wiley-VCH, 2001.
[26] M. Irie, Mol. Cryst. Liq. Cryst. **227**, 263 (1993).
[27] V. Balzani, A. Credi, F. M. Raymo, J. F. Stoddart, Angew. Chem. Int. Ed. **39**, 3349 (2000).
[28] M. Asakawa, P. R. Ashton, V. Balzani, A. Credi, C. Hamers, G. Mattersteig, M. Montalti, A. N. Shipway, N. Spencer, J. F. Stoddart, M. S. Tolley, M. Venturi, A. J. P. White, D. J. Williams, Angew. Chem. Int. Ed. **37**, 333 (1998).
[29] C. P. Collier, G. Mattersteig, E. W. Wong, Y. Luo, K. Beverly, J. Sampaio, F. M. Raymo, J. F. Stoddart, J. R. Heath, Science **289**, 1172 (2000).
[30] L. A. Bumm, J. J. Arnold, M. T. Cygan, T. D. Dunbar, T. P. Burgin, L. Jones II, D. L. Allara, J. M. Tour, P. S. Weiss, Science **271**, 1705 (1996).
[31] M. A. Reed, C. Zhou, C. J. Muller, T. P. Burgin, J. M. Tour, Science **278**, 252 (1997).
[32] C. Kegueris, J. P. Bourgoin, S. Palacin, D. Esteve, C. Urbina, M. Magoga, C. Joachim, Phy. Rev. B **59**, 12505 (1999).
[33] J. Reichert, R. Ochs, D. Beckmann, H. B. Weber, M. Mayor, H. v. Löhneysen, Phys. Rev. Lett. **88**, 176804 (2002).
[34] J. K. Gimzewski, E. Stoll, R. B. Schlittler, Surf. Sci. **181**, 267 (1987).
[35] D. L. Allara, T. D. Dunbar, P. S. Weiss, L. A. Bumm, M. T. Cygan, J. M. Tour, W. A. Reinerth, Y. Yao, M. Ko-zaki, and L. Jones, Ann. N. Y. Acad. of Sci. **852**, 349 (1998).
[36] P. S. Weiss, L. A. Bumm, T. D. Dunbar, T. P. Burgin, J. M. Tour, and D. L. Allara, Ann. N. Y. Acad. of Sci. **852**, 145 (1998).

[37] J. Chen, M. A. Reed, C. L. Asplund, A. M. Cassell, M. L. Myrick, A. M. Rawlett, J. M. Tour, and P. G. Van Patten, Appl. Phys. Lett. **75**, 624 (1999).

[38] J. K. Stroscio, D. M. Eigler, Science, **254**, 1319 (1991).

[39] R. S. Becker, J. A. Golovenchko, B. S. Schwarzentruber, Nature, **325**, 419 (1987).

[40] G. Dujardin, R. E. Walkup, P. Avouris, Science, **255**, 1232 (1992).

[41] A. A. Gewirth, B. K. Niece, Chem. Rev. **97**, 1129 (1997).

[42] C. Joachim, J. Gimzewski, Chem. Phys. Lett. **265**, 353 (1997).

[43] E. Scheer, N. Agraït, J. C. Cuevas, A. L. Yeyati, B. Ludolph, A. M. Rodero, G. R. Bollinger, J. M. van Ruitenbeek, C. Urbina, Nature **394**, 154 (1998).

[44] H. B. Weber, J. Reichert, F. Weigend, R. Ochs, D. Beckmann, M. Mayor, R. Ahlrichs, H. v. Löhneysen, Chem. Phys. **281**, 113 (2002).

[45] H. Park, A. K. L. Lim, A. P. Alivisatos, J. Park, P. L. McEuen, Appl. Phys. Lett. **75**, 301 (1999).

[46] J. Park, A. N. Pasupathy, J. I. Goldsmith, C. Chang, Y. Yaish, J. R. Petta, M. Rinkoski, J. P. Sethna, H. D. Abruna, P. L. McEuen, D. C. Ralph, Nature **417**, 722 (2002).

[47] H. Park, J. Park, A.K.L. Lim, E.H. Anderson, A.P. Alivisatos, P.L. McEuen, Nature **407**, 57 (2000).

[48] W. Liang, M. P. Shores, M. Bockrath, J. R. Long, H. Park, Nature **417**, 725 (2002).

[49] C. M. Fischer, M. Burghard, S. Roth, K. von Klitzing, Appl. Phys. Lett. **66**, 3331 (1995).

[50] R. M. Metzger, B. Chen, U. Höpfner, M. V. Lakshmikantham, D. Vuillaume, T. Kawai, X. Wu, H. Tachibana, T. V. Hughes, H. Sakurai, J. W. Baldwin, C. Hosch, M. P. Cava, L. Brehmer, G. J. Ashwell, J. Am. Chem. Soc. **119**, 10455 (1997).

[51] R. M. Metzger, B. Chen, D. Vuillaume, M. V. Lakshmikantham, U. Höpfner, T. Kawai, J. W. Baldwin, X. Wu, H. Tachibana, H. Sakurai, M. P. Cava, Thin Solid Films **327**, 326 (1998).

[52] O. Kwon, M. L. McKee, and R. M. Metzger, Chem. Phys. Lett. **313**, 321 (1999).

[53] C. Zhou, M. R. Deshpande, M. A. Reed, L. Jones II, and J. M. Tour., Appl. Phys. Lett. **71**, 611 (1997).

[54] J. Chen, W. Wang, M. A. Reed, A. M. Rawlett, D. W. Price, J. M. Tour, Appl. Phys. Lett. **77**, 1224 (2000).

[55] J. M. Seminario, A. G. Zacarias, J. M. Tour, J. Am. Chem. Soc. **122**, 3015 (2000).

[56] C. P. Collier, E. W. Wong, M. Belohradsky, F. M. Raymo, J. F. Stoddart, P. J. Kuekes, R. S. Wlliams, J. R. Heath, Science **285**, 391 (1999).

[57] S. N. Yaliraki, M. Kamp, M. Ratner, J. Am. Chem. Soc. **121**, 3428 (1999).

[58] R. Landauer, IBM J. Res. Dev. **1**, 223 (1957).

[59] A. Nitzman, Ann. Rev. Phys. Chem. **52**, 681 (2001).

[60] S. Datta, *Electronic transport in mesoscopic systems*, Cambridge University Press, 1995.

[61] J. M. Tour, L.R. Jones II, D. L. Pearson, J. J. S. Lamba, T. P. Burgin, G. M. Whitesides, D. L. Allara, A. N. Parikh, S. V. Atre, J. Am. Chem. Soc. **117**, 9529 (1995).

[62] J. Heath, P. Kuekes, G. Snider, S. Williams, Science **280**, 1716 (1998).

[63] Y. Luo, C. P. Collier, J. O. Jeppesen, K. A. Nielsen, E. Delonno, G. Ho, J. Perkins, H.-R. Tseng, T. Yamamoto, J. F. Stoddart, and J. R. Heath, Chem. Phys. Chem **3**, 519 (2002).

[64] P. J. Kuekes, R. S. Williams, J. R. Heath, Patent No. US 6,314,019 B1, Nov. 6, 2001; P. J. Kuekes, R. S. Williams, Patent No. US 6,256,767 B1; see also: http://www.nanotechweb.org/articles/news/1/9/8/1

[65] J. C. Ellenbogen and J. C. Love, Proc. IEEE, 386, March 2000.

[66] S. Ami and C. Joachim, Phys. Rev. B **65**, 155419 (2002).

[67] J. Heurich, J. Cuevas, J. C. Wenzel, G. Schön, Phys. Rev. Lett. **88**, 256803 (2002).

Random Access Memories

Contents of Part IV

Introduction to Part IV	527
21 High-Permittivity Materials for DRAMs	537
22 Ferroelectric Random Access Memories	563
23 Magnetoresistive RAM	589

Introduction to Part IV

Contents

1 Definition of Random Access Devices   527
2 Physical Storage Principles   528
3 Timing schemes   531
4 General Scaling Trends for Future Memory Generations   532

1 Definition of Random Access Devices

Present-day digital information storage devices are commonly grouped into random access devices and sequential access devices. In **random access devices** the storage cells are organized in a *matrix*. Sometimes, the memory matrix is also called a memory cell array, a cell array, or simply an array. This structure facilitates short access times which are independent of the location of the data. Random access devices are used as the main (or: primary) computer memory to store instructions and data for fast access by the processing unit(s). In **sequential access devices**, the access time depends on the physical location of the data relative to the position of read/write unit(s). In computer systems, these devices are used as a large and permanent secondary memory for information archiving and retrieval purposes. They are termed mass storage devices and will be covered in Block V.

The earliest random access devices were **magnetic-core memories** (Figure 1a) which found widespread application in the 1960s. The memory cells consist of tiny wire-threaded ferrite toroids. The two states of remanent magnetization in the toroid represent the binary states, "0" and "1". For a write operation, current pulses are passed through the selected row and column wires (Figure 1b). Only at the crossing point, i.e. the cell at the addressed node, is the current large enough to switch the magnetization of the toroid (Figure 1c). For a read operation, a "1" is written into the addressed element. In the case of a "0" stored earlier in this element, the magnetization change induces a current pulse in the read line. This line runs through all toroids of the memory matrix and is connected to the input of a sense amplifier. The appearance or absence of a pulse

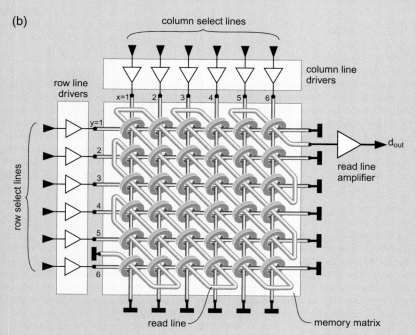

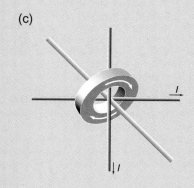

Figure 1: Magnetic core memory.
(a) Photograph of an historical 1kb ferrite-based magnetic-core memory. One side of the square has a length of 10 cm.
(b) Schematics of the magnetic-core memory.
(c) A ferrite core at the intersection (node) of the row and column lines used for writing the magnetization direction which represents the binary logic states.

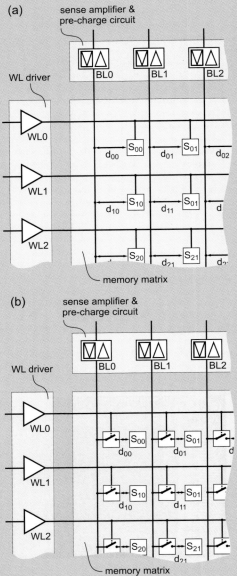

Figure 2: Configuration of matrix-based memories including the word line (WL) drivers located at the rows of the matrix and sense amplifier / pre-charge circuit units driving and sensing the bit lines (BL) located at the columns of the matrix. The sketch is generalized and simplified. The actual configuration will depend on the type of memory and design.
(a) Passive matrix.
(b) Active matrix.
S_{ik} storage elements, d_{ik} data signals
(i: WL number, k: BL number).

on the read line is interpreted as a "0" and "1", respectively. Obviously, the read operation destroys the information in the addressed core (destructive read-out, DRO) and requires a subsequent write-back operation. Magnetic-core memories are (1) true *random-access memories, RAM,* for full read and write access in contrast to read-only memories, ROM, (2) they are *non-volatile* because they do not lose their information if the power supply to the system is turned off, and (3) they are based on a *passive memory* matrix.

In a **passive matrix** (Figure 2a), the amplifying and actively switching cells are only located at the periphery of the storage matrix acting as line drivers or sense amplifiers. In this matrix, all non-addressed cells in a selected row or column experience a fraction (e. g. half) of the signal, even if they are not located at the addressed node. This requires the physical storage function to exhibit a very precisely defined threshold value for switching between the two binary states, and a high quality and reproducibility of the signals of the line drivers. The demands become more challenging with an increasing size of the matrix. In an **active matrix** (Figure 2b), an active switch, typically a select transistor, is located at every node, which significantly relaxes this threshold requirement because non-addressed cells do not experience any (disturb) signal. This benefit is purchased at the expense of additional elements, because for each storage element a switching element is needed. Usually, the rows are called word lines (WL) and the columns are called bit lines (BL) or data lines.

The **storage capacity** of matrix-based memories is determined by the product of the column number and row number. Since in digital computers the addresses are binary coded, the number of columns and rows are powers of two, e. g. 2^m and 2^n, resulting in a matrix capacity of 2^{m+n} bits (Figure 3). For example, in a chip with n + m = 20 one can store 1 Mbit (or simply Mb) of binary information, where M denotes 1024 k, with k = 1024 (which is a slight deviation from the usual definition, M = 1000 k and k = 1000). A 1 Gb RAM is established by n + m = 30, where G = 1024 M. In general, the matrix memory organization was introduced to reduce the number of lines and line drivers compared to a linear array. Obviously, the corresponding optimum is at n = m. Real RAM organizations may deviate from this condition because of different optimization criteria for WLs and BLs. While a small memory (e. g. a 64 kb DRAM chip) is typically organized in a simple matrix as sketched in Figure 3, a large capacity memory (e. g. a 64 Mb DRAM chip) comprises several hierarchical levels of subarrays and banks of matrix units on a chip in order to boost the access speed. In addition, the width of the data bus is increased. A 4 Gb RAM chip may, for instance, contain 16 x 256 Mbit banks for which read/write operations on the data are performed on a word (e.g. 16 bit) base. Details are described in Ref. [1].

2 Physical Storage Principles

For all matrix-based memories, the storage principle is based on physical states which can be read electrically by addressing the matrix element. Matrix-based memories (Figure 4) are based today on semiconductor chips and can be grouped into read & write **random access memory (RAM)**, typically used for data storage, and **read only memory (ROM)**, typically used for instruction storage. The latter group can be further divided into **once-programmable ROM** and **re-programmable ROM**. The programming of the first group is either performed during fabrication of the chip by an appropriate layout of the last metallization mask (**mask-based ROM**), which is cost-effective only in very large numbers (> 10 000 pcs.) or, in the case of programmable ROMs (**PROMs**), by the customer in a first programming step in which, for example, tiny metal bridges at the matrix nodes are either fused or left intact to represent the binary information. In the case of re-programmable ROMs, MOSFETs with an additional *floating gate* are used as storage elements. The floating gate is charged to different voltages during the programming sequence to open or close the MOSFET channel in a non-volatile fashion. Before a new programming sequence, the information must be erased by discharging the floating gates of all memory cells. For erasable PROMs (EPROMs), this discharge is performed by UV light while for electrically erasable PROMs (EEPROMs) it is conducted by an enhanced programming voltage, typically generated on the chip. Flash EEPROMs show a very simple, cost-effective architecture and short re-programming times. The re-programming needs to be applied to the whole chip as for EPROMs, in contrast to standard EEPROMs which can be reprogrammed blockwise. In all types of re-programmable ROM, the re-writing is far more time-consuming than the reading. Hence, they are used only when re-writing is rarely required.

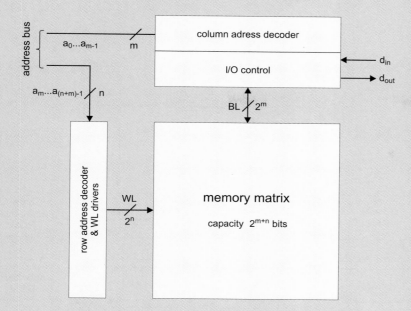

Figure 3: General memory chip configuration comprising the row address decoder and WL drivers, the column address decoder and I/O control at the bit lines including the data I/O, and the memory matrix. The address bus comprises m+n lines. The lower m address bits are used to select a column for addressing a specific BL, the higher n address bits are used to select a row for addressing a specific WL.

RAM devices are classified into volatile RAM and non-volatile RAM. The **volatile RAM** types comprise **static RAM (SRAM)** and employ flip-flop-based latches (see Chapter 6) as storage elements, while the so-called **dynamic RAM (DRAM)** uses a tiny capacitor with two different charge storage levels to represent the binary information. Due to the unavoidable self-discharge of capacitors, the information needs to be refreshed periodically. The refreshing period is a fraction of a second. Since the storage cell capacitor and the single select transistor of a DRAM cell require a much smaller area on a Si chip than the six transistors needed to make up the cell select and flip-flop of an SRAM cell, DRAMs are employed for the main primary memory in common personal computer systems. On the other hand, SRAM has a faster access time than DRAM and, hence, is used as cache memory to temporarily store frequently used instructions and data for quicker access by the processing unit(s) of a system. **Nonvolatile RAM** technologies are typically based on novel electronic materials, which are not utilized in classical semiconductor technology.

The following **physical storage principles** have been employed or are suggested for future matrix-based memory devices:

- **interlocked state of logic gates** realized as flip-flops are used for *Static RAM (SRAM)*; short description in Chapter 7.
- **magnetization** of ferromagnetic materials utilized in the *magnetic-core memories*; see Figure 1.
- **a capacitor with a (non-hysteretic) dielectric** is used for temporary charge storage in *Dynamic RAM (DRAM)*; new dielectric materials for DRAMs: see Chapter 21.
- **a capacitor with a hysteretic dielectric,** i. e. a ferroelectric, is employed in the nonvolatile *Ferroelectric RAM (FeRAM)*; Chapter 22.
- **electrical resistance** of a metal bridge (fuse) on the chip is utilized in *PROMs*.
- charge on an additional **floating gate of a FET** providing a hysteresis to the channel resistance is used in *EEPROMs* and *flash memories*,
- **polarization charges of ferroelectric gate oxides** of a FET are employed in *ferroelectric FET (FeFET)*; see Chapter 14.
- **charge confined into a nano-dot** in which the confinement leads (1) to a discretisation of the energy levels by quantum mechanics (in resonant tunnelling devices [10]; see Chapter 15) and/or (2) to a discretisation of the energy levels due to Coulomb blockade effects (in single electron devices [8], [11], [12]; see Chapter 16). These quantum devices may gain importance as the feature size F is decreasing into the deep sub-100 nm regime. Even traditional concepts such as the Flash memory principle can be modified utilizing nano-dots for further miniaturization [9].
- **electrical resistance of a magnetoelectronic tunneling junction** in which the tunneling probability depends on the direction of the magnetization is utilized in *magnetoresistive RAM (MRAM)*; see Chapter 23.

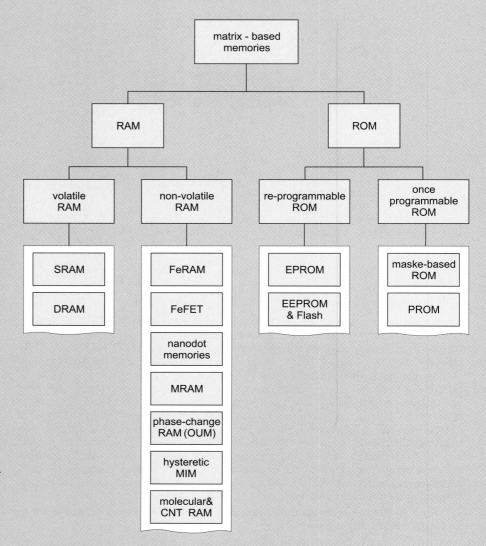

Figure 4: Categories of matrix-based memory chips showing the most relevant types. The abbreviations are explained in the text. The boundary between non-volatile RAM and re-programmable ROM is not sharp. Today, it is given by the write times, which are in the ns to µs range for the non-volatile RAMs and in the seconds range for the re-programmable ROM.

- **electrical resistance of a phase-change material** which can be thermally switched between a crystalline and an amorphous phase is used by a novel non-volatile RAM concept called OUM® (Ovonic unified memory) [2], based on the same writing principle as the rewritable DVDs; see Chapter 26.
- **electrical resistance of metal-insulator-metal (MIM) elements**, in which the insulator denotes e.g. an oxide [13]. Numerous mechanisms of the resistance hysteresis are discussed such as redox processes, ferroelectric effects, electrode and interface reactions, etc.
- **electrical resistance of a carbon nanotube (CNT)** switchable by external electrostatic potentials are described in Chapter 19.
- **electrical resistance of organic molecules,** which depends on a presettable polarization or configuration state; Chapter 20.

Other variants and combinations have indeed been reported in the literature. Still, this list provides the relevant devices realized today or showing a high potential for future systems. In most cases, the physical storage principles for RAMs are based either on the charge which is stored at the node (charge-based RAMs) or the node shows a different resistance (resistance-based RAMs) for the logic states. In some case, the physical storage principle is combined in the gate of the select transistor, for which the channel resistor is read-out (e. g. in the case of EEPROM or FeFET). An general overview of such devices is given in Ref. [7].

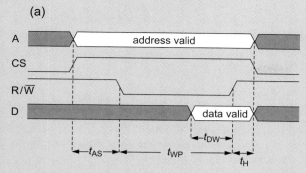
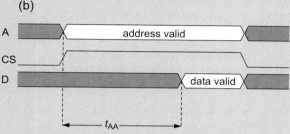

Figure 5: Timing scheme for (a) write operations and (b) read operations as seen at the external contacts of the chip. The scheme is simplified to show the operating principles.
t_{AS}: Address Setup Time
t_{WP}: Write Pulse Width
t_{DW}: Data Valid to End of Write Time
t_H: Hold Time
t_{AA}: Address Access Time

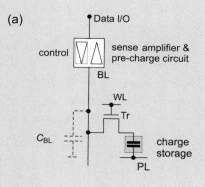

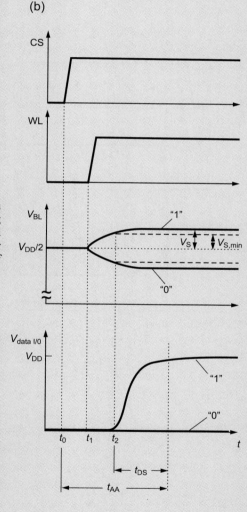

Figure 6: Charge-based RAM.
(a) Bit-line configuration showing the memory cell at one node of the matrix and the schematic sense amplifier & pre-charge circuit. The control line is driven by the R/$\overline{W}$ line.
(b) Internal read-out scheme. The symbols and signals are explained in the text.

3 Timing schemes

The simplified **timing of read and write operations** of a RAM chip are illustrated in Figure 5 [3]. The operations are initiated by selecting the chip (CS high) and applying the address A. To allow a settling of the address lines, the write signal (R/$\overline{W}$ low) should be set after a (minimum) delay time t_{AS} (Figure 5a). The write pulse width is t_{WP}. After the settling time t_{DW}, the data D are written into the addressed cells until the edge of R/$\overline{W}$, starts rising. For some RAM types, the data need to be applied for a hold time, t_H, after the rising edge of R/$\overline{W}$. Hence, the total write cycle time is:

$$t_W = t_{AS} + t_{WP} + t_H \qquad (1)$$

The read operation is shown in Figure 5b. After selecting the chip and applying the address, a (minimum) time t_{AA} is needed, including a (minimum) time t_{DS} for settling the data, before the data are valid at the data bus. t_{AA} is identical to the read access time. The timing schemes in Figure 5 are significantly simplified. In order to save I/O pins at RAM chips, the address bus is usually multiplexed, e. g. the row address is first read and latched, followed by the column address. In addition, there are many time-saving modes permitting fast access to large blocks of data which are stored in a coherent address range. Details can be found in the data sheets of the RAM producers. It should be noted that the time of the first access (so-called latency) to data in a RAM is always longer than for access to subsequent data. For example, by sophisticated timing and parallel access to many banks, data rates up to 10 Gbit/s for large data blocks can be achieved for modern DRAM modules although the (single) access time to a single DRAM cell is still in the order of 10 ns, see e. g. [4].

For a discussion of the internal **read-out scheme of charge-based RAMs**, a generalized and simplified circuit diagram of a node is shown in Figure 6a. The figure also includes the bit line capacitance, C_{BL}, the sense amplifier, and the precharge circuit. Figure 6b shows the essential part of the timing. A short time after applying the external address and the chip select CS (at time t_0), the word line WL becomes active and opens the select transistor Tr (at time t_1). Consequently, the charge is shared between C_S and C_{BL}. Hence, the voltage on the bit-line V_{BL}, which has been precharged to $V_{DD}/2$, will approach the steady state voltage

$$V_{BL} = \frac{V_{DD}}{2}\left[1 \pm \frac{C_S}{C_{BL}+C_S}\right] = V_{DD}/2 \pm V_S \qquad (2)$$

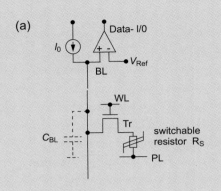

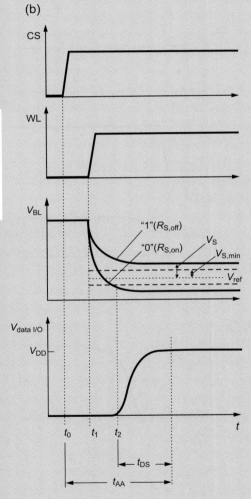

Figure 7: Resistance-based RAM.
(a) Bit-line configuration showing the memory cell at one node of the matrix and the schematic current source and the sense amplifier.
(b) Internal read-out scheme. The symbols and signals are explained in the text.

where "+" leads to V_{BL}("1") and "-" leads to V_{BL}("0"). The time constant for this process is given by R_L and $(C_{BL} + C_S)$. The resistance R_L comprises the channel resistance of Tr and the interconnect resistance, which is usually negligible. Due to tolerances and noise, the sense amplifier needs a minimum difference $V_{S,min}$ to discriminate between the two logic states, i. e. C_S must be sufficiently large to guarantee

$$V_S > V_{S,min} \tag{3}$$

where V_S is given by Eq. (2).

At time t_2, the output signal of the sense amplifier starts to approach the full data signal. After an additional delay t_{DS}, which accounts for the data settling and signal transfer, the stable data signal is available at the data output contacts of the chip. The overall read-access time t_{AA} has been introduced in Figure 5. The detailed circuitry, signals, and timing will depend on the type of charge-based RAM (e. g. DRAM or FeRAM) and specific aspects of the technical realization.

For **reading out a resistance-based RAM** (such as MRAM, FeFET, etc.), typically a constant current source is utilized. A simplified circuit diagram of a node is shown in Figure 7a, where R_s represents the switchable resistor. The timing is illustrated in Figure 7b. When the write line WL becomes active and, hence, the select transistor Tr opens, the bit line voltage drops from a precharge value (which depends on the details of the current source and the circuit) to a value which is determined by

$$V_{BL} = I_0 \cdot (R_S + R_L) \tag{4}$$

where $R_S = R_{S,OFF}$ represents "1" and $R_S = R_{S,ON}$ represents "0". The time constant for settling V_{BL} is given by C_{BL} and the sum of R_S and the line resistance R_L. The latter comprises the channel resistance of Tr and the - typically much smaller - ohmic resistance of the bit line. The reference voltage of the sense amplifier is set in the middle between the voltages V_{BL} for ON and OFF situations. As described above, the sense amplifier needs a minimum difference $V_{S,min}$ to discriminate between the two logic states, i. e. the resistance difference $\Delta R_S = R_{S,OFF} - R_{S,ON}$ must be sufficiently large to guarantee condition (3). Here, V_S is given by

$$V_S = \frac{I_0}{2} \cdot \Delta R_S \tag{5}$$

The rest of the timing is identical to the scheme shown in Figure 6.

4 General Scaling Trends for Future Memory Generations

The future scaling of matrix-based memory chips is determined by economic, technological, and physical boundary conditions. The historical development of memories in the past decades and the International Technology Roadmap for Semiconductors (ITRS) [5] provide reasonable guidelines for further evolution. Here, we will only consider general aspects while the specific topics are discussed in the following chapters. In contrast to the ITRS, we will not give any information on the estimated year of introduction but only on trends concerning the geometrical and electrical specifications. The data for the predictions are mainly taken from [5], [1], and [6].

First, we introduce definitions to describe the geometry scaling shown in Fig. 8 vs. the minimum feature size, F. The die size, A_{DS}, is the total area of a memory chip which has grown in the past, but is expected to grow only slightly during the coming memory generations for economic reasons. Large DRAMs and processors in the year 2002 had die sizes of approx. 250 mm². The fraction of the storage matrix area on the chip is given by X_{Matrix}. For the 256 Mb DRAM generation, the storage matrix area is approx. 55 % of the total die area, i. e. $X_{Matrix} = 0.55$. During the next generations, this value may increase to 0.75 ... 0.85 (Figure 8). The area of an individual storage cell at a node of the matrix is denoted A_{CA}, which is given by the square of the feature size, F^2, times the cell area factor, X_{CA}, which describes how many F^2 are needed to realize the cell:

$$A_{CA} = X_{CA} F^2 \tag{6}$$

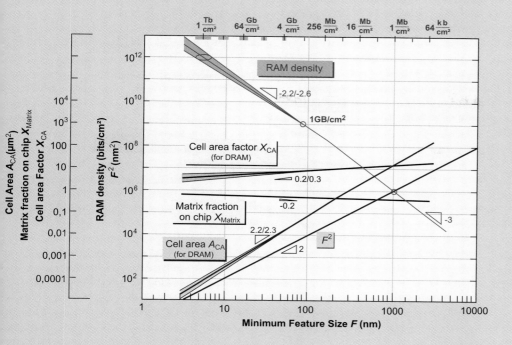

Figure 8: Roadmap trends of the chip and cell geometry. The symbols are explained in the text. The upper axis shows the RAM density which results from the calculations given in the text.

For example, on a 256 Mb DRAM chip a cell area factor $X_{CA} = 8$ is required to realize a cell. According to the ITRS, it is expected that it will be possible to reduce X_{CA} to 4 in the long run (Figure 8). From these considerations one can calculate the total storage capacity of a chip, i. e. the RAM capacity (bits/chip) by

$$\text{RAM Capacity} = \frac{X_{\text{Matrix}} A_{\text{DS}}}{X_{CA} F^2} \qquad (7)$$

Figure 8 indicates, that a feature size in the range of 5 to 8 nm is needed in order to manufacture an 1 Terabit (Tb) memory within 1 cm². This is true of all types of matrix-based memories which are able to follow the extrapolations for X_{Matrix} and X_{CA} sketched in the diagram. The storage density in bits/cm² shows the progress in technology more clearly than the total RAM capacity.

$$\text{RAM Density} = \frac{X_{\text{Matrix}}}{X_{CA} F^2} \qquad (8)$$

In fact, in the long run the ITRS predicts a decreasing chip size for economic reasons which will delay the introduction of RAM generations on a single chip while the functional density is scaled aggressively. In this time frame, also 3D integration techniques (see Chapter 29) may become relevant which would again accelerate the introduction of RAM generations per package.

In addition to the geometrical aspects, there are general trends for the electrical specifications which are independent of the specific type of memory (Figure 9). The sense amplifier margin $V_{S,\text{min}}$ is expected to decrease slightly. This is also true of the bit line capacity C_{BL} due to the shorter line for connecting the same number of cells. The product reveals the minimum charge Q_S which is needed to be stored in a cell

$$Q_S = V_{S,\text{min}} C_{BL} \qquad (9)$$

In order to reduce Q_S significantly, the BL may be divided into sub-BLs to reduce C_{BL}. This may be very helpful since the major challenge for future charge-based RAMs lies in the area needed for the storage capacitor which is proportional to Q_S. This topic will be discussed briefly in Chapter 22. In addition, the operating voltage V_{DD} will continue to decrease (Figure 9). According to the ITRS roadmap, V_{DD} will scale in the range between F^1 and $F^{1/2}$, depending on an optimization for minimum power or for maximum speed. In conjunction with Q_S, this gives the scaling of the storage cell capacitor C_S for DRAM (see Chapter 21)

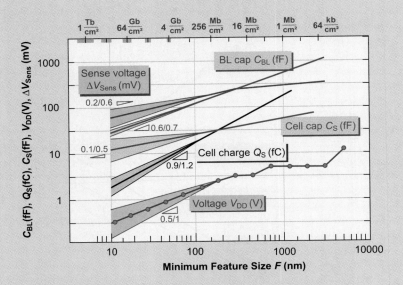

Figure 9: Roadmap trends of electrical specifications such as the BL capacitance, the minimum signal voltage and the operating voltage. The BL capacitance is shown for 512 storage cells per BL which is typical of a 64 Mb DRAM. The cell capacitance is relevant for DRAMs. The symbols are explained in the text.

$$C_S = \frac{Q_S}{V_{DD}} \tag{10}$$

while for FeRAM a different relationship applies (see Chapter 22).

For resistance-based memories, there is a boundary condition on the product of I_0 and $R_{S,OFF} + R_L$ according to Eq. (4), since $V_{BL}("0")$ may not exceed V_{DD}. Hence, using Eq. (4) reveals

$$I_0 \cdot (R_{S,OFF} + R_L) = V_{BL}("0") \leq V_{DD} \tag{11}$$

Insertion Eq. (11) into Eq. (5) gives

$$\frac{\Delta R_S}{2(R_{S,OFF} + R_L)} = \frac{V_S}{V_{DD}} > \frac{V_{S,min}}{V_{DD}} \tag{12}$$

From the scaling of $V_{S,min}$ and V_{DD}, discussed above, one reads the required scaling of the relative change of the resistance between the two logic states, expressed by Eq. (12) and shown Figure 10. This applies in general to all resistance-based memories which are read out by the current sensing scheme described above.

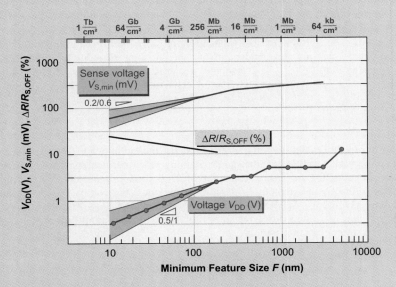

Figure 10: Roadmap trend for the required minimum resistance change of active matrix resistance-based RAMs.

Acknowledgements

The editor is grateful for critical reviews and fruitful suggestions by Jürgen Rickes (Agilent Technologies), Wolfgang Hönlein and Werner Weber (Infineon Technologies). He thanks Fotis Fitsilis for technical support in the compilation.

References

[1] K. Itoh, VLSI Memory Chip Design, Springer series in advanced microelectronics, Springer, Berlin, Heidelberg, New York, 2001.

[2] S. R. Ovshinsky, Phys. Rev. Lett. 36, 1469 (1968). RAM based on phase-change materials have been introduced by Ovonyx Inc., http://www.ovonyx.com, http://www.ovonic.com; IEDM 2001 Techn. Digest, Session 36, Dec. 2001; G. Wicker, SPIE vol. 3891, pp. 2-9, Oct. 1999.

[3] U. Tietze, Ch. Schenk, E. Schmid, Electronic Circuits, 2nd internat. edition, Springer, New York, 1993.

[4] http://www.rambus.com/

[5] International Technology Roadmap for Semiconductors (ITRS), 2001 Edition, http://public.itrs.net/Files/2001ITRS/Home.html

[6] A. Nitayama, Y. Kohyama, and K. Hieda, IEDM 1998 Techn. Digest, Dec. 1998.

[7] D. Goldhaber-Gordon, M.S. Montemerlo, J. C. Love, G. J. Opiteck, J. C. Ellenbogen, IEEE Proc. **85** (4), 521-40, 1997.

[8] H.Okada, H. Hasegawa, Jpn. J. Appl. Phys. **40**, 2797-2800, 2001.

[9] S. Tiwari, F. Rana, H. Hanafi, A.Hartstein, E. F. Crabbe, K. Chan, Appl. Phys. Lett. 68 (10), 1377-79, 1996.

[10] L. Montes, G.F. Grom, R. Krishnan, P.M. Fauchet, L. Tsybeskov, B.E. White, Mater. Rec. Soc. Proc. **638**, pp. F 2.3.1-6, 2001.

[11] K. Yano, T. Ishii, T. Hashimoto, T. Kobayashi, F. Murai, K. Seki, IEEE Electr. Dev. **40**, 1628-38, 1994.

[12] K. Hofmann, B. Spangenberg, and H. Kurz, Microel. Eng. **57-8**, 851 (2001).

[13] www.itrs.net/Common/2004Update/2004_05_ERD.pdf

High-Permittivity Materials for DRAMs

Herbert Schroeder, Department IFF, Research Center Jülich, Germany

Angus Kingon, North Carolina State University, Raleigh, USA

Contents

1 Introduction	539
2 Basic Operation of DRAM Cells	540
3 Challenges for Gb DRAM Capacitors	541
4 Properties of High-Permittivity Dielectrics	543
4.1 Requirements for DRAM	543
4.2 Field Dependence	543
4.3 Temperature Dependence	544
4.4 Microstructure Dependence	544
4.5 Thickness Dependence	545
4.6 Frequency Dependence	547
5 Stability of Capacitor Charge and Reliability of the Cell	547
5.1 Relaxation Currents	548
5.2 Leakage Current Mechanisms	549
5.3 Resistance Degradation and Breakdown	550
6 Integration Aspects	551
7 High-Permittivity Materials in DRAM	554
7.1 Status of Implementation of New Dielectric Materials	554
7.2 Trends for the End of the Roadmap	556
7.3 Alternative High-Permittivity Materials	558

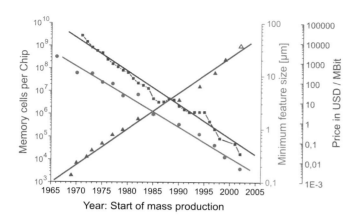

Figure 1: Exponential growth of number of memory cells per chip, "*Moore's law*". (Moore was one of the founders of Intel Corporation.) This was the result of the exponential decrease of the minimum feature size and a similar decrease of the price per Mb of memory.

High-Permittivity Materials for DRAMs

1 Introduction

Effective and cheap information storage in semiconductor memory cells is one of the primary reasons for the success of Si-based information technology in the last three decades. Furthermore, the most prominent representative memory type, the dynamic random access memory (DRAM) patented in 1967 [1] and introduced into the market by Intel Corporation in 1972, was the driving force for the exponentially growing large-scale integration in memory chips and promoted similar advances in logic chips. Besides necessary improvements in lithography and dry-etching technology the simplicity of the 1T-1C DRAM cell, one transistor as a switch and one capacitor as charge storage element (see Figure 2), as well as the "simple", Si-based dielectrics used in the capacitor, Si oxide (SiO_2) and nitride (SiN_X), guaranteed the tremendous and long-lasting success of DRAMs.

The DRAM success story is demonstrated in Figure 1. For more than three decades an exponential growth of the number of memory cells per chip has been observed and described by Moore´s law. This evolution started with a few kb per chip. Mass production of 256 Mb chips was introduced in 2000. On the average, this is an increase by a factor of four – this corresponds to a new chip generation – every 3 to 4 years. The reasons for this success story are manifold: Besides the mentioned simple structure of the DRAM cell two other reasons are shown in Figure 1. One is the exponential decrease of the minimum feature size, F, structured by lithography and (dry) etching, enabled by continuous improvements in these technologies. For example, for the 256 Mb generation F is 250 nm or 180 nm. The state-of-the-art lithography methods are presented in Chapt. 9. There is also another closely related reason: The exponential decrease of the prices for memory chips (here plotted as price in US$ per 1 Mb of memory), modulated by several short-term variations due to production shortages in view of high demand (high prices) or overproduction with low prices. The nearly constant chip prices (compared to the previous generation) for new DRAM generations with four times higher capacity are due to the tremendous improvement of the fabrication productivity: Besides the mentioned decrease of F, which decreases the area consumed by one DRAM cell, the increase of wafer size from 10 mm (4") diameter in 1990 to 30 mm (12") diameter in new production lines since 2002 has contributed to this economically important fact.

Up to now, the necessarily increased charge storage density of the capacitor with decreasing cell area of new DRAM generations has been achieved by thinning the dielectric film and enlarging the capacitor area using the third dimension (3-D trench or stacked capacitors). (Note: We will use the term '3-D folding' to describe the use of these three dimensional structures.) This 3-D folding has been employed since the 4 Mb generation. As the technology enters the Gb DRAM generations, these solutions are reaching their limits: the thickness of the dielectric film (< 4 nm) because of unacceptable charge losses due to tunneling leakage currents, and the 3-D folding because of production complexity and, thus, reliability, yield and unacceptable costs. Therefore, for the first time since the introduction of the DRAM, new dielectric materials with higher effective permittivity compared to that of Si oxide/nitride composites, $\varepsilon_r \approx 7$, have to be considered for higher density DRAM generations.

Individual companies plan to introduce new high permittivity materials in sequence into the future DRAM generations. These plans loosely follow the roadmap published by the Semiconductor Industry Association [2], [3] although company-to-company variations are already apparent. For example, Samsung introduced Ta_2O_5 ($\varepsilon_r \approx 22$ on poly-Si [4]) into its 256 Mb generation, and has reported a 1 Gb demonstration chip based on Ta_2O_5 [5]. Eventually it is expected that very high permittivity materials such as complex oxides with the perovskite structure will be required, although there is no consensus at the present time as to the technology node where this will occur.

In this chapter we focus on the complex oxide $(Ba,Sr)TiO_3$ (or "BST") as the vehicle to discuss the relationships between high permittivity material properties and DRAM performance. Although BST is clearly not the first high permittivity material to be incorporated into DRAMs, this material has been rather extensively studied, more so than the simpler materials that will be incorporated first. It therefore provides greater insights into the issues involved in the integration of high permittivity materials into high density DRAMs. Towards the end of the chapter we return to the simpler materials, and discuss progress in the integration of these materials with advancing technology generation.

2 Basic Operation of DRAM Cells

The 1T-1C DRAM cell has quite a simple structure (see Figure 2). The access transistor, Tr, acts as a switch and is addressed by the word line, WL, controlling the gate. The memory capacitor, C_S, represents the charge storage element for the information and is connected to the bit line, BL, via the transistor. When the switch is closed the information, the voltage levels $+V_{DD}/2$ or $-V_{DD}/2$ are applied to C_S via BL. The corresponding charge on C_S represents the binary information, "1" or "0". After this write pulse (e.g. for the 256 Mb generation this pulse is less than 10 ns at $V_{DD}/2 = 1.25V$) the capacitor is disconnected by opening the transistor switch.

The memory state is read by closing the switch and sensing the charge on the capacitor via the bit line, which is usually pre-charged to $V_{DD}/2$. The cell charge is redistributed between the cell capacitance, C_S, and the bit line capacitance, C_{BL}, leading to a voltage change in the bit line. This voltage change is detected by the sense amplifier in the bit line and amplified for the subsequent use of the read information as described in Sect. 3 of the Introduction to Part IV. As a read pulse destroys the charge state of the capacitor, it has to be followed by a re-write pulse to maintain the stored information. The plate line, PL, is kept at $V_{DD}/2$ in order to reduce the electric voltage stress on the capacitor dielectric, which is charged to $\pm V_{DD}/2$ instead of being discharged to 0 V and charged to full V_{DD}.

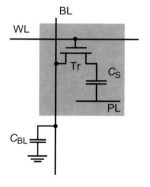

Figure 2: Schematic representation of a DRAM memory cell.

The DRAM is volatile, i.e. the information, the stored capacitor charge, is lost after the supply voltage is cut. In addition, the stored charge decreases with time because of leakage currents through both, the capacitor and the transistor. In order to guarantee a certain charge and, thus, a corresponding minimum voltage change in the BL detectable by the sense amplifier during the read operation, the charge of each memory capacitor has to be refreshed periodically. For the 256 Mb generation, the refresh time is about 64 ms [6]. In general, by selecting a WL all the select-transistors in this row are active so that all cells in this row (1024 or 2048, depending on organization) can be read simultaneously. As mentioned above, those cells are then refreshed by the necessary re-write. A special built-in counter takes care of the refresh for those rows which are not selected by the working program during the given refresh period.

High-Permittivity Materials for DRAMs

The value of the cell capacitance, C_S, varies only slowly with DRAM generation due to the same trends for the cell charge, Q_S, and the operating voltage, V_{DD}. For the 256 Mb chip, C_S is approx. 25 fF. The geometrical dimensions can be approximated very well by using the equation for the parallel plate capacitor:

$$C_S = \varepsilon_0 \varepsilon_r \frac{A_S}{t_{phys}} = \varepsilon_0 \varepsilon_{r,SiO_2} \frac{A_S}{t_{eq}} \quad \text{with} \quad t_{eq} = \frac{\varepsilon_{r,SiO_2}}{\varepsilon_r} t_{phys} \quad \text{and} \quad \varepsilon_{r,SiO_2} = 3.9. \quad (1)$$

A_S is the total area of the capacitor, (defined by the surface area of the bottom electrode), t_{phys} is the physical thickness of the dielectric, and ε_r its relative permittivity, while ε_0 is that of vacuum. Eq. (1) also defines the equivalent dielectric thickness, t_{eq}, with respect to the relative permittivity of SiO_2, $\varepsilon_{r,SiO2} = 3.9$, as this dielectric was used at the beginning of DRAM history. With higher integration density, it was replaced by a mixture of Si oxide and nitride layers ($\varepsilon_{r,SiNx}$ approx. 11) with higher effective dielectric constant, the so-called ON dielectric. Starting with the 4 Mb generation, the required capacitor area, A_S, was too large for the cell area, A_{CA}, and the cell capacitance could not be achieved with a planar geometry. Consequently, the third dimension had to be used in the form of trenches (see Figure 3) and later also of posts in a stacked geometry (see Figure 4).

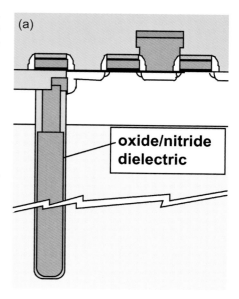

3 Challenges for Gb DRAM Capacitors

Entering the Gb era, the capacitor of the DRAM cell in the conventional design is approaching its limits: (i) The thickness of the (Si) oxide/nitride (ON) compound layer allows no further thinning beyond approx. 5 nm because of unacceptably high tunneling leakage currents (see Sect. 5.2) reducing the stored capacitor charge to too low values, and (ii) the 3-D geometry is already very complicated and, hence, rather expensive due to the high number of processing steps. Some examples of very advanced 3-D structures are shown in Figure 3 (trench capacitor) and Figure 4 (stacked capacitor). Both have been produced with the 250 nm technology, i.e. the minimum feature size, F, was 250 nm, which was used when this generation was introduced into the market. The DRAM cell area, A_{CA}, was approx. $12 F^2 = 0.75 \mu m^2$ and the equivalent thickness, $t_{eq} = 5$ nm (see Eq. (1)), which corresponds to a physically thicker layer using ON dielectrics [9]. The challenges of the capacitor for future Gb DRAM generations will be demonstrated with the help of a detailed description of Figure 3 and Figure 4. Some important characteristic numbers for these capacitors will be summarized in Table 1 together with the respective ones for higher permittivity materials such as Ta_2O_5 ($\varepsilon_r \approx 22$) or $(Ba,Sr)TiO_3$ (BST) ($\varepsilon_r = 200$).

The most impressive feature of the trench capacitors shown in Figure 3b is their depth, more than 7 μm with 250 nm width at the top resulting in an aspect ratio (i.e. depth/width) of near 30. Thus, the necessary capacitor area, $A_S = 5.1 \mu m^2$ corresponding to a capacitance of 35 fF, can be supplied. The projected area of the trench capacitor in the cell area is only $2 F^2 = 0.125 \mu m^2$ or 16 % of A_{CA} (see Figure 5a). The reason for this small fraction can be realized from the schematic representation of a trench cell in Figure 3a. The cell area has to be shared with the select transistor and the local wiring, the capacitor is in an offset position similar to the planar capacitors in early DRAM generations.

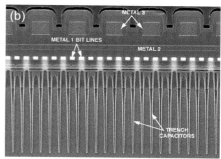

Figure 3: Example of a 3-D capacitor: (cross section) deep trench with oxide/nitride dielectric ($\varepsilon_r \approx 7$).
(a) Schematic, from [8];
(b) Scanning electron micrograph for Toshiba/Infineon 64 Mb chip. The design is similar in 256 Mb and 1 Gb chips with much higher aspect ratios (height/ depth).

Figure 4: Example of a 3-D capacitor (cross section). Stack with oxide/nitride dielectric ($\varepsilon_r \approx 7$) with COB design (capacitor over bit-line).
(a) Schematic disk-type capacitor [8].
(b) Scanning electron micrograph: Stacks with roughened surface of the bottom electrode (HSG-Si) for increased effective area. (Mitsubishi 64 Mb) [7].
(c) Scanning electron micrograph: stacks as disks made of SiO_2 as base for capacitor (Mitsubishi 256 Mb) [9].

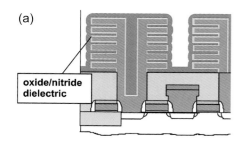 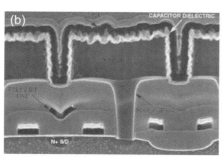 

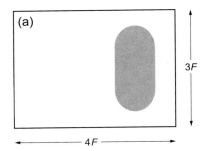
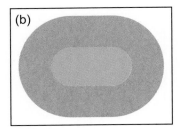
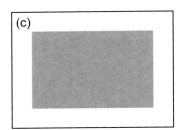

Figure 5: Projected areas of cell capacitors in a 256 Mb DRAM cell in 250 nm technology with a cell area of $12\,F^2$.
(a) Trench ($2\,F^2$)
(b) Stacked disc ($6\,F^2$)
(c) stacked post or planar ($6\,F^2$)

For higher density generations the decreasing lateral dimensions can be compensated by thinning the ON dielectric (a factor of two at most!), or by increasing the trench area. The latter can be achieved by even deeper trenches at reduced width, leading to aspect ratios beyond 50, a real challenge for the etching technology. This can be somewhat relaxed by a "bottle" trench geometry, i.e. the cross section is widened like a bottle by a special etch technique beginning at a depth not occupied by the transistor structure. Nevertheless, in the long run a higher permittivity material has to be introduced as a dielectric for the trench. The deposition of high-permittivity oxides such as BST onto the trench walls will be extremely challenging since the problems are comparable to the task of the introduction of high-k gate oxides in MOSFETs (see Chapt. 13).

The main disadvantage of the trench, the small fraction of the cell area available to be assigned to the capacitor, is eliminated by a different cell architecture, the stacked cell, i.e. the capacitor is stacked on top of the select transistor (see Figure 4). Especially if – as shown – the local wiring is below the capacitor structure (capacitor over bitline – COB), then the dielectric layer as well as the top electrode layer of the capacitors can be made as continuous films covering the complete cell area. The capacitor area is defined by the bottom electrode which can occupy a much larger portion (about 50 %) of the cell area, as can be seen from Figure 5b. It shows the projected area of an (oval) disc-type bottom electrode, similar to that shown in Figure 4c, formed around a cylindrical structure as shown in Figure 4a. This complicated, three-dimensional structure is necessary to supply the 3.6 µm^2 area for the 25 fF capacitance, again using an equivalent dielectric thickness of 5 nm. For Mitsubishi's 256 Mb chip shown in Figure 4c each ring segment had a total area of about 1 µm^2 and a height of 100 nm (rings of 50 nm height and 50 nm apart) so that four segments with a total height of 0.4 µm were needed for the cell capacitance [9].

The complexity of the 3D stacked capacitor can be reduced by using higher permittivity materials, e.g. Ta$_2$O$_5$ with $\varepsilon_r \approx 22$. Using the same t_{eq} as above does not seem to be a good choice because it leaves the most problematic number, the large capacitor area, unchanged. Instead, as an example, using a simple post structure (as shown in Figure 4b) with reduced surface area ($A_S = 1.38$ µm^2) as the bottom electrode with a projected area of $2\,F$ times $3\,F$ (see Figure 5c) and a height of 0.4 µm (as above) results in a reduced equivalent thickness of $t_{eq} = 1.8$ nm, corresponding to a physical thickness of 10 nm Ta$_2$O$_5$.

Using a high-permittivity material such as BST ($\varepsilon_r = 200$) the capacitor of the 256 Mb DRAM cell could be realized in a planar structure, e. g. with an area $A_S = 6\,F^2 = 0.38$ µm^2 and a physical thickness of 26 nm corresponding to $t_{eq} = 0.5$ nm. For higher density Gb DRAM generations the thickness of the BST dielectric can be decreased and the bottom electrode geometry has to become three-dimensional again to supply the necessary capacitance.

This comparison nicely demonstrates the large capabilities using high dielectric constant materials for Gb DRAM generations, if the high ε_r can be retained under working conditions and if these materials can be successfully integrated into the CMOS technology. These issues will be covered in the following sections. This comparison also demonstrates that the equivalent thickness as defined in Eq. (1) is not a good descriptor for a DRAM generation as it depends not only on the dielectric constant and the physical thickness of the chosen material, but also on capacitor geometry.

Material	Relative Permittivity	Cap. Area A_S µm^2	t_{phys} nm	t_{eq} nm	Proj electrode area µm^2
ON Trench (C_S=35fF)	7	5.1	9	5	0.13
ON Stacked (C_S=25fF)	7	3.6	9	5	0.38
Ta$_2$O$_5$ Stacked (C_S=25fF)	22	1.38	10	1.8	0.38
BST Stacked (C_S=25fF)	200	0.38	26	0.5	0.38

Table 1: Characteristic numbers for a cell capacitor of a 256 Mb DRAM in 250 nm technology with a cell area ACA = 12 F^2 = 0.75 µm^2.

4 Properties of High-Permittivity Dielectrics

There are numerous materials with higher permittivity than ON films, but only a few of them are serious candidates for use in an ultra-large scale integrated DRAM cell as the capacitor dielectric. We will first define a requirement catalogue for the properties of that dielectric. Then we will review the properties of the most thoroughly investigated material, barium strontium titanate, $Ba_xSr_{1-x}TiO_3$ (abbreviation: BST), in the light of these demands.

4.1 Requirements for DRAM

The most relevant requirements are listed for a thin film dielectric with high permittivity to be integrated into a capacitor of a DRAM 1T-1C cell in a CMOS chip of an advanced Gb generation:

- Most important is a high permittivity ($\varepsilon_r \geq 200$) to guarantee simpler electrode geometries, i.e. low 3-D folding factor, for several DRAM chip generations. A charge of about 25 fC at voltages ≤ 1 V has to be loaded to the cell capacitor. This corresponds to a usable cell capacitance $C_S \geq 25$ fF.
- The material has to be homogeneously deposited as a thin film with thickness of ≤ 30 nm over large areas (12" wafer size). The deposition method has to ensure a conformal coverage in order to coat 3-D structures. The thermal budget should be as low as possible. For these reasons, the metal-organic chemical vapor deposition (MOCVD) technique is most suited for the deposition (see also Chapt. 8).
- As the read and write times will approach 1 ns, the dielectric behavior should not show a significant dispersion up to frequencies of a few GHz.
- Refresh times of the order of 1 s are desirable. Within the refresh time, the charge loss due to polarization relaxation and leakage currents has to be smaller than 10%.
- Long-term stability of the properties (10 years is the projected lifetime) is indispensable.
- All the processes for the dielectric material itself as well as for the electrodes and the possibly necessary diffusion barrier and adhesion layers have to be compatible with the CMOS process technology.

These requirements will be discussed in the context of one of the most promising candidate materials, BST, which is a mixed oxide of the ABO_3 type with a perovskite crystal structure. It is derived from the ferroelectric $BaTiO_3$, but it does not show a ferroelectric hysteresis as a thin film on a Si substrate because (i) the solid solution of strontium on the barium "A" site shifts the Curie point to lower temperatures and (ii) the geometrical confinement as a thin film on a thick Si substrate leads to the typical biaxial tensile stress conditions on cooling from high processing temperatures (550°C to 750°C) to ambient temperatures if its thermal expansion coefficient is larger than that of Si. Rather, it behaves as incipient ferroelectric or superparaelectric film [10]. The properties of BST depend on the relative concentration of the components, especially on the ratio of Ba and Sr (the Ba:Sr ratio is about 70:30 in the films discussed in this section), and on the conditions of deposition and recrystallization treatment. Therefore, we report the properties from only a few sources to assure comparability of the different dependences.

4.2 Field Dependence

Inherently, high-permittivity materials such as BST are non-linear dielectrics, i.e. the induced polarization, $P(E)$, is non-linearly dependent on the applied field, E = applied voltage (bias)/film thickness (Figure 6a). At low fields, the polarization increases at a higher rate and approaches a linear relation at higher fields. This can be described by the thermodynamic Landau-Ginzburg-Devonshire (LGD) theory of ferroelectrics and non-linear paraelectrics above the transition temperature (see Chapt. 2). As mentioned, bulk BST is a ferroelectric material with a transition temperature around 0°C (for the Ba:Sr ratio 70:30), but in thin films the phase transition from paraelectric to ferroelectric is suppressed (see Sect. 4.3 and Figure 7b). The small signal capacitance, $C(V)$, can be measured directly using a small ac amplitude superimposed on a dc bias voltage, V. Alternatively, it can be calculated from the slope of the curve electrical displacement, $D(V)$, vs. applied voltage, V. Please note that for high values of ε_r, i.e.

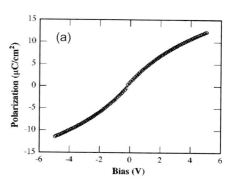

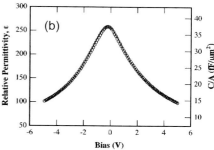

Figure 6: Field dependence of (a) polarization and (b) relative permittivity ε_r (capacitance area density C/A) for a 61 nm thick MOCVD-BST film between Pt electrodes measured at 300 K and 1 kHz (after Basceri et al. [11]).

$\chi_r = \varepsilon_r - 1 \gg 1$, the following approximations hold: $\varepsilon_r(V) \approx \chi_r(V)$ and $D(V)$ (which is measured!) $\approx P(V)$ (which is usually plotted). Both properties, $\varepsilon_r(V)$ and the area capacitance, C/A are plotted in Figure 6b vs. the applied voltage.

It should be pointed out that the relevant property for the DRAM cell switching from the state "0" to the state "1" (or vice versa) by changing the voltage from $V_{"0"} = -V_{DD}/2$ to $V_{"1"} = +V_{DD}/2$ is the total charge difference, ΔQ_S, on the cell capacitor which is the integral of $C_S(V)$ between the respective voltages:

$$\Delta Q_S = \int_{V_{"0"}}^{V_{"1"}} C(V) \mathrm{d}V = \frac{A_S \varepsilon_0}{t} \int_{-V_{DD}/2}^{+V_{DD}/2} \varepsilon_r(V) \mathrm{d}V \qquad (2)$$

i.e. it corresponds to an average permittivity obtained by integration.

4.3 Temperature Dependence

The temperature dependence of the permittivity for a BST thin film prepared by MOCVD is shown in Figure 7. The relative permittivity vs. applied nominal field is plotted for various temperatures in Figure 7a. The field is corrected to the field at the maximum permittivity. This maximum (see e.g. Figure 6b) is usually not exactly at the zero field indicating an internal field which might be induced e.g. by asymmetrical electrodes and interfaces. The maximum of the relative permittivity decreases from about 290 at 192 K to about 190 at 473 K while the tails at larger fields are independent of temperature.

This dependence on temperature is also demonstrated in Figure 7b, in which the inverse values at the maximum, $(\varepsilon_r(E \approx 0))^{-1}$, are plotted vs. the measuring temperature. For temperatures higher than 373 K, the behavior can be described by the Curie-Weiss law (see Chapt. 2):

$$\varepsilon_r(T) = \frac{C}{T - T_0} \qquad (3)$$

The extracted Curie temperature, $T_{0,\text{film}}$, of the film is about 110 K, much lower than the indicated Curie temperature of the bulk for the same composition ($T_{0,\text{bulk}} = 300$ K). The reason for this may be a temperature-independent, finite interfacial capacitance, possibly caused by a "dead layer", i.e. a reduced permittivity layer, as discussed in Sect. 4.5. In contrast to the bulk, for which the phase transition from paraelectric to ferroelectric behavior is characterized first by an increase of ε_r with decreasing temperature down to the transition temperature and then by a subsequent decrease of ε_r in the ferroelectric phase with further decreasing temperature, the latter is suppressed in this thin film. Instead, in the low temperature regime, the permittivity stays at high values nearly independent of temperature. The absolute permittivity values are much lower than for bulk ceramics. Several reasons may contribute to this behavior. Besides the interfacial capacitance, effects of the biaxial stress of the films and the microstructure have been proposed. In addition, the (Ba,Sr) to Ti stoichiometry shows an influence. For a detailed discussion see Refs. [12], [13].

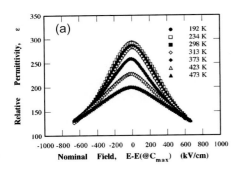

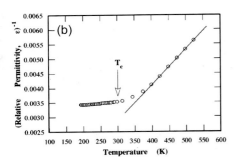

Figure 7:
(a) Field dependence of relative permittivity ε_r at different temperatures.
(b) Temperature dependence of inverse relative permittivity ε_r at corrected zero field. (53 nm thick MOCVD-BST film between Pt electrodes at 1 kHz) (after Basceri et al. [11]).

4.4 Microstructure Dependence

As mentioned above, the microstructure has also some effect on the permittivity value in thin films, similar to its grain size dependence investigated in BaTiO$_3$ (BTO) bulk ceramic [14], which shows decreasing ε_r values with decreasing grain size for ceramics with a grain size below 700 nm. This was modeled by a different behavior of the near grain boundary layer compared to the intragranular bulk (brick wall model). It was found to extend this relationship for BTO polycrystalline films with grain sizes from 30 nm to 100 nm [15]. In contrast, films of same thickness with columnar grains, i.e. only grain boundaries perpendicular to the substrate, have much higher permittivities [16]. A similar increase of the maximum dielectric constant was observed in 20 nm-thick epitaxial BST films with all-oxide electrodes compared to the same MIM capacitor with polycrystalline microstructure [17]. In addition, some influence of the top electrode, platinum or SrRuO$_3$, respectively, was observed in this investigation with lower permittivities for Pt top electrode.

4.5 Thickness Dependence

The field dependence of capacitance area density, C/A, is shown in Figure 8a for different thicknesses ranging from 24 nm to 160 nm all measured at 300 K. As expected, the capacitance increases with decreasing thickness. But the increase at the maximum is much less than proportional to the inverse thickness as is expected from simple theory (see Eq. (1a)). This is demonstrated in Figure 8b. The relative permittivity, ε_r, at maximum decreases significantly with thickness. At a film thickness of 160 nm, ε_r is about 280 while at 24 nm it is only about 170, i.e. ε_r is not independent of thickness. This indicates that the permittivity measured in these films is an *effective* one, dependent on thickness and decreased by factors such as the already mentioned finite interfacial capacitance, the importance of which obviously increases with decreasing film thickness. This interfacial capacitance may result from dielectric "dead layers" [18] with much reduced permittivity at surfaces and interfaces. Possible reasons for their occurrence are discussed below.

The plot in Figure 9 supports this interpretation. The thickness dependence of $(C/A(E \approx 0))^{-1}$ is shown for different temperatures. Theoretically it should result in a linear dependence with no offset, i.e. all the lines should go through the origin, which is obviously not the case. Instead, all the extrapolated lines intercept at a common positive value of about 0.008 μm²/fF. The simplest model to describe such behavior would be some additional capacitance (in series with the film), which is sketched in the inset of Figure 9. It assumes some interfacial capacitances at the bottom (BI) and at the top (TI) electrode of the film with certain thickness, (C_{BI}, t_{BI}) and (C_{TI}, t_{TI}), respectively. The measured value $(C/A)^{-1}$ can be written as:

$$\left(\frac{C}{A}\right)^{-1} = \frac{t}{\varepsilon_0 \, \varepsilon_{r,\text{eff}}} = \frac{t_{BI}}{\varepsilon_0 \, \varepsilon_{r,BI}} + \frac{t - (t_{BI} + t_{TI})}{\varepsilon_0 \, \varepsilon_{r,BST}} + \frac{t_{TI}}{\varepsilon_0 \, \varepsilon_{r,TI}}$$

$$\text{for} \quad t \gg t_{BI}, t_{TI} \quad \approx \frac{1}{\varepsilon_0 \, \varepsilon_{r,BST}} t + \left(\frac{A}{C}\right)_{BI} + \left(\frac{A}{C}\right)_{TI} \quad (4)$$

Using the approximate Eq. (4) for evaluation, the relative permittivity of the "bulk" BST can be calculated from the slope of the lines in Figure 9 and the total interface capacitance (C_{BI+TI}) from the intercept. Assuming the identical capacitance at the bottom and top electrode, only the quotient of $(\varepsilon_r / t)_{BI,TI}$ of the interface layers can be determined. As an example, some numbers are extracted from the data of Figure 9. From the slope, the relative permittivity of the BST at zero field, $\varepsilon_{r,BST}$, is determined as 324 at 25°C, compared to the effective value of 285 from the corresponding curve in Figure 8b with a sample thickness of 160 nm. From the mean intercept value, 0.008 μm²/fF, and due to the condition $\varepsilon_{r,I} \geq 1$ the minimum thickness of one interface layer, $t_{BI,TI}$, can be calculated: $t_I \geq 0.035$ nm, about a tenth of a lattice constant in BST. For a more realistic thickness, $t_{BI,TI}$ between 0.4 nm and 2 nm (corresponding to about 1 and 5 lattice constants), the respective values for $\varepsilon_{r,I}$ are between 11 and 55.

The total interface capacitance from both interfaces determined from the intercept in Figure 9 is 127 fF/μm². As long as this interface capacitance is large compared to that of the high permittivity BST, the latter dominates the measured capacitance although some influence is seen even at larger thicknesses (e.g. about 10% decrease of $\varepsilon_{r,\text{eff}}$ compared to $\varepsilon_{r,BST}$ for a thickness of 200 nm at 300 K). The crossover, at which both components are equal, is at a BST thickness of 23 nm for 300 K and $\varepsilon_{r,BST} = 324$. Other reported values of the interface capacitance for thin films are in the same range, e.g. 136 fF/μm² for columnar CSD-BST [19], 160 fF/μm² [20] and 230 fF/μm² [12] for fiber-textured MOCVD-BST, and 220 fF/μm² for epitaxial $SrTiO_3$ [21].

The microscopic reason for the effect leading to this thickness dependence of the permittivity is not yet understood but several different mechanisms have been proposed in the literature, all (except the last one) leading to an interface-related capacitance with reduced permittivity. Worth noting is the experimental observation that the "dead layer" effect, decreasing $\varepsilon_{r,\text{eff}}$ with decreasing dielectric film thickness, is reduced by using conducting oxides as the electrode material instead of Pt [22]. The most important mechanisms will be mentioned here:

- As the lattice periodicity is broken at any surface or interface a changed phonon spectrum in the dielectric at or near the interface may be responsible for inhibition of the "soft" phonon mode. This mode is thought to be induced by a cooperative mechanism ("long range" dipole-dipole interaction) leading to ferroelectricity or high dielectric constants (Thomas theory [18], [23]). The range of the disturbed interaction

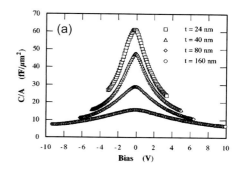

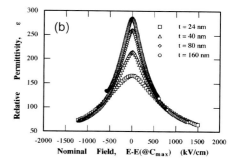

Figure 8:
(a) Field dependence of capacitance area density C/A (proportional to the quotient of effective relative permittivity and thickness t, $\varepsilon_{r,\text{eff}} / t$) at different thicknesses;
(b) Field dependence of effective relative permittivity ε_r at different thicknesses; (MOCVD-BST films between Pt electrodes at 300 K and 1 kHz) (after Basceri et al. [11]).

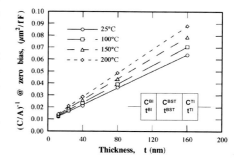

Figure 9: Thickness dependence of inverse capacitance area density, $(C/A)^{-1}$, for zero bias (proportional to the quotient of effective relative permittivity and thickness t, $\varepsilon_{r,\text{eff}} / t$) at different temperatures; (MOCVD-BST films between Pt electrodes at 1 kHz) (after Basceri et al. [11]).
Note: If the total film thickness, t, approaches the thickness of the interfacial layers, $t_I = t_{BI} + t_{TI}$, (i.e. $t_{BST} \to 0$), the slope of the curves should increase to value proportional to $(\varepsilon_{r,I})^{-1}$ instead of $(\varepsilon_{r,BST})^{-1}$ as at larger thickness. As a result the curves should bend to the origin point.

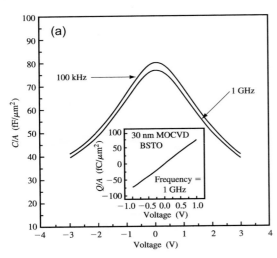
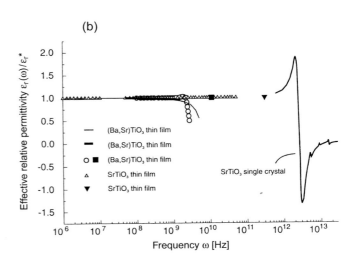

Figure 10:
(a) Capacitance area density (C/A) (proportional to the effective relative permittivity $\varepsilon_{r,\text{eff}}$ at constant thickness) at 300 K for a 30 nm thick MOCVD-BST film between Pt electrodes vs. applied voltage measured at 100 kHz and 1 GHz;
(Inset: charge density (Q/A) vs. voltage at 1 GHz) (after Kotecki et al. [38]).
(b) Frequency dependence of effective relative permittivity $\varepsilon_{r,\text{eff}}$ normalized to the value ε_r^* at low frequencies [39].

(i.e. thickness of the "dead layer") was estimated by Zhou and Newns [18] as 1 to 3 nm. As this effect is a feature of the interface it may also be influenced by the electrode material.

- A similar disturbance of the soft mode may be expected by defects related to the interface, i.e. missing ions (oxygen vacancies) [17] or local in-diffusion of electrode material [24], [25]. Such extrinsic effects may be controlled by the choice of electrode material and its treatment.

- An obvious changed permittivity is expected for a chemically different (reaction) layer at the interface [26]. This may also be influenced by the production and treatment of the films. However, the few HRTEM studies do not show any indications of a secondary phase for high-quality films [27], [28].

- Another reason for an interface capacitance is electronic surface or interface states with energies within the gap of the dielectric. These states can change their charge with a variation of applied voltage. The density and properties of these states are dependent on many extrinsic parameters (point defects, lattice mismatch, treatment) [29]. A special kind of an interface state with intrinsic character is the metal-induced gap state (MIGS), which is induced in the surface of the insulating dielectric by the electrode metal electronic wave function decaying into the insulator (or semiconductor) gap [30], [31]. These MIGS are somewhat dependent on the materials forming the interface, but they cannot be avoided completely.

- Black and Welser [32] have suggested another possibility. They have adapted an idea of Ku and Ullmann [33] who pointed out that the displacement charge on the electrodes couldn't be an ideal surface charge but occupies a finite spatial extent in the electrodes, i.e. the applied field penetrates also the electrodes. From Gauss' law they calculated a potential drop in the electrodes decreasing the applied voltage in the dielectric to an effective voltage. This results in an effective capacitance of the dielectric in the same form as Eq. (4) substituting $\varepsilon_{r,I}$ by an effective permittivity of the metal, $\varepsilon_{M,\text{eff}}$ and t_I by a characteristic (screening) length in the metal. As an important consequence this idea would result in an intrinsic, electrode determined upper limit for the capacitance of a high permittivity thin dielectric film. As there is still a discussion about the correct values of $\varepsilon_{M,\text{eff}} (= 1 \ldots 100)$ and t_I ($=$ Thomas-Fermi electron screening length) [32], [34] this limit (and hence the model) is very uncertain.

- The hardening of the soft phonon modes may also be homogeneous throughout the whole film [35]. If the extent of this hardening increases with decreasing thickness, the effective permittivity will be thickness-dependent.

In general, this three-capacitor model is a simple and, therefore, a coarse description, as discussed below. It should be most applicable for "dead layers", for which the two conditions $t_I \ll t$ and $\varepsilon_{r,I} \ll \varepsilon_{r,\text{Bulk}}$ hold. On the other hand, both t_I and $\varepsilon_{r,I}$ have lower limits: $\varepsilon_{r,I} \geq 1$ due to basic physics and $t_{BI,TI}$ larger than at least one lattice constant in order to apply continuum theories such as the concept of the permittivity. If the reason for the "dead layer" interface capacitance is intrinsic in nature it will be hard or even impossible to increase the area capacitance of high dielectric constant materials such as BST by decreasing its thickness further and further. This limit is not yet well defined.

Consider further the limitations of the simple model. It should be noted that in the "dead layer" model of Zhou and Newns [18], those authors assumed a very simple form of the permittivity distribution versus thickness, i.e. simply a step function with only two values of permittivity (bulk and interface). Additionally, the permittivity of the interfacial dielectric was assumed to be temperature and field independent. (These assumptions are also consistent with the term "dead layer"). These simple assumptions were not a consequence of the physics of the Thomas theory of ferroelectrics [23] from which the model was derived, but rather a simplifying convenience for finding an analytical solution. Note again that the primary experimental observations for thin films have been the intercept of the 1/C versus film thickness plots (such as Figure 9). However, the simple assumptions are inadequate *if* a wider range of experimental phenomena need to be included, for example the thickness dependencies of the permittivity-temperature maxima, and the diffuseness of the phase transitions, in addition to the electrode effects, as discussed recently by Parker et al. [36]. If these experimental data must be included, or microscopic origins investigated, a model with a more realistic form of the polarizability versus distance from the interface is required, such as that provided by Kretschmer and Binder [37]. Further advantages of the Binder model are that it accommodates depolarization fields, and finite field penetration into the electrodes could also be included.

(a)

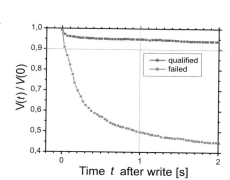

(b)

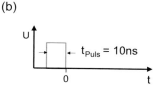

4.6 Frequency Dependence

For the 256 Mb DRAM generation the write time for the capacitor charge is of the order of 10 ns. Therefore, it is important that the high permittivity is also retained at high frequencies up to the GHz range. Figure 10b presents a collection [39] of the effective relative dielectric constants at zero field measured for different BST and STO thin films up to the THz range. The data are normalized to values in the same films measured at low frequencies. Generally, for high quality films the dispersion is negligible. The decreasing values at the high frequency end of some data sets are due to limits of the measuring set-ups. The resonance at frequencies beyond 1 THz is due to the ionic polarization (phonons).

In Figure 10a the area capacitance of a 30 nm thick MOCVD-BST film between platinum electrodes is shown versus the applied voltage measured at room temperature for two frequencies, 100 kHz and 1 GHz, respectively. The result is the same; the difference in the capacitance and thus in effective permittivity is quite small, about 5 % at the peak. Therefore, the frequency dependence does not seem to represent a limit for the use of BST films in Gb DRAM capacitors even if the write time approaches 1 ns. It should be mentioned that there is a relationship between the exponent $(-\alpha)$ of the decrease of the relaxation current vs. time, which usually is very close to -1 (see Sect. 5.1), and the exponent $(\alpha-1)$ (≈ -0) of the dispersion of the susceptibility $\chi_r(\omega)$ $(\approx \varepsilon_r(\omega))$, which may be deduced with the help of a Fourier transformation (see Chapt. 1).

Figure 11:
(a) Improvement of charge loss of metal-insulator (BST)-metal capacitor structures due to improved MOCVD-deposition technique. Shown is the time dependence of capacitor voltage $V(t) = Q(t)/C$, normalized to the value immediately after charging ($t = 0$).
(b) Schematic representation of a "write"-pulse.

5 Stability of Capacitor Charge and Reliability of the Cell

As described in Sect. 2, the stored charge in DRAM cells is unstable, i.e. it decreases with time. The reasons are a non-ideal and leaky capacitor, and a leaky transistor. As the sense amplifier cannot identify the state of a cell below a certain limit the charge state is refreshed periodically. For future DRAM generations the refresh times have to be in the several hundred ms regime so that 1s as an upper limit seems to be a good choice. Within this time the capacitor charge should not decrease by more than 10 %.

Examples of two films are presented in Figure 11a. The voltage of a Pt/MOCVD-BST/Pt capacitor normalized to the voltage immediately present after charging is plotted vs. time for different BST films. The charging pulse of ≤10 ns is shown schematically in Figure 11b. While the BST film of the lower curve has an unacceptably large charge loss during the first second, the film of the upper curve satisfies the requirement for DRAM: less than 10 % loss within the first second.

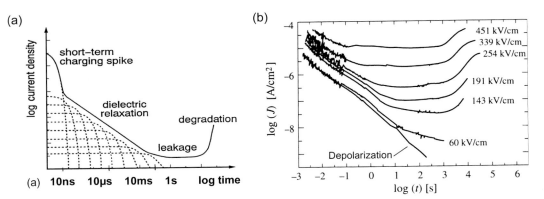

Figure 12:
(a) Schematic representation of current density j vs. time t for DC-loading of a MIM-capacitor showing different regimes.
(b) Experimental loading curves at different DC-fields at 425 K for MIM-capacitor (Pt/250 nm CSD STO/Pt). Also shown is one depolarization curve (after Dietz [40]).

5.1 Relaxation Currents

In contrast to the low-permittivity materials such as ON, for high-permittivity materials such as BST it is not the leakage current which dominates the charge loss but the polarization relaxation current. This is true especially at short times as is demonstrated schematically in Figure 12a. Generally, the relaxation currents in insulators such as BST follow the empirical Curie-von Schweidler relation with a time dependence of the relaxation current density $j_{relax}(t)$:

$$j_{relax}(t) = j_0 \cdot \left(\frac{t}{t_0}\right)^{-\alpha} \; ; \quad 0.5 \leq \alpha \leq 1. \tag{5}$$

Usually α is very close to 1 (see Table 2). Such a time dependence as in Eq. (5) can be generated e.g. by a sequence of a large number of "Debye" relaxations ($j_i(t) \propto \exp(-t/\tau_i)$) with very different relaxation times τ_i, as indicated in Figure 12a (dashed lines), and is discussed in more detail in Chapt. 1. There is still no microscopic identification of such a sequence of polarization relaxations for the high-permittivity materials.

This relaxation regime is followed by the leakage regime of steady-state current until the current increases at larger times by resistance degradation, i.e. the high resistivity of the insulator decreases gradually (soft breakdown) or nearly instantaneously by several orders of magnitude (hard breakdown), which both are unacceptable for DRAM operation because of possible lifetime restrictions.

In Figure 12b several experimental curves [40] are plotted which generally follow the described behavior. They were measured on Pt/SrTiO$_3$/Pt capacitors of 250 nm thickness at 425 K for different, steadily applied DC fields as indicated. The log (j) – log (t) plot shows that the relaxation currents at small times decrease as predicted by Eq. (3) and their variation with field is not very large. In contrast, the constant leakage currents are very dependent on the applied field. They change by many orders of magnitude, from nearly 10^{-9} A/cm^2 at 60 kV/cm (1.5 V) to 10^{-5} A/cm^2 at 451 kV/cm (11.3 V). At longer times, as a function of the applied field resistance degradation occurs. For the lowest field value the depolarization current is also shown (capacitor shorted after field removed). Its magnitude is equal to the polarization relaxation current at the same field.

Thickness [nm]	Exponent α of j_{relax}	C/A [fF/μm^2]	$Q_{lost\,by\,relax.}$ [%] at 1 V	Limit for $j_{leakage}$ [A/cm^2] ($Q_{lost\,by\,leak.} < 10\%$ at 1 V)
100	0.99	40.4	10.9	$4 \cdot 10^{-7}$
30	0.99	80.0	10.3	$8 \cdot 10^{-7}$
15	0.99	92.4	7.3	$9.2 \cdot 10^{-7}$

Table 2: Charge loss by relaxation and leakage currents in MOCVD BST capacitors up to 1 s (partly after [38]).

Usually the relaxation currents are regarded as the main cause for the charge loss in BST. Note that this is in marked contrast with the ON-dielectric most familiar to the semiconductor industry. Table 2 gives some values calculated from experimental data at room temperature [38] assuming that the measured time dependence of the relaxation

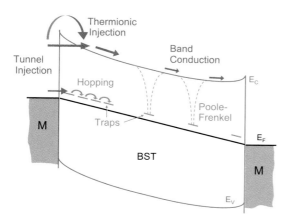

Figure 13: Schematic representation of leakage conduction mechanisms.

current at larger times is also valid for very short times. The corresponding limits for the leakage current to keep this contribution to the charge loss below 10 % are also given. As the relative importance of leakage currents increases with increasing refresh times and probably with decreasing film thickness a brief overview of the main mechanisms of leakage currents will be given in the next section.

For the DRAM community it is important also to consider the optimum method of characterizing the relaxation currents in a quality control environment. Measurement of the frequency dependence of the permittivity is not convenient as a routine quality measure. The measurement of current versus time is even more time consuming. As a result, the preferred procedure is the measurement of dielectric loss tangent, which has been shown to correlate with the relaxation behavior [41].

5.2 Leakage Current Mechanisms

The electronic conduction in dielectrics, which are insulators or wide-band gap semiconductors (the energy band gap in BST is about 3.2 eV at 300 K [43]), is complex and far from being well understood, especially in thin films. Besides electronic band conduction usually applied to metals and semiconductors also other models are applicable to account for the generally low carrier concentration of insulators and the low carrier mobility (about 1 cm^2/Vs for BST at 300 K [44]). The most important transport mechanisms are represented schematically in Figure 13 and described briefly below; more details are given in various textbooks [45], [46].

The current through the dielectric film has to be supplied by the injecting electrode, cathode for electrons and anode for holes. As the Fermi energy levels of the metal electrode and the insulator have to equilibrate in contact, there is generally a barrier at the interface for the injection of electronic carriers into the conduction band or valence band, or into gap states. The height of this barrier is an interface property. It has to be overcome by (1) thermionic emission, (2) field-enhanced thermionic emission ("Schottky" effect due to the Coulomb mirror potential), or by (3) quantum-mechanical tunneling through a (thin) barrier (see Figure 13). Therefore, in the literature the leakage current mechanisms are usually divided into "bulk limited" or "interface limited". For the latter, there are only very few exotic examples of *true* interface limited currents, e.g (1) tunneling through a very thin (< 3 nm) dielectric film from one electrode to the other or (2) "quasi-ballistic" flight of injected electronic carriers through the dielectric, i.e. without any significant (scattering) interaction. Therefore, the mean free path in the dielectric has to be larger than its thickness. Such conditions are not realized in insulating dielectrics, so that the injected current is generally decreased by a factor smaller than one due to the limitation of the bulk conduction mechanism. Simulation calculations of the steady state leakage current density show that only for rather extreme conditions the current is approaching the injection current limit, e.g. the Schottky limit of field enhanced thermionic emission at very high applied fields ($V/t > 1$ MV/cm) [42] and the tunnel injection limit for tunneling through a thin dead layer at the electrode interface [49]. All the injection currents are additive, tunneling can be dominant at lower temperatures while thermionic emission usually dominates at temperatures higher than 300 K.

Electronic bulk conduction mechanisms discussed for leakage current are as follows: (1) Band conduction, i.e. drift and diffusion of non-localized, quasi-free electronic carriers with an effective mass, as known from semiconductors. (2) Hopping conduction, a diffusion-like motion of electronic carriers using overlapping localized gap states without using the bands. (3) Poole-Frenkel conduction, a combination of trapping in deep localized gap states and band conduction after release from the traps. This release can be modeled similar to the Schottky effect at the interface. (4) Space-charge limited currents (SCLC). For this mechanism deviations from an ohmic current are obtained if the density of injected carriers (dependent on the applied voltage) is larger than the internal carrier density of the insulator. As the injected carriers represent space charges they create internal fields limiting the injection und, thus, the current. Depending on trap densities and energy levels, very different current-voltage characteristics are possible.

Experimental data of leakage current density, j, in BST (as well as other titanates) quite often show a linear behavior in a "Schottky plot", i.e. $\ln j$ vs. square root of applied voltage, V, or of mean applied field, $E = V/t$, as can be realized in Figure 15. Most authors interpret such a behavior as electrode limited leakage current. Fitting the equation for the (saturation) current of (field enhanced) thermionic emission to the data, unfortunately, in many cases the optical dielectric constant extracted from the slope of the curves is too small and often unphysical, (i.e. $\varepsilon_{r,opt} < 1$ instead of 5.6 for BST), e.g. $\varepsilon_{r,opt} \approx 0.16$ for the different annealing conditions shown in Figure 15 (Baniecki et al. [47]).

This can be corrected to some extend using the 3-capacitor model with low dielectric constant interface layers as discussed in Sect. 4.5, as it was done by Shin et al. [48] and in the re-interpretation of the leakage currents for the samples annealed in forming gas (see Figure 15) by Baniecki et al. [49]. In both papers the *injection saturation current* (neglecting any influence of the dielectric film conduction) is used to describe the experimental results. Although they achieved excellent description of their data and therefore interpreted their currents as electrode injection limited the extracted parameters for their similar Pt/BST/Pt film systems are very different, e.g. the effective electron mass. In contrast, Schroeder et al. [42], [50] were able to describe successfully experimental leakage currents (field, temperature, thickness dependence) in BST at higher temperatures ($T > 100°C$) by a combined injection/bulk conduction/recombination model without tunneling contributions, but also using dead layers at the electrode interfaces. This model is a modified version of the thermionic emission-diffusion theory [46], similar to that used in [49]. But still, further work is required to understand leakage current mechanisms in thin dielectric films.

5.3 Resistance Degradation and Breakdown

The steady-state leakage current regime is generally followed by a period of resistance degradation, i.e. a decreasing resistance with time (resulting in increasing leakage currents) and subsequently by a soft (slow) or hard (abrupt) breakdown, i.e. the complete loss of the insulating properties (see Figure 12a). These phenomena determine the lifetime and thus the reliability of the capacitor. The lifetime can be defined as the time for which the leakage current has reached 10x that of the steady state minimum. The benchmark for the lifetime of capacitors to be used in future Gb DRAM generations is 10 years at 1 V (= $V_{DD}/2$) and 125°C [51]. There are only a very few systematic investigations available on thin films [51]-[54].

Generally, the data for the lifetime have to be collected in accelerated tests (in order to avoid extremely long testing times), i.e. at temperatures and/or voltages much higher than the benchmark values. The temperature dependence of the lifetime can be described by Arrhenius behavior, i.e. the controlling mechanism is thermally activated. The extracted activation energies for BST thin films vary between 0.7 eV and 1.6 eV [55], [56] and they are dependent on various parameters. In particular, they decrease with increasing electrical field, with increasing film thickness and with a decrease of the titanium content from 53 at% to 51 at% [52]. A power law can describe the thickness dependence of the lifetime best but the exponents are dependent on temperature and field and stoichiometry. As an example, they are 1.85, 1.48, and 1.47 for measuring temperatures 200°C, 225°C, and 250°C, respectively, for BST films with thickness from 20 nm to 100 nm at a field of 750 kV/cm and 53 at% Ti [52]. A power law can also fit the voltage/field dependence with high exponents (approx. –5), which in turn are dependent on temperature, thickness, and stoichiometry. But also other dependences on voltage have been suggested, e.g. $\exp(1/V)$ or $\exp(-V)$. Additional dependences of the lifetime on doping and electrode material have been reported [53]. Due to the complex functional dependence of the lifetime on many parameters reliable extrapolations to the benchmark conditions are impossible, only rough estimates can be given, indicating that the lifetime of thin BST films may be satisfactory [52].

achievements have been made in process integration. For example, several manufacturers have been investigating high aspect ratio cylindrical designs [72]-[75]. The Fujitsu/Toshiba consortium has developed an innovative geometry, which they entitle the "line supported cylinder" [72]. This involves the deposition by MOCVD of a sacrificial, conformal TiN layer into a cylindrical well formed in SiO_2. The MOCVD Ru electrode is then deposited onto the TiN, followed by wet etch removal of both the TiN and the SiO_2. This leaves a freestanding Ru cylinder, with impressively large aspect ratio, and which is mechanically stable during subsequent processing. The Ta_2O_5 dielectric is then deposited on both inside and outside of the cylinder, increasing the effective surface area. The TiN etch also results in a clean Ru surface, which promotes local epitaxy of Ta_2O_5 and thus the high permittivity crystal structure. A number of integration issues relating to oxidation, adhesion, thermal budget, and CMP have been resolved by innovative approaches [72], [73], although some problems remain. Other companies are investigating closely related structures [74], [75]. A study of the mechanical stability of the electrode cylinder by the Fujitsu/Toshiba consortium suggests that this $Ru/Ta_2O_5/Ru$ technology could be scaled beyond the 100 nm technology node. For example, the existing achievable cylinder aspect ratio of 8 (internal diameter to height ratio) would allow the 65 nm technology node (4 Gb) to be achieved [72].

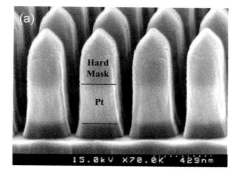

Another dielectric that has received attention over the past few years is Al_2O_3 [77]-[79]. This material may appear a surprising candidate, as it has a lower permittivity than Ta_2O_5, around 9 to 10. However, there are two factors favoring the material, namely its low leakage current density, and also the availability of high reliability layer-by-layer MOCVD growth (termed "atomic layer deposition" or ALD), along with the ALD equipment suitable for a manufacturing environment. Samsung have demonstrated a fully functional MIS 1 Gb DRAM using Al_2O_3 deposited by ALD on HSG Si with TiN plate (top) electrodes [78]. The low process temperature of 350°C is attractive to minimize the formation of a low permittivity SiO_2 layer at the Si interface. No predictions are yet being made regarding the scalability to further generations. However, it should be noted that the high quality conformal coverage makes the material attractive for high aspect ratio geometries, in particular deep trenches. This is demonstrated in a recent publication from Infineon, which demonstrates the use of ALD Al_2O_3 in trenches for sub-100 nm technology [80]. This work additionally uses rough polysilicon (HSG) in the trench, and uses a 'bottle' geometry to maximize the surface area.

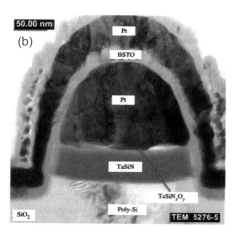

Recently, DRAM manufacturers have been developing layered or 'laminate' dielectrics, in order to gain maximum advantage from each of two different dielectric materials. For example, Al_2O_3/HfO_2 laminate structures have been demonstrated by Samsung [81]. The apparent advantage is that Al can be deposited with minimal accompanying growth of a deleterious SiO_2 interfacial layer, while HfO_2 has a higher dielectric constant (around 22, depending on structure).

The difficulties of processing of BST have also caused manufacturers to study the far end of the (Ba,Sr)-perovskite solid solution series, namely $SrTiO_3$. This material can be processed at a significantly lower temperature than BST, thus lessening the severe problems of stoichiometry control and conformal coverage, which has dogged the MOCVD of BST (see earlier section). For example, a $Ru/SrTiO_3/Ru$ MIM cell structure has been reported with an extremely high capacitance density [82]. It should be noted that this dielectric, particularly in conjunction with electrodes which form a conducting oxide interface, display a smaller dependence of permittivity upon thickness than BST. As a result, for later generations, and thinner dielectrics, the permittivity advantage of BST over $SrTiO_3$ is reduced.

Figure 16: Electron micrographs.
(a) SEM micrograph of post or pillar structure for highly integrated capacitor structures [8].
(b) TEM cross sectional micrograph of a prototype of Pt/BST/Pt post capacitor suitable for integration in 1 Gb DRAM generation (feature size =0.2 μm) [38].

It is worth emphasizing the importance of the geometry of the electrode or post structure on which the dielectric can be successfully deposited. The high aspect ratio Ru cylinders on which Ta_2O_5 can be deposited have already been mentioned. Considerable effort is required to process these cylinders, and the higher permittivity of BST is therefore an advantage. For comparable technology nodes, the posts required for BST (actually solid cylinders called "pedestals" in the roadmap [3]) have considerably smaller aspect ratios than those for Ta_2O_5. Posts fabricated of Pt bottom electrode that were achieved a number of years ago are shown in Figure 16a [9]. The area and pitch are approximately suitable for the 130 nm technology node [3]. The aspect ratio, after removing the hard etching mask (see Figure 16a), is around 1.3. Using these posts, and values for capacitance density for a 15 nm BST film taken from Table 2 [38], it is possible to calculate a probable BST cell capacitance. This estimated value is around 18 fF, which is a little less than the required 25 fF/cell. This calculation is consistent with the data calculated by Melnick [8], which is presented in Table 6. It can be seen that Melnick similarly calculates that a somewhat larger post aspect ratio, around 2.2, is required

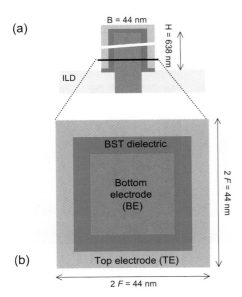

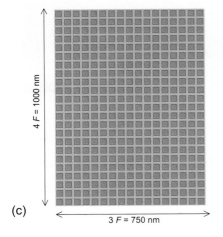

Figure 17: Comparison of DRAM cells: (a) 64 Gb cell (COB) with (b) projected areas of BE and layers of BST and TE; (c) 256 Mb cell with 387 cells of the 64 Gb DRAM.

for the Gb generation with feature size of 0.13 μm. However, the Table 6 also emphasizes the advantage of the higher permittivity of BST, in that the post aspect ratios required with the alternative dielectrics are considerably larger.

Dielectric material	SiO$_2$	SiN$_x$	Ta$_2$O$_5$ (MIS)	Ta$_2$O$_5$ (MIM)	BST* (MOCVD)
Capacitance area density [fF / μm^2]	7	12	17	37	85
Bottom electr. aspect ratio ($F = 0.18$ μm)	14.8	8.5	5.9	2.5	1.0
Bottom electr. aspect ratio ($F = 0.13$ μm)	28.7	16.6	11.6	5.1	2.2

* An additional 10 % increase in required capacitance area density was assumed for BST to account for charge losses due to relaxation currents.

Table 6: Bottom electrode pedestal aspect ratios for different dielectrics for 180 and 130 nm DRAM generations. Adapted from [8].

Figure 16b shows the cross-section of a successfully integrated BST capacitor, on a similar Pt post to that discussed above [38]. This post consists of a TaSiN diffusion barrier between Pt electrode and the poly-Si plug. The cross-section also displays one of the problems of BST integration discussed previously, namely the imperfect conformality of the BST. This problem increases at higher levels of integration where the aspect ratios will necessarily increase.

In summary, for the next few Gb generations of DRAMs, high permittivity materials will be utilized. The initial higher permittivity material is Ta$_2$O$_5$ in an MIS configuration. MIM configurations will follow, with BST or SrTiO$_3$ on simple posts competing with Ta$_2$O$_5$ on high aspect ratio cylinders, and ALD Al$_2$O$_3$ deposited in deep trenches.

7.2 Trends for the End of the Roadmap

Some aspects of the stacked capacitor with a high dielectric constant material such as BST have already been discussed in Sect. 3 and Figure 5. The 2001 ITRS roadmap [3] gives estimates of long-term trends all the way to the 64 Gb generation. A minimum feature size, F, of 22 nm and a cell area, A_{CA}, of $4 F^2$ are suggested for this generation. In addition, the roadmap suggests that higher permittivity materials, with $\varepsilon_r = 3000$, will be required [3]. Alternatively, is it still reasonable to consider BST at this technology node? To address this question, let us assume a minimum thickness of 5 nm can be achieved (with satisfactory leakage properties), and a dielectric constant of $\varepsilon_r = 200$ maintained, for the capacitor of this generation. Helpfully, the cell capacitance will decrease slightly to 20 fF (see Introduction to Part IV, Figure 9).

A cross section of such a COB capacitor layout for the 64 Gb generation is shown in Figure 17a, following the design guidelines of the roadmap. We assume a 6 nm-thick top (plate) electrode layer. The projected area of bottom electrode, which acts as a post (or pedestal), together with the BST layer and the top electrode layer, are plotted into the $4 F^2$ cell area in Figure 17b. This results in a cross section area of $1 F^2$ for the post. In order to illustrate the large integration factor for the 64 Gb cell in comparison to the 256 Mb cell (see Figure 5), such a cell is filled in Figure 17c with approximately 400 small cells of the 64 Gb generation. As the layer thickness of both dielectric and top electrode are a large fraction of the minimum feature size, complicated geometrical structures such as fins or even a rugged surface cannot be adopted in order to contribute a significant surface area increase. Therefore, the bottom electrode has to be assumed to be the simple pedestal with a quadratic or circular cross section, probably getting somewhat narrower at the top due to a practical etching angle less than 90°. Thus, a pedestal with quadratic cross section provides the upper limit for the calculation of the capacitor surface area, A_S.

The BST film parameters given above correspond to an equivalent SiO$_2$ thickness $t_{eq} = 0.098$ nm. The capacitor area can now be calculated by Eq. (1), yielding $A_S = 0.057$ μm^2, which corresponds to $117 F^2$. After subtracting $1 F^2$ for the top surface, we can calculate very simply that the side walls of the pedestal have to be at least $29 F$ ($= 638$ nm) tall! Is this reasonable or possible? It is an aspect ratio of 29, which is much greater than has been achieved for stack structures to date, but is similar to the aspect ratio of current trenches (see Sect. 3). It is not reasonable to expect that such structures could be etched directly into a noble metal, or that such stable, high aspect ratio pedestals could be achieved. However, deposition of the metal layers and BST into

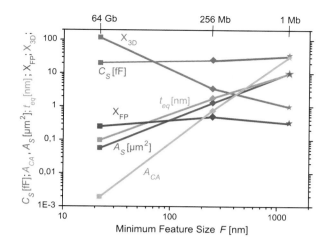

Figure 18: DRAM trends for capacitance C_S, cell area A_{CA}, capacitor area A_S, equivalent oxide thickness t_{eq}, footprint area fraction X_{FP} in A_{CA}, and 3-D folding factor $X_{3D} = A_S / X_{FP} A_{CA}$.
1 Mb: planar, SiO_2; 256 Mb: Stacked post, Ta_2O_5 (see Table 1); 64 Gb: Stacked post, BST.

a deep trench may be feasible if advances are made in layer-by-layer deposition technology. Some encouragement can be gained from the initial announcements of an atomic layer deposition technique for BST [66]. This scenario is actually less aggressive than the roadmap for trench DRAMs, which assumes a capacitance density that is three times lower than the above, and results in trench aspect ratio requirements around 100 !

The most important parameters for the extrapolated 64 Gb DRAM cell are plotted in Figure 18, together with those of the 256 Mb cell with Ta_2O_5 dielectric (see Table 1), and of the planar 1 Mb cell with SiO_2 dielectric, vs. the respective minimum feature size, F. These cells are each four DRAM generations apart corresponding to an integration factor of 256. From the different distances of the 1 Mb and 64 Gb data from 256 Mb data on the logarithmic abscissa it is concluded that F will shrink faster in the future than in the past, i.e. tremendous and accelerated improvements of lithographic and etching technologies are expected. On the other hand this trend is necessary to keep the die size, A_{DS}, within practical limits (see Figure 8, Introduction to Part IV). In general, most of the dependencies are smoothly extrapolated to the 64 Gb generation. The DRAM cell size decreases with decreasing F as $F^{2.4}$, the capacitance very slowly as $F^{0.1}$, t_{eq} shrinks as $F^{1.2}$, and hence the capacitor area shrinks as $F^{1.3}$ (see Eq. (1)). The only parameters that change differently in the extrapolation than in the past are the footprint area fraction, X_{FP}, and the 3D-folding factor, X_{3D}. The first, X_{FP}, increased from 1 Mb to 256 Mb because of the position change of the capacitor from the "offset" to the "stacked", but it will decrease in the future as $F^{0.3}$ because the dielectric and TE layers need an increasing fraction of the cell area, namely 75 % for the 64 Gb generation (see Figure 17b). Therefore, the 3D-folding factor, X_{3D}, defined as

$$X_{3D} = A_S / (X_{FP} \cdot A_{CA}), \tag{6}$$

increases as $F^{-1.4}$ to 117 for the 64 Gb generation, which results in the enormous technological problems discussed above.

The discussion shows that 64 Gb DRAM chips on a chip area of approx. 500 mm² (see Figure 8, Introduction to Part IV) are still conceivable on the basis of *geometric* considerations, although technologically huge challenges (such as dielectric deposition down deep features) would have to be met. Independent of these technological issues, the 64 Gb cell is close to a basic geometrical limit for this design. As can be recognized from an inspection of Figure 17 the layers for the dielectric and the top electrode cannot be smaller than $F/2$. As an example, using a required minimum layer thickness of 5 nm for the dielectric (to avoid unacceptably high leakage due to tunneling currents) and an assumed minimum thickness of 3 nm for the top electrode layer, the limit in the minimum feature size, F, would be 16 nm, close to the projected 22 nm to be used for the 64 Gb generation.

It should also be noted that the decrease of the dimensions of the local wiring to a few nanometers does not significantly affect the specific resistance of the electrode material at room temperature, due to the predominant surface scattering (i.e. size effect), because the mean free path length of the bulk resistivity of e.g. Pt, Ru or RuO_2 electrodes is of the same order. As a result, the RC delay time of the local wiring is still negligible compared to that of the bitline for the 64 Gb generation. Depending on the length of the bitline, this RC delay may approach the nanosecond regime. If necessary, the bitline may be divided into appropriate parts connected by "repeaters".

For the generations in the lower Gb range (i.e. the highest densities), the technological requirements discussed above could be partially released by (1) dielectrics with still higher permittivities and/or (2) a sub-bitline circuit that reduces the BL capacitance and, hence, decreases the required cell capacitance, C_S. This option is discussed in the context of FeRAMs (Chap. 22), while the material option will be sketched in the next section.

7.3 Alternative High-Permittivity Materials

In order to relax the high folding factor, only an increase of the dielectric constants is suitable, as the BST cannot be thinned beyond about 5 nm due to tunneling limits. As mentioned previously, the 2001 ITRS assumes that permittivities around 3000 will be required at the highest density generations at the end of the roadmap. Is this reasonable? There are several other mixed oxides with permittivities of several thousands to ten thousands in bulk ceramics, e.g. in the lead/titanium system such as $(Pb,La)(Zr,Ti)O_3$ (PLZT) and $Pb(Mg,Nb)O_3$-$PbTiO_3$ (PMN-PT) [83]. Such materials are ferroelectric in general, but for special stoichiometries [83] they are paraelectric, or their hysteresis loop is very narrow, i.e. the coercive field is sufficiently small that it can be used as dielectric in a DRAM capacitor. Permittivities of > 4000 have been demonstrated for films of PMN-PT of a thickness of 700 nm [84]. Of course, it has to be proven that these materials can retain a substantial part of the high ε_r as the thickness is reduced into the 5 to 10 nm range. In addition, it is known that the permittivity is field-dependent, decreasing strongly with increasing applied field (as shown in Figure 6). Hence, the gain predicted by Eq. (2) will be somewhat less than the permittivity improvement suggests. At the same thickness of 5 nm (as the BST) any increase of ε_r beyond 200 would linearly decrease the height of the pedestal bottom electrode. E.g. for $\varepsilon_r = 1000$ the height of the BE decreases from 29 F to 5.8 F = 128 nm, which significantly increases the probability for practical realization. However, it is expected that the relaxation, leakage, and resistance degradation issues will become more significant for materials of increasing permittivities. For these reasons, and because of the issues associated with the integration of Pb-based materials, there are no detailed investigations into Pb-based ultrahigh-permittivity materials for DRAM applications.

Another option would be the use of epitaxial thin films deposited on single-crystalline substrates and epitaxially grown BE, usually metallically conducting oxides. An example of such a system is SRO/20 nm epi-BST/SRO (SRO = $SrRuO_3$). Permittivities of up to 900 have been measured [85]. Still, for multi-Gb capacitors with non-planar geometries it would appear to be extremely difficult to develop an epitaxial layer-based complex oxide capacitor cell.

For DRAM generations at the advanced technology nodes there may be reasons other than a higher ε_r for replacing BST by alternative high permittivity materials. An example is $Ba(Zr,Ti)O_3$ (BZT) [13], films of which have ε_r similar to BST, but improved properties with respect to the dielectric losses, leakage, and resistance degradation.

In general, the materials having very high permittivities as bulk ceramics represent a large source of potential dielectric thin films in an advanced DRAM capacitor cell although they have not been investigated thoroughly with respect to their properties.

Finally, it must be pointed out that, faced with difficult processing, integration, and materials obstacles, alternative and competitive approaches will always be considered by the industry. To mention just a few examples: a DRAM architecture has been reported in which DRAM cells are stacked vertically upon one another [86]; and a stacked-surrounding gate transistor based DRAM has been reported, also consisting of vertically integrated cells [87]. An even more dramatic development is the announcements of capacitor-less DRAMs by several groups [88]-[90]. In one variant, the body-charging of partially depleted SOI-based transistors are used to store the binary logic states [88].

Acknowledgements

The authors gratefully acknowledges Cem Basceri, Dan Gealy Tom Graettinger, and Kunal Parekh (Micron Inc.) for a careful review of this chapter and they would like to thank Sam Schmitz (Caesar, Bonn) for checking the symbols and formulas in this chapter.

References

[1] R.H. Dennard, United States Patent 3387286, July 14, 1967.
[2] http://www.itrs.net/1999_SIA_Roadmap/Home.
[3] http://public.itrs.net/ International Technology Roadmap for Semiconductors, 2001 Edition: Front End Processes.
[4] K. Kishiro, N. Inoue, S.-C. Chen, and M. Yoshimaru, Jpn. J. Appl. Phys. **37**, 1336 (1998).
[5] www.samsung.com/news Sept. 24, 1998 ; (New) address (possibly changed): http://samsungelectronics.com/semiconductors/DRAM/product_news/olddata_993788883281_101.html.
[6] NEC 256 Mb SDRAM (μPD45256441) Datasheet (1998).
[7] H. Watanabe, T. Tatsumi, S. Ohnishi, T. Hamade, I. Honna and T. Kikkawa, IEDM Techn. Dig. 259 (1992).
[8] B.M. Melnick, Intern. Symp. on Integrated Ferroelectrics 1999.
[9] T. Morihara, Y. Ohno, T. Eimori, T. Katayama, S. Satoh, T. Nishimura and H. Miyoshi, Jpn. J. Appl. Phys. **33**, 4570 (1994).
[10] R. Waser and O. Lohse, Integrated Ferroelectrics **21**, 27 (1998).
[11] C. Basceri, S.K. Streiffer, A.I. Kingon and R. Waser, J. Appl. Phys. **82**, 2497 (1997).
[12] S.K. Streiffer, C. Basceri, C.B. Parker, S.E. Lash, and A.I. Kingon, J. Appl. Phys. **86**, 4565 (1999).
[13] S. Hoffmann and R. Waser, Integrated Ferroelectrics **17**, 141 (1997).
[14] G. Arlt, D. Hennings, and G. de With, J. Appl. Phys. **58**, 1619 (1985).
[15] R. Waser and S. Hoffmann, J. Kor. Phys. Soc. **32**, S1340 (1998).
[16] S. Hoffmann and R. Waser, J Eur. Ceram. Soc. **19**, 1339 (1999).
[17] M. Ihazu, K. Abe, N. Fukushima, Jpn. J. Appl. Phys. **36**, 5866 (1997).
[18] C. Zhou and D.W. Newns, J. Appl. Phys. **82**, 3081 (1997).
[19] U. Ellerkmann, R. Liedtke, and R. Waser, Ferroelectrics **271**, 315 (2002).
[20] T.M. Shaw, Z. Suo, M. Huang, E. Liniger, R.B. Laibowitz, and J.D. Baniecki, Appl. Phys. Lett. **75**, 2129 (1999).
[21] H.-M. Christen, J. Mannhart, E.J. Wiliams, and Ch. Gerber, Phys. Rev. **B49**, 12095 (1994).
[22] C.S. Hwang et al., J. Appl. Phys. **85**, 287 (1998).
[23] E.g. M.E. Lines and A.M. Glass, *Principles and Applications of Ferroelectrics and Related Materials,* Gordon and Breach, New York, 1976.
[24] I. Stolichnov et al., Appl. Phys. Lett. **75**, 1790 (1999).
[25] D. Choi et al., J. Appl. Phys. **86**, 3347 (1999).
[26] V. Craiciun and R.K. Singh, Appl. Phys. Lett. **76**, 1932 (2000).
[27] S. Stemmer, G.R. Bai, N.D Browning, and S.K. Streiffer, J. Appl. Phys. **87**, 3526 (2000).
[28] C.L. Jia, K. Urban, S. Hoffmann, and R. Waser, J. Mater. Res. **13**, 2206 (1998).
[29] E.H.Rhoderick and R.H. Williams, *Metal Semiconductor Contacts,* Oxford University, New York, 1988.
[30] A.W. Cowley and S.M. Sze, J. Appl. Phys. **36**, 3212 (1965).
[31] W. Mönch, Phys. Rev. Lett. **58**, 1260 (1986); Surf. Sci. **299**, 928 (1994).
[32] C.T. Black and J.J. Welser, IEEE Transactions on Electron Devices **46,** 776 (1999).
[33] H.Y. Ku and F.G. Ullman, J. Appl. Phys. **35**, 265 (1964).

[34] M. Dawber, L.J. Sinnamon, J.F. Scott, and J.M. Gregg, Ferroelectrics **268**, 455 (2002).
[35] A.A. Sirenko et al., Nature **404**, 373 (2000).
[36] C.B. Parker, J.-P. Maria, and A.I. Kingon, Appl. Phys. Lett. **81**, 340 (2002).
[37] R. Kretschmer and K. Binder, Phys. Rev. **B 20**, 1065 (1979).
[38] D.E. Kotecki et al., IBM J. Res. Develop. **43**, 367 (1999).
[39] R. Waser, Integrated Ferroelectrics **15**, 39 (1997) and references therein.
[40] G. Dietz, Fortschritt-Berichte VDI-Verlag, Düsseldorf, Reihe 21, Nr. 215, p. 77 (PhD Thesis RWTH Aachen, (1997).
[41] J.D. Baniecki, R.B. Laibowitz, T.M. Shaw, P.R. Duncombe, D.A. Neumayer, D.E. Kotecki, H. Shen, and G.M. Ma, Appl. Phys. Lett. **72**, 498 (1998).
[42] H. Schroeder, S. Schmitz, and P. Meuffels, Integrated Ferroelectrics **47**, 197 (2002); Appl. Phys. Letters **82**, 781 (2003).
[43] D. Goldschmidt and H.L. Tuller, Phys. Rev. **B35**, 4360 (1987).
[44] R. Moos and K.H. Härdtl, J. Am. Ceram. Soc. **80**, 2549 (1997).
[45] J.J. O'Dwyer, *The Theory of Electrical Conduction and Breakdown in Solid Dielectrics*, Clarendon Press, Oxford, 1973.
[46] S.M. Sze, *Physics of Semiconductor Devices*, 2nd Ed., John Wiley & Sons, New York, 1981.
[47] J.D. Baniecki et al., J. European Ceramic Soc. **19**, 1457 (1999).
[48] J.C. Shin, J. Park, C.S. Hwang, and H.J. Kim, J. Appl. Phys. **86**, 506 (1999).
[49] J.D. Baniecki et al., J. Appl. Phys. **89**, 2873 (2001); J. Appl. Phys. **94**, 6741 (2003).
[50] S. Schmitz and H. Schroeder, Integrated Ferroelectrics **46**, 233 (2002); H. Schroeder and S. Schmitz, Appl. Phys. Letters **83**, 4381 (2003).
[51] S. Zafar et al., Appl. Phys. Lett. **73**, 175 (1998).
[52] C. Basceri et al., Mat. Res. Soc. Proc. **493**, 9 (1998).
[53] M. Grossmann et al., Integr. Ferroel. **22**, 83 (1998).
[54] R. Kita, Y. Matsu, Y. Masuda, and S. Yano, Mod. Phys. Lett. **B13**, 983 (1999).
[55] R. Waser, in *Ferroelectric Ceramics*, N. Setter and E.L. Colla, eds., Birkhäuser-Verlag, 1993.
[56] G. Dietz, R. Waser, S.K.Streiffer, C. Basceri, and A.I.Kingon, J. Appl. Phys. **82**, 2359 (1997).
[57] R. Waser, T. Baiatu, and K.-H. Härdtl, J. Am. Ceram. Soc. **73**, 1645 (1990); ibid. 1654.
[58] C.B. Parker, Ph.D. Thesis, Department of Materials Science and Engineering, North Carolina State University, Raleigh, North Carolina (USA), September 2002.
[59] B.R. Chalamala et al., Appl. Phys. Lett. **74**, 1394 (1999).
[60] A.J.Hartmann, M. Neilson, R.N. Lamb, K. Watanabe, and J.F. Scott, Appl. Phys. **A70**, 239 (2000).
[61] S. Summerfelt, in *Thin Film Ferroelectric Materials and Devices*, R. Ramesh, ed., Kluwer Academic Publ., 1997.
[62] F. Fitsilis et al., Integrated Ferroelectrics **38**, 211 (2001).
[63] M. Yoshida et al., J. of Electroceramics **3**, 123 (1999).
[64] T. Horikawa et al., Mat. Res. Soc. Proc. **541**, 3 (1999).
[65] J. Park, C.S. Hwang, and D.Y. Yang, J. Mater. Res. **16**, 1363 (2001).
[66] M. Leskela and M. Ritala, Thin Solid Films **409**, 138 (2002).
[67] M. Hiratani, T. Nabatame, Y. Matsui, K. Imagawa,and S. Kimura, J. Electrochem. Soc. **148**, C524 (2001).
[68] S.R. Summerfelt, Intern. Symp. on Integrated Ferroelectrics (2000).
[69] For information see Internet pages of companies.
[70] Y. Park and K. Kim, Proceedings of the IEEE IEDM, 391 (2001).
[71] M. Hiratani et al., Symposium on VLSI Technology Digest of Technical Papers, 41 (2001).
[72] Y. Fukuzumi et al., Proceedings of the IEEE IEDM, 793 (2000).
[73] J. Lin et al., Proceedings IEEE, 183 (2002).
[74] W.D. Kim et al., 2000 Symposium on VLSI Technology Digest of Technical Papers, 100 (2000).
[75] Y. Nakamura, I. Asano, M. Masahiko, T. Saito, and H. Goto, Symposium on VLSI Technology Digest of Technical Papers, 39 (2001).

[76] M. Takeuchi et al., Symposium on VLSI Technology Digest of Technical Papers, 29 (2001).

[77] Y.K. Kim et al., Proceedings of the IEEE IEDM, 369 (2000).

[78] I.-S. Park et al., Symposium on VLSI Technology Digest of Technical Papers, 42 (2000).

[79] Y.K. Kim, S.M. Lee, I.S. Park, C.S. Park, S.I. Lee, and M.Y. Lee, Symposium on VLSI Technology Digest of Technical Papers, 52 (1998).

[80] J. Lutzen et al., Symposium on VLSI Technology Digest of Technical Papers, 178 (2002).

[81] J.-H. Lee, Y.-S. Kim, H.-S. Jung, J.-N.-I. Lee, L.-K. Kang, and K.-P. Suh, Symposium on VLSI Technology Digest of Technical Papers, 114 (2002).

[82] C.M. Chu et al., Symposium on VLSI Technology Digest of Technical Papers 43 (2001).

[83] R.C. Buchanan (ed.), *Ceramic Materials for Electronics: Processing, Properties, and Applications*, 2nd ed., Marcel Dekker, Inc., 168 New York, 1991.

[84] Z. Kighelman, D. Damjanovic, and N. Setter, J. Appl. Phys. **89**, 1393 (2001).

[85] N. Fukishima, K. Abe et al., Mat. Res. Soc. Proc. **493**, 3 (1998).

[86] F. Masuoka, T. Endoh, and H. Sakuraba, Proceedings of the Fourth IEEE International Caracas Conference on Devices, Circuits, and Systems, (Aruba,, April 17-19, 2002). Published by IEEE, p. CO 15-1 to CO 15-6 (2002).

[87] T. Endoh, M. Suzuki, H. Sakuraba, and F. Masuoka, IEEE Trans. on Electron Devices **48**, 1599 (2001).

[88] S. Okhinin, M. Nagoga, J.M. Sallese, and P. Fazan, IEEE Electron Device Letters **23**, 85 (2002).

[89] H.-J. Wann and D. Hu, IEDM Tech. Dig. 635 (1993).

[90] C. Kuo, T.-J. King, and C. Hu, IEEE Electron Device Letters **23**, 345 (2002).

Ferroelectric Random Access Memories

Ulrich Böttger, Institute of Electronic Materials, RWTH Aachen University, Germany

Scott R. Summerfelt, Texas Instruments, Dallas, USA

Contents

1 Introduction	565
1.1 Ferroelectric Non-Volatile Memory	566
1.2 State of the Art	566
2 FeRAM Circuit Design	567
2.1 Writing Schemes	568
2.2 Read Schemes	568
2.2.1 Cell Plate Step-Driven Read Scheme (CPSD)	568
2.2.2 Cell plate Pulse-Driven Read Scheme (CPPD)	569
2.3 Reference Voltages	569
2.4 Memory Architecture	571
2.4.1 Cell Plate Nondriven Read Scheme (NDP)	571
2.4.2 Divided Cell Plate Architecture	571
2.4.3 Chain FeRAM (NAND Architecture)	572
3 Ferroelectric Thin Film Properties	573
3.1 General Aspects	573
3.2 Ferroelectric Switching	573
3.3 Thickness Effects	575
4 Thin Film Integration	576
4.1 General Aspects	576
4.2 Electrodes	577
4.3 Special Integration Topics	578
4.3.1 Oxygen Barrier	578
4.3.2 Etching	579
4.3.3 Sidewall Diffusion Barrier	579
4.3.4 Back-End Processing	579
5 Failure Mechanisms	580
5.1 Polarization Fatigue	580
5.2 Retention Loss	581
5.3 Imprint	582
5.4 Lifetime Extrapolation	582
6 New Challenges	583
6.1 Polymer Ferroelectric RAM	583
6.2 Future Trends	584
7 Summary	586

Ferroelectric Random Access Memories

1 Introduction

Non-volatile memories are needed for high-performance digital cameras, digital audio appliances, digital video cameras and other popular multimedia applications. They can also be used with portable products such as mobile phones, PDAs, Palm PCs, and sub-notebook PCs as a high-density storage device, in place of a hard disk drive. Due to their higher power-consumption, larger size and higher sensitivity to shock and vibration, hard disk drives are typically not used in these handheld products.

Besides the desktop computation power and electrical power, the mobilization of large memory sizes plays a key role in the facilitating expansion of functions in mobile equipment and other future applications. A current trend in chip design is reflected by the increasing use of **embedded memories**, where memory is included on-chip. The embedded concept has the potential as low power and high performance device because the memory is directly connected to the logic circuits and analog components via on-chip bus (enhanced parallel processing). A further benefit is the reduction of the number of chips by a stringent integration approach, resulting in lower package cost and smaller number of pins per chip.

For **stand-alone memories** as well as embedded memories, the ideal characteristics are low power operation, fast write/read times, a near infinite number of write/read cycles, compatibility with Si process, non-volatility and minimum added process cost. The non-volatility is particularly helpful in reducing standby memory power. For high-density standalone memory applications a small cell size is indispensable to create a low cost memory. Whereas, for some embedded memories applications, it is more relevant to maintain acceptable electrical properties than to realize a high density of memory cells [1]. In order to impact the much larger embedded memory market it is necessary to reduce the embedded memory cost which is mostly a combination of small cell size and minimum added process cost.

Conventional non-volatile memories such as Flash devices only partially fulfil these demands. Flash belongs to the family of EEPROM (electrically erasable programmable read-only memory) which is a special type of PROM that is capable of erasure by exposing it to a large electrical voltage (~12 V) compared to ~1 V of scaled CMOS. Like other types of PROM, EEPROM retains its contents even when the power is turned off. However, this memory (or specific sectors) must be erased once to be rewritten and the number of writing cycles of EEPROM is limited to 10^6. Also, the writing cycle is rather ineffective compared to that of RAMs (random access memory) because the duration, as well as the power consumption, for writing exceeds RAM by some orders of magnitudes. The relatively large write energy consumption is one negative for this technology.

The most common embedded memory is SRAM because it has no added process cost although it has a large cell size. This volatile memory has a scaling problem with standby power. DRAM is also currently available as an embedded memory and it potentially has a much lower cell size than SRAM but it also has a significant added process cost. DRAM is also a volatile memory with a significant standby power, especially when embedded. The smallest embedded DRAMs require 6 – 8 added masks and while other flavors are available with fewer (~4) added masks they also have a significantly larger cell size.

Flash and EEPROM have also been developed as an embedded memory. Because of the performance limitations with these memories they can not replace standard main memory (currently SRAM) and are used in combination with SRAM. In particular, embedded Flash has a very small cell size compared to SRAM but unfortunately a significant added process cost (6 – 8 added masks is typical) plus added area for charge pump and other high voltage transistors.

Because of the limitations of the standard embedded memories, there is a search for a new "ideal" embedded memory.

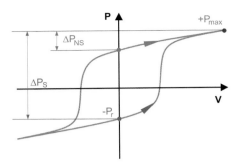

Figure 1: Non-switching and switching of the polarization of a ferroelectric capacitor

1.1 Ferroelectric Non-Volatile Memory

A recent type of memory device is the ferroelectric random access memory (FeRAM) incorporating a ferroelectric film as a capacitor to hold data [2]. The ferroelectric film has the characteristic of a remanent polarization, which can be reversed by an applied electric field, which gives rise to a hysteretic *P-E* loop (see Chapter 2). By using thin film technologies, capacitors of thickness at the sub-micron scale can be prepared so that operation voltages are reduced to a level below standard chip supply voltages. FeRAM use the *P-E* characteristic to hold data in a non-volatile state and allows data to be rewritten fast and frequently. In other words, a FeRAM has the advantageous features of both, RAM and ROM.

Voltage pulses are used to write and read the digital information. If the electrical field of the applied pulse is in the same direction as the remanent polarization, no switching occurs (Figure 1). The change of polarization ΔP_{NS} is due to the dielectric response of the ferroelectric material. If the initial polarization is in the opposite state due to the field, the polarization reverses giving rise to an increased switching polarization change ΔP_S.

The different states of the remanent polarization ($+P_r$ and $-P_r$) cause different transient current behavior of the ferroelectric capacitor to an applied voltage pulse, as shown in Figure 2. By integrating the current, the **switched charge** ΔQ_S and the **non-switched charge** ΔQ_{NS} can be determined. The difference in charge $\Delta Q = A \, \Delta P$ (area of capacitor A) enables to distinguish between the two logic states.

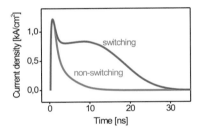

Figure 2: Current response of non-switching and switching case of the ferroelectric polarization.

1.2 State of the Art

A recently introduced product based on the use of a non-volatile ferroelectric random access memory is the RF-operated **smart card**. These smart cards do not only store ordinary credit card information but also process data by an embedded micro controller. The current standard FeRAM size of this embedded application is 1 to 4 kB, today. Smart cards are anticipated for use in public transportation and traveling, electronic banking, and sales, as well as in intelligent production and logistics with an expected high impact on daily life in the future [3]. Figure 4 shows the functional principle of a smart card in a block diagram and an embedded FeRAM chip. A reader provides an RF connection, which fulfils two purposes: to feed electrical energy into the card, since the card itself has no battery, and to perform data exchange by modulation of the RF signal. The RF signal is rectified and the energy is stored in a capacitor. The voltage control module supplies the dc power to all other modules of the card, i.e. the modulator/demodulator for the RF signals, the micro controller, some control logic, and the non-volatile memory.

Property	SRAM	eFlash	eDRAM	eFeRAM (projected)
Min. Voltage	> 0.5 V	> 12 V (± 6 V)	> 1 V	> 1 V
Write Time	< 10 ns	100 µs / 1 s	< 20 ns	< 20 ns
Write Endurance	> 10^{15}	< 10^5	> 10^{15}	> 10^{15}
Read Time	< 10 ns	20 ns	< 20 ns	< 20 ns
Read Endurance	> 10^{15}	> 10^{15}	> 10^{15}	> 10^{15}
Nonvolatile	no	yes	no	yes
Cell Size (F=half metal pitch)	~ 80 F^2	~ 8 F^2	~ 8 F^2	~ 15 F^2
Mask Count Adder	0	5 – 8	5 – 9	2

Table 1: Comparison of different memory technologies.

A comparison between the various embedded memories is shown in Table 1. The embedded FeRAM or eFeRAM is based on projected characteristics at leading edge technology nodes (for example 0.13 µm) [5] – [7]. SRAMs have a relative large cell size, which make this concept less suitable for high-density applications. DRAMs need self-refreshing cycles resulting in additional energy consumption. In addition, FeRAMs have fast write times ($\approx$ ns) and fast read times (down to 20 ns) as well as low write voltage

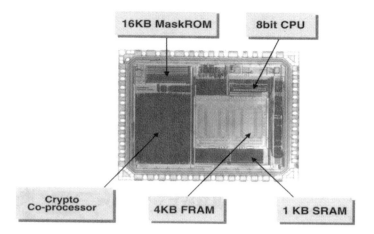

Figure 3: Embedded chip for smart card based on 4 kB FeRAM [4].

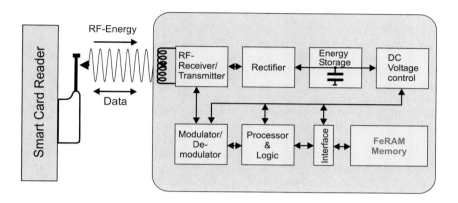

Figure 4: Block diagram of the functional principle of a smart card [3].

(0.9 – 3.3 V) which are similar to those of conventional memories. Therefore, it is evident that a FeRAM has the potential for portable equipment products as well as to become an universal memory type. From the table it is evident that FeRAM not only has performance advantages over current embedded memories but also cost advantages (smaller cell size than SRAM and fewer added masks than (eFlash and eDRAM). However, the performance and the reliability of FeRAMs should be verified in competition with other novel concepts for non-volatile memories as ferroelectric field transistor (FeFET), ferromagnetic tunnel junctions (MRAMs) or molecular devices. Key parameters are a number of 10^{15} write / read cycles and a data retention of longer than 10 years. If the number of write / read cycles is less than this, then some applications will not be able to use FeRAM. The larger the cycle number, the larger the applications which can be served.

The memory market for stand-alone devices is driven by the ongoing progress with standard semiconductor memories such as DRAMs (see Introduction to Part IV). The current task of the semiconductor industry is the introduction of the 1 Gb DRAM. The development of ferroelectric high-density memories is still about 3 to 4 generations behind that of DRAMs. The current stand-alone FeRAM devices are instead trying to compete with low power SRAM and EEPROM (not Flash) [7]. These markets are $\sim 10^9$ \$/year which is much less than DRAM or Flash but still significant. The top of the current development has been a 4 Mb FeRAM in 0.13 µm technology with an access time of 15 ns which was introduced by Texas Instruments, Agilent and Ramtron [5].

2 FeRAM Circuit Design

Four kinds of cell configurations, differing by the number of transistors and capacitors, have been introduced as ferroelectric memory cells: the **2T-2C cell**, the **1T-2C cell**, the **1T-1C cell** and the **1T cell**. The last, also referred as FeFET cell, is discussed in detail in Chapter 14 of this book. The 2T-2C cell has two ferroelectric capacitors where one is

IV Random Access Memories

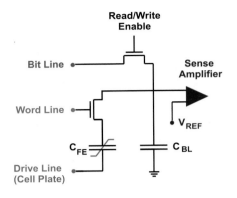

Figure 5: Simplified ferroelectric memory circuit.

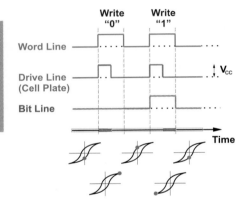

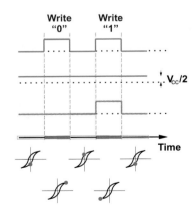

Figure 6: Simplified writing schemes and polarization states of the ferroelectric capacitor at short pulse scheme (top) and at fixed pulse level at the cell plate line (bottom) [10].

acting as a reference for the other [8]. This configuration enables a stable operation, but its cell size is twice that of 1T-1C. Therefore, the smaller 1T-1C is the target of present developments [9]. A detailed review on ferroelectric circuits was given in Ref. [10].

The 1T-1C ferroelectric memory cell is quite similar to a 1T-1C DRAM cell (see Chapter 21). Both cells are addressed by **wordlines** and **bitlines**. The only difference between the two cells is the **driveline** (or **cell plate** or **plateline**, see Figure 5) in the ferroelectric memory cell which needs to be switched in one of the writing an dread schemes dicussed below. This scheme does not exist in the DRAM cell. The function of each line during the write / read operations of the 1T-1C memory cell is described in the following section.

2.1 Writing Schemes

In order to write a binary digit 0 to a cell corresponding to the positive polarization state, a positive voltage V_{CC} is applied to the cell plate while the bitline is grounded and the wordline is addressed, see Figure 6 (top). The word line is raised to $V_{CC}+V_T$, where V_T is the threshold voltage of the access transistor. The applied voltage is known as **boosted** V_{CC} [11]. This allows a full V_{CC} to appear across the ferroelectric capacitor. The state of the capacitor is independent of the initial state. In writing a binary digit 1 that corresponds to the negative polarization state, a positive voltage V_{CC} is applied to the bitline while the cell plate is grounded and the wordline is addressed. It should be noted that a negative voltage would be required on the bitline to produce the same voltage across the ferroelectric capacitor if the cell plate was grounded. As shown in Figure 6 (top), it is standard practice to start the writing cycle with a short pulse on the cell plate line. Then, writing a "0" or a "1" is assigned by the level "low" or "high" at the bitline.

A different writing mode is given when the cell plate is fixed at $V_{CC}/2$ [12]. Writing the negative polarization state is caused by applying the V_{CC} voltage level to the bitline while the wordline is at the high level. When the bitline is grounded and the wordline is addressed the positive polarization state is written (Figure 6 (bottom)). This scheme is rather simple compared to the first one. However, the voltage level across the ferroelectric capacitor can only be $V_{CC}/2$, which could be too low to drive the polarization completely to saturation, i.e. to write the binary information definitely. Or, V_{CC} has to be set to twice the value. In any case, the equivalent circuit of the write operation is represented by the ferroelectric capacitor in series with the resistance of the access transistor. Consequently, there is an effective time constant that limits the writing speed.

2.2 Read Schemes

2.2.1 Cell Plate Step-Driven Read Scheme (CPSD)

A cell can be read by floating the bitline, i.e. write disabled, and applying a positive voltage V_{CC} to the driveline while asserting the wordline [9]. This establishes a capacitor divider consisting of C_{FE} (**ferroelectric capacitance**) and C_{BL} (**bitline capacitance**) between the driveline and the ground. The **sense amplifier** is off. The voltage at the driveline V_{CC} is divided between C_{FE} and C_{BL}, the parasitic capacitance of the bitline, according to their relative capacitance. Depending on the data stored, the capacitance of the FE capacitor can be approximated by C_0 or C_1, as shown in Figure 7. Therefore, the voltage developed on the bitline can be one of the two values $V_{BL}^{(0)}$ or $V_{BL}^{(1)}$.

$$V_{BL}^{(0)} = \frac{C_0}{C_0 + C_{BL}} V_{CC} \quad \text{if "0" is stored}$$
$$V_{BL}^{(1)} = \frac{C_1}{C_1 + C_{BL}} V_{CC} \quad \text{if "1" is stored} . \tag{1}$$

The voltage of the bitline corresponds to the possible polarization stages $+P_r$ or $-P_r$ as follows, where A is the FE capacitor area, P_S is the polarization at V_{CC} and P_r denotes the remanent polarization.

$$V_{BL}^{(0)} = \frac{A}{C_{BL}}(P_s - P_r)$$
$$V_{BL}^{(1)} = \frac{A}{C_{BL}}(P_s + P_r).$$
(2)

At this point, the sense amplifier is activated to drive the bitline to full V_{CC} if the voltage developed on the bitline is $V_{BL}^{(1)}$, or to 0 V if the voltage on the bitline is $V_{BL}^{(0)}$ as discussed in Section 2.3.1. Since this operation is destructive, the data must be written back into after each read, i.e. data restore. The wordline is kept activated until the sensed voltage on the bitline restores the original data back into the memory cell.

This method is called cell plate step-driven (CPSD) or "read-on-pulse" since the voltage on the cell plate line is a step voltage during the sensing operation, see Figure 8.

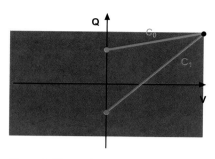

Figure 7: Hysteresis loop is approximated by two linear capacitors C_0 and C_1.

2.2.2 Cell plate Pulse-Driven Read Scheme (CPPD)

An alternative read procedure is the CPPD or **read-after-pulse scheme** [12] ensuring a stable operation of a FeRAM cell even at low voltages V_{CC} or at a small capacitance ratio of the bitline to the ferroelectric capacitor C_{BL}/C_{FE}. In both cases, the effective available voltage V_{FE} across the ferroelectric capacitor could be too small to switch the polarization of the capacitor completely. According to the equivalent circuit of the read operation the ferroelectric capacitor is in series with the bitline capacitor and V_{FE} is given by:

$$V_{FE} = \frac{1}{1 + C_{FE}/C_{BL}} V_{CC}.$$
(3)

A ratio C_{BL}/C_{FE} of 2:1 is normally suggested for 5 V power supply operation. Higher ratios will provide larger voltage for the polarization switching but, at the same time, reduces the voltage available for sensing. In the CPPD scheme, the sensing operation is activated after the positive voltage on the cell plate line returned back to zero, see Figure 8. The bitline voltage, in this case, is proportional to $2P_r$ for a digital 1 data, and 0, for a digital 0 data.

$$V_{BL}^{(1)} = \frac{A}{C_{BL}} 2P_r$$
$$V_{BL}^{(0)} = 0.$$
(4)

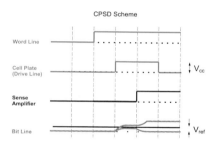

Comparing the CPSD and CPPD sensing schemes, they both develop the same voltage difference between digital 1 and 0 ($V_{BL}^{(1)} - V_{BL}^{(0)}$), which is proportional to $2P_r$ with the above mentioned factor. However, the **common mode voltage** of the CPSD scheme is different from that of the CPPD scheme.

$$\overline{V}_{CPSD} = \frac{V_{BL}^{(1)} + V_{BL}^{(0)}}{2} \propto P_s$$
$$\overline{V}_{CPPD} = \frac{V_{BL}^{(1)} - V_{BL}^{(0)}}{2} \propto P_r.$$
(5)

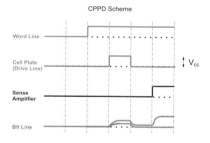

Figure 8: The reading schemes, cell plate step-driven (CPSD) and cell plate pulse-driven (CPPD), differ from the sensing operation which is activated before or after the positive voltage on the cell plate line returned back to zero [10].

Depending on the FE capacitor characteristics, P_S can be several times larger than P_r. Therefore, the step sensing scheme provides a higher common mode voltage which simplifies the sense amplifier design when a bias voltage is required. The pulse sensing scheme is normally used in conjunction with fully differential cell structures and direct bitline sensing technique.

2.3 Reference Voltages

The sense amplifier determines if the voltage that develops on the bitline is equal to V_0 or V_1 by comparing it against a reference voltage V_{REF} which is ideally the half between V_0 and V_1. In case of V_1 on the bitline the difference between V_1 and V_{REF} is amplified to V_{CC} otherwise it is shortened.

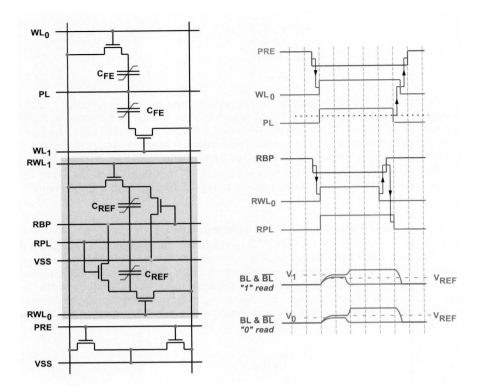

Figure 9: Cell design of *One oversized reference capacitor per column* [10].

There are some challenges in generating a reference voltage. It is very difficult to assume fixed values of V_0 and V_1 which are determined by C_{BL}, C_0 and C_1 (see Eq. (1)). C_0 and C_1 are cell dependent and can vary across the memory array, and change with time under cycling, fatigue, or without cycling, as imprint, see Sec. 5. The fatigue behavior implies that often accessed cells degrade faster than less accessed cells. In any case no fixed value of the reference voltage can be used across the whole chip, a variable V_{REF} is required to accurately track the process variation and the ferroelectric material degradation.

The cell design and the **timing diagram** for a read operation is shown in Figure 9. The reference circuit consists of two reference cells, i.e. one per bitline (BL), with its dedicated wordlines RWL_0 and RWL_1. When RW_0 and WL_0 are activated, the voltage on BL is compared with the reference voltage on $\overline{BL}$. The reference capacitor C_{REF} is always driven in the non-switching stage ("0" stored), but it is sized larger than C_{FE} in order to evoke a signal on the bitline between V_0 and V_1. After precharging the bitlines, WL and RWL are activated together, followed by a simultaneous step voltage on the PL and the RPL. The $\overline{BL}$ is raised to V_{REF}, while the BL is raised to either V_0 or V_1, depending on the storing stage. Since V_{REF} is between V_0 and V_1, the sense amplifier drives the bitline of higher voltage ($\overline{BL}$ or) to V_{CC} and the bitline of lower voltage to 0 V. In order to avoid a cycling of the reference capacitor that could cause fatigue, only a positive or zero voltage should act on the reference. This is realized by the timing of RBP relatively to those of RWL and RPL.

In addition, the reset transistor of the reference cell pushes the storage node to 0 V to avoid a voltage built up across C_{REF} which could otherwise cause contribution to a next immediate read (before the storage node leaks its charge to the substrate).

There are other reference schemes using the 1T-1C cell as *Two Half-Sized Reference Capacitors Per Column* (2 0.5C/BL), *One Half-Sized Reference Cell Per Half-Columns* (0.5C/0.5BL), *Two Full-Sized Reference Capacitors Per Two Columns* (2C/2BL) or *Adding Reference Cells to Rows* (2C/WL). A review is given in [10]. The 2C/2BL and 2C/WL schemes have superior sensing complexity and fatigue immunity, respectively.

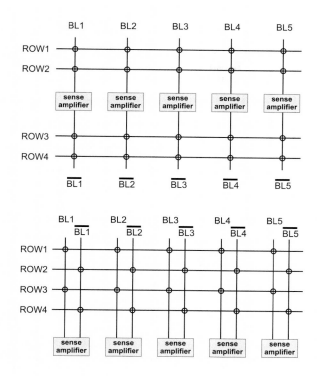

Figure 10: Block diagram of a ferroelectric memory with
(top) an open-bitline architecture and
(bottom) a folded-bitline architecture.

2.4 Memory Architecture

Two principle architectures of ferroelectric memories are available similar to DRAM cells: the **folded-bitline** and the older **open-bitline** architecture [13]. The first is now well adopted in FeRAM [14].

In that case the bitlines are folded to lie on the same side of a sense amplifier, as shown in Figure 10, instead of lying open on opposite sides of the sense amplifier, to reduce chances of any bitline mismatch that could occur due to process variations.

On the other hand, the requirement for pulsing the plateline in a FeRAM has called for original circuit techniques that were not required in a DRAM. There are various memory architectures that have been developed for a FeRAM with moving plateline.

2.4.1 Cell Plate Nondriven Read Scheme (NDP)

A plateline is slow to move due to its relatively high capacitance. This has motivated the researchers to innovate a constant-plateline architecture. The cell plate nondriven read scheme is useful for **high-speed operation** of FeRAMs [15]. In a conventional FeRAM cell-access operation, the cell plate is pulsed so as to apply the V_{CC} voltage level across the ferroelectric capacitor. However, the large capacitance of the cell plate slows down the cell access operation. In the "nondriven scheme", the cell plate line voltage is fixed at $V_{CC}/2$ during the read operation as shown in Figure 11. This leads to a faster access time as compared to that of DRAMs. For a given technology, the delay from addressing of the wordline to the activation of the sense amplifier can be shortened to 13 ns, while the access time of the NDP scheme is about 60 ns [15]. Besides the above mentioned disadvantage of low operation voltages in the NDP mode, additional refresh cycles are required. Because of the constant voltage level $V_{CC}/2$ of the cell plate line leakage current flow from the capacitor storage nodes to the substrate through p-n junctions occurs. In this way, the memory information could be destroyed.

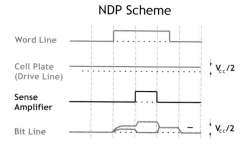

Figure 11: Reading scheme: nondriven cell plate (NDP).

2.4.2 Divided Cell Plate Architecture

In conventional cell architecture, the cell plate is realized by **global cell plate lines** covering the total area of the memory. Such a global cell plate line has a high capacitance. However, additional parasitic capacitances formed by the bottom electrode of the ferroelectric capacitors, also called the **local cell plate line**, enhance the effective capacitance

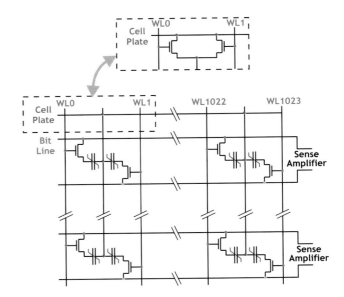

Figure 12: Transition from conventional cell plate configuration to divided cell plate configuration by insertion of transistors [9].

substantially. Thus, this architecture leads to large power consumption and slow access times. Following the concept of the divided cell plate design [9] shown in Figure 12, transistors are inserted between the local cell plate lines and the global cell plate lines in order to separate the local cell plate lines from each other and reduce the effective capacitance [9]. The gate of the transistors is linked to the wordline. While the wordline is addressed, only the corresponding local cell plate line is connected to the global cell plate line. e.g. in a 256-kbit SBT-FeRAM, the parasitic capacitance is decreased by about 80 % and the current is reduced from 0.52 mA to 0.12 mA (at $V_{DD} = 3$ V and $t_{cycle} = 200$ ns).

2.4.3 Chain FeRAM (NAND Architecture)

In contrast to conventional FeRAM, the transistor and capacitor of the Chain-Type Ferroelectric Random Access Memory (CFeRAM) cell are connected in parallel instead of in series, based on a suggestion by Toshiba [16]. Single cells are lined up to a chain enabling a very compact layout sharing a single contact to the bitline at one end and a single contact to the plateline at the other end. The combined effect of reducing the number of contacts on the bitline and the plateline is reduced access time and chip area. This architecture is similar to a NAND Flash memory architecture [17], in which a number of memory cells are grouped in series to share a single contact to the bitline. As cell size is a key parameter for high-density semiconductor memory, CFeRAM may have some advantages for high memory densities.

The circuit diagram and a cross-section layout of two so-called cell blocks is depicted being terminated by a BL at one end and a PL at the other end (Figure 13). In the active operation, a cell is accessed by grounding its corresponding WL and raising the Block-Select (BS) signal. All other wordlines remain high. The BL voltage as well as the PL voltage reach the selected cell via the chain of access transistors. Both, the driven plateline and the non-driven plateline read schemes, can be incorporated in this technique. Because all the FE capacitors are short-circuited during a standby operation via their parallel transistors, no refreshing cycles are required in the non-driven mode.

By the CFeRAM approach the **bitline capacitance** is decreased compared to the 1T-1C architecture since only one cell per cell block requires a direct contact to the bitline. The other cells contribute less capacitance to the bitline as they share diffusion area with neighboring cells. So, a higher number of cells per bitline is possible from the design point of view. However, the increase leads to higher feed resistances and parasitic capacitances causing a higher access time of the memory cell. Therefore, a balance should be found among the bitline capacitance (which could be decisively determined by the parasitic capacitance of the transistors in series), the readout delay, and the chip size. It was shown that for 1024 cells per bitline, and 16 cells per cell block, the bitline capacitance of the CFeRAM is comparable to that of the 1T-1C architecture while the total chip area is reduced to 63 % of that of the 1T-1C architecture [16].

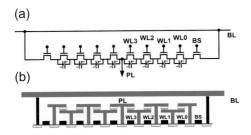

Figure 13:
(a) Circuit diagram of two cell blocks and (b) its corresponding cross-section layout in a chain FERAM architecture [10].

3 Ferroelectric Thin Film Properties

3.1 General Aspects

The most promising materials for FeRAM application are lead zirconate-titanate, $Pb(Zr_x,Ti_{1-x})O_3$ (PZT) and strontium bismuth tantalate, $SrBi_2Ta_2O_9$ (SBT), see also Chapter 2. Besides the ferroelectric properties, the compatibility with the silicon CMOS technology, the endurance of the ferroelectric capacitor (more than 10^{15} read / write cycles) as well as the device concept are playing decisive roles.

With respect to future high-density applications and for enhanced reliability the value of the **remanent polarization** should be as high as possible. In a simple consideration for **planar ferroelectric capacitors** [18], as depicted in Figure 14, it is assumed that the area of the capacitor takes a third of the space of the total cell area and that the polarization is not changed by size effects. When reducing the memory cell size, if the switched charge $\Delta Q = A\,\Delta P$ falls below a threshold value of about 30 fC a read failure will result. For a switched polarization $\Delta P = 20\ \mu C/cm^2$, i.e. a remanent polarization of $P_r = 10\ \mu C/cm^2$, the critical charge is crossed already when a 0.15 µm process technology corresponding to a cell size of about 0.45 µm is introduced.

In order to fulfil the demand of low power memory devices, a **low coercive voltage** is required of the ferroelectric capacitor. As shown in Chapter 2, SBT has lower coercive fields compared to PZT. In a first approach, this disadvantage could be overcome by reducing the film thickness, however, it should be taken into account that in general by decreasing the film thickness the coercive field is increased, and the remanent polarization may be reduced. Thus, for each individual case, the thickness scaling has to be verified with respect to the figure of merit. The thickness dependence of the coercive field as well as of the switching properties in ferroelectric thin films has been explained by the assumption of the existence of a (non-ferroelectric) layer at the interface between the electrodes and the film. This will be discussed later (see Sections 3.3).

As mentioned in Chapter 2, reversible as well as irreversible contributions to the polarization process exist in ferroelectrics. Even if the irreversible part is decisive, since the reversible process cannot be used to store information, both of the contributions are of importance to the design of external circuits of FeRAMs. Using the cell plate pulse-driven read scheme, the reversible parts do not play any role because the level on the bit-line is measured after the voltage step ($V_{BL} \propto P_r \pm P_r$), where P_r represents the irreversible polarization part, see Eq. (4). In contrast to the CPPD scheme, the cell plate step-driven read scheme (CPSD) leads to an offset voltage ($V_{BL} \propto (P_S-P_r)+P_r \pm P_r$) while the voltage step is applied caused by the reversible contribution (P_S-P_r), as shown by Eq. (2). This might change the requirements to the sense amplifier. In any case the difference between the switching and the non-switching case is the same: $\Delta V_{BL} \propto 2P_r$

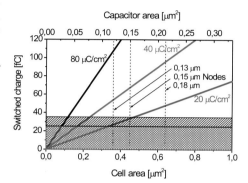

Figure 14: Correlation between charge and critical memory size with respect to different process nodes for $\Delta P = 20\ \mu C/cm^2$, $\Delta P = 40\ \mu C/cm^2$ and $\Delta P = 80\ \mu C/cm^2$ for planar geometry. Non-planar capacitors are discussed in Section 6.1.

3.2 Ferroelectric Switching

From the device point of view it is very important to guarantee that the switching process of the polarization in ferroelectric thin films is complete in order to obtain unambiguously the digital information. This requires on the one hand a sufficiently high external field, and on the other hand a sufficient pulse width longer than the ferroelectric switching time.

As mentioned in Chapter 2, the **switching time** depends on many factors, the domain structure, the nucleation rate of opposite domains, the mobility of the domain walls, and so on. The lower limit is given by the time for a domain wall to propagate from the top to the bottom electrode (or reversal) in a capacitor film with the thickness d:

$$t_0 = \frac{d}{c}. \tag{6}$$

Assuming that the upper limit of the domain wall velocity corresponds to the sound velocity $c \approx 4000$ m/s, the theoretical switching time t_0 is about 50 ps for a 200 nm-thick ferroelectric capacitor. By current investigations of the ultra-fast switching dynamics with the means of a novel femtosecond laser technique, a switching time for PZT thin films between 100 – 200 ps was found, which is very close to the theoretical limit.

Applying voltage steps on capacitors of PZT thin films up to voltages above the coercive voltage, the **current response** could behave differently, even if the switching is complete. In Figure 15 the responses of two $PbZr_{0.3}Ti_{0.7}O_3$ thin films (undoped and

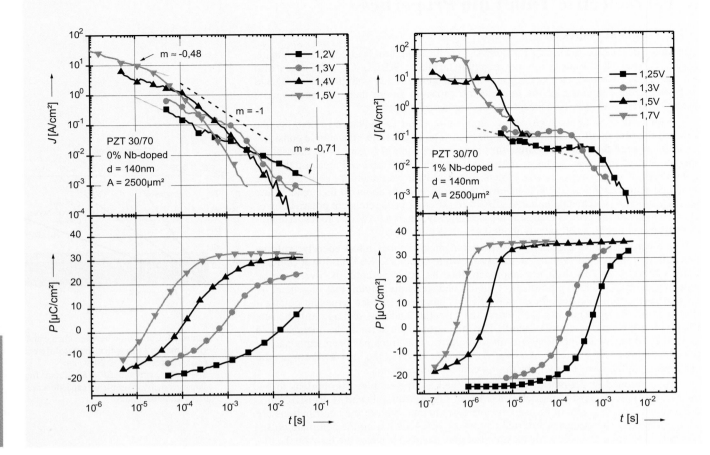

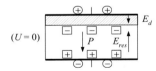

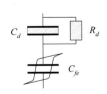

Figure 15: Polarization switching current and calculated polarization at different electrical fields for PZT thin film, undoped (left) and 1 mol % Nd-doped (right) [23].

Figure 16: Simplified scheme of a real ferroelectric capacitor and equivilant circuit.

1 % Nb-doped) are depicted. It depends on the squareness of the hysteresis loop whether the current shows the typical ferroelectric behavior with maximas $i_{max}(E)$ as known from single crystals (see Chapter 2) or not. This becomes important from the device point of view, because a ferroelectric capacitor with such characteristics could lead to read error in memory cells since the collected charges does not distinguish between the both switching states $-P_r$ and $+P_r$, i.e. the two logic states.

During polarization switching, the current response of the undoped film clearly shows a typical Curie-von-Schweidler behavior. It can be described by the following law, where κ is a constant and has a value of less or equal to one.

$$J \propto t^{-\kappa}. \tag{7}$$

The detection of this kind of current response leads to a different view of the polarization reversal since the **Curie-von-Schweidler behavior** is found in all thin films, in dielectric [19] as well as in ferroelectric thin films [20]. Two main models are discussed to be responsible for the Curie-von-Schweidler relaxation, firstly a many-body interaction model which is based on the fact that the hopping motion of a charged particle always affects the motion of the neighboring charges which can lead to a Curie-von-Schweidler law [21].

Secondly, there is the distribution of relaxation times (DRT) model which is based on a superposition of Debye-type relaxations with a large distribution of relaxation times which may be caused by a variation in charge transport barrier, for example at the grain boundaries [22].

The physical origin is not further discussed, but all evidence suggests that dielectric polarization processes are responsible. The Curie-von-Schweidler law in non-ferroelectric materials leads to a linear dispersion in the frequency regime, which reads as

$$\varepsilon = \varepsilon_\infty + k_0\, \omega^{\kappa-1}. \tag{8}$$

where ε_∞ is the permittivity at very high frequencies, k_0 a constant, and κ being the same constant as in (7).

Using a simplified picture of a real ferroelectric capacitor the switching behavior is modeled by a simulation tool for integrated circuits as e. g. SPICE. The capacitor consists of an ideal ferroelectric and a non-ferroelectric **interface layer**. The interface layer implies defects in the ferroelectric, amorphous grain boundaries as well as the real interface between electrode and ceramic film. It is well known that these regions have a strong dielectric dispersion, i. e. dielectric losses. The behavior of the dielectric interface is described by a parallel circuit of linear RC-terms, as shown in Figure 16. Lohse added to the ideal non-dispersive ferroelectric capacitor the dielectric RC-behavior in the above mentioned concept of the distribution of relaxation times (DRT) model on a superposition of Debye-type relaxations [24]. The switching is implemented by the Kolmogorov-Avrami process of the nucleation of domains and the subsequent domain wall motion.

As a result, it can be shown that the dielectric interface has a strong influence on the switching process. The delay of the polarization reversal is determined rather by dielectric dispersive polarization mechanisms than by the real ferroelectric switching (Figure 17).

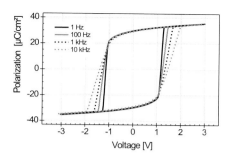

Figure 17: Simulated frequency behavior of ferroelectric thin film.

These interface effects could also provide an explanation for the **frequency dependence** of the coercive field in ferroelectric thin films which have a different origin compared with single crystals. Therefore, the frequency dependence of the coercive field depends on the ferroelectric itself and the electrode material. In $SrBi_2Ta_2O_9$ with Pt electrodes the exponential factor β of the empirical relation [25]

$$E_C \propto f^{-\beta} \qquad (9)$$

is about 0.1 in a raw approximation, while in $PbZr_{0.3}Ti_{0.7}O_3$ with Pt electrodes it is about an order of magnitude less, see Figure 18.

An alternative approach to the Kolmogorov-Avrami model is given in [26]. It is assumed that the film consists of many areas, which have independent switching dynamics. The switching in an area is considered to be triggered by an act of the reverse domain nucleation. The switching kinetics is described in terms of the distribution function of the nucleation probabilities in these areas.

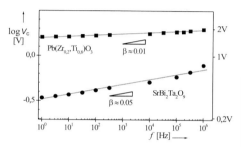

Figure 18: Frequency dependence of the coercive voltage in ferroelectric thin films of $SrBi_2Ta_2O_9$ and $PbZr_{0.3}Ti_{0.7}O_3$ [24].

3.3 Thickness Effects

As mentioned, there is a demand for the decrease of the film thickness in order to reduce the operating voltage of ferroelectric memory cells. It should be taken into account that the coercive field may be increased when the film thickness is decreased. The degree depends on the ferroelectric-electrode system and the ferroelectric itself. Three classical types are discussed: Pt/PZT/Pt, Pt/SBT/Pt and oxide/PZT/oxide (like RuO_2, $(La,Sr)CoO_3$ and so on). As summarized in Figure 19 the **size effect** is present when a ferroelectric thin film of PZT is combined with Pt or other noble metal electrodes. By the use of oxide/PZT/oxide or SBT a weak or a negligible dependence is observed. In addition, changes of the substrate as well as the composition (e.g. from rhombohedral to tetragonal PZT) or the measruing frequency, as shown in Eq. (9), lead to variations of the coercive field.

An explanation for the dependence of the coercive field on the film thickness was given on the assumption of the existence of a (non-ferroelectric) **interface layer** between the electrode and the ferroelectric with different dielectric properties [34]. These authors suggested that due to the field in the dielectric layer, charge carriers will be injected from the electrode to the interface between the non-ferroelectric layer and the ferroelectric thin film, thus screening the polarization of the ferroelectric. The effect of the near-by-electrode **charge injection** leads to the thickness dependence of the coercive field of a ferroelectric capacitor.

In Figure 20, a model of a real ferroelectric capacitor with a ferroelectric thin film, a non-ferroelectric interface layer and electrodes is shown. By a simple electrostatic consideration the electrical fields follow the equation below, where E is the applied field and σ the surface charge density at the interface:

$$E_f = E - \frac{d}{L}\frac{P-\sigma}{\varepsilon_0 \varepsilon_d} = E - \frac{d}{L}E_d. \qquad (10)$$

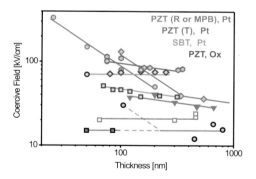

Figure 19: Thickness dependence of the coercive field for different ferroelectric-electrode systems summarized of [27] – [33].

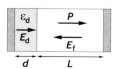

Figure 20: Model of a real ferroelectric capacitor with a ferroelectric (thickness L), a non-ferroelectric interface layer (thickness $d \ll L$) and electrodes (gray).

Taking into account that during cycling the field in the dielectric layer can reach values high enough to provoke electronic injection ($E_d > E_{th}$), the switching behavior is analyzed in terms of the relation between the threshold field E_{th} and the maximal polarization P_m on the hysteresis loop. Three different cases are distinguished.

If $\varepsilon_d E_{th} > P_m$, no injection occurs, the field in the surface layer is always smaller than the threshold field of the injection so that the injection is off during the cycling and the surface layer behaves as an ideal insulator. No thickness effects are expected, the measured coercive field E_C is equal to the coercive field of the film without interface layer $E_C^{(0)}$.

$$E_C = E_C^{(0)}. \tag{11}$$

If $\varepsilon_d E_{th} < P_m < 2\varepsilon_d E_{th}$, the field in the surface layer E_d will reach the threshold field during cycling and injection will be on after the polarization has changed its sign. For $P_m = 2\varepsilon_d E_{th}$, the injection starts, when the polarization is zero at the coercive field.

$$E_C = E_C^{(0)} + \frac{d}{L}\left(\frac{P_m}{\varepsilon_0 \varepsilon_d} - E_{th}\right). \tag{12}$$

If $P_m > 2\varepsilon_d E_{th}$, the injection is already on, before the coercive field is reached during cycling.

$$E_C = E_C^{(0)} + \frac{d}{L} E_{th}. \tag{13}$$

Following the conclusions of Tagantsev, the different behavior of the various ferroelectric-electrode systems is explained as follows. For Pt/PZT/Pt, case (ii) is valid and a strong influence of the film thickness on the coercive field is expected. With the decrease of the thickness L the „charge injection controlled" term of (13) becomes larger and increases the measured coercive field E_C. Considering an SBT thin film capacitor, the polarization is significantly lower than that of PZT. Therefore the electrical field of the interface layer will not reach the threshold field, and injection will be absent during the entire cycling. This case corresponds to (i). It is known that oxide electrodes enhance the leakage current of ferroelectric capacitors. The threshold field is very low ($E_{th} \approx 0$) or, in other words, the interface layer is shortened by the use of oxide electrodes. Therefore, case (iii) becomes independent of the film thickness.

4 Thin Film Integration

4.1 General Aspects

This section addresses the integration of ferroelectric capacitors in FeRAMs. However, some of the issues can be generally applied to other ferroelectric thin film applications in semiconductor circuits as well, e.g. pyroelectric and infrared sensors, piezoelectric-acoustic components, or electro-optic devices. The solutions to the process-related issues are critical for making ferroelectric thin film memories commercially successful and the efforts are subject of an intense industrial competition.

Generally, the fabrication process of the ferroelectric memories can be divided into three subsequent processing steps, the order of which is determined by decreasing processing temperatures. Firstly, the silicon devices such as transistors are formed by a **CMOS process** referred to as the *CMOS front-end* in the literature. Then, the ferroelectric capacitor with its electrodes is deposited and finally, the interconnect metallization and protection layers are processed, referred to as *back-end processing*. The process flow is shown in Figure 21. The **thermal budget** for the ferroelectric crystallization lies between the higher temperature treatment at 850 – 900°C needed for the current CMOS technology which varies significantly with the technology node and the lower temperature treatment of about 450°C used for the interconnection process [35].

The upper limit is required in order to activate the transistor's source/drain after the implantation of dopants. It is obvious that smaller nodes demand lower activation temperatures for a better control of the diffusion process. For the future sub-100 nm technology, the front-end process temperatures are about 650 – 700°C [35]. The future move from cobalt silicide to nickel silicide might further reduce maximum temperature that a

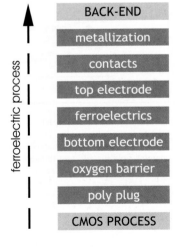

Figure 21: Process flow of a ferroelectric stacked cell.

logic front end can tolerate to ~500 – 600°C. The maximum temperature is of course lower for logic based processes (i.e. embedded memories) than for commodity memory processes (i.e. stand-alone memories). In addition, a certain thermal energy is needed to initiate the flow of phosphorus silicate glass (PSG) or boron phosphorus silicate glass (BPSG). This problem could be overcome by the means of alternative techniques such as Chemical Mechanical Planarization (CMP).

The lower limit is given by the final forming gas anneals in N_2/H_2, used to avoid trap levels in the transistor channel and hillocking of aluminium interconnects [37]. The maximum temperature at the sub-100 nm node will be lowered to 400°C. The requirement to lower the temperature is caused by the use of copper as interconnect metal and by the use of low-K materials as substitution of PSG or BPSG as interlevel dielectric.

The simplest way to integrate a ferroelectric capacitor in a 1T-1C cell is to contact top and bottom electrodes via metal lines resulting in a so called **offset cell**, see Figure 22 (top) and Figure 23 (top). A smaller cell size resulting in a **stacked cell** structure as shown in Figure 22 (bottom) and Figure 23 (bottom) can be obtained when the ferroelectric capacitor is arranged on top of the transistor and is connected by a conducting plug. For high density applications as required for standalone memories the stacked cell is indispensable, but introduces additional process complexity.

However, even the low-density memory cell design has the potential to be viable in the commercial market. Embedded memories where memory is included on-chip can tolerate lower packing densities corresponding to larger cell sizes, if the ferroelectric memory can be easily adapted to the other on-chip CMOS processes (i.e. logic circuits, analog components etc.). Since it offers a less stringent integration approach, it is therefore more likely to maintain acceptable electrical properties of the capacitor [38].

For both designs, notice that an interlevel dielectric, e.g. BPSG, is used between the CMOS front-end and the back-end processes. The purpose is to isolate the Si devices from the processing of the ferroelectric capacitor, i.e. the film as well as the electrodes.

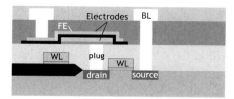

Figure 22: Schematic view of an offset cell (top) and a stacked cell (bottom) after [36].

4.2 Electrodes

Because the ferroelectric oxides are deposited in an oxygen-containing atmosphere and at relatively high temperatures (usually about 550 – 800°C), the electrodes must be resistant to oxidation, i.e. they may not form an insulating layer, and they must be able to withstand the processing conditions of the oxide. For this reason, candidates are oxidation-resistant **noble metals**, such as Pt and Pd, **metals** such as Ru and Ir and their **oxides** which have a high conductivity or form a conductive oxide such as RuO_2 [39] or IrO_2 [40] as well as multicomponent conductive oxides such as $SrRuO_3$ [41] and many others. The choice of electrodes for FeRAM applications and their deposition technique depend on device architecture, performance, reliability, processing and processing-related control requirements, environmental issues, cost, and the material of the ferroelectric capacitor.

Platinum electrodes are deposited by techniques as sputtering and e-beam evaporation (see Chapter 8). In the FeRAM cells poor adhesion prevents the deposition of Pt directly onto the interlevel dielectric (ILD). To improve adhesion, an intermediate layer of Ti is often deposited onto the ILD prior to Pt deposition, often. Ti is only used in offset cells, see Figure 23 (bottom). The thickness of the Pt and Ti films are typically 50 – 200 nm and 5 – 20 nm, respectively. However, under thermal cycling the combination Pt/Ti promotes the **formation of Hillocks** due to Pt stress relief [42]. The height of the hillocks varies, though in the worst cases, it has been observed to be greater than the ferroelectric film thickness, thus shorting the capacitor electrically [43]. Changing the Pt deposition technique, deposition conditions, annealing atmosphere, and film thickness the hillock formation can be avoided.

During the ferroelectric deposition and crystallization at high temperatures interdiffusion of Pt and Ti can occur [44]. If the Ti diffuses to the Pt/ferroelectric interface, TiO_x may form while the electrode is in an oxygen environment. The presence of TiO_x at the Pt surface promotes the formation of PZT. According to Hase et al. [45], the formation of TiO_x on the surface and in the grain boundaries of the Pt is a nucleation site for the perovskite PZT phase. With the Ti layer, PZT shows a homogeneous (single-phase) microstructure. Without the Ti layer, PZT exhibits an inhomogeneous microstructure, which consisted of perovskite regions surrounded by a pyrochlore (non-ferroelectric) matrix most likely due to the lack of appropriate nucleation sites.

In contrast to PZT, TiO_x on the top surface of a Pt bottom electrode impairs the ferroelectric properties of SBT resulting in a reduction of the remanent polarization of SBT. P_r changes from 13 to 4 $\mu C/cm^2$, when a Ti adhesion layer of 20 nm thickness is

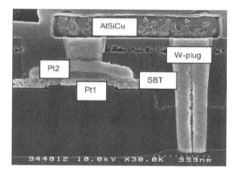

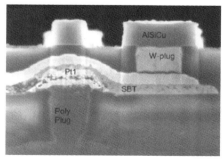

Figure 23: Cross section of a ferroelectric offset cell (top) and stacked cell (bottom) [36].

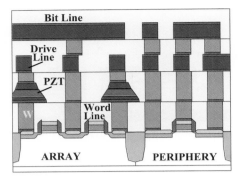

Figure 24: Cross-sectional schematic diagram of embedded FRAM. For clarity, only the first two of five Cu metal interconnect levels are shown [5].

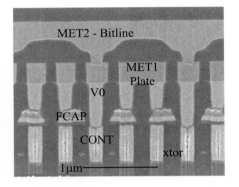

Figure 25: Cross-sectional scanning-electron micrograph showing the eFRAM module integrated between CONT and MET2 [5].

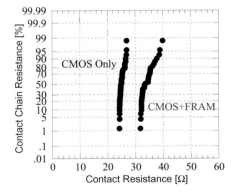

Figure 26: Measured 220 kΩ contact chain resistance from MET1 to NMOAT. The chain resistance is higher for the wafers with the FRAM module because of the additional VIA [5].

deposited. The reason is the diffusion of Ti and Si through the platinum to the SBT/bottom electrode at elevated temperatures. At high temperatures (> 700°C), i.e. high diffusion rate of Ti atoms, the incorporation of Bi into the film was negligible, because the bonding of the oxygen to Ti and Si species compared to the Bi species, is favoured. This results in Bi evaporation due to its high vapor pressure. It should be noticed that Sr and Ta were always readily incorporated into the film [46]. In order to overcome the difficulties the deposition temperature of SBT can be reduced below 700°C and the incorporation of Bi into the films was ensured by using Bi excess [46].

A different approach is the use of alternative diffusion barriers or adhesion layers such as TiAlN, TaSiN or in situ formed Ir/IrO_2. Ir/IrO_2 especially shows excellent oxygen barrier characteristics which is needed for the formation of capacitors in stacked cells, see Sec. 4.3. Because of these properties, for PZT as well as SBT capacitors bottom electrodes based on Ir have been introduced. The only other electrodes that are being pursued are $SrRuO_3$.

In any case, it should be kept in mind that changes in the integration and process parameters will also significantly affect the electrical properties of the capacitor such as the remanent polarization, the saturation polarization, the coercive field, the fatigue behavior, or the leakage current. In PZT based thin films, for example, several groups [47] have shown that various electrical properties are affected when Pt or RuO_x ($0 \leq x \leq 2$) electrodes are used. It was observed that the use of a bottom electrode of Ru/RuO_2 reduces significantly the value of the remanent polarization and increases slightly the average coercive voltage compared to Pt top and bottom electrode. It was suggested that this could be due to a better crystallization or a more favorable orientation of the PZT grown on Pt as compared to RuO_2. Furthermore, the asymmetrical charge-voltage hysteresis loop of the PZT with Ru/RuO_2 electrodes is indicative of a different top and bottom electrode.

4.3 Special Integration Topics

Besides the ferroelectric itself and the electrodes other integrations aspects such as miniaturization, oxygen diffusion as well as sidewall diffusion barriers, etching technologies and metallization connections play a decisive role for a successful integration of a ferroelectric capacitor into a memory device. The state of the art of the integration is shown in Figure 24 and Figure 25 [5]. A 1.5 V embedded ferroelectric memory process has been created using 130 nm, 5 metal layer Cu/FSG (Copper with Fluorosilicate glass with low-k dielectric) logic process and only 2 additional masks. Both FeRAM memory (0.58 μm^2 cell size) and SRAM memory (1.95 μm^2 cell size) have been created on the same wafers [5]. The eFeRAM density (~900 kb/mm^2) demonstrated here surpasses SRAM density by roughly 2.5 x and has been the most compact FeRAM in 2002. For an eFeRAM to be successful, it is essential that the CMOS characteristics are unaffected by the insertion of the eFeRAM module. The only significant impact on the CMOS characteristics is that the additional VIA increases the apparent contact resistance to MET1 by roughly 50 % (~24 Ω to 35 Ω as shown in Figure 26). Almost no difference in standard transistor characteristics has been observed. Fully functional 4 Mb SRAMs without repairs have been demonstrated on same wafers with a 4 Mb FeRAM test chips. The critical FeRAM processes used here are MOCVD PZT, Ir/TiAlN lower electrode, IrO_x top electrode, hardmask, one mask stack etch, and AlO_x H_2 diffusion barrier. There is starting too emerge a consensus on these processes for high density FeRAM as shown by recent conference publications [5] – [7]. The average properties of the ferroelectric are shown in Figure 27 [5] and the bit distribution shown in Figure 28 [48].

4.3.1 Oxygen Barrier

After the standard CMOS process **plugs** are formed to connect the transistor to the capacitor in the stacked cells. For commodity processes poly-Si typically is used, for logic based processes (i.e. embedded memory) tungsten is convenient. Then, an oxygen barrier is introduced in order to maintain a stable contact between the transistor and the bottom electrode after the anneal of the ferroelectric layer. This barrier layer is supposed to prevent the oxidation of the plug material and has to inhibit reactions between the bottom electrode material and the plug material which could lead to an increase of the resistance of the electrical connection [36]. Additionally, all reactions between the barrier material itself and either the plug or the bottom electrode have to be avoided. An overview of different oxygen barriers proposed in the literature was given in [36].

As mentioned above, the excellent oxygen barrier characteristics of Ir / IrO$_x$ are the main reason for the use as electrode materials of capacitors directly over contacts. IrO$_2$ based barriers are especially used, if the annealing temperature is higher than 600 °C in oxygen ambience. In the case of IrO$_2$ based barriers, IrO$_2$ has the function of oxygen barrier and Ir prevents the reduction of IrO$_2$ by the plug material during annealing. For structures on plugs some groups use different approaches of barrier-adhesion layer. These materials include TiN, TaSiN, TiAlN, or Pt-Rh alloys, see Table 2.

Top electrode	Pt	IrO$_2$	IrO$_2$/Pt	Pt	Pt
Barrier / Bottom electrode	Poly-Si/TaSiN / Ir/IrO$_2$/Pt	W/TiAlN/Ir	Ir/IrO$_2$/Pt	Ir/IrO$_2$/Pt	Poly-Si/Pt-Rh-O$_x$ / Pt-Rh/Pt-Rh-O$_x$
Temperature [°C]	700	< 600	700	700	650
Contact resistance [Ω]	200 – 400	100 – 200	100 – 200	500 – 700	1500
Plug diameter [µm]	0.5	0.5	0.5	0.6	0.6

Table 2: Overview of different oxygen barriers given in [36]

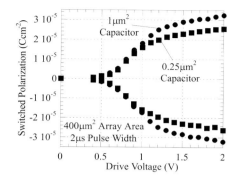

Figure 27: Measured switched polarization (Pp-Pu, Pn-Pd) for integrated capacitors (4×10^{-6} cm^2) after the first Cu interconnect layer using a resistor-based (200 Ω) PUND technique (2 µs pulse width) [5].

4.3.2 Etching

Reactive ion etching (RIE) or **reactive ion beam etching** (RIBE) are suitable tools for structuring integrated ferroelectric capacitors. For a general overview of these etching technologies, the reader is referred to Chapter 10.

The main topic for the etching technologies of ferroelectric capacitors is the reduction of the number of masks. There is a need for 1-mask etching processes of the total structuring which can produce ferroelectric capacitor fence slope better than 80 °C. It was shown that multiple masks not only are expensive but they greatly increase the cell size without increasing the capacitor size. For small feature sizes i.e. small capacitor to capacitor spacing of 0.18 µm or less, the only way to etch these structures is with hardmasks, see Chapter 8. Adding hardmasks has a big impact on integration flow since they also need to be deposited, etched and either removed again or kept [49].

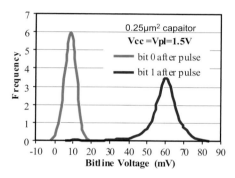

Figure 28: Measured bit distribution histogram for a 512 kb segment of the 4 Mb embedded FeRAM demonstrating bit operation for this device. A 100 ns plate line pulse duration was used for this measurement [5].

4.3.3 Sidewall Diffusion Barrier

A sidewall diffusion barrier is absolutely essential for integration. The primary reason is that a barrier against **hydrogen diffusion** is required. Hydrogen is used in many semiconductor backend processes, e.g. the final forming gas annealing with N$_2$/H$_2$, and it will impair or even destroy the ferroelectric properties [50]. The hydrogen sensitivity of ferroelectric films is one of the most important integration challenges. In order to save sufficient ferroelectric properties after back-end-of-line processing (BEOL) the barrier layers have to prevent contact of ferroelectric with hydrogen during subsequent temperature steps. The principal material that has been used for this purpose is AlO$_x$, firstly reported by [51].

An alternative barrier material proposed in the literature is TiO$_2$. In Ref. [36], the authors used a SiO$_2$ buffer layer and a SiN layer by low pressure chemical vapor deposition (LPCVD) to a SBT capacitor. The reason for this approach is that LPCVD SiN deposition contains hydrogen compounds like NH$_3$ and SiH$_4$ by itself. The presented hysteresis loops show only a slight change after encapsulation and forming gas annealing. In case of PZT, however, the AlO$_x$ barrier is preferred because Pb reacts with SiO$_2$ or SiN.

4.3.4 Back-End Processing

An important challenge of back-end processing is the **metallization connection** to the Si through the FeRAM layer. There are a number of varied integration approaches. For example, the concepts of offset cell as well as stacked cell uses contacts with different depths. Another approach is shown in Figure 29, where the cell area is reduced by means of stacked vias [52]. Each approach creates new problems for standard processes which causes development problems and potential yield impacts.

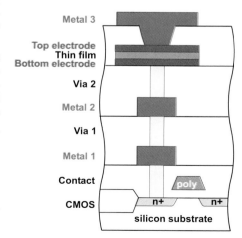

Figure 29: Cross section of a ferroelectric memory technology that uses three metal layer and allows stacked vias to minimize the memory cell area [10].

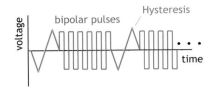

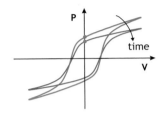

Figure 30: Pulse schemes for measuring fatigue behavior (top) and fatigue effect of hysteresis of ferroelectric capacitors (bottom).

Another topic of back-end processing is the ferroelectric cross contamination [53]. New metal contaminants, such as Pb, Zr, and Ir, introduced during FeRAM fabrication are of great concern because of their potentially adverse effects on complementary metal oxide semiconductor device characteristics. Numerous routes for **cross-contamination** exist in a Si wafer fab.

Contamination becomes possible by transfer of elements between the wafer handling system of a typical shared tool and the back side of clean wafers. Specifically, the potential for transferring Pb, Zr, Ti, and Ir from a contaminated surface to clean wafers could lead to contaminations. Typically, cycling a series of clean wafers through the tool led to a rapid decrease in the quantity of transferred contaminants, and is an effective method for eliminating these elements altogether. Moreover, the Pb, Zr, Ti, or Ir contaminants transferred in this way are easily removed using a traditional surface cleaning process.

5 Failure Mechanisms

The reliability of ferroelectric capacitors as well as CMOS circuits is one of the most crucial issues in the commercialization of FeRAMs, as both ferroelectric capacitors and CMOS circuits need to be highly reliable for at least 10 years and to allow a number of read/write cycles larger than 10^{15}. It is known that the reducing atmospheres used in the back-end processing lead to degradation of the properties of the ferroelectric capacitors. The ferroelectric processes which often use an oxidative atmosphere affect the reliability of the CMOS circuits. Furthermore, as non-volatility is a specific function of FeRAMs, novel reliability test methods have to be developed.

In the case of the ferroelectric memory cell, three different failure mechanisms have been reported in the literature which can affect the operation.

5.1 Polarization Fatigue

Upon continuous cycling the hysteresis loop to the positive and negative saturation, which emulates continuous read and write operation of the same cell, the P-V hysteresis loop may become flatter and the remanent polarization may decrease (see Figure 30, [2]). Hence, the difference between the switching and the non-switching charge becomes smaller which can lead to a failure of the device if this difference is too small to be detected by the sense amplifier. The decrease of the remanent polarization depends on the material properties of the ferroelectric capacitor as well as the test conditions such like pulse amplitude and width.

The constituents of a ferroelectric capacitor, the electrodes as well as the ferroelectric itself, are decisive for the fatigue behavior. A summary of some results of different ferroeolectric-electrode systems, as shown in Figure 31, was given in [54], [55]. In PZT-based thin films with standard Pt top and Pt bottom electrodes the value of the remanent polarization is reduced by about 50 % already after 10^{10} cycles. Using oxide electrodes, ferroelectric PZT films exhibit less significant fatigue. For IrO_2 and an optimized PZT composition with different dopants, a fatigue stability larger than 10^{15} was obtained.

Figure 31: Fatigue behavior of the normalized remanent polarization $P_r(N)/P_r(0)$ of SBT and PZT thin films with different electrodes on
(a) Ti/SiO$_2$/Si,
(b) BTO/YSZ/Si,
(c) Ru/SiO$_2$/Si and poly-Si/SiO$_2$/Si [55].

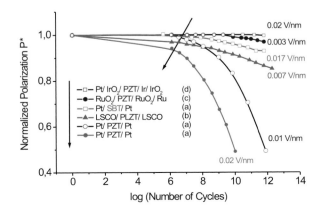

The physical origin of fatigue is still under discussion [54]. One model explains the fatigue phenomena to be caused by **defect traps** at the interface between the electrode and the ferroelectric film [56]. Assuming that the defects are oxygen vacancies, these can be eliminated at the oxide electrodes but not between PZT and the Pt interface. Another approach is that oxygen vacancies order into two-dimensional planar arrays that are capable of pinning domain wall motion [57].

Analytical analysis such as TEM or SEM confirms the different behavior of the PZT-Pt and the PZT-RuO$_2$ interfaces. On the one hand, for PZT-Pt capacitors the interface layer has an initial thickness less than 1 nm which increases up to 5 – 10 nm under fatigue treatment. On the other hand, Ru migrates from the top electrode into the PZT, forming channel-like structures. For Pt, no electromigration into PZT has been observed.

SBT thin films with Pt electrodes show little fatigue behavior and low leakage levels. These facts are the major motivation to introduce SBT as a capacitor material even though its remanent polarization is only about 30 % of that of PZT. Switching experiments with variable width and amplitude of the write pulse on PbZr$_{0.3}$Ti$_{0.7}$O$_3$ thin films show that small amplitudes in combination with a small pulse width of the write pulse does not provide sufficient switching of the ferroelectric film, thus wrongly suggesting a good fatigue behavior [58]. For fatigue measurements the degree of switching caused by the fatigue excitation signal strongly influences the fatigue results. In the case of complete switching, the fatigue behavior is found to be independent of the fatigue frequency and only the number of switching cycles is decisive for the polarization decrease.

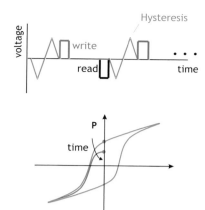

Figure 32: Pulse scheme for measuring retention loss in ferroelectric capacitors (top) and hysteresis loop (bottom).

5.2 Retention Loss

After establishing a remanent polarization, the ferroelectric cannot ideally retain its remanent polarization because of depoling or backswitching. The polarization slightly decreases with time which is called retention loss (Figure 32, [2]). Similar to fatigue, the difference between switching and non-switching charge becomes smaller. The retention loss is observed as a long-term relaxation of the remanent polarization of the hysteresis. It is determined by measuring the retained charge, e.g. by means of a negative read pulse, after a certain period of time when the state was stored by a positive write pulse.

FeRAMs should keep the digital information longer than 10 years. However, it is not possible to verify this requirement directly. One approach is to test the reliability at elevated temperatures in order to accelerate the retention loss. The retention loss at room temperature can then be extrapolated to 10 years assuming that the rate of decrease of the remanent polarization is **thermally activated** with a certain activation energy [59]. It has been empirically found that the polarization follows a logarithmic decay that begins at t_0 with the decay rate m:

$$P(t) = P(t_0) - m \log\left(\frac{t}{t_0}\right) \qquad (14)$$

with

$$m \propto \exp\left(-\frac{W}{kT}\right) \qquad (15)$$

When the remanent polarization falls below a critical threshold value P_c at $t = t_c$, the sense amplifier can no longer distinguish between the switching and non-switching case. Combining Eqn. (14) and (15) results in:

$$\log\log\left(\frac{t_c}{t_0}\right) \propto \frac{W}{kT}. \qquad (16)$$

In Figure 33 (top) the cumulative failures measured by a retention test of a 288-bit FeRAM at different temperatures is depicted [60]. The critical time t_c is reached when the cumulative failure crosses a certain value. Replotting the results of Figure 33 (top) for a cumulative failure of 50 % as an Arrhenius plot (Figure 33 (bottom)) yields an activation energy of 350 meV. From (16) it can easily be shown that the retention loss will not cause a device failure after 10 years at room temperature.

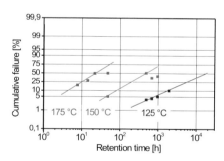

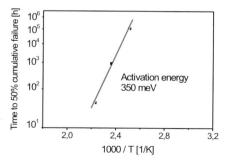

Figure 33: Cumulative failure in 288-bit SBT-FeRAM as function of time at different temperatures (top) and Arrhenius plot for a cumulative failure of e.g. 50 % (bottom) [60].

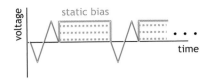

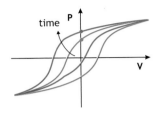

Figure 34: Pulse schemes for measuring imprint behavior (top) and imprint in ferroelectric capacitors (bottom).

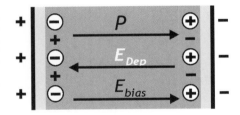

Figure 35: Charge separation in the interior due to E_{Dep} (caused by the dielectric gap at the electrode-thin film interface) resulting in trapping screening at the surface.

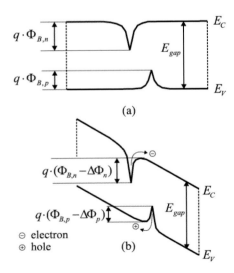

Figure 36: Sketch of the Frenkel-Poole emission within the surface layer.

5.3 Imprint

In literature, imprint is referred to as the preference of one polarization state over the other or the inability to distinguish between the two different polarization states [61]. Imprint affects the ferroelectric behavior of the thin films in two ways. On one hand, a **shift of the ferroelectric hysteresis loop** on the voltage axis is observed and on the other hand, imprint also leads to a **loss of remanent polarization**. Establishing and maintaining a positive state of polarization leads to a shift of the hysteresis loop on the voltage axis to the left (i.e. to negative voltages) and, additionally, to a loss of the negative state of polarization, i.e., the state opposite to the established one (see Figure 34).

Large signal aging of ferroelectric bulk ceramics and single crystals is very similar to the imprint behavior of ferroelectric thin films. The aging effect manifests itself as a shift (poled state) or a constriction (unpoled state) of the P-E hysteresis loop. From a physical point of view, different mechanisms like the reduction of domain wall mobility, the defect dipole alignment, or grain boundary effects are discussed [62]. A quantitative model of aging in acceptor-doped $BaTiO_3$ bulk ceramics was given by Arlt et al. [63]. By acceptor doping (e.g. with nickel) the Ni^{2+}-ions occupy the sites of the titanium ions (Ti^{4+}). In order to maintain charge neutrality, an oxygen vacancy is created in the lattice cell. The Ni ion and the oxygen vacancy represent a non-ferroelectric **permanent dipole**. With the beginning of the aging process these dipoles start to align with the direction of the surrounding spontaneous polarization of the ferroelectric. In poled ceramics the alignment of the microscopic dipoles stabilizes the macroscopic polarization and evokes an internal bias field E_{bias} that manifests itself as a shift of the hysteresis loop. With increasing the temperature, the imprint effect is accelerated because of the thermal activated mobility of oxygen vacancies.

In ferroelectric thin films, an additional imprint mechanism due to the screening of the ferroelectric polarization near the electrode-thin film interface by electronic charges was found. Assuming the existence of a (non-ferroelectric) interface layer [14] between the electrode and the ferroelectric with different dielectric properties, the screening charges on the electrodes are spatially separated from the polarization charges. This gives rise to an internal field [64]. The residual **depolarizing field** E_{Dep} is responsible for charge separation in the interior of the film (see Figure 35). The electronic charge carriers become trapped (the nature of the traps is still unclear) near the electrode-thin film interface resulting in a partial internal screening of the polarization. The de-trapping time constant of these charges exceeds the switching time of the ferroelectric polarization by several orders of magnitude. As a result, after switching the ferroelectric polarization, the trapped charges remain in their position leading now to an internal bias field E_{bias}, which causes the shift of the hysteresis loop. In accordance with this model Dimos et al. [65] could explain an enhancement of the imprint behavior in PZT thin films under illumination with band gap light, i.e. at a higher density of free electronic charge carriers.

A slightly modified model has been proposed in Ref. [66]. As a result of the different dielectric properties in bulk and interface, a large electric field E_{if} arises within the surfae layer pointing in the direction of the polarization. The field E_{if} is responsible for **charge separation** within the surface layer according to a Frenkel-Poole effect (see Figure 36). In Figure 37 experimental results are compared to simulation results based on the interface screening model. With the interface screening model also the impact of other experimental results are compared to simulation results based on the interface screening model. With the interface screening model the impact of other experimental parameters can also be successfully simulated, such as time, temperature, film thickness, and the improvement due to the use of oxide electrodes.

5.4 Lifetime Extrapolation

Imprint can lead to two different failure modes of a memory cell. Firstly, the voltage shift might become too large so that the programming voltage V_{cc} cannot switch the ferroelectric capacitor (**write failure**) anymore. Secondly, if the loss of polarization is dominant the sense amplifier cannot distinguish between the two logic states (**read failure**). To switch a ferroelectric, a sufficiently high voltage V_{min} has to be applied so that the resulting remanent polarization is high enough to meet the circuit design requirements. Hence, the failure point due to the write failure depends on the difference between the minimum required voltage V_{min} of the thin film material and the programming voltage V_{cc} of the memory device. A write failure occurs, when the **voltage shift** becomes

$$V_{c,\text{shift}} \geq V_{cc} - V_{\min} \tag{17}$$

The time dependence of the voltage shift follows in good approximation a logarithmic law: $|V_{c,\text{shift}}| \propto t^a$. For SBT thin films, the voltage shift during an imprint treatment is plotted in Figure 38 as function of the time. The lifetime of the ferroelectric capacitor limited by the write failure can be easily estimated by extrapolating this logarithmic dependence of the voltage shift $V_{c,\text{shift}}$ to the failure criterion $V_{cc} - V_{\min}$.

The read failure occurs when one relaxed remanent polarization drops to zero. Hence, the failure criterion for the read failure can be defined as the point when the loss of polarization equals the value of the relaxed polarization state of the initial hysteresis loop. However, for the read failure no simple logarithmic dependence of the polarization loss versus time is observed (see Figure 38).

However, a more detailed investigation reveals a strong correlation between the loss of the polarization, the voltage shift and the quasi-static hysteresis loop [67]. Using this correlation each value of the loss of polarization can be assigned to a certain voltage shift and the non-logarithmic time dependence of the polarization loss can be transformed to the logarithmic time dependence of the voltage shift. The failure criterion for the read failure ($P_{\text{rel}} = 0$) reads as:

$$V_{c,\text{shift}} \geq V_{c,\text{static}} \tag{18}$$

Assuming characteristic values of a SBT film (190 nm), i.e. $V_{\min} \approx 1$ V and $V_{c,\text{stat}} = 0.25$ V, it can be shown by Eqn. (17) and (18) that the write failure is only decisive for very low voltage operations $V_{cc} < 1.25$ V, whereas the read failure determines the lifetime of the device for more realistic values of the operation voltage $V_{cc} > 1.25$ V.

In Figure 39 the lifetime extrapolation for an SBT capacitor at an elevated temperature of 85°C is shown. A unipolar rectangular excitation signal has been applied to the ferroelectric which emulates a permanent read and write without changing the polarization state. It can be seen that even for unipolar stress a lifetime of well over ten years can be achieved.

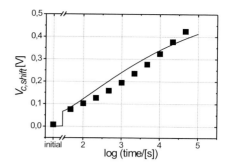

Figure 37: Experimental data and numerical simulation according to the interface screening model.

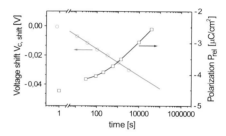

Figure 38: Logarithmic decrease of the voltage shift $|V_{c,\text{shift}}| \propto t^a$, which is responsible for a write failure, and time dependence of the relaxed remanent P_{rel}, which is responsible for a read failure, of SBT thin film [67].

6 New Challenges

6.1 Polymer Ferroelectric RAM

Driven by cost reduction combined with fast access and storage times and an infinite number of write cycles, new ferroelectric materials are suggested in thin film capacitors for non-volatile memory applications. One of the promising candidates are polymer chains with a dipole moment. The data are stored by changing the polarization of the polymer between metal electrodes. This class of memories is called polymeric ferroelectric RAM (PFeRAM) [68]. It should be noted in advance that because of the access and storage times these memories compete rather with Flash memories (mass data) than with conventional FeRAMs.

Intel has proposed such a device, which uses two electrode layers of metal lines running perpendicular to each other separated by a thin polymer layer, as shown in Figure 40. The complete film can be prepared by simple processes as printing or spinning. The intersection of the cross-points of the metal lines represents a memory cell. The matrix architecture is not restricted to read-only, it allows to write, to read and to erase data according to the applied voltage level.

The polymer material is sandwiched between two metal lines, which expose the material to a measured voltage. This arrangement eliminates the 1T storage cell, as there are no transistors per cell, i.e. it is a **passive matrix memory**. The individual polymer addressed bits are activated by word and bit lines. A common set of sense amplifiers in the CMOS base wafer sense the memory bit values. Moreover, these polymer layers can be stacked similar to multilayer capacitors with a polymer layer between different bit layers (Figure 40). In that way, real three-dimensional memories are created.

Relating to the absence of any transistor per cell, the PFeRAM has a potential for high density applications which could be higher than that of current NAND and NOR Flash memories as well as conventional FeRAMs. Nevertheless, the passive concepts make high demands on the ferroelectric properties of the polymer. Especially a sufficient squareness and a high stability of the hysteresis loop are required.

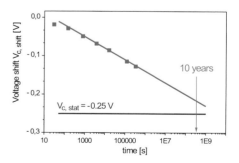

Figure 39: Lifetime extrapolation at 85 °C under unipolar rectangular cycling ($V_{cc} = 3$ V, 100 kHz) of a SBT film (190 nm). The extrapolated voltage shift will not cross the read failure criterium, $V_{c,\text{shift}} > V_{c,\text{static}}$, within 10 years [67].

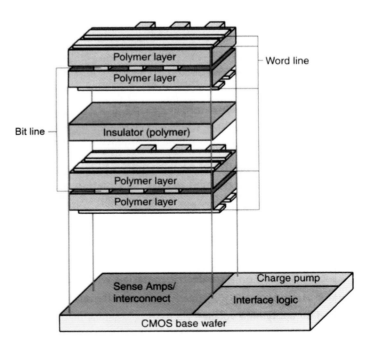

Figure 40: Details of polymeric memory being developed by Intel in collaboration with Opticom.

In that case polymer memory could provide a very low cost per bit with a high chip capacity. The process would be simple and could be easily integrated with standard CMOS processes. It uses a small cell size ($4\,F^2$, feature size F) and can stack up to eight layers for a high chip bit density. The cells do not require any standby power or any refresh cycles, but polymer memory is not a fast access memory. Write speeds exceed those of Flash memories, read times are on the order of 10 μs (for destructive mode) or less than 50 ns (for non-destructive mode), which is adequate for disk-storage applications [69]. At the publishing date of the book, this concept still has to prove its suitability as commercial products.

6.2 Future Trends

As discussed in the Introduction to Part IV, there are some general aspects about the ongoing integration density of random access memories. Key parameters are the shrinking of the feature size, the reduction of the operation voltage and the enhancement of the voltage sensing. For the future scaling of FeRAMs, a possible scenario is shown in Figure 41.

A higher integration level needs a more complicated process technology with larger number of masks and additional single process steps. The approach has been initiated, as mentioned in Sec. 4.1, by the transition from the offset cell to a stacked cell. In the future, by the means of multi-stacked or trench cells as well as by the means of folded ferroelectric films, the effective capacitor area will be enhanced at a fixed cell size. In the following, the feature size is estimated for which a **folded 3-D structure** will become necessary.

The area of an individual storage cell A_{CA} is given by the square of the feature size F and an area factor X_{AF} describing how many F^2 are needed to realize the cell. Assuming a scaling dependence of $X_{AF} \propto F^{\,0.2-0.4}$ the cell area becomes $A_{CA} \propto F^{\,2.2-2.4}$.

The capacitor footprint area is the projection of the capacitor on the cell area A_{CA}. The factor X_{FP} indicates how much of cell area is covered by the ferroelectric capacitor: $A_{FP} = X_{FP}\, A_{CA}$. When this factor is enhanced with further miniaturization ($X_{FP} \propto F^{-0.1-0.3}$), A_{FP} results in

$$A_{FP} \propto F^{1.9-2.3} \qquad (19)$$

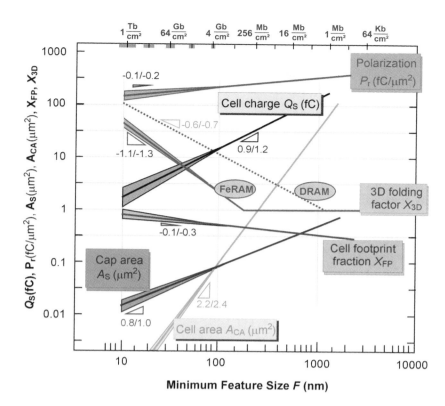

Figure 41: Roadmap of future FeRAM.

The limitations of the actual area of the ferroelectric capacitor A_S are determined by the switched cell charge and the remanent polarization: $A_S = Q_s / 2 P_r$. A reduction of A_S is possible as far as the charge is high enough to be detected by the sense amplifier. As shown in the Introduction of Part IV, it is expected that Q_s decreases with $F^{0.9-1.2}$, i.e. at feature size of 70 nm, a charge of 10 fC should be detectable.

In addition, scaling effects on the material properties of the ferroelectric which are independent of all design aspects should be taken into account. Ideally, the remanent polarization P_r should be independent of the capacitor area. In reality, a certain decrease of P_r with decreasing the area will occur, e.g. because of etching effects). Therefore, it is assumed that P_r follows: $P_r \propto F^{0.1-0.2}$, so that even at feature size of 10 nm a remanent polarization is still larger than 10 $\mu C/cm^2$.

Combining the estimations for Q_s and P_r results in the following relation between capacitor area and feature size:

$$A_S \propto F^{0.8-1.0} \tag{20}$$

When an actual area of the ferroelectric capacitor is required which is larger than the footprint area, 3-D capacitor structures have to be introduced. This requirement is expressed by the three-dimensional folding factor $X_{3D} = A_S / A_{FP}$ and corresponds to the relation $X_{3D} > 1$. As shown in Figure 41, it is expected that memory generations above 256 Mb will base on non-planar technologies, in contrast to DRAMs where these techniques are established since the introduction of 1 Mb chip (at 1 μm feature size). This is caused by the fact that the detected charge of a ferroelectric capacitor is higher than the charge of a dielectric one at the same capacitor area. Taking into account Eqn. (19) and (20) the folding factor depends on F with:

$$X_{3D} \propto F^{-1.0--1.5} \tag{21}$$

Long-term extrapolation to high density FeRAMs with 64 Gb with a feature size of 30 nm the folding factor is about 10. This value is achievable by current techniques, but it requires production at low tolerances.

7 Summary

Generally, the FeRAM can be used for data storage applications which have traditionally used DRAM. At the same time, it has the high-speed processing capability of SRAM and the non-volatility of Flash. Since all these advantages are featured in a single device, it is being hailed as the ultimate memory component of the next millennium.

Discussing the ferroelectric material aspect of a FeRAM, there are some benefits of PZT compared to SBT. The most important parameters for PZT and SBT are summarized in Table 3.

A decisive property of PZT is the lower crystallization temperature. It is expected that the thermal budget of future silicon 0.13 μm-technologies is below 650°C, which is less than the crystallization temperature of current SBT thin film deposition processes. Furthermore, the higher remanent polarization of PZT delivers the potential for a higher density of memory cells.

Parameter	SBT	PZT	Comment
Stand-alone FeRAM	yes	yes	DRAM market about US$14.6 B (2002)
Embedded FeRAM	no	yes	ASIC market about US$15.3 B (2002)
Crystal. Temperature [°C]	> 700	550 – 600	Primary integration issue
Volatile element	Bi	Pb	lower temp. for Pb process
Cap. @ 0.18 μm	3-D	Planar	Planar: simpler process, fewer masks
Endurance R/W	>10^{14}	>10^{14}	For PZT with metal /oxide electrodes
Voltage @ 100 nm	1.2	1.7	MOCVD 50 nm PZT
Electrode structure	Pt	IrO_2/Ir or RuO_2/Ir	Pt vr. Ir availability
Switched polarization[C/m²]	20	40 – 80	Varies with Zr/Ti ratio

Table 3: Summary of the key parameters for PZT and SBT.

Acknowledgements

The authors gratefully acknowledge Jürgen Rickes (Agilent Technologies) and Angus Kingon (NCSU) for a review of this chapter and Rainer Bruchhaus (Infineon AG) for fruitful discussion. They would like to thank Peter Schorn (RWTH Aachen) for checking the symbols and formulas in this chapter.

References

[1] P. Zurcher et al., IEEE Trans. Comp. Pack. & Manufact. Techn. A **20**, 175 (1997).
[2] J.F. Scott and C.A. Paz de Araujo, Science **246**, 1400 (1989).
[3] R. Waser, J. Euro. Ceram. Soc. **19**, 655 (1998).
[4] Fujitsu Ltd., *FRAM Guidebook* (2000).
[5] T. Moise, et. al, IEDM 2002, session 21 (2002).
[6] Y. Horii, et. al., IEDM 2002, session 21 (2002).
[7] S.Y. Lee and K. Kim, Future 1T1C, IEDM 2002, session 21 (2002).
[8] J.T. Evans and R. Womack, IEEE Jour. Solid-State Circuits **23**, 1171 (1988).
[9] T. Sumi et al., IEEE Intern. Solid-State Circ. Conf., Dig. Techn. Papers, 268 (1994).
[10] A. Sheikholeslami and P.G. Gulak, Proc. IEEE **88**, 667 (2000).
[11] W. Kraus et al., Symp. VLSI Circuits Dig. Tech. Papers, 242 (1998).
[12] K. Asari et al., IEICE Transactions on Electronics **E81-C**, 488 (1998).
[13] R. C. Foss, ISSCC Dig.Tech. Papers, 140 (1979).
[14] W. Känzig, Phys. Rev. A **98**, 549 (1955).
[15] H. Koike et al., IEEE International Solid-State Circuits Conference. Dig. Techn. Papers, ISSCC, 368 (1996).
[16] D. Takashima and I. Kunishima, IEEE J. Solid-State Circuits, **33**, 787 (1998).
[17] B. Ricco, G. Torelli, M. Lanzoni, A. Manstretta, H. Maes, D. Montanari, and A. Modelli, Proc. IEEE **86**, 2399 (1998).
[18] S. Summerfelt. Plenary talk, 12th ISAF 2000, Hawaii, USA, 2000.
[19] M. Schumacher, G. W. Dietz, and R. Waser, Integr. Ferroelectr. **10**, 231 (1995).
[20] X.Chen, A. I. Kingon, L. Mantese, O. Auciello, and K. Y. Hsieh, Integr. Ferroelectr. **3**, 355 (1993).
[21] L. A. Dissado and R. M. Hill, J. Mater. Sci. **16**, 1410 (1981).
[22] H Kliem, IEEE Trans. Electr. Insul. **24**, 185 (1989).
[23] O. Lohse, M. Grossmann, D. Bolten, U. Boettger, R. Waser, Mat. Res. Soc. Symp. **655**, CC 7.6.1 (2001).
[24] O. Lohse, PhD Thesis, RWTH Aachen (2001).
[25] J. F. Scott, Ferroelectrics Review **1**, 1 (1998).
[26] A. Tagantsev, I. Stolichnov, N, Setter, will be published in Phys. Rev. B
[27] J. F. Scott et al., J. Appl. Phys. **64**, 767 (2001).
[28] J. F. M. Cillessen, M. W. J. Prins, and R. M. Wolf, J. Appl. Phys. **81**, 2779 (1997).
[29] J. Zhu, X. Zhang, Y. Zhu, and S. B. Desu, J. Appl. Phys. **83**, 1610 (1998).
[30] C. H. Lin, P. A. Friddle, C. H. Ma, A. Daga, and Haydn Chen, J. Appl. Phys. **90**, 1509 (2001).
[31] P. K. Larsen, J. M. Dormans, D. J. Taylor, and P. J. van Veldhoven, J. Appl. Phys. **76**, 2405 (1994).
[32] D. J. Wouters, G. J. Norga, and H. E. Maes, Mater. Res. Soc. Symp. Proc. **541**, 381 (1999).
[33] Joo-Dong-Park, Tae-Sung-Oh, Jae-Ho-Lee, Joon-Yeol-Park, Thin Solid Films **379**, 183 (2000).
[34] A. Tagantsev, I. Stolichnov, App. Phys. Lett. **74**, 1326 (1999).
[35] International Technology Roadmap for Semiconductors (ITRS), 2001 Edition, http://public.itrs.net/Files/2001ITRS/Home.html
[36] N. Nagel et al., MRS Fall Meeting, Boston, USA, 2000.
[37] H. Achard, H. Mace, and L. Peccoud, Microelectr. Eng. **29**, 19 (1995).
[38] P. Zurcher et al., IEEE Trans. Comp., Pack. & Manufact. Technology Part A **20**, 175 (1997).
[39] H.N. Al-Shareef, K.R. Bellur, A.I. Kingon, and O. Auciello, Appl. Phys. Lett. **66**, 239 (1995).
[40] T. Nakamura, Y. Nakao, A. Kamisawa, and H. Takasu, Appl. Phys. Lett. **65**, 1522 (1994).
[41] W. Bensch, H. W. Schmalle, and A. Reller, Solid State Ionics **43**, 171 (1990).
[42] G.A. Spierings, J.B. Von Zon, M. Klee, and P.K. Larsen, Integr. Ferroelectr. **3**, 283 (1992).

[43] D.J. Eichorst, T.N. Blanton, C.L. Barnes, and L.A. Bosworth, Integr. Ferroelectr. **4**, 239 (1994).
[44] K. Sreenivas, I. Reaney, T. Maeder, N. Setter, C. Jagadish, and R.G. Elliman, J. Appl. Phys. **75**, 232 (1994).
[45] T. Hase, T. Sakuma, K. Amanuma, T. Mori, A. Ochi, and Y. Miyasaka, Ferroelectrics **8**, 89 (1994).
[46] J. Im, A.R. Krauss, A.M. Dhote, D.M. Gruen, O. Auciello, R. Ramesh, and R.P.H. Chang, Appl. Phys. Lett. **72**, 2529 (1998).
[47] D.J. Taylor, J. Geerse, and P.K. Larsen, Thin Solid Films **263**, 221 (1995).
[48] S. R. Summerfelt, et. al., ISIF 2003 extended abstracts (2003).
[49] Y.S. Lee et al., Symposium on VLSI Technology. Dig. Techn. Papers, 111 (2001).
[50] T. Sakoda et al., Jpn. J. Appl. Phys. **40**, 2911 (2001).
[51] I.S. Park et al., Int. Electron Device Meeting, IEDM Technology. Digest, 617 (1997).
[52] K. Amanuma et al., Tech. Dig. IEEE Int. Electron Devices Meeting, 363 (1998).
[53] S.R. Gilbert et al., Journal of Electrochem. Soc. **148**, 195 (2001).
[54] A.K. Tagantsev, I. Stolichnov, E. Colla, and N. Setter, J. Appl. Phys. **90**, 1387 (2001).
[55] W. Wersing, 12th ISIF, Aachen, Germany, 2000.
[56] S.B. Desu and I.K. Yoo, Integr. Ferroelectr. **3**, 365 (1993).
[57] J. F. Scott and Matthew Dawber, Appl. Phys. Lett. **76**, 3801 (2000).
[58] M. Grossmann et al., Appl. Phys. Lett. **77**, 1894 (2000).
[59] J.M. Benedetto, R.A. Moore, and F.B. McLean, Appl. Phys. Lett. **75**, 460 (1994).
[60] T. Otsuki in *Handbook of Thin Film Devices, Vol. 5*, ed. D. Taylor, Academic Press, San Diego, USA (2000)
[61] W.L. Warren et al., Jap. J. Appl. Phys. **35**, 1521 (1996).
[62] W.A. Schulze and K. Ogino, Ferroelectrics **87**, 361 (1988).
[63] R. Lohkämper, H. Neumann, and G. Arlt, J. Appl. Phys. **68**, 4220 (1990).
[64] U. Robels, J.H. Calderwood, and G. Arlt, J. Appl. Phys. **77**, 4002 (1995).
[65] D. Dimos, W.L. Warren, and B.A. Tuttle, Mater. Res. Soc. Symp. Proc. **310**, 87 (1993).
[66] M. Grossmann, PhD Thesis RWTH Aachen (2001).
[67] M. Grossmann et al., Appl. Phys. Lett. **76**, 363 (2000).
[68] R. Weiss, Electronic Design **49**, 56 (2001).
[69] G. Marsh, Materialstoday Sep./Oct. 2001, 34 (2001).

Magnetoresistive RAM

Jon M. Slaughter, Motorola Labs, Tempe, USA

Mark DeHerrera, Motorola Semiconductor Products Sector, Tempe, USA

Hermann Dürr, BESSY, Berlin, Germany

Contents

1 Introduction   591

2 Implementation of MRAM Devices   592
2.1 Single Transistor/ Single TMR (1T1TMR) MRAM Cell   593
2.2 Other MRAM Cell Architectures   593
2.3 MRAM Memory Circuit Architecture   594

3 Magnetic Stability of MRAM Devices   596
3.1 Thermally-Activated Reversal of a Magnetic Element   596
3.2 Switching Induced Demagnetization   599
3.2.1 Solution: Exchange Biasing   600
3.2.2 Additional Considerations   602

4 Ultrafast Magnetization Reversal   602

5 Summary and Outlook   603

Magnetoresistive RAM

1 Introduction

The development of modern computers is intimately linked to data storage technology. In present-day computers the information being processed is temporarily stored in fast dynamic random access memory (DRAM) devices (see Chap. 21) with data access times of typically 10 ns. DRAMs are fast but have the disadvantage that they are volatile, i.e. the stored data need to be constantly refreshed. This becomes evident when a computer is switched on. There is a time lapse in which the data (operating system, etc.) have to be loaded into the DRAM from permanent but relatively slow mass storage devices such as hard disc drives. Data access times for hard disc drives are typically in the msec range and mainly determined by the positioning speed of the read/write head.

Faster magnetic data storage was, in a sense, realized in early computers. In the 1960s, before integrated electronic circuits became available, so-called magnetic core memory was used as described briefly in the Introduction to Part IV. Within the novel magnetic RAM (MRAM) concept, the bulky ferrite cores were replaced by integrated thin magnetic layers elements. Writing the bits into MRAM is similar as in the original core memory devices. For reading the bit state, more sensitive magnetoresistance effects are employed. This allows a highly increased miniaturization of the magnetic data storage devices which have been developed in recent years and today follow a Moore's law type evolution. In addition to being non-volatile, fast read/write access times can be realized approaching those of present day random access memories (RAMs).

The earliest random access memories based on magnetoresistance in thin ferromagnetic films were proposed and developed in the 1980s [1]-[3]. These memories exploited the fact that the resistance of a ferromagnetic conductor depends on the angle between the magnetization and the current, an effect known as anisotropic magnetoresistance (AMR). This type of MRAM has found applications in the radiation-hard memory area but not in the commercial memory market. Its competitiveness with other commercial memories is limited by the rather small resistance change (< 2%) of AMR memory cells which leads to low memory density and high current demands.

In the late 1980s, much larger magnetoresistance values were observed in multilayered metallic films and recognized to be a new effect termed giant magnetoresistance (GMR) as described in Chap. 4. The new technology was quickly adopted for commercial hard disk read heads and spawned new research efforts in GMR-based MRAM circuits designed to exploit the higher signal [4]-[6]. Although the GMR effect provides a larger resistance change than AMR, in the range of 4–8% for typical memory cell designs, the low resistivity inherent to these metallic films results in a small signal at the low read currents desired for semiconductor applications. Marginally higher resistance devices are possible by using GMR cells that pass the current perpendicular to the film plane. In such perpendicular devices the resistance can be scaled to higher values via their smaller area. Still the resistance of such devices using current-generation microelectronic lithography dimensions is quite low, thus, limiting their application potential for the present.

The development of magnetic tunnel junctions (MTJ) with high tunnelling magnetoresistance (TMR) in 1995 [7], [8] (Chap. 4) offered MRAM technology a much improved signal due to its higher resistance and improved resistance change. In recent years the techniques and materials for producing MTJs have been dramatically improved so that TMR changes in the 30-50% range are now widely reported [9]-[12]. Several properties make MTJs ideal for MRAM circuits: the resistance can be adjusted to match the circuit, the TMR is several times higher than the GMR effect, and the flow of current perpendicular to the plane allows fabrication of small area memory cells.

In this chapter, the characteristics of TMR-based MRAM devices will be presented. The next section deals with the principle and technical implementation of MRAM devices and circuits. In the following sections, we will identify the main technological challenges and outline how modern scientific techniques can highlight the underlying microscopic processes. Magnetic stability issues associated with MRAM materials and possible solutions will be described in Section 3. Section 4 will be concerned with ultrafast magnetic switching processes that determine the writing speed of MRAMs. The chapter will conclude with a summary and outlook.

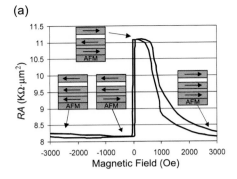

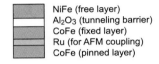

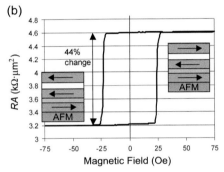

Figure 1: Resistance area product vs. external magnetic field of TMR stacks. The AFM layer pins the magnetization of the lowermost CoFe layer (*pinned layer*). The pinned layer is strongly antiferromagnetically coupled to the second CoFe layer (*fixed layer*) through a non-magnetic Ru layer (red). On top of the Al_2O_3 tunnelling barrier there is a NiFe layer that can be switched relatively easily. The magnetic layers show a thickness of 2 to 6 nm, the Ru layer of 0.7 to 0.9 nm.
(a) Characteristics of the TMR material (10 x 10 μm^2 bit).
(b) Characteristics of a 0.6 x 1.2 μm^2 TMR device.

2 Implementation of MRAM Devices

At the heart of the TMR device is a set of ferromagnetic layers separated by an insulating oxide layer (typically Al_2O_3). The electrical resistance of this arrangement depends on the magnetization direction of the two magnetic layers with respect to one another. Additional ferromagnetic and antiferromagnetic layers may be added to fix the magnetization of one of the ferromagnetic layers in one direction and provide stability to the magnetic material. The resistance change, $\Delta R/R$, is also known as the magnetoresistance ratio (MR), and has been shown to be as high as 50% for ferromagnetic electrodes such as NiFe, NiFeCo, and CoFe. The resistance of the barrier is exponentially dependent upon barrier thickness and the RA product can be varied from several $\Omega\mu m^2$ (material resistance, R, times patterned cell area, A) to more than one $M\Omega\mu m^2$ [13]-[14]. The TMR effect has been described in Chap. 4 in more detail.

The Al_2O_3 tunnel barrier can be engineered by optimizing the Al thickness and oxidation time [11]-[15] to have TMR ratios of $\Delta R/R = 40\% - 50\%$ for the entire range of $RA > 200$ $\Omega\mu m^2$. [12] Material used in the first TMR-based MRAM devices was in the 1 to 20 $k\Omega\mu m^2$ range to produce a reasonable device resistance for devices with an area on the order of 1 μm^2. However a bit fabricated in sub-0.1 μm lithography having an area of only 0.01 μm^2 would require material with $RA = 100$ $\Omega\mu m^2$ to have a reasonable resistance of 10 $k\Omega$. Much work is already underway to improve the performance of low-RA TMR material, including development of alternative barrier materials. Progress in this area would ensure resistance scaling to deep sub-0.1 μm lithography.

The resistance versus field response of such a material is shown in Figure 1a. Since MRAM devices must operate at relatively low magnetic fields due to power consumption considerations, the low field resistance versus field response of patterned TMR materials is shown in Figure 1b. The insets show the layers of the MTJ stack with the direction of magnetization for each ferromagnetic layer indicated by an arrow. The top layer is the free magnetic layer, typically a soft magnetic alloy such as NiFe, that reverses easily when a (relatively small) external magnetic field is applied. The purpose of the bottom four layers is to provide a fixed magnetic layer that does not move when a (relatively small) field is applied to switch the free layer. The two green ferromagnetic layers (typically CoFe alloy) separated by a red spacer layer form a *synthetic antiferromagnet* (SAF), described in more detail in Section 3.2.2. The ferromagnetic layers of the SAF are strongly coupled antiferromagnetically, remaining antiparallel until a field much larger than the free-layer switching field is applied. An antiferromagnetic exchange layer (AFM) is provided at the bottom to pin the SAF in the desired direction. When a positive field is applied, the free magnetic layer (top NiFe layer) becomes oriented antiparallel to the fixed magnetic layer (middle CoFe layer) across the tunneling barrier and the resistance is high. Conversely, when a negative field is applied, the free magnetic layer becomes oriented parallel to the fixed magnetic layer and the resistance is low. The hysteresis of the material ensures that once a change of state is obtained and the magnetic field is removed, the bit will remain in that state until a sufficient field of the opposite polarity is applied. Since properly designed TMR devices exhibit two stable resistance states in the absence of magnetic field, the memory is nonvolatile.

Figure 2: TMR bit stack placed at the crossing point of a bit line and a digit line. The magnetic moment of the fixed layer (blue) does not move in the applied field while the free layer (green) can be switched between two states.
(a) Single MRAM cell in "Write" Mode. Current in the bottom conductor (digit line) provides a field in the hard axis direction leading to a titling of free layer magnetization, as shown in the inset. This hard axis field lowers the easy axis field, provided by the top conductor (bit line), required to switch the bit. Only the bit in the array that experiences both hard and easy axis fields will switch. The isolation transistor is turned off during the write operation. Inset: Without a digit line current, the magnetization (red and blue arrows) is along the long axis of the rectangular shaped free layer (green), while a digit line current leads to a tilting of the magnetization directions which assists the switching procedure.
(b) Single MRAM cell in "Read" Mode. The top conductor is in contact with the top electrode of the TMR device. When the isolation transistor is turned on, a small sense current may flow from the top conductor, through the bit, and to ground.

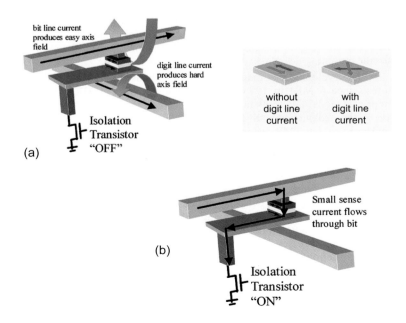

2.1 Single Transistor/ Single TMR (1T1TMR) MRAM Cell

In the most basic implementation of TMR-based MRAM devices, orthogonal conductors are placed above and below each patterned TMR bit stack as shown in Figure 2. An electrical connection is established from the base contact of the bit to ground via an isolation transistor in the underlying CMOS. In order to introduce a shape anisotropy, the bit stack is rectangularly shaped with the long side parallel to the digit line. Passing current through the conductors while the isolation transistor is in the "off" state generates the magnetic fields necessary to change the resistance state of the bit (Write operation, Figure 2a). In this implementation, the *bit line* positioned above the bit produces switching fields along the easy axis of the bit. The direction of the current flow through the bit line determines the direction in which the free layer of the bit is switched. The *digit line* positioned below the bit produces magnetic fields along the hard axis of the bit, i.e. perpendicular to the easy axis. An applied hard axis field tilts the magnetization direction in the free layer (Figure 2) and, hence, reduces the necessary easy axis field for switching the bit (green curve in Figure 3). This is necessary to distinguish between the bit to be written and the other bits along the same bit line which should remain unaffected by the write operation (black curve in Figure 3). The digit line current is unipolar as the hard axis response of the bit is independent of the sign of the field. In order to read the bit, the isolation transistor is turned on, and a small current of about 10 μA (depending on the resistance of the material, size of the bit, and desired bias condition) is passed through the bit to sense the resistance as seen in Figure 2b.

Within the MRAM active matrix array core, a parallel set of bit line conductors are arranged orthogonal to a parallel set of digit line conductors. A TMR device residing at each bit and digit line intersection along with its isolation transistor defines a single memory cell as depicted in Figure 4. This is known as the 1T1TMR cell. As a consequence of the orthogonal array architecture, the current pulses of several ns duration flowing through a given bit and digit line can be adjusted so that only the bit at the intersection of the two lines sees the field from both pulses. All other bits along either line are *half-selected* by a field which is insufficient to switch its state [16]-[17]. In addition, parallel processing is possible by pulsing current through a digit line while simultaneously addressing several bit lines. For the reading process the individual bits are selected by the bit lines and additional word lines which run parallel to the digit lines. A cross-sectional view of a MRAM process stack for a 1T1TMR cell architecture is shown in Figure 4b. Because the MRAM module is inserted in the back end of the process flow after Metal 3, MRAMs may be ideal for embedded applications as they cause little or no perturbation to the CMOS process.

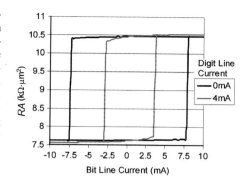

Figure 3: Resistance versus current response of a 0.6 x 1.2 μm² TMR device. When digit line current is applied, the hysteresis loop collapses, so less bit line current is required to change the resistance state of the bit. A bit line current of ± 5 mA will write the selected bit (green curve) but not the other bits along the same bit line (black curve).

2.2 Other MRAM Cell Architectures

Other possible implementations of MRAM include the **twin-cell arrays** (2T2TMR), the **diode cell arrays** and the **transistorless arrays**.

In the *twin-cell* approach, each cell is defined by two TMR devices and an isolation transistor [18]. During the write process, one TMR device is written high, and one is written low. The memory state is determined by which device is high and which is low. The advantage of this approach is that the read process is less sensitive to resistance variation and the effective signal of the cell is doubled. This potentially makes the circuit faster than the 1T1TMR implementation. The disadvantage is that the cell size is doubled, making it less attractive on a cell density and cost basis.

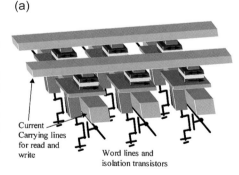

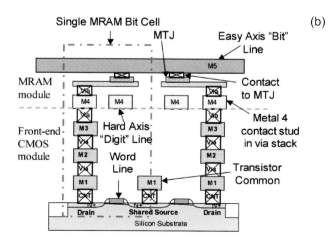

Figure 4:
(a) MRAM cross-point architecture with bits between orthogonal conductors. Each cell is defined by a TMR device at the intersection of a conductor above and below, along with one isolation transistor.
(b) Cross-sectional view of MRAM process stack. The MRAM module is inserted within the last two metallization layers of the process flow.

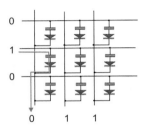

Figure 5: Schematic diagram of an array of TMR cells using diodes as isolating devices. The red line shows the path of a read operation.

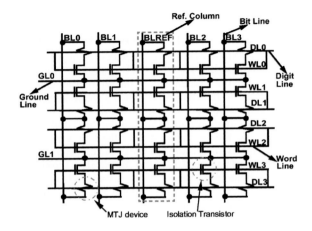

Figure 6: MRAM memory core block with midpoint reference column.

In the *diode cell* approach, a diode may be used as the isolating device instead of a transistor as shown in Figure 5. The tunnel junction acts as a load resistance for the diode and sets its operation point, i.e. it determines the actual voltage division between TMR element and diode. The advantage of this approach is that it may eventually enable the MRAM module to be decoupled from the Silicon substrate by using a poly-Si diode in a silicon-on-insulator (SOI) process. This would be immensely valuable to embedded applications as it would allow the integration of a high performance non-volatile memory with most circuits without sacrificing valuable silicon wafer surface area. Unfortunately, SOI diodes suitable to the application have yet to be developed.

Another architecture that is being explored is *transistorless array*, where all of the bits are connected in parallel between the digit and bit lines. If feasible, this architecture would offer large advantages in reduced cell area, as well as leaving large blocks of substrate under the arrays available for the driver circuitry and perhaps even other circuitry in embedded applications. Furthermore, a 3-D stacking of the arrays would be conceivable. Unfortunately, there are significant unsolved issues associated with this technique. For instance, since all of the devices are in parallel and the memory is based on resistors, the circuitry required to read the bit state would be very complex and the read operation might be quite slow. Also, during the write process small amounts of the write current would be shunted through the bits. Therefore, the write current pulse amplitude would attenuate as it moves from one end of the array to the other.

2.3 MRAM Memory Circuit Architecture

Figure 6 is an example of a 1T1TMR circuit schematic. Shown in the figure is a reference column made up of TMR elements similar to those in the nearby memory arrays. The bits in this reference column are set to a known resistance state and are then used by the read circuitry as a comparison to the target bits along the same row to determine the resistance state of a given target bit. The actual reference cells can be made up of a single TMR device or a series-parallel combination of four devices. If a series-parallel combination of four devices is used in each cell and two are programmed high and two are programmed low, then the resulting resistance is:

$$(R_{max} + R_{min}) \| (R_{max} + R_{min}) = \tfrac{1}{2}(R_{max} + R_{min}) \qquad (1)$$

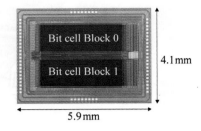

Figure 7: Microphotograph of Motorola's 1 Mb MRAM circuit.

or the midpoint between R_{max} and R_{min}. From here, the read circuitry determines whether the target bit is above or below the midpoint to determine the memory state. The close proximity of the reference cells to the bits, as well as the fact that TMR devices similar to the array devices are used in the reference cells, ensures that the reference cells closely track the active elements and are insensitive to variations in processing, temperature, etc.

Figure 7 is a photograph of Motorola's 1 Mb MRAM circuit [19]. This circuit has measured read access and write cycle times of 50 ns at 3 V operation and can be used as a non-volatile drop-in replacement for a SRAM chip. There are onboard bias, reference and clock generation sub-circuits, so no external reference voltage, bias voltage or clock signal is required. There are three modes available: active, sleep and standby. Active mode has full power consumption and the circuit is ready for random access. In sleep mode, random access and other functions are disabled, but power consumption is reduced. There is no power-up procedure required from sleep mode to active mode.

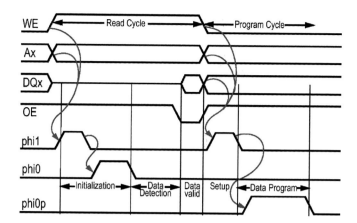

Figure 8: Timing diagram of program operation for MRAM circuit.

There is almost no power consumption when the chip is in standby mode, but a power-up sequence is required to put the chip into active mode. This circuit was processed using single damascene Cu interconnect technology in a 0.6 μm CMOS process utilizing five layers of metal and two layers of polysilicon. The 1 Mb MRAM circuit is arranged into two 524 kb banks. Each bank contains sixteen 32 kb blocks, with each block containing a mid-point reference generator column. Each mid-point generator cell services sixteen bits to its left and sixteen bits to its right.

The read operation for the 1 Mb MRAM is as follows: a memory cell is selected by driving a word line / digit line to V_{DD}, selecting a column, and turning on all ground switches. A current conveyor is shared by every 32 bit lines and every reference bit line has its own current conveyor. Once the current conveyors are turned on, they clamp the target bit lines and reference bit lines to their respective voltages, take the resulting target and reference bit line currents and convert them to a voltage signal with substantial boost. The target and reference current conveyors form a differential pair and their outputs are fed into a two stage differential comparator followed by a regenerator, which again boosts the signal. The read circuitry has been optimized to achieve high bandwidth, maintain offset insensitivity and consume minimal silicon area [20].

The timing sequence for *reading* and *writing* bits is shown in Figure 8. The read cycle is initiated by making write enable (WE) high (the signal is active low). The address of the bit to be read is stored on the 16-bit Ax bus. The active high transition of the phi1 clock starts the initialization sequence by discharging the data (DQx) lines, followed by an active high transition of the phi0 clock which precharges the data lines. A data detection delay following the low transition of phi0 allows the comparator signals to settle and precedes the active low assertion of the output enable (OE) clock. While the OE clock is active, the data bits are available on the DQx lines.

During the program mode sequence, the Ax bus contains the address to be written and the DQx bus contains the data to be written. The write sequence is initiated with the WE pin active low and a high transition of the phi1 clock. After a short setup delay to allow addressing to complete, phi1 transitions low, triggering the phi0p high. The write pulses are active on the lines as long as phi0p is high, then shut off when phi0p transitions low.

  WE = write enable (low for write, high for read)
  Ax = Address (16 lines)
 DQx = output data bus and input data to be written (16 bits)
  OE = output enable
 phi1 = clock signal to discharge the read lines and delay the write cycle
 phi0 = clock signal to precharge the read lines
phi0p = signal to enable write pulse

3 Magnetic Stability of MRAM Devices

Two fundamental effects that have consequences for the stability and scalability of MRAM devices are: the energy barrier of thermally-activated reversal of the free layer and switching induced demagnetization of the hard layer in hard/soft devices. The former effect can increase the error rate and ability of the device to retain information over time. The latter is manifested as a loss of read-out signal progressing with usage and is, thus, an analog of "wear-out" in other memory technologies.

3.1 Thermally-Activated Reversal of a Magnetic Element

When considering the operation of very small magnetic devices, it is important to consider thermal effects. We begin by describing the operation of the device in terms of an energy barrier, where the two stable magnetic states are potential wells separated by an energy barrier that prevents reversal of the free layer. The programming operation relies on the magnetic field to reduce the energy barrier to magnetization reversal of the free layer. The magnetic shape anisotropy of the elongated bit creates an energy barrier W_b to magnetization reversal, the energy barrier being critical for the nonvolatility of the cell. The size of W_b can be reduced through the application of a magnetic field along the easy-axis (parallel to the long axis of the bit) or hard-axis directions (transverse to the long-axis of the bit). As shown schematically in Figure 9, W_b is a maximum with no field applied. With easy-axis (H_{easy}) or hard-axis fields (H_{hard}) applied separately, W_b is reduced but still finite. This case corresponds to the ½-selected bits that are only exposed to fields from one line. With both easy and hard-axis fields applied, W_b is reduced to zero and the bit located at the intersection of the current carrying lines is programmed. The easy-axis field direction determines the written bit state depending on the current polarity. The hard-axis field, however, can be unidirectional, since its only function is to symmetrically reduce the energy barrier to allow the intersecting easy-axis line to program the bit.

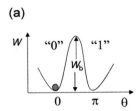
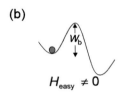

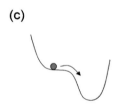

Figure 9: Schematics of energy barrier W_b that separates state 0 from state 1 in a MRAM cell.
(a) In zero magnetic field, W_b is maximum.
(b) For non zero H_{hard} or H_{easy}, W_b is reduced but finite.
(c) For nonzero H_{hard} and H_{easy}, W_b is reduced to 0 and the cell is programmed.

The energy barrier can be probed experimentally by measuring the reduction in the field required to switch the bit as the length of the field pulse is increased or by carefully measuring the change in time to switch as the magnitude of the pulse is varied [21], [22]. Multiple barriers or a distribution of barrier heights may be present if there are multiple reversal modes available to the bit. However if the bit switches nearly coherently, or at least by the same mode each time, one expects agreement with the predictions of the Arrhenius-Néel law for thermal activation over a single energy barrier: the switching probability is well described by $P(t, H) \propto \exp(-t/\tau)$, where τ is a characteristic reversal time dependent on the size of W_b. Because thermally activated magnetization reversal is a random process, only the probability of reversal can be defined for a given set of parameters such as time, magnetic field and temperature. The Arrhenius-Néel theory of thermally activated magnetization reversal assumes that for a given bit, the probability per unit time of reversal over W_b is $1/\tau$ with $\tau = \tau_0 \exp(W_b/k_B T)$ [23], [24]. Here τ_0 is the time for a thermally activated reversal and the exponential is the Boltzmann factor, where k_B is Boltzmann's constant and T is the temperature. If the same bit has its reversal time measured many times, the number of times N the bit has not reversed in time t will be given by the rate equation $dN/dt = -N/\tau$. Its solution is an exponential so that $N(t) = N_0 \exp(-t/\tau)$, where N_0 is the number of times the experiment is performed. Single energy barrier behavior implies exponential switching probabilities as well as Poisson statistics for the counting noise in the measurement [25].

Figure 10a shows the results of experiments in which the amplitude I of a current pulse generating a magnetic field was fixed and t_p was varied to give the probability of not switching as function of time [22]. In this experiment the device was a 0.45 × 1.13 μm² bit with a free layer thickness of 4 nm. The figure shows multiple data sets, each one taken with the easy axis field generated by current pulses of different amplitude I ranging from 9 mA to 7 mA. A constant hard axis field of magnitude 40 Oe was applied externally to this bit for the data shown. The solid lines are fits to an exponential function $\exp(-t/\tau)$ as predicted by the Arrhenius-Néel theory. The data are plotted with a logarithmic time axis so that exponentials with different τ have the same shape but are shifted along the t_p axis. The semilogarithmic axis also allows the exponential behavior to be observed over a much larger range of t_p than if a linear axis was used. For smaller values of I, τ is larger since the W_b is larger and $\tau = \tau_0 \exp(W_b/k_B T)$. As I increases, W_b is reduced and consequently τ is reduced as observed. The reduced χ^2 of the fits were between 1 and 2, indicating good agreement between theory and experiment.

Figure 10b shows the values of τ extracted from exponential fits to the data shown in Figure 10a as well as for hard axis fields ranging from 0 to 40 Oe. The exponential dependence of τ on I is clear. The exact functional dependence of τ on I requires knowing how W_b depends on I. This dependence can be derived for the case of a single domain Stoner-Wohlfarth particle undergoing coherent magnetization rotation with the result

$$W_b = \frac{M_s V_{\text{eff}} H_{\text{sw}}}{2}\left(1 - \frac{I}{I_{\text{sw}}}\right)^2 \qquad (2)$$

where M_s is the saturation magnetization of the free layer, V_{eff} is the effective magnetic volume participating in the reversal, and H_{sw} (I_{sw}) is the intrinsic switching field (current) in the absence of thermal effects [26], [27]. A hard axis field has the effect of reducing W_b primarily by reducing H_{sw} [28].

The solid lines in Figure 10 are fits to the data assuming such a quadratic dependence of W_b on I and the exponential dependence of τ on W_b, assuming $\tau_0 = 1$ ns. Using the fits, one can extract the energy barrier at zero I (or H), $\alpha = W_b/k_B T$ which is a critical parameter for long term data storage. For nonvolatility and reasonable error rates, it is required that the characteristic reversal time $\tau \gg 10$ yr. Since $\tau = \tau_0 \exp(\alpha)$, this implies $\alpha > 50$ at a minimum. For this particular bit, α ranges from 200 to 100 over the range of applied hard axis fields.

The results and discussion above show the importance of characterizing the energy barrier to switching for MRAM bits. These measurements show fundamental tests that must be passed for the technology to reach is full potential of nonvolatility and functionality. For a non-volatile memory the ratio of the energy barrier to $k_B T$ must be large (> 50) except when the bit is being intentionally written. As the lithography dimensions used in semiconductor technology become smaller, a simple scaling of bit size leads to a reduced bit volume. Since the Stoner-Wohlfarth model predicts that the energy barrier is proportional to the volume of coherently rotating magnetic particle, it will be important to demonstrate that the new bit designs can meet the requirements for data retention.

The demonstration of single-energy-barrier switching in patterned bits is a critical result for the proper functioning of MRAMs. A single energy barrier for each individual bit ensures the narrowest possible distribution is attained. Since switching field distributions must be minimized for practical operation of large memory arrays, this single-bit criterion is necessary but not entirely sufficient. In addition to narrow single-bit distributions, the behavior of all the bits in the array must be very nearly identical to ensure narrow distributions across the array. For bits with switching field determined largely by their shape anisotropy, this requirement leads to the necessity of having very good shape fidelity in the patterned devices.

Alternative approaches

Other approaches to MRAM switching have been proposed which would change the scaling properties of the free layer. One approach is to replace the free layer with a **synthetic antiferromagnet** (SAF) composed of two or more ferromagnetic layers that are strongly coupled antiferromagnetically by a nonmagnetic spacer layer such as Ru. A SAF exploits the strong antiferromagnetic coupling that can be achieved with certain non-magnetic spacer layer materials that exhibit the oscillatory interlayer exchange coupling phenomenon described in detail in Chapter 4. The antiferromagnetic coupling

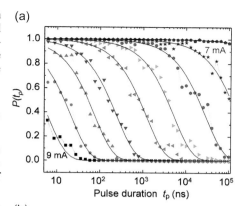

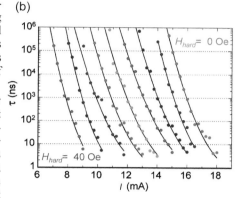

Figure 10:
a) Probability of not switching $P(t_p)$ as a function of pulse duration t_p for the free layer of 0.45 x 1.13 μm^2 magnetic tunnel junction. Each data point is the result of 1000 pulses at that t_p. Each set of data is for a different pulse current amplitude I ranging from 9 to 7 mA. The solid lines are exponential fits to the data. An exponential switching probability is expected for thermal activation over a single energy barrier.
b) Characteristic reversal time τ vs. switching current I, where τ is derived from the exponential fits of $e^{-t/\tau}$ to the data shown in Figure 1 and other data not shown. Each set of data is for a different hard axis field ranging from 0 to 40 Oe. The solid lines are fits of $\tau(I) = \tau_0 \exp(W_b(I)/k_B T)$ where $W_b(I) \propto (1-I/I_{sw})^2$, and where we have assumed $\tau_0 = 1$ ns.

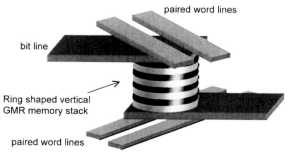

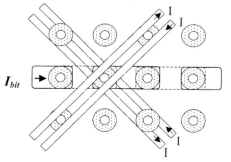

Figure 11: A proposed vertical MRAM cell based on GMR multilayer material patterned into a ring [30]. The magnetization of the layers forms stable circular states that can be parallel or antiparallel.

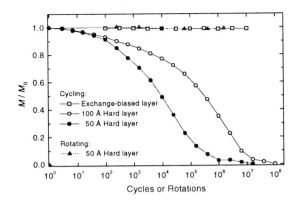

Figure 12: Change of magnetic moment of the hard layer with the number of field cycles [33].

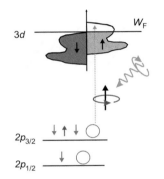

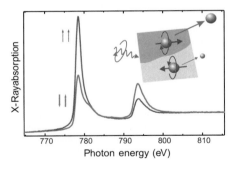

Figure 13: Element specific magnetic contrast can be achieved by using the X-ray equivalent of the magneto-optical Kerr effect (top). When the photon energy is tuned into an absorption edge of magnetic materials the X-ray absorption cross section depends sensitively on the relative alignment of sample magnetization and X-ray helicity vector (indicated by the arrows). The spectra (bottom) show the effect for a ferromagnetic Co sample when X-ray helicity and sample magnetization are parallel (blue) and antiparallel (bottom). This difference in X-ray absorption results in an intensity variation of emitted secondary electrons for different magnetic domains as illustrated in the inset.

mediated by the spacer layer causes the ferromagnetic layers to orient with antiparallel magnetic moments. Since the net moment of a SAF free layer is the difference of the moments of the two ferromagnetic layers, one can design a free layer with very low moment but a high magnetic volume. The high magnetic volume adds to thermal stability of the magnetic state and the low moment results in lower switching fields and reduced fringing fields. The use of a pinned SAF for the fixed layer is described in Section 3.2.2.

Another MRAM approach is shown in Figure 11 which illustrates a **vertical MRAM** (VMRAM) memory cell [30]. The *GMR memory stack* is composed of ferromagnetic layers separated by a metallic spacer layer such as copper (see Chap. 4). The current passes perpendicular to the layers, taking advantage of the large current-perpendicular-to-plane GMR effect [31]. Because the GMR material is patterned into a ring, i.e. a circle with a hole in the middle, the stable magnetization states are circular. Each magnetic layer forms a flux-closed circle, essentially a vortex with the center singularity missing due to the hole in the center. Adjacent layers have antiparallel magnetization if they are oriented in the opposite sense, clockwise and counterclockwise, and parallel magnetization if they are oriented in the same sense. Since the magnetic flux is fully enclosed inside the rings, such bits are predicted to be magnetically stable down to sizes of ~10 nm [29]. In addition, the reduced magnetic stray fields avoid cross talk between neighboring cells.

The ferromagnetic layers of the VMRAM cell are alternating soft/hard layers that switch at low/high field. This stack design allows information to be stored by using a low/high field to write the alternate magnetic layers into antiparallel/parallel states similar to a conventional GMR MRAM design [5]. In the high-resistance state the magnetization of the soft layers is antiparallel to that of the hard layers, resulting in a structure with alternating antiparallel moments separated by the nonmagnetic spacer layers. In the low-resistance state, the magnetic moments of all the ferromagnetic layers are parallel.

To efficiently write the VMRAM magnetic states, the cell must generate circular fields with clockwise or counter clockwise direction. Such a field can be generated by passing a large current through the stack itself. It is possible to switch bits using only this current-generated self-field [32], but selecting bits for writing in a memory array would require a very large transistor for each cell. The addition of paired word lines above and below the stack, as shown in the figure, have been proposed as a way to select a bit in an array. It is proposed that the approximately radial field generated at the cross-point cell will substantially lower the bit current needed to switch the element. The bit current could then be applied to an entire row with the write only occurring at the cross-point. Although no test VMRAM test circuits have been produced, it is clear from the proposed cell design that the advantage of magnetic stability comes at the price of cell size and array efficiency. The double word lines combined with the ring shape of the bit dictate a minimum cell size of 32 F^2, where F is the minimum feature size of the patterning process, and the bit line driver transistors must be large enough to source the large current needed for switching. Whether such engineering tradeoffs are necessary at some future lithography generation remains to be seen.

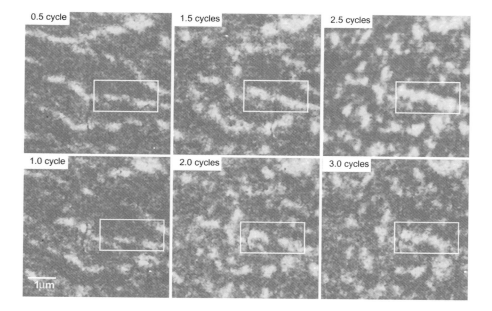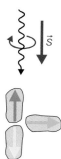

Figure 14: PEEM images of magnetic domains in the $Co_{75}Pt_{12}Cr_{13}$ hard layer after various switching cycles of the $Co_{84}Fe_{16}$ soft layer of a 50 Å $Co_{75}Pt_{12}Cr_{13}$/15 Å Al_2O_3/100 Å $Co_{84}Fe_{16}$ TMR structure. In the experiment the projection of the sample magnetization onto the light helicity direction, S, is measured. As shown on the right red areas correspond to regions with the magnetization pointing upwards. In the yellow regions the magnetization is oriented downwards. Orange indicates that the magnetization points either to the left or to the right [34].

3.2 Switching Induced Demagnetization

An important consideration for applications is the magnetic stability of a TMR junction after repeated switching of the soft magnetic layer. Figure 12 shows how the total magnetization of the hard layer develops as a function of the number of switching cycles. While for some TMR structures (using exchange-biased hard layers as explained below) remain unaffected, the magnetization of the hard layer of devices discussed in the last section decays to zero after about 10^7 reversals. This corresponds roughly to one write cycle per second for one year and is clearly unacceptable for commercial devices. In the next two sub-sections we will discuss the microscopic mechanisms for this behavior and possible solution.

In order to 'harden' a magnetic layer it is necessary to increase its magnetic anisotropy. This is the energy barrier that a magnetic moment has to overcome during reversal. Materials with a large magnetic anisotropy can often only be manufactured with a granular structure. A commonly used hard magnetic material is, for instance, $Co_{75}Pt_{12}Cr_{13}$ which consists of CoPt grains embedded in a Cr matrix [34]. The individual CoPt grains have a very high anisotropy barrier and represent ferromagnetic particles that each behave as one single large magnetic moment. These particles are coupled to their neighbors by only relatively weak magneto-static interactions. Although the particles themselves are difficult to resolve the magnetic domain structures that are formed by magnetically aligning neighboring particles have been imaged as shown in Figure 13. These images of the magnetic microstructure reveal insight into the demagnetization of the hard layer.

The images were obtained by using a so-called photoemission electron microscope (PEEM). Its general principle of operation is briefly described in Chap. 11. The photoelectron distribution in the illuminated spot is magnified onto a screen via several electrostatic lens systems [35]. It mainly contains information about the topology of the sample surface and the three dimensional surface structure. However, also magnetic sensitivity can be obtained when the energy of *polarized* X-rays is tuned into an absorption edge. In this case core electrons are resonantly excited into unoccupied valence levels. This process is schematically depicted in the left panel of Figure 13. In ferromagnetic materials there are more spin down (red) than spin up (blue) valence electrons. Consequently the core electrons are excited mainly into the unoccupied spin up (blue) states. Since the electron spin is conserved in such transitions the strong spin-orbit coupling in the core shell leads to different X-ray absorption cross sections when the X-ray polarization is changed. This is nothing but the X-ray equivalent of the magneto-optical Kerr effect. The well-known Kerr rotation of the polarization plane for linearly polarized light depending on the sample magnetization can be translated directly into different absorption cross-sections when *circularly polarized X-rays* are used. This can be seen clearly in the right panel of Figure 13 where typical X-ray absorption spectra for ferromagnetic Co metal are shown. The two peaks visible in the spectra are caused by exciting electrons from the spin-orbit split $2p$ core levels into the Co $3d$ valence shell. Using

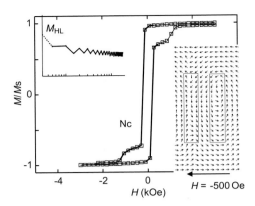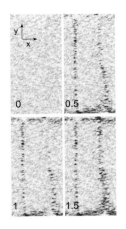

Figure 15: Micro-magnetic calculations of the magnetization dynamics in a 50 Å $Co_{75}Pt_{12}Cr_{13}$/15 Å Al_2O_3/100 Å $Co_{84}Fe_{16}$ TMR structure [34]. The left panel shows a hysteresis loop of the total magnetization as a function of applied magnetic field. The $Co_{75}Pt_{12}Cr_{13}$ hard magnetic layer switches at ±1 kOe, the soft magnetic $Co_{84}Fe_{16}$ layer at much smaller fields. The top inset shows how the magnetization of the hard layer, MHL, decays with increasing number of switching cycles. The bottom right inset shows the calculated spin orientations of the soft layer during magnetization reversal. The vortices indicated by the rectangles cause magnetic stray fields that leave behind trails of areas with reversed magnetization in the hard magnetic layer (panels on the right).

left and right circularly polarized X-rays or alternatively changing the sample magnetization direction while keeping the X-rays helicity fixed results in the red and blue spectra shown in the figure. If the photon energy is set to 778 eV, for instance, the X-ray absorption yield varies very strongly when the sample magnetization is reversed as it might happen in different magnetic domains on the sample. This is shown schematically in the inset of Figure 13. The X-ray absorption can be monitored by measuring the emitted secondary electrons which are generated when the $2p$ core hole are filled again. This will result in different electron intensities for different magnetic domains which can be monitored with the PEEM instrument. Since the core electron binding energies vary strongly for the different elements this method also allows element resolved magnetic imaging.

Figure 14 shows typical PEEM images of the magnetic domain configurations in the hard magnetic $Co_{75}Pt_{12}Cr_{13}$ of a 50 Å $Co_{75}Pt_{12}Cr_{13}$/15 Å Al_2O_3/100 Å $Co_{84}Fe_{16}$ TMR structure [34]. The soft magnetic $Co_{84}Fe_{16}$ layer was cycled through various magnetization reversals as indicated on the individual images. Starting from an almost homogeneously magnetized $Co_{75}Pt_{12}Cr_{13}$ layer (red color in Figure 14) areas of reversed magnetization start to appear (yellow color) and become more pronounced with increasing number of switching cycles. This is very surprising since for the applied magnetic field of ±500 Oe only the soft layer magnetization should become reversed and the hard magnetic layer should remain unaffected. This can be seen from the hysteresis loop in Figure 15 which demonstrates that fields above 1 kOe are needed to switch the hard layer.

The results of micro magnetic calculations shown in Figure 15 demonstrate that this phenomenon is a consequence of the complex magnetization reversal in the soft layer. The ensemble of spins shown in the center of Figure 15 represent a snapshot of the soft layer spin configuration during a magnetization reversal. Initially all spins point to the right. Then a magnetic field of –500 Oe is applied. The magnetization reversal process starts at the top where spins are flipped to the right and proceeds downwards. The magnetization reversal is characterized by magnetic vortices propagating from top to bottom (highlighted by the rectangles). These structures generate large stray fields in the adjacent hard magnetic layer leaving behind a trail of areas with reversed magnetization (black regions in the right panels of Figure 15). Initially the hard layer is homogeneously magnetized (gray colored areas in Figure 15). The black regions in Figure 15 correspond to the yellow areas in Figure 14. They clearly increase with the number of switching cycles. This demonstrates how stray fields resulting from the magnetization reversal of the soft layer can demagnetize a granular hard magnetic layer.

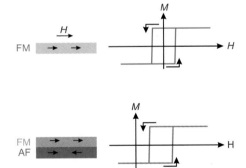

Figure 16: Schematic illustrations of hysteresis loops $M(H)$ of a ferromagnetic (FM) layer (top). In exchange biased FM/AF (antiferromagnetic) systems (bottom) the FM hysteresis loop is shifted to one side by the exchange bias field.

3.2.1 Solution: Exchange Biasing

As shown in Figure 16 exchange biasing is a way to avoid switching-induced magnetization decay. The exchange bias effect is based on bringing a ferromagnetic layer in contact with an antiferromagnet. A shift of the hysteresis loop (see Chap. 4) is observed and is schematically reproduced in Figure 16. This effect is rather surprising since at the surface of an ideal antiferromagnet (red layer in Figure 16) equal amounts of atomic magnetic moments pointing *left* and *right* should exist (indicated by arrows in Figure 16). The microscopic basis of this effect is still debated. It is presumably the defect structure (atomic steps etc.) at real interfaces that leads to uncompensated spins, i.e. spins that have no counterpart with opposite spin orientation. These uncompensated spins can introduce a preferential orientation of the magnetization in an adjacent magnetically coupled ferromagnetic layer (purple layer in Figure 16). The advantage of such an exchange biasing approach is the possibility to generate more homogeneous hard ferromagnetic layers, which drastically reduces their susceptibility to local magnetic stray fields in TMR devices.

In this section we will describe an experiment that enabled the first insight into the underlying microscopic mechanism of exchange biasing. The idea of this experiment is rather simple. If the exchange bias is caused by uncompensated spins located in the antiferromagnetic interface layer then exchange biasing should be especially pronounced in small magnetic domains. This has statistical reasons. In large domains there is a higher chance for an uncompensated spin of finding a suitable partner somewhere in the domain which has an opposite spin orientation. The experimental challenge so far has been to image antiferromagnetic domains on a μm scale. We have seen in the last section that the magneto-optical Kerr effect in the soft X-ray range can be utilized to image ferromagnetic domains. This effect is based on the different X-ray absorption cross sections for circularly polarized X-rays when the sample magnetization direction is reversed. We can, therefore, say that this effect is linear in the magnetization M [35], [36]. There is, however, also a higher-order and, thus, weaker magneto-optical effect which is quadratic in M. It can be measured with linearly polarized soft X-rays and results in different X-ray absorption cross sections when the linear X-ray polarization is oriented parallel or perpendicular to the magnetization direction [35], [36].

In Figure 17 PEEM images are shown of a ferromagnetic Co film deposited onto an antiferromagnetic LaFeO$_3$ substrate [34]. The images consist of 15 μm x 15 μm areas and show the magnetic domain structure of the ferromagnetic Co and antiferromagnetic Fe spin configurations. At the Co $2p$ absorption edges ferromagnetic contrast is observed with circularly polarized X-rays (top image in Figure 17). Since the X-rays where incident onto the surface along the vertical direction in Figure 17 the experiment could separate magnetic domains with spins pointing upwards (blue areas) from those with spins pointing downwards (yellow areas). No separation could be made between spins pointing to the left and to the right. Both types of domains give the same contrast (green areas). At the Fe $2p$ absorption edge the antiferromagnetic domain structure is resolved with linearly polarized X-rays (Figure 17 middle panel). In this case one can distinguish between the domains with spins oriented horizontally (along the X-ray polarization vector) from those with the spin oriented vertically (perpendicular to the X-ray polarization vector).

The importance of Figure 17 is the almost 1:1 correspondence of Fe antiferromagnetic and Co ferromagnetic domains. Blue areas in the Co layer of Figure 17 where spins are pointing upwards (an external magnetic field was applied to almost completely remove any domains with spins pointing downwards as indicated by the yellow areas) correspond to dark blue areas in Figure 17 where the antiferromagnetic domains are oriented vertically (Fe spins are either pointing upwards or downwards). A similar situation occurs for horizontally oriented spins. It is now possible to measure the spatially resolved hysteresis loops for each individual domain. A magnetic field was applied vertically in Figure 18. The figure shows the determined shift of the hysteresis loop which corresponds directly to the local strength of the exchange-bias field (see Figure 16). The solid black domains contain spins that are oriented horizontally. The green areas show domains with spins pointing upwards while the spins are oriented downwards in the purple domains. The distribution of local bias fields which is displayed in the right panel of Figure 18 shows that there is a net zero macroscopic exchange bias. However, it is evident from Figure 18 that there is a local exchange bias. Moreover, the strength of the exchange bias clearly increases as the domains become smaller.

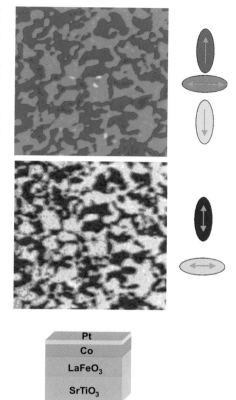

Figure 17: PEEM images (15 μm x 15 μm size) of the magnetic domain configurations of (top) a ferromagnetic 20 Å Co film deposited onto (bottom) an antiferromagnetic 400 Å LaFeO$_3$ film. The magnetization orientation is indicated at the right of each image. A schematic cross section of the whole sample is depicted in the bottom panel [35].

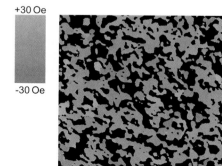
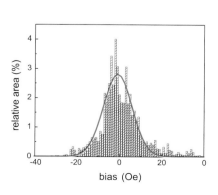

Figure 18: Map of vertically oriented exchange bias (left) field measured in the Co layer of Figure 17. In the solid black domains the spins are oriented horizontally. The distribution of local bias fields (right) result in a net zero macroscopic exchange bias [35].

3.2.2 Additional Considerations

The observations of switching induced demagnetization have been made in blanket films or very large devices. As magnetic bits become smaller, their reversal mode becomes more like coherent rotation of the magnetic moments. In such cases the reversal no longer requires the movements of a domain wall or a vortex and the associated large stray fields. Thus it remains to be seen if the demagnetization problem persists at small dimensions.

In addition to using an exchange bias layer to pin the fixed magnetic layer, current MRAM structures typically use a synthetic antiferromagnet (SAF) structure to further improve the stability of the fixed layer. A SAF exploits the strong antiferromagnetic coupling that can be achieved with through certain non-magnetic spacer layer materials, such as Ru or Rh, that exhibit the oscillatory coupling phenomenon [37]. A typical SAF consists of two ferromagnetic layers separated by a nonmagnetic metal spacer layer that causes them to orient their moments antiparallel. Because the antiferromagnetic coupling strength is large, the pinned SAF tends to be magnetically very stable. This structure adds additional stability to prevent demagnetization of the fixed magnetic layer. An additional benefit of the antiparallel orientation is that these structures have little or no net moment to create stray fields that can shift the hysteresis loop of the free layer.

4 Ultrafast Magnetization Reversal

One of the central issues in data storage is the read/write access time. The read-out of an MRAM involves determining the magneto-resistance of the element. This essentially reduces to measuring the voltage drop across the memory cell. The level of integration with existing CMOS technology and the available clock speed, therefore, presently limits it to several nanoseconds. The writing of a MRAM involves a magnetization reversal. Limitations of this process will be explored in this section.

For the study of the magnetization reversal a sub-ns temporal resolution is mandatory. There are several ways to achieve this. With parallel imaging methods such as PEEM it is in principle possible to obtain a movie of a single reversal process. However, the data acquisition time for each image limits the achievable temporal resolution severely. A different and more promising approach is to image the reversal process stroboscopically. This means that images are averaged over many individual reversal cycles. Such a so-called pump-probe measurement is shown schematically in Figure 19. A pump signal triggers a magnetic field pulse. The switching of an MRAM structure is then probed at various pump-probe time delays. The use of fs-laser pulses allows synchronization of pump and probe signals. Usually each fs-laser pulse is split into two. One pulse triggers a current pulse by illuminating a photo-conductive switch. This generates a magnetic field at the MRAM structure that induces the magnetization reversal. The second laser pulse is delayed relative to the other and reaches the sample after the pump pulse. It can probe the magnetic status of the MRAM by using for instance the magneto-optical Kerr effect.

The switching characteristics of a magnetically soft permalloy ($Ni_{80}Fe_{20}$) microstructure (20 μm x 2 μm size, 15 nm thickness) was studied by Choi et al. [38]. Figure 20 (left) shows schematically the experimental setup. A current pulse through a gold transmission line generates a magnetic field pulse (H_s = 24 kA/m) of about 10 nsec duration. This is used to switch the magnetization of the permalloy sample. An applied static longitudinal bias field of about the same magnitude assures that the sample spins return to the initial orientation when the current pulse has passed. In addition a static transverse bias field of various strength could be applied along the transmission line. The x-component of the sample magnetization, M_x, was measured using the magneto-optical Kerr effect in the optical spectral region. Images of the whole microstructure were obtained by scanning the sample under the focused spot of a fsec-laser. The same fsec-laser pulse synchronously triggered the current pulses and the images shown in Figure 20 were obtained stroboscopically for various time delays.

The magnetization reversal shown in Figure 20 is clearly different with or without an applied transversal bias field, H_t. For H_t = 0 the areas with reversed magnetization start to nucleate significantly about 1.3 nsec after the field pulse was turned on. Only after about 5 nsec the sample magnetization is homogeneously reversed. A similar delayed magnetic switching is observed after the field pulse has passed. Much faster (< 1 nsec) reversal is observed when a transverse bias field is applied. In this case the

Figure 19: Schematics of a pump-probe experimental setup to measure the magnetic switching characteristics of a MRAM device. Ultrafast electron pulses are generated by illuminating a photo-conductive GaAs switch with a fs laser pump pulse. The MRAM is situated on the electrical biasing lines (yellow). Magnetization reversal occurs by the magnetic field (red arrow) co-propagating with the electron pulse. A time delayed probe laser pulse is used to measure the magnetic state of the MRAM via the magneto-optical Kerr effect.

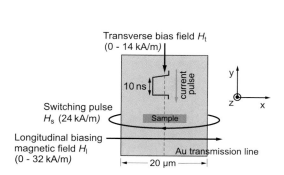 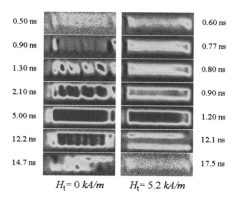

Figure 20: Schematic experimental layout for 180° dynamic reversal experiments as described in the text (left). Spatial profile of the M_x component as a function of time after the magnetic pulse was applied (right) [38].

sample spins are already partially rotated away from the easy magnetization direction (along the long side of the sample). It is, therefore, easier to completely overcome the anisotropy barrier between the two opposite magnetization directions. In addition the images reveal a more homogeneous magnetization of the sample consistent with a magnetization reversal via domain wall movement. These measurements demonstrate that sub-nanosecond magnetic switching can be achieved by optimizing the MRAM geometry. It is worthwhile to note that the rather large electrical power needed for the magnetic switching process can be substantially decreased with the system dimensions.

5 Summary and Outlook

The memory state in a MRAM device is maintained by the direction of the magnetic moments and not by power. This non-volatile character alone makes MRAM an attractive alternative to the established DRAM. The discovery of high MR ratios from magnetic tunnel junctions has spurred the development of MRAM based on the integration of TMR material with CMOS circuits, targeted at commercial memory markets. Key attributes of MRAM technology are nonvolatility and unlimited read and write endurance. In addition, it is anticipated that MRAM could operate at high speed and low voltage, with competitive densities.

Although the initially achievable storage densities are expected to be in the Mbit range, there seems to be potential for continuing increases as the technology matures. Controlling the tunneling resistance uniformity, achieving single-energy-barrier switching behavior of magnetic bits, and integrating TMR material with CMOS are some of the key challenges that have been overcome to successfully demonstrate feasibility of this technology. Challenges in scaling MRAM technology to smaller dimensions include: reducing the resistance-area product of the TMR material while maintaining the MR ratio (signal), generating the magnetic fields needed for switching the bits using shrinking metal lines, maintaining magnetic stability and reproducibility as described in Section 3.1, and accommodating the increased magnetostatic fields generated when magnetic elements are defined at reduced dimensions. The latter concern is addressed by the synthetic antiferromagnetic free layer and VMRAM designs described in Section 3.1.

The performance of prototype devices demonstrated to date is very encouraging and is anticipated to improve significantly as the technology matures and moves to competitive lithography dimensions. These results show that MRAM is a unique high-speed, nonvolatile memory with the potential to become a universal memory for a variety of applications.

Acknowledgements

Saied Tehrani, Brad Engel, Nick Rizzo, Mark Durlam contributed to this chapter. We thank Thomas Mikolajick for reviewing this Chapter and his helpful suggestions. The editor gratefully acknowledges Thomas Mikolajick (Infineon Technologies) for a detailed review of this chapter, and he would like to thank Simon Stein (FZ Jülich) for checking the symbols and formulas in this chapter.

References

[1] L. J. Schwee, P. E. Hunter, K. A. Restorff, M. T. Shephard, J. Appl. Phys **53**, 2762 (1982).

[2] C. W. Baugh, J. H. Cullom, E.A. Hubbard, M. A. Mentzer, R. Fedorak, IEEE Trans. Magn. vol. MAG-18, **6**, 1782 (1982).

[3] A. V. Pohm, J.S.T. Huang, J.M. Daughton, D.R. Krahn, V. Mehra, IEEE Trans. Magn. **24**, 3117 (1988).

[4] J. M. Daughton, Thin Solid Films **216**, 162 (1992).

[5] S. Tehrani, E. Chen, M. Durlam, T. Zhu, and H. Goronkin, International Electron Devices Meeting. Technical Digest, 193, IEEE, New York, NY, 1996.

[6] S. Tehrani, E. Chen, M. Durlam, M. DeHerrera, J. M. Slaughter, J. Shi, G. Kerszykowski, J. Appl. Phys., **85**, 5822 (1999).

[7] T. Miyazaki and N. Tezuka, J. Magn. Magn. Mater., **139**, L231 (1995).

[8] J. S. Moodera, L. R. Kinder, T. M. Wong, and R. Meservey, Phys. Rev, Lett. **74**, 3273 (1995); J. S. Moodera and L. R. Kinider, J. Appl. Phys. **79**, 4724 (1996).

[9] S. S. P. Parkin, R. E. Fontana, and A. C. Marley, J. Appl. Phys. **81**, 5521 (1997).

[10] R. C. Sousa, J. J. Sun, V. Soares, P. P. Freitas, A. Kling, M. F. da Silva, J. C. Soares, Appl. Phys. Lett. **73**, 3288 (1998).

[11] S. Tehrani, B. Engel, J. M. Slaughter, E. Chen, M. DeHerrera, M. Durlam, P. Naji, R. Whig, J. Janesky, J. Calder, IEEE Trans. Magn. **36**, 2752 (2000).

[12] J. M. Slaughter, Renu Whig Dave, M. DeHerrera, M. Durlam, B. N. Engel, N. D. Rizzo, and S. Tehrani, J. Superconduct. Incorp. Novel Magnetism, **15**, 19 (2002).

[13] S. S. Parkin, et al., J. Appl. Phys. **85**, 5828 (1999).

[14] E. Y. Chen, R. Whig, J. M. Slaughter, D. Cronk, J. Goggin, G. Steiner, S. Tehrani, J. Appl. Phys. **87**, 2000.

[15] J. M. Slaughter, E. Y. Chen, R. Whig, B. N. Engel, J. Janesky, and S. Tehrani, JOM-e, **52** (6) (2000), available online at www.tms.org/pubs/journals/JOM/0006/Slaughter/Slaughter-0006.html.

[16] J. Shi, T. Zhu, M. Durlam, E. Chen, S. Tehrani, Y. F. Zheng, and J.-G. Zhu, IEEE Trans. Magn., **34**, 997 (1998).

[17] S. Tehrani, B. Engel, J. M. Slaughter, E. Chen, M. DeHerrera, M. Durlam, P. Naji, R. Whig, J. Janesky, J. Calder, IEEE Trans. Magn. **36**, 2752 (2000).

[18] R. Scheuerlein, W. Gallagher, S. Parkin, A. Lee, S. Ray, R. Robertazzi, W. Reohr, IEEE ISSCC Dig. of Tech. Papers **43**, 128 (2000).

[19] M. Durlam, P. Naji, A. Omair, M. DeHerrera, J. Calder, J. M. Slaughter, B. Engel, N. Rizzo, G. Grynkewich, B. Butcher, C. Tracy, K. Smith, K. Kyler, J. Ren, J. Molla, B. Feil, R. Williams, S. Tehrani, *A low power 1 Mbit MRAM based on 1T1MTJ bit cell integrated with Copper Interconnects*, in press.

[20] P. K. Naji, M. Durlam, S. Tehrani, J. Calder, M. DeHerrera, IEEE ISSCC Dig. of Tech. Papers, **438**, 22 (2001).

[21] R. H. Koch et al., Phys. Rev. Lett. **84**, 5419 (2000).

[22] N. D. Rizzo, M. DeHerrera, J. Janesky, B. Engel, J. Slaughter, S. Tehrani, Appl. Phys. Lett. **80**, 2335 (2002).

[23] S. Arrhenius, Z. Phys. Chem. **4**, 226 (1889) [Selected Readings in Chemical Kinetics, edited by M. H. Back and K. J. Laidler, Pergamon, Oxford, 1967].

[24] L. Néel, Ann. Geophys. **5**, 99 (1949).

[25] C. W. Helstrom, *Probability and Stochastic Processes for Engineers*, Macmillan Publishing, New York, 1984.

[26] P. Gaunt, J. Appl. Phys. **59**, 4129 (1986);

[27] C. P. Bean and J. D. Livingston, J. Appl. Phys. **30**, 120S (1959).

[28] B. D. Cullity, *Introduction to Magnetic Materials,* Addison-Wesley, Reading, 1972.

[29] J.-G. Zhu, Y. Zheng, G. A. Prinz, J. Appl. Phys. **87**, 6668 (2000).

[30] Jian-Gang Zhu, Youfeng Zheng, and Gary A. Prinz, J. Appl. Phys. **87**, 6668 (2000).

[31] W. P. Pratt Jr., S.-F. Lee, J. M. Slaughter, R. Loloee, P. A. Schroeder, and J. Bass, Phys. Rev. Lett. **66**, 3060 (1991).

[32] K. Bussmann, G. A. Prinz, R. Bass, and J.-G. Zhu, Appl. Phys. Lett. **78**, 2029 (2001).

[33] S. Gider, B.-U. Runge, A. C. Marley, S. S. P. Parkin, Science **281**, 797 (1998).

[34] L. Thomas, et. al., Phys. Rev. Lett. **84**, 3462 (2000).

[35] W. Swiech, et. al., J. Electron Spectr. & Rel. Phenom. **84**, 171 (1997).

[36] F. Nolting, et. al., Nature **405**, 767 (2000).

[37] S. S. P. Parkin, Phys. Rev. Lett. **67**, 3598 (1991).

[38] B. C. Choi, M. Below, W. K. Hiebert, G. E. Ballentine, M. R. Freeman, Phys. Rev. Lett. **86**, 728 (2001).

Mass Storage Devices

Contents of Part V

Introduction to Part V	607
24 **Hard Disk Drives**	615
25 **Magneto-Optical Discs**	631
26 **Rewritable DVDs Based on Phase Change Materials**	643
27 **Holographic Data Storage**	657
28 **AFM-Based Mass Storage – The Millipede Concept**	685

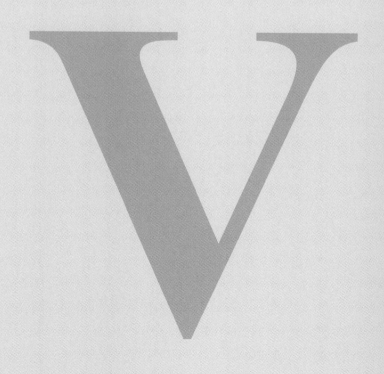

Introduction to Part V

Contents

1 Definition ... 607
2 Physical Storage Principle ... 608
3 Distributed Storage ... 612

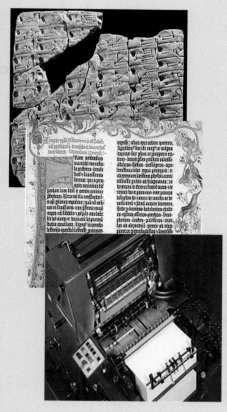

Figure 1: Impressions of the development of the storage of written information. Approx. 5000 years ago, the Sumerian developed the cuneiform [1]. The picture (top) shows the table of Uruk. In the late 15th century, printing on movable letters led to a revolution in reproducing book (center). In the second half of the 20th century, the electronic offset printing (bottom) started to supersede the traditional printing.

1 Definition

In order to pass on human experience and knowledge among contemporaries and from generation to the next, it was advantageous to store information external to human memory. In fact, the invention of writing can be regarded as the most important ingredient in the cultural evolution of mankind because it was the prerequisite for the accumulation of experience, knowledge, impressions and ideas. The speed of this evolution is correlated to the ease in which information storage and retrieval has been possible.

It started with the Sumerian cuneiform, impressed by a stylus into clay, followed by pen writing on parchment and paper, accelerated through the invention of printing from movable type, to the electronic reproduction of books (Figure 1). Similar developments took place for the storage of images, from prehistoric cave drawings to the accurate capturing of complex images of the real world by photography – first photochemically and, nowadays, electronically. Early storage systems for music were various types of musical-clockworks invented in the 14th to 16th century, followed by the phonograph (later called record player), magnetic tapes, and digital compact disc. Moving pictures (movies) started on celluloid film, followed by magnet tapes and discs again. All these examples represent information storage systems which can be directly accessed. The counterparts for the access by electronic information systems are called **mass storage devices** (MSD), which are typically used for the non-volatile storage of large amount of data.

The table in Figure 2 summarizes a comparison between mass storage devices and random access devices. The main difference is the relationship between the storage media and the way to access the information. In random access devices (Part IV), the storage media is distributed at the nodes of a memory matrix and the information can be accessed electronically in a random fashion by the application of the corresponding column and row addresses. Mass storage devices are composed of one (or more) access

	Mass Storage Devices	Random Access Devices
System	• access unit (e.g. R/W head) • storage medium	• matrix of conductor lines • storage elements at nodes
Addressing of Data	mutual positioning of the R/W unit & storage medium	application of column & row address signals
Data exchange	• mechanical • optical • magnetic fields • electric fields	electronic access via matrix (voltage mode or current mode)

Figure 2: Comparison of the concepts of mass storage devices and random access memories.

V Mass Storage Devices

Figure 3: Historical and modern mass storage devices and their underlying physical access principles.

Physical Access Principles

Mechanical
- record player
- punch card / tape reader
- SPM-based Milliped /concept (*Chap. 28*)

Optical
- CD/DVD-ROM system
- CD/DVD Recordable
- MO Discs (*Chap. 25*)
- phase-change based CD/DVD-RW (*Chap. 26*)
- Holographic 3-D system (*Chap. 27*)

Magnetic Field
- magnetic / tape / systems
- Floppy Disc system
- Hard Disc system (*Chap. 24*)

Electric Field
- SPM based ferroelectric thin film R/W system (future?)

unit(s), such as read heads or read/write heads and a storage medium. The information is modulated onto the medium. The information is accessed by a mutual positioning of the access unit and the storage medium and a sequential transfer of the information to or from the medium. The writing operation corresponds to a modulation of the medium, the reading is based on a demodulation. The data exchange between the storage medium and the access unit may be by various physical interactions (Sec. 2). This is, in principle, irrespective of the coding of the data in an analog or digital form.

2 Physical Storage Principle

This section provides a brief survey of the physical principles which have been and are used typically for accessing the storage media. The development of the mechanically and magnetically media is sketched and the basic information on compact disc based media is outlined.

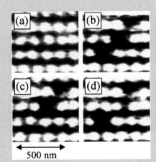

Figure 4: Information storage on a ferroelectric $Pb(Ti,Zr)O_3$ thin film of 100 nm thickness [2]. The brightness is the piezoresponse AFM image of the direction of the ferroelectric polarization and may be used for binary data storage. Switching is performed by positive and negative voltage pulses at the AFM tip. Pictures (a) to (d) show different data bits erased and rewritten. The storage density is approx. 6 Gb/cm^2.

2.1 Classifications

One of the most relevant classifications for the user of MSD systems concerns the write procedure:
- **pre-written media** are modulated during the fabrication process (such as standard CDs made by music companies).
- **write-once media** may be once written by the user. They are often called WORM media (write once read many). A typical example is the CD-R (CD Recordable).
- **read/write** (short: R/W media) can be re-written multiple times by the user. Magnetic tapes and discs as well as magneto-optical and phase-change discs are common R/W media.

A variety of physical access principles have been developed. The table in Figure 3 lists some examples of MSD systems based on access by mechanical forces, by optical means, by magnetic fields, and by electrical fields. Some of the systems will be described in more detail below. The utilization of ferroelectric films which may be accessed by electrical fields are only in a research stage. Ferroelectric thin films can be polarized locally by voltage pulses which are applied to conducting SPM tips. The read operation is based on the electric field microscope mode or the piezo response mode of

SPM systems (Chap. 12). Very high storage densities of up to 6 GB/cm² and good data retention have been demonstrated (Figure 4) [2]. To overcome the relatively slow read / write rate of a single SPM tip, this principle could be combined with the multi-tip concept described in Chap. 28.

Another classification concerns the exchange of media. In some cases, such as magnetic hard discs, the access heads and the media are manufactured as an integrated unit. In the case of hard discs, this is required because head and media need to be mutually calibrated with extreme precision. In operation, the head is flying approximately ten nanometers above the disc surface. In addition, the disc has to be isolated from dust particles by a perfectly sealed case. If dust is of lower concern and the operating distance between the access head and medium is signficantly larger, **exchangeable media** systems such as floppy disc systems or CD/DVD systems are built [3].

2.2 Dimensionalities

The speed of access to an arbitrary position in a MSD depends on the dimensionality of mutual motion of the access unit and the medium:

- **1-D access**
  All tape-based media (magnetic tapes, punch tape) are lead in an 1-D motion along the access head. Accessing a new position means winding the tape until the position is reached.

- **2-D access**
  Any modem disc (magnetic hard discs, optical discs) is rotated and, in addition, the head is positioned in an radial direction. This represents a full 2-D access.

- **3-D access**
  Information stored in a volume medium can be accessed optically, e.g. by the depth of focus or the viewing angle. This approach is exploited in the holographic systems described in Chap. 27.

Figure 5: The Jacquard loom using punched cards to control the weaving of pattern into fabrics.
(a) drawing showing the loom with the Jacquard system on top.
(b) a more detailed view of the Jacquard system and the chain of punched cards [3],[4].

This classification can only be regarded as a first approximation. In many cases, a kind of fractioned dimensionality is realized. For example, the magnetic recording in the Digital Audio Type (DAT) and in video tape systems utilizes the width of the tape to expand the area for information storage by inclined recording. Similarly, the high-density DVDs have two layers which are accessed by different depths of focus.

2.3 Early Mechanical Recording Media

The first data storage media in history for direct process control were the **punched cards** developed by Joseph-Marie Jacquard in 1805. The Jacquard system enabled looms to produce fabrics with individual woven patterns (Figure 5). These patterns were encoded on the exchangeable punched cards using designated perforation positions. Later in the 19th century, Charles Babbage developed a (never realized) plan for an Analytical Engine, the forerunner of the modern digital computer, and he envisioned punched cards for the input of instructions and data.

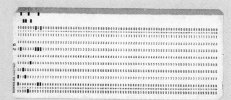

Figure 6: Picture of a punched card for feeding computer systems [5]. The total width of the card is 185 mm

The electrical reading and sorting of punched cards was introduced by Herman Hollerith in machines for automating the tabulation work of the census in the United States in 1890. Hollerith founded the Tabulating Business Machine Company, to manufacture these machines. This company later merged with other firms to become IBM in 1924. Punched cards were used as a storage media for computers until the 1970s (Figure 6) [6]. They have been complemented by punched (or perforated) paper tapes since the 1950s (Figure 7).

For many decades of the 20th century, record players were the most popular systems for sound reproduction. Record players consist of motor-driven turntables for carrying the record disc and the pick-up hold by a tone-arm. The analog sound information is encoded in a groove impressed into the disc. It is read by moving the needle of the pick-up as it is moving along the groove. The mechanical motions of the needle are electromagnetically or piezoelectrically converted into electric signals in the pick-up head and passed on to the amplifier.

The AFM based Milliped concept (Chap. 28) can be regarded as a possible successor of these early systems. It uses a digital coding similar to the punched cards and tapes - now on a nm-scale instead of a mm-scale, in order to obtain a very high density data storage.

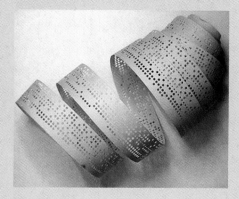

Figure 7: Perforated (or punched) tape [7]. The width of the tape is 25 mm.

Mass Storage Devices

2.4 Magnetic Tapes and Discs

Compared to the mechanical storage principle, magnetic recording provided the inherent advantage of a read/write system [8][9][10]. The principle of magnetic recording was first demonstrated by Valdemar Poulsen in 1900, when he invented a machine that recorded sound on a steel wire. Subsequent development led to a plastic film ribbon coated with magnetic oxides or metal. A wide variety of magnetic tapes have been introduced for recording sound (audio tapes), motion pictures (video tapes), and data (Figure 8).

While most of the systems are based on analog recording, i.e. the magnitude of the magnetization of the tape represents the amplitude of the signal, there is also a range of digitally coded tape systems, such as the Digital Audio Tape (DAT) and digital video formats. These systems were also used in data storage applications. Many Tb of data are stored in **magnetic cartridge archives** which can be accessed automatically through robot systems (Figure 9).

Magnetic disc systems have been introduced for data storage in the 1950s. The media are metal or polymer discs coated on one or both sides with magnetic material. R/W heads "fly" over the disc surface at close distance. In hermetically sealed **hard disc** drives up to 20 discs are mounted on the spindle of a drive unit. The present evolution of hard discs is described in Chap. 24. In the 1970s, the **floppy disc** was introduced. The medium is a flexible disc and represents a typical exchangeable medium. Because of the geometrical tolerances required by the exchangeability and the lower protection against dust, the storage capacity is much less than in hard discs.

Figure 8: Magnetic tapes for data storage.
(a) Example for tape machines: Honeywell system, 1963 [11].
(b) selection of tapes [12].

2.5 Compact Disc Based Systems

The Compact Disc (CD) was commercially introduced in the early 1980s by Philips as a digitally coded medium for high-quality audio information. It is the root of a variety of optical storage systems including the CD-ROM for data storage, the CD-Recordable (CD-R) as a WORM medium, the CD-R/W, as well as the Digital Versatile Disc (DVD) which has a much higher storage density. In this section, the basic operation will be described [14]. The physical background of the CD-R/W and DVD-R/W concept based on phase-change material is given in Chap. 26.

The standard CD disc is 120 mm in diameter and 1.2 mm thick. The geometry is shown in Figure 10. The disc body is made from optical grade polycarbonate. In the data area of the disc there is a single spiral track of pits which contain the digital information.

In the standard pre-recorded CD these pits are generated by molding the polycarbonate onto an appropriate metal master. The vicinity of the pits is called land. The track distance is 1.6 µm, the minimum pit or land distance along a track is 0.83 µm, the maximum distance is 3.3 µm. A comparison of these spacings for a CD and a DVD is given in Figure 11. A 50 to 100 nm thin layer of metal (Al, Au or Ag) is deposited onto the modulated surface of the disc. This is followed by a polymer protection layer (10 to 30 µm) and the screen-printed label.

Figure 9: At the Fermilab, access to huge amounts of data is provided by robotic systems retrieving magnetic tape cartridges [13].

The disc is read from the bottom by focussing a laser beam through the polycarbonate body onto the track. The incident spot entering the polycarbonate of approx. 800 µm in diameter is focused down to about 1.7 µm at the metal interface. It is a major advantage of the CD concept that the large incident diameter greatly reduces the effect of dust and scratches on the polycarbonate surface (Figure 12). The reading head of the CD drive consists of an infrared AlGaAs laser with a wavelength of 780 nm (Figure 13). Due to the refractive index $n = 1.55$, the wavelength within the polycarbonate is reduced to approx. 500 nm. The pits are fabricated to a quarter of this wavelength (approx. 125 nm). If the focused laser beam hits a pit area, it will cover the pit and – because its diameter is larger than the pit width – some of the surrounding land. The light reflected by the land is delayed $2 \times \frac{1}{4} = \frac{1}{2}$ of a wavelength; i.e. it is out of phase with the light reflected by the pit (Figure 12). These two waves interfere destructively and, hence, the total intensity of the reflected light is significantly reduced compared to a region without a pit. The polarizer shown in Figure 13, is needed to reflect the light into the photo detector on its

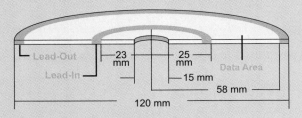

Figure 10: Geometrical format of the Compact Disc (CD).

Introduction

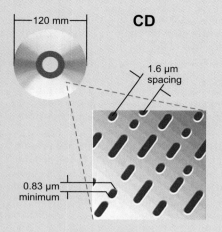
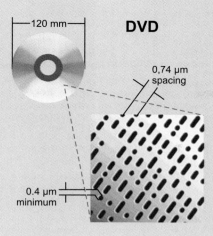

Figure 11: Track distances and pits sizes for the CD and the DVD.

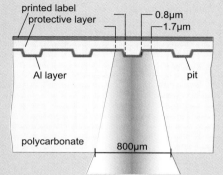

way back. Any change from pit to land and vice verse is interpreted as a logical "1", no change means "0". In order to achieve a constant speed of reading data, the linear velocity is kept constant at 1.3 m/s. This means that the angular velocity is reduced from 500 rpm at the lead-in to 200 rpm at the lead-out.

The read system requires at least two and at most eleven "0"s between two "1"s. For this reason the byte (8bit) data are expanded into a 14 bit code plus a 3 bit separator between the byte data. 24 of these 17-bit units together with three bytes for error detection and correction plus some overhead are called a *frame*. Since the audio signal on a standard CD is resolved by 16 bit, a frame contains six audio values per channel:

- 1 frame: 24 byte = 6 audio values per channel
- 1 sector = 98 frames: 6 x 98 = 588 values per section
- 1 second = 75 sectors: 588 * 75 = 44100 values

Figure 12: Cross section of a CD showing the pit and land area at the reflective metal layer and the interaction of the laser spot.

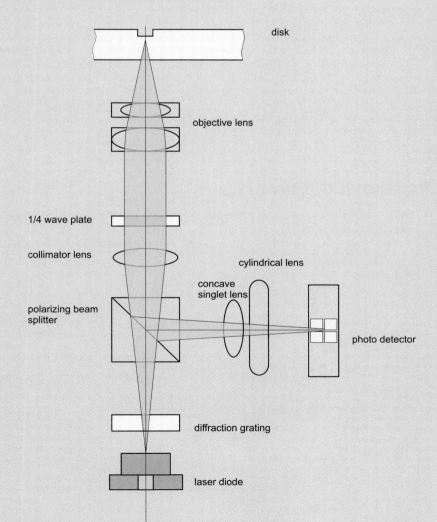

Figure 13: Optical read system of a CD. The polarising beam splitter leads to a linear polarization of the light, the ¼ wave plate turns this into a circular polarization. On its way back through the ¼ wave plate the light gets linearly polarized again, however, with a perpendicular direction compared to the incident light. This leads to a reflection into the photodetector array. The autofocus system uses the cylindrical lens and the four segments of the photodetector array.

	CD	DVD
Diameter	120 mm	120 mm
Thickness	1.2 mm	2 x 0.6 mm or 4 x 0.3 mm joint layers
Track distance	1.6 μm	0.74 μm
Width of pits	0.83 μm	0.40 μm
Laser wavelength	780 nm	650/635 nm
Data layers / sides Capacity	1/1 ≙ 680 MB	1/1 ≙ 4.7 GB 2/2 ≙ 8.5 GB 1/2 ≙ 9.4 GB 2/2 ≙ 17 GB

Figure 14: Comparison of technical data for a CD and different DVD formats.

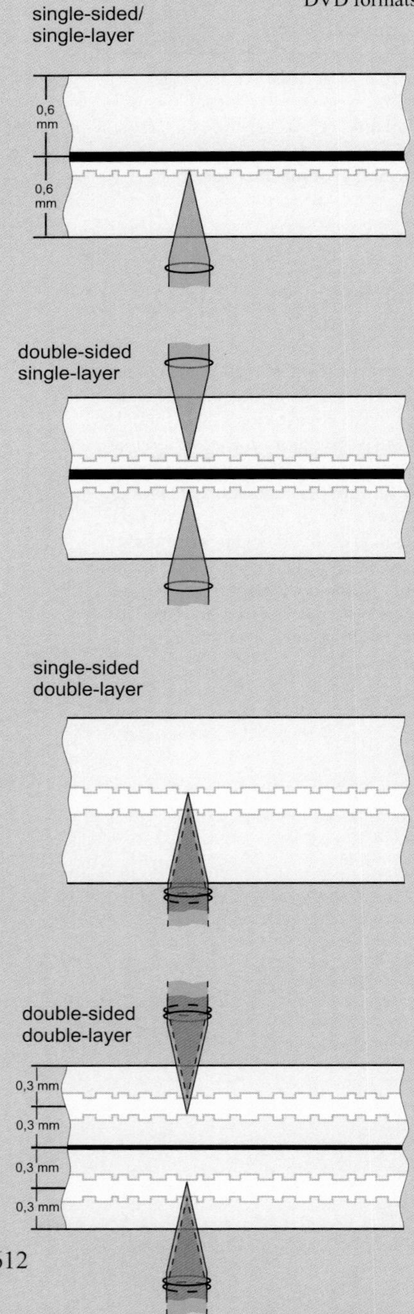

Figure 15: DVD formats.

single-sided/ single-layer

double-sided single-layer

single-sided double-layer

double-sided double-layer

This reveals the standard sampling rate of 44.1 kHz used for audio CD systems. In the case of the CD-ROM, some of the data bytes per sector are used for an addition error-correction, in order to increase the data security.

DVDs use reduced spacing (Figure 11) and a shorter laser wavelength (650 or 635 nm). The relevant specifications of CD and DVDs are compared in Figure 14. In order to enhance the capacity further, there are DVDs formats with two layer, with two sides, and with both (Figure 15). The reading of two layers is achieved by semitransparency of the reflection data layer and by setting the focus on the layer to be read.

Regular CDs and DVDs are not recordable. As described, their information is transferred from a master onto the disc during the fabrication process. **CD/DVD-Recordable** (CD-R, DVD-R) are WORM media which are fabricated with an additional light-sensitive organic dye layer. Depending on the type of organic dye and the metal (Au or Ag) of the reflective layer the disc exhibit different base colors. Commonly, cyanine, phthalocyanin and azo compounds are employed as dye. During fabrication the polycarbonate disc receives a *pregroove* at the track position with pits representing a frequency of 22.05 kHz. This pregroove is required for keeping the track during writing. Subsequently, the thin dye layer is deposited on the disc, followed by the metal reflective layer and the protection layer, as for regular CDs and DVDs. The writing is performed by a laser of high power density which locally pyrolysizes the dye at the designed positions of the data pits (Figure 16). The pyrolysis residues are opaque and/or absorptive. In addition, the interfaces of the dye layer at metal layer side and the polycarbonate side get modified. As a consequence, such a modified pit shows a significantly reduced light reflection during the read operation.

Two principles, on which optical R/W discs are based will be described comprehensively in Chap. 25 and Chap. 26.

3 Distributed Storage

With increasing transfer speed of data communication system (see Part VI), the requirement that MSDs are a physically integral part of a computer system becomes more and more relaxed. In typical client/server systems, the major storage capacity is provided by MSDs which are part of the network. The MSDs may be distributed among the servers or they may be separate systems within the network. If a suitable hardware and software concept is used, distributed storage systems offer advantages such as very high access speed, data redundancy for fault tolerance, topological flexibility and expandability, and the opportunity of a remote management [15],[16]. It is expected that distributed storage in local area networks (LANs) as well as in wide area networks (WANs, Internet) will have a growing relevance in future.

Acknowledgements

The editor gratefully acknowledges Christoph Buchal (FZ Jülich) for a careful review of this introduction.

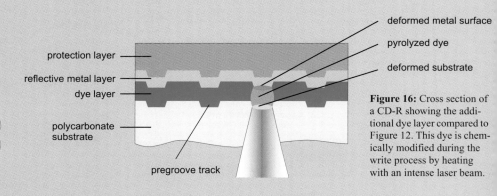

Figure 16: Cross section of a CD-R showing the additional dye layer compared to Figure 12. This dye is chemically modified during the write process by heating with an intense laser beam.

References

[1] A. Cavigneaux, *Uruk*, Mainz, 1996.
[2] P. Paruch, T. Tybell, and J.-M. Triscone, Appl. Phys. Lett. **79**, 530 (2001).
[3] A. Khurshudov, *The Essential Guide to Computer Data Storage: From Floppy to DVD*, Prentice Hall PTR, 2001.
[4] http://www.weller.to/his/h04-automatischen-rechenmaschinen.htm
[5] http://www.jgiesen.de/Divers/Rechner/GifJpg/lochkarte.jpg
[6] W. J. Eckert and J. C. McPherson, *Punched Card Methods in Scientific Computation*, MIT Press, 1984.
[7] http://www.funet.fi/index/FUNET/history/internet/en/reika.html
[8] C. D. Mee, E. D. Daniel (Eds.), *Magnetic Storage Handbook*, McGraw Hill, 1996.
[9] J. C. Mallinson, *The Foundations of Magnetic Recording*, Academic Press, 1993.
[10] H. N. Bertram, *Theory of Magnetic Recording*, Cambridge University Press, 1994.
[11] National Museum of American History.
[12] http://artemis.rrze.uni-erlangen.de/iser/pictures/I0387_02.jpg
[13] http://www.fnal.gov/pub/inquiring/physics/technology/00-572-8.jpg
[14] L. Purcell, *CD-R/DVD Disc Recording Demystified*, McGraw Hill, 2001.
[15] T. Clark, *Designing Storage Area Networks: A Practical Reference for Implementing Fibre Channel SANs*, Addison Wesley Longman, 1999.
[16] M. Farley, *Building Storage Networks*, McGraw-Hill Professional Publishing, 2nd ed., 2001.

Hard Disk Drives

Andreas Dietzel, Micro- and Nano- Scale Engineering, TU Eindhoven, Netherlands

Contents

1 Introduction	617
2 Magnetic Hard Disk Drives	617
3 Inductive Write Head	618
3.1 Construction	618
3.2 Write Field Calculation	618
3.3 Write Head Materials	620
4 Magnetic Recording Media	620
4.1 Thin Film Disks	620
4.2 Macroscopic Properties	620
4.3 Microscopic Properties	621
4.4 Thermal Stability	622
4.5 Antiferromagnetic-Coupled (AFC) Media	622
5 Magnetic Read Head	623
6 Head-Disk Interface	625
7 Future Trends	626
7.1 Perpendicular Recording	626
7.2 Patterned Media	626
7.3 Self-Organized Particle Media	627
7.4 Thermally-Assisted Recording	628
7.5 Tunneling Magnetoresistive Sensors	628
7.6 Contact Recording	628
7.7 Small-Form-Factor Drives	628
8 Summary	628

Hard Disk Drives

1 Introduction

Hard disk drives (HDDs) provide data storage at a cost per bit that is more than ten times lower than for DRAM's. On the other hand HDDs feature high data rates compared to other common mass storage device systems. Magnetic data storage is currently the dominant form of secondary non-volatile storage used in various kinds of computers. It is foreseeable that HDDs will continue to be further developed and used as mass storage devices throughout this decade. The progress in magnetic recording technology, particularly that for HDDs has been as fascinating as that in semiconductor memory. The figure of merit for the technological progress in HDDs is the areal density. It indicates a faster pace since the early 1990s mainly due to the introduction of magnetoresistive (MR) sensors for the read-back of magnetically stored data. In the last years, driven by the introduction of giant magnetoresistive (GMR) sensors, storage density growth has accelerated again and the annual compound growth rate is close to 100 % and significantly surpasses the growth rate for semiconductor memory. In 2001, a new type of media with antiferromagnetic coupled layers (AFC media) was realized in current mass-fabricated products with areal densities of > 30 Gbit/inch^2 which corresponds to a bit cell size of about 350 nm × 45 nm. Figure 1 shows a magnetic force micrograph of recorded bit patterns.

Figure 1: Magnetic force micrograph (MFM) of recorded bit patterns. Track width is 350 nm. The tracks are written at a frequency (linear bit density) slightly lower than used in the actual HDDs and every second track is skipped in order to give a clearer image.

2 Magnetic Hard Disk Drives

The history of the HDD began in 1956 [1]. The very first hard disk drive was called the random access method of accounting and control (RAMAC) and was introduced by IBM. Since then a lot of technological improvements have been achieved but the basic concept remained unchanged. Digital data are stored along concentric tracks in the magnetic medium on a spinning disk. Figure 2 shows an up to date recording scheme. The information is written into the magnetic medium by the magnetic field from an inductive write element. The write field is spreading from the gap between the magnetic poles (P_1 and P_2) into the recording medium. Even today a longitudinal recording scheme is used which means that the magnetization lies in the plane of the thin film medium. The stray fields arising at the transitions between areas of opposing magnetization direction can be read back by a magnetic sensor such as a GMR element which is shielded between two soft magnetic layers to avoid interferences with stray fields from neighbouring transitions. P_1 may also serve as one of the sensor shields.

As displayed in Figure 3 an HDD is mainly composed of the following parts:

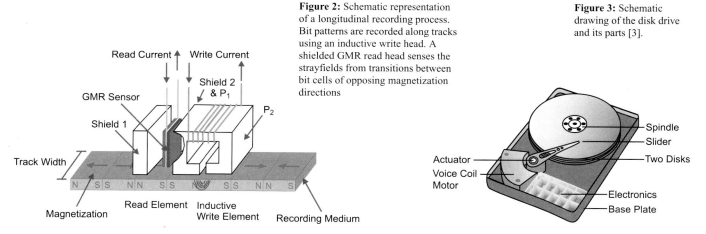

Figure 2: Schematic representation of a longitudinal recording process. Bit patterns are recorded along tracks using an inductive write head. A shielded GMR read head senses the strayfields from transitions between bit cells of opposing magnetization directions

Figure 3: Schematic drawing of the disk drive and its parts [3].

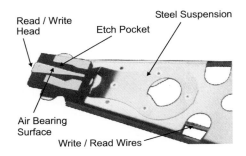

Figure 4: Micrograph of the head-gimbal assembly (HGA).

1. **The slider carrying the magnetic read/write heads.** Data are written onto the disk with the inductive write element and can be read back with the MR or GMR read element. The slider is mounted onto the end of a stainless steel gimbal-suspension, forming a so called head-gimbal assembly (HGA) which is displayed in Figure 4. The slider has a profiled surface facing the medium that forms an air-bearing surface (ABS) allowing the head to 'fly' at a close distance above the medium. In todays HDDs the fly height is about 10 nm.

2. **The magnetic disks** (up to 10). Todays disks are between 1 inch and 5.25 inches in diameter. The disk substrates are made of aluminum or glass. Both surfaces are covered with thin sputtered magnetic layers. The disk typically rotates at 5,400 – 15,000 RPM giving a relative velocity of the slider of more than 100 km/h.

3. **Electronics for data detection and write circuit.** A channel converts the digital data to be stored into write currents and supplies them to the head coil and on the other hand receives signals from the read-element and translates them to digital data.

4. **Mechanical servo and control system** including the spindle, actuator and voice coil motor. It provides a means of moving the head/slider to the desired track and also fine adjustments to retain the head in the center of the track.

The following sections will discuss the main components and their functionality. However, the electronics and the read/write channel functionality will not be outlined here. A detailed review of hard disk drive technology is given in [2], [3], [4].

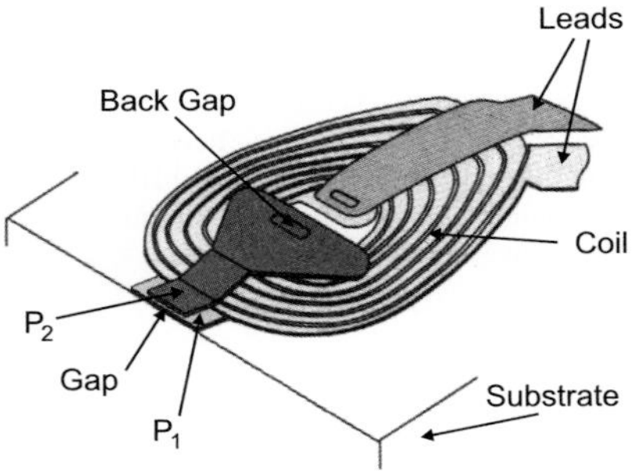

3 Inductive Write Head

3.1 Construction

Already in the 1980s the conventional electromagnet with coiled copper wires was replaced by a thin film head comprising a stack of thin films structured by lithographical methods. Since separate read elements in the form of MR sensors were introduced in the early 1990s inductive heads are only used for the writing and no longer for the reading operation. Figure 5 shows the principal parts of a state-of-the-art thin film write head. The yoke consists of structured $Ni_{81}Fe_{19}$ (permalloy) or $Ni_{45}Fe_{55}$ films P_1 and P_2. These films are all deposited on top of a substrate that is commonly a compound of alumina (Al_2O_3) and titanium carbide (TiC) which is carrying already the read element layers. The width of the trailing layer (P_2) is below 1 μm and much smaller than the leading layer. The P_2 width is defined by photo lithography and determines the gap length and thereby the writing track width of the head (see also Figure 6). A planar thin film copper coil of serveral turns passes between the permalloy sheets P_1 and P_2. A number of insulation layers separate the metallic layers from each other. The gap width which is also a critical dimension for high density recording is defined by the thickness of an Al_2O_3 insulation layer between P_1 and P_2 which is below 100 nm. The P_2 layer is deposited over the coils structure and comes into contact with P_1 at the back gap (see Figure 5) to form a complete magnetic circuit enclosing the coils. Therefore, it is not a entirely planar layer which aggravates the photo lithography process.

Figure 5: Thin film head coil and yoke construction [2].

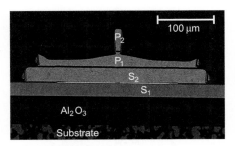

Figure 6: Micrograph of the read/write head taken with a scanning electron microscope from the ABS side. The write lement layers P_1 and P_2 and also the read element shields S_1 and S_2 are marked. The write gap is located between P_1 and P_2, the read element between S_1 and S_2.

3.2 Write Field Calculation

The field produced by the head must at least exceed the coercivity H_c of the medium. However, applying a field equal to H_c to a typical medium will reverse only half the magnetization. Due to the nonsquare nature of the media hysteresis loops (as discussed in the following section), a field equal to two or three times H_c is needed to fully reverse the magnetization. In the following the field from the write element shall be estimated by using simplifying models. All the formulas and quantities will be reflected in SI units. Detailed deductions of the presented formula are given in [2], [3], [4] in which also tables for conversion into CGS units can be found.

Even though the yoke of a thin film head has a complex geometric structure it shall be approximated by a ring head structure as shown in Figure 7. The magnetic flux Φ penetrating through an area A is given by the scalar product of the magnetic induction (or flux density) B and A. In analogy to an electric circuit the model of a magnetic circuit can be applied (see Table 1).

Electric Circuit	Magnetic Circuit
Current (I)	Flux ($\Phi = \mathbf{B} \cdot \mathbf{A}$)
Voltage (V)	Magnetic Potential (nI)
Resistance (R)	Reluctance ($\mathscr{R} = l/(\mu\mu_0 A)$)
Conductivity (σ)	Permeabilty ($\mu\mu_0$)

Table 1: The analogy between electric and magnetic circuit.

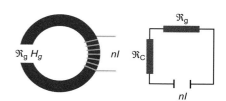

Figure 7: Schematic of a magnetic ring head and its corresponding magnetic circuit diagram

Using this model the ring head can be described by two reluctances in series, that of the head gap and that of the head core $\mathscr{R}_g$ and $\mathscr{R}_c$ respectively. The fraction of the magnetic potential nI (number of turns in the coil times the coil current) that is effective at the head gap is then given by the ratio of the reluctances. Applying Ampere's law reveals the field H_g within the gap of width g as:

$$H_g = \frac{\mathscr{R}_g}{\mathscr{R}_g + \mathscr{R}_c}\frac{nI}{g} = \frac{nI}{g + (l_c\mu_0/\mu)(A_g/A_c)} \quad (1)$$

The reluctances $\mathscr{R}_g$ and $\mathscr{R}_c$ are defined by the geometry of the ring head (g = gap width, l_c = length of the ring head core, A_g = gap cross-section, A_c = core cross-section) and the head material permeability μ. The model of a magnetic circuit used for this equation can only describe the linear part of the core magnetization curve. If the magnetization of the yoke reaches its saturation value M_s, H_g attains its maximum value which is not reflected in Eq. (1).

In a second step, the field at any position x, y in front of the gap shall be calculated for a given field inside the gap H_g. An analytic head model was given by Karlqvist [5], which assumes an idealised inductive head:
a. The permeability of magnetic core or film is infinite. For $\mu \to \infty$ the head has an ideal efficiency and Eq. (1) simplifies as $H_g = nI/g$.
b. The magnetic core is much wider than the head gap in x direction, and the whole head is infinite in z direction. Therefore, the problem becomes two-dimensional in the x-y plane (see Figure 8).
c. The magnetic field H_g within the head gap ($y < 0$) is constant.

In this context we will concentrate on longitudinal recording in which the magnetic medium is magnetized along the data track direction, so that the x component of the head field is relevant.

For a Karlqvist head it can be shown that the magnetic field component H_x at the distance y (>0) from the head pole pieces is with good approximation given by:

$$H_x = \frac{H_g}{\pi}\theta, \quad \theta = \left[\arctan\left(\frac{x+g/2}{y}\right) - \arctan\left(\frac{x-g/2}{y}\right)\right] \quad (2)$$

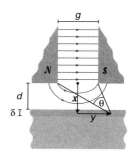

Figure 8: Schematic view of the head gap region indicating the head field and the angle θ used for the Karlqvist field calculation.

Figure 8 illustrates how the angle θ can be geometrically derived for any position (x,y) in front of the gap. For very small gap width g the field has circular contours comparable to the magnetic field surrounding a straight conductor. Figure 9 shows plots of H_x for increasing y/g ratios, i.e. for increasing distance from the gap. The x-axis values are normalized to g and the head field values are normalized to H_g to make the plots universally applicable. For magnetic recording with high density it is favourable to have a write field distribution with sharp contours. Figure 9 shows that this can be achieved by reducing the parameter y/g. For a given head this can be realised by reducing the distance between the storage medium and the write head y. At the same time also the peak height of the write field increases. For a calculation of the write field within the magnetic medium, y can be taken as $d+\delta/2$, where d is the magnetic spacing (see also paragraphs 5 and 6) and δ is the thickness of the magnetic medium as illustrated in Figure 8.

Calculation of the required write current for a given head structure (example):
To properly write a medium with coercivity $H_c = 200$ kA/m a field of $H_c \times 2.5 = 500$ kA/m is necessary. Assuming a medium thickness $\delta = 10$ nm and a head medium spacing of $d = 20$ nm the parameter y is given by $d + \delta/2 = 25$ nm. For $x = 0$ the required minimum field in the head gap H_g of width $g = 100$ nm can then be determined using Eq. (2) or Figure 9 with $y/g = 0.25$ as $H_g = 710$ kA/m. For an ideal Karlqvist head Eq. (2) simplifies as noted above. If the write head coil consists of 10 turns the current necessary for proper writing results in $I = 7$ mA.

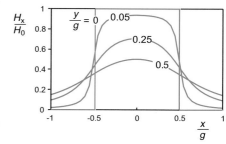

Figure 9: Normalized head field contours at various distances from the head y/g.

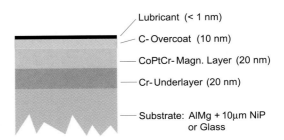

Figure 10: Schematic cross section of a thin film hard disk.

3.3 Write Head Materials

In order to achieve high data rates, high areal densities as well as reliable performance suitable magnetic materials for inductive write heads have to fulfill a list of requirements:

a. High saturation magnetization is necessary because it defines the maximum head field. High fields are necessary to write magnetic media with high coercivity that are suitable for high density storage.

b. The soft magnetic materials must have a large permeabilty over a wide frequency range to achieve sufficient head efficiency at high data rates. A permeability loss at higher frequencies due to eddy currents can be suppressed by highly resistive materials or laminated materials with insulating layers. Laminated structures require dry deposition processes such as sputtering.

c. The yoke materials must be magnetically soft in order to minimize hysteresis losses.

d. The head materials have to withstand high temperatures during fabrication and have to be mechanically and chemically stable also during operation within the HDD.

Because they satisfy the requirements to large extend, NiFe alloys such as permalloy ($Ni_{81}Fe_{19}$) and $Ni_{45}Fe_{55}$ are widely used head materials. With the demand for materials with higher saturation magnetization also NiFe alloys with higher Fe content and Co based alloys are applied. The process commonly used for forming the magnetic poles is electroplating because it offers an economic deposition processs. Nevertheless dry deposition processes have more flexibility and may replace electrodeposition in the future. Electroplated copper is used as coil material. Sputtered Al_2O_3 is as well as hard baked-photoresist between the coil and yoke films are used as insulation layers.

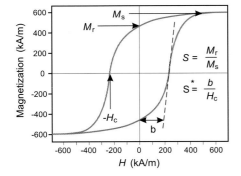

Figure 11: Typical hysteresis of a thin film disk sample indicating the macroscopic magnetic parameters which are important for the recording behaviour.

4 Magnetic Recording Media

4.1 Thin Film Disks

Thin-film disks are complex multilayer structures as schematically shown in Figure 10. In case of magnesium doped aluminium as substrate material the disk is plated with an amorphous NiP undercoat to make the disk smooth and hard. A chromium underlayer is often used to control magnetic properties and microstructures of the magnetic recording layer. The magnetic layer (typically sputtered CoPtCr doped with B) is covered by a carbon overcoat layer and lubricant. The last two layers are necessary for the tribological performance of the head-disk interface and for the protection of the magnetic layer. The fly height is the distance between the ABS at the trailing edge (see also Figure 19) and the disk top surface, while the magnetic spacing d is the distance between head pole tips and the magnetic layer. The latter is most relevant in the write and read processes.

Rigid-disk media magnetic properties can be conveniently subdivided into macroscopic and microscopic properties. An understanding of each is needed to design thin film media as required by the magnetic recording system.

4.2 Macroscopic Properties

For high density recording the macroscopic properties such as coercivity (H_c), remanence (M_r), coercive squareness (S^*) and remanence squareness (S) determine read-back signal variables such as pulse shape, amplitude, and resolution arising from magnetic transitions. These disk parameters are determined by means of VSM (vibrating sample magnetometer). A magnetization curve of a thin film disk as can be measured with a VSM is shown in Figure 11.

Achieving the narrowest possible transitions allows placing recorded bits close together and hence results in high storage density. In an ideal situation, in the absence of demagnetizing fields arising from adjacent bit cells, the transition would be abrupt. A more realistic analytical model that can be used for head-disk-studies is given by Williams and Comstock [6]. A magnetic transition of finite width shall be described by an arctangent function:

$$M(x) = \frac{2}{\pi} M_r \arctan\left(\frac{x}{a}\right) \qquad (3)$$

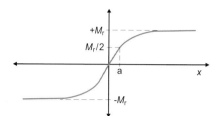

Figure 12: Arctangent form of a magnetic transition. The transition parameter a is indicated.

This function, plotted in Figure 12 is smooth, i.e. microstructural inhomogeneities are neglected, and has even symmetry. The parameter a describes the width of the transition from $-M_r$ to $+M_r$.

Assuming that contributions to the transition parameter from characteristics of the write head are negligible and that the recording medium has a squareness S^* close to one a simplified relation can be derived from the Williams Comstock model [4]:

$$a = \sqrt{\frac{M_r \delta (d + \delta/2)}{\pi H_c}} \qquad (4)$$

A general design rule for magnetic media is to minimize the transition parameter a. Eq. (4) indicates that this can be achieved by decreasing the magnetic spacing d, increasing the coercivity H_c of the medium, and decreasing the thickness δ of the magnetic medium. Smaller M_r could also reduce a, but as will be described in Eq. (7) in Section 5, a smaller M_r would reduce the magnetic field at the read element and thereby the signal voltage. As the linear recording density is scaled up, the remanence thickness product $M_r\delta$ of the magnetic layer needs to be scaled down. Therefore δ was continuously reduced on the way to higher storage densities. Today, 20 nm (see magnetic layer thickness in Figure 10) is typical value for δ.

The addition of alloying elements helps to enhance and control medium coercivity, to minimize the intergranular exchange coupling, and to increase corrosion resistance. H_c is largely determined by magnetocrystalline anisotropy and to lesser extent by shape anisotropy of the magnetic grains. But coercivity is also effected by film stress, crystal defects, grain size, grain orientation and grain boundaries. The remanence squareness $S = M_r/M_s$ describes the squareness of the magnetization curve. Coercive squareness S^* is related to the slope of the magnetization curve and can be determined as indicated in Figure 11. A perfect square hysteresis loop is described by $S = S^* = 1$. S and S^* values depend on the grain interaction within the film. High S^* values (close to one) lead to narrow switching field distribution and sharply recorded transitions.

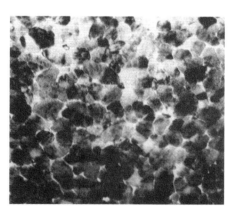

Figure 13: Transmission Electron (TEM) micrograph of a storage medium thin film structure at 200,000x magnification. The grain structure is noticeably voided, leading to reduced magnetic interactions and lower transition noise [3].

4.3 Microscopic Properties

Microscopic properties such as grain size, grain coupling, and grain crystallographic orientation mainly determine the signal-to-noise ratio (S/N) in the read-back signal. The magnetic grains within a magnetic thin film are so close that magnetostatic and intergranular exchange coupling are relatively strong. Highly intergranular coupled magnetic thin films with long correlation lengths tend to form zigzag domain walls in recorded transitions which result in excessive transition noises. Therefore the reduction of intergranular exchange coupling in thin-film media becomes important for medium design and fabrication. There are three major approaches to noise reduction: physical grain segregation, compositional segregation, and small grains with very narrow size distributions. Physical grain segregation has been observed in TEM micrographs in low-noise media as shown in Figure 13.

Sputtering at conditions where the sputtered atoms have low mobility during film growth enables the formation of voided structures. The model of Thornton [7] (see also Chap. 8) describes the formation of voided columnar films at low substrate temperature and high sputter pressure. In addition low sputter rates and the absence of bias voltage lead to a preferred columnar growth, essential for the fabrication of low-noise media. In addition magnetic films tend to adapt to the underlayer structure below. Thus forming well defined segregated underlayer structures leads to segregated magnetic film structures. Compositional segregation of grains in thin film disks is another effective way to minimize interactions. Higher chromium content and higher sputter temperatures are considered as to accelerate the chromium segregation to grain boundaries, forming non-magnetic phases. Sputtering quaternary alloys such as CoPtCrTa and CoPtCrB is also an effective way to reduce the noise. The low solubility of boron in the cobalt alloy leads to a compositional segregation [9].

To summarize, an ideal thin-film magnetic recording layer with high coercivity and low noise would be composed of grains with high-anisotropy, which are much smaller than the recording bit cell, uniform in size, and magnetically isolated.

4.4 Thermal Stability

For high density recording the grains of a magnetic disk have to be small compared to the bit cell in order to get acceptable transition signals. In a simplified model assuming isolated grains the thermally induced switching of magnetization can be described as indicated in Figure 14.

For the reversal of its magnetization the grain has to overcome an energy barrier. If the energy barrier is small a spontaneous reversal may occur due to thermal energy fluctuations. In this model the switching probability f is given by an Arrhenius equation:

$$f = f_0 \exp\left(-\frac{\Delta W}{kT}\right), \quad \Delta W = K_u V \tag{5}$$

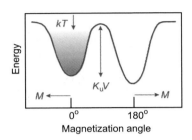

Figure 14: Schematic representation of a simplified model for the thermal instability of magnetization. It illustrates the energy of a small isolated particle as a function of magnetization orientation. The energy barrier between the two local minima separated by 180° rotation is in the order of thermal energy fluctuations. The energy levels for the local minima may be different in the presence of an external field.

Where f_0 is about 10^9 Hz [2], k is the Boltzmann constant, T is the absolute temperature, and ΔW is the energy barrier for the switching. For an isolated grain the energy barrier is $K_u V$, where K_u is the uniaxial anisotropy constant and V is the volume of the grain. If the grains become very small, the magnetization will switch very easily. This phenomenon is called the *superparamagnetic effect*. High anisotropy magnetic media giving increased thermal stability to avoid spontaneous loss of data are therefore necessary for increasing the storage densities in HDDs. However this approach is limited by the write capabilities of inductive write heads.

Estimation of minimum grain size (example): For currently used thin disk materials K_u is about 2×10^5 J/m^3. If we claim that the bits shall be safely stored for ten years at room temperature ($f < 3.33 \times 10^{-9}$ Hz at $T = 300$ K), then the diameter of grains assumed to be spheric should be greater than 9 nm. In current thin film disks the grains have already a typical diameter of around 10 nm.

4.5 Antiferromagnetic-Coupled (AFC) Media

In order to delay the problem of thermal instability arising with increasing areal density, multilayer magnetic media employing antiferromagnetic coupling have been recently developed [10]. An AFC medium is schematically represented in Figure 15 in which two magnetic layers are separated by a thin layer of nonmagnetic ruthenium. A precise control of the Ru thickness allows to establish an anti-parallel (antiferromagnetic) coupling between the two magnetic layers. It was discussed in previous sections that the $M_r\delta$ value has to be scaled down to achieve sharper transitions and that grains have to be small to achieve good S/N for single-layer media. In AFC media an effective $M_r\delta$ value which is the difference of $M_r\delta$ values for each of the two magnetic layers is sensed by the read element and is defining the sharpness of the transitions:

$$M_r\delta(\text{eff}) = M_r\delta(\text{top}) - M_r\delta(\text{bottom}) \tag{6}$$

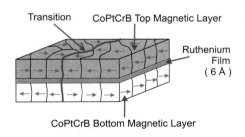

Figure 15: Schematic cross-section of a AFC storage medium indicating also the grain structure and a single magnetic transition separationg grains of opposing mag netization.

Therefore the $M_r\delta$ value for the top magnetic layer can be relatively large in comparison to single layer media designed for permitting the same areal density. In other words, the grains in AFC media can be larger (in the vertical direction). This means that the thermal stability can be increased without affecting transition sharpness, transition noise or required write fields.

It is worth mentioning that AFC media can be manufactured using the same production equipment as for conventional media. Even though the GMR output pulse from a transition in an AFC medium is a superposition of signals from both, top layer and bottom layer transition, it can be detected as a single pulse. The recording heads and data channel electronics in the HDD do not need to be modified for AFC media and existing disk manufacturing equipment can still be used. Therefore AFC media could be rapidly introduced and are found in HDD products with storage densities beyond 25 Gbit/in^2.

5 Magnetic Read Head

The signal output for anisotropic magnetoresistive (AMR) heads is directly proportional to the MR ratio ($\Delta R/R$) of the sensor material (NiFe) and therefore limited to about 2 – 3 %. A little bit more than ten years ago the giant magnetoresistive effect (GMR) was first discovered by P. Grünberg [11] and was also reported by M. N. Baibich et al. [12]. It is impressing, that in less than a decade this scientific breakthrough transformed into commercial products. An introduction to GMR is given in this book in Chapter 4. For the application in HDDs a GMR sensor structure was developed, which can be economically fabricated, which is sensitive to local fields from recorded media, and which operates at ambient temperature, the so-called 'spin-valve'[13].

A spin-valve (see Figure 16) achieves a $\Delta R/R$ of typically about 10 %. Its sandwich structure consists of two uncoupled ferromagnetic films separated by a nonmagnetic film. The preferred ferromagnetic layers are NiFe and CoFe. The nonmagnetic layer which is typically Cu is thin compared to the mean free path of the electrons that can pass from one ferromagnetic layer to the other. An antiferromagnetic layer of PtMn, NiMn, FeMn or NiO is used to pin the magnetization of the CoFe-layer. The NiFe *free layer* can rotate its magnetization in the presence of an external magnetic field, e.g. emanating from the storage disk. Figure 16 reflects a sensor with Current In Plane (CIP) geometry where the voltage drop occurs in the plane of the GMR films in contrast to a Current Perpendicular Plane (CPP) geometry, where the contacts are formed below and on top of the GMR stack and the voltage drop occurs perpendicular to the film plane.

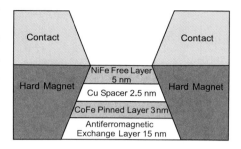

Figure 16: Schematic cross-section of a spin-valve sensor in CIP geometry as used in recording heads.

The difference of available states for electrons with spin orientation parallel and antiparallel to the magnetization causes spin dependent scattering in the free and the pinned layer. The current through the spin-valve stack is formed by electrons of both kinds passing back and forth between the ferromagnetic layers via the nonmagnetic interlayer. It turns out that in the case of parallel magnetization of the ferromagnetic layers one electron species has a low resistivity path whereas in case of opposite oriented magnetization both species suffer from strong scattering in either of the layers. Therefore the total resistance of the spin-valve stack is increased for antiparallel magnetizations. In the presence of an external field the free layer can gradually rotate its magnetization within the film plane and the resistance of the structure increases as:

$$\frac{\Delta R}{R_0} = \frac{\Delta R_{\max}}{R_0}\left(\frac{1-\cos(\Delta\vartheta)}{2}\right) \quad (7)$$

$\Delta\vartheta$ is the angle between the magnetizations of the free and the pinned layer. R_0 is the resistance at parallel magnetization and $\Delta R_{\max}$ is the difference in resistance between parallel and anti-parallel magnetization. $\Delta\vartheta$ can be influenced by the flux induced by the storage disk and is determined by minimizing the total energy of the free layer. If $\Delta\vartheta$ is 90° in the quiescent state the read-back voltage from a spin-valve GMR head V_{GMR} is given by:

$$V_{\text{GMR}} = \frac{I\Delta R_{\max}}{2\mu_0 tw M_s^f} E_{\text{GMR}} \Phi_{\text{sig}} \quad (8)$$

Where I is the current through the sensor, E_{GMR} is the efficiency of the sensor accounting for the fact that the signal flux decays from the ABS to the top of the free layer and Φ_{sig} is the signal flux injected into the free layer. w is the width (read track width) and t the thickness of the sensor. M_s^f is the saturation magnetization of the spin-valve free layer. For calculating the signal flux arising from a recorded transition a theorem called *reciprocity principle* can be used. It states that the mutual inductance between any two objects is one quantity and therefore the same in both directions. This principle can be directly applied between the writing and reading if both processes are using the same inductive head structure. Even though the GMR sensor is only a reader and the inductive write head is a separate structure in modern HDDs a shielded GMR head still allows to apply a conceptual reciprocity between the reading and a hypothetical writing process as indicated in Figure 17. One of the soft magnetic sensor shields is viewed as one of the pole pieces in Figure 8 and the free layer of the GMR sensor is viewed as the other pole piece.

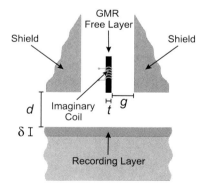

Figure 17: Shielded GMR sensor showing an imaginary coil used for the calculation of the sensor voltage.

The field created by an imaginary coil surrounding the GMR sensor could in principle magnetize the medium. The two-gap GMR structure can therefore be treated as a summation of two inductive heads. Therefore the signal voltage V_{GMR} at the sensor resulting from a single transition can similarly to Eq. (2) be deducted from the Karlqvist field and is then given by:

$$V_{GMR}(x) = \frac{E_{GMR} I \Delta R_{max} M_r \delta}{\pi M_s^f t} \frac{g+t}{g} \left[\arctan\left(\frac{x+g/2}{a+d}\right) - \arctan\left(\frac{x-g/2}{a+d}\right) \right] \quad (9)$$

The sensor is assumed to be placed in the middle between the two magnetic shields separated by $2g+t$. x is the centre of a magnetic transition. In Eq. (9) an arctangent magnetic transition of width a as defined in Eq. (3) is assumed. Therefore the distance d between the magnetic medium and the head can be replaced by $d + a$ which is one motivation for using the arctangent transition form. Eq. (9) is useful for understanding the influences of head/media parameters on signal output. First, a small transition parameter a and a small head/disk spacing d are desirable for high signal outputs. Second, if $M_r\delta$ is reduced to achieve storage media allowing smaller transitions and higher storage densities as discussed in 4.2 this reduces the flux in the sensor and has to be compensated by a higher GMR sensitivity which is defined by the quantities E_{GMR}, I, ΔR_{max} and M_s^f.

Calculation of GMR output voltage (example): If M_r is 500 kA/m and the recording layer thickness is 20 nm the medium is characterized by $M_r\delta$ = 10 mA. Let us further assume that the free layer of the GMR sensor has a saturation magnetization M_s^f of 700 kA/m and a thickness of $t = 10$ nm and that the sensor is placed centric between two shields defining a read gap width of $2g = 100$ nm. Due to losses in the leads and a gradient of sensor magnetization E_{GMR} is reduced to 0.1. The sensor current is typically 5 mA and ΔR_{max} typically 2.5 Ω. If the head medium spacing is $d = 20$ nm and the transition is characterized by $a = 20$ nm, the readback peak to peak voltage is given as $V_{p-p} = 2 \times V_{GMR} (x = 0) = 1.5$ mV.

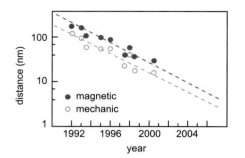

Figure 18: Distance of the read-write-head from the hard disk until year 2000 and extrapolated to 2005. Shown are both the mechanical distance as well as the magnetic distance. The latter is a few nanometers larger owing to non-magnetic protection layers [15].

A more adequate description of the situation in an HDD is given if we consider a dense sequence of transitions in the medium instead of a single arctangent transition. To simplify the model the magnetization can be assumed to have the form of a sinus wave $M(x) = M_0 \sin(kx)$ reflecting a sequence of bit cells with alternating magnetization. The bit length (distance between transition centres) is given by π/k. Under this assumption three contributions to the loss of signal can be identified which are of general importance for magnetic recording:

$$V_{GMR} \propto \left[e^{-kd} \right] \left[\frac{1-e^{-k\delta}}{k\delta} \right] \left[\frac{\sin\left(\frac{kg}{2}\right)}{\frac{kg}{2}} \right] \quad (10)$$

Each of the three terms in square brackets is smaller than unity and reduces V_{GMR} by a corresponding fraction. The first term describes the losses due to the spacing. The signal decays exponentially with the magnetic spacing d. This is the reason for the continuous reduction of head/disk spacing which will be discussed in the following section. The second term describes the losses due to the thickness δ of the magnetic layer. For a thinner magnetic layer the bottom of the medium is closer to the head and the loss reduces. The third term describes the losses related to the gap dimension. The first zero of this oscillating term appears having the gap length of twice the bit length and therefore equal positive and negative magnetization is located under the gap. Therefore g should be kept sufficiently below $2\pi/k$.

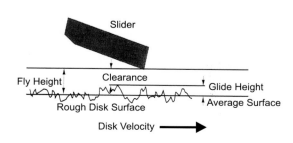

Figure 19: Schematic of the head-disk-interface [2].

6 Head-Disk Interface

The head/disk interface is crucial to the performance of the HDD. Eq. (10) can also be written as:

$$V_{GMR} = J_{medium} S_{sensor} e^{-kd} \qquad (11)$$

S_{sensor} stands for the sensitivity of the read sensor and J_{medium} is a term describing the magnetic medium properties. If the magnetic spacing d is too high, the read-back signal is reduced and the probability of data errors increases. Also the write field at the medium is to weak to write the disk properly. On the other hand very low head/disk spacing may lead to mechanical wear and can substantially reduce the lifetime of the HDD. Nevertheless the evolution of areal densities demands a continuous decrease of the head/disk spacing (see Figure 18). To achieve very low magnetic spacing not only the fly height but also the thickness of the non-magnetic protection layers of heads and disks has to be reduced [14].

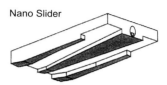
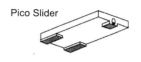

Figure 19 shows a schematic of the head-disk interface. It outlines disk roughness and asperities. The nominal fly height is considered as the average mechanical distance between the disk surface and the trailing edge of the slider. As the head disk spacing is reduced, at some point the head will come into contact with asperities. This spacing is called the glide height and defines lowest possible fly height for a given head/disk combination. As discussed in previous sections the magnetic spacing or magnetic fly height d is a very relevant parameter for the read-write performance. There are several factors that influence this parameter:

a. **The structure of the ABS.** Slider ABS can be designed to reduce variations in the fly height. Such variations occure for example due to different relative velocities between slider and spinning disk at inner and outer disk diameters.

b. **The roughness of the disk surface and asperities.** To reduce both a burnishing procedure is applied to the disks, to get a very smooth surface .

c. **Tribological coatings,** such as carbon coating and lubricant.

d. **Radial curvature and radial slope.** They contribute to variations in the fly height especially when the thickness of the substrate is reduced. Glass substrates provide higher rigidity and surface flatness than Al substrates.

e. **The load force** defined by spring constant of the stainless steel suspension.

Figure 20: A nano Al_2O_3-TiC slider ($2.05 \times 1.6 \times 0.45$ mm) with etched rails and a pico slider ($1.25 \times 1 \times 0.3$ mm) with a 'tripad' ABS [2].

The air-bearing surface of the slider is structured by ion milling and photolithographic processing. With such processes the sliders surface (see Figure 4, Figure 20 and Figure 21) can be shaped to tailor positive and negative pressure zones. The structure of the ABS, along with the disk rotational speed and the load-force, determine the physical spacing or fly height above the disk. For proximity recording where the slider flies near the glide height of the disk, so-called "tripad" nano sliders (see Figure 20) were developed. Smaller slider dimensions not only make possible that more sliders can be produced on a single wafer but are also beneficial to their tribological performance and position control.

As already discussed the roughness of the disk has to be much lower than the fly height. Because super-smooth disk surfaces increase the stiction forces between slider and disk a dynamical load/unload mechanism is implemented in today's HDDs. With this technique the slider starts from and lands on a ramp instead of the disk surface.

The magnetic recording layer cannot withstand the friction and wear forces from head slider contact. Therefore an overcoat and a lubricant are applied. The most widely used overcoat material is sputtered carbon, which features diamond-like properties such as high hardness and good corrosion resistance. In current disk drives these layers are only a few nanometers thick to obtain a low d. The most widely used lubricants are perflouropolyethers (PFPE) which are applied by dip coating. They are highly chemically and thermally stable, have a low vapour pressure, low surface tension and low viscosity. The lubricant layer is less than one nm thick.

Figure 21: Scanning electron micrograph of a Pico slider. The ABS profile as well as the four bonds at the trailing edge for read and write element are displayed.

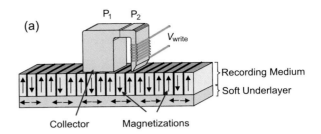
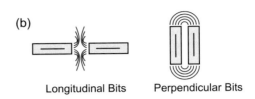

Figure 22: Schematic illustration of the perpendicular recording principle (a) and a schematic indicating demagnetizing fields arising from adjacent bit cells comparing longitudinal with perpendicular arrangement of bit cells (b).

7 Future Trends

A continuous progress in areal storage densities can only be achieved by improvements of all components of a hard disk drive. In the following some of the current research topics are briefly discussed. Many researchers believe that the progress in areal densities may be limited by the *superparamagnetic phenomenon* therefore attention is given to approaches to delay the problem of superparamagnetism. It is worth noticing that it is difficult to foresee whether the discussed approaches will be implemented in future mass produced HDDs.

7.1 Perpendicular Recording

In contrast to longitudinal recording (see Figure 2) where the magnetization lies in the plane of the recording layer information can also be stored using a perpendicular (or vertical) recording scheme as indicated in Figure 22 where the medium is magnetized perpendicular to the film plane. In longitudinal recording the demagnetizing fields between adjacent magnetic bits tend to separate the bits, making the transition parameter a large. Perpendicular recording bits do not face each other and can therefore be written with higher density (see Figure 22b). In comparison to a longitudinal medium, a perpendicular medium supporting the same areal density can be thicker and problems with superparamagnetism can be delayed. As illustrated in Figure 22a) a narrow write gap is formed between P_2 and a soft magnetic under-layers which is serving as a flux return path towards the much wider P_1 pole piece. A lot of research and development work on perpendicular recording, including suitable media and head designs, has been undertaken during the last three decades. Up to now no HDD product employing this technology has ever been on the market but in order to achieve the ultimate high densities it may be necessary to implement this technology in future products. The principal problems of suitable media materials with perpendicular anisotropy and suitable write head geometries have already been solved [2].

7.2 Patterned Media

In today's HDDs the information is stored as a magnetization pattern within a film consisting of weakly coupled grains. In order to get reasonable signal-to-noise ratios every bit covers an area containing hundreds of grains. Further increase in storage density would therefore require a reduction of grain volume V leading to superparamagnetic behaviour in which thermal energy can alter magnetization of single grains. According to Eq. (5) this could in principal be compensated by increasing K_u which would lead to media with higher coercivity H_c. But write elements are limited by material properties and supply current which does not allow arbitrary high magnetic fields to be generated.

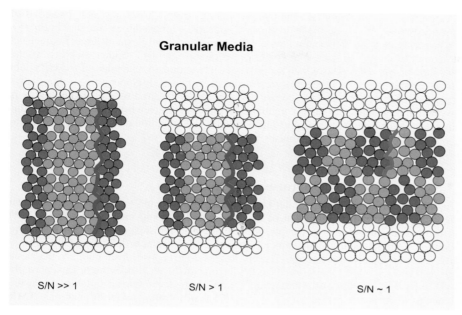

As previously discussed the grain volume V decrease can be delayed to some extent by using AFC media. Nevertheless, for acceptable S/N in the read back signal from media with storage densities of several hundred Gbit/in^2 grain sizes with inherent thermal instability will be needed. A possible solution for this dilemma is the use of a patterned magnetic medium. Such a medium consists of a regular pattern of isolated nanomagnets each of them consisting of a single domain and carrying the information of a single bit. Because edges of such bit areas are well defined by the patterning process such media do not suffer from the noise generated by zigzag transitions in continuous granular media as depicted in Figure 23.

Such large-scale periodic nanomagnet arrays can for instance be realized using photoresist patterning by laser interference followed by etching and deposition steps [16]. A suitable fabrication process for the pre-patterning of such domains of about 50 nm size has to fulfil special requirements such as facilitating patterns with rotational symmetry and preserving the ultra smooth disk surface. A potential industrial fabrication process is the resistless ion projection direct structuring (IPDS) which makes use of a direct conversion of magnetic properties through ion irradiation [17]. Magnetic islands of any shape and in arbitrary arrangements can be defined in a stencil mask and are projected with reduced dimensions onto the recording medium without affecting its surface topography. For example CoPt multilayer films which are a suitable medium for perpendicular storage can be locally intermixed by ion irradiation. The intermixing of the Co/Pt interfaces reduces the interfacial perpendicular anisotropy within the small islands to a level that allows to write within but not outside these islands.

Figure 23: Schematic of bit cells generated in continous granular media compared to predefined bit cells in patterned media. Indicated are also the transitions between the bit cells. The zigzag shape of the transitions gives rise to the signal noise. The improved signal-to-noise ratio *S/N* at a given grain size and bit size favours a patterned media concept.

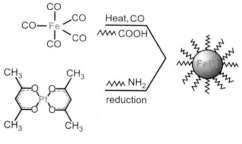
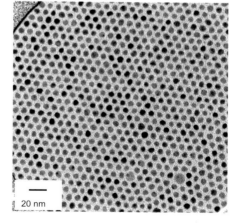

7.3 Self-Organized Particle Media

Recently, a new recording material approach based on self-assembling FePt nanoparticles has been reported [18]. The nanoparticles are chemically synthesized as indicated in the insert of Figure 24 by the reduction of platinum acetylacetonate and decomposition of iron pentacarbonyl. Their size and composition can be precisely controlled. When the particles are spread out on a substrate the carrier solvent can evaporate and the particles self-assemble into an ordered lattice with controlled spacing as is demonstrated by the micrograph in Figure 24. An annealing process transfers the internal particle structure into a chemically ordered face-centered tetragonal phase whereby ferromagnetic nanocrystals are formed. The annealed nanocrystal assemblies are smooth ferromagnetic films which can be considered as recording medium. The particle size is tunable from 3 to 10 nm with a standard deviation below 5 %. Compared to standard recording media the grain size distribution is very narrow. Therefore the number of grains per bit could be significantly reduced and optimized ferromagnetic nanocrystal superlattices can be considered as potential future storage media.

Figure 24: TEM micrograph of self assembled FePt nanoparticles (bottom). The schematic illustrates the fabrication roots for this self-assembled material (top) [18].

7.4 Thermally-Assisted Recording

A different approach to attack the thermal stability problem has not been discussed so far. Besides K_u and V also the temperature T appears as controllable parameter in Eq. (5). It suggests that the information should be recorded at an elevated temperature level whereas it should be stored at lower temperatures giving a higher stability. If the medium has a high H_c (i.e. a high stability) at room temperature the necessary write field is lowered if the write process is carried out at an elevated temperature. In thermally-assisted recording the medium coercivity is thermally lowered by a heat supply that is limited in time and space [19]. Potential heat sources are given by small focus laser illumination or near field resistive heating elements. The operation of the heat source has to be synchronized with the write pulses.

7.5 Tunneling Magnetoresistive Sensors

In order to read smaller and smaller bits the read sensitivity has to be improved. This can be achieved by the development of new sensors. Tunneling magnetoresistive (TMR) sensors as discussed also in Chapter 4 are made of a ferromagnetic metal / insulating barrier / ferromagnetic metal sandwich with electrical contacts defining a CPP geometry. Electrons can tunnel through the very thin (< 2 nm) barrier from one ferromagnetic electrode to the other. The electrons in the ferromagnetic electrodes are at least partially spin polarized. Therefore the tunneling current through the barrier is larger when the magnetization in the two electrodes is parallel than for antiparallel orientations. TMR ratios $\Delta R/R$ of more than 40 % at room temperature have been reported making such sensors operating in CPP geometry very attractive for applications in future magnetic hard disk drives [20].

7.6 Contact Recording

As outlined in previous sections the maximum performance with an available head/media combination will be achieved when the *magnetic spacing* is as close as possible to zero. In tape storage systems contact recording regularly occurs, but for disk drive technology the goal is more challenging mainly because of higher relative velocities. Developing reliable designs of head and media and suitable lubricants are the main challenges.

7.7 Small-Form-Factor Drives

For mobile applications such as laptop computers, notebooks, digital cameras, and multipurpose mobile phones smaller drives are gaining more interest. The special requirements for such applications are small size, light weight, robustness against external shock, and low power consumption. Currently there are 2.5", 1.8" and 1" format mobile disks on the market. Because magnetic storage for the same amount of data is significantly cheaper than semiconductor memory and hand held devices as mobile internet phones and cameras require more and more storage capacity a new mass market field is opening for this technology.

8 Summary

The continuous scaling of magnetic recording is more and more facing scientific and technological challenges. But within this decade there will probably be no alternative technology available that can replace magnetic hard disk storage. In addition the internet, mobile computing and digital photography are already beginning to expand the market for hard disk storage. Therefore, the industrial investments in research and development are huge and will lead to further significant improvements in all components which will make it difficult for alternative technologies to compete with cost and performance of advancing HDDs.

Acknowledgements

Special thanks are due to R. Berger, H. Grimm and F. Voges who provided some of the figures and gave helpful suggestions for the preparation of the manuscript. The editor would like to thank Simon Stein (FZ Jülich) for checking the symbols and formulas in this chapter.

References

[1] E. D. Daniel, C. D. Mee, M. Clark, *Magnetic Recording: The first 100 years*, IEEE Press, New York, 1998.
[2] S. X. Wang, A. M. Taratorin, *Magnetic Information Storage Technology*, Academic Press, New York, 1999.
[3] K. G. Ashar, *Magnetic Disk Drive Technology*, IEEE Press, New York, 1997.
[4] C. D. Mee, E. D. Daniel, *Magnetic Recording Technology*, Mc-Graw Hill, New York, 1987.
[5] O. Karlqvist, Trans. Roy. Inst. Technol., Stockholm, **86**, 3 (1954).
[6] M. L. Williams, R. L. Comstock, AIP. Conf. Proc. **5**, 738 (1972).
[7] J. A. Thornton, J. Vac. Sci. Technol. **A4**, 3059 (1986).
[8] C. Arnoldussen, L. L. Nunnelly, *Noise in Digital Magnetic Recording*, World Scientific, London, 1992.
[9] C. R. Paik, I. Suzuki, N. Tani, M. Ishidawa, Y. Ota, K. Nakamura, IEEE Trans. Magn. **26**, 256 (1990).
[10] E.E. Fullerton et al., Appl. Phys. Lett. **77**, 3806 (2000).
[11] G. Binasch, P. Grünberg, F. Saurenbach, W. Zinn, Phys. Rev. **B 39**, 4828 (1989).
[12] M. N. Baibich et al., Phys. Rev. Lett. **61**, 2472 (1988).
[13] B. Dieny, V. Speriosu, S. Parkin, B. Gurney, D. Wilhoit, D. Mauri, Phys. Rev. B**43**, 1297 (1991).
[14] A. Leson, H. Hilgers, Physikalische Blätter **55**, 63 (1999).
[15] D. Speliotis, Data Storage Jan. 1999, 25 (1999).
[16] A. Carl, S. Kirsch, J. Lohau, W. Weinforth, E. F. Wassermann, IEEE Trans. Magn. **35**, 3106 (1999).
[17] A. Dietzel et al., Adv. Mater. **15**, 1152 (2003).
[18] S. Sun, C. B. Murray, D. Weller, L. Folks, A. Moser, Science **287**, 1989 (2000).
[19] J. J. M. Ruigrok, R. Coehoorn, S. R. Cumpson, H. W. Kesteren, J. Appl. Phys. **87**, 5398 (2000).
[20] S. S. P. Parkin et al., J. Appl. Phys. **85**, 5828 (1999).

Magneto-Optical Discs

Klaus Röll, Experimental Physics, Kassel University, Germany

Contents

1 Introduction — 633
2 Principle of Magneto-Optical Data Storage — 633
3 Material Properties — 635
4 Application of Exchanged Coupled Layers — 638
   4.1 Basic Principles — 638
   4.2 Laser Intensity Modulation for Direct Overwrite (LIM-DOW) — 638
   4.3 Magnetically Induced Super Resolution (MSR) — 639
5 Summary and Outlook — 640

Magneto-Optical Discs

1 Introduction

Optical data storage devices are available at present in various types as outlined in the Introduction to Part V. The success of these storage techniques started in 1982 with the introduction of the audio compact disc (CD). At the same time, the erasable magneto-optical (MO) disc was announced but did not become available before 1988, although the physical principles of MO recording were discussed as early as 1970. The magneto-optical MiniDisc (MD) is now a common product on the consumer market along with the audio CD, the computer CD-R, and the digital versatile disc DVD. MO discs with capacities of several GBytes are available for computer applications [1] – [4].

In the recordable optical media, the information is generally written in a thin layer on one surface of the disc by means of local heating by a laser pulse. In the case of CD-R, the radiation is absorbed by a dye layer and this process leads to local variations of the reflectivity. In CD-RW discs changes of reflectivity are created by transitions between amorphous and crystalline states of the information layer. A reversible transition can be obtained by different pulse lengths and power levels. In magneto-optical (MO) discs, the binary information is stored by small magnetic domains which are formed (or erased) in a magnetic layer by the laser pulses in the presence of a magnetic field (thermomagnetic writing). Read-out is obtained from intensity modulations of the reflected laser light caused by the written domains.

2 Principle of Magneto-Optical Data Storage

The **thermomagnetic writing process** used in the MO media is shown in Figure 1. A magnetic layer, which is homogeneously magnetized in one direction perpendicular to the surface, is heated by the laser pulse to a sufficiently high temperature where the coercivity H_c of the magnetic layer becomes smaller than a simultaneously applied bias field H_b. By this process a small domain is created whose magnetization is opposite to the original magnetization direction. As a result small bubble domains with magnetization down are formed in an environment with magnetization up. The two different magnetizations can be attributed to the binary numbers 1 or 0.

For read-out, the laser is operated with a sufficiently low power so that it does not change the domain structure. The polarization of the reflected light is then influenced by the magnetizations of the different domains due to the **magneto-optical Kerr effect (MOKE)**. When linearly polarized light is reflected from a sample with magnetization M perpendicular to the surface the polarization becomes elliptical with a small tilt (Kerr rotation) $\theta_K \leq 0.5$ degree, which is proportional to the magnetization (Polar Kerr effect, Figure 2a). In particular, $\pm \theta_K$ is obtained for opposite magnetization directions. By

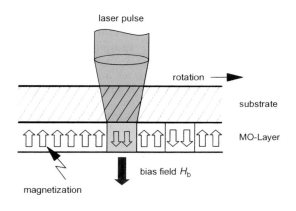

Figure 1: Principle of magneto-optical recording [3].

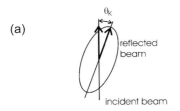
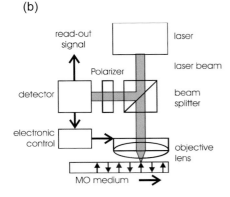

Figure 2: Read-out by the magneto-optical Kerr effect (MOKE).
(a) Linear and elliptical polarization of the incident and the reflected beam, respectively.
(b) Optical paths and electronic control of the optical write and read head.

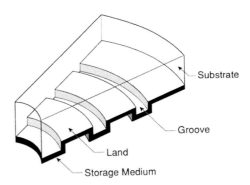

Figure 3: Cross section of a magneto-optical disc [1].

means of a polarizer the Kerr rotation can be transformed into an intensity modulation of the reflected light, which corresponds to the binary information represented by the bubble domain structure. The optical path and the electronic control of the magneto-optical write and read head are shown schematically in Figure 2b. For the writing and read-out processes the laser is operated at a high or low power level, respectively. Focusing and tracking are performed by the electronically controlled positioning of the objective lens.

The overwriting of given information by new information can be performed by **magnetic field modulation (MFM)** when the bias field is appropriately switched from one direction to the other between sequential laser pulses during the writing process. This method is applied in the well-known MiniDisc (MD), which operates at a frequency of 44.1 kHz. For computer applications, however, the required high frequency is not compatible with the MFM technique. In the first step, therefore, the information is erased along the track by continuous laser heating in the presence of a constant bias field. In the next step new information is written by the thermomagnetic process and, finally, read out for verification. The overwrite process, therefore, requires at least two rotations of the disc which reduces the average access time. Attempts have been made to obtain a direct overwrite (DOW) by appropriate layer systems (see Sec. 4).

A cross section of the **magneto-optical disc** is shown in Figure 3. A polycarbonate substrate 1.2 mm thick with grooves on one side is formed by injection moulding. The pregrooved surface is then coated with the storage medium by sputter deposition. Finally a protective layer about 10 μm thick is prepared by the spin coating of a lacquer. In order to obtain double capacity often two discs are glued together with the storage layers in the middle of the disc. The *land and groove structure* is necessary for focusing and tracking of the optical head during disc rotation. The magnetic information is normally written on the land areas of the medium. The track pitch in the present MiniDisc, for instance, is 1.6 μm and the width of the land area is 1.2 μm. The depth of the groove is approximately 100 nm. As the light is focused through the substrate onto the track, the diameter of the laser beam at the disc surface is of the order of 1 mm. Small contaminations on the substrate surface, for instance dust, particles or scratches, do normally not disturb the signal. The MO discs, therefore, only need to be protected mechanically by a cartridge and can be handled in a normal atmosphere as easily as a CD or CD-R.

An example of a **storage medium** is shown in more detail in Figure 4, [2]. The essential part is a very thin magneto-optical layer (thickness typically 25 nm) of TbFeCo, where the magnetic domains are formed. Because the rare earth Tb is very sensitive to corrosion and oxidation the MO layer is protected from both sides by layers of Si_3N_4 which prevent diffusion of water vapour from the polymer substrate or from the lacquer into the magneto-optical layer. In addition, the SiN layers act as antireflective coatings and improve the read-out signal. A further increase of the signal is achieved by an Al reflecting layer because the reflected beam can pass through the MO layer twice.

The layer stack is prepared by **sputter deposition** (Figure 5). The injection-moulded substrate is loaded into a vacuum chamber and transported sequentially to different sputtering positions. The layer stack may therefore consist of up to twelve different layers. The deposition of thicker layers, for instance Si, is often performed in more than one chamber in order to keep the cycle time constant for each position. The productivity of this industrial sputtering system is of the order of 500 discs per hour. The concept of sequential sputtering provides a high flexibility for the design of the multilayered film system. In the arrangement of Figure 5, for instance, a layer sequence different from that of Figure 3 is produced. It contains exchange coupled double layers of TbFeCo and GdFeCo, which enable an optical resolution beyond the diffraction limit (see Sec. 4.3).

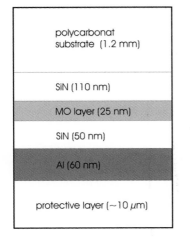

Figure 4: Layer stack of a typical MO disc [2].

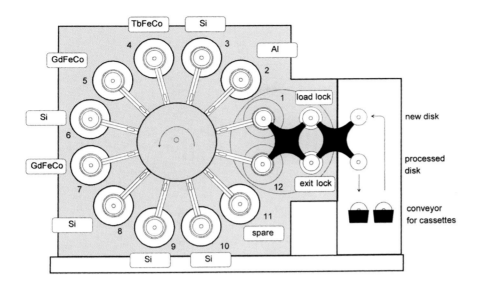

Figure 5: Sputtering system for the deposition of magneto-optical layer stacks (courtesy of Balzer's Process Systems).

3 Material Properties

The *polar Kerr effect* used for the read-out procedure is sensitive to the magnetization component perpendicular to the film plane. In magnetic thin films, however, the magnetization M_s is usually parallel to the film plane due to the high s*hape anisotropy* $(1/2)\,\mu_0 M_s^2$. In order to obtain a complete perpendicular orientation an intrinsic *perpendicular anisotropy*, characterized by the anisotropy constant K_u, must exist which is larger than the shape anisotropy:

$$K_u \geq (1/2)\,\mu_0 M_s^2 \tag{1}$$

The intrinsic perpendicular anisotropy is a property of only a selected group of magnetic materials, for instance some Co alloys, whereas the magnetization M_s can be reduced by the composition. An appropriate combination of both properties has to be found to fulfil Eq. (1).

This can be achieved by amorphous films consisting of **rare earth (RE)** and **transition metal (TM)** components (see article by Gambino in [2]). A typical example is $Tb_{75}Fe_{25}$ where the Tb and the Fe elements can be partially substituted by other RE or TM atoms, for instance Gd, Dy… or Co, Ni…, respectively. Because the films have an antiferromagnetic coupling between the RE and the TM components, the *net magnetization* M_s may be small in spite of the fact that the magnetizations of the RE and TM subnetworks are high. As the temperature dependencies of the two subnetwork magnetizations are different often a *compensation temperature* exists where the net magnetization is even zero (Figure 6). At this compensation temperature T_{comp} the coercivity of the material becomes infinite because an external field cannot turn the magnetization in any given direction. This is important for the stability of the domains. At a high temperature near the *Curie temperature* T_C the coercivity is sufficiently small to form a domain by the external bias field H_b. During cooling to room temperature the coercivity increases rapidly and, as a consequence, a written domain will not be disturbed by any external stray field.

An example for different TbFeCo films is shown in Figure 7, [6]. Within a small range of composition the compensation temperature changes from zero to a value above the Curie temperature. For the stability of the domains a compensation temperature near room temperature is favourable, which can be obtained by a Tb content of approximately 25 %. The composition also influences the Curie temperature T_C but in a much smaller range. A Curie temperature near 200 °C is convenient for the writing process. For a given Tb content the Curie temperature can be slightly shifted to higher values by replacing Fe by Co [5].

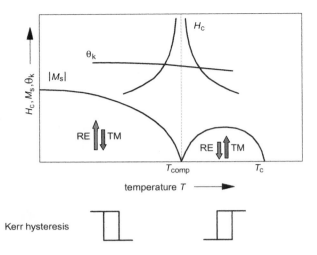

Figure 6: Magneto-optical properties of amorphous rare earth (RE) and transition metal (TM) thin films (M_s: saturation magnetization; θ_K: Kerr rotation; H_c: coercivity) [3].

The Kerr effect required for read-out is related to interband and intraband transitions of the RE/TM alloy. For wavelengths between 620 and 780 nm, which are commonly used for read-out, the main contribution comes from the 3d electrons of the transition metals. The *Kerr rotation* θ_K, therefore, is mainly proportional to the magnetization of the TM subnetwork. In the temperature range near the compensation temperature, it varies only slightly in spite of the fact that the net magnetization changes dramatically. An external field, however, acts on the net magnetization, which changes from a RE-dominated behaviour below the compensation temperature to a TM-dominated behaviour above the compensation temperature. Hysteresis loops determined by MOKE, therefore, change their sign, but not the magnitude of θ_K when the temperature passes the compensation temperature T_{comp} (Figure 6).

The *amorphous film structure* is necessary to keep the noise level low during rotation of the disc. A very important contribution to noise comes from irregularities of the written domains. They depend on the composition as well as on the writing conditions. In Figure 8 this is demonstrated for some test structures observed in Tb/Fe multilayers. Domains which are formed spontaneously normally have an irregular shape (Figure 8a). By using laser pulses of 20 μs pulse lengths and various power levels circular domains can be written in a film having $H_c = 170$ kA/m. Depending on the power level the diameter can be varied between 5 μm and 1 μm (Figure 8b). The regularity of the domains also depends on the magnitude of the bias field during the writing process (Figure 8c). For a pulse length of 10 μs and a power level of 1.8 mW the field must be above 15 kA/m otherwise a granular substructure is observed within the domains. In Figure 8d a domain structure is shown which is obtained by *magnetic field modulation*. A circular domain can be partially overwritten when the laser spot is slightly shifted and the bias field is reversed from one direction to the opposite one. The result is a track of crescent-like domains. It is obvious that a high recording density along the track can be obtained in spite of a relatively large diameter of the individual domain. This type of recording is applied in the present generation of the MiniDisc.

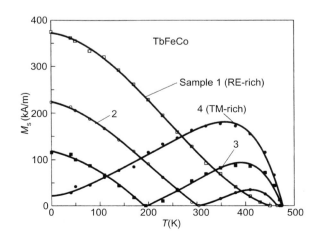

Figure 7: Magnetization M_s of four TbFeCo films with different compositions; sample
(1): $Tb_{29,9} Fe_{62,6} Co_{7,5}$;
(2): $Tb_{27,2} Fe_{65,5} Co_{7,3}$;
(3): $Tb_{23,6} Fe_{67,6} Co_{8,8}$;
(4): $Tb_{21,2} Fe_{71,9} Co_{6,9}$ [6].

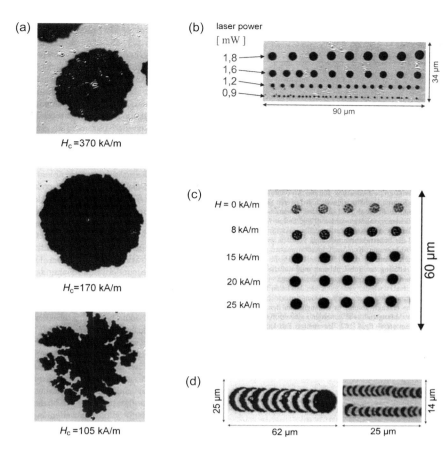

Figure 8: Magnetic domains in Tb/Fe multilayers observed by Kerr microscopy (λ = 680 nm) [7].
(a) Spontaneously formed domains in films with different coercivities H_c.
(b) Thermomagnetic writing of domains in the film with H_c = 170 kA/m by variation of the laser power at a pulse length of 20 µs.
(c) Thermomagnetic writing of domains by variation of the bias field H_b at a laser power of 1.8 mW and a pulse length of 10 µs.
(d) Demonstration of domains written by magnetic field modulation.

A basic limitation for the recording density is the **optical diffraction**. The diameter D of the diffraction pattern produced by the objective lens during writing and reading is given by

$$D = 1.22 \, \lambda / NA \qquad (2)$$

where λ is the wavelength of the radiation (at present between 780 nm and 620 nm) and NA is the numerical aperture of the objective lens of the optical head. The writing process is not so critical because the size of the written domain depends mainly on the relation between the applied bias field H_b and the temperature distribution of H_c produced by the laser pulse (cf. Figure 8b). By appropriate conditions domains smaller than the optical limit D can be created. The main problem comes from the reading process because it is not possible to distinguish very small domains at a distance below ($D/2$). It is assumed that the problem can be overcome in the near future by magnetically induced superresolution (MSR) or related techniques (see Sec. 4). A more straightforward solution would be the application of a green or blue laser instead of the present red and infrared lasers. In the near future semiconductor lasers will be available with wavelengths of 428 nm (blue) which would allow recording density to be increased by a factor 3.8 by a linear reduction of the dimensions. However, for the presently available RE/TM media the figure of merit drops rapidly with decreasing wavelengths. The most promising candidates for green and blue laser radiation are *Co/Pt or Co/Pd multilayer structures* with a typical composition 25 × (0.4 nm Co + 1.9 nm Pt) [2] – [4]. Although laboratory experiments have been successfully performed it is an open question as to how a large number of discs with such a complicated structure can be produced on an industrial scale.

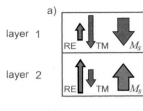
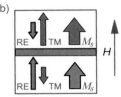

Figure 9: Subnetwork magnetizations of the RE and TM components and net magnetization M_s of exchange coupled double layers (ECDLs) with antiparallel coupling;
(a) ground state without external field;
(b) saturation in an external field H [3].

4 Application of Exchanged Coupled Layers

4.1 Basic Principles

As already explained the standard MO discs have two drawbacks: The limited access time due to the erasing process and the limited recording density due to the optical diffraction. In both cases exchange coupled layer systems can overcome the restrictions. They allow laser intensity modulation for direct overwrite (LIM-DOW) and magnetically induced superresolution (MSR) [2] – [4], [7] – [9]).

An exchanged coupled double layer (ECDL) with *antiparallel coupling* is shown in Figure 9. It consists of two layers with different compositions where layer 1 is TM-dominated and layer 2 is RE-dominated. In the ground state the exchange coupling of the two RE subnetworks, as well as of the two TM subnetworks, is very strong. As a consequence the net magnetizations are opposite to each other (antiparallel coupling). In a sufficiently high saturation field H the net magnetizations become parallel and the sub-network magnetizations are opposite to each other. Therefore, an *interface wall* with *wall energy* σ_w is formed. When the coercivity of both layers is sufficiently high this situation may also be stable in an opposite external field. By switching of one layer the wall disappears and the wall energy is released.

Switching of a single uncoupled layer would take place at the coercivity field H_c, and the energy per area would change by the amount $2 M_s H_c d$, where d is the thickness of the film. In the exchanged coupled double layer, in addition, the wall energy σ_w is released by or required for the switching process. The energy for switching one layer in a double layer stack, therefore, is given by $W_S = 2 M_s H_c d \pm \sigma_w$. The switching does not take place at the coercivity H_c but at the *switching fields*

$$H_s = H_c + \sigma_w/(2 \cdot M_s \cdot d) \quad \text{or} \quad H_s = H_c - \sigma_w/(2 \cdot M_s \cdot d) \tag{3}$$

The sign (+) is related to the creation of a wall, and the sign (–) to wall annihilation. Figure 10 shows the switching fields of an ECDL with antiparallel coupling measured by MOKE from both sides of the film. Layer (1) is TM-dominated and layer (2) is RE-dominated. For the measurement, the external field was increased from negative saturation (initial state) to positive saturation (final state) at different temperatures. Note that below a transition temperature T_s firstly layer (2) switches and then layer (1) whereas above T_s the switching takes place in the opposite sequence.

4.2 Laser Intensity Modulation for Direct Overwrite (LIM-DOW)

A behaviour as shown in Figure 10 can be used to obtain LIM-DOW in ECDLs, which was first proposed by Saito (see article by Saito in reference [2]). An appropriate film system consists of a TM-dominated layer (1), which acts as a memory layer, and an RE-dominated reference layer (2) (Figure 11). By means of this combination new information can be written into the memory layer simply by applying *laser pulses with a high or low power level*. The energy for switching comes partially from the release of the wall energy, which is triggered by the laser pulse.

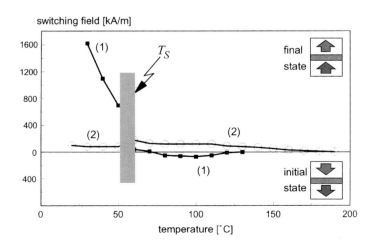

Figure 10: Switching fields of an ECDL with antiparallel coupling as a function of temperature [3].

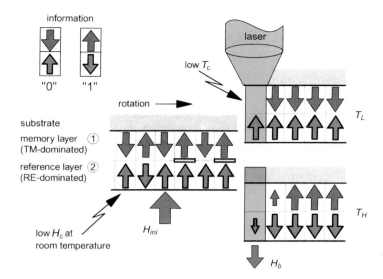

Figure 11: Principle of direct overwrite by laser intensity modulation (after Saito [2]).

In the ground state there are two magnetic configurations which can be attributed to the information "0" or "1" (Figure 11). During disc rotation the domain structure first runs through an area where an initializing field H_{ini} is present at room temperature. In this area the RE-dominated reference layer (2) can be saturated by the magnetic field because of the lower switching field at room temperature (cf. Figure 10). This process creates an interface wall in those domains which carry the information "1". A low-level laser pulse then leads to the temperature T_L, which is near to the Curie temperature T_C of the TM-dominated layer (1). As a consequence, the exchange energy forces layer (1) into a magnetic configuration which represents the information "0".

The high-power laser pulse, on the other hand, leads to a temperature T_H near to the Curie temperature of the layer (2), which is higher than that of layer (1) (cf. Figure 10). Therefore, the information in both layers is destroyed and first formed again in layer (2) by the bias field H_b. During further cooling the exchange coupling leads to an orientation of layer (1) which corresponds to the information "1".

As a consequence, the low-power process leads to the new configuration "0" and the high-power process to the information "1", irrespective of the former configuration. Switching field diagrams as shown in Figure 10 can be considered as phase diagrams which define stable magnetization configurations of the ECDL. They can be used to find appropriate bias fields and temperature ranges to perform a reliable high-power or low-power process [9].

4.3 Magnetically Induced Super Resolution (MSR)

The resolution of conventional optical instruments is limited by diffraction as explained by Eq. (2). However, this limit can be overcome by apertures, the diameter of which is much smaller than the optical wavelength. When the aperture is shifted over the sample in a very close distance the transmitted intensity allows the detection of details below the optical limit. This technique, for instance, is applied in the scanning near field optical microscope (SNOM). ECDLs allow the principle of *near-field optics* to be also transferred to the magneto-optical read-out process. This is the basic idea of magnetically induced super resolution (MSR).

As an example MSR by **central aperture detection (CAD)** is shown in Figure 12. The information is stored in the memory layer TbFeCo by magnetic domains much smaller than the diffraction spot of the laser. A read-out layer consisting of GdFeCo has an in-plane magnetization at modest temperatures, but may switch to a perpendicular magnetization in a temperature range above T_1. The domain structure of the memory layer is masked by the read-out layer in the area where the temperature is below T_1 because the polar Kerr effect is only sensitive for perpendicular magnetization. If the maximum temperature T_2 is adjusted appropriately only a very small area of the read-out layer has a perpendicular anisotropy. Within this area the information of the

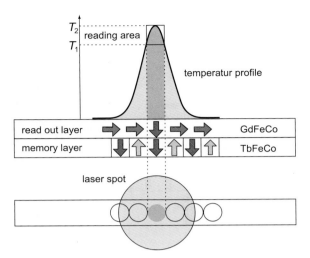

Figure 12: Principle of magnetically induced superresolution (MSR) by central aperture detection (CAD) [7].

TbFeCo film is copied into the read-out layer by exchange coupling and can be detected by the reflected light. The central part of the laser spot, therefore, forms a very small aperture which is shifted during the disc motion over the information stored in the memory layer. The intensity modulation of the reflected light corresponds to structures which are smaller than the diffraction limit of the laser beam.

In practice, the MSR technique explained in Figure 12 has to be modified due to the fact that the temperature distribution is elliptical rather than circular because of the motion of the disc. This leads to *front aperture detection (FAD)* or *rear aperture detection (RAD)*, which were initially proposed by Japanese authors in 1991 (see articles by Kaneko in [2], [4]; qualitative descriptions are given in [3], [8]).

5 Summary and Outlook

Magneto-optical recording plays an important role among the various optical storage techniques. The storage media are thin layers of amorphous ferrimagnetic alloys of rare earth (RE) and transition metal (TM) materials. A typical composition is $Tb_{25}Fe_{75}$, where Tb and Fe can be partially replaced by other RE and TM materials. The magneto-optical film system normally consists of several different layers, the composition and the thickness (typically 10 to 100 nm) of which have to be adjusted to the requirements for optimum recording, such as high density, thermal stability, stability to corrosion etc. The layer stack is prepared by sputter deposition on grooved polymer substrates. Exchange coupled layers (ECDLs) as part of the film system can help to improve the access time by direct overwrite (DOW) or increase the recording density beyond the optical diffraction limit by means of magnetically induced superresolution (MSR).

At present a 5.25 inch double-sided disc has a capacity of 5.2 GBytes. Compared with the first generation put on the market in 1988 this is the eightfold capacity. Major progress was achieved by a systematic reduction of the linear dimensions of the bit area. The track pitch, for instance, was reduced from 1.6 μm to 0.85 μm. Other contributions came from better coding methods and slightly smaller wavelengths. Further improvements can be expected from land and groove recording (factor 2) and, in particular, from the various MSR techniques (factor $\geq$ 2). In addition, there are other concepts for improving the resolution above the optical standard, for instance by *magnetic amplifying MO systems (MAMOS)* or by *solid immersion lenses (SIL)* and *near-field recording (NFR)* [2], [4], [7], [8]. A significant improvement step will be made when the blue laser is finally available and the magneto-optical materials are adjusted to new specifications required by the small wavelength.

Acknowledgements

The editor would like to thank Simon Stein (FZ Jülich) for checking the symbols and formulas in this chapter.

References

[1] M. Mansuripur, *The Physical Principles of Magneto Optical Recording,* Cambridge Univ. Press, Cambridge, 1995.
[2] R. J. Gambino and T. Suzuki, *Magneto Optical Recording Materials,* IEEE Press, New York, 1999.
[3] K. Röll, in *Magnetic Multilayers and Giant Magnetoresistance*, ed. U. Hartmann, Springer Verlag, Berlin Heidelberg, 2000.
[4] M. Kaneko, in *Magneto-Optics,* eds. S. Sugano, N. Kojima, Springer Verlag, Berlin Heidelberg, 2000.
[5] P. Hansen, in *Handbook of Magnetic Materials*, ed. K. H. J. Buschow, Elsevier, Amsterdam, 1991.
[6] F. J. A. M. Greidanus, B. A. J. Jacobs, J. H. M. Spruit, and S. Klahn, IEEE Trans. Magn. MAG **25**, 3524 (1989).
[7] S. Knappmann, *Untersuchung von Schichtsystemen für die magnetooptische Datenspeicherung mit Hilfe der Kerr-Mikroskopie,* PhD thesis, Kassel, 2000.
[8] S. Becker, c't **25**, 190-195 (1998).
[9] S. Becker, H. Rohrmann and K. Röll, J. Magn. Magn. Mater. **171**, 225 (1997).

Rewritable DVDs Based on Phase Change Materials

Matthias Wuttig

Institute of Physics, RWTH Aachen University, Germany

Contents

1 Introduction and Principle of Phase Change Media   645

2 The Velocity of Phase Transformations   647
2.1 The Driving Force of Phase Transformations   647
2.2 Nucleation and Growth   647

3 Requirements for Phase Change Media   649

4 Present Status of Phase Change Materials   652
4.1 Write Process   652
4.2 Erase Process   652

5 Concepts to Improve Storage Density   653

6 Phase Change Random Access Memory (PCRAM)   654

7 Summary   656

Rewritable DVDs Based on Phase Change Materials

1 Introduction and Principle of Phase Change Media

Optical data storage is believed to replace or complement magnetic data storage in several important areas. For example, it is expected that within a short period of time the DVD (Digital Video Disc or Digital Versatile Disc) will replace the VCR (videocassette recorder) since it offers higher data transfer rates, i.e. better video quality, faster access and lower storage space requirements combined with better longevity. Nevertheless the DVD does not yet have the full functionality of the magnetic hard disk, i.e. it still lacks fast rewritability, i.e. the potential to read, write and erase information at very high speeds. In industry labs the next generation of optical drives is already under development. The next formats presumably will be DVRs, with a storage capacity of 9.5 and 22.5 GB, respectively, employing red and blue laser diodes. In such drives the minimum bit length will be around 0.25 μm while only a few nanoseconds have to be sufficient to read, write or erase a bit. It is clear that this imposes very stringent requirements on possible materials that could be used for rewritable optical data storage.

In this manuscript we will address the question whether rewritable optical storage will be able to fulfil all requirements of modern mass storage devices and replace magnetic storage media. At present there are two different optical storage technologies which enable rewritable optical data storage. These are magneto optic media (see Chap. 25) and phase change media, the subject of this chapter. The principle of rewritable storage media based upon phase change recording is displayed in Figure 1.

Phase change recording [1] is based upon the reversible transformation of small regions of an active layer between the crystalline and amorphous state (Figure 1). The crystalline film is heated rapidly by a focussed laser beam and melts locally. After the laser beam is turned off, the film is cooled rapidly (cooling rate > 10^9 K/s). This leads to a drastic reduction of atomic mobility which is negligible at room temperature. Hence the atoms do not return to the stable, crystalline state but are trapped in a metastable, amorphous state. Figure 2 shows the result of a structure investigation of the crystalline and amorphous state [2].

While a majority of materials can be amorphized by a short, intense laser pulse only few materials show a very pronounced difference in optical properties between the amorphous and crystalline state. This is depicted in Figure 3. Since the optical properties of the amorphous and crystalline state differ significantly (see Figure 3), different material states can be detected by a weak laser beam.

Figure 3 shows the index of refraction n and the extinction coefficient k as a function of wavelength. It is clearly visible from this figure, that both quantities differ significantly for the amorphous and crystalline state. While the crystalline state behaves more

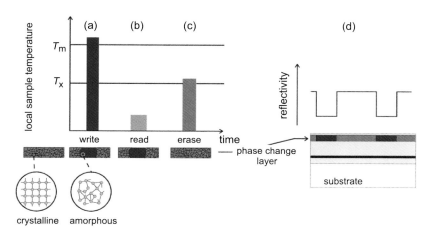

Figure 1: Principle of data storage with phase change media: To write a bit, the crystalline layer is locally molten. Recrystallization is prevented by the very high cooling rate (10^9 K/s), 'freezing' the disorder of the melt, which leads to an amorphous region. The amorphous and the crystalline areas have rather different optical properties, leading to a difference in reflectance ((b) and (d)). By a laser beam of intermediate intensity the amorphous region is heated above the glass transition temperature, so that recrystallization can take place and the bit is erased (c).

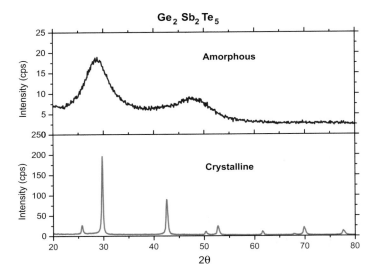

Figure 2: x-ray diffraction spectrum of a thin film of phase change material ($Ge_2Sb_2Te_5$) after deposition (amorphous) and after heating to 200°C (crystalline).

like a semi-metal and has only a very small band gap, the amorphous phase has a much larger band gap and hence a lower extinction coefficient. These different optical properties lead to a strong contrast in both reflection and transmission. Only a limited number of materials shows such a pronounced difference between the amorphous and crystalline state. Materials which show this peculiar feature will be thoroughly discussed later.

The large difference in optical properties is an essential material characteristic since it enables the error-free reading of information. The information is encoded in the length of amorphous regions which are located on a concentric spiral. To read, write and erase bits the intensity of the laser diode has to be adjusted. This is displayed in Figure 1, where the resulting local sample temperature is denoted. The sample temperature is highest when data are written (amorphization), while much lower intensity illumination is sufficient to read the information. The amorphous regions of the material can be recrystallized upon laser illumination. The raise in temperature increases the mobility of the atoms and allows structural rearrangements into the stable crystalline phase. The thin film will recrystallize if the mobility of the atoms is increased sufficiently upon annealing. This erases previously written bits.

As described in the introduction, the main challenges for phase change recording are ongoing improvements in data transfer rate and storage density. Hence the next chapter is devoted to a discussion of the velocity of phase transformations. Subsequently the materials will be introduced which enable fast phase transformations. Finally technologies will be presented allowing to achieve optical data storage with very high storage densities.

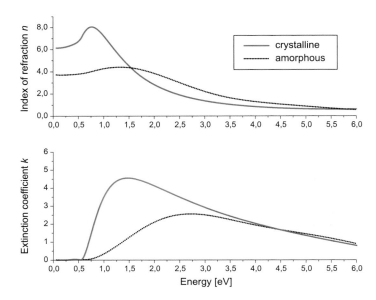

Figure 3: Index of refraction n and extinction coefficient k for amorphous (dashed line) and crystalline (solid line) $Ge_2Sb_2Te_5$.

2 The Velocity of Phase Transformations

2.1 The Driving Force of Phase Transformations

To achieve high data transfer rates, materials need to be identified which show fast phase transitions. This is one of the major conditions phase change materials have to fulfill. A phase transformation can only proceed if this reduces the free enthalpy G (or Gibbs free energy) of the system. The driving force of a phase transition is the difference ΔG of the free enthalpy of the two phases. The velocity of the phase transformation, however, will be strongly influenced by the height of the activation barrier W_A, which has to be surmounted for the transformation to proceed (see Figure 4).

The free enthalpy will be determined from the free energy H, temperature T and entropy S:

$$G = H - TS \qquad (1)$$

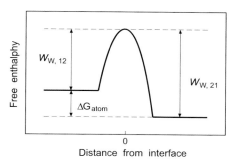

Figure 4: Change of the free enthalpy G of the system upon a phase transition from phase 1 to 2. While $W_{W,12}$ denotes the activation barrier for a transition from phase 1 to phase 2. $W_{W,21}$ characterizes the activation barrier for the reverse transformation, i.e. a transition from phase 2 to phase 1. Both activation barriers differ by ΔG, the free enthalpy difference between phase 1 and 2.

In a first approximation we will neglect the temperature dependence of S and H of a given phase. Then $G(T)$ is a straight line with slope $-S$. Figure 5 shows the free enthalpy of a material with liquid, crystalline and amorphous phase. T_m denotes the melting point and T_G the glass transition temperature. At this temperature the viscosity of the glass is 10^{13} $Pa\,s$ and a transition between an undercooled liquid and a solid amorphous state proceeds.

At the melting point, $\Delta G = 0$ and hence $\Delta S = \Delta H / T_m$, where ΔH denotes the latent heat of melting. This leads to a driving force for the transition crystalline → liquid:

$$\Delta G = \Delta H \frac{T_m - T}{T_m} \qquad (2)$$

For the transition amorphous → crystalline one obtains:

$$\Delta G = \begin{cases} \Delta H \dfrac{T_m - T}{T_m} & T > T_G \\ \Delta H_{ac}\left[1 - \dfrac{T}{T_G}\left(1 - \dfrac{\Delta H}{\Delta H_{ac}}\dfrac{T_m - T_G}{T_m}\right)\right] & T \leq T_G \end{cases} \qquad (3)$$

H_{ac} describes the exothermic energy of the transformation from the amorphous to the crystalline state [13]. The equation for temperatures $T < T_G$ is of no relevance in the following, since at these low temperatures the transformation proceeds extremely slowly. This is desirable since amorphous bits could otherwise already be erased by thermal fluctuations without laser irradiation.

2.2 Nucleation and Growth

The preceding paragraph has discussed the kinetics of phase transformations based upon very general concepts. The specific process that needs to be optimized in phase change materials employed for optical data storage is the recrystallization, i.e. the formation of crystallites in a previously amorphous surrounding. Crystallization consists of nucleation and growth. If we want to achieve high crystallization rates we hence need to understand the relevance of both processes for the class of materials studied here.

To form a nucleus with a different phase not only the energy gain ΔG but also the energy to produce the phase boundary ΔG_{IF} and the elastic energy ΔG_E have to be considered. The total energy change upon the formation of a new phase is then:

$$\Delta G_{total} = \Delta G + \Delta G_{IF} + \Delta G_E \qquad (4)$$

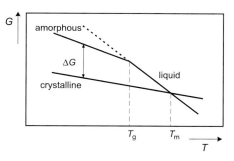

Figure 5: Temperature dependence of free enthalpy G.

In the following we will neglect the elastic energy since this term will not affect the general shape of the phase transition.

For a nucleus with volume V and area A, ΔG_{total} can be written as:

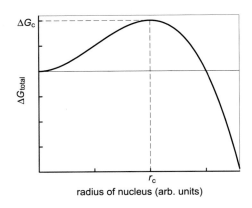

Figure 6: Energy required to form a nucleus with radius r.

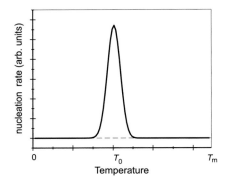

Figure 7: Temperature dependence of the nucleation rate.

$$\Delta G_{\text{total}} = \Delta G_V \cdot V + \gamma \cdot A \tag{5}$$

where G_V is the free enthalpy per volume and γ the specific interface energy. For a spherical nucleus with radius r we then obtain

$$\Delta G_{\text{total}}(r) = \Delta G_V \cdot \frac{4\pi}{3} r^3 + \gamma \cdot 4\pi \, r^2 \tag{6}$$

The dependence of ΔG_{total} upon the radius of the nucleus is displayed in Figure 6. For radii smaller than the critical nucleus r_c a further nucleus growth is unfavorable. For $r \geq r_c$ a further growth is energetically more favorable than the decay. Nuclei with $r < r_c$ are denoted as embryos in the following, nuclei with $r > r_c$ as nuclei and nuclei with $r = r_c$ as critical nuclei.

For the critical radius r_c ΔG_{total} reaches a maximum. From Eq. (6) one obtains:

$$r_c = -\frac{2\gamma}{\Delta G_V} \tag{7}$$

and

$$\Delta G_C = \frac{16\pi}{3} \cdot \frac{\gamma^3}{\Delta G_V^2} \tag{8}$$

The negative sign in Eq. (7) reflects the fact that only negative values of ΔG_V lead to a phase transition. ΔG_C represents an energy barrier that first has to be overcome for critical nuclei to form. ΔG_C is essential for the rate with which critical nuclei are formed.

By thermal fluctuations embryos constantly form and decay leading to an equilibrium nucleus size distribution

$$N(r) = N_0 \, e^{-\frac{\Delta G_{\text{total}}(r)}{k_B T}} \tag{9}$$

where $N(r)$ describes the number of nuclei with radius r, N_0 the total number of possible nucleation sites and k_B the Boltzmann constant. The number of critical nuclei is then

$$N_C = e^{-\frac{\Delta G_C}{k_B T}} \tag{10}$$

This number density is crucial for the nucleation rate, since the capture of one additional atom by a critical nucleus leads to a stable nucleus. The probability of an atom to jump across the interface and adhere to the critical nucleus is proportional to $\exp(-W_K/k_B T)$, where W_K is the activation barrier to cross the interface. According to the nucleation theory by Vollmer and Weber [3] and by Becker and Döring [4], the nucleation rate is given by $I = \alpha \, \exp[-(W_K + \Delta G_C)/k_B T]$, where α is a proportionality constant. While α and W_K are, in a first approximation, independent of temperature, ΔG_C is decreasing with increasing distance to the melting temperature T_M. This leads to a pronounced maximum of nucleation rate with temperature. The position and width of the maximum depend upon melting temperature, latent heat of melting, interface energy γ and activation energy W_K (Figure 7).

Once nucleation has taken place two phases coexist separated by an interface between them. The transfer of atoms through this interface is determined by the difference in free enthalpy of the two phases and the activation energy to overcome the interface. This is schematically depicted in Figure 4.

A rigorous derivation [4] of the growth rate leads to the following formula:

$$U = \xi \cdot n \cdot \nu \cdot p \cdot B \cdot V_{\text{atom}} \, e^{-\frac{W_{W21}}{k_B T}} \left(1 - e^{-\frac{\Delta G_{\text{atom}}}{k_B T}}\right) \tag{11}$$

where n denotes the areal density at the interface, p the probability of a place change to the other phase, B is the probability for an atom to remain in the new phase. ξ considers the fact that not all areas of the interface can adhere atoms, ν describes the attempt frequency, while ΔG_{atom} is the difference of the free enthalpies (per atom) between the two phases. The expression $\nu \cdot \exp(-W_W/k_BT)$ is proportional to $1/\eta$, hence the growth rate is related to the viscosity η. A schematic presentation of the temperature dependence of the growth rate is depicted in Figure 8. Again a maximum is clearly visible, but now this occurs at a higher temperature than the maximum for nucleation. Finally we want to add that for the materials discussed here the enthalpy difference ΔG_{atom} is much smaller than $W_{W,21}$ and $W_{W,12}$. They can hence be described by one quantity W_W which is identical to the activation energy for nucleation W_K.

Various models have been proposed to describe the interplay between nucleation and growth for isothermal phase changes. Most frequently a model by Johnson and Mehl [5] and Avrami [6] is used, which describes the transformed volume fraction χ as a function of time. Often experimental data are presented in the form of an exponential

$$\chi(t) = 1 - e^{-(Kt)^{n_A}} \qquad (12)$$

The parameters n_A and K are characteristic for the transformation considered. K is a rate constant while n_A is the Avrami exponent. Often the value of the Avrami coefficient already gives helpful information on the mechanism of the transformation, i.e. whether the transformation is diffusion controlled or not.

A compact presentation of the time and temperature dependence of isothermal phase transformations are TTT diagrams, where TTT stands for time, temperature and transformation. An example is shown in Figure 9 which displays the typical behavior. The nose shaped behavior is caused by the combined effect of the temperature dependent mobility and the temperature dependent driving force. At low temperatures the driving force is high but the atomic mobility is low, leading to a slow crystallization. At temperatures close to the melting temperature the mobility is high but the driving force is low. Again, this leads to a slow crystallization. Only at intermediate temperatures are the driving force and the mobility high enough to cause fast crystallization. The optimum temperature T^* lies between the temperature of maximum nucleation rate and the temperature of maximum growth rate.

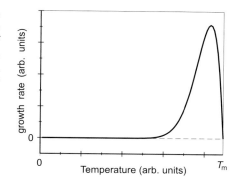

Figure 8: Temperature dependence of the growth rate.

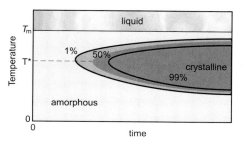

Figure 9: Schematic presentation of a TTT-diagram.

3 Requirements for Phase Change Media

Rewritable storage media have to fulfill five main data storage requirements. They have to enable writing of data (writability). The stored information has to be stable (archival storage) and easy to read (readability). Then the information should also be erasable (erasability) and the storage medium should allow numerous write/erase cycles (cyclability). These data storage requirements can be translated to media requirements (Table 1).

storage requirement	materials requirement	material property
Writability	easy glass former	melting point/layer design
archival storage	stable amorphous phase	high activation energy
Readability	large s/n ratio	high optical contrast
Erasability	fast recrystallization	simple crystalline phase, low viscosity
Cyclability	stable layer stack	low stresses

Table 1: Requirements for phase change media.

Writability implies that the material is an easy glass former. Desirable is a material with a relatively low melting point of around 500 °C. In order to reach such temperatures easily in the active layer it is desirable to incorporate the storage layer in a stack of layers that ensures that a maximum amount of light is absorbed in the phase change film. This layer design should also help to convert the power absorbed in the storage layer into heat. An important aspect of this layer design is the optimization of thermal properties. It is desirable, for example, to surround the active layer by dielectrics with medium or small thermal conductivity to obtain sensitive media where less than 5 mW/μm² are suf-

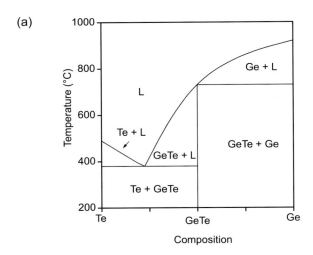
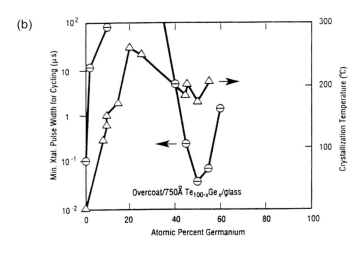

Figure 10:
(a) Phase diagram and
(b) minimum laser time for recrystallization and minimum temperature for recrystallization as a function of Ge content for Ge_xTe_{1-x} [1].

ficient to write a bit. For the stored information to be stable, we need to have a stable amorphous state, i.e. a high activation energy for recrystallization so that amorphous bits are stable for more than 10 years for media at room temperature. This implies that the activation energy for recrystallization is higher than 1 eV/atom. On the other hand fast erasure at elevated temperatures should be possible. This necessitates a better understanding of the recrystallization process and means to improve it. Such improvements should lead to an erasure time per bit of less than 10 ns. Such short times are essential for applications where high data transfer rates are mandatory. This holds for the recording of high definition TV signals as well as for computer based mass storage applications. The stored information should also be read easily. This is achieved if the material has a large signal-to-noise ratio. A large optical contrast and a low noise level help achieve this goal. Finally a large number of write and erase cycles should be possible.

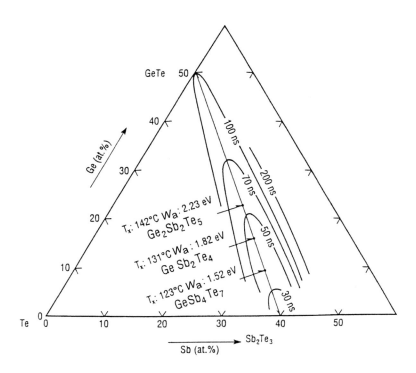

Figure 11: Minimum laser recrystallization time for various Ge-Sb-Te films with a thickness of 100 nm employing an 8 mW laser pulse [7].

The goal is to reach 10^7 cycles. This can be achieved by a stable layer design. The main problem are mechanical stresses linked with the phase transformation and the related volume change. It is important to note that the requirements of archival storage and fast erasibility are to some extent contrary requirements. Archivability requires a high activation energy for recrystallization at room temperature. For fast erasibility at elevated temperatures a high mobility is required. The best strategy to enable fast recrystallization is to ensure a short diffusion length. This is most likely found in systems which show no phase separation upon melting and crystallize into a single simple structure. Such a behavior is found in stoichiometric systems with low anisotropy, such as cubic crystal structures, for example.

Already in the seventies first experiments where performed to achieve optical data storage with tellurium compounds. Te was chosen since it is an easy glass former. Table 2 displays the history of phase change recording materials. The most favorable systems crystallize into a single phase and are stoichiometric. Table 2 shows that frequently Ge-Te compounds are used. This is illustrated in Figure 10a, where the phase diagram for GeTe is shown on the left hand side. The corresponding crystallization temperature as a function of composition and the minimum pulse length for laser recrystallization is shown as well (Figure 10b). Stable storage is only achieved if the crystallization temperature is above 150 °C. This value is reached for a Ge content of more than 20 %. However, only around 50 % Ge do we also see a short laser recrystallization. Even small deviations from the ideal composition of 50 % Ge lead to a much longer recrystallization time [1]. The reason becomes clear if we analyze the phase diagram. For compositions other than 50 % Ge, phase separation proceeds upon melting. Recrystallization is then connected with a long diffusion length and hence long recrystallization times. Hence the precise preparation of the right composition of $Ge_{50}Te_{50}$ is crucial. A less sensitive system, which is hence easier to produce, is found in the ternary GeSbTe system (Figure 11). The best properties are found along the pseudo-binary line of GeTe-Sb_2Te_3 [7]. Here small deviations from the ideal composition do not lead to such a pronounced increase in the minimum recrystallization time. Furthermore the atoms crystallize in a simple NaCl lattice where one sublattice is occupied by Te atoms and the second sublattice by Ge and Sb atoms. This pseudo-binary system is most frequently used for present optical data storage applications.

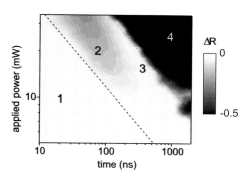

Figure 12: Power-time-effect diagrams for the amorphization process for a 80 nm thick $Ge_2Sb_2Te_5$ film. One can clearly distinguish the amorphization zone (zone 2), the melt crystallization zone (zone 3) and the ablative hole formation zone (zone 4).

Year	Composition
1971	Te-Ge-Sb-S
1974	Te-Ge-As
1983	Te-Ge-Sn-O
1985	Te-Sn-Se, Ge-Se-Ga
1986	Te-Ge-Sn-Au, Sb_2Se, In-Se, GeTe, Bi-Se-Sb, Pd-Te-Ge-Sn
1987	GeTe-Sb_2Te_3, ($Ge_2Sb_2Te_5$, $GeSb_2Te_4$), In-Se-Tl-Co
1988	In-Sb-Te, In_3SbTe_2
1989	GeTe-Sb_2Te_3-Sb, Ge-Sb-Te-Pd, Ge-Sb-Te-Co, Sb_2Te_3-Bi_2Se_3
1991	Ag-In-Sb-Te

Table 2: History of materials development for phase change media.

This table demonstrates that all phase change media employ Te or Se alloys. Te and Se alloys are well-known glass formers. Yet, in the last decade scientist have realized that while glass forming properties are an essential prerequisite for storage applications the crucial property of the media can best be described as 'bad' glass forming. To achieve fast crystallization rates materials have to be chosen that already recrystallize upon slow cooling. This requirement immediately excludes a vast number of typical glass formers.

4 Present Status of Phase Change Materials

4.1 Write Process

Using a focussed laser beam amorphous bits are written in a crystalline matrix [8]. Figure 12 shows the corresponding PTE-diagram (power, time, effect). P is the power of the laser beam, T is the pulse length of the laser and E is the effect of the laser pulse given as a normalized reflectance change $\Delta R_C = (R_A - R_C)/R_C$, where R_C is the reflectance of the crystalline material prior to the modification pulse and R_A is the reflectance after the modification pulse. Figure 12 shows that for low pulse powers no modification of the crystalline region is obtained, i.e. the material does not melt (region 1). Higher pulse powers and/or longer pulse lengths lead to a local melting and subsequent quenching into the amorphous state (region 2). The dashed line denotes the onset of amorphization at short times. A detailed data analysis shows that the straight line can be fitted by a power law with the pulse power necessary to write proportional to $t^{-1/2}$ [14]. Finite element calculations show that this power law describes an isotherm [14]. This implies that the material melts as soon as the melting temperature is reached. Superheating is apparently not observed down to 10 nanoseconds. In region 4 ablation is observed, which leads to a strong reduction in reflectance. Again the isoeffect line for ablation can best be described by a power law of the form $P \propto t^{-1/2}$. Again, this implies that instantaneous evaporation proceeds once the evaporation temperature is reached. Interesting is also the behavior in region 3. Here no change in optical properties is observed, even though the material has locally been molten. The reason for this behavior is melt-crystallization. If high powers are absorbed in the storage layer and its environment, fast heat diffusion and hence rapid cooling after the laser is turned off is no longer possible. Upon cooling the amorphous bit already recrystallizes and no change of optical properties is observed. Hence only a small parameter window of pulse powers and pulse times can be used for amorphization. Important for the fast storage of data is the fact that 10 ns are already sufficient to melt the storage layer. Further increases in the available pulse power and/or the sensitivity of the storage layer and media stack should even reduce the required powers further.

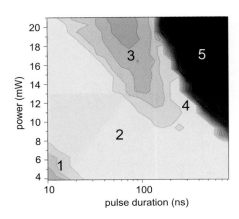

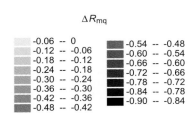

Figure 13: PTE-diagram for the crystallization of melt-quenchend amorphous bits (erase process). Complete erasure (zone 2) is possible in 10 ns.

4.2 Erase Process

PTE-diagrams can also be used to characterize the erasure process, i.e. the crystallization of melt quenched amorphous bits in a crystalline film. Figure 13 shows the change in reflectance after the partial erasure of amorphous bits as a function of power and duration of the erase pulse. The fixed write parameters for the amorphous bits led to a 60 % reduction of reflectance relative to the maximum possible reduction which would be reached if the whole area probed had become amorphous. The partial erasure of the bits leads to a partial recovery of reflectance. The difference of the reflectance before the write pulse and after the erase pulse characterizes the efficiency of the erase pulse. The PTE-diagram of the erase process (Figure 13) exhibits 5 zones. Erase pulses in zone 1 are insufficient to completely crystallize the amorphous bit. The reflectance after the erase pulse (R_3) is therefore smaller than the original reflectance (R_1). Thus $\Delta R_{mq} = R_3 - R_1$ is negative which is represented by increasingly darker gray values. Complete erasure of the bits is achieved in zone 2. The gray value of this zone corresponds to $\Delta R_{mq} = 0$, i.e. $R_3 = R_1$. In zone 3 the erase pulses lead to a local melting. Subsequent quenching results in a reamorphization of the film. R_3 is smaller than R_1 (negative ΔR_{mq}, darker gray value). Zone 4 corresponds to melt crystallization with $R_3 = R_1$ and hence $\Delta R_{mq} = 0$ as in zone 2. The pulses in zone 5 finally lead to ablative hole formation with large negative values for ΔR_{mq}.

From the PTE-diagram in Figure 13 it is clear that complete erasure of melt quenched amorphous bits (zone 2) can be achieved with pulses as short as 10 ns. This is considerably shorter than the minimum incubation time (100 ± 10) ns which was found for the crystallization of the as deposited amorphous film. Since the incubation time is needed for the production of critical nuclei it is clear that the nucleation process is not important for the crystallization of melt quenched amorphous bits. There are two possible reasons for this finding [9]. Either the crystalline phase starts to grow from the crystalline rim of the amorphous bits towards the center or the crystalline phase starts to grow from crystalline nuclei in the volume of the amorphous bits.

In a first attempt to answer this questions the efficiency of a certain erase pulse as a function of the size of the amorphous bit has been examined (data not shown). No dependence on the bit size was found which indicates that the crystalline phase does not start to grow from the rim of the bit [9]. Moreover the erasure of melt quenched amorphous bits always leads to the formation of the rock salt structure independent of the actual structure of the surrounding crystalline $Ge_2Sb_2Te_5$ which can either show hexagonally close packed (hcp) or also rock salt structure, as shown previously [2]. This finding also does not support the scenario of the growth of the crystalline phase from the rim towards the middle, since then an hcp-surrounding should lead to the growth of an hcp-phase. For $Ge_2Sb_2Te_5$ the scenario of the growth of quenched-in crystalline nuclei is much more likely. The preferred formation of nuclei with rock salt structure during the quenching of the melt should be due to the higher symmetry and smaller complexity of the unit cell of the rock salt structure as compared to the hcp phase.

For the application of $Ge_2Sb_2Te_5$ films for data storage applications it can hence be concluded that writing and erasure is possible within 10 ns. Experiments are under way to demonstrate if even faster data transfer rates are feasible.

5 Concepts to Improve Storage Density

Besides the improvement in data transfer rates an increase of storage density is crucial to ensure the future success of optical recording. Storage densities are rapidly approaching 50 Gbit/in². This is a serious challenge for optical recording. In the optical far-field no structures can be read or written which are considerably smaller than $1.22 \cdot \lambda / 2NA$ where λ is the wavelength of the laser and NA the numerical aperture of the lens used to focus the laser onto the sample [11]. At present objectives with a numerical aperture of 0.5 and 0.6 are employed. In the near future optical heads with an aperture up to 0.8 will find usage. However this would not allow to create structures much smaller than ¾ λ. With the solid state lasers presently used, which have a wavelength of 780 nm, storage densities of approximately 1 bit/µm² can be reached only. This is presently state of the art. A further increase in storage density can then be reached only if the wavelength of the laser diode is reduced further. The next generation of optical storage systems will employ InGaAlP laser diodes with a wavelength of 635 nm. A further step in storage density will be enabled by blue laser diodes working at around 400 nm. However, even with such a laser diode, which is not yet available with sufficient power and long lifetime for an acceptable price, storage densities of around 4 bit/µm² can be reached at best. Hence new concepts need to be developed to reach higher storage densities.

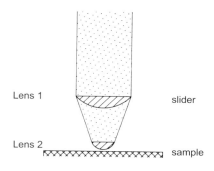

Figure 14: Optical head using a solid immersion lens [10] which focuses a laser beam onto the sample surface.

One such concept employs an immersion lens. The principle is schematically depicted in Figure 14. A collimated laser beam is focussed by a first lens. The light is then directed into a second lens with high index of refraction. In the close vicinity of the second lens the evanescent field of light still has components with the wavelength the light had inside the lens, which is λ/n, where n is the index of refraction. This enables a reduction in bit dimension by a factor of n and an increase in storage density by a factor of n^2. For dielectric materials a high index of refraction of 2.5 can be reached. This would enable a more than 6-fold increase in storage density.

Another interesting approach to achieve higher resolution is near-field microscopy. In this technique a light source with an aperture which is smaller than the wavelength λ is brought into close proximity of the sample (~λ/10). Scanning of the light source over the sample allows to image the sample [11]. Using this concept which is called scanning near-field optical microscopy (NSOM or SNOM) a resolution of 12 nm has been achieved. As a light source a tapered fiber tip was used. The tip of the fiber is coated by an opaque metal such that light can only leave the tip through a small aperture at the tapered tip. Such fiber tips have already been employed to write small domains into magneto-optic films. While the spatial resolution of the fiber tips is very high the transmission through the aperture is very low (10^{-4}). The intensity of the light coupled into the fiber is limited since absorption and hence heating of the metal film could lead to a melting of the metal and hence destruction of the fiber tip. Therefore only a small photon flux is available for the characterization and modification of material. The light intensity in the near field should however be higher by two orders of magnitude. This improves the chances to modify phase change media by near field techniques. Nevertheless the distance regulation of a near-field microscope is rather slow. This is a further obstacle for efficient and fast storage and manipulation schemes employing such light sources. Both drawbacks of conventional near field microscopes can however be circumvented by a different near-field light source design [12]. This light source which has been developed by Bell Labs, Lucent Technologies, is presently commercialized. The basic principle of the light source is displayed in Figure 15. An edge emitting laser

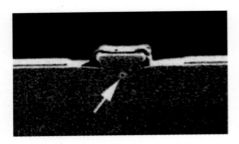

Figure 15: Focussed ion-beam (FIB) image of the metallized front facet of the very-small-aperture laser, VSAL. The white arrow shows a 200-nm-square aperture etched into the metal layer by the FIB [12].

diode is coated by a metallic film which acts as the mirror for the laser resonator. With a focussed ion beam a small aperture is etched into the metal mirror. Light will leave the diode through this aperture. If one brings such a light source in close proximity of the sample, material modification is possible. The main advantage of this concept are the high output powers compared with optical fibers. The far field powers of such light sources are up to 4 orders of magnitude larger than the ones from tapered fiber tips. Optimized designs and shorter wavelength laser diodes should enable a further increase in output powers.

A further advantage of the very-small-aperture laser (**VSAL**) concept is that the laser diode can even read bits in phase change media. The storage layer is practically part of the resonator mirror. A change in reflectance of the phase change film will change the resonator properties and lead to a voltage change across the laser diode. Hence a fast electronic circuit can quickly pick up this signal.

6 Phase Change Random Access Memory (PCRAM)

Optical non volatile data storage is based on the optical contrast between the amorphous and the crystalline state. Another not surprising, but remarkable feature is the pronounced difference in resistivity between the amorphous and the crystalline state. This feature is mainly based on the lower carrier-mobility and the smaller carrier concentration in the amorphous phase.

Due to the fact that the electrical resistance is a feature which can be read out very easily electronically, this makes phase change media a prime candidate for a Random Access Memory (RAM) as described in Part IV of this Textbook.

Figure 16 shows a van der Pauw Measurement of the sheet resistance of a Phase Change film which is heated in a furnace. After heating the film above the crystallisation temperature the resistance changes by three orders of magnitude.

Heating the phase change media can also be done by a current flow, which makes it possible to write information electronically into a bit. The principle is the same as the one used for optical data storage, a short strong pulse to amorphise the bit, and a longer less intense pulse to crystallise it.

The main problem for amorphisation is to obtain high cooling rates of 10^9 K/s. To achieve this the surface to volume ratio has to be high enough, which is achieved by making the switched volume very small.

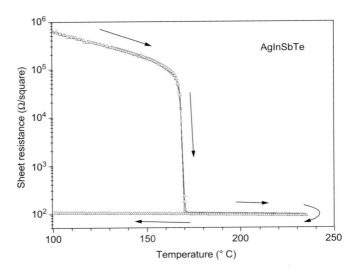

Figure 16: Sheet resistance of a PCM film at different temperatures before and after crystallisation [15].

Figure 17 shows the structure of a basic cell design. The Phase Change Material is stacked between two metallic contacts. It is noteworthy that the contact area to the bottom electrode has to be quite small in order to minimize the switched volume. In this picture only the part of the phase change media which is close to the bottom electrode is switched, due to the higher current densities there.

In Figure 18 the resistance of such cell is shown as a function of the current of the programming pulse. By applying a current below 0.1 I/I_{reset} the resistance of the memory cell does not change. This enables a **non destructive read** (a feature which is not found in DRAM).

After a current pulse above 0.6 I/I_{reset} the resistance of the cell will independently of its initial state end up in the low resistive state. By applying pulses above I_{reset} the system always changes to the high resistive state. So independently of the starting resistance the end state only depends upon the applied current: This enables a direct overwrite and is superior to the switching of FLASH memory where one has to delete the memory before it can be rewritten.

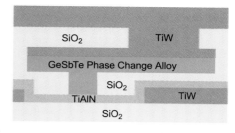

Figure 17: Simple planar structure to investigate basic device physics [16].

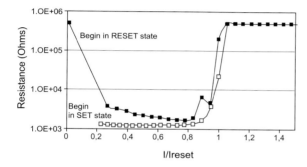

Figure 18: Device resistance vs. applied programming current pulse [16].

To produce a complete memory chip with these cells, only few extra electronics are required. In Figure 19 a schematic diagram of a memory array is shown. Each cell can be read or written by setting the corresponding bitline (BL n) to 0 V (selecting all transistors of this bitline) and sending a read, set or reset current through the corresponding wordline (WL m).

Within a memory chip this design is expected to have fast read and write speeds of ~10 ns at an operating voltage of <1V.

In contrast to other non volatile memory like FLASH where extra charge pumps on the chip are required, the PCRAM needs only very few extra electronics on the chip. Another positive feature is the high storage density, which in conjunction with the easy integration into CMOS manufacturing makes PCRAM a cost-effective technology. At the publishing date of this Textbook, > 10^{12} SET/RESET cycles have been demonstrated without any significant degradation [17].

PCRAM provides a *non volatile memory* with a high signal to noise ratio. It is an easy to manufacture, low cost storage technology with a high storage density. This makes PCRAM a promising candidate for future applications in new PDAs (Personal Digital Assistants) and new UMTS Applications. Therefore today several major chip manufacturers are investigating this data storage technique. Even though the concept is quite simple, there still remain some challenges. First of all there is the quest for faster materials with low writing powers and suitable resistances. Fortunately universal scaling laws predict, that with decreasing cell size, e.g. going from a 180 nm process to a 130 nm process, the cell will become faster and the switching power will diminish. However, at the same time the transistor size decreases which reduces the available switching current. And last but not least there is the question of the life time of such a cell: the frequent switching leads to a build up of stress which might destroy the cell. Here the electrode / phase change media combination is of major interest to scientists and engineers. For materials scientists this is a tremendous opportunity to combine interesting basis research with potentially very rewarding applications.

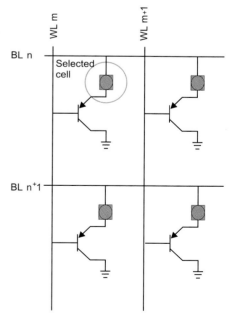

Figure 19: Simple array structure to use PCM-cells in a memory chip.

7 Summary

Phase change materials at present are very promising materials for optical data storage in multimedia applications. They allow inexpensive, reliable and fast storage of vast amounts of data. Nevertheless considerable improvements in materials understanding and optimization as well as improvements in optical recording technology are necessary to compete with ingoing improvements of storage rate and density in magnetic hard disks.

In addition, phase change materials are utilized for non-volatile random access memories based on the difference in resistivity between the crystalline and amorphous phase.

Acknowledgements

I am indebted to Henning Dieker, who wrote the paragraph about the PCRAM. Furthermore, it is a pleasure to acknowledge my co-workers which include R. Detemple, I. Friedrich, J. Kalb, W. Njoroge, D. Wamwangi, V. Weidenhof, H.W. Wöltgens and S. Ziegler. The editor would like to thank Ralf Liedtke (RWTH Aachen) for checking the symbols and formulas in this chapter.

References

[1] M. Libera and M. Chen, MRS Bulletin, 40 (1990).
[2] I. Friedrich, V. Weidenhof, W. Njoroge, P. Franz and M. Wuttig, J. Appl. Phys. **87**, 4130 (2000).
[3] M. Vollmer and A. Weber, Z. Phys. Chem. **119**, 277 (1925).
[4] R. Becker and W. Döring, Ann. Phys. **24**, 719 (1935).
[5] W.A. Johnson and R.F. Mehl, Trans. A.I.M.E. **135**, 416 (1939).
[6] M. Avrami, J. Chem. Phys. **7**, 1103 (1939).
[7] N. Yamada, MRS Bulletin, 48 (1996).
[8] V. Weidenhof, N. Pirch, I. Friedrich, S. Ziegler and M. Wuttig, J. Appl. Phys. **88**, 657 (2000).
[9] J.H. Coombs, A.P.J. Jongemelis, W. van Es-Spiekman and B.A.J. Jacobs, J. Appl. Phys. **78**, 4906 (1995).
[10] S.M. Mansfield and G.S. Kino, Appl. Phys. Lett. **57**, 2615 (1990).
[11] E. Betzig et al., Appl. Phys. Lett. **61**, 142 (1992).
[12] A. Partovi et al., Appl. Phys. Lett. **75**, 1515 (1999).
[13] C. Peng, L. Cheng and M. Mansuripur, J. Appl. Phys. **82**, 4183 (1997).
[14] V. Weidenhof, N. Pirch, I. Friedrich, S. Ziegler and M. Wuttig, J. Appl. Phys. **88**, 657 (2000).
[15] W. K. Njoroge and M. Wuttig, J. Appl. Phys. **90**, 3816 (2001).
[16] www.ovonyx.com .
[17] M. Gill, T. Lowrey, and J. Park, TD 12.4, Proc. ISSCC, February 4-6, 2002.

Holographic Data Storage

Mirco Imlau, Physics Department, University of Osnabrück, Germany

Thomas Bieringer, Central Research, Bayer AG, Germany

Serguey G. Odoulov, Institute of Physics, National Academy of Sciences, Kiev, Ukraine

Theo Woike, Institute of Mineralogy, University of Cologne, Germany

Contents

1 Introduction — 659

2 Fundamentals of Holographic Information Storage — 659

3 Optical Processes — 662
3.1 Writing of a Holographic Grating — 662
3.2 Read-Out of a Holographic Grating — 663
3.3 Mismatch of Bragg Incidence — 664
3.4 Concepts for Multiplexing — 665
3.5 Geometries for Writing — 666
3.6 Bit Error Rate (BER) — 667
3.7 M# Number — 668

4 Inorganic Materials — 669
4.1 Fundamentals — 669
4.2 Iron-Doped Lithium Niobate ($LiNbO_3$:Fe) — 670
4.3 Photorefractive Effect — 670
4.3.1 One-Center Model — 670
4.3.2 Charge Transport Mechanism — 670
4.3.3 Space Charge Formation and Refractive Index Modulation — 671
4.3.4 Thermal Fixing — 673
4.3.5 Holographic Data Storage in $LiNbO_3$:Fe — 674
4.4 Outlook: Pulse Holography — 675

5 Photoaddressable Polymers — 676
5.1 Introduction — 676
5.2 Photochemistry of Azobenzene — 676
5.3 Azobenzene-Containing Polymers — 677
5.4 Liquid Crystalline Side Chain Polymers — 677
5.5 Photoaddressable Polymers for Optical Storage — 677
5.6 Holographic Measurements with Photoaddressable Polymers — 681

6 Outlook — 682

Holographic Data Storage

1 Introduction

As mentioned before, many research activities focus on the development of new concepts and materials to increase the storage density and the data transfer rates. In 1997, read-only media like CD-ROM and CD-AUDIO (650 MByte, which is sufficient for 75 minutes of music) held a market share of 90 % of the exchangeable optical storage media. Now they are being replaced by two new media: the CD-R (recordable) and CD-RW (rewritable), which permit one or many writing cycles. Both disks incorporate several additional functional recording layers.

The digital versatile disc (DVD) with its increased capacity of 4.7 GBytes (which corresponds to a 135 min Hollywood movie) will probably soon reach a level of acceptance similar to the CD today. "Rewritable DVD is shaping as the pervasive PC storage system of the next three or four years", as David Firth from Hewlett Packard stated in September 1999. Nevertheless, more powerful storage technologies than the DVD are already under development.

One new technology is based on volume holographic data storage. Recently, significant improvements have been demonstrated in the development of the hardware for holographic data storage devices, including low-cost components [1] and new storage materials [2]. Further progress in the field of holographic data storage is expected from the development of superior hardware, as fast electronic drivers for spatial light modulators, and from new recording materials. New results have been presented by the National Storage Industrial Consortium (NSIC), funded by the Defense Advanced Research Projects Agency (DARPA). Finally we may say that all the hardware needed for holographic data storage has been demonstrated – more than 50 years after the discovery of the principles of holography by Dennis Gabor. Gabor was far ahead of his time, indeed.

The most important requirements for the storage materials are good optical quality, high light sensitivity at the desired light wavelength, preferentially at the Nd:YAG laser wavelength of 532 nm, wide dynamic range, high spatial resolution, ease of processing, physical and chemical stability under operating conditions. In addition, for a rewritable memory, the storage process must be reversible. Later in this chapter, two distinct classes of holographic storage materials will be discussed: the inorganic materials and the organic photoadressable polymers. In the first sections of this chapter, an introduction is given to the basics of holography and holographic storage.[1]

1) An excellent extensive description of holographic data storage is given by H. J. Coufal, D. Psaltis, G. T. Sincerbox [3].

2 Fundamentals of Holographic Information Storage

History

The principles of holography were discovered in 1948. Dennis Gabor found that both phase and amplitude of a wavefront can be recorded by the interference of two coherent waves [4]. The first application of his discovery aimed to improve the quality of X-ray images. However, the widespread application of recording and retrieving images only became possible after the development of lasers in the 1960s. A first system describing holographic information storage was established by van Heerden in 1963 [5]. Van Heerden discussed the recording of multiple holograms within the same volume element and several holograms were stored by Leith in a photographic plate in 1966 [6]. In 1973, the angular multiplexing technique was applied to holographic information storage for the first time [7].

Today, 40 years later, a holographic information storage system is not yet commercially available. Most of the necessary technological components, as lasers, cameras and two-dimensional liquid crystal displays, have been around for a long time, but only in the last few years has a breakthrough in performance and cost been attained. Still, there is an ongoing search for better storage materials, since they have to fulfill many conflicting properties, as for example high optical sensitivity together with high optical transparency. During the last four decades, a very diverse spectrum of developments has been promoted [8] - [23].

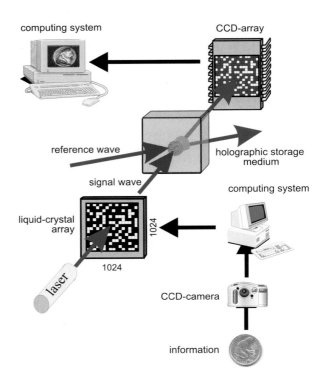

Figure 1: Basic setup of a system for holographic information storage in transmission geometry. The information is transferred into a two-dimensional bitmap, which is projected onto the holographic storage medium via a laser system. Superposition with the reference wave leads to an interference pattern which can be stored. The reconstruction of the information is realized by illumination of the stored interference pattern. The holographic picture is restored. The bitmap on the CCD array is read out by a computing system and the stored information is retrieved.

Basic setup

Figure 1 shows the basic setup of a holographic information storage system and its components: charge coupled device (CCD), computing system, 2D-liquid crystal array, laser and holographic storage medium. The green arrows denote the direction of the laser beams, the black ones the logical flow.

As an example see the hologram of the German "50 pfennig" coin. It is already a historical document after the introduction of the "euro" in 2002. First the picture of the coin is digitized. This can be done with a standard camera device (CCD) producing a TIFF file. The file is read into a computing system, where it is transferred into a format which can be processed by the liquid crystal display (LCD): the file is divided into parts of about 1 Mbit and the bits are arranged into arrays of 1024×1024. The arrays, or "data pages", can be reproduced on the liquid crystal by setting the pixels to transparent for the "1"s or opaque for the "0"s.

The coherent light from a laser system projects each array onto a charge coupled device (CCD) having the same resolution as the LCD. The holographic storage medium is placed at the focus of the optical path of this beam, called the signal wave. A second laser beam with a homogeneous intensity profile, called the reference beam, is projected onto the storage medium having a different angle of incidence but overlapping with the first beam within the medium. Due to the coherence of the two beams they interfere with each other, which leads to an interference pattern of dark and light areas. This interference pattern has characteristic dimensions of a few micrometers and is unique for each data page. It is used to expose the photosensitive storage medium. The storage of the interference pattern is called "writing of the hologram".

The writing process has to be repeated for each new data page. Holography in 3D recording materials allows further holograms to be written into the same volume element after a slight rotation of the storage medium. This procedure is called "angular multiplexing" and is the key to achieve large storage capacities. Additionally, one can successively select and address different nonoverlapping or partially overlapping volume elements of a larger block of the storage medium. This is called "spatial multiplexing". Further types of multiplexing are described below.

The illumination of the written hologram by the reference beam leads to the restoration of the image and is used as the read-out process. The signal wave is diffracted by the hologram. The array of high and low bits re-appears on the CCD. For the read-out of specific data pages the angular and spatial positions have to be exactly identical to the writing configuration. All images are reconstructed in the same way by controlled rotation and translation of the storage medium. The final bitmaps are transferred from the CCD into a computer, where the graphic picture is restored – the "50 pfennig" coin is back.

A schematic setup of the optical system for holographic storage is shown in Figure 2. It uses a storage medium in the form of a flat disc, e.g. a photoadressable polymer. This configuration is of great interest, since many of the technological developments already established in the field of CDs and DVDs can be used.

System components and specifications

The necessary components for such a holographic storage system are: light source, holographic storage medium, rotation and translation stage, liquid crystal display, camera system and a computer. The light source and the holographic storage medium have to fulfill special requirements: as light source a laser is chosen, because an optical coherence length from at least centimeters up to meters is available. The choice of the laser depends on the output power and the wavelength needed for recording. Both are determined by the holographic storage material used. Argon ion lasers, frequency-doubled Nd-YAG lasers or diode-pumped frequency-doubled DBR lasers are common in most of the setups covering the blue-green spectral range with an output power of up to 5 W. Materials for holography have to satisfy many demands. Besides the compulsory requirement that a material has to be sufficiently transparent for the wavelength used, a high spatial resolution is needed. Compared to normal photographic emulsions with approximately 100 lines/mm, holographic films need a resolution of 5000 lines/mm because the modulation of the interference pattern is of the order of the wavelength of the light, i.e. a few hundred nanometers. Further, high optical sensitivity and a large dynamic range of the holographic storage media are important for the storage of a large amount of data. For each writing of the many holograms a short exposure time is desired. The dynamic range is the basis for a large storage density and will be introduced by the M#-number. Finally, the spectral photosensitivity of the material is an important parameter, since there are great differences in the operation and in the costs of the laser systems.

Today, liquid crystal displays and cameras are readily available. Liquid crystal displays are standard components in video beamers with constantly improving contrast, speed and resolution, now reaching 2048 × 2048 pixels. Important for the quality of the hologram is the contrast ratio between 0 and 1, which is typically in the range of 1:300 or more. CCD-arrays are available with up to 4096 × 4096 pixels.

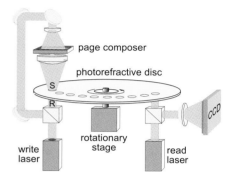

Figure 2: Schematic arrangement of a holographic storage system using a disc-like storage medium. The use of "holo-discs" is advantageous, since many developments in the field of CDs and DVDs can be used for this kind of holographic storage system. The page composer is also called a spatial light modulator, SLM.

Storage Capacity and Density

The ultimate density for optical recording at wavelength λ is given by $D = 1/\lambda^3$. However, the actual storage density of a holographic system will be lower and has to be estimated. Let us consider some common components and assume a transparent storage medium with dimensions of $50 \times 50 \times 3$ mm^3 and a diameter of the reference or signal beam of 1 mm. Each hologram contains 1 Mbit of data using a liquid crystal display with a resolution of 1024×1024 pixels. For angular multiplexing, the crystal is rotated by about 0.01 degrees after recording of each hologram. Therefore 1000 holograms can be written within a range of $\pm 5°$. Consequently 1000 Mbit of raw data can be stored within a volume element of about 2.4 mm^3. In addition, spatial multiplexing may be used. If the holograms are recorded with a spatial separation of 2 mm (from center to center), an array of 25×25 positions is possible. Recording 1000 holograms at each of these 625 positions will finally result in a capacity of 625×1000 Mbit, which is approximately 78 Gbyte. With respect to the complete volume of the storage medium we get a storage density of 10.4 Mbyte/mm^3. This realistic estimate is two orders of magnitude smaller than the ultimate density. In summary, the storage capacity C can be estimated by:

$$C = N_{LCD} \times N_{ROT} \times N_{SPAT} \qquad (1)$$

and the storage density by:

$$D = \frac{C}{w \times h \times d} \qquad (2)$$

whereby N_{LCD} denotes the number of pixels of the LCD, N_{ROT} the number of holograms recorded by angular multiplexing, N_{SPAT} the number of holograms recorded by spatial multiplexing, w, h are the width and height of the storage medium and d is the hologram thickness. For a further increase of the storage capactiy into the Tbyte regime one has to increase the resolution of the liquid crystal display, increase the angular range for recording holograms, reduce the angular distance between the holograms, reduce the beam diameter of the reference and signal waves, reduce the spatial distance between the holograms, use a gray scale LCD (not only 1 and 0 bits) and enlarge the size of the crystal. It has to be noted that N_{ROT} depends on the recording parameters of the storage system and the thickness of the storage medium, whereby N_{SPAT} depends on the size of the storage medium and the beam diameter. The total density of stored data pages will be reduced significantly, if there is an interference between different exposures ("hologram cross-talk") or diffuse light scattering from defects.

In conclusion, holographic systems provide two crucial advantages which lead to high storage density:

- **parallel storage**:
  the information is stored in parallel, not serially
- **multiplexing**:
  a multitude of data pages can be stored within one volume element, which is not possible using conventional storage techniques.

Writing and Reading Speed

Let us now estimate the writing and reading velocity of a holographic storage system in which the writing process is performed by a cw laser. For **writing** we assume a writing time of 1 ms for each hologram. We have to add the time which is necessary to rotate the crystal to the next angular position: about 5 ms for each movement. Therefore, we need $1000 \times (1 \text{ ms} + 5 \text{ ms}) = 6$ seconds for one volume element. To go to the next volume element we have to move the crystal or the reference and signal waves. In both cases we again need e.g. about 5 ms before we reach the desired position. In our previous example we estimated 625 spatial positions, so we have to additionally consider 625×5 ms for the positioning and 625×6 s for the writing. In sum, we get a total time of about 3753 seconds for the recording of about 78 Gbyte. This is equal to a total writing velocity of approximately 20 Mbyte/s. We can estimate the writing velocity by:

$$v_{\text{WRITE}} = \frac{C}{N_{\text{SPAT}}\left[N_{\text{ROT}} \times (t_{\text{HOLO}} + t_{\text{ANG}}) + t_{\text{POS}}\right]} \tag{3}$$

whereby $t_{\text{HOLO}}, t_{\text{ANG}}, t_{\text{POS}}$ denote the recording time of one hologram and the angular and spatial positioning time. To increase the writing velocity one has to decrease the recording time, increase the data content of each hologram and decrease the angular and spatial positioning times. The decrease of the recording time can be realized by illumination with short laser pulses.

For the **read-out velocity** we have to consider the time needed for the movement of the crystal to the desired angular and spatial position. Assuming the same values as used for the writing process and a camera system with about 100 frames per second, we can read one page within 10 ms. More than two and a half hours are necessary for the read-out of 78 Gbyte. This is equal to a read-out velocity of 8.3 Mbyte/s. We estimate the read-out velocity by:

$$v_{\text{READ}} = \frac{C}{N_{\text{SPAT}}\left[N_{\text{ROT}} \times (t_{\text{CCD}} + t_{\text{ANG}}) + t_{\text{POS}}\right]} \tag{4}$$

whereby t_{CCD} denotes the time needed to read out one hologram with the CCD array.

In conclusion, the main time demand for the read-out is given by the mechanical positioning of the equipment, and for the writing, in addition, by the time needed by the LCD array (SLM), given by t_{HOLO}. Obviously the processing capacity of the computer has to match the high data flow.

3 Optical Processes

3.1 Writing of a Holographic Grating

The simplest form of a hologram is a sinusoidal intensity distribution. Physically, this process corresponds to the creation of a diffraction grating, which is called an elementary holographic grating. By diffraction of a lightwave from this holographic grating, both the amplitude and phase of one of the waves used for recording are reconstructed [24].

As an instructive and important example, this section explains the writing of an elementary holographic grating. The elementary grating is recorded by the interference of two electromagnetic waves within the volume of the holographic storage medium of thickness d. The reference and signal waves with wavelength λ, intensities I_R and I_S and wavevectors $\mathbf{k}_i$ and $\mathbf{k}_j$, are superposed under the Bragg angle θ_B, as shown schematically in Figure 3:

The sinusoidal intensity modulation inside the volume of the recording medium is given by:

$$I(x) = (I_R + I_S)\left[1 + m\cos(|\mathbf{K}|x)\right] \quad (5)$$

In a photosensitive recording material, this results in a modulation of the refractive index $n(x)$ or the absorption coefficient $\alpha(x)$:

$$n(x) = n_0 + \Delta n \cdot \cos(|\mathbf{K}|x + \varphi_n) \quad (6)$$

$$\alpha(x) = \alpha_0 + \Delta\alpha \cdot \cos(|\mathbf{K}|x + \varphi_\alpha) \quad (7)$$

Δn and $\Delta \alpha$ denote the amplitude of the light-induced refractive index and the absorption change, respectively. Depending on the detailed microscopic processes, phase shifts φ_n and φ_α between the light intensity modulation and the refractive index or the absorption modulation will occur. The periodic variations of the refractive index and the absorption constant given by Eq. (6) and Eq. (7) are in fact the phase and amplitude hologram, respectively.

An elementary holographic diffraction grating with the grating spacing Λ is formed by $n(x)$ and $\alpha(x)$, respectively. The grating spacing Λ is connected to the grating vector $\mathbf{K}$ via the relation $|\mathbf{K}| = 2\pi/\Lambda$ and $\mathbf{K} = \mathbf{k}_i - \mathbf{k}_j$. The grating spatial frequency $|\mathbf{K}|$ depends on the wavelength of the writing beams and their relative angle. With respect to Bragg's law $\lambda = 2\Lambda\sin\theta_B$, we get $|\mathbf{K}| = 4\pi\sin\theta_B/\lambda$. The grating vector or hologram spatial frequency can be varied by the Bragg angle or by the laser wavelength. Typical values for Λ are in the range of 10^{-6} m. (Note that all angles and wavelengths are considered with their actual values within the storage material. The numbers for propagation in vacuum are not to be used.)

The diffraction of an unmodulated reference wave from the holographic grating reconstructs the signal wave. The amplitude of the reconstructed wave depends on the amplitude of the refractive index modulation Δn and the modulated absorption $\Delta \alpha$, respectively.

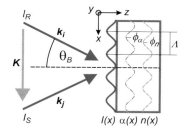

Figure 3: Reference and signal beam with intensities I_R and I_S and wavevectors $\mathbf{k}_i$ and $\mathbf{k}_j$ superpose in the volume of the storage medium under the angle θ_B. The resulting sinusoidal intensity modulation $I(x)$ is transferred into a modulation of the refractive index $n(x)$ and absorption coefficient $\alpha(x)$. Phase shifts ϕ_n and ϕ_α with respect to the incoming light interference pattern, respectively, occur depending on the physical process. K is the grating vector of the resulting grating with grating spacing Λ.

3.2 Read-Out of a Holographic Grating

The diffraction efficiency describes the intensity of the reconstructed signal wave. To measure this parameter quantitatively, the hologram is illuminated with the reference beam with the intensity I_i. In the case of a transmission hologram a part of the transmitted reference beam with the intensity I_0 and a diffracted beam with the intensity I_1 is observed behind the hologram, as shown in Figure 4.

The diffraction efficiency is defined as the ratio of the diffracted intensity to the incoming intensity:

$$\eta = \frac{I_1}{I_i} \quad (8)$$

The diffraction efficiency depends on both external parameters, e.g. the Bragg angle, and on material parameters, as the amplitude of the modulation of the dielectric constant. An expression for the diffraction efficiency as a function of these parameters was derived by Kogelnik in 1969 for thick holographic gratings that obey the Bragg condition [25], [26]. From this "coupled wave theory" the diffraction efficiency η for a mixed (phase and amplitude) holographic grating in an absorbing material is given by:

$$\eta = \exp\left(\frac{-\alpha \cdot z}{\cos\theta_B}\right) \times \left[\sin^2\left(\frac{\pi \cdot \Delta n \cdot z}{\lambda \cos\theta_B}\right) + \sinh^2\left(\frac{\Delta\alpha \cdot z}{4\cos\theta_B}\right)\right] \quad (9)$$

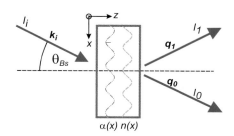

Figure 4: Diffraction of the incoming wave I_i incidenting under the Bragg angle θ_B from the sinusoidal grating $n(x)$ and $\alpha(x)$ leads to the diffracted wave I_1 besides the transmitted part I_0. The diffraction efficiency is defined as the ratio of the diffracted intensity to the incoming intensity.

We discuss the diffraction efficiency in order to evaluate the parameters which result in large intensities of the diffracted wave. The formula itself consists of two terms in square brackets, one describing the contribution to the diffracted beam from the phase grating (refractive index grating) and the other to the diffraction from the amplitude grating (absorption coefficient grating). A common exponential factor in front of the square brackets gives the exponential decrease of the beam intensity in the material which is due to absorption. It can be seen that both too large a thickness or too high an absorption negatively influences the diffraction efficiency. For a visualization of this important equation we have plotted the diffraction efficiency for a pure refractive index ($\Delta\alpha = 0$) and a pure absorption grating ($\Delta n = 0$) in Figure 5, using the material parameters of iron-doped lithium niobate (LiNbO$_3$:Fe).

Characteristic of the pure refractive index grating is the sinusoidal behavior of $\eta_{\Delta n}$. Note that a large index modulation Δn does not inevitably lead to a large diffraction efficiency $\eta_{\Delta n}$. In contrast, the diffraction efficiency of an absorption grating shows only one maximum, resulting from the competition between the sinus hyperbolicus and the exponential term in Eq. (9). In general, the ultimate diffraction efficiency of phase gratings is large compared to amplitude gratings and can reach 100 % in the limit of vanishing absorption. Equation (9) permits the material properties Δn and $\Delta\alpha$ to be calculated by a direct measurement of the incoming and the diffracted intensities for a pure phase and a pure amplitude grating, respectively.

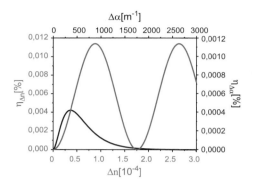

Figure 5: Diffraction efficiencies of a pure refractive index grating ($\Delta\alpha = 0$) and a pure absorption grating ($\Delta n = 0$) as a function of the amount of the refractive index and absorption modulation, Δn and $\Delta\alpha$, respectively. The diffraction efficiency of the refractive index grating shows a sinusoidal behavior, i.e. several maxima and minima are reached with increasing refractive index modulation. In contrast, the diffraction efficiency of the absorption grating clearly shows one maxium. Material parameters of iron-doped lithium-niobate were used.

3.3 Mismatch of Bragg Incidence

In the previous considerations we have assumed that the hologram is read out with the reference beam having the same angle of incidence and wavelength as used for the writing process, so that the Bragg condition $q_l = k_i - K$ is exactly fulfilled (see Figure 4). If the angle or the wavelength is varied by $\Delta\Theta_B$ or $\Delta\lambda_B$, the corresponding diffraction efficiencies $\eta(\Delta\theta), \eta(\Delta\lambda)$ of the thick holographic grating decrease dramatically. This mismatch of the Bragg condition is exploited for angular and wavelength multiplexing. A general solution of the wave equation, which includes the angular and wavelength mismatch of the incident beam with the wavevector k_i, leads to the following expression for the diffraction efficiency of a pure refractive index grating ($\Delta\alpha = 0$) [25]:

$$\eta = \frac{\sin^2\sqrt{\left(\frac{\pi \cdot \Delta n \cdot z}{\lambda \cdot \cos\theta_B}\right)^2 + \left(\frac{\vartheta \cdot z}{2 \cdot \cos\theta_B}\right)^2}}{1 + \left(\frac{\vartheta \cdot z}{2 \cdot \cos\theta_B}\right)^2 \bigg/ \left(\frac{\pi \cdot \Delta n \cdot z}{\lambda \cdot \cos\theta_B}\right)^2} \quad (10)$$

with the "off-Bragg" parameter:

$$\vartheta = \Delta\theta_B |K| \cos\theta_B + \frac{\Delta\lambda \cdot |K|^2}{4\pi n} \quad (11)$$

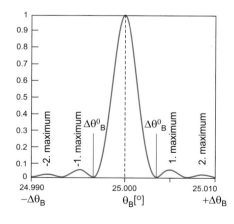

Figure 6: Angular dependence of the diffracted intensity with the first and second maxima and the first nulls. The maximum corresponds to Bragg incidence.

The "off-Bragg" parameter describes the angular deviation $\Delta\theta_B$ from the Bragg angle plus the offset of the read-out wavelength $\Delta\lambda_B$. Note that the diffraction efficiency as given by Eq. (9) is included in Eq. (10) for Bragg incidence ($\vartheta = 0$).

Now consider the mismatch of the Bragg condition, either by a variation of the read-out angle $\Delta\theta_B$ or the read-out wavelength $\Delta\lambda_B$ with respect to the initial Bragg values θ_B, λ_B. The dependencies $\eta(\Delta\theta_B)$ and $\eta(\Delta\lambda_B)$, respectively, are called the "rocking curve". As an example $\eta(\Delta\theta_B)$ is plotted in Figure 6 with the values $\theta_B = 25^0$, $\lambda = 532$ nm and $d = 3$ mm. The diffraction efficiency is normalized to the maximum at Bragg incidence $\eta(\theta_B)$.

Besides the strong maximum from zero-order diffraction at $\Delta\theta_B = 0$ (Bragg incidence), further side maxima appear due to higher order diffraction. The mismatch $\Delta\theta_B$ can be adjusted in such a way that the diffraction efficiency decreases to zero. In this case the illumination of the holographic material does not give any information about the already written hologram. In the present example this first null is reached at $\Delta\theta_B^0 = 0.0038^0$. An analogous behavior of the diffraction efficiency is revealed for a mismatch of the wavelength $\Delta\lambda_B$. The first nulls of the diffraction efficiency are reached at:

$$\Delta\theta_B^0 = \frac{\lambda}{2 \cdot z \cdot \pi \cdot \sin(2\theta_B)} \quad \text{and} \quad \Delta\lambda_B^0 = \frac{\lambda^2}{2 \cdot z \cdot \pi \cdot \sin^2\theta_B} \quad (12)$$

respectively. The quantities $\Delta\theta_B$ and $\Delta\lambda_B$ are called the angular selectivity and spectral selectivity of the holographic grating respectively. The values of the first null are strongly dependent on the wavelength used, the hologram thickness and the Bragg angle. In Figure 7, $\Delta\theta_B^0$ and $\Delta\lambda_B^0$ are plotted as a function of the Bragg angle θ_B for $\lambda = 532$ nm and $d = 3$ mm. The smallest value of the position of the first null is reached for an angular mismatch at $\theta_B = 45^o$ with an amount of $\Delta\theta_B^0 = 0.0016^o$ and for a wavelength mismatch at $\theta_B = 90^o$ with an amount of $\Delta\lambda_B^0 = 0.015$ nm.

3.4 Concepts for Multiplexing

In most cases, the maximum diffraction efficiency is reached if the written grating is read under identical angle θ_B and wavelength λ_B as used during writing. Then the Bragg condition is exactly fulfilled. If the read-out angle or the read-out wavelength is changed, the diffracted intensity decreases until the holographic image vanishes.

On the other hand, the loss of the diffraction efficiency resulting from an angular or wavelength mismatch is a most useful feature for dense holographic information storage, since it allows a multitude of holograms to be written within the same volume element: After having written one grating, the mismatch of the Bragg condition is optimized by varying θ_B or λ_B until the diffraction vanishes. Now another hologram can be written without any crosstalk from the previous one. This procedure is called "multiplexing". In this chapter different multiplexing techniques are presented. Multiplexing increases the storage density dramatically.

The **angular multiplexing** technique is based on the high angular selectivity of 3D gratings [6], [27], [28]. A multitude of holograms is written within the same volume element by a slight rotation of the storage material. After the recording of each hologram the storage medium is rotated by a specific angle. The angle is given by the first minimum of $\eta(\Delta\theta_B)$, as illustrated for two elementary holographic gratings in Figure 8.

Two maxima are visible referring to the two written elementary gratings #1, #2. If a read-out angle of θ_B^1 is chosen the diffraction efficiency is maximum for hologram #1 and zero for hologram #2 and vice versa. There will be no crosstalk between the holograms reading-out #1 or #2 in the maximum of the diffraction efficiency. The number of holograms N_{ROT} which can be written by angular multiplexing at a single location of the medium is given by a) the first null $\Delta\theta_B^0$ (Eq. (12)) and b) the maximum possible rotation range of the holographic material $\pm\theta_{max}$:

$$N_{ROT} = \frac{2 \cdot |\theta_{max}|}{\Delta\theta_B^0} + 1 \qquad (13)$$

Coming back to the example ($\lambda = 532$ nm, $\theta_B = 25°$, d = 3 mm) with $\Delta\theta_B^0 = 0.0038°$, one finds $N_{ROT} \approx 2632$ within a range of $\pm 5°$. Depending on the dynamic range of the photorefractive material, the ultimate number of multiplexed holograms is limited by the amount of diffraction efficiency of each hologram. The intensity of the diffracted beam has to be sufficient for an error-free read-out.

The angular multiplexing technique can be realized in different directions of rotation. The storage medium or the writing beams can be rotated. Three angular multiplexing techniques are shown in Figure 9: a) rotation around an axis orthogonal to the plane of incidence (in-plane), b) rotation around an axis parallel to the plane of incidence (out-of-plane) and c) rotation around an axis normal to the storage medium (peristrophic). However the rotations in b) and c) do not have a high selectivity compared with the rotation shown in a), especially in the case of inorganic non-centrosymmetric crystals.

The wavelength multiplexing technique is based on the wavelength mismatch of the diffraction efficiency [6]. A multitude of holograms is written into the same volume, using a slightly changing wavelength. After the recording of one hologram the wavelength is tuned by a specific amount, until the next hologram is recorded. The wavelength is tuned such that the first minimum of $\eta(\Delta\lambda_B)$ is reached, see Eq. (12). As in angular multiplexing, there is no crosstalk between the holograms. The number of holograms N_{WAVE} that can be written with this technique at a single location is given by the grating spectral selectivity $\Delta\lambda_B^0$ and the maximum range of tuning the wavelength $\pm\lambda_{max}$ of the laser:

$$N_{WAVE} = \frac{2 \cdot |\lambda_{max}|}{\Delta\lambda_B^0} + 1 \qquad (14)$$

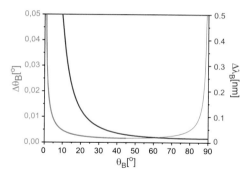

Figure 7: The first null of the angular and wavelength mismatch as a function of the Bragg angle.

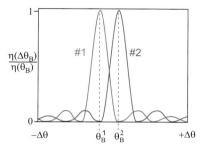

Figure 8: Rocking curve of two holographic elementary gratings #1, #2 with Bragg angles θ_B^1 and θ_B^2, respectively. The diffraction efficiencies of both curves are normalized to their values at Bragg incidence. It can be seen that there is no crosstalk between the gratings, if the angular distance between the gratings is chosen such that the Bragg angle of grating #2 is selected at the angle of the first null of grating #1 and vice versa.

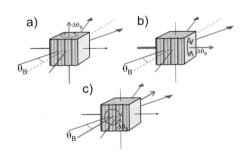

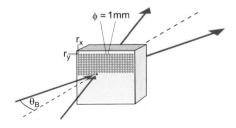

Figure 9: Three different rotation directions of the angular multiplexing technique:
a) rotation about an axis orthogonl to the plane of incidence,
b) rotation around an axis parallel to the plane of incidence (out-of-plane) and
c) rotation around an axis normal to the storage medium (peristrophic).

Figure 10: Spatial multiplexing technique. The crystal is divided in to volume elements, which do not overlap. In the present example the beam diameters of the reference and signal waves are chosen as 1 mm so that holograms can be written with a distance of 2 mm measured from center to center.

2) Geometries for retrieval: "image-plane geometry", "Fourier-plane geometry", "Fresnel-zone-geometry", "van der Lugt geometry" and geometries using phase-conjugated signal beams are summarized in [3].

Using the previous values ($\lambda = 532$ nm, $\theta_B = 25°$, $d = 3$ mm) for a reflection grating recording, i.e. $\theta_B = 90°$, where the minimum of $\Delta\lambda_B^0 = 0.015$ nm appears within the range of ± 10 nm, one finds: $N_{WAVE} \approx 1334$. There is no need for any mechanical rotation using the wavelength multiplexing technique. Therefore, this technique enables a storage system with fast write and read-out cycles. Unfortunately, tunable lasers with sufficient intensity, tunable spectral range and resolution are not commonly available. The prospects for wavelength multiplexing will therefore depend on the development of small tunable lasers.

Beside an angular and a wavelength detuning, the diffraction efficiency decreases by a spatial displacement of the read-out beam, which is used for the **spatial multiplexing** technique [29]. Different positions of the holographic medium are used, without overlap. This allows a large volume of the storage medium to be used, as shown in Figure 10:

It can be seen that the number of positions N_{SPAT} depends on the area of the medium (width w and height h) and on the dimensions of the reference and signal beams:

$$N_{SPAT} = \frac{w \cdot h}{4 \cdot r_x \cdot r_y} \qquad (15)$$

whereby r_x and r_y denote the radii of the beams in x- and y-directions of the medium. This means that the number of holograms increases quadratically by spatial multiplexing with increasing area and additionally quadratically with decreasing beam diameter. In comparison to angular multiplexing technique, where the number of holograms increases linearly with decreasing Bragg selectivity $\Delta\theta_B^0$, see Eq. (12), spatial multiplexing leads to especially high storage capacites.

The **shift multiplexing** method is similar to the spatial multiplexing method with the difference that the distance between two spatial positions is smaller than the beam diameter of the writing beams [30]. Shift multiplexing is possible within and orthogonal to the plane of the reference and the signal beam, conforming to "in-plane" and "out-of-plane" phase shifting. It can be realized using a high-resolution linear drive. To reduce the crosstalk betweeen the holograms, shift multiplexing is commonly combined with the angular multiplexing technique, i.e. every shift is combined with a slight rotation of the crystal.

A more complicated multiplexing technique is **phase coded multiplexing** [31]. This technique makes use of a spatial phase distribution of the reference beam. Although mostly electromagnetic waves with planar wave fronts of constant phases are assumed, because they allow a simple mathematical description of writing and reading-out of holograms, phase-coded multiplexing assumes a two-dimensional coded phase front for the writing of each hologram. The phase of the reference beam is planar, but spatially coded in values from 0 to 2π. As a consequence, the hologram can only be reconstructed if the read-out wave has the same two-dimensional coded phase front. For sufficient multiplexing the different phase patterns have to belong to a set of orthogonal codes. The limit of this multiplexing technique is given by the spatial resolution of the phase pattern and its phase resolution given by the amount of the set of orthogonal phase codes. The phase multiplexing technique can be realized by inserting a spatial light modulator into the reference beam.

In fact, a combination of several multiplexing techniques may be possible and will lead to the highest storage capacities.

3.5 Geometries for Writing[2]

The geometrical arrangements of the reference and signal beams with respect to the storage medium can be rather different. The basic arrangements result from the directions of the reference and signal beam impinging onto the storage medium. They are divided into: transmission geometry, 90° geometry and reflection geometry, as shown schematically in Figure 11.

In **transmission geometry** (Figure 11a) the reference and signal waves enter the storage material from the same side. The resulting interference pattern is characterized by the wavevector **K** parallel to the entrance plane. Bragg angles can be chosen in the range of $0° < \theta_B < 90°$. In order to get a high diffraction efficiency combined with a sharp angular selectivity one should use thick storage media and large Bragg angles. In **90° geometry** (Figure 11 b) both beams enter the storage medium at two orthogonal faces. Therefore, the grating vector is tilted by 45°. In this geometry the storage medium has to have a cubic form. The highest angular selectivity of the Bragg condition is realized and due to the beam diameters the thickness of the medium is large enough to get efficient gratings with a sharp selectivity. In **reflection geometry** (Figure 11c) both beams penetrate the storage material in counterpropagation, so that the grating vector is orthogonal to the entrance face of the storage medium. The Bragg angles lie in the range of $0° < \theta_B < 90°$. The reflection geometry provides optimal Bragg selectivity for wavelength multiplexing, especially for $\theta_B = 90°$. The thickness of the storage medium enhances the selectivity (see Eq. (12)).

The usable angular range of the different geometries is limited by the refractive indices of the storage medium. This is significant for materials with high refractive indices. Table 1 lists the accessible Bragg angles inside iron-doped $LiNbO_3$ ($n = 2.235$), which is an inorganic storage material.

	External Bragg angle	Internal Bragg angle
transmission geometry	$0° < 2\theta_B < 180°$	$0° < 2\theta_B < 53.16°$
$90°$ geometry	$-90° < 2\theta_B < 270°$	$36.84° < 2\theta_B < 143.16°$
reflection geometry	$0° < 2\theta_B < 180°$	$126.84° < 2\theta_B < 180°$

Table 1: Internal and external Bragg angles for iron-doped $LiNbO_3$ in the three main geometries.

Table 1 shows, that efficient recording of a multitude of holograms is restricted to transmission and 90° geometry for angular, and to reflection geometry for the wavelength multiplexing technique.

3.6 Bit Error Rate (BER)

In the first sections, the storage capacity of a standard holographic storage system has been estimated with regard to the writing process. This capacity gives the amount of data written into the medium, the so-called raw data capacity. However, for a fair comparison with the more common storage systems the actually usable storage capacity is of interest. That is the amount of data which can be retrieved from the storage system with tolerable errors. Several sources of errors have to be considered:

- pixel errors from the LCD
- pixel errors from the CCD
- misalignment of the optical systems between the reconstructed image of the LCD and the CCD (spatial or rotational)
- optical errors from the optical systems
- optical errors from the storage medium
- crosstalk between holograms
- noise from light scattering.

To estimate all read out errors, the bit error rate (*BER*) is introduced. The bit error rate is defined as the ratio of the number of errors $\#_{errors}$ to the total number of bits n in the limit of a very high number n:

$$BER := \lim_{n \to \infty} \frac{\#_{errors}}{n} \qquad (16)$$

In practice, the measure of the *BER* seems to fail since there is almost only a finite number of bits available. In fact, this error counting method is only useful if a sufficient number of errors can be detected within the finite amount of bits, i.e. the *BER* is large. However, the *BER* should be small in a storage system containing a large capacity, so that there is need for an estimation of the *BER* leading to a comparable result with respect to the error-counting method, but with a smaller amount of data. Such an estimation seems possible since some of the error sources include statistical errors, e.g. probability distributions of the 0 s and the 1 s resulting from the CCD. An inspection of the distribution of the revealed data from one hologram containing one data page makes this very impressive. Figure 12 shows the picture of one holographic data page stored in a crystal of $LiNbO_3$:Fe. The data page contains of 320×256 pixels of the LCD, whereby the frequency of high and low bits is equal. Additionally, an array of 22×22 pixels is shown in an enlargement.

The picture shows a two-dimensional distribution of pixels. The intensities of the 0 s and 1 s are not zero and 100 %, as might be expected. In contrast, it seems that there is a gray-scaling of the pixels. For the reconstruction of the data from the photograph the

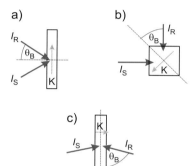

Figure 11: Geometries for the writing of holograms.
a) Transmission geometry – reference and signal beams irradiate the storage medium from the same side. The grating vector lies parallel to the surface of the medium.
b) 90° geometry – reference and signal beams enter the storage medium at two orthogonal entrance faces of the storage medium, commonly with an angle of 90° to each other. The grating vector encloses an angle of 45° with the surfaces of the medium.
c) Reflection geometry – reference and signal beams penetrate the storage medium in counterpropagation. The grating vector is orthognal to the entrance face of the medium.

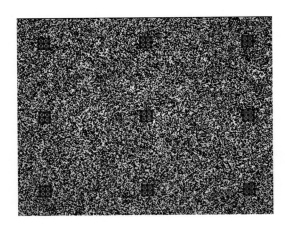

Figure 12: Photograph of a data page containing 320×256 pixels: whole page (left), magnification of 23×23 pixels (right). The page is taken with a CCD that has a resolution four times larger. High and low bits are equally distributed.

spatial distribution of the pixel intensities has to be assigned unambiguously to the spatial distribution of the low and high bits. This is quite doubtful since the contrast of the reconstructed image is lower, if compared to the initial page. However, the CCD allows the intensity of each individual pixel to be measured with 8 bit resolution, so that it is possible to plot the entire intensity distribution of all pixels measured. This is shown in Figure 13 for an area of 128 × 96 pixels.

This histogram shows impressively that the high and low bits are not reconstructed with an unambigous intensity, i.e. we get a statistical distribution with a characteristic maximum and width. For comparison the intensity distribution of the inital data page gives two characteristic delta functions for the 0 s and 1 s. This statistical distribution is the basis for the estimation of the *BER* with a finite number of data points, i.e. errors can be regarded as random events in a sufficiently large amount of data [32]. With this result the *BER* can then be approximated by a) a fit of two Gaussian functions to the distributions of the 0 s and the 1 s, and b) the determination of bits within the region of overlap between the two distributions, i.e. bits, which cannot be assigned unambiguously. The *BER* is then given by:

$$BER \approx \frac{1}{2\sqrt{2\pi}\sigma}\left[\int_0^{I_c} A_1 \exp\left(-\frac{(I-I_c)^2}{2\sigma^2}\right)dI + \int_{I_c}^{\infty} A_0 \exp\left(-\frac{(I-I_c)^2}{2\sigma^2}\right)dI\right] \quad (17)$$

where 2σ denotes the half width at full maximum and A_0, A_1 the amplitudes of the Gaussian functions. In conclusion, the *BER* can be estimated via the fit of two Gaussian functions to the flanks of the pixel distribution. From this, holographic storage in e.g. iron-doped lithium niobate gives a *BER* of $\approx 2.4 \cdot 10^{-6}$ due to its high optical quality [33].

3.7 M# Number

Recording of a multitude of holographic gratings is possible within the same volume element of the storage material by various multiplexing techniques. The upper limit for the number of holograms within one volume element is given by the dynamic range of the storage medium and the diffraction efficiency of each hologram. The dynamic range is described by the M number *M#* [34] taking various storage processes, techniques and materials into account. For equal time constants for the writing and erasure processes *M#* is expressed for pure phase gratings by:

$$M\# = \frac{\pi \cdot \Delta n \cdot d}{\lambda \cdot \cos\theta_B} \exp\left(-\frac{\alpha \cdot d}{2}\right) \quad (18)$$

Δn denotes the saturation value of the light-induced change of the refractive index, d the hologram thickness, λ the recording wavelength, θ_B the Bragg angle, and α the material absorption. Therefore, *M#* defines the largest possible light-induced modulation in the recording medium. It is limited by the particular features of the recording process (e.g. space-charge limitations in oxidic photorefractive materials) and the thickness of the

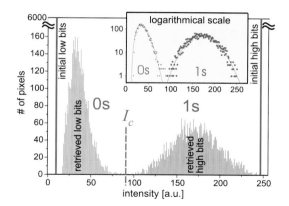

Figure 13: Number of pixels of intensity I as a function of the intensity. The distribution can be divided into two parts, belonging to the 0 s and 1 s of the data page, with the threshold set to I_c. An unambiguous assignment of a pixel intensity to a high or low bit in the surrounding of I_c is doubtful. In comparison, the bold red and green lines denote the delta-like distribution of the initial data page.

sample. This phase modulation needs to be distributed between all information-bearing holograms. Every particular hologram will profit from a part of this ultimate phase modulation. The smaller this part is, the larger the number of holograms that can be recorded. As for small phase modulation, the diffraction efficiency is proportional to the square of the phase modulation (according to Eq. (9)) we get an expression for the dependence of the maximum number of holograms with diffraction efficiency $\eta(N)$ on $M\#$:

$$\eta(N) = \frac{(M\#)^2}{N^2} \qquad (19)$$

The optimum value of the $M\#$ number is reached if $\alpha \cdot d = 2$. Typical values of $M\#$ lie in the range of 0.01 - 10. As an example, up to 1000 holograms can be stored assuming a storage material with $M\# = 1$ and a diffraction efficiency for all holograms of $\eta(N) = 1 \cdot 10^{-6}$, i.e. the intensity of each reconstructed data page is 1 μW if the intensity of the read-out beam is 1 W. Recording of a larger number of holograms is possible taking into account a diffraction efficiency of less than $1 \cdot 10^{-6}$. In conclusion, storage materials with a large $M\#$ are desirable for large storage densities.

4 Inorganic Materials

4.1 Fundamentals

Crystals, like $LiNbO_3$, $LiTaO_3$, $BaTiO_3$, $KNbO_3$, $Sr_xBa_{1-x}Nb_2O_6$, $Bi_{12}TiO_{20}$, $Ba_{1-x}Ca_xTiO_3$, $KTa_{1-x}Nb_xO_3$, are primarily used as holographic storage media in the holographic demonstration platforms since they provide important features such as the possibility for multiple recording and erasure. The materials themselves are:

- insensitive to external electric fields
- insensitive to external magnetic fields
- insoluble in organic liquids
- insoluble in most inorganic liquids
- insensitive to temperature.

However, the recorded holograms can be sensitive to external electric and magnetic fields, thus changing the Bragg condition. If the phase transition is near room temperature, the polarization will be destroyed, as in SBN and $BaTiO_3$.

Inorganic electro-optic crystals were first recognized for holographic data storage in 1968 by Chen et al. [35]. In 1975 Staebler et al. reported the recording of a multitude of holograms in lithium niobate [36]. In this chapter, inorganic materials, especially oxide crystals, are introduced with respect to their properties and applications as holographic storage media. It is the presence of impurities that leads to a photorefractive behavior,

3) A complete description of the photorefractive effect in electro-optic crystals can be found in e.g. [37].

i.e. the refractive index is changed with light illumination.[3] The cooperation of impurities, charge transport mechanism and electro-optic effect is explained in detail for the example of lithium niobate doped with iron.

4.2 Iron-Doped Lithium Niobate (LiNbO$_3$:Fe)

The most extensively studied oxide crystal is lithium niobate (LiNbO$_3$) [38]. We summarize the most important crystallographic results: The structure of lithium niobate is related to the ABO$_3$ perovskites and crystallizes in the quasi-perovskite structure, space group R3c. The environments of Li$^+$ and Nb^{5+} are similar because of their nearly identical radii. They are surrounded by distorted octahedra formed by six O^{2-} ions. The Nb^{5+}-O^{2-} bond is stronger than the Li$^+$-O^{2-} bond, so lithium niobate is normally grown in the non-stoichiometric composition with a lithium to niobium ratio of [Li]/[Nb] = 0.942. The composition is [Li$_{1-5x}$Nb$_{5x}$]Nb$_{1-4x}$O$_3$, with $x = 0.0118$ for the congruent composition [39], so that Li is missing from 5.9 % of all sites. Here, Nb^{5+} is incorporated on a Li$^+$ position as an intrinsic defect. The charge compensation is achieved by 4.7 % of Nb vacancies.

The flexibility of the structure allows doping with different types of elements, such as Fe, Cr, Cu, Mg or Zn. Figure 14 shows the absorption spectra of an undoped in comparison to an iron-doped crystal of lithium niobate.

The iron-doped crystal shows a broad absorption band centered at 500 nm, which is used for the recording of holograms. The broad absorption band can be attributed to the Fe^{2+} dopants and is responsible for the orange-brown color of the iron doped LiNbO$_3$ crystals. In contrast, undoped LiNbO$_3$ crystals are transparent. The position and intensity of the absorption band can be adjusted by the doping element and doping concentration. Doping with further elements is also possible, e.g. LiNbO$_3$:Cu:Mn or LiNbO$_3$:Zn:In [40].

4.3 Photorefractive Effect

4.3.1 One-Center Model

In 1966 Ashkin et al. observed a change of the refractive index in crystals of lithium niobate during illumination with light [41]. The light-induced charge transport mechanism is described best in lithium niobate doped with iron or copper by the "one center model" at standard light intensities [42], [43]. A look at the band scheme of lithium niobate in Figure 15 clearly demonstrates the decisive role of the impurities:

The impurity energy levels lie within the band gap between the valence and conduction band. Optical, Mössbauer and EPR measurements at lithium niobate crystals demonstrated that iron ions are present in the valence states Fe^{2+} or Fe^{3+}. Irradiating with light of an energy equal to or larger than the energy separation between this deep trap and the conduction band leads to the excitation of electrons from Fe^{2+} into the conduction band. The state of the deep trap is changed from Fe^{2+} to Fe^{3+}. A reverse change from Fe^{3+} to Fe^{2+} will occur if an excited electron is trapped.

Besides one impurity in two valence states (one-center model), the existence of two impurities in two valence states (two-center model [44]) and the existence of one impurity in three valence states are reported (three-valence model [45]). A detailed summary of these models is given by Buse [46].

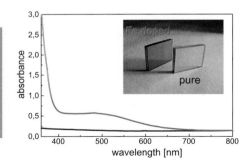

Figure 14: Absorption spectra of pure LiNbO$_3$ (blue line) and of LiNbO$_3$ doped with 0.05 wt. % Fe (red line). The iron-doped crystal shows a characteristic broad absorption band centered at about 500 nm. As a result the crystal is orange.

4.3.2 Charge Transport Mechanism

Free charge carriers migrate in the conduction band because of diffusion, drift or the photovoltaic effect leading to a total current:

$$\boldsymbol{j} = \boldsymbol{j}_{\text{diff}} + \boldsymbol{j}_{\text{drift}} + \boldsymbol{j}_{\text{phv}} \qquad (20)$$

The **diffusion current** $\boldsymbol{j}_{\text{diff}}$ results from an inhomogeneous illumination of the material leading to a spatially varying concentration of the charge carriers in the conduction band. It can be expressed by:

$$\boldsymbol{j}_{\text{diff}} = -Q \cdot \hat{D} \cdot \nabla N_{\text{e,h}} = -\frac{Q \cdot \hat{\mu}_{\text{e,h}} \cdot k_B \cdot T}{e} \nabla N_{\text{e,h}} \qquad (21)$$

with $Q = -e$ for electron and $Q = +e$ for hole conductivity, $\hat{D}$ the diffusion tensor, $\hat{\mu}_{e,h}$ the mobility tensor, k_B the Boltzmann constant, T the temperature, e the elementary charge, and $N_{e,h}$ the density of movable electrons or holes, respectively.

The **drift current** j_{drift} results from the Coulomb interaction of an externally applied electric field with the free charge carriers and follows Ohm's law for photorefractive oxide crystals [47], [48], [49]:

$$j_{drift} = \hat{\sigma} \cdot E = e \cdot \hat{\mu}_{e,h} \cdot N_{e,h} \tag{22}$$

The **photovoltaic current** j_{phv} in the volume or bulk photovoltaic current was found in a lithium niobate crystal in 1974 by Glass et al. [50]. Illumination causes stationary currents, which are proportional to the light intensity and to the absorption of the material and sesquilinear dependence on the light polarization. The photovoltaic current is described by [51], [52]:

$$j_{phv}\big|_i = \frac{1}{2}\left(\beta_{ijk} E_j^* E_k + \text{c.c.}\right) \tag{23}$$

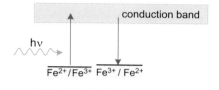

Figure 15: Energy position of Fe^{2+} and Fe^{3+} ions within the gap between the valence and conduction band of iron-doped lithium niobate. Electrons can be excited by light illumination from Fe^{2+} in the conduction band and are trapped by Fe^{3+}.

with the third-rank photovoltaic tensor β_{ijk} and the electric field components E_j and E_k of the light wave. The photovoltaic effect is the dominating driving force in iron-doped lithium niobate crystals [53], [54]. However, in contrast to the diffusion and drift currents a clear visual explanation of the photovoltaic effect is missing. The most simple description is that the photovoltaic current is based on the excitation of free charge carriers from noncentrosymmetric positions. As a consequence, such carriers might be forced to move in a certain direction. Attempts at a more detailed understanding of this effect have been made by e.g. [55] - [58]. A detailed phenomenological description of the photovoltaic effect is given by Sturman [52]. This theory predicted that some photovoltaic currents can be excited with linearly polarized eigenwaves of the crystal, ordinary or extraordinary (identical indices j = k in Eq. (23)). The others can be excited only if two orthogonally polarized waves propagate inside the sample simultaneously (j ≠ k in Eq. (23)). Nonvanishing currents of the second type enable the grating recording with two orthogonally polarized waves, one ordinary and the other extraordinary. Surprisingly, in such an interference pattern the intensity is distributed uniformly and only the polarization of the light field is spatially modulated.

4.3.3 Space Charge Formation and Refractive Index Modulation

The modulation of the refractive index due to illumination can be explained by the one-center model and a charge transport mechanism. We consider the illumination of a lithium niobate crystal by a sinusoidal light interference pattern. The situation is sketched in Figure 16.

In the bright regions of the light interference pattern $I(x)$ charge carriers are excited into the conduction band. They migrate due to drift, diffusion or the photovoltaic effect. Figure 16 illustrates the space charge formation which is due to the intensity-dependent linear photovoltaic current. If the charge carriers are trapped in the dark regions of the interference pattern, they will remain there because there is no light to reexcite them. This leads to a charge separation, as shown in Figure 16 a. The resulting photovoltaic current in Figure 16b is transferred in the steady state to the charge density $\rho(x)$ in Figure 16c via the continuity equation:

$$\dot{\rho} = -\text{div}\, j \tag{24}$$

The charge density generates an electric field

$$\text{div}(\hat{\varepsilon} \cdot E) = \frac{\rho}{\varepsilon_0} \tag{25}$$

This sinusoidally modulated electric field is called the space charge field $E_{SC}(x)$ in Figure 16d. This field couples to the linear electro-optic coefficient (Pockels effect) leading to a modulation of the refractive index Δn in Figure 16e:

$$\Delta n(x) = -\frac{1}{2} n^3 r_{eff} E_{SC}(x) \tag{26}$$

Figure 16:
a) Reference and signal waves superpose within the crystal volume and produce a sinusoidal light interference pattern $I(x)$. Electrons are excited in the bright regions and are trapped in the dark regions.
b) Excited electrons migrate due to the bulk photovoltaic effect. The photovoltaic current is sinusoidally modulated.
c) The photovoltaic current is transferred to a sinusoidal modulation of the charge density via the continuity equation. The modulation is phase-shifted by $\pi/2$ with respect to the incoming interference pattern.
d) The charge density is transferred to a sinusoidal modulation of the electric field via the Poisson equation. Again, there is a phase shift of $\pi/2$ with respect to the modulation of the charge density. Therefore, the electric field is phase-shifted by π with respect to the incoming light interference pattern.
e) The space charge field generates a modulation of the refractive index via the electro-optic effect (Pockels effect).

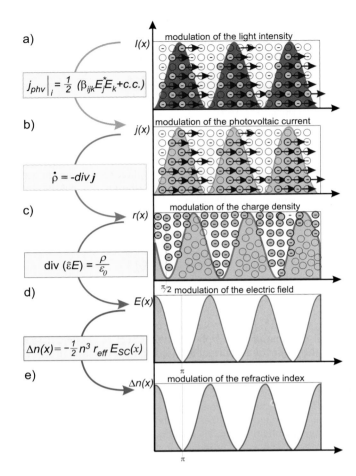

whereby $\Delta n(x)$ denotes the amplitude of the light-induced refractive index change, n the refractive index of the storage material and r_{eff} the effective electro-optic tensor component. As can be seen in Figure 16, the resulting sinusoidal refractive index modulation is π-phase-shifted with respect to the incoming light intensity modulation. In the case of diffusion this phase shift is $\pi/2$. Depending on the field polarity and field strength the phase shift may be either 0 or π (small fields) or approach $+\pi/2$ and $-\pi/2$ (strong fields). In summary, the photorefractive effect in electro-optic materials consists of the following succession of physical processes:

$$I(x) \Rightarrow j(x) \Rightarrow \rho(x) \Rightarrow E_{\text{SC}}(x) \Rightarrow \Delta n(x),$$

describing the transformation of an incoming light interference pattern into a modulation of the refractive index. A feature of the photorefractive effect in inorganic crystals is that it depends on the number of free charge carriers and the charge transport mechanisms, on the electrooptic coefficients, the dielectric constant and the refractive index of the material. Since the electrooptic tensor r_{ijk} is a third-rank tensor, this photorefractive process can only occur in materials which do not have a center of inversion. Therefore, this effect is observed almost exclusively in crystalline materials.

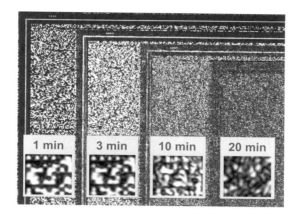

Figure 17: Decrease of the quality of a reconstructed data page by continuous light illumination for 1, 3, 10 and 20 minutes duration of light exposure. With increasing exposure the quality of the data page decreases dramatically so that the bit-error rate increases. A reconstruction of the data is not possible after 10 minutes.

4.3.4 Thermal Fixing

A long lifetime of a recorded hologram is of the utmost interest for many archival applications. In inorganic materials there are two processes which limit the useful lifetime due to a loss of refractive index modulation:

a **without further illumination**: after shutting the recording beams, the photovoltaic current vanishes and the space charge field remains forming the refractive index modulation via the electrooptic effect. However, this space charge field is not stable for an infinite time. It decreases, since the electrons migrate to their former places due to the dark conductivity σ_{dark}. E.g. the dark conductivity of lithium niobate is in the order of 10^{-15} $\Omega^{-1}m^{-1}$, so that a lifetime of up to a few years can be guaranteed if the hologram is kept in the dark [59].

b **with further illumination:** the lifetime of holograms in inorganic materials is extremely short if the holograms are illuminated. Of course, illumination is necessary for the retrieval of the stored data. In this case not the dark conductivity, but the photoconductivity σ_{photo} of the material is relevant for the lifetime. Since $\sigma_{photo} > \sigma_{dark}$ the lifetime of the stored hologram drops dramatically to a few minutes or even to a few seconds: the stored data are erased during read-out. Figure 17 shows this effect for the read-out of a holographically recorded data page in iron-doped lithium niobate. The hologram is continuously illuminated with the reference beam for up to 20 minutes. It can be seen that the quality of the data page decreases with increasing exposure, i.e. the bit error rate increases. A reconstruction of the data becomes impossible after 10 minutes of illumination.

To prevent this loss of stored information, the written holograms have to undergo a "thermal fixing" treatment. In 1971 a thermal fixing process was discovered for lithium niobate [60], which uses a heating procedure during or after the recording process of the hologram.

For the process of thermal fixing [61], the written holograms are heated up to temperatures above 150°C. At this high temperature, ions become mobile in the crystal. These ions travel through the crystal, driven by the space charge field with the aim of compensating the internal electric field modulation. Protons play the major role in this thermal fixing process in lithium niobate [62], [63]. If the crystal is cooled down to room temperature, the ions lose their mobility again, so that electronic and ionic gratings coexist. To get a refractive index modulation the electronic grating has to be removed. This development of the ionic grating mirrors the inital space charge field. For the development of the hologram, the sample is homogeneously irradiated with white light of high intensity. This excites the electrons from the Fe^{2+} ions into the conduction band. The homogeneous light illumination leads to a homogeneous redistribution of the electrons, so that the ionic grating remains. The refractive index is again modulated by the modulation of the ionic space charge field via the electro-optic effect. In comparison to the initial electronic grating, the ionic grating is stable at room temperature. This is due to the fact that ions are immobile at low temperatures and cannot be influenced with light. Erasure of the ionic grating is possible if the crystal is heated to above 200°C, where the ionic mobility is large enough for a redistribution. In summary, the electronic grating is transferred into an ionic grating by the thermal fixing process. The lifetime of the thermally fixed holograms is larger since the decay times of the ionic and electronic gratings are different [63].

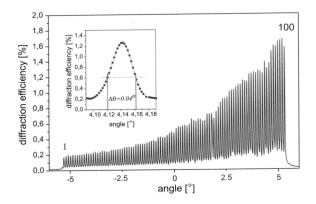

Figure 18: Diffraction efficiency as a function of the angle (rocking curve) of 100 holograms recorded in lithium-niobate. The holograms were written up to an efficiency of 1.6 % using transmission geometry: $\Theta_B = 26.6°$, $\lambda = 532$ nm. The angular distance between two holograms is 0.1°. The inset shows the rocking curve for one hologram at an angle 4.14°. The 1/e width of the rocking curve is 0.04°.

4.3.5 Holographic Data Storage in LiNbO$_3$:Fe

Recording of 100 holograms

In this chapter the recording of a multitude of holograms is demonstrated in iron-doped lithium niobate using the angular multiplexing technique. The processes of recording, thermal fixing and development of the holograms are shown in detail.

100 elementary holographic gratings were recorded in transmission geometry at $\lambda = 532$ nm and a Bragg angle of $\theta_B = 26.6°$ within an angular range of $\pm 5°$. The angular distance between two recording positions was chosen as 0.1°. The recording process of each hologram was controlled by measuring the diffraction efficiency as a function of the light exposure. It was interrupted at a diffraction efficiency of 1.6 % to prevent an overmodulation or optical damage. Figure 18 shows the rocking curve of 100 recorded holograms within the range of $\pm 5°$, i.e. the diffraction efficiency is plotted as a function of the angle. The inset exemplarily shows the single rocking curve of a written hologram. The 1/e width of the rocking curve of all holograms is measured to 0.04°.

Although each hologram was recorded up to the same diffraction efficiency of 1.6 %, this maximum is only reached for the holograms recorded last. In contrast, the diffraction efficiency of the first hologram is 0.2 %. Therefore, the holograms recorded first are partially erased by the subsequent recording of further holograms. Eq. (18) considers this effect for the diffraction efficiency via the M#number. In conclusion, if there is the need for a constant diffraction efficiency of a multitude of recorded holograms, the recording procedure has to be adapted. For example, the first gratings have to be recorded up to higher values of the diffraction efficiency.

Thermal fixing of 100 holograms

After recording, the holograms were thermally fixed by heating the crystal up to 150° to ensure a sufficiently long lifetime. The development was realized by homogeneous illumination at room temperature with white light. The diffraction efficiency of all thermally fixed holograms is shown in Figure 19. Again the rocking curve of a single hologram is shown in the inset.

After the procedure of thermal fixing is completed, the diffraction efficiency of the holograms is similar to that of the initially recorded patterns. All 100 holograms can be read out with sufficient diffraction efficiency, however, it is smaller by a factor of 100.

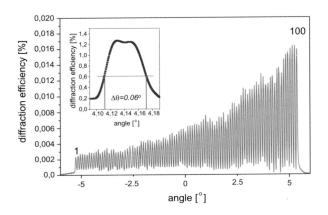

Figure 19: Diffraction efficiency as a function of the angle of 100 holograms recorded and thermally fixed in iron-doped lithium niobate. The inset shows the rocking curve for one hologram.

An interesting point is that the width of the rocking curve increased to 0.06° and its maximum is slightly lower. To improve the long-time stability of the fixed holograms, the crystal was illuminated with the reference beam for 200 hours. A decrease of the diffraction efficiency was not observed.

The experimental results demonstrate that a multitude of holograms can be recorded and thermally fixed in iron-doped lithium niobate using the angular multiplexing technique. Several characteristic processes are observed with recording, which have to be taken into account and can be optimized [64] for any technological application. In conclusion, thermal fixing allows the long-term storage of several thousands of holograms. The successful process of writing and subsequent thermal fixing of up to 10,000 holograms has been reported [65].

4.4 Outlook: Pulse Holography

To increase the recording velocity and mechanical stability of holographic storage systems, the intensity for the writing process has to be increased to get a short recording time. One possibility is to use pulsed laser systems, which provide both short laser pulses in the range of nanoseconds and large intensities. Additionally, inorganic materials present unique effects if illuminated with short laser pulses, and can be used for an efficient recording of holograms. Most interesting is the excitation of small polarons in oxide photorefractive crystals, which provide a two-step recording process.

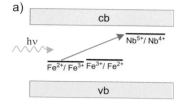

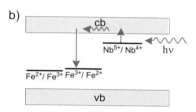

Figure 20: a) Excitation of Nb^{4+} polarons from the deep trap Fe^{2+} by illumination with high light intensities in the blue-green spectral range. b) Writing of holograms by excitation of electrons from the Nb^{4+} polarons into the conduction band in the infrared spectral range. Electrons are redistributed by the photovoltaic effect and are trapped by Fe^{3+} ions.

Two-step recording

As sketched in Figure 20, besides the photorefractive center Fe^{2+}/Fe^{3+}, the niobium ions lie energetically near to the conduction band in the valence states Nb^{4+}/Nb^{5+}. Electrons can be excited directly from the deep trap Fe^{2+} to the shallow trap Nb^{5+} ion by illumination with high light intensities (Figure 20a). This means that the charge transport described above via the conduction band in the one-center model fails for high light intensities.

The excited electrons are energetically stabilized at the Nb^{5+} ion by a deformation of the crystal lattice. Thus, these bound electrons influence a large area around the Nb^{5+} ion, so that in principle a Nb^{4+} ion is present. The resulting combination of electron and crystal lattice deformation is a quasi-particle known as a polaron. The lifetime of a polaron is generally in the range of micro- to milliseconds at room temperature and depends, among other factors, on the number of deep traps (e.g. Fe^{3+}), the stoichiometric composition of the crystal and the temperature.

The excited polarons can be used to write holograms with light in the infrared spectral range, as shown in Figure 20 b. After excitation of the Nb^{4+} polaron electrons can be excited from this shallow trap into the conduction band, where they migrate and are trapped by the deep photorefractive center Fe^{3+}. This redistribution of charges leads to the build-up of a space-charge field, which modulates the refractive index via the electro-optic effect.

In iron-doped lithium niobate Nb^{4+} polarons can be excited with a laser pulse in the blue-green spectral range. Electrons can be excited from these shallow traps into the conduction band with laser light in the near infrared spectral range. Using a frequency-doubled Nd-YAG pulse laser, excitation of polarons is possible with a laser pulse of 532 nm, followed by pulses of 1064 nm for holographic recording within the lifetime of the polarons. The most interesting feature of this two-step recording process is that holograms are written with light of 1064 nm. This means that the erasure of the holograms during read-out with light in the infrared spectral range does not occur, since the space charge field is formed by a modulation of Fe^{2+}/Fe^{3+} and the energetic distance of the deep traps to the conduction band is too large to allow excitation of electrons with light of 1064 nm wavelength. Therefore, the written holograms are stable during the read-out process without any further need of fixation. The lifetime of the hologram is limited by the dark conductivity only.

Using the technique of two-step recording by excitation of polarons, about 1 Mbit of data can be written into and read out of a crystal of $LiNbO_3$:Fe or $LiTaO_3$:Fe with a single laser pulse of 10 ns duration. The effective writing velocity is then limited by the repetition rate of the laser, the time constant of the page composer and the velocity of the angular and translational motion of the crystal. If we assume that these components work at a frequency of several kHz, a writing speed of more than 1 Gbit/s becomes possible.

5 Photoaddressable Polymers

5.1 Introduction

Polymers are perfect materials for a diversity of applications. Due to their macromolecular architecture there is an enormous variety to modify and control the synthesis of polymers. Thus, they can be tailored to even quite difficult demands. Since a whole industry deals with the processing of polymers, efficient production lines have been developed for almost every polymer. Not only the molecular composition, but also the cost of polymers has been optimized. Now these materials can be considered even for highly sophisticated applications, as optical and holographic data storage.

Looking at the development of optical media, it is obvious that polymeric materials play a major role for the compact disc (CD): Important for the success of the CD-ROM and CD-AUDIO was the development of a specific polymer – polycarbonate – which serves as substrate material, carrying the digital information in the form of tiny pits. In the ROM the information is imprinted during production and cannot be erased.

In photopolymers the optical information is stored by changing the optical properties of the material by a light-induced polymerization or a ring opening reaction. These materials turned out to be the most suitable holographic media for write-once-read-many (WORM) applications [66]. Since especially the optical requirements for holographic data storage are very stringent, there are some drawbacks that hinder the introduction of those polymers into mass production. So far, the shrinkage during illumination is the most severe problem for photopolymers.

As far as rewritable holographic storage is concerned, photorefractive crystals like iron-doped lithium niobate ($LiNbO_3$) are mostly used for rewritable (R/W) laboratory demonstrations [67]. In the following chapter, a polymeric class of materials, the so-called photoaddressable polymers [68] - [72] and the results of holographic measurements with these materials will be presented. These polymers show no shrinkage effects and react in a reversible way to the illumination of light and are therefore able to meet the requirements for R/W holographic data storage.

In principle, all materials that react to light with a change of specific properties can be described as photoaddressable polymers. Since the change of material properties is the basic condition for storing information in a material, we have to deal with a reversible modification in order to fulfill the requirements of an R/W material. In the following section the name photoaddressable polymer (PAP) is used as a synonym for a class of polymers with azo dyes as reversible antennae for the incident light. Those PAPs are basically azobenzene-containing side-group polymers. The aim of this chapter is to give an overview of the optical and photophysical properties of those polymers.

5.2 Photochemistry of Azobenzene

The photophysical reactions of azobenzene have been well studied. It can be demonstrated that due to controlled light-induced reactions of azo chromophores, properties of the whole system incorporating the dye can be modified. Those properties include for example viscosity, solubility, mechanical parameters, bioactivity, and optical constants [73]. Additionally azobenzene molecules can be used as a probe for their molecular surroundings. In this way a detailed study of polymeric parameters can be performed by monitoring the photochemical behavior of the azo dyes. The success of azo dyes in all these applications can be explained by detailed knowledge of their light-induced photoreactions: Azobenzene chromophores exist in two isomeric states. The rodlike long shaped trans form and the bent cis configuration. The isomerization can be induced by light in both directions, from trans to cis and from cis to trans, whereas the cis-isomer can also undergo a thermal back relaxation to the thermodynamically more stable trans-isomer (Figure 21).

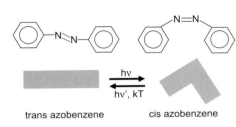

Figure 21: Trans-cis isomerization of azobenzene.

A look at a typical absorption spectrum (Figure 22) of unsubstituted azo chromophores (Figure 21) shows two absorption bands: The so-called $\pi\pi^*$ band with a maximum absorbance at $\lambda_{max} \sim 360$ nm and the so-called $n\pi^*$ band with $\lambda_{max} \sim 460$ nm. The spectral position of the absorption maximum of the $\pi\pi^*$ band can be shifted by chemical modification of the dye, namely by the replacement of donor/acceptor substituents. Those substituents do not influence the spectral position of the $n\pi^*$ band. The isomerization cycles have a distinct influence on absorption and optical index, which can be used for data recording.

The absorption bands define the laser wavelength needed: Azo-dyes undergo their isomerization cycles by illumination with light of wavelengths from the UV to the green/yellow range of the optical spectrum.

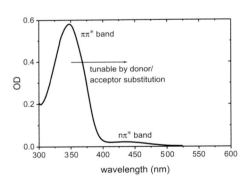

Figure 22: Absorption spectrum of unsubstituted azobenzene in solution.

5.3 Azobenzene-Containing Polymers

There are three ways to incorporate chromophores into a polymer:

- In guest-host-systems chromophore guests are doped into a polymeric host. In these systems the chromophore concentration cannot exceed a specific value, because highly concentrated chromophores tend to phase separation and crystallization.
- This disadvantage can be avoided by attaching the chromophores to the backbone as side chains.
- Or the chromophores are fixed as part of the main chain.

The first investigations of the isomerization kinetics of azo chromophores in polymers were performed in 1972 [74] and mainly focused on spectroscopic measurements [75] – [78]. Those experiments can be summarized by the experimental observation that the isomerization of azo dyes in a polymeric environment is possible, even at temperatures below the glass transition temperature. This result applies to side chain as well as to main chain polymers.

Since the isomerization results in a change of the molecular shape of the azo chromophores, as described above, there is a specific demand for free volume to enable this reaction[4]. In solution this free volume condition is fulfilled. In polymers, however, there might be some sterical constraints due to the inhomogeneous free volume distribution. For some chromophores there are local environments with large free volume, where the isomerization can be performed in a similar way as in solution. But there are also configurations where the free volume is not sufficient for the isomerization of the dyes. Therefore strong deviations of the photochemical reactions in polymers occur, if compared to the reactions in a solution.

However, experiments performed in polymeric systems demonstrate that azo dyes are able to undergo isomerization reactions in this environment.

4) In the range of 10^{-1} nm^3 in the case of unsubstituted azos [74].

5.4 Liquid Crystalline Side Chain Polymers

Liquid crystalline side chain polymers are polymers with different types of functional side groups. Figure 23 shows a sketch of a typical polymer under investigation: A chromophoric and a mesogenic side chain are attached to the polymer backbone via flexible spacers. Azobenzene or chemical modifications, so-called derivatives of azobenzene are used as chromophores, acting as antennae for the incident light.

When illuminated with visible light, a typical mesogenic side chain does not show any absorbance. Mesogens of the rigid-rod class are characterized by their long-shaped molecular geometry and their tendency to a spontaneous self-organization in a specific temperature range: Like wooden logs floating in a river, the mesogenic molecules tend to align side by side. The temperature range where this alignment occurs is located between the solid and the liquid phase, the corresponding phase is therefore named the mesogenic[5] phase. For liquid crystalline polymers, the transition temperature between the solid and the liquid crystalline phase is called the glass transition temperature T_g; the temperature between the liquid crystalline and the liquid phase is called the clearing temperature T_{cl}. This nomenclature results from the fact that in the liquid crystalline phase the molecules tend to form polydomains, acting as scattering centers for the incident light. At the clearing temperature the polymers become liquid and the polydomain structure breaks down. The result is a transparent, clear film. In contrast to a crystal with a long-range order even in the position of the single components, the molecules in the liquid crystalline phase only follow an order principle in the orientation. In the case of nematic liquid crystals, for example, the majority of molecules are aligned along a special direction. Details about the liquid crystalline phases can be found in [79]. We conclude that the liquid-crystalline molecules tend to align along one or more common directions.

5) Greek: in between.

5.5 Photoaddressable Polymers for Optical Storage

The motivation of the development of polymers following the scheme of Figure 23 can be summarized as follows:

- The dye acts as an antenna for the incident light. In order to optimize the response to light, adequate dye systems must be used. Azo dyes are the perfect chromophores for this purpose because of their well-known photochemistry.
- It is the task of the mesogenic groups to stabilize and to amplify the reorientation of the chromophores. The physical reason for this can be explained by the intrinsic tendency of the mesogens to spontaneously organize in domains.

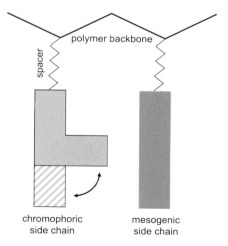

Figure 23: Scheme of a azobenzene containing liquid crystalline copolymer.

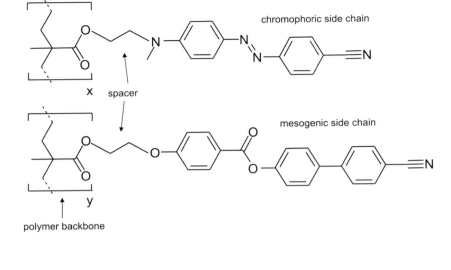

Figure 24: Typical chemical structure of PAP.

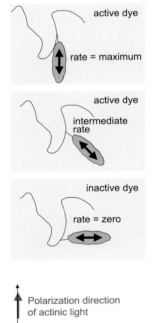

Figure 25: Orientational hole burning by irradiation of azo dyes with incident (actinic) polarized light. (For details see the text).

In the past decade, the synthesis of polymers with different optical and thermodynamical properties revealed details about the structure-property-relations of the polymers under laser illumination: Polymers with a strong tendency to form a liquid crystalline phase are difficult to prepare for optical applications, since a polydomain texture has to be suppressed in order to avoid diffuse scattering. To achieve sufficient thermal stability of the light-induced molecular reorientation, the temperature of the glass transition must be well above room temperature. In this way the different relaxation modes are frozen out. For the first time, these requirements have been fulfilled in amorphous polymers with high T_g [80]. Even though those purely amorphous high T_g polymers show an excellent light-induced birefringence [80], they have problems in the long-term stability of the light-induced reorientation. Therefore Bayer focuses on systems with a liquid crystalline phase. It is a challenge to produce polymers with an intrinsic mesogenic potential without any degradation of the optical properties, as the formation of light-scattering polydomains. Furthermore, the preparation should be adaptable to future mass production. It can be shown that it is possible to produce photoaddressable liquid crystalline copolymers where the liquid crystalline phase can be quenched by rapid cooling of the polymer melt or by other techniques as spin coating [81], [82].

Figure 24 shows a typical chemical structure of the polymers under investigation. The fraction x of the chromophoric and y of the mesogenic component can be varied over a wide range.

As mentioned above, the polymeric architecture allows many degrees of freedom for tailoring the polymers to specific applications. Now we discuss the physical principle of storing data in PAP and give an overview of the impact of molecular modifications on the recording behavior of the polymer system.

- The movement of the polymer main chain is mostly caused by the tendency to form a configuration, where the entropy reaches a maximum, e.g. where the system is mostly disordered. This competes with the tendency of the mesogens to order along a common direction, as described above. The movement of the backbone can be decoupled from the movement of the side chains by attaching chromophores and mesogens to the backbone via flexible **spacer groups**. Methylene-spacers, for example supply the chromophores and the mesogens with the flexibility needed to undergo the light-induced orientation. Furthermore, even large structural changes of side-groups in the glassy state of the polymer can be achieved by guaranteeing enough flexibility due to flexible spacer attachment of these side groups [83].

- The attachment of the azo dyes via flexible spacer groups can be considered as a boundary condition for the dye that leads to an effect which is very important for the application in optical data storage, an effect called **orientational hole burning.** In order to take advantage of this effect, photoaddressable polymers have to be illuminated with **polarized light**.

This polarized illumination leads to a formation of an optical anisotropy and therefore to birefringence of the illuminated sample. Figure 25 explains the occurrence of the optical axes during illumination. In the following, the quantum-mechanical description of the excitation will be performed in the dipole approximation. The isomerization probability

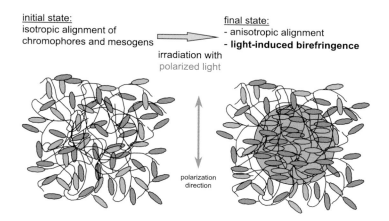

Figure 26: Formation of light-induced birefringence with photoaddressable polymers.

of an azo dye is a function of the angle Θ between the polarization direction $\vec{\varepsilon}$ of the incident (actinic) light and the dipole momentum $\vec{\mu}$ of the chromophore[6] [84]:

$$W \propto |<|\vec{\mu}\vec{\varepsilon}|>|^2 \propto \cos^2\Theta. \qquad (27)$$

6) The expression $<|\vec{\mu}\vec{\varepsilon}|>$ is the transition dipole momentum of the excitation from the initial electronic state to the excited state.

In the $n\pi^*$ band (see Figure 22) the dipole momentum of the trans azo dye lies parallel to the rod-like axis of the dye, for cis molecules $\vec{\mu}$ lies almost parallel to the N = N double bond. Since all isomerization cycles finally result in the thermodynamic more stable trans form, the following explanation can be restricted to the trans form of the dye: Dyes with their long axis parallel to the polarization direction of the incident light will preferably undergo an isomerization ($\cos^2\Theta = 1$). Dyes with an intrinsic orientation parallel to $\vec{\varepsilon}$ (Figure 25, top picture) will lose this orientation after a trans-cis-trans isomerization cycle[7]. As long as the angle between the polarization direction and the long axis of the dye in the resulting orientation does not reach 90°, this azo dye will undergo further isomerization cycles (center picture of Figure 25, $\cos^2\Theta \neq 0$). If an isomerization results in a dye orientation perpendicular to $\vec{\varepsilon}$ (Figure 25, bottom picture), this dye cannot be addressed anymore, since the excitation probability is equal to zero ($\cos^2\Theta = 0$). As every dye with an orientation perpendicular to the polarization direction is trapped and will not be able to take part in further excitation steps, illumination with polarized light results in an increased number of chromophores oriented in this direction. After a series of cycles the macroscopic orientational distribution function in the direction parallel to the polarization is decreased (orientational hole burning).

7) This behavior is, on the one hand, a result of the flexible attachment of the dyes via the spacers. On the other hand, the chromophore loses the memory to the starting configuration after a whole isomerization cycle.

A measure of the efficiency of this orientational hole burning is the light-induced birefringence (Figure 26). One starts from an isotropic distribution of the chromophores and dyes. After illumination with polarized light, the dyes will be preferentially oriented in a direction perpendicular to the polarization[8]. The different molecular configurations parallel and perpendicular to the polarization direction result in a different refractive index along these directions: the medium has become birefringent.

8) In Figure 26 a system is sketched, where the mesogenic (gray) side chains follow the orientation of the (red) chromophores (cooperative motion).

As mentioned above, there are several ways to optimize the polymers for the light-induced reorientation processes. The following modifications are used to change the photochemical and photophysical properties:

- A variety of **azo substitution** reactions can be found in literature. Chemical modifications of the **azo groups** have an influence on the absorption characteristics of the dye and therefore on the optical properties of the whole polymer system. By changing the substitution of the chromophore one can modify the photostationary equilibrium and the time constants of the isomerization steps and therefore the photochemical response time of the polymer. The polarity of the substitute and the anisotropy of the resulting azo group influence the light-induced birefringence. Long-shaped azo groups are able to provide a stronger birefringence than unsubstituted short chromophores.

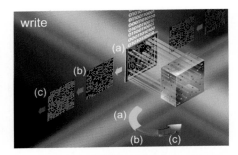

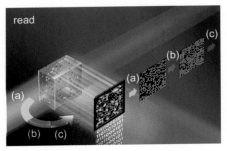

Figure 27: Principles of holographic multiplexing experiments. The three colors yellow, blue and red represent three incident angles of the reference beam (description, see text).

- The **mesogenic side groups** are used to stabilize and amplify the orientation of the chromophores without impairing the optical quality of the polymer (non-scattering samples). The choice of an adequate liquid crystalline neighbor depends on the chromophore under investigation. Since the mesogenic partner has to follow the reorientation of the dye, there has to be an interaction between both groups either by steric or, more efficiently, by dipolar interaction. Therefore the substitution of the mesogenic unit has to be adapted to the chromophoric group. The most effective cooperative reorientation between chromophores and mesogens is observed in systems containing groups of similar molecular shape. Following this molecular architectural concept, Bayer was able to synthesize copolymers with birefringence values of up to $\Delta n = 0.5$ [85], whereas typical homopolymers, containing solely chromophoric groups, reach values of only about $\Delta n \sim 0.1$.

- The **composition of the copolymer** determines the intrinsic tendency of the liquid to form oriented domains. A pure liquid crystalline polymer with no azo groups attached to the backbone would form a turbid sample of highly ordered domains. The thermodynamic tendency towards mesogenic order can be gradually suppressed by incorporating chromophoric co-monomers. This is due to the fact that most chromophores are not able to build up a liquid crystalline phase and therefore act as a disturbance for oriented domains. At a specific azo concentration this disturbance reaches a point where the formation of a mesogenic phase is no longer possible.

It has been demonstrated that polymers with a borderline concentration between the liquid crystalline and the amorphous phase show the best orientational properties. They are amorphous or display a weak liquid crystalline phase only. No diffuse light-scattering is observed. However, they retain the tendency to form an aligned liquid phase. Therefore the largest birefringence can be achieved with these polymers.

- The choice of the **main chain** has a major influence on the glass transition temperature T_g. Polyacrylates and -methacrylates (PMMA) [82], [86]-[93], polysiloxanes [94], polycarbonates [68], polyurethans [95], polyimides [96], and aliphatic polyesters [69], [97]-[102] have been investigated. The mechanical and thermodynamic properties of these systems can be changed over a very wide range. The rigidity of a PMMA backbone favors the long-term stability of the polymer. T_g values well above 100°C can be realized, whereas the light-induced reorientation steps are induced at room temperature. At room temperature, the main chain is locked into position and the backbone constitutes a rigid matrix, stabilizing the molecular reorientation processes.

The systematic optimization of PAP led to a polymeric system with outstanding properties, which can be summarized as follows:

PAP show the **highest photoinducable Δn** values reported for **amorphous** polymers. We measured resonance enhanced Δn values of more than 0.5 at reading wavelengths between 633 and 670 nm. This promises high dynamic ranges for PAP storage materials for holographic multiplexing. Other outstanding properties of PAP are the **temporal and thermal optical stability**. We investigated the orientational stability of light-induced birefringence values of PAP films with thicknesses in the range of micrometers from room temperature up to temperatures of 160°C. The relative Δn changes were less than 5 % under these conditions within a time span of four weeks, which confirms the long stability of the PAP systems and their usefulness for archival storage.

Besides these features, PAP materials fulfill the additional requirements for **rewritable** storage media, namely: The rewritability is provided by the reversible molecular orientation process. After a change of the polarization direction of the incident light, the optical axis of the polymer will follow accordingly. Furthermore, there is **no evidence for** any **shrinkage** of the PAP materials in holographic storage experiments.

PAP materials are amorphous at operating temperatures. The amorphous phase provides a **high optical quality** which is essential for all holographic applications.

Bulk samples are **easy to prepare** from the amorphous polymers. Widely used low-cost fabrication methods as injection molding are applicable. Finally, the **information storage density** is very high, since the photophysical processes are directly based on molecular reorientation.

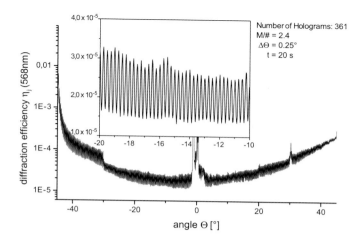

Figure 28: Rocking curve of 360 holographic gratings, written into a 1 mm thick polymer sample by angular multiplexing. To write a new hologram the sample was rotated by 0.25°. The illumination parameters for writing one holographic grating were: wavelength = 568 nm, intensity = 100 mW/cm², writing time = 20 s.

5.6 Holographic Measurements with Photoaddressable Polymers

As described before, illumination of photoaddressable polymers with polarized light leads to a light-induced birefringence. By changing the polarization direction of the incident light, the polymer can be switched **reversibly** between states, which differ strongly in their molecular orientation. This effect can be used to erase the birefringence and to record new information into the same volume.

In holographic experiments advantage is taken of this effect.

Figure 27 (top) gives an example of a holographic storage device based on PAP as storage material: A hologram is written into the PAP in the two-beam geometry. The signal beam carries the optical information in form of a two-dimensional pattern, provided by a spatial light modulator, SLM, see Figure 1 and Figure 2. The PAP storage material is placed into the intersection of the signal and the reference beam, generating birefringence according to the interference intensity. The written hologram can be reconstructed from the diffracted intensities (bottom picture of Figure 27). The high data rates of the holographic data storage are due to the parallel optical processing – all bits of a page are written and read simultaneously.

Figure 27 also demonstrates the angular multiplexing technique to increase the data storage density: After a small change of the incident angle of the reference beam a new hologram can be stored.[9] The angular shift between two consecutive holograms can be reduced if the PAP thickness is increased. This is a consequence of the Bragg diffraction. Thus one aims for thick polymer layers in order to increase the maximum number of holograms using the multiplexing technique. For applications, a PAP thickness in the range of millimeters is feasible. Another important goal is a high optical transparency of the storage material in order to enable the laser beams to penetrate the medium unattenuated.

PAP films with thicknesses ranging from 5 to 200 μm exhibit volume phase holograms with large diffraction efficiencies of more than 90 % [102]. To further demonstrate the feasibility in formats close to the technical requirements, holographic discs with thicknesses of 1 - 2 mm were prepared. The molecular structure of PAP was improved by incorporating new chromophores with a modified absorption behavior. Injection molding was used as the preparation method. The resulting thick polymer discs show high Bragg selectivity. The low angular width of $\Delta\theta < 0.2°$ allowed several hundred holographic gratings to be written into the same 3×3 mm² area of the PAP sample, using angular multiplexing (see Figure 28). The dynamic range of those PAP samples is in the range of M# number= 2 - 5, measured in holographic grating experiments.

Holographic recording of two-dimensional data masks has been performed at the Almaden IBM Research Center. The hardware (IBM PRISM tester) is described in detail elsewhere [1]. We stored a 256 kbit mask (512 × 512 pixel) and a megapixel mask (1024 × 1024 pixel) as holograms in a PAP sample. Due to the optical quality of the films, the measured bit error rate is very low: For the 256 kbit mask it turned out to be in the range of $BER = 1.2 \times 10^{-6}$, which is a very encouraging result and underlines the potential of PAP for holographic data storage.

9) A well known example for angular multiplexing can be found on credit cards: The small image changes its shape with the perspective of the viewer.

6 Outlook

The two main advantages of holographic data storage have been explained:

1. Data can be stored as a volume diffraction grating within the recording medium, enabling the storage of data in the range of terabytes.

2. The data transfer is parallel, permitting a simultaneous fast recording and read-out of complete data pages.

Inorganic materials offer themselves as a promising holographic medium for guaranteed long-term archivation, especially if pulse holography is considered. Pulse holography is not impaired by mechanical vibrations, both the writing and the read-out process are fast. An additional fixing of the holograms is not necessary for short storage times (approx. 1 year). It can be added after writing whenever sufficient time is available.

Under favorable illumination conditions the writing time of holograms in PAP materials lies in the range of several 100 ms, in some systems even in the range of several seconds. The writing speed is therefore at least one order of magnitude too long for technical applications. Since the chromophoric subsystem is the antenna for the incident light, it is responsible for the writing speed. Therefore these dyes need further improvement.

It remains a major challenge to find one material which meets all requirements for holographic storage – either an inorganic crystal or a photoaddressable polymer. Many answers will be given by experimental results in the years to come.

Acknowledgements

Financial support by the Deutsche Forschungsgemeinschaft (DFG) and the Ministerium für Schule und Bildung des Landes Nordrhein Westfalen is gratefully acknowledged. The authors thank Prof. Dr. E. Krätzig for helpful discussions. The editor gratefully acknowledges Christoph Buchal (FZ Jülich) for his comprehensive editorial and technical assistance in the compilation of this chapter. The editor would like to thank Sam Schmitz (CAESAR, Bonn) for checking the symbols and formulas.

References

[1] J. Ashley, M.-P. Bernal, G. W. Burr, H. Coufal et al., IBM J. Res. Develop. **44**, 341 (2000).

[2] S. Xie, A. Natansohn, P. Rochon, Chem. Mater. **5**, 403 (1995).

[3] H.J. Coufal, D. Psaltis, G.T. Sincerbox, *Holographic Data Storage*, Springer Verlag, Berlin, 2000.

[4] D. Gabor, Nature **161**, 777 (1948).

[5] P.J. van Heerden, Appl. Opt. **2**, 393 (1963).

[6] E.N. Leith, A. Kozma, J. Upatnieks, J. Marks and N. Massey, Appl. Opt. **5**, 1303 (1966).

[7] L. d'Auria, J.P. Huignard E. Spitz, IEEE Trans. Magn., MAG-**9**, 83 (1973).

[8] L. K. Anderson, Bell Labs Record **45**, 319 (1968).

[9] J. Lipp, J. Reynolds, in *Applications of Holography*, E. S. Barakette et al. eds. New York: Plenum Press, 1970.

[10] K. K. Stherlin, J. P. Lauer, R. W. Olenick, Appl. Optics **13**, 1345 (1974).

[11] W. C. Steward, R. S. Mezrich, L. S. Cosentino, E. M. Nagle, F. S. Wendt, R. D. Lohmann, RCA Rev. **34**, 3 (1973).

[12] L. D. Auria, J. P. Huignard, V. Slezak, E. Spitz, Appl. Optics **13**, 808 (1974).

[13] N. Nishida, M. Sakaguchi, F. Saito, Appl. Optics **12**, 1663 (1973).

[14] A. Bardos, Appl. Optics **13**, 832 (1974).

[15] Y. Tsunoda, K. Tatsuno, K. Kataoka, Y. Takeda, Appl. Optics **15**, 1398 (1976).

[16] A. Mikaeliane, in *Optical Information Recording 2*, E. S. Barrekette et al. eds., Plenum Press, New York, 1978.
[17] K. Kubota, Y. Ono, M. Kondo, S. Sugama, N. Nishida, M. Sakaguchi, Appl. Optics **19**, 944 (1980).
[18] I. Sato, M. Kato, K. Fujito, F. Tateishi, Appl. Optics **28**, 2634 (1989).
[19] S. Redfield, in *Holographic Data Storage*, H. J. Coufal et al. eds., Springer Verlag, 2000.
[20] M. P. Bernal, H. Coufal, R. K. Grygier, J. A. Hoffnagle, C. M. Jefferson, R. M. Mcfarlane, R. M. Shelby, G. T. Sincerbox, G. Wittmann, Appl. Optics **35**, 2360 (1996).
[21] J. Heanue, M. Bashaw, L. Hesselink, Science **265**, 749 (1994).
[22] I. Michael, W. Christian, D. Pletcher, T. Y. Chang, J. H. Hong, Appl. Optics **35**, 2375 (1996).
[23] G. W. Burr, J. Ashley, H. Coufal, R. K. Grygier, J. A. Hoffnagle, C. M. Jefferson, B. Marcus, Opt. Letters **22**, 639 (1997).
[24] L.Solymar, D.J. Cooke, *Volume Holography and Volume Gratings*, Academic Press, London, 1981.
[25] H.Kogelnik, Bell Syst. Tech. J. **48**, 2909 (1969).
[26] G. Montemezzani, M. Zgonic, Phys. Rev. E **55**, 1035 (1997).
[27] H. Lee, X.-G. Gu, D. Psaltis, J. Appl. Phys. **65**, 2191 (1989).
[28] K. Curtis, A. Pu, D. Psaltis, Opt. Lett., **19**, 993 (1994).
[29] D. Psaltis, Byte **17**, 179 (1992).
[30] G. Barbatathis, M. Levene, D. Psaltis, Appl. Opt. **35**, 2403 (1996).
[31] C. Denz, G. Pauliat, G. Roosen, Opt. Commun. **85**, 171 (1991).
[32] J.A.Hoffnagle, C.M.Jefferson in H.J. Coufal, D. Psaltis, G.T. Sincerbox, *Holographic Data Storage*, Springer Verlag Berlin, 2000.
[33] R.M. Shelby, J.A. Hoffnagle, G.W.Burr, C.M. Jefferson, H.Coufal, R.K. Grygier, H.Günther, R.M. Macfarlane, G.T. Sincerbox, Opt.Lett. **22**, 1509 (1997).
[34] F. H. Mok, G. W. Burr, D. Psaltis, Opt. Lett. **21**, 96 (1996).
[35] F. S. Chen, J. T. LaMacchia, D. B. Fraser, Appl. Phys. Lett. **13**, 223 (1968).
[36] D. L. Staebler, W. J. Burke, W. Phillips, J. J. Amodei, Appl. Phys. Lett. **26**, 182 (1975).
[37] P. Günther, *Physics Report*, North Holland Publish. Company, 1982.
[38] Y. S. Kuz'minor, *Lithium-Niobate-Crystals*, Cambridge International Science Publishing, 1999.
[39] S. C. Abrahams, P. Marsh, Acta. Cryst. B **42**, 61 (1986).
[40] K. Kasemir, K. Betzler, B. Marzas, B. Tiegel, T. Wahlbrink, M. Wöhlecke, J. Appl. Phys. **84**, 5191 (1998).
[41] A. Ashkin, G. D. Boyd, J. M. Diedzic, R. G. Smith, A. A. Ballmann, J. J. Levinstein, Appl. Phys. Lett. **13**, 223 (1968).
[42] E. Krätzig, R. Orlowski, Ferroelectrics **27**, 241 (1980).
[43] E. Krätzig, O.F. Schirmer, in *Photorefractive Materials and their Aapplications I*, P. Günther, J.-P. Huignard eds., Topics Appl. Phys. Vol. 61, Springer, Berlin, Heidelberg, 1988.
[44] G. C. Valley, Appl. Opt. **22**, 3160 (1983).
[45] K. Buse, E. Krätzig, Appl. Phys. B **61**, 27 (1995).
[46] K. Buse, Appl. Phys. B **64**, 273 (1997).
[47] P. Günther, Ferroelectrics **22**, 671 (1978).
[48] E. Krätzig, Ferroelectrics **21**, 635 (1978).
[49] K. Buse, H. Hesse, U. van Stevendaal, S. Loheide, D. Sabbert, E. Krätzig, Appl. Phys. A **59**, 563 (1994).
[50] A.M.Glass, D. von der Linde, T.J. Negran, Appl. Phys. Lett. **25**, 233 (1974).
[51] V. I. Bilinicher, B. Sturman, Sov. Phys. Usp. **23**, 199 (1980).
[52] B.I. Sturman, V.M. Fridkin, *The Photovoltaic and Photorefractive Effects in Noncentrosymmetric Materials*, Gordon and Breach Science Publishers S.A, 1992.
[53] E. Krätzig, Ferroelectrics **21**, 635 (1978).
[54] A. M. Glass, D. von der Linde, T. J. Negran, Appl. Phys. Lett. **25**, 233 (1974).
[55] W.Ruppel, R. von Baltz, P.Würfel, Ferroelectrics, **43**, 109 (1982).
[56] H.Presting, R. von Baltz, phys. Stat. Sol. (b), **112**, 559 (1982).
[57] N.Kristoffel, R. von Baltz, D. Hornung, Z. Phys. B **47**, 293 (1982).
[58] V.N.Novikov, B.I. Sturman, Ferroelectrics **75**, 199 (1987).

[59] I. Nee, M. Müller, K. Buse, E. Krätzig, J. Appl. Phys. 88, 4282 (2000).

[60] J. J. Amodei, D. L. Staebler, Appl. Phys. Lett. **59**, 256 (1991).

[61] K. Buse, S. Breer, K. Peithmann, S. Kapphan, M. Gao, E. Krätzig, Phys. Rev. B **56**, 1225 (1997).

[62] H. Vormann, G. Weber, S. Kapphan, E. Krätzig, Solid State Commun. **40**, 543 (1981).

[63] Y. Yang, I. Nee, K. Buse, D. Psaltis, Appl. Phys. Lett. **78**, 4076 (2001).

[64] C.R. Hsieh, et al., Appl. Optics **38**, 6141 (1999).

[65] X. An, D. Psaltis, G.W. Burr, Appl. Optics **38**, 386 (1999).

[66] V. L. Colvin, R. G. Larson, A. L. Harris, M. L. Schilling, J. Appl. Phys. **81**, 5913 (1997).

[67] K. Buse, A. Adibi, D. Psaltis, Nature (London) **393** (6686), 664 (1998).

[68] M. Eich, J. H. Wendorff, B. Reck, H. Ringsdorf, Makromol. Chem. Rapid Commun. **8**, 59 (1987).

[69] S. Hvilsted, F. Andruzzi, P.S. Ramanujam, Opt. Lett. **17**, 1234 (1992).

[70] A. Natansohn, P. Rochon, J. Gosselin, S. Xie, Macromolecules **25**, 2268 (1992).

[71] Wiesner, N. Reynolds, C. Boeffel, H. W. Spiess, Liq. Cryst. **11**, 251 (1992).

[72] D.Y. Kim, L. Li, J. Kumar, S.K. Tripathy, Appl. Phys. Lett. **66**, 1166 (1995).

[73] S. Barley, A. Gilbert, and G. Mitchell, J. Mater. Chem., **1**, 481 (1991).

[74] C. S. Paik and H. Morawetz, Macromolecules, **5**, 171 (1972).

[75] C. D. Eisenbach. Ber. Bunsenges. Phys. Chem., **84**, 680 (1980).

[76] E. Dubini-Paglia, P. L. Beltrame, B. Marcandalli, P. Carniti, and L. Seves, A. and Vicini, J. Appl. Polym. Sci., **31**, 1251 (1986).

[77] I. Mita, K. Horie, and K. Hirao, Macromolecules **22**, 558 (1989).

[78] T. Naito, K. Horie, and I. Mita, Macromolecules, **24**, 2907 (1991).

[79] S. Chandrasekhar, Liquid Crystals, Cambridge University Press (1992).

[80] A. Natansohn, S. Xie and P. Rochon, Macromolecules **25**, 5531 (1992).

[81] T. Bieringer, R. Wuttke, D. Haarer, U. Geßner, J. Rübner, Macromol. Chem. Phys. **196**, 1375 (1995).

[82] S. Zilker, T. Bieringer, D. Haarer, R. S. Stein, J. W. van Egmond, Adv. Mat. **10**, 855 (1998).

[83] H. Ringsdorf and R. Zentel. Makromol. Chem. **183**, 1245 (1982).

[84] B. H. Bransden and C. J. Joachain, *Physics of Atoms and Molecules*, Wiley and Sons, 1991.

[85] V. Cimrová, D. Neher, S. Kostromine, T. Bieringer, Macromolecules **32**, 8496 (1999).

[86] M. Eich and J. Wendorff, Makromol. Chem., Rapid Commun. **8**, 467 (1987).

[87] K. Anderle, R. Birenheide, M. Eich and J. Wendorff, Makromol. Chem., Rapid Commun. **10**, 477 (1989).

[88] V. Shibaev, S. Kostromine, S. Ivanov, L. Lasker, T. Fisher, J. Stumpe, Macromol. Symp **96**, 157 (1995).

[89] J. Eickmans, T. Bieringer, GIT Labor-Fachz. **43**(10), 1108 (1999).

[90] S. Zilker, M. R. Huber, T. Bieringer, D. Haarer, Appl. Phys. B **68**, 893 (1999).

[91] J. Eickmans, T. Bieringer, S. Kostromine, H. Berneth and R. Thoma, Jpn. J. Appl. Phys. **38**, 1835 (1999).

[92] F. L. Labarthet, J.-L. Bruneel, T. Buffeteau, C. Sourisseau, M. R. Huber, S. J. Zilker, T. Bieringer, Phys. Chem., **2**, 5154 (2000).

[93] R. Wuttke, K. Fischer, T. Bieringer, D. Haarer, Polym. Mater. Sci. Eng. **72**, 520 (1995).

[94] R. Ortler, C. Braeuchle, A. Millera and G. Riepl, Makromol. Chem., Rapid Commun. **10**, 189 (1989).

[95] O. Watanabe, M. Tsuchimori, A. Okada, and H. Ito, Appl. Phys. Lett. **71**, 750 (1997).

[96] Z. Sekkat, J. Wood, W. Knoll, W. Volksen, R.D. Miller, A. Knoesen, J. Opt. Soc. Am. B **14**, 829 (1997).

[97] S.Hvilsted, F. Andruzzi, C. Kulinna, H. Siesler and P.S. Ramanujam, Macromolecules **28**, 2172 (1995).

[98] N. Holme, P.S. Ramanujam and S. Hvilsted, Appl. Opt. **35**, 4622 (1996).

[99] N.Holme, P.S. Ramanujam and S.Hvilsted, Opt. Lett. **21**, 902 (1996).

[100] S. Hvilsted and P.S. Ramanujam, Current Trends in Polymer Science **1**, 53 (1996).

[101] L. Nikolova, T. Todorov, M. Ivanov, S. Hvilsted and P.S. Ramanujam, Appl. Opt. **35**, 3835 (1996).

[102] R.H. Berg, S. Hvilsted and P.S. Ramanujam, Nature **383**, 505 (1996).

AFM-Based Mass Storage – The *Millipede* Concept

Peter Vettiger, Michel Despont, Urs Dürig, Mark A. Lantz,
Hugo E. Rothuizen and *Gerd K. Binnig*
IBM Research, Zurich Research Laboratory, Switzerland

Contents

1	Introduction, Motivation, and Objectives	687
2	The Millipede Concept	688
3	Thermomechanical AFM Data Storage	689
4	Polymer Medium	690
5	Array Design, Technology, and Fabrication	692
6	Array Characterization	695
7	First Write/Read Results With the 32 × 32 Array Chip	696
8	Discussion of Possible Millipede Applications in Data Storage	696
9	Summary and Outlook	697

AFM-based Data Storage – The *Millipede* Concept

1 Introduction, Motivation, and Objectives

An emerging technology being considered as a serious candidate to replace an existing but limited technology must offer long-term perspectives. For instance, the magnetic disc storage industry is huge and requires correspondingly enormous investments, which makes it long-term-oriented by nature. The consequence for storage is that any new technique with better areal storage density than today's magnetic recording [1] should have long-term potential for further scaling, desirably down to the nanometer or even atomic scale.

The only available tool known today that is simple and yet provides these very long-term perspectives is a nanometer-sharp tip. Such tips are now used in every atomic force microscope (AFM) and scanning tunneling microscope (STM) for imaging and structuring down to the atomic scale. The simple tip is a very reliable tool that concentrates on one functionality: the ultimate local confinement of interaction.

In the early 1990s, Mamin and Rugar at the IBM Almaden Research Center pioneered the possibility of using an AFM tip for readback and writing of topographic features for the purposes of data storage. In one scheme developed by them [2], reading and writing were demonstrated with a single AFM tip in contact with a rotating polycarbonate substrate. The data were written thermomechanically via heating of the tip. In this way, densities of up to 30 Gb/inch^2 were achieved, representing a significant advance compared to the densities of that day. Later refinements included increasing readback speeds to a data rate of 10 Mb/s [3] and implementation of track servoing [4].

In making use of single tips in AFM or STM operation for storage, one must deal with their fundamental limits for high data rates. At present, the mechanical resonant frequencies of the AFM cantilevers limit the data rates of a single cantilever to a few Mb/s for AFM data storage [5], [6], and the feedback speed and low tunneling currents limit STM-based storage approaches to even lower data rates.

Currently a single AFM operates at best on the microsecond time scale. Conventional magnetic storage, however, operates at best on the nanosecond time scale, making it clear that AFM data rates have to be improved by at least three orders of magnitude to be competitive with current and future magnetic recording.

The objectives of our research activities within the Micro- and Nanomechanics Project at the IBM Zurich Research Laboratory are to explore highly parallel AFM data storage with areal storage densities far beyond the expected superparamagnetic limit (60 - 100 Gb/inch^2) [7] and data rates comparable to those of today's magnetic recording.

The "Millipede" concept presented here is a new approach for storing data at high speed and with an ultrahigh density. It is not a modification of an existing storage technology, although the use of magnetic materials as storage media is not excluded. The ultimate locality is given by a tip, and high data rates are a result of massive parallel operation of such tips. This chapter will be focused on demonstrating the Millipede concept with areal densities up to 500 Gb/inch^2 and parallel operation of very large 2D (32 × 32) AFM cantilever arrays with integrated tips and write/read storage functionality. The chapter is an updated and extended version of Ref. [8].

The fabrication and integration of such a large number of mechanical devices (cantilever beams) will lead to what we envision as the VLSI age of micro- and nanomechanics. It is our conviction that VLSI micro/nanomechanics will greatly complement future micro- and nanoelectronics (integrated or hybrid) and may generate applications of VLSI-Nano(Micro) ElectroMechanical Systems [VLSI-N(M)EMS] not conceived of today.

2 The Millipede Concept

The 2D AFM cantilever array storage technique [9], [10] called "Millipede" is illustrated in Figure 1. It is based on a mechanical parallel x/y scanning of either the entire cantilever array chip or the storage medium. In addition, a feedback-controlled z-approaching and -leveling scheme brings the entire cantilever array chip into contact with the storage medium. This tip-medium contact is maintained and controlled while x/y scanning is performed for write/read. It is important to note that the Millipede approach is not based on individual z-feedback for each cantilever; rather, it uses a feedback control for the entire chip, which greatly simplifies the system. However, this requires stringent control and uniformity of tip height and cantilever bending. Chip approach and leveling make use of four integrated approaching cantilever sensors in the corners of the array chip to control the approach of the chip to the storage medium. Signals from three sensors (the fourth being a spare) provide feedback signals to adjust three magnetic z-actuators until the three approaching sensors are in contact with the medium. The three sensors with the individual feedback loop maintain the chip leveled and in contact with the surface while x/y scanning is performed for write/read operations. The system is thus leveled in a manner similar to an antivibration air table. This basic concept of the entire chip approach/leveling has been tested and demonstrated for the first time by parallel imaging with a 5 × 5 array chip [11]. These parallel imaging results have shown that all 25 cantilever tips have approached the substrate within less than 1 μm of z-actuation. This promising result has led us to believe that chips with a tip-apex height control of less than 500 nm are feasible. This stringent requirement for tip-apex uniformity over the entire chip is a consequence of the uniform force needed to minimize or eliminate tip and medium wear due to large force variations resulting from large tip-height nonuniformities [4].

During the storage operation, the chip is raster-scanned over an area called the storage field by a magnetic x/y scanner. The scanning distance is equivalent to the cantilever x/y pitch, which is currently 92 μm. Each cantilever/tip of the array writes and reads data only in its own storage field. This eliminates the need for lateral positioning adjustments of the tip to offset lateral position tolerances in tip fabrication. Consequently, a 32 × 32 array chip will generate 32 × 32 (1024) storage fields on an area of less than 3 mm × 3 mm. Assuming an areal density of 500 Gb/inch^2, one storage field of 92 μm × 92 μm has a capacity of about 6.6 Mb, which can be increased to about 10 Mb after application of a code that requires at least two "0"s between each "1", and the entire 32 × 32 array with 1024 storage fields has a capacity of about 10 Gb on 3 mm × 3 mm. As shown in Sec. 7, the storage capacity scales with the number of elements in the array, cantilever pitch (storage-field size) and areal density, and depends on the application requirements. Although not yet investigated in detail, lateral tracking will also be performed for the entire chip, with integrated tracking sensors at the chip periphery. This assumes and requires very good temperature control of the array chip and the medium substrate between write and read cycles. For this reason the array chip and medium substrate should be held within about 1 °C operating temperature for bit sizes of 30 to 40 nm and array chip sizes of a few millimeters. This will be achieved by using the same material (silicon) for both the array chip and the medium substrate in conjunction with four integrated heat sensors that control four heaters on the chip to maintain a constant array-chip temperature during operation. True parallel operation of large 2D arrays results in very large chip sizes because of the space required for the individual write/read wiring to each cantilever and the many I/O pads. The row and column time-multiplexing addressing scheme implemented successfully in every DRAM is a very elegant solution to this issue. In the case of Millipede, the time-multiplexed addressing scheme is used to address the array row by row, with full parallel write/read operation within one row.

The current Millipede storage approach is based on a new thermomechanical write/read process in nanometer-thick polymer films. As previously noted, thermomechanical writing in polycarbonate films and optical readback were first investigated and demonstrated with a single cantilever by Mamin and Rugar [2]. Although the storage density of 30 Gb/inch^2 obtained originally was not overwhelming, the results encouraged us to use polymer films as well to achieve density improvements.

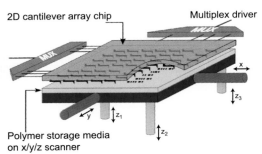

Figure 1: "Millipede" concept. Reproduced with permission from [23](a), © 1999 IEEE.

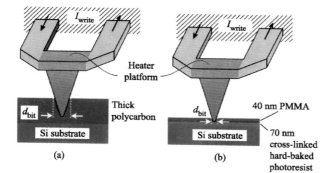

Figure 2:
(a) Earlier storage medium consisting of a bulk polycarbonate layer.
(b) New storage medium for smaller bit sizes, consisting of a thin writable PMMA layer on top of Si substrate separated by a cross-linked film of photoresist.
From [13], reproduced with permission.

3 Thermomechanical AFM Data Storage

In recent years, AFM thermomechanical recording in polymer storage media has undergone extensive modifications, primarily with respect to the integration of sensors and heaters designed to enhance simplicity and to increase data rate and storage density. Using cantilevers with heaters, thermomechanical recording at 30 Gb/inch^2 storage density and data rates of a few Mb/s for reading and 100 Kb/s for writing have been demonstrated [2], [3], [12]. Thermomechanical writing is a combination of applying a local force by the cantilever/tip to the polymer layer and softening it by local heating. Initially, the heat transfer from the tip to the polymer through the small contact area is very poor, improving as the contact area increases. This means that the tip must be heated to a relatively high temperature (about 400 °C) to initiate the melting process. Once melting has commenced, the tip is pressed into the polymer, which increases the heat transfer to the polymer, increases the volume of melted polymer, and hence increases the bit size. Our rough estimates indicate that at the beginning of the writing process only about 0.2% of the heating power is used in the very small contact zone (10 - 40 nm^2) to melt the polymer locally, whereas about 80% is lost through the cantilever legs to the chip body and about 20% is radiated from the heater platform through the air gap to the medium/substrate. After melting has started and the contact area has increased, the heating power available for generating the indentations increases by at least ten times to become 2% or more of the total heating power. With this highly nonlinear heat-transfer mechanism, it is very difficult to achieve small tip penetration and thus small bit sizes, as well as to control and reproduce the thermomechanical writing process.

This situation can be improved if the thermal conductivity of the substrate is increased, and if the depth of tip penetration is limited. The use of very thin polymer layers deposited on Si substrates has been explored to improve these characteristics [13], [14], as illustrated in Figure 2. The hard Si substrate prevents the tip from penetrating farther than the film thickness allows, and it enables more rapid transport of heat away from the heated region because Si is a much better conductor of heat than the polymer. Si substrates were coated with a 40-nm film of polymethylmethacrylate (PMMA) and achieved bit sizes ranging between 10 and 50 nm. However, increased tip wear was noticed, probably caused by the contact between Si tip and Si substrate during writing. Therefore, a 70-nm layer of cross-linked photoresist (SU-8) was introduced between the Si substrate and the PMMA film to act as a softer penetration stop that avoids tip wear but remains thermally stable.

Using this layered storage medium, data bits 40 nm in diameter have been written, as shown in Figure 3. These results were obtained using a 1-μm-thick, 70-μm-long, two-legged Si cantilever [12]. The cantilever legs are made highly conducting by high-dose ion implantation, whereas the heater region remains low-doped. Electrical pulses 2 μs in duration were applied to the cantilever with a period of 50 μs. Figure 3(a) demonstrates that 40-nm bits can be written with 120-nm pitch, or very close to each other, without merging (Figure 3(b)), implying a potential bit areal density of 400 Gb/inch^2. Subsequently, we have demonstrated single-cantilever areal densities up to 1 Tb/inch^2, although currently at a somewhat degraded write/read quality [15].

Imaging and reading are done using a new thermomechanical sensing concept [16]. The heater cantilever originally used only for writing was given the additional function of a thermal readback sensor by exploiting its temperature-dependent resistance. The resistance (R) increases nonlinearly with heating power/temperature from room temperature to a peak value at 500 - 700 °C. The peak temperature is determined by the doping

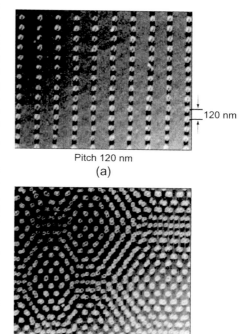

Figure 3: Series of 40-nm data bits formed in an uniform array with
(a) 120-nm pitch and
(b) variable pitch ($\geq$ 40 nm), resulting in bit areal densities of up to 400 Gb/inch^2. Images obtained with a thermal readback technique.
Adapted from [13], with permission.

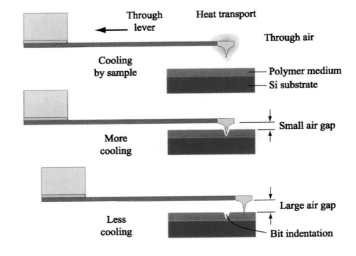

Figure 4: Principle of AFM thermal sensing. The heater cantilever is continuously heated by a dc power supply while it is being scanned and the heater resistivity measured.
Adapted from [23](a), with permission.

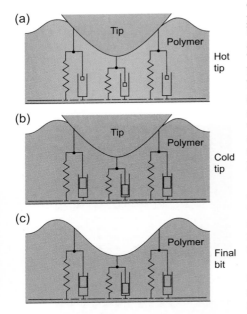

Figure 5: Visco-elastic model of bit writing:
(a) The hot tip heats a small volume of polymer material to more than T_g: the shear modulus of the polymer drops drastically from GPa to MPa, which in turn allows the tip to indent the polymer. In response, elastic stress (represented as compressed springs) builds up in the polymer. In addition, viscous forces (represented as pistons) associated with the relaxation time for the local deformation of molecular segments limit the indentation speed.
(b) At the end of the writing process, the temperature is quenched on a microsecond time scale to room temperature: The stressed configuration of polymer is frozen-in (represented by the locked pistons).
(c) The final bit corresponds to a metastable configuration. The original unstressed flat state of the polymer can be recovered by heating the bit volume to more than T_g, which unlocks the compressed springs.
Reproduced with permission from [15], © 2002 IEEE.

concentration of the heater platform, which ranges from 1×10^{17} to 2×10^{18}. Above the peak temperature, the resistance drops as the number of intrinsic carriers increases because of thermal excitation [17]. For sensing, the resistor is operated at about 350 °C, a temperature that is not high enough to soften the polymer, as is necessary for writing. The principle of thermal sensing is based on the fact that the thermal conductance between the heater platform and the storage substrate changes according to the distance between them. The medium between a cantilever and the storage substrate – in our case air – transports heat from one side to the other. When the distance between heater and sample is reduced as the tip moves into a bit indentation, the heat transport through air will be more efficient, and the heater's temperature and hence its resistance will decrease. Thus, changes in temperature of the continuously heated resistor are monitored while the cantilever is scanned over data bits, providing a means of detecting the bits.

Figure 4 illustrates this concept. Under typical operating conditions, the sensitivity of thermomechanical sensing is even better than that of piezoresistive-strain sensing, which is not surprising because thermal effects in semiconductors are stronger than strain effects. The good $\Delta R/R$ sensitivity of about 10^{-5}/nm is demonstrated by the images of the 40-nm-size bit indentations in Figure 3, which were obtained using the thermal-sensing technique described.

In addition to ultradense thermomechanical write/read, we have also demonstrated the erasing and rewriting capabilities of polymer storage media [13]. Thermal reflow of storage fields is achieved by heating the medium to about 150 °C for a few seconds. The smoothness of the reflowed medium allowed multiple rewriting of the same storage field. This erasing process does not allow bit-level erasing; it will erase larger storage areas. However, in most applications single-bit erasing is not required anyway, because files or records are usually erased as a whole.

The erasing and multiple rewriting processes, as well as bit-stability investigations, are topics of ongoing research.

4 Polymer Medium

The polymer storage medium plays a crucial role in Millipede-like thermo mechanical storage systems. The thin-film-sandwich structure with PMMA as active layer (see Figure 2) is not the only choice possible, considering the almost unlimited range of polymer materials available. The ideal medium should be easily deformable for bit writing, yet written bits should be stable against tip wear and thermal degradation. Finally, one would also like to be able to repeatedly erase and rewrite bits. In order to be able to scientifically address all important aspects, some understanding of the basic physical mechanism of thermomechanical bit writing and erasing is required.

In a gedanken experiment, we visualize bit writing as the motion of a rigid body, the tip, in a viscous medium, the polymer melt. For the time being, the polymer, i.e. PMMA, is assumed to behave like a simple liquid after it has been heated above the glass-transition temperature in a small volume around the tip. As viscous drag forces must not exceed the loading force applied to the tip during indentation, we can estimate an upper bound for the viscosity η of the polymer melt using Stokes' equation

$$F = 6\pi\eta Rv. \qquad (1)$$

In actual Millipede bit writing, the tip loading force is on the order of $F = 50$ nN, and the radius of curvature at the apex of the tip is typically $R = 20$ nm. Assuming a depth of the indentation of, say, $h = 50$ nm and a heat pulse of $\tau_h = 10$ µs duration, the mean velocity during indentation is on the order of $v = h/\tau_h = 5$ mm s^{-1} (note that thermal relaxation times are on the order of microseconds [18], and hence the heating time can be equated to the time it takes to form an indentation). With these parameters we obtain $\eta < 25$ Pa s, whereas typical values for the shear viscosity of PMMA are at least 7 orders of magnitude larger even at temperatures well above the glass-transition point [19].

This apparent contradiction can be resolved by considering that polymer properties are strongly dependent on the time scale of observation. At time scales on the order of 1 ms and below, entanglement motion is in effect frozen in, and the PMMA molecules form a relatively static network. Deformation of the PMMA now proceeds by means of uncorrelated deformations of short molecular segments rather than by a flow mechanism involving the coordinated motion of entire molecular chains. The price one has to pay is that elastic stress builds up in the molecular network as a result of the deformation (the polymer is in a so-called rubbery state). On the other hand, corresponding relaxation times are orders of magnitude smaller, giving rise to an effective viscosity at Millipede time scales on the order of 10 Pa s [19] as required by our simple argument (see Eq. (1)). Note that, unlike the normal viscosity, this high-frequency viscosity is basically independent of the detailed molecular structure of the PMMA, i.e. chain length, tacticity, poly-dispersity, etc. In fact, we can even expect that similar high-frequency viscous properties are found in a large class of other polymer materials, which makes thermomechanical writing a rather robust process in terms of material selection.

We have argued above that elastic stress builds up in the polymer film during indentation, creating a corresponding reaction force on the tip on the order of $F_r \sim 2\pi GR^2$, where G denotes the elastic shear modulus of the polymer [20]. An important property for Millipede operation is that the shear modulus drops by orders of magnitude in the glass-transition regime, i.e. for PMMA from ~ 1 GPa below T_g to ~ 0.5 ⋯ 1 MPa above T_g [19]. (The bulk modulus, on the other hand, retains its low-temperature value of several GPa. Hence, in this elastic regime, formation of an indentation above T_g constitutes a volume-preserving deformation.) For proper bit writing, the tip load must be balanced between the extremes of the elastic reaction force F_r for temperatures below and above T_g, i.e. for PMMA $F \ll 2.5\,\mu N$ to prevent indentation of the polymer in the cold state and $F \gg 2.5$ nN to overcome the elastic reaction force in the hot state. Unlike the deformation of a simple liquid, the indentation represents a metastable state of the entire deformed volume, which is under elastic tension. Recovery of the unstressed initial state is prevented by rapid quenching of the indentation below the glass temperature with the tip in place. As a result, the deformation is frozen in because below T_g motion of molecular-chain segments is effectively inhibited (see Figure 5).

This mechanism also allows local erasing of bits – it suffices to locally heat the deformed volume above T_g whereupon the indented volume reverts to its unstressed flat state driven by internal elastic stress. In addition, erasing is promoted by surface tension forces, which give rise to a restoring surface pressure on the order of $\gamma(\pi/R)^2 h \approx 25$ MPa, where $\gamma \sim 0.02$ Nm^{-1} denotes the polymer-air surface tension.

One question immediately arises from these speculations: If the polymer behavior can be determined from the macroscopic characteristics of the shear modulus as a function of time, temperature, and pressure, can then the time-temperature superposition principle also be applied in our case? The time-temperature superposition principle is a very successful concept of polymer physics [21]. It basically says that the time scale and the temperature are interdependent variables that determine the polymer behavior such as the shear modulus. A simple transformation can be used to translate time-dependent into temperature-dependent data and vice versa. It is not clear, however, whether this principle can be applied in our case, i.e. under such extreme conditions (high pressures, short time scales and nanometer-sized volumes, which are clearly below the radius of gyration of individual polymer molecules).

To test this, the heating time, the heating temperature, and the loading force were varied in bit-writing experiments on a standard PMMA sample. The results are summarized in Figure 6. The minimum heater temperature at which bit formation starts for a

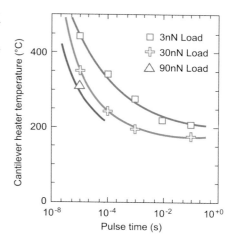

Figure 6: Bit-writing threshold measurement. The load was controlled by pushing the cantilever/tip into the sample with a controlled displacement and a known spring constant of the cantilever. When a certain threshold is reached, the indentations become visible in subsequent imaging scans. The solid lines are guides to the eye. Curves of similar shape would be expected from the time-temperature superposition principle.
Reproduced with permission from [15], © 2002 IEEE.

given heating-pulse length and loading force was determined. This so-called threshold temperature is plotted against the heating-pulse length. A careful calibration of the heater temperature has to be done to allow a comparison of the data. The heater temperature was determined by assuming proportionality between temperature and electrical power dissipated in the heater resistor at the end of the heating pulse when the tip has reached its maximum temperature. An absolute temperature scale is established using two well-defined reference points. One is room temperature, corresponding to zero electrical power. The other is provided by the point of turnover from positive to negative differential resistance (see Figure 10), which corresponds to a heater temperature of 550 °C. The general shape of the measured threshold-temperature versus heating-time curves indeed shows the characteristics of time-temperature superposition. In particular, the curves are identical up to a load-dependent shift with respect to the time axis. Moreover we observe that at constant heater temperature the time it takes to write a bit is inversely proportional to the tip load. This property is exactly what one would expect if internal friction (owing to the high-frequency viscosity) is the rate-limiting step in bit writing (see Eq. (1)).

The time it takes to heat the bit volume of polymer material to more than the glass-transition temperature is another potentially rate-limiting step. Here the spreading resistance of the heat flow in the polymer and the thermal contact resistance are the most critical parameters. Simulations suggest that equilibration of temperature in the polymer occurs within less than 1 μs [18].

Very little is known, however, on the thermal coupling efficiency across the tip-polymer interface. There are several indications that the heat transfer between tip and sample plays a crucial role, one of them being the asymptotic heater temperature for long writing times, which according to the graph (Figure 6) is approximately 200 °C. The exact temperature of the polymer is unknown. However, the polymer temperature should approach the glass-transition temperature (around 120 °C for PMMA) asymptotically. Hence, the temperature drop between heater and polymer medium is substantial. Part of the temperature difference is due to a temperature drop along the tip, which according to heat-flow simulations [18] is expected to be on the order of 30 °C at most. Therefore, a significant temperature gradient must exist in the tip-polymer contact zone. Further experiments on the heat transfer from tip to surface are needed to clarify this point.

One of the most striking conclusions of our model of the bit-writing process is that it should in principle work for most polymer materials. The general behavior of the mechanical properties as a function of temperature and frequency is similar for all polymers [21]. The glass-transition temperature T_g would then be one of the main parameters determining the threshold writing temperature.

5 Array Design, Technology, and Fabrication

As a first step, a 5 × 5 array chip was designed and fabricated to test the basic Millipede concept. All 25 cantilevers had integrated tip heating for thermomechanical writing and piezoresistive deflection sensing for read-back. No time-multiplexing addressing scheme was used for this test vehicle; rather, each cantilever was individually addressable for both thermomechanical writing and piezoresistive deflection sensing. A complete resistive bridge for integrated detection has also been incorporated for each cantilever.

The chip has been used to demonstrate $x/y/z$ scanning and approaching of the entire array, as well as parallel operation for imaging. This was the first parallel imaging by a 2D AFM array chip with integrated piezoresistive deflection sensing. Details of the all-dry micromachining process for the fabrication of the 5 × 5 array and parallel imaging results are described in [11]. The imaging results also confirmed the global chip-approaching and -leveling scheme, because all 25 tips approached the medium within less than 1 μm of z-actuation. Unfortunately, the chip was not able to demonstrate parallel writing because of electromigration problems due to temperature and current density in the Al wiring of the heater. However, we learned from this 5 × 5 test vehicle that 1) global chip approaching and leveling is possible and promising, and 2) metal (Al) wiring on the cantilevers should be avoided to eliminate electromigration and cantilever deflection due to bimorph effects while heating.

Encouraged by the results of the 5 × 5 cantilever array, a 32 × 32 array chip was designed and fabricated. With the findings from the fabrication and operation of the 5 × 5 array and the very dense thermomechanical writing/reading in thin polymers with

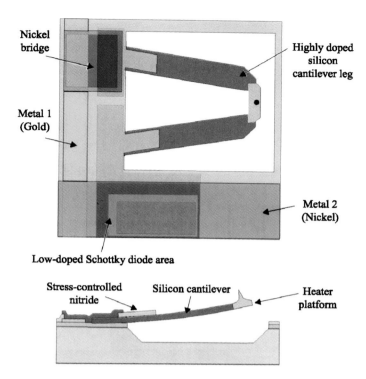

Figure 7: Layout and cross section of one cantilever cell. Reproduced with permission from [23](a), © 1999 IEEE.

single cantilevers, we made some important changes in the chip functionality and fabrication processes. The major differences are 1) surface micromachining to form cantilevers at the wafer surface, 2) all-silicon cantilevers, 3) thermal instead of piezoresistive sensing, and 4) first- and second-level wiring with an insulating layer for a multiplexed row/column-addressing scheme.

Because the heater platform functions as a write/read element and no individual cantilever actuation is required, the basic array cantilever cell becomes a simple two-terminal device addressed by multiplexed x/y wiring, as shown in Figure 7. The cell area and x/y cantilever pitches of 92 µm × 92 µm result in a total array size of less than 3 mm × 3 mm for the 1024 cantilevers. The cantilever is fabricated entirely of silicon for good thermal and mechanical stability. It consists of the heater platform with the tip on top, the legs acting as a soft mechanical spring, and an electrical connection to the heater. The legs are highly doped to minimize interconnection resistance and to replace the metal wiring on the cantilever to eliminate electromigration and parasitic z-actuation of the cantilever due to the bimorph effect. The resistive ratio between the heater and the silicon interconnection sections should be as high as possible; currently the highly doped interconnections are 400 Ω and the heater platform is 5 kΩ (at 3 V reading bias).

The cantilever mass must be minimized to obtain soft (flexible), high-resonant-frequency cantilevers. Soft cantilevers are required for a low loading force in order to eliminate or reduce tip and medium wear, whereas a high resonant frequency allows high-speed scanning. In addition, sufficiently wide cantilever legs are required for a small thermal time constant, which is partly determined by cooling via the cantilever legs [12]. These design considerations led to an array cantilever with 50-µm-long, 10-µm-wide, 0.5-µm-thick legs, and a 5-µm-wide, 10-µm-long, 0.5-µm-thick platform. Such a cantilever has a stiffness of 1 N/m and a resonant frequency of 200 kHz. The heater time constant is a few microseconds, which should allow a multiplexing rate of 100 kHz.

The tip height should be as small as possible because the heater platform sensitivity depends strongly on the distance between the platform and the medium. This contradicts the requirement of a large gap between the chip surface and the storage medium to ensure that only the tips, and not the chip surface, are making contact with the medium. Instead of making the tips longer, we purposely bent the cantilevers a few micrometers out of the chip plane by depositing a stress-controlled plasma-enhanced chemical vapor deposition (PECVD) silicon-nitride layer at the base of the cantilever (see Figure 7). This bending as well as the tip height must be well controlled in order to maintain an equal loading force for all cantilevers of an array.

Figure 8: Photograph of fabricated chip (14 mm × 7 mm). The 32 × 32 cantilever array is located at the center, with bond pads distributed on either side. Reproduced with permission from [23](a), © 1999 IEEE.

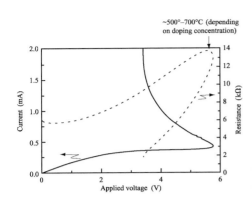

Figure 10: *I/V* curve of one cantilever. The curve is nonlinear owing to the heating of the platform as the power and temperature are increased. For doping concentration between 1×10^{17} and 2×10^{18} at./cm^3, the maximum temperature varies between 500°C and 700°C.
Reprinted from [23](b), by permission of Elsevier Science.

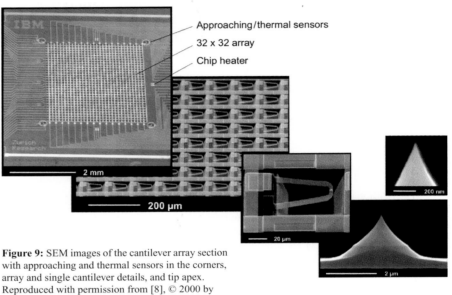

Figure 9: SEM images of the cantilever array section with approaching and thermal sensors in the corners, array and single cantilever details, and tip apex. Reproduced with permission from [8], © 2000 by International Business Machines Corporation.

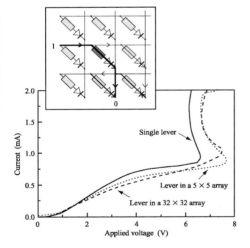

Figure 11: Comparison of the *I/V* curve of an independent cantilever (solid line) with the current response when addressing a cantilever in a 5 × 5 (dotted line) or a 32 × 32 (dashed line) array with a Schottky diode connected serially to the cantilever. Little change is observed in the *I/V* curve between the different cases. Also shown in the inset is a sketch representing the direct path (thick line) and a parasitical path (thin line) in a cantilever-diode array. In the parasitical path there is always one diode in reverse bias that reduces the parasitical current. Reprinted from [23](b), by permission of Elsevier Science.

Cantilevers are released from the crystalline Si substrate by surface micromachining using either plasma or wet chemical etching to form a cavity underneath the cantilever. Compared to a bulk-micromachined through-wafer cantilever-release process, as performed for our 5 × 5 array [11], the surface-micromachining technique allows an even higher array density and yields better mechanical chip stability and heat sinking. Because the Millipede tracks the entire array without individual lateral cantilever positioning, thermal expansion of the array chip must be either small or well-controlled. Because of thermal chip expansion, the lateral tip position must be controlled with better precision than the bit size, which requires array dimensions as small as possible and a well-controlled chip temperature. For a 3 mm × 3 mm silicon array area and 10-nm tip-position accuracy, the chip temperature has to be controlled to about 1 °C. This is ensured by four temperature sensors in the corners of the array and heater elements on each side of the array. Thermal expansion considerations were a strong argument for the 2D array arrangement instead of 1D, which would have made the chip 32 times longer for the same number of cantilevers.

The photograph in Figure 8 shows a fabricated chip with the 32 × 32 array located in the center (3 mm × 3 mm) and the electrical wiring interconnecting the array with the bonding pads at the chip periphery.

Figure 9 shows the 32 × 32 array section of the chip with the independent approach/heat sensors in the four corners and the heaters on each side of the array, as well as zoomed scanning electron micrographs (SEMs) of an array section, a single cantilever, and a tip apex. The tip height is 1.7 μm and the apex radius is smaller than 20 nm, which is achieved by oxidation sharpening [22].

The cantilevers are interconnected by integrating Schottky diodes in series with the cantilevers. The diode is operated in reverse bias (high resistance) if the cantilever is not addressed, thereby greatly reducing crosstalk between cantilevers. More details about the array fabrication are given in Ref. [23].

6 Array Characterization

The array's independent cantilevers, which are located in the four corners of the array and used for approaching and leveling of chip and storage medium, are used to initially characterize the interconnected array cantilevers. Additional cantilever test structures are distributed over the wafer; they are equivalent to but independent of the array cantilevers. Figure 10 shows an *I/V* curve of such a cantilever; note the nonlinearity of the resistance. In the low-power part of the curve, the resistance increases as a function of heating power, whereas in the high-power regime, it decreases.

In the low-power, low-temperature regime, carrier mobility in silicon is affected by phonon scattering, which depends on temperature, whereas at higher power the intrinsic temperature of the semiconductor is reached, resulting in a resistivity drop due to the increasing number of carriers [17]. Depending on the heater-platform doping concentration of 1×10^{17} to 2×10^{18} at./cm^3, our calculations estimate a resistance maximum at temperatures of 500 °C and 700 °C, respectively.

The cantilevers within the array are electrically isolated from one another by integrated Schottky diodes. Because every parasitic path in the array to the addressed cantilever of interest contains a reverse-biased diode, the crosstalk current is drastically reduced, as shown in Figure 11. Thus, the current response to an addressed cantilever in an array is nearly independent of the size of the array, as demonstrated by the *I/V* curves in Figure 11. Hence, the power applied to address a cantilever is not shunted by other cantilevers, and the reading sensitivity is not degraded – not even for very large arrays (32 × 32). The introduction of the electrical isolation using integrated Schottky diodes turned out to be crucial for the successful operation of interconnected cantilever arrays with a simple time-multiplexed addressing scheme. Figure 12 shows the measured cantilever-resistance uniformity across the array for both forward- and reverse-biased diodes. Good uniformity is essential for a robust write/read electronics design.

The tip-apex height uniformity within an array is very important because it determines the force of each cantilever while in contact with the medium and hence influences write/read performance as well as medium and tip wear. Wear investigations suggest that a tip-apex height uniformity across the chip of less than 500 nm is required [4], with the exact number depending on the spring constant of the cantilever. In the case of the Millipede, the tip-apex height is determined by the tip height and the cantilever bending. Figure 13 shows the tip-apex height uniformity of one row of the array (32 tips) due to tip height and cantilever bending. It demonstrates that our uniformity is of the order of 100 nm, thus meeting the requirements.

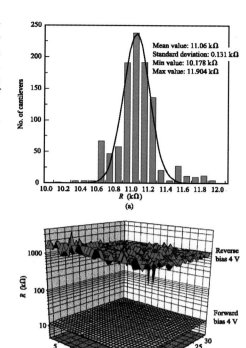

Figure 12: Cantilever resistance uniformity for forward- and reverse-biased diodes: (a) Resistance histogram (4 V forward bias); (b) resistance mapping in 4 V forward and reverse bias. Reproduced with permission from [8], © 2000 by International Business Machines Corporation.

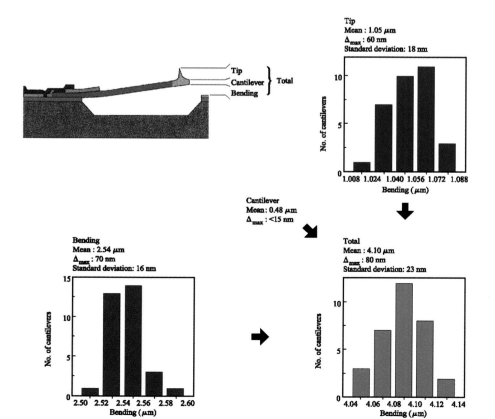

Figure 13: Tip-apex height uniformity across one cantilever row of the array, with individual contributions from the tip height and cantilever bending. Reproduced with permission from [8], © 2000 by International Business Machines Corporation.

7 First Write/Read Results With the 32 × 32 Array Chip

Two $x/y/z$ scanning approaching schemes were explored to operate the array for writing/reading. The first one is based closely on the Millipede basic concept shown in Figure 1. A 3 mm × 3 mm silicon substrate is spin-coated with the SU-8/PMMA polymer medium structure described in Sec. 3. This storage medium is attached to a small magnetic $x/y/z$ scanner and approaching device. The three magnetic z-approaching actuators bring the medium into contact with the tips of the array chip. The z-distance between the medium and the Millipede chip is controlled by the approaching sensors (additional cantilevers) in the corners of the array. The signals from these cantilevers are used to determine the forces on the z-actuators and, hence, also the forces of the cantilever while it is in contact with the medium. This sensing and actuation feedback loop continues to operate during x/y scanning of the medium. The PC-controlled write/read scheme addresses the 32 cantilevers of one row in parallel. Writing is performed by connecting the addressed row for 20 µs to a high, negative voltage and simultaneously applying data inputs ("0" or "1") to the 32 column lines. The data input is a high, positive voltage for a "1" and ground for a "0". This row-enabling and column-addressing scheme supplies a heater current to all cantilevers, but only those cantilevers with high, positive voltage generate an indentation ("1"). Those with ground are not hot enough to make an indentation, and thus write a "0". When the scan stage has moved to the next bit position, the process is repeated, and this is continued until the line scan is finished. In the read process, the selected row line is connected to a moderate negative voltage, and the column lines are grounded via a protection resistor of about 10 kΩ, which keeps the cantilevers warm. During scanning, the voltages across the resistors are measured. If one of the cantilevers falls into a "1" indentation, it cools, thus changing the resistance and voltage across the series resistor. The written data bit is sensed in this manner.

The SEM and the AFM image in Figure 14 show our first parallel writing/reading results. Figure 14 (a) shows a SEM image of a large area of the polymer medium, in which the many small bright spots indicate the location of storage fields with data written by the corresponding cantilevers. The data written consisted of an IBM logo composed of indentations ("1"s) and clear separations ("0"s). Figure 14 (b) shows magnified images of two storage fields. The dots are about 50 nm in diameter, which results in areal densities of 100 - 200 Gb/inch^2, depending on the ability to separate the bit indentations by a distinguishable amount. A first successful attempt demonstrates the read-back of the stored data by the integrated thermomechanical sensing: Figure 14 (c) shows the raw read-back data of two different storage fields with areal density similar to that in Figure 14 (b). Figure 15 is a photograph of the complete experimental Millipede setup.

The second $x/y/z$ scanning and approaching system explored makes use of a modified magnetic hard-disk drive. The array chip replaced the magnetic write/read head slider and was mechanically leveled and fixed on the suspension arm. The z-approaching and -contacting procedure was performed by a piezoelectric actuator mounted on top of the suspension, which brought the array chip into contact with the medium and maintained it there. More details are given in Ref. [8].

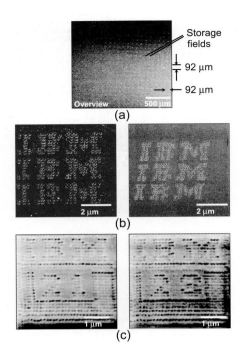

Figure 14: SEM images of $x/y/z$ scanner Millipede writing results:
(a) SEM image of many storage fields.
(b) Magnified views of two individual storage fields with IBM logo represented by bit indentations/separations equivalent to a storage density of 100 - 200 Gb/inch^2 (50-nm-thick PMMA medium).
(c) Thermomechanical read-back signals of two storage fields demonstrating areal density similar to those in (b).
Reproduced with permission from [8], © 2000 by International Business Machines Corporation.

8 Discussion of Possible Millipede Applications in Data Storage

Our current 32 × 32 array chip is just one example of the many possible designs of a data-storage system; the design and concept strongly depend on the intended application. Figure 16 shows the storage capacity and access time as a function of array size and cantilever pitch/scan range of a Millipede data-storage system according to the basic concept of Figure 1, assuming an areal density of 500 Gb/inch^2. It demonstrates the great potential for low- to high-end applications, of which the current 32 × 32 array is but one case. The two shaded circles in Figure 16 represent the storage capacity and access time for the current 32 × 32 array. It is important to note that the same data capacity can be achieved, for example, using large arrays with small cantilever pitch/scan range or, conversely, using small arrays with a larger scan range. In addition, terabit data capacity can be achieved by

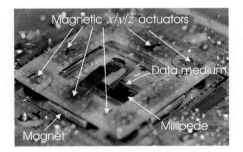

Figure 15: Photograph of experimental Millipede setup with $x/y/z$ magnetic actuators.
Reproduced with permission from [8], © 2000 by International Business Machines Corporation.

one large array, by many identical small ones operating in parallel, or by displacing a small array on a large medium. Out of this wide range of design and application scenarios, we would like to explain two cases of particular interest.

- *Small-form-factor storage system (Nanodrive)*

IBM's miniaturized harddrive, the Microdrive, represents a first successful step into miniaturized storage systems. As we enter the age of pervasive computing, we can assume that computer power is available virtually everywhere. Miniaturized and low-power storage systems will become crucial, particularly for mobile applications. The availability of storage devices with gigabyte capacity having a very small form factor (in the range of centimeters or even millimeters) will open up new possibilities to integrate such "Nanodrives" into watches, cellular telephones, laptops, etc., provided such devices have low power consumption.

We have recently demonstrated that the magnetic $x/y/z$ scanner can be micromachined and miniaturized [24], [25] with potential for even further miniaturization. Figure 17(a) shows the basic principles of the integrated micromagnetic scanner concept, whereas Figure 17(b) is a photograph of the first micromachined silicon scanner. Further details on the design and fabrication have been published in Refs. [24], [25].

The array chip with integrated or hybrid electronics and the micromagnetic scanner are key elements demonstrated for a Millipede-based device called Nanodrive, which is of course also very interesting for audio and video consumer applications. All-silicon, batch fabrication, low-cost polymer media, and low power consumption make Millipede very attractive as a centimeter- or even millimeter-sized gigabyte storage system.

- *Terabit drive*

The potential for very high areal density renders the Millipede also very attractive for high-end terabit storage systems. As mentioned above, terabit capacity can be achieved with three Millipede-based approaches: 1) very large arrays, 2) many smaller arrays operating in parallel, and 3) displacement of small/medium-sized arrays over large media.

Although the fabrication of considerably larger arrays (10^5 to 10^6 cantilevers) appears to be possible, control of the thermal linear expansion will pose a considerable challenge as the array chip becomes significantly larger. The second approach is appealing because the storage system can be upgraded to fulfill application requirements in a modular fashion by operating many smaller Millipede units in parallel. The operation of the third approach was described above with the example of a modified hard disk. This approach combines the advantage of smaller arrays with the displacement of the entire array chip, as well as repositioning of the polymer-coated disk to a new storage location on the disk. A storage capacity of several terabits appears to be achievable on 2.5- and 3.5-inch disks. In addition, this approach is an interesting synergy of existing, reliable (hard-disk drive) and new (Millipede) technologies.

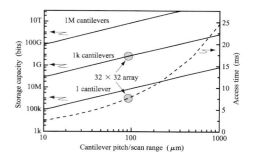

Figure 16: Storage capacity and access time vs. cantilever pitch/scan range for various array sizes, assuming 500 Gb/inch² storage density. Reproduced with permission from [8], © 2000 by International Business Machines Corporation.

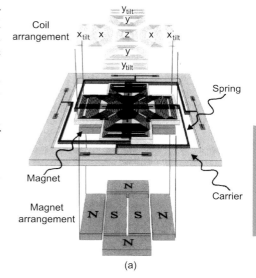

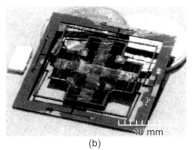

Figure 17:
(a) Diagram illustrating basic concept of integrated micromagnetic $x/y/z$ scanner.
(b) Photograph of micromachined $x/y/z$ scanner.
Reproduced with permission from [8], © 2000 by International Business Machines Corporation.

9 Summary and Outlook

In this chapter, the fabrication and operation of large 2D AFM arrays for thermomechanical data storage in thin polymer media are described and the key milestones of the Millipede storage concept are outlined. The 400 - 500-Gb/inch² storage density demonstrated with single cantilevers is among the highest reported so far. The initial densities of 100 - 200 Gb/inch² achieved with the 32 × 32 array are very encouraging, with the potential of matching those of single cantilevers. Well-controlled processing techniques have been developed to fabricate array chips with good yield and uniformity. This VLSI-NEMS chip has the potential to open up new perspectives in many other applications of scanning-probe techniques as well. Millipede is not limited to storage applications or polymer media. The concept is very general if the required functionality can be integrated on the cantilever/tip. This of course applies also to any other storage medium, including magnetic ones, making Millipede a possible universal parallel write/read head for future storage systems. Besides storage, other Millipede applications can be envisioned for large-area, high-speed imaging and high-throughput nanoscale-lithography [26], [27], as well as for atomic and molecular manipulation/modification.

The current Millipede array chip fabrication technique is compatible with CMOS circuits, which will allow future microelectronics integration. This is expected to produce better performance (speed) and smaller system form factors, as well as lower costs.

Although the authors have demonstrated the first high-density storage operations with the largest 2D AFM array chip ever built, a number of issues must be addressed before the Millipede can be considered for commercial applications; a few of these are listed below:

- Overall system reliability, including bit stability, tip and medium wear, erasing/rewriting.
- Limits of data rate (S/N ratio), areal density, array and cantilever size.
- CMOS integration.
- Optimization of write/read multiplexing scheme.
- Array-chip tracking.

Our near-term future activities are focused on these important aspects.

Finally, we would like to conclude this chapter with some thoughts and visions on the long-term outlook for the Millipede concept.

The highly parallel nanomechanical approach is novel in many respects. There is at least one feature of the Millipede that we have not yet exploited, and we might ask whether a new device of a yet inconceivable level of novelty could possibly emerge from the Millipede. With integrated Schottky diodes and the temperature-sensitive resistors on the current version of the Millipede array chip, we have already achieved the first and simplest level of micromechanical/electronic integration, but we are looking for much more complex ones to make sensing and actuation faster and more reliable. However, we envision something very much beyond this. Whenever there is parallel operation of functional units, there is the opportunity for sophisticated communication or logical interconnections between these units. The topology of such a network carries its own functionality and intelligence that goes beyond that of the individual devices. It could, for example, act as a processor. For the Millipede this could mean that a processor and VLSI-nanomechanical device may be merged to form a "smart" Millipede. We do not completely foresee what this could ultimately mean, but we do have a vision of the direction in which to go. If the Millipede is used, for example, as an imaging device, let us say for quality control in silicon chip fabrication, the amount of information it can generate is so huge that it is difficult to transmit these data to a computer to store and process them. Furthermore, most of the data are not of interest at all, so it would make sense if only the pertinent parts were predigested by the specialized smart Millipede and then transmitted. For this purpose, communication between the cantilevers is helpful because a certain local pattern detected by a single tip can mean something in one context and something else or even nothing in another context. The context might be derived from the patterns observed by other tips. A similar philosophy could apply to the Millipede as a storage device. A smart Millipede could possibly find useful pieces of information very quickly by a built-in complex pattern recognition ability, e.g., by ignoring information when certain bit patterns occur within the array. The bit patterns are recognized instantaneously by logical interconnections of the cantilevers. Even with this somewhat vague vision, we are very confident that the "smart" Millipede will have interesting long-term prospects in many application fields, possibly in fields that we cannot even envision today.

Acknowledgements

The authors acknowledge the invaluable technical contributions of our team mates U. Drechsler, B. Gotsmann, W. Häberle and R. Stutz, the technical assistance of R. Beyeler and D. Caimi, as well as the encouraging support of P. F. Seidler.

Special thanks and appreciation go to our colleagues T. Albrecht, T. Antonakopoulos, P. Bächtold, G. Cherubini, A. Dholakia, E. Eleftheriou, D. Jubin, T. Loeliger, H. Pozidis, D. Wiesmann, H. Dang, A. Sharma, and S. Sri-Jayantha for their contributions to our current storage system prototyping effort.

In addition, it is our pleasure to acknowledge J. Frommer, C. Hawker, V. Lee, J. Mamin, and R. Miller of the IBM Almaden Research Center for their motivated collaboration on alternative and modified polymer materials.

The editor gratefully acknowledges Andreas Roelofs (RWTH Aachen) for his editorial and technical assistance in the compilation of this chapter.

References

[1] E. Grochowski and R. F. Hoyt, IEEE Trans. Magn. **32**, 1850 (1996).

[2] H. J. Mamin and D. Rugar, Appl. Phys. Lett. **61**, 1003 (1992).

[3] R. P. Ried, H. J. Mamin, B. D. Terris, L. S. Fan, and D. Rugar, J. Microelectromech. Syst. **6**, 294 (1997).

[4] B. D. Terris, S. A. Rishton, H. J. Mamin, R. P. Ried, and D. Rugar, Appl. Phys. A **66**, S809 (1998).

[5] H. J. Mamin, B. D. Terris, L. S. Fan, S. Hoen, R. C. Barrett, and D. Rugar, IBM J. Res. Develop. **39**, 681 (1995).

[6] H. J. Mamin, R. P. Ried, B. D. Terris, and D. Rugar, Proc. IEEE **87**, 1014 (1999).

[7] D. A. Thompson and J. S. Best, IBM J. Res. Develop. **44**, 311 (2000).

[8] P. Vettiger, M. Despont, U. Drechsler, U. Dürig, W. Häberle, M. I. Lutwyche, H. E. Rothuizen, R. Stutz, R. Widmer, and G. K. Binnig, IBM J. Res. Develop. **44**, 323 (2000).

[9] G. K. Binnig, H. Rohrer, and P. Vettiger, *Mass-Storage Applications of Local Probe Arrays*, U.S. Patent 5,835,477, November 10, (1998).

[10] P. Vettiger, J. Brugger, M. Despont, U. Drechsler, U. Dürig, W. Häberle, M. Lutwyche, H. Rothuizen, R. Stutz, R. Widmer, and G. Binnig, J. Microelectron. Eng. **46**, 11 (1999).

[11] M. Lutwyche, C. Andreoli, G. Binnig, J. Brugger, U. Drechsler, W. Häberle, H. Rohrer, H. Rothuizen, and P. Vettiger, Proceedings of the IEEE 11th International Workshop on Micro Electro Mechanical Systems (MEMS '98), Heidelberg, Germany, 1998, M. Lutwyche, C. Andreoli, G. Binnig, J. Brugger, U. Drechsler, W. Häberle, H. Rohrer, H. Rothuizen, P. Vettiger, G. Yaralioglu, and C. Quate, Sensors & Actuators A **73**, 89 (1999).

[12] B. W. Chui, H. J. Mamin, B. D. Terris, D. Rugar, K. E. Goodson, and T. W. Kenny, Proceedings of IEEE Transducers '97, Chicago, IL, (1997).

[13] G. Binnig, M. Despont, U. Drechsler, W. Häberle, M. Lutwyche, P. Vettiger, H. J. Mamin, B. W. Chui, and T. W. Kenny, Appl. Phys. Lett. **74**, 1329 (1999).

[14] G. K. Binnig, M. Despont, W. Häberle, and P. Vettiger, *Method of Forming Ultrasmall Structures and Apparatus Therefor*, filed at the U.S. Patent Office, March 17, 1999, Application No. 147865.

[15] P. Vettiger, G. Cross, M. Despont, U. Drechsler, U. Dürig, B. Gotsmann, W. Häberle, M. A. Lantz, H. E. Rothuizen, R. Stutz, and G. K. Binnig, IEEE Trans. Nanotechnol. **1**, 39 (2002).

[16] G. K. Binnig, J. Brugger, W. Häberle, and P. Vettiger, "Investigation and/or Manipulation Device," filed at the U.S. Patent Office, March 17, 1999, Application No. 147867.

[17] S. M. Sze, *Physics of Semiconductor Devices*, John Wiley, New York, 1981.

[18] W. P. King, J. G. Santiago, T. W. Kenny, and K. F. Goodson, in *ASM MEMS*, Vol. 1, 583 (1999); W. P. King, T. W. Kenny, K. E. Goodson, G. L. W. Cross, M. Despont, U. Dürig, H. Rothuizen, G. Binnig, and P. Vettiger, J. Microelectromech. Sys. (submitted).

[19] K. Fuchs, Chr. Friedrich, and J. Weese, Macromolecules **29**, 5893 (1996).

[20] The estimate is based on a fluid dynamic deformation model of a thin film. U. Dürig and B. Gotsmann, unpublished.

[21] J. D. Ferry, *Viscoelastic Properties of Polymers*. Wiley and Sons, New York, 1980.

[22] T. S. Ravi and R. B. Marcus, J. Vac. Sci. Technol. B **9**, 2733 (1991).

[23] (a) M. Despont et al., Technical Digest, 12th IEEE International Micro Electro Mechanical Systems Conference (MEMS'99), Orlando, FL, 1999, (b) idem, Sensors & Actuators A **80**, 100 (2000).

[24] M. Lutwyche et al., *Magnetic Materials, Processes, and Devices: Applications to Storage and Microelectromechanical Systems (MEMS)*, L. T. Romankiw, S. Krongelb, and C. H. Ahn, eds., The Electrochemical Society, Pennington, NJ, 1999.

[25] H. Rothuizen, U. Drechsler, G. Genolet, W. Häberle, M. I. Lutwyche, R. Stutz, R. Widmer, and P. Vettiger, Microelectron. Eng. **53**, 509 (2000).

[26] S. C. Minne, G. Yaralioglu, S. R. Manalis, J. D. Adams, A. Atalar, and C. F. Quate, Appl. Phys. Lett. **72**, 2340 (1998).

[27] N. C. MacDonald, Digest of Technical Papers, 7th International Conference on Solid-State Sensors and Actuators, Yokohama, Japan, 1993.

Data Transmission and Interfaces

Contents of Part VI

	Introduction to Part VI	703
29	Transmission on Chip and Board Level	713
30	Photonic Networks	727
31	Microwave Communication Systems – Novel Approaches for Passive Devices	753
32	Neuroelectronic Interfacing: Semiconductor Chips with Ion Channels, Nerve Cells, and Brain	777

Introduction to Part VI

Contents

1 Signal Transmission . . . 703
2 Types of Signals and Limits to Transmission . . . 704
3 Unmodulated Transmission – Transmission Lines . . . 705
4 Modulated Transmission - Communication Systems . . . 707
   4.1 Shannon Communication Model . . . 707
   4.2 Channel Capacity and Carrier Modulation . . . 707
   4.3 Source Encoding . . . 708

1 Signal Transmission

As described in the General Introduction, information is always bound to a physical medium in the form of signals in order to be processed, stored, or transmitted. The transmission of signals can be classified into a transmission across interfaces and a transmission along channels as shown in Figure 1. An entire communication system typically involves both. The transmission along channels is further subdivided into a group which operates with a modulation of the signal onto a carrier and a group which operates without modulation. The later is typically used for short and medium distances. Examples range from signal transfer on the interconnects of integrated chips over distances of micrometers to millimeters, data buses within systems, to serial and parallel connections between systems, including serial busses and local area networks (LAN). Modulated transmission can be divided into wire-bound and wireless systems. Examples of wire-bound transmission are rf-carrier based transmission through coaxial wires, transmission by light through optical fibers, and the transmission of action potential spikes along the axon of biological neurons. Wireless communication is typically based on modulated electromagnetic radiation. It is either omnidirectional (broadcasting) or point-to-point (e. g. by microwave or laser beam links).

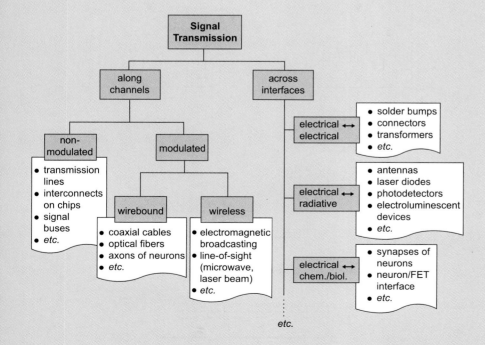

Figure 1: Categories of the signal transmission showing the most relevant classes.

VI Data Transmission and Interfaces

Interfaces are required whenever signal parameters (e. g. voltage levels) and/or the medium for signal transmission are changed in a communication system. Often, this involves a change of the information code too. Electrical/electrical interfaces include connectors on the board and system level, transformers, bus transceiver etc.. Interfaces between electrical conductors and electromagnetic radiation include all types of antennas, components of the optical data communication such as laser diodes to feed optical fibres, photodetectors to convert optical signals into electrical signals. Conceptionally, one can regard displays (Part VIII) as systems which represent an electrical $\rightarrow$ optical interface for information transfer. In the same manner, devices with an non-electrical $\rightarrow$ electrical interface can be regarded as electrical sensors (Part VII).

A more detailed overview of the theory and methods of information transmission is compiled in references such as [1], [2], [3].

2 Types of Signals and Limits to Transmission

Depending on the alphabet used, information is represented by either discrete or continuous values, which are coded as signals onto a physical medium either as a continuous-time stream or as a sequence of discrete sampling times. Signals which are continuous-value and continuous-time are called *analog*, while those which are discrete-value and discrete-time are called *digital* [4]. In the contemporary communication technology typically binary coded, digital signals are transmitted because they are suitable for numerical processing. This processing is used, for example, to include redundant code and to facilitate detection and correction of error which entered the signal during the transmission (see Sec. 4.2). If the information to be transmitted is not already binary coded, as in data communication, there is a need to convert analog signals (e. g. speech, audio, video) into digital signals. Analog-to-digital (A/D) conversion comprises sampling and quantizing as illustrated in Figure 2. The sampling process substitutes the continuous signal $s(t)$ by

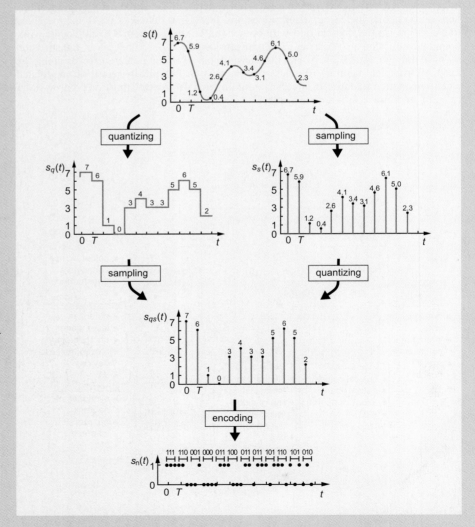

Figure 2: Types of signals and illustration of the pulse code modulation (PCM), adapted from Ref. [5] with modifications. The analog signal is sampled and quantized first (analog-to-digital conversion) and then encoded into the binary code. In this illustration, the digital signal has a resolution of $2^3 = 8$. To convert into binary signals, the sampling time T has to be reduced to $T/3$.
$s(t)$ continuous-value and continuous-time signal (analog signal)
$s_q(t)$ discrete-value and continuous-time signal
$s_s(t)$ continuous-value and discrete-time signal
$s_{qs}(t)$ discrete-value and discrete-time signal (digital signal)
$s_n(t)$ binary coded signal

$$s_s(t) = s(t) \sum_{n=-\infty}^{\infty} \delta(t - nT) \qquad (1)$$

where δ denotes the delta function. For transmission, the signal is encoded from the multivalued digital form into the binary coded form. A resolution of the digital signal of 2^N requires a decrease of the sampling time T to T/N.

The analog signal $s(t)$ to be sampled by an A/D converter needs to be bandwidth limited, i. e. it may not contain spectral components with frequencies higher than a frequency f_g, where

$$2 f_g \leq \frac{1}{T} \qquad (2)$$

and $1/T$ is the sampling rate. Eq. (2) is called the **Nyquist criterion**. The rational for the Nyquist criterion becomes obvious in the frequency regime. $S(f)$ is supposed to be the spectral distribution of an original signal $s(t)$ (Figure 3 a). The substitution of $s(t)$ by $s_s(t)$ according to Eq. (1) leads to a change of the spectral distribution as shown in Figure 3 b. Any selected spectral component at a frequency f_i of the original signal is reproduced at frequencies

$$f_i^* = \frac{m}{T} \pm f_i \quad \text{with m = 1, 2, 3, ...} \qquad (3)$$

i. e. the original spectrum $S(f)$ is repeated periodically with the period m/T. As long as the Nyquist criterion is fulfilled, these additional spectral components cause no problem since the original signal can be recovered by appropriate low-pass filtering with a cut-off frequency at f_g. However, if the sampling rate is chosen too low, the spectra caused by the sampling and the original spectrum overlap (Figure 3 c). As a consequence, spectral components are produced by the sampling which cannot be eliminated any more.

Independent of the type of signal, there are additional influences which degrade signals and limit the information which can be transmitted via a given channel:

- **Noise** denotes any undesired, unpredictable component added to the signal by the environment. For example, it may be generated by the thermal energy in a resistor or other components which the signal has to pass. Another typical source are electric signals in the vicinity to the signal channel which induce noise by electromagnetic emissions.
- **Distortion** is any undesired modification of the signal amplitude or of its phase and spectral composition, e.g. by a *non-linear characteristics*. It includes *dispersion* of transmitted signals because of a frequency dependent signal velocity.
- **Attenuation** is the reduction of the power of a signal while it travels along a channel or passes an interface. The signal attenuation is usually expressed in decibels (dB) per unit distance. Its frequency dependence is described by the spectral transfer function of the system.

The implication of these general degradation effects will be discussed throughout the following text. In addition, signal degradation effects will be mentioned which are specific for one or another system to be discussed.

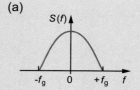
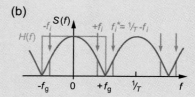
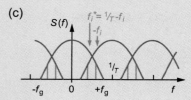

Figure 3: Illustration of the effect of sampling. (a) Spectrum $S(f)$ of an analog signal $s(t)$. (b) Spectrum of the signal after sampling at rate $1/T$. The orignal spectrum is repeated at frequencies m/T (m = 1, 2, 3, …). Since the Nyquist criterion for the sampling rate $1/T \geq 2f_g$ is fulfilled, the additional spectra can be eliminated by a low-pass filter $H(f)$. (c) Spectrum for sampling at a rate too low. Components of the original spectrum at frequency f_i generate artificial components at frequencies f_i^* which are in interval of the original spectrum and, hence, cannot be eliminated.

3 Unmodulated Transmission – Transmission Lines

The transmission properties of electrical wires and cables and the corresponding theory can be subdivided into three categories, depending on the wavelength λ_g corresponding to the highest frequency f_g which may occur in the signal to be transmitted and the geometry of the line (Figure 4):

- **Large wavelength $\lambda_g \gg l$:**

  If the wavelengths are large compared to length of the line under consideration, a *low-frequency* description of the system is appropriate. The ohmic resistances dominate the overall impedance, while line inductances play a minor role. The large wavelength condition is fulfilled, for example, in wiring a home audio equipment or for the interconnects of integrated circuits which are operated at moderate frequencies.

- **Intermediate wavelength $d \ll \lambda_g \ll l$:**

  In this regime, the capacitances and inductances distributed along the line dominate the impedance consideration. Wires and cables under these conditions are treated as *transmission lines*. The boundary conditions and implications are sketched below.

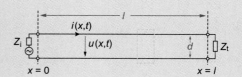

Figure 4: Illustration of a system for the transmission of signals along wires and cables. From left to right, there is a signal generator with a source impedance Z_i, the cable sketched by two conductors at distance d (e. g. the core and shield of a coaxial cable) and a length l, and a termination impedance Z_t. The dashed rectangle represents an element of the line.

- **Short wavelength** $d \geq \lambda_g$:

   This is the *waveguide* regime. For these high frequencies, the formation of modes of electromagnetic waves have to be considered along the line. A short introduction into the theoretical treatment is given in Chapter 1. It is typically applied in signal transmission using high-frequency carriers, for example, in the fiber-based optical communication (Chapter 30) and some aspects of the microwave-based communication (Chapter 31). Frequencies which can occur in the unmodulated transmission, are too low to require a waveguide description of the lines.

Wires and cables between systems (local-area networks, LAN), within systems (system-wide buses), and on the board level have always been treated as *transmission lines*. Since the operating frequencies of digital electronics have increased beyond one GHz and the signal rise times go below approx. 150 ps [6], transmission line considerations are needed for the global interconnects (Chapter 29) on the chip level as well.

A transmission line is described as a distribution of equivalent circuits along the line (Figure 5). The **characteristic impedance** of the line is given by

$$Z_0 = \sqrt{\frac{R' + j\omega L'}{G' + j\omega C'}} \approx \sqrt{\frac{L'}{C'}} \qquad (4)$$

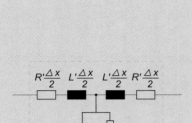

Figure 5: Equivalent circuit of a transmission line segment of length Δx. The elements are explained in the text.

where R', G', L', and C' denote the conductor resistance per unit length, the conductance of the dielectric insulator per unit length, the inductance per unit length, and the capacitance per unit length, respectively. For a derivation of Eq. (4), the reader is referred to standard textbooks, e. g. [1], [2]. For not too long distances and reasonably good conductors and dielectrics, R' and G' can be neglected and one arrives at the simplified term on the right. The value of Z_0 for typical cables is in the range of a few ten to a few hundred Ohms.

The propagation velocity v of a signal wave is

$$v = \frac{1}{\sqrt{L'C'}} = \frac{c_0}{\sqrt{\mu_r \varepsilon_r}} \qquad (5)$$

where c_0 represents the speed of light and the denominator in the right term is not much larger than 1 for typical materials used in cables and lines.

The transmission line description leads to several implications. The impedance of the transmitting circuit should be low enough to drive the characteristic impedance of the line. Furthermore, the line needs to be terminated by a matching load impedance $Z_t = Z_0$, which can be either an ohmic resistance or a subsequent transmission line. For such a termination, the wave energy is either dissipated in the ohmic terminal or guided away by the subsequent line. Without impedance matching, the signal wave is (partially or completely) reflected back. The reflection factor r represents the ratio of the signal voltage of the reflected and incident wave at the termination position:

$$r = \frac{Z_t - Z_0}{Z_t + Z_0} \qquad (6)$$

For an open termination ($Z_t \to \infty$), $r = 1$, for a short-circuited termination ($Z_t = 0$), $r = -1$. While the termination of buses and cables by a matching impedance is relatively simple, the situation on an integrated chip is much more delicate [6]. This is because in contrast to, e. g., a coaxial cable consisting of a core and shield, an interconnect line on a chip typically has no parallel counter line to establish an uniform characteristic impedance. Due to discontinuities of Z_0, caused by a changing environment (neighbouring interconnects) along a global interconnect, there are partial reflections which must be kept at a minimum. Furthermore, resistive terminations which are the simplest and most efficient method, lead to power dissipation and are, therefore, replaced by non-linear and active elements. For on-chip interconnects, Eq. (4) has to be used in its full form because the line resistances are not negligible. Often, the contribution of R' is dominant, requiring that the description has to be modified. Reflections are no issue anymore. However, the signal velocity is smaller than stated by Eq. (5), as described in the Introduction of Part III.

Another important issue that becomes relevant at much lower frequencies (above approx. 10 MHz) is the capacitively induced noise, which is induced by the **capacitive coupling** of signal-carrying lines in the vicinity of a line under consideration.

All these signal degradation effects must be kept below a critical level by a careful chip design and layout [6].

4 Modulated Transmission - Communication Systems

4.1 Shannon Communication Model

According to Shannon [8], a general communication system can be represented by an abstract model as sketched in Figure 6. A *source* transfers an information to an encoder which adopts it for the transmission along a *channel*. At the end of the channel, the information is decoded and transferred to the *receiver*. A simple example is the conversation of two persons by an analog telephone line. One person (information source) speaks into the telephone receiver (encoder), which converts the acoustic signal into an electrical signal of appropriate impedance. The electrical signal is transmitted over the telephone line (channel) to the other end of the line, where it is converted back into an acoustic signal to be listened by the second person (information receiver). The concept can be further generalized from a communication in the *time domain* to a communication in the *space domain* which then represents information storage systems.

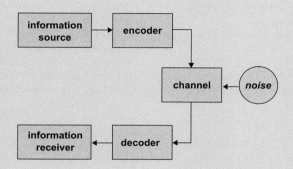

Figure 6: Shannon's model of a communication system.

In the case of high-rate digital communication, the encoder in Figure 6 has to be subdivided as shown in Figure 7:

- the **source encoder** extracts redundancies from the information, in order to increase the amount of information which can be transmitted through a given communication system,
- the **channel encoder** adds redundant information again, to allow errors to be detected or/and corrected, depending on the coding
- the **line encoder** adapts the signal to the requirements of the physical channel. In carrier-based transmission, it performs the modulation of the carrier.

The decoder comprises the appropriate stages, to recover the information. We will briefly describe the function of the three encoder stages for a modulated transmission.

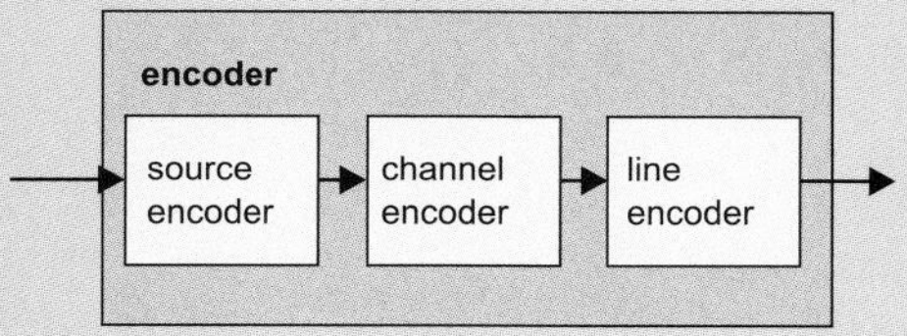

Figure 7: Functional blocks of the encoder for digital communication systems.

4.2 Channel Capacity and Carrier Modulation

As described in the General Introduction, the average information content (per position) is measured by the average information H. With r as the rate at which information is transferred we arrive at the *information flux*

$$H^* = rH \qquad (7)$$

which has the unit bit/s.

These parameters, average information and information flux, are helpful to answer the question of the maximum rate of information transfer without loosing information or with a given information loss across a given channel. This value is the channel capacity.

Given parameters to derive the channel capacity are the entropy and the transfer rate. Also, the transition failure rate from a defined input value to a changed output value is known from observation of the channel.

The information flux at the output of the channel, is defined as

$$H^*_{Output} = H^* - H^*_{loss} \qquad (8)$$

H^* represents the information flux of the source, H^*_{loss} is the loss of information due to the non-ideality of the channel. Now, the channel capacity C^* is defined as the maximum of the information flux for all possible sources,

$$C^* = \max\{H^*_{Output}\} \qquad (9)$$

This means a certain source information can be transferred across a given channel without loss of information if the information flux is smaller or equal to the channel capacity.

Most physical carriers are utilizing continuous signals. The line encoder has to convert the digital signal into a signal which is suitable for transmission, either modulated or non-modulated.

For the Gauss-Channel which is a good approximation for most of the transmission channels, the channel capacity is described by the following equation

$$C^* = f_B \, \mathrm{ld}\left[1 + S/N\right] \qquad (10)$$

where S denotes the power of the signal and N is the power of the noise. In the case of ideal low-pass systems (e. g. all unmodulated transmission systems), f_B represents the maximum signal frequency f_g. In the case of ideal band-pass systems, f_B is the bandwidth Δf.

Thus, a channel capacity can be realized with different combinations of the parameters bandwidth and signal to noise ratio. If we increase the bandwidth of the channel, we can achieve a transmission with the same error probability at a smaller signal to noise ratio. But there is a limit in reducing the signal to noise ratio at the expense of the bandwidth, when f_B approaches infinity. In this case, we find the value of the **channel capacity**

$$C_\infty^* = 0.72 \frac{S}{N} \qquad (11)$$

From this equation we derive the minimum power required for transmission $S = C_\infty^* N/0.72$. At a transmission time of a single binary value $T = 1/C_\infty^*$ the minimum energy for the transfer of a single bit can be derived as $W_{min} = S \times T = N/0.72$ and finally the Shannon limit, the minimum signal to noise ratio for error free data transmission, is found. For smaller values, the error probability increases drastically. The Shannon limit is shown in Figure 8. Using a channel encoding, like block coding, the discrepancy between the non-coded transmission and the theoretical limit can be abated (see Section 4.3).

$$\left.\frac{S}{N}\right|_{min} = 1.39 = 1.42 \, dB \qquad (12)$$

4.3 Source Encoding

For an optimized transmission in view of speed, failure probability and requested energy per bit, a coding of the source data is very helpful. Hereby, we can distinguish between source encoding, channel encoding and line encoding.

Source encoding methods are applied to reduce the redundancy R by increasing H and bringing it closer to the maximum possible value H_{max}. In the case of pulse code modulation (PCM), source encoding may be based on generating and transmitting a differential signal (differential pulse code modulation, DPCM). In the case of text messages, this can be facilitated by taking the natural occurrence probability of the letters into account. H can be further increased if the statistical interdependencies of the letters in a word, a word in a sentence, etc. are considered. In other cases, the source encoder searches for repeating blocks of information and sends only a short code, if a repetition is encountered. In the case of music or pictures, those parts of the entire information are eliminated which are physiologically not or less relevant. These compression algorithms can be very effective and are able to enhance H by factors up to ten and more. Examples are the MPEG3 standard for music and the MPEG4 standard (MPEG: Motion Pictures Expert Group). The MPEG3 coding for example is based on psycho-acoustic effects, which take into consideration the cognition of ear and brain. The frequency spectrum of the source signal is divided into 32 subbands. Those parts which are very sensitive to the ears, are set into narrow bands. Afterwards a masking effect eliminates parts which are not recognized by the ear. These methods help to compress the source signal by a factor of 12. To achieve an optimal data compression, often several methods are combined.

4.4 Channel Encoding

The purpose of **channel encoding** is to introduce additional information which can be used to detect and correct errors caused by noise during the transmission. Since these additional bits "dilute" the original information, they increase its redundancy. One method of channel encoding is to calculate redundant bits from the contents of a block of information in a way that a corresponding calculation in the channel decoder can locate and correct any erroneous bit in the block. The possibility of error detection or correction closely relates to the difference or distance between all valid codes, called **Hamming distance**. 100, 010 and 001 might be valid codes to be used for information transfer across a channel. Since in each case 2 digits have to be changed to convert one valid code into another, the Hamming distance amounts to 2. To detect an n bit error, a Hamming distance of $n+1$ is required, in case of the correction of n bit errors, a Hamming distance of $2n+1$ is required. In our example, just a one bit error ($n=1$) can be detected (e.g. 110 or 000). A restoration of the transmitted information is not possible. A code with a Hamming distance of 3 that also enables error correction, might consist of the two valid digit sequences 111 and 000. If, due to a perturbed signal transmission, one bit of the sent digit sequence flips (received signal 101, 110, ...), the signal can be reconstructed without loss of information. In case of 2 incorrect bits, at least the occurrence of a transmission fault can be detected.

In general, a Hamming distance larger than the minimum distance between the code words is chosen in order to ensure low error probability. Another method of channel encoding is often used in bi-directional communication systems. Here, error detection is realized, e.g. by the use of parity bits and parity checks. In case of an error, the channel decoder simply calls for a repetition of the transmission.

Figure 8 displays the Shannon limit as well as the characteristics for transmission without channel encoding (unipolar signal) and a channel encoded signal (Bose-Chaudhuri-Hocquenghem-block code, BCH). By the use of channel encoding, this limit can be approached much better. The BCH-block code consists of 127 bits in length. 92 bits source signals are extended by 35 bits redundant information, which enables the correction of up to five bits within one block. The distance to the Shannon limit can be further reduced by more tricky coding methods.

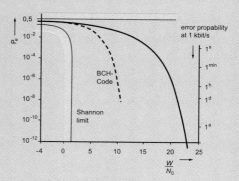

Figure 8: Error probability of different transmission methods, dashed line: BCH-code represents a transmission method including channel encoding. Solid line: unipolar transmission and the theoretical limit for error free transmission, the Shannon limit (redrawn after Ref. [5], with modifications).

4.5 Line Encoding and Modulation

After source and channel encoding, the signal to be transmitted has to be transformed into a suitable form with respect to the physical properties of the channel, e.g. for fiber optic communication, an electrical signal has to be converted into an optical signal. In general, it can be distinguished between **line encoding** and **modulation**. *Line encoding* denotes the *base band transmission* of the respective signal. Base band transmission is characterized by a conservation of the frequency spectrum of the original signal into the encoded signal. Transmission of digitized signals via base band transmission is also possible. For example, the digits can be encoded by unipolar or bipolar signals. Both methods differ in the error probability. For a given signal to noise ratio, bipolar transmission shows a smaller error probability than the unipolar transmission. Depending on the physical channel properties, a variety of further digital line encoding methods can be used.

Signal modulation

Signal modulation is required, if the information has to be transmitted across a *band pass transmission* channel. Reasons might be the physical restrictions of the channel that does not allow for a transmission of low frequency signals or the aim to transmit more than one signal across the some channel, e.g. by frequency multiplexing. An example of frequency multiplexing will be given below.

Analogue and digital signal encoding for band pass transmission

For the band pass transmission a sinusoidal *carrier* signal such as $\cos(2\pi f_0 t)$ is needed. The carrier signal frequency f_0 is chosen to meet the requirements for transmission through the band pass channel. In case of an analogue input signal, amplitude modulation (AM), frequency modulation (FM), or phase modulation (PM) of the carrier signal can be performed. All modulation methods only cause weak changes of the carrier signal frequency that does not affect the band pass transmission. Binary signal carrier encoding can be done by corresponding modulation methods which are called amplitude shift keying (ASK), phase shift keying (PSK), frequency shift keying (FSK), or minimum shift keying (MSK). Figure 9 illustrates the different digital modulation methods.

VI Data Transmission and Interfaces

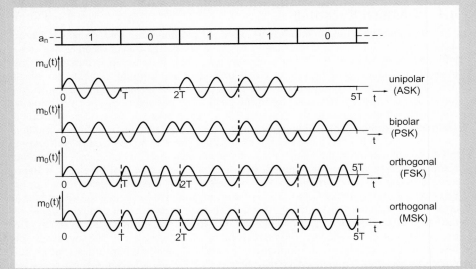

Figure 9: Modulation methods for digital band pass transmission (redrawn after Ref. [5]).

Demodulation of band pass signals

To restore the information transmitted across a band pass channel via a modulated carrier signal, the incoming signal has to be demodulated. Modulation and demodulation will be demonstrated on the example of an AM transmission, because it can be understood easily and intuitively. The demodulation of binary coded carrier signals follows similar principles.

As an example, let us consider $f(t)$ as an analogue base band signal in the time domain characterized by its highest frequency f_g. Fourier transforming the signal into the frequency domain, we obtain the signal $F(f)$ as shown in Figure 10. According to the definition of a band pass channel to be impermeable for low frequency signals, $f(t)$ can not be transmitted directly, but a frequency shift of the low frequency signal must be performed. Multiplying $f(t)$ with the carrier signal causes the required frequency shift. The modulated carrier signal is then given by $m(t) = f(t) \cos(2\pi f_0 t)$. The amplitude of the fast oscillating sine wave changes with respect to the source signal. In the frequency domain, $M(f)$ is shown in Figure 10. If f_0 is in the transmission frame of the channel, the signal can be transmitted, e.g. propagated by an electromagnetic wave.

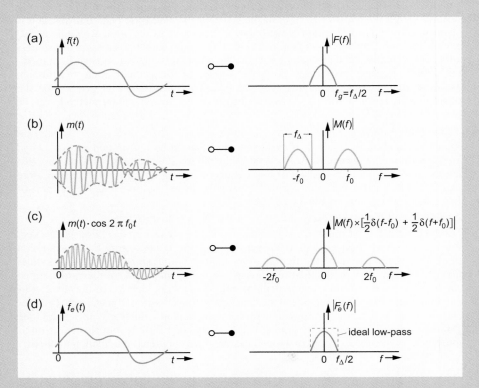

Figure 10: Signal forms in the time and frequency regime of the amplitude modulation method (redrawn after Ref. [5])..
(a) Analog source signal and corresponding frequency spectrum.
(b) Modulated carrier transformation of base band pass.
(c) Demodulation by multiplying modulated signal with the carrier signal.
(d) Extraction of the base band by a low-pass filter.

After receiving the transmitted signal, the low frequency contribution containing the transferred information has to the extracted. For the demodulation, the incoming signal is again multiplied by a sinusoidal signal of the carrier frequency. According to an addition theorem given by

$$(\cos x)^2 = \frac{1}{2}(1+\cos 2x) \qquad (13)$$

a low frequency part representing the base band signal $F(f)$ and a high frequency part with a center frequency of twice the carrier frequency is found in the frequency domain. After low pass filtering, the original signal is recovered.

Frequency and time multiplexing

Similar to the transmission of one signal, several signals can be transmitted through a single channel by using *different carrier frequencies*, if an overlapping of the modulated carrier signals in the frequency domain is prevented. Figure 11 shows a transmission of different signals with the carrier frequencies f_0, f_1 and f_2. In principle, multiplying the received information with the respective carrier frequency allows a selective reconstruction of the information of each signal without any use of additional band pass filters. Since a synchronization of sampling times is not required, this method is, in contrast to the time multiplex method mentioned below, simple to realize. In practice, e.g. interference problems are known to occur due to non-ideal characteristics of the preamplifier in the receiver. Therefore, bandpass filters are used in technical applications, if a large number of signals has to be transferred across a single channel. In order to further increase the information density, improving the steepness of filters is of major importance. A detailed description of novel filter approaches is given in Chapter 31.

Time multiplexing describes an alternative method to transmit a number of signals across a single channel. In this method each source signal is related to a certain time slot. The receiver must be synchronized to these time slots, otherwise the transferred information will be scrambled.

Attenuation

The attenuation of the signal on the transmission line is also of importance. For long distance transfers a refresh of the signal is needed. As a rule, the refresh procedure is complex and slows down the transmission speed. Especially for optical signal transmission, the refresh procedure has been the speed limiting factor for a long time. Optical signals had to be converted into electrical signals and converted back into optical signals after amplification. Nowadays, this bottle neck is bypassed by optical transceivers based on Erbium-doped optical fiber amplifiers. Further detailed information will be given Chap. 30.

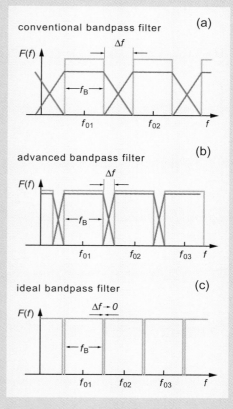

Figure 11: Band pass transmission of several channels and influence of the filter steepness on the number of carrier frequencies on a single channel.
(a) Real,
(b) Optimized,
(c) Ideal.

Acknowledgements

The editor gratefully acknowledges René Meyer (RWTH Aachen) and Stephan Tiedke (aixACCT Systems) for writing and compiling Section 4 of this Introduction and for stimulating discussions. Thanks are due to Stefan Tappe (RWTH Aachen) for corrections.

References

[1] R. Simon, J. R. Whinnery, T. Van Duzer, *Fields and Waves in Communication Electronics*, 3rd ed., John Wiley & Sons, 1994.
[2] N. Gershenfeld, *The Physics of Information Technology*, Cambridge University Press, 2000.
[3] L. P. Hyvärinen, *Information Theory for Systems Engineers*, Springer, 1970.
[4] S. Haykin and B. Van Veen, *Signals and Systems*, John Wiley & Sons, 1999.
[5] J.-R. Ohm and H. D. Lüke, *Signalübertragung* (in Ger.), 7th ed., Springer, 2002.
[6] H. B. Bakoglu, *Circuits, Interconnections, and Packaging for VLSI*, Addison-Wesley, 1990.
[7] W. J. Duffin, *Electricity and Magnetism*, 4th ed., McGraw-Hill, 1990.
[8] C. E. Shannon, *A Mathematical Theory of Communication*, Bell System Tech. J., **27**, 379-423, 623-656 (1948).

Transmission on Chip and Board Level

Wilfried Mokwa,
Institute of Electronic Materials, RWTH Aachen University, Germany

Contents

1	**Introduction**	715
2	**On-Chip Interconnection Technology**	716
3	**Chip to Substrate Interconnection**	718
3.1	Chip-and-Wire Techniques	718
3.2	Flip-Chip Bonding	721
3.3	Tape Automated Bonding	722
3.4	Comparison	723
4	**Ball-Grid-Array**	723
5	**Multi Chip Modules**	723
6	**Three Dimensional Packaging**	724
7	**Summary**	726

Transmission on Chip and Board Level

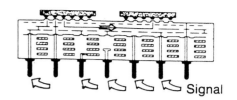
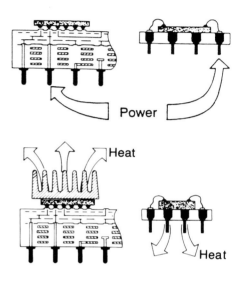
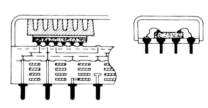

Figure 1: Four major functions of a package.

1 Introduction

Transmission on chip and board level requires certain interconnection technologies. On chip level mainly transistors, diodes, resistors and capacitors have to be interconnected, which today is done by a multi level metallization. Then these integrated circuits (ICs) have to be packaged which is as important as chip fabrication itself. A package consists of a certain number of elements, which build a unit. The elements of an electronic package are electrical components like resistors, coils, capacitors, diodes, transistors and integrated circuits (ICs). To combine these components to circuits, interconnections are required. Some circuits also need to be connected to each other to generate functional units, in which the signal distribution has to be ensured. The number of inter-circuit connections in ICs steadily increases. Such sensitive ICs and their interconnections need – besides mechanical support – an appropriate environmental protection. Electrical circuits need to be supplied with electrical energy. The energy is consumed and transformed into thermal energy (heat). Because all circuits operate best within a limited temperature range, packaging must offer an adequate means for removal of heat [1]. These major functions of a package are illustrated in Figure 1.

The type of interconnection depends on the packaging level. Zero level packaging includes all interconnections on chip level. The chip metallization belongs to this level. First level packaging means the assembly of bare dies into packages. Usually plastic or ceramic packages are used. The result of first level packaging is called a single chip module (SCM). The second level packaging leads to cards that are equipped with several SCMs together with discrete devices like resistors or capacitors (e.g. a sound card for a PC). The third level packaging combines several cards to a main board (e.g. PC main board). The fourth level packaging means the installing of boards into a rack (Figure 2).

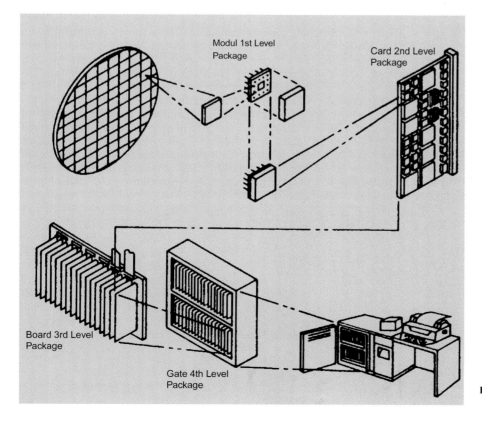

Figure 2: Packaging hierarchy.

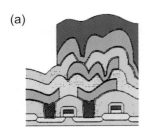

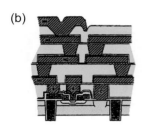

Figure 3:
(a) Cross section of a conventional multi level CMOS-metallization.
(b) Cross section of a CMP planarized multi level CMOS-metallization.

Besides these more conventional techniques, for sophisticated high frequency applications like in the field of cellular phones an increasing number of companies are assembling bare dies directly to a common substrate like printed circuit boards or ceramic boards. This is the so called chip on board (COB) technique. Several chips on one substrate or board form a multi-chip-module (MCM). This board can be mounted now into a plastic or ceramic package or directly into a device. Especially for high frequencies (100 MHz and higher) MCMs show less signal distortions and losses due to signal propagation than the conventional integrated circuit packages. The connecting structures of unpackaged chips are usually shorter and thinner. From this fewer parasitic effects follow. Investigations showed that the signal transmission rate rises by the use of unpackaged chips by around 30 % to 80 % [1].

In the following sections an overview of the most important microelectronic interconnection and packaging technologies is presented including the demands on the materials used.

2 On-Chip Interconnection Technology

On-chip interconnection is the starting point within the packaging hierarchy. The International Technology Roadmap for Semiconductors projects that by 2011 over one billion transistors will be integrated into a single monolithic die [9]. The wiring system of this billion-transistor will deliver power to each transistor, provide a low-skew synchronizing clock to latches and dynamic circuits, and distribute data and control signals throughout the chip. So new concepts are required to reach this goal.

A few years ago only aluminum was used for on chip interconnection. Figure 3a shows a cross section of a standard multilevel metallization. The metallization process starts with etching of contact holes in the intermediate oxide to the source drain regions or to the polysilicon gate of a MOSFET. Advances in interconnection technology have played a key role in the continued improvements in integrated circuit density, performance, and cost per function. One main step forward was the use of chemical mechanical planarization (CMP) that provides global and local planarization.

This enables the manufacturing of chips with multilevel metallization in a planar manner. In Figure 3b the cross section of a CMP based multi-level metallization is shown.

The other important step was the replacement of aluminum by copper. At the same time the subtractive metal etching process was replaced by a damascene (metal inlay) process. In the damascene process, wire patterns (trenches and via holes) are etched into an insulator and then filled electrolytically with copper. The excess copper is then removed by CMP. The minimum wiring pitch today is about of 0.6 μm. The height-to-width ratios of the combined via and trench structures are in the order of 3 to 4 [8]. Copper interconnections offer advantages in performance, cost, and reliability over existing aluminum wiring processes. Performance is gained because the resistivity of copper is approximately 40 % lower than that of aluminum, so that copper wires exhibit approximately 40 % lower RC delay than aluminum wires of the same cross section. Reliability is improved because the electrolytically deposited copper, when compared to aluminum, exhibits far less electromigration and far less stress migration. For example IBM uses this technology in its CMOS 7s process. A representative cross section of the CMOS 7s wiring is shown in Figure 4.

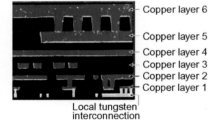

Figure 4: Cross section of the IBM CMOS 7s wiring [8].

In this figure perhaps the most striking feature is the dramatic difference in size between the bottom and top levels of the wiring stack. Minimum contacted pitch is 0.63 μm at the high-density first copper level, 0.81 μm at succeeding levels, and the pitch and thickness of the fifth and sixth levels can be 2 times larger, as shown in the figure, to obtain a low RC delay. CMOS 7s is thus an example of a hierarchical wiring system in which successive wire levels at increasing thickness and width enable long wire runs with low RC delay. If the height and width of a wire and the thickness of surrounding insulators are all increased by a factor of λ, the capacitance (C) per unit length remains in a first order unchanged, while the resistance (R) per unit length is reduced by a factor of $1/\lambda^2$ (the scaling factor in Figure 4 is 2). In principle the RC delay can be reduced to arbitrarily low values by implementation of such *fat wires*. This is sometimes referred to as *reverse scaling*.

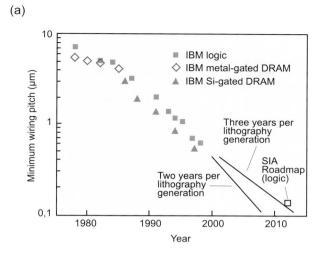
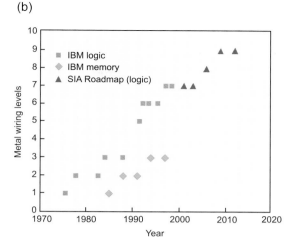

Figure 5:
(a) Scaling of the minimum wiring pitch.
(b) Development of the metal wiring levels [8].

To determine the total RC delay in a logic circuit the effective internal resistance, R_t, of a driver transistor and the input capacitance, C_t, of the transistors that form the load, as well as the lumped resistance, R_w, and lumped capacitance, C_w, of the connecting wire have to be taken into account. The total delay of the circuit is approximated by a weighted sum of delay terms of the form $R_w C_w$, $R_w C_t$, $R_t C_w$, and $R_t C_t$, with coefficients that depend on the circuit that is modeled [7]. Thus, enlarging wire cross-sectional dimensions by a factor of λ while leaving wire length and transistor dimensions fixed causes $R_w C_w$ and $R_w C_t$ to decrease as $1/\lambda^2$ while the third wire-related delay term, $R_t C_w$, is unchanged. This last term, if significant, can be reduced by the use of a "wide" driver transistor, which costs little in terms of additional chip area. Thus all wire-related RC delays in this simple model circuit can easily be reduced. In general, the combination of reverse scaling and appropriate circuit design techniques prevents interconnection delays from overwhelming transistor delays in current microprocessor designs. Since only a relatively small fraction of the wires represent long or critical paths, the cost of the additional metal area required to implement fat wires has so far been acceptable.

For future developments, the minimum wiring pitch must continue to scale with transistor dimensions. The scaling rate can be estimated from historical data, as shown in Figure 5.

At the same time, an ever-increasing fraction of the total wires on chip will have to be implemented with larger pitches and thicknesses in order to meet RC constraints. The following example shows that this is true. In a circuit all transistor and wire dimensions are shrunk by a factor of $1/s$. Since the length of every wire in the circuit shrinks as $1/s$, R_w for any wire increases as s, while C_w decreases as $1/s$, and $R_w C_w$ remains constant. However, in the simplest transistor scaling scenario, R_t is constant, C_t decreases as $1/s$, and thus $R_t C_t$ decreases as $1/s$. So some fraction of the wire runs which had been acceptable in terms of RC delay before scaling will become unacceptable (compared to transistor switching delays) after scaling. To maintain parity with the improved transistor performance, such wires must be moved up the wiring hierarchy to the next larger pitch. This means that an existing wiring layout cannot be simply shrunk, but must instead be redesigned. Furthermore, the area required for the new wiring layout does not scale as $1/s^2$, so either the circuit area also fails to scale as $1/s^2$, or additional wiring levels must be added to contain those wires with pitches that do not scale. Since these fat wires require more area than traditional wires implemented at, or close to, minimum lithographic width, aggressive implementation of hierarchical wiring could drive the increase of wiring levels at a rate above the historical norm and the current industry projections, which are shown in Figure 5 [8].

Of course, RC constraints are not the only factor driving the evolution of wiring systems. Wires for power distribution must be scaled to limit voltage drops. Electromigration constraints must also be avoided. Long wires that operate as transmission lines must scale in width as the square root of the clock frequency for wires of constant length; the thickness of transmission lines need not to be increased at all with clock frequency once the thickness significantly exceeds the skin depth. Such considerations are additional reasons why, at this time, the metal thickness and minimum pitch of last or global wiring levels are no longer decreasing and will increase in the future. The combined effects of shrinking the minimum pitch, adding intermediate metal levels, and increasing the pitch of global wiring levels are schematically illustrated in Figure 6. Wiring systems will become increasingly hierarchical and increasingly three-dimen-

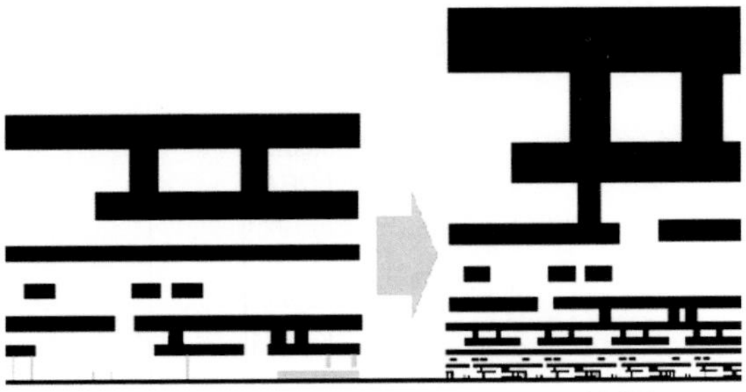

Figure 6: Development of the minimum pitch and the pitch of the global wiring [8].

1998   2012?

sional, with an increasing disparity between minimum and maximum wire dimensions [8], [9].

RC values can also be decreased by using interlayer materials with a lower dielectric constant ε_r. Therefore, the semiconductor industry is moving towards the replacement of the traditional SiO_2 interlayer dielectric ($\varepsilon_r \approx 4$) with one or more materials having a lower dielectric constant. Among these are fluorinated SiO_2 ($\varepsilon_r = 3.2$-3.6) and polymers such as PTFE with $\varepsilon_r = 1.9 - 2.5$. Even lower effective values ($\varepsilon_r < 2$) are realized by introducing nanoscaled porosity into these materials, e.g. porous silica (so-called *Nanoglass*) or porous polyimide (so-called *Nanofoam*). In some of these materials mechanical stability and/or thermal conductivity are critical issues. Another approach is the use of *air* gaps (or bridges), i.e. the intentional introduction of holes of similar dimensions as the interlayer dielectric material. Thus, the effective dielectric constant can be tailored depending on the volumetric ratio between holes and dielectric material frame. Due to their good mechanical stability the well-known SiO_2 or SiN_x dielectrics are preferred candidates for the frame material. [10] and [11] give a good overview of low dielectric constant materials (**low-k dielectrics**) under discussion.

3 Chip to Substrate Interconnection

3.1 Chip-and-Wire Techniques

The standard method for connecting a bare die to a board is the chip-and-wire technique. This technique includes two processes, namely die bonding and wire bonding.

Die bonding serves for the mechanical attachment of the chip to the substrate. In addition thermal and in some cases electrical conductivity is required by the chip substrate connection in order to create back side contacts and to dissipate energy. A special problem of the die bonding is the adjustment to the different coefficients of thermal expansion (CTE) of the substrate materials. The greater the difference of the temperature expansion, the greater the area of the chip surface and the higher the processing and operating temperature are, the larger are the mechanical stresses which can occur in the connection. As a result of large stresses the connection or the silicon chip is destructed. The linear CTEs of some materials are specified in Table 1.

There are basically three possibilities of reducing the stress in the connection:

- the use of substrates with coefficients of expansion, that come close to that of the silicon
- indirect assembly with spans, e.g. from molybdenum, that compensates large differences in the expansion coefficients
- manufacturing of the connection with an additional flexible material, e.g. epoxy resin adhesive

Three techniques are applied, in order to die bond chips to carrier strips or substrates:
- soldering
- eutectic soldering
- adhesive joining

material	CTE [10^{-7}/K]
Al_2O_3	75
BeO	85
AlN	34
FeNi42	66
copper	178
epoxy adhesive	200 – 400
solder (SnPb36Ag2)	240
Si	25

Table 1: Linear coefficients of thermal expansion (CTEs) of some substrate materials.

Adhesive joining of the chips is world wide mostly used (99 % of all cases). The two other techniques are used, if large energy dissipation is necessary or if high operation or process temperatures do not permit the use of adhesive joining [2].

Wire bonding is a technique for the production of discrete electrical connections, generally from a chip on a substrate (Figure 7) [4].

For wire bonding the components, which have to be connected, must have suitable contact areas (so-called *pads*). The wires used consist predominantly of gold or aluminum alloys with diameters down to below 10 µm. All wire bond techniques have in common that they break the oxide films of the wires and the pads by application of pressure, heat and ultrasonic power. The wires are not melted, but they come in such a close contact with the pads, that van-der-Waals forces become effective and provide a permanent connection [3]. In practice three bonding methods are used (Table 2):

- thermo compression (TC) welding
- ultrasonic (US) welding
- thermosonic (TS) welding

Table 2 shows all methods in comparison.

Figure 7: Close proximity down bonding.

	US welding	TS welding	TC welding
forms of energy used	ultrasonic pressure	ultrasonic pressure heat	pressure heat
welding temp. [°C]	25	120 – 200	280 – 380
pressure [mN]	300	300 – 900	300 – 900
welding time [ms]	below 25	5 – 20	up to 1000
frequency [kHz]	60 – 120	60 – 120	–
US power [mW]	100 – 500	100 – 500	-
wire materials	Al, AlSi1, Au	Au, Cu	Au
wire diameter [µm]	7 – 500	17 – 100	12 – 100

Table 2: Comparison of wire bonding methods.

These different methods are implemented in the form of two techniques:
- ball bonding
- wedge bonding

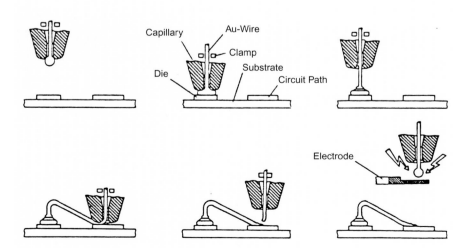

Figure 8: Process steps in ball bonding.

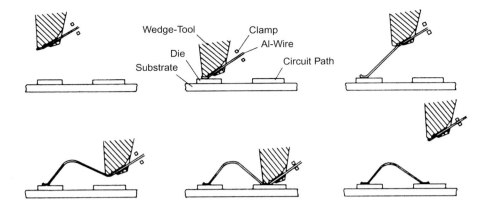

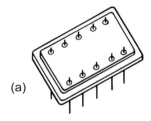

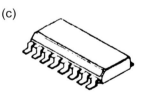

Figure 9: Wedge bonding process.

Figure 10: Package types,
(a) metal,
(b) ceramic,
(c) plastic.

Ball bonding is the most frequently applied technique. The different steps of the ball bonding process are shown in Figure 8.

A wire (preferably gold wire) is guided through a tool with a centric drilling, called "capillary". The end of the wire is melted by a small gas burner flame or by an electrical discharge. Due to the surface tension the end of the wire deforms to a drop having a two or three times larger diameter than the wire. The drop is pressed with the capillary onto the pad and welded there with the surface. Afterwards the capillary takes off. Subsequently, the capillary is driven over the second pad. There the tool is again lowered. With the edge of the capillary the wire is pressed on the pad, deformed plastically and welded. At the same time a break section is brought into the bonding wire by the special forming out of the capillary. When taking the capillary off the wire breaks off at this point, and the connection steps are finished. The end of the wire at the capillary is again melted, and the bonder is prepared for a further duty cycle. The advantage of this technique is, that when setting the first bonding site the wire is perpendicularly on the surface. Therefore the second bond can be set in any direction. A disadvantage of this technique consists of the fact that only gold wire can be used. Other wire materials have to be protected by a flow of nitrogen in order to avoid a strong oxidation during the melting process.

Within the **wedge bonding** process both bonds are set with the wedge-shaped point of the tool used for this technique (Figure 9). Since the wire of the first bond is not perpendicular to the surface, the direction of the second bond is determined. Therefore the substrate has to be aligned according to the bond direction before the bond process is started. An advantage of this technique is, that also aluminum wires can be used, because the wire does not need to be melted. Further advantages are that one gets along with smaller bond space and shorter wire loops. That is very favourable with very many junctions on a chip and with applications in the high frequency region [3].

The substrate used for bonding is dependent on the type of package used. Three different types of packages are normally used. There are packages from metal, ceramic and plastic (Figure 10). In contrast to the plastic package metal and ceramic packages are sealed with a cover. Plastic packages are sealed via injection moulding. Before molding the chip is mounted to a lead frame in chip-and-wire technique. After the injection moulding the plastic package is punched out at the end of the leads. Finally these are bent, so that they can be attached to a substrate by surface mount technology (SMT).

Table 3 contains as examples the values of relative dielectric constants of some general substrate materials.

material	ε_r
epoxy fiberglass	4,8
polyimid fiberglass	4,8
epoxy aramid fiber	3,9
polyimid aramid fiber	56
fiberglass/teflon laminates	2,3
alumina-beryllia ceramic	8,0

Table 3: Relative dielectric constants of some substrate materials.

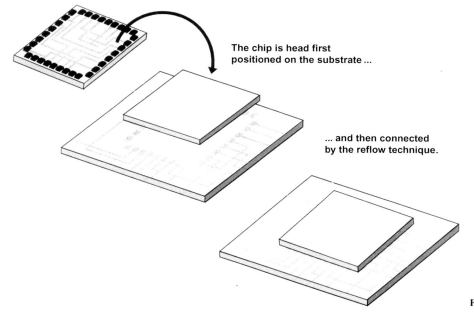

Figure 11: Process steps in flip chip technology.

3.2 Flip-Chip Bonding

Flip-Chip Bonding is the oldest procedure for simultaneous contacting. With this technique the maximal component density on a substrate can be achieved, since the space requirement reduces the size of the chip. Prior to flip-chip bonding bumps have to be grown on to the bond pads of the chip. These bumps serve for increasing the height of the contact. The bumps are fabricated after the normal production process of a chip is finished. First in a sputtering or an evaporating system, intermediate layers called under bump metallization (UBM; e.g. Cr, Ti, Ni, Pt, Cu, Ti/W) are grown onto the pads, which serve as an adhesion layer and at the same time as a diffusion barrier. There are several methods for the production of the bumps: electroplating, evaporating, printing or dispensing. Typical bump heights are between 50 and 100 µm.

The principle of the flip chip technology is represented in Figure 11. The flip chip process can therefore be divided into four steps:

- uptake of the chip face down
- aligning the mirror-symmetric pad grids of chip and substrate to each other
- putting down the chip onto the substrate
- simultaneous production of all mechanical and electrical connections. For the connection procedure three different procedures are possible. Which procedure is applied depends on the type of the bumps:
- soldering with PbSn-, SnAg-, AuSn- or In-bumps
- adhesive joining with Au-, Ni/Au or polymer-bumps
- thermocompression with Au-bumps

With the flip chip technology the smallest distance between contacts (pitch) can be achieved. For example the minimal pitch between solder bumps with a height of 65 – 95 µm is 150 µm.

Thermally induced mechanical stresses caused by different CTEs of chip and substrate have to be compensated by an underfilling. The underfill is a particle filled adhesive that is applied by dispensing. The application of an underfill has the following effects:

- The stress on the bumps is reduced by the shear stiffness of the underfill layer. The larger the shear modulus of the underfill, the smaller the load on the bumps.
- Interfacial and die stresses increase with increasing shear modulus of the underfill.

Because the shear stiffness of the underfill layer increases with chip size, larger chips get more CTE mismatch reduction than smaller chips. The CTE mismatch reduction is best, if the CTE of the underfill is approximately equal to that of the bumps.

3.3 Tape Automated Bonding

To increase performance of wire bonding, engineers have replaced the wire with etched copper leads plated with gold. The leads are supported with polyimide plastic film. It was once predicted that tape automated bonding (TAB) would replace wire bonding, but it has not, due to higher costs and inflexibility in design associated with the production and inventory of the tape. As a result, TABs are only rarely used for digital logic chips like this one.

In TAB technique the connection to the chip is performed by metal strips, the leads. These metal stripes are made photolithographically on a flexible polyimide plastic film that has a shape like a movie film tape (Figure 12). Three types of tapes can be used. They differ as far as the number of layers to produce the tape is concerned. A one-layer tape only consists of the metal structure. In a two-layer tape the metal is directly attached to a polyimide carrier. In the case of a three-layer type the metal is glued on a polyimide carrier by means of an adhesive layer. The tape has an opening that has the size of the chip that is to be connected to the tape. The metal stripes are looking into the opening just above the bond pads of the chip. The first joining step is the so called inner lead bonding (ILB). In this step the inner ends of the leads are connected with the pads of the chip simultaneously via thermo-compression or soldering (Figure 13). For soldering the leads are coated with a thin tin layer. As in the case of flip chip bonding bumps are necessary. They can be either placed on the chip or on the inner ends of the leads.

After joining testing is performed. then the chips are encapsulated in such a way that the outer parts of the leads are left free. The encapsulated chips are separated from the tape by punching them out. Thus the outer ends of the leads are separated from the film carrier. After a second test the encapsulated chip is mounted by simultaneous bonding of all outer ends of the leads to a ceramic substrate or to a printed wiring board (PWB) in a soldering process. This second joining is called outer lead bonding (OLB). The whole manufacturing process is shown in Figure 13.

The resulting assembly has better electrical performance than could be reached with wire bonds, and can achieve a higher density of interconnections. It also has a lower overall height than other methods of assembly. For these reasons, it is the dominant method of packaging chips for driving liquid crystal displays. It is also used for some very cost-sensitive high volume consumer electronics products like digital watches and calculators.

Figure 12: Different TAB tape formats.

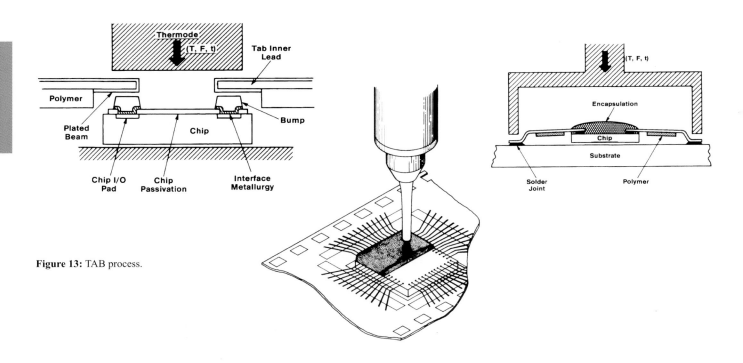

Figure 13: TAB process.

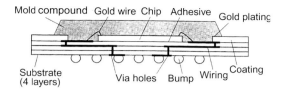

Figure 14: BGA cross section.

3.4 Comparison

Flip chip technology shows the best electrical properties that is mainly due to the absence of longer connection lines. In addition it allows for the smallest miniaturisation. The rectangular conductive stripes of TAB have larger cross section geometry and thereby show reduced inductances in comparison to the wire bond technology. TAB lines are also shorter than wire bond lines and thereby enable smaller signal delay and distortion, low characteristic impedances and fewer reflections in relation to the wire bond technique. By use of multi-layer tapes the couple effects are further reduced. In comparison to Flip Chip technology the TAB lines are longer than the flip chip connections. A substantial advantage of the TAB technique in relation to flip chip technology is the possibility for the electrical test on the tape before contacting [5].

4 Ball-Grid-Array

A BGA is a highly integrated package for the surface mount technology (SMT). Figure 14 shows the cross section of a typical structure. With the area array of robust solder balls this form of package represents an alternative to the common package types with peripheral arranged leads. The BGA-substrates take over the function of contact redistribution and expansion on a larger, planar arranged grid.

The solder balls are brought on the lower surface of the package. The ICs can be mounted both on the lower surface (cavity down) or on the top side (cavity up). With upside equipped BGAs plated through hole connections to the lower side are necessary. In the cavity down arrangement the surface, which is needed for chip contacting, is lost for the solder ball assembly.

In ceramic package versions (CBGA) the ICs are usually contacted by means of wire bonding, flip chip bonding or TAB. Contrary to other package types CBGAs can be sealed hermetically. In newer plastic BGAs (PBGA) the chips typically are wire-bonded on an organic PWB. Afterwards they are overmolded, sealed with a Glop-Top or provided with a cap made of plastic or metal.

Due to the good electrical characteristics PBGA packages are used for fast SRAM. PBGA indicate smaller values of inductance, capacity and electrical resistance compared with package types as for example quad flat packages. With an optimised voltage supply and grounding in the BGA a low inductance is achieved. Flip chip versions in BGA packages increase the electrical performance because of the short signal paths. The BGAs can be used as single chip or multi-chip packages (compare Section 5) [5].

5 Multi Chip Modules

A multi chip module (MCM) is a device consisting of two or more ICs electrically connected to a common circuit base and interconnected by conductors in that base. The base is a substrate, that can be mounted into a package or can be part of a package itself.

The basic idea behind developing MCM technology is to decrease the average spacing between ICs in an electronic system. Therefore, the fundamental aspect of MCM technology is chip interconnection, which includes connecting I/O conductors on a chip to an MCM substrate. The goals are higher performance resulting from reduced signal delays between chips, improved signal quality between chips, reduced overall size and reduced number of external components.

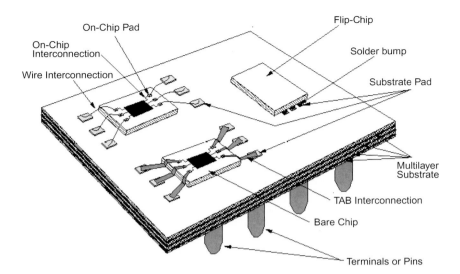

Figure 15: Schematic diagram of a MCM.

An MCM substrate can be composed of different layers and their number depends on the MCM technology used. These layers provide all the interconnections between the different mounted ICs in addition to the interconnections needed to interface for second level packaging [6].

There are three types of MCMs. The substrate of an MCM-L is a multilayer PWB. A ceramic substrate is used with an MCM-C. An MCM-D is built by deposition of metals and dielectrics as thin films on top of a polished ceramic substrate or onto silicon.

Figure 15 shows how bare dice are interconnected to an MCM multilayer substrate using different interconnection technologies.

The basic architecture of an MCM is composed of:

- integrated circuits: packaged and bare chips, mounted on/in the surface of the substrate.
- first level packaging interconnections: wire bonding, TAB, flip chip bonding
- substrate: the common base that provides all the signal interconnections and the mechanical support for all chips
- MCM seal: provides a degree of protection to the circuits in addition to heat removal and interconnections
- second level packaging interconnections: provides the necessary interface to the second level of interconnection [6].

6 Three Dimensional Packaging

There are several possibilities for 3-D integration. Within the current silicon technology there are certain devices where the third dimension is used like in trench or stacked capacitors, vertical transistors or multilevel integration. The active devices are still in a plane.

The most sophisticated 3-D integration would have arbitrary 3-D location of active structures and equal interconnect densities in all three dimensions. The technology used for the realization of such devices is the recrystallization of amorphous or polycrystalline silicon.

Another possibility is the laminated 3-D approach that seems technologically more feasible. In this technique layers of fully processed packaged or unpackaged chips or even wafers are stacked above each other. Before stacking, the wafer/chips have to be thinned and holes for inter-layer contacts have to be etched. For an "ideal" laminated 3-D integration there should be no modification of the base process, only finished wafers/chips should be used and only „known good dies" should be used. A large

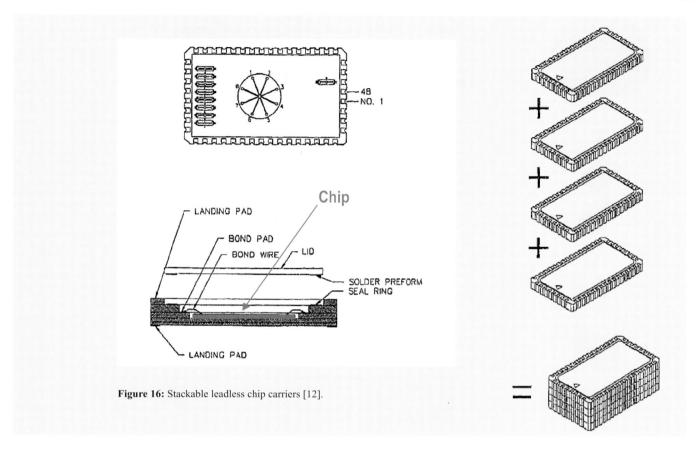

Figure 16: Stackable leadless chip carriers [12].

number of stacked layers and a high density of inter-layer contacts that could be located arbitrarily should be possible. The chip size should be arbitrarily and it should be possible to mix different technologies. Of course the technology has to be simple and cheap.

There are the following key questions to be technologically solved for 3-D integration:

- How to provide an electrical contact from "top" to "bottom" layer?
  - Etching of through-layer vias
  - metallization of through-layer vias
  - Lateral isolation of metal pin to Si
- How to bond "top" and "bottom" layers together?
- How to reduce the thickness of the " top" wafer/chip
- How to handle thin wafers/chips?

One example for stackable leadless chip carriers is shown in Figure 16. Two chips are bonded in one leadless carrier. Up to eight carriers are stacked together. The interconnections are realised by soldering. Memory stacks up to 1 Gb are in production [12].

Another example shows the stacking of chips (Figure 17). The interconnection of the stacked chips is done by chip-edge metallization. Figure 18 shows a 20 chip stack. In production today are stacks with up to 40 layers [13].

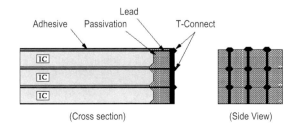

Figure 17: Interconnection of stacked chips by chip-edge metallization.

Neo-Wafer Fabrication and Thinning

Layer Dicing

Stack Lamination

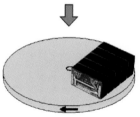

Lap / Lead Exposure

Metallization

Segmentation

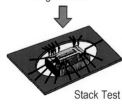

Stack Test

Figure 18: 20 chip stack [13].

7 Summary

In the given Chapter an overview of the most important microelectronic interconnection and packaging technologies is presented. The driving force for all of these developments is the increasing number of active elements on one chip and the decrease in switching time for the transistors, which leads to very high clock frequencies. Therefore the number of interconnections is increasing very fast and the wiring delay times play an important rule.

In on-chip connection technology this will lead to a hierarchical wiring system with up to nine metallization layers. Reverse scaling has to be applied to hit the needs in wiring delay.

In chip to board interconnection Ball-Grid-Arrays and TAB are used to meet the increasing number of input/output connections.

The requirements for minimum assembly area are reached with flip-chip-technology together with MCMs. They lead to very compact assemblies with a high number of input/output connections together with short wiring lengths.

A further integration is only possible if one goes to a three dimensional packaging. This can be achieved by 3-D integration on chip as well as in a laminated 3-D approach.

Acknowledgements

The editor thanks Gernot Steinlesberger (Infineon Technologies) for providing a compilation on 3-D integration techniques. He would also like to thank Carsten Kügeler (RWTH Aachen) for checking the symbols and formulas in this chapter.

References

[1] R.R. Tummala, E.J. Rymaszewski, A. G. Klopfenstein, *Microelectronics Packaging Handbook*, Part I, Chapman & Hall, 2. Edition, New York, 1997.
[2] H. Reichl, *Hybridintegration*, Hüthig, Heidelberg, 1988.
[3] W. Menz, J. Mohr, *Mikrosystemtechnik für Ingenieure*, VCH, Weinheim, 1997.
[4] AVT Report, Heft 4/91, *Technologie des Drahtbondens*, VDI/VDE, Berlin, 1991.
[5] H. Reichl, *Direktmontage*, Springer, Berlin Heidelberg, 1998.
[6] S. F. Al-sarawi, *3D VLSI Packaging Technology*, The Centre for High Performance Integrated Technologies and Systems, Department of Electrical & Electronics Engineering, University of Adelaide, 1997.
[7] H. B. Bakoglu, *Circuits, Interconnections, and Packaging for VLSI*, Addison-Wesley Publishing Co., Inc., Reading, MA, 1990.
[8] T.N. Theis, IBM Research Journal **9** (1999).
[9] J. A. Davis et al., Proc. of the IEEE **89**, 305 (2001).
[10] Proc. of Mat. Res. Soc. Symp. *Low-Dielectric Constant Materials* I, **381** (1995), to V, **565** (1999); Mat. Res. Soc. Bull. **22** (Oct. 1997).
[11] M.J. Loboda, R. Singh, A.S. Ang, and H.S. Rathore, Proc. The Low and High Dielectric Constant Materials: Materials Science, Processing and Reliability Issues, Electrochemical Society, New York, 2000.
[12] www.dense-pac.com.
[13] www.irvine-sensors.com.

Photonic Networks

Christoph Buchal
Department ISG, Research Center Jülich, Germany

Contents

1 Introduction 729
1.1 A General Overview 729
1.2 The Materials Arsenal of Lightwave Systems 730

2 Guiding Photons in Optical Fibers 733
2.1 Optical Waveguides 733
2.2 Optical Fibers 735
2.3 Wavelength Division Multiplexing 737

3 Light Sources 739

4 Photodetectors 743

5 Optical Amplifiers 745

6 Switches and Modulators 746
6.1 The Two Basic Functions of Switching Devices 746
6.2 MEMS - Microelectromechanical Systems 747
6.3 Optical Ferroelectrics 747
6.4 Acousto-Optic Devices 749

7 Summary 750

Photonic Networks

1 Introduction

1.1 A General Overview

Today optical communication and lightwave systems form the backbone for the high-speed data traffic [1] – [4]. In particular, long-distance ground-based systems rely on optical fibers. On land, fiber cables are installed, wherever the right of way permits: they are buried in the ground, laid along railroad tracks or on top of high voltage transmission lines, inside gas pipelines or sewage systems. They also follow rivers and coastlines under water and cross the oceans.

Many business and research facilities already have direct fiber connections and the fiber to the home (FTTH) is not too far in the future. These developments are driven especially by the high demand for bandwidth necessary for the numerous computers participating in internet traffic. Lightwave systems are one of the fastest growing industrial branches. This is due to a few important inventions and a huge amount of research and development by physicists and engineers.

The key components of a long-distance lightwave communication system are
- low-loss glass fibers
- semiconductor lasers
- erbium-based all-optical amplifiers
- photodetectors .

The concept of guiding light in optical fibers has been known for a long time, but the fibers were not pure and transparent enough to permit long-distance connections until approximately 1970. The fabrication process had to be continuously improved, until the glass was sufficiently free of impurities. Traces of H_2O, which strongly absorbs infrared photons, are especially detrimental. At present, optical signal transmission is possible over a distance of up to 100 km without intermediate amplification.

Around 1970 the semiconductor industry developed reliable heterostructure lasers. These form the only adequate light source for launching optical signals into thin fibers.

Four different fiber links are shown in Figure 1. The simple fiber link (Figure 1a) uses just one diode laser, which converts the electronic data into optical pulses. These pulses are transmitted through a fiber and detected by a photodetector, then converted back to electrical signals. This type of optical link uses light of a single wavelength and spans a maximum distance of 80 to 100 km. The capacity of a single fiber can be increased by transmitting several different carrier wavelengths simultaneously (Figure 1b). This is called wavelength division multiplexing, WDM, or even DWDM, where the first D denotes "densely packed" optical frequencies. A separate diode laser is needed for each wavelength λ_i. The different signals are combined in a multiplexer (MUX), transmitted through one fiber and separated by a demultiplexer (DEMUX). After separation they are fed into individual receivers. More details are given in Sec. 2.3.

The transmission span of a fiber link can be increased to global distances, if intermediate signal amplification is provided. The first optical long-distance cables used one single wavelength per fiber. These long-haul systems are essentially formed by a repetition of simple links, as shown in Figure 1a. Each connecting point forms a repeater. A repeater consists of a receiver, an electronic amplifier and a laser. A submarine transatlantic cable needed a sequence of up to 60 repeaters per fiber.

Establishing a long-haul WDM link with electronic repeaters is prohibitively difficult and expensive, because at each repeater one has to provide stable demultiplexing, detection, electronic amplification, optical retransmission by individual lasers and remultiplexing again. Instead, several parallel fibers with simple repeaters were more practical and represented a cheaper solution. It was the invention of the erbium-doped all-optical fiber amplifier (EDFA) which permitted a significant improvement of the long-haul systems. The EDFA completely avoids the conversion from the optical to the electronic regime at each repeater. It uses a directly stimulated photon emission to refresh the optical signal. For each arriving photon, the EDFA generates typically 100 new photons of exactly identical wavelength, phase and polarization. Within its bandwidth, the EDFA is able to amplify signals of several wavelengths simultaneously. It is called "wavelength-transparent". Only ten years after its first demonstration in 1987, the EDFA was implemented by telecommunication companies. This is an unusually short time span, but the EDFA provided the "missing link" for successfully operating WDM long distance systems, which are now installed worldwide.

Besides these key elements, there are numerous additional functions needed and available to modulate, switch, combine and analyze the optical signals. Also the network traffic management and the switching, routing and distribution systems are essential. The reader will find up-to-date information and further references concerning these issues in [3].

This chapter will provide an introduction to the following subjects:
- Optical fibers – guiding the photons
- Semiconductor lasers – efficient light sources
- Detectors – converting photons to electrical signals
- Optical amplifiers – directly refreshing the photonic signals
- Switches and modulators – rerouting the photons.

All these topics are part of the field of "Photonics". Photonics covers linear and non-linear optics, waveguides, semiconductor and solid-state lasers, amplifiers, light sources, optical amplifiers, photon detectors, photon switching and manipulation, electro-optics and micro-optics as well as optical storage and holography. For a more comprehensive coverage, the reader is referred to the literature [1] – [4].

1.2 The Materials Arsenal of Lightwave Systems

We start with a first glance at the materials which provide the powerful basis for modern lightwave systems.

Hyperpure SiO_2 (doped with GeO_2 in the core) is the material most modern fibers are made of. Typically, the core diameter is 8 μm and the cladding diameter 125 μm. The outside is protected by several layers of polymer coating. These extremely pure fibers transmit light at 1.5 μm for 100 km with a total attenuation of only 20 dB.

Direct bandgap semiconductors provide the best material for the efficient generation of photons in light-emitting diodes (LED) or laser diodes (LD). Photons are generated by the recombination of electrons and holes in *pn*-junctions, operated in the forward direction. The bandgap is responsible for the photon energy. Semiconductor lasers span a huge range from the $1-laser for the CD player to the most expensive communication network lasers (> $ 10000).

Silicon or direct bandgap semiconductors are the material of choice for photodetectors (PD), which convert optical pulses to electrical signals. For each absorbed photon, an electron-hole pair is generated and separated in the depletion zone of a *pn*- or *pin*-junction in reverse bias. Avalanche photodetectors (APD) provide additional signal amplification by avalanche carrier multiplication. The avalanche effect leads to more than one electron-hole pair per photon. For very high bit rate signals, the detectors have to be designed with very short carrier drift path lengths (< 1 μm).

Highly transparent ferroelectric crystals such as $LiNbO_3$ are another important material class. In these materials the optical index is changed if electrical fields are applied. Therefore $LiNbO_3$ is used for high-speed waveguide modulators and switches. Acoustic waves can also be easily generated in $LiNbO_3$. Because the density pattern of the acoustic wave acts as a grating for light waves, a large family of acousto-optic devices has been developed.

Passive devices are frequently planar waveguide structures, made from glass or semiconductors. Simple passive devices are signal splitters. A more sophisticated design is needed for wavelength-sensitive devices such as diffraction gratings or phased arrays. They combine or separate the different wavelengths to be guided in one fiber. This is called multiplexing or demultiplexing.

Micromechanical devices are beginning to play an important role. A good example is Lucent's Lambda Router, which uses micromechanically tilted mirrors to switch the signals between different input and output fibers. This is one form of an all-optical network switch. It should be noted that the network switching is currently performed by huge electronic computers. The optical input signals are received by photodetectors and converted to electrical signals. The entire signal handling is provided by these computers. They feed the outgoing electrical signals into laser diodes, which launch optical output pulses into the outgoing fibers.

Photonic Networks

(a)

Direct Fiber Link

(b)
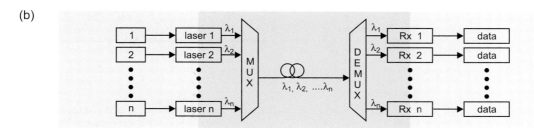
Wavelength Division Multiplexing: WDM

(c)
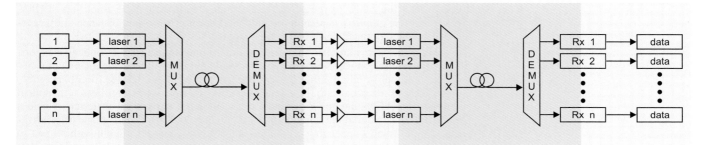
Electronic repeater for long haul

(d)
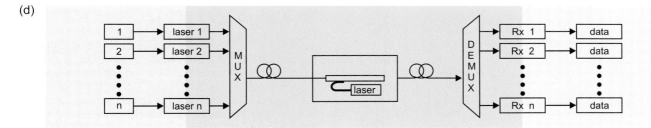
Erbium Doped Fiber Amplifier: EDFA

Figure 1: The development of optical fiber links; (a) Direct fiber link between one transmitter (laser diode) and one receiver. The direct link operates at one wavelength λ and has a maximum span of approximately 80 km.
(b) Wavelength division multiplexing (WDM) permits the transmission of several closely spaced wavelengths $\lambda_1, \lambda_2, \ldots \lambda_n$. A multiplexer MUX combines the signals. A demultiplexer DEMUX separates the signals and feeds them into n separate receivers Rx. This technique increases the capacity of each fiber.
(c) Long haul transmission needs amplification approximately every 80 km. Before the invention of the EDFA an electronic repeater was the only solution. The insertion of the demultiplexers and the numerous lasers makes this a difficult and expensive technique.
(d) The all-optical erbium-doped fiber amplifier uses stimulated emission to restore the signal strength. It is "wavelength-transparent" over its bandwidth. Therefore it can amplify different signal wavelengths simultaneously without the need for a demultiplexer.

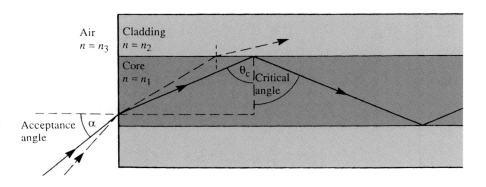

Figure 2: Critical angle and numerical aperture NA of a step index fiber. The ray outside the acceptance cone is lost. Typical values for a monomode fiber are $n_2 = 1.44$ at 1.5 µm (SiO_2-cladding). The numerical aperture NA = $\sin \alpha$; $\sin \alpha = (n_1^2 - n_2^2)^{1/2}$; $\Delta = (n_1 - n_2)/n_1 = 0.3\%$; $\theta_c = 85.7°$, $\alpha = 6°$; NA ≈ 0.11.
Note that the core diameter is 8 µm. The small acceptance angle together with the tiny core spot demand high precision alignment for the coupling of an optical beam.

An especially advanced device uses a combination of $LiNbO_3$ optical waveguides and the interaction of acoustic waves in $LiNbO_3$ with the propagating optical signal. In this way it is possible to realize wavelength-dependent add-drop switches, which can add or withdraw optical signals of selectable wavelengths from one fiber or a fiber ring.

Erbium-doped fibers or erbium-doped waveguides in planar structures use the stimulated emission of Er to refresh optical signals. These all-optical amplifiers (EDFA, Er-doped fiber amplifier) are an important component of very-long-distance lines, especially the intercontinental submarine lightwave cables.

Magneto-optic materials play an important and unique role as optical isolators. They transmit the optical power in one direction only. They allow the launching of a signal from a laser or an amplifier into a fiber, but protect these devices from backreflections. This is essential for the stable operation of lasers or amplifiers. An optical "one-way street" is unknown in standard ray optics of transparent materials, but in a magneto-optic material the Lorentz force acts on the electrons, and the plane of polarization of light rotates depending on the propagation direction of the light with respect to the direction of the applied magnetic field. With two polarizers and a 45° rotation within the magneto-optical material, one beam may pass unattenuated, but the reverse beam is completely blocked.

Polymers play important roles for optical storage (CD, DVD), for optical waveguides on a chip and – most recently – in the form of polymer LED displays ("Poly-Light").

Today's lightwave systems are incredibly powerful. In a laboratory demonstration, one single strand of glass fiber already carried a data stream of more than 5 terabit/s (5000 Gbit/s). This number is very difficult to visualize. For a first impression, consider 30 volumes of an encyclopedia, each consisting of 1000 pages. In total, these books contain 1 Gbit of information. This information can be stored in optical form on one standard CD or on a small fraction of a new DVD (Digital Versatile Disc). One DVD-9 holds 68 Gbit of data, which is sufficient for a two-hour high-definition video film. This huge amount of data (68 Gbit) may be transmitted over one single fiber within approximately $1/100$ of a second.

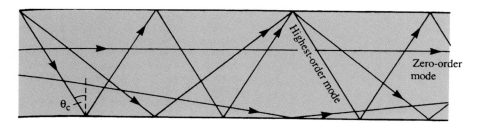

Figure 3: Visualization of a multimode guide with many "ray angles". The smaller the core, the lower the number of allowed angles, see Figure 4. The ray picture is very simplistic. Solving Maxwell's equations for the appropriate boundary condition is the only correct approach.

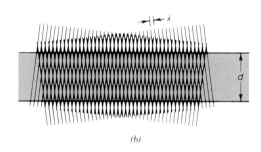

Figure 4: The number of allowed modes is limited. This is visualized by a considering the interference process.
(a) Condition of self-consistency: as a wave reflects twice it duplicates itself.
(b) At angles for which self-consistency is satisfied, the two waves interfere and create a pattern that does not change with z [1].

2 Guiding Photons in Optical Fibers

2.1 Optical Waveguides

Guiding photons within optical fibers relies solely on optical index contrast, which acts on the phase propagation of the wave, well known from refraction and total reflection. The reader is reminded, that the process of refraction or total reflection between two transparent media takes place without any loss of energy - unless imperfections such as scattering centers are present. Index-induced reflection is very different from reflection at a metal-coated mirror, where the unavoidable absorption by the metal's free electrons results in an energy loss of at least 1 % from each reflection process. This is no problem for most optical instruments. In contrast, it would be disastrous if light had to be guided in a mirror waveguide, which can be formed by two closely spaced metallized mirrors: After 1000 reflections, the light intensity would be attenuated to approximately 10^{-5}. In a typical waveguide, 1000 reflections span a length of only a few cm.

As shown in Figure 2, a highly transparent core has to be encapsulated by a cladding of identical optical transparency and slightly lower optical index, which provides total reflection. Although most of the wave's energy is guided within the core, a significant amount of the wave extends into the cladding.

The critical angle θ_c, the numerical aperture NA and the acceptance angle α of a fiber are depicted in Figure 2. Typically the optical index of the material is $n \approx 1.44$ at $\lambda = 1.5$ µm. The index contrast δ between core (n_1) and cladding (n_2) is small:

$$\Delta = (n_1 - n_2)/n_1 = 0.3\%.$$

This leads to an acceptance angle α of only 6°. All rays entering at an angle larger than α are lost into the cladding and are not guided. The two-dimensional ray picture (Figure 2, 3) might suggest that all angles smaller than α correspond to allowed beams in the fiber. Surprisingly, this is not the case. Only very few discrete ray angles $\theta > \theta_c$ correspond to propagating waves in the guide. These stable solutions follow directly from the solutions of Maxwell's equations [1]. For a two-dimensional slab waveguide they may be visualized by the interference picture shown in Figure 4. The allowed solutions are called modes. Each mode has a different transverse field profile, resulting in a different fraction of the energy being guided in the cladding as compared to the core. Since there is an index gradient between cladding and core, this results in a different effective propagation constant for each mode. This is the reason for the observed modal dispersion. If several modes are guided in a fiber, the fiber is called a multimode fiber (Figure 5a, c).

In practical fibers, there is always an energy exchange amongst different modes. Therefore modal dispersion will lead to a temporal spreading of short optical pulses, while the pulses propagate through the fiber. Since monomode fibers sustain only one

Figure 5: Geometry, refractive-index profile, and typical rays in:
(a) a multi-mode step-index fiber,
(b) a single-mode step-index fiber, and
(c) a multimode graded-index fiber [1].

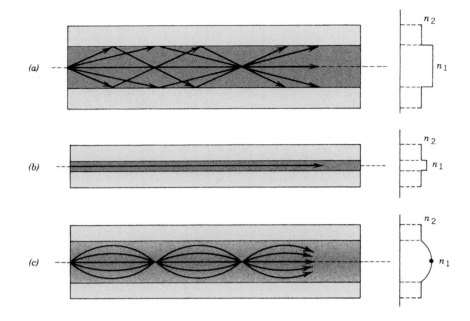

propagating mode (Figure 5b), they are free of modal dispersion and are therefore preferred for long-distance communication. The core of a standard monomode fiber has a diameter of 8 µm. The resulting effective diameter of the optical mode amounts to 10 µm, since the mode extends into the cladding. The strongest modal dispersion is found in step-index fibers (Figure 5a). For intermediate distances, the graded-index fiber (Figure 5c) provides a good compromise, because modal dispersion is mitigated and less pronounced. This is due to the fact that in the case of a graded index fiber the more oblique modes, which zigzag along at greater angles, also travel longer distances within the cladding, where the phase velocity is slightly higher, because the index is lower than in the center of the core. A well-designed graded index fiber may show a very small modal dispersion.

Naturally, the detailed analysis of light propagation within a fiber has to rely on Maxwell's equations and the appropriate boundary conditions at all interfaces [1]. Frequently numerical solvers have to be applied. Even if only the simplest stationary solutions are discussed, the electrical and the magnetic field both must satisfy Maxwell's equations. The stationary, time-independent spatial form is called the Helmholtz equation [1]. Δ denotes the Laplace differentiation operator and U stands either for the magnetic or the electric field:

$$\Delta U + n^2 k_0^2 U = 0 \tag{1}$$

The respective indices are inserted for the core ($n = n_1$) and the cladding ($n = n_2$ in the case of a step, or a variable $n = n_2(r)$ for a graded index profile). The vacuum wavevector is denoted by k_o:

$$k_0 = 2\pi/\lambda_0 \tag{2}$$

As pulse propagation is due to the influence of both core and cladding, one defines an effective index n_{eff} and a corresponding propagation constant β:

$$\beta = 2\pi n_{eff}/\lambda_0 \tag{3}$$

For a fiber of cylindrical symmetry, all stationary solutions can be factorized using cylindrical coordinates:

$$U(r,\varphi,z) = U(r)e^{-il\varphi}e^{-i\beta z} \quad \text{with } l = 0, \pm 1, \pm 2, \ldots \tag{4}$$

The solutions for the radial part $U(r)$ are given by a set of Bessel functions [1]. For each integer l, Eq.(4) has multiple solutions $\beta_{l,m}$, which represent the multitude of m modes and the phase propagation factor $e^{-i\beta z}$ varies for each mode. It reflects a propagation with a speed in between the values for pure core (n_1) or pure cladding (n_2):

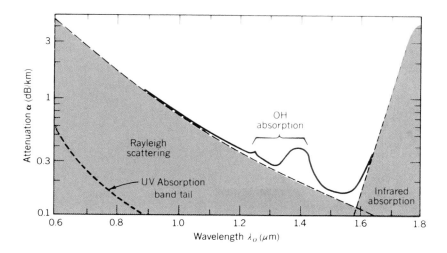

Figure 6: Dependence of the attenuation coefficient α of silica glass on the wavelength λ_0. There is a local minimum at 1.3 µm ($\alpha \approx 0.3$ dB/km) and an absolute minimum at 1.55 µm ($\alpha \approx 0.16$ dB/km) [1]. The magnitude of the OH-absorption double peak could be further reduced by technological improvements, but the shaded area is intrinsic to SiO_2.

$$n_1 k_0 > \beta_{l,m} > n_2 k_0 \quad (5)$$

Only the fundamental mode (l = 0) has a radial shape similar to a Gaussian intensity profile. All other modes have nodes in their field profile. For monomode fibers, only the fundamental mode has to be considered. This mode also provides the strongest intensity confinement. It should be noted that a monomode fiber is free from modal dispersion, but the material dispersion $n = n(\lambda)$ remains. Material dispersion becomes noticeable for the propagation of very short pulses, because these pulses contain a spectrum of frequencies.

In a monomode fiber, the light is confined to a very small cross-section. If powerful laser diodes (20 mW) are used for long-distance traffic, the optical intensities become high (0.3 GW/m²). The problem is even more serious for multi-wavelength WDM traffic, because the added intensities of several transmitters will eventually lead to a nonlinear optical response of the SiO_2, resulting in pulse distortion and frequency conversion. To avoid additional coupling losses and reflections, all junctions and splices of single mode fibers have to be aligned with submicrometer precision.

The power or the photon flux launched into a fiber is attenuated according to an exponential law:

$$P(z) = P(0) \cdot 10^{-\alpha z/10} \cong P(0) \cdot e^{-0.23 \alpha z} \quad (6)$$

with α in dB/km and z in km.

A sequence of lossy elements leads to a product of the individual transmission factors, therefore power losses and amplifier gains in dB units are additive. For example, if a fiber connection has a transmission of 10^{-2} (equivalent to –20 dB), an amplifier gain of 100 (+20 dB) will restore the original signal strength (0 dB, factor 1).

2.2 Optical Fibers

For long-distance data traffic, all fiber links are operated in the near-infrared at a wavelength of 1.3 µm or 1.5 µm, because SiO_2 is most transparent at these wavelengths, see Figure 6. Rayleigh scattering at intrinsic structural inhomogeneities drops with increasing wavelength, until at 1.6 µm absorption due to higher harmonics of optical phonon frequencies sets in and dominates the losses. Additional energy losses would be encountered if optically active impurities were excited within the SiO_2. In the past, H_2O has been an especially serious problem for fiber transmission, because the OH-bond of H_2O impurities in SiO_2 has a vibrational resonance at approx. 0.45 eV (λ = 2.75 µm). Harmonic and anharmonic overtone excitation leads to absorption peaks at λ = 1.38, 1.24 and 0.92 µm. This is a very strong effect, since 1 ppm of H_2O leads to a loss of 40 dB/km at 1.38 µm. Therefore any water concentration has to be reduced to the low ppb level during fiber manufacturing. Today an OH concentration of less than 10 ppb is standard. Of course, the installed fibers must be carefully protected from any possible water diffusion into the cladding. Modern fibers now practically reach the theoretical absorption minimum of < 0.2 dB/km or 20 dB/100 km. 20 dB absorption corresponds to

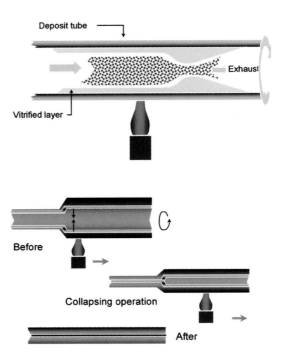

Figure 7: Fabrication of the preform. The inside of a hyperpure SiO$_2$ tube is coated with the glass composition of the core by a CVD process. In a second step the tube is collapsed to the completely bubble-free and uniform perform of 80 mm diameter [ALCATEL].

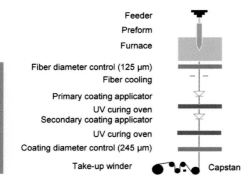

Figure 8: The fiber pulling tower is more than 10 m high. A preform rod of approximately 80 mm diameter and 2 m length is heated at its lower end. The glass becomes so soft that a fiber of 125 µm diameter can be drawn, cooled and coated with one or two layers of plastic coating in a continuous process. Typical data are:
Preform speed: 2.75 mm/min = 16.5 cm/h
Fiber pulling speed: 1100 m/min = 66 km/h
Outer fiber diameter: 125 µm ⌀
Variation of fiber diameter: ± 0.1 µm (over 400 km)
Fiber core: 8 µm ⌀
Optical mode diameter: 10 µm ⌀ at λ = 1.5 µm
Fiber length on one bobbin: 400 km
[ALCATEL].

10^{-2} of the launched photons reaching the detector after travelling 100 km through the fiber. This presently constitutes the practical limit for a transmission span without intermediate amplification.

Modern single mode fibers permit bit rates of 20 Gbit/s at one wavelength for transoceanic distances. The present record value is 40 Gbit/s at one wavelength [6]. In addition, many wavelengths may be transmitted at a spacing of 0.4 nm, leading to bit rates exceeding 1 Tbit/s for one fiber.

A modern way to fabricate single-mode fibers starts from a 1 m long tube of ultrapure SiO$_2$. The inside core is formed by a chemical vapor deposition (CVD) process, which adds GeO$_2$ and possibly P$_2$O$_5$, Na$_2$O$_3$, Al$_2$O$_3$ and F to the SiO$_2$ in order to increase the index by 0.3 to 1 %, see Figure 7 [5]. By varying of the flow rate of the different CVD-gases, a diameter-dependent index profile or a step profile can be designed. This allows an optimization of the total fiber dispersion. During the CVD process, the chemical elements are introduced into the core in the form of SiCl$_4$, GeCl$_4$, PCl$_3$ and SF$_6$. Amongst competing methods, the CVD method is chosen because it provides a core glass quality which contains the lowest concentration of detrimental metal impurities. Impurity concentrations of less than 1 ppb are achieved. Finally, the tube is collapsed to become the preform. The preform is an extremely pure, compact and bubble-free glass rod with an outer diameter of 8 cm up to 2.5 m length. In a pulling tower, this rod is heated at its lower end to 2000 – 2100 °C. The glass rod is lowered very slowly with a typical speed of a few mm per minute. At the very tip of the softened rod a thin fiber forms. The viscous glass at the fiber forming point is accelerated ("drawn") to 1100 m/min. The following process is very fast. As soon as it has melted away from the preform rod, it takes only a fraction of a second for any (tiny) glass volume element to become part of the finished plastic-coated fiber on the bobbin. In a feedback loop, controlled by optical sensors, the force and speed of the pulling machine are regulated to provide the exact outside diameter. For a fiber length of 400 km, the diameter is kept constant at ±0.1 µm. To avoid any contact of the glass with dust or moisture, the fiber is immediately protected by two layers of plastic coatings (Figure 8). A coated fiber with a core/cladding diameter of 8/125 µm is now typically 400 µm thick. The coating material is very rapidly cured and hardened by UV radiation. The finished fiber is coiled up onto a large bobbin, which holds a total length of 400 km. From one meter of preform rod, 80 mm in diameter, a total fiber length of 400 km will be drawn.

For an extensive discussion of the processes for fiber and cable fabrication, the reader is referred to [5]. A completed fiber cable for land installation may contain up to 120 fibers in different bundles, Figure 9. If one considers the enormous bandwidth of a single fiber, one might doubt, whether so many fibers per cable will ever be needed. Although certainly not needed for bandwidth, a high number of fibers per cable is an

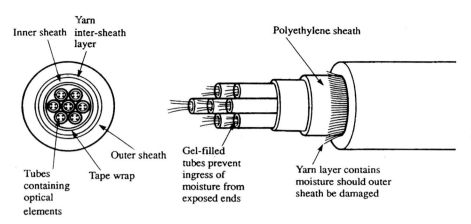

Figure 9: A typical land cable may contain up to 120 fibers. As many as 600 fibers per cable have been fabricated. The high number of fibers is needed because many professional users prefer to establish their own optical network, which they do not want to share with anybody [5].

economic alternative to expensive multiplexing or add-drop devices. It also enables many customers to operate their own fiber networks without sharing hardware with third parties. The bandwidth advantage of the monomode fiber is illustrated in the Table below.

	Twisted pair (telephone)	**Coaxial (broad band)**	**Monomode fiber**
Material	Cu	Cu/insulator	SiO_2/GeO_2
Diameter (mm)	2	10	0.2
Loss (dB/km)	20 dB/km @ 4 MHz	20 dB/km @ 60 MHz	20 dB/100 km @ 10000 MHz
Bandwidth	6 MHz	500 MHz	10 GHz to many THz using DWDM
Repeater span	1 – 2 km	1 – 2 km	~ 50 km

2.3 Wavelength Division Multiplexing

It is possible to send several different wavelengths in parallel through one fiber – just think of them as several different colors. A 1.5 µm SiO_2/GeO_2 monomode fiber will carry optical signals in the wavelength range from 1.525 µm to 1.610 µm. Presently it is only the central segment of this bandwidth which is used:

- Center band (C-band): 1538 nm – 1570 nm
- Total available band width: 4 THz = 4000 GHz
- Carrier frequency spacing: 50 GHz (= 0.4 nm) at 80 channels
- Expected total bitrate for 80 channels (C-band): 80 · 10 Gbit/s = 0.8 Tbit/s.

The entire usable bandwidth of the fiber of around 1500 nm includes shorter wavelengths (S-band: 1525 – 1538 nm) and longer wavelengths (L-band: 1570 – 1610 nm). This more than doubles the available bandwidth. A basic diagram for wave division multiplexing (WDM) or dense WDM (DWDM) is shown in Figure 1. For every wavelength a separate, extremely stable transmitter laser is needed. Each laser runs on one wavelength, which differs by 0.4 nm from the next laser, for instance: No. 1: 1538.0 nm, No. 2: 1538.4 nm, No. 3: 1538.8 nm and so on up to No. 80: 1570.0 nm.

Stabilizing these lasers requires laser cavities of a high Q and careful temperature regulation. Then their signals are combined by a multiplexer and launched into a single fiber. Presently, research attempts are being made to reduce the cost of the expensive temperature-controlled laser modules by stabilizing the laser frequency using externally added fiber components. These are Bragg gratings within the fibers (FBG).

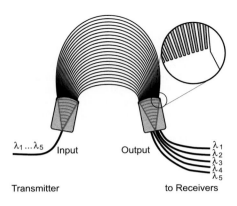

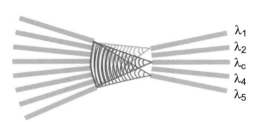

Figure 10: Phased array. Light of different wavelengths λ_i from a transmitting station is uniformly coupled into a staggered array of monomode guides. After passing through the array, the focusing into the output guides is wavelength selective: demultiplexing is achieved.

Figure 11: Focusing condition at the receiver side of the phased array. Due to the tilted phase front different wavelengths are focused onto different exit spots.

If many laser signals are coupled into one fiber, the optical intensity in the fiber is accordingly higher than in a standard application. As a consequence of the high transmitted power, intensity-dependent nonlinear effects have to be considered. It is necessary to limit the total power at the transmitter end and to lower the repeater spacing from 90 km to 40 km.

In order to get some idea of the optical intensities: a good semiconductor laser launches up to 20 mW into a monomode fiber. The mode has an effective area of 50 μm². For $P = $ 20 mW the intensity is $I = 2 \cdot 10^{-2}$ W / $50 \cdot 10^{-12}$ m² = 0.4 gigawatt/m².

If the effect of 80 channels DWDM within one monomode fiber were included, the unrealistic number of 32 gigawatt/m² would be reached. Even minute absorption in the fiber will lead to local heating and nonlinearity. (Remember that a glowing red hotplate in your kitchen runs at approx. 100 kW/m²).

The WDM scheme needs dispersive devices to combine and separate the channels. A very elegant method is the addition or separation of one single wavelength from a wavelength spectrum by add/drop waveguiding devices, as described in Sec. 6.4 on surface acoustic wave (SAW) components. The SAW devices have the great advantage that they can be tuned instantaneously by changing the electrical drive frequency, but SAW add/drop systems become very expensive if many channels have to be operated in parallel because each channel needs a separate device. Gratings or prisms are well known dispersive devices, but neither lends itself readily to waveguiding geometries. For instance, a vertically etched grating works perfectly well for integrated optics (IO), but is expensive to fabricate.

Instead, phased array devices have become the MUX/DEMUX solution for WDM transmission [7]. They consist of an array of single-mode strip waveguides of staggered length, spread out on a plane. Their fabrication relies entirely on standard IO patterning techniques. Figure 10 shows the principle of a phased array. A single waveguide is coupled into a planar waveguide section. In this section, the light is no longer confined laterally and spreads out. The emerging light reaches the input apertures of the arrayed waveguides and excites propagating modes. These modes start with identical phase and propagate through the array. They reach the output, after having passed through guides of *different* lengths l_i. The length values l_i are staggered in such way that the optical path length between the guides varies by multiple integer wavelengths:

$$l_{i+1} - l_i = m \cdot \lambda_c / n_{\text{eff}} \quad m = 1, 2, \ldots \quad (7)$$

As a consequence, the staggering seems to be compensated and has no direct effect on interference. As input and output apertures are designed symmetrically, a new phase front forms, which focusses the emerging light (for $\lambda = \lambda_c$) onto one spot in the image plane corresponding to the input. Essentially the divergent beam from the input is "mirrored back" to a convergent beam at the output. However, Eq. (7) holds only for one wavelength, called the central wavelength λ_c. A phased array is always dispersive, because any variation of the wavelength λ leads to a different phase distribution after the staggered guides. For a wavelength $\lambda \neq \lambda_c$ the phase front of the emerging light will be tilted and focused onto a different spot in the image plane, see Figure 11. This is exactly analogous to a Rowland-type grating with a circle mounting.

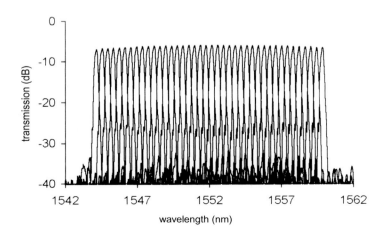

Figure 12: 40 optical channels demultiplexed by a phased array. The channel spacing is 50 GHz ≈ 0.4 nm [ALCATEL].

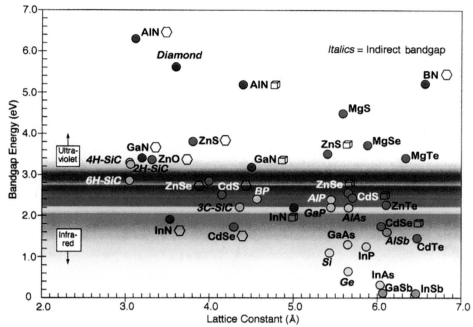

Figure 13: The bandgap energy and lattice constant of many semiconductors important for optoelectronics. In the case of SiC and III-N compounds, both cubic (zinc blende) and hexagonal crystal polytypes are shown. For the hexagonal materials, the basal-plane lattice constant is used. The colored regions approximate the band-to-band photoemission corresponding to the bandgap energy.

Figure 12 shows a demultiplexed WDM spectrum containing 40 channels spaced 0.4 nm apart. The light can be generated by the same semiconductor laser, and tuned to 40 different wavelengths by changing the temperature. Typically a temperature step of 1 °C results in a shift in the emission wavelength of 0.2 nm to 0.3 nm.

3 Light Sources

The great success of optocommunication is crucially enabled by mastering the epitaxial growth of compound semiconductor structures. All available **semiconductor laser diodes** (LD) are made of compound semiconductors, because only these semiconductors provide a direct bandgap. A direct bandgap permits electrons from the conduction band and holes from the valence band to recombine "directly" in a two-particle process. No third particle such as a phonon is needed for momentum conservation. Note that all kinds of defects or impurities form centers for nonradiative recombination. Therefore the material quality and purity is crucial. The electron-hole-recombination energy is emitted in the form of a photon. The energy of the photon is equal to the bandgap energy. A wide bandgap semiconductor will emit in the blue, while lower bandgaps will lead to red or infrared radiation. Simple binary semiconductors as GaAs, InP, InAs or GaN provide characteristic photons energies, see Figure 13.

A well-controlled variation of the bandgap becomes possible by using ternary compounds. The most famous system is $Al_xGa_{1-x}As$, which permits the adjustment of the bandgap in the range 1.4 – 2.2 eV. Excellent epitaxial growth of layers of different bandgap on top of each other becomes possible, because for all concentrations x the alloy possesses the same crystal lattice structure and lattice spacing. A defect density below 10^4 defects per cm^2 has become a minimum requirement.

As shown in Sec. 2, the transparency of quartz glass fibers is highest at 1.3 or 1.5 µm, see Figure 6. For laser diodes operating at these wavelengths, the significantly more complex quaternary compounds are best suited:

$\lambda = 1.3$ µm (= 0.95 eV) : $In_{0.73}Ga_{0.27}As_{0.58}P_{0.42}$

$\lambda = 1.53$ µm (= 0.8 eV) : $In_{0.58}Ga_{0.42}As_{0.9}P_{0.1}$

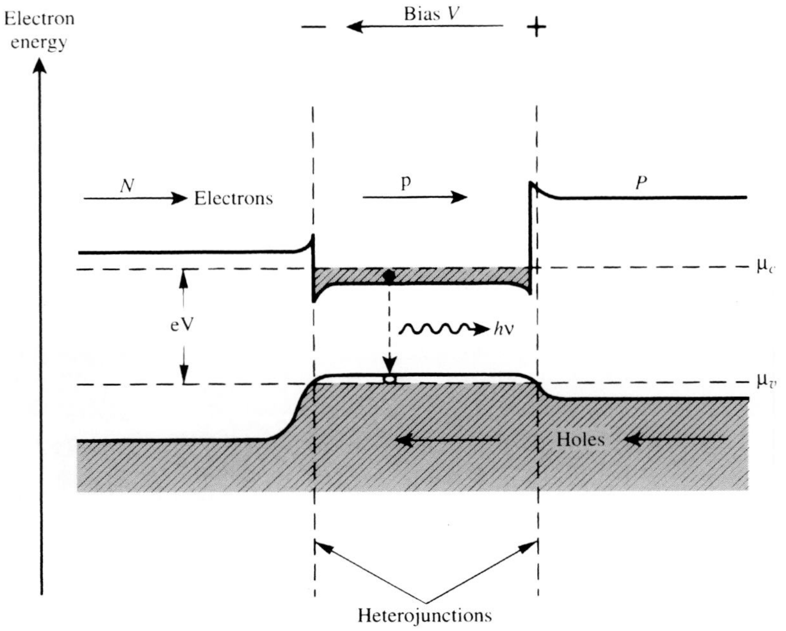

Figure 14: Schematic band diagram of a forward biased N-p-P double heterostructure in which the P region is heavily doped. Electrons flowing from the N region and holes flowing from the P region are trapped in the p region potential well creating a population inversion [8].

The design and fabrication of semiconductor lasers is complex. It is always the *pn* transition which forms the key functional structure. The other key feature is the optical cavity, which is normally realized in the form of an optical waveguide, and which has to be matched and connected to the optical fiber. Since the electron-hole recombination is a band-to-band process, it has a typical linewidth of 50 nm, which results in a correspondingly wide emission spectrum. This has to be compared to the WDM wavelength spacing of 0.4 nm. It is the optical cavity which defines the exact lasing wavelength within this broad peak by forming a narrow bandwidth resonator. On the other hand, the wide emission spectrum of a semiconductor LD is not only a disadvantage. It allows the operation of the LD together with a tunable resonator or the addition of external fiber Bragg gratings for wavelength selection.

Within light emitting diodes (LED) or diode lasers (LD), photon emission originates from the region where electrons and holes meet under high forward current condition. In a simple *pn* transition, the carriers traverse the space charge region and electrons are injected deep into the p-doped region, while holes are injected into the n-doped volume. Both recombine as minority carriers within the large volume, which is defined as the minority carrier diffusion length. In other words: In order to achieve strong inversion with a high electron-hole density within a small volume, the simple *pn* transition has to be modified.

The density of electrons and holes is strongly increased if an additional semiconducting layer with a lower bandgap is added into the central region of the *pn*-junction. This heterostructure is used for all LDs, see Figure 14 and Figure 15. Under forward bias, electrons will be injected from the heavily doped N region (at the left side) into the central region, which is only weakly doped. They cannot cross the heterojunction into the P region due to the band offset. At the same time holes are easily injected into the central region (from the right). These holes are also unable to cross the heterojunction into the N regime due to the built-in band offset in the conduction band. Therefore a high density of electrons and holes is confined to the central part of the heterostructure, leading to a very strong inversion and accordingly to a very strong stimulated emission of photons.

In addition, the center layer with the lower bandgap automatically has a higher index of refraction, see Figure 15. (This is a free gift from Mother Nature). Therefore it may be used as an optical waveguide, which is needed for the formation of a laser cavity. Although this cavity has the form of a strip waveguide, it still has to be adapted to match the optical fibers. The epitaxial heterostructures of the LD are thin, a typical vertical extension amounts to $1 - 2$ μm. A fiber has a core diameter of 8 μm. Therefore coupling tapers or lenses are needed.

While the vertical confinement is provided by the heterostructure growth, the lateral confinement needs an additional processing step. In the simplest case, a stripe would be defined by narrow electrode stripes. Only optical modes under the electrodes are amplified. This is called *gain guiding.* Better designs use either a laterally defined introduc-

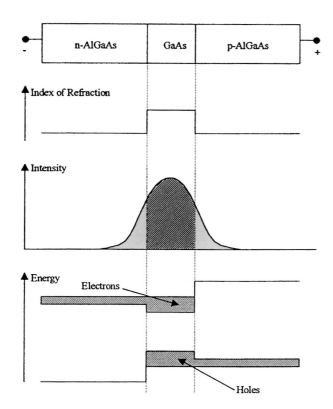

Figure 15: Index of refraction, optical intensity and carrier densities in a heterostructure laser.

tion of a different material composition, an insulator, or an ion implantation step. In III/V semiconductors, ion implantation is an efficient way to form deep traps, which render the material insulating. In all cases, the optical index of the material at the sides of the waveguide is lowered. This leads to *index guiding*, which provides a stable and well-defined optical waveguide. Also mesa structures have been formed for lateral confinement. For a communication laser, an index-defined waveguide width of 5 μm is a typical value.

A complete optical cavity also requires mirrors. The form and quality of the cavity determines the spectral properties of the emitted light. In the simplest cases, such as the LDs for CD players or the CD-ROM drive of a PC, no special mirrors are added. Instead, the laser end facets are cleaved. This automatically forms parallel surfaces at both ends. Due to the high index of the semiconductor ($n \cong 3.5$) and the associated Fresnel reflections, the free end faces form mirrors with 30% reflectivity. This is sufficient for a low quality cavity, if the spectral purity is not critical. In contrast, all LDs for communication must incorporate carefully designed waveguides with integrated reflectors. These lasers are called "distributed feedback" (DFB) or "distributed Bragg" reflectors (DBR), see Figure 16. In these devices, a corrugated guide forms an extended mirror. A grating of period d selects a set of m well-defined wavelengths. Only one wavelength λ_0 will be included in the laser's gain curve:

$$m \cdot \lambda_0 = 2d \cdot n_{\text{eff}} \qquad (8)$$

m is typically 1 or 2, higher orders are not practical. The corrugation shown in Figure 16 is designed to form a weak periodic disturbance of the waveguide, leading to a weak reflection of the guided mode at each step. Constructive interference, as known from a dielectric multilayer mirror, leads to the final reflectivity result. Bragg reflectors may be conveniently incorporated into optical fibers and the corrugation periodicity may be varied over the length of the Bragg reflector. This allows the design of spectral characteristics tailored for transmission and reflection.

Tuning of a semiconductor laser can be achieved by an external movable grating, by replacing external Bragg reflectors (most conveniently integrated into a glass fiber) or by changing the temperature of the entire assembly.

VI Data Transmission and Interfaces

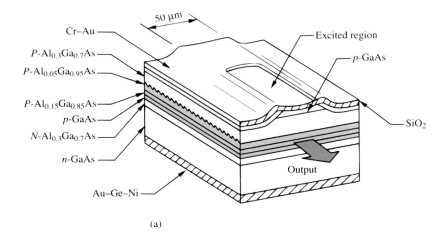

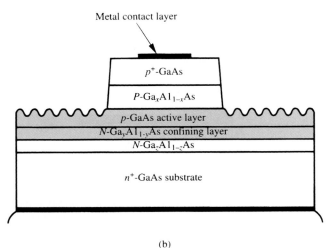

Figure 16:
(a) Double heterojunction distributed feedback laser. The corrugated growth of the top semiconductor layer has the effect of a continuous optical reflector, selecting the lasing wavelength: DFB-laser,
(b) a distributed Bragg reflector outside the cavity: DBR-laser [8].

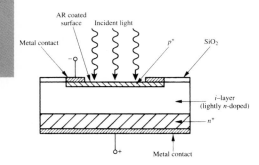

Figure 17: *PIN* diode. The incident light is absorbed in the intrinsic layer, which provides a sufficiently large photosensitive volume for good photon detection. The i-layer is depleted of free carriers due to the applied reverse voltage between p^+ and n^+.

All lasers are known to be disturbed by foreign light being injected into the cavity. In the case of fiber communication, these may be backreflected photons returning from the depths of the fiber thus having a different phase, or photons of a different wavelength from another WDM transmitter. Therefore an optical isolator made of a magnetooptic material must be installed at the output of each laser.

A major advantage of the electrically driven LDs results from the fact that they can be turned on and off by switching the external current. This provides a simple method for on/off light pulse generation. Unfortunately, this is only practical for frequencies up to approximately 2 GHz. At higher bit rates the energy decay time constant of the cavity and the required frequency stability prevent the direct modulation of the laser. In this case, the LD is operated in a well-stabilized cw mode. For modulation a fast external modulator is added. Interferometric electro-optic modulators from $LiNbO_3$ or electro-absorption modulators are mounted after the LD. There are modern commercial devices fabricated on the basis of InP, which incorporate the laser and the electro-absorption modulator on a single substrate.

The development of Si-based **optical emitters** is highly attractive, because of the process compatibility with the silicon CMOS technology. In fact, several light sources on the basis of doped silicon devices have been demonstrated [26] – [28]. The most efficient design is based on the electroluminescence of rare earth fluorescent centers embedded in thin SiO_2 layers. They are excited by injected electrons. As pure SiO_2 degrades under the influence of an electron current, a matrix of silicon nanoparticles embedded within SiO_2 or within silicon-rich oxides seems most promising. As a first complete device, an on-chip optocoupler has been demonstrated, which transmits information from the CMOS logic part to the high power area of the same Si chip. The logic circuits control the power regulator without a galvanic electrical link, being well isolated and protected from high voltage transients during power switching. Silicon-based emitters may not only be used for on-chip communication, but could find an important use for the data transfer between different boards or chips within a computer.

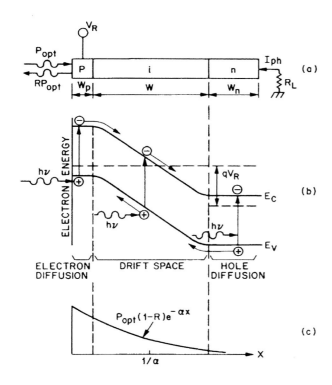

Figure 18: *PIN* diode in reverse bias,
(a) schematic; photons enter from the left,
(b) under reverse bias each absorbed photon creates an electron-hole pair, which leads to an external current I_{ph},
(c) absorption of the optical power as a function of the junction depth [9].

4 Photodetectors

At the receiving end of optical transmission systems the light pulses have to be converted to electrical signals for further processing. This is always done by semiconducting photodetectors (PD). Mostly a *pn*-junction under reverse bias is used. In the depletion region of a *pn*-junction there is a strong electrical field, but no free carriers. If a photon is absorbed in the valence band in this region, an electron-hole pair is created which is immediately separated by the strong electrical field in the junction. Since a fast response is essential, communication detectors are always designed for operation with an applied external field, which accelerates the carriers to saturation drift velocity v_s (typically $v_s \approx 0.1$ µm/ps).

Detection sensitivity is an important issue. If light passes through matter, the optical power is absorbed. This has been discussed for the pulse propagation along a fiber, Eq. 6. In contrast to the highly transparent fiber, α must be ten orders of magnitude larger for a detector:

$$P(x) = P(0) \cdot e^{-\alpha x} \tag{9}$$

For direct bandgap semiconductors and photon energies $h\nu$ exceeding the bandgap value, α is typically larger than 10^4/cm or 1/µm. Therefore an absorption length of several micrometers is appropriate. For silicon, an indirect bandgap material, which is

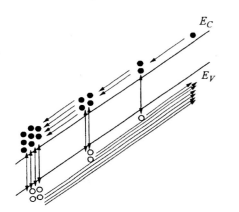 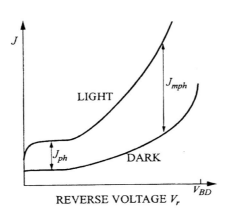

Figure 19: The avalanche photodiode (APD). Left: electrons in the intrinsic region are strongly accelerated by the reverse applied bias. Whenever the kinetic energy exceeds the bandgap energy, a scattering process creates another electron-hole-pair. Right: current characteristics of an APD. The dark current is due to thermally generated carriers and leakage. The illuminated current starts with the normal photocurrent, at higher voltages carrier multiplication sets in.

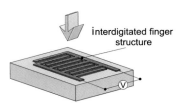

A. planar structure

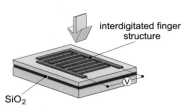

B. planar structure on SOI substrate

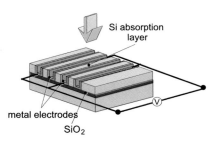

D. trench structure on SIMOX substrate

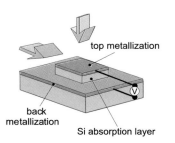

D. vertical structure

Figure 20: Four different designs for fast detectors;
(a) simple planar finger structure;
(b) additional insulating layer reduces carrier diffusion from the depth of the substrate and increases the device speed;
(c) trenches provide a uniform electrical field distribution;
(d) a parallel plate capacitor structure is the fastest and best design, because it combines a uniform electrical field for maximum carrier acceleration with a short vertical carrier drift length. In the case of edge or waveguide illumination, also a favorable long optical absorption length is provided.

transparent for $\lambda > 1.1$ μm, the value $\alpha = 10^4$ is reached at the visible wavelength $\lambda = 0.6$ μm (2 eV) [1]. For the standard IR communication wavelengths, generally III/V semiconductors are preferred. More details can be found in [3], [7] – [12].

Typically the depletion layer width of a semiconductor does not exceed 1 μm, which is insufficient for the complete absorption of the optical signal [9]. The sensitive volume of a PD is further increased if an intermediate intrinsic ("undoped") layer is introduced: *pin* detector, see Figure 17 and Figure 18. The *pin* photodiode has become a standard communication detector. The electrical signals are even stronger if the applied reverse bias leads to carrier multiplication within the intrinsic region by the avalanche effect, see Figure 19 [9]. In this case the photogenerated carriers are strongly accelerated by the applied bias. They reach kinetic energies exceeding the bandgap value. As a consequence the fast carriers scatter inelastically (with atoms) and are able to generate more electron-hole pairs. Typical multiplication factors are 5–10 and avalanche photodiodes (APD) deliver a total signal of more than one electron-hole pair per absorbed photon.

Another important issue is the device speed. The maximum carrier speed v_s under the influence of a strong field typically is in the order of $v_s = 10^7$ cm/s. The electrical response speed of a detector is limited by the carrier transit time t_{tr}

$$t_{\text{tr}} = d / v_{\text{s}} \tag{10}$$

For an electrode spacing $d = 5$ μm this results in $t_{tr} = 50$ ps. If shorter transit times are needed, different geometries have to be chosen, for instance a narrow waveguide between two electrodes attached to both sides. This is sometimes called "edge illumination" of the detector. It combines a short transverse carrier drift length and with a long optical absorption length [3], [13]. Fast detectors have to rely on drift processes in strong fields. Diffusion of photogenerated carriers, which constitutes the dominant mechanism for photovoltaic devices, must be avoided. The diffusion time t_{diff} is given by

$$t_{\text{diff}} = length^2 / 2D \tag{11}$$

For a length of 5 μm and a typical value of $D = 10$ cm^2/s, this results in $t_{\text{diff}} \cong 10$ ns. Diffusion of carriers out of low field areas considerably lengthens the electrical response time.

An ultrashort electrical response is observed if a metal-semiconductor-metal (MSM) detector can be used. Schematic designs are shown in Figure 20. In this case, the Schottky barriers between the semiconductor and the two metal electrodes provide the diode function. The semiconductor remains the photosensitive part, where electron-hole-pairs are generated. It may be fully depleted of free carriers by the applied field, as long as the optical signal is turned off. In this case, the applied field extends across the entire semiconductor and leads to saturation carrier velocities whenever carriers are generated. In addition, Schottky diode based detectors are unipolar devices in the sense that no electron-hole recombination inside the semiconductor is involved. This gives another speed advantage, because the recombination time is avoided. For a silicon-based MSM detector, pulse response times of 3.2 ps FWHM have been demonstrated [13], [14].

Finally it is worth mentioning that the "internal" photoeffect at Schottky barriers can be used for photodetection. In this case, photons are absorbed in the metal and carriers are emitted out of the metal over the Schottky barrier into the semiconductor, accelerated and collected at the counter electrode. This process has a low quantum efficiency, because most carriers do not cross the barrier into the semiconductor and thermalize inside the metal. On the other hand, Schottky barriers are strongly dependent on the materials (metals) involved and a wide spectrum of barrier heights can be fabricated. For instance, the barrier between Si and PtSi is very low (0.26 eV) and offers a corresponding responsivity to long wavelength infrared photons. At present, cooled Schottky barrier detectors are only used for IR imaging.

Typical materials for optocommunication detectors are InGaAs and InP for wavelengths between 0.9 and 1.7 μm. Silicon has been used sucessfully for 0.8 μm. In either case the photon energy has to be larger than the bandgap energy. Due to the processing compatibility with microelectronics, silicon-based optoelectronic components have advantages wherever they can be employed. There is a huge arsenal of photodetectors on the market. Frequently the PD package includes an integrated preamplifier stage for low noise performance. For a detailed analysis of photocarrier dynamics see Refs. [3], [9], [13], [15].

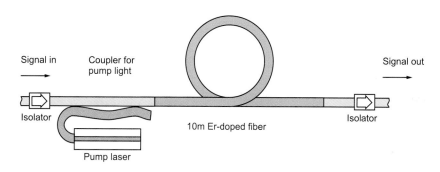
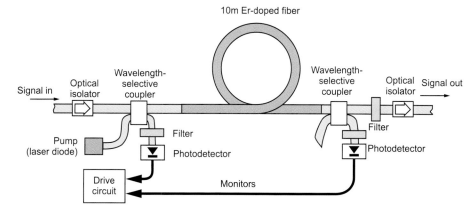

Figure 21: Erbium-doped fiber amplifier layout; a strand of erbium-doped fiber (green) is spliced in line with the transmission fiber (red). A wavelength-selective coupler combines the pump light and signal light. The pump light, from a high-power semiconductor laser diode, has a wavelength of 810, 980 or 1480 nm. Optical isolators prevent unwanted reflections. Optoelectronic feedback loops (not shown) control amplifier gain and pump power.

5 Optical Amplifiers

In an erbium-doped optical fiber amplifier (EDFA) for 1.5 µm light, a fiber of up to 30 m length is doped with laser-active Er ions in the center of core, see Figure 21, [16], [17].

The optical amplifier operates like a laser, but it lacks the mirrors, which form the laser cavity. As in a laser, a population inversion has to be generated by a pumping mechanism. An EDFA is optically pumped. The process of stimulated emission creates new photons, which are completely identical to the first photon in energy, phase and polarization.

In addition to the stimulated emission, there are also spontaneously emitted photons. For instance, all lasers start by emitting some photons spontaneously. These photons are reflected inside the cavity and create their stimulated "siblings". This process rapidly multiplies the photon density in the cavity. Since an amplifier lacks the cavity, all spontaneously emitted photons are sent into the signal line, where they form an unwanted optical "noise background". Therefore an amplifier will not reach the lasing condition. Instead, a population inversion remains until an external signal arrives. This signal will be amplified. As an optical signal pulse passes through the doped region of the fiber, the process of stimulated emission creates many new identical signal photons. A signal photon multiplication factor of 1000 is achieved. In this case the amplifier is said to have a gain of 30 dB. If an optical pulse is attenuated by 20 dB after passing through 100 km of fiber, the EDFA is easily able to refresh the pulse to its full starting power.

The great success of EDFAs is due to several facts:

- Highly reliable laser diodes are available running at $\lambda = 1.48$ µm or 0.98 µm to pump the EDFA.
- The level spacing ($^4I_{13/2} - ^4I_{15/2}$) of the Er^{3+} ion (Figure 22) exactly matches the energy required for long-distance traffic at $\lambda = 1.5$ µm.

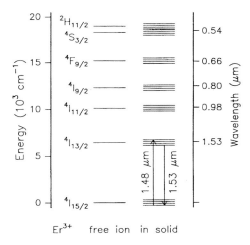

Figure 22: Energy level diagram of Er^{3+}. In the free ion the energy levels are sharp; in a solid the levels are split due to the Stark effect. Pump (1.48 µm) and signal (1.53 µm) wavelengths for an Er-doped amplifier are indicated.

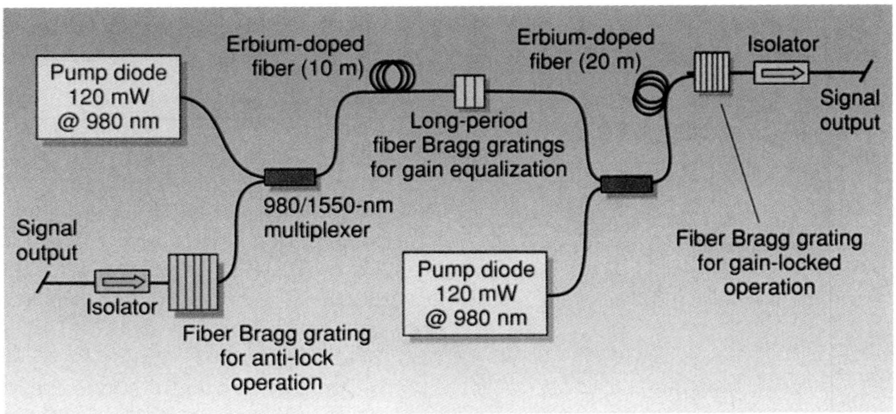

Figure 23: Gain-locked, gain-flattened EDFAs can support the addition of more channels to DWDM systems because of their tolerance for signal variations. In this double-forward pumping scheme, two pumps emit 120 mW at 980 nm; fiber Bragg gratings and long-period fiber Bragg gratings provide gain flattening and equaliziation [18].

- The level spacing between $^4I_{13/2}$ and $^4I_{15/2}$ is broadened by Stark splitting, Figure 22. Depending on the detailed local environment of each Er^{3+} ion, somewhat different energy levels are observed ("inhomogeneous level broadening"). This permits pulse amplification for wavelengths ranging between 1.525 µm and 1.610 µm. The amplifier is called "color transparent" for this wavelength range and it is able to amplify the many wavelengths of a DWDM multiplexed signal. This is extremely useful if already deployed fiber links are to be upgraded later from one to many signal wavelengths. In fact, the amplification window of $\Delta\lambda = 85$ nm corresponds to a total bandwidth of $\Delta\nu = 196721$ GHz $- 186335$ GHz $= 10386$ GHz. This explains why the bandwidth of the EDFA is one of its outstanding advantages which cannot be reached by any single electronic repeater/amplifier.
- In addition, the EDFA is insensitive to the polarization of the incident light.
- EDFAs work in both directions. This is potentially dangerous, because any small backreflection will also be amplified. Therefore EDFAs are always installed between two optical isolators.
- Each EDFA constantly emits a small amount of spontaneously emitted photons, unfortunately exactly at the signal wavelength. These photons are amplified as well and contribute to the signal noise. Amplified spontaneous emission (ASE) determines the maximum length of an EDFA refreshed fiber link. The maximum length of a fiber link with EDFAs installed every 50 km is estimated to run "once around the globe", which fortunately seems sufficient.
- Of course, the EDFA will only work at $\lambda = 1.5$ µm, not in the 1.3 µm window. There are many researchers working on different ions, for instance Pr^{3+}, to establish optical amplifiers for 1.3 µm as well.
- In addition, there are significant efforts under way to flatten the gain characteristics of the EDFA in order to make its full bandwidth usable for practical applications. At present, the deployed EDFAs only run in the center band of $1.530 - 1.560$ µm. A very recent approach to flatten the gain and make a wider DWDM amplification window practical is shown in Figure 23 [18].

6 Switches and Modulators

6.1 The Two Basic Functions of Switching Devices

Two different classes of switching devices can be distinguished:
- Slow switches reroute the entire optical pulse train from one fiber or waveguide to another fiber or waveguide. These devices may have "slow" operating characteristics, because it may take up to 10 ms or even longer to go through a switching step. On the other hand, the most advanced all-optical switches are completely color- and bit-rate-transparent. Therefore they are able to transfer the entire bit rate transported

through one fiber. These switches are called network switches or routers. The micro-mechanical devices (Sec. 6.2) are used as routers, but the electro-optic materials (Sec. 6.3) can be used for "slow" switching as well.

- Ultrafast switches are operated in conjunction with a continuously running cw-LD. They are designed to turn the optical signal off and on for fast pulse generation. They are sold for operation rates of up to 40 GHz, and higher rates have been demonstrated in the laboratory. These switches are called modulators. Ultrafast modulation is the domain of electro-optics (Sec. 6.3) or of the electro-absorption effect within semiconductors [19].

Please note, that the term "switch" is also used for the huge servers and computers, which operate network nodes and perform network management. The above mentioned "slow" all-optical switches are intended to replace a few of the routing functions of the electronic node management.

6.2 MEMS - Microelectromechanical Systems

Silicon technology has not only led to the well known superb electronic devices, but has also opened a "side road" to micromechanics, because lithographically defined etched structures may be used for creating microscopic mechanical parts from Si. One example is an all-optical switch, which consists of an array of 256 small mirrors on one Si wafer [20]. The mirrors shown in Figure 24 have a diameter of 0.5 mm and a spacing of 1 mm. They are mounted on silicon hinges and may be tilted in any direction up to 5°. The deflection of the mirrors is controlled by electrostatic forces, which result from tiny capacitor structures and applied voltages of up to 150 V. Each mirror is used to reflect light from one fiber output into another fiber input. Additional mirrors and microlenses are part of the optical path. With this arrangement, it is possible to achieve a complete reconfiguration of a network node with 256 inputs and 256 outputs within 50 ms. Of course, this mirror array based on a silicon MEMS layout is completely color and bit-rate-transparent.

Figure 24: Micromechanical mirror (one of 256) from the WaveStar LambdaRouter, a trademark of Lucent Technologies [20].

6.3 Optical Ferroelectrics

Transparent ferroelectrics change their refractive index in the presence of an applied electrical field. If a light beam passes through the material, the high frequency of the carrier wave ($\lambda = 1.5$ µm corresponds to 200 THz) solely interacts with the electrons of the material. (The refractive index of matter is due to this interaction of light with electrons). A less rapidly varying field is also able to shift the ions. For ionic displacements, the material's high frequency limit is given by the optical phonon frequencies. Displacements of the ions deform the electronic orbitals from their equilibrium configuration. This modifies the polarizability of the electrons and is the reason for the field-induced index change, called the electro-optical effect, see Chap. 1. Important materials are $LiNbO_3$, $LiTaO_3$, $BaTiO_3$ as well as the III-V-semiconductors. The semiconductors are not ferroelectric, but their crystalline structure provides ordered electric dipoles (for example: Ga^--As^+) within each unit cell. Within a ferroelectric domain, the dipoles are oriented parallel. In this case, the sign of the index change depends on the sign of the applied external field with respect to the ferroelectric axis.

As the ferroelectrics are anisotropic, their optical index is described by the indicatrix (index ellipsoid). In the absence of an external field, the indicatrix is oriented along the main crystallographic directions [1], [8], [19], [21]

$$x^2/n_x^2 + y^2/n_y^2 + z^2/n_z^2 = 1 \qquad (12)$$

An applied electrical field tilts and deforms the indicatrix. This demands the introduction of an electro-optic tensor containing 18 components. Frequently, the crystal symmetry and the device structure reduce the complexity and lead to simple expressions. For example, if the electric field is applied along the c-axis of the $LiNbO_3$ crystal, n_x and n_y are identical and are called n_o, while n_z propagates the extraordinary beam and is called n_e. In this case, the principal axes of the indicatrix only change their length, but the indicatrix is not rotated. This leads to the well-known form [1], [8], [19], [21]:

$$n_0(\boldsymbol{E}) = n_0 - (1/2)n_0^3 r_{13} \cdot \boldsymbol{E} \quad \text{for } \boldsymbol{E} \text{ parallel } z : \boldsymbol{E} = (0,0,E) \qquad (13)$$

VI Data Transmission and Interfaces

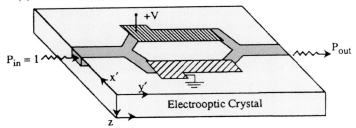

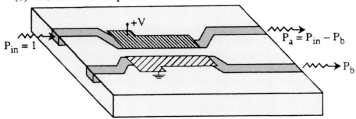

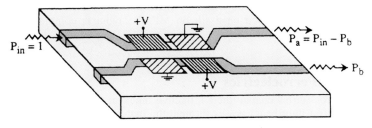

Figure 25: Three different electro-optic waveguide devices
(a) Mach-Zehnder modulator with push-pull arrangement of the electrodes,
(b) electrically controlled directional coupler,
(c) coupler with phase-reversal electrodes [1].

$$n_e(E) = n_e - (1/2)n_e^3 r_{33} \cdot E \quad (14)$$

The observed index changes are generally small. For $LiNbO_3$, the largest coefficient is r_{33}:

$$r_{33} = 31 \cdot 10^{-12} \, m/V = 31 \, pm/V \quad \text{(for } LiNbO_3\text{)} \quad (15)$$

Electrical breakdown limits the usable field to approximately 10 V/μm. This results in an index change of $\Delta n = 1.65 \cdot 10^{-3}$. Therefore a long path is required to induce a phase change of π for a wave propagating in a waveguide. A typical path length is 10^3 to 10^4 wavelengths. This adds up to a total typical length of 1 cm.

For optocommunication, where the light is guided in fibers, only planar structures with strip waveguides are appropriate for building modulators or switches, see Figure 25. The top device is an interferometric Mach-Zehnder waveguide modulator [1], [21]. It is realized on a $LiNbO_3$ chip with waveguides, which are formed by the in-diffusion of lithographically defined narrow Ti stripes. A typical stripe width is 3 μm. Titanium locally raises the optical index and permits the fabrication of waveguide patterns on the surface of $LiNbO_3$, while the electro-optical properties of the $LiNbO_3$ remain unchanged. The light propagates with very low losses through these waveguides. The input guide is split into two equal branches which recombine at the output. An electrical field can be applied to each branch, but, due to the geometry, the field will have an opposing sign with respect to the ferroelectric axis, increasing the optical index in one branch, and lowering it in the other. This leads to a mismatch of the optical path length between the two arms. When the optical signal is recombined at the end of the two, constructive or destructive interference is observed, see Figure 26 and Figure 27. In the case of zero-phase mismatch, interference leads to a complete reconstruction of the input signal – the modulator is in the "on" state. If the phase mismatch amounts to π, interference leads to an antisymmetric optical mode profile. This belongs to a mode of higher order, which is not supported by the *monomode* waveguide, but radiated into the substrate – the modulator is in the "off" state. Further increasing the field leads to a periodic repetition of the modulation curve, see Figure 27.

The directional coupler shown in Figure 25b relies on the fact that there is a coupling of energy between two very closely spaced guides. It depends on the extension of the evanescent tail of one guided mode into the other guide. The effect is frequently compared to the physics of two coupled resonators (two coupled pendulum oscillators). The application of a field in an electro-optic material modifies the index of the material *between* the guides, changing the coupling strength. This may lead to a *complete* transfer of the signal from one guide into the other after a certain propagation length. In fact, a further change of the field leads to a subsequent retransfer back into the first guide, and even many oscillations back and forth are possible [1], [21]. Again, a periodic response of the device as a function of the applied field is observed, which also depends on the wavelength of the transmitted signal. Therefore a coupler may be designed as a chromatic switch, used for switching (demultiplexing) one single wavelength. A coupler may also be used as a fast "on"-"off" modulator, similar to the Mach-Zehnder device, or as a router.

At high modulation rates, the dynamic behavior of the system becomes an important issue. The material itself is able to respond to frequencies in the 1000 GHz range. Only when the resonances of the intrinsic oscillators of the crystal are reached will the electro-optic effect vanish. These frequencies are the frequencies of the optical phonons, which are in the low THz range. Nevertheless, it is extremely difficult to reach modulation rates of 100 GHz, because the electrical signal has a limited propagation speed and the electrical drive circuit has to cope with the RC time constants of the electrode configuration. The optical signal moves with a phase velocity of $v_0 = c/n$ with $n \sim 2$. The microwave phase velocity is $v_m = c/\sqrt{\varepsilon}$, with $\varepsilon \sim 40$. Therefore a 100 GHz microwave has a vacuum wavelength of $\lambda_{vac} = 3$ mm but of 0.47 mm in the presence of the ferroelectric. As a typical modulator is 10 mm long, this corresponds to more than 40 half-waves. At a first glance, this device will cancel its modulation effect completely to zero.

How is it at all possible to design 100 GHz modulators [22]?

- The electrical signal propagates partly in air. This reduces the effective microwave index and brings it closer to the optical index.
- Very thick metallic electrodes (up to 30 μm Au) provide low ohmic resistance and lower RC constants.
- A favorable design uses a coplanar ridge waveguide for the microwave, which travels parallel to the optical signal. As has been shown by [22], this structure allows a good match between microwave and light signal. In this case, microwaves and light travel with nearly the same phase velocity and strong modulation is achieved.
- Some authors have designed electrode configurations, which reverse the sign of the modulating field along the path of interaction: "phase-reversal electrodes". This is shown in Figure 25c for a coupler where phase reversal is used to match the phase between microwave and optical pulse. The polarity reversal of the electrodes compensates the sign change of the modulating microwave due to the phase velocity mismatch.

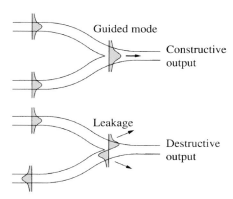

Figure 26: Illustration of the constructive and destructive interference pattern in a Mach-Zehnder waveguide modulator. Only if the outgoing waveguide is of a "single mode" type will the resulting antisymmetric mode be completely radiated into the substrate. A multimode waveguide cannot be used for an efficient Mach-Zehnder modulator.

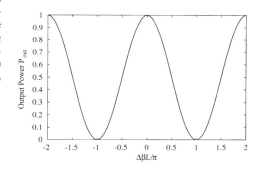

Figure 27: The observed $\cos^2$(voltage) signal from a Mach-Zehnder modulator. A phase mismatch of π gives one full modulation amplitude. The pattern is completely periodic for higher values of the phase mismatch, i.e. higher voltages. A well-designed device achieves an on/off ratio exceeding 30 dB.

6.4 Acousto-Optic Devices

Although the detailed physics of acousto-optic devices for communications is fairly involved, the basic principles are easily understood [3], [21], [23]-[25].

Consider a sound wave travelling through the bulk of a material or along the surface. In the simplest case, a sound wave consists of a periodic pressure variation and a periodic density variation. In a crystalline material, there is also stress and strain, which leads to the transverse sound waves. At the surface, the crystalline symmetry is broken and another wave mode, travelling along the surface, becomes possible: a surface acoustic wave (SAW). In the simplest case, the density variation within a wave has an increased optical index at the density maxima and a reduced index at the density minima. In an optical ferroelectric, piezoelectricity also plays a role. Therefore the stress

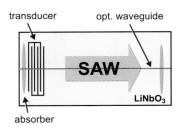

Figure 28: Basic structure for copropagating a SAW parallel to an optical waveguide. The long interaction path permits efficient coupling. The interdigitated electrode transducers generate the SAW, the absorbers are a sticky polymer which dampens the SAWs [25].

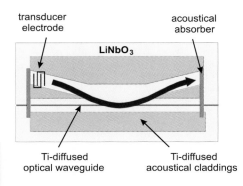

Figure 29: Advanced design of copropagating SAW: The SAW is guided in an acoustic waveguide, the optical signal is guided in an optical waveguide. In both cases Ti is diffused into $LiNbO_3$. The Ti increases both the optical index and the mechanical stiffness of $LiNbO_3$ as well [25].

modulates the optical index via the electro-optic effect and in summary the index variations are stronger than in non-polar media. In the following, SAWs in an optical ferroelectric will be considered.

The most important material for SAW devices, especially optical SAW functions, is $LiNbO_3$. In this material, a number of useful properties are combined:

- In the ferroelectric $LiNbO_3$ SAWs are easily excited by interdigitated metallic finger electrodes, deposited on the surface, see Figure 28. (Note that the conversion of electrical signals to SAWs and the reverse process of picking up a SAW as an electrical signal in $LiNbO_3$ are so efficient that there is a huge market for SAW electronic filters and SAW delay lines for TV and mobile communication applications).

- The phase velocity of the SAW is $c_s \approx 3500$ m/s. The wavelength λ_s of the SAWs can be changed over a very wide range by varying the frequency ν_s:

- $\lambda_s = c_s/\nu_s$

- A frequency $\nu_s = 1$ GHz corresponds to $\lambda_s = 3.5$ µm. Note that the wavelength of a SAW may become comparable to an optical IR wavelength. Therefore some interaction of a SAW and an optical wave is expected.

- Diffusing Ti into $LiNbO_3$ forms optical waveguides, as has been mentioned in the previous section. But it also stiffens the elastic constants and increases c_s. This makes it possible to fabricate acoustic waveguides which guide the SAW, see Figure 29. This is especially useful, if a SAW has to be guided parallel or antiparallel to an optical signal in an optical waveguide.

- Although the optical index modulation due to an SAW is only in the order of $\Delta n \sim 10^{-3}$, the SAW may propagate parallel or antiparallel to an optical waveguide, especially if the SAW is guided in a coplanar acoustic guide. It is useful to consider the SAW as a distributed Bragg reflector (DBR), which has been discussed in Sec. 4. Although the index modulation is small, the extended interaction over many mm along a waveguide assures efficient coupling of the SAW and the optical signal. The two important differences between a guided SAW and a stationary DBR are:
    - the index grating of the SAW moves along with c_s and
    - its wavelength is tunable by changing the microwave drive frequency.

- An optical wave may be diffracted or partially reflected by an SAW. Note that the diffraction of light (frequency ν_0) on a moving diffraction grating (frequency ν_s) involves a Doppler shift, which generates sidebands at $\nu_0 \pm m\,\nu_s$ (m = 1, ...) Only the first sideband (m = 1) is important. The effect can be used to frequency shift a carrier signal by adjustable amounts or to filter at an adjustable frequency by Bragg diffraction of the optical signal at the SAW, as shown in Figure 30.

- In $LiNbO_3$, the TE and TM polarization modes have different phase velocities. With an SAW running parallel to a TE and a TM mode, the mismatch in the propagation vector can be compensated and an effective energy exchange between TE and TM mode becomes possible. Therefore the phase matching between TE and TM with the help of an acoustic mode enables the TE-TM-mode conversion. Since the SAW frequency can be changed, the mode conversion becomes frequency selective. This allows one color from a DWDM pulse train to be selected by adjusting the SAW frequency, this signal to be converted to another polarization state and guided out of the primary optical path. This is called "frequency-selective signal drop".

- The tuning range of such SAW devices is typically limited by the bandwidth of the finger electrodes. A tuning range of 50 nm (spanning the entire C-band at 1.5 µm) has been demonstrated, see Figure 31 and Figure 32.

7 Summary

Worldwide, more than 100 000 km of optical fibers are installed daily. The volume of optically transported data has been growing at a rate of 60 % per year and probably will continue to do so for at least one decade.

ALCATEL has demonstrated a single fiber DWDM link that operates at 5.12 Tbit/s over 300 km. It uses 128 DWDM channels at 40 Gbit/s each. A data rate of 5 Tbit/s exceeds our imagination. The text of the entire Encyclopaedia Britannica is just 1 Gbit. In terms of the old telephone lines (64 kbit/s), this leads to the rather absurd and amusing picture of 80 million calls in parallel through the same glass fiber without cross-interfer-

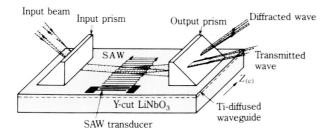

Figure 30: Acousto-optic Bragg cell. The optical input beam is coupled through a prism into a planar waveguide in LiNbO$_3$. A surface acoustic wave SAW is generated by the interdigitated electrodes. The SAW forms an index grating for diffracting the optical beam. Changing the SAW frequency changes the SAW periodicity [21].

ence. Of course, the potential of modern optocommunication lies in its effortless and rapid use for browsing through remote servers, searching and using programs, graphics, news and data pools, video conferencing, remote video-supported process control, video on demand, and many other possibilities. It permits the worldwide distributed, connected and interlaced operation of engineers and scientists on a common task.

The industries which provide fiber communication devices and systems report a total sales volume of more than $ 30 billion per year. This does not include the revenue made by cable operators and telecom companies. Singapore Telecom plans the installation of a PanAsia Submarine link of 17 000 km total length and 7.5 Tbit/s capacity. This demonstrates that more than any other system optical communication systems are empowering the information exchange, which is one of the cornerstones of the information society. (The other cornerstones are information storage and information processing).

Surprisingly, the gigantic improvements in the worldwide communication network are not very conspicuous, because the numerous fiber links are mostly hidden. Optocommunication is replacing long-distance electric telephone lines everywhere, but it is a silent revolution. While the wireless mobile and satellite systems have to fight for frequencies and pay outrageous sums for the limited UMTS bandwidth licenses, the fiber optic systems really offer virtually unlimited bandwidth, satisfying even the most demanding perspectives for internet traffic and live video transfer protocols. It is fiber communication more than any other technology which promotes the effective operation within today's and tomorrow's *Global Village*.

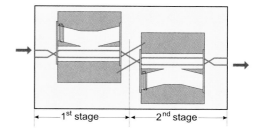

Figure 31: Polarization-independent double stage tunable SAW filter. The input is coupled to a polarization splitter ("first crossing"). After that, a highly wavelength selective TE → TM conversion takes place. Only the converted signal wavelength enters the second stage. The second stage reconverts TM → TE. By this process the frequency shift of the first stage is completely compensated. With a slightly more complex architecture, tunable add-drop WDM multiplexers can be fabricated [25].

Acknowledgements

I would like to thank Dr. Jürgen Rosenkranz and Dr. Wilhelm Reiners at ALCATEL Optical Fibers, Mönchengladbach and Dr. Harald Herrmann, Dr. Hubertus Suche and Prof. Wolfgang Sohler at Universität Paderborn for cooperation and the generously offered information and support. The editor would like to thank Stefan Tappe (RWTH Aachen) for checking the symbols and formulas in this chapter.

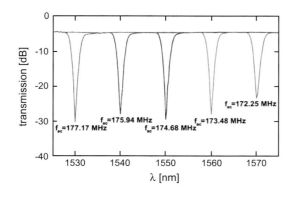

Figure 32: Response of the tunable two-stage filter shown in Figure 30. Tuning is achieved by simply changing the frequency of the SAW [25].

References

[1] B. E. A. Saleh and M. C. Teich, *Fundamentals of Photonics*, J. Wiley, New York, 1991.

[2] R.Menzel, *Photonics,* Springer, Berlin, 2001.

[3] N.Grote, H.Venghaus (eds.), *Fibre Optic Communication Devices,* Springer, Berlin 2001.

[4] J. Wilson and J. F. B. Hawkes, *Optoelectronics*, Prentice Hall, New York, 1989.

[5] H. Murata, *Handbook of Optical Fibers and Cables*, Marcel Dekker, Inc., New York, 1996.

[6] Lucent Technologies, *Wave Star System* (www.bell-labs.com/news/1999/march/9/1).

[7] M.K.Smit, *PHASAR-Based WDM-Devices,* IEEE J Selected Topics in Quantum Electronics 2, 236 (1996).

[8] C. C. Davis, *Lasers and Electro-optics*, Cambridge Univ. Press, Cambridge, 1996.

[9] S. M. Sze, *Physics of Semiconductor Devices*, Wiley, New York, 1981.

[10] K. W. Böer, *Survey of Semicondcutor Physics*, Van Nostrand Reinhold, New York, 1990.

[11] J. I. Pankove, *Optical Processes in Semiconductors*, Dover Publ., New York, 1971.

[12] K. J. Ebeling, *Optoelectronics*, Springer, Berlin, 1993.

[13] Ch. Buchal in: 31. Ferienkurs, Femtosekunden und Nano-eV, IFF-Forschungszentrum Jülich, 2000.

[14] Ch. Buchal, M.Löken, Th.Lipinsky, L.Kappius, S.Mantl, J Vac Sci Technol A18, 630 (2000).

[15] H.Zimmermann, *Integrated Silicon Optoelectronics,* Springer, Berlin, 2000.

[16] E. Desurvire, Physics Today 47, 20 (1994).

[17] M. J. F. Digonnet, *Rare Earth Doped Fiber Lasers and Amplifiers*, M. Dekker, New York, 1993.

[18] V. Morin and E. Taufflieb, Laser Focus World 35, 127 (1999).

[19] Ch.Buchal, M.Siegert in O.Bisi (ed), *Silicon-Based Microphotonics,* IOS Press, Amsterdam, 1999.

[20] www.lucent-optical.com/products/lambdarouter.

[21] H. Nishihara, M. Haruna and T. Suhara, *Optical Integrated Circuits*, McGraw-Hill, New York, 1989.

[22] K. Noguchi, H. Miyazawa and O. Mitomi, Electr. Lett. 34, 661 (1998).

[23] H. Herrmann, K. Schäfer and Ch. Schmidt, IEEE Photonics Techn. Lett. 10, 120 (1998).

[24] F. Wehrmann, Ch. Harizi, H. Herrmann, U. Rust, W. Sohler and S. Westenhöfer, IEEE J. Selected Topics in Quantum Electr. 2, 263 (1996).

[25] H. Herrmann, U. Rust and K. Schäfer, J. Lightwave Technol. 13, 364 (1995).

[26] Ch. Buchal, S. Wang, F. Lu, R. Carius, S. Coffa, Nucl. Instr. and Meth. in Phys. Res. B 190, 40 (2002).

[27] S. Wang, A. Eckau, E. Neufeld, R. Carius, Ch. Buchal, Appl. Phys. Lett. 71, 2824 (1997).

[28] ST Microelectronics news release Oct. 29, 2002, see: http://eu.st.com.

Microwave Communication Systems – Novel Approaches for Passive Devices

Norbert Klein
Department ISG, Research Center Jülich, Germany

Contents

1 Introduction	755
2 Some Important Aspects of Microwave Communication Systems	756
2.1 Cellular Communication	756
2.2 Satellite and Point-to-Point Communication	757
3 Basic Properties of Resonant Microwave Devices	758
3.1 Electromagnetic Resonators	758
3.2 Multipole Filters	761
4 Microwave Properties of Metals, Superconductors, and Dielectric Materials	762
4.1 Surface Impedance of Normal Metals	762
4.2 Surface Impedance of High-Temperature Superconductor Films	763
4.3 Microwave Properties of Dielectric Single Crystals, Ceramics, and Thin Films	764
5 Novel Passive Devices for Microwave Communication Systems	769
5.1 Dielectric Filters	769
5.2 Planar High-Temperature Superconducting Filters and Subsystems	770
6 Micromechanics for Microwaves: RF MEMS and FBARs	771
7 Photonic Bandgap Structures	773
8 Summary	774

Microwave Communication Systems – Novel Approaches for Passive Devices

1 Introduction

Microwave communication has turned out to be one of the most rapidly growing fields of our information society. Wireless communication based on individual handsets has revealed breathtaking growth rates with a dramatic increase of transferred date rates and number of subscribers. In particular, wireless multimedia applications are considered to enforce a dramatic increase of wireless system capability in the near future.

In contrast to optical communication links, where the data rate capability appears to be as infinite from the point of view of physical limitations, microwave communication systems are intrinsically limited by the availability of bandwidth. Therefore, modern systems have to utilise the available bandwidth most efficiently. In addition, the cost issue is rather important, for example due to strong competition between different service providers in mobile communication.

Current system specifications including those for the 3rd generation mobile telephone systems (UMTS in Europe) are based on the available performance of current devices and subsystems. However, novel devices with improved performance are likely to increase the capacity of the systems or may become even a necessity in order to operate a system properly under any circumstances, e.g. in mobile communication at locations and in situations with strong interference problems. Therefore, any new material and device development leading to a performance increase or / and to a cost reduction is worth to be investigated.

This chapter is focussing on novel approaches to passive microwave devices and related material properties with emphasis on the analogue part of communication systems. Novel approaches in improving the digital part of communication systems are discussed in other chapters of the book. In most cases passive devices are frequency selective devices, like filters or frequency determining elements in oscillator circuits. In addition, flexible utilisation of bandwidth in the future may require switchable and tuneable passive devices.

The performance of passive devices has a strong impact on relevant system properties like receiver sensitivity, interference capability, required handset power, relative fraction of usable bandwidth and system costs. In addition to the passive devices to be discussed in this lecture, novel active devices with relevance to the analogue part of communication systems have emerged. Prominent examples are gallium nitride field effect transistors for high frequencies which exhibit a high potential to surpass the performance of other III/V field effect transistors in terms of noise and high-frequency amplification [1]. In addition, SiGe gets more and more important for the construction of integrated micro- and millimetre wave oscillators [2].

The chapter is organised as follows. First, a very basic introduction into the architecture of modern wireless communication systems with emphasis on how the performance of passive microwave devices defines their properties will be given (Sec. 2). Sec. 3 is devoted to a selection of relevant basic properties of electromagnetic resonators and filters. After discussing the physical aspects of microwave properties of metals, superconductors and dielectrics (Sec. 4), different types of novel devices based on these materials will be presented. In Sec. 5, cryogenically cooled subsystems based on high-temperature superconducting devices will be discussed. Sec. 6 provides a brief overview about micromechanical microwave switches and a novel type of bulk acoustic resonators for the GHz range. Finally, in Sec. 7 photonic bandgap structures and their potential use in millimeter wave systems will be discussed.

VI Data Transmission and Interfaces

2 Some Important Aspects of Microwave Communication Systems

It is beyond the scope of this chapter to describe the various aspects of modern microwave communication systems. However, in order to get a basic understanding of how device performance enters the system performance some system properties are worth discussing here. For a more comprehensive introduction to microwave communication systems the reader is referred to [3].

2.1 Cellular Communication

Cellular communication means that a certain available range of frequency bandwidth can be used simultaneously in different cells, i.e. areas within the reach of one (or several) base stations. Typically, each base station consists of a receiving unit which receives signals being transmitted from handsets of users inside the cell (uplink). For the most common European systems the signals are within frequency windows of 890 to 950 MHz for uplink and 935 to 960 MHz for downlink (GSM 900, in Germany D1 and D2) and at around 1800 MHz (DCS 1800, in Germany referred as "E-Netz"), respectively [4]. The receiving unit of a base station consists of segments as the one sketched in Figure 1. The main parts of it are an antenna, an input bandpass filter, a low noise preamplifier and a frequency converter which transfers the incoming microwave signal to the base band frequency. The base band signal is further processed employing an A/D converter. For the GSM (Global System of Mobile Communication) standard the downlink frequency band is well separated from the uplink window (usually located above the downlink interval). The received digital information is employed to modulate the signal of a microwave oscillator, which is fed back to the antenna after amplification (transmitter unit). Data transmission to other basestations or to the wire-based telephone network is provided either by optical links or by microwave point-to-point or point-to-multipoint microwave links usually operating at higher frequency bands above 20 GHz.

Usually, the frequency windows of each base station have to be shared between different service providers. This can result in interference problems due to intermodulation distortion by the preamplifier [7]: Suppose that two strong signals at frequencies f_1 and f_2 (assume $f_1 < f_2$) are received within one subband, the nonlinearities of the preamplifier creates so-called intermodulation signals at frequencies $2f_1 - f_2$ and $2f_2 - f_1$, i.e. unwanted signals at frequencies shifted by an amount $\Delta f = f_2 - f_1$ below f_1 and above f_2. Therefore, small uplink signals in adjacent subbands may interfere with such intermodulation signals leading to bit errors beyond the bit-error-compensation ability of the software employed for digital signal processing. Therefore, analogue bandpass filters in the receiver frontend are important – not only to isolate the downlink signals from the receiver frontend, but also to protect certain subbands from intermodulation signals occuring in adjacent subbands. An ideal bandpass filter has a transmission-frequency characteristic like a Θ-function, i.e. infinite off-band rejection, infinite steepness of skirts at the passband edges and zero signal attenuation within the passband. It will be discussed in Subsec. 3.2 how close real filters come to this ideal situation.

Finally, it is worth mentioning that interference problems may become even more severe for the upcoming 3rd generation systems. In UMTS ("Universal Mobile Telecommunications Systems" = European standard of 3rd generation) there exist bands for FDD (frequency domain duplexing) with well separated uplink (1920 to 1980 MHz) and downlink windows (2110 to 2170 MHz) used for voice communication and TDD (time domain duplexing) bands for unbalanced internet communication with frequency bands at 1900 to 1920 MHz and 2010 to 2030 MHz [5]. For the TDD bands the two frequency windows can be used for uplink and downlink signals simultaneously. Since in some cases base stations for TDD and FDD are directly adjacent to each other, strong TDD downlink signals may cause intermodulation signal in the FDD receiver units, if no sufficient filtering is accomplished.

For UMTS the FDD receiving interval of 60 MHz width is subdivided in 12 frequency windows of 5 MHz width for each. In Germany, two adjacent windows will be used by one service provider. This channel subdivision is different to GSM where the channel bandwidth is in the kHz range. The UMTS system architecture allows voice messages and high-data rate messages to be dealt with in the same way, which results in a much more efficient utilisation of available bandwidth. However, the extremely narrow

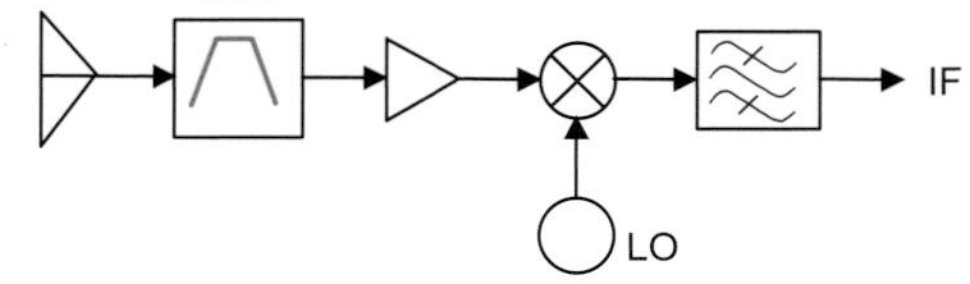

Figure 1: Block diagram of a part of a receiver unit as being employed in base stations for modern digital wireless communication systems (LNA: low noise amplifier, LO: local oscillator, IF: intermediate frequency).

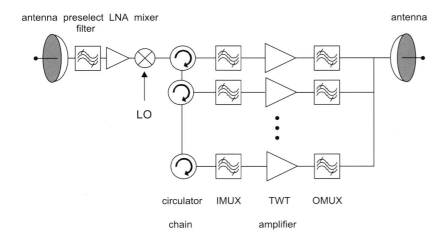

Figure 2: Block diagram of a satellite transponder consisting of a preselect filter, a preamplifier and frequency transforming unit, an input multiplexer (IMUX), a switch matrix, high-power TWT amplifiers, and an output multiplexer (OMUX) [6], further explanation see text).

subbands of just 0.3% relative bandwidth may enforce extremely narrow bandwidth filters for the case of strong interference situations. As discussed within Subsec. 5.1 and 5.2, this situation is very challenging for the development of new filter technologies.

Beyond 3rd generation, in future communications systems the allocation of frequency bandwidth may become even more flexible than for the 3rd generation systems: For example, in case of mobile internet traffic the downlink datarate is expected to be much higher than the uplink rate. In this case, the strict separation between uplink and downlink bands of equal widths may be replaced by a most variable up- and downlink allocation. Such a situation would imply even stronger requirements for analogue filtering both in the base station and in the mobile terminal, because strong transmit signal may interfere with weak receive signals.

2.2 Satellite and Point-to-Point Communication

In contrast to cellular communication, satellite and point-to-(multi)point microwave communication links often rely on terminals at fixed locations. This holds true for geostationary satellites (GEO), while low earth orbit (LEO) satellites represent moving transponder stations which may even be linked with mobile terminals on the ground. In the case of stationary links there is no upper frequency limit due to shadow effects like in cellular communication. The physical limitation for the use of very high frequencies occurs due to attenuation of electromagnetic waves in the atmosphere and scattering by rain and fog, which is extremely crucial for satellite-earth links.

Current geostationary satellite links operate from L-band frequencies (around 2 GHz) over C-band (3.7 to 4.2 GHz for downlink) and Ku-band (10.95 to 12.2 GHz for downlink) towards Ka-band frequencies (19.7 to 20.2 GHz for downlink) [6]. The uplink frequency intervals are above the downlink frequencies for each band, but all the filtering takes place within the downlink intervals. Typically, L-band is used for GSM and Meteosat, C-band for communication links, Ku-band for European satellite TV, and Ka-band will become very important for multimedia applications in the future. In addition, for intersatellite links V-band frequency (around 60 GHz) transponders are under development, but optical links represent a challenging alternative.

In contrast to base stations in mobile communications, the typical transponder architecture being used in satellite communication (see block diagram in Figure 2) is much simpler and usually requires no on-board digital processing (on board processing is supposed to become more important in the future, but simplicity and robustness of the payload hardware sets strong limits on complexity). Basically, a satellite transponder receives the uplink signal, transfers its carrier frequencies (usually more than one) to the downlink frequency band, and sends the signal back to earth after amplification. Due to intermodulation distortion in the amplifiers, each of the N channels has to be amplified separately. According to Figure 2, a multicarrier signal, usually composed of N modulated signals with a bandwidth of 36 MHz for each (representing e.g. one TV program) is de-multiplexed in the IMUX (input multiplexer) after pre-amplification and transformation from the uplink to the downlink frequency interval. The IMUX consists of a circulator chain directing the multicarrier signal to N individual bandpass filters (a circulator is a standard three port devices based on ferrites which allows for transmission from port i to $i+1$ ($i = 1, 2, 3$) but *not* via versa [7]). Each filter transmits one of the N carriers, rejects the remaining ones and hands them to the next filter of the multiplexer chain. After channel switching the signal of each channel is amplified to power levels of typically several ten watts using travelling wave tube (TWT) amplifiers. Switching is required e.g. to replace a damaged TWT by a redundant amplifier). In order to avoid interference between different channels (usually, TWT amplifiers generate lots of harmonics), the amplified signal of each channel has to pass through a second bandpass filter with sufficiently high power handling capability. The filtered signals are combined (employing a passive combiner network usually consisting of waveguide sections) and fed back to the antenna. The hardware unite composed of output filters and passive combiner network is called "OMUX" (output multiplexer).

In contrast to cellular communication the specifications for the filters can be fulfilled by standard resonator filters, which are composed of coupled sections of coaxial transmission lines, hollow waveguide sections or dielectric resonators (see Subsec. 3.2). Since a typical TV satellite transponder consists of 64 channels, the weight of the IMUX and OMUX represent a significant contribution to the satellite payload. Therefore, new filter technologies with the potential of weight reduction by matching or surpassing the performance of existing filter technologies are important. In addition to filters, a further reduction of payloads can be achieved by replacing motor-driven mechanical waveguide switches by micromechanical switches (MEMS, see Sec. 6). In addition to the passive components, on the active side the high-power TWT amplifiers may be replaced by modern solid-state amplifiers (e.g. GaN), if there power efficiency will approach the one of tube amplifiers.

For higher frequencies like Ka-band the lack of performance of devices is a general problem, and any improvement, e.g. phase noise reduction of the on-board local oscillator, is of particular importance. Upon approaching the millimetre wave range (frequencies above 100 GHz) to-be-developed resonator and filter technologies based on photonic bandgap materials may play a role in the future (see Sec. 7). Besides communications, this frequency range is of particular relevance for imaging systems and radar.

3 Basic Properties of Resonant Microwave Devices

3.1 Electromagnetic Resonators

In microwave technology, the basic element of a filter and the frequency stabilising element of an oscillator circuit is an electromagnetic resonator. With respect to such applications, an electromagnetic resonator is characterised by its resonant frequency f_0, its unloaded quality factor Q_0 of the selected resonant mode(s) and its spectrum of spurious modes. In order to measure the resonator properties or to use a resonator as part of a filter structure or another microwave circuit, the resonator needs to be equipped with one or two coupling ports, which are characterised by external quality factors Q_{ext1} and Q_{ext2}. The loaded quality factor Q_L, which determines the 3dB bandwidth or resonant halfwidth $\Delta f_{3\text{dB}}$ by $Q_L = f_0 / \Delta f_{3\text{dB}}$ depends on Q_0 and the coupling strength [8]:

$$\frac{1}{Q_L} = \frac{1}{Q_0} + \frac{1}{Q_{\text{ext1}}} + \frac{1}{Q_{\text{ext2}}} \tag{1}$$

Consider a two-port resonator with equal input and output coupling strength, i.e. $Q_{\text{ext}} \equiv Q_{\text{ext1}} = Q_{\text{ext2}}$. Such a device has a Lorentzian shaped transmission behaviour characterised by its 3dB bandwidth and its minimum insertion loss (at the resonance frequency). According to Eq. (2) the ratio of transmitted to incident power (referred to by the absolute value of the element S_{21} of the scattering matrix describing a two port device) is given by [34]:

$$|S_{21}| = \frac{1}{\left[1 + \frac{Q_{\text{ext}}}{2Q_0}\right]^2} \approx 1 - \frac{Q_{\text{ext}}}{Q_0} = 1 - \frac{1}{Q_0}\frac{f_0}{\Delta f_{3\text{dB}}} \tag{2}$$

The approximation given in Eq. (2) is valid for strong coupling ($Q_{ext} \ll Q_0$), which is the typical situation for bandpass filters. According Eq. (2) the insertion loss $1-|S_{21}|$ of such a one-pole bandpass filter in the centre of the passband increases with decreasing bandwidth and decreasing unloaded quality factor. This indicates, that high quality factors become very important to obtain sufficiently low values of the in-band insertion loss. As to be discussed in Subsec. 3.2, for a multipole filter the obtainable steepness of filter skirts at the passband edges is also limited by the value of Q_0.

For the microwave frequency range there exist a variety of resonators with different values of Q_0. With regard to recent developments of novel structures, dielectric resonators, planar resonators, bulk acoustic resonators (see Sec. 6) and defect resonators see Sec. 7) will be discussed in some detail. For the discussion of more conventional resonators like cavity or coaxial resonators the reader is referred to standard microwave engineering textbooks [7].

Dielectric resonators – In general, a dielectric resonator consists of one (sometimes more than one) piece(s) of dielectric material characterised by its relative permittivity $\varepsilon_r \equiv \mathrm{Re}\{\varepsilon\}$ and its loss tangent $\tan\delta \equiv \mathrm{Im}\{\varepsilon\}/\mathrm{Re}\{\varepsilon\}$, with $\varepsilon(\omega,T)$ representing the complex dielectric function of the dielectric medium (see Sec. 4). For many applications, the dielectric body is of cylindrical shape with circular cross section, but square rectangular and hemispherical shapes are currently under development (see Subsec. 5.1). The dielectric resonator is arranged inside a metallic shielding cavity machined from a highly conducting metal like copper or lightweight aluminium based alloys with silver coating, but wall segments based on high-temperature superconducting thin films can also be used. The unloaded quality factor of such a shielded dielectric resonator is given by [8]

$$\frac{1}{Q_0} = \kappa \tan\delta + \frac{R_s}{G}; \quad \kappa = \frac{\varepsilon_r \int_{DR} E^2 dV}{\int_C E^2 dV + \varepsilon_r \int_{DR} E^2 dV}; \quad G = \frac{\omega\mu_0 \int_{DR+C} H^2 dV}{\int_C H^2 dA} \quad (3)$$

with R_s representing the surface resistance of the shielding cavity material (see Sec. 4), κ the filling factor indicating the fraction of electric field energy stored in the dielectric resonator (DR = "dielectric resonator", C = "shielding cavity filled with air / vacuum") and G a geometric factor (in units of "Ohms") representing the ratio of cavity volume to

Table 1: Modes in cylindrically shaped dielectric resonators

Mode / Properties	$HE_{11\delta}$	$TE_{01\delta}$	"whispering gallery"
Electric field distribution			
Degeneracy	dual mode	monomode	dual mode
Quality factor	κ = 0.8-0.9 G = 500 – 1500 maximum Q: 30.000 @ 4 GHz with BMT (see section 4.3)	κ = 0.9 G = 2000 maximum Q: 50.000 @ 10 GHz with alumina (see section 4.3)	κ = 0.95 G = 10^4 maximum Q: 10^7 @ 10 GHz and 77 K for sapphire (see section 4.3)
Spurious density	low (fundamental mode)	medium	very high
Most relevant applications	multipole bandpass filters for satellite communication	oscillators, microwave characterisation of dielectrics and superconducting films, filters	low-phase noise oscillators, microwave frequency standards

surface area which increases with increasing cavity size for a fixed frequency. For a given resonator geometry, the quantities κ and G can be calculated for each eigenmode by solving the integrals in Eq. (3), if the distribution of the electric field $E(x)$ and the magnetic fields $H(x)$ have been calculated either analytically or – nowadays very conveniently with commercial state-of-the art eigenmode solver programs – numerically based on finite element methods.

The first term in Eq. (3) represents the dielectric losses, the second one the losses associated with the metallic walls of the shielding cavity. In general, the quantities G and κ depend on the geometry of the cavity and the DR, on the permittivity of the DR, and on the employed resonant mode.

Among all possible geometries the cylindrical geometry is the most commonly used one. This is because of the simplicity of fabrication by pressing pellets from ceramic powder and subsequent sintering. The other reason is that the resonant modes of a cylindrically shaped DR can be solved either analytically (in case of an infinitely extended cylinder or a piece of cylinder clamped between two metallic walls) or by mode matching techniques [9]. However, due to recent achievements in numerical simulation techniques there are novel approaches based on different geometries.

Table 1 summarises the relevant properties of the most commonly used modes of cylindrically shaped dielectric resonators. Depending on the values of G, which is a measure for the loss contribution of the metallic shielding cavity (see Eq. (3)), a reduction of the loss tangent leads to an increase of the resonator Q value. In particular, for the whispering-gallery modes the loss contribution of the metal walls becomes negligible, which results in extremely high Q values if a high-purity single crystalline dielectric like sapphire is employed. Whispering gallery modes (named after the "whispering gallery" in St. Pauls Cathedral in London) are modes with high azimuthal mode number (typically $n \geq 7$) corresponding to a standing wave propagating along the circumference of the dielectric cylinder with $2n$ field maxima / minima per circumference. According to geometrical optics, total reflection occurs at the circumference upon transition from a dense medium (dielectric puck) to the surrounding air, leading to a strong confinement of electromagnetic field energy in the DR.

In table 1 the possible applications of the different modes are given. For most filters, only modes of low order (in many cases only the fundamental mode can be used) because of specifications with regard to the off-band rejection properties of the filter.

Among other geometries which have been investigated recently, it is worth mentioning that the fundamental dual mode of a dielectric hemisphere arranged on a cylindrically shaped dielectric socket (mushroom configuration, see [10]) gives a somewhat higher Q. However, machining of such "mushrooms" is difficult and expensive, in particular due to the hardness and brittleness of most low loss dielectric ceramics. Another attractive approach is the fundamental dual mode of a square shaped dielectric disc. In a recently developed filter geometry such a disc is fixed by soldering inside a cylindrically shaped metal housing at its edges, resulting in reasonable high Q values and very good thermal contact between the dielectric ceramic and the metal housing [11]. The latter is important for output filters operating at elevated power levels in order to reduce heating (resulting in detuning due to the temperature coefficient of its dielectric constant).

Eq. (3) is not only valid for dielectric resonators. Any other type of electromagnetic resonator employing dielectric parts, like metal ceramic coaxial-type resonators (e.g. used as filters in mobile phones) and microstrip or coplanar resonators (used in microwave integrated circuits) have a Q-contribution due to dielectric losses. For the latter type of resonator the dielectric losses are negligible in comparison to metallic losses, unless high temperature superconducting metallization layers are applied.

Planar resonators – Figure 3 shows typical examples of planar resonators which have been used as building blocks of multipole filter structures (see Subsec. 3.2) or can be used as frequency stabilising element of an integrated oscillator circuit. Planar resonators are either segments of TEM (transverse electromagnetic) transmission lines with length L of half a wavelength corresponding to the resonances to be at frequencies $f_n \approx n / 2 L$ ($n = 1, 2, ...$) (b,d) or lumped element resonators each composed of a discrete inductor L and capacitor C with its resonance at a frequency $f = 1 / [2\pi (LC)^{1/2}]$. Pure lumped element resonators are mostly based on interdigital capacitors and small segments of metallization defining the circuit inductance (a). In many cases hybrids of lumped elements microstriplines are used, e.g. folded microstriplines disrupted by a capacitive gap (c).

Planar resonators and microstrip and coplanar transmission lines represent the passive elements in almost any integrated circuit technology: at first, on chip integration in so called MMICs (monolithic microwave integrated circuits) with SiGe or III/V semiconducting active devices has been the most important technological effort with regard to the commercial exploitation of the microwave frequency range. In addition, so-called LTCC (Low-Temperature Co-fired Ceramics) technology based on stacked ceramic plates with integrated metallization layers has become extremely important to achieve reasonable Qs in mobile communication units (see Subsec. 4.3). On the high-performance end, planar resonators based on high-temperature superconducting (HTS) thin films to be operated at cryogenic temperatures between about 50 and 80 K exhibit Q values even higher than those for cavity or dielectric resonators. As shown in Subsec. 5.2, this technology currenty allows for the highest filter performance attainable for the microwave range.

According to Eq. (3) the unloaded Q of a microstrip resonator can be approximated by $Q_0 \approx \omega \mu_0 t / R_s$ (neglecting substrate losses), if the stripline width is large in comparison to the substrate thickness t. In the case of a lumped element resonator with feature size small in comparison to the wavelength the unloaded Q can be approximated by $Q_0 \approx \omega L / R$ with L being the inductance of the resonator and R its resistance. In the case of ring of diameter d $R \approx R_s \pi d / w$ and $L \approx \mu_0 d / 2$ [12] resulting in $Q_0 \approx f \mu_0 w / R_s$. In general, the Q of a planar resonator needs to be calculated by numerical simulation tools.

As mentioned before, the highest Q values can be achieved with HTS thin film technology. In this case also ground plane metallization needs to be formed by a HTS film, e.g. in general double sided films are required for ultimate performance. As an alternative, coplanar resonators have been used (d) with some compromise on Q in comparison to microstrip lines of the same size but with the advantage of single sided coating. For the highest Q and ultimate power handling capability circular 2D microstrip resonators excited in the TM_{010}-mode have been used (e, [34]), also requiring double sided coating.

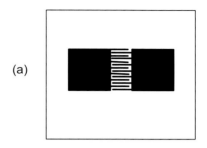
(a)

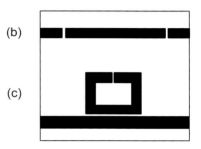
(b)

(c)

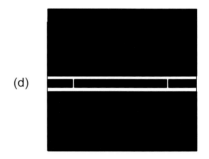
(d)

3.2 Multipole Filters

For most cases, the Lorentzian shape of the transmission characteristic of a single resonator is not very useful for filters in microwave communication systems. Ideally, one would like to have a flat passband response, i.e. low insertion loss over the whole passband. In addition, the out-of-band rejection below and above the passband should become very high as close as possible to the passband edges. In order to approximate such a rectangular shaped characteristic, multipole filters composed of mutually coupled resonators are commonly used in microwave technology.

In general, such a multipole filter can be constructed as a combination of single resonators. The number of poles of the multipole filter corresponds to the number of resonances. For single mode resonators, the number of modes is identical with the number of resonators, i.e. a N-pole filter consists of N resonators. For dual-mode resonators, the required number of resonators is $N / 2$.

The most simple approach of an N-pole filter is a forward-coupled Chebyshev filter consisting of a chain of N resonators (for simplicity let us assume single mode resonators) with coupling between adjacent resonances, i.e. between resonators i and $i+1$ ($i = 1....N - 1$). The 1st and the Nth resonators are coupled to an external port, which usually is a coaxial cable, a waveguide port or a microstripline within an integrated circuit. For the sake of simplicity, in most cases symmetric filters are used with coupling coefficients $k_{i, i+1} = k_{N-i, N-i+1}$ and input- and output coupling of identical strength.

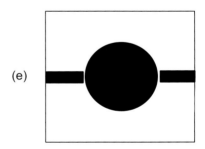
(e)

For such a filter design, the steepness of filter skirts increases gradually with increasing number of N up to a limit determined by the unloaded Q-values of the resonators. This means, that even for an infinite number of poles a rectangular filter characteristic cannot be obtained due to the finite value of Q_0. For given values of N and Q_0 (in the following assumed to be equal for all N resonators), the steepness of skirts can be improved drastically by introducing damping poles in the filter characteristic [6], [7]. One common way to accomplish damping poles is to introduce cross couplings (mostly negative, i.e. there exists an overlap between positive and negative electric field amplitudes at the location of the coupling port, examples are shown in Sec. 5) between non-adjacent resonators. The maximum number of cross couplings for an N pole filter with even value of N is $m = N - 2$. These cross couplings generate $m/2$ damping poles arranged symmetrically (ideally) on both sides of the passband (see Figure 4). Such filters are called quasielliptic filters (real elliptic filters are filters with an odd number of poles of N poles and $m = (N - 1)/2$ damping poles on each side, but cannot be constructed from forward coupled resonators with cross couplings). As a drawback of elliptic and quasielliptic filters, the out-of-band transmission is higher than for Chebyshev filters of the same order.

Figure 3: Typical planar resonators being used as building blocks for filters: lumped element (a), microstrip (b), folded microstrip with integrated capacitors (c), coplanar (d), and 2D microstrip resonator. Omitting the capacitive gap in the folded microstrip design (e) leads to a ring resonator (square if circular shaped), which also represents a quite commonly used microstrip resonator design.

Similar to a single resonator (Eq. (2)), the in-band insertion loss of an N-pole filter of fractional bandwidth BW is proportional to $N/(Q_0 \cdot BW)$ and differs slightly for different filter topologies. The Q-limitation of the skirt steepness S (theoretically only achievable for an infinite number of poles) is given by S [dB/MHz] $= 4 \cdot 10^{-3} \cdot Q_0 \cdot \xi / f$ [GHz] with $\xi = 0.33$ for Chebyshev and $\xi = 1$ for elliptic filters (the values for quasielliptic filters are in between, but much closer to the values for the elliptic filter) [6]. This indicates that (quasi)elliptic filters composed of resonators with Q_0 values of several 10^4 are quite attractive for applications in base stations and satellite transponders.

4 Microwave Properties of Metals, Superconductors, and Dielectric Materials

For the physical description of the microwave properties of dielectric materials and metals we consider Maxwell´s equations in the absence of localised charges:

$$\nabla \times \boldsymbol{E} = -\frac{\partial \boldsymbol{B}}{\partial t} \;;\quad \nabla \times \boldsymbol{H} = \frac{\partial \boldsymbol{D}}{\partial t} + j \;;\quad \nabla \boldsymbol{B} = 0 \;;\quad \nabla \boldsymbol{E} = 0 \tag{4}$$

The universal treatment of dielectrics and metals including superconductors relies on the fact that for a harmonic time dependence according to $\exp(i\omega t)$ one can describe the displacement current $\partial \boldsymbol{D}/\partial t$ and the Ohmic current density j by a generalised complex dielectric function $\underline{\varepsilon}$ or a generalised complex conductivity $\underline{\sigma}$ (assuming that the current density obeys a generalised Ohms law with complex conductivity, i.e. nonlocal effects can be neglected):

$$\underline{\sigma} = i\omega\underline{\varepsilon}; \quad \underline{\sigma} = \sigma' - i\sigma''; \quad \underline{\varepsilon} = \varepsilon_0 \varepsilon_r (1 + i\tan\delta) \tag{5}$$

The quantity ε_r is the rael part of the relative dielectric constant or permittivity of a dielectric medium, $\varepsilon_0 = 8.85 \cdot 10^{-12}$ As/Vm. The quantity $\tan\delta$ is the loss tangent of a dielectric medium (see also Chap. 1). Metals in the microwave range are usually described by a complex conductivity with dominant real part for normal metals and dominant imaginary part for superconductors.

It is useful to consider the solution of Maxwell´s equation (Eq. (4)) for plane electromagnetic waves in the absence of boundary conditions, which can be written as $\exp[i(\underline{\beta}z - \omega t)]$ assuming propagation in z-direction of cartesian coordinates. The quantity $\underline{\beta}$ is the complex propagation constant of the medium with dominant real part for dielectrics and dominant imaginary part for metals. The impedance of the medium, Z, defined as ratio of electric to magnetic field is related to $\underline{\beta}$ by $\underline{Z} = \omega\mu_0/\underline{\beta}$ with $\mu_0 = 1.256 \cdot 10^{-6}$ Vs/Am. As it can be derived from Maxwell´s equation, the impedance is related to the conductivity / dielectric function by the following expression:

$$\underline{Z} = \sqrt{\frac{i\omega\mu}{\underline{\sigma}}} = \sqrt{\frac{\mu_0}{\underline{\varepsilon}}} \tag{6}$$

In case of vacuum ($\varepsilon = \varepsilon_0$) the impedance is real and its absolute value is 377 Ω (wave impedance of vacuum). In case of a metal an incident wave decays rapidly from the surface, thus the impedance of a metal is called "surface impedance" $Z_s = R_s + i X_s$. The real part of Z_s is called surface resistance (R_s), the imaginary part surface reactance (X_s). As described in the Subsec. 4.2, the surface resistance is an important figure-of-merit of superconducting films.

4.1 Surface Impedance of Normal Metals

For a normal metal at microwave frequencies the imaginary part of the conductivity can be neglected and the real part is equal to the dc conductivity σ_{DC}. In this case a simple expression for the surface impedance follows from Maxwell´s Equation and Ohm´s law:

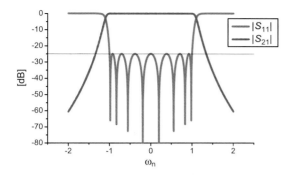
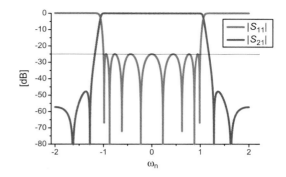

Figure 4: Calculated filter characteristic (without losses) of an 8-pole Chebyshev filter (left) and a quasielliptic 8-pole filter with 4 negative cross coupling between non-adjacent resonators (right) leading to two symmetric damping poles on each side of the filter characteristic. The quantities $|S_{21}|$ and $|S_{11}|$ (in dB) represent the normalised transmitted and reflected power, respectively, ω_n frequency deviations from the mid band frequency normalised to half of the filter bandwidth. The damping poles results in a significant increase of steepness of filter skirts.

$$Z = \sqrt{\frac{\omega\mu_0}{2\sigma_{DC}}}(1+i) = \frac{1}{\delta\sigma_{DC}}(1+i); \qquad \delta \equiv \sqrt{\frac{2}{\omega\mu_0\sigma_y}} \text{ (skin depth)}. \qquad (7)$$

Eq.(7) is valid as long as the skin depth is large in comparison to the mean free path of the electrons in the metal. This holds true in the microwave range at room temperature, for cryogenic temperature the surface resistance lies above the values predicted by Eq. (7) and exhibits a $f^{2/3}$ rather than a $f^{1/2}$ frequency dependence (anomalous skin effect [8]).

As an example, for copper with a room temperature conductivity of $5.8 \cdot 10^7$ $(\Omega m)^{-1}$ the surface resistance at 10 GHz is 26 mΩ, the skin depth is 0.66 µm. Therefore, the Q of a cavity resonator with a geometric factor of several hundred is in the 10^4 range. However, for planar resonators like the ones shown in Figure 3 the G values are only a few Ohms leading to Q values of only a few hundred. This is too small for many filter and oscillator applications.

4.2 Surface Impedance of High-Temperature Superconductor Films

As discussed in Chap. 3, the microwave response of a superconductor can be understood by two physical phenomena: First, by the Meissner effect, which causes any magnetic field applied to the surface of a superconductor drops exponentially to zero inside the superconductor on the lengthscale of the London penetration depth λ_L, which is about 160 nm for the most relevant high-temperature superconductor (HTS) compound YBa$_2$Cu$_3$O$_7$ (called "YBCO") at $T \rightarrow 0$ and increases with temperature according to $(\lambda_L(0)/\lambda_L(T))^{-2} \approx 1 - (T/T_c)^2$ (T_c = transition temperature = 92 K for YBCO). This shielding behaviour is almost frequency independent up to the THz range for YBCO leading to a frequency independent skin depth equal to λ_L. The second fact is that at finite temperatures below T_c Cooper pairs (corresponding to "ballistic" charge carriers without dissipation) and quasiparticles (corresponding to normal conducting charge carriers with a temperature dependent density) coexist. Therefore, the complex conductivity of a superconductor is composed of two parts:

$$\sigma = \sigma'(T) - i\sigma''(T) = \sigma'(T) - \frac{i}{\omega\mu_0\lambda_L^2(T)} \qquad (8)$$

The real part describes the conductivity of the quasiparticles, which are assumed to behave like conventional electrons, i.e. obey Ohm´s law. The imaginary part of the conductivity corresponds to the inductive response of the Cooper pairs. The explicit relation between σ_2 and λ_L follows from the London equations (see Chap. 3 or any textbook about superconductivity, e.g. [13]). At microwave frequencies $\sigma' \ll \sigma''$ holds true resulting in a simple expression for the surface resistance and reactance of a superconductor (by a Taylor expansion of Eq. (6) using Eq. (8)):

$$R_s = \frac{1}{2}\omega^2\mu_0^2\sigma'(T)\lambda_L^3(T); \qquad X_s(T) = \omega\mu_0\,\lambda_L(T). \tag{9}$$

In contrast to a normal metal, R_s exhibits a quadratic frequency dependence. Figure 5 shows the measured frequency dependence R_s at $T = 77$ K for state-of-the thin films of YBCO.

In contrast to normal metals, the surface resistance exhibits a quadratic frequency dependence. According to these data, there is a clear advantage by orders of magnitude upon using HTS for the whole range of microwave communications bands.

Above 77 K, R_s increases strongly towards the critical temperature due to the strong increase of the London penetration depth. Below 77 K, R_s decreases gradually towards zero temperature by less than one order of magnitude [8], [14].

The surface resistance data discussed so far correspond to the linear regime corresponding to low levels of microwave power. For higher power levels corresponding to higher values of the high frequency magnetic fields nonlinearities occur resulting either in an increase of R_s or / and intermodulation distortion [8]. This is of particular importance, because – as a result of the Meissner effect – the current density across a superconducting microstrip line is strongly enhanced by about one order of magnitude at the edges on a lengthscale of the London penetration depth (this effect can only be avoided in the so called "edge current free geometries" like the one shown in Figure 3e. Typically, intermodulation distortion as well as a field dependent surface resistance occurs at power levels which depend strongly on the film quality. As an example, misoriented grains in the strongly anisotropic materials and grain boundaries give raise to strong nonlinearities [8]. In general, one can claim that planar resonators made from high quality HTS can handle power levels in the milliwatt range, which is sufficient for most of the receiver applications. In contrast, applications in the transmit circuits of communications systems, i.e. power levels above several watts, can only be handled in "edge current free" geometries (Figure 3e).

The nonlinear effects in HTS films are strongly temperature dependent, in particular close to T_c [8]. Therefore, operation temperature of 50 to 65 K are more favourable than 77 K. As discussed in Subsec. 5.2, temperatures in this range can be attained by low-power closed-cycle cryocoolers.

Apart from YBCO, thin films with reasonable microwave properties have been prepared from the thallium-based compounds $Tl_2Ba_2CaCu_2O_8$ ($T_c \approx 105$ K) and $Tl_2Ba_2Ca_2Cu_3O_{10}$ ($T_c \approx 115$ K) [8], [14]. For the thallium-based and the more recently discovered mercury-based compounds with T_c values up to 140 K epitaxial growth is very difficult because of the volatility of thallium and mercury, respectively. By growth of amorphous thallium free precursor films at room temperature and subsequent annealing in thallium atmosphere thin films of the thallium-based compounds (2212 and 2223) with a high degree of crystalline orientation with respect to the substrate orientation can be grown.

HTS films with reasonable and qualified microwave properties nowadays can be grown on wafers up to more than 4" in diameter, the most common size are 2" and 3" for microwave applications [15]. A very important step was the preparation of double-sided coating, which have turned out to be essential for planar microwave devices, where the metal ground plane needs to be superconducting in order to achieve high quality factors.

4.3 Microwave Properties of Dielectric Single Crystals, Ceramics, and Thin Films

Until now, the physical understanding of the origin of microwave losses of dielectrics is very rare. So far, only for single crystals a complete microscopic understanding has been achieved. This is because single crystals of very high purity are at their theoretical limit of absorption by the phonon system. In contrast, for dielectric bulk ceramics and thin films the losses are mostly determined by defects. There is a large variety of defects and possible loss mechanism associated with defects like acoustic phonon excitation by point defects, point defect Debye relaxation (see Chap. 1), unwanted doping and losses initiated by unwanted, possibly conducting phases. In this section, such mechanisms are discussed briefly and the state-of-the art of existing materials will be presented. Apart from linear microwave dielectrics, nonlinear dielectrics will be discussed with respect to frequency agility.

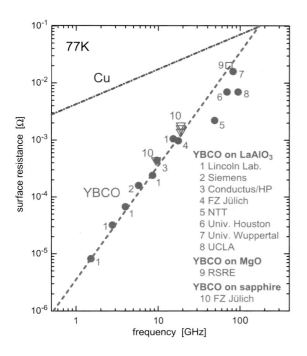

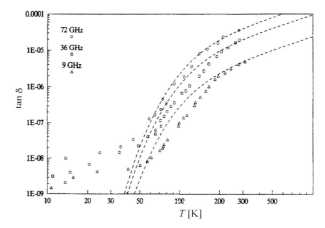

Figure 5: Measured surface resistance of epitaxially grown YBCO films grown at different laboratories at $T = 77$ K in comparison to that of copper.

Figure 6: Measured temperature dependence of the loss tangent of sapphire and theoretical calculations based on the SKM model [21].

As discussed in Chap. 1, the most simple model of dielectrics in the microwave and near infrared regime is based on classical harmonic oscillators with damping [16]:

$$\underline{\varepsilon}(\omega) = \varepsilon_\infty + \sum_{j=1}^{N} \frac{S_j \omega_j}{\omega_j^2 - \omega^2 - i\omega\Gamma_j}. \qquad (10)$$

The quantity ε_∞ represents the optical permittivity, which is determined by the electronic polarizability. The second term represents the ionic crystal lattice as a sum of N classical harmonic oscillators with eigenfrequencies ω_j, damping constants Γ_j and oscillators strengths S_j. In order to fit Eq. (10) to the observed far infrared spectra, these parameters are used as fit parameters. For the microwave regime, the frequency ω is small in comparison to the phonon frequencies and simplified expressions for ε_r and $\tan\delta$ can be derived (Eq.(5)):

$$\varepsilon_r = \frac{1}{\varepsilon_0}\left[\varepsilon_\infty + \sum_{j=1}^{N} S_j\right]; \qquad \tan\delta = \frac{\omega}{\varepsilon_0 \varepsilon_r} \sum_{j=1}^{N} \frac{S_j \Gamma_j}{\omega_j^2}. \qquad (11)$$

In agreement with experimental observations, most ionic crystals exhibit no significant frequency dependence of the permittivity and a linear frequency dependence of the loss tangent in the microwave regime. However, measured absolute values of the loss tangent and its temperature dependence differ significantly from Eq. (11) using parameters determined by infrared measurements. The theoretical problem relies on a proper quantum mechanical calculation of the relaxation rate $\Gamma_{TO}(\boldsymbol{q}=0,j,\omega,T)$ of transverse optical phonons of branch j at a wave vector $\boldsymbol{q}=0$. Since the microwave frequencies are much smaller than ω_{TO}, multiphonon absorption and emission processes will determine the loss tangent because of energy conservation. In any case, one expects a monotonous increase of the loss tangent with temperature due to the fact that the occupation of phonon states is determined by the Bose statistics. Therefore, for $T \to 0$ the intrinsic phonon losses disappear.

Sparks, King and Mills [17] and Gurevich and Tagentsev [18] investigated theoretically two- and three phonon difference absorption processes and calculated the temperature dependence of the loss tangent of various single crystals. Figure 6 shows experimental results on the temperature dependence of the loss tangent of high-purity single crystals of sapphire ($\varepsilon_r = 9.2 - 11.4$) [19] and calculations based on the SKM model [21]. Apart from some deviations below 50 K, which are due to extrinsic losses, there is a good agreement between theory and experiment. It is worth mentioning that sapphire is the microwave dielectric with the lowest losses at all. In particular, at cryogenic temperatures, tanδ values in the 10^{-8} range allow for Q values as high a several 10^7.

A similarly good agreement between theory and experiment has been achieved for single crystals of rutile with a permittivity of about 100 [21].

However, the situation becomes already more complicated for ternary single crystals like lanthanum-aluminate (LaAlO$_3$, ε_r = 23.4). The temperature dependence of the loss tangent depicted in Figure 7 exhibits a pronounced peak at about 70 K, which cannot be explained by phonon absorption [22]. Typically, such peaks, which have also been observed at lower frequencies for quartz, can be explained by defect dipole relaxation (see Chap. 1). The most important relaxation processes with relevance for microwave absorption are local motion of ions on interstitial lattice positions giving rise to double well potentials with activation energies in the 50 to 100 meV range and colour-centre dipole relaxation with activation energies of about 5 meV.

Figure 7 shows the measured temperature dependence of the loss tangent of LaAlO$_3$ single crystals and a fit employing the SKM model plus a Debye relaxation term (see Chap. 1). The best fit was achieved for an activation energy of 31 meV, the estimated defect concentration is only 10^{16}/cm^3. It was suggested that aluminium atoms on interstitial lattice positions are responsible for the observed relaxation phenomena. This indicates, that the dielectric losses are extremely sensitive to very small concentrations of point defects [21], [22].

Charged point defects on regular lattice positions can also contribute to additional losses: the translation invariance, which forbids the interaction of electromagnetic waves with acoustic phonons, is perturbed due to charged defects at random positions. Such single-phonon processes are much more effective than the two- or three phonon processes discussed before, because the energy of the acoustic branches goes to zero at the Γ point of the Brillouin zone. Until now, only a classical approach to account for these losses exists, which has been successfully used to describe the losses in garnets, glasses and in doped sapphire at low temperatures [21]. At $T \rightarrow 0$ this mechanism may become dominant, because intrinsic phonon losses and thermally activated Debye losses disappear.

Finally, any doping leading to a finite conductivity σ will lead to a loss contribution according to tan$\delta = \sigma/(\omega\varepsilon_0\varepsilon_r)$ (see Eq.(5)), which is the dominant mechanism for semiconductors. However, different conduction mechanism like hopping conductivity may give rise to complicated temperature and frequency dependences, which cannot be discussed in detail here.

Most of the materials discussed so far (sapphire, lanthanum aluminate, magnesium oxide,...) are common substrate materials for HTS films, thus their low dielectric losses are essential to utilise the low conduction losses of HTS films in planar microwave circuits. For dielectric resonators operating at room temperature they are not very attractive, first, because single crystals are quite expensive, and secondly, because they exhibit a relatively strong temperature dependence of the dielectric constant leading to a temperature coefficient of the resonant frequency of a dielectric resonator of about 100 ppm / K (typically for sapphire) and up to more than 1000 ppm / K for rutile. For HTS or cryogenically cooled dielectric resonator devices this is not very important, because the temperature has to be controlled anyway (because of the temperature dependent London penetration depth of a superconductor), but for room temperature devices, e.g. in mobile phones, base stations or satellites the temperature can vary between typically –40 and +50°C. In addition, devices operating at high power levels might heat up due to power dissipation in the dielectric material.

Fortunately, they are several species of low-loss dielectric ceramics with tailored temperature coefficient of dielectric constant, which can be made lower than 1 ppm/K for a certain temperature window around room temperature. Physically, this can be accomplished either by intrinsic compensation of the temperature dependence of thermal volume expansion $V(T)$ and lattice polarisability $\alpha(T)$ via the Clausius-Mossotti relation (see Chap. 1):

$$\frac{\varepsilon_r(T)-1}{\varepsilon_r(T)+1} = \frac{4\pi}{3}\frac{\alpha(T)}{V(T)} \qquad (12)$$

For a suitable doping, both effects compensate each other for some materials and lead to a broad turning point TP (ε_r^{-1} dε_r/d$T|_{T=TP}$ = 0) in the temperature dependence of $\varepsilon_r(T)$. As a second approach, polycrystalline mixtures of grains of different materials with opposite temperature coefficient of ε_r lead to an effective medium with zero temperature coefficient.

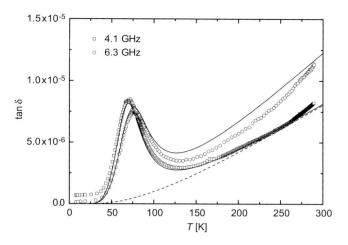

Figure 7: Measured temperature dependence of the loss tangent of LaAlO$_3$ single crystals and a theoretical fit employing the SKM model and defect dipole relaxation [18] (from [22]).

Table 2 summarises the most relevant commercially available microwave dielectrics with compensated temperature coefficients. Usually, the quantity $Q \cdot f$ is quoted for the microwave losses: Assuming that $Q = 1 / \tan\delta$ for a dielectric resonator with filling factor $\kappa = 1$ (Eq. (3)) and $\tan\delta \propto f$ according to Eq. (11) $Q \cdot f$ is a measure for the loss tangent. However, for most dielectrics $Q \cdot f$ drops below a certain frequency gradually to lower values indicating that the frequency dependence of the loss tangent is weaker than linear. As an example, for high-quality BMT-ceramic $Q \cdot f = 350$ THz at 10 GHz and only about 200 THz at 4 GHz [23]. This behaviour is not yet fully understood.

So far, most of the material development has been performed in companies, indicating that the market for microwave dielectrics has been grown tremendously over the last years (as an example, each commercial satellite receiver contains a small microwave ceramic disk as frequency stabilising element for the local microwave oscillator). In fact, the optimisation of microwave ceramics with respect to low losses has been pursued by empirically improving the preparation conditions (with particular emphasis on the sintering procedure) rather than due to a systematic study of physical loss mechanisms (which now is appreciated as a challenge within the microwave ceramic community). As a tendency, the general requirements for low losses are:
- dense materials with large grains
- small amount of impurity phases
- high structural order
- low level of free charge carriers.

For sure, this list is incomplete and may have neglected important details.

The materials listed in table 2 are also not complete but the selection represents the most relevant composites which are of commercial interest. In addition to the performance issue, there is a strong tendency in research and development to reduce the costs. For that reason, a lot of research is devoted to the niobates as possible replacement for the tantalates, because niobium is cheaper than tantalum. In addition, compensated materi-

MATERIAL	ε_r	τ_f	$\tan\delta^{-1}$	f [GHz]	
Al$_2$O$_3$	10	-60	50,000/ 100,000	10	BC SC
LaAlO$_3$	24	-60	40,000	10	SC
Ba(Mg$_{1/3}$Ta$_{2/3}$)O$_3$	24	0	26,000	10	BC
Ba-Zn-Ta-O	29	-3…3	50,000	2	BC
Zr-Sn-Ti-O	38	-3…+3	8000	7	BC
Ca-Ti-Nd-Al	47	20	6000	6.5	BC
Ba-Nd-Ti-O	80	90	2500	5.5	BC
TiO$_2$	100	450	16,000	3	BC, SC
CaTiO$_3$	155	800	5000	2.3	TF
SrTiO$_3$	270	1200	1500	2	TF
Ba$_x$Sr$_{1-x}$TiO$_3$	420	?	1500	1.5	TF

Table 2: Microwave properties of the most important microwave dielectrics.
SC = bulk single crystals
BC = bulk ceramics
TF = thin films
τ_f = temperature coefficient of resonant frequency
The materials marked with the grey box are tuneable dielectrics.

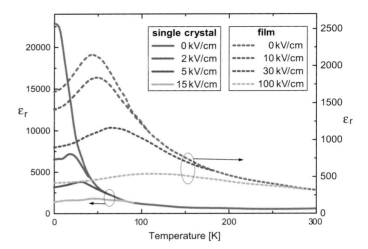

Figure 8: Temperature dependence of the dielectric constant of thin films and bulk single crystals of SrTiO$_3$ for different values of the applied dc electric fields (from [28]).

als with very high values of the permittivity are currently under development. Recently, for the compound Ag(Nb$_{1-x}$Ta$_x$)O$_3$ with $0.35 < x < 0.65$ ε_r values of 450 were achieved for potential use as filters (to replace the surface-acoustic-wave devices) and planar antennas in mobile phones [24]. In addition, a novel high Q material (Qf = 30,000 at 1 GHz) for base-station filters with ε_r = 60 is currently under development [25].

The microwave ceramics discussed so far have been prepared by sintering pellets of pressed powder. Usually, the sintering temperatures are very high (typically around 1200 to 1500°C) in order to obtain large and highly ordered grains, which is essential for low values of the loss tangent.

For planar single- and multilayer microwave circuits dielectric ceramics are important as low-cost substrates. One very cost effective method of manufacturing integrated microwave circuits is the so-called LTCC process. LTCC stands for "Low Temperature Co-fired Ceramics" and is based on unsintered ceramic foils (in most cases composed of aluminium oxide with organic additives), where metal circuits and vias connections through a foil are prepared by printing, stamping and metal thick film deposition processes. Several (typically up to 30) of such foils are stacked together and subsequently the entire stack is sintered at relatively low temperatures (typically around 850 °C). The LTTC technology allows for a very cheap and versatile manufacturing of multilayer circuits with complex wiring. The permittivity values of LTCC ceramics are between 4 and 10. Due to the low temperatures sintering temperatures loss tangent value are in the range of about $5 \cdot 10^{-4}$ at 10 GHz, which is higher than for high-temperature ceramics [26].

In summary, the field of microwave ceramics has been continuously progressing over the last ten years. More recently, a strong interest has come up to utilise the non-linear dielectric properties of thin or thick films of ferroelectrics (or incipient ferroelectrics) above the Curie-temperature to build tuneable microwave devices (see also Chap. 1). Figure 8 shows the temperature dependence of the permittivity at different applied voltages for thin films and bulk single crystals of SrTiO$_3$. Whereas SrTiO$_3$ exhibit a maximum tuneability at about 80K and therefore ideally suited for planar tuneable HTS devices, Ba$_x$Sr$_{1-x}$TiO$_3$ with $x = 0.5 - 0.7$ is ideal for room temperature applications (voltage dependence of permittivity is shown in Chap. 1). However, due to the high values of the static electric field required for a significant change of ε_r only thin-film based electrode designs with micrometer-sized capacitive gap areable to accomplish significant tuning ranges at practicable voltages of several ten volts. This fact limits ferroelectric tuning mostly to planar integrated microwave circuits.

Currently, the most important aspect of research on these materials is the improvement of the loss tangent of about 10^{-3} to 10^{-2} [29] towards the theoretical values of several 10^{-4} (achieved in single crystals of SrTiO$_3$) [27]. The best results obtained until now are already competitive (or even better) with commercial semiconducting varactor diodes. As a clear advantage, ferroelectric thin film varactors are easy to integrate in MMIC or LTCC devices. Finally, the question of intermodulation distortion caused by the nonlinear dielectric constant is still open. Moreover, new materials with tailored properties and low losses are highly desired, but the number appears to be quite limited.

5 Novel Passive Devices for Microwave Communication Systems

In this chapter some examples of new devices based on bulk dielectrics and high-temperature superconducting thin films will be presented. The selection is quite subjective and does not account properly for the enormous amount of work which has been done in this rapidly growing field. However, the selection should illustrate the ongoing progress in this field and the challenges for future research and development work.

5.1 Dielectric Filters

As discussed in Sec. 3.2, high-Q values are essential to achieve low values of the in-band insertion loss and steep skirts at the passband edges. Among the different types of resonators for room temperature operation, dielectric resonators possess the highest Q values. Therefore, they do represent the devices of choice for high-performance communication hardware such as satellite transponders and base stations.

According to Sec. 3.2, bandpass filters are composed of mutually coupled resonators. As an example, typical filters for satellite output multiplexers (Figure 2) require steep filter skirts which can be achieved for filters with at least four poles and one damping pole on each side ($N = 4$ and $m = 1$ according to Sect. 3.2). Figure 9a) shows a possible arrangement of monomode dielectric resonators (TE$_{01\delta}$ from Table 1). The input signal ("in") from a waveguide excites the first resonator ("M"ode 1) via an overlap of waveguide fields and evanescent dielectric resonator fields (in an equivalent circuit this would correspond to a capacitor C_{kin} between the input port and a parallel LC resonance circuit composed of an inductance L_0 and a capacitance C_0 with $k_{in}=C_{kin}/C_0$). Inter-resonator coupling between mode i and $i+1$ (see Sec. 3.2) is accomplished by coupling holes or slits in the housing wall between resonator i (M$_i$) and $i+1$ (M$_{i+1}$). According to Sec. 3.2 this arrangement corresponds to a Chebyshev filter. The most simple way to accomplish a damping pole in the characteristic of a four pole filter is given by an additional negative coupling between resonator 1 and 4. In Figure 9a a negative coupling is realised by an additional waveguide segment connecting M1 and M2 via coupling holes or slits. Negative coupling means that the field amplitude in this waveguide has to be of opposite sign at the positions of the coupling hole, i.e. the length of the guide segment has to be adjusted properly. This difficulty can be avoided by the use of dual-mode resonators (HE$_{11\delta}$ in Table 1).

In the arrangement shown in Figure 9b mode 1 of resonator 1 is excited by a coaxial antenna entering the filter housing from underneath ("in"). The polarisation of M1 indicated by the arrow is enforced by the antenna itself, because it represents the strongest perturbation of rotational symmetry of the dielectric resonator. Coupling between M1 and M2 of resonator 1 is accomplished by a dielectric or metallic rod entering the filter housing from underneath. Similarly, M2 and M3 are mutually coupled. Coupling between M2 of resonator 1 and M3 of resonator 2 is accomplished similar to Figure 9a by a coupling slit between the neighbouring resonator subhousings. It is important that the coupling slit is mode selective, i.e. there is no parasitic coupling between other possible mode combinations (e.g. M1 to M4). In the case of Figure 9b this is fulfilled due to the fact that the electric field of the HE$_{11\delta}$ mode is strongly polarised along the direction indicated by the arrows (there are other possible designs like dielectric cylinders stacked along the rotational axis being coupled via cross shaped apertures employing one bar of the cross for k_{23} and the other one for k_{14}).

The most obvious advantage with respect to monomode resonators is that the required negative coupling between M4 and M1 is simply attained by placing the k_{34} rod in resonator 2 at an angular position rotated by 90 degree with respect to the position of the corresponding k_{12} coupling rod in resonator 1. This trick causes the amplitude of M4 to be negative with respect to the amplitude of M1 at the position of the coupling slit for k_{14}.

One common way of building 8, 12 and even 16 pole filters is the so called "cascaded quadruplet" design: M4 in Figure 9 is coupled to M1* of a second identical quadruplet (M1* to M4* corresponding to M5 to M8 of the resulting eight pole filter) by a single coupling slit (k_{45}).

The conventional dual-mode filter design is based on dielectric cylinder excited in the HE$_{11\delta}$ - mode (see Table 1) [9]. More recently, a novel concept based on dielectric hemispheres has resulted in a world record for highest Q of a C - band ($f = 3.8$ GHz) dual-mode filter (about 35000 at room temperature employing BMT and 300 K and 100000 at 77 K employing single crystalline LaAlO$_3$) (Figure 10) [6], [30], [31].

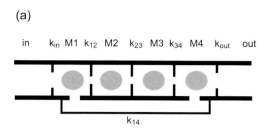

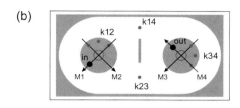

Figure 9: Filter topology of
(a) a four-pole filter composed of monomode resonators and
(b) a quasielliptic dielectric filter developed at Juelich research centre [6], [30], [31].

Figure 10: Three channel prototype output multiplexer (OMUX) based on the hemiospherical dual mode filter shown in Figure 9 [6].

The potential use of high-Q dielectric filters is for base stations and IMUX and OMUX in communication satellites. Recently at Alcatel in France, a novel-type of quasielliptic dielectric filter based on square shaped dielectric plates has been developed with lower Q_0 but of smaller size and weight [11]. This is a particular important argument for existing satellites, where filter Q_0s of 10000 are sufficient to fulfil the filter specifications. However, for future satellites operating at higher frequencies (e.g. at Ka-band) one may utilise higher Q-values and even cryogenic operation temperatures.

5.2 Planar High-Temperature Superconducting Filters and Subsystems

Over the past few years, tremendous progress has been achieved in the development of HTS planar filters and their integration in subsystems, mostly for base stations in mobile communication. Currently, the commercialisation of such subsystems is in good progress: Meanwhile, in the US about 1000 base stations are equipped with HTS subsystems. Ongoing field trials have shown significant improvements in terms of number of dropped calls, maximum number of calls per time interval and other quantities related to benefits for the service provider [32].

Filters: For HTS planar filters the resonator types shown in Figure 3 and others have been used. The main emphasis of the recent years was to develop filters for base stations with steep skirts coming as close as possible to the Q-limitation described in Sec. 3.2. In most cases, these are (quasi)elliptic filters with 8 to 17 poles or Chebyshev filters with up to 30 poles.

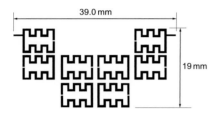

Figure 11: Example of an HTS-planar filter with 8 poles and quasielliptic characteristic [35].

Figure 11 shows an example of an 8-pole quasielliptic filter based on folded microstrip resonators similar to Figure 3 c. In this case of monomode resonators, negative cross coupling was achieved using electric field rather than magnetic field coupling (used for the k_i, k_{i+1} forward coupling elements). Recently, for a 17-pole elliptic filter at 1.8 GHz with 5% relative (resonator Q_0 = 50000 at 65 K) the steepness of skirts of 85 dB/MHz was demonstrated (Figure 12). Such a high performance cannot be achieved with any other filter technology.

In addition, high quality filters based on microstrip resonators have been developed for C-band satellite transponders. Figure 13 shows a three channel IMUX test module developed at Bosch SatCom GmbH in Germany. This test module, which is considered to be part of a space experiment, consists of 3 HTS quasielliptic eight pole channel filters, cryogenic circulators, a cryogenic preamplifier and a wide band HTS input filter.

Cryogenics and subsystems: The most important "enabling technology" for applications of HTS in telecommunication is cryotechnology. Of course, cooling by liquid nitrogen is unacceptable, because the amount of maintenance for each base station should be as small as possible. The situation for satellite communications is even more severe, because the lifetime of a geostationary satellite is at least 10 years and any failure of cryogenic cooling would be a disaster.

The most convenient way of cooling are Stirling-type closed-cycle refrigerators consisting of a small stainless steel compressor with helium gas pressure of a few bar and a cold finger with a displacer inside, both connected by a thin stainless steal tube. Typically, a compressor with an AC 50 Hz power consumption of 100 - 200 watts with size of one or two beer cans provides a cooling power of 3 to 10 watts at 77 K, which is the typical amount of cooling power needed for an HTS subsystems consisting of about

Figure 12: Measured characteristic of a HTS planar 17-pole elliptic filter [33].

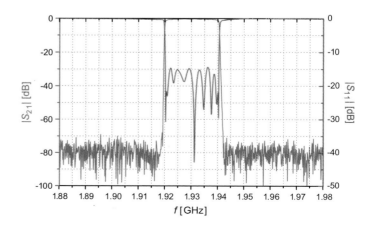

three filters and one cryogenic LNA. For the sake of filter performance the base station subsystems are operating between 60 and 70 K, the satellite systems at 77 K because cooler efficiency is a crucial parameter with respect to system mass reduction [6].

Figure 14 shows a commercial system for potential use in base stations. The filters are assembled inside a vacuum dewar (left side), the metal cylinder in the right part of the assembly is the compressor of the Stirling type refrigerator.

Currently, companies are focussing on improving the coolers with respect to lifetime (currently about 3 - 5 years for continuous operation) and cost reduction (currently about 5000 to 10000 Euro per unit).

6 Micromechanics for Microwaves: RF MEMS and FBARs

The abbreviation MEMS stands for "microelectromechanical systems", which – in general – represents an extensive field of research and development. For high-frequency applications there are two types of MEMS devices which are currently under development. At first, MEMS switches allow for fast switching of electromagnetic power passing through a planar microwave transmission line. Devices based on such switches are phase shifters, as e.g. used in radar systems. Secondly, microbridge acoustic resonators based on piezoelectric thin films (FBAR = frequency bulk acoustic resonators) have approached the Gigahertz frequency range. FBARs are considered to be used to build extremely miniaturised filters for mobile wireless communications units.

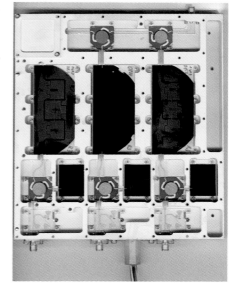

Figure 13: Photograph of a three-channel IMUX module based on HTS planar filters as part of 3 cryogenic C-band satellite transponder. The HTS filter are indicated as large black squares [6].

MEMS switches and switching devices: In Figure 15 (a), a cross section view (b) and an equivalent circuit (c) of a capacitive MEMS shunt switch are shown. The most commonly used technology to fabricate such bridges with typical dimensions in the range if several 100 μm is the so-called "sacrificial layer technology" based on a photoresist pad prepared at the location of the free standing bridge. After deposition of a metallic layer (e.g. gold or aluminium) on the patterned photoresist layer, the latter is removed ("sacrificial layer") by chemical solvents. The pull-down electrode consists of a metal pad covered by a thin dielectric layer. This dielectric layer avoids a short circuit between the bridge and the pull-down electrode. In this particular design, the dielectric layer also defines the high-frequency capacitance in the down-state of the switch. There exist other designs with separate switch contact and pull-down electrode (for a recent review see [36]).

Switching of the microbridge from the up-state (as depicted in Figure 15) to the down-state (membrane or cantilever in contact with the pull-down electrode) is provided by the electrostatic force F due to a dc voltage U applied across the air-gap g of about 2 - 3 μm. In order to switch the microswitch to the down state, the electrostatic force has to be larger than the repelling force of the spring k ($g - g_0$):

Figure 14: Commercial HTS subsystem for potential use in base stations. The filters and the LNA are inside a vacuum dewar (left side), the metal cylinder in the right part of the assembly is the compressor of the Stirling type refrigerator [33].

$$\frac{\varepsilon_0 A U^2}{2g^2} \geq k(g - g_0). \tag{13}$$

In Eq. (13) A represents the area of the microbridge, g_0 the equilibrium position at zero voltage, and $\varepsilon_0 = 8.85 \cdot 10^{-12}$ As/Vm. The spring constant k depends on the bridge dimensions and the Young modulus of the bridge material, typical values are 5 - 30 N/m

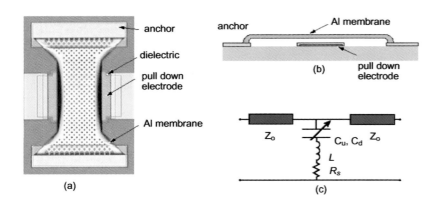

Figure 15: Picture of a capacitive MEMS switch (a), cross section view (b) and equivalent circuit (c) with C_u and C_d representing the value of the capacitance in the "up" and "down" state (from [36]).

[39]. According to Eq. (13), the total force on the spring (electrostatic force minus spring repelling force) has a minimum at $g = 2g_0/3$. Consequently, the pull-down voltage U_p at which the bridge collapses is given by:

$$U_p = \sqrt{\frac{8kg_0^3}{27\varepsilon_0 A}} \qquad (14)$$

As an example, for $g_0 = 2.5$ μm, $A = 100$ μm^2 and $k = 10$ N/m the pull-down voltage comes out to be 23 V. Such values are typical for electrostatic microwave MEMS switches. Since the electrostatic force increases upon decreasing the gap, the voltage to keep the switch in the down state is lower than the switching voltage.

The fundamental mechanical resonance frequency of a double-clamped flexural beam of length l and thickness d in direction of motion is given by the expression [37]

$$f_r = \frac{1}{2\pi}\sqrt{\frac{k}{m}} = 1.03\sqrt{\frac{E}{\rho}}\frac{d}{l^2} \qquad (15)$$

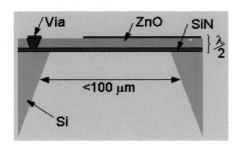

Figure 16: Schematic of an FBAR consisting of a free standing metal-piezoelectric multilayer membrane [40].

with E representing Youngs modulus and ρ the density of the beam material (Au: $E \approx 80$ GPa, $\rho = 19280$ kg/m^3) [39]. For typical dimensions of MEMS switches ($l \approx 300$μm, $d \approx 2$ μm) this value is in the order of 50 kHz. Typically, the resonance is damped from typical Q-values in vacuum of about 50 – 1000 to values of only $0.3 \leq Q \leq 5$ by operating the switch at atmospheric pressure (damping by removing the air underneath the bridge). Due to the sensitivity to contamination, packaging of the MEMS switches has to be performed under clean room conditions. With such a design, switching times in the order of 2 - 50 μs and several billions of switching cycles can be achieved.

Possible application areas of rf-MEMS are switched filter banks for wireless communications systems (both for base stations and portable units), phase shifters for radar systems switching networks, e.g. for communication satellites.

FBARs (Film Bulk Acoustic Resonators): Today, acoustic resonators based on bulk single crystals of quartz are all-present as time defining (i.e. frequency stabilising) element in almost any commercial electronic device. For the microwave frequency range, surface rather than bulk acoustic resonators are most commonly used, i.e. as filters in mobile phones (SAW = Surface Acoustic Wave devices, see [38]).

Due to the possibilities associated with micromachining bulk accoustic resonators for the microwave regime have become feasible, representing ultimate performance with respect to their miniaturisation and on-chip integration potential. As an example, for silicon a doubly–clamped flexural beam of $l = 10$ μm and $d = 1$ μm would resonate at about 50 - 100 MHz according to Eq. (15) ($E = 160$ GPa and $\rho = 2330$ kg/m^3 for silicon [39]). In order to achieve frequencies in the GHz range submicron lithography is required. For such dimensions the issues of resonance excitation, precise control of resonance frequency and small values of resonant energy need to be addressed in order build devices for practical use in telecommunication.

As an alternative, free-standing bridges of piezoelectric materials of 100 μm length can be used to build bulk acoustic resonators for the GHz range. Figure 16 from [40] shows the schematic of an FBAR (Film Bulk Acoustic Resonator) structure consisting of ZnO (or AlN) membrane with thin-film metallic electrodes (deposited on a silicon wafer with subsequent underetching). The fundamental resonance occurs when the film thickness is half the wavelength. For the fundamental mode being at $f = 2$ GHz, the corresponding thickness is 2.6 μm (sound velocity of AlN: 10,400 m/s).

FBAR filter devices for mobile phones are currently under development and have already surpassed the performance of state-of-the art SAW devices. Similar to MEMS switches, the issue of packaging is rather crucial, since high Qs above 1000 can only be achieved at reduced atmospheric pressure.

Finally, piezoelectric thin films on silicon membranes can also be used as switching devices [41]. In contrast to the electrostatically driven MEMS, lower switching voltages are required.

7 Photonic Bandgap Structures

Photonic bandgap (PBG) structures are periodic dielectric structures in one, two or three dimensions –ideally infinitely extended – in reality large in comparison to the free space wavelength [42]. By analogy to Bloch waves, which represent the eigenstates of the electronic wave function of a real crystal (see Chap. 3), the electromagnetic fields (for convenience solutions are expressed in terms of the magnetic field H) can be represented as a Fourier series of Bloch states $h(x)$ which are periodic in space with a representing the unit vectors of the Bravais lattice:

$$H(x) = \sum_{\vec{G}} \sum_{\lambda=1}^{2} h_{G,\lambda} e_\lambda \exp i[k+G]x \qquad (16)$$
$$h_{G,\lambda}(x) = h_{G,\lambda}(x+a).$$

As for real crystals, the quantity G represents the reciprocal lattice vectors of the PBG lattice. In contrast to real crystals, the summation over λ accounts for the two possible orthogonal polarisation states of plane electromagnetic waves.

By analogy to the time independent Schrödinger equation, the eigenvalues ω/c (angular frequency / vacuum velocity of light) of the electromagnetic field can be calculated by the eigenvalue equation

$$\nabla \times \left[\frac{1}{\varepsilon(x)} \nabla \times H(x) \right] = \frac{\omega^2}{c^2} H(x) \qquad (17)$$

which can be derived from Maxwell's equation (Eq. (4)). In Eq. (17) the specific role of the potential $V(x)$ in Schrödinger's equation is taken over by the spatial distribution of the dielectric constant $\varepsilon(x)$, which typically is composed of θ-step functions. Similar to Schrödinger's equation, the differential operators on the left side of Eq. (17) are hermite, i.e. the eigenvalues are real and positive.

By analogy to the electronic bandgap of an insulator or a semiconductor there exist certain frequency intervals where electromagnetic waves cannot propagate. This frequency intervals are called "photonic bandgaps". In one dimension an alternating sequence of parallel dielectric slabs represents a rather trivial PBG structure. In contrast to the electronic bandgap in a real crystal, a photonic bandgap depends on the polarisation of the electromagnetic waves.

Figure 17 shows the calculated band structure of a hexagonal lattice of dielectric rods for TM waves. Typically, the wave vector k is given in normalised units: at the M point (direction parallel to a row of rods) $k = 2\pi/a$. The Γ-M direction is rotated by 30 degree with respect to the Γ-K direction. The shaded green areas correspond to the bandgaps of this structure.

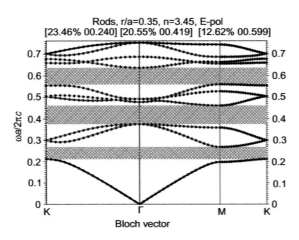

Figure 17: 2D hexagonal lattice of dielectric rods (left) and calculated band structure for TM waves (electric field parallel to the rod axis) for $\varepsilon_{r,rod} = 10$ and rod radius $r_{rod} = 0.35\,a$. Both frequencies and k vectors (Bloch vectors) are normalised to the lattice constant a.

VI Data Transmission and Interfaces

Possible novel types of high Q resonators can be created by local modifications of the PBG lattice. As an example, a simple point defect is generated by removing one rod from the lattice. Similar to localised states in a real crystal, such defects cause a localisation of electromagnetic field at the position of the defect. Figure 18 shows the calculated electromagnetic field distribution of an extended defect in a hexagonal lattice of air holes in a dielectric disc. In contrast to the structure shown in Figure 17, this so-called "inverse hexagonal structure" supports primarily bandgaps for TE waves (magnetic field oriented parallel to the hole axis). The particular extended defect depicted in Figure 18 is a hexagonally shaped air hole.

The numerically calculated field distribution visualizes the strong confinement of electromagnetic field energy in the defect area. This localisation corresponds to a resonant mode. Since the filling factor of this mode is only 0.35 (see Eq. (3)), the Q-factor caused by dielectric losses from the disc material is about three times smaller than for a dielectric resonator machined from the same material.

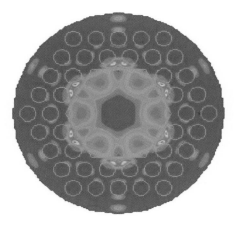

Figure 18: Calculated distribution of electric field energy for a defect resonance in a two dimensional PBG structure. The extended defect is formed by a hexagonal shaped hole in an alumina ($\varepsilon_r = 10$) disk, the PBG structure is given by a hexagonal arrangement of holes in the disk. The value of k corresponds to the quanitity κ in Eq. (3). The lattice period is 3 mm.

However, the size of PBG structures is relatively large (lattice period is about one third of the free space wavelength) and thus their application potential may be limited to high frequencies above 30 to 40 GHz. The particular emphasis of PBG structures is supposed to be for the millimetre and submillimetre frequency range (100 GHz to 1 THz), were conventional waveguide techniques are at their limits. In this frequency range high-resistive silicon represents an almost ideal dielectric material (see Subsec. 4.3). In addition, silicon is very attractive because of its micromachining capability. Beyond the millimetre wave frequency range, photonic bandgap structures are considered to become most relevant for optical communications bands (wavelength 1.5 μm).

In addition to defect resonators, waveguides can be generated by a linear defect in the hexagonal / inverse hexagonal PBG structure (Figure 19). Besides low losses, such waveguides are attractive because they can bend around sharp corners without radiation losses. This becomes possible because Bragg reflection (in contrast to total reflection at the transition form a high-ε_r to a low-ε_r region) occurs at any incidence angle.

The PBG structures discussed so far are two dimensional structures. In order to utilise the potential of hexagonal PBG structure, the field confinement in the third dimensions needs to be provided by other means. As shown recently, dielectric slabs of certain thickness being of the order of the wavelength allow for 3D confinement by Bragg reflection in the lateral direction and total reflection in the direction normal to the slab [44]. Such a configuration is of particular importance for optical communication bands where such a slab can be created by thin film technologies. For the visible range propagation of light through a bent optical waveguide as observed recently in an epitaxially grown AlGaAs/GsAs structure [45].

For the millimetre wave range stacked wafers could be used to create real three dimensional PBG structures. Recently, a stacked 3D PBG structure composed of a sequence of hexagonal and inverse hexagonal 2D layers has been suggested (Figure 20). One crystallographic unit cell in z-direction is composed of a stacked sequence of six alternating hexagonal /inverse hexagonal slabs, which are stacked in a repeating three layer sequence along the crystallographic (111) direction [46]. The most important feature of this structure is that defect resonators and waveguides in one of the 2D sublayers (both "rod" and "hole" layer) exhibit electromagnetic properties resembling that of the 2D structure, but providing strong field confinement within the particular slab containing the defect structure [46], [47]. However, manufacturing of this structure has still not been demonstrated, but laser machining appears to be one possible approach.

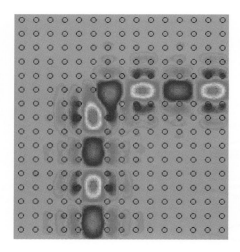

Figure 19: Linear-defect TM-type waveguide in a PBG structure consisting of a rectangular lattice of dielectric rods in [43].

8 Summary

It was shown that microwave communication systems can strongly benefit from new and improved materials and novel devices. In this chapter, an overview about recent results of research and development of passive devices and related materials was given. From the commercial point-of-view, the lost cost-and medium-performance devices like FBARs, LTCC-circuits and MMICs are expected to represent the largest market contribution. The applications include all kind of end-user mobile communications terminals (e.g. mobile phones) and commercial radar applications like collision avoidance radar systems. On the other hand, high-performance devices like high-Q dielectric filters and high-temperature superconducting filters are expected to have strong impact on the capability of future communication systems. Since this capability currently represents the bottleneck of information processing in general, there is a fair chance for new technologies which allow to utilise the available frequency bands in the most efficient way. The dream of purely digital processing (i.e. a receiver consisting of an antenna and an A/D converter) is by far not realistic and analogue processing of microwave signals will become even more sophisticated in future.

The spectrum of millimetre waves (above about 40 GHz) is expected to become more important for the future. The driving technological force are active semiconducting devices (HBTs, HEMTs), which are currently developing quite rapidly with respect to a continuous increase of the maximum operation frequency. Since planar microstrip circuits are already at their upper frequency limits and metallic waveguide structures are too expensive, photonic bandgap structures based on silicon or ceramics are a great challenge for the future. For such structures and also for fast switches and novel type of acoustic resonators, micromachining will become a key technology for future development of microwave communication systems.

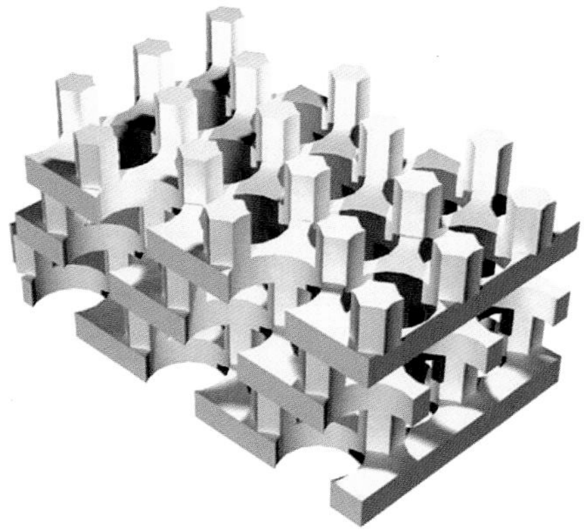

Acknowledgements

The editor would like to thank Stefan Tappe (RWTH Aachen) for a careful review of this chapter and for checking the symbols and formulas.

Figure 20: 3 dimensional PBG structure consisting of stacked layers of the hexagonal and inverse hexagonal structure [46].

References

[1] J.A. Garrido et al., Appl. Phys. Lett. **76**, 3442 (2000), and L.F. Eastman, Cornell University, private communication.
[2] Further information and references about SiGe are available in the Internet: http://ctd.grc.nasa.gov/5620/SiGe.html
[3] F. Losee,23 *RF systems, components and circuits handbook*, Artech House, Boston, 1997.
[4] Internet: http://ccnga.uwaterloo.ca/~jscouria/GSM/gsmreport.html.
[5] Internet: http://www.umts-forum.org/information.html.
[6] M. Klauda et al., IEEE Transactions on Microwave Theory and Techniques **48**, 1227 (2000).
[7] See e.g. D. M. Pozar, *Microwave Engineering*, John Wiley & Sins Inc., 1998.
[8] M. Hein, *High-temperature-superconductor thin films at microwave frequencies*, Springer Tracts in Modern Physics, 155, Springer Verlag, 1999.
[9] D. Kajfez and P. Guillon, *Dielectric resonators*, Artech House Inc., 1986.
[10] N.Klein et al., IEEE Transactions on Applied Superconductivity **9**, 3573 (1999).
[11] Y. Latousche et al., IEEE MTT-S Digest, 1607, (2001).
[12] see e.g. http://emcsun.ece.umr.edu/new-induct/circular.html or F.W. Grover, *Inductance Calculations*.
[13] see e.g. J.R. Waldram, *Superconductivity of metals and cuprates*, Institute of Physics Publishing, Bristol and Philadelphia, 1996.
[14] N. Klein, *Electrodynamic properties of oxide superconductors*, Juelich report Jül-**3773**, Habilitationsschrift (1997).
[15] http://www.theva.com/ (German commercial vendor of qualified HTS films).
[16] see e.g. N.W. Ashcroft and N.D. Mermin, *Solid State Physics*, Sunders College, Philadelphia, 1976.
[17] M. Sparks et al., Phys. Rev. B. **26**, 6987 (1982).
[18] V.L. Gurevich and A.K. Tagantsev, Adv. Phys. **40**, 719S (1991).
[19] V.B. Braginsky, V.S. Ilchenko, Kh. S. Bagdassarov, Phys.Lett. *A* **120**, 300 (1987).
[20] N. Klein et al., J. Appl. Phys. **78**, 6683 (1995).
[21] C. Zuccaro, Mikrowellenabsorption in Dielektrika und Hochtemperatursupraleitern für Resonatoren hoher Güte, Juelich report Jül-**3631**, Dissertation (1998).
[22] C. Zuccaro et al., J. Appl. Phys. **82**, 5695 (1997).

[23] N. Klein, M. Winter, H.R. Yi, Superc. Science and Technology **13**, 527 (2000).
[24] M. Valant et al., J. Europ. Ceram. Soc. **21**, 2647 (2001).
[25] M. Matsui, Kyocera, private communication.
[26] see e.g. http://www.imst.de/.
[27] J. Krupka et al., IEEE Trans. on Microwave Theory and Techniques **42**, 1886 (1994).
[28] R. Ott und R. Wördenweber, "Tunable Ferroelectric Capacitors for Microwave Applications", to be published in Appl. Phys. Lett.
[29] K. Bouzehouane et al., Appl. Phys. Lett. **80**, 109 (2001).
[30] H.R. Yi, N. Klein, IEEE Transactions on Applied Superconductivity **11**, 489 (2001).
[31] N. Klein et al., IEEE Transactions on Applied Superconductivity **9**, 3573 (1999).
[32] B. Willemsen, IEEE Transactions on Applied Superconductivity **11**, 60 (2001).
[33] Cryoelectra GmbH, technical data sheet.
[34] B.A. Aminov et al., IEEE Transactions on Applied Superconductivity **9**, 4185 (1999).
[35] Jia-Sheng Hong et al., IEEE Transactions on Applied Superconductivity **9**, 3893 (1999).
[36] G. M. Rebeiz, J.B. Muldavin, "RF MEMS switches, switch circuits, and phase shifters", submitted HF review (2001), and http://www.eecs.umich.edu/rebeiz/Current_Research.html.
[37] A.N. Cleland, M.L. Roukes, Sensors and Actuators **72**, 256 (1999).
[38] http://www.sawtek.com/techsupport.htm.
[39] http://www.memsnet.org/material .
[40] http://www.mst-design.co.uk/cs/fbar.html.
[41] http://www.mst-design.co.uk/cs/mswitch.html.
[42] J.D. Joannopoulos et al., *Photonic Crystals*, Princeton University Press (1995).
[43] http://ab-initio.mit.edu/photons/bends.html.
[44] S.G. Johnson et al., Phys. Rev. B **62**, 8212 (2000).
[45] S. Yamada et al., J. Appl. Phys. **89**, 855 (2001).
[46] S.G. Johnson and J.D. Joannopoulos, Appl. Phys. Lett. **77**, 3490 (2000).
[47] M.L. Povinelli et al., Phys. Rev. B **64**, 075313 (2001).

Neuroelectronic Interfacing: Semiconductor Chips with Ion Channels, Nerve Cells, and Brain

Peter Fromherz,
Max Planck Institute for Biochemistry, Department of Membrane and Neurophysics, Martinsried, Germany

1 Introduction		779
2 Iono-Electronic Interface		780
2.1	Planar Core-Coat Conductor	780
2.2	Cleft of Cell-Silicon Junction	782
2.3	Conductance of the Cleft	785
2.4	Ion Channels in Cell-Silicon Junction	788
3 Neuron-Silicon Circuits		790
3.1	Transistor Recording of Neuronal Activity	790
3.2	Capacitive Stimulation of Neuronal Activity	793
3.3	Circuits with Two Neurons on Silicon Chip	795
3.4	Towards defined Neuronal Nets	799
4 Brain-Silicon Chips		800
4.1	Tissue-Sheet Conductor	801
4.2	Transistor Recording of Brain Slice	802
4.3	Capacitive Stimulation of Brain Slices	803
5 Summary and Outlook		804

Neuroelectronic Interfacing: Semiconductor Chips with Ion Channels, Nerve Cells, and Brain

1 Introduction

Computers and brains both work electrically. However, their charge carriers are different – electrons in a solid ion lattice and ions in a polar fluid. Electrons in silicon have a mobility of about 10^3 cm^2/Vs, whereas the mobility of ions in water is around 10^{-3} cm^2/Vs. That enormous difference of mobility is at the root for the different architecture of the two information processors. It is an intellectual and technological challenge to join these different systems directly on the level of electronic and ionic signals as sketched in Figure 1.

In the 18th century, Luigi Galvani established the electrical coupling of inorganic solids and excitable living tissue. Now, after fifty years of dramatic developments in semiconductor microtechnology and cellular neurobiology, we may envisage such an integration by far more complex interactions, right on the level of individual nerve cells and microelectronic devices or even on the level of biomolecules and nanostructures. Today, however, we are not concerned whether brain-computer interfacing can be really implemented in the forseeable future, with neuronal dynamics and digital computation fused to thinking-computing systems. The issue is an elucidation of the fundamental biophysical mechanisms on the level of nanometers, micrometers and millimeters, and the development of a scientific and technological culture that combines the theoretical concepts and experimental methods of microelectronics, solid state physics, electrochemistry, molecular biology and neurobiology. If we succeed in that endeavour, then we shall be able to fabricate ionoelectronic devices to solve problems in molecular biology, to develop neuroelectronic devices for an experimental physics of brain-like systems, and to contribute to medicine and information technology by creating microelectronic neuroprostheses and nerve-based ionic processors.

Having worked for some time with artificial biomembranes on semiconductor electrodes, I wrote in 1985 a note "Brain on line? The feasibility of a neuron-silicon-junction" [1]. The idea of brain-computer interfacing was scaled down to the level of a real project: "The utopian question may be shaped into a proper scientific problem: How to design a neuron-silicon junction?" I outlined the mechanism of neuron-semiconductor interfacing in both directions. On that basis, the first experimental results were reported in 1991 and 1995 with nerve cells of the leech on open transistors and on capacitive stimulation spots of silicon chips [2], [3]. After those elementary steps, two directions were followed: (i) Downwards, the microscopic nature of the cell-semiconductor contact was investigated with respect to its structure and electrical properties [4] – [23]. The goal is a physical rationalization of the junction in order to have a firm basis for a systematic optimization of neuron-silicon interfacing [24] – [29]. (ii) Upwards, hybrid systems were assembled with neuronal networks joined to microelectronic circuits [30] – [41]. Here the goal is a supervision of numerous neurons in a network by noninvasive contacts to a semiconductor substrate as required for long term studies of dynamical processes such as learning and memory.

The present article relies on own publications [2] – [41] and reviews [42] – [45]. It discusses the physics of the cell-silicon junction, the electronic interfacing of individual neurons by transistor recording and capacitive stimulation, and first steps towards a connection of silicon chips with neuronal networks and with brain slices. Literature on the background of the field is found in the reference lists of the original publications.

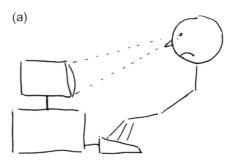

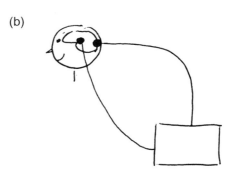

Figure 1: Cartoon of brain-computer interfacing.
(a) Communication through the macroscopic optical and mechanical pathways screen-eye and finger-keyboard.
(b) Hypothetical microscopic interfacing of a computer with the visual and motor cortex [1].

2 Iono-Electronic Interface

A neuron-silicon chip with an individual nerve cell from rat brain and a linear array of transistors is shown in Figure 2. A nerve cell (diameter about 20 μm) is surrounded by a membrane with an electrically insulating core of lipid. That lipid bilayer (thickness about 5 nm) separates the environment with about 150 mM (10^{20} cm^{-3}) sodium chloride from the intracellular electrolyte with about 150 mM potassium chloride. Ion currents through the membrane are mediated by specific protein molecules, ion channels with a conductance between 10 pS and 100 pS. Silicon is used as an electronically conductive substrate for three reasons: (i) Coated with a thin layer of thermally grown silicon dioxide (thickness 10 - 1000 nm), silicon is a perfect inert substrate for culturing nerve cells. (ii) The thermally grown silicon dioxide suppresses the transfer of electrons and the concomitant electrochemical processes that lead to a corrosion of silicon and to a damage of the cells. (iii) A well established semiconductor technology allows the fabrication of microscopic electronic devices that are in direct contact to the cells, shielded by the inert oxide layer.

In principle, a direct coupling of ionic signals in a neuron and electronic signals in the semiconductor can be attained by electrical polarization. If the insulating lipid layer of the neuron is in direct contact to the insulating silicon dioxide of the chip, a compact dielectric is formed as sketched in Figure 3a and Figure 3b. An electrical field across the membrane – as created by neuronal activity – polarizes the silicon dioxide such that the electronic band structure of silicon and an integrated transistor is affected (Figure 3a). Vice versa, an electrical field across the silicon dioxide – as caused by a voltage applied to the chip – polarizes the membrane in a way that conformations of field-sensitive membrane proteins such as voltage-gated ion channels are affected (Figure 3b).

However, when a nerve cell grows on a chip as illustrated in Figure 2, we cannot expect, that the lipid layer of the cell and the oxide layer of silicon form a compact dielectric. Cell adhesion is mediated by protein molecules that protrude from the cell membrane (integrins, glycocalix) and that are deposited on the substrate (extracellular matrix proteins). These proteins keep the lipid core of the membrane at a certain distance from the substrate, stabilizing a cleft between cell and chip that is filled with electrolyte as indicated in Figure 3c and Figure 3d. The conductive cleft shields electrical fields and suppresses a direct mutual polarization of silicon dioxide and membrane.

The cell-silicon junction forms a planar electrical core-coat conductor: the coats of silicon dioxide and membrane insulate the core of the conductive cleft from the conducting environments of silicon and cytoplasm. The first step of neuroelectronic interfacing is determined by the current flow in that core-coat conductor [10], [26]: (i) The activity of a neuron leads to ionic and displacement currents through the membrane (Figure 3c). The concomitant current along the core gives rise to a *Transductive Extracellular Potential* (*TEP*) between cell and chip. (ii) A voltage transient applied to silicon leads to a displacement current through the oxide coat (Figure 3d). Again a *Transductive Extracellular Potential* appears between chip and cell due to the concomitant current along the cleft. In a second step of interfacing, the *Transductive Extracellular Potential* in the core-coat conductor is detected by voltage-sensitive devices in the chip or in the cell: (i) The *TEP* induced by the neuron gives rise to an electrical field across the silicon dioxide that is probed by a field-effect transistor (Figure 3c). (ii) The *TEP* induced by the chip gives rise to an electrical field across the membrane that is probed by voltage-gated ion channels (Figure 3d).

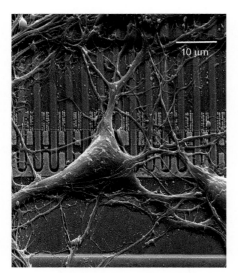

Figure 2: Nerve cell from a rat brain on a silicon chip [22]. Colored electron micrograph, scale bar 10 μm. The surface of the chip consists of thermally grown silicon dioxide (green). The metal free gates of a linear array of field-effect transistors are visible as dark squares. The neuron (blue) is cultured on the chip for several days in an electrolyte.

2.1 Planar Core-Coat Conductor

The *Transductive Extracellular Potential* mediates the coupling of neurons and silicon. It is determined by the current balance in the core-coat conductor of the junction [10]. To describe current and voltage, we use the two-dimensional area-contact model or the zero-dimensional point-contact model as sketched in Figure 4.

Area-contact model. We describe the current in each area element of the junction by the area-contact model symbolized by the circuit of Figure 4a [10], [14]. The current along the cleft is balanced by the displacement current through silicon dioxide and by the ionic and displacement current through the attached membrane. The conservation of electrical charge per unit area of the junction is expressed by (1) where the left hand side refers to the balance of current per unit length in the cleft and the right hand side to the current per unit area through membrane and oxide with the electrical potential V_M in the cell (membrane potential), the potential V_S in the substrate, the *Transductive Extracellular*

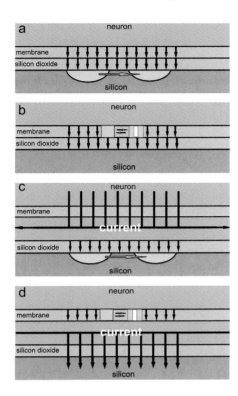

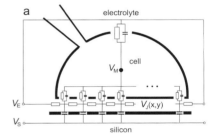

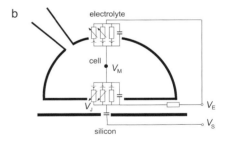

Figure 3: Iono-electronic interfacing. Schematic cross sections, not to scale.
(a) and (b) direct polarization of cell and chip. In (a) the electrical field in the membrane of an excited neuron polarizes silicon dioxide and modulates the source-drain current of a transistor (yellow: source and drain). In (b) an electrical field in silicon dioxide polarizes the membrane and opens ion channels (yellow: closed and open conformations).
(c) and (d) neuron-silicon coupling by electrical current. In (c) current through the membrane of an excited neuron leads to an *Transductive Extracellular Potential* in the cleft between cell and chip which polarizes the oxide and modulates the source-drain current. In (d) capacitive current through the oxide gives rise to a *Transductive Extracellular Potential*, which polarizes the membrane and opens ion channels.

Potential V_J in the junction and the two-dimensional spatial derivative operator ∇. If the bath electrolyte is kept on ground potential ($V_E = 0$), V_M, V_S and V_J are the voltages between cell, silicon and junction and the bath.

$$-\nabla\left(\frac{1}{r_J}\nabla V_J\right) = c_S\left(\frac{\partial V_S}{\partial t} - \frac{\partial V_J}{\partial t}\right) + c_M\left(\frac{\partial V_M}{\partial t} - \frac{\partial V_J}{\partial t}\right) + g_{JM}(V_M - V_J) \quad (1)$$

Parameters are the sheet resistance of the cleft, the area specific capacitances c_M and c_S of membrane and substrate and an area specific leak conductance g_{JM} of the attached membrane. Voltage-dependent ion conductances are not included in (1), for sake of clarity. The specific capacitance c_M in the attached membrane is assumed to be the same as in the free membrane. The sheet resistance r_J can be expressed by the width d_J and the specific resistance ρ_J of the cleft r_J with $r_J = \rho_J/d_J$.

Point-contact model. For many applications it is convenient to describe the core-coat conductor by an equivalent circuit shown in Figure 4b [9], [10], [26]. The conductive cleft is represented by a global Ohmic conductance G_J, attached membrane and silicon dioxide by the global capacitances C_{JM} and C_S. We take into account global ion specific conductances G_{JM}^i in the attached membrane. The reversal voltages V_0^i originate in the concentration differences of the ions between cell and environment, which flow through the conductances G_{JM}^i. They are assumed to be the same as in the free membrane. When we define area specific parameters with respect to the area A_{JM} of the attached membrane as $c_S = C_S/A_{JM}$, $c_M = C_{JM}/A_{JM}$, $g_{JM}^i = G_{JM}^i/A_{JM}$ and $g_J = G_J/A_{JM}$, Kirchhoff's law is expressed by (2) where V_J and V_E are the potentials in the junction and in the bulk electrolyte.

$$g_J(V_J - V_E) = c_S\left(\frac{dV_S}{dt} - \frac{dV_J}{dt}\right) + c_M\left(\frac{dV_M}{dt} - \frac{dV_J}{dt}\right) + \sum_i g_{JM}^i(V_M - V_J - V_0^i) \quad (2)$$

Electrodiffusion. The area-contact and the point-contact model as expressed by (1) and (2) imply that the ion concentrations in the narrow cleft between cell and chip are not changed with constant r_J, constant g_J and constant V_0^i. A change of the ion concentrations in the cleft may become important when the density of ion channels in the junction is high and when these channels are open for an extended time interval. An electrodiffusion version of the area-contact and of the point-contact model accounts for these effects [11].

Figure 4: Core-coat conductor of cell-semiconductor junction [10], [26]. The heavy lines indicate silicon dioxide, cell membrane and micropipette. The cross sections are not to scale: the distance of membrane and chip is between 10 nm and 100 nm, the diameter of a cell is between 10 μm and 100 μm.
(a) AC circuit of area-contact model. The infinitesimal elements of oxide, membrane and electrolyte film in the junction are represented as capacitors and Ohmic resistances.
(b) DC circuit of point-contact model with voltage-dependent ion conductances. Oxide, membrane and electrolyte film of the junction are represented by global capacitances and resistances. V_M is the electrical potential in the cell, V_J the *Transductive Extracellular Potential* in the junction, V_S the potential of the substrate and V_E the potential of the bath.

Area-contact vs. point-contact. A comparison of (1) and (2) shows that the Laplace operator in a homogeneous area-contact model is replaced by a constant in the point-contact model with $-\nabla^2 \to r_J g_J$. To match the two models, we must express the area specific conductance $g_J = G_J/A_{JM}$ by the sheet resistance r_J. Various averaging methods lead to a relation $G_J^{-1} = r_J/\theta\pi$ between global resistance and sheet resistance with a scaling factor $\theta = 4 - 6$ [10], [12], [13]. For a circular junction of radius a_J with $A_{JM} = a_J^2\pi$, we obtain (3) with $r_J = \rho_J/d_J$.

$$g_J = \theta \frac{1}{r_J a_J^2} = \theta \frac{d_J}{\rho_J a_J^2} \tag{3}$$

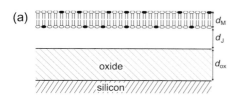

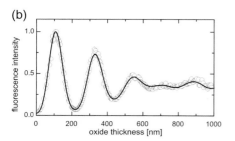

Figure 5: Fluorescent lipid membrane on silicon [5]. (a) Schematic cross section of lipid bilayer with incorporated dye molecules on oxidized silicon. The distance of membrane and chip is d_J, the thickness of the oxide is d_{ox}.
(b) Experimental fluorescence intensity of a bilayer with the cyanine dye DiI versus oxide thickness. The data are fitted by the electromagnetic theory of dipole radiation with a single free parameter, the scaling factor of intensity.

Area specific parameters are preferred in the point-contact model, because a single parameter g_J combines three unknown properties of the junction – the specific resistance ρ_J, the width d_J and the radius a_J, because the area specific capacitances c_M and c_S of membrane and chip are usually known and because area specific membrane conductances g_{JM}^i are common in the neurophysiological literature.

Intracellular dynamics. The *Transductive Extracellular Potential* V_J is determined by the current in the junction alone (1), (2), if the potentials V_M and V_S in cell and chip are under external control. Usually that condition holds for the chip, where V_S is held constant or is determined by a waveform $V_S(t)$ of stimulation. For the cell, V_M is held constant in voltage-clamp situations when the intracellular space is controlled by a micropipette (Figure 4). In situations of noninvasive extracellular recording and stimulation by a chip, the intracellular potential $V_M(t)$ obeys an autonomous dynamics, governed by the balance of ionic and displacement currents through the free and attached membrane as indicated in Figure 4.

For the point-contact model we obtain (4) using Kirchhoff's law, where the left hand side describes the outward current through the free membrane, and the right hand side refers to the inward current through the attached membrane with the area specific ion conductances g_{JM}^i and g_{FM}^i of attached and free membrane and with the ratio $\beta_M = A_{JM}/A_{FM}$ of attached and free membrane area.

$$c_M \left(\frac{dV_M}{dt} - \frac{dV_E}{dt} \right) + \sum_i g_{FM}^i \left(V_M - V_E - V_0^i \right)$$
$$= -\beta_M \left[c_M \left(\frac{dV_M}{dt} - \frac{dV_J}{dt} \right) + \sum_i g_{JM}^i \left(V_M - V_J - V_0^i \right) \right] \tag{4}$$

(2) and (4) together describe the coupled dynamics of the intracellular and extracellular potentials $V_M(t)$ and $V_J(t)$ for the point-contact model. In analogy, the area-contact model has to be amended by the intracellular dynamics. There, Kirchhoff's law for the cell is given by the outward current through the free membrane as given by the left hand side of (4) and by an integral over all local inward currents through the attached membrane area [13].

Conclusion. The interfacing of neuron and semiconductors is mediated by a *Transductive Extracellular Potential*. A large *TEP* results from high currents through membrane and silicon dioxide, and from a low conductance of the junction. Recording and stimulation of neuronal activity are promoted by a small distance d_J, a high specific resistance ρ_J, and a large radius a_J of the cell-chip junction. Efficient recording requires high ion conductances g_{JM}^i in the attached membrane, efficient stimulation a high area specific capacitance c_S of the chip.

2.2 Cleft of Cell-Silicon Junction

The distance d_J between a cell membrane and a silicon chip is a fundamental parameter of cell-silicon junctions. The distance is measured by the method of fluorescence interference contrast (*FLIC*) microscopy which relies on the formation of standing modes of light in front of the reflecting surface of silicon.

Fluorescence on silicon. We consider a lipid bilayer on oxidized silicon as sketched in Figure 5a. The membrane is labelled with amphiphilic dye molecules with transition dipoles in the membrane plane. Upon illumination, light is reflected at all interfaces, in particular at the interface silicon to silicon dioxide. Also the fluorescence light emitted by the dye molecules is reflected. Due to interference effects the excitation and the fluorescence of the dye depend on the distance between membrane and silicon [4].

The electrical field of a light wave has a node in the plane of an ideal mirror. For normal incidence of light, the probability of excitation of a membrane-bound dye is described by the first factor of (5) with a thickness d_{ox} and a refractive index n_{ox} of silicon dioxide, with a width d_J and refractive index n_J of the cleft between membrane and chip and with a wavelength λ_{ex}. An analogous interference effect occurs for light that is emitted from the dye directly and with reflection. The probability of fluorescence at a wavelength λ_{em} in normal direction is described by the second factor of (5). The detected stationary fluorescence intensity $J_{fl}(d_J, d_{ox})$ is proportional to the product of excitation and emission probability according to (5) [4] which can be read as a function $J_{fl}(d_J)$ at constant d_{ox} or as a function $J_{fl}(d_{ox})$ at constant d_J.

$$J_{fl}(d_J, d_{ox}) \propto \sin^2\left[\frac{2\pi(n_{ox}d_{ox} + n_J d_J)}{\lambda_{ex}/2}\right] \cdot \sin^2\left[\frac{2\pi(n_{ox}d_{ox} + n_J d_J)}{\lambda_{em}/2}\right]. \quad (5)$$

For a cell on silicon, the complete electromagnetic theory of dipole radiation has to be applied. It leads to a more involved function $J_{fl}(d_{ox}, d_J)$ which takes into account the layered optical structure, the aperture of a microscope, the spectral bandwidth of illumination and detection and the nearfield interaction of dye and silicon [5].

For an experimental test of the modulated fluorescence on silicon, we attach ($d_J \approx 0$) a pure lipid bilayer with the cyanine dye DiI to a silicon chip with 256 terraces of silicon dioxide. The observed fluorescence intensity $J_{fl}(d_{ox})$ is plotted in Figure 5b. We observe a damped periodic variation of the intensity which levels out above 600 nm due to the large aperture and the wide spectral bandwidth of detection. The experiment is perfectly fitted with the relation $J_{fl}(d_{ox})$ of the complete electromagnetic theory using a single free parameter, the scaling factor of the intensity [5].

FLIC microscopy. The modulation of fluorescence on silicon is the basis of *FLIC* microscopy which allows to determine the distance between a chip and a cell. A direct evaluation of d_J from the measured fluorescence intensity and the theoretical function $J_{fl}(d_{ox}, d_J)$ at a given value d_{ox} is not possible (i) because we cannot measure absolute intensities [4] and (ii) because there is a background fluorescence from the upper membrane of the cell out of focus [6]. To overcome that problem, the intensity $J_{fl}(d_{ox}, d_J)$ of the membrane is measured on several oxide layers of different height d_{ox} at a certain unknown value of d_J. Usually 4 or 16 quadratic terraces are fabricated in a 10 μm × 10 μm unit cell of the silicon surface [6], [7]. The data are fitted by a function $\tilde{J}_{fl}$ according to (6) with three parameters, a scaling factor a, a background b and the optical width of the cleft $n_J d_J$.

$$\tilde{J}_{fl} = aJ_{fl}(d_{ox}, d_J) + b \quad (6)$$

It is a main advantage of *FLIC* microscopy that the theoretical function $J_{fl}(d_{ox}, d_J)$ is dominated by the optics of the well defined interface of silicon and silicon dioxide. Not well known optical parameters of the cell – the thickness of the membrane including protein complexes and the refractive indices of membrane and cytoplasm – play almost no role. Prerequisite of *FLIC* microscopy is a similar geometry of cell adhesion on the different terraces and a homogeneous staining of the membrane.

Astrocyte on laminin. A fluorescence micrograph of a glia cell from rat brain (astrocyte) on 16 different terraces is shown in Figure 6a. The chip is coated with a protein from the extracellular matrix (laminin) with a thickness of 3 nm in its dry state. The checkerboard pattern of fluorescence matches the oxide terraces [6]. Two features of the picture are important: (i) The intensity is rather homogeneous on each terrace. (ii) The intensity is periodic with the unit cells of 4 x 4 terraces. These observations indicate that the membrane is stained homogeneously and that a well defined distance of membrane and chip exists on all terraces.

The fluorescence intensity $J_{fl}^{exp}(d_{ox})$ on 16 terraces is plotted in Figure 6b versus the height of the terraces. It is highest on the thinnest oxide, drops and increases again on higher terraces. That result is quite in contrast to the model experiment of Figure 5b. For comparison, the result of a control experiment is plotted in Figure 6b where a stained vesicle made of a pure lipid bilayer is attached to the same microscopic terraces with polylysine and observed under the same optical conditions [16]. There the fluorescence starts with a minimum on the thinnest oxide as in the model experiment with a

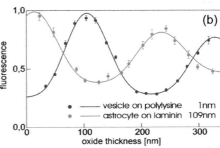

Figure 6: *Fluorescence interference contrast (FLIC) microscopy of astrocyte [7].*
(a) Fluorescence micrograph of the adhesion region of a rat astrocyte on a silicon chip with quadratic 2.5 μm x 2.5 μm terraces of silicon dioxide. Scale bar 10 μm. The chip is coated with laminin. The membrane is stained with the dye DiI.
(b) Fluorescence intensity versus height of the terraces for the astrocyte (red dots) and a lipid vesicle on polylysine (blue dots). The lines are computed by an electromagnetic theory assuming a water film between oxide and membrane of 109 nm thickness for the astrocyte and of 1 nm for the lipid vesicle.

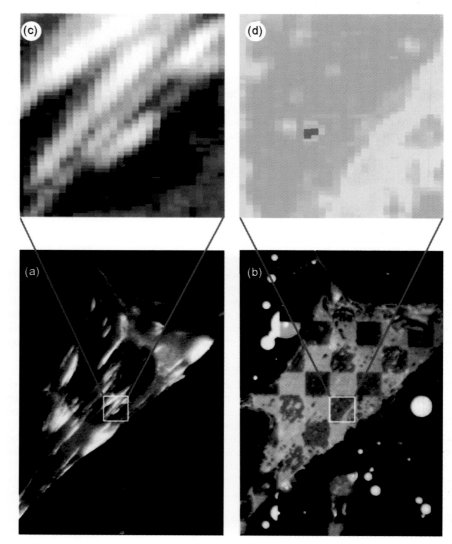

Figure 7: Focal contact of fibroblast on fibronectin [8].
(a) Fluorescence micrograph in the light of *GFP* (*green fluorescent protein*) fused to vinculin showing elongated focal contacts.
(b) *FLIC* micrograph in the light of the cyanine dye *DiI*. The size of the four terraces of different height is 5 μm x 5μm.
(c) Blow up of a terrace of the vinculin picture.
(d) Color coded map of the distance between cell and chip obtained by *FLIC* microscopy on the selected terrace. Within the lateral resolution of about 400 nm, there is no close contact in correlation to the areas of focal adhesion.

planar lipid bilayer Figure 5. A fit of the data according to (6) leads to $d_J = 1$ nm assuming a refractive index of water. On the other hand, a fit of the data for the astrocyte membrane on laminin leads to $d_J = 109$ nm.

Focal contact. For comparison we consider fibroblast cells on the extracellular matrix protein fibronectin. Their special cellular structures promote strong adhesive forces such that we may expect a particularly close distance of cell and chip. We visualize these focal contacts by fusing *green fluorescent protein* (*GFP*) to vinculin, one of their protein components, as shown in Figure 7a [8]. Choosing a different illumination we perform a *FLIC* experiment with the cyanine dye DiI on a chip with four different terraces as shown in Figure 7b. For a selected terrace depicted in Figure 7c, we compute a distance map $d_J(x,y)$. Figure 7d shows that even at focal contacts with their strong adhesion, the separation of the lipid core of the cell membrane and the chip is 50 nm within the lateral resolution of the microscope.

Conclusion. The lipid core of a cell membrane and the silicon dioxide layer of a silicon chip are not in close contact. The large distance is caused by dangling polymer molecules that protrude from the membrane (glycocalix) and that are deposited on the chip (laminin) [17]. They give rise to repulsive entropic forces that balance the attractive forces of cell adhesion between the integrins in the membrane and laminin molecules. It will be an important task to lower the distance of cells and chips by special treatments of the chip surface and by genetic modifications of the membrane without impairing the viability of the cells.

2.3 Conductance of the Cleft

Given a cleft between cell and chip, we have to ask for the sheet resistance r_J of the junction in the area-contact model or for the area specific conductance g_J in the point-contact model. Various approaches can be chosen to obtain r_J or g_J from measurements of the voltage transfer in the junction, considering the circuits of Figure 4: (i) We may apply a voltage $V_M - V_E$ between cell and bath [9] - [15] or a voltage $V_S - V_E$ between chip and bath [15], [16], [21]. (ii) We may probe the voltage drop $V_J - V_S$ across the oxide with field-effect transistors [9] - [14], [15], [16] or the voltage $V_M - V_J$ across the membrane with voltage-sensitive dye [16]. (iii) We may use ac voltages [9] - [16] or voltage steps [21] for stimulation. (iv) We may use the point-contact model [9], [10], [15] or the area-contact model [10], [15] to evaluate the data. After considering the nature of transistor recording, we discuss here an intracellular ac stimulation with transistor recording evaluated by the point-contact model, and an extracellular ac stimulation with transistor recording evaluated by the area-contact model. Finally, we describe an extracellular pulse stimulation with optical recording by a voltage-sensitive dye.

Transistor recording. In a p-type *metal oxide silicon field effect transistor* (*MOSFET*) the source-drain current I_D is controlled by the voltage V_{DS} between drain and source and the voltage V_{GS} between metal gate and source. Above the threshold $V_{GS} > V_T$ of strong inversion and below pinch-off, the current is described by (7) where the proportionally constant depends on the length and width of the channel, the mobility of the holes and the capacitance of the gate oxide.

$$I_D \propto \left[V_{DS}(V_{GS} - V_T) - V_{DS}^2/2 \right] \quad (7)$$

An electrolyte replaces the metal gate in an *electrolyte oxide silicon field effect transistor* (*EOSFET*). It is joined to an external metallic contact by a Ag/AgCl electrode that transforms ionic into electronic current. The source-drain current is controlled by the voltage $V_{ES} = V_E - V_S$ applied to the electrolyte. In (7) we substitute $V_{GS} \to V_{ES}$ and $V_T \to V_T^{(E)}$ where the threshold $V_T^{(E)}$ is determined by the work function of silicon, the redox potential of Ag/AgCl, the contact potential of the Ag/AgCl electrode and the electrical double layer at the interface electrolyte / silicon dioxide.

When we probe the electrical effect of a cell, the source-drain current of an *EOSFET* is modulated by the voltage $V_{JS} = V_J - V_S$ in the cell-silicon junction, of course. The cell affects the voltage drop $V_{JE} = V_J - V_E$ between junction and bulk electrolyte at a constant external voltage V_{ES}. With $V_{JS} = V_{JE} + V_{ES}$ we obtain for transistor recording (8).

$$I_D \propto \left[V_{DS}\left(V_{JE} + V_{ES} - V_T^{(E)}\right) - V_{DS}^2/2 \right] \quad (8)$$

The change of the extracellular potential occurs in the cleft between cell and chip, far beyond the electrical double layer which has a thickness of 1 nm in 100 mM NaCl. Thus we are dealing with a genuine modulation of the gate voltage. Local voltage recording by an *EOSFET* has to be distinguished from the application of an *EOSFET* as an *ion-sensitive transistor* (*ISFET*). There molecular interactions of protons and other ions in the electrical double layer modulate the threshold voltage $V_T^{(E)}$.

The characteristics $I_D(V_{DS}, V_{ES})$ of an *EOSFET* is measured in a calibration experiment by variation of the bath potential without cell. The transconductance $\left(\partial I_D / \partial V_{ES}\right)_{V_{DS}}$ is determined at a working point defined by the potentials V_E, V_D and V_S. When we assume that the transconductance of the calibration experiment is valid for the local recording of a potential, we obtain from the experimental ΔI_D the extracellular potential V_J with $V_E = 0$ according to (9).

$$\Delta I_D = \left(\frac{\partial I_D}{\partial V_{ES}}\right)_{V_{DS}} V_J \quad (9)$$

We use p-type transistors where all parts of the silicon chip are held at a positive voltage with respect to the bath with $V_{ES} = V_E - V_S < 0$, $V_{DS} > V_{ES}$, and with bulk silicon on source potential $V_B = V_S$. The bias voltages at the working point prevent cathodic corrosion of the chip and an invasion of sodium ions into the transistors. The thickness of the gate oxide is around 10 nm. In arrays with close spacing, the transistors are placed in a large area with a common gate oxide where they are separated from each other by local field oxide made by a *LOCOS* (*local oxidation of silicon*) process [14], [15].

VI Data Transmission and Interfaces

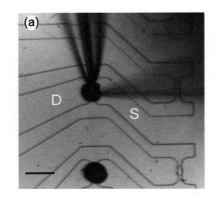
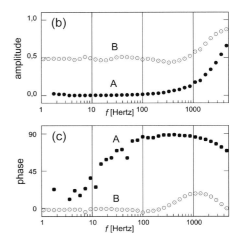

Figure 8: Intracellular ac stimulation and transistor recording [9], [10].
(a) Micrograph of leech neuron on *EOSFET* with source S and drain D. The cell is contacted with a patch pipette. From the right a second pipette is impaled to measure the actual voltage $\underline{V}_M$ in the cell.
(b) Amplitude of voltage transfer $\underline{V}_J/\underline{V}_M$ from cell to junction versus frequency f.
(c) Phase of the voltage transfer. The dots mark the A-type spectrum, the circles the B-type spectrum.

Cell stimulation with transistor recording. We apply an intracellular ac voltage with an amplitude $V_M(\omega)$ at an angular frequency ω to a cell using a patch-pipette in whole cell configuration. The bath is held on ground potential $V_E = 0$. We record the complex response $V_J(\omega)$ in the junction with a transistor [9], [10]. A leech neuron on a transistor is shown in Figure 8a.

Amplitude and phase of the transfer spectrum $\underline{V}_J/\underline{V}_M$ are plotted in Figure 8b and Figure 8c versus the frequency $f = \omega/2\pi$. We find two types of spectra: (i) The A-type spectrum has a small amplitude at low frequencies, an increase of the phase around 10 Hz and an increase of the amplitude above 1000 Hz. (ii) The B-type spectrum has a high amplitude at low frequency and a further increase at 1000 Hz. There is only a minor change of the phase around 1000 Hz. Similar measurements can be made with an array of transistors beneath a single leech neuron as illustrated in Figure 9. In that case the voltage transfer as a function of frequency and space coordinate is evaluated with the area contact model [14].

We evaluate the experiment of Fig. 8 with the point-contact model. We insert in (2) an intracellular stimulation $V_M = \underline{V}_M \exp(i\omega t)$ with a complex response $V_J = \underline{V}_J \exp(i\omega t)$. When we take into account a leak conductance g_{JM} in the attached membrane we obtain (10) at $dV_S/dt = 0$ and $V_E = 0$ with the time constants τ_J and τ_{JM} of the junction and the attached membrane.

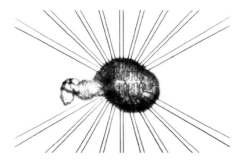

Figure 9: Leech neuron on array of field-effect transistors (diameter of the cell about 60 μm). The array consists of two rows with eight transistors shining through the cell body. The drain contacts are radially directed upwards and downwards. The contact of the common source is at the left and right of the array. The transistors and contacts are separated by local field oxide (*LOCOS* process). The cell is connected with a patch-pipette and the transistor array is used to measure the profile of the extracellular voltage in response to applied intracellular ac voltage [14].

$$\frac{\underline{V}_J}{\underline{V}_M} = \frac{c_M}{c_M + c_S} \cdot \frac{\tau_J/\tau_{JM} + i\omega\tau_J}{1 + i\omega\tau_J}, \quad \tau_J = \frac{c_S + c_M}{g_J + g_{JM}}, \quad \tau_{JM} = \frac{c_M}{g_{JM}}. \quad (10)$$

The high frequency limit of the amplitude $|\underline{V}_J/\underline{V}_M|_\infty = c_M/(c_M+c_S)$ is determined by the capacitances, the low frequency limit $|\underline{V}_J/\underline{V}_M|_0 = g_{JM}/(g_{JM}+g_J)$ by the conductances. There is no phase shift in the limits of low and high frequency. If an intermediate frequency range exists with $\omega\tau_{JM} \gg 1$ and $\omega\tau_J \ll 1$, a phase shift of $\pi/2$ appears where the current is determined by the membrane capacitance and the junction conductance in series.

We interpret the spectra of Figure 8 in terms of (10) using a membrane capacitance $c_M = 5$ μF/cm² of leech neurons and a stray capacitance $c_S = 0.3$ μF/cm² of the chip. In the A-type spectrum the small amplitude at low frequencies indicates a low membrane conductance g_{JM}. Concomitantly, the increase of the phase at a rather low frequency reflects a large time constant τ_{JM} of the membrane. The increase of the amplitude at a high frequency indicates a small time constant τ_J and a large conductance g_J. When we fit the data we obtain $\tau_{JM} = 14$ ms and $\tau_J = 25$ μs and the conductances $g_{JM} = 0.36$ mS/cm² and $g_J = 217$ mS/cm². In the B-type spectrum, the enhanced amplitude at low frequencies indicates a large membrane conductance g_{JM}, the further increase at a high frequency is due to a large conductance g_J. The minor change of phase suggests that a range with $\omega\tau_{JM} \gg 1$ and $\omega\tau_J \ll 1$ does not exist, i.e. that the two time constants are similar. When we fit the data we obtain $\tau_{JM} = 130$ μs and $\tau_J = 66$ μs and the conductances $g_{JM} = 38.5$ mS/cm² and $g_J = 40.8$ mS/cm².

The crucial difference of A-type and B-type junctions is the leak conductance of the attached membrane [9], [10]. Whereas in an A-type contact the membrane conductance is normal, it is enhanced in B-type junction by two orders of magnitude. From the specific junction conductances $g_J = 217$ mS/cm^2 and $g_J = 40.8$ mS/cm^2 we obtain with (3) at $\theta = 5$ and with an estimated contact area $A_{JM} = 1000$ μm the sheet resistances $r_J = 7.7$ MΩ and $r_J = 41$ MΩ. If the cleft is filled with bulk electrolyte ($\rho_J = 100$ Ωcm), its width is $d_J = 130$ nm and $d_J = 24.4$ nm.

Bath stimulation with transistor recording. In a second experiment we apply an extracellular ac stimulation and map the response of the junction with a transistor array [15], [16]. The experiment is performed with a pure lipid membrane [16]. A giant vesicle is sedimented onto the chip and attached by polylysine as shown in Figure 10a.

An ac voltage $\underline{V}_E$ is applied to the electrolyte with respect to ground potential and the modulation of the extracellular voltage $\underline{V}_J$ with respect to ground is observed in amplitude and phase. The amplitude of voltage transfer $\underline{V}_J/\underline{V}_E$ is plotted in Figure 10b versus the position of the transistors and the frequency f. At low frequencies, the cleft perfectly follows the voltage in the bath. That coupling is mediated by the conductance of the cleft considering Figure 4a. Already around $f = 2$ Hz the voltage transfer drops in the center of the junction and a hammock-like profile appears. There the capacitive current through membrane and oxide begins to contribute. At high frequencies where the capacitive current dominates, a plateau is observed again.

We use the area-contact model of (1) to evaluate the profile of the transfer function $\underline{V}_J/\underline{V}_E$ [16]. We do not consider explicitly the current balance of the intracellular space for the area-contact model, but assign the area elements of the free membrane serially to the area elements of the attached membrane. Assuming that the properties of the free and attached membrane are identical, we define an effective area specific capacitance and conductance $\tilde{c}_M = c_M/(1+\beta_M)$ and $\tilde{g}_{JM} = g_{JM}/(1+\beta_M)$. This "local approximation" avoids an integration over the attached membrane [13], [16]. The voltage transfer from the electrolyte to the junction in a circular junction with a radius a_J is given by (11) as a function of the radial coordinate a and the angular frequency ω with the modified Bessel function I_0 and the time constants τ_{JM} and $\tilde{\tau}_{JM}$ and the complex reciprocal length constant $\tilde{\gamma}_J$ of the core-coat conductor.

$$\frac{V_J(a,\omega)}{V_E(\omega)} = \frac{I_0(\tilde{\gamma}_J a)}{I_0(\tilde{\gamma}_J a_J)} + \frac{i\omega\tau_{JM}}{1+i\omega\tilde{\tau}_{JM}}\left[1 - \frac{I_0(\tilde{\gamma}_J a)}{I_0(\tilde{\gamma}_J a_J)}\right] \quad (11)$$

$$\tau_{JM} = \frac{c_M}{g_{JM}}, \rightarrow \quad \tilde{\tau}_{JM} = \frac{\tilde{c}_M + c_S}{\tilde{g}_{JM}}, \rightarrow \quad \tilde{\gamma}_J^2 = r_J \tilde{g}_{JM}(1+i\omega\tilde{\tau}_{JM})$$

For a radius $a_J = 25$ μm and an area ratio $\beta_M = 0.7$ estimated from the shape of the vesicle, with the capacitance $c_M = 0.6$ μF/cm^2 for solventfree lipid bilayers, we obtain a perfect agreement of theory and experiment, when we assume a sheet resistance $r_J = 130$ GΩ and a membrane conductance $g_{JM} < 1$ μS/cm^2. The low conductance reveals the perfect quality of the lipid bilayer. The sheet resistance is surprisingly high. With $d_J = 1$ nm measured by *FLIC* microscopy we obtain from $r_J = \rho_J/d_J$ a specific resistance $\rho_J = 13000$ Ωcm which is far higher than the specific resistance $\rho_E = 250$ Ωcm of the bulk electrolyte. The discrepancy can be assigned to a lowered concentration of ions in the narrow cleft, caused by the image force near the oxide and the membrane with their low dielectric constants.

Chip stimulation with optical recording. In a third experiment a voltage $V_S - V_E$ is applied between chip and electrolyte and the response of the voltage $V_M - V_J$ across the attached membrane is observed with a voltage-sensitive dye [21]. We use cells of the line *HEK 293* (*human embryonic kidney cells*) on a chip coated with fibronectin, a protein from the extracellular matrix. The outer surface of the cell membrane is stained with the voltage-sensitive dye *diButyl-Naphtylamine-Butylsulfonato-IsoQuinolinium* (*BNBIQ*) [19]. At selected wavelengths of excitation and emission, the dye responds with a decrease of fluorescence when a positive voltage is applied to the cytoplasm [18]. Voltage pulses with a height V_{SE}^0 are applied to a highly p-doped silicon chip and the fluorescence change is recorded by signal averaging. A rather thick oxide ($d_{ox} = 50$ nm) is chosen to get high fluorescence intensity in front of the reflecting silicon. Optical transients in the attached and free membrane are depicted in Figure 11. After a negative voltage step, the fluorescence transient is negative in the adhesion region indicating a positive change of the membrane voltage $V_M - V_J$. For a positive voltage step, the change of the membrane voltage $V_M - V_J$ is negative. The data are fitted with exponentials. For the attached membrane the time constant is 2.9 μs.

We evaluate the experiment with the point contact model [21]. A step stimulation with an amplitude V_{SE}^0 applied to the chip with respect to the bath is inserted into (2) with $c_S dV_S/dt = c_S V_{SE}^0 \delta(t)$ When we neglect all ionic currents we obtain from (4) an

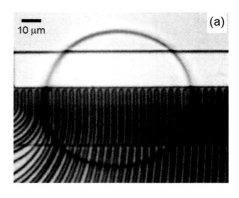

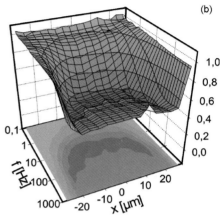

Figure 10: Membrane-silicon junction probed by extracellular stimulation and transistor recording [16].
(a) Micrograph of giant lipid vesicle on a linear transistor array. Scale bar 10 μm. The gates are between the ends of the dark lanes of local field oxide.
(b) Amplitude of voltage transfer $\underline{V}_J/\underline{V}_E$ (ratio of voltage amplitude in the cleft and voltage amplitude in the bath with respect to ground) versus position x and frequency f.

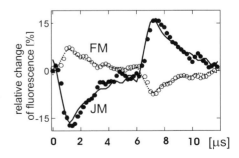

Figure 11: Membrane-silicon junction probed with voltage-sensitive dye *BNBIQ* [21]. A negative voltage of –6 V pulse is applied to the chip from 0 μs to 6 μs, a positive pulse of +6 V from 6 μs to 12 μs. The transient change of fluorescence is plotted for the attached (JM) and free membrane (FM) of a *HEK293* cell. The data are fitted with exponentials convoluted with the transfer function of the chip and the response function of the photomultiplier.

exponential response of the voltage across the attached membrane according to (12) with a time constant $\tilde{\tau}_J$ where the effective capacitance per unit area is $\tilde{c}_M = c_M/(1+\beta_M)$

$$\frac{V_M - V_J}{V_{SE}^0} = -\frac{1}{1+\beta_M}\frac{c_S}{\tilde{c}_M + c_S}\exp\left(-\frac{t}{\tilde{\tau}_J}\right), \quad \tilde{\tau}_J = \frac{\tilde{c}_M + c_S}{g_J} \quad (12)$$

From the experimental time constant $\tilde{\tau}_J = 2.9$ μs with $c_M = 1$ μF/cm², $c_S = 0.07$ μF/cm² and $\beta_M = 0.4$ we obtain a specific conductance $g_J = 270$ mS/cm² of the junction. Using (3) with $\theta = 5$ and a contact area $A_{JM} = 725$ μm², a sheet resistance of $r_J = 8$ MΩ is evaluated. That result is similar to the A-type junction of leech neuron. For the *HEK293* cells, however, we are able to measure the width of the cleft by *FLIC* microscopy. We find $d_J = 50$ nm. From $r_J = \rho_J/d_J$ we obtain a specific resistance $\rho_J = 40$ Ωcm in the cleft. This value is quite similar to the surrounding bath with a specific resistance 74 Ωcm. We conclude: the cleft between a cell and a silicon chip is filled with bulk electrolyte. Whether the difference of 40 Ωcm and 74 Ωcm is significant has to be checked by more detailed experiments.

Conclusion. The cleft between neuronal cells and chips has an electrical resistance that corresponds to a thin film of bulk electrolyte. The sheet resistance is in the order of $r_J \approx 10$ MΩ with a global resistance around $G_J^{-1} \approx 1$ MΩ. There is no gigaohm seal between neuronal cells cultured on a chip. It should be noted that the width of the cleft is far larger than the thickness of the diffuse electrical double layer at the silicon dioxide and at the membrane with a Debye length around $\kappa_D^{-1} \approx 1$ nm in 100 mM NaCl and also far larger than the Bjerrum length $l_B \approx 0.7$ nm of Coulombic interactions which governs the interaction with image charges in membrane and silicon dioxide. It will be a difficult task to enhance the sheet resistance by lowering the width or by enhancing the specific resistance of the cleft.

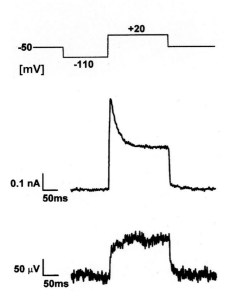

Figure 12: Rat neuron on a transistor under voltage-clamp [22]. The protocol of the intracellular voltage V_M is shown at the top. In the center the total membrane current I_M is plotted, at the bottom the extracellular voltage V_J recorded by a transistor obtained by averaging 30 records.

2.4 Ion Channels in Cell-Silicon Junction

During neuronal excitation, the *Transductive Extracellular Potential* $V_J(t)$ depends on the current through ion conductances in the attached membrane. During capacitive stimulation of neurons the primary target of the *Transductive Extracellular Potential* $V_J(t)$ are the ion conductances in the attached membrane. Thus we have to ask: (i) Are there functional ion channels in the contact region at all? (ii) Is the density of ion channels in the contact the same as in the free membrane? We consider two systems, intrinsic potassium channels in rat neurons and recombinant potassium channels in *HEK293* cells.

Rat neurons. Neurons from rat hippocampus are cultured on a chip with transistors as shown in Figure 2. The intracellular voltage of a cell is varied by the whole-cell patch-clamp technique. Simultaneously, we measure the current I_M through the total membrane with the micropipette and the extracellular voltage in the contact area V_J with a transistor, holding the bath at ground potential [22]. The sodium current is inhibited by tetrodotoxin. The intracellular voltage V_M, the outward current I_M and transistor record V_J are plotted in Figure 12.

At a depolarization $V_M = 20$ mV there is a stationary current $I_M = 0.25$ nA and a superposed transient current. These two current components are due to two different potassium conductances, a K-type conductance and an inactivating A-type conductance. The extracellular voltage V_J detected by the transistor shows a stationary response that matches the stationary K-type current, but no component corresponding to the A-type current [22]. The slow relaxation of the transistor signal after the depolarizing and hyperpolarizing step is due to electrodiffusion effects.

We discuss the result in terms of the point-contact model using (2) and (4) [22], [23]. At a constant voltage V_M, the membrane current I_M for a single ion conductance with an average area specific conductance g_M^i in the whole cell membrane is given by (13) with the total membrane area A_M, assuming that the extracellular voltage is small with $V_J \ll V_M - V_0^i$. The extracellular voltage V_J is described by (14) with the specific conductance g_{JM}^i in the attached membrane.

$$I_M = A_M g_M^i (V_M - V_0^i) \quad (13)$$

$$V_{\rm J} = \frac{1}{g_{\rm J}} \overline{g}_{\rm JM}^{\rm j} \left(V_{\rm M} - V_0^{\rm i} \right) \tag{14}$$

If the channels in the attached and free membrane have the same functionality, the relative conductances $g_{\rm M}^{\rm i}/\overline{g}_{\rm M}^{\rm i}$ and $g_{\rm JM}^{\rm j}/\overline{g}_{\rm JM}^{\rm j}$ follow the same voltage-dependence where $\overline{g}_{\rm M}^{\rm i}$ and $\overline{g}_{\rm JM}^{\rm j}$ are the maximum conductances with open channels. Considering (13) and (14), the transistor record $V_{\rm J}$ and the pipette record $I_{\rm M}$ are proportional to each other for all voltages $V_{\rm M}$ according to (15).

$$\frac{V_{\rm J}}{I_{\rm M}} = \frac{1}{g_{\rm J} A_{\rm M}} \frac{\overline{g}_{\rm JM}^{\rm j}}{\overline{g}_{\rm M}^{\rm i}} \tag{15}$$

Considering Figure 12 with (15) we conclude: (i) The absence of a transient in the transistor response indicates that there is no A-type potassium conductance in the attached membrane with $\overline{g}_{\rm JM}^{\rm A} << \overline{g}_{\rm M}^{\rm A}$. (ii) The visible response of the transistor shows that functional K-type channels exist in the junction. To evaluate the ratio $\overline{g}_{\rm JM}^{\rm K}/\overline{g}_{\rm M}^{\rm K}$ from the experimental $V_{\rm J}/I_{\rm M} = 480$ kΩ with (15) we need a value of the scaling factor $(g_{\rm J} A_{\rm M})^{-1}$.

$g_{\rm J}$ and $A_{\rm M}$ are obtained by an ac measurement when the channels are closed. From (2) and (4) we obtain for the complex response of the current $\underline{I}_{\rm M}$ and of the extracellular voltage $\underline{V}_{\rm J}$ (16) and (17) at $g_{\rm JM}^{\rm j} = 0$ and $\omega \tau_{\rm J} << 1$. The scaling factor is given by the ratio $\underline{V}_{\rm J}/\underline{I}_{\rm M}$ according to (18).

$$\underline{I}_{\rm M} = i\omega c_{\rm M} A_{\rm M} \underline{V}_{\rm M} \tag{16}$$

$$\underline{V}_{\rm J} = \frac{i\omega c_{\rm M}}{g_{\rm J}} \underline{V}_{\rm M} \tag{17}$$

$$\frac{\underline{V}_{\rm J}}{\underline{I}_{\rm M}} = \frac{1}{g_{\rm J} A_{\rm M}} \tag{18}$$

From the response by ac stimulation at $\omega = 200$ Hz, we obtain with (16) and (17) a membrane area $A_{\rm M} = 1100$ µm^2 and a junction conductance $g_{\rm J} = 1000$ mS/cm^2 at $c_{\rm M} = 1$ µF/cm^2. With the resulting scaling factor $(g_{\rm J} A_{\rm M})^{-1} = 91$ kΩ and with $V_{\rm J}/I_{\rm M} = 480$ kΩ the ratio of the maximum conductances in the attached and total membrane is $\overline{g}_{\rm JM}^{\rm K}/\overline{g}_{\rm M}^{\rm K} = 5.3$ using (15). We conclude: the K-type potassium channels are significantly accumulated in the junction.

Recombinant channels. A more detailed investigation is possible with recombinant *hSlo* potassium channels that are overexpressed in *HEK293* cells. No signal averaging is required, and the voltage-dependent gating can be analyzed in detail. The cells are cultured on a transistor array as depicted in Figure 13a.

The voltage $V_{\rm M}$ in the cell is changed step by step with a patch-pipette, and the total membrane current $I_{\rm M}$ and the extracellular voltage $V_{\rm J}$ are simultaneously recorded [23]. Examples are shown in Figure 13b for a high extracellular potassium concentration. A depolarization leads to an enhancement of the current and a correlated enhancement of the transistor signal. Thus, functional *hSlo* potassium channels must exist in the junction. (With low extracellular potassium concentrations, slow transients appear in the transistor signal that are due to electrodiffusion effects.)

In that experiment we are able to check whether the functionality of the channels in the attached and free membrane is the same [23]. We plot $V_{\rm J}$ versus $I_{\rm M}$ in Figure 13c for all voltages $V_{\rm M}$ and obtain a strict linear relation over the whole range of gating. The result shows that (15) is valid with a constant ratio $V_{\rm J}/I_{\rm M} = 73$ kΩ. We measure the scaling factor $(g_{\rm J} A_{\rm M})^{-1}$ with an ac stimulation at $\omega = 6 - 20$ Hz. From the current and the transistor signal we obtain with (16) and (17) a membrane area $A_{\rm M} = 2360$ µm^2 and a junction conductance $g_{\rm J} = 1960$ mS/cm^2 at $c_{\rm M} = 1$ µF/cm^2. The high junction conductance indicates a small effective radius $a_{\rm J}$ of the junction (3). It is due to a peripheral location of the transistor in the area of cell adhesion. With the scaling factor $(g_{\rm J} A_{\rm M})^{-1} = 22$ kΩ and with $V_{\rm J}/I_{\rm M} = 73$ kΩ, the ratio of the maximum area specific conductance in the attached and total membrane is $\overline{g}_{\rm JM}^{\rm hSlo}/\overline{g}_{\rm M}^{\rm hSlo} = 3.3$ using (15). Thus the recombinant *hSlo* potassium channels are accumulated in the attached membrane.

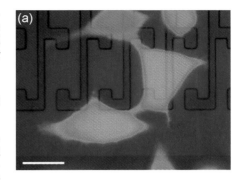
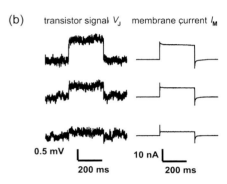
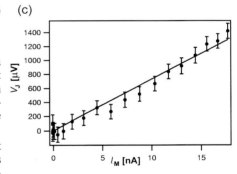

Figure 13: Recombinant *hSlo* potassium channels on a transistor [23].
(a) *HEK293* cells on linear transistor array. With a blue illumination the transfected cells appear in the color of the fluorescence of *GFP* (*green fluorescent protein*) used as a marker.
(b) Iono-electronic coupling at three intracellular voltages $V_{\rm M} = 30, 45, 58$ mV. Left: extracellular voltage $V_{\rm J}$ in the cell-silicon junction. Right: current $I_{\rm M}$ through the total cell membrane. Intracellular potassium concentration 152 mM, extracellular concentration 100 mM.
(c) Extracellular voltage $V_{\rm J}$ versus membrane current $I_{\rm M}$. The regression line has a slope $V_{\rm J}/I_{\rm M} = 73$ kΩ.

VI Data Transmission and Interfaces

Sensorics. An overexpression of ligand-gated channels and transistor recording is a promising approach to develop cell-based biosensors. In such systems the intracellular voltage is not controlled. The *TEP* is obtained by combining (2), (3) and (4) for small signals with $V_J \ll V_M - V_0^i$ and $dV_J \ll dV_M$ at $V_E = 0$ according to (19): A cellular electronic sensor relies on an accumulation or depletion of channels with $g_{JM}^i - g_{FM}^i \neq 0$, a high driving voltage $V_M - V_0^i$ and a junction with a large radius a_J, a high specific resistance ρ_J and a small distance d_J.

$$V_J = \frac{\rho_J a_J^2}{5 d_J (1 + \beta_M)} \left(g_{JM}^i - g_{FM}^i \right) \left(V_M - V_0^i \right) \tag{19}$$

Conclusion. The combination of transistor recording with whole-cell patch-clamp shows that functional ion channels exist in the area of cell adhesion. Important for neuronal interfacing and cellular biosensorics is the observation, that ion channels are selectively accumulated and depleted. A control of the expression and sorting of ion channels is an important task to optimize cell-chip contacts.

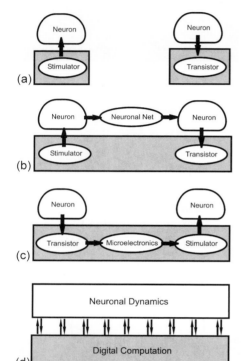

Figure 14: Neuroelectronic hybrids.
(a) Capacitive stimulation and transistor recording of individual neurons from semiconductor.
(b) Two-neuron pathway with capacitive stimulation, signal transmission through neuronal network and transistor recording at a second neuron.
(c) Two-neuron pathway with transistor recording, signal processing by microelectronics on the chip and capacitive stimulation of a second neuron.
(d) Integration of neuronal dynamics and digital electronics by bidirectional signaling on a microscopic level.

3 Neuron-Silicon Circuits

Cell-silicon junctions are the basis for an integration of neuronal dynamics and digital electronics. The first step is an interfacing of individual nerve cells and silicon microstructures (Figure 14a) with (i) eliciting neuronal activity by capacitive stimulation from the chip and (ii) recording of neuronal activity by a transistor. On a next level, pairs of nerve cells are coupled to a chip with two fundamental pathways: (i) Stimulation of a neuron, signal transfer through a neuronal network with synapses to a second neuron and recording of neuronal activity there by a transistor (Figure 14b). (ii) Recording activity of one neuron by a transistor, signal transfer through the microelectronics of the chip and capacitive stimulation of a second neuron (Figure 14c). In a further step, defined neuronal networks are created on the chip such that an intimate communication of network dynamics and computation can be envisaged (Figure 14d).

Identified neurons from invertebrates are preferred in these experiments because they are large and easy to handle, because they form strong neuroelectronic junctions, and last but not least, because small neuronal networks have a distinct biological function in invertebrates and may be reconstituted and studied on a chip.

3.1 Transistor Recording of Neuronal Activity

The activity of a nerve cell – an action potential – consists in a fast opening of sodium channels with a concomitant current into the cell and a delayed opening of potassium channels with a compensating outward current. The ionic currents are coupled by the membrane potential $V_M(t)$ which controls the opening and closing of the channels. The voltage transient $V_M(t)$ is confined by the reversal voltages of sodium and potassium channels. The neuronal excitation drives ionic and capacitive current through the membrane attached to a chip. That current is squeezed through the cleft between cell and chip and gives rise to a *Transductive Extracellular Potential* $V_J(t)$ that is recorded by a transistor. First we derive $V_J(t)$ expected for neuronal excitation using the point-contact model. Then we consider transistor recordings of leech and rat neurons.

Small signal approximation. The *Transductive Extracellular Potential* is determined by the coupled dynamics of the extracellular and intracellular potentials V_M and V_J according to (2) and (4). We assume that the extracellular potential is small with $V_J \ll V_M - V_0^i$ and $dV_J \ll dV_M$, and that the capacitive current to the chip is negligible. With $V_E = 0$ we obtain (20) and (21) with the current j_{INJ} injected by a pipette per unit area of the cell membrane [24]: The extracellular potential $V_J(t)$ is determined by the capacitive and ionic current through the attached membrane. The intracellular potential $V_M(t)$ is governed by the currents through attached and free membrane.

$$g_J V_J = \sum_i g_{JM}^i \left(V_M - V_0^i \right) + c_M \frac{dV_M}{dt} \tag{20}$$

$$(1+\beta_M)\, c_M \frac{dV_M}{dt} = -\sum_i \left(g^i_{FM} + \beta_M g^i_{JM}\right)\left(V_M - V^i_0\right) + (1+\beta_M) j_{INJ} \qquad (21)$$

A-, B- and C-type response. When the attached membrane contains no voltage-gated conductances, we obtain (22) with a leak conductance g_{JM} in the attached membrane using (20).

$$g_J V_J = g_{JM} V_M + c_M \frac{dV_M}{dt} \qquad (22)$$

For negligible leak conductance, the capacitive current through the attached membrane dominates. Then the *TEP* is proportional to the first derivative of the intracellular voltage with $V_J \propto dV_M/dt$ [2]. This situation corresponds to the A-type junction observed in ac measurements. We call it an A-type response. For a dominating ohmic leak conductance, the *TEP* reflects the intracellular waveform itself with $V_J \propto V_M$ [9], [10], [25]. We call it a B-type response in analogy to the B-type junction found with ac experiments.

When we insert (21) into (20), the capacitive current is expressed by the ionic current through the free membrane and we obtain (23). The *Transductive Extracellular Potential* is determined by the differences $g^i_{JM} - g^i_{FM}$ of the area specific ion conductances in the attached and free membrane [24].

$$g_J V_J = \frac{1}{1+\beta_M} \sum_i \left(g^i_{JM} - g^i_{FM}\right)\left(V_M - V^i_0\right) + j_{INJ} \qquad (23)$$

This striking relation shows that the *TEP* of an action potential relies on an inhomogenity of the membrane. In particular (23) reveals, that a selective accumulation or depletion of voltage-gated channels can give rise to a wide spectrum of waveforms $V_J(t)$. We call them C-type responses [26]. Details must be treated by numerical simulation. A special signal is expected when all conductances are accumulated by the same accumulation factor $\mu^i_J = g^i_{JM}/g^i_{FM}$. For $\mu_J > 1$ the response is proportional to the negative first derivative of the intracellular voltage according to (24).

$$V_J = \frac{1-\mu_J}{1+\mu_J \beta_M} \frac{c_M}{g_J} \frac{dV_M}{dt} \qquad (24)$$

Leech neuron. Transistor records of leech neurons are shown in Figure 15. Two positions of the cells are illustrated in Figure 15a and Figure 15b, with the cell body right on a transistor [25] and with the axon stump placed on a transistor array [26]. The cells are impaled with a micropipette and action potentials are elicited by current injection. The intracellular potential $V_M(t)$ is measured with the pipette. The response of the transistors is calibrated in terms of the extracellular potential $V_J(t)$ on the gate. Three types of records are depicted in Figure 15c: (i) With a cell body on a transistor, $V_J(t)$ resembles the first derivative of the intracellular voltage $V_M(t)$. (ii) With a cell body on a transistor, $V_J(t)$ resembles the intracellular voltage $V_M(t)$ itself. (iii) With an axon stump on a transistor, $V_J(t)$ resembles the inverted first derivative of $V_M(t)$.

The results perfectly match the A-type, the B-type and the special C-type response considered above. The C-type record is observed only when the axon stump is placed on the transistor (Figure 15b). That region is known for its enhanced density of ion channels. The small amplitude is due to the small size of the axon stump with a high junction conductance g_J according to (3). The density of channels is depleted in the cell body where we observe A-type and B-type records.

Rat neuron. Neurons from rat hippocampus are cultured for seven days in neurobasal medium on a chip coated with polylysine [27], [29]. Selected cells, such as depicted in Figure 2, are contacted with a patch pipette and action potentials are elicited by current pulses. Transistor records are obtained by signal averaging, locking the transistor signal to the maximum of the intracellular transient. A result obtained with 63 sweeps is shown in Figure 16.

The amplitude of the extracellular potential $V_J(t)$ is around 0.15 mV. The action potential $V_M(t)$ give rise to two positive transients in $V_J(t)$, one in the rising phase and one in the falling phase. The upward and downward jumps in the record match the upward and downward steps of the injection current (23). The small amplitude of the transistor record is a consequence of a high junction conductance g_J which is expected for the small size of rat neurons (3). From an ac measurement we estimate $g_J = 600$ - 700 mS/cm². The shape of the transistor response can be interpreted in terms of (23): (i)

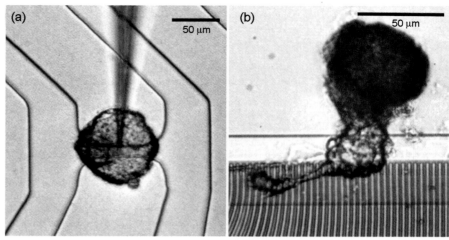

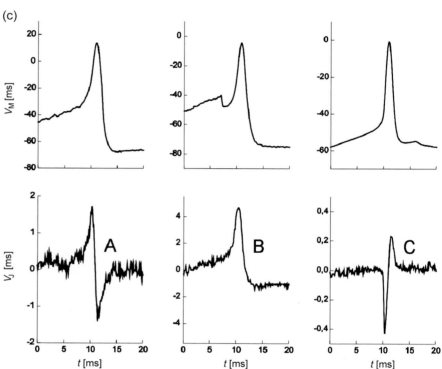

Figure 15: Transistor recording of neuronal excitation in leech neurons [25], [26].
(a) Cell body of a neuron on the open gate oxide of a field-effect transistor. Scale bar 50 μm. The cell is impaled with a micropipette.
(b) Axon stump of a neuron on a linear array of field-effect transistors. Scale bar 50 μm.
(c) A, B and C-type coupling. The upper row shows the intracellular voltage $V_M(t)$, the lower row the extracellular voltage $V_J(t)$ on the gate oxide. A and B-type couplings are observed for arrangement a), C-type couplings for arrangement b).

The positive peak in the rising phase is related with the sodium current. Considering (23) with $V_M - V_0^{Na} < 0$, it must be connected with $g_{JM}^{Na} - g_{FM}^{Na} < 0$, i.e. a depletion of sodium channels in the junction. In other words, the sodium inward current through the free membrane gives rise to a capacitive outward current through the attached membrane. (ii) The positive peak in the falling phase can be assigned to a potassium outward current through the attached membrane, considering an accumulation of potassium channels with $g_{JM}^K - g_{FM}^K > 0$ and $V_M - V_0^K > 0$.

Conclusion. Neuronal activity is detected by field-effect transistors. The response is rationalized by a *Transductive Extracellular Potential* in the cell-chip contact that plays the role of a gate voltage. There is no unique response to action potentials, the shape of the extracellular record depends on the cell type and the cell area attached to the chip. The amplitude of the extracellular records is small, because the junction conductance is high compared to the effective ion conductances in the contact with $g_J \gg g_{JM}^i - g_{FM}^i$. The signals are particularly weak for mammalian neurons due to their small size. We may attempt to optimize recording (i) by improving the cell-chip contact - reducing the width of the cleft or enhancing its specific resistance, (ii) by enhancing the inhomogeneity of channel distribution using recombinant methods or (iii) by lowering the noise of the transistors choosing improved design and fabrication.

3.2 Capacitive Stimulation of Neuronal Activity

A changing voltage $V_S(t)$ applied to a stimulation spot beneath a neuron leads to a capacitive current through the insulating oxide. The concomitant current along the cleft between cell and chip gives rise to a *Transductive Extracellular Potential* $V_J(t)$ beneath the neuron. As a result voltage-gated ion channels may open in the membrane and an action potential $V_M(t)$ may arise. We consider first the extracellular and intracellular voltage of the point-contact model after stimulation with a voltage step. Then we discuss experiments with a voltage step and with a burst of voltage pulses.

A-type stimulation. Stimulation starts from the resting state of a neuron with low ion conductances. For the initial phase, we completely neglect the ion conductances in the attached and free membrane. Considering the coupling of the intracellular and extracellular potentials $V_M(t)$ and $V_J(t)$ with (2) and (4), we obtain (25) and (26) for $V_E = 0$ with $\tilde{c}_M = c_M/(1+\beta_M)$ [3], [21].

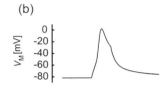

$$(c_S + \tilde{c}_M)\frac{dV_J}{dt} + g_J V_J = c_S \frac{dV_S}{dt} \qquad (25)$$

$$\frac{dV_M}{dt} = \frac{\beta_M}{1+\beta_M}\frac{dV_J}{dt} \qquad (26)$$

For a voltage step of height V_S^0 at time $t = 0$ the perturbation of the junction is $c_S dV_S/dt = c_S V_S^0 \delta(t)$. The voltages across the attached and free membrane $V_M - V_J$ and V_M respond with exponentials for $t > 0$ according to (27) with the time constant $\tilde{\tau}_J$.

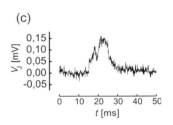

$$V_M - V_J = -\frac{1}{1+\beta_M}\frac{c_S}{\tilde{c}_M + c_S} V_S^0 \exp\left(-\frac{t}{\tilde{\tau}_J}\right)$$

$$V_M = \frac{\beta_M}{1+\beta_M}\frac{c_S}{\tilde{c}_M + c_S} V_S^0 \exp\left(-\frac{t}{\tilde{\tau}_J}\right) \qquad \tilde{\tau}_J = \frac{c_S + \tilde{c}_M}{g_J} \qquad (27)$$

For a positive voltage step $V_S^0 > 0$, the voltage drop is negative (hyperpolarizing) across the attached membrane and positive (depolarizing) across the free membrane. For $V_S^0 = 5V$, $c_S = 0.35 \mu F/cm^2$, $c_M = 5 \mu F/cm^2$ and $\beta_M = 1/6$ as estimated for leech neurons, the voltage amplitude is $|V_M - V_J|_0 \approx 300$ mV across the attached membrane and $|V_M|_0 \approx 50$ mV across the free membrane. With $g_J \approx 200$ mS/cm^2 of an A-type junction of a leech neuron, we expect a time constant $\tilde{\tau}_J \approx 25\mu s$. For mammalian neurons, the time constants are even shorter due to the larger g_J and smaller c_M as verified by optical recording (Figure 11). The crucial question is, how such short transients can affect the ion conductance of a membrane.

B-type stimulation. We consider the role of a leak conductance g_{JM} in the attached membrane. To avoid complicated equations, we assume that the intracellular voltage is small compared to the extracellular voltage with $V_M \ll V_J$ and $dV_M \ll dV_J$. From (2) we obtain (28) for the extracellular voltage at $V_E = 0$ with the exponential solution of (29) after a voltage step of height V_S^0.

Figure 16: Transistor record of neuronal excitation in a rat neuron [29].
(a) Injection current applied by a patch pipette.
(b) Intracellular potential $V_M(t)$.
(c) Transistor record scaled as an extracellular potential $V_J(t)$ on the gate (63 averaged signals).

$$(c_S + c_M)\frac{dV_J}{dt} + (g_{JM} + g_J)V_J = c_S \frac{dV_S}{dt} \qquad (28)$$

$$V_J = \frac{c_S}{c_M + c_S} V_S^0 \exp\left(-\frac{t}{\tau_J}\right), \quad \rightarrow \quad \tau_J = \frac{c_M + c_S}{g_{JM} + g_J} \qquad (29)$$

The upward jump of the extracellular transient $V_J(t)$ injects a charge pulse into the cell by capacitive polarization. That charge is withdrawn during the decaying exponential. Additional charge is injected during the exponential transient through a leak conductance in the attached membrane. Using (2) we obtain the injected current per unit area according to (30).

$$j_{JM} = V_S^0 c_S \left\{ \frac{c_M}{c_M + c_S} \left[\delta(t) - \frac{\exp(-t/\tau_J)}{\tau_J} \right] + \frac{g_{JM}}{g_J + g_{JM}} \frac{\exp(-t/\tau_J)}{\tau_J} \right\} \quad (30)$$

Considering (2), the perturbation of the intracellular potential $V_M(t)$ is obtained from (31).

$$c_M \frac{dV_M}{dt} + \sum_i g_{FM}^i (V_M - V_0^i) = \beta_M j_{JM} \quad (31)$$

When we neglect the voltage-gated conductances g_{FM}^i in the free membrane in the stimulation phase, the intracellular potential $V_M(t)$ is given by (32) with the resting potential V_M^0.

$$V_M = V_M^0 + V_S^0 \beta_M \frac{c_S}{c_M} \left[\left(\frac{c_M}{c_M + c_S} - \frac{g_{JM}}{g_{JM} + g_J} \right) \exp\left(-\frac{t}{\tau_J}\right) + \frac{g_{JM}}{g_{JM} + g_J} \right] \quad (32)$$

There is a jump of the intracellular voltage at $t = 0$ due to the capacitive effect. The subsequent relaxation within the time τ_J levels out at a potential determined by the leak conductance. There is a stationary change of the intracellular potential (33)

$$V_M - V_M^0 = V_S^0 \beta_M \frac{c_S}{c_M} \frac{g_{JM}}{g_{JM} + g_J} . \quad (33)$$

If that potential change is above a threshold, an action potential is elicited, if it is below the threshold, it relaxes with the time constant of the total cell. For a positive voltage step $V_S^0 = 5V$ applied to the B-type junction of a leech neuron with $\beta_M = 1/6$, $c_S = 0.35 \, \mu F/cm^2$, $c_M = 5 \, \mu F/cm^2$, $g_{JM} \approx 40 \, mS/cm^2$ and $g_J \approx 40 \, mS/cm^2$ we expect a depolarization of $V_M - V_M^0 = 25 \, mV$.

C-type stimulation. When we take into account voltage-gated conductances in the attached and free membrane, different neuronal responses are expected to a positive or negative voltage step, depending on channel sorting. Such junctions must be treated by numerical simulation on the basis of an assumed dynamics of voltage-gated channels [28].

Step stimulation of leech neuron. Neurons from the leech are stimulated by a single positive voltage step applied to a capacitive stimulation spot on a silicon chip as illustrated in Figure 17a [3]. The height of the steps is $V_S^0 = 4.8, 4.9, 5.0$ V. The intracellular response $V_M(t)$ is shown in Figure 17b. When the height exceeds a threshold, an action potential is elicited.

A positive step in an A-type contact leads to a hyperpolarizing effect on the attached membrane. No sodium channels can open there to induce an action potential. The free membrane is affected by a depolarizing transient with a small amplitude $|V_M| \approx 50$ mV and a time constant $\tau_J \approx 25 \, \mu s$. It is difficult to imagine, how sodium channels with a time constant in the millisecond range are able to respond to such short transients. Thus it is likely that a B-type junction exists. The *Transductive Extracellular Potential* $V_J(t)$ injects Ohmic current into the cell through a leak conductance g_{JM} according to (30). A resulting quasi-stationary depolarization of about 25 mV is sufficient to elicit an action potential.

Burst stimulation of snail neuron. In many junctions of neurons from leech and snail, excitation is achieved only when a burst of voltage pulses is applied to a chip [30], [34], [35]. An example is shown in Figure 18. A snail neuron is attached to a two-way junction made of a stimulation area and a transistor (Figure 18a). When a burst of voltage pulses is applied to the stimulation area, the intracellular voltage $V_M(t)$ responds with short capacitive transients at the rising and falling edge of each pulse and a stationary depolarization during the pulses as illustrated in Figure 18b. After the third pulse, the intracellular potential rises such that an action potential is elicited [34]. The transistor allows us a look into the junction. The rising and falling edges of each pulse lead to capacitive transients $V_J(t)$. At the rising edge an additional negative transient $V_J(t)$ is initiated that slowly decays during the pulse and during the subsequent pulse interval.

We conclude that the positive capacitive transients in the cleft at the onset of the pulses induce an ionic inward current through the attached membrane. That conductance decays slowly and is not affected by the negative capacitive transient. It is responsible for the intracellular depolarization $V_M(t)$. However, the positive extracellular transients

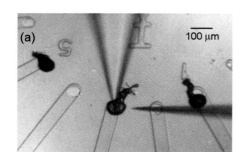
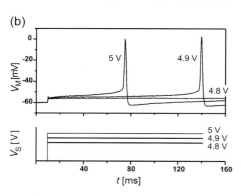

Figure 17: Capacitive stimulation by silicon chip [3].
(a) Leech neurons are attached to circular stimulation spots (from 20 μm to 50 μm diameter) covered by 10 nm silicon dioxide. The rest of the chip is insulated by a 1 μm thick field oxide. One neuron is impaled with a micropipette electrode. An additional electrode (right) measures the local bath potential. Scale bar 100 μm.
(b) Top: intracellular voltages $V_M(t)$. Bottom: voltage steps $V_S(t)$ applied to the chip (not on scale).

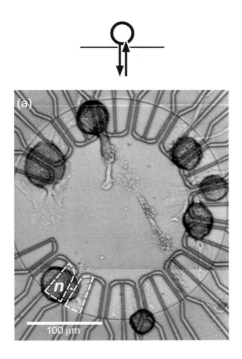
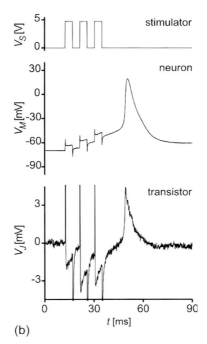

Figure 18: Capacitive stimulation of neuron by a burst of voltage pulses [34].
(a) Micrograph of snail neurons on a chip with a circular arrangement of two-way contacts. The stimulation area with two wings under neuron n is marked with a dashed line, the transistor is located between the two wings. Scale bar 100 μm.
(b) Voltage $V_S(t)$ applied to the stimulation area.
(c) Intracellular voltage $V_M(t)$ measured with an impaled pipette.
(d) Extracellular voltage $V_J(t)$ measured with the transistor.

has a hyperpolarizing effect on the attached membrane and cannot open voltage-gated channels there. Maybe the stimulus with an amplitude $|V_M - V_J|_0 \approx 300$ mV is sufficient to induce a transient electroporation of the membrane. The resulting inward current is recorded by the transistor and depolarizes the cell. Possibly the negative capacitive transient is not sufficient to induce electroporation, because its electrical field is opposite to the given electrical field in the resting state.

Conclusion. The concept of a core-coat conductor can guide a rationalization of capacitive stimulation of neuronal activity on silicon chips. But the situation is less clear than with transistor recording. Optical recording directly reveals that fast voltage transients actually exist in the attached and free membrane. But how those transients affect the cell is uncertain. Current injection through a leaky membrane, capacitive gating of ion channels and transient electroporation are difficult to distinguish. Further studies on neuronal excitation are required with a recording of the local voltage by transistors or voltage-sensitive dyes, comparing voltage-clamp and current-clamp. Detailed studies on the capacitive gating of ion channels and on electroporation will be most helpful. An optimized stimulation may be achieved (i) by lowering the junction conductance, (ii) by inserting recombinant ion channels into the junction and (iii) by fabricating stimulation spots with an enhanced specific capacitance.

3.3 Circuits with Two Neurons on Silicon Chip

We consider two hybrid circuits with two neurons on a silicon chip, the signal transmission from a neuron through a chip to another neuron and the signal transmission from the chip through two synaptically connected neurons back to the chip (Figure 14). With respect to the second device, the formation of electrical synapses and the immobilization of neurons are addressed.

Neuron-chip-neuron. The equivalent circuit of recording and stimulation of two neurons on a chip is shown in Figure 19: A field-effect transistor probes the extracellular voltage V_J in the junction between the first cell and the chip as caused by the membrane currents of an action potential. Capacitive stimulation induces a voltage V_J in the junction of the second cell to elicit neuronal excitation by activating membrane conductances. The processing unit accomplishes five tasks (Figure 19): (i) The source-drain current of the transistor is transformed to a voltage and amplified. (ii) The response to an individual action potential is identified by a threshold device (Schmitt trigger). (iii) A

VI Data Transmission and Interfaces

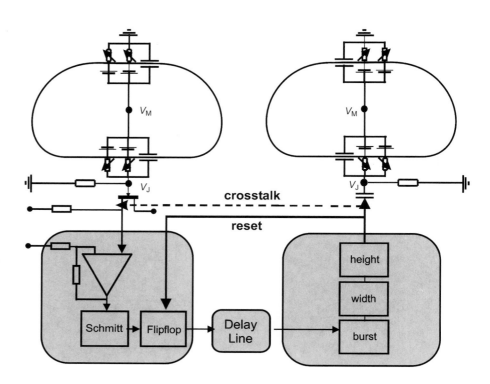

Figure 19: Electronic coupling of two disconnected neurons [31], [32]. In the upper part the interfacing between the neurons and the semiconductor is represented as an equivalent circuit with the point-contact model. In the lower part the function of the interneuron unit is sketched as a block circuit with three stages: (i) amplification and signal recognition, (ii) delay line and (iii) pulse generator. The cross talk from stimulation to recording is marked by a dashed line, the reset of the flip-flop after stimulation by a drawn line.

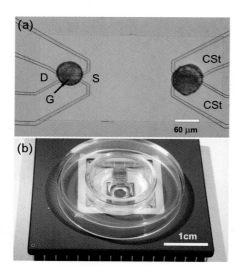

Figure 20: Neurochip.
(a) Micrograph of a silicon chip with an all-silica surface. Two snail neurons are attached to two two-way interface contacts. The source (S), drain (D) and gate (G) of a transistor, and the wings of a capacitive stimulator (CSt) are marked [32]. The surface of the chip is made of silicon dioxide. The bright rectangle is the area of thin oxide.
(b) Perspex chamber with chip. The quadratic interface unit forms the bottom of the circular perspex chamber in contact to the culture medium. The interneuron unit is bonded to it side by side, shielded from the electrolyte.

delay line is started. (iv) A train of voltage pulses is generated and applied to a capacitive stimulator. (v) The crosstalk from stimulator to transistor in the chip is eliminated by a refractory circuit: the delay line is not started directly by the output of the Schmitt-trigger, but by the onset of a flip-flop as triggered by an action potential. The flip-flop is reset after stimulation.

In a first implementation, the chip consists of two parts, an interface unit with transistors and stimulation spots and an interneuron unit implemented as a conventional integrated circuit for discrimination, delay line and pulse shaping [31], [32]. Figure 20a shows two snail neurons attached to two-way contacts of the interface unit. The two silicon chips of the interfacing and the interneuron unit are bonded side by side with the interface unit exposed to the culture medium in a chamber as illustrated in Figure 20b.

The connection from a spontaneously firing neuron A along the chip to a separated neuron B is shown in Figure 21. Both neurons are impaled by micropipette electrodes to observe their intracellular voltage V_M. On the left we see four action potentials of neuron A and the response of the transistor after amplification. The negative transient of the source-drain current is caused by an outward current through the cell membrane. On the right we see four delayed bursts of voltage pulses (17 pulses, height 2 V, width 0.3 ms, interval 0.3 ms) that are applied to the stimulator beneath neuron B. The firing of neuron B is in strict correlation to the firing of neuron A.

The first step of processing on the chip is the assignment of a digital signal to each action potential of neuron A (Figure 21). However, in addition the transistor records strong short perturbations in coincidence with the stimulation pulses at neuron B. They are caused by capacitive crosstalk on the chip. A digital signal is assigned to these artifacts, too, as shown in Figure 21. A second processing step on the chip prevents these artifacts from feeding back into the pulse generator: the response to each action potential activates a flip-flop (Figure 21). The onset of the flip-flop starts the delay line that triggers the voltage generator. The flip-flop is reset after completed stimulation. Thus the stimulation artifact meets an activated flip-flop and cannot interfere.

The pathway neuron – silicon – neuron demonstrates that single action potentials from individual nerve cells can be reliably fed into a digital electronic processor, and that after computations a single action potential in an individual nerve cell can be reliably elicited, all on a microscopic level. The study relies on the established physiology of

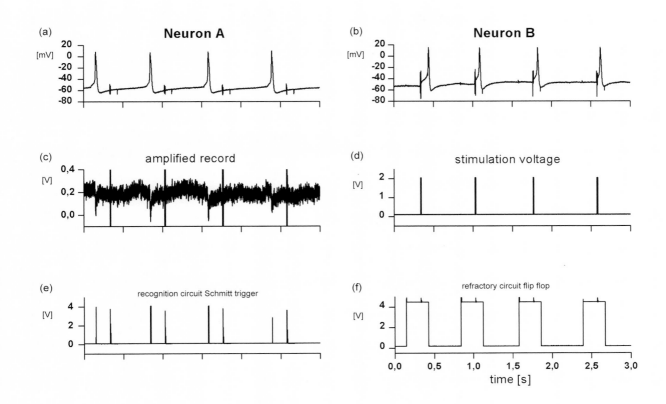

Figure 21: Pathway neuron-silicon chip-neuron [32].
(a) Intracellular voltage of neuron A.
(b) Intracellular voltage of neuron B which fires with a constant delay after neuron A.
(c) Response of the transistor beneath neuron A after current-voltage conversion and amplification. Weak downward signals of the action potentials and strong perturbations due to cross talk of the stimulation voltages at neuron B.
(d) Bursts of voltage pulses at the stimulator beneath neuron B. Each burst of 10 ms duration consists of 17 voltage pulses (width and separation 0.3 ms).
(e) Output of the Schmitt trigger. Signals are assigned to the action potentials and to the crosstalk perturbations.
(f) State of the refractory flip-flop. The flip-flop is set by an action potential. This transition triggers the delay line that elicits the stimulation voltages. It is reset after the end of the stimulation burst.

neurons, on the known physics of interfacing and on standard electronics. It is a fundamental exercise in neuroelectronic engineering. The crosstalk may be avoided (i) by an enhanced capacitance of the stimulators and a lower amplitude of the stimulation pulses, (ii) by transistors with lower noise to provide better defined waveforms of the records such that they can be distinguished from artifacts and (iii) by blocking devices on the chip to reduce the interferences.

Electrical synapses. Neuronal networks rely on synaptic connections. In a first stage we use nerve cells from the snail *Lymnaea stagnalis* which form strong electrical synapses [33]. In these contacts electrical current flows from the presynaptic to the postsynaptic cell through the gap junctions that bridge both cell membranes. An electrical synapse transmits current of hyperpolarisation. The transmission of a stationary presynaptic hyperpolarization $\Delta V_M^{(pre)}$ to a postsynaptic hyperpolarization $\Delta V_M^{(post)}$ defines the coupling coefficient $k_{pre,post}$ according to (34).

$$k_{pre,post} = \frac{\Delta V_M^{(pre)}}{\Delta V_M^{(post)}} \qquad (34)$$

A pair of snail neurons with a coupling coefficient $k_{pre,post} = 0.29$ is shown in Figure 22 for two neurons grown on a linear pattern [33]. The left neuron is stimulated by an impaled micropipette, the pre- and postsynaptic membrane potentials are measured with impaled pipettes. Apparently, the synapse transmits a hyperpolarization. When the left neuron is stimulated by depolarizing current, a train of action potentials is elicited. Each action potentials injects current into the postsynaptic cell such that after four spikes the threshold is reached of an action potential in the postsynaptic neuron.

Immobilization of neurons. An electronic supervision of neuronal nets requires a precise placement of the cells on two-way contacts and a "wiring" by neurites that form synaptic connections. However, when the snail neurons are attached to defined sites of a

VI Data Transmission and Interfaces

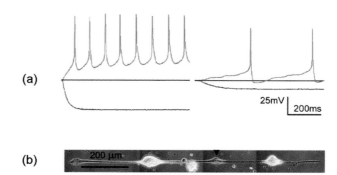

Figure 22: Electrical synapse between two neurons [33].
(a) Electrical signal transfer from the left to the right neuron. In blue, transmission of hyperpolarization, in red, transmission of neuronal activity.
(b) Micrograph of two snail neurons grown on a linear lane forming an electrical synapse (marked with a triangle).

chip, the sprouting neurites exert strong forces on the cell bodies. As a result the cells are displaced from their contact sites [34]. An example is shown in Figure 23. Several neurons are attached to two-way contacts in a circular arrangement. Within two days a network of neurites is formed in the central area of the chip where the neurons are connected by electrical synapses. But the cell bodies are removed from their two-way contacts. The mechanical instability of the arrangement dramatically lowers the yield of a simultaneous interfacing of two neurons.

The problem of displacement is overcome by a mechanical fixation of the cell bodies. Picket fences are fabricated around each two-way contact by photolithography of a polyimide [35]. Neuronal cell bodies are inserted into the cages as illustrated in Figure 24a. They are immobilized even after extensive outgrowth. A chip with a network on a circular array of two-way contacts is shown in Figure 24b.

Mechanical immobilization allows an interfacing of more than one neuron in a net with sufficient probability. Two aspects of the technology are crucial: (i) The fabrication of the cages is a low-temperature process and does not interfere with the semiconductor devices. (ii) The neurites grow on the same surface where the cell bodies are attached. No forces arise that lifts the cell bodies form the two-way contacts.

Signaling chip-neuron-neuron-chip. An experiment with a signaling pathway silicon-neuron-neuron-silicon (Figure 14) is shown in Figure 25 using a network of immobilized snail neurons [35]. Two neurons are selected which are connected by an electrical synapse and which are placed on two-way junctions. A burst of seven pulses is applied to excite neuron 1 as checked with an impaled micropipette. In neuron 2 we observe a subthreshold postsynaptic depolarization. A second burst of voltage pulses elicits another action potential in neuron 1 and leads to a further depolarization of neuron 2. After the third burst, which fails to stimulate neuron 1, the fourth burst gives rise to an

Figure 23: Displacement of neuronal cell bodies by neuronal outgrowth [34]. Snail neurons are placed on a chip with a circular array of two-way contacts. The chip is coated with polylysine and the culture medium is conditioned with snail brains.
(a) 4 hours after mounting.
(b) 44 hours after mounting.

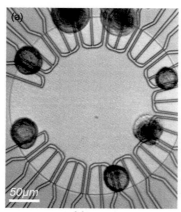
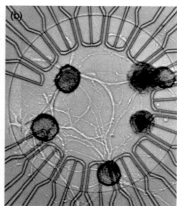

4 hours　　　　　44 hours

 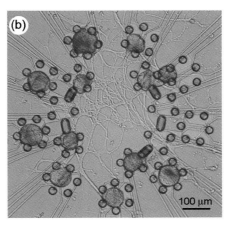

Figure 24: Mechanically stabilized network of neurons on silicon chip [35].
(a) Electron micrograph of a snail neuron immobilized by a picket fence on a two-way contact after three days in culture. Scale bar 20 μm.
(b) Micrograph of neuronal net with cell bodies (dark blobs) on a double circle of two-way contacts with neurites grown in the central area (bright threads) after two days in culture. Scale bar 100 μm. Pairs of pickets in the inner circle are fused to bar-like structures.

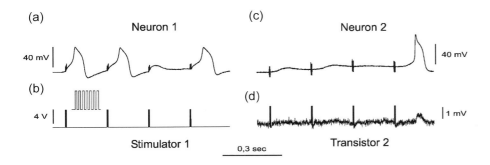

Figure 25: Electronic interfacing of two synaptically connected snail neurons [35].
(a) Intracellular voltage of neuron 1.
(b) Bursts of voltage pulses applied to the stimulator (seven pulses, amplitude 5 V, duration 0.5 ms).
(c) Intracellular voltage of neuron 2.
(d) Transistor record of neuron 2.

action potential in neuron 1 which finally leads to a postsynaptic excitation in neuron 2. That postsynaptic action potential is recorded by the transistor underneath neuron 2, completing an electronically interfaced monosynaptic loop.

Correlated with the burst of stimulation pulses, we observe perturbations of the transistor records and of both microelectrode signals. These perturbations do not reflect actual changes of the voltage on the gate or of the intracellular voltage, respectively. Control experiments without neurons reveal a direct pathway of capacitive coupling through the chip from stimulators to transistors. Control experiments with an open stimulator and a neuron far away reveal a capacitive coupling to the micropipette through the bath. A subsequent depolarization of the cell is not observed in that case. On the other hand the shape, the delay and the temporal summation of the postsynaptic signals correspond to the experiments with intracellular presynaptic stimulation (Figure 22). We conclude that the depolarization of the postsynaptic neuron in Figure 25 is induced by synaptic transmission and not by direct chip stimulation.

3.4 Towards defined Neuronal Nets

The function of neuronal networks is generally based on two features: (i) A mapping of a set of neurons onto another set or a mapping of a set of neurons onto itself as in the symmetrical Hopfield net illustrated in Figure 26. (ii) Hebbian learning rules with an enhancement of synaptic strength as a consequence of correlated presynaptic and postsynaptic activity. Systematic experiments on network dynamics require (i) a noninvasive supervision of all neurons with respect to stimulation and recording to induce learning and to observe the performance of the net on a long time scale, and (ii) a fabrication of neuronal maps with a defined topology of the synaptic connections. To achieve the second goal we have to control neuronal outgrowth, the direction and bifurcation of neurites and the formation synapses.

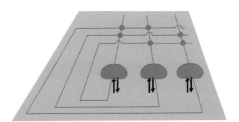

Figure 26: Defined neuronal network (red) with symmetrical connections of axons, synapses and dendrites supervised from a semiconductor chip (blue) by two-way interfacing (black).

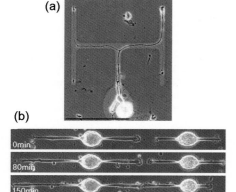

Chemical guidance. The motion of neuronal growth cones is guided by chemical patterns. Defined arborizations of leech neurons are achieved by chemical guidance with lanes of extracellular matrix protein [37], [38]. They are fabricated by UV photolithography of a homogeneous film of extracellular matrix protein using metal masks. When a cell body is placed on the root of a tree-like pattern with orthogonal branchings, the growth cone perfectly follows the lanes and is perfectly split at the branch points into daughter cones as illustrated in Figure 27a.

Using linear chemical patterns, we are able to guide the outgrowth of two neurons such that their growth cones are forced to collide and to form a synapse. That experiment is performed with snail neurons [33]. The chip is coated with polylysine and incubated with dissected snail brains. Secreted proteins are adsorbed on the chip and patterned by UV photolithography. Three stages are illustrated in Figure 27b. After the encounter of the growth cones, synapse formation is checked as discussed above.

Though the process of chemical guidance is rather perfect, there are two problems: (i) The stability of neuritic trees is limited. Neurites have an tendency to shorten. When they are guided around a corner, they dissociate from the guiding lane and cross the nonguiding environment. (ii) A patterned substrate provides a restricted area of growth, but it does not guide a neurite in a certain direction on a branched pattern; with several neurons it does not guide a certain neuron into a certain direction. These two problems of chemical guidance are illustrated in Figure 28 with leech neurons on a hexagonal pattern [36].

Apparently, neuronal growth cones are perfectly guided by the lanes of extracellular matrix protein. But, grown neurites dissociate from the bent lanes such that the defined shape of the neuritic tree is lost. Different neurites grow on the same lane, such that the neuritic tree is not defined by the pattern.

Figure 27: Guided outgrowth by chemical patterns [33], [37].
(a) Defined bifurcations of a leech neuron. The pattern with orthogonal branchings is made by UV photolithography of extracellular matrix protein. Scale bar 100 μm.
(b) Controlled formation of a synapse between two snail neurons. The lanes are made by UV photolithography of brain derived protein adsorbed to the substrate. Scale bar 100 μm.

Topographical guidance. The instability of grown neuritic trees is overcome by topographical guidance. There the grown neurites are immobilized by microscopic grooves which are used as cues for the guidance of the growth cones. Microscopic grooves are fabricated from a polyester photoresist on the chip [38]. An example of a topographical structure obtained by photolithography with five pits and connecting grooves is depicted in Figure 29a. The resin is compatible with cell culture and the low temperature process does not damage microelectronic devices of the chip. The insert of Figure 29a shows the perfectly vertical walls of the pits and grooves.

The chip is coated with polylysine. Secreted proteins from excised snail brains in the culture medium are adsorbed and render the surface of silicon dioxide on the bottom of grooves and pits as well as the surface of the polyester equally suitable for outgrowth. Cell bodies of snail neurons are placed into the pits of the polymer structure. Neurites grow along the grooves and are split at bifurcations of the groove to form neuritic trees as shown in Figure 29b. Because a micrograph does not allow an assignment of the neurites to the different neurons, we sequentially stain individual neurons by injection with the dye Lucifer Yellow which cannot pass the electrical synapses of snail neurons. The synaptic connections are checked by impaling with micropipettes as discussed above.

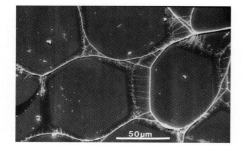

Figure 28: Incomplete control by chemical guidance [36]. Electronmicrograph of the neurites of leech neurons. Scale bar 50 μm. A hexagonal pattern of extracellular matrix protein is made by UV photolithography. Grown neurites dissociate from the guiding lanes. Neurites grow without control on all given lanes of the pattern.

Electrical guidance. Topographical guidance does not solve the problem, that a neuritic tree is not uniquely defined by the guiding pattern. We must combine it with control signals that are only effective at certain positions and certain times to promote or inhibit the growth at a crossing or a branch points and that induce or prevent synapse formation at certain places. Electrical manipulation of growth cones may be a tool to accomplish that task [39].

Conclusion. In a next step, we can envisage joining small networks with defined geometry - made by chemical, mechanical or electrical guidance - to silicon chips with two-way contacts. It remains unclear, however, whether large neuronal nets with hundreds and thousands of neurons can be joined in a defined way by the various methods of guidance. An alternative strategy may be kept in mind, the outgrowth of a disordered neuronal net on a chip with thousands of closely packed interface structures. There most neurons would be on a contact at any time and the rearranging network would be continuously supervised in its structure and dynamics.

4 Brain-Silicon Chips

Instead of culturing defined neuronal nets by controlled outgrowth, we may use neuronal nets given by brains. Considering the planar nature of semiconductor chips, planar networks are preferred in order to attain an adequate supervision by the chip. Organotypic brain slices are particularly promising as they are only a few cell layers thick and conserve major neuronal connections. A cultured slice from rat hippocampus on a chip with a linear transistor array is shown in Figure 30. When we succeed in coupling a brain-grown net to numerous, closely packed transistors and stimulation spots, we are able to study the distributed dynamics of the neuronal network.

An interfacing of a transistor or stimulation spot with an individual neuron in neuronal tissue can hardly be achieved. Thus with brain slices, we have to consider the stimulation and recording of local populations of neurons. The concept of a core-coat conductor of individual cells is not adequate. To guide the development of appropriate chips and to evaluate experimental data, we discuss an approach that relates (i) neuronal currents in a slice to the extracellular potential that is recorded by transistors and (iii) the capacitive stimulation currents from a chip to the extracellular potential that elicits neuronal excitation of the tissue [40]. On that basis, we consider first experiments with transistor recording in brain slices [41] and the implications of capacitive stimulation.

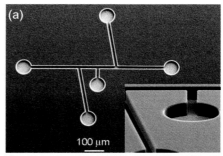

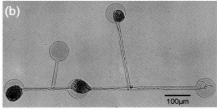

Figure 29: Topographical guidance [38].
(a) Polyester structure on silicon chip with pits and grooves. Electronmicrograph, scale bar 100 μm.
(b) Micrograph of a network of three snail neurons formed by topographical guidance. The neurons are connected by electrical synapses.

4.1 Tissue-Sheet Conductor

In an organotypic brain slice, the neurons are embedded in a tissue of about 100 μm thickness between the insulating silicon dioxide of a chip and an electrolytic bath on ground potential as illustrated in Figure 31a. Excited neurons are local sources or sinks of current that flows to adjacent regions of the tissue layer and to the bath. As a consequence, an extracellular potential appears in the tissue that may be recorded by transistors in the substrate. On the other hand, capacitive contacts in the substrate may locally inject current into the tissue layer, that flows to adjacent regions of the slice and to the bath. The resulting extracellular potential may elicit neuronal excitation. In contrast to the studies with individual neurons, we have to deal here with the stimulation and recording of neuronal populations.

Volume conductor. In a tissue of densely packed neurons, we do not consider the true extracellular voltage on a submicrometer level between the cells, but an average field potential V_{field} that arises from currents per unit volume j_{source} of cellular sources or j_{stim} due to stimulation electrodes. The continuity relation of current in three dimension leads to (35) with the three-dimensional spatial derivative operator ∇ assuming an isotropic average specific resistance ρ of the tissue.

$$-\nabla\left(\frac{1}{\rho}\nabla V_{\text{field}}\right) = j_{\text{stim}} + j_{\text{source}} \qquad (35)$$

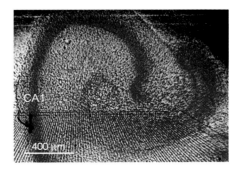

Figure 30: Organotypic slice from rat hippocampus on silicon chip [41]. Nissl staining of a slice cultured for 14 days. Scale bar 400 μm. The dots are neuronal cell bodies. A linear array of field-effect transistors is aligned perpendicular to the CA1 region through the stratum pyramidale and the stratum radiatum until the gyrus dentatus (see text).

We have to consider two boundary conditions [40]: (i) The slice is on ground potential at its surface. With the z-direction normal to the layer plane and a height h of the slice, the constraint is $V_{\text{field}}(h) = 0$ at $z = h$, when we neglect voltage drops in the bath. (ii) At the substrate, the capacitive current per unit area $j_{\text{stim}}^{(2)}$ determines the gradient of the potential with $-\left(dV_{\text{field}}/dz\right)_0 = \rho j_{\text{stim}}^{(2)}$ at $z = 0$. In principle, the field potential can be computed from (35) with the boundary conditions for an arbitrary pattern of stimulation electrodes and an arbitrary distribution of neuronal excitation. However, to get a simple picture of the electrical properties of a slice, we consider a two-dimensional model based on the volume conductor theory.

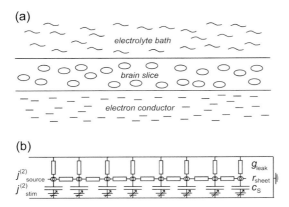

Figure 31: Brain slice on silicon substrate [40].
(a) Geometry of tissue layer between electron conductor and electrolyte bath.
(b) Sheet conductor model. The neurons in the slice give rise to a current-source density per unit area $j_{\text{source}}^{(2)}$. The current flows along the slice (sheet resistance r_{sheet}) and to the bath (g_{leak}, leak conductance). The slice is stimulated by a current density $j_{\text{stim}}^{(2)}$ from the chip by capacitive contacts (specific capacitance c_S).

VI Data Transmission and Interfaces

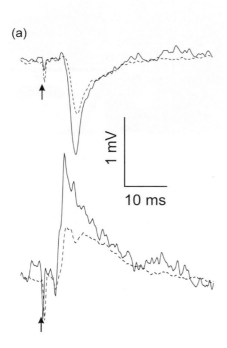

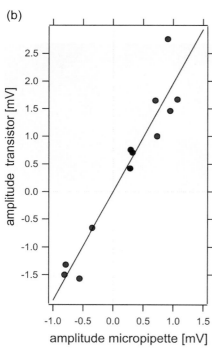

Figure 32: Transistor recording in organotypic slice of rat hippocampus [41].
(a) Evoked field potentials in the stratum radiatum (top) and stratum pyramidale (bottom) of the CA1 region. The arrows mark the stimulation artifact. The dashed lines are the simultaneous records by micropipette electrodes.
(b) Amplitude of transistor records versus amplitude of microelectrodes. The regression line has a slope of 1.96.

Sheet-conductor model. A brain slice between chip and bath is a planar conductor with a capacitive bottom and a leaky cover. Neglecting the z-dimension normal to the plane, we describe the thin tissue by a sheet resistance $r_{sheet}(x,y)$, the shunting effect of the bath by an ohmic conductance per unit area $g_{leak}(x,y)$, and the substrate by a capacitance per unit area $c_S(x,y)$ as illustrated by the circuit of Figure 31b. The neuronal current sources per unit area $j^{(2)}_{source}(x,y)$ and the stimulation current due to a changing voltage V_S at the capacitive contacts are balanced by the current along the sheet and by ohmic and capacitive shunting to bath and substrate according to (35) with the potential $V_{field}(x,y)$ and the 2-D spatial derivative operator ∇.

$$-\nabla\left(\frac{1}{r_{sheet}}\nabla V_{field}\right) + g_{leak}V_{field} = j^{(2)}_{source} + c_S\left(\frac{\partial V_S}{\partial t} - \frac{\partial V_{field}}{\partial t}\right) \quad (36)$$

The sheet conductor model formulated by (36) describes (i) the extracellular field potential that arises from neuronal activity with $\partial V_S/\partial t = 0$, as it may be recorded with transistors, and (ii) the extracellular field potential that is caused by capacitive stimulation with $j^{(2)}_{source}(x,y) = 0$, as it may elicit neuronal excitation.

Note the similarity and the difference between Figure 31b with (36) and Figure 4a with (1): In both cases the circuits and the differential equations describe the continuity relation of electrical current in a twodimensional system. However, Figure 4a and (1) refer to the junction of an individual cell and a chip, whereas Figure 31b and (36) describe a tissue sheet with numerous neurons on a chip.

4.2 Transistor Recording of Brain Slice

We culture a slice from rat hippocampus on a silicon chip with an all-oxide surface and a linear array of transistors as shown in Figure 30. The slice has a thickness $h = 70$ μm. It is stimulated with a tungsten electrode in the gyrus dentatus. A profile of evoked field potentials is recorded across the CA1 region [41].

Field potential. Two transistor records from the stratum radiatum (layer of dendrites) and from the stratum pyramidale (layer of cell bodies) are shown in Figure 32a. Excitatory postsynaptic potentials of neuronal populations are observed. There is a negative amplitude in the region of the dendrites where current flows into the cells and a positive amplitude in the region of the cell bodies with a compensating outward current. For comparison the records of micropipette electrodes are plotted in Figure 32a. The transistor records have an identical shape, but a higher amplitude than the micropipette records. The amplitudes are proportional to each other as shown in Figure 32a. The identical shape and the similar and proportional amplitude validate the approach of transistor recording. Larger amplitudes are expected for a measurement near the insulating substrate where the shunting effect of the bath is smaller as compared to a measurement near the surface where the microelectrodes are positioned.

Potential profile across CA1. The amplitudes of the evoked field potentials across the CA1 region are plotted versus the position of the transistors in Figure 33 [41]. The region of negative potentials matches the stratum radiatum, the region of positive potentials the stratum pyramidale.

We evaluate the experimental field potential with the sheet-conductor model. Along the CA1 layer of the hippocampus the electrical activity is usually assumed to be constant. In that case (36) can be reduced to a one-dimensional relation across the CA1 region along the soma-dendrite direction x. Without stimulation we obtain (37) with the length constant $\lambda_{sheet} = 1/\sqrt{g_{leak}r_{sheet}}$ when we neglect the capacitive current with $c_S dV_{field}/dt \ll g_{leak} V_{field}$.

$$-\lambda_{sheet}^2 \frac{d^2 V_{field}}{dx^2} + V_{field} = \frac{j^{(2)}_{source}}{g_{leak}} \quad (37)$$

We express the sheet resistance as $r_{sheet} = \rho/h$ by the specific resistance of the slice and its thickness and the leaks as $g_{leak} = 2/\rho h$ by the conductance from the center of the slice to the bath. The length constant is $\lambda_{sheet} = h/\sqrt{2}$.

The simplest current profile that is physiologically meaningful is a constant density of synaptic inward current in the stratum radiatum that is balanced by a constant outward current density in the stratum pyramidale. For a constant current-source density $j^{(2)}_{source}$ in a range $-x_0 < x < x_0$, the field potential is given by (38).

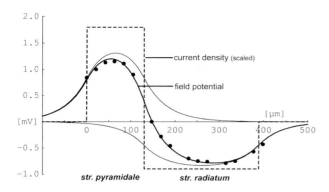

Figure 33: Profile of evoked field potentials across the CA1 region [41]. The amplitude of voltage transients is plotted versus the position of the transistors (black dots). The data are fitted with the field potential V_{field} computed by the sheet conductor model with a constant negative current density in the stratum radiatum and a balancing constant positive current density in the stratum pyramidale. The contributions from the two strata are drawn as thin lines. The scaled profile of current density $j^{(2)}_{\text{source}}/g_{\text{sheet}}$ is indicated (dashed line).

$$V_{\text{field}}(x) = \frac{j^{(2)}_{\text{source}}}{g_{\text{leak}}} \begin{Bmatrix} 1 - \exp\left(-\frac{x_0}{\lambda_{\text{sheet}}}\right)\cosh\left(\frac{|x|}{\lambda_{\text{sheet}}}\right) & |x| < x_0 \\ \sinh\left(\frac{x_0}{\lambda_{\text{sheet}}}\right)\exp\left(-\frac{|x|}{\lambda_{\text{sheet}}}\right) & |x| > x_0 \end{Bmatrix} \quad (38)$$

The width of the stratum radiatum is about 260 μm, of the stratum pyramidale about 130 μm. Using (38) we compute the potential profile of the two regions and superpose them. We obtain a perfect fit of the experimental data with $\lambda_{\text{sheet}} = 50$ μm choosing scaled current densities $j^{(2)}_{\text{source}}/g_{\text{leak}} = -0.9$ mV in the stratum radiatum and $j^{(2)}_{\text{source}}/g_{\text{leak}} = 0.9$ mV (260 μm/130 μm) = 1.8 mV in the stratum pyramidale as shown in Figure 33. In the center of the stratum radiatum the curvature of the field potential $V_{\text{field}}(x)$ is small. Considering (37), there the potential reflects the synaptic current density with $V_{\text{field}} \propto j^{(2)}_{\text{source}}$. This result is in contrast to the volume conductor theory for bulk brain where the curvature is proportional to the local current source density according to (36).

4.3 Capacitive Stimulation of Brain Slices

In a next step we will implement the stimulation of brain slices from silicon chips. The sheet conductor model is a useful guide to develop optimal capacitive contacts.

Circular contact. We consider the capacitive stimulation through a circular stimulation spot. From (36) we obtain (39) for a homogeneous slice in cylinder coordinates with the radius coordinate a when we disregard neuronal activity and assume $dV_{\text{field}}/dt \ll dV_S/dt$ [40].

$$-\lambda^2_{\text{sheet}}\left(\frac{\partial^2 V_{\text{field}}}{\partial a^2} + \frac{1}{a}\frac{\partial V_{\text{field}}}{\partial a}\right) + V_{\text{field}} = \frac{c_S}{g_{\text{leak}}}\frac{\partial V_S}{\partial t} \quad (39)$$

For a stationary stimulation dV_S/dt = const. with a capacitive contact $c_S \neq 0$ for $a < a_0$ we obtain (40) with the modified Besselfunctions I_0, I_1, K_0, and K_1.

$$V_{\text{field}}(a) = \frac{c_S}{g_{\text{leak}}}\frac{dV_S}{dt}\begin{Bmatrix} 1 - \frac{a_0}{\lambda_{\text{sheet}}}K_1\left(\frac{a_0}{\lambda_{\text{sheet}}}\right)I_0\left(\frac{a}{\lambda_{\text{sheet}}}\right) & a < a_0 \\ \frac{a_0}{\lambda_{\text{sheet}}}I_1\left(\frac{a_0}{\lambda_{\text{sheet}}}\right)K_0\left(\frac{a}{\lambda_{\text{sheet}}}\right) & a > a_0 \end{Bmatrix} \quad (40)$$

The field potential for a radius $a_0 = 25$ μm is drawn in Figure 34 with $\lambda_{\text{sheet}} = 35$ μm, $r_{\text{sheet}} = 60$ kΩ and $g_{\text{leak}} = 13.3$ nS/μm². These parameters correspond to a thickness $h = 50$ μm and a specific resistance $\rho = 300$ Ωcm. We apply a voltage ramp for 100 μs with an amplitude of 1 V to a contact with a capacitance $c_S = 30$ μF/cm².

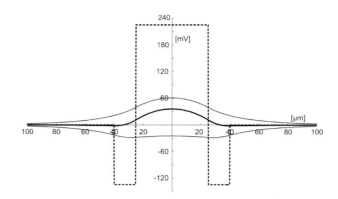

Figure 34: Theory for capacitive stimulation of brain slice [40]. Field potential versus electrode radius computed for a length constant $\lambda_{sheet} = 35$ μm. The central area has a radius of 25 μm, a ring-shaped surround a width of 15 μm. The scaled stimulation current per unit area $(c_S/g_{leak})dV_S/dt$ is indicated as a dashed line. The potential profile $V_{field}(a)$ due to center-surround stimulation is plotted (heavy line), as well the profiles due to positive stimulation in the center (upper thin line), and negative stimulation in the surround (lower thin line).

The field potential is not localized but spreads far beyond the contact area. The maximum amplitude in the center of the contact is around 60 mV, far lower than the amplitude $(c_S/g_{leak})dV_S/dt = 225$ mV that would be attained for an infinitely large electrode. The localization of the field potential can be improved by a negative stimulation in a circular surround as illustrated in Figure 34. The improved localization, however, leads to a further reduction of the amplitude. On the basis of these computations, adequate capacitive contacts on silicon chips have to be developed.

5 Summary and Outlook

The present paper shows that the basic questions on the electrical interfacing of individual nerve cells and semiconductor chips are fairly well answered, the nature of the core-coat conductor, the properties of the cleft, the role of accumulated ion channels, the mechanism of transistor recording and capacitive stimulation. With respect to the latter issue, however, studies on the capacitive gating of ion channels and on electroporation on planar stimulation contacts are required. In the near future, we are faced with important steps of optimization: (i) On the side of the semiconductor, the capacitance of the stimulation contacts must be enhanced and the noise of the transistors has to be lowered. (ii) On the side of the neurons, the structural and electrical properties of the cell membrane (glyocalix, ion channels) in the neuron-semiconductor junction must be studied and optimized by recombinant methods.

With respect to systems of neuronal networks and digital microelectronics, we are in a rather elementary stage. Two directions may be envisaged in the future: (i) Small defined networks of neurons from invertebrates and mammals must be created with learning chemical synapses and with a defined topology of synaptic connections. (ii) Large neuronal nets may be grown on closely packed arrays of two-way interface contacts, such that the rearranging structure and dynamics of the net is under continuous control of the chip. An adaptation of the industrial standard of *CMOS* technology will be crucial.

The interfacing of brain slices is in its infancy. In the near future, the two-way interfacing of groups of neurons in a tissue has to be studied in detail. Then two directions must be followed: (i) Arrays of two-way contacts will lead to a complete spatiotemporal mapping of brain dynamics. (ii) Learning networks on a chip will be implemented and will allow systematic studies of memory formation.

The availability of involved integrated neuroelectronic systems will help to unravel the nature of information processing in neuronal networks and will give rise to new and fascinating physical-biological-computational questions. Of course, visionary dreams of bioelectronic neurocomputers and microelectronics neuroprostheses are unavoidable and exciting, but they should not obscure the numerous practical problems.

Acknowledgements

The work reported in this chapter was possible only with the cooperation of numerous students who contributed with their skill and enthusiasm. Most valuable was also the help of Bernt Müller (Institute for Microelectronics, Technical University Berlin) who fabricated several chips and gave his advice for our own chip technology. The neurochip project was generously supported by the University Ulm, the Fonds der Chemischen Industrie, the Deutsche Forschungsgemeinschaft, the Max-Planck-Gesellschaft and the Bundesministerium für Bildung und Forschung. The editor acknowledges Christiane Hofer (RWTH Aachen) for checking the comprehensive set of symbols and formulas in this chapter. Furthermore, he would like to thank Matthias Schindler (RWTH Aachen) for checking the clarity and consistency of this chapter.

References

[1] P. Fromherz, 20th Winterseminar "Molecules, Memory and Information", Klosters 1985. (Available at http://www.biochem.mpg.de/mnphys/.)
[2] P. Fromherz, A. Offenhäusser, T. Vetter, J. Weis, Science **252**, 1290 (1991).
[3] P. Fromherz, A. Stett, Phys. Rev. Lett. **75**, 1670 (1995).
[4] A. Lambacher, P. Fromherz, Appl. Phys. A **63**, 207 (1996).
[5] A. Lambacher, P. Fromherz, J. Opt. Soc. Am. B**19**, 1435 (2002).
[6] D. Braun, P. Fromherz, Appl. Phys. A **65**, 341 (1997).
[7] D. Braun, P. Fromherz, Phys. Rev. Lett. **81**, 5241 (1998).
[8] Y. Iwanaga, D. Braun, P. Fromherz, Eur. Biophys. J. **30**, 17 (2001).
[9] P. Fromherz, C.O. Müller, R. Weis, Phys. Rev. Lett. **71**, 4079 (1993).
[10] R. Weis, P. Fromherz, Phys. Rev.E **55**, 877 (1997).
[11] P. Fromherz, unpublished results.
[12] V. Kiessling, P. Fromherz, unpublished results.
[13] Biophys. J. **87**, 631 (2004)
[14] R. Weis, B. Müller, P. Fromherz, Phys. Rev. Lett. **76**, 327 (1996).
[15] V. Kiessling, B. Müller, P. Fromherz, Langmuir **16,** 3517 (2000) .
[16] P. Fromherz, V. Kiessling, K. Kottig, G. Zeck, Appl. Phys. A **69**, 571 (1999).
[17] Langmuir **19**, 1580 (2003)
[18] J. Phys. Chem. B **107**, 2445 (2004)
[19] H. Ephardt, P. Fromherz, J. Phys. Chem. **97,** 4540 (1993).
[20] A. Lambacher, P. Fromherz, J. Phys. Chem. **105**, 343 (2001).
[21] D. Braun, P. Fromherz, Phys. Rev. Lett. **86**, 2905 (2001).
[22] S. Vassanelli, P. Fromherz, J. Neurosci. **19**, 6767 (1999).
[23] B. Straub, E. Meyer, P. Fromherz, Nature Biotech. **19**, 121 (2001).
[24] P. Fromherz, Eur. Biophys. J. **28**, 254 (1999).
[25] M. Jenkner, P. Fromherz, Phys. Rev. Lett. **79**, 4705 (1997).
[26] R. Schätzthauer, P. Fromherz, Eur. J. Neurosci. **10**, 1956 (1998).
[27] S. Vassanelli, P. Fromherz, Appl. Phys. A **65**, 85 (1997).
[28] C. Figger, A. Stett, P. Fromherz, unpublished results.
[29] S. Vassanelli, P. Fromherz, Appl. Phys. A **66**, 459 (1998).
[30] A. Stett, B. Müller, P. Fromherz, Phys.Rev.E **55**, 1779 (1997).
[31] M. Ulbrich, P. Fromherz, Adv. Mater. **13**, 344 (2001).
[32] P. Bonifazi, P. Fromherz, Adv.Mater. (2002) in press.
[33] A. A. Prinz, P. Fromherz, Biol. Cybern. **82**, L1-L5 (2000).
[34] M. Jenkner, B. Müller, P. Fromherz, Biol. Cybern. **84**, 239 (2001).
[35] G. Zeck, P. Fromherz, Proc. Natl. Acad. Sci. USA **98**, 10457 (2001).
[36] P. Fromherz, H. Schaden, T. Vetter, Neurosci. Lett. **129**, 77 (1991).
[37] P. Fromherz, H. Schaden, Eur. J. Neurosci. **6**, 1500 (1994).
[38] M. Merz. P. Fromherz, Adv. Mater. **14**, 141 (2002).
[39] S. Dertinger, P. Fromherz, unpublished results.
[40] P. Fromherz, Eur. Biophys. J. **31**, 228 (2002).
[41] B. Besl, P. Fromherz, Eur. J. Neurosci. **15**, 999 (2002).
[42] P. Fromherz, Ber. Bunsenges. Phys. Chem. **100**, 1093 (1996).

[43] P. Fromherz, in: *Electrochemical Microsystems Technology*,. Eds. J.W.Schultze, T.Osaka, M.Datta., Taylor and Francis, London, 2002.
[44] P. Fromherz, Physik. Blätter **57**, 43 (2001).
[45] P. Fromherz, ChemPhysChem **3**, 276 (2002).

Sensor Arrays and Imaging Systems

Contents of Part VII

Introduction to Part VII	809
33 Optical 3-D Time-of-Flight Imaging System	817
34 Pyroelectric Detector Arrays for IR Imaging	829
35 Electronic Noses	847
36 2-D Tactile Sensors and Tactile Sensor Arrays	861

Introduction to Part VII

Contents

1 **Classification and Physical Principles of Sensors**   809
2 **Electronic Sensor Arrays**   811
3 **Biological Sensor Arrays**   812
    3.1 Visual Sense   812
    3.2 Olfactory Sense   814
    3.3 Sense of Touch   815

1 Classification and Physical Principles of Sensors

Transducers establish a class of components, which convert signals of one kind of energy (e. g., thermal, mechanical, magnetic) into signals of another type of energy (e. g., electrical, optical, mechanical). For a classification, one can distinguish between six classes of input or output signals [8]:

    mechanical, **thermal**, **electrical**, **magnetic**, **radiant**, **chemical**

The particular types of transducers that output an electrical signal, are called **electronic sensors** or simply sensors. They pick up information about physical or chemical quantities from the outside world and forward a correlated electronic quantity (voltage, current or charge) into a system for information processing, transmission, and/or storage.

Sensors can be classified into two basic groups, self-generating elements and modulating elements. *Self-generating sensors*, sometimes also denoted as *active sensors*, directly transform a non-electrical energy (thermal energy, chemical energy, etc.) into electrical energy. Typical examples are thermocouples based on the Seebeck effect, Nernst type oxygen sensors, and piezoelectric pressure sensors. In all cases, the non-electrical energy is used to separate charges, i. e. to generate electrical energy. In *modulating* or *passive sensors*, the non-electrical energy modulates an electric signal. Typical examples are thermoresistive sensors, ion-sensitive field effect transistors, and piezoresistive pressure sensors. In these cases, an electrical current conducted through the element is modulated by the non-electrical energy. A classification taking into account the nature of the input, modulating and output energy is illustrated in Figure 1 [1]. The disposition in the *sensor effect cube* allows a finer subdivision between different transducer principles. A compilation of physical and chemical transducer principles linking the input quantity with the output quantity is given in Figure 2 [10].

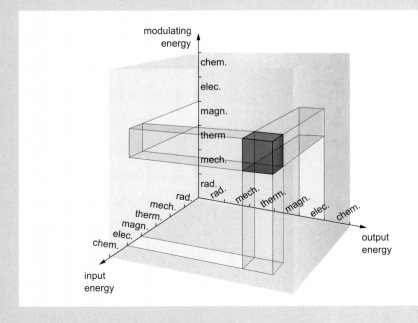

Figure 1: Middlehoek-Noorlag diagram for classifying transducers [1]. As an example, a resistive sensor for magnetic fields is shown (input energy: electrical, output energy: electrical, modulating energy: magnetic).

VII Sensor Arrays and Imaging Systems

According to the given definition, electronic sensors represent a subgroup of transducers. In Figure 2, they are highlighted in the white column. Since a large number of sensor principles exists at present and new sensors and sensing principles are being developed, only a collection of some effects can be given here. Comprehensive surveys on electronic sensors are compiled, for example, in Refs. [2] and [3].

Output Signal / Input Signal	Mechanical	Thermal	Electrical	Magnetic	Radiant	Chemical
Mechanical	• (Fluid) Mechanical and Acoustic Effects e.g. Diaphragm, Gravity Balance, Echo Sounder	• Friction Effects • Cooling Effects	• Piezoelectricity • Piezoresistivity • Resistive, Capacitive and Inductive Effects	• Magnetomechanical Effects e.g. Piezomagnetic Effect	• Photoelastic Systems • Interferometers • Sagnac Effect • Doppler Effect	
Thermal	• Thermal Expansion • Radiometer Effect		• Seebeck Effect • Thermoresistance • Pyroelectricity • Thermal (Johnsen) Noise		• Thermooptical Effects • Radiant Effects	• Reaction Activity e.g. Thermal Dissociation
Electrical	• Electrokinetic and –mechanical Effects: e.g. Piezoelectricity, Electrometer Ampere's Law	• Joule (Resistive) Heating • Peltier Effect	• Charge Collectors • Langmuir Probe	• Biot-Savart's Law	• Electrooptical Effects, e.g. Kerr-Effect, Pockels Effect, Electroluminescense	• Electrolysis • Electromigration
Magnetic	• Magnetomechanical Effects: e.g. Magnetostriction, Magnetometer	• Thermomagnetic Effects • Galvanomagnetic Effects	• Thermomagnetic Effects e.g. Ettingshausen -Nernst Effect • Galvanomagnetic Effects, e.g. Hall Effect Magnetoresistance		• Magnetooptical Effects, e.g. Faraday Effect Cotton-Mouton Effect	
Radiant	• Radiation Pressure	• Bolometer • Thermopile	• Photoelectric Effects, e.g. Photovoltaic Effect, Photoconductive Effect		• Photorefractive Effects • Optical Bistability	• Photosynthesis, -dissociation
Chemical	• Hygrometer • Electrodeposition Cell • Photoacoustic Effect	• Calorimeter • Thermal Conductivity Cell	• Potentiometry, Conductimetry, Amperometry • Flame Ionization • Volta Effect • Gas Sensitive Field Effect	• Nuclear Magnetic Resonance	• Emission and Absorption Spectroscopy • Chemiluminescense	

Figure 2: Table of transducer principles. Electronic sensors are highlighted.

2 Electronic Sensor Arrays

To scan a certain area with a single sensor, the detector has to be moved to any position, where data acquisition is desired. Consequently, sequential recording and a realization with mechanical elements is necessary. The requirement of fast acquisition of information (parallel data recording), positioning problems (precision), the aim of a non-mechanical device realization (cost reduction), but also the need for different kinds of sensing elements integrated on a single chip promote the use of sensor arrays instead of single sensor devices. Two kinds of sensor arrays can be distinguished:

First, sensor arrays with identical kinds of sensing elements: Here, fast data acquisition is made possible by the parallel recording of a spatially resolved signal of one kind. Well defined distances between the sensing elements on the sensor array guarantee a minimal linear distortion of the recovered signal. The spatial resolution of the sensor is limited by the number of sensing elements. The sensitivity of the sensor is proportional to the area of the sensing element. Using an identical fabrication process, an increase of resolution is therefore usually accompanied with a decrease of sensitivity.

A typical example of this type of sensor array consisting of a large number of identical elements is the *Charge Coupled Device* (CCD) sensor used for digital imaging. Todays CCD sensor array generations found in commercial digital cameras consists of up to more than 10 million identical light sensitive elements. Each of these elements might be realized by a MOS-structure. A sketch of the structure and the operating principle of a single MOS element is given in Figure 3 [9]. Electron-hole pairs created by absorption of incoming photons are separated by an internal electric field formed by the space charge region of the MOS element. According to the sign of the field, holes are driven towards the interior of the element, where charge neutrality is restored by compensation with electrons from the substrate electrode. Affected by the inner field, electrons migrate towards the semiconductor oxide interface. The information is stored in the amount of negative charge that is created during the illumination period and collected in the near-interface potential well. Since all sensing elements are exposed to light at the same time, the data recording is a parallel process.

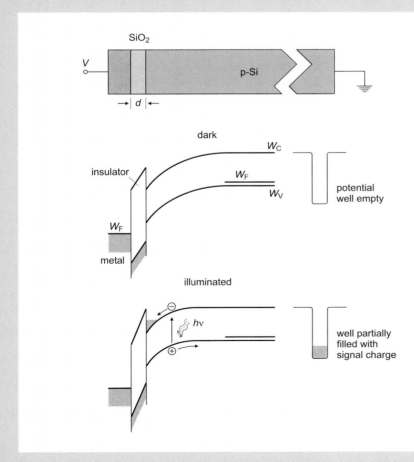

Figure 3: MOS structure and band diagram of a single CCD element. Electron hole pairs created by photon collisions with valence electrons are separated by the inner electrical field in the space charge region. The optical information in form of negative charge is stored in a potential well at the oxide semiconductor interface.

Figure 4: Principle of the charge shift to an adjacent potential well. The right barrier height of the potential well is lowered by a positive voltage applied to the right gate electrode.

In contrast, reading out the stored information is made possible by a combination of parallel and serial processes. Applying a defined voltage sequence to adjacent gates, the stored charge can be shifted by modulation of the height of the potential barrier on one side of the potential well [9]. Figure 4 illustrates this shifting process. Applied to each line of the sensor array, a parallel shift of the charge of all lines across the sensor chip is performed. In the last row, the line information is read-out by an output amplifier after serial shifting the line elements. Sensitivity to colors is made possible by the use of color filters or rotating color wheels. Advanced concepts to improve signal quality are antiblooming, antivignetting, shuttering, the use of lenslets or frontside and backside illumination.

Most sensor arrays are built up of similar sensing elements. In the framework of this book, a 3-D optical imaging technique (Chap. 33), a pyroelectric sensor array for IR imaging (Chap. 34), as well as a tactile sensor array (Chap. 36) are discussed in detail as further examples.

Second, arrays with different kinds of sensing elements: They can be characterized by simultaneous sensing of different quantities. For example, that sensor array converts a chemical information into an electrical signal. Here, it is essential to analyze the chemical composition of a complex environment, consisting of a large number of different chemical compounds. The sensor array consists of different sensor elements, where each sensor is ideally just sensitive to one substance or one group of substances. In practice, each sensing area of an array consisting of identical sensor elements is coated with an individual reactive surface to obtain a selective sensitivity. In contrast to the first type of sensor arrays, where a spatial resolution of the signal is essential, the recognition of a large number of different substances is desired in this case.

As a rule, calibration of chemical sensor arrays is needed. In the simplest case of a linear sensor response of each sensing element, calibration constants have to be determined. Due to the nature of chemical reactions, most sensor arrays obey a non-linear sensor response. For data acquisition, information can only to be obtained by the use of look-up tables. Additionally, cross sensitivities and stability problems caused by long term drift or contamination require complex analysis and limit the number of detectable substances of the sensor array. The last two aspects can be regarded as current challenges for research and development. Electronic noses as one example of sensor arrays with different sensor elements are introduced in Chap. 35.

3 Biological Sensor Arrays

For a comparison with the technical sensor arrays covered in this Part, a brief description of three types of biological sensor arrays of vertebrates will be given: the visual sense, the olfactory sense, and the sense of touch. These and all other types of biological senses (such as hearing, balance, taste) rely on a basic set of sensory neurons, such as photoreceptors, chemoreceptors, and mechanoreceptors.

3.1 Visual Sense

Photoreceptors represent the innermost layer of the **retina** (Figure 5). There are approx. 120 million *rods* and 6 million *cones*. The rods are 30 – 100 times more light sensitive then the cones and – when adapted to darkness – may be stimulated by a single photon. The cones come in three types, being sensitive to light of short, intermediate, and long wavelengths and, hence, make it possible to distinguish colors. Rods and cones are dis-

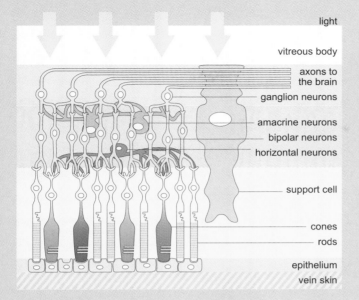

Figure 5: Schematic cross-section of the retina. The photoreceptors are located at the pigment cell layer on the inner part of the eye. The first layer of synapses connects to the bipolar cells. The second layer of synapses connects to the ganglion cells. Within the layers of synapses there are horizontal cells and amacrine cells, which perform additional lateral interconnections in the layers. Only the ganglion cells create action potentials. (Also shown: supporting cells.)

tributed differently across the retina. While the center of the retina consists almost exclusively of cones, the periphery contains primarily rods. The overall dynamic intensity range of the human vision is approximately 10^{12}.

A layer of bipolar cells connects the photoreceptors with approximately 1 million ganglion cells; the axons form the optic nerve that projects to higher centers of the brain. A ganglion cell in the visual center is typically connected to a single cone, via an appropriate bipolar cell, resulting in a very high spatial resolution. In the outer area, more than a hundred rods are connected via bipolar cells to one ganglion cell. This increases the light sensitivity, but decreases the spatial resolution. The area of photoreceptors connected to a ganglion cell is called the receptive field of this ganglion cell. Because of lateral connections, the receptive fields of neighbouring ganglion cells overlap.

The absorption of light in the photoreceptor causes a graded change of the membrane potential of the cell. The light-sensitive pigment of rods is called **rhodopsin** (or: visual purple), which is a molecule consisting of the chromophore *retinal* and the protein *opsin*. Upon excitation by light, the retinal undergoes an isomerisation reaction from the 11-*cis*-form to the all-*trans*-form (Figure 6). The excited rhodopsin triggers a cascade of enzymatic reactions which, on a time scale of approx. 10 ms, leads to the closing of ion channels, which are open in the dark. The closing of ion channels result in a hyperpolarisation of the membrane potential. This signal is transferred to bipolar cells through synapses. The amplification factor of this photoelectrical transduction process is high, since the photo-excitation of one rhodopsin causes a large number of ion channels to close. The retinal is restored to its original form through a complex sequence of chemical reactions which include detaching the all-trans-retinal from the opsin, transport, redox reactions, and re-attachment of the 11-cis-retinal to opsin. The overall regeneration process takes several seconds to minutes. This long duration is one of the reasons for the slow dark adaption of the eye. The cones also contain rhodopsin, which is different for the three different light–sensitive cone photoreceptors.

The excitation conveyed to the ganglion cell of a given receptive field may be excitatory in the center of the field and inhibitory around the center area, or it may be vice versa. In any case, this mechanism leads to a high sensitivity to detect sharp contrast.

In many cases, blindness is caused by defective photoreceptors, while the ganglion cells are still operational. In these cases, retina implants are developed which are based on a miniaturized camera outside the eye, e. g. worn in glasses, and a chip implanted onto the retina. The image information as well as energy for operating the **retina implant** are transmitted to the eye and onto the implant by light and microwave radiation, respectively. Within the retina implant the information is sent to an electrode array which stimulates the ganglion cells within a certain area of the retina. Since the artifical image created by this technique is very different to the natural image, image processing (outside the eye) as well as a learning phase of the patient is required to regain some sight. At the publishing date of this textbook, this area is still in the early stages of development. Especially, life time and microelectronic / cell contact issues need to be addressed. For more details, the reader is referred to Ref. [4].

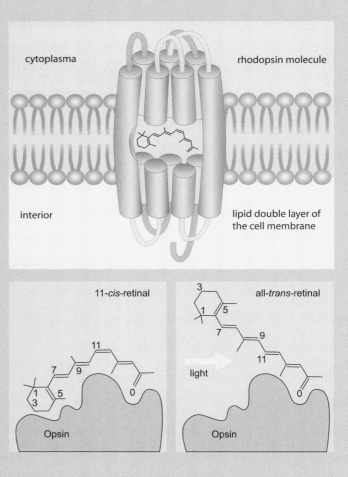

Figure 6: Structure of rhodopsin embedded in the membrane (top) and of the retinal in the non-excited and illuminated form (bottom).

3.2 Olfactory Sense

The human nose comprises approx. 10 million chemoreceptor cells, the olfactory sensory neurons, spread over an area of 3 to 5 cm^2. These cells, together with supporting cells etc. are embedded in a cell layer, called epithelium (Figure 7). This cell layer is covered by a mucous layer which is actively kept wet and is exposed to the air stream from which odours are to be detected. An olfactory neuron has a single dendrite which extends to the apical surface of the epithelium. Ten to fifty thin fibers, called olfactory cilia, are located at the tip of the dendrite and extend into the mucous layer. The membrane of the cilia contains proteins which bind odour molecules (odorants). The binding of an odorant to the corresponding protein triggers a chain of enzymatic reactions which leads to a depolarization of the olfactory neuron and eventually generates an action potential. Its axon transmits the signal to a so-called glomerulus in the olfactory bulb. These glomeruli collect signals from the olfactory neurons which specifically express one kind of odorant receptor. From the glomerulus, the signal is distributed within another layer of interconnecting neurons (so-called mitral cells) and then guided to central areas of the brain.

More details about the molecular mechanism of odorant receptors and a comparison to photoreceptors can be found in Ref. [6].

The human nose comprises approx. 300 different types of olfactory neurons. There are a few thousand cells of each type, connected to the same glomerulus, of which approx. 2000 are contained in the olfactory bulb. The reason why much more than 300 odorants can be distinguished (up to several thousands by trained humans) lies in the fact, that the neural processes in the olfactory region of the brain are working with weighted signals from several olfactory neurons which are typically activated for a given smell. This principle is much the same as that used in electronic noses (Chap. 35).

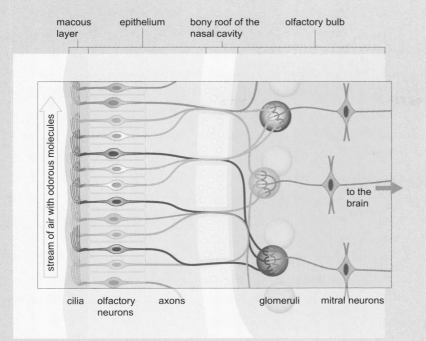

Figure 7: Schematic cross section of the olfactory nervous system of vertebrates. The cilia of the olfactory neurons float in the wet mucous top layer of the epithelium. Olfactory neurons which are specific for a given type of odorant molecule are sketched in color. This illustrates, that olfactory neurons of the same specifity extend to the same glomerulus.

3.3 Sense of Touch

The sense of touch in the skin is mediated by mechanoreceptors. Together with other sensory neurons (thermoreceptors, nociceptors, etc.), they are distributed over the entire surface of the body, at very different area densities. The highest density of 17000 cm^{-2} is encountered at the fingertips [7], while at the lower arm, for example, the density is decreased to approx. 20 cm^{-2}. There are six different types of mechanoreceptors, at different depths below the skin surface. Some types are specialized for the glabrous skin (Figure 8), others innervate the hairy skin, and react to different kinds of touch. The free receptors (so-called Ruffini endings), for example, penetrate the entire dermis and detect the intensity and duration of a touch. The Pacinian corpuscles are located in the deeper parts of the dermis and are specifically sensitive to vibration. The Meissner corpuscles are relatively near the surface and provide information about the velocity of touch. The Merkel´s disk are found in the epidermis and react on pressure.

The biomolecular mechanism of mechanoreceptors is based on (fast) direct effects of the mechanical stress on the opening of ion channels and, consequently, on a change of the membrane potential. A sufficiently large stimulus triggers action potentials, which are sent to the central nervous system.

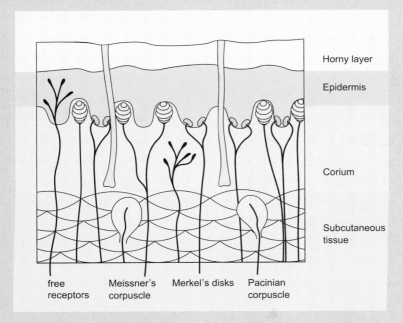

Figure 8: Schematic cross section of the glabrous skin embedding different types of tactile sensors.

Acknowledgements

The editor gratefully acknowledges Benjamin Kaupp (FZ Juelich) for carefully reviewing and correcting Section 3 as well as René Meyer (RWTH Aachen) for writing Section 2 and assisting in the compilation of this overview.

References

[1] S. Middlehoek and D. W. Noorlag, Sensors and Actuators, **2**, 29 (1981).
[2] W. Gopel, J. Hesse, and J. N. Zemel, Sensors, (9 volumes), John Wiley and Sons, 1995.
[3] J. Fraden, *Handbook of Modern Sensors: Physics, Designs, and Applications*, Springer, 2nd ed., 1997.
[4] J.-U. Meyer, Digest of Techn. Papers Transducer '01, Munich, 2001.
[5] E. R. Kandel, J. H. Schwartz, and T. M. Jessell, *Principles of Neural Science*, 4th ed., McGraw-Hill/Appleton & Lange, 2000.
[6] U. B. Kaupp and R. Seifert, Physiol. Rev., **82**, 769 (2002).
[7] R. S. Johansson et al., Trends Neurosci., **56**, 550 (1984).
[8] S.M. Sze, *Physics of Semicanductor Devices*, John Wiley and Sons, 2nd ed. edition, 1981.
[9] K.S. Lion, *IEEE Transactions*, **IECI-16**, 2-5 (1969)
[10] T. Grandke, J. Hesse in Sensors, A Comprehensive Survey, ed. by W. Göpel, J. Hesse, J.N. Zemel Vol. 1, Fundamentals and General Aspects, VCH Verlagsgesellschaft mbH, 1989.

Optical 3-D Time-of-Flight Imaging System

Bedrich J. Hosticka, *Werner Brockherde*, and *Ralf Jeremias*
Fraunhofer Institute of Microelectronic Circuits and Systems, Duisburg, Germany

Contents

1 Introduction — 819
2 Taxonomy of Optical 3-D Techniques — 820
3 CMOS Imaging — 820
4 CMOS 3-D Time-of-Flight Image Sensor — 822
5 Application Examples — 826
6 Summary — 826

Optical 3-D Time-of-Flight Imaging System

1 Introduction

In numerous application areas there is a need for nontactile (i.e. noncontact) image acquisition of spatial objects. These are distinguished from their environment by differences in density and/or composition. In this chapter, we will limit ourselves to a specific imaging technique of condensed matter objects in a non-condensed environment (which is typically the ambient atmosphere).

The imaging techniques range from 1-dimensional acquisition (e.g. rangefinders) to 3-dimensional acquisitions (3-D imaging). They are usually either based on [1] – [3]

- sound waves (i.e. ultrasound) or
- electromagnetic waves (microwaves, infrared radiation, visible radiation).

Which technology is the best for each particular application depends on several factors, among others on object size and range, on material properties of the object and the medium, and the ambient conditions. An additional classification of the techniques concerns the source of the radiation used for image acquisition:

- active emitter techniques depend on the (usually modulated) emission of the radiation and the detection of the back-scattered waves or
- passive techniques are based on the detection of naturally emitted radiation.

The technique described in this chapter employs an active emitter.

Since **ultrasound** technologies employ the longest wavelength, they are quite robust with respect to adverse ambient conditions, e.g. dirty environment. However, they are affected by properties of the propagation medium (thus air temperature, air streams, and humidity affect the performance) and properties of the object surfaces and object size [4] – [5]. In addition, emitted beams are very wide. Hence, the directivity can be implemented only by using complex beamforming techniques, which require costly sensor arrays. Multiple reflections from near objects make it difficult to acquire more complex images and moving objects cause the well-known Doppler effect.

Microwave **radar** is frequently employed to image large objects (airplanes, ships, etc.) located at a considerable distance. At a close distance (< 100 m), however, its spatial resolution is rather poor and, hence, radar is here used only as proximity or velocity sensor.

Since **optical** technologies exhibit the shortest wavelength with respect to ultrasound and radar, they show a completely different behavior: they enable high directivity, but perform poorly in dirty environment. For those techniques, which use radiation sources, the range depends on emitted power, which is in practice usually limited by eye safety regulations. Microwave radar can be *located* somewhere in the middle: which technology is better depends on the specific application.

As far as resolution is concerned, optical technologies are clearly the winner due to their shortest wavelength. Hence, this is the only technology that can be used to investigate microscopic object geometry, e.g. roughness of surfaces. In this contribution, however, we will discuss only the acquisition of macroscopic geometry, i.e. shape of objects. For discussion of acquisition of 3-D microscopic geometry, the reader is referred elsewhere [6]. Also, we will concentrate on optical technologies only and omit the discussion of ultrasound and microwave radar technologies. Infrared imaging that employs uncoiled pyroelectric detectors is discussed in Chap. 34 of this book.

2 Taxonomy of Optical 3-D Techniques

Optical techniques for nontactile acquisition of macroscopic geometries can be divided into five main categories: triangulation, interferometry, time-of-flight methods, tomography, and shape from shading [7].

Triangulation methods are either based on active or passive approach [8] – [9]. Active triangulation evaluates beam geometry generated by a structured light source and reflected from an object using either point, line, or grid projection, while passive triangulation uses mostly stereoscopic vision. Autofocus technique, frequently used in still imaging cameras, also belongs to the passive triangulation category. Interferometry is either based on acquiring Moiré fringes or holography [6], [10]. Time-of-flight methods employ active sources (e.g. lasers) that are either pulsed or emit continuous-wave modulated signals and evaluate the delay or the phase of the reflected signal, respectively, which depends on object distance [11] – [13]. Tomography and shape from shading methods are trying to reconstruct information about the volume of objects [7]. While tomography relies on computerized 2-D slice analysis, shape from shading is based on reflectometry.

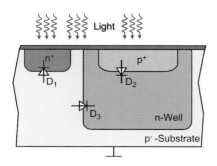

Figure 1: Cross-sectional view of a CMOS chip.

All these methods have their advantages and disadvantages. Active 3-D triangulation tends to be slow and expensive, as it requires mechanical scanning, while passive 3-D triangulation based on stereoscopic vision suffers from infamous correspondence problems. This problem stems from the fact that the depth information is extracted from the difference between two corresponding image coordinates. Nevertheless, it is difficult in practice to find such corresponding coordinates and ensure their uniqueness. In addition, each stereo camera can "see" only part of the image: this is called aperture problem. Interferometry, on the other hand, requires highly complex hardware. Tomography and shape from shading also require complex hardware and, in addition, abundant computing power and thus still elude low-cost implementation. Note that 3-D imaging acquisition for tomography and shape from shading also requires mechanical scanning around the object. Time-of-flight methods are often used for 1-D rangefinders, but classical 3-D imaging requires again mechanical scanning and is thus expensive. Nevertheless, at least two *scannerless* time-of-flight methods have appeared recently that have the potential to implement low-cost optical 3-D imagers. They employ silicon-based imagers, i.e. either CCD- or CMOS-based [14] – [16]. Especially, the use of CMOS imaging is of high interest as it enables a co-integration of electronics on the same chip as the 3-D image sensor then leading potentially to a single chip 3-D imager [17] – [18]. In contrast with this, the CCD technology does not have the potential to co-integrate electronics, it can realize solely the sensor array. Thus when using CMOS, only the active source *eludes* the monolithic integration. Hence ideally a low-cost 3-D camera implemented in CMOS technology would consist of only 2 parts: an active source and a single chip 3-D imager with extremely simple optics and mechanics.

In the following we will discuss one of these methods more in depth, namely a time-of-flight method based on a pulsed laser. In contrast with this, time-of-flight approaches based on continuous-wave modulated laser beam rely on synchronous demodulation of the beam that has been reflected from a distant object.

3 CMOS Imaging

The 3-D imager we shall demonstrate below relies on CMOS imaging which represents the cheapest imaging hardware today. This stems from the fact has already the most basic CMOS processes, which can be employed for monolithic integration of logic devices, and memories contain parasitic light-sensitive devices, such as pn-junctions. As imager signals are of analog nature it is advantageous, however, to use CMOS processes that offer the capability to handle analog signals: in practice, this means that the CMOS process to be used should have linear capacitors. The coexistence of imaging devices and analog and digital circuits on a single CMOS chip is implying that circuits carrying out readout, amplification, image processing, storage, and A/D-conversion can be now all integrated on a CMOS chip together with imaging, control, and interfacing. This feature means that assembly and packaging costs can be also greatly reduced while reliability is increased in rugged environments. Also, this concept is implying that a certain amount of computational intelligence can be located on the imager chip - this is then often called "smart" imager.

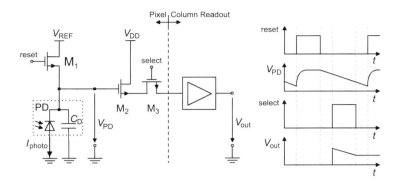

Figure 2: Typical three-transistor pixel circuit.

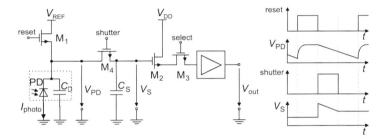

Figure 3: Pixel circuit with on-chip shutter.

As a matter of fact in most modern standard CMOS processes there is a variety of light-sensitive devices available such as pn-diodes, photogate devices, parasitic bipolar transistors, and MOS transistors [18]. While all these devices differ somewhat in their structure, sensitivity, and noise, they all exhibit sensitivity in the visible part of the light spectrum up to near-infra red (NIR) wavelength although the practical use is limited up to about 950 nm.

The simplest imaging device in CMOS technology is represented by the pn-diode formed either between a drain or a source diffusion and the substrate or a well, or between a well and the substrate. These diodes can be easily identified as D_1, D_2, and D_3, respectively, in Figure 1, which shows the cross-section of a CMOS chip. Obviously, CMOS imaging utilizes parasitic devices that are present in some or other form in all CMOS processes and, hence, no special processing steps for diode fabrication are needed. Let us now turn to the photodiode operation.

When illuminated, photons incident to light generate electron-hole pairs in the depleted space-charge region of the pn-junction which gives rise to electrical charge flow. The conversion into a voltage can be realized either using a continuous-time diode loading, a transimpedance amplification, or a discrete-time charge integration. The load and the transimpedance can be both, linear or nonlinear. The pn-diode, called now photodiode, together with the readout circuit forms a *picture element* or *pixel*. In image sensors the ratio of the light sensitive area to the total pixel area is called a fill factor. A low fill factor implies a low pixel photoresponsivity and a wide spatial frequency response. The latter can induce undersampling of fine structures in images. This phenomenon frequently referred to as *aliasing* is visible as Moiré patterns in the images and can degrade imager performance. The pixel spacing called *pixel pitch* determines the pixel count for a given lens size. Note that the pixel count basically defines the spatial resolution of the imager. Another important factor that greatly affects imager performance is the noise. Imagers can generate *noisy* images due to statistical fluctuations of their properties varying in time and space: in practice we speak of temporal *random noise* and spatial *fixed-pattern noise* (FPN), respectively. The noise then determines other important parameters such as signal-to-noise ratio (SNR) and dynamic range (DR) of the imager.

A typical pixel circuit with its clock timing is shown in Figure 2. This is a so-called *three-transistor* pixel circuit with a photodiode PD, formed e.g. by an n-diffusion in the p-substrate. The photodiode capacitance C_D is periodically charged to the reference voltage V_{REF} using the reset device M_1. The discharge voltage due to the integration of the photocurrent I_{photo} at C_D after a defined time (= integration time) now corresponds to the irradiance measured by the pixel. It is buffered by the source follower M_2 and it can be accessed by activating the select switch M_3 so that its value can be passed to the readout

VII Sensor Arrays and Imaging Systems

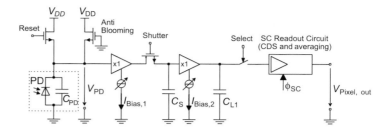

Figure 4: Pixel circuit diagram of the 3-D imager.

circuitry. Typically, an analog amplifier based on switched-capacitor (SC) technique is used for the readout of the pixel circuit. The pixel circuit of Figure 2 is not the only possible pixel circuit design as the CMOS technology allows realization of variety of different pixel structures.

Another CMOS pixel circuit is depicted in Figure 3. This circuit contains in addition a so-called shutter switch (transistor M_4) which controls the integration time of the photocurrent at the storage capacitor C_S. At the beginning of each integration the reset and the shutter switches are thrown on to precharge the capacitance C_D of the photodiode and the storage capacitance C_S to the reference voltage V_{REF}. After the reset switch has been thrown off the integration starts as the capacitances are discharged due to the photocurrent generated by the incident light. This process is stopped at C_S when the shutter switch M_4 is thrown open. The resulting voltage stored at C_S can be accessed via the source follower M_2 by activating the select switch M_3.

This pixel circuit can be used either to capture images containing fast motion, or – as in our application – it can be used together with an active illumination. In the latter case, the electronic shutter operation has to be synchronized with a pulsed active light source, e.g. emitting light in the near-infrared (NIR) range where it is invisible to human eye but silicon is still light sensitive. As the eye safety limit is basically defined by the energy received at the retina we can increase the light source power if sufficiently short pulses are used. This is a particular advantage when CMOS imagers are employed as they can be easily equipped with on-chip synchronous shutters operating even at speeds below 50 ns (note that the term "shutter speed" has been borrowed from photography and it simply means shutter opening time).

4 CMOS 3-D Time-of-Flight Image Sensor

The 3-D imager presented here is based on the time-of-flight method and relies on the measurement of elapsed time that emitted light pulses need to travel when reflected by an object. It has been pointed out that the classical time-of-flight realizations require 3-D laser scanners, i.e. combination of lasers, photodiodes, and rotating mirrors. The 3-D imager demonstrated here is completely *scannerless* and can be – with the exception of the light source – completely integrated on a single CMOS chip [15].

The imager chip contains a photodiode array, fast synchronous electronic shutter, analog switched-capacitor (SC) readout amplifier, multiple integration, clocking generator, pulse synchronization, and control. The circuit schematic of the pixel circuit is illustrated in Figure 4. The circuit operates similarly as the circuit in Figure 3, but there is an additional source follower circuit to buffer the photodiode. The resulting photodiode

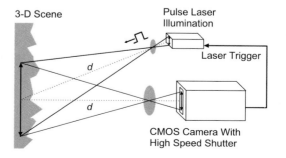

Figure 5: Principle of 3-D CMOS camera.

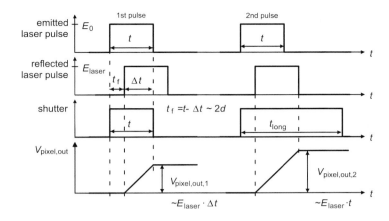

Figure 6: Pulse and shutter timing.

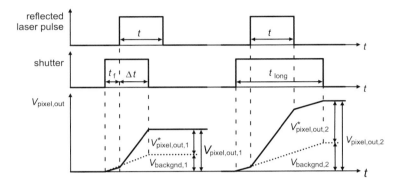

Figure 7: Effect of background illumination.

voltage stored at C_S is transferred to an inverting SC readout amplifier. Note that the pixel contains a so-called *antiblooming* MOS transistor. This ensures that the photodiode always operates in reverse-bias mode (a forward-biased photodiode could cause *blooming* effects known from CCD imagers).

The principle of the time-of-flight measurement with CMOS-cameras can be found in Figure 5. A few nanoseconds long light pulse generated by an NIR laser diode illuminates the entire imager field of view. This is accomplished by employing a simple optics to widespread (i.e. to defocus) the laser beam so that the entire field of view is illuminated. After the pulse has been emitted the electronic shutter switch is thrown on so that the photocurrent due to the reflected pulse can start discharging the photodiode capacitance C_{PD} after its arrival. The amount of the received light (and thus the amount of the discharge) depends on synchronous timing of the laser diode, reflectance of the object in the scene, travel time of the pulse (and, hence, the distance of the object in the field of view), and the shutter switch timing. Note that the received light contains also background illumination of the captured scene, not only the reflected laser pulse. When the shutter switch is switched off, the voltage at the storage capacitor C_S, remains *frozen* (see Figure 4). This voltage contains, among others, the information about range. In a second cycle this measurement is repeated at a different integration time t_{long}, i.e. the shutter switch timing is different, though the laser pulse length is the same. The second cycle is necessary in order to be able to separate the range information from the other factors, such as laser power, background illumination, and object reflectance. This is shown in the following calculations.

To start with, consider the first pulse in Figure 6, which shows the pulse and shutter timing for the circuit in Figure 4 (note that the reset and select switch timing have been omitted). The time t_f elapsed between the emission and reception of the pulse depends on the travel distance d as $t_f = 2d/c_0$ where c_0 is the velocity of light. This yields a pixel output voltage $V_{pixel,\,out,\,1} = R\,E_{laser}\,\Delta t$, where R is the sensor responsivity, E_{laser} is the irradiance measured at the sensor, and $\Delta t = t - t_f$. The shutter time is in the case the same as the pulse length t. Note that the amount of received light, i.e. E_{laser}, depends on emitted irradiance E_0, reflectance of the object, and its distance, if we disregard any background irradiance, e.g. due to effects of other light sources present (this will be considered below).

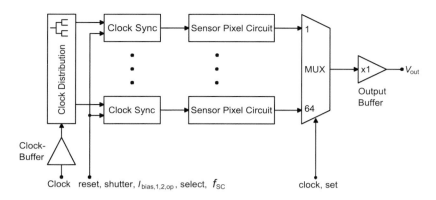

Figure 8: Block diagram of the 3-D imager.

After the first pulse a second pulse is fired and the above procedure is repeated, however, with shutter opening time greatly exceeding the pulse length t. This yields another pixel voltage, namely $V_{\text{pixel, out, 2}} = R\, E_{\text{laser}}\, t$. Forming a quotient from these two pixel voltages shows that the effects of E_{laser} and R can be completely eliminated, since

$$\eta = \frac{V_{\text{pixel,out,1}}}{V_{\text{pixel,out,2}}} = 1 - \frac{2d}{t \cdot c_0} \tag{1}$$

and, hence, the distance d is given by

$$d = \frac{t \cdot c_0}{2} \cdot (1 - \eta) \tag{2}$$

Figure 7 shows the same procedure, but this time including the effects of background illumination. Performing the two cycles as above but each time executing two measurements per cycle (i.e. one with and one without laser illumination) and subtracting these measurements, the effect of background illumination can be cancelled. We simply form the differential output pixel voltages $V^*_{\text{pixel, out, 1}} = R\,(E_{\text{laser}}\, \Delta t + E_{\text{backgnd}}\, t) - R\, E_{\text{backgnd}}\, t$ and $V^*_{\text{pixel, out, 2}} = R\,(E_{\text{laser}}\, t + E_{\text{backgnd}}\, t_{\text{long}}) - R\, E_{\text{backgnd}}\, t_{\text{long}}$. Creating the quotient $\eta = V^*_{\text{pixel, out, 1}} / V^*_{\text{pixel, out, 2}}$ yields exactly the same formula for the distance as above, so that the effect of background illumination which generates the voltage V_{backgnd} has no effect at all. Thus just two differential images are needed to form a 3-D image.

Note that this applies individually to all photodiodes if a photodiode array is used. To summarize our discussion we can state that the individual irradiance intensity measured at each pixel depends not only on the range of its grid point in the observed scene space, but also on the laser power, the background illumination, and on the object reflectance. However, with the above approach these effects are cancelled. Optionally, each of the two measurement cycles may be repeated n times using laser pulse bursts thus yielding multiple integration. The resulting voltages are accumulated in the SC readout circuit operated in accumulation mode. This allows signal-to-noise ratio improvement by sqrt(n) and extends the sensor dynamic range.

For laser pulses lasting tens of nanoseconds, a high power laser can be used while still remaining eye-safe under laser class 1 regulations. It is mandatory that the imager can be operated with as short shutter opening times as possible to minimize the effect of the background illumination for short laser pulses. Too long shutter times decrease the

Figure 9: 3-D imager chip micrograph.

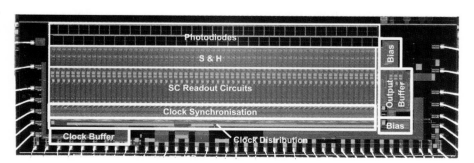

dynamic range of the imager because the background illumination generates large offsets. The most important sensor parameter is the light sensitivity of the photodiode array since the reflected laser light is low for distant objects. Therefore, in the design of the sensor, the noise analysis is of paramount importance, because the noise defines the smallest signal to be detected.

Shutter clocking is another important issue since shutter phase noise of 1 ps results in a 0.15 mm range inaccuracy. To obtain high-speed synchronous shutter clocking the imager employs synchronization circuit design to ensure minimum skew (see Figure 8 for block diagram of the sensor chip).

The CMOS 3-D imager described above has been fabricated in a standard 0.5 μm CMOS technology featuring p-substrate, n-well, and three metal and a single polysilicon layer [15]. The photodiodes are formed between the n-well and the p-substrate. The chip contains two linear photodiode arrays, each consisting of 32 pixels, readout electronics, and clock generation (see Figure 9). The chip area is 42 mm² including all peripherals and bonding pads, while the area of a single photodiode is $(260\ \mu m)^2$.

The chip has been tested using a camera system developed especially for this sensor [19]. It employs sensor and laser control as well as image preprocessing. Both shutter and accumulation times are user programmable.

Measurements of background light sensitivity show no observable dependence up to $\cong$ 10 klux. The sensitivity to reflectances of various objects is below 1 %. Due to noise optimization, the light sensitivity of the sensor, i.e. measured noise-equivalent-power NEP, is 5.2 W/m² at 30 ns shutter speed (technical data are given in Table 1).

Figure 10 depicts a calibrated distance measurement showing linearity error of < 5 % for different pulse counts. An acquired 3-D image exhibiting ± 1 cm accuracy is shown in the insert. As indicated above, two images ("capture 1" and "capture 2") are needed to form a 3-D image, as shown in Figure 11 [20]. As the photodiode array yields only two pixel points in one lateral dimension, linear mechanical scan has been used for 3-D image recording in the test system described here. The measurement range is up to 10 m, depending on optics used and number of pulses employed resolution required (see Table 1). An area imager that allows 3-D imaging without mechanical scan is under development.

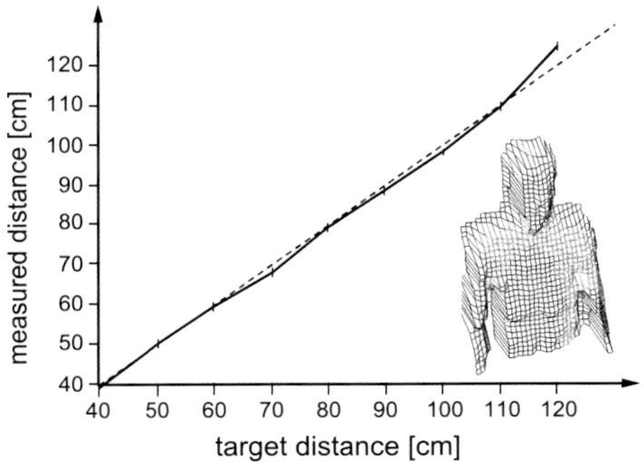

Figure 10: Calibrated distance measurement (insert: 3-D image).

Parameter	Data	Remark
Pixel count	32 x 2	
Power dissipation	330 mW	single 3.3 V supply
Max. shutter speed	30 ns	
Measurement range	> 5 m	depending on optics
Range resolution (1 pulse)	< 5 cm	
Range resolution (100 pulses)	< 1 cm	burst operation
Max. clock skew	< 100 ps	
Sensitivity (NEP)	5.2 W/m²	@ 30 ns shutter speed
Linearity	< 5 %	
Max. frame rate	20 000 frames/s	3-D images @ 66MHz clock
Frame rate @ 100 pulses	4 000 frames/s	3-D images @ 66MHz clock
Frame rate @ 100 pulses	4 000 frames/s	3-D images @ 66MHz clock

Table 1: Technical data of the 3-D imager.

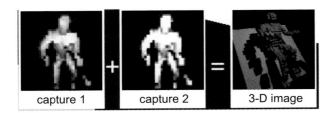

Figure 11: Formation of a 3-D image based on the two differential images (Courtesy of Siemens).

5 Application Examples

There are numerous applications for nontactile 3-D imaging. Examples include automotive, multimedia, medical, safety, security, surveillance, robotics, machine vision, visual inspection, and remote manipulation applications. In order to limit the scope of our discussion we will discuss only the first application, i.e. the automotive.

There are two areas where 3-D imaging can be used in a car: in the interior compartment and in the exterior area. The interior applications include, among others, seat occupancy, intrusion, child restraint, and passenger out-of-position detection. Head position detection is also useful, e.g. for drivers' fatigue recognition. Exterior applications cover parking control, precrash detection, "blind spot" observation, and obstacle and pedestrian detection. These applications require low-cost, easy to install and maintain 3-D cameras. These cameras must exhibit high dynamic range with respect to scene brightness in order to ensure reliable operation even under the most extreme driving and environment conditions.

6 Summary

In this contribution we have presented optical imaging systems for nontactile acquisition of 3-D objects. We have briefly discussed main technologies and different approaches to 3-D imaging. As an example, a simple CMOS 3-D camera has been presented and various applications have been discussed.

Acknowledgements

The 3-D imager has been developed in collaboration with Siemens AG, Munich, and Siemens VDO Automotive AG, Regensburg, both in Germany, which have also provided part of the funding. The editor would like to thank Stephan Tiedke (aixACCT Systems GmbH, Germany) for checking the clarity and consistency of this chapter. Furthermore, he would like to thank Peter Gerber (RWTH Aachen) for checking the clarity and consistency of this chapter.

References

[1] J. Fraden, *AIP Handbook of Modern Sensors*, American Inst. of Physics, New York, 1993.
[2] I. Skolnik, Proc. of the IEEE **73**, 182 (1985).
[3] G. R. Osche and D. S. Young, Proc. IEEE **84**, 103 (1996).
[4] G. Lawitzky and W. Feiten, KI: Zeitschrift des Fachbereiches 1 (Künstliche Intelligenz) der GI, 1994.
[5] R. Dillmann, F. Wallner, and F. Wechesser, Proc. of the 2nd Conf. of Mechatronics and Robotics, Duisburg/Moers (Germany), 1993.
[6] H. Höfler and M. Seib; *Sensors-A Comprehensive Survey,* eds. W. Göpel, I. Hesse, and J. N. Zemel, Vol. 6, VCH-Weinheim, 1992.
[7] H. Maître, *Prof. EUSIPCO-88, Signal Processing IV: Theories and Applications,* eds. J. L. Lacoume, A. Chehikian, N. Martin, and J. Malbos, Elsevier Science Publ. (North-Holland), 1988.
[8] A. Gruss, L. R. Carley, and T. Kanade, IEEE Journal of Solid-State Circuits **26**, 184 (1991).
[9] J. Kramer, P. Seitz, and H. Baltes, Sensors and Actuators A**31**, 241 (1992).
[10] J. W. Goodman, Proceedings of the IEEE **59**, 1292 (1971).
[11] R. J. Wangler and R. A. Olson, Laser Focus World, 105 (1993).
[12] R. Miyagawa and T. Kanade, IEEE Journal of Electron Devices **44**, 1648 (1997).
[13] K. Osugi, K. Miyauchi, N. Furumi, and H. Miyakoski, JSAE Review **20**, 549 (1999).
[14] R. Schwarte, B. Buxbaum, H. Heinol, Z. Xu, J. Schulte, H. Riedel, P. Steiner, M. Schever, B. Schneider, and T. Ringbeck, Proc. Advanced Microsystems for Automotive Applications, Berlin, 2000.
[15] R. Jeremias, W. Brockherde, G. Doemens, B. Hosticka, L. Listl, and P. Mengel, Digest of Techn. Papers IEEE International Solid-State Circuits Conf. 2001, San Francisco, 2001.
[16] R. Lange and P. Seitz, IEEE Journal of Quantum Electronics **37**, 390 (2001).
[17] E. R. Fossum, Digest of Techn. Papers IEEE International Electron Devices Meeting, Washington, 1995.
[18] M. Schanz, W. Brockherde, R. Hauschild, B. J. Hosticka, and M. Schwarz, IEEE Trans. on Electron Devices **44**, 1699 (1997).
[19] P. Mengel, G. Doemens, and L. Listl, Proc. IEEE International Conference in Image Processing (ICIP2001), Thessaloniki (Greece), 2001.
[20] B. Röthlein, New World – Das Siemens Magazin, 2001.

Pyroelectric Detector Arrays for IR Imaging

Paul Muralt
Swiss Federal Institute of Technology EPFL, Lausanne, Switzerland

Howard R. Beratan
Raytheon Systems, Texas, USA

Contents

1 Introduction   831

2 Operation Principle of Pyroelectric IR Detectors   831
2.1 The Pyroelectric Response   832
2.2 Noise and Detectivity   834
2.3 Numerical Values   836
2.4 Thermal Wavelength Effects   836
2.5 Characterization of Focal Plane Arrays for Thermal Imaging   837

3 Pyroelectric Materials   838

4 Realized Devices, Characterization, and Processing Issues   839
4.1 Fabrication Technology   839
4.2 Read-Out Integrated Circuits (ROIC)   842

5 Summary   844

Pyroelectric Detector Arrays for IR Imaging

1 Introduction

The radiation maximum of a black body at 300 K is in the infrared (IR) regime at a wavelength of approx. 10 μm. To obtain electronic images in the IR regime, two different physical principles are frequently employed:

1. quantum detectors turn absorbed photons into the generation of electrical carriers,
2. thermal detectors work on the temperature increase upon the absorption of radiation and a consequent electrical sensing of the temperature change.

Quantum detectors are based on the internal photoelectric effect, i.e. they are utilizing the electron-hole pair generation by absorbing photons in the space charge region of a semiconductor junction. The photon energy hc/λ (where h denotes Planck's constant, c is the velocity of light, and λ the wavelengths) needs to exceed the characteristic energy barrier ΔW_b for the carrier generation. For intrinsic semiconductors, which are frequently employed in IR detectors, ΔW_b is the band gap. For extrinsic semiconductors, it would be the ionization energy of a dopant. Or it can be the band gap between two sub-bands of a multiquantum well system. The most important quantum IR detector material for IR imaging arrays is $(Cd_{1-x}Hg_x)Te$. In this intrinsic compound semiconductor, the bandgap energy and hence the spectral range of the sensitivity can be tuned between from 1.5 eV ($\lambda = 0.8$ μm) for x = 0 (CdTe) to zero for x = 1 (HgTe). While quantum IR detectors exhibit high temperature resolution and can be fabricated with sensitivities almost up to the theoretical quantum efficiency, they show two strong disadvantages: firstly, their production is very costly, and secondly, they need to be cooled (e.g. to 77 K) during operation. For this reason, their spread has been limited mainly to night-vision systems in military applications [35], [15].

Figure 1: IR-image of a house with major thermal insulation damage [33].

However, there is a large potential for civilian and commercial applications ranging from the fire detection, night-driving assistance, surveillance, and building diagnosis. For this market, **thermal detectors** based on pyroelectric detector arrays are perfectly suited. They operate without cryogenic cooling and can be fabricated by standard MEMS and microelectronic technology. Figure 1 shows a pyroelectric infrared image, while typical infrared imaging cameras for civilian and commercial applications are shown in Figure 2.

Pyroelectricity is an extraordinarily sensitive principle to detect temperature changes. Although the first pyroelectric material was discovered some 2300 years ago by the greeks, it is not until the 1950's when pyroelectricity was also employed for technical use [34]. In the last decades, pyroelectric detectors have been mainly built as single element systems made from small bulk single crystals of pyroelectric materials. These sensors react to very slight but sudden changes of the temperature within their sight field. They are typically applied for intruder and fire alarm systems and, for example, as contactless lighting switches in buildings. In addition, they are used as temperature sensors for air-conditioning systems, for biomedical systems, flow meters, etc..

Figure 2: Commercial system for a police car (top), an automotive driving aid (center), and a general-purpose camera (PalmIR)(bottom) [32].

2 Operation Principle of Pyroelectric IR Detectors

The basic principle of pyroelectric materials is briefly described in Sec. 2.4 of Chap. 2. This section treats the device physics of pyroelectric infrared detectors, including some model calculations for typical thin film device geometries and dimensions used in IR imaging arrays. A general description of pyroelectric detectors can be found in reviews [1], [2] and textbooks [3] that have been written for single crystal and bulk ceramic devices. The aim of this chapter is to elucidate the specific behavior of thin film devices, which arises due to much different quantitative values, such as the heat capacity, as compared to bulk devices. A more comprehensive survey emphasizing specific topics on the thin film integration has been given in Ref. [4].

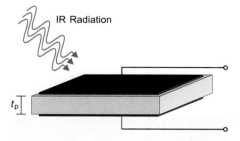

Figure 3: Schematic drawing of a pyroelectric element. Depending on the design and fabrication technique these are either bulk elements (typically $t_p > 30$ mm) or thin film elements (typically $t_p < 1$ mm), representing a pixel in an IR imaging array.

2.1 The Pyroelectric Response

The pyroelectric charge ΔQ generated by a temperature change ΔT of a pyroelectric element (see Figure 3) with electrode surface A amounts to

$$\Delta Q = p \cdot A \cdot \Delta T, \tag{1}$$

where p is the pyroelectric coefficient. For infrared radiation detection, ΔT is the result of the absorbed radiation power during a certain time interval. The thermal properties of the device have to be suitably optimized by design and meterials choice for obtaining a maximal ΔT at the required response frequency. Thin film devices exhibit small heat capacities. This has to be compensated by a good thermal insulation in order to obtain large enough thermal time constants. Micromachining is the main tool to reduce thermal conduction. Micromachining pays off especially well in case of arrays, for which the thermal crosstalk can be very much reduced. If the device is not operated in vacuum, air convection to the surrounding walls (housing, infrared optics) is also important. The temperature change is obtained from the balance of the heat flows, as sketched in Figure 4. First of all, some fraction η of the infra-red radiation (power P) falling onto the element is absorbed, i.e., transformed into heat. The temperature change at a given heat input $\eta \cdot P \cdot \Delta t$ depends on the heat conductance G to the surrounding heat reservoir (heat sink) at temperature T_0 and the heat capacity H of the element. Denoting by $\theta = T - T_0$ the temperature increase above the constant heat sink temperature T_0, the increase of the internal energy of the element by $H \cdot \Delta \theta$ is equal to the heat input ($\eta \cdot P \cdot \Delta t$) reduced by the amount of heat flowing to the heat sink of $G \cdot \theta \cdot \Delta t$. This yields the following differential equation (see, e.g., [3]):

$$G \cdot \theta + H \cdot \frac{\partial \theta}{\partial t} = \eta \cdot P \tag{2}$$

In the case of a stationary, sinusoidally modulated radiation power $P = P_\omega \cdot e^{i\omega t}$, the stationary solution is readily obtained, yielding a temperature modulation amplitude of:

$$\theta_\omega = \frac{\eta \cdot P_\omega}{G + i\omega \cdot H}, \quad |\theta_\omega| = \frac{\eta \cdot P_\omega}{G\sqrt{1 + \omega^2 \cdot \tau_{th}^2}} \tag{3}$$

The thermal time constant $\tau_{th} = H/G$ has been introduced. There are thus two frequency domains separated by the inverse thermal time constant. On the low frequency side, the temperature modulation amplitude is governed by the heat loss to the surrounding; on the high frequency side the heat capacity determines the thermal behavior. Combining Eqs. (1) and (3), the pyroelectric current with frequency ω is derived as:

$$J_\omega = p \cdot A \cdot \omega \cdot |\theta_\omega| \tag{4}$$

The current **responsivity** (i.e. the specific response) is the current per P of radiation power falling onto the detector element:

$$\Re_J(\omega) = \frac{J_\omega}{P_\omega} = \frac{p \cdot \eta \cdot \omega \cdot A}{G\sqrt{1 + \omega^2 \cdot \tau_{th}^2}} \tag{5}$$

Alternatively, it is also possible to measure the voltage across a parallel resistor R_p. In bulk devices, a value of 10 GΩ is typically applied (R_p should not exceed the gate impedance of the amplifier.) In thin film devices, it might be possible to avoid this parallel resistor, because thin film capacitors exhibit larger leakage currents than bulk capacitors. In the following we will consider R_p as an effective value, composed of the film dc resistance and the eventually mounted external parallel resistor. At higher frequencies the dielectric loss tangent $\tan\delta$ comes also into play. The conductance Y of the complete element (thus including the parallel resistance) is obtained as:

$$Y(\omega) = \frac{1}{R_p} + \omega C \cdot \tan\delta + i \cdot \omega C, \quad |Y(\omega)| = \frac{1}{R_p}\sqrt{(1 + \omega \tau_{el} \tan\delta)^2 + \omega^2 \cdot \tau_{el}^2} \tag{6}$$

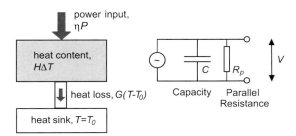

Figure 4: Schematic of thermal and electrical circuit of a pyroelectric element.

The electrical time constant $\tau_{el} = R_p \cdot C$ has been introduced. The voltage response is proportional to the impedance $Z = Y^{-1}$. Above the angular frequency τ_{el}^{-1}, the impedance decreases as ω^{-1}. The pyroelectric element thus works like a low pass filter, cutting off the high frequency response. Omitting the small term in $\tan\delta$, the complete voltage responsivity is written as:

$$\Re_V(\omega) = |Z(\omega)| \cdot \Re_J(\omega) = \frac{p \cdot \eta \cdot \omega \cdot A \cdot R_p}{G \cdot \sqrt{1 + \omega^2 \cdot \tau_{th}^2} \cdot \sqrt{1 + \omega^2 \cdot \tau_{el}^2}} \tag{7}$$

The general trends of the two responsivity functions can now easily be sketched as a function of the frequency, provided we know the relative size of the two time constants. In **bulk devices** one generally encounters $\tau_{el} < \tau_{th}$ [1], [41]. For thin film devices the opposite is true. The down scaling i.e. the transition from the bulk to the thin film element, consists essentially in a shrinkage of the thickness of the pyroelectric material. The parallel resistor is adapted to the input impedance of the amplifier and typically amounts to 10 GΩ for thin film and bulk materials. The heat conduction to the heat sink (surrounding) is difficult to decrease by the same amount as the thickness, because bulk pyroelectric materials are good thermal insulators. So we may conclude that the ratio of the time constants scales roughly as:

$$\frac{\tau_{el}}{\tau_{th}} = \frac{R_p C}{H/G} \propto \frac{C}{H} \propto (t_p)^{-2} \tag{8}$$

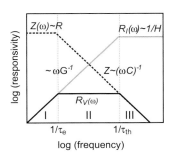

Figure 5: Log-log scheme of current and voltage response, together with the impedance for the typical thin film situation where the thermal time constant is much shorter than the electrical one. Thermal wave length effects are neglected (from ref [5], [6]).

This strong dependence on the element thickness t_p makes the ratio switch from less than one to larger to one when scaling down from a single crystal to a thin film.

The resultant frequency behavior is schematically shown in Figure 5. The voltage response for **thin film devices** may look similar as for bulk sensors. However, the time constants have the opposite order (i.e., $\tau_e > \tau_{th}$). As one consequence, the voltage response in this intermediate region (region II of the figure) is determined by other parameters. Approximative formulas for this region are obtained by the appropriate approximations in the voltage responsivity:

$$1 + \omega^2 \cdot \tau_{th}^2 = \begin{cases} 1 & \text{thin film} \\ \omega^2 \cdot \tau_{th}^2 & \text{bulk} \end{cases} \qquad 1 + \omega^2 \cdot \tau_{el}^2 = \begin{cases} \omega^2 \cdot \tau_{el}^2 & \text{thin film} \\ 1 & \text{bulk} \end{cases} \tag{9}$$

thin film:
$$R_\nu(II) \cong \frac{\eta \cdot p \cdot A}{C \cdot G}$$

bulk:
$$R_\nu \cong \frac{\eta \cdot p \cdot A \cdot R_p}{H}$$

In thin film devices the parallel resistance is thus not important for the voltage response. This is very advantageous, as the film resistivity is not very precisely controllable. The mounting of parallel resistors could thus be avoided. In case of the thin film sensor, Eq. (9) can be further modified by observing that $C = \varepsilon\varepsilon_0 A / t_p$:

$$R_\nu(II) \cong \frac{\eta \cdot p \cdot t_p}{\varepsilon\varepsilon_0 \cdot G} \tag{10}$$

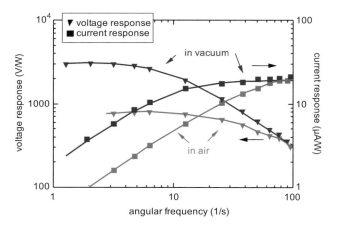

Figure 6: Voltage and current responsivity of a linear array element (1x12) measured in air and in vacuum. One element was 0.4 x 0.9 mm large. The curves are calculated, the points are measured [7].

This shows that the responsivity is essentially independent of the surface A. (The same can also be concluded for the bulk behavior, since the heat capacity is proportional to A.) An optimization of the voltage response requires rather thick films and a very good thermal insulation. The figure of merit of the pyroelectric thin film material for the voltage response is therefore:

$$F_V^* = \frac{p}{\varepsilon\varepsilon_0} \qquad (11)$$

This assumes that the limiting capacitance is provided by the material, and not the read-out circuit. In case of bulk detectors this value is often devided by the specific heat capacity: $F_V = p/(c_p\varepsilon\varepsilon_0)$. Although not relevant for thin films, it might be useful to use the latter figure of merit in order to be compatible with the more frequently applied definition.

The upper frequency region (III) appears to be the best for current detection. Applying the appropriate approximations for $\omega > 1/\tau_{th}$, and $\omega >> 1/\tau_e$, the current response is obtained as:

$$R_J(III) \cong \frac{\eta \cdot p \cdot A}{H} \qquad (12)$$

Here a small thickness, giving a small heat capacity, and a large surface seem to be the best choice. However, one has to keep in mind that the heat conduction has to be small enough for obeying $\omega > 1/\tau_{th}$. One observes again that the responsivity does not depend on A, if the design is such that the volume defining the heat capacity has the same area as the absorbing layer. If a device is scaled down in surface, the current *responsivity* does not change. (The current *response* may be reduced, because the power falling onto the element might decrease. In fact, if the optical part is maintained, the power density stays constant and the power falling onto one element, and hence also the response, decreases as A^{-1}).

The impact of the heat conductance G in region II, and its unimportance in region III, according to Eqs. (10) and (12), is nicely demonstrated in Figure 6, which depicts the response of a pyroelectric detector in air and in vacuum, i.e. once with air convection (large G), and once without air convection (small G).

2.2 Noise and Detectivity

The ultimate detection limit of a sensor is given by the intrinsic noise of the detector element. In order to reach this limit with the complete sensor set-up, the noise contribution from the amplifier and from external sources of the environment have to be smaller than the intrinsic noise. Using low-noise amplifiers like JFET transistors, the theoretical limit of the intrinsic noise can indeed be reached [9]. Pyroelectric sensors have two major contributions to the noise: thermal noise and Johnson noise [1], [8]. The thermal noise power

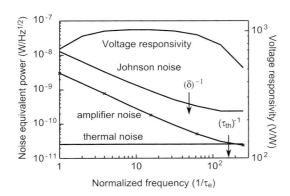

Figure 7: Calculated noise equivalent powers (NEP) of different sources for a 1 mm² element of PZT 15/85 in the frequency range between $1/\tau_e$ and $1/\tau_{th}$ (region II). The displayed frequency is normalized to $1/\tau_e$ [5], [6].

$$P_{thn} = \sqrt{4kT^2GB} \qquad (13)$$

is proportional to the square root of the heat conduction G, and is very small for optimized thin film devices. The Johnson noise is normally dominating. The contributing noise sources are the parallel resistance and the dielectric loss, i.e., the real part g of the admittance Y. The noise current J_n for a bandwidth B is obtained as [8]:

$$J_n = \sqrt{4kTgB} \quad \text{with} \quad g = R_p^{-1} + \omega C \tan\delta$$

$$J_n = \left(4kTB\left(R_p^{-1} + \omega C \tan\delta\right)\right)^{1/2} = \left(\frac{4kTB}{R_p}\left(1 + \omega\tau_{el}\tan\delta\right)\right)^{1/2} \qquad (14)$$

The minimal detectable power is the noise equivalent power defined by $NEP = J_n/\Re_J$. The different noise contributions to the NEP are shown in Figure 7. The amplifier noise was calculated according to typical J-FET performance.

As can be seen in Figure 7, the resistor noise (Johnson noise) – decreasing as ω^{-1} – yields the largest contribution in the frequency range II, before the dielectric noise (as $\omega^{-1/2}$) becomes dominant above $\omega > 1/(\tau_e \tan\delta)$. At the low frequency side of region II approximate formulas can thus be obtained by considering the resistor noise only. Very often, the detectivity D^* is given instead of the NEP:

$$D^* = \frac{A^{1/2}B^{1/2}}{NEP} = \frac{\Re_J(\omega) \cdot A^{1/2}B^{1/2}}{J_n} = \frac{p \cdot \eta \cdot \omega \cdot A^{3/2} \cdot R_p^{1/2}}{G\left(4kT \cdot \left(1 + \omega^2 \cdot t_{th}^2\right) \cdot \left(1 + \omega\tau_{el}\tan\delta\right)\right)^{1/2}} \qquad (15)$$

The multiplication with $A^{1/2}B^{1/2}$ cancels *area* and *bandwidth* contribution from the noise. As seen above, the current response in region III is independent of A. In this frequency range, D^* is thus a parameter that is independent of the arbitrary detector surface and detection bandwidth. Despite of a limited universality (no imaging properties are in fact measured for deriving D^*), D^* is frequently used as a measure of performance to compare different detectors.

The noise introduces a third time constant: $\tau_{el}\tan\delta$. Typically, this value should lie between τ_{th} and τ_{el}. We thus have 3 regions of different frequency behavior:

$$(1) \quad \tau_{el}^{-1} < \omega < \left(\tau_{el}\cdot\tan\delta\right)^{-1} \quad D^* \cong \frac{p\cdot\eta\cdot A^{3/2}\cdot R_p^{1/2}}{G(4kT)^{1/2}}\cdot\omega$$

$$(2) \quad \left(\tau_{el}\cdot\tan\delta\right)^{-1} < \omega < \tau_{th}^{-1} \quad D^* \cong \frac{p\cdot\eta\cdot A^{3/2}}{G(4kT\cdot C\tan\delta)^{1/2}}\cdot\omega^{1/2} \qquad (16)$$

$$(3) \quad \tau_{th}^{-1} < \omega \quad D^* \cong \frac{p\cdot\eta\cdot A^{3/2}}{H(4kT\cdot C\tan\delta)^{1/2}}\cdot\omega^{-1/2}$$

The best operating conditions are found near the inverse thermal time constant, where the signal to noise ratio is peaking, i.e. between regions 2 and 3. Ideally τ_{th} should be matched to $\tau_{el} \tan\delta$. The materials figure of merit for the detectivity is best defined as

$$F_{D^*} = \frac{p}{(\varepsilon\varepsilon_0 \tan\delta)^{1/2} c_p} \tag{17}$$

which applies in region 2 as well as region 3. This definition is also used for bulk materials.

2.3 Numerical Values

The calculated curves for the responses as a function of frequency follow quite well the experimental behavior (see Figure 6) as long as the dielectric, pyroelectric, and thermal properties do not vary with frequency. The numerically simulated thermal behavior, as obtained from finite element calculations, correspond well to the thermal parameters H and G derived from the pyroelectric measurements (responsivity of Figure 6) [7]. The responsivities and the detectivity of a focal plane element can be predicted by scaling down heat capacity and the thermal conductance to the dimensions needed for thermal imaging. Below, an example is given with a 100 x 100 μm area and absorptivity of 1. The heat capacity and the heat loss have been estimated for an element on a membrane operated in air, according to the experimental values found for the responsivities in Figure 6. For such a small surface, the air convection is less important. This results in a higher voltage responsivity. Peak values of roughly 40,000 V/W, 50 μA/W and 6.4 x 10^8 cmHz$^{1/2}$/W are obtained (see Figure 8).

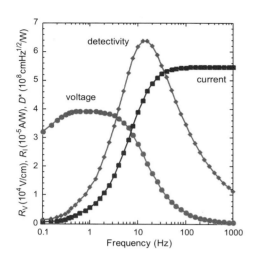

Figure 8: Calculated responsivities and detectivity for a 0.01mm² large PLT element, taking typical published parameters, as given in Table 1.

Pyroelectric coefficient	p = 500 μC/m²/K	Pyroel. Film thickness	t_p = 1 μm
Dielectric constant	ε = 250	Heat capacity	$H = 6.4 \cdot 10^{-8}$ J/K
Dielectric loss	$\tan\delta$ = 0.01	Heat conductance	$G = 4 \cdot 10^{-6}$ W/K
Specific resisitivity	$\rho = 1.0 \cdot 10^9$ Ωm	→ Parallel resistor	R_p = 100 GΩ
Heat convection coeff.	γ = 50 W/m²/K	→ Thermal time const.	τ_{th} = 16 ms
Specific heat capacity	c_v = 3.2 MJ/K/m³	→ Capacity	C = 22 pF
Absorption coefficient	η = 1.0	→ Electrical time const.	τ_e = 2.2 s

Table 1: Numerical values for a 0.01 mm² model element on a thin ceramic membrane, and some cooling by air convection (after ref. [10]).

2.4 Thermal Wavelength Effects

The simple theory presented above neglected any time dependent propagation effects of heat conduction. In reality, there is some time needed to distribute the heat inside the pyroelectric element. If the surface of a body is heated up by a modulated radiation source (angular modulation frequency ω), the temperature change that is in phase with the excitation frequency ω decreases inside the body with a decay length of

$$\lambda_{th} = \sqrt{\frac{2\kappa}{\omega c_p}} \tag{18}$$

where κ is the thermal conductivity, c_p is the heat capacity per volume. This decay length is also addressed as thermal wavelength. In PbTiO₃ with c_p = 3.2 MJ/m³/K and κ = 3.8 W/m/K, λ_{th} amounts to 600 μm at 1 Hz and 20 μm at 1 kHz. In the frequency range of interest, pyroelectric thin films are thus heated up uniformly throughout the thickness. In Si (1.7 MJ/m³/K, 148 W/m/K) λ_{th} is quite large: 5.3 mm and 166 μm for the respective frequencies. In fused silica (1.65 MJ/m³/K, 1.4 W/m/K) 520 μm and 16 μm are obtained. When pyroelectric films are operated directly on substrates, without care for thermal insulation, as is especially the case for early work with thin films [11],

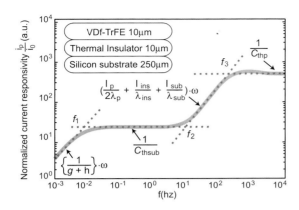

Figure 9: Current response as a function of frequency for a pyroelectric film separated by an insulating layer from the Si substrate (from [13]).

the relevant heat capacity decreases with increasing frequency like the thermal wavelength does. An increasing current response (as $\omega^{1/2}$) is measured in this case, as the current response is proportional to $1/H$ (Eq. (12)). At low frequencies, the complete substrate below the element is heated up synchronously and a response plateau at $1/H$ (substrate) is seen. At higher frequencies the current increases to another plateau defined by the smaller H (p-film), or more complicated for more layers (see Figure 9). Much work has been done in this field for polymer pyroelectrics (for a review see ref. [12]).

2.5 Characterization of Focal Plane Arrays for Thermal Imaging

The detectivity D^* is used to characterize the sensitivity of an IR detector independently of size and detection bandwidth, within the limits described earlier. D^* is a poor descriptor for pyroelectric sensors. However, for thermal imaging applications, D^* alone does not sufficiently characterize the complete device, which also contains an optical part. The weakness is not only in the optical part of the formulation. The quantity of interest is the minimal temperature difference of a black body target object with respect to the background temperature one still can detect. It is called the "noise equivalent temperature difference", NETD. In order to determine this quantity one has to measure the noise voltage V_n of the detector when the target temperature is in equilibrium with the background temperature, then to increase the target temperature by ΔT and to measure the signal voltage $V_s(\Delta T)$. The NETD is derived as [14]:

$$\text{NETD} = \frac{\Delta T}{V_s / V_n} \quad (19)$$

ΔT is equal to the NETD if the signal is just equal to the noise voltage. Of course, the NETD can be related to the NEP and D^* of the detector knowing the optical properties of the imaging system. The temperature increase ΔT has to be such that the increase of the power falling onto one pixel of the focal plane array (FPA) equals the NEP. The power increase is proprtional to $dP/dT \Delta T$ in the wavelength intervall $\Delta \lambda$ detected, where P is the power density emitted by the target surface. It is further proportional to the FPA pixel area A_D, and the transmissivity τ of the optics. The reader is referred to special literature on this topic (ref. [14]) for the derivation of general formulas. The formula given below finds frequent use [15]:

$$\text{NETD} = \frac{(4F^2+1)}{A_D \cdot \tau \cdot (dP/dT)_{\Delta\lambda}} \cdot \text{NEP} = \frac{(4F^2+1)}{A_D^{1/2} \cdot \tau \cdot (dP/dT)_{\Delta\lambda}} \cdot \frac{B^{1/2}}{D^*} \quad (20)$$

F is the f-Number of the optics. For black body radiation at room temperature and for a detection intervall from 8 to 14 µm, the factor $(dP/dT)_{\Delta\lambda}$ is calculated as 2.6 W/m²/K from Plank's radiation law. Taking an optic with $F = 1$, a pixel area of 10^{-8} m², a transmissivity of 0.5, and a bandwidth B of 100 Hz (> frame rate), the NETD is obtained from D^* by the formula:

$$\mathrm{NETD}[K] = \frac{2.1 \cdot 10^7}{D^*[cmHz^{1/2}W^{-1}]} \tag{21}$$

Hence our model detector from above with $D^* = 3.5 \times 10^8$ cmHz$^{1/2}$W^{-1} would yield a NETD of 100 mK. That corresponds to the target value of 0.1 K, which has been set for uncooled imaging systems ([15], [16]). The NETD does still not sufficiently describe the final performance of a focal plane array. In this case one is additionally interested in the spatial resolution, i.e. the minimum resolvable temperature difference (MRT) and the modulation transfer function (MTF, ability to detect small features see, e.g., Hanson [42], and the dynamic range Butler et al. [43]) . Generally, micromachined detectors offer better MTF's as compared to reticulated ceramic versions, due to a lower thermal cross talk between pixels. It turned out that AC coupled devices (pyroelectric FPA with chopper) and DC coupled devices (as e.g., VO$_x$ bolometer) cannot simply be compared using NETD figures [44]. DC coupled devices are more prone to degrade MTF and MRT due to 1/f noise at less than 5 Hz and fixed pattern noise (drifts that are systematically linked to pixels). It is also clear that the spatial resolution depends on the frequency in case of DC coupled devices. Using a chopper makes all signals appear with the same frequency of 30 to 100 Hz. The white noise related to such frequencies is more averaged out by the human eye than the low frequency noise seen and generated by DC coupled devices. Hanson et al. [44] conclude that DC coupled devices need a NETD of about a factor 5 better than of AC coupled devices in order to give to human eyes the impression of equal image quality.

3 Pyroelectric Materials

A review on bulk materials can be found in an article of Whatmore [8], a review on thin film materials is given by Muralt [45]. Properties of thin film materials differ from those of bulk materials in as much as microstructure and substrate influence are of importance. In contrast to bulk ceramics, thin films can be grown textured or even completely oriented (epitaxy). The optimal texture is achieved when the polar axis (polarization) stays perpendicular to the electrodes everywhere in the film. In this case, a performance similar to the one of single crystal materials is obtained. This leads to a considerable improvement of true pyroelectric properties in the case of substances that only exist as polycrystalline ceramics. A good demonstration of this case is epitaxial PbTiO$_3$, whose figure of merit F_V was measured as 291 kVm^{-1}K^{-1} [36] at thin films, whereas only 107 kVm^{-1}K^{-1} is reached in bulk ceramics. Operating in the dielectric bolometer mode near a phase transition (cubic/tetragonal for example), the dielectric tensor is nearly spherical, and ceramic performance should be virtually the same as the single crystal. In practice, the ceramic is better due to the limitations found in preparing uniform single crystals from multi-component materials.

There is a trade off between temperature stability and size of the pyroelectric effect. If a cheap, simple and reliable device is the goal, materials with high critical temperatures (T_c) such as LiTaO$_3$ and PbTiO$_3$ are more adequate. If the operating temperature can be held precisely, close to or exactly at the critical temperature, materials with near room temperature T_c, with ideal first order or relaxor-type behavior, yield much larger responses [37]. The materials choice taken in the published thin film work indicates that PbTiO$_3$ derived compounds are the clear favorites until now, and much less work has been spent on the second category of materials. Pure PbTiO$_3$ has been mostly abandoned because of too high dielectric losses and difficulties to pole. The most advanced integration work on Si has been achieved with PZT (15 to 30 %) Zr films [26]. This is a kind of spin-off from the work done in ferroelectric memories and actuators. In general, films grown in epitaxial quality on single crystal substrates such as MgO, SrTiO$_3$ or sapphire exhibit figures of merit that are roughly two times higher than the ones of films deposited onto Si [38]. The earlier work on relaxor-type materials and lead scandium tantalate (PST) dealt with depositions on inert substrates only. Only recent efforts have been successful using ordinary platinized Si substrates. The properties are, however, by a factor 3 to 10 lower. Main problem are the higher nucleation and growth temperatures as compared to lead titanate family, resulting in a degradation of the platinum bottom electrode and its interfaces. Suitable seeding layers are applied to lower the nucleation temperature ([46], [47]). Improvements in sol-gel chemistry still seem to be possible. Nevertheless it looks difficult to synthesize PST without second phases and with B-site ordering below 700 °C. Low deposition temperatures have been achieved by MOCVD [48], however, no material properties were reported.

Thin film materials cannot be considered independently from the substrate on which they are grown. The substrate may well influence microstructure and properties of the ferroelectric thin films. Three types of **substrate effects** can be identified (see Ref. [45] for details):

1. Growth phenomena related to the surface on which the film is grown (usually the bottom electrode).
2. Thermal strains imposed by the bulk of the substrate during cool down from the growth temperature (the substrate during growth is typically is 100 to 1000 times thicker than the film).
3. Thermal expansion mismatch with thin film support during operation of the pyroelectric film yielding a piezoelectric contribution to the pyroelectric effect.

Beside oxide pyroelectric materials there exists **polymer materials** which reveal pyroelectric properties. Due to their commercial availability at a very low cost in streched-sheet form pyroelectric PVDF films as thin as 6 μm are interesting for low-cost detectors. Their low thermal conductivity (λ = 0.13 W/Km) makes a self-supporting and thermally well-insulated mounting possible and multidetectors can be produced together on a PVDF foil [39]. In the future, modern micromachining and thin-film techniques for the integration of pyroelectric thin films deposited on highly isolating membranes with semiconductors wil also lead to a tremendous change of the structures of single-element detectors and detector arrays. Beside pyroelectric ceramic thin films, pyroelectric films of the copolymer polyvinylidene fluoride tri-fluorethylene (P(VDF/TrFE)) deposited by spin-coating may become considerably attractive for low-cost detectors [40].

Material	Substrate	p [μCm^{-2}K^{-1}]	e	tan δ	F^*_V [kVm^{-1}K^{-1}]	F_D [10^{-5}Pa$^{-0.5}$]	Lit.
PbTiO$_3$ [1]	MgO	250	97	0.006	291	3.4	[36]
PbTiO$_3$ [2]	-	180	190	0.01	107	1.5	[1]
Pb(Sc,Ta)O$_x$ [1]	Sapphire	6000	6500	0.02	104	6-9	[37]
PZT 30/70 [1]	Pt	200	340	0.01	66	1.3	[26]
PZT 45/55 [3]	MgO	420	400	0.013	119	2.0	[38]
PVDF	-	30	12	0.015	282	1.4	[39]

1) Thin film 2) bulk 3) epitaxial

Table 2: Published data of pyroelectric materials. F_V and F_D have been evaluated according Eqs. (11) and (17).

4 Realized Devices, Characterization, and Processing Issues

4.1 Fabrication Technology

There are various fabrication concepts for pyroelectric devices. Si is not the best suited substrate for pyroelectric thin film growth. Still, Si is the best suited (and cheapest) substrate for all other fabrication steps, such as semiconductor processing and micromachining. For this reason some concepts have been developed to grow the pyroelectric film on an other substrate and later glue it onto Si structures (so-called hybrid assembly), instead of directly depositing onto Si wafers. Because of the requirements on thermal insulation discussed in the theory Sec. 2, the substrate material underneath the pyroelectric pixel is removed and a cantilever or membrane is formed using micromachining techniques. There are two different approaches to micromachining.

VII Sensor Arrays and Imaging Systems

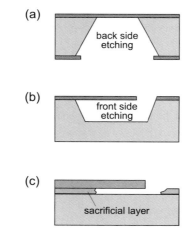

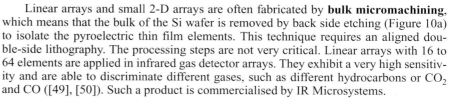

Figure 10: Si wafer cross section to demonstrate different micromachining techniques for thermal insulation of suspended structures:
(a) bulk micromachining,
(b) surface micromachining by front-side etching of the Si substrate,
(c) surface micromachining by sacrificial layers.

Figure 11: Typical structure of pyroelectric elements of a linear array on a thin membrane fabricated by means of micromachining [23]. The elements are contacted to pads in the other direction than the one seen in this cut.

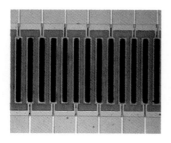

Figure 12: Top view on 50 element array with 200 μm period obtained with bulk micromachining, membrane size: 2 x 11 mm. The black platinum absorbers, the Cr-Au contact lines, the membrane layers between the elements, and the SiO$_2$ layer for reduction of parasitic capacitance are well visible (from [24], [25]).

Linear arrays and small 2-D arrays are often fabricated by **bulk micromachining**, which means that the bulk of the Si wafer is removed by back side etching (Figure 10a) to isolate the pyroelectric thin film elements. This technique requires an aligned double-side lithography. The processing steps are not very critical. Linear arrays with 16 to 64 elements are applied in infrared gas detector arrays. They exhibit a very high sensitivity and are able to discriminate different gases, such as different hydrocarbons or CO$_2$ and CO ([49], [50]). Such a product is commercialised by IR Microsystems.

Bulk mircomachining is less suitable for *large* 2-D arrays due to the brittleness of the membrane. Additionally, no transistors of the read-out circuit can be placed underneath the pixel cell, since there is no material left.

For this reason, large arrays have to be manufactured by **surface micromachining** techniques. This term comprises processing steps which act on the surface of the device side of the wafer, again to thermally isolate the pyroelectric elements (Figure 10 b,c). This approach allows to integrate the complete microelectronic read-out circuit into the wafer underneath the pyroelectric detector array. Although double-side lithography is not required, surface machining processes are more challenging. For example, a protection of the devices during Si etching is necessary. This problem can solved by the use of easily removable sacrificial layers. They can be an additional layer (typically BPSG), as indicated in Figure 10, or they can be obtained by a local transformation of Si to porous Si, which can be removed selectively afterwards [17], [18]. With the sacrificial layer techniques, only small distances to the Si substrate are achieved, making the use of vacuum imperative. Two dimensional modeling of thermal properties of small gap structures is treated in ref. [19].

Example 1: Membrane structure obtained by bulk micromachining for 1-D arrays

A stress compensated membrane layer of Si$_3$N$_4$/SiO$_2$ [20] or low stress nitride is coated on both sides of a double side polished wafer. This coating fulfills the following functions: It serves as a mask for back-side etching in KOH or in an equivalent base [21]. Second it serves as a support of the pyroelectric elements (membrane) exhibiting a low thermal conductivity. Bottom electrode and pyroelectric film (PZT15/85) are deposited by sputtering and CSD, respectively. The top electrode is deposited and patterned by a lift-off technique before a quartz layer is sputter deposited for reduction of parasitic capacity below the contact pads [22]. Windows down to the top electrodes are opened by a CF$_4$ reactive ion etching. The PZT elements on the membrane part are etched free in a HCl:HF solution, leaving only narrow bridges between the elements and the bulk Si part, as needed for separation of bottom and top conductor. The platinum bottom electrode is removed between the elements by electrochemical etching. This etching technique does not attack the membrane material. After deposition and patterning of the conductor lines, pads (Au/Cr) and absorbing layer, the Si is removed below the elements by back-side etching, as defined by a window in the back-side nitride layer, in order to obtain the result shown in Figure 11 and Figure 12. The 0.9 μm thick membrane with a specific conductivity of 2 W/m/K gives a fairly good thermal insulation, which allowed to obtain rather high

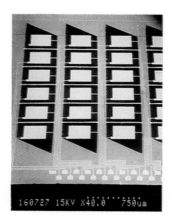

Figure 13: View on 11 x 6 array of Siemens [28].

voltage responses at 1 Hz of 800V/W in air [23]. The membrane roughly doubles the heat capacity of the pyroelectric element. Rather long thermal time constants of 28 ms in air, and 104 ms in vacuum have been obtained with such devices [7].

Siemens has developed a 11 x 6 array for thermal imaging (see Figure 13). Sputter deposited PZT 20/80 films were applied and D^* of 3×10^8 cmHz$^{1/2}$/W is reported for 10 Hz. The suspended structures have been obtained by bulk micromachining of (110) silion wafers. This Si cut allows fabrication of narrower Si bridges, which are needed in this design for carrying the electrical connection lines to the pads to which the read-out electronics is connected [28].

Example 2: Surface micromachining for 2-D arrays

For large 2-D arrays bulk micromachining is not practicable. Suitable surface micromachining processes are thus required. The example presented here bases on 0.5 μm Si nitride layer on a 1 to 2 μm thick sacrificial layer. The Si nitride is the mechanical supporting layer for the pyroelectric capacitor stack. After removal of the sacrificial layer, the suspended structure is kept by two bridges (see Figure 14). The geometrical arrangement combines in an optimal way the need for long bridges on one hand and the need for an as large as possible detector area. The extremities of the bridges rest on previously prepared via hole studs, which also serve to contact bottom and top electrode. The CMOS processing for the readout electronics, including the wiring, must be performed before the micromachining and ferroelectric film processing is carried out. Main problem is the high processing temperature for the pyroelectric film, which does not allow to wire with standard CMOS-technolgy, i.e. with Al conductor lines.

Raytheon-TI Systems reported the successful development of a monolithically integrated focal plane array based on a pyroelectric PLZT thin film, and containing 320x240 pixels of 48.5 x48.5 μm size each [29], [30], [31]. The pyroelectric detector array is built on top of the Si-CMOS read-out circuit (ROIC) chip. Crucial to this achievement was the balancing of the thermal budget of the chemical solution deposition (CSD) process to allow perovskite PLZT crystallization on the one hand, and to avoid degradation of the aluminum metallization on the other hand. A subset of such an array is shown in Figure 15.

The critical temperature for the CMOS metallization was evaluated to be 550 °C during short anneals. The sensor structure consisted of suspended pyroelectric thin film plates obtained by a sacrificial layer technique (see Figure 16, similar to the structure shown in Figure 14), with the difference that the gap is used to trap the IR radiation.

For this purpose, the gap was adjusted to a quarter of the IR wavelength to be detected. The IR radiation which is not adsorbed upon first incidence and, hence, passing through the pixel is reflected by a mirror on top of the ROIC. Due to the ¼ wavelength wide gap, the incoming and reflected light show a distructive interference in the plane of the pyroelectric pixel which converts the IR energy into heat.

Thermal imaging with a NETD of around 200 mK has been demonstrated. This value is higher than expected from basic pyroelectric and thermal properties. Major problems in transferring processes from the laboratory stage to a complex device fabrication have been encountered. Film stresses, cracking, warping, metal oxidation and interdiffusion reactions resulted in a lowering of pyroelectric properties and absorption coefficient.

Figure 14: Example of surface micromachining for focal plane arrays in monolithic technology, as designed for pyroelectric thin film focal plane array by GEC-Marconi (from [26]).

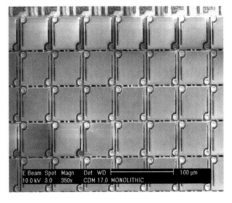

Figure 15: Partial array of thin film ferroelectric (TFFE) pixels on 48.5 μm centers [28].

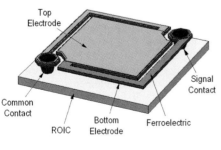

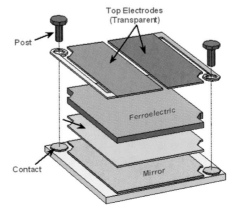

Figure 16: Drawing of a monolithic TFFE detector pixel [29].

Figure 17: Explode view of pyroelectric TFFE pixel.

As an example for a process flow, we will describe the fabrication of the Raytheon Commercial Infrared pyroelectric imaging chip in somewhat more detail. In the first step, the ROIC chip is manufactured in a standard Si-CMOS line. Subsequently, a metal reflector layer is deposited and patterned. In Figure 17 an exploded view of the pyroelectric pixel is shown, which is a modified structure of Figure 16. Here the top electrode is split. This simplifies device processing by enabling both electrical contacts to the pixel to be made from the upper surface. The bottom electrode serves as an equipotential plane, so the equivalent circuit is two capacitors in series with each other.

The approach that is most compatible with Si wafer foundry practice relies on a sacrificial layer to separate the micromachined device from the ROIC. An example of a device in production prepared using processes similar to those employed to micromachine pyroelectric arrays is the DLP engine manufactured by Texas Instuments.

As said, the pyroelectric pixel is separated from the ROIC by a distance of about 1/4 wavelength (~2 μm). The spacing is determined by a sacrificial layer, which in conjunction with the reflector, shown in Figure 18 (a) and (b), is part of an optical interference cavity that efficiently absorbs energy in the 7.5 to 14 micron waveband. The reflector prevents radiation from being absorbed by the ROIC protective overcoat. Device response is directly proportional to absorption efficiency, so this factor should be greater than 0.8 for a well designed structure. The sheet resistance of the semi-transparent electrodes separated by the ferroelectric film as shown in (c)-(e) taken in parallel combination, about 350 Ω/square, is selected for high absorption efficiency.

The bottom electrode (Figure 18c) must be compatible with device thermal budget of about 550°C. Conductive oxides like $(La,Sr)CoO_3$ are candidate materials. The pyroelectric layer may be prepared by any method that results in good properties at reasonable thermal budget (Figure 18d). The thermal budget is largely determined by compatibility with the ROIC and sacrificial layer. Growth techniques like CSD are well suited for investigating the compositional space of the pyroelectric material, while MOCVD and sputter deposition are more appropriate for production. The top electrode (Figure 18e) and arm metals are selected for their sheet resistance and low thermal conductivity. Metals like NiCr are appropriate for this application, while Pt is too reflective at practical thicknesses to tune the cavity to absorb radiation efficiently. The Wiedmann-Franz formalism illustrates the connection between electrical and thermal conductivity. The top electrode metal may also serve as the arm metallization, and connect the top electrode to the ROIC.

Pixel body and arms, shown in Figure 17, are formed by wet etch. The dimensions are selected to result in a thermal time constant of about 16.7 ms to optimize device performance. The limiting thermal conductance is defined by the pixel arm and its metallization, shown in Figure 18 (right, top-view).

The pixel is connected to the ROIC by coating a via, shown in Figure 18 (f), with thin film metal, shown in Figure 18 (g). The via metal does not limit thermal conduction, so any easily etchable metal, like Al, is suitable. After removing the sacrificial layer, the ferroelectric is poled at 150 °C by biasing one of the split electrodes positive, and the other negative, relative to each other.

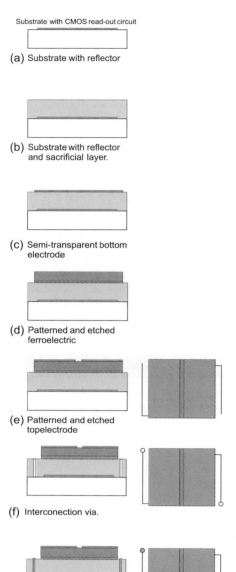

Figure 18: Sketch of the process flow, (left) cross-section and (right) top-view.

4.2 Read-Out Integrated Circuits (ROIC)

If it is the primary goal to determine the slowly fluctuating temperature of an object, as, e.g. for detecting badly insulated parts of a house, a chopper needs to be installed. The chopper blades define the reference temperature, and the signal output is proportional to $\{T^4(object)-T^4(chopper)\}$. For security applications and imaging of moving objects (or with moving camera) the chopper might not be necessary. A suitable signal treatment may be helpful in this case. As concerns the point detector of Figure 19 employed, e. g., for intruder detection, the image of an intruder walking by is first projected onto one and then onto the other element. Because of series connection of the elements, the signals from the two elements differ in the sign. The signal thus is first positive, then negative, or vice versa. The electronics catches this feature to avoid false alarms that could be otherwise caused by a heating up of the environment.

The preamplification stage of the read-out electronics is the most crucial part. Special components or designs need to be selected for lowering the electronics noise contribution to below the dielectric noise of the pyroelectric element. The classical solution with discrete electronics is the use of a junction field effect transistor (JFET) in a source follower circuit (see Figure 20 (a) [1]). JFET's exhibit smaller noise levels than MOSFET's at the low frequencies required (1/f noise). For voltage amplification, the effective parallel resistor in schematic (a) needs to be smaller than the input resistor of the JFET, which amounts typically to 10^{11} Ω. The R_p of the thin film element is often much below this value and so the external R_{pa} can be omitted.

For thermal imaging, the measurement of the pyroelectric current is more adequate than the measurement of the voltage, because the voltage decreases as $1/\omega$ in the frequency range of interest, whereas the current stays stable (see Sec. 2). The current can be measured by an operational amplifier (see Figure 20 (b)) operated like a charge amplifier (integrator) with a feedback capacitor. The feedback resistor must be rather high in order to have the cut-off frequency $1/R_f C_f$ below the measuring frequency. In integrated circuits R_f is often replaced by a switch allowing resetting the integration of charge. Modern CMOS designs combined with filtering techniques allow the use of CMOS technology. For arrays, the read-out IC is also equipped with a sample and hold, multiplexing unit, and analog to digital conversion. One of the designs developed for 2D arrays is known under the name of LAMPAR (Low-noise Arrays of Mosfets for Pyroelectric Array Readout) [27]. If a chopper is used, it is possible to improve the signal to noise ratio by applying the lock-in technique, at the price of a lower response time.

The ROIC of the Raytheon Commercial Infrared system is a CMOS device fabricated using about 0.8 micron design rules. The unit cell shown in Figure 21 contains a high-pass filter, a gain stage, a tunable low-pass filter, a buffer, and an address switch. The high-pass filter consists of the detector capacitance and a feedback resistor. The feedback resistance is typically about 300 GΩ, which gives a characteristic frequency of about 10 Hz for the high pass filter. The preamp is a simple CMOS inverter. A variable resistor tunes the low-pass filter. A diode is an appropriate device to use to create a tunable resistor. The high-pass capacitor is the gate of a metal oxide semiconductor (MOS) transistor biased to accumulation. Together, the high-pass and low-pass filters result in a bandwidth of about 120 Hz. The near unity-gain output buffer provides the ability to drive the relatively high capacitance load of the column-address lines and column amplifiers. Multiplexing the outputs of the column amplifiers provides sequential external access to the outputs. Thus the array output is compatible with standard TV formatting.

Figure 19: Top view on a single-element detector, mounted on TO-39 header. The dice is 4.5 x 4.5 mm large. Two rectangular pyroelectric elements are sitting on a transparent membrane. A JFET is glued onto the dice at the upper left corner, (from Ref [7]).

Optimizing performance involves compromises between possible operating modes. Generally, thin-film ferroelectric devices are designed to operate poled, well below the ferroelectric transition temperature. This offers the possibility of temperature independent operation, and minimizes the risk that device leakage current will saturate the amplifier. The ROIC contributes a parasitic capacitance C_s that is large compared to detector capacitance C_d (see Figure 21). The stray capacitance $C_s \sim 12$ pF is large compared to C_d, which is usually less than 3 pF. C_s is the effective capacitance of the integrating capacitor in the feedback loop of a high-gain amplifier circuit. Its effective capacitance is large because it is modified by the Miller effect.

This results in an attenuation factor of:

$$A_S = \frac{C_d}{C_d + C_s} \qquad (22)$$

The parameter A_s attenuates the detector signal, but not all noises are reduced. Temperature fluction noise, and detector Johnson noise are attenuated, while system noise is not. Consequently, the selection of detector material, operating temperature and biasing mode is not straightforward.

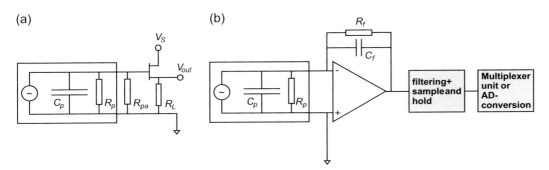

Figure 20: Typical concepts for read-out electronics with voltage measurement (a) and current measurement (b). C_p and R_p stand for the capacity and resistance of the pyroelectric thin film element. R_{pa} is the external parallel resistor, C_f and R_f are feedback capacity and resistor respectively.

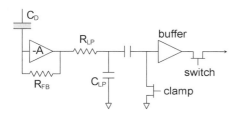

The ROIC output is sychronized with a 30 Hz chopper to create a periodic temperature change. The chopper is an Archimedes spiral that approximates a square wave thermal input. The chopper also AC couples the detector to the scene. This allows for the real-time calibration of offset non-uniformity. Consequently, AC coupled sensors work well in dynamic thermal environments.

In Figure 22 the chopper is shown. One half of the chopper is transparent, and the other half opaque or diffusing which gives the chopper the Archimedes spiral look.

5 Summary

This chapter reported the status of research in pyroelectric thin films, of pyroelectric thin film devices and applications. The sensitivity limit of pyroelectric IR detectors was derived and several ways of improving the present performance for approaching the theoretical limits were pointed out. It has been shown that materials engineering combined with thin film processing skills help to improve the relevant figures of merit. This includes not only an increase of the pyroelectric coefficient, but also a decrease of dielectric constant and dielectric loss. On the device side, integration technology is of paramount importance for achieving the optimal combination with the required thermal properties and the needs for electrical read-out. Good solutions for integration issues are the keys for obtaining competitive pyroelectric devices. Prototype devices of pyroelectric thin film detectors show already now high quality IR images. This is demonstrated in Figure 23 showing IR images shot with a monolithic thin film 320x240 array.

- 0,8-μm CMOS
- Preamp per pixel
  - Gain (400x)
  - High-pass filter (10 Hz)
  - Low-pass filter (150 Hz)
  - Buffer
  - Automatic offset elimination
- Row-address shift register
- Multiplexer
- Programmable
  - Temperature control
  - Low-pass filter

Figure 21: Readout IC Unit Cell of the Raytheon-TI system.

Acknowledgements

The editor gratefully acknowledges Ralf Liedtke (RWTH Aachen) for his editorial and technical assistance in the compilation of this chapter.

Figure 22: Chopper (Archimedes spiral) for the IR light.

References

[1] S. G. Porter, Ferroelectrics **33**, 193 (1981).
[2] R. W. Whatmore, A. Patel, et al., Ferroelectrics **104**, 263 (1990).
[3] A. J. Moulson and J. M. Herbert, Electroceramics. London, Chapman & Hall, 1990.
[4] D. J. Taylor and M. H. Francombe, Ferroelectric Film Devices, Academic Press, 2000
[5] P. Muralt, 10th IEEE Symp. Appl. Ferroelectrics, East Brunswick (USA), IEEE, 1996.
[6] P. Muralt, Revue de l'électricité et de l'électronique, 56 (1996).
[7] M. Kohli, C. Wüthrich, K. Brooks, M. Forster, P. Muralt, N. Setter, P. Ryser, Sensors and Actuators A **60**, 147 (1997).
[8] R. W. Whatmore, Reports on progress in physics **49**, 1335 (1986).
[9] M. Kohli, Y. Huang, T. Maeder, C. Wüthrich, A. Bell, P. Muralt, N. Setter, Microelectronic Engineering **29**, 93 (1995).
[10] C. H. Kohli, P. E. Schmid, F. Levy, IMF-9, Séoul, 1997.
[11] H. Vogt, P. Würfel, U. Hetzler, W. Ruppel, Ferroelectrics **33**, 243 (1981).
[12] S. Bauer and S. B. Lang, IEEE Trans. Dielectrics Electr. Insulation **3**, 647 (1996).
[13] J. J. Simone, F. Bauer, L. Audaire, Ferroelectrics **171**, 239 (1995).
[14] J. M. Lloyd, Thermal imaging systems, Plenum Press, 1975.
[15] P. W. Kruse, Infrared Physics & Technology **36**, 869 (1995).
[16] P. W. Kruse, SPIE 2552, 556 (1995).

[17] T. Bischoff, G. Müller, W. Welser and F. Koch, Sensors and Actuators A **60**, 228 (1997).
[18] C. Dücsö, E. Vazsonyi, M. Adam, I. Szabo, I. Barsony, J. G. E. Gardeniers and A. V. d. Berg, Sensors and Actuators A **60**, 235 (1997).
[19] B. Ploss, D. Lienhard, F. Sieber, Microelectronic engineering **29**, 75 (1995).
[20] A. Bell, Y. Huang, O. Paul, Y. Nemirovsky and N. Setter, Int. Ferroelectrics **6**, 231 (1995).
[21] H. Seidel, A. Heuberer, H. Baumgärtel, J.Electrochem.Soc. **137**, 3612 (1990).
[22] M.-A. Dubois and P. Muralt, IEEE Trans. UFFC (1998).
[23] P. Muralt, K. Brooks, M. Kohli, T. Maeder and C. Wüthrich, Ferroelectric thin films for microsystems, ECASIA, Montreux (Switzerland), John Wiley & Sons, 1995.
[24] B. Willing, M. Kohli, P. Muralt, N. Setter and O. Oehler, Transducers - International Conf. on Solid- State Sensors and Actuators, Chicago, 1997.
[25] B. Willing, M. Kohli, P. Muralt, N. Setter and O. Oehler, Sensors and Actuators A **66**, 109 (1998).
[26] N. M. Shorrocks, A. Patel, M. J. Walker and A. D. Parsons, Microelectronic Eng. **29**, 59 (1995).
[27] R. Watton, P. A. Manning, M. J. Perkins and J. P. Gillham, Infrared technology and applications XXII, Orlando, SPIE, 1996.
[28] R. Bruchhaus, D. Pitzer, R. Primig, W. Wersing and Y. Xu, Integrated Ferroelectrics **14**, 141 (1997).
[29] H. R. Beratan, C. M. Hanson, J. F. Belcher and K. R. Udayakumar, International Symposium on Integrated Ferroelectrics (ISIF), Monterey, 1998.
[30] C. M. Hanson, H. R. Beratan and J. F. Belcher, Infrared detectors and focal plane arrays VI, San Jose, SPIE, 2000.
[31] C. M. Hanson, H. R. Beratan, J. F. Belcher K. R. Udayakumar and K. L. Soch, Infrared detectors and focal plane arrays V, Orlando, Florida, SPIE, 1998.
[32] C. M. Hanson, H. R. Beratan, J. F. Belcher K. R. Udayakumar and K. L. Soch, Next-Generation ferroelectric uncooled IR detectors, Raytheon Systems Company, 1998.
[33] H. R. Tränkler, E. Obermeier, Sensortechnik, Springer-Verlag, 1998.
[34] S. B. Lang, Sourcebook of pyroelectricity, London, Gordon and Breach Science Publishers, 1974.
[35] A. Rogalsky and J. Piotrowsky, Progr. Quant. Electr. **12**, 87 (1988).
[36] K. Iijima, Y. Tomita, J. Appl. Phys. **60**, 361 (1986).
[37] R. Watton and M. A. Todd, Ferroelectrics **118**, 279 (1991).
[38] R. Takayama and Y. Tomita, J. Appl. Phys. **65**, 1666 (1989).
[39] G. Mader and H. Meixner, Sensors and Actuators **A21-A32**, 503 (1990).
[40] N. Neumann, R. Köhler and G. Hofmann, Ferroelectrics **118**, 319 (1991).
[41] E.H. Putley, Semicond.Semimet. **5**, 259-285 (1970).
[42] C. M. Hanson, Semiconductors and Semimetals **47**, 123 (1997).
[43] N.R. Butler, Infrared detectors and focal plane arrays VI, Orlando, Florida, SPIE, 2000.
[44] C. M. Hanson, H. R. Beratan and J. F. Belcher, Infrared detectors and focal plane arrays VI, San Jose, SPIE, 2000.
[45] P. Muralt, Rep. Progr. Phys. **64**, 1339 (2001).
[46] P. Muralt, T. Maeder, L. Sagalowicz, S. Hiboux, S. Scalese, D. Naumovic, R. G. Agostino, N. Xanthopoulos, H. J. Mathieu, L. Patthey and E. L. Bullock, J. Appl. Phys. **8**, 3835 (1998).
[47] Z. Kighelman, D. Damjanovic, A. Seifert, L. Sagalowicz and N. Setter, Appl.Phys.Lett. **73**, 2281 (1998).
[48] D. Liu and H. Chen, Mat. Lett. **28**, 17 (1996).
[49] B. Willing, M. Kohli, P. Muralt and O. Oehler, Infrared Physics and Technology **39**, 443 (1998).
[50] B. Willing, P. Muralt, T. Reimann and O. Oehler, Proc. Transducers'99, Sendai, 1999.

Figure 23: Images formed with first monolithic thin-film ferroelectric 320 × 240 array installed in a PalmIR commercial camera [29].

Electronic Noses

Claus-Dieter Kohl

Physics Department, Justus Liebig University Giessen, Germany

Contents

1 Introduction   849

2 Operating Principles of Gas Sensor Elements   849
   2.1 Calorimetric Sensors   850
   2.2 Electrochemical Cells   850
   2.3 Surface and Bulk Acoustic Wave Devices   851
   2.4 Gas-Sensitive FETs   851
   2.5 Resistive Semiconductor Gas Sensors   852

3 Electronic Noses   854

4 Signal Evaluation   854

5 Dedicated Examples   855
   5.1 Identification of Hazardous Solvents and Gases   855
   5.2 KAMINA – A Semiconductor Sensor Array   856
   5.3 A Study on Smelling the Aroma of Food   856
   5.4 Monitoring the Roasting Process of Food   857

6 Summary and Outlook   858

Electronic Noses

1 Introduction

The types of gas sensor elements found in electronic noses are described first and it is shown that by the impact of microstructuring technology and the demand of multisensor systems new elements are developed. Within the last decade the availability of micro controller chips with low prices has stimulated more complex gas sensor systems, usually termed as electronic noses. One of the first successful applications, the identification of solvents by a set of electrochemical cells, in 1984, is still available as an optional module of a commercial electronic nose [1]. Besides the description of this historical landmark more recent results as the German KAMINA, the English eNose, the Swedish Nordic Nose, and others are covered. The survey will be restricted to sensors for online analysis. The traditional chemical gas analysis based on mass spectrometers, gas chromatographs, conventional optical spectrometers etc. will not be discussed [41].

2 Operating Principles of Gas Sensor Elements

A wide variety of physico-chemical principles are employed to detect gases and their chemical nature. In the following section, the most relevant principles are briefly introduced and explained.

For all types of sensors, there are basic sensor properties:

- The **sensitivity**, which gives the amplitude of the electrical output signal y for a given concentration x of the specific gas to be detected. The *absolute* sensitivity is defined by

$$S = \frac{dy}{dx} \qquad (1)$$

while the *relative* sensitivity is given by

$$\alpha = 100\% \frac{S}{y} = 100\% \frac{dy}{y\,dx} \qquad (2)$$

- The **resolution, which** defines the smallest increment Δx which leads to a detectable change Δy. The resolution is usually limited by noise, by hysteresis effects in the sensor or by the resolution of AD converters employed in the signal processing.
- The **selectivity** of a gas sensor for a specific gas species is defined by the cross sensitivity to other gases, which occurs with a concentration x'.

$$Q(\%) = 100 \frac{dy/dx'}{dy/dx} \qquad (3)$$

- The **response time,** which describes the time until the sensor output signal has reached $1-1/e$ (i.e. approx. 63%) of its saturation value after applying the stimulation (i.e. the gas concentration to be detected) in a step-function. In the frequency domain, this corresponds to a bandwidth. This definition of the response time assumes low-pass transfer characteristics of first-order, although in many cases it is more complex.

The selectivity is of vital importance for the realization of electronic noses. The selectivity is controlled by tailoring the sensitivity with respect to the properties (adsorption coefficient on a specific sensor surface, reactivity concerning combustion, molecule mass, etc.) of the gas molecules and to employ an array of sensors for an electronic nose. This will be explained in Sec. 3.

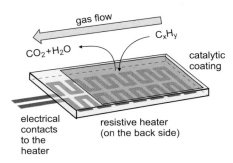

Figure 1: Sketch of a pellistor with platinum-filament, which was sintered together with the surrounding ceramic.

2.1 Calorimetric Sensors

Microcalorimetric gas sensors, so-called pellistors, burn explosible gases (e.g. hydrocarbons C_xH_y) with the oxygen in the surrounding air according to

$$C_xH_y + (x+\frac{y}{4})O_2 \longrightarrow xCO_2 + \frac{y}{2}H_2O \qquad (4)$$

on the surface of a small ball or film of a catalytically active metal [2] on a sensor system (e.g. a cantilever or membrane) of a given (small) heat capacity H (Figure 1). The catalyst, e.g., Pt, Pd or Rh, is kept at 500-600°C. The heat of combustion (oxidation reaction) W_R in the presence of a gas is balanced by a reduction of electrical heating energy, W_{el}. The change of the electrical power consumption, dW_{el}/dt, serves as signal indicating the concentration of flammable gases. The differential equation for describing this process has also to take into account the heat conductivity G between the sensor element and its ambience. Denoting $\theta = T - T_0$ as the difference between the operating temperature of the sensor, T, and the ambient temperature, T_0, the equation reads

$$G \cdot \theta + H \cdot \frac{\partial \theta}{\partial t} = \frac{\partial W}{\partial t} \qquad (5)$$

where $W = W_R + W_{el}$ is the total energy heating up the sensor. As already said, the principle is based on isothermal operation of the sensor, i.e.

$$\partial \theta / \partial t = 0 \qquad (6)$$

which reveals

$$\frac{\partial W_{el}}{\partial t} = G \cdot \theta - \frac{\partial W_R}{\partial t} \qquad (7)$$

The relation between dW_R/dt and the concentration of the gas is given by the specific heat of oxidation and the diffusion rate of the molecules onto the sensor surface. The electric power dW_{el}/dt is determined by the current, I, and the electrical resistance, R, of the heating element of the sensor by Ohms' law:

$$I = \sqrt{\frac{\partial W_{el}}{R \partial t}} \qquad (8)$$

This type of sensor is the current standard for the detection of explosives in plants. One electronic nose utilizes this sensor principle [1].

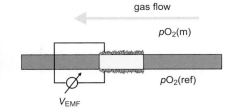
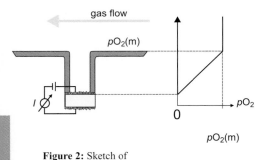

Figure 2: Sketch of
(a) a potentiometric cell and
(b) of an amperometric cell.
The signal of the potentiometric cell is the voltage VEMF, calculated in equation (9), the signal of the amperometric cell is proportional to the pressure difference of oxygen.

2.2 Electrochemical Cells

Electrochemical gas cells ionize gas molecules at a three-phase boundary layer (atmosphere, electrode of a catalytically active material (e.g. Pt), electrolyte) at the surface of the sensor. One type of ions, e.g. O^{2-}, H^+, Cl^-, which is involved in the reaction on the surface, can be specifically conducted in the electrolyte. There are two principles of operation, the potentiometric and the amperometric principle.

In an **amperometric cell**, see Figure 2 (a), an external voltage is applied and all ions generated at the surface are driven through the electrolyte to the counter electrode. The current is a measure of the rate of generation of new ions, and this, in turn, depends on the concentration of the involved gas and the rate of transfer of gas molecules to the surface, e.g. by diffusion.

In **potentiometric cells**, Figure 2 (b), the potential difference U_{EMK} generated between the electrodes by different partial pressures on both sides ($p_{O_2}^{(m)}$, $p_{O_2}^{(ref)}$) of the cell serves as signal, according to the Nernst equation:

$$V_{emf} = \frac{k_B T}{4e} \ln \frac{p_{O_2}^{(m)}}{p_{O_2}^{(ref)}} \qquad (9)$$

Where T denotes the temperature, k_B Boltzmann's constant and e the charge of an electron. Here $p_{O_2}^{(ref)}$ is the oxygen partial pressure of the reference atmosphere. Most lambda probes in car exhausts are of this type [3]. Again one electronic nose utilizes this sensor principle [1]. Details on electrochemical cells are described in [42].

2.3 Surface and Bulk Acoustic Wave Devices

Mass sensitive sensors detect a weight change of an adsorptive layer. The simplest device of that type utilizes a piezoelectric quartz disc oscillator. The shift Δf of the frequency f of the oscillator is given by

$$\frac{\Delta f}{f} = -C_f \cdot \frac{f \cdot \Delta m_{ad}}{A} \qquad (10)$$

where m_{ad} denotes the adsorbed mass, A is the coated area of the disc oscillator and C_f the mass sensitivity. By coating such a quartz disc with, e.g., a polymer absorbing the gas of interest a concentration can be monitored [4]. The added mass of an adsorbed gas lowers the resonance frequency. Since these sensors utilize piezoelectrically stimulated acoustic waves in the bulk of the crystal, they are called Bulk Acoustic Wave devices (BAW). The coatings are often the same as in gas chromatography, so the existing knowledge can be used [5]. A closer look reveals that not only the mass increase changes the resonance frequency but also stress evoked by swelling of the polymer layer. Further on conductivity changes of the material can contribute with up to 50% to the signal. Nevertheless such devices are usually named microgravimetric sensors. The sensitivity increases with the frequency, typically BAWs with 10 to 30 MHz are in use, 50 MHz are also realized. Lower sensitivity limits are around 50 ppm. Higher frequencies are attainable by Surface Acoustic Wave devices (SAW). In this case one uses the filter devices of cellular phones, consisting of a silicon substrate with piezoelectric, e.g. ZnO, layers covered by interdigital structures to generate an acoustic surface wave, which travels along the silicon surface (coated by an adsorptive polymer layer) to a detector structure (same structure as the generator).

Figure 3: Array of an 8-microcantilever sensors imaged by scanning electron microscopy. Length and width of these silicone tongues amount to 500 μ x 100 μ. Thickness 8.6 μm. Resonance frequency 50 kHz. Each cantilever is coated by an adsorptive polymer. Applications e.g. in solvent vapor detection [6].

Frequencies up to more than 1 GHz were realized and sensitivities below 1 ppm were attained. To bring this type of devices closer to micromachining standards, recently also cantilevers with adsorptive coatings have been realized [6], see Figure 3.

Each of the microcantilevers is coated on one side by 2 to 3 μm of a polymer layer (carboxymethylcellulose, polyvinylalcohol, polyvinylpyridine, polyvinylchloride, polyurethane (2 tongues), polystyrene, polymethylmethacrylate). Five different perfume oils could be separated using the shift of the resonance frequency of the 8 microcantilevers as a pattern. Such a device can be the core element of an electronic nose. However, due to its frequency of only 50 kHz its use is restricted to nearly saturated vapors. The first electronic nose with BAWs is still on the market [7]. Other manufacturers followed, e.g. [1].

2.4 Gas-Sensitive FETs

Gas-FET`s are devices consisting of a field effect transistor with a gate metallization exposed to the surrounding atmosphere (open gate MOSFET). The source-drain current at a fixed (regulated) overall gate voltage is taken as signal. Hydrogen or hydrogen containing gases dissociate or decompose on the surface; the protons diffuse to the metal/insulator interface. The adsorbed protons produce a dipole layer with a certain dipole moment p which result in a change of threshold voltage ΔV of MOS transistor:

$$V_T = V_{T0} - \Delta V \qquad (11)$$

where

$$\Delta V = n_i \cdot p / \varepsilon_0 \qquad (12)$$

n_i is the number of hydrogen atoms (per unit area) at the inner interface. In the saturation regime of a MOS transistor the drain current I_D depends quadratically on the threshold voltage

$$I_{DS}(\Delta V) = K_d \frac{(V_G - V_{T0} - \Delta V)^2}{2} \qquad (13)$$

Here K_d is a specific transistor parameter and V_G is the applied gate voltage. The measured drain current is the sensor signal for detecting adsorbed gases.

VII Sensor Arrays and Imaging Systems

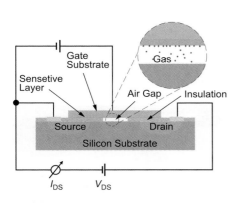

Figure 4: Field effect type gas sensor device [10].

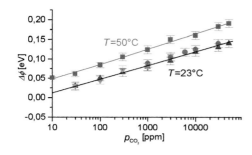

Figure 5: CO_2 sensitivity of the field effect type gas sensor device, shown in the last figure [10].

Common gate metallizations are Pd or Pt, the operation temperature is limited to about 200°C on silicon FET`s, to about 300°C on FET`s with special barrier layers and to about 600° C for silicon carbide FET`s [8]. Gas-FET arrays are found in one electronic nose [9].

A few years ago the Siemens research laboratories [10] in Munich presented a sensor for CO_2 operating at room temperature using the principle of work function measurement without the moving electrode of the Kelvin probe. The sensor consists of a special FET structure (see Figure 4).

The micromachined suspended gate is covered by $BaCO_3$, whose change of work function acts the same way as the conventional external bias voltage. The first contact of the gaseous CO_2 molecules is with an adsorbed water film (thickness about 5 monolayers) from the ambient humidity forming on top of the $BaCO_3$ gate. $(CO_3)^-$ ions together with the $BaCO_3$ form a double layer. The potential drop across the double layer varies logarithmically with the CO_2 concentration according to Nernst`s law (see Figure 5).

A corresponding logarithmic dependence on the gas concentration is observed in the source-drain conductance. Since the devices are made by silicon microstructuring, it is possible to manufacture the sensor very cost efficiently. The sensors can be part of MOS circuitry with neglectable power consumption. A possible application is fire detection as an add-on feature of a cellular phone.

2.5 Resistive Semiconductor Gas Sensors

Since nearly no sensor based electronic nose lacks the implementation of homogeneous semiconductor sensors this type shall be discussed in more detail. Homogeneous semiconducting oxides operated at elevated temperatures, between 500° and 750° C, serve as resistive oxygen sensors [3]. In this temperature range the concentration of oxygen vacancies in the bulk is in thermodynamic equilibrium with the oxygen pressure outside. As an example, the formation of point defects such as oxygen vacancies $V_{\ddot{O}}$ in an ideal, undoped oxide sensor material requires energy ($\Delta H > 0$) and increases the entropy ($\Delta S > 0$). The latter results from an increased disorder of the previously perfect lattice in the absence of defects. Because of the energy required to produce point defects, e.g. in the reaction

$$O_0 \rightleftharpoons 1/2 O_2 + V_{\ddot{O}} + 2e' \tag{14}$$

the point defect concentration is negligible at low temperatures and increases at higher temperatures. In Eq. (14) O_0 denotes lattice oxygen, $V_{\ddot{O}}$ positively charged oxygen vacancies, and e' free electrons in the oxide. The equilibrium constant for this reaction

$$K = \frac{\sqrt{p_{O_2}}[V_{\ddot{O}}][e']^2}{[O_0]} \tag{15}$$

is influenced by the oxygen partial pressure p_{O_2} in the gas phase, and by the concentrations of both, oxygen vacancies [$V_{\ddot{O}}$] and electrons [e']. For low defect concentrations, [O_0] is constant. Changes in p_{O_2} may be detected sensitively by conductivity changes because of changes in

$$[e'] = 2[V_{\ddot{O}}] \sim p_{O_2}^{-1/6} \tag{16}$$

If charged acceptors are incorporated in the oxide lattice the situation might be more complex. In general the conductivity of the material, which determines the sensor signal is a function of the oxygen partial pressure. For a two different oxygen partial pressures $p_{O_2,1}$ and $p_{O_2,2}$ and distinct change of conductivity follows the law

$$\frac{\sigma_1}{\sigma_2} \sim \left(\frac{p_{O_2,2}}{p_{O_2,1}}\right)^{\alpha} \tag{17}$$

Here α is the exponent, which for the simple case of formula (16) is 1/6.

A completely different mechanism is the change of resistance in materials with grain boundaries due to different adsorbed gases. Here the bulk resistance is used as a sensor signal. At temperatures below about 500°C the adsorbed molecules can form ionic bonds and transfer electric charge into the semiconductor grains of a sensitive layer [12]. For these polycrystalline layers the current transport through Schottky depletion layers at the boundaries between neighbored grains dominates the sensor resistance (cp. Figure 6). The basic effect can be described as follows. A donator doped, polycrystalline material may contain grain boundaries, which are characterized by trapped electrons in the disturbed crystal region. Due to this local charge with a charge area density n_{t0} the conduction band is raised and potential energy barrier of the magnitude (height)

$$\phi_{B0} = \frac{e^2 n_{t0}^2}{8\varepsilon\varepsilon_0 N_d} \quad (18)$$

and depletion width

$$d = \frac{n_{t0}}{2N_d} = \left[\frac{2\varepsilon\varepsilon_0 \phi_{B0}}{N_d e^2}\right] \quad (19)$$

is developed. Here ε_0 is the permittivity of the free space, ε is the relative permittivity, N_d is the density of trapped electrons at the boundary (at equilibrium). If a small voltage bias is applied the current above the barrier might be described due to Schottky emission in the simplest case as

$$J = AT^2 \exp\left(-\frac{\phi_{B0}}{k_B T}\right) \quad (20)$$

Here A is the Richardson constant, T the temperature and k_B Boltzmann's constant. Depending on temperature the bonds between the gas molecules and the sensor surface are broken at a certain rate. For a given concentration of gas, the adsorbed molecules are incorporated at the grain boundary and change the density of trapped electrons n_{t0} producing a new steady state situation.

$$n_{t0} = n_{t0}(c_{ad}) \quad (21)$$

The correlation of the adsorbed molecule density c_{ad} and the variation in the density of trapped electrons is molecule specific and very complicated process, which will not be described here.

In more complicated cases a sequence of surface reactions is involved. [13]. The resistance becomes a reversible function of the gas concentration (Figure 7). The material most frequently used is SnO_2, but Ga_2O_3, WO_3, ZnO, In_2O_3, TiO_2 and Fe_2O_3 are also employed. A survey of materials for oxidic gas sensors is presented in [15].

The sensitivity and the specificity (ability to respond to a certain species in a gas mixture) of the device can be controlled to some extent by selecting the temperature at which the metal oxide is hold. Different types of surface oxygen differ in reactivity: the "nucleophilic" O^{-2} ions bound within the lattice at the surface react with hydrogen or dehydrogenate hydrides and hydrocarbons. With respect to oxidation reactions the adsorbed superoxide $(O_{2\,ads})^-$ and $(O_{ads})^-$ species and also exposed oxygen atoms at steps are classified as "electrophilic" reactants which preferentially attack the $C=C$ double bond of adsorbates abstracting electrons [16]. Exposure to hydrogen and to hydrogen containing gases at elevated temperatures produces surface oxygen vacancies. They are well known as donors for example on TiO_2 (110) faces [17].

Figure 6: Schematic diagram of a grain boundary. In equilibrium the electrons are depleted in the grain boundary area with a typical depletion layer thickness d. The trapped electrons within the grain boundary n_{t0} produce a symmetric electrostatical potential barrier of energy ϕ_{B0}.

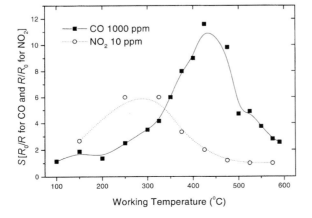

Figure 7: Sensitivity for carbon monoxide and for nitrogen dioxide versus working temperature of a nanocrystalline tin oxide sensor. Grain size 6-34 nm made by a pyrolytic reaction $SnCl_4 \cdot 5(H_2O) \rightarrow SnO_2 + 4\,HCl + 3\,H_2O$. 800°C growth temperature [14].

The decomposition of hydrogen- and hydrocarbon molecules on catalytically active metals is used in semiconductor-type sensors with metal-cluster deposits. The sensitivity can be enhanced by partial oxidation of the gases on the active metal deposit and subsequent spillover of hydrogen to the semiconductor substrate. The activity of a catalyst depends on its oxidation state. Palladium oxide oxidizes the adsorbed species thereby preventing spillover of hydrogen. The d-orbitals of metallic Pd transfer electronic charge into antibonding levels of the adsorbate facilitating dissociation or dehydrogenation with subsequent spillover. A more extended discussion is given in [13]. After a prolonged stay at operation temperature semiconductor gas sensors for reducing gases in the absence of target gas can „fall into sleep", caused by the formation of palladium oxide. Exposure to a reducing gas reduces the palladium and re-establishes its spill over function.

3 Electronic Noses

Whilst analytical techniques rely on the separation of an aroma into its constituent volatile components, electronic noses ([11], [30], [40]) detect the overall aroma across an array of gas sensor elements. Data from the sensor array provide a fingerprint identifying the sample. Sample aromas are recorded in real-time providing an instantaneous graphical representation. This multidimensional fingerprint may then be used as a comparison against a standard [38], [39]. Measurement of the Euclidian distance or more sophisticated procedures of comparison between the aroma pattern of one sample against another provides a quantitative indication of the difference between their aromas, compare Figure 8.

An electronic nose complements human sensory panels and chemical component analysis techniques by providing real-time evaluation of a sample's overall aroma. Thereby the opportunity for more frequent aroma management at critical points in industrial processes is given. Such electronic noses are more or less compact machines comprising up to some tens of sensor elements and electronics recognizing a single gas out of a given set of gases, or complex gas mixtures, e.g. the aroma of a certain type of food. These systems were proposed for the identification of solvents, for routine quality control of materials and other purposes. Sometimes there exists a special interest in off-flavors. A (not complete) survey of activities in Europe, including future conferences, is given by the EU funded NOSE - Network on artificial Olfactory Sensing, Network of Excellence (http://nose.uia.ac.be and http://www.nose-network.org).

4 Signal Evaluation

If more than two linear sensors are used to characterize a sample, a multivariate evaluation is usually applied to visualize and evaluate the data [18]. There are many multivariate techniques in use. In electronic noses pattern recognition, PARC, techniques are generally applied. Unsupervised PARC techniques try to cluster the results, which are usually visualized in a two-dimensional plot. This allows classifying off odors of unknown origin for example. Such results form a base for further search to the underlying reasons for the differences. More often a supervised technique is used to find out if an identification of a set of materials with different smells is possible. During a learning process a set of descriptors is generated for identification of future samples. A side effect of these models is, that one can find out how much a single sensor element contributes to a reliable identification. Thereby the initial number of sensors can be reduced without significant loss of accuracy.

Sensors with linear response, electrochemical cells or microgravimetric sensors are usually evaluated by Principal Component Analysis (PCA). PCA involves a mathematical procedure that transforms a number of (possibly) correlated variables into a (smaller) number of uncorrelated variables called principal components. The first principal component accounts for as much of the variability in the data as possible, and each succeeding component accounts for as much of the remaining variability as possible. To illustrate it by an example, imagine five linear gas sensors exposed to a varying mixture of two gases A and B. Each sensor is sensitive to at least one of the gases, some to both

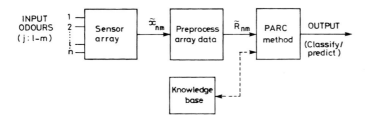

Figure 8: Scheme of an electronic nose [43]. An array of gas sensor elements is exposed to an odor or smell composed of one or many chemical compounds. After preprocessing by preamplifiers the signal pattern is compared to prerecorded odor signal samples by a PAttern ReCognition unit (PARC).

of the gases. The sensitivities to A and B of each sensor are different. Then PCA forms from the five output signals two new signals, one proportional to gas A and the other proportional to gas B.

The linear output of the mass spectrometer based Smart Nose, mentioned below, is also evaluated by PCA. The characteristics of semiconducting sensors may be linear, if an initial increase is used for evaluation. If, however, saturation values are evaluated, nonlinear methods have to be applied to separate the effects of the single gases. Artificial Neural Networks, ANN, and Polynome Networks, PN, can be used, a comparison of both for sensor applications is found in reference [19]. The mathematical basics of PCA and ANN can be found in [20], [21]. The PCA stastical mathematics can be found online, too [22].

5 Dedicated Examples

5.1 Identification of Hazardous Solvents and Gases

Already in 1980 Stetter [23], [24] addressed the problem of identification and quantification of compounds for emergency response personnel in field situations by the use of arrays of electrochemical gas sensors. The US Coast Guard as a part of the US Department of Transportation was responsible for spills of hazardous chemicals in waterways and on land. The chemicals of concern ranged from petrochemicals for plastics and fuels to those used in agriculture and pharmacological processes. A representative list of compounds was selected for the initial evaluations. This list of compounds, reproduced in the table, includes aromatic, aliphatic, as well as substituted hydrocarbons and inorganic compounds.

Compounds selected for initial sensor array development	
Acetic acid	Formic acid
Acetone	Hydrogen sulphide
Acrylonitrile	Nitric oxide
Ammonia	Nitrobenzene
Benzene	Nitrogen dioxide
Carbon monoxide	Nitromethane
Carbon tetrachloride	Pyridine
Chlorine	Sulphur dioxide
Chloroform	Sulphuryl fluoride
Cyclohexane	Tetrachloroethylene
Ethyl acrylate	Tetrahydrofuran
Formaldehyde	Toluene Vinyl acetate

Table 1: Compounds selected for initial sensor array development.

VII Sensor Arrays and Imaging Systems

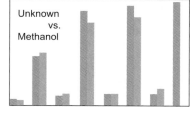

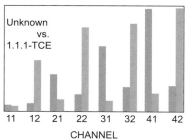

Figure 9: Channel by channel comparison of unknown vector with two known library vectors. After Stetter [23].

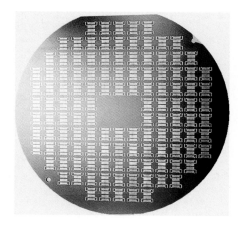

Figure 10: 3" silicon waver with 26 chips of the Karlsruhe micro nose, KAMINA [25].

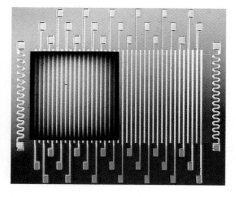

Figure 11: KAMINA silicon chip 8 mm x 9 mm covered for the specific detection of organic vapors. Left-hand are 20 WO_3 elements, right-hand are 20 SnO_2 elements to be seen [25].

Two filaments were operated as crackers before the gas reached an array of four electrochemical cells. The cells obtain their specificity from the choice of an electrocatalyst and the electrochemical potential. Since the cell currents vary linearly with concentrations, the 16 channel signals could be normalized and taken as patterns in a library. The application was restricted to cases with one predominant gas or vapor. Unknown gases are identified by the minimum of the sum of the quadratic deviations relative to all the stored patterns. After identification a calibration factor was applied to calculate the concentration. The described battery operated sensor system equipped with a micro controller was installed in 1984 by the U.S. Coast Guard and performed well.

Figure 9 shows the comparison of an unknown volatile to two different patterns. By the similarity to the methanol pattern it is identified as methanol or at least as an alcohol. The results were crosschecked by gas chromatographic analysis. A set of 4 electrochemical cells, made by Stetter, is now available as optional EC-module for MOSES II [1], a modular electronic nose. This instrument can be equipped with an oxide sensor (8x SnO_2), a quartz microbalance (8x BAW) and the EC-module. A microcalorimetric module contributed by Freiberg university is also announced.

5.2 KAMINA – A Semiconductor Sensor Array

For the detection of organic vapours KAMINA (abbreviation for <u>Ka</u>rlsruhe <u>mi</u>kro nose, in German nose = <u>Na</u>se) [25] is equipped with special semiconductor oxide sensor elements. From a 3" silicon waver (Figure 10) 26 KAMINA chips, sized 8 mm x 9 mm, are taken. Each chip is equipped with 2 x 20 single sensing elements and two temperature sensors on the front side (Figure 11). The backside bears four separate heating elements to generate a temperature gradient (cp. Figure 12). Two different materials can be combined on one chip. Figure 13 shows such a multisensor element in its housing. Three different materials, tin oxide (with or without platinum addition) tungsten oxide (without or with gold addition) and iron oxide, can be deposited by chemical vapour deposition. The elements are covered with a thin ceramic membrane (a few nanometers) of selective permeability, either SiO_2 or Al_2O_3, to tune the specificity [26], [27]. The resulting signal patterns allow a highly sensitive detection of single compounds in complex gaseous atmospheres. The following applications are claimed by the developers [25], [28].

- Quality control in food production
- Development of an intelligent extractor hood
- Analysis of breath and body odor
- Leak detection of pipes of natural gas
- Environmental monitoring, e.g. contaminated soil

5.3 A Study on Smelling the Aroma of Food

In a study ([29], [32]), a sensor based electronic nose has been tested to discriminate four ripening stages (ages) of Emmental cheese, 1, 21, 98, and 180 days. The cheese has been produced at four different sites. The instrument, an Marconi Applied Technologies eNose 5000 (Essex, UK), is equipped with semiconducting sensors (8 tin oxide elements). Via an autosampler the samples are placed in an oven. 5 ml are taken with a syringe and mixed with nitrogen (carrier gas) at a defined flow rate and fed along the sensor array. The instrument stores relative values of the sensor resistance taken at a preset time or the maximum response for further evaluation. When the oxide sensors were continuously exposed to measurement cycles with 1000 ppm ethanol the response stayed rather stable. However, after a break of about a week the response decreased for all but two sensor elements by about a factor of two (this observation may be due to a slight oxidation of the noble metal clusters on the oxides which is reverted by interaction with reducing gases, as mentioned above). Cheese ripened for 1, 21, and 98 days could be very well separated in a multiple discriminant analysis from the oxide element signals. The production sites of the cheese brands could be identified in a reliable way, too. A further study [29] performed a comparison to a mass-spectrometer based electronic nose. The conclusion was that the semiconductor based eNose showed a significantly better selectivity.

The characteristic flavors of foods, such as bread or coffee, are often generated during processing by a sequence of chemical reactions from odorless precursors. When the human nose is used as a „sensor array" it has been shown, e.g. in aroma extract dilution analysis [33] for a number of foods [34], that of the hundreds of volatiles present, only a small number is needed to generate an overall food aroma. That means the human nose needs only a limited number of compounds to create an odor impression. Therefore, food processing can be more efficiently controlled by sensor elements optimized to detect the volatiles relevant for the human nose.

5.4 Monitoring the Roasting Process of Food

Metal oxide sensors used were either prepared as described in reference [36] or supplied by UST (Umweltsensortechnik, Geschwenda, Germany). The sensor materials, including the addition of noble metal clusters such as palladium and platinum, as well as the temperature of the sensitive layer influence the responses of the sensor elements [13]. N-heterocycles formed during processing of foods, such as bread or coffee via the Maillard reaction, were used to choose sensor materials sensitive to these compounds [35], [37]. A model mixture of four pyrazines 2,3,5-trimethyl-, 2-ethyl-3,6-dimethyl-, 2-ethyl-3,5-dimethyl- and 2,3-diethyl-5-methyl-pyrazine, 2-acetylthiazol and 2-acetyl-2-thiazoline was analysed. Structural formulas of the sulphur compounds, 2-acetylthiazol and 2-acetyl-2-thiazoline, are sketched in Figure 14.

At a temperature of 450°C, about 2-5 ng of these volatiles could be detected by a ZnO/Pd sensor. A ZnO/Pt sensor showed a slightly lower overall sensitivity, whereas a SnO_2 sensor was unable to detect these volatiles. Decreasing the temperature of the sensors to 350°C increased the sensitivity of the sensors for 2-acetylthiazol and 2-acetyl-2-thiazoline.

At 350°C the ZnO sensor doped with platinum (upper trace in Figure 15) showed a distinct specificity for 2-acetyl-2-thiazoline. The higher oxidized molecule, 2-acetylthiazol causes only a small elevation and the four pyrazines do not show up. The specificity towards 2-acetyl-2-thiazoline could be further improved by lowering the operating temperature to 180°C. 2-acetyl-2-thiazoline belongs to the key odorants of roast meat [87]. Therefore, the ZnO/Pt sensor operated between 180°C and 350°C might be useful in the control of thermal processing. The results obtained for the pair 2-acetylthiazol and 2-acetyl-2-thiazoline suggest that highly oxidized molecules do not react with ZnO/Pt at lower temperatures. This assumption was checked by the pair 2-acetylpyrrol and 2-acetyl-1-pyrroline, the structural formulas of these compounds are sketched in Figure 16.

Figure 17 shows the check of this assumption. The ZnO/Pt sensor, kept at 350°C, responds only to 2-acetyl-1-pyrroline. Only at 500°C also the higher oxidized 2-acetylpyrrol shows up, too. But the signals are much smaller. ZnO with palladium addition was found to be not specific to the different oxidation states of 2-acetylpyrrol and 2-acetyl-1-pyrroline. The 2-acetyl-1-pyrroline (Figure 16) and 2-acetyltetrahydropyridine are key odorants in wheat bread crust and freshly popped corn. ZnO/Pt sensors are candidates to monitor these reactions, too. Both, 2-acetyl-2-thiazoline and 2-acetyl-1-pyrro-

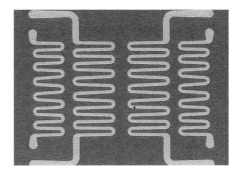

Figure 12: Four heating elements on the back of the KAMINA chip to get a temperature gradient along the waver [25].

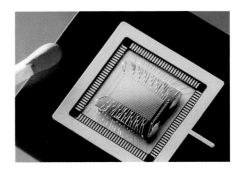

Figure 13: KAMINA chip with opened housing [25].

Figure 14: Structural formula of 2-acetyl-2-thiazoline (left side) and the higher oxidized 2-acetylthiazol (right side). 2-acetyl-2-thiazoline belongs to the key odorants of roast meat [37].

Figure 15: A flame ionization detector (FID) and a ZnO/Pt sensor are exposed to a mixture of 2-acetylthiazol and 2-acetyl-2-thiazoline. The relative conductance increase of the sensor and the FID signal are sketched. The peak at 950 s belongs to 2-acetyl-2-thiazoline, a key odorant of roast meat which is specifically detected by the ZnO/Pt sensor kept at 180°C (middle trace) and at 350°C (upper trace).

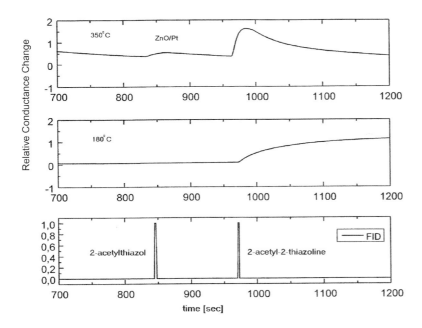

line add strongly to the smell of baked or roasted food. During baking and cooking also the more strongly oxidized species, 2-acetylthiazol and 2-acetylpyrrol, featuring only one double bond less are formed.

The oxidized species are characterized by higher activation energy for further oxidation. At least for the two pairs investigated, the higher oxidation energy results in a smaller signal for ZnO/Pt at low and intermediate temperature. Since the human nose is less sensitive to higher oxidized volatiles emitted during food processing it is important to select sensors (and their operating conditions) responsive to the volatiles strongly influencing the smell. Higher oxidized species present in higher concentration do not influence the human nose significantly. Therefore, an electronic nose should mimic this behaviour.

The toasting of white bread was shown to be controllable by key aroma compounds, e.g. 2-acetyl-1-pyrroline (typical popcorn note), using a simple electronic nose with two oxide sensor elements (no preselection by a chromatographic column necessary) [37].

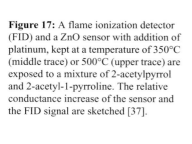

Figure 16: Structural formula of 2-acetyl-1-pyrroline (left side) and the higher oxidized 2-acetylpyrrol (right side).

6 Summary and Outlook

The first decade of electronic noses development has shown that this type of instruments complements the established laboratory techniques intended to identify distinct chemical compounds in an atmosphere. Electronic noses do not rely on separation of compound constituents. Instead a certain smell is characterized by the relative signals of a set of sensors, which are specific only to a certain degree. A smell is recognized by the similarity of a signal pattern to prerecorded sample patterns. Such a procedure is much faster than a chemical analysis using a gas chromatograph or a spectrometer because the time consuming separation does not have to be performed. So these instruments allow a fast process control, e.g. of roasting processes, and a fast inspection of potential hazards, e.g. in contaminated soil.

Most of the present sensor elements were not originally developed for the inclusion in multisensor arrays. Applications in electronic noses afford to reinspect specificity, sensitivity and stability to optimize them for electronic noses. Another task for the future is the development of recalibration procedures. Standards for smells are more difficult to stabilize than single compound calibration samples.

Low power sensor elements, e.g. the electrochemical cells, allow the construction of truly portable instruments with rechargeable batteries for on site inspections outside of laboratories. Only suspicious samples have to be further investigated by certified analytical instruments in the laboratory.

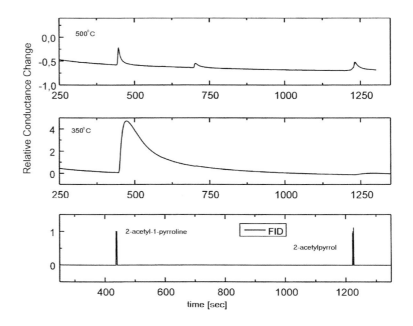

Figure 17: A flame ionization detector (FID) and a ZnO sensor with addition of platinum, kept at a temperature of 350°C (middle trace) or 500°C (upper trace) are exposed to a mixture of 2-acetylpyrrol and 2-acetyl-1-pyrroline. The relative conductance increase of the sensor and the FID signal are sketched [37].

Acknowledgements

The editor gratefully acknowledges Ralf Liedtke (RWTH Aachen) for his editorial and technical assistance in the compilation of this chapter.

References

[1] Lennartz electronic GmbH, Bismarckstrasse 136, D-72072 Tübingen, http://www.lennartz-electronic.de/, EC-modul: http://www.ipc.uni-tuebingen.de/weimar/nose/gassensors/transducers/amperometric.html

[2] H. Debéda, D. Rebière, J. Pistré and J. Ménil, Sensors and Actuators **B 27**, 297 (1995).

[3] A. D. Brailsford, M. Yussouff and E. M. Logothetis, Technical Digest of the 4th International Meeting on Chemical Sensors, Tokyo, ed. N. Yamazoe, Japan Association of Chemical Sensors, 1992.

[4] R. Lucklum and P. Hauptmann, Sensors and Actuators, **B 70**, 30 (2000).

[5] W. Grate and H. A. Abraham, Sensors and Actuators **B 3**, 85 (1991).

[6] F. M. Battiston, J.-P. Ramseyer, H.P. Lang, M. K. Baller, Ch. Gerber, J. K. Gimzewski, E. Meyer, H.-J. Güntherodt, Sensors and Actuators B77, 122 (2001). Compare also http://www.chem.ucla.edu/dept/Faculty/gimzewski/id11.htm

[7] Perkin-Elmer Corporation, 761 Main Ave., Norwalk, CT 06859-0010, http://www.perkin-elmer.com, http://www.hkr-sensor.de/start.htm

[8] P. Tobias, P. Mårtensson, A. Baranzahi, P. Salomonsson and I. Lundström, Sensors and Actuators **B 47**, 125 (1998).

[9] AppliedSensor Sweden AB, Teknikringen 6, SE-583 30 Linköping, Sweden and AppliedSensor GmbH, Aspenhausstraße 25, 72770 Reutlingen, Germany, http://www.appliedsensor.com/

[10] B. Ostrick, M. Fleischer, H. Meixner and D. Kohl, Sensors and Actuators **B 68**, 197 (2000).

[11] Aromascan electronic nose company portrait http://www.aromascan.com/products/core.html

[12] T. Ochs, W. Geyer, C. Krummel, M. Fleischer, H. Meixner and D. Kohl, Advances in Solid State Physics **38**, 623 (1999).

[13] D. Kohl, in *Handbook of Biosensors and Electronic Noses: Medicine, Food & the Environment*, ed. E. Kress-Rogers, CRC Press Inc., Boca Raton, 1996.

[14] A. Cirera, A. Diéguez, R. Diaz, A. Cornet, J. R. Morante, Sensors & Actuators B **58**, 360 (1999).

[15] U. Lampe, M. Fleischer, N. Reitmeier, H. Meixner, J. B. McMonagle and A. Marsh, *Sensors Update*, eds. H. Baltes, W. Göpel and J. Hesse, VCH, Weinheim, 1996.

[16] J. Haber, Proc. 8th International Congress on Catalysis, Berlin, FRG, Verlag Chemie, Weinheim, 1984.

[17] S. Munnix and M. Schmeits, J. Vac. Sci. Technol. **A 5**, 910 (1987).

[18] J. W. Gardner and P. N. Bartlett, in *Sensors and Sensory Systems for an Electronic Nose*, eds. J. W. Gardner and P. N. Bartlett, Dordrecht, Kluwer Academic Publishers, 1992.

[19] V. Sommer, P. Tobias, D. Kohl, H. Sundgren, I. Lundström, Sensors & Actuators: B. Chemical **28**, 217 (1995).

[20] O. Erkki, International Journal of Neural Systems, **1**, 61 (1989).

[21] C. Rafael, E. Gonzalez and R. E. Woods, Digital image processing, Addison Wessley Publishing Company, 1992.

[22] http://www.cis.hut.fi/~jhollmen/dippa/node1.html.

[23] J. R. Stetter, in *Sensors and Sensory Systems for an Electronic Nose*, ed. J. W. Gardner and P. N. Bartlett, NATO ASI Series E: Applied Sciences, Kluwer Academic Publishers Dordrecht, 1992.

[24] J.R. Stetter, S. Zaromb, W. R. Penrose, M. W. Findlay, Jr., T. Otagawa, Hazardous materials conference proceedings, April 1984, Nashville, Tennessee, ISBN 0-86587-064-0, (1984).

[25] http://irchsurf5.fzk.de/mox-sensors/Default_Eng.htm

[26] P. Althainz, A. Dahlke, M. Frietsch-Klarhof, J. Goschnick and H.J. Ache, Sensors and Actuators B **24-25**, **366** (1995).

[27] P. Althainz, A. Dahlke, M. Frietsch-Klarhof, J. Goschnick and H.J. Ache, Phys. Stat. Sol. **145**, 611 (1994).

[28] S. Ehrmann, J. Jüngst, and J. Goschnick, Technical Digest of the 7th International Meeting on Chemical Sensors, July 27-30, 1998, Beijing, CHINA, 861-863 (ISBN 7-80003-423-2),.

[29] Emmanuelle Schaller, Thesis "Applications and limits of electronic noses in the evaluation of dairy products", ETH Zürich, (2000).

[30] EM Microelectronic-Marin SA, Laboratory Dr. Zesiger, Fleur-de-Lys 9, CH-2074 Marin-Epagnier, http://www.smartnose.com/

[31] E. Schaller, S. Zenhäusern, T. Zesiger, J.O. Bosse, F. Escher, Analysis, **28**, 743-749 (2000). Available via http://www.edpsciences.org/articles/analusis/abs/2000/08/an2009/an2009.html

[32] R. T. Marsili, J. Agriculture and Food Chemistry **47**, 648 (1999).

[33] W. Grosch, Trends Food Sci. Technol. **4**, 68 (1993).

[34] P. Schieberle in Characterization of Foods: Emerging Methods, Edt. A. G. Gaonkar, Elsevier, Amsterdam, 1995.

[35] L. Heinert, Doctoral thesis, Systematic Structure-Effect Investigations between Semiconducting Oxide Sensors and Hydrocarbons, Gießen, 2000.

[36] T. Hofmann, P. Schieberle, C. Krummel, A. Freiling, J. Bock, L. Heinert and D. Kohl, Sensors & Actuators B **41** 81 (1997).

[37] P. Schieberle, T. Hofmann, D. Kohl, C. Krummel, L. Heinert, J. Bock and M. Traxler, *Flavor Analysis, Developments in Isolation and Characterization*, eds. C. J. Mussinan and M. J. Morello, American Chemical Society, Washington DC, 1998.

[38] Method and semiconductor gas sensor device for detection for explosive hazards, EU-Patent EP 0 608 483 B1 25.11.1998, Siemens AG München together with RWE Energie AG Essen, inventors: J. Hellmann, H. Petig, D. Kohl, J. Kelleter and O. Kiesewetter, 1998.

[39] A Cooker Hood: European patent WO 00/66950, Electrolux, Stockholm 29.4.1999. Inventors Häusler, Jens; Kohl, Dieter; Eskildsen, Christian; Ovenden, Neil.

[40] Osmetech plc, Electra House, Electra Way, Crewe CW1 6WZ, UK, http://www.osmetech.plc.uk, http://www.aromascan.com/acrobats/osmebroch.pdf, http://www.aromascan.com/acrobats/osmetechnology.pdf

[41] R. L. Grob, Modern Practice of Gas Chromatography, Wiley, NY, 1995.

[42] H.-D. Wiemhöfer and K. Camman, *Sensors, A comprehensive Survey*, ed.. W. Göpel, J. Hesse and J.N. Zemel, VCH, Weinheim, 1991.

[43] J. W. Gardner and E. L. Hines, *Handbook of Biosensors and Electronic Noses: Medicine, Food & the Environment*, ed. E. Kress-Rogers, CRC Press Inc., Boca Raton, 1996.

2-D Tactile Sensors and Tactile Sensor Arrays

Katsuyuki Machida
NTT Microsystem Integration Laboratories, Atsugi, Japan

Joel Kent
Elo TouchSystems Inc., Fremont, California, USA

Contents

1	Introduction	863
2	Definitions and Classifications	863
3	Resistive Touchscreens	866
4	Ultrasonic Touchscreens	867
5	Robot Tactile Sensors	868
6	Fingerprint Sensors	869
6.1	Introduction	869
6.2	Basic Principles of Capacitive Fingerprint Sensors	870
6.3	Single Chip Fingerprint Sensor Array with Integrated Fingerprint Identification Processing	870
6.4	Sensor Structure and Fabrication Process	871
7	Summary and Outlook	873

2-D Tactile Sensors and Tactile Sensor Arrays

1 Introduction

This chapter reviews commercially developed tactile sensor technologies used in touchscreens and fingerprint sensors. Tactile sensors for robots, such those that used in dexterous robotic end effectors, will also be briefly considered. Supporting this commercial activity is a significant body of engineering experience and expertise. This knowledge may also prove useful in emerging and future applications of tactile sensors.

Touchscreens are transparent input devices that are placed in front of displays. When touched by a finger or a stylus, two-dimensional coordinates are reported to a host computer. Touchscreens are widely used in a variety of applications including handheld computers, train ticketing machines, restaurant cash registers, information kiosks, etc. Responding to marketplace competition, touchscreen technology provides a tactile sense at low cost per unit area.

Compared to touchscreens, fingerprint sensors are more sophisticated tactile sensors. A touchscreen often replaces a mouse and typically produces little more information than the (x,y) coordinate of a single touch position. In contrast, fingerprint sensors generate high-resolution two-dimensional images of fingerprints. Fingerprint sensors provide excellent examples of tactile sensor arrays that generate a large quantity of data.

The commercial development of fingerprint sensors and touchscreens provides a solid foundation of knowledge and experience, which may lead to new applications. Perhaps such new applications will be found in the field of robotics. Robotics has its own rich body of experience and research on the topic of tactile sensors. Nevertheless, for certain robot sensor applications, robotics may well have something to learn from the low-cost-per-unit-area sensors of touchscreen technology as well as from the cost-effective high-density tactile sensor arrays with integrated local signal processing developed for fingerprint sensors.

2 Definitions and Classifications

A wide variety of sensor technologies exist, each with its own characteristics. Some care is required to select the best sensor technologies for specific applications. For this purpose, Figure 1 introduces a classification scheme. Also, let us define some terminology.

Here we define a *touch* as any contact between the sensor system and an object external to the system. For a touchscreen interface at an interactive museum exhibit or a railroad-ticketing kiosk, the *touch* is due to a human finger. For a touchscreen in a handheld computing device, the touch is typically from a small stylus. In the artificial hand of a robot, many individual sensor elements may be *touched* thus providing means to analyze shape and surface structure much like a human hand. Fingerprint sensors divide the sensing area into many pixels [1], each of which may or might not have a touch. Tactile sensors detect *touches*.

Here we limit the term *tactile sensor* or *tactile sensor arrays* to sensors that not only detect the presence of a touch, but also detect the position of a touch or touches. This excludes, for example, a simple elevator button that is either "ON" or "OFF". Touchscreens and fingerprint sensors are two examples of tactile sensors for which industry and academia have developed a considerable body of knowledge and experience.

The most basic information provided by a tactile sensor is the location of touches. The complexity of this position information can vary greatly depending on the sensor technology. A fingerprint sensor represents one extreme. Such sophisticated sensors generate complete high-resolution two-dimensional images of fingerprints. At the other extreme is the resistive touchscreen of a handheld computer. A typical resistive touchscreen generates only a single (X,Y) coordinate pair. It will respond to simultaneous

VII Sensor Arrays and Imaging Systems

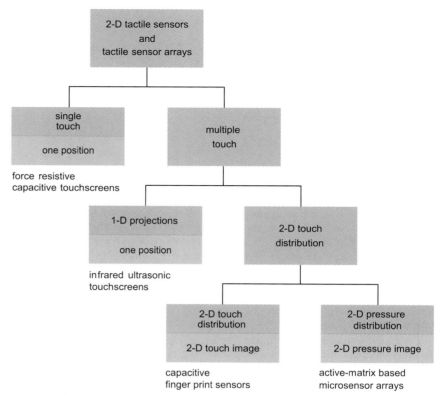

Figure 1: Hierarchy of Tactile Sensor Arrays.

Figure 2: Force-based sensor.

touches at several locations with a single intermediate (X,Y) position. Other tactile sensors are between these two extremes. Figure 1 presents a hierarchy of tactile sensor types based on the nature of the position information.

Some tactile sensors can only generate a single touch coordinate. This is clearly illustrated by the force-based touchscreen system illustrated in Figure 2. When the plate is touched, pressure or strain sensors at the corners 1, 2, 3, and 4 measure the forces F_1, F_2, F_3, and F_4. If only a single localized touch of force F_A is applied, normalized (X,Y) coordinates of the touch are easily determined by the measured forces as follows.

$$X = \frac{(F_3 + F_4)}{(F_1 + F_2 + F_3 + F_4)} \qquad (1)$$

$$Y = \frac{(F_2 + F_3)}{(F_1 + F_2 + F_3 + F_4)} \qquad (2)$$

However, if a second for force F_B is simultaneously applied, the detected forces F_1, F_2, F_3, and F_4 will be identical to a single localized touch of force $F_A + F_B$ at an intermediate location as shown in Figure 2.

Typical resistive touchscreens are also single touch tactile sensors. Mechanically, a resistive touchscreen is a variation of membrane switch. As indicated in Figure 3, it is electronically a variable resistor where V_0 is the applied voltage and V_{ADC} is the voltage measured by an analogue-to-digital converter. The touch position is related to the measured voltage by the by $X = V_{ADC}/V_0$. Even if electrical contact is made between the flexible membrane and the base plate at several locations, there will still be only one measured voltage and hence only one intermediate touch location reported. Section 3 on Resistive touchscreens discusses the generalization from 1-D to 2-D position measurements.

Capacitive touchscreens are another example of tactile sensors of the *single touch* category (Figure 4). A capacitor is formed when a user's finger touches a capacitive touchscreen. One electrode is a resistive layer in the substrate and the other electrode is the flesh of the user's finger. This capacitance is typically detected by observing a shunt to ground of current at frequencies in the kHz range. While an insulating stylus is not

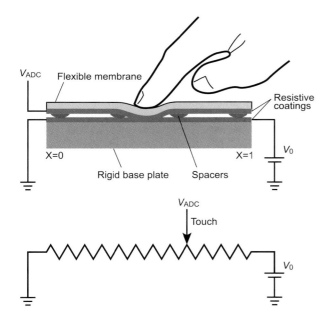

Figure 3: Resistive touchscreen mechanism.

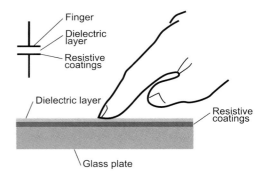

Figure 4: Capacitive touchscreen mechanism.

detected, such systems can provide excellent touch sensitivity for ungloved human fingers. The current shunted through the finger is supplied by circuitry connected to the four corners of the touchscreen. Position information obtained by observing how this current is divided between the four corners. These four corner currents are analysed much like the distribution of forces at the four corners of the force-based touchscreen discussed above.

We consider a tactile sensor to have *multiple touch* capability if it can distinguish between a single localized touch and simultaneous multiple touches. Infrared touchscreens as illustrated in Figure 5 have limited multiple-touch capability. A grid of infrared beams is formed, typically in a sequential scan. Each infrared beam starts at an LED and ends at a phototransistor. The electronics observes which beams are interrupted by touches. The basic data is in the form of one-dimensional (1-D) projections. If two touches are simultaneously present, each will cast its own shadow. There may be an ambiguity regarding which shadow in X corresponds to which shadow in Y; see dotted circles in Figure 5. Often timing information will resolve such ambiguities. The raw measurements of infrared touchscreens contain more complex touch position information than force, resistive and capacitive touchscreens.

Section 4 considers ultrasonic touchscreens. Ultrasonic touchscreens, like infrared touchscreens, also generate raw data, which is in the form of 1-D projections, and hence are also an example of tactile sensors with multiple touch capability.

Commercial ultrasonic and infrared touchscreen systems typically do not make full use of the multiple touch capability of the sensor. Touchscreen systems generally present touch information to the host computer's operating system in mouse format. The operating system expects only a single (x,y) position at a time from mouse-like drivers such as touchscreen drivers. Due to limitations of the host computer's operating system, infrared and ultrasonic touchscreens typically generate only *one position* at a time as indicated in Figure 1.

Fingerprint identification requires a high-resolution two-dimensional (2-D) image. Fingerprint sensors, such as described in Section 6, provide a complete *2-D touch distribution*, which is used to produce a black-and-white *2-D touch image*. Like a display device, a two-dimensional area is finely divided into pixels. Each pixel provides, at a minimum, one bit of information on the presence of absence of touch contact. If more than one bit of information is provided per pixel, the sensor provides a *2-D pressure distribution* and the system can create a grey-scale *2-D pressure image*. Here *pressure* may be a direct measurement of physical pressure, or simply any variable measured quantity related to the degree of touch.

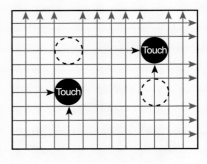

Frame with IR emitters and receivers

Figure 5: Infrared touch mechanism.

3 Resistive Touchscreens

Referring to Figure 3, now let us consider resistive touchscreens in more detail.

The flexible membrane of a resistive touchscreen is typically constructed of a polyethylene terephthalate (PET) film. A thin hard coating is applied to the exterior film surface to improve scratch resistance. A resistive coating of indium-tin oxide (ITO) is sputtered on the film's interior surface. Variations and refinements to this basic membrane construction have been developed. The rigid base plate is typically ITO coated glass. Alternately a polymer base plate may be used. The spacers are dots of transparent and insulating material bonded either to the flexible membrane or the base plate. The desired two-dimensional array of dots may be, for example, screen-printed and then cured using ultraviolet light. In commercial products, spacer dots heights are typically less than 25 μm and separated by at most a few millimetres.

Touchscreens must be transparent. This constrains the choice of materials and fabrication methods. If resistive touchscreen technology is applied, for example, to the design of artificial skins for robots, transparency is no longer required and more material and fabrication options become available.

Figure 3 presents the basic mechanism and equivalent electronic circuit for coordinate measurement. Commercial products use two very different approaches to extend this basic principle of operation to two dimensions: *4-wire* and *5-wire* resistive touchscreens. These descriptions refer to the number of electrical connections between the sensor and associated electronics.

Hand held personal digital assistants (PDA) typically use 4-wire resistive touchscreens. Figure 6 illustrates this type of touchscreen.

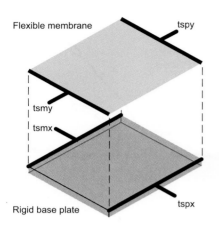

Mode	tspx	tsmx	tspy	tsmy
X	power	ground	ADC	ADC
Y	ADC	ADC	power	ground
Detect	power	power	ground	ground

Figure 6: 4-wire resistive touchscreen.

In the Figure 6, *tspx* labels the wire connected to a bus bar on the resistive coating of touchscreen's positive X side [2]. During an X coordinate measurement a voltage gradient is created in the resistive coating of the base plate by applying power to *tspx* while grounding *tsmx*. Wires *tspy* and *tspx* from the membrane's coating transmit the touchpoint voltage to circuitry including an analogue-to-digital converter (ADC). During Y coordinate measurement, the roles of the base plate and flexible membrane are swapped.

To save power, resistive touchscreens have a *detect* mode. In detect mode, the sensor draws no current and functions as a simple "ON/OFF" membrane switch. Before a touch, the sensor *sleeps* in detect mode. When a touch is detected, the touchscreen system measures two-dimensional touch coordinates by sequentially alternating between the X and Y modes.

Figure 7 illustrates the operating principles of a *5-wire* resistive touchscreen. In a 5-wire touchscreen, the membrane's resistive coating is always connected to the voltage sensing circuitry. Both X and Y coordinate measurements use the base plate's resistive coating as the resistor in the equivalent circuit of Figure 3. The only function of the membrane's resistive coating is to electrically transmit the base plate voltage at the touch point to the ADC.

ITO is a brittle ceramic material. Touchscreen use causes flexing of the membrane, which in turn can cause degradation in the electronic properties of the resistive coating. 4-wire resistive touchscreens depend on uniform resistivity of the membrane's ITO coating. Wear may lead to distorted coordinate measurements. In contrast, 5-wire resistive touchscreens have no requirement for the resistivity of the membrane coating to be stable or uniform. The performance of 5-wire touchscreens does not degrade with such wear. Resistive touchscreens that must be reliable despite heavy use, such as in restaurant cash registers, are typically 5-wire resistive touchscreens.

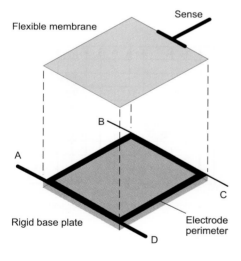

Mode	A	B	C	D	Sense
X	ground	ground	Power	Power	ADC
Y	ground	power	power	ground	ADC
Detect	power	power	power	power	ADC

Figure 7: 5-wire resistive touchscreen.

The *electrode perimeter* schematically shown in Figure 7 typically includes a complex geometry of conductive regions, resistive regions and insulating regions [3]. This is the topic of much of the patent literature for 5-wire touchscreens. The challenge for the electrode perimeter is to provide boundary conditions resulting in linear voltage gradients for both X and Y measurement modes.

The 5-wire patent literature also considers an interesting alternate approach [4]. The design of the electrode perimeter can be greatly simplified if one is willing to accept equipotential lines that are no longer straight and equally spaced. In this sense, raw measured coordinates become *non-linear*. In fact, if resistive touchscreen technology is applied to the design of an artificial skin covering a section of a curved non-Euclidian surface, it is mathematically impossible to avoid such non-linear equipotential lines. Such non-linear raw position measurements can be mapped back to any preferred coordinate system by applying appropriate mapping algorithms in downstream software.

After calibration, commercial resistive touchscreens typically generate measured touch coordinates within about 1 % of the touchscreen's diagonal dimension from the actual touch position. This is more than enough accuracy for typical touch applications. Greater accuracy is possible if tighter tolerances are placed on the uniformity of the ITO coatings and/or more complex calibration schemes are tolerated. However, for typical touchscreen applications, that would add unnecessary cost and complexity.

Relative position resolution is often much better than absolute accuracy. PDAs often take advantage of this in handwriting input systems.

Touch sensitivity is largely determined by the membrane and spacer dot parameters shown in Figure 8. Typically, a finger touch covers many spacer dots. In such cases, it is the applied pressure (force per unit area), not the total touch force itself that determines if the membrane flexes enough to make electrical contact with the base plate's coating. If the spacer dots have height h and spacing s, and the membrane has Young's modulus E and thickness d, then the pressure to activate, p_{activate}, scales as follows.

$$p_{\text{activate}} = \text{constant} \frac{E \, d^3 \, h}{s^4} \qquad (3)$$

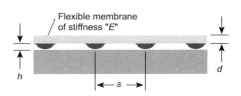

Figure 8: Parameters affecting sensitivity.

The constant depends on the geometry of the two-dimensional array of spacer dots. The pressure to activate depends on the inverse fourth power of the dot spacing s. The engineer can greatly vary the touch sensitivity with even modest changes in dot spacing. If small dot spacing is desired, for example to achieve a smoother writing surface, lower dot heights h and/or thinner membranes are required.

For a given force, a pointed object like a stylus results in a much higher applied pressure (force per unit area) than a human finger. For handwriting applications, it is possible to design the touch pressure threshold to respond easily to a stylus and at the same time tend to ignore accidental finger and palm contact with the writing surface.

4 Ultrasonic Touchscreens

The characteristics of ultrasonic touchscreen technology may be of particular interest to the development of alternate applications of touchscreen technology. At a low cost per unit area, it may be possible to give a sense of touch to hard exposed surfaces of robot shells of non-planar geometry. Furthermore, multiple touch capability is possible. This motivates us to choose ultrasonic touchscreens as a second touchscreen technology to consider in more detail.

Most commercial ultrasonic touchscreens sense touches with Rayleigh waves. The Rayleigh wave is an example of a surface acoustic wave (SAW). Rayleigh waves can be visualized as miniature ocean waves propagating on the surface of an elastic material. In touchscreens, Rayleigh waves propagate in glass. Glass has the advantages of transparency, low cost, and a relatively low attenuation rate for ultrasonic waves. Wavelengths are typically a fraction of a millimetre and correspond to an operating frequency in the neighbourhood of 5 MHz.

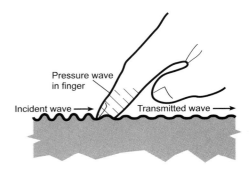

Figure 9: Ultrasonic touch mechanism.

The basic touch sensing mechanism is illustrated in Figure 9. When a finger makes contact with the glass surface, it absorbs power from the incident Rayleigh wave. The dominant absorption mechanism is radiation damping. The Rayleigh wave velocity in glass (≈ 3160 m/sec) is supersonic with respect to the speed of sound in water (≈ 1500 m/sec) and finger flesh. This radiation of sound into the finger rapidly absorbs Rayleigh-wave power much like a supersonic aircraft looses energy to radiated sound. The result is that even a light touch of a finger casts a shadow in Rayleigh waves. The shadow in turn reveals one coordinate of the touch position.

Note the simplicity of the touch sensitive substrate. Unlike resistive and capacitive touchscreens (Figure 3 and Figure 4), no coatings or layered structures are required. Even window glass from a house will be touch sensitive if means are provided to excite and detect Rayleigh waves. Ultrasonic Rayleigh waves also propagate very nicely in metals such as aluminium and steel, but of course such materials are not transparent and cannot be used for touchscreens. In contrast, polymer materials are less attractive as they more rapidly attenuate ultrasonic waves.

Like infrared touchscreens (see Figure 5), ultrasonic touchscreens also detect shadows produced by touches. However, in this case, the shadow corresponds to attenuation of a surface acoustic wave such as a Rayleigh wave.

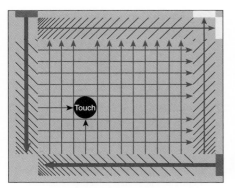

Figure 10: Ultrasonic touchscreen.

Figure 10 is a schematic drawing of an ultrasonic touchscreen. A transmit transducer (dark rectangle) emits a beam of surface acoustic waves. Arrays of 45° reflectors scatter the waves across the desired touch sensitive area. A second array redirects the waves to a receive transducer (lighter rectangle). Position information is determined by the delay time at which the received signal is attenuated. These components are duplicated to provide both X and Y coordinate measurements.

Typically, commercial touchscreen products are flat as required to match the geometry of flat rectangular liquid-crystal displays. However, ultrasonic touchscreen technology is also well suited to curved surfaces. Rayleigh waves simply follow the curvature of the surface much like ocean waves follow the curvature of the Earth. One variation of ultrasonic touchscreen technology is to use the curved glass surface of a CRT display itself as the touch sensitive surface [5]. The generalization of ultrasonic touch sensor geometry to rather arbitrary curved surfaces has been considered in the patent literature [6].

As shown in Figure 1, ultrasonic touchscreens are tactile sensors with inherent multiple touch capability of the 1-D projection variety. To date, commercial touchscreen have made little use of the capability. Menu driven software applications, whether activated via a touchscreen or a mouse, generally assume only one touch or *click* location at a time. Perhaps this multiple-touch capability will have value in future applications of ultrasonic tactile sensor technology.

The activation characteristics of an ultrasonic sensor may be modified with the addition of a coversheet [7]. In this case, the sensor would respond to force or pressure in a manner similar to resistive touchscreens. For touchscreen applications there is little commercial interest in this possibility as one can simply use a resistive touchscreen. Perhaps for other tactile sensor applications it may be of interest to combine the resistive touchscreen's activation characteristics with other features of ultrasonic touchscreens such as multiple-touch capability.

5 Robot Tactile Sensors

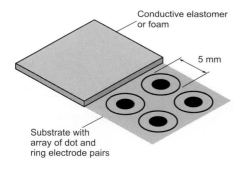

Figure 11: Conductive elastomer or foam based robot sensor.

This section touches only briefly on the subject of robot tactile sensors. See [8] and [9] for a more complete discussion.

A key motivation for the development robot tactile sensors has been the desire for dexterous end effectors (*robot hands*) to perform delicate manipulations of work pieces. The work piece must be gripped with sufficient force to be firmly held and at the same time not subjected to excessive forces that may cause damage. If the surfaces of the robot that are in contact with the work piece are provided with tactile sensors, then feedback control can assure that applied forces are in the desired range. The needs of such applications have guided much of the development of robot tactile sensors.

Because quantitative monitoring and control of applied forces is often the purpose of robot tactile sensors, the fundamental touch activation mechanism is typically force based. Note the contrast with touchscreen sensors whose detection mechanisms may be based not only on applied forces, but also on ultrasonic damping, on capacitive coupling and on the interception of infrared beams.

While robot tactile sensors typically interact with the external environment via forces, within the sensor itself, a wide variety of mechanisms may be used to convert the applied forces to electronic signals. For example, each sensor element may contain a resistor formed of a piezoresistive conductive polymer. When a force is applied to such a sensor element, a change in resistance is detected. Alternately, each sensor element may contain a capacitor with area A and gap distance d. An applied force will induce a measurable change in capacitance $C = \varepsilon A/d$ if there is a force induced change in either A or d. For example, if the gap distance is filled with a compressible dielectric material, the gap d will decrease when a force is applied. Sensing may also be based on magnetic field effects. An electronic signal is generated if an applied force changes the distance between a small magnet and a Hall probe. Another magnetic approach is to electronically monitor changes of inductance of transformers or inductors for which associated magnetic field lines pass through a magnetoelastic material subject to external forces. Optical approaches have also been developed. If an applied force deforms the geometry of the optical path between an optical source and receiver, then the applied force changes the received optical intensity. Likewise, if applied forces modulate the transmission characteristics of an optical wave-guide such as an optical fibre, received light intensity is again a measure of applied forces. Yet another possibility is that each sensor element contains a strain guage attached to a surface that is distorted when a force is applied. In the design of robot tactile sensor arrays, there are indeed many options for converting applied forces into electronic signals.

To date, the majority of tactile sensors used in many forms of robots have been based on observing resistance changes associated with the compression of a conductive elastomer or foam. See Figure 11 for a representative example. Each element of the sensor array has a dot and a ring electrode pair in contact with a layer of conductive foam. The entire sensitive area of the tactile sensor is covered with this layer, but in the drawing it is partially removed to reveal some of the dot and ring electrodes on the supporting substrate. The conductive elastomer or foam may be, for example, rubber loaded with fine carbon particles. The resistivity of such materials decreases when compressed. A map of the applied forces over the area of the tactile sensor is obtained by monitoring the resistance between the electrodes of each dot and ring pair.

Figure 12 shows an example of a dexterous end effector [11]. A number of tactile sensors are embedded in the fingers. The embedded force sensors are based on a piezoresistive polymer material [10]. Feed-back from these sensors plays a key role in the grasping process. As more sophisticated tactile sensor technology becomes available, the capabilities of such dexterous end effectors will increase.

Development of MEMS (micro electro-mechanical systems) technology creates new possibilities for the design of robot tactile sensor arrays. The promise of a high density of sensor elements in an integrated package is enticing. This is an active area of research and development that may in time lead to innovative new commercial tactile sensor products.

As tactile sensors are developed with increasing numbers of sensor elements, it becomes increasingly difficult to provide the necessary interconnections and real-time signal processing. For example, an n by n sensor array generally requires a minimum of $2n$ wire connections and generates n^2 force measurements to be processed and interpreted. Researchers have shown the feasibility of addressing these problems by integrating the sensor array with LSI (large scale integration) circuitry that not only measures the force applied to each sensor element, but also provides for parallel processing of the measured data. Such technology still needs to mature before it is ready for large-scale commercialisation, but nevertheless holds great promise for the future of robot tactile sensors.

Interestingly, while of great interest to the field of robotics, technology that integrates tactile sensor arrays and LSI circuitry for sensor read out and parallel signal processing is likely to first mature and be commercialised as a mass-produced product for an entirely different application: fingerprint ID sensors. Fingerprint ID sensors address an immediate and significant need within our expanding information-age infrastructure for improved identification and security solutions. The field of robotics may well benefit from relevant design and manufacturing technology developed in response to marketplace demand for improved fingerprint ID sensors. The next section presents an interesting example of such a fingerprint ID sensor.

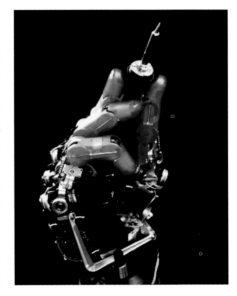

Figure 12: Example of a dexterous end effector. The photograph shows the DLR Hand II holding a solder pen, made by DLR, German Aerospace Center (DLR), Institute of Robotics and Mechatronics (from [11]).

6 Fingerprint Sensors

6.1 Introduction

Fingerprint ID sensors are another important example of tactile sensors. In the near future, fingerprint sensors are likely to become mass-produced commercial products much like touchscreens are today. As the underlying sensor technology and manufacturing processes matures, they may well find new applications beyond fingerprint identification.

Conventional methods for gaining access to electronic systems use secret passwords or codes, which unfortunately can often be easily compromised. On the other hand, techniques based on biometrics, that is the quantitative analysis of biological characteristics of the user, hold promise for much more secure access control of electronic information technology. Biometrics includes analysis of facial features, speech, the iris scans, signatures, hands, or fingerprints. Perhaps fingerprint analysis is the simplest biometrics method and may well have the greatest commercial potential.

VII Sensor Arrays and Imaging Systems

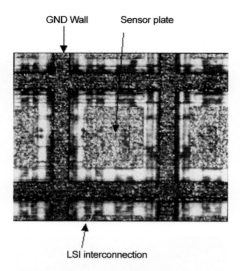

(a) Photograph

(b) After USPCT

Figure 13: A principle of capacitive fingerprint sensor LSI.

Identification by fingerprint has a long and interesting history [12]. While fingerprints in clay appear to have been used as a kind of signature as far back as ancient China and Babylon, a 1684 publication by the Englishman Nehemiah Grew is credited as the start of the modern study of fingerprints. In recent times, many systems based on various techniques have been developed for fingerprint identification. Many of today's fingerprint identification systems [13] use optical sensors. Such optical systems include many components such as sensors, prisms, and a computer. Recent technology advances have enabled sensors of much smaller size including micro machined pressure sensors [14], thermal sensing sensors [15] and capacitive sensors based on thin-film transistor (TFT) technology [16]. To further reduce the size and number of parts in the system, a semiconductor capacitive sensor has also been developed and implemented with CMOS LSI electronics [17], [18].

From a commercial perspective, such compact and highly integrated capacitive fingerprint ID sensors show great promise. Furthermore, such sensors provide an excellent example of a *2-D touch distribution* tactile sensor of Figure 13. Below we describe such a capacitive fingerprint ID sensor in some detail.

6.2 Basic Principles of Capacitive Fingerprint Sensors

Figure 13 shows the sensing principle of a capacitive sensor array fabricated using LSI technology. Such semiconductor sensor chips have been developed for fingerprint identification systems using LSI multilevel interconnection process technology and a conventional plate structure [17], [18]. When a finger makes contact with the sensor surface, capacitances are formed between the finger and the sensor plates. By detecting such added capacitances at the sensor plates, a fingerprint image is directly captured. The sensor contains a large array of pixels, each containing a sensor plate and sensing circuit. When a ridge of the fingerprint contacts the surface of the pixel, the sensed capacitance is high, and when a valley between ridges prevents contact, the measured capacitance is low. The capacitance for each pixel is converted to one bit of digital data, thus creating a black-and-white image of the fingerprint.

Once a digital image of a fingerprint has been generated, it may be transferred outside the chip for processing. However, for a high level of security, it is best to avoid digital transmission of fingerprint image files. The security of a fingerprint sensor system is greatly improved if no image files are transmitted and all fingerprint image processing and identification is done locally within the fingerprint sensor system. This is the case for the example discussed in detail in this section. While the special security needs of fingerprint ID systems motivate such integration, methods to integrate sensors arrays and associated signal processing is of general interest to the field of tactile sensors.

6.3 Single Chip Fingerprint Sensor Array with Integrated Fingerprint Identification Processing

Figure 14 shows a block diagram of a single-chip fingerprint sensor/identifier fabricated using 0.5-μm CMOS LSI technology (FIL-chip) [19]. After standard LSI processing steps, additional novel processing steps described in Section 6.4 create a sensor array on top of the CMOS circuitry. Digital processing of fingerprint images is done in parallel. Under each sensor pixel, the associated electronics for signal detection and imagine processing is located. In addition to the pixel array, the chip also includes a small-embedded controller to orchestrate the parallel image processing in the pixel array and communicate results of the fingerprint matching algorithms to the host computer.

Figure 15 shows the basic elements contained in each pixel. The sensor is a conducting plate that forms a capacitance with a finger when touched. The sensing circuit represented by the triangle measures the sensor's capacitance and generates a digital 1 if the capacitance is sufficiently large. The 1-bit memory represented by the small square stores a template of the user's fingerprint captured when the computer was registered. The large square represents a processing circuit that compares the fingerprint images from the sensor and the template in the memory. Figure 16 shows the elements of the processing circuit in more detail. In order to correct for translations and distortions of the sensed fingerprint image with respect to the template stored in memory, the controller may instruct the pixel processors to shift sensor data to neighboring pixel processors. The pixel processor informs the controller whether or not the bit in the pixel memory agrees with the sensor bit (perhaps shifted from another pixel). All these elements are provided for each pixel of the array.

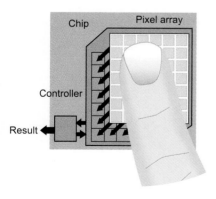

Figure 14: A block diagram of single chip architecture.

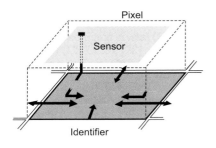

Figure 15: The concept of identifying pixel.

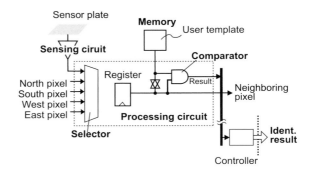

Figure 16: Processing circuit of the pixel.

Fingerprint identification is achieved by the following pixel-parallel image processing steps:
1. Store the user's fingerprint image template in pixel memory as 1-bit per pixel.
2. Capture fingerprint image with sensors and digitize as 1-bit datum per pixel for digital processing.
3. Shift digitized image between pixel processors to allow for variations in finger position.
4. Compare the shifted sensed datum with the template datum, and transmit the results from each pixel to the controller.

The first step is performed once when the computer is registered. Subsequent fingerprint identification involves steps 2, 3, and 4. The fingerprint verification algorithm is based on thinned-image pattern matching [20]. The algorithm reduces the complex task of fingerprint verification into simple parallel image processing steps that are well matched to the limited capabilities of the pixel processors. As a result, the system provides fingerprint identification at high speed and low power.

For dexterous end effectors of robots, this type of parallel processing of tactile-sensor-array data may be of interest for slip sensing.

6.4 Sensor Structure and Fabrication Process

To be of commercial value, a fingerprint sensor must be durable and reliable. The sensor must not break when subjected to a broad range of applied finger pressures. The circuit elements must not be damaged by electrostatic discharge (ESD) generated by static electricity through the user's finger. Moisture and salt residues from finger touches must not corrode the sensor. For capacitive sensor arrays built on LSI silicon chips, these reliability issues require special attention. Product reliability must be a high priority for sensor design and fabrication methods, as is well illustrated by the fingerprint ID sensor discussed in this section. Extensive testing (published elsewhere [21]) has confirmed that highly reliable capacitive fingerprint sensors can indeed be designed and built.

Figure 17 shows the sensor structure. This structure simultaneously meets the need for a reliable product design and a low-cost manufacturing process. The layers labeled 0.5 μm *CMOS LSI* contain the sensing and logic circuits and are fabricated using standard LSI processes. The upper layers labeled *Capacitive Sensor* contains the array of sensor plates and involves innovate designs and processing as described in detail below.

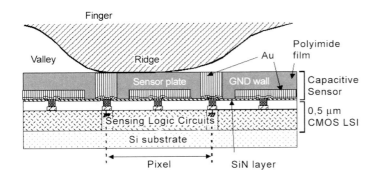

Figure 17: Sensor structure.

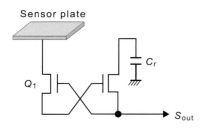

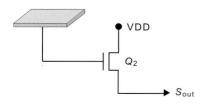

Figure 18: Circuit configuration. C_r is the reference capacitance. S_{out} is the sensing signal output.

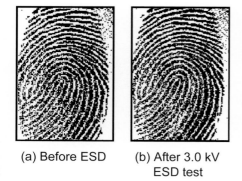

Figure 19: Sensed image before and after ESD test.

Ground walls (GND wall) surrounding each sensor plate provide tolerance to ESD. Unlike the sensor plates, the ground walls provide a conductive path to the exposed surface of the sensor array. Static electricity from a user's finger will discharge through the ground walls rather than through the sensor plates and associated circuitry. Thus the ground walls protect the sensors and electronics from ESD much like a lighting rod protects a house from lighting.

Improved design of the electronic interface between the sensors plates and the sensing circuits provides further ESD protection. Figure 18 illustrates both conventional and the improved interface design. Conventionally the sensor circuit is connected to the gate electrode as shown in Figure 18b. Here ESD tolerance determined by the gate breakdown voltage. In Figure 18a, to avoid the gate breakdown of conventional circuits like gate electrode of MOSFET Q_2, and the sensor plate is connected to the source electrode of MOSFET Q_1. Because the junction breakdown voltage is higher than the gate breakdown voltage, ESD protection is improved [22]. Figure 19 shows fingerprint images captured before and after ESD testing, thus demonstrating immunity to ESD.

The *Capacitive Sensor* layer shown in Figure 17 is also designed to protect underlying structures from moisture and salt residues. The grounded walls are fabricated as gold electrodes to prevent oxidation. The other exposed material at the surface is a thick polyimide film that is also resistant to moisture and salt. The SiN provides a moisture barrier to further protect the underlying electronics.

Testing has shown that the approximately 2-μm thick polyimide film also effectively protects sensor and LSI structures from mechanical damage. Furthermore, testing has also shown that even with applied finger forces well above the pain threshold, the 675-μm thick silicon substrate will not crack.

Figure 20 shows the sensor fabrication process flow. Note the integration of the sensor array with associated LSI electronics. The sensor is fabricated after the 0.5-μm CMOS LSI and three-metal interconnection processes. The sensor process temperature is below 350 °C. 350 °C is below all LSI processing temperatures and is sufficiently low to prevent diffusion of moisture and sodium contaminants into sensitive LSI circuitry. The sensor plate and GND wall are 1.0- and 3.0-μm thick, respectively. They are made of gold and are fabricated by electroplating, which is an easy and simple way to form a thick film. The seed layers for electroplating are deposited by evaporating Au/Cr on the SiN film (Figure 20a). The SiN film is deposited at a substrate temperature of 300 °C, using $SiH_4/NH_3/N_2$ gas by plasma CVD (chemical vapor deposition). The Au, Cr and SiN films are 0.1-μm thick. After resist patterning and electroplating, the sensor plate and GND wall are patterned by wet etching (Figure 20b). Photosensitive polyimide film is applied to the thick, hard passivation layer. Polybenzoxazole is spin-coated onto the sensor plate and the GND wall and polymerized to form polyimide. The polyimide on the GND walls is exposed to UV irradiation in order to planarize the surface (Figure 20c). After that, the polyimide film is annealed at 310 °C in nitrogen atmosphere. Finally, the planarization structure of the sensor surface is obtained (Figure 20d). This fabrication process provides a simple and cost effective way to make a single-chip fingerprint sensor/identifier LSI chip (FIL-chip).

Figure 21 shows a scanning electron micrograph (SEM) and a focused ion beam (FIB) photograph of the sensor structure. The SEM shows the sensor/identifier array at the FIL-chip surface. One can see that the planarization structure of the sensor surface is clearly achieved. Each pixel is 81.6 μm × 81.6 μm. The FIB photograph shows that the sensor plate is fabricated on the MOSFETs and the GND wall is exposed at the sensor surface.

The FIL-chip fabricated by NTT in this manner also includes an embedded controller and program memory. The controller operates at 1 MHz and consumes 187 μW of power at a supply voltage of 3.3 V. The FIL-chip has 20,584 pixels, and each pixel contains 158 MOSFETs. It operates at the practical sensing and identifying time of 102 ms. The characteristics of NTT's FIL-chip are summarized in Table 1.

	Die size		15 mm × 15 mm
Identifying array	area		10.1 mm × 13.5 mm
	pixels		20,584 (124 × 166)
	density		311.3 dpi
Pixel	size		81.6 μm × 81.6 μm
	Tr. count		158
Time	Sensing & binarizing		2 ms/Image
	identifying		100 ms/Image
Supply voltage			3.3 V

Table 1: FIL-chip characteristics.

The above capacitive sensor array with integrated image processing meets the need for an inexpensive, thin and mechanically compact, tamper proof and reliable tactile sensor for fingerprint identification. Furthermore, it provides an interesting example of the general concept of a tactile sensor array with integrated LSI circuitry for sensor read out and parallel signal processing.

7 Summary and Outlook

Touchscreens, robot sensors and fingerprint sensors are examples of tactile sensors. These tactile sensors are the focus of significant commercial activity and provide a body of knowledge and industrial infrastructure that may be of interest to future tactile sensor applications.

In the field of robot sensors, there is increasing interest in *smart sensors* in which sensing and LSI circuitry are integrated to provide local signal processing. The fingerprint sensor system described in this chapter is an example of such a *smart* tactile sensor. The commercial demand for small, low-cost, reliable, and secure fingerprint ID systems may well lead to technology and industrial infrastructure development of relevance to the field of robotics.

As discussed in the robotic literature, tactile sensing is a key component in many robotic systems, ranging from moving parts to assisting surgeons in orthopaedics procedures. While detailed sensing requirements such as dynamic range and resolution will depend on the application, all tactile sensors in the field need to exhibit reliability

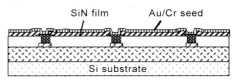
(a) Seed layer

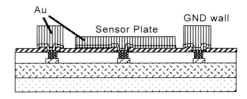
(b) AU/Cr wet etching

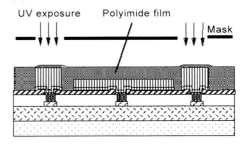
(c) Photosensitive film

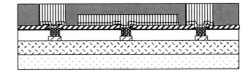
(d) Surface planarization

Figure 20: Sensor fabrication process.

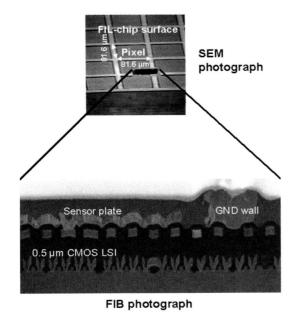

Figure 21: FIL-chip surface of SEM (scanning electron microscope) and the cross section of FIB (focused ion beam) photographs.

through robustness of design. To some extent the move from the research laboratory to the production line has proved to be a significant hurdle for the developers of tactile sensors. It is interesting to note that in many robotic applications vision sensors are more developed than tactile sensors. Indeed at the conference on biologically inspired robotics [23] it was noted by Rodney Brooks of MIT, that the most significant challenge in robotics was in the development of sensor that would give robots a sense of touch that will in some form approach that of a human. It is hoped that some of the high-resolution sensors discussed in this chapter, will in some way go towards meeting this goal.

As robot applications expand from highly controlled factory environments to a broader range of less controlled applications, existing touchscreen technology may become the basis of future sensors for robots. In particular, for applications involving large area tactile sensors, touchscreen technology provides interesting examples of sensor designs with a low cost per unit area. Touchscreen technology may provide external shells of moving robot parts with a tactile sense as a safety feature for collision detection. More generally touchscreen technology may enable robots to respond in a more sophisticated manner to touches. For example, imagine a toy robot dog whose back is simultaneously an ultrasonic touch sensor and a force-based touch sensor. With such touch sensor input, the robot dog could be programmed to wag its tail when petted with a human hand (both ultrasonic and force-based touch systems have large signals), to whine when its back is poked with a sharp object (strong force signal but little ultrasonic signal), and to shake when the dog's back is sprinkled with water (strong ultrasonic signal but little force signal).

Current touchscreen, robot, and fingerprint sensor technology provides a solid foundation for the field of tactile sensors. Time will tell how the next generation of engineers will build on this foundation to further innovate the technology and applications of tactile sensors.

Acknowledgements

The authors wish to thank members of the NTT research team that developed the FIL-chip. In particular thanks go to H.Kyuragi for his helpful support and encouragement. Thanks also go to S. Shigematsu, who proposed the single chip, H. Morimura, who proposed the sensing circuit, H. Ishii, H. Unno, Y. Tanabe, K. Sakuma, S. Yagi, N. Sato, Y. Okazaki for their helpful discussions and support, and M. Yano, K. Kudou, T. Kumasaki for the chip fabrication and measurement.

The authors thank Elo TouchSystem, Inc. for its enthusiastic support. Despite numerous other competing activities of more immediate business concern, high priority was given to the preparation of this chapter.

We most gratefully acknowledge Prof. Richard M. Crowder of the University of Southampton (UK), an expert in the field of robotics, for his critical review and excellent editing suggestions.

The editor would like to thank Peter Gerber (RWTH Aachen) for checking the symbols and formulas in this chapter.

References

[1] Strictly speaking, *pixels* are *picture elements* of a display image. An alternate term such as *taxels* for *tactile elements* would more accurately describe elements of a tactile sensor array.

[2] The notation used here is similar that used in the data sheet for the Philips Semiconductors UCB1300 advanced modem/audio analog front-end chip. This chip is intended for PDA applications and includes a flexible resistive touchscreen controller.

[3] Examples of electrode perimeter designs used in commercial products are shown in Figs. 1 and 2 of US patent 5,045,644 of David J. Dunthorn as well as and Figure 7 of US patent 4,198,539 of William Pepper Jr.

[4] For example, see the PCT patent application WO 98/19283 of TopoTec, Inc. (bwarmack@comcast.net).
[5] Elo TouchSystems, Inc.'s iTouch™ product line; see www.elotouch.com.
[6] US patents 5,854,450 (1998) and 6,091,406 (2000).
[7] US patent 5,451,723 of Jianming Huang and Terence J. Knowles.
[8] R.M. Crowder, in *Handbook of Industrial Automation,* ed. by Richard L. Shell and Ernest L. Hall, Marcel Dekker, Inc., 2000.
[9] N.I. Glossas, N.A. Aspragathos, Mechatronics **11**, 899 (2001); C. Melchiorri, IEEE-ASME Trans. Mechatronics. **5**, 235 (2000); M.H. Lee, Int. J. Robot. Res. **19**, 636 (2000).
[10] H. Liu, P. Meusel, and G. Hirzinger Proc. of the 4th International Symposium on Measurement and Control in Robotics (ISMCR'95), June12-16, 1995, Slovakia.
[11] http://www.robotic.dlr.de/mechatronics/ .
[12] H. Lee and R. Gaensslen, *Advances in Fingerprint Technology*, Elsevier, 1991.
[13] R. Sandage and J. Connelly, in Proceedings of the International Electron Device Meeting, 171 (1995).
[14] P. Rey, P. Charvet, M. T. Delaye, S. Abou Hassan, Transducers '97, Chicago, 1453 (1997).
[15] K. Sato, T. Kadowaki, H. Jisong, and M. Shihida, International Conference on Solid-State Sensors and Actuators, Digest of Technical Papers, LN.7 (1999).
[16] N.D. Young, G. Harkin, R. M. Bunn, D. J. McCuulloch, R. W. Wilks, and A. G. Knapp, IEEE Electron Device Lett. **18**, 19 (1997).
[17] M. Tartagni and R. Guerrieri, IEEE Tran. Solid-State Circuits **33**, 133 (1998).
[18] D. Inglis, L. Manchanda, R. Comizzoll, A. Dickinson, E. Martin, S. Mandis, P. Silveman, G. Weber, B. Ackland, and L. O. Gorman, ISSCC Dig. Tech. Papers, 284 (1998).
[19] S. Shigematsu, H. Morimura, Y. Tanabe, T. Adachi, and K. Machida, IEEE Tran. Solid-State Circuits. **34**, 1852 (1999).
[20] T. Kobayashi, Proceeding of 4th International Conference, Computing and Information, 341 (1992).
[21] K. Machida et al., IEEE Transactions on Electron Devices **48**, 2273 (2001).
[22] H. Morimura, S. Shigematsu, and K. Machida, IEEE Journal of Solid-State Circuits **35**, 724 (2000).
[23] WGW'02 EPSRC/BBSRC International workshop on Biologically Inspired Robotics- The legacy of W. Grey Walter, Bristol, 2002.

Displays

Contents of Part VIII

	Introduction to Part VIII	879
37	Liquid Crystal Displays	887
38	Organic Light Emitting Devices	911
39	Field-Emission and Plasma Displays	927
40	Electronic Paper	953

Introduction to Part VIII

Contents

1 **Definition**   879
2 **Photometry**   880
3 **Sensitivity of Human Eye**   881
4 **Color Theory**   882
5 **Display Concepts and Addressing**   883
6 **3-D Display Concepts**   883
   6.1 Aided Viewing   883
   6.2 Free Viewing   884
   6.3 Further Approaches   885

1 Definition

One of the most powerful means of making electronic information intelligible to humans is the visible display of this information, since the eye accepts very high amounts of data in parallel. A variety of emerging information appliances is available from very large public messaging screens (> 100 inches diagonal), high definition television screens (HDTV, 100-30 inches), table top monitor screens (30-15 inches), screens for portable devices with a size below 15 inch, and finally down to displays on SmartCards and tickets with only one inch in size.

Through many decades, the cathode ray tubes (CRTs) [1] developed by F. Braun in 1897 was the only means of visualisation of complex information electronically. Its image performance is highly functional, although it has one decisive drawback for the mobile society: it is hardly a portable device because of the high operation voltage, the large weight, and the high sensitivity to vibrations. The CRT is lined with a phosphorous material that glows when it is struck by a stream of electrons which is produced by a set of electron guns at the back of the monitor, see Figure 1 [2]. This phosphorous material is arranged into an array of millions of tiny cells, called dots. To produce a picture on the screen, the electron beam scans the luminescent screen row by row by the means of the deflection plates. The video signal controls the intensity of the electron beam at each dot and, therefore, the color and brightness of each pixel on the screen. On a color monitor, there are three electron guns in order to display red, green, and blue light respectively. The surface of the CRT is arranged to have these dots placed adjacently in a specific pattern. By varying the intensity of the red, green, and blue streams, the full spectrum of colors is accessible. The surface of the CRT only glows for a small fraction of a second before beginning to fade. This means that the monitor must redraw the picture very quickly in order to avoid flickering of the screen. Current CRTs operate at horizontal scanning frequencies up to 140 kHz and a vertical scanning frequencies above 100 Hz.

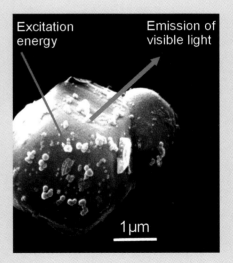

Figure 1: Exciting radiation is absorbed in a phosphor particle of a few micrometers in diameter, and converted into visible light and heat, since the conversion efficiency is < 100 % [2].

Today, a variety of display realizations and concepts have been introduced. There is no mainstream path, many very different examples of combination of engineering and materials science have been created. An overview is given in Ref. [3], where the display approaches are classified by their physical mechanisms which is applied to convert an electronic into a visible signal, see Figure 2. **Light generating displays** such as light emitting diodes (LEDs), including organic light emitting diodes (OLEDs) or cathode ray tubes (CRTs) emit light itself, i.e. generate photons. The contrast strongly depend on the surrounding light, and they have a high power consumption because of the low efficiency of optoelectronic transducers. **Light controlling displays** such as liquid crystal displays (LCDs) or electronic paper diffuse, polarize, absorb or depolarize the incident light, i.e. interact with ambient photons. The advantages are low power consumption and a high contrast in bright ambiences.

In the framework of this textbook, the content of Part VIII focuses on liquid crystal displays (Chapter 37), organic light emitting devices (Chapter 38), plasma displays and field emission displays (Chapter 39), and electronic paper (Chapter 40).

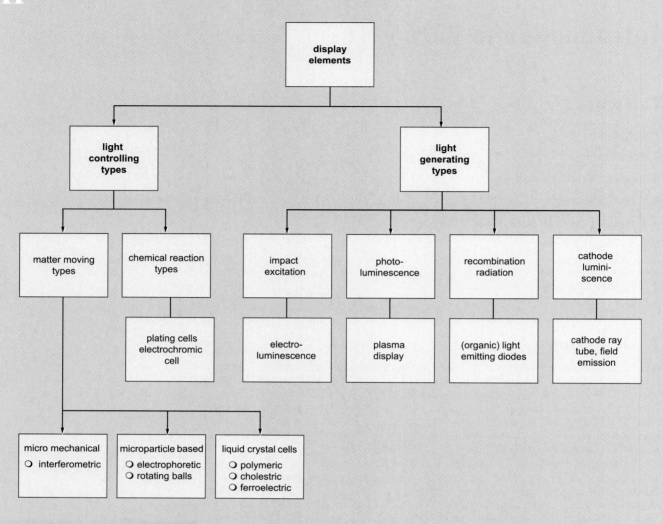

Figure 2: Overview and classification of the different families of displays based on Ref. [3]

2 Photometry

Photometry deals with the measurement of light, i.e. optical radiation in the visible spectral range. The measurement is executed just as the human eye perceives brightness and color, using specific photometric units and nondimensional numbers for characterizing colors. The properties of light sources, measurement devices and illumination are specified with photometric quantities and quantities for color characterization [4].

The **luminous intensity** I is the light emitted by a light source in a given space angle. It is one of the basic units of the International System of Units, and is measured in candela. The **luminous flux** ϕ (unit lumen) corresponds to the total visible spectrum of a point light source emitted in all space directions perceived by the eye. It describes the performance of lamps without taking into account any dependence on the direction or distance of the observation position. The **illuminance** E rates the illumination of areas (e.g. working environments). It is given by the luminous flux per illuminated area. The unit of the illuminance is lux. The **luminance** or **brightness** L is the physical value of a illuminated area of a light source or a lighted object perceived by the eye. Depending on the view angle and the illumination scene an illuminated area will be seen with different brightness. The unit is candela/m². The luminous flux in a given time is described by the **luminous energy** Q in lumen h. It is used to specify the lifetime of light sources. The **luminous exposure** H is the product of illuminance and duration in lux×s.

The **relations** between the various photometric values are shown below, e.g. the derivation of the luminous flux ϕ with respect to the steradian Ω leads to the luminous intensity I. It should be distinguished between the area of a receiver A_R and an emitter A_E.

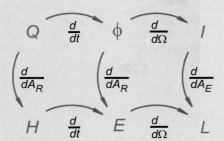

In photometry, all values are **physiological**, i.e. they depend on the sensitivity of the human eye. The luminous energy follows from the physical radiation energy Q_e with K_m=683 lm/W by:

$$Q = K_m \int_{\text{visible}} \nu(\lambda)\, Q_e(\lambda)\, d\lambda \tag{1}$$

3 Sensitivity of Human Eye

The curves in the Figure 3 show the normalized response of an human eye to various intensities of ambient light [5]. The shift in sensitivity occurs since two types of photoreceptors, cones and rods, are responsible for the eye's response to light, see also Introduction to Part VII. Under normal lighting conditions, the **photopic response** of the eye is acting. This curve peaks at 555 nm which means that under these light conditions, the eye is most sensitive to a greenish yellow color. In that case, the cones, which are composed of three different photo pigments that enable color perception, respond to the light. When the light levels drop to near total darkness, the sensitivity of the eye is shifted to shorter wavelength (**scotopic response**). Then, the rods are most active. Rods are highly sensitive to light but are comprised of a single photo pigment, which accounts for the loss in the ability to discriminate color. At this very low light level, sensitivity to any amount of light is present, but is less sensitive to the range of color. Blue, violet, and ultraviolet is rather perceived than yellow and red. The bold curve in the middle represents the eye's response at ambient light level found in an intermediate range.

The perception of the human as a function of illumination is illustrated in Figure 4 [6]. The human eye typically exhibits an instantaneous **dynamic range** of less than 200:1. There is a threshold for this range depending on the illumination conditions which separates the stochastic behavoir (in the mesopic region and probably in the scotopic region) from the deterministic one (photopic region). At 10^{-4} Lamberts (= 0.318 cd/m^2) the transition from mesopic to scotopic vision takes place. All chromatic response is lost. As the scotopic region is approached, the human reports some residual chromatic sensations in the blue green after all response in the red region is lost. The chromatic response of the human eye is also degraded at very high levels. The hyperopic region is characterized by a loss in chromatic saturation that accentuates the perceived sensitivity in the region near 487 and 580 nm. A subject generally reports a yellowing of the scene.

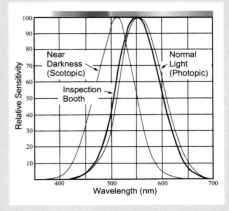

Figure 3: Normalized response of an average human eye as function of wavelength at normal daylight, near darkness and an intermediate range [5]

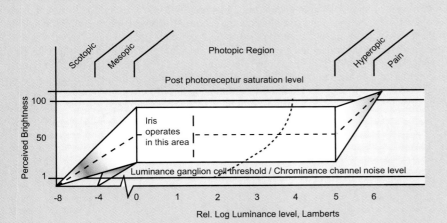

Figure 4: Luminous transfer function of human vision for an uncontrolled sky source of unknown color temperature [6]

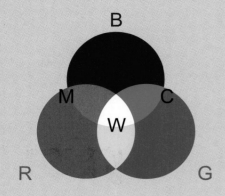

Figure 5: Additive colour mixing. Illustration of the colour impression resulting from a superposition of red, green and blue primary colours

The average luminous transfer function between object space and perceptual space in the absence of any large signal information is indicated by the dashed line. A combined effect of the iris and the adaptation amplifiers is to maintain a constant level of perceived response over a wide illumination range (photopic region). Within this range, the eye reacts as if the average input illumination is fixed [6].

4 Color Theory

There are two basic ways colors can be mixed to produce other colors. One is **additive color mixing**, in which the combined colors are formed by adding light from two or more light sources. If colored light is mixed, the brightness of the colored lights are added also. The three primary colors of light are red, green and blue. White light is formed by adding all three primary colors, see Figure 5.

The other way is **subtractive color mixing** by means of no-light generating elements as filters or reflecting objects. A filter absorbs the light of certain wave lengths. Only the color, which is evoked by the superpositon of the colors passing the filter, will be observed. The absorption and reflecting of light is also responsible for the perception different colors of objects.

Colorimetry is the science of measuring colors [5]. Because each individual perceives colors slightly differently, the Commission Internationale d'Eclairage (CIE) has defined a **standard observer**. A set of standard conditions for performing color measuring experiments has also been established by CIE. These experiments consist of choosing three particular light sources, that emit light on the white screen, where three projections overlap and form an additive mixture. On the other side of the screen a target color is projected, and an observer tries to match the target light by altering the intensities of the three light sources. The weights of the light sources are in the range [-1,1]. A negative weight means adding this color to the target color. In order to avoid negative values a linear transformation of the matching functions was introduced by CIE resulting in x, y and z matching functions, as shown in Figure 6 [5].

If the surface reflectance, and the light source distribution are known, their product defines color as $C(\lambda)$. The weights, X, Y and Z define a color in the CIE XYZ space which is a 3-D linear color space.

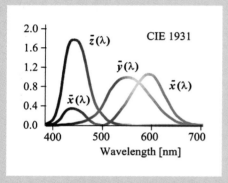

Figure 6: The x, y, and z colour-matching functions of a standard observer as defined by the Commission Internationale de l'Eclairage (CIE) 1931

$$X = \int C(\lambda)\,\bar{x}(\lambda)\,d\lambda \qquad Y = \int C(\lambda)\,\bar{y}(\lambda)\,d\lambda \qquad Z = \int C(\lambda)\,\bar{z}(\lambda)\,d\lambda \qquad (2)$$

It is common to project this space onto a 2-D space known as the CIE chromaticity diagram. The coordinates in this diagram are usually called x and y and they are derived from XYZ using the following equations:

$$x = \frac{X}{X+Y+Z} \qquad y = \frac{Y}{X+Y+Z} \qquad z = 1-x-y \qquad (3)$$

Each distribution function of the light source results precisely in a definite xy point of the chromaticity diagram. However, each *xy* point could be caused by many arbitrary distribution functions (Figure 7).

The border of the colored area in the CIE chromaticity diagram represents the color coordinates of monochromatic radiation, marked with its wavelength in nm, i.e. the whole visible spectrum. The straight line connecting the lowest wavelength blue and the highest wavelength red is called the *purple line* and does not represent spectral colors. The small quadrilaterals indicate chromaticity standards as defined by the European Broadcasting Union (EBU) for colour television for red (R), green (G) and blue (B) primary emission. In the centre, the so-called Black-Body line indicates the colour points for emission of a Planckian black body radiator, with different color-temperatures (CT) indicated. D65 is the color point of daylight at CT = 6500 K which has the coordinates $x = 0.313$ and $y = 0.329$, while A corresponds to an incandescent lamp emission with CT = 2850 K.

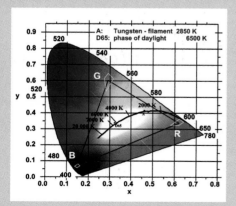

Figure 7: CIE chromaticity diagram

5 Display Concepts and Addressing

Every display consists of one, few or many single illuminated **elements**. Corresponding to their number the displays are based on one icon, few simple segments or many pixels. The differences become clear in Figure 8. Only in the case of simple alphanumerical displays or in the case that a linear relation exists between luminous intensity and operation voltage, it is applicable to address every single one separately. Otherwise, the pixel have to be organized in a **matrix** (see also Introduction to Part IV). The pixels on a screen are addressed one row at a time via the row electrodes. A single pixel is switched on or off since an appropriate voltage is applied to the column electrode at the same time. That way, the number of addressed lines of a (n,m) matrix is reduced to $2(n\,m)^{1/2}$. An important challenge of the matrix concept is to avoid crosstalk with non-addressed elements. This requires a large difference between the states *on* and *off* (contrast). A further reduction of the addressed lines can be achieved by a dynamic mode (**multiplex addressing**). In addition, this technique allows to vary the operation voltages of different matrix elements. An important criterion for the suitability of multiplexing a display is the dynamic response of the single pixel. The raise time has to be short, the decay has to be long enough, in order to avoid any flickering. In Chapter 37, a detailed description of the multiplex addressing scheme is given.

In a **passive matrix display**, as described above, the visible information represented by a row of pixels fades during the period of time needed to address all the other rows in the display (Figure 9). A non-flickering image requires a balance of the display fade rate and the persistence of vision in the human eye. In an **active matrix display**, each pixel is connected to its corresponding row and column electrodes by additional transistors keeping the pixel on or off, even when the row in which the pixel resides is not being addressed. Flickering is not a problem since the row of pixels remains static, Figure 9.

Figure 8: From top to bottom: Schematic presentation of icon-based, segment-base, and pixel-based displays. The display *flexibility* increases from top to bottom.

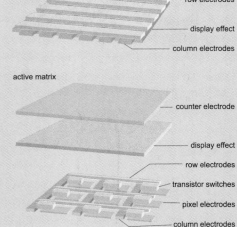

Figure 9: Concepts of passive and active matrix

6 3-D Display Concepts

There is a manifold of areas in which 3-D display are desirable. These areas comprise the field of medicine such as telematic operations, diagnostics or training as well as the fields of accident simulations, prototyping, and entertainment.

Depth perception, i.e. 3-D viewing, is primarily based on the binocular information from the differences between the retinal images in the left and right eyes. When the eyes converge on an object at a certain distance, the image projects to corresponding places in the two eyes, but the images of objects at other distances will project to different places, as shown in Figure 10. This separation between the images of these objects in the two eyes is known as **binocular disparity** [7]. The human brain analyzes both images and synthesizes a single 3-D image. This effect is called **stereoscopy**. Additional information sources of depth perception are the physiological information from the eye muscles, monocular information from complex patterns, and dynamic information from motions.

Stereoscopy is by far the most important element of 3-D perception. To display a 3-D image, it is necessary to feed a specific image with different perspective to the left as well as to the right eye, i.e. a display system has to handle two image informations. The common tasks of every concept are collecting, processing, transporting, (and as the case may be superposition, separating) and feeding the images to the eyes.

6.1 Aided Viewing

This class of 3-D displays use stereoscopic means to separate the superposed image informations [8]. The **Anaglyph process** requires two images taken from two slightly different vantage points. One image is colored in blue, the other one in red. Both images are superposed and merged by glasses with a blue and a red filter. That way, a 3-D illusion is produced. This is the oldest 3-D technique, and was introduced through movies in the 1950's. A similar approach is the variation of light projection with two projection objectives with **polarizers** crossed to each other. A 3-D effect is evoked by the uses of glasses with an analog pair of polarizers. Synchronized **shutter glasses** fall in the same category. Two shutters from liquid crystals alternately open and close the left and the right eye. If the shutter frequency is higher than 25 Hz, the brain senses the imaging at

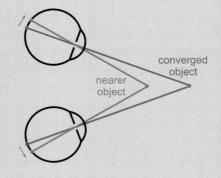

Figure 10: Binocular disparity

VIII Displays

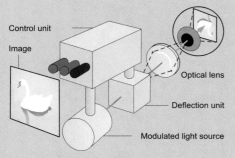

Figure 11: Schematic view of a retina projection in the Virtual Retinal Display of Microvisions Inc. [9]

the same time. **Head Mounted Displays (HMD)** such as glasses or helmet also use the procedure of stereoscopy. Here, separated images are generated by two small liquid crystal displays, directly to each eye.

In current developments, the projection of images takes place directly on the retina of the human eye, e.g. in the **Virtual Retinal Display** (VRD), see Figure 11 [9]. A control unit processes a video signal, manages the deflection unit, and writes the image by a single stream of pixels in the eye. As the beam is rapidly scanned across the eye, the human visual system completes the function which is otherwise performed by the matrix array of a flat panel, visually integrating the pixel elements into a stable, coherent image. The intensity of the light source makes it possible to adapt the intensity of the image to the environment. The power consumption is very low because all the light is directly fed onto the retina. This technology can eliminate any screen outside of the eye, and avoid some of the screen performance and cost inefficiencies.

6.2 Free Viewing

Free Viewing, also referred to as auto-stereoscopy, is another category of 3-D concepts. These systems are only capable for single user because the display is related to the position of the observer. The different informations for the left and the right eye are superposed on a screen. They are separated without the use of any external means, and appropriately merged to the eyes. The image seems to be before or behind the screen.

Figure 12: Idealised 3-D image of a tracked stereo display of SeeReal Technologies [10]

A common feature of the most auto-stereoscopy displays is the **interlaced mode** for the different eyes, e.g. the even columns and the odd columns of pixels are guided to the left and the right eye, respectively. Various techniques are used in order to guide the information to the corresponding eye. One development is based on the refraction by a backside illuminated prisms mask. A tracking system determines the position of the observer (*eye finder camera*) and aligns the orientation of the mask. That way, a movement of the observer becomes possible, see Figure 12 [10].

A different display approach operates in an interlaced mode row by row which is realized by a matrix of holographic elements being illuminated by two light sources (Figure 13). The imaging is displayed on a liquid crystal monitor. Again, a tracking system for the alignment of the light sources guarantees the independence of the observer position .

Related to the imaging of a **mirror**, a new 3-D display concept was introduced . The light beams come out of the mirror plane simultaneously. Since it is not possible to synthesize all light beams of a virtual 3-D reality at the same time, a serial procedure of processing steps is adopted. If the processing rate is high enough, the brain senses a total image. The image is produced by a CRT. The column raster of a light modulator which is placed in front of the CRT allows the corresponding perspective to be passed to each eye. Ten sub-images showing different perspectives are generated so that a pronounced 3-D effect with continuous transitions between the sub-images is achieved. Thus, different images depending on the position are observed.

Figure 13: Operation principle of 3-D display using holographic elements for light deflection row by row

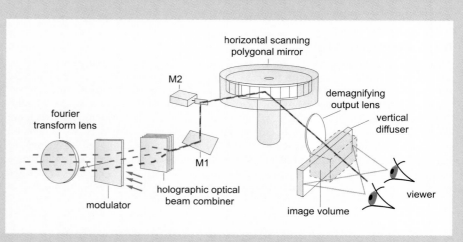

Figure 14: Realization of an holographic video display ot Spatial Imaging Group at the MIT Media Lab [12]

884

6.3 Further Approaches

Holography is a photographic procedure for 3-D imaging by laser light. A splitted beam of coherent laser light falls on a recording medium as well as on the object to be imaged. The unmodified beam and the reflected beam generate an interference pattern on a plate. The developed plate is the hologram. When light falls on the hologram, a 3-D image of the object is generated by the recorded interference pattern. An **holographic video** becomes possible using a real-time imaging system that can render and display computer generated holograms at near-video rates. In 2000, developments were demonstrated by the MIT [12], see Figure 14. The display is capable of rendering full color $25\times25\times25$ mm^3 images with a 15 degree view zone at rates around 20 frames per second. The greatest challenge to the system is the processing of a vaste amount of data which needs further research in order to reduce the computation time.

Spatially Immerse Displays (SID) render pronounced 3-D images by the projection on hemispherical or cylindrical screens. Additional stereoscopic effects are achieved by the combination of these displays with 3-D glasses. With **Virtual Model Displays** (VMD), the *pure display* functionality of rendering is expand by interaction by the means of 3-D pointers or data gloves.

Acknowledgements

The editor gratefully acknowledges Ulrich Böttger (RWTH Aachen) for writing and compiling this overview.

References

[1] L.F. Weber, in *Flat panel Displays and CRTs*, L.E. Tannas, ed., Van Nostrand Reinhold Company, New York, (1985).

[2] With the courtesy of Philips Research Laboratories, Aachen, Germany.

[3] E.O. Johnson, *Electronic Displays*, in Topics in Applied Physics, ed. J. I. Pankove, **40**, 235 (1980).

[4] Peter A. Keller, *Electronic Display Measurement*, John Wiley & Sons, New York, Chichester (1997).

[5] G. Wyszecki, W. S. Stiles, *Color Science, Concept and Methods, Quantitive Data and Formulae*, John Wiley & Sons, 1982.

[6] J. T. Fulton, Processes in Animal Vision, Vision Concepts, Available on the Internet: URL:http://www.4colorvision.com (2000).

[7] G. Wyszecki and W. S. Stiles, *Color Science: Concepts and Methods, Quantitative Data and Formulae*, John Wiley & Sons, 2000.

[8] Three-dimensional Sterographic Visual Displays in Marketing and Consumer Research, M. B. Holbrook, Columbia Univ. 1997.

[9] Homepage of Microvision Inc., USA, http://www.mvis.com/home.htm

[10] Homepage of SeeReal Technologies, Dresden, Germany, URL: http://www.dresden3d.com/default.de.htm

[11] Homepage of VisuReal Diplaysysteme, Oelsnitz, Germany, URL: http://www.visureal.de

[12] Homepage of Spatial Imaging Group at the Massachussets Institute of Technology, Media Lab, Boston, USA, URL: http://spi.www.media.mit.edu/groups/spi/holoVideoAll.htm

Liquid Crystal Displays

Reiner Zorn, Department IFF, Research Center Jülich, Germany

Shin-Tson Wu, School of Optics, University of Central Florida, USA

Contents

1 Introduction	889
2 Liquid Crystal Materials	890
2.1 Physical Properties of Liquid Crystals	891
2.2 Chemistry of Liquid Crystals	893
3 Twisted Nematic Cell	894
4 Addressing of Liquid Crystal Displays	896
4.1 Dynamic (Multiplex) Addressing	896
4.2 Grey Scale and Colour Generation	898
5 Cells for High-Resolution Displays	899
5.1 Supertwisted Nematic Cells	899
5.2 Active Matrix Displays	900
6 Backlighting	901
7 Reflective Liquid Crystal Displays	902
7.1 Two-Polarizer R-LCDs	902
7.2 Zero-Polarizer R-LCDs	903
7.3 Single-Polarizer R-LCD	903
7.4 Reflectors	904
8 Transflective Displays	905
9 Projection Displays	905
9.1 Transmissive Displays	905
9.2 Reflective Displays	905
10 New Liquid Crystal Display Principles	906
10.1 Ferroelectric Liquid Crystals	906
10.2 Polymer Dispersed Liquid Crystals	908
11 Summary	909

Liquid Crystal Displays

1 Introduction

Today, liquid crystal displays can be found in numerous applications from simple wristwatches to complex computer displays. They offer many advantages: They are flat and compact (in comparison to cathode ray tubes). Their power consumption is low (compared to light-emitting diodes or plasma displays). Many of them can be used with ambient light (in contrast to light-emitting devices which lose contrast in broad daylight). At least the low resolution liquid crystal displays are cheap, so that they can be used in low-cost applications. For many applications like hand-held computers or cellular phones it is even difficult to imagine how they could be realised without such displays.

The market volume of liquid crystal displays has been approx. 20 M$ in the year 2000 and is showing an annual rise of well above 10%. 60 - 70% of the *value* of this market consist of computer displays. Nevertheless, consumer applications (e.g. displays in household appliances) are the most important sector (approx. 80%) if one considers the *numbers* of units sold.

The history of liquid crystals is a good example of the fact that scientific discoveries which are considered for a long time to be of only "scholastic interest" can suddenly gain an enormous importance in technology. The first observations of liquid crystals date back to the middle of the 19th century when the first materials were studied which we now can identify as liquid crystals.

Nevertheless, it took nearly half a century until Reinitzer and Lehmann in 1888 correctly identified the liquid crystalline state as a phase of its own occurring between the liquid and crystalline state. These two researchers were also the first to use the term "liquid crystal" for this new state of matter. Their opinion was not undisputed at that time and even well-known researchers such as Nernst strongly opposed it. Eventually, the upcoming methods for studying the molecular structure by x-ray scattering revealed the intermediate symmetry and thus proved Reinitzer and Lehmann right.

The next steps important for the current applications of liquid crystals were done by Friedel in the 1920s. Not only did he establish the classification of liquid crystals into nematic, smectic, and cholesteric dependent on their symmetry properties. He also discovered that liquid crystals can be oriented by external (electric or magnetic) fields, an effect which is the basic principle of all display applications of liquid crystals.

The first theory of liquid crystal phases was established by Maier and Saupe in 1958 [1]. In a mean field approximation the temperature dependence of the order parameter S (Eq. (1)) was derived and the first order phase transition nematic-liquid is explained by the theory.

Although the first ideas of liquid crystal displays came up during the sixties the final breakthrough towards a technological use was triggered by the work of Schadt and Helfrich [2]. The main advantage of the "twisted nematic (TN) cell" invented by them is the low control voltage. This feature made it possible to use liquid crystal displays in battery powered units. The basic principle is the use of the birefringence of a liquid crystal which can be controlled by an electrical field combined with a "frustrated" structure generated by technically imposed boundary conditions (twist). This structure is less stable and can therefore be more easily influenced than the native structure of a liquid crystal. In combination with optical polarisers control of the birefringence changes the attenuation of transmitted light. Interestingly, up to now all commercially successful liquid crystal displays are based on this principle.

As soon as the feasibility of application was proven several improvements of the original design were developed:

Contrast, viewing angle, and addressability were highly improved by introducing "supertwisted" nematic (STN) liquid crystal displays [3]. In these displays the basic principle of a structure near instability is improved to its limits allowing the control of the display by voltage changes in the millivolt range.

The problem of cross-talk in high resolution displays was solved by using an active matrix (AM), i.e. thin film transistors (TFT) which are integrated into the display rather than completely external driver circuits. Interestingly, the idea of an active matrix was

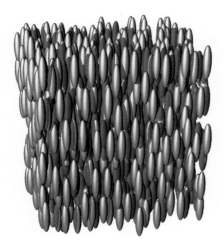

Figure 1: Nematic Phase. Molecules are shown as uniaxial prolate ellipsoids. To show the type of ordering more clearly a perfect directional order (S = 1) has been assumed. Note that in real systems the order parameter is much lower, i.e. the molecules are not arranged as parallel as in this schematic representation (If no other reference is given the 3D figures were generated by POV-Ray™ [7]).

Figure 2: Smectic Phase. As in the preceding figure perfect directional and positional order has been assumed. In reality the molecules will be less parallel and the layers less well defined. Also the gap between the layers is enhanced with respect to a real thermotropic liquid crystal to show the separation more clearly.

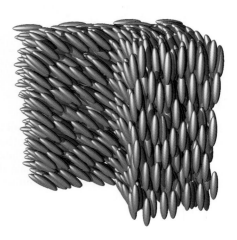

Figure 3: Chiral Nematic (Cholesteric) Phase. To show the twist along the depth axis the front-left quarter of the cube has been cut out. The total twist is 90° and the cut reveals the middle plane in which the molecules are inclined by 45°.

established already in 1971 [4] but became commercially interesting not before the nineties when computer notebook and terminal applications called for high resolution and high quality displays.

The development of liquid crystal displays is still going on. On one hand the existing technology is developed towards larger, higher resolution, better response displays for video and computer applications. On the other hand new technologies (as ferroelectric liquid crystals) are studied which may offer completely new qualities.

Firstly the liquid crystal materials are introduced in this chapter. Then, the physics and electronics of current display technology are explained. Finally, an outlook on future technologies is given.

For a general and more detailed review of liquid crystals the reader is referred to [5], [6].

2 Liquid Crystal Materials

A common way to categorise the aggregate states of matter is by translational symmetry into *crystalline* and *amorphous* states. In crystals translations by a discrete set of lattice vectors transform the solid into itself. In amorphous materials (gases, liquids, and glasses) this symmetry is not present.

The fact that there are three space co-ordinates and in case of molecules without rotational symmetry three angles of orientation makes it possible that states of matter exists where the discrete translational symmetry is only present in some of these co-ordinates, e. g. in the case of the smectic phase SmA (Figure 2) in one space co-ordinate and two angles.

Such states are called *liquid crystalline* because they show properties of both liquids and crystals. In most of them the molecules are directionally ordered. Therefore they show crystal-like anisotropy of susceptibilities. On the other hand they do not form rigid bodies but appear mechanically as viscous liquids.

2.1 Physical Properties of Liquid Crystals

The physical properties of liquid crystals are largely determined by their molecular anisotropy. The molecules are rod- or disk-like shaped, similar to ellipsoids with one axis strongly differing from the other two. This anisotropy causes a tendency of the molecules to arrange in parallel which is often enhanced by electrical dipolar interactions. In the most simple approximation of the molecules as uniaxial ellipsoids the average orientation is described by a vector **n** called "*director*". The degree of orientation is expressed by the order parameter

$$S = \frac{1}{2}\left(3\,\overline{\cos^2\theta} - 1\right) \text{ with } \overline{\cos^2\theta} = \frac{1}{2}\int_0^\pi p(\theta)\cos^2(\theta)\sin(\theta)d\theta \quad (1)$$

where θ is the angle between the molecule axes and the director and $p(\theta)$ its distribution. In addition to the directional ordering one often finds positional order. This ordering concerns one or more co-ordinates in space, i.e. the molecules are arranged in layers, columns or macrocrystalline structures.

The type of symmetry characterising the liquid crystalline order is the (nowadays used) basic classification property. The lowest symmetry is that of the *nematic* (N) phase (Figure 1). In this phase a (long-range) order exists only for the orientation of the molecules. Their positions only show a local order similar to that of liquids but are uncorrelated over large distances.

The next higher symmetry is that of *smectic* (Sm) liquid crystals in which also one spatial co-ordinate is ordered (Figure 2). This means that the molecules arrange in layers. The smectic class of liquid crystals is divided into many subclasses SmA-SmO which differ in the order of molecules within the layers. Especially it is not necessary that (as in Figure 2) the director is identical to the layer normal. The molecules may be tilted resulting in the SmC phase. Higher spatial symmetries are e.g. columnar (Col) or cubic (Cub). But those materials are not (yet?) technologically important.

An important variation of these phases results from the use of *chiral* molecules. These are molecules which are not mirror symmetrical. They are intrinsically optically active. Such molecules build chiral phases, i.e. phases in which the arrangement of molecules is not identical to its mirror image. This fact is expressed by an asterisk appended to the abbreviation of the phase name. Figure 3 shows the most frequent of these phases, chiral nematic N*. Due to its widespread occurrence it was considered a class of its own in the old Friedel nomenclature, the *cholesteric* phase. The cut in Figure 3 shows that the director rotates if one proceeds in one spatial direction perpendicular to the molecules. By this rotation a helicity is defined which corresponds to the chirality of the molecule. If one would build the same structure from mirrored molecules the director would spiral in the opposite way.

Another chiral phase of technological interest is the SmC* (Figure 4). Each of the layers is identical to a SmC layer (similar to SmA, Figure 2, but with the director inclined to the layer normal) but the directors of different layers are not the same. Rather the director rotates around the layer normal when proceeding from one to the next layer. The spiral of directors again defines a helicity of the phase.

Sometimes different phases may be observed for the same material. In the case of the materials discussed here the phase state depends on the temperature. For this reason they are called "*thermotropic*" liquid crystals [8]. As one would expect from the principles of thermodynamics the sequence of phases goes from most ordered to least ordered when raising the temperature: crystalline --> smectic (H -->A) --> nematic --> isotropic liquid.

The anisotropy of the molecules combined with their ordering causes an anisotropy of many macroscopic physical properties, e.g. index of refraction (birefringence), dielectric constant, and elastic constants. All these observables reflect in their temperature dependence the underlying order parameter which in the nematic phase usually can be described by

$$S(T) = \left(1 - yT/T_{NI}\right)^b \quad (2)$$

where T_{NI} is the nematic-isotropic phase transition temperature and y and b are empirical parameters (Figure 5). This means that in the liquid crystalline phase the order decreases when the temperature is raised causing the anisotropy to reduce. At the phase transition it still has a final value before it vanishes upon entering the liquid isotropic phase. This temperature dependence has consequences for technical applications: Often the desired properties of a liquid crystal cell degrade when the temperature is raised.

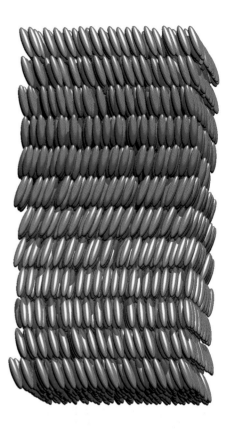

Figure 4: Smectic C* Phase. The height of the picture corresponds to the "pitch" of the helical structure, i.e. the director is rotated by 360° when going from the bottom to the top layer.

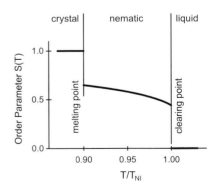

Figure 5: Temperature dependence of the order parameter S(T) defined by Eq. (1). In the crystal the molecules are near-perfectly oriented (S(T) ≈ 1), in the liquid-isotropic phase they are completely disordered (S(T) = 0). In the intermediate nematic phase the behaviour of S(T) is empirically described by Eq. (2).

Figure 6 shows the temperature dependence of the birefringence, the different indices of refraction for the ordinary beam (polarisation perpendicular to the long axis, low molecular polarisability) and the extraordinary beam (polarisation parallel, high molecular polarisability). The behaviour of the indices of refraction can be calculated using the Lorentz-Lorenz equation [10]

$$\frac{n_i^2 - 1}{\overline{n}^2 + 2} = \frac{N}{3\varepsilon_0} \langle \alpha_{ii} \rangle \tag{3}$$

where n_i is the index of refraction for a specific polarisation, with subscript 1 (or e) for the extraordinary beam and 2 or 3 (or o) for the ordinary beam. $\overline{n}$ is its directional average. $\langle \alpha_{ii} \rangle$ are the averaged molecular (anisotropic) polarisabilities. N is the number of molecules per unit volume. They can be calculated from the individual molecular polarisabilities by orientational averaging:

$$\langle \alpha_{11} \rangle = \frac{1}{4\pi} \int_0^\pi \sin\theta d\theta \int_0^{2\pi} d\phi \; p(\theta)\left(\alpha_{11}\cos^2\theta + \alpha_{22}\sin^2\theta\cos^2\phi + \alpha_{33}\sin^2\theta\sin^2\phi\right)$$
$$= \frac{1}{2} \int_0^\pi \sin\theta d\theta p(\theta)\left[\alpha_{11}\cos^2\theta + \frac{1}{2}\alpha_{22}\sin^2\theta + \frac{1}{2}\alpha_{33}\sin^2\theta\right]. \tag{4}$$

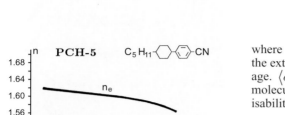

Figure 6: Temperature dependence of the birefringence of the liquid crystal PCH-5. The upper curve shows the extraordinary index of refraction n_e the lower curve shows the ordinary n_o. At the nematic-liquid transition around 54 °C both merge into one value n_i.

Here we assume that the minor axes 2 and 3 of the molecule are randomly distributed. If we compare the last expression with the definition of the order parameter, equation (1), we obtain

$$\langle \alpha_{11} \rangle = \frac{2S+1}{3}\alpha_{11} + \frac{1-S}{3}\alpha_{22} + \frac{1-S}{3}\alpha_{33}$$
$$= \overline{\alpha} + \frac{2}{3}\left(\alpha_{11} - \frac{1}{2}(\alpha_{22} + \alpha_{33})\right)S(T) \tag{5}$$

with $\overline{\alpha} = (\alpha_{11} + \alpha_{22} + \alpha_{33})/3$ being the average polarisability. In a similar way one obtains

$$\langle \alpha_{22} \rangle = \overline{\alpha} - \frac{1}{3}\left(\alpha_{11} - \frac{1}{2}(\alpha_{22} + \alpha_{33})\right)S(T) . \tag{6}$$

From formulae (5) and (6) it becomes clear that the birefringence expressed by $\Delta n = n_1 - n_2$ is a monotonous function of the order parameter. It decreases with temperature and vanishes in the liquid isotropic phase.

The importance of birefringence for the technical application is immediately clear because it allows to control light transmission through an liquid crystal cell via the change of polarisation. The second important physical quantity is the dielectric anisotropy which is related to the possibility to orient liquid crystal molecules by an external field. In a nematic phase it is usually defined by the difference of dielectric constants parallel and perpendicular to the liquid crystal director [9]:

$$\varepsilon_\| - \varepsilon_\perp = NFh\left(\Delta\alpha + \frac{p_e^2 F}{2k_B T}(3\cos^2\beta - 1)\right)S(T) \tag{7}$$

In the large parenthesis one can recognise two terms: $\Delta\alpha = \alpha_{11} - (\alpha_{22} + \alpha_{33})/2$ describes the influence of the anisotropic polarisability on the dielectric constants. Here, F and h are reaction field and cavity field factor which account for the field dependent interaction of a molecule with its environment; for their definition see Chap. 4 of [1]. In the large parenthesis one can recognise two terms: $\Delta\alpha = \alpha_{11} - (\alpha_{22} + \alpha_{33})/2$ describes the influence of the anisotropic polarisability on the dielectric constants. The second term where p_e is the electric dipole moment of the molecule and β its angle with the molecule's long axis describes the effect of rotation of the molecules by imposing an external electrical field. Again the total value is proportional to the order parameter.

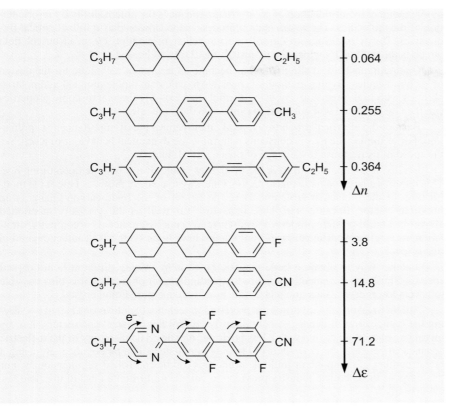

Figure 7: Effect of variation of the chemical structure on the physical properties of liquid crystals. The upper figure shows a variation of birefringence due to the length of the conjugate electron system. The lower figure shows the influence of polar residues on the anisotropy of the dielectric constant.

2.2 Chemistry of Liquid Crystals

The liquid crystal molecules which form the technologically important phases mentioned in the last section are invariably rod-shaped, i.e. their extension in one spatial direction is significantly larger than in the other two. They usually consist of a central part connected to two (possibly substituted) hydrocarbon chains. The central part is a concatenation of ring-shaped groups; these can by aliphatic (e.g. cyclohexane), aromatic (e.g. benzene), or heterocyclic (e.g. pyrimidine). The rings are either linked directly or by short functional groups as ether or ester bridges. The cyclic groups may also have substituents, mostly halogen atoms. The hydrocarbon chain can differ in length and can be either saturated or unsaturated. They are if at all substituted in the end position by halogens in order to increase the dipole moment. It is clear that there is a vast number of possible chemical structures for liquid crystal molecules even under these "construction rules". This allows to vary the physical properties of liquid crystals over a wide range.

As an example the upper part of Figure 7 shows the birefringence of structurally similar but chemically different liquid crystal molecules. The difference in polarisability $\Delta\alpha$ is mostly a consequence of the presence of conjugated electron systems, i.e. alternating sequences of single and double bonds. In the topmost structure the central part consists only of cyclohexane rings, there is no aromatic character. Therefore the polarisability is low arising from displacements of electrons in the restriction of σ bonds only. The lowest molecule in contrast has a π electron system which extends over three benzene rings and the connecting triple bond. In this structure the electrons can move over a much larger distance if an electrical field is applied. Therefore the polarisability is about six times higher.

The lower part of Figure 7 shows that introduction of polar residues increases the dielectric anisotropy. Orientation of such polar molecules in an external field leads to an additional contribution to the dielectric constant. The effect can be enhanced by substitution of an aromatic group. The electronegativity shifts the electron distribution in the π electron system creating a large permanent dipole moment. The lowest structure of Figure 7 shows an extreme application of this principle. The dielectric constant of this liquid crystal in parallel direction is larger than that of water.

Although the effect of the anisotropy of the dielectric constant expressed by Eq. (7) is similar, the variations described in the two parts of Figure 7 have different effects for application. Increasing the number of conjugated double bonds enhances the electronic polarisability and thus only affects the optical properties (birefringence). Increase of the

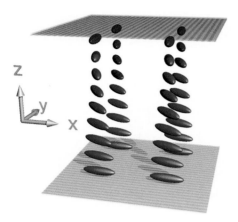

Figure 8: Molecular structure in a twisted nematic liquid crystal display cell. The top and bottom planes are the cell walls of the display. The stripes symbolise the orientation of the surface coating. In operation the two planes will also be the control electrodes and light will go through the cell vertically.

dipole moment by introduction of polar groups enhances the polarisability by re-orientation of the molecules. The re-orientation process is too slow to play a role at optical frequencies. But it is desired to control the molecular orientation by an electrical field which is the method by which the majority of LCDs is addressed.

Quantum mechanical simulation has become an important technique for the design of such molecules. By this method it is possible to calculate dipole moment and polarisabilities with high precision. By using formulae as (3) - (7) the macroscopic properties can then be predicted without need to synthesise the molecules.

Besides the electrical properties several other features are of importance for application. For example, the elastic constants determine the dynamical behaviour of a cell, i.e. the speed of its response to a change of the electrical field.

Finally, it is important to have the properties stable over a large temperature range. This requires a large nematic phase region. One way to achieve this is as before the modification of the molecular structure. Here the length of the hydrocarbon chains is an important factor. Because the phase behaviour (as well as the viscosity mentioned above) is a multi-particle property it can (currently) not be predicted by computer simulation. In lengthy experiment series homologues have to be synthesised and compared in that respect.

Another way to obtain a large nematic range is by mixing different liquid crystal species. Similar to the eutectic range in the phase diagram of an alloy it is thus possible to have a lower crystal-nematic phase transition than in a pure liquid crystal.

Another reason why mixtures are preferred in technical applications is that by varying the composition it is easy to continuously control the physical parameters. E.g. by mixing a polar liquid crystal with a less polar one any desired anisotropy of the dielectric constant in between the $\Delta\varepsilon$ values of the pure components can be obtained. By using a larger number of components multiple properties can be controlled simultaneously. For this purpose the chemical industry provides so-called "multi-bottle kits" of miscible liquid crystals.

3 Twisted Nematic Cell

The core of the twisted nematic liquid crystal display cell is shown in Figure 8. It is a liquid crystal having a structure which is similar to that of the chiral nematic (cholesteric) phase. The main difference is that the structure is not self-assembled but maintained by minute forces exerted by the upper and lower cell walls. The cell walls are coated with polyimide which is "brushed" or "buffed", i.e. rubbed along one direction with a piece of fabric. By this rubbing process the polymer chains are microscopically aligned. Liquid crystal molecules close to these surface order parallel to the polymer molecules. By having the rubbing direction of the top and bottom plate perpendicular to each other one can force the liquid crystal into a twisted structure.

The cell walls are made from glass so that light will be transmitted through the cell. If this light is initially polarised parallel to the molecular director the polarisation will be turned (approximately) with the twist of the structure (Figure 9, top). From the parallel change of molecule and polarisation direction this is called the "waveguide mode" of a liquid crystal cell. The emerging light will be polarised at an angle of about 90° with respect to the incoming light. Therefore, if we put the cell between two parallel polarisers no light will be transmitted because at the second polariser the light will just have the wrong polarisation to be transmitted (normally-black (NB) cell). Conversely, if the polarisers are crossed light will be transmitted through the arrangement which otherwise would be blocked by the second polariser (normally-white (NW) cell).

Figure 9: Function of the TN cell (schematic) in waveguide mode. The ellipsoids indicate the orientation of the liquid crystal molecules. The blue and red ends mark the dipole moment of the molecules. The yellow band shows the polarisation of light traversing the cell from left to right. The lower part shows the effect of an external field: the molecules are oriented parallel to the light and the polarisation-rotation effect vanishes.

The point of this arrangement is now that one can switch off the rotation of the light's polarisation plane by applying an electric field. This is done by coating the glass walls of the cell with indium-tin-oxide (ITO). ITO is an electrically conducting compound which in thin layers is nearly ($\approx 95\%$) transparent. If a voltage is applied to these electrodes there will be a torque on the liquid crystal molecules trying to orient them parallel to the field, i.e. perpendicular to the cell walls (Figure 9, bottom). Although the molecules close to the wall will retain their parallel order the majority of the molecules in the centre will turn parallel to the light path. This means that a large part of the polarisation-rotation will vanish and the cell will go to the opposite transmission state – i.e. black for an NW cell, clear for an NB cell. Figure 10 shows the transmission-voltage characteristics of the first TN cell constructed by Schadt and Helfrich [2] in 1971.

In order to calculate the change of polarisation of light going through a liquid crystal cell exactly it is necessary to take into account the birefringence which changes the incoming linear polarised light into elliptically polarised [12]. Elliptically polarised light can in general be described by the time-dependent field in the direction parallel to the long axis of the molecule (x', ordinary beam, see Figure 11 for definition) and perpendicular to it (y', extraordinary beam):

$$E_{x'}(z) = A(z)\sin\omega t + B(z)\cos\omega t$$
$$E_{y'}(z) = C(z)\sin\omega t + D(z)\cos\omega t. \qquad (8)$$

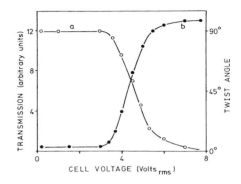

Figure 10: Transmission-voltage characteristics of a twisted nematic cell. The right ordinate shows
(a) the angle of rotation of polarisation effected by the cell. The left axis shows
(b) the intensity of transmitted light in the case of parallel polarisers (normally-black mode). From [2].

The twisted nematic liquid crystal is now considered as composed of infinitesimal layers of thickness dz. If the total twist of the structure is ϑ on a thickness d each layer causes a trivial rotation of $d\vartheta = -(\vartheta/d)dz$ in the molecules' co-ordinate system. In addition the birefringence causes a phase shift $d\varphi = (2\pi\Delta n/\lambda)dz$ for a wavelength λ. Expanding the trigonometric functions of (8) and those arising from the rotation by $d\vartheta$ we obtain a set of linear differential equations for the amplitude factors A–D:

$$\frac{dA}{dz} = -\frac{\vartheta}{d}C \qquad \frac{dB}{dz} = -\frac{\vartheta}{d}D$$
$$\frac{dC}{dz} = \frac{\vartheta}{d}A - \frac{2\pi}{\lambda}\Delta n D \qquad \frac{dD}{dz} = \frac{\vartheta}{d}B + \frac{2\pi}{\lambda}\Delta n C. \qquad (9)$$

These differential equations can be combined into one fourth order differential equation for each amplitude, e.g. for A:

$$\frac{d^4 A}{dz^4} + \left(\frac{2\vartheta^2}{d^2} + \frac{4\pi^2 \Delta n^2}{\lambda^2}\right)\frac{d^2 A}{dz^2} + \frac{\vartheta^4}{d^4} = 0. \qquad (10)$$

As boundary conditions one has the four amplitudes upon entry into the cell at $z = 0$. In the case of the TN cell the light enters polarised parallel to the molecules: $A(0) = 1$, $B(0) = C(0) = D(0) = 0$. The respective solutions for the z dependence of the amplitudes are:

$$A = \frac{q^2}{1+q^2}\cos\frac{\vartheta z}{qd} + \frac{1}{1+q^2}\cos\frac{q\vartheta z}{d}$$
$$B = \frac{-q^2}{1+q^2}\sin\frac{\vartheta z}{qd} + \frac{1}{1+q^2}\sin\frac{q\vartheta z}{d}$$
$$C = \frac{q}{1+q^2}\sin\frac{\vartheta z}{qd} + \frac{q}{1+q^2}\sin\frac{q\vartheta z}{d} \qquad (11)$$
$$D = \frac{q}{1+q^2}\cos\frac{\vartheta z}{qd} - \frac{q}{1+q^2}\cos\frac{q\vartheta z}{d}$$

with

$$q = \sqrt{1 + 2u^2 + 2u\sqrt{1+u^2}} \quad \text{and} \quad u = \frac{\pi d \Delta n}{\vartheta \lambda} \qquad (12)$$

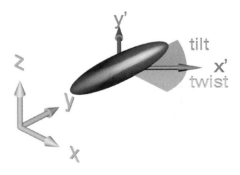

Figure 11: Definition of the local co-ordinates x' and y' attached to the liquid crystal molecule (green), the twist angle (pink) and tilt angle (light blue).

From Eq. (11) one can see that in general the polarisation will be elliptic when the light leaves the cell at $z = d$. Only in the case of very thick cells, $\lambda \ll d\Delta n$, the hand-waving argument that the polarisation of the light stays parallel to the molecule axis is

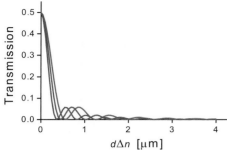

Figure 12: Transmission of a normally-white TN cell with 90° twist. The curves are calculated for red (650 nm), green (530 nm), and blue (425 nm) light.

approximately correct. This is usually called the Mauguin limit. But also for special values $d\Delta n/\lambda = \sqrt{m^2 - (\vartheta/\pi)^2}$ with integer m the polarisation will be elliptic only within the cell but linear (and perpendicular to the initial polarisation) upon exit. Figure 12 shows the transmission expected from the calculations for the normally-black case for different wavelengths and different cell thicknesses.

For economic reasons cells are normally not built in the Mauguin limit but at one of the transmission minima. But as Figure 12 shows at a certain thickness only one wavelength will be completely extinguished. This causes the common problem of "off-state colour" of normally-black devices.

The problem is reduced for normally-white cells because then it just means a slight nuance of white which is more tolerable. In addition the normally-white state offers theoretically unlimited contrast. By raising the voltage arbitrarily high extinctions can be achieved.

Unfortunately, these advantages of the normally-white state are more than compensated by its smaller viewing angle. The range of viewing angles is a common weak point of liquid crystal displays. The problem arises from the different action of the birefringence on rays with oblique incidence. The calculation done above is only valid if the light direction is perpendicular to the molecules. For a beam which is tilted with respect to the cell normal this is not true. Therefore, the contrast may be lower or even inverted when a display is observed under an angle. Figure 13 shows the contrast as the ratio of transmissions of the "white" and "black" states. It can be seen that for the normally-white state the contrast depends strongly on the azimuthal angle what is usually not desirable.

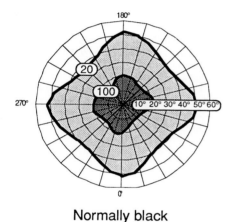

Normally black

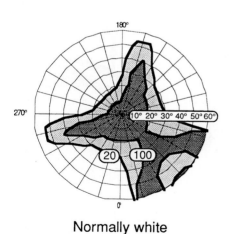

Normally white

Figure 13: Dependence of contrast of a liquid crystal on the viewing angle (calculation). The figures show the contrast of the display when viewed under a polar angle of up to 60° from all azimuthal directions. The iso-contrast line of 20 is usually considered as the tolerable limit. From [5].

4 Addressing of Liquid Crystal Displays

For simple devices (e.g. electronic watch displays) each segment is represented by an individual electrode on one cell wall and the other cell wall is completely coated by ITO. This is called "static" addressing because in principle DC voltages would be sufficient to control the display [11]. For static addressing the number of lines contacting the segments and the number of drivers controlling the voltages is equal to the number of segments.

4.1 Dynamic (Multiplex) Addressing

It is evident that for a laptop computer display with $768 \times 1024 \approx 800000$ pixels it is impossible to realise such a number of leads and drivers. Clearly, the pixels have to be addressed as a matrix through rows and columns. Front and back wall bear electrodes formed as stripes whose intersections define the pixels. This kind of addressing is only possible because the voltage-transmission characteristics (as in Figure 10) show a threshold voltage V_{th} below which the cell stays in the off-state. If we want to address a single pixel we can put a voltage $-V$ to the row electrode and $+V$ to the column. Then only at the part of the cell in between the selected row and column electrode a voltage $2V$ occurs (Figure 14). For all pixels where only one electrode is activated the voltage is V, for the rest of the panel it is zero. If we chose $V < V_{th} < 2V$ only one pixel will go into the on-state.

In order to create an arbitrary pattern of on-states it is now necessary to scan the display row by row similar to a cathode ray tube. That is why this form of matrix addressing is usually called "dynamic". All dynamic addressing schemes rely on the fact that only the root-mean-square voltage $V_{rms} = \sqrt{\overline{V^2}}$ determines the state of the cell if the frequency is sufficiently high.

Figure 15 shows a commonly used addressing scheme called "3:1" (from the number of voltages used). In the example we have 3 rows by 4 columns. It can be seen that during the first three time intervals the three rows are addressed one after the other by a voltage $+2V$. When a row is selected the desired on-state pixels get a voltage $-V$ on the column electrode, those which should be off a voltage $+V$. In this way only for the selected pixels a voltage difference of $2V - (-V) = 3V$ lies across the cell. For all other pixels it is $|0 - (+V)| = |0 - (-V)| = |2V - (+V)| = V$. Because a pixel is selected only during one time interval of three the r.m.s. voltage on the once-selected pixel is

$\sqrt{(V^2+V^2+(3V)^2)/3} = \sqrt{11/3}V = 1.91 \cdot V$. For all pixels which are never selected the r.m.s. voltage is simply V. After the first three time intervals the sequence is repeated invertedly to ensure a zero DC component and then periodically continued.

So by using the 3:1 scheme it is possible to put a r.m.s. voltage which is 91% higher than the "off" voltage onto an arbitrary pattern of selected pixels for a three row panel. It is easy to see that for a panel having n rows the "selected" voltage V_s decreases with n as

$$V_s = \sqrt{1+8/n}\ V. \tag{13}$$

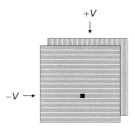

Therefore the relative voltage difference $\Delta V/V$ becomes smaller and smaller if the number of rows increases. This makes dynamic addressing finally impossible because the effected contrast becomes too small. A significant improvement above the 3:1 scheme was introduced by Alt and Pleshko [13]. They simply suggested to change the levels of the row and column voltages to values which optimise $\Delta V/V$:

$$2V \rightarrow n^{3/4}/\sqrt{2(\sqrt{n}-1)}\ V \tag{14}$$

$$\pm V \rightarrow \pm n^{1/4}/\sqrt{2(\sqrt{n}-1)}\ V. \tag{15}$$

Figure 14: Matrix addressing of a liquid crystal cell. The front and back cell wall are covered with stripes of conducting material (indium tin oxide). In order to bring a selected pixel into the on-state a voltage is put on one of the row electrodes and a voltage of opposite sign onto the column. If the voltages are appropriately chosen only the pixel at the crossing of the electrodes (black) will be selected.

These values now depend on the row number n; only for $n = 4$ this scheme is identical to the 3:1. But there are still only three distinguished voltages to be set by the driver circuit so it is electronically easy to realise. By this simple modification leaving the principle of row-wise addressing unchanged the select voltage can be raised to

$$V_s = \sqrt{\frac{\sqrt{n}+1}{\sqrt{n}-1}}\ V. \tag{16}$$

As Figure 16 shows a substantial improvement of the relative voltage difference is possible for high row numbers as they occur in computer display applications. Interestingly, it could be proven that the absolute limit for matrix addressing schemes is not much higher than what can be achieved by the Alt-Pleshko scheme [14]. That is why eq. (16) is still called the "iron law" of dynamic addressing.

Comparing the select voltage ratios in Figure 16 with the voltage-transmission characteristics (Figure 10) one can see that dynamical addressing of TN displays quickly reaches its limits when it comes to high resolution. The select and unselect voltages can not be placed far enough to produce a sufficient contrast. To make it worse the characteristics depend on the viewing angle and the operation temperature. Therefore, an optimal setting of V and V_s may be bad for oblique observation or at different ambient temperature. This usually makes a "contrast control knob" necessary with which the user manually adjusts the voltages to the optimum. Here, the situation can be improved substantially only by the use of supertwisted nematic liquid crystals or an "active matrix" which will be described below.

Another problem related to dynamic addressing results from the response time of the liquid crystal. If the liquid crystal reacts too fast scanning the rows will become visible as a flicker. If it is too slow the display of motions (e.g. a mouse cursor dragged across the screen) will smear out. This problem can nowadays be solved by multi-line

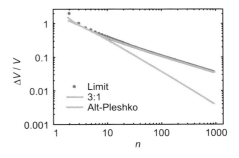

Figure 16: Relative voltage difference between selected and unselected pixels which can be achieved by different addressing schemes. The blue curve represents the 3:1 scheme the green curve the Alt-Pleshko scheme [13]. The red dots show the fundamental limit derived by Nehring and Kmetz [14].

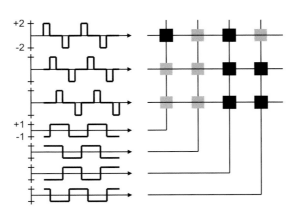

Figure 15: 3:1 addressing scheme. The top three graphs on the left side show the time dependence of the row electrode voltages which are the same for any selected pattern. The bottom four graphs contain the column voltages which are necessary to generate the select pattern displayed on the right.

addressing. The dynamic addressing schemes shown here can be generalised in a way that the row voltages activate several rows at the same time. Then the column voltages do not depend on the intended pattern in a single row but a linear combination of several rows. The calculations of the column voltages are more complicated but can be done by state-of-the-art display controller circuits.

4.2 Grey Scale and Colour Generation

Up to here, only addressing schemes were discussed which generate a pattern of black and white pixels. Of course for video or computer applications also grey levels are necessary. Different principles have been exploited to achieve this:

The most straightforward way is to use analog voltages instead of the finite number of voltage levels described before. Because the voltage differences between "black" and "white" are already very low in highly multiplexed displays the accuracy of voltage generation becomes a problem here. Also a deeper analysis shows that if one uses interpolated voltages, e.g. between the positive and negative value of (15), the voltages on the non-select pixels are not equal among each other. This means that addressing one pixel causes others in the same row or column to change their grey levels. This phenomenon is called "cross-talk" and is one of the major problems of liquid crystal displays.

This can be avoided by switching between the select and unselect voltage during the time interval a pixel is addressed (pulse width modulation). In principle this resolves also the problem to generate exact-valued analog voltages. On the other hand pulses can become very short for "light grey" or "dark grey" situations. This means that the limiting frequency of the driver circuit has to be high and due to capacity between the electrodes new cross-talk can emerge.

A similar possibility is the *frame rate modulation* where a pixel is addressed as black in one period (which would be a "frame" in video applications) and white in the next. Provided that the slow response of the liquid crystal and/or the averaging by the cognitive system of the user is sufficient an impression of "grey" will result. Otherwise frame modulation leads to a flickering display.

Finally, a pixel can be divided into sub-pixels of different area (*spatial dithering*). Putting together black and white pixels in a distance which can not be resolved by the eye creates the impression of one grey pixel. The technical disadvantage is here that the number of addressing lines increases.

In general different principles will be combined. E.g. by a frame rate modulation over two refresh periods one can get three grey levels. If these are distributed onto four sub-pixels with the area ratio 1 : 3 : 9 : 27 a total of 81 grey levels can be synthesised which is sufficient for standard applications.

Similar problems are caused by the demand for colour displays. Here the approach is in nearly all applications spatial dithering as it is for colour cathode ray tubes. Three very small (≈ 100 μm) sub-pixels are combined for each pixel of the display. Each sub-pixel is equipped with a colour filter which can be made of dye or be an interference filter. Interference filters are preferred for projection devices where the energy of the absorbed light would lead to excessive heating of the device. Because each of the sub-pixels only uses a narrow wavelength band the problem of incomplete blocking of light in the off-state can be solved in a simple way here: By matching the gap of the cell to the wavelength of light for each pixel one can go exactly to the minimum positions of Figure 12. The disadvantage of such arrangements is that the colour filters reject about 80% of the white light. For reflective applications where the light goes twice through the cell this is not tolerable. Therefore current colour liquid displays for computer applications have to be operated with backlighting also if ambient light is present.

In a few cases the birefringence colours of a cell which is not operated at a transmission minimum are directly used as the display colours (*electronically controlled birefringence*). In applications where only a limited number of colours have to be represented (e.g. cellular phone displays) this principle has been successfully used. It offers the possibility to build "transflective" displays, which can be operated with ambient light and/or backlight. But if one wants to address the whole colour space – what is expected from a computer display – it is necessary to stack several such devices to create colours by subtractive mixing. Due to the high cost this principle has never been realised in commercial computer display applications although it would offer a higher transparency than common colour liquid crystal displays.

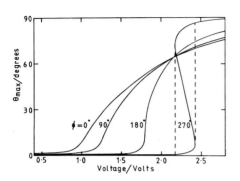

Figure 17: Dependence of the mid-cell tilt angle (see definition in Figure 11) of a (super)twisted nematic structure on the applied voltage. The curves represent different total twist angles ϕ. The mid-cell tilt angle is also the highest value, called θ_{max} here. From [15].

5 Cells for High-Resolution Displays

5.1 Supertwisted Nematic Cells

As already pointed out the problem of losing contrast in highly multiplexed displays can not be solved satisfactorily with 90° twisted nematic liquid crystals. Fortunately, the situation can be improved by increasing the twist angle. Figure 17 shows the calculated dependence of the mid-cell tilt angle on the applied voltage. For small twist angles a curve similar to that of Figure 10 is obtained; up to the threshold voltage all molecules remain more or less parallel to the cell walls, above it the tilt rises continuously. But for higher twist the dependence becomes steeper and finally, beyond about 240°, even a bistable situation occurs. From this microscopic behaviour of the molecules it is evident that also the electro-optic characteristics improve with increasing twist.

In practice additional efforts are necessary to maintain the "supertwisted" nematic (STN) structure with a twist angle $\vartheta > 90°$. An ordinary nematic liquid crystal would optimise its elastic energy by "flipping back" to $\vartheta - 180°$ which has the smaller absolute value. Therefore it is necessary to "pre-twist" the material. This is done by admixing a small amount of a chiral nematic (cholesteric) liquid crystal. In this way a intrinsic twist with arbitrary pitch between that of the chiral component and infinity can be created. The only limitation arises from the fact that for too small pitch the structure may change to one which is periodic in a direction parallel to the cell walls. This "stripe structure" was discovered earlier by Chigrinov [16]. This effect limits the twist to $\approx 240°$, the bistable situation (which would be desirable for many applications) can not be used. The twist angles in commercial displays are in the range 180 - 240°.

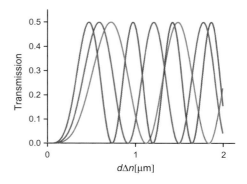

Figure 18: Transmission of a normally-black TN cell with 240° twist and polarisers under 45° angle with the liquid crystal molecules. The curves are calculated for red (650 nm), green (530 nm), and blue (425 nm) light.

Another major difference of STN displays is that the polarisers on both sides are placed in a non-standard way. They are not parallel to the respective liquid crystal molecule orientations as in the TN cell. This becomes necessary because of the residual twist in the select state which makes the standard way of operation (usually called wave-guide mode) impossible. Instead the *interference mode* is used where one has to accept a highly coloured off-state. Figure 18 shows the transmission curve calculated for the 240°/45°/45° (twist angle/first polariser angle/second polariser angle) in a similar way as for the 90°/0°/0° case in Sec. 3. It shows periodic zeros for the different colours as before but not the general decaying envelope. Therefore the transmission for "wrong" colours is much higher here than in the waveguide mode. The resulting colour is chosen in a way that the subjective contrast is still acceptable – usually yellow or blue. Therefore, in the context of STN displays one speaks about normally-yellow and normally-blue displays.

Several principles are used to improve this situation: Using coloured backlight one can achieve colours of the transmitted light which may be just the intended ones (e.g. in car displays). Replacing the polarisers by dichroitic devices one can let the opposite colour of the off-state through so that additive mixing results approximately in white. Extremely thin cells do not show this coloration ($d\Delta n \to 0$ in Figure 18) but have a low transmission even in the "white" state.

The best but also most expensive solution is the optical compensation. This principle is most easily explained from the double supertwisted nematic (DSTN) cell which for a long time was the standard in laptop display applications. There, one combines two layers of STN liquid crystal with opposite twist. This is done by mirroring the direction of the orienting coatings on the cell walls and using the enantiomeric (mirrored chemical structure) chiral component in the liquid crystal mixture. Without any calculation symmetry tells that a light beam crossing both structures will remain unchanged in its polarisation. All effects of birefringence of the first layer are identically compensated by the second. This means one can get a white off-state by parallel polarisers and a black one by perpendicular polarisers. To get the device into the on-state only on one of the liquid crystals, the "active" layer, a voltage is applied. In this way the symmetry is destroyed and the cell switches to the opposite colour.

Because the construction of the second mirrored liquid crystal layer is very expensive, devices have been developed where the compensation is accomplished by a solid foil (FSTN). The foil is made from a polymer which contains liquid crystalline sidegroups. Here, the disadvantage with respect to the DSTN principle is that the optical compensation will be perfect only for one wavelength and one temperature. Therefore, a slight coloration of the off-state results which increases for "abnormal" ambient temperatures.

VIII Displays

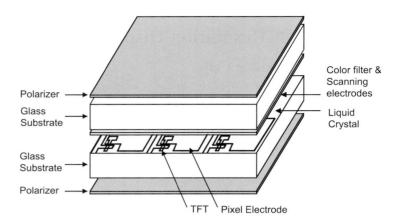

Figure 19: Basic structure of a TFT display.

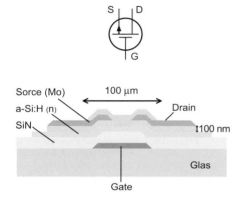

Figure 20: Construction of a thin-film transistor. The lower figure shows the material layers after etching, from bottom to top: glass substrate, gate electrode, silicon nitride insulation, amorphous silicon semiconductor, highly doped semiconductor layer, source and drain contacts, protection insulator. Above this drawing the common symbol for a metal-insulator field-effect transistor is shown. Note that to be on scale with the thickness the horizontal extension would have to be ten times larger.

5.2 Active Matrix Displays

The current standard for liquid crystal computer displays is the active matrix (AM) or thin-film transistor (TFT) principle. Instead of solving the problems of highly multiplexed addressing by chemistry here the solution is an electronic one: Each display pixel is equipped with its own driver. This is only possible by putting the driving transistor close to the pixel into the display's matrix. The basic structure of a TFT display is shown in Figure 19.

One of the glass substrates contains a matrix of thin film transistors whose gates are connected to the row lines and source electrodes are connected to the column lines (see Figure 21). The pixels are not directly constructed as the intersection of row and column lines (as for the passive matrix addressing in Figure 14). They are rather created by individual electrode patches connected to the drains of the TFTs. The counterelectrode extends over the whole area of the opposed glass substrate.

This integration of a transistor onto the glass matrix was only made possible by an invention made much earlier (Lilienfeld, 1933!), the thin film transistor. In principle it is a field-effect transistor similar to the well-known MOS (metal-oxide-silicon) FET. But instead of the conventional way to make the transistor from bulk material, here it is made by coating (vapour deposition, sputtering) the glass cell wall with semiconductor and conductor layers and etching. Figure 20 shows the final structure which is obtained after about a dozen processing steps.

The source and drain electrodes are connected by a layer of n-type amorphous hydrogen-doped silicon (a-Si:H) [17]. Below the semiconductor layer, insulated by silicon nitride the metal gate electrode is placed. Because the carriers in the semiconductor are electrons a positive voltage on the gate enhances source-drain conductivity and a negative voltage suppresses it. So the TFT can be viewed as an electronically controllable switch.

Figure 21 shows the addressing scheme which is very simple. The TFTs simply act as switches. By applying row-by-row a positive voltage the thin-film transistors of the rows are opened sequentially. When the source-drain line is open the voltage on the column lines is forwarded to the liquid crystal cell. After a row is addressed the TFT will close again and shield the pixel from the voltages applied for the next row. The electric field on the liquid crystal is kept by its own capacity and in most applications by an additional capacitor (not shown in the figure).

TFT active matrix displays are currently the best solution for high resolution displays. Cross-talk can be avoided except for capacitive effects. There is no limitation in the select/non-select voltages. Therefore normal TN liquid crystals can be used and analog grey scale addressing is possible.

On the other hand TFT displays are three times more expensive than DSTN ones. Therefore, this technology is currently reserved to applications which are already costly, e.g. computer and video displays. The market fraction (value, not device numbers) of TFT displays was 74% in 1999. It is expected to increase further, mostly at the expense of DSTN devices which will sooner or later disappear from the market of high-resolution displays.

During the years since its discovery TFT technology has made an enormous progress. The image resolution is already approaching that of printers a decade ago. As an example Figure 22 shows a TFT display manufactured by IBM. It boasts a so-called QUXGA resolution of 3840×2400 pixels. At a diagonal size of 22.2" (564 mm) this corresponds to a pixel distance of 0.1245 mm (to be compared to 0.26 mm of standard cathode ray tube displays) or a pixel density of 204 per inch. The retail price of the monitor based on this display has dropped from 15000 US$ in 2001 to 3000 US$ in 2004.

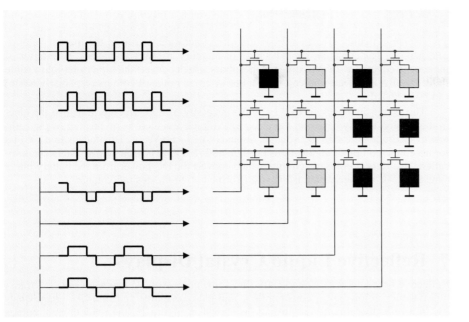

Figure 21: Active matrix display circuit diagram. The right hand drawing shows the schematic circuit diagram: Each liquid crystal cell pixel (grey or black square) is equipped with a thin-film transistor. The row lines are connected to the gates of the TFTs, the column lines to their sources. One liquid crystal cell electrode is connected to the TFT's drain the other is a grounded plate. On the left the addressing voltages are shown.

6 Backlighting

As mentioned in the preceding sections some LCD display principles require backlighting. In other cases the application forces employing backlight in addition to the ambient light. Therefore, design of the light source is often an important part of the construction of an LCD display module. This is also due to the fact that about 90% of the electrical power is consumed by the backlight (if present). Therefore, its power consumption is the major factor determining the operation time of battery powered devices.

Figure 22: Top-end TFT LCD display TF 221 (IBM). The display has 3840×2400 pixels and a diagonal size of 22.2" (564 mm). Picture courtesy of IBM.

Source:	Filament Lamp (Halogen)	Electroluminiscent Foil	Light Emitting Diode	Cold Cathode Fluorescent Lamp
Power Consumption	1.5 W	0.5 - 1.5 W	1 - 3 W	0.5 W
Intrinsic Colour	white	orange–blue (white possible)	red–green (blue and white expensive)	Different phosphors (white possible)
Life Span	2000 h	3000–5000 h	100000 h	10000 – 20000 h
Remarks	many shapes, simple power supply	very flat, high AC supply voltage (inverter)	many shapes, mounting on circuit board possible	high AC supply voltage (inverter)

Table 1: Comparison of typical LCD backlighting sources. The power consumption is stated for a standard application at 100 cd/m^2 from a surface of 100 cm^2 (approx. 3 lumen).

Table 1 shows a comparison of the most popular currently used backlighting sources. It can be seen that with respect to the power consumption the cold cathode fluorescent lamp is the best available source. That is why it is used in notebook computers without exception. The filament lamp and electroluminiscent foil are used in small scale applications as wristwatches and cellular phone displays. Because there the display lighting is only intermittently activated the life span is not an issue but rather the cost of production. Light emitting diodes have a limited application (e.g. in car displays) if a coloured backlighting is desired.

Because of its widespread application the cold cathode fluorescent lamp (CCFL) will be presented in more detail: CCFLs are what is commonly called "neon" lamps [18]. Their use dates back to 1910 when they were first employed in advertising signs. Because of the high voltage necessary the use of CCFLs was mostly restricted to this "neon sign" application until their renaissance as LCD backlighting.

A CCFL consists of a tube which is filled with a mixture of mercury vapour and "penning gas" (a noble gas mixture, often containing neon) at about 7 hPa pressure. The gas discharge mainly creates a 253.7 nm ultraviolet radiation from the mercury atoms. The penning gas enhances the ionisation of mercury at lower voltages. The interior of the tube is covered with a mixture of phosphors which convert the ultraviolet into visible light. By an appropriate choice of phosphors a broad range of colours (especially white) can be obtained.

The major drawback of CCFLs is the necessity of high voltage. Because the electron emission is only based on the electrical voltage and there is no heating of the cathode to facilitate emission the voltage amounts to 25 - 50 V/cm. This means that CCFLs for notebook computer applications need an 800 - 1000 V power supply. This makes so-called "inverters" necessary, circuits which convert the DC power from the battery into a sinusoidal AC voltage which is then transformed up to the desired voltage.

7 Reflective Liquid Crystal Displays

Reflective liquid crystal displays [20] offer some competitive advantages over the transmissive ones in low power consumption, sunlight readability, and film-like image quality. In a reflective direct-view display as shown in Figure 23, the readout light comes from ambient. Because no backlight is used, its power consumption and panel weight are both reduced. In a bright outdoor environment, the images of a transmissive display could be washed out by sunlight. Since reflective displays utilize ambient light as reading source, the brighter the ambient light, the more vivid the displayed images. In most reflective LCDs, the thin-film-transistors (TFT) are imbedded underneath the reflector. Therefore, the aperture ratio as large as 90% can still be obtained even for a high-resolution (1024 × 768) device. High aperture ratio makes the displayed images like film quality. The screen door effect as observed in some transmissive LCDs no longer exists.

Besides LCD, several reflective display technologies have been developed, e.g., digital micromirror device, [21] grating light valve, [22] interferometric modulation, [23] electrophoretic display, [24] and rotating ball display [25]. Each technology has its own merits and demerits. Some devices are more suitable for direct-view and others are preferred for projection displays.

Within the liquid crystal family, several light modulation mechanisms have been discovered. These include phase retardation, polarization rotation, absorption, light scattering, and Bragg reflection [20]. Phase retardation and polarization effects need to incorporate at least one polarizer. A reflective direct-view display employing a polarizer can achieve 30 - 40 : 1 contrast ratio and good color saturation, except that the light throughput is sacrificed. On the other hand, the display mechanisms involving absorption, light scattering, and Bragg reflection do not require any polarizer. High brightness and wide view angle are the inherent advantages of such displays. However, their contrast ratio is usually limited to 5 - 10 : 1.

Several types of reflective direct-view liquid crystal displays have been developed. They can be roughly categorized by the number of polarizers employed, e.g., 2, 1 and 0. The benchmark for these types of reflective displays is white paper. A white paper and newspaper respectively has reflectivity of 80 and 55% and contrast ratio of 12:1 and 6:1. All the electronic displays developed so far still cannot match the performance of a white paper in reflectivity and viewing angle.

7.1 Two-Polarizer R-LCDs

Wristwatch displays use two polarizers. A LC cell is sandwiched between two crossed linear polarizers. The 90° twisted nematic (TN) cell is used as light switch. Behind the second polarizer, a diffusive reflector reflects the incident light back to the viewer. The 90° TN cell has been widely used for transmissive display due to its high efficiency and weak color dispersion, high contrast ratio, and low operation voltage. However, when a 90° TN cell is used for reflective display, two crossed polarizers are required in order to achieve high contrast ratio. The second polarizer (~ 200 μm thick) sitting between the back substrate and diffuse reflector not only reduces brightness but also causes parallax (double image) that limits the device resolution. Thus, the two-polarizer R-LCD is not suitable for high resolution displays.

7.2 Zero-Polarizer R-LCDs

Reflective LCDs with no polarizer offer high brightness and wide viewing angle. Four types of reflective LCDs that do not require a polarizer have been actively pursued: polymer-dispersed liquid crystal (PDLC), [26] liquid crystal gels, [27] cholesteric liquid crystal display (Ch-LCD), [28] and guest-host (GH) display [29]. PDLC and LC gels are light scattering devices. Ch-LCD utilizes Bragg reflection, and guest host display takes advantages of the anisotropic absorption of dichroic dyes. Among them, Ch-LCD and guest host displays gain more momentum than PDLC and LC gels.

An important advantage of the Ch-LCD is that the reflected light is already colored. Therefore, a color display without using color filters can be achieved. Since no polarizer is needed, the reflectivity of a Ch-LCD is at least 6 times higher than a reflective display employing a polarizer and color filters. Without using a polarizer, the viewing cone of a Ch-LCD is wider than 60° except that its contrast ratio is in the 5 - 10 : 1 range. To achieve multiple colors, one approach is to stack 2 or 3 RGB panels together [30]. A potential problem of the stacked approach is parallax, i.e., the incident light and reflected light pass through different pixels. Parallax leads to mixed color which becomes a serious problem for a high resolution LCD. The operation voltage of a Ch-LCD is inversely proportional to the pitch length. Owing to the short pitch involved for visible displays, the operation voltage of a Ch-LCD is in the vicinity of 20 V_{rms}. Ch-LCD is a bistable device. Its low power consumption is particularly suitable for electronic book application.

The guest-host display [29] can be driven by TFT arrays, however, its contrast ratio is limited to ~5:1. The operation principle of a guest-host display is to utilize the absorption anisotropy of dichroic dye molecules doped in an aligned LC host. Normally, a guest-host display system consists of 1 - 5% of absorbing dye molecules dissolved in a liquid crystal host. The host LC is highly transparent in the visible spectral region and the guest dyes strongly absorb one polarization of the incoming light and transmits the other. To avoid using any polarizer in a guest-host display, special device configurations have been developed. For instance, the Cole-Kashow cell uses a homogeneous GH cell with a quarter-wave film sitting between the rear substrate and the reflector. The White-Taylor cell uses chiral dopant to form twisted GH layers so that both polarization of the incident light can be absorbed while traversing through the GH LC layer twice. The double cell uses two orthogonal homogeneous cells; the first GH layer is responsible for one polarization and the second layer for the other. All the three above-mentioned cells have their own pros and cons. The typical performance of a GH LCD is ~ 50 - 60% reflectivity, ~ 5:1 contrast ratio, ~ ±60° viewing angle, and ~ 50 - 80 ms response time. Low power consumption is the major attraction for the guest-host display.

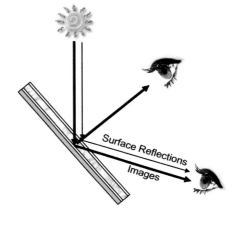

Figure 23: Reflective LCD using ambient light for displaying images.

7.3 Single-Polarizer R-LCD

For a reflective LCD using one polarizer, the light still traverses the polarizer twice so that the maximum light throughput is limited to ~ 40%. However, as compared to the two-polarizer LCD, the single-polarizer LCD offers two advantages. Firstly the removal of the second polarizer sitting between back glass substrate and reflector eliminates the parallax, and secondly the elimination of the second polarizer enhances the reflectivity by ~ 25%. On the other hand, the single-polarizer LCD exhibits a higher contrast ratio than a non-polarizer LCD, although its reflectivity is lower. Contrast ratio is essential for color displays. If the contrast ratio is 5-10:1, light leakage in a color pixel is relatively large resulting in a mixed color. If the contrast ratio exceeds 30:1, the color saturation is much improved, although still far from perfect. In a reflective display, due to surface reflections, it is not easy to get contrast ratio higher than 50:1 under ambient light condition. The single-polarizer LCD offers a compromise between display brightness and contrast ratio, and has gradually become the mainstream approach.

Several cell configurations have been developed for single-polarizer reflective LCDs [20]. For example, the mixed-mode TN (called MTN) cell, super-twist TN (STN) cell, homeotropic cell, film-compensated homogeneous cell and π-cell all have their own merits. For passive matrix display, STN with 220 - 240° twist angle has been considered for achieving high duty ratio. For active matrix display, MTN cells offer a large cell gap tolerance to accommodate the bumpy reflector surfaces so that a good contrast ratio can still be obtained. For reflection type projection displays, MTN and homeotropic cells are favored choices due to their high contrast ratio.

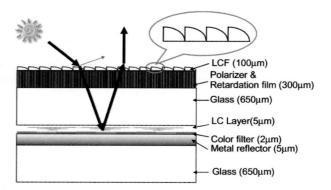

Figure 24: Reflective LCD based on a light control film and a reflector.

The viewing angle of a reflective cell is equivalent to a two-domain transmissive cell. This is because the mirror image effect experienced by the incident and reflected beams. For computer monitor applications, such ±45° viewing angle is insufficient. But for handheld displays, the panel can be conveniently adjusted to fit the viewer's position. Therefore, viewing angle is not a critical issue for most handheld reflective LCDs.

7.4 Reflectors

For reflective direct-view displays, the reflector design and fabrication is probably the most critical issue. An ideal reflector needs to meet two criteria: 1. It steers the displayed images to the viewer without overlapping with surface reflections. If the images mix with glares, the contrast ratio is substantially reduced. 2. It provides sufficiently wide viewing cone and optical gain. A specular mirror has too narrow viewing angle and cannot be used alone for reflective displays. By contrast, a Lambertian reflector has uniform scattering all over the angles so that the light intensity in the preferred viewing cone is reduced.

Extensive efforts have been put into reflector design and fabrication with acceptable viewing angle while providing optical gain and preserving good contrast ratio. So far, four types of reflectors have been investigated: (1) Light control films, [31] (2) Rough surface reflectors, [32] (3) Holographic reflectors, [33] and (4) Cholesteric reflectors [34]. A light control film is designed to steer the images to the observer at approximately normal viewing direction which is well separated from the specular angle, as illustrated in Figure 24. This film has been applied to STN, PDLC and cholesteric displays. The display brightness and contrast ratio are greatly enhanced. On the other hand, the bumpy reflector reflects and diffuses the incident light to a preferred viewing cone to avoid overlapping with the specular reflection. It involves a structural change. The holographic reflector can be either laminated to the backside of the display or integrated with color filters. Although the former approach is simple, it creates undesirable parallax so that its usefulness is limited to low resolution displays. The internal holographic reflector design is challenging, but holds promise for displays with high brightness, high contrast, and excellent chromaticity. Lastly, cholesteric liquid crystal layer has also been used as diffusive reflector for some birefringence color reflective LCDs.

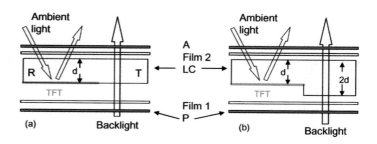

Figure 25: Device configurations of transflective LCDs: (a) single cell gap approach and (b) double cell gap approach. d = cell gap, P = polarizer, A = analyzer, Films 1 and 2 are phase retardation films.

8 Transflective Displays

A reflective LCD relies on ambient light to read out the displayed information. When the ambient light is dim, the display is not readable if no built-in light is available. To overcome this problem, a transflective display has been developed [35]. Figure 25 shows two types of transflective LCDs that have been devised: single cell gap and double cell gap. In a transflective LCD, each pixel is split into R (reflective) and T (transmissive) sub-pixels. Usually, the R and T area ratio is 4:1, in favor of reflection. The backlight is used for dark ambient only in order to conserve power.

In the single cell gap approach, the cell gap (d) for R and T modes is the same and is optimized for R-mode. As a result, the light transmittance for the T-mode is lower than 50% because the light only passes the LC layer once. In the double cell gap approach, the cell gap is d and 2d for the R and T pixels, respectively. In this approach, both R and T have high light efficiency. However, the T mode has four times slower response time than R mode. A common problem for most transflective displays is that R and T pixels have different color saturation. For R pixels, the ambient light passes the color filter twice, but for T pixels the backlight only passes the color filter once. In order to achieve equal color saturation, the pigment concentration or the color filter thickness of the transmissive pixels should be twice that of the reflective ones. For simplicity, most transflective displays simply ignore this problem.

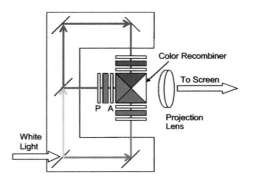

Figure 26: Projection display using three transmissive LCD panels. Each LCD is sandwiched between a polarizer (P) and an analyzer (A).

9 Projection Displays

According to today's technology, to fabricate 40-inch direct-view LCDs is feasible, but not very economical. To overcome the panel size limit, projection is an alternative way for achieving large screen displays, such as high definition TVs, boardroom presentation and electronic cinema. Two types of projection displays are often configured: front and rear projections. Front projection configurations utilize a distant screen to view the magnified image. Rear projection configurations enclose the magnification optics in back of an imaging screen to produce a self-contained system. Both transmissive and reflective LCDs have been considered for projection displays [36].

9.1 Transmissive Displays

A LCD projector using three transmissive poly-silicon thin-film-transistor (TFT) panels is shown in Figure 26. The incoming white light from a lamp is divided into three channels by dichroic mirrors. In each channel, a LCD panel is sandwiched between two linear polarizers. The 90°-TN and vertical-aligned (VA) LC cells are two popular choices because the TN cell is known to have weak wavelength dependency and VA cell possesses an unprecedented contrast ratio (>1000:1). The X-cube recombines the RGB beams and the projection lens throws the images to screen.

Most LC devices require linearly polarized light. To convert unpolarized light into linear, a few polarization conversion devices have been invented. Conversion efficiency reaches 70 - 80%. Moreover, in order to enhance aperture ratio of a high resolution TFT-LCD panel, a microlens array is proven useful. The effective aperture ratio is improved to ~ 80% for a XGA device. Overall speaking, the power efficiency of ~ 10-15 lm/watt in the projection system using three transmissive LCD panels has been routinely obtained. Under such circumstance, a 100 W short-arc lamp would produce 1000 - 1500 lumens on the screen.

Figure 27: Liquid crystal microdisplay devices based on silicon wafer. (Courtesy of Dr. M. Bone, Aurora Systems).

9.2 Reflective Displays

Liquid crystal on silicon (LCOS) [37] has been developed as the optical engines for projection and virtual display systems. LCOS also called microdisplay, is a reflective image transducer that is capable of accepting video signals and converting it into a high-brightness, high-contrast, large-screen images. In the silicon backplane, as shown in Figure 27, a CMOS transistor and a capacitor are fabricated and connected. The CMOS transistor could support high data rate owing to the high electron mobility of single-crystal silicon. The electron mobility of c-Si is about two orders of magnitude higher than that of amorphous silicon (a-Si). The reflective aluminum electrode is connected to

drain of the transistor and capacitor through a light blocking metal layer. After smoothing, the aluminum pixel mirror approaches a reflectivity of 91%. The dimension of the electrode is 13x13 µm and the gap between electrodes is 0.5 µm. Thus, the aperture ratio is about 93%. For a 1365 (H) x 1024 (V) imager in a 4:3 aspect ratio, the diagonal is 23 mm.

Two types of projection displays using LCOS have been actively developed: rear projection TVs and near-eye projection. Figure 28 shows a 58-inch rear projection TV developed by Three-Five Systems. Three LCOS panels with 1280×720 resolution elements are used. Owing to the reflective optics employed, the display depth is only 20 inches.

Virtual projection is a display that uses optics to create virtual magnified images. A user looks into a small viewfinder, as shown in Figure 29, and sees a projected image that appears much larger and floating some distance from the viewer. For the interest of reducing weight and cost, a single high- resolution and high speed LCOS is commonly used as optical engine. Three sequential color LEDs are used to display full-color images. LEDs are known to have fast response time, light weight, high brightness, low power consumption, and long lifetime. Thus, they are ideal light sources for virtual projection displays.

A critical element for virtual display is light-weight and high-magnification projection optics. If a 12.7 mm diagonal LCOS imager (800×600 pixels) is used as the object, the required magnification is about 15x in order to view an equivalent 19" display at about 2 meters away. Simple optics such as a magnifying glass or an eyepiece are often limited to a maximum magnification of about 10x. Compound optical systems such as microscopes, permit much higher magnification, but are too bulky to be wearable. To achieve a large magnification, a folded compound magnifier has been developed [38]. The eye relief is about 20 mm and exit pupil is 10 mm.

In addition to optics, a major challenge for color sequential virtual display is LC response time. For a frame rate of 85 Hz, the LC frame (rise + decay) time needs to be shorter than 3.9 ms in order to avoid color breakup. To achieve such a fast response, LC cell gap as thin as 1 µm has been considered and color sequential operation from –10 to 70°C has been demonstrated [39]. Although thin cell is favorable for speeding up LC response, the manufacturing is more difficult.

Figure 28: A rear projection TV using three LCOS panels (Courtesy of Dr. R. L. Melcher, Three-Five Systems).

10 New Liquid Crystal Display Principles

10.1 Ferroelectric Liquid Crystals

The ferroelectric liquid crystals used in technical applications are not nematic but of the SmC* type shown in Figure 4. By applying symmetry considerations it can be proven that such a phase always has a spontaneous electric polarisation of the smectic layers [19]. That is because the dipole moment is connected to the molecular orientation and it points into the same direction for every molecule in the ordered layers of a SmC* phase. If all layers would have the same preferred orientation of the molecules (director) this would add up to a macroscopic polarisation like that of an electrete. Of course, such a structure has a high self-energy. But because in the smectic liquid crystal the orienting forces between different layers are small this situation can be avoided by rotating the director around the layer normal from one layer to the next (Figure 30). In this way the helical structure shown in Figure 4 arises. In this structure the polarisation of each layer is also changing from layer to layer and the total polarisation is kept zero.

If such a liquid crystal is put into a cell which is thinner than the helical pitch (≈ 2 µm) the helix structure is lost [41]. This is so because the tendency to arrange parallel to a surface becomes stronger than the electric forces. Then, there are only the two energetically equivalent ways to orient the molecules shown in Figure 31. The polarisation which is now always perpendicular to the cell walls can either point "upwards" or "downwards". By applying an external field one can switch between the two energy minima. The polarisation then follows the typical hysteresis behaviour of ferroelectric or -magnetic materials (Figure 32). Because this type of structure only occurs between close walls the correct name is "surface stabilised ferroelectric liquid crystal" (SSFLC).

Figure 33 schematically shows the display devised by Clark and Lagerwall based on SSFLCs. The left part shows the orientational state of the molecules corresponding to the "black" optical state. The light enters from the front and is firstly polarised parallel

Figure 29: A near eye virtual projection display using a color sequential LCOS.

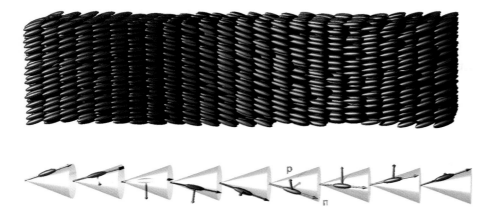

Figure 30: Layer-dependence of director and electric polarisation in a SmC* liquid crystal. The top picture shows the molecular arrangement, the red and blue sides of the molecule indicating the partial charges. The lower drawing indicates the director **n** (grey arrow) and the molecular polarisation $\mathbf{p}_e$ within the respective layer (red arrow).

to the molecule direction (symbolised by the change from a cylinder to a flat band in the figure). For this situation there will be only an ordinary beam which keeps its linear polarisation. The polariser on the exit side is perpendicular to the other and therefore blocks the light nearly completely. If the cell is switched to the other state (right part of the figure) the molecules are not any more parallel to the incoming polarisation. Therefore, an extraordinary beam emerges in addition which propagates with different wavelength and a phase difference. The light changes from linear to elliptical polarisation. Upon leaving the cell the parallel component is transmitted through the analyser. Because the polarisation is not completely twisted as in the TN cell but only partially changed the transmission is not complete. Therefore the open state is called "grey" here.

The main advantage of the SSFLC cell is its faster switching behaviour. Because the field couples directly to the molecules' dipole moment switching times of ≈ 10 µs can be achieved. This is more than 1000 times faster than the switching time for nematic devices. Therefore, this type of cell is predestined for video application where it is important to reproduce fast changing patterns.

Also the problems resulting in nematic displays from a loss of orientation in between the row addressing strobes are not present here. Due to the bistable ferroelectric characteristic the cell keeps its state also when the external field is switched off.

An important quality improvement results from the much larger viewing angle. Because the molecules are never (as for the nematic types) tilted out of the cell plane there is virtually no difference in the birefringence properties for an oblique beam.

A further advantage is the small pixel size which is possible mainly because of the thin cells. With a size of ≈ 5 µm SSFLC pixels can be made a factor ten smaller than those in nematic displays. Although this is much better than what is needed in computer displays (E.g. in an XGA, 768×1024 display of $14''$ diagonal the subpixel representing a single colour is 90 µm wide.) it makes this display type useful for microdisplay applications, i.e. light-valves for projectors or head-mounted displays.

The performance disadvantages are only small: (1) Because of the bistability in principle no grey scales can be displayed. But already for nematic cells techniques have been developed to produce grey levels in a digital way, e.g. frame rate modulation and spatial dithering (Sec. 4.2). The latter is even easier to use here because of the small pixel size. (2) Because in such a thin cell no complete (i.e. 90° linear-linear) change of the polarisation direction can be achieved as in TN cells the transmitting state is never completely "white". This makes a strong backlight necessary.

On the other hand, the production of the ferroelectric cell is more complicated. While the twisted nematic structure emerges spontaneously between the orienting surfaces the SSFLC structure has to be generated by a definite phase sequence: The isotropic liquid crystal filled into the cell is firstly cooled into the nematic phase which is oriented parallel to the brushing direction of the cell walls. Upon further cooling the liquid crystal enters the SmA phase with the molecules still parallel and layers perpendicular to the brushing direction. Further cooling leaves the molecules arranged in the same layers but the orientation changes to the oblique one of the SmC* phase. This phase sequence (isotropic→N→SmA→SmC*) is prerequisite for obtaining the correct macrostructure but it is not ensured by every type of liquid crystal. Small irregularities during the cooling process can cause domains with wrong orientation making the display unusable.

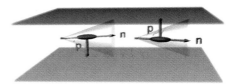

Figure 31: Surface stabilised ferroelectric liquid crystal. The cone shows the orientations of the director possible in the bulk state. Between close walls only the two orientations shown by the arrows are possible with the corresponding polarisations pointed up- and downward respectively.

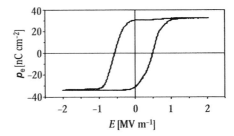

Figure 32: Hysteresis of a surface stabilised ferroelectric liquid crystal. From [40].

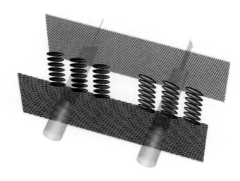

Figure 33: Surface stabilised ferroelectric liquid crystal display. The left part shows the "black" state, the right part the partially transmitting "grey" state. The hatchings on the front and back glass plate indicate the orientation of the polariser and analyser. The polarisation of the light propagating from front to back is shown by the yellow band or elliptical cylinder. After [42].

But the main drawback of ferroelectric liquid crystal displays is that the structure may also be lost by mechanical stress during the use of the display. If an ordinary nematic display is gently pressed it will lose its function for a short moment because the twisted structure is distorted. This may lead to unwanted colours of even a complete "blackout". But after a fraction of a second the structure will be restored and the display will again function properly. Unfortunately, this is not true for a ferroelectric display. As soon as domains are built which do not have the right orientation (parallel to the polarisation if the incident light) they will remain stable. This seemingly small flaw has (until now) obstructed commercial use of this display type on a larger scale. Only for specialised applications where small size is important and it is not possible for the user to touch the display (e.g. head-mounted displays, projection displays) such displays are commercially available [43].

10.2 Polymer Dispersed Liquid Crystals

Here, a technology is presented which is rather different from the preceding. The devices discussed here work in "scattering mode". The displays presented before all work by a liquid crystal cell which changes the polarisation. With a second polariser this change of polarisation is converted into a controllable *absorption*. The displays ideally switch between transparent and black. Scattering mode displays in contrast switch between a transparent state and a (milky white) state where light is diffused. Interestingly, the first experimental displays were of that type [44] but they were not commercially successful.

While these first displays used an intrinsically homogeneous liquid crystal in which turbulences are created by the electric field polymer dispersed liquid crystals are inhomogeneous materials (Figure 34). Small liquid crystal droplets of $\approx 2\,\mu$m size are imbedded in a solid polymer matrix. The liquid crystal is chosen so that the ordinary index of refraction n_o is identical to the (isotropic) index of refraction of the polymer. If there is no field applied the directors in the liquid crystal droplets have a random orientation. Therefore the effective index of refraction being an orientation dependent mixture of the ordinary and extraordinary will in general be different from that of the polymer. In consequence a beam of light will be scattered by the droplets. If a field parallel to the direction of the light is applied the directors will orient. Then the ordinary index of refraction of the liquid crystal will strongly dominate. This means that the index difference vanishes and so does the scattering.

There are several different ways to produce such materials [45]:

The monomer can be mixed with the liquid crystal before polymerisation. When the polymerisation starts the growing chains favour a two-phase situation for entropic reasons. Therefore, a spontaneous demixing can take place which causes the liquid crystal to segregate into small droplets.

The same can be done after polymerisation by cooling a polymer melt/liquid crystal mixture if the phase diagram contains an upper consolute temperature.

If polymer and liquid crystal are immiscible but dissolve in a common solvent such a solution can be dried. When the solvent content is lowered phase separation will occur.

Finally, micro-encapsulation techniques known from pharmacology can be applied to firstly create encapsulated liquid crystal materials which then will be mixed into a polymer melt.

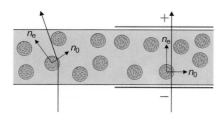

Figure 34: Polymer dispersed liquid crystal cell. The darker spheres represent the liquid crystal droplets. The lines in "gooseberry arrangement" indicate the nematic directors. The left shows the off-state scattering the light coming from the bottom. On the right side the liquid crystal molecules are aligned and transmitted light is not scattered.

All of these methods are comparatively inexpensive. Therefore, it is possible to produce polymer dispersed liquid crystals in large quantities, e.g. as foil materials of square-meter size.

Because of this low price as a current application "switchable windows" are produced. Those windows can be switched electrically from a transparent state to opaque. But also the use in display devices is under consideration. Especially, for projection devices it does not matter whether a beam is absorbed or scattered out of its path. The latter may be even more desirable because the energy is not deposited within the cell.

The main drawback preventing a more widespread use of scattering mode devices is the incomplete orientation if a field is applied. As indicated in Figure 34 the liquid crystal molecules close to the droplet surfaces tend to orient parallel to the surface and not the field. This causes a residual scattering which is seen as smokiness in window application and low transmissions for projection devices. In addition the voltages to be applied are in general higher that those for (super)twisted nematic type displays.

11 Summary

Liquid crystals are known and studied since the 19th century. Nevertheless, they did not gain their enormous technological importance until about 30 years ago. The features of liquid crystal molecules which enable them to be used for displays are their optical anisotropy and dipole moment. The anisotropy allows the control of transmitted light by means of birefrigence. The dipole moment of the molecules allows the control of their orientation by an electric field. The first successful "light valve" cells constructed on grounds of these principles used the twisted nematic (TN) phase. In this structure the nematic order is artificially distorted to a helical one. The structure is then able to rotate the plane of polarisation by an amount which can be controlled by an electrical field.

As the demands on displays (resolution, grey scales, colours...) grew mainly because of applications in information technology the limits of simple TN cells became obvious. Although elaborate multiplex addressing schemes were developed it could be proven that there are mathematical limits on them for the number of pixels of a display. Therefore, the basic principle of TN cells was improved along two different lines: (1) The twist angle was increased in order to obtain an intrinsically less stable structure which is easier to control electrically (supertwisted nematic, STN). A smaller difference between the "on" and "off" voltage is possible which in turn enables addressing of larger arrays of pixel. This comparably less expensive solution is mostly used in intermediate quality displays (e.g. for mobile phones). (2) A transition from "passive" displays to "active matrix" (AM) was done by integrating the final stage of the addressing electronics, a thin-film transistor (TFT), onto the cell wall. This approach allows a theoretically unlimited increase of size and resolution and has proven to be the most successful for high-end applications (e.g. for computer monitors).

Besides these technical, quantitative improvements also fundamentally different principles are explored. Ferroelectric liquid crystal structures could resolve several drawbacks of traditional TN type cells. They can be controlled faster, have a larger viewing angle, and smaller pixel size. Polymer dispersed liquid crystals allow the construction of large area devices at low cost. Nevertheless, until now none of these radically different technologies could gain an important market share [46].

Acknowledgements

We thank Optrex Europe GmbH for providing valuable information. The editor would like to thank Carsten Kügeler (RWTH Aachen) for checking the symbols and formulas in this chapter.

References

[1] W. Maier, A. Saupe, Z. Naturforschg. **14a**, 882 (1959); W. Maier, A. Saupe, Z. Naturforschg. **15a**, 287 (1960).
[2] M. Schadt, W. Helfrich, Appl. Phys. Lett. **18**, 127 (1971).
[3] T. J. Scheffer, J. Nehring, Appl. Phys. Lett. **45**, 1021 (1984).
[4] B. J. Lechner, F. J. Marlowe, E. O. Nester, J. Tults, Proc. IEEE 59, 1566 (1971).
[5] B. Bahadur, *Liquid Crystals, World Scientific*, Singapore, 1990.
[6] D. Demus, J. Goodby, G. W. Gray, H.-W. Spiess, V. Vill, *Handbook of Liquid Crystals, Fundamentals,* Wiley-VCH, Weinheim, 1998.
[7] The POV-Ray Team™, www.povray.org.
[8] In contrast to lyotropic liquid crystals which contain a solvent and where the concentration of the liquid crystal material in the solvent determines the phase. Lyotropic liquid crystals are important in biology, e.g. lipids which constitute the cell membrane which in our nomenclature would be a single-layer smectic liquid crystal.
[9] L. Pohl, U. Finkenzeller, *Physical Properties of Liquid Crystals*, in [5].
[10] J. D. Jackson, *Classical Electrodynamics*, Wiley, New York, 1975.

[11] In practice, even for "static" addressing one has to use AC. A persistent voltage of same sign will damage the liquid crystal by electrolysis. Therefore one always has to use an alternating voltage with zero average value.

[12] C. H. Gooch, H. A. Tarry, J. Phys. D: Appl. Phys. **8**, 1575 (1975).

[13] P. M. Alt, P. Pleshko, IEEE Trans. Electron Devices ED-21, 146 (1974).

[14] J. Nehring, A.R. Kmetz, IEEE Trans. Electron Devices ED-26, 795 (1979).

[15] E. P. Raynes, Mol. Cryst. Liq. Cryst. Lett. **4**, 1 (1986).

[16] V. G. Chigrinov, V. V. Belyaev, S. V. Belyaev, M. F. Grebenkin, Sov. Phys. JETP 50, 994 (1979).

[17] Currently, this is the only possible material which can be applied as such a thin film. The hydrogen doping restores the semiconductor properties which were otherwise destroyed by the amorphisation. Research with the aim of using polycrystalline silicon instead are currently underway.

[18] This statement refers to the US-american use of the term "neon". In Europe "neon lamp" denotes lamps which are larger in size, based on the same discharge principle, but have filament (i.e. hot) electrodes.

[19] R. B. Meyer, L. Liebert, L. Strzelecki, P. Keller, J. de Phys. Lett. **36**, L69 (1975).

[20] S. T. Wu and D. K. Yang, *Reflective Liquid Crystal Displays*, Wiley, 2001.

[21] J. B. Sampsell, SID Tech. Digest **25**, 669 (1994).

[22] D. T. Amm and R. W. Corrigan, SID Tech. Digest 29, 29 (1998).

[23] M. W. Miles, SID Tech. Digest **31**, 32 (2000).

[24] S. A. Swanson, M. W. Hart, and J. G. Gordon, SID Tech. Digest **31**, 29 (2000).

[25] N. K. Sheridon et al, J. SID 7, 141 (1999).

[26] P. Drzaic, *Liquid Crystal Dispersions*, World Scientific, Singapore, 1995.

[27] R. A. M. Hikmet, J. Appl. Phys. **68**, 4406 (1990)

[28] D. K. Yang, J. W. Doane, Z. Yaniv and J. Glasser, Appl. Phys. Lett. **65**, 1905 (1994).

[29] B. Bahadur, *Liquid Crystals Applications and Uses*, Vol. 3 and references therein, World Scientific, Singapore, 1992.

[30] D. Davis et al, J. SID 7, 43 (1999).

[31] Y. B. Huang, et al, SID Tech. Digest (paper 26.3, 2002).

[32] Y. Itoh et al, SID Tech. Digest **29**, 221 (1998).

[33] A. G. Chen et al, SID Tech. Digest **26**, 176 (1995).

[34] R. van Asselt, R. A. W. Van Rooij, and D. J. Broer, SID Tech. Digest **31**, 742 (2000).

[35] M. Kubo et al, US patent 6,295,109 B1 (2001).

[36] E. H. Stupp and M. S. Brennesholtz, *Projection Displays*, Wiley, New York, 1999.

[37] F. Sato, Y. Yagi and K. Hanihara, SID Tech. Digest **28**, 997 (1997).

[38] N. Bergstrom et al, SID Tech. Digest **31**, 1138 (2000).

[39] D. J. Schott, EuroDisplay'99 (1999).

[40] I. Dierking, Phys. Blätter **56**, 53 (2000).

[41] J. Dijon: Ferroelectric LCDs, in [5].

[42] N. A. Clark, S. T.Lagerwall, Appl. Phys. Lett. **36**, 356 (1984).

[43] Displaytech Inc., www.displaytech.com.

[44] R. Williams, J. Chem. Phys. **39**, 384 (1963); R. Williams, G. H. Heilmeier, J. Chem. Phys. **44**, 638 (1966).

[45] J. W. Doane, *Polymer Dispersed Liquid Crystals*, in [5].

[46] It seems that once a technology is established on the market it tends to fend off innovations because of its economic advantage. Since the infrastructure for its production already exists less economic risk is involved in improving the current technology gradually than changing to an innovation which may yield a better result in the end but implies the risk of complete failure. In this way a "conventional" technique may also be "pressed" beyond limits which once were considered insurmountable (e.g. pixel sizes less than 100 μm for TFT displays). Therefore, new technologies as FLCD will probably not get a chance before the development potential in traditional TN-based devices is completely exhausted.

Organic Light Emitting Devices

Martin Pfeiffer
Institute for Applied Photophysics, University of Technology Dresden, Germany

Stephen R. Forrest
Department of Electrical Engineering, Princeton University, USA

Contents

1 Introduction — 913

2 Organic Semiconductors — 914
2.1 Isolated Molecules and Molecular van der Waals Crystals — 914
2.2 Absorption, Exciton Diffusion and Fluorescence — 915
2.3 Charge Carrier Transport — 916

3 Organic Light Emitting Diodes — 916
3.1 Prerequisites for an Efficient OLED — 916
3.2 Single and Double Layer OLEDs — 917
3.3 Improving the Device Efficiency Using Fluorescent Dopants — 918
3.4 High Efficiency Phosphorescent OLEDs — 918
3.5 Reduced Voltages by Doped Charge Transport Layers — 920

4 Organic Displays — 923
4.1 Passive and Active Matrix Displays — 923
4.2 Patterning Techniques — 924
4.3 Outlook: Transparent, Stacked and Flexible OLEDs — 924

Organic Light Emitting Devices

1 Introduction

Organic dyes with conjugated π-electron systems have attracted a large and rapidly increasing interest in the last decade as possible materials for various electronic and optoelectronic devices because of their advantageous semiconductor properties. For application as photoactive materials in photocopiers, they have already reached maturity and replaced inorganic semiconductors on a large scale. Efficient electroluminescence (EL) from an organic solid was first demonstrated in large (50 μm to 1 mm thick) single crystals of anthracene [1], [2]. Although these devices showed a high EL quantum efficiency of up to 8 %, they were impractical due to the large applied voltage (>1000 V) required to inject electrons and holes into the crystal. More recently, practical organic light-emitting devices (OLEDs) consisting of sequentially vacuum-deposited layers of hole- and electron-transporting molecular materials [3], [4], [5], [6] or of spin-coated thin polymer films [7], [8] have been demonstrated with active device thicknesses of only a few hundred Ångstroms. Applying about 3 to 5 V, they are already brighter than a conventional TV screen with much higher efficiencies, brilliant colors, large viewing angle, switching times fast enough for video real time image displays and lifetimes well above 10,000 hours. The first displays based on organic semiconductors have become commercially available and companies like Pioneer and Philips and many others are currently developing this promising technology. Small independent firms, such as Universal Display Corporation in the U.S. and Cambridge Display Technology in the U.K., have been created solely to turn these innovations into commercial reality.

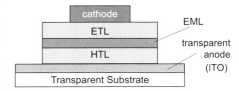

Figure 1: Schematic cross section of an organic light emitting device (OLED) showing the contacts, the electron transport layer (ETL), the light-emitting layer (EML) and hole transport layer (HTL). The total thickness of the organic layers is typically around 100 nm.

The large interest is not only caused by technological aspects such as the low costs, the possibility to prepare flexible large area devices at low process temperatures on polymer foils, and the almost unlimited variety of organic compounds that allow tuning of e.g. energy levels and emission colors. They also have some unique physical properties and advantages compared to inorganic semiconductors [9]. Due to their low dielectric constants, the reflection losses at interfaces are lower than for III-V-semiconductors. The internal fluorescence efficiency is close to unity for some organic dyes with their emission peaks strongly red shifted to the absorption edge thus avoiding re-absorption losses in the devices. For this reason, it is possible to prepare efficient sandwich structures of several OLEDs with different emission wavelengths [10]. The electronic structure of organic semiconductors is largely determined by the individual molecules and only weakly modified by solid state effects. Therefore, interface states play a minor role. Also, amorphous layers grown on cheap, flexible substrates exhibit attractive electrical properties to allow for efficient electroluminescence. The amorphous matrices also enable the incorporation of a wide range of dopants, if necessary in high concentrations, to tune both the conduction type [5], [11] and the emission wavelength [12]. For instance, the admixture of about 7 % of phosphorescent dyes has led to OLEDs that convert 100 % of the electrically excited molecular states into visible light with internal efficiencies close to unity [4], [13], [14].

Another striking feature of some organic dyes is their high absorption coefficients (up to 2×10^5 cm^{-1}) in the visible, making them attractive for cheap and extremely thin photovoltaic cells [15]. The use of organic materials requires low energy consumption in the production process. However, the efficiencies are still too low for large scale power generation applications [16].

A schematic cross section of a conventional, small molecule based OLED with three organic layers (a double heterostructure) is shown in Figure 1. The top, ohmic, electron-injecting electrode consists of a low work function metal alloy, typically Mg-Ag or Li-Al, deposited by vacuum evaporation. The bottom, hole-injecting electrode is typically a thin film of the transparent semiconductor indium tin oxide (ITO), deposited onto the substrate by sputtering or electron beam evaporation. Light is emitted through this electrode when the device is operated in *forward bias*, i.e., with the ITO biased positive with respect to the top electrode. The next layer deposited is a hole transporting layer (HTL) of a material such as TPD. This is followed by the light emitting (EML) layer consisting of, for example, Alq$_3$ for green light or a host material doped with fluorescent or phosphorescent dyes and optionally an additional electron transport layer (ETL). The chemical structures of N,N'-diphenyl-N,N'-bis(3-methylphenyl)1-1' biphenyl-4,4' diamine (TPD) and tris (8-hydroxyquinoline) aluminum (Alq$_3$) are shown in Figure 2. Before we discuss the details of the device architecture in Section 3, we will briefly introduce some basic properties of organic semiconductors. The Chapter is concluded in Section 4. where we consider the state of the art and specific techniques of organic full color displays.

Organic light emitting devices can be prepared either from **small molecules** (c.f. Section 2.1) or from **polymers**. While the basic physics of both classes of materials is similar, the fabrication processes are very different. Small molecules are usually deposited by thermal evaporation in high or ultrahigh vacuum. Alternatively, the molecules can be evaporated into a stream of a hot inert gas directed to the substrate [17]. This so-called organic vapor phase deposition (OVPD) may provide a means for inexpensive and large-scale manufacturing. The typical sublimation temperatures are between 200 and 400°C. The molecules often have a thermal decomposition temperature much higher than the sublimation temperature. In this case, materials can be purified by repeated sublimation in vacuum or in an inert gas [18]. To separate components with different sublimation temperatures, long glass tubes inside a furnace with a temperature gradient are used. Such high purity materials are not only needed for good device performance, but are essential for long device operational lifetimes. Thermal deposition in vacuum or in the gas phase also allows for easy fabrication of multilayer devices. With sophisticated multilayer architectures, the distribution of charge carriers and excitons can be tuned (cp. e.g. Section 3.2 and 3.5). Small molecular weight devices with long lifetimes have been demonstrated to emit across the visible spectrum [19]. The highest emission efficiencies have been reported for small-molecular weight devices using phosphorescent emitters [13].

Polymers, on the other hand, cannot be deposited or purified by thermal evaporation because their decomposition temperature is generally lower than their sublimation temperature. Devices are accordingly prepared from solution, e.g. by spin coating. A particular advantage of polymers is their adaptability to potentially low cost large-scale display production by ink jet printing [20]. With optimized chemical synthesis and purification procedures, green and blue polymers with high fluorescence efficiencies have been obtained. Green polymeric OLEDs have also reached operational lifetimes in excess of 10,000 h. Polymer devices also have lower operating voltages than devices made from small molecules. However, using controlled doping, small molecular weight devices can compete here as well [5].

In this chapter, we will restrict the discussion to small molecule based devices. A review on polymer OLEDs can be found in Ref. [8].

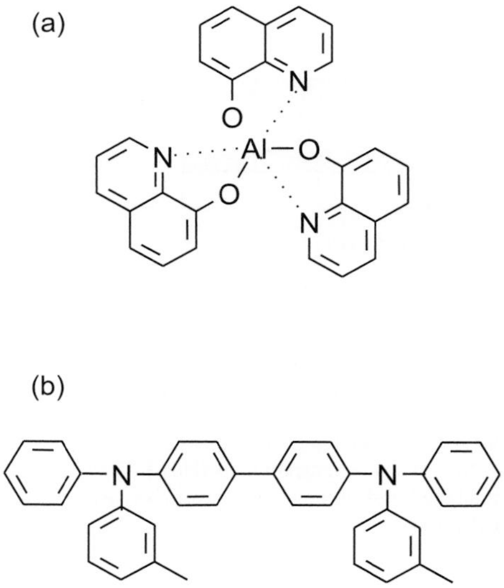

Figure 2: Chemical structures of (a) Alq$_3$, an electron transport and green fluorescent emitting material and (b) TPD, a hole transport material.

2 Organic Semiconductors

2.1 Isolated Molecules and Molecular van der Waals Crystals

The basic electronic structure of the class of organic molecules typically used for OLEDs is similar to the benzene discussed in Chapter 5. They feature a π−electron system delocalized over the entire molecule and characterized by frontier orbitals know as the HOMO (highest occupied molecular orbital) and LUMO (lowest unoccupied molecular orbital). As discussed below, these two orbitals largely determine the electrical and optical properties of both the single molecules and the molecular solids. The HOMO is a fully occupied bonding π−orbital, and the LUMO is an antibonding unoccupied π^*−orbital. Thus, the π-electron-system is typically saturated (i.e. fully occupied with no upaired spins); hence intermolecular covalent bonds cannot form. Consequently, the solids formed of such molecules are only bonded by weak van der Waals forces [9].

The overlap of the π-electron-system of adjacent molecules is small, leading to an energetic splitting of the order of tens to a few hundred meV. Nevertheless, this splitting is sufficiently large for ordered molecular crystals to show band-like charge transport at low temperatures, i.e. the HOMO levels form a narrow hole transport band and the LUMO levels split into a narrow electron transport band. Note that the overlap of the π−orbitals of neighbouring molecules is generally much smaller than the *intra*molecular overlap of the p_z−orbitals of neighbouring C-atoms inside the individual molecules, leading to the splitting between the different π−electron-orbitals of the molecule. Accordingly, the narrow bands that arise from each of the molecular π−orbitals exhibit only a small *inter*molecular overlap. Such systems are often called *small molecular* systems or *low molecular weight* systems even though the total number of atoms forming such small molecules can be on the order of fifty or more.

The situation is somewhat different in polymeric systems where the large number of C-atoms taking part in the π−conjugation leads to broad quasi-continuous energy bands of π−electron states along the molecular chains. However, this does not imply that polymers have larger charge carrier mobilities than small molecular weight solids because the interchain overlap in polymers tends to be small and mobilities are always limited by the most energetically costly hopping steps – in this case a hop between chains. In fact, the highest mobilities are achieved for highly ordered crystals of small, flat molecules that form closely packed stacks with high *inter*molecular π−electron overlap [21], [22].

2.2 Absorption, Exciton Diffusion and Fluorescence

In the framework of the Hückel-approximation introduced in Chapter 5, the HOMO energy levels support the conduction of holes, the LUMO conducts electrons, and the lowest energy excitation, i.e. the absorption edge, is given by a HOMO-LUMO transition. However, this is the point were the limitations of the Hückel-approximation become obvious. Hückel theory is based on an LCAO-approach (linear combination of atomic orbitals), i.e. it is a one electron picture that does not account for electron-electron interaction energies for (i) an electron promoted from the HOMO to the LUMO leading to an excited neutral molecule or (ii) an electron brought from infinity onto the LUMO leading to a negatively charged (anion) molecule. Self-consistent Hartree-Fock calculations reveal that the difference between the electronic energy gap (equal to the ionization energy minus the electron affinity) and the optical gap (the energy of the first excited singlet state) can be several electron volts for single molecules. However, the difference becomes smaller in the solid state (see Figure 3) where the ionization energy is lower and the electron affinity is higher by 1-2 eV because both the anionic and the cationic molecular states are energetically stabilized by polarizing the surrounding solid [23].

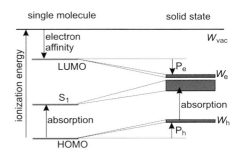

Figure 3: Ionization energy, electron affinity and first optical transition (S_0 to S_1) in a single molecule and formation of charge carrier and exciton bands in molecular solids from the respective molecular orbitals. The ionic energy levels are energetically stabilized in the solid state by the polarization energies P_e and P_h for electrons and holes, respectively. Note that an energetic stabilization of a hole leads to an upward shift in this electron energy picture.

In the solid state, the optical gap is 0.5 to 1 eV smaller than the electronic gap [24] due to the Coulomb energy gained when a free electron and hole meet at a molecule to form an excited state or *exciton* [9]. Molecular excitons are different from hydrogenic Wannier-excitons found in inorganic semiconductors. In a molecular *Frenkel exciton*, the electron and the hole are localized on the same molecule, typically on the same ligand. In ordered crystals, bands of Frenkel excitons can be formed by wave-type linear combination of molecular excitations, i.e. the exciton may be delocalized even though the electron and the hole are coulombically bound. For closely packed materials of flat molecules, the energy of these delocalized states may differ from an excitation on an individual, *free* molecule by several hundred meV. This leads to the formation of broad exciton bands, and the absorption of the solid differs significantly from single molecules in solution [9]. Accordingly, the excitons couple strongly the phonon system, which can lead to exciton self-trapping by excimer formation [25]. Here, an excimer is an excited state shared by two, adjacent molecules, causing them to *bind* somewhat more tightly than in their ground state, thus giving rise to a new, lower energy excitation than that afforded by a simple Frenkel state. Excimers provide additional channels for non-radiative decay (e.g. aggregate quenching). Therefore, non-planar molecules that form amorphous, loosely packed solids are typically chosen for OLEDs. In this latter case, the optical absorption spectrum is similar in solution and in the solid state since the Frenkel exciton is localized on a single molecule. This excitation can diffuse around the solid by hopping transport. For Alq$_3$, e.g., the exciton lifetime is 16 ns [26], leading to a diffusion length of 10 – 20 nm [26], [27].

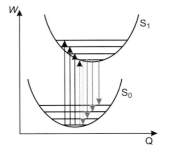

Figure 4: Potential energy lines and vibronic levels for the ground state S_0 and the first excited state S_1 as a function of a configuration coordinate Q reflecting the intramolecular atomic distances. The electronic transitions without change in Q are always most pronounced in the spectra. Clearly, the absorption wavelengths (black) are larger, in general, than the emission wavelengths (green), causing a red shift of the emission spectrum with respect to the absorption spectrum.

As mentioned in Section 1, another important feature of organic semiconductors is the red shift of the main fluorescence peak with respect to the first absorption peak (the *Stokes shift*) observed for many materials due to intramolecular reorganization upon excitation. As the HOMO is a bonding and the LUMO is an antibonding orbital, the equilibrium distances of the atoms comprising a molecule is different in the ground and in the excited states. Therefore, fast electronic excitation leads to an excited state that relaxes into thermal equilibrium within a small fraction of its lifetime (Figure 4) [9]. Similarly, the relaxed excited state transition can make a transition to a vibronically excited ground state (i.e. one in which the molecular bonds are thermally excited above their energies at 0 K). Both effects add up to the red shift of fluorescence as compared to absorption. The transition probabilities from a given electronic state into the different vibronic sublevels of another electronic state are expressed by *Franck-Condon factors*. Accordingly, the typical absorption and fluorescence spectra of organic molecules in solution and of weakly coupled solids show electronic transitions with a series of side bands reflecting the vibronic sublevels of the molecular species.

2.3 Charge Carrier Transport

As in the case of conventional crystalline semiconductors such as Si, perfect molecular crystals show band-like charge carrier transport at low temperatures even though the bands are significantly energetically narrower in organic materials [21]. However, at room temperature, carriers are often localized on individual molecules [21] since the energy bands narrow at higher temperature along with increasing thermally induced molecular vibrations and lattice vibrations (phonons) that reduce the intermolecular π-electron overlap. Eventually, the electron-phonon interaction energies exceed the electronic bandwidth and charge carriers become trapped on individual molecules. In contrast, wider bandwidths exist in conventional semiconductors due to the considerably tighter covalent interatomic bonds. In this case, lattice vibrational energies are much smaller than the electronic bandwidth, significantly reducing the probability of electronic localization (i.e. self trapping) on an individual constituent lattice atom. These effects lead to typical room temperature hopping mobilities of ~1 cm^2/Vs for most organic single crystals, which is two to three orders of magnitude lower than observed for conventional semiconductors. Furthermore, the charge carrier mobility decreases by another four to six orders of magnitude due to structural disorder in amorphous or polycrystalline thin films. Here, a distribution in the trapping energy (the polarization energy) due to the distribution in relative molecular orientations characteristic of disordered solids, leads to a distribution of the energetic density of transport states (DOS) [23]; consequently electron transport can be described by variable-range hopping in the tails of these distributions [28]. In such a situation, the effective mobility increases with the density of injected carriers or the electron donor or acceptor dopant density, respectively: With increasing occupation, the highly localized states deep in the tail of the distribution saturate while states close to the center of the DOS become occupied. Here, the density of states is high, the mean hopping distance is correspondingly reduced, and the hopping rate increases. This leads to a superlinear increase of conductivity with the doping concentration [29] and, for undoped samples, to a current that increases rapidly with the applied voltage [30] (c.f. Section 3.5).

3 Organic Light Emitting Diodes

3.1 Prerequisites for an Efficient OLED

For an efficient fluorescent OLED, an electron and a hole should form a singlet molecular excited state, this exciton should recombine radiatively and the energetic losses throughout the process should be low. Accordingly, the total external power efficiency η_P of an OLED is given by:

$$\eta_P = \gamma r_{st} \Phi_r \Phi_{out} \frac{\overline{\hbar\omega}}{eV} . \qquad (1)$$

Here, γ, also called the charge-balance factor, is the ratio of the number of excitons formed to the number of electrons flowing in the external circuit, r_{st} is the fraction of excitons which are formed as singlets, Φ_r is the efficiency of radiative decay, Φ_{out} is the light outcoupling efficiency, $\hbar\omega$ is the average energy of the emitted photons, e the elementary charge and V the applied voltage.

Singlet excitons refer to those states that are spin antisymmetric (with a total spin quantum number $S=0$). Singlet exciton transitions to the ground state (also in a singlet configuration) are quantum mechanically allowed, and hence are fast (~1 – 10 ns) and efficient, giving rise to fluorescence. In contrast, triplets have even symmetry ($S=1$). Their transition to a singlet ground state does not conserve spin angular momentum, and hence are disallowed. For this reason, triplet transitions are slow (100 μs – 10 s) and highly inefficient, leading to phosphorescence. From the multiplicity of angular momemtum states, there is one $S=0$ state ($m_s=0$ is the quantum number for the projection of the spin on the z-axis), and three $S=1$ states ($m_s=\pm 1,0$); hence the names singlet and triplet. Due to the random nature of spin production via electron injection in electrically driven devices such as OLEDs, simple statistical arguments suggest that only 25 % of the injected charge will combine into emissive fluorescent states. In Section 3.2, we will discuss device architectures that ensure that γ is close to unity, and that by employing heavy metal containing phosphor molecules, all electrically excited molecules can be made efficiently emissive from their triplet state.

The efficiency of radiative decay, Φ_r, is given by

$$\Phi_r = \frac{\kappa_r}{\kappa_r + \kappa_{nr}} \qquad (2)$$

where κ_r and κ_{nr} denote the rate constants for radiative and non-radiative decay of an exciton, respectively. The radiative decay constant is largely determined by the strength of the transition dipole moment [25], and is also influenced by the proximity of the radiating molecule to a reflecting surface, such as a metal cathode. Here, the transition dipole moment is proportional to the overlap of the wavefunctions in the molecular excited and ground states: the larger overlap leading to increased probability for the electron to transit from one state to the other, emitting light when this transition releases energy. The rate of non-radiative decay is a function of the coupling of the electronic excitation to intramolecular vibration modes. However, in molecular solids, further effects like traps and aggregate formation can further increase the non-radiative decay rate [31]. These losses can be minimized by using mixed emitter layers as described in Section 3.3. A successful strategy to reduce the driving voltage is by doping the transport layers by electronic acceptor or donor molecules (cp.3.5). Finally, light outcoupling is ~20 % [13] when glass substrates are used due to total reflection at the interface between the substrate and the ambient. To maximize Φ_{out}, the emission region should be removed by approximately one quarter of a wavelength from the reflecting top contact to provide constructive interference for emission normal to the substrate. Otherwise, the probability increases that the emitted light is coupled into waveguide modes in the substrate. Furthermore, various techniques have been reported to reduce waveguiding losses by structuring the substrate surface, e.g. by employing an array of microlenses [32] or by using microcavities.

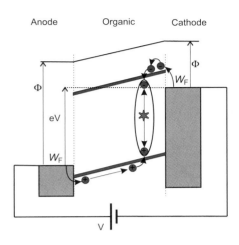

3.2 Single and Double Layer OLEDs

The operation principle of a single layer OLED is schematically shown in Figure 5. For such devices, the exciton recombination region is near the middle of the device if injection of holes from the anode and electrons from the cathode into the organic material is equally efficient, and the mobilities for both carrier types are also equal. Otherwise the active region where excitons are formed moves close to one of the electrodes which leads to a reduced efficiency: The type of carrier which is more mobile or more efficiently injected has a higher probability of reaching the opposite electrode without being captured by the opposite charge type (thereby reducing γ). Moreover, excitons formed close to an electrode may be quenched by defects or diffuse to the electrode where they non-radiatively recombine.

Typical OLEDs therefore consist of at least two or three layers of different organic materials (c.f. Figure 1), i.e. forming single or multi-heterojunction devices. The single heterojunction device of Tang and van Slyke [3] consisting of TPD and Alq$_3$ sandwiched between ITO and Al, marked a breakthrough in low voltage, high efficiency device performance. In their OLED, the energy levels were chosen such that there was only a small barrier to hole injection from the hole transport layer (HTL) into the light emission layer (EML), while electrons from the EML met a high barrier so they were not able to penetrate into the HTL. Consequently, there are almost no loss by electrons reaching the anode. On the other hand, the hole mobility in the EML was much lower than the electron mobility. Therefore, the active region is close to the heterointerface in the EML, i.e. roughly in the middle of the OLED structure, thus avoiding the problems encountered in single layer devices (see, for example, Ref. [33]).

Even in the double layer OLED, losses by holes reaching the cathode or by quenching at the cathode can be significant, especially if the thickness of the EML is low, i.e. on the order of the exciton diffusion length. However, thin layers of these low mobility (and hence high resistance) layers are desirable to minimize the operating voltage. Consequently, optimized devices comprise a second heterojunction: An electron transport layer (ETL) is inserted between the EML and the cathode. It should have a wider HOMO-LUMO gap and a larger ionization potential than the EML so that neither excitons nor holes from the EML can penetrate into the ETL [34] (c.f. Figure 5). Therefore, this layer is variously referred to as the hole-blocking or exciton blocking layer (HBL or EBL, respectively). Such heterojunction architectures are self-balancing due to their internal barriers, i.e. they guarantee $\gamma \to 1$ without requiring either the mobilities or contact injection barriers to be equal.

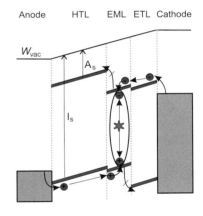

Figure 5: Energy level scheme for a single layer (top) and double heterojunction OLED (bottom) with an applied bias voltage V showing the vacuum energy level W_{vac}, the Fermi levels W_F and workfunctions Φ of the metallic contacts and the hole and electron transport levels of the organic layers. The level offsets at the organic heterojunctions are determined by the different ionization energies I_s and electron affinities A_s of the adjacent layers.

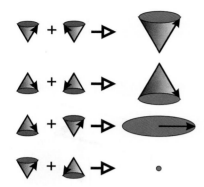

Figure 6: Schematic representation for the formation of singlet and triplet excitons (green) from electrons (red) and holes (blue) with different spin-polarization. The spins are symbolized by vectors in a vertical magnetic field.

3.3 Improving the Device Efficiency Using Fluorescent Dopants

There are organic dyes with photoluminescence (PL) quantum yield Φ_r close to unity in the gas phase or in solution. In the solid state, collisions between excitons and phonons open additional non-radiative decay channels. Additionally, the PL yield in organic solids is sensitive to impurities, as excitons can diffuse into low energy, defect-related traps [35]. Consequently, the PL quantum yield is often considerably lower in the solid state. This problem can be overcome by doping a small amount (~ 1 mol %) of a fluorescent dye into a conductive host material forming the EML [12]. The concentration must be low to ensure that the dopants are separated from each other to avoid non-radiative aggregate traps.

Excitons in doped systems can be formed directly on the conductive host molecules by combination of a free hole and electron [33]. Subsequently, this excited host molecule can transfer its excitation to the dopant by a Förster mechanism [36]. Here, Förster transfer is a non-radiative process, involving the excitation of the dipole transition of the dopant molecule by the simultaneous de-excitation of the host molecule. Förster transfer requires that the excited state energy of the host be equal to (i.e. resonant with) the unexcited HOMO-LUMO gap of the receiving, or guest molecular dopant. This is apparent when the host fluorescence spectrum overlaps the absorption spectrum of the dopant.

An alternative exciton formation mechanism entails the trapping of a hole on the dopant prior to recombining with an electron on the host (or vice versa), thereby forming the radiative exciton.

An additional benefit of doping by fluorescent molecules is that the roll-off of quantum efficiency at very high emission brightness is less pronounced for doped emission layers than for undoped (neat) layers. If the concentration of freely moving excitons is high in a single component material, two excitons might collide leading to exciton-exciton annihilation [27]. If the excitons are rapidly trapped on dopants, the concentration of freely moving excitons, and thus their annihilation probability, is significantly reduced.

Finally, doping a host with emissive guest molecules provides an effective means to tune the OLED emission color. Using, for example, a blue fluorescent host, the entire visible spectrum is accessible simply by using an admixture of different guests into the emission layer [37].

3.4 High Efficiency Phosphorescent OLEDs

The internal quantum efficiency of fluorescent OLEDs cannot exceed 25 % because of the singlet-triplet-ratio r_{st} discussed above. As shown in Figure 6, there are three ways to form a triplet and only one way to form a singlet. This intrinsic limitation can be overcome by incorporating *phosphorescent guest molecules* into a host matrix material to generate light from both the triplet and the singlet excitons, and achieve internal quantum efficiencies close to unity [13], [14].

Figure 7: Chemical structure and energy level scheme for the metalorganic phosphor, Ir(ppy)$_3$ (from [38]). The ligand singlet state (^{1}ligand) and metal-to-ligand charge-transfer singlet state (^{1}MLCT=S$_1$) were determined by the absorption peaks in toluene solution (10^{-5} M). Also, the triplet MLCT state (^{3}MLCT=T$_1$) was estimated from the phosphorescence peak. κ_{NP}, κ_P, and κ_{NP} are quantum yields for nonemissive transitions from ^{1}MLCT intrinsic phosphorescent transitions, and nonemissive transitions from ^{3}MLCT, respectively. Also, Φ_{ISC} is the yield for intersystem crossing from ^{1}MLCT to ^{3}MLCT.

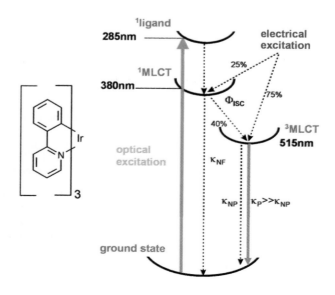

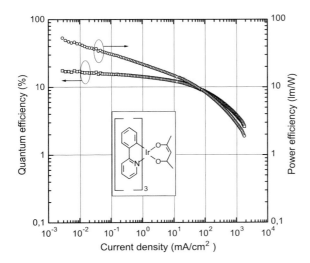

Figure 8: External quantum and power efficiencies of an ITO/HMTPD (60 nm)/12 %-(ppy)$_3$-Ir(acac):TAZ(25 nm)/Alq$_3$ (50 nm)/Mg:Ag OLED. A maximum external quantum efficiency of 19 % and power efficiency of 60 lm/W were obtained. Inset: The chemical structure of (ppy)$_3$Ir(acac) (from [13]).

As noted above, radiative triplet exciton transitions are quantum mechanically forbidden due to the necessity for spin conservation. This quantum mechanical selection rule can be broken by spin-orbit coupling provided by a heavy metal atom incorporated into the emitting organic molecule: If the exciton spin couples to the orbital angular momentum of the metal electrons, the exciton spin is no longer a good quantum number. The selection rule of spin conservation is therefore weakened and intersystem crossing from a singlet into a triplet state becomes possible. The spin-orbit coupling is especially strong if the lowest singlet state is a metal-to-ligand charge transfer (MLCT) state as is the case e.g. for fac-tris(2-phenylpyridine) iridium (III) (Ir(ppy)$_3$) (Figure 7), a material used as a high efficiency phosphorescent emitter in green OLEDs [4], [38].

As a rule, the energy of the lowest triplet exciton state, T_1, is lower than the lowest singlet exciton state, S_1, by typically 0.5 to 1.5 eV (0.86 eV for Ir(ppy)$_3$ [38]). The reason is that the electrons forming a triplet have the same spin symmetry. The Pauli exclusion principle requires that the total electronic state be antisymmetric. Hence, a triplet has an odd spatial symmetry, resulting in a lower exciton energy as compared to the spatially symmetric singlet state by approximately 0.5 to 1.0 eV (the "exchange" energy). The photoluminescence process in phosphorescent molecules is as follows: An electron is excited from the ground state S_0 into the first state S_1 (or any other singlet state from where it rapidly thermalizes into S_1, Figure 7). From there, it undergoes intersystem crossing (ISC) to the triplet state T_1 with a probability Φ_{isc}. Finally, it can recombine either radiatively or non-radiatively from T_1 to S_0 with a rate κ_p and κ_{np}, respectively. Accordingly, the efficiency for phosphorescent photoluminescence is given by

$$\Phi_p = \Phi_{isc} \frac{\kappa_p}{\kappa_p + \kappa_{np}} . \qquad (3)$$

In electroluminescence, the situation is different. Under electrical excitation, both singlet and triplet excitons are directly created on either the guest or host molecules with a statistical splitting of $r_{st} \sim 25\%$ singlets and $(1-r_{st}) \sim 75\%$ triplets. For sufficiently high concentrations of phosphorescent guest molecules in a host molecular matrix with a suitable match of the energy levels to enable energy transfer, we can assume that all excitons generated on matrix molecules will completely transfer their energy to the guests within their lifetime. Thus follows:

$$\eta_{int} = \left[(1-r_{st}) + r_{st}\Phi_{isc}\right] \frac{\kappa_p}{\kappa_p + \kappa_{np}} \qquad (4)$$

Consequently, the internal quantum efficiency η_{int} will be at least 75 % if only the term outside the brackets, i.e. the probability of radiative decay of a triplet exciton, is close to unity. This is the case for instance in Ir(ppy)$_3$, where the PL efficiency of the molecule in solution is only 40 % due to a limited efficiency of inter-system crossing into the triplet state [38] (c.f. Figure 7). Nevertheless, phosphorescent OLEDs with internal quantum yields of nearly 90 % have been demonstrated using Ir(ppy)$_3$ [14] or (ppy)$_2$Ir(acac) [13].

A general problem of phosphorescent OLEDs is that the efficiency can decrease more severely at higher brightness levels compared with in a fluorescent OLEDs (Figure 8). This is due to the longer lifetime of triplet excitons (typically 500 ns [38] to

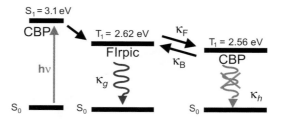

Figure 9: Energy-level diagram of singlet and triplet levels of the CBP host and the FIrpic guest in a blue phosphorescent OLED. Due to the energy lineup of CBP and FIrpic triplet levels, endothermic transfer is dominant. Here, κ_g and κ_h are the radiative decay rates of triplets on the guest and host molecules; κ_F and κ_B are the rates of exothermic (forward) and endothermic (reverse) energy transfers between CBP and FIrpic.

100 μs [39] in phosphorescent materials) as compared with singlet excitons (~10 ns), leading to triplet-triplet annihilation or quenching of triplet excitons by charge carriers at high brightness [40].

To reduce the effects of annihilation processes, it is necessary to design phosphorescent molecules with short lifetime, i.e. with a large spin-orbit coupling. An alternative approach is to co-dope a matrix with a phosphorescent sensitizer and a fluorescent emitter [41]. Here, long-lived triplet states are avoided by fast and efficient resonant Förster transfer from the triplet state of the sensitizer to the singlet state of the emitter. Generally, the long lifetime and accordingly long diffusion lengths of triplets require device architectures that include exciton blocking layers to avoid quenching at contacts or at interfaces [4], [14].

The price one has to pay for the increased efficiency in phosphorescent OLEDs is the energy loss in the transition from singlet to triplet state. For a given emission wavelength, this requires host materials with a correspondingly wider band gap.

In the case of green phosphorescence, high efficiencies are obtained by resonant, exothermic energy transfer from both the host singlet and triplet states to the phosphor. As the triplet energy of the phosphor increases, it becomes less likely to find an appropriate host with a suitably high-energy triplet state. Thus, one route to efficient blue electrophosphorescence involves the endothermic energy transfer from a near-resonant excited state of the host to the higher triplet energy of the phosphor [42]. Provided that the energy required in the transfer is not significantly greater than the thermal energy, this process can be very efficient. It might appear surprising that recombination takes place predominantly from a high energy phosphor state in quasi-equilibrium with a low energy host singlet state. This occurs since the triplet lifetime on the phosphor dopant is significantly shorter than the forbidden triplet transition of the fluorescent conductive host (Figure 9). With 4,4'-N,N' dicarbazole biphenyl (CBP) as a host (triplet energy 2.56 eV) and FIrpic as guest (triplet energy 2.62 eV), a maximum external electroluminescent quantum efficiency of ~5.7 % and a luminous power efficiency of ~6.3 lm/W were obtained [42], representing a significant improvement over efficiencies of blue fluorescent emitters reported to date.

3.5 Reduced Voltages by Doped Charge Transport Layers

To obtain a high power efficiency and low driving voltage, efficient charge injection at interfaces and low ohmic losses in the transport layers are required. These conditions are fulfilled in conventional inorganic light-emitting diodes (LEDs) by using heavily n-and p-doped electron and hole transport layers, leading to efficient tunneling injection and flat energy band conditions. The operating voltages of such LEDs are thus usually close to the thermodynamic limit, i.e., the photon energy divided by the elementary charge.

Organic light-emitting diodes contain nominally undoped layers, requiring high internal fields to overcome contact barriers and to drive carriers into the emission region. In this case, the operating voltage for high brightness significantly exceeds the thermodynamic limit.

There have been a number of attempts to overcome these problems: one method to improve hole injection is to insert buffer layers between the anode and hole transporting layer [43], [44]. To improve electron injection, one uses low work function metal cathodes, or inserts a thin energy-barrier-reducing layer between cathode and electron transporting layer [45], [46]. However, even with optimized contacts, there remains the fundamental problem of space-charge-limitation of the current whenever undoped transport layers are used. When carriers are injected into a material that does not contain any equilibrium carriers, the material will be charged [23], [47]. As like carriers are coulombically repelled, this space charge limits further injection. In the absence of traps, the current density j follows

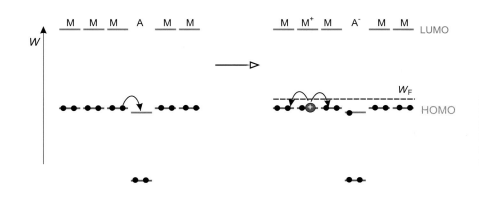

Figure 10: Charge transfer from a host (M) to an acceptor-type dopant (A) leading to the formation of a free hole on the host and a negatively charged acceptor (A⁻).

$$j = \frac{9}{8}\varepsilon\mu\frac{V^2}{d^3} \qquad (5)$$

where ε is the permittivity, μ is the carrier mobility, V is the applied voltage and d is the thickness of the sample. If there is a distribution of traps in the material that is subsequently filled with increasing voltage, the current increases with a higher power of the voltage and thickness. Consequently, a strategy to increase the current density and thus the brightness at a given applied voltage is to minimize the thickness of the *undoped* layers in the OLED and to ensure that transport in the *doped* layers is ohmic with a negligible voltage drop [5], [48] – [53].

We briefly introduce the concept of doping for the case of p-type organic semiconductors [11], [52]. When an acceptor molecule A is brought into a host matrix of molecules M, the unoccupied LUMO of A, is close in energy to the HOMO of M (Figure 10). The energetically favorable electron transfer from M to A then leads to a negatively charged acceptor and a positively charged host molecule, i.e. a hole that can move relatively freely throughout the film. The material is electrically neutral with

$$p = n + N_A^- , \qquad (6)$$

where p and n are the densities of free holes and electrons, respectively, and N_A^- denotes the density of negatively charged acceptors. Generally, the minority carrier density, n, can be neglected as compared to p in p-type semiconductors. In doped materials, charge transport does not imply the formation of space charges. The charge carriers in a doped material move with a drift velocity proportional to V, according to Ohm's law.

An example of a p-doped hole transport material is 4,4',4''-tris(3-methylphenylphenylamino)triphenylamine (MTDATA) doped with F$_4$-TCNQ (Figure 11). The electron affinity of F$_4$-TCNQ is comparable to the ionization energy of MTDATA, implying

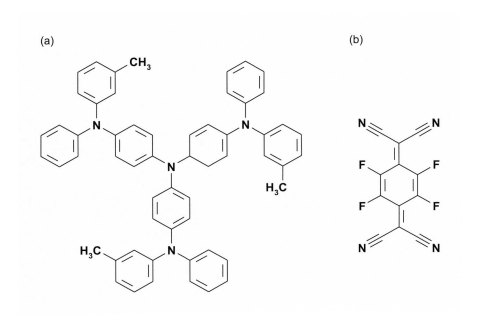

Figure 11: Chemical structures of (a) MTDATA (host) and (b) F$_4$-TCNQ (acceptor), a model system for p-doped hole transport materials.

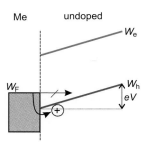

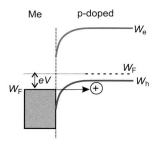

Figure 12: Hole injection from a metallic anode into an undoped and a p-doped organic semiconductor. The energy level schemes are shown for an applied bias voltage V.

complete charge transfer, i.e. all dopants carry exactly one negative charge [11]. Here, a doping ratio of 1:50, realized by co-evaporation of the two materials, leads to a conductivity of ~5×10^{-7} S/cm at room temperature [53]. Accordingly, the voltage drop across a 50 nm thick hole transport layer of doped MTDATA at 10 mA/cm^2, a typical value for OLEDs employed displays, is only 0.1 V.

A further advantage of doped transport layers is that they can enable efficient charge injection from contacts even over a high energy barrier. If we contact a p-type semiconductor with a low work function metal, a rectifying junction is formed, thereby preventing efficient hole injection into the semiconductor. However, the width of the depletion region at the interface scales as $N_A^{-1/2}$. For very high doping levels, the depletion region becomes sufficiently thin that carriers can tunnel through the barrier (Figure 12). It has been shown by thickness dependent photoelectron spectroscopy that this principle also applies to organic semiconductors [54].

As an example of the effects of doping, we consider ITO, a transparent conductive oxide that is commonly used as an anode in OLEDs. Due to the high work function of ITO (~4.5 eV), it cannot efficiently inject holes into many organic transport materials because the barrier exceeds 1 eV without a modifying surface treatment such as exposure to ozone or an oxygen plasma. However, efficient injection of both holes into highly p-doped hole transport materials [48], [49] and electron injection into highly n-doped electron transport materials [55] is possible even from untreated ITO.

From this, it is obvious that an OLED can benefit from a p-i-n structure. The emission region is undoped (intrinsic) as dopants tend to provide non-radiative exciton recombination centers. Thin undoped exciton blocking layers are inserted between the emission layer and the doped charge transport layers to keep the efficiency high. The total thickness d of the undoped region is kept low since space charge limited currents scale with d^{-3}.

The energy level scheme of a p-i-n OLED [5] is shown in Figure 13. The device has a high luminance at low operating voltages, where 100 cd/m^2 is reached at 2.5 (close to the thermodynamic limit for green emission); and 1000 cd/m^2 at 2.9 V. In this device, the total thickness of undoped layers is only 35 nm.

Recently, the approaches of using phosphorescent emitters combined with the doping of the charge transport layers have resulted in low voltage, high efficiency p-i-n OLEDs. These devices attain a luminance of 1000 cd/m^2 with a power efficiency of 28 lm/W at 3 V, and 10,000 cd/m^2 with 17 lm/W at 4 V [56].

The performance of phosphorescent OLEDs of different colors is summarized in the following table. Here, Φ is the conversion factor from light power (in Watt) to brightness (in lumen), the latter unit corresponding to the spectral sensitivity of the eye, η_P is the power efficiency, V_λ is the photon energy at the peak of the emission spectrum divided by e, and V is the applied voltage to achieve a given current density. An optical microcavity model [13] was used to calculate the internal quantum efficiency η_{Qint} from the measured external quantum efficiency η_{Qext}. For phosphorescent green emitters, a current density of 1mA/cm^2 corresponds to a brightness of 300 – 500 cd/m^2, which is 3 to 5 times brighter than a television monitor.

Figure 13: Luminance-voltage characteristics and energy level scheme for a p-i-n OLED under applied forward bias (from [5]).

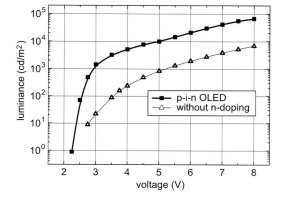

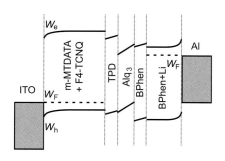

phosphor	host	color	Φ (lm/W)	1μA/cm² η_{Qint}	1μA/cm² V_λ/V	η_P (lm/W)	1mA/cm² η_{Qext}	1mA/cm² η_{Qint}	1mA/cm² V_λ/V
(ppy)$_2$Ir(acac) [13]	TAZ	green	530	0.87	0.60	20	0.15	0.68	0.25
btpIr(acac) [57]	CBP	red	170	0.65	0.34	2	0.10	0.44	0.22
FIrpic [42]	CBP	blue	260	0.27	0.83	5	0.06	0.23	0.34
Ir(ppy)$_3$ (p-i-n OLED) [56]	CBP	green	530	0.25	1.02	29	0.09	0.45	0.86

Table 1: Efficiencies of several different electrophosphorescent OLEDs.

4 Organic Displays

4.1 Passive and Active Matrix Displays

Figure 14 shows a typical layout of a passive matrix (PM) OLED display. On the surface of a transparent substrate (which can be glass or plastic) are patterned columns of transparent, conducting anode contacts. Onto their surface is deposited the full organic light emitting structure, with the hole transporting layer in contact with the anodes. The display is completed by depositing rows of metal cathode contacts. To address a particular OLED *picture element* or pixel, a potential is placed across the appropriate row and column contacts, and current flows across the organic layers at the intersection of those contacts, thereby lighting up that device. A full image is produced by rapidly scanning through the row lines while individually energizing the appropriate column lines using *line driver* electronics connected to the rows and columns at the edge of the PM display. Typical handheld video image display pixels must be less than 250 μm across, and must be spaced by less than 50 μm.

To fabricate a full color display, each row must be divided into three, with each of the three row lines contacting an OLED which emits a different color of the primary triad: red, green and blue. The entire 3-color triad must fit into the same space as that of a single pixel in the monochrome display, making the fabrication of a full color display far more complex. A drawback of passive matrix displays is that the individual pixels are active only for a small fraction of the viewing time since each row must be scanned sequentially during a frame period (typically ~1/30 s), and must therefore be driven at a very high brightness to achieve a reasonable time averaged brightness. However, the power conversion efficiency of typical OLEDs decreases with brightness due to an increase in driving voltage. Here, the use of doped charge transport layers (c.f. Section 3.5) can reduce these losses. Also the quantum efficiency tends to decrease with brightness due to multiple exciton annihilation or exciton-polaron quenching. The required brightness increases linearly with the number of pixels as does the capacitive loss during the rapid switching of the pixels. Passive matrix displays are therefore limited in size, with practical dimensions not exceeding 5 cm x 5 cm.

Fabrication complexity can be reduced, and display performance can be increased by placing a transistor *switch* at each contact intersection. These switches are set by the line driver electronics once per each display frame, resulting in an active matrix (AM) OLED display. Here, all pixels emit simultaneously and continuously during each display frame, so that the maximum brightness of a particular pixel does not have to be multiplied by the number of display rows to achieve its desired *viewing brightness*, as is the case for PMOLED displays. This type of display is enabled by polysilicon thin film transistor technology. Very attractive active matrix full color OLED displays (13 inch and 17 inch diagonal) have already been demonstrated using this approach. Their power consumption is a factor of 2-3 lower than for comparable liquid crystal displays [58].

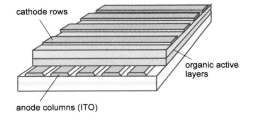

Figure 14: Layout of a passive matrix OLED display.

4.2 Patterning Techniques

A significant limitation to the realization of advanced organic electronic devices is the lack of a simple and low-cost means for patterning devices which contain fragile organic thin films. In displays, the issue of patterning is focused primarily on confining the cathode metal to the region occupied by a particular display pixel. There are several methods one considers in the contact patterning process. For example, conventional semiconductors are patterned using standard photolithography. Unfortunately, this technique depends on wet chemical processing which can degrade most small molecular weight and polymer organic materials. Attention has, therefore, been focused primarily on using shadow-masking during metal deposition to pattern the OLED contacts. Such a mask must be sufficiently thick to reduce bowing or wrinkling of the mask during use when it is held in very close (~1 μm) proximity to the substrate during deposition to maintain a small feature size. As a general rule, the mask thickness must be equal to or greater than the aperture (e.g. contact) dimension, limiting contact diameters to >70 μm. Further, mask cleaning after frequent use is required to prevent clogging of the small apertures by the deposited material. Several alternative patterning methods such as the use of a shadow mask prepatterned on the substrate using photoresist features, excimer laser ablation of organic materials, direct patterning of polymer pixels employing ink-jet printing, and conformal masking using elastomeric membranes have been demonstrated (c.f. [59] and references therein). A promising new approach is direct patterning of the metal cathode by cold welding of the metal on the organic surface to a "stamp", followed by lift-off of the metal in unwanted areas on the substrate [59]. In this process, a prepatterned, metal-coated stamp composed of a rigid material such as Si is pressed onto an unpatterned film consisting of the organic device layers coated with the same metal as that used to coat the stamp. When a sufficiently high pressure is applied, an intimate metallic junction is formed between the metal layers on the stamp and the film, leading to a cold-welded bond. When the stamp and film are separated, the metal cathode fractures at the edges of the stamp. The stamp is then removed from the substrate, removing the metal with which it came into contact from the wafer surface, thus forming a well-defined patterned electrode. The resolution of this process has been shown to be ~10 nm [61].

4.3 Outlook: Transparent, Stacked and Flexible OLEDs

In many respects, OLEDs are an alternative approach providing the same functionality with higher image quality, higher power efficiency or lower costs than existing display technologies such as liquid crystal or plasma displays. Here, we will introduce a few examples showing that OLEDs also offer new possibilities not accessible by these other approaches.

Transparent OLEDs

The Transparent OLED (TOLED) uses a transparent cathode, e.g. ITO sputtered on top of the active organic layers, to create displays that can be made both top and bottom emitting on an ITO coated substrate. When placed on a light absorbing surface (such as an active transistor back plane in an AMOLED display), TOLEDs can greatly improve contrast and aperture ratio (i.e. the ratio of emitting to non-emitting display areas) making displays easier to view in bright ambient lighting. Because TOLEDs are >70 % transparent when turned off, they may be integrated onto car windshields, architectural windows, and into helmet-mounted or "head-up" systems for virtual reality applications. The use of a transparent top contact furthermore enables the use of metals, foils, silicon wafers and other opaque substrates for top-emitting devices. Low voltage high efficiency phosphorescent TOLEDs reaching 1000 cd/m^2 at 3.4 V with 7 % quantum efficiency have recently been demonstrated using doped transport layers [60]. Doping is a key technology for TOLEDs since it enables efficient injection of both electrons and holes if ITO is used as an anode or as a cathode contact material. Moreover, the sensitive emission region of an OLED can be protected from sputtering damage during ITO deposition by using thick, doped transport layers.

Stacked OLED displays

Transparent OLEDs are the basic building block for a novel pixel architecture that is based on stacking the red, green, and blue subpixels on top of one another instead of side by side positioning used in conventional full color displays. This improves display resolution up to three-fold as each individual stacked pixel can provide full color. In large screen displays, individual pixels are frequently large enough to be resolved at short range. With side-by-side positioning of the sub-pixel color elements, the eye may perceive the individual red, green and blue instead of the intended color mixture. With stacked OLEDs, each pixel emits the desired color, and thus is perceived correctly from any viewing distance.

Flexible OLED displays

Flat panel displays have traditionally been fabricated on glass substrates because of structural and/or processing constraints. However, organic layers can easily be deposited onto a wide variety of substrates that range from optically clear plastic films to reflective metal foils. These materials provide the ability to conform, bend or roll a display into any shape (see Figure 15).

Figure 15: A flexible 0.18 mm thick passive matrix OLED display fabricated by Universal Display Corporation.

Acknowledgements

The editor would like to thank Fotis Fitsilis (FZ Jülich) for checking the symbols and formulas in this chapter.

References

[1] M. Pope, H. Kallman, P. Magnante, J. Chem. Phys. **38**, 2042 (1963).
[2] W. Helfrich, W.G. Schneider, Phys. Rev. Lett. **14**, 229 (1965).
[3] C.W. Tang, S.A. VanSlyke, Appl. Phys. Lett. **51**, 913 (1987).
[4] M.A. Baldo, S. Lamansky, P.E. Burrows, M.E. Thompson, S.R. Forrest, Appl. Phys. Lett. **75**, 4 (1999).
[5] Jingsong Huang, M. Pfeiffer, A. Werner, J. Blochwitz, Shiyong Liu and K. Leo, Appl. Phys. Lett. **80**, 139 (2002).
[6] S. Miyata, H.S. Nalwa, eds., *Organic electroluminescent materials and devices*, Gordon and Breach Science Publishers, Amsterdam, 1997.
[7] J.H. Burroughes, D.D.C. Bradley, A.R. Brown, R.N. Marks, K. Mackay, R.H. Friend, P.L. Burns, A.B. Holmes, Nature **347**, 539 (1990).
[8] R.H. Friend, R.W. Gymer, A.B. Holmes, J.H. Burroughes, R.N. Marks, C. Taliani, D.D.C. Bradley, D.A. Dos-Santos, J.L. Bredas, M. Löglund, W.R. Salaneck, Nature **397**, 121 (1999).
[9] M. Pope, C.E. Swenberg, *Electronic processes in organic molecular crystals*, Oxford University Press, New York, 1982.
[10] P. E. Burrows, S. R. Forrest, S. P. Sibley, M. E. Thompson, Appl. Phys. Lett. **69**, 2959 (1996).
[11] M. Pfeiffer, T. Fritz, J. Blochwitz, A. Nollau, B. Plönnigs, A. Beyer, K. Leo, Advances in Solid State Physics **39**, 77 (1999).
[12] C.W. Tang, S.A. VanSlyke, C.H. Chen, J. Appl. Phys. **65**, 3610 (1989).
[13] C. Adachi, M. A. Baldo, M. E. Thompson, S.R. Forrest, J. Appl. Phys. **90**, 5048 (2001).
[14] M. Ikai, S. Tokito, Y. Sakamoto, T. Suzuki, Y. Taga, Appl. Phys. Lett. **79**, 156 (2001).
[15] C.W. Tang, Appl. Phys. Lett. **48**, 183 (1986).
[16] C. J. Brabec, N. S. Sariciftci, J. C. Hummelen, Adv. Funct. Mater. **11**, 15 (2001).
[17] M.A. Baldo, V.G. Kozlov, P.E. Burrows, S.R. Forrest, V.S. Ban, B. Kroene, M.E. Thompson, Appl. Phys. Lett. **71**, 3033 (1997).
[18] R.A. Laudise, Ch. Kloc, P.G. Simpkins, T. Siegrist, J. Cryst. Growth **187**, 449 (1998).
[19] P.E. Burrows, S.R. Forrest, T.X. Zhou, L. Michalski, Appl. Phys. Lett. **76**, 2493 (2000).
[20] T.R. Hebner, C.C. Wu, D. Marcy, M.H. Lu, J.C. Sturm, Appl. Phys. Lett. **72**, 519 (1998).
[21] J. H. Schön, Ch. Kloc, B. Batlogg, Phys. Rev. B - Condensed Matter **63**, 245201 (2001).
[22] S.R. Forrest, Chem. Rev. **97**, 1793 (1997).
[23] E.A. Silinsh, Springer Series in *Solid-State Sciences*, Springer-Verlag, Berlin, Heidelberg, New York, 1980.

[24] I.G. Hill, A. Kahn, Z.G. Soos, R.A. Pascal Jr., Chem. Phys. Lett. **327**, 181 (2000).
[25] J.B. Birks, *Photophysics of Aromatic Molecules*, John Wiley and Sons, London, (1970).
[26] C.W. Tang, S.A. VanSlyke, C.H. Chen, J. Appl. Phys. **65**, 3610 (1989).
[27] I. Sokolik, R. Priestley, A.D. Walser, R. Dorsinville, C.W. Tang, Appl. Phys. Lett. **69**, 4168 (1996).
[28] M.C.J.M. Vissenberg, M. Matters, Phys. Rev. B - Condensed Matter **57**, 12964 (1998).
[29] B. Maennig, M. Pfeiffer, A. Nollau, X. Zhou, P. Simon, K. Leo, Phys. Rev. B - Condensed Matter 64, 195 (2001).
[30] P.E. Burrows, Z. Shen, V. Bulovic, D.M. McCarty, S.R. Forrest, J.A. Cronin, M.E. Thompson, J. Appl. Phys. **79**, 7991 (1996).
[31] K. Puech, H. Fröb, M. Hoffmann, K. Leo, Opt. Lett. **21**, 1606 (1996).
[32] S. Möller and S. R. Forrest, J. Appl. Phys. **91**, 3324 (2002).
[33] J. Staudigel, M. Stössel, F. Steuber, J. Simmerer, J. Appl. Phys. **86**, 3895 (1999).
[34] J. Kido, C. Ohtaki, K. Hongawa, K. Okuyama, K. Nagai, Jpn. J. Appl. Phys. Pt. **2**, 32 (7A), L917 (1993).
[35] N. Karl, Mat. Sci., **10**, 365 (1984).
[36] T. Förster, Discuss. Faraday Soc. 27,7 (1959).
[37] C. Hosokawa, M. Eida, M. Matsuura, K. Fukuoka, H. Nakamura, T. Kusumoto, Synthet. Metal. **91**, 3 (1997).
[38] C. Adachi, M.A. Baldo, S.R. Forrest, M.E. Thompson, Appl. Phys. Lett. **77**, 904 (2000).
[39] D.F. O'Brien, M.A. Baldo, M.E. Thompson, S.R. Forrest, Appl. Phys. Lett. **74**, 442 (1999).
[40] M.A. Baldo, C. Adachi, S.R. Forrest, Phys. Rev. B - Condensed Matter **62**, 10967 (2000).
[41] B.W. D'Andrade, M.A. Baldo, C. Adachi, J. Brooks, M.E. Thompson, S.R. Forrest, Appl. Phys. Lett. **79**, 1045 (2001).
[42] C. Adachi, R. C. Kwong, P. Djurovich, V. Adamovich, M. A. Baldo, M. E. Thompson and S. R. Forrest, Appl. Phys. Lett. **79**, 2082 (2001).
[43] S.A. VanSlyke, C. H. Chen, and C. W. Tang, Appl. Phys. Lett. **69**, 2160 (1996).
[44] Y. Shirota, Y. Kuwabara, H. Inada, Appl. Phys. Lett. **65**, 807 (1994).
[45] L.S. Hung, C.W. Tang, M.G. Mason, Appl. Phys. Lett. **70**, 152 (1997).
[46] C. Ganzorig, M. Fujihira, Jpn. J. Appl. Phys. Pt. **2**, 38, L1348 (1999).
[47] M.A. Lampert, Phys. Rev. **103**, 1648 (1956).
[48] J. Blochwitz, M. Pfeiffer, T. Fritz, K. Leo, Appl. Phys. Lett. **73**, 729 (1998).
[49] X. Zhou, M. Pfeiffer, J. Blochwitz, A. Werner, A. Nollau, T. Fritz, K. Leo, Appl. Phys. Lett. **78**, 410 (2001).
[50] X. Zhou, J. Blochwitz, M. Pfeiffer, A. Nollau, T. Fritz, K. Leo, Adv. Funct. Mater. **11**, 310 (2001).
[51] J. Blochwitz, M. Pfeiffer, M. Hofman, K. Leo, Synthet. Metal. **127**, 169 (2002).
[52] M. Pfeiffer, A. Beyer, T. Fritz, K. Leo, Appl. Phys. Lett. **73**, 3202 (1998).
[53] J. Drechsel, M. Pfeiffer and K. Leo, Synthet. Metal. **127**, 201 (2002).
[54] J. Blochwitz, T. Fritz, M. Pfeiffer, K. Leo, D.M. Alloway, P.A. Lee, N.R. Amstrong, Organic Electronics **2**, 97 (2001).
[55] X. Zhou, M. Pfeiffer, J. Blochwitz and K. Leo, Appl. Phys. Lett, **81**, 922 (2002).
[56] M. Pfeiffer, K. Leo, S. R. Forrest, and M. E. Thompson, Adv. Mater. **14**, 1633 (2002).
[57] C. Adachi, M.A. Baldo, S.R. Forrest, S. Lamansky, M.E. Thompson, R.C. Kwong, Appl. Phys. Lett. **78**, 1622 (2001).
[58] G. Rajeswaran, M. Itoh, M. Boroson, S. Barry, T.K. Hatwar, K.B. Kahen, K. Yoneda, R. Yokoyama, T. Yamada, N. Komiya, H. Kanno, H. Takahashi, SID 00 Digest **40**, 1 (2000).
[59] Changsoon Kim, Paul E. Burrows, and Stephen R. Forrest, Science **288**, 831 (2000).
[60] M. Pfeiffer, X. Zhou, K. Leo and S.R. Forrest, Organic Electronics, submitted.
[61] C. Kim, M. Shtein and S. R. Forrest, Appl. Phys. Lett. **80**, 4051 (2002).

Field-Emission and Plasma Displays

Peter K. Bachmann, *Helmut Bechtel,* and *Gerd Spekowius*
Philips Research Laboratories, Aachen, Germany

Contents

1 Introduction	929
2 Field Emission Displays (FEDs)	929
2.1 Basic Principles of FEDs	929
2.2 FED Cathodes and Cathode Materials	930
2.2.1 Spindt-Type Tip Emitters	930
2.2.2 Carbon Nanotube Emitters	931
2.3 Phosphor Screens for FEDs	932
2.3.1 Phosphor Efficiency and Emission Properties of Cathode-Ray Phosphors	932
2.3.2 Low Voltage Excitation of Phosphors	933
2.3.3 Electron Penetration Depth	933
2.3.4 Voltage Dependence of the Phosphor Efficiency	934
2.3.5 Luminance Limitations under Display Operating Conditions	935
2.4 Matrix Addressing of an FED	936
2.5 FED-Performance and Prospects	936
3 Plasma Display Panels (PDPs)	937
3.1 PDP Discharge	938
3.2 Operation Principles of PDPs	939
3.3 PDP Design and Manufacturing	940
3.4 High Gamma Coatings in PDP Cells	940
3.4.1 General	940
3.4.2 Ion-induced Electron Emission Mechanism and Data Analysis	942
3.4.3 Parameters Influencing g-Emission: Some Examples	943
3.5 Phosphors for PDPs	944
3.5.1 Phosphor Efficiency	945
3.5.2 Electronic Transitions Involved in Europium Luminescence	945
3.5.3 Colour Point and Efficiency of the Red Phosphors	945
3.5.4 Stability and Colour Point of the Blue Emitting BAM Phosphor	946
3.6 PDP Performance and Prospects	947
4 Summary	948

Field Emission and Plasma Displays

1 Introduction

Displays are a key technology of the information age acting as the final interface between information providers, like, e.g., computers, the internet or broadcasting stations and the human visual system. If we look into an electronic shop today, we probably recognise at first two different types of displays, namely cathode-ray tube (CRT)-based television sets and conventional computer monitors, and a rapidly increasing number of liquid crystal display (LCD)-based computer monitors. Secondly, we will probably see myriads of small LCDs implemented in mobile phones, watches and other instruments of any kind. Moreover, we find flat displays with very large diagonals of about 100 cm or more, so-called plasma display panels (PDPs). With the PDP the dream of the hang-on-the-wall TV has become a reality. Analysts are predicting stunning growth for PDPs in the next few years. Other so-called flat panel display technologies, like field emission displays (FEDs) and electro luminescence (EL) displays today predominatly can be found in markets such as the automotive and medical industries.

In this contribution, we concentrate on the discussion of FEDs and PDPs as role models for novel flat emissive display technologies. Both display principles rely on matrix-addressed arrays of picture elements (pixels) and, contrary to LCDs, in both cases phosphors are excited by either electrons (FEDs) or UV photons (PDPs) to generate the picture. After a brief introduction into the general display principles, the physical aspects of light generation (field emission of electrons, generation of plasma discharges, phosphor emission) are discussed. Special attention is paid to the functional materials used inside the display (protective materials, cathodes and phosphors). Finally, current display performance data are briefly reviewed and some of the trends in the display industry are illustrated.

2 Field Emission Displays (FEDs)

2.1 Basic Principles of FEDs

Field emission displays (FEDs) are basically flat cathode-ray tubes. Electrons are emitted from the surface of a cathode material and are accelerated in vacuum towards an anode by applying an electrical field. This field determines the energy of the electron upon arrival at the phosphor-coated anode surface. The electron energy is, like in CRTs, used to excite a luminescent phosphor particle layer, which, upon de-excitation, emits visible light. Despite these similarities, there are also distinct differences between CRTs and FEDs. CRTs are based on rather bulky, heavy, three-dimensional, thick-walled evacuated glass tubes that contain 1000 °C hot filaments as continuously operating electron sources and magnetic coils to deflect electron beams in order to address and light up individual picture elements (pixels) on the phosphor-coated front plate opposing the electron source. FEDs consist of two glass plates at a distance of only a few millimetres. One of the plates carries an array of electron emitters that are switched on individually for a short period of time only when irradiation of the corresponding phosphor pixel that resides on the anode plate is triggered by the driver electronics. In FEDs, the electron emitters are at a much lower temperature, even down to room temperature. Emission occurs because of high electrical fields and enhancement of such fields at sharp edges and tips, rather than the high temperature of the emitter material. In Figure 1, the cross section of a very simple, diode-type variant of an FED structure, consisting of the emitter-bearing cathode plate, and the phosphor-bearing anode plate is depicted.

The potential energy $W(z)$ of an electron at a distance z from the cathode surface is the sum of four terms

$$W(z) = W_F + \Phi - \frac{e^2}{(4\pi\varepsilon)4z} - eEz \qquad (z > 0) \qquad (1)$$

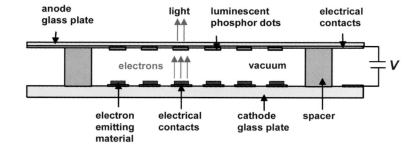

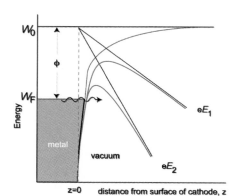

Figure 1: Cross section of a very simple variant of an FED structure, consisting of an emitter-bearing cathode plate and a phosphor-bearing anode plate. Electron extraction is induced by applying a high voltage V between cathodes and anodes using a matrix addressing scheme. Emitted electrons are accelerated towards the phosphor-coated anode, excite individual pixels of luminescent material, thus stimulating the emission of visible light.

Figure 2: Surface potential barrier experienced by electrons in a metal at two different applied electrical fields $E_2 > E_1$. The barrier is composed of four separate terms (see Eq. (1)). Tunneling of electrons from filled electronic states up to the Fermi level W_F is fostered by lowering the potential barrier, i.e. by increasing the strength of the applied field [1].

where W_F is the Fermi energy, Φ is the work function, $-e^2/4z$ represents the image force that an electron of charge e experiences when leaving the solid, and $-eE$ is the energy contribution from the applied electrical field. Figure 2 depicts the contributions from each of the four terms and the total potential energy of the electron for two applied fields E (with $E_1 < E_2$). Obviously, there are two ways to lower the energy barrier for electrons to leave the solid and to be emitted into the vacuum. Firstly, by lowering the work function Φ to values near the Fermi level, electrons will be able to leave the surface more easily. Secondly, as the applied field increases, the shape of the potential barrier changes and the probability that an electron can tunnel from the solid through the energy barrier into the vacuum gets higher. In this case, the density of the current emitted by the cathode depends on the tunneling probability, the number of electrons near the interface and the density of states.

The relationship between these terms can be evaluated at room temperature to give what is known as Fowler-Nordheim equation:

$$j(E) = 6.2 \cdot 10^6 \frac{(W/\Phi)^{1/2}}{W_F + \Phi} E^2 e^{-\frac{6.8 \cdot 10^7 \cdot \Phi^{3/2}}{E}} \qquad (2)$$

2.2 FED Cathodes and Cathode Materials

Deformation of the potential barrier by increasing the local field strength (see Figure 2) is used to increase the tunnelling probability of electrons into the vacuum, thus increasing the emission current. Likewise, the emission probability is increased by lowering the work function of the emitter material. Both principles of emission enhancement were investigated in the past (see, e.g., [3] – [10]) and the first FED prototypes and products appeared at trade shows and on the market in the year 2000 (see Sect. 2.5).

2.2.1 Spindt-Type Tip Emitters

The principle of enhancing the field at an emitter surface locally by tipped structures in order to facilitate emission was first investigated by C. Spindt [3] in 1976 and now starts to yield the first products [10], [12]. Figure 3 depicts a schematic view of the cross section of a Spindt-tip type FED structure outlining the basic principles of the corresponding displays.

Rather than the diode-type structure shown in Figure 1, the structure in Figure 3 comprises of three electrodes, namely the cathode, the gate electrode and the anode. In this case, the cathodes are grown as large arrays on the contact-bearing back plate and

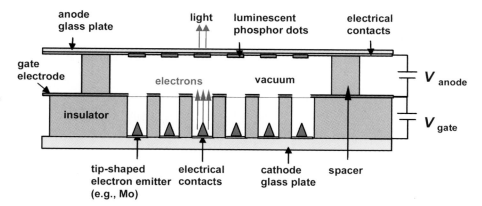

Figure 3: Schematic cross sectional view of a triode-type, so-called Spindt-tip emitter field emission display arrangement. A small gate voltage (5 – 10 V) is controlling the electron emission. With a high anode voltage (3 – 5 kV) electrons are accelerated towards to phosphor screen.

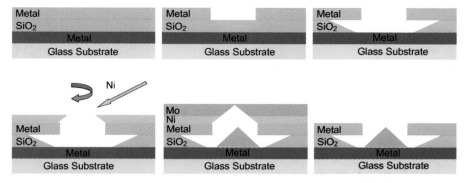

Figure 4: Fabrication of Spindt-tip emitter arrays. On the supporting glass structure (yellow), a metal cathode contact layer (blue) is deposited and coated with an insulator (red), e.g., spin-on glass. On top, the gate metal electrode is grown (green). Holes are generated in the gate and in the insulator by photolithography and etching. A sacrificial Ni layer (orange), evaporated under an angle with rotating substrate, protects the structure during subsequent deposition of the Mo, thus creating tips inside the holes. The Ni and the Mo layer are electrochemically removed [2].

consist of tip-shaped metal structures. Nowadays Molybdenum is the material of choice to fabricate such tips, because it turned out to be sufficiently sputter-resistant not to turn blunt (thus loosing field enhancement properties) too early by ion bombardment. Moreover, it is sufficiently resistant to poisoning by materials that evaporate into the rather small volume gas phase of the display. Emission is governed by Eqs. (1) and (2) and the fact, that with a radius of curvature in the order of nanometers, the field enhancement leads to sufficient deformation of the tunnelling barrier and thus to emission currents of some mA/cm^2. This is sufficient to excite a phosphor dot that is located on the opposing anode plate. Thousands of tips are used to excite a single pixel. Between cathode and anode plate, two additional elements are visible when comparing Figure 1 and Figure 3. An insulator separates the cathode contacts from the gate electrode. The gate voltage V_{gate} is used to extract electrons from the Spindt tips. They propagate through holes in the gate and are accelerated towards the phosphor-bearing anode plate by applying a second voltage to the system, the anode voltage V_{anode}. This structure, which is common to most prototype displays that appear today, allows to partially decouple electron generation from phosphor excitation. While electron extraction is facilitated at low voltages, the additional anode voltage allows exciting phosphors with accelerated 1–10 keV electrons, thus partly reducing phosphor efficiency problems and phosphor degradation (see Sect. 2.3) The fabrication steps that lead to such structures are outlined in Figure 4. Figure 5 depicts a 3-D view of a Spindt-tip based FED.

Although Spindt-tip based emitters came a long way, remaining emitter- and phosphor-related problems (lifetime, poisoning, burn-out, degradation) fully justify thorough investigation of alternative materials and structures such as flat emitters based on the negative electron affinity of diamond [5] or, especially, emission from carbon nanotubes and other carbonaceous structures ([4], [6] – [9]). Sect. 2.2.2 describes selected examples of such emitter structures, their performance and their preparation.

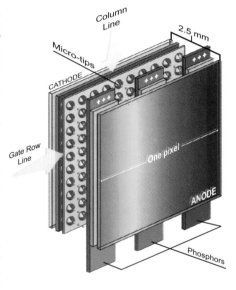

Figure 5: 3-D view of a Spindt-tip-based FED. The structure shown is used by Pix-Tech Inc. to fabricate small (5.2") monochrome and also full colour field emission displays for medical systems, car systems and military applications (see also Figure 17) [10]. The anode voltage used is 500 V, switched between red, green und blue emitting phosphor lines, which are addressed sequentially.

2.2.2 Carbon Nanotube Emitters

Over the past years, much effort in research was devoted to the evaluation of carbon-based electron emitter structures. Initially, CVD diamond films were expected to be a good candidate for electron emission because of the negative electron affinity (i.e. a vacuum level below conduction band minimum) of hydrogen-terminated diamond surfaces ([5], [31]). Diamond turned out to be a good photo emitter, however, as a field emitter, diamond is neither sufficiently conductive nor is it possible to populate the conduction band sufficiently to take advantage of the low electron affinity. In the course of related studies it turned out, that mixed phase carbon materials containing diamond and non-diamond carbon phases are superior to pure diamond as field emitters ([9], [14], [15]) and ultimately carbon nanotubes (CNTs) were found to emit electrons even better (see, e.g., [4][6][7][8]).

Figure 6 illustrates the excellent field emission properties of such oriented nanotubes by showing an *I-V* curve and an emission site map of oriented carbon nanotubes grown by microwave plasma chemical vapour deposition from methane-hydrogen mixtures at 600 °C and 40 mbar.

Patterning of carbon nanotube (CNT) arrays is feasible by simply patterning the transition metal catalyst (e.g., Fe or Ni) that is needed to initiate CNT growth. An excellent example of highly oriented, patterned nanotubes is shown in Figure 7 [6]. The task of present R&D efforts is to replace the Mo-based emitter tips in Figure 3 or Figure 17 by such patterned arrays of simple-to-manufacture, high performance carbon nanotube

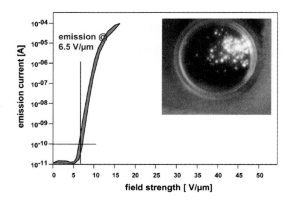
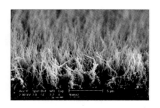

Figure 6: Emission current vs. field strength (left) measured at the nanotube array (right). The insert depicts an emission site map based on the blue luminescence emitted by a transparent ITO anode when irradiated by electrons emitted from a nanotube-covered surface during the *I-V*-measurements. The material starts emitting at fields of around 7 V/μm and is capable of current loads in the order of A/cm² [15].

emitters. Samsung of Korea recently announced success in developing the first prototype 15" full colour display based on such emitters and device structures similar to those shown in Figure 3 [8].

2.3 Phosphor Screens for FEDs

Light generation in a field emission display (FED) is almost identical to light generation in a classical cathode-ray tube (CRT). As depicted in Figure 8 a layer of phosphor powder is deposited onto the inner surface of a glass tube. Patterns of red, green and blue emitting phosphors, when hit by high-energy electrons, are fabricated by UV photolithography. Charge accumulation on the surface is prevented through secondary electron emission or suitable conductive coatings. If the electron energy is high enough, a thin conductive aluminium layer is deposited evenly on top of the phosphor layer. This aluminium layer of about 100 nm thickness is sufficiently conductive to prevent charging and also reflects light emitted towards the inside of the display to increase the light output by a factor of about two.

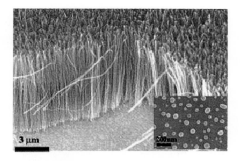

Figure 7: Patterned, highly oriented carbon nanotube arrays grown by microwave plasma CVD [6].

Phosphors consist of inorganic powders with particle sizes of 3 to 8 μm. This implies that layers scatter visible light and the amount of light emitted from the screen decreases with increasing thickness of the phosphor layer. On the other hand, the layer has to be thick enough to cover the glass surface completely and to absorb the full energy of the incident electrons. Both requirements are met when the phosphor layer thickness is about twice the median grain size of the phosphor ([32], [33]).

In the following text, the quantification of light and color is needed. The corresponding description is given in the Introduction to Part VIII.

2.3.1 Phosphor Efficiency and Emission Properties of Cathode-Ray Phosphors

When fast electrons are absorbed in solids they create electrons and holes by ionisation of lattice atoms as shown in Figure 10. In addition, the primary and the secondary particles loose also energy by excitation of lattice vibrations. Generally, the energy W to generate one electron-hole pair with the band gap energy W_g of a semiconductor can be written as:

$$W = W_i + W_{op} + 2W_f = \beta \cdot W_g \quad (3)$$

W_i is the ionisation threshold energy, W_{op} is the average energy lost to optical phonons and W_f is the energy below threshold before thermalisation.

Robbins has performed a quantitative treatment of the optical phonon generation in 1980. He related the constant β to the high frequency and static dielectric constants of the host lattice ε_∞ and ε_0 and the optical phonon frequencies [35]. Robbins found a convincing agreement between his predicted and experimental phosphor efficiency. β ranges from 2.9 for ZnS to about 4 for the oxysulfides to 7 for oxide materials.

The full expression given by Robbins for the energy efficiency η of a phosphor under electron excitation is

$$\eta = (1 - r_b) \cdot h\nu_{em} \cdot \eta_t \eta_a \eta_{esc} / \beta \cdot W_g \quad (4)$$

where r_b is the fraction of back-scattered electrons, $h\nu_{em}$ is the mean energy of emitted photons, η_t, η_a and η_{esc} denote the efficiency of electron-hole pair transfer to the luminescence centers, radiative recombination at the luminescence center, and escape of visible

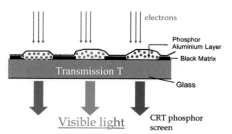

Figure 8: Cross sectional view of the phosphor screen of a colour cathode-ray tube (CRT). Patterns of phosphor powder layers emit red, green and blue light when irradiated with high-energy electrons. A thin aluminium layer, which is almost transparent for high-energy electrons and optically reflective, conducts incident electrons and increases the light output through the glass plate. A so-called Black Matrix, consisting of a thin carbon layer separating different phosphor patterns, combined with a tinted glass minimises the reflectivity of the screen for ambient light.

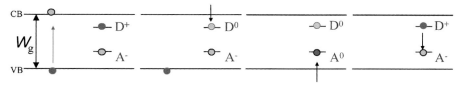

Figure 9: Energy level and excitation and emission process for the donor acceptor pair recombination of ZnS:Ag. The valence band VB and the conduction band CB with the band gap energy W_g are indicated. After excitation of an electron to the conduction band a hole is left in the valence band. Both, electrons and holes are captured by the donor and acceptor levels, respectively, and recombine under emission of a photon.

light. Phosphors developed for cathode-ray excitation operate at the physical limit with η_a and η_{esc} equal to one. Thus, Eq. (4) simplifies to:

$$\eta = (1-r_b) \cdot \eta_t \frac{h\nu_{em}}{\beta \cdot W_g} \qquad (5)$$

The most efficient cathode-ray phosphors are the zinc sulfides and their derivatives. They possess a relatively low band gap energy of about 3.8 eV and a small β-factor. The energy efficiency of the blue ZnS:Ag phosphor is about 20 percent, which is by far the most efficient blue emitting cathode-ray phosphor. The ZnS host material is doped with about 0.03 mole percent of Ag, which leads to formation of an acceptor level within the ZnS band gap. For charge compensation, aluminium or chlorine can be added, to form shallow donor levels. Figure 9 displays the energy level scheme and the electron – hole generation and recombination process.

Since the emission wavelength for the ZnS:Ag phosphors is determined by the band gap energy of the lattice, the emission band can easily be shifted with the addition of Cd into the green and red spectral region. This phosphor system has been used for both colour cathode-ray tubes and field emission displays.

Today, Cadmium has become unacceptable for environmental reasons, and ZnS:Cu, Au is applied as the green emitting phosphor. The red emitting phosphor is nowadays $Y_2O_3S:Eu^{3+}$, where emission occurs in narrow lines from inner f states of the Eu^{3+} ion. Although the energy efficiency of this phosphor is only 13 percent, the lumen efficiency is far superior to wide band emitters and the Eu^{3+} ion is used almost exclusively as emitter for red phosphors in display applications. Figure 11 depicts emission spectra for the red, green and blue emitting cathode-ray phosphors.

In the Introduction to Part VIII, the colour coordinates of the phosphor emission spectra of Figure 11 are shown together with the European Broadcasting Union (EBU) chromaticity standards for television. According to the rules of additive colour mixing, all colours within the colour gamut, i.e., a triangle formed by the three primary colour coordinates for red, green and blue, can be displayed.

2.3.2 Low Voltage Excitation of Phosphors

A remarkable difference of CRTs and FEDs lies in the flatness of the display. CRTs are operated with anode voltages of 20 to 30 kV. With the advent of FEDs, a strong desire for phosphors that can be operated at voltages even below 1 kV arose, in order to minimise the risk of arcing and leakage currents in the flat FED devices. Research and development, also for the application in vacuum fluorescent displays (VFD), is ongoing. A 20-year chronology of that work is found in reference [36].

Most of the work on low voltage operation of phosphors concentrates on the phosphor efficiency decrease. However, as discussed below, not only the efficiency of the phosphors deteriorates when the operating voltage is decreased. Also the magnitude of the luminance that can be generated in a short amount of time decreases due to efficiency saturation phenomena, and the phosphor lifetime decreases as well.

2.3.3 Electron Penetration Depth

Bohr and Bethe derived the basic theories for the stopping power of solids for impinging particle energies much greater than the energies of the atomic electrons. For kilo electron volt (keV) energies, the electron penetration depth λ_e was investigated intensively by energy loss measurements of electrons in thin foils. There is general agreement that the expression relating the λ_e and the energy W of impinging electrons has the form:

$$\lambda_e = bW^n \qquad (6)$$

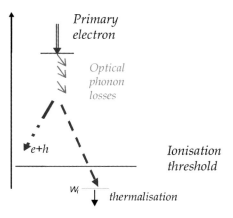

Figure 10: Schematic representation of energy loss processes of fast electrons in a solid. From one fast electron a large number of electrons and holes (e+h) are produced via lattice ionisation. In addition, energy is lost to optical phonons from fast electrons and to phonons during a thermalisation process when the ionisation threshold energy is passed.

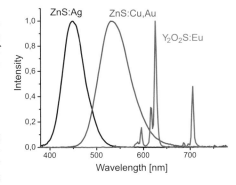

Figure 11: Emission spectra of the red, green and blue emitting cathode-ray phosphors used in today's CRTs and FEDs.

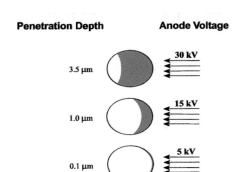

Figure 12: Scheme of electron penetration depths (shaded area) at different anode voltages for a 5 μm ZnS phosphor particle.

where b is a constant related to the material and n is a constant assumed to be independent of the material. A discussion of the underlying theories and comparison with experiments and Monte Carlo calculations can be found, e.g., in an article by Shea and the references therein [37]. A practical formula we used throughout our work giving the electrons range in (μm) with the voltage in (kV), is

$$\lambda_e = \frac{0.046}{\rho} \cdot V^{1.67} \qquad (7)$$

where ρ is the density of the solid in units of g/cm³. Figure 12 displays the electron range in a 5 μm diameter particle for anode voltages of 5, 15 and 30 kV, respectively. Only at the high operating voltage of a CRT of about 30 kV, the incoming energy is distributed over the entire particle. Even at 5 kV, which is still a rather high operating voltage for an FED, only the surface of a phosphor grain is used for light generation.

2.3.4 Voltage Dependence of the Phosphor Efficiency

Figure 13 shows the measured luminance (right axes) as a function of the operating voltage for a blue emitting ZnS:Ag phosphor screen without aluminium backing layer. From the deviation between measured data and the line indicating strict proportionality, it appears that the phosphor efficiency decreases with decreasing operating voltage, as can be seen more clearly from the efficiency plot (left axes).

From the discussion of the cathode-ray efficiency of phosphors in Sect. 2.3.1, it is obvious that there are hardly any factors depending on the electron energy in the range of a few hundred electron volts to a few kilo electron volts. Both, βW_g and $h\nu_{em}$, are independent of the incident energy of the exciting electrons. At normal incidence, the back-scattering coefficient r_b increases with atomic number Z. For $Z = 15$ to 30 (the typical atomic numbers for atoms of phosphors) r_b is nearly independent of the energy of the incident electrons in the range of 1 to 100 keV and amounts to 20 – 30% [5]. In conclusion, the only energy dependent number in Eq. (5) is the transfer efficiency η_t of electrons and holes to the activators.

Successful models for the description of the voltage-dependence of the phosphor efficiency were developed by various authors, considering non-radiative luminescence at the phosphor surface in combination with the so-called ambipolar diffusion of charge carriers ([38], [39]). Assuming the crystal dimensions to be large compared to the electron penetration depth λ_e and the diffusion length L of the charge carriers, the problem can be treated one-dimensionally. With the electron penetration depth λ_e defined above (see Eqs. (6) and (7)), Gergeley solved the diffusion equation for a uniform energy distribution over the total range, and obtained for the voltage dependent (transfer) efficiency, normalised to unity [39]

$$\eta = 1 - Q \cdot \left[\frac{1 - \exp\left(-\frac{\lambda_e}{L}\right)}{\left(\frac{\lambda_e}{L}\right)} \right] \qquad (8)$$

where Q is a surface recombination loss parameter and L is the ambipolar diffusion length.

With the voltage range relation of Eq. (7), Eq. (8) can be expressed in terms of the anode voltage V and a characteristic voltage V_0 by

$$\eta = 1 - Q \cdot \left[\frac{1 - \exp\left(-\frac{V}{V_0}\right)^{1.67}}{\left(\frac{V}{V_0}\right)^{1.67}} \right] \qquad (9)$$

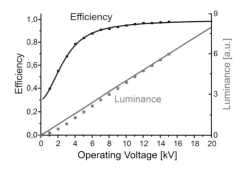

Figure 13: Luminance and efficiency of a ZnS:Ag phosphor screen (without aluminium backing layer) as a function of the operating voltage. Symbols are measured data. The blue line is a least square fit to Eq. (9). The red line indicates the expected proportional increase of the luminance with the operating voltage.

The efficiency, i.e. the measured luminance divided by the power input, is displayed in Figure 13 (left axes) as a function of the operating voltage. The line is a least square fit to Eq. (9).

The ambipolar diffusion length determined from voltage dependent phosphor efficiency measurements is in the range of 15 to 60 nm, decreasing with an increase of the activator concentration of the phosphor [40]. The dependence of the phosphor efficiency, as depicted in Figure 13 is typical for all phosphor materials. A minimum value of 0.5 found in our laboratory for the surface recombination parameter of all phosphor materi-

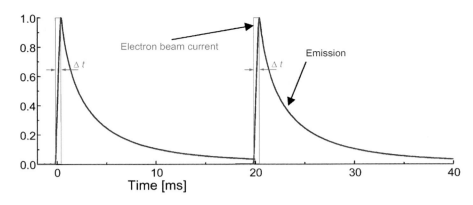

Figure 14: Electron beam current and emitted light from a phosphor layer for a single pixel of a display operated at an image repetition time of 20 ms (50 Hz repetition rate).

als imposes a practical limit for the low voltage intrinsic phosphor efficiency. Moreover, it is obvious that all kinds of non-luminescing materials coated onto the phosphor (e.g., the aluminium layer) further decrease the low voltage efficiency.

For anode voltages larger than about 4 kV, aluminium coating of the phosphor screens has a positive effect on the screen luminance as proven experimentally and theoretically.

2.3.5 Luminance Limitations under Display Operating Conditions

We have seen so far that a phosphor layer emits more light if more energy is dissipated. If the energy input is increased by increasing the operating voltage this is always true and the increases can even be higher than linear. For current increase this is not always the case. It is a general experience in phosphor operation that the phosphor efficiency decreases with increasing current input.

For a display operated at 50 Hz, the image is refreshed every 20 ms. Depending on the number of lines that have to be addressed and on driving considerations of the electronics, a phosphor pixel element will be addressed for a time Δt = 20 ms divided by the number of lines of the display and divided by 3 for colour displays. For a VGA display with 480 lines, the excitation time is then

$$\Delta t = \frac{20\text{ms}}{3 \cdot 480} = 13.9\ \mu\text{s} \tag{10}$$

For most phosphors, this time is small compared to the decay time of the phosphor. Typical shapes of the time dependent beam currents and emitted light vs. time for a single pixel in a display are depicted in Figure 14.

Since the number of activators that can be excited in a phosphor is limited (energy dissipation and transfer of electrons and holes to the activators are much faster than the luminescence decay), electron-hole pairs recombine non-radiatively and the efficiency of the phosphor decreases (the phosphor saturates).

Bril developed an activator ground-state depletion model to describe the efficiency saturation of phosphors ([41], [42]). The basic equation describing the phosphor efficiency η/η_0 is

$$\frac{\eta}{\eta_0} = \frac{N}{N+g\tau} + \frac{N\,g\tau^2}{\Delta t(N+g\tau)^2}(1-e^{-(g/N+1/\tau)\Delta t}) \tag{11}$$

with N the concentration of luminescence centers, τ the decay time, and Δt the excitation time, respectively. g is proportional to the volume rate for center excitation. In addition to the rate of center excitation, the duration of the excitation pulse Δt relative to the phosphor decay time τ plays a dominant role for saturation behaviour. For $\Delta t/\tau \ll 1$, efficiency saturation η/η_0 is only a function of $g\Delta t$, which is proportional to the so-called excitation energy density per pulse j. In this case, Eq. (11) in a linear approximation of the exponential term reads:

$$\frac{\eta}{\eta_0} = \frac{1}{1+\dfrac{j}{j_{50\%}}} \tag{12}$$

Here, $j_{50\%}$ is the energy density per pulse for which the phosphor efficiency is reduced by 50%. A more complete discussion of the phosphor saturation and timing aspects is given in [40].

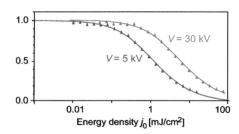

Figure 15: Phosphor efficiency as a function of the energy density per pulse for a red emitting Y_2O_2S:Eu phosphor at two different operating voltages. j_0 is the energy density in the center of a Gaussian electron beam profile. The lines are least square fits to Eq. (13), averaged over the Gaussian beam profile (see reference [40] for details).

Figure 15 shows experimental data of phosphor efficiency saturation for the red emitting Y_2O_2S:Eu phosphor measured at two different operating voltages. As expected from the decreasing excitation volume with decreasing operating voltage, phosphor saturation is stronger with decreasing operating voltage. The measurement is done with a pulsed stationary Gaussian electron beam. Therefore Eq. (12) has to be averaged over the x and y coordinates of the beam profile resulting in [40]

$$\frac{\eta}{\eta_0} = \frac{j_{50\%}}{j_0} \ln\left(1 + \frac{j_0}{j_{50\%}}\right) \quad (13)$$

where j_0 is the energy density in the center of the beam.

The luminance L of a phosphor screen can be easily calculated from the deposited power density P, multiplied by the efficency of the phosphor screen and divided by π, i.e.,

$$L[\text{cd/m}^2] = \frac{1}{\pi} \cdot \eta_{\text{phosphor}}(V; j; \frac{\Delta t}{\tau}) \cdot P[\text{W/m}^2] \quad (14)$$

The energy density j governing phosphor saturation is given by

$$j = \frac{P_{\text{local}}}{R_{\text{im}}} \quad (15)$$

Here R_{im} is the image repetition rate (50 Hz in the above example) and P_{local} is the power density at the phosphor pixel. In general, P_{local} is larger than the power averaged over the whole display screen area, since phosphor pixels usually do not cover the entire screen (see Figure 8). As can be seen from Eq. (15), increasing the image repetition rate reduces saturation and can increase the luminance of the display.

2.4 Matrix Addressing of an FED

As described in Sect. 2, a field emission display (FED) consists of a number of phosphor dots located on transparent anode electrodes (e.g. Indium Tin Oxide (ITO)). Light is generated when electrons from electron emitting cathode dots are accelerated in an electrical field and impinge on the phosphor dot. Video information is displayed when the electrical energy of the electrons interacting with the phosphor dots is modulated. Energy modulation can be done in various ways. In any case, pixel addressing is achieved by employing a matrix of electrodes. This allows addressing of one row (or one line) at a time Δt (see also Sect. 3.2) at the intersection of two electrodes. The maximum time Δt available for deposition of the electrical energy required to generate the desired luminance is given by the frame time (time to display one image) divided by the number of lines of the display. Fast switching of high voltages is difficult. Therefore, FEDs often employ a so-called triode configuration (cf. Figure 3). The matrix formed by the cathode and gate electrodes is utilized for addressing. Emitted electrons are then accelerated by a stationary anode voltage of 3 to 10 kV between gate and anode, i.e. a continuous ITO electrode below the phosphor dots. For an anode voltage higher than approx. 4 kV, a thin Aluminium layer (50 nm thick) on top of the phosphor dots (cf. Figure 8) can replace the ITO anode. Light intensity modulation can be achieved either with current amplitude modulation at a constant time Δt, or with modulation of the on-time of one pixel at a fixed current, so-called pulse-width-modulation. Combinations of both principles are also possible. As described in Sect. 2.3.4, the electrical energy needs to be stored in the phosphor material prior to conversion into light. This severely affects the achievable efficiency. In any case, in FEDs, light generation is done during addressing, while in PDPs addressing and light generation are separated (see Sect. 3.2).

Figure 16: A 13" SVGA-FED from Sony/Candescent [12].

2.5 FED-Performance and Prospects

A considerable number of people expect FEDs to be the biggest threat to the LCD dominance in the flat display market. FEDs can be build to approximately the same size as LCD screens and exhibit some intrinsic advantages related to using emissive phosphor screens. One advantage is the absence of the backlights in FEDs. Consequently, the power consumption depends mainly on the image content and is, therefore, lower than for LCDs In addition, viewing angle, colour rendering and appearance of moving pictures are superior to LCDs. As thousands of electron emitters are used to illuminate a single pixel, FEDs possess an inherent emitter redundancy. In LCDs, however, a single failing transistor results in a dead pixel and thus may render the display useless.

For anode voltages of 5 – 10 kV the feasibility of up to 15" diagonal FED colour displays was recently demonstrated. Figure 16 depicts a photograph of a 33 cm (13") diagonal prototype FED exhibited at the year 2000 symposium of the Society for Information Display (SID) [12]. Today substantial research and development effort is directed towards trying to identify and validate industrially viable FED emitter technologies, including diamond-based emitters, corralline-type carbon, screen-printed carbon emitters and emitters based on multiwalled carbon nanotubes (MWCNTs) [49].

Currently, several companies supply first FED models to the market. In Figure 17, a range of monochrome as well as full colour FEDs developed and marketed by PixTech Inc. [10] is shown. These displays are 5.2" in diagonal and less than 1 cm thick. They consist of 320 x 240 pixels. The power consumption (50% white) is less than 10 W and the brightness can be as high as 1200 cd/m^2 (see Introduction to Part VIII for a detailed explanation of display-related units). While monochrome displays are rugged and exhibit a lifetime that is sufficient for practical use, full colour displays still suffer from lifetime limitations due to phosphor degradation and poisoning of the Mo-tips during operation.

3 Plasma Display Panels (PDPs)

Displays based on plasma discharges are known since the early 1950s [51]. In principle, all plasma displays employ a simple neon glow-discharge. A sealed glass envelope is filled with neon or a rare gas mixture (e.g., 5% Xe/Ne) and applying a sufficiently high voltage between two electrodes ignites a plasma discharge. Both, DC- and AC-discharges were used in PDPs [50]. Initially, the orange/red light generated in the Neon glow discharge was used directly for monochrome display purposes. A breakthrough for this display technology was achieved by applying the so-called AC-surface discharge principle illustrated in Figure 18. A dielectric barrier AC-discharge is sustained on the surface of a front-glass plate of a display by applying a high frequency voltage of several hundred volts between two transparent electrodes embedded in the glass. Typically, a mixture of 3 – 10% Xenon in Neon is used at a pressure of about 600 mbar. In the discharge, Xe atoms are excited to generate vacuum UV (VUV) radiation. This radiation is converted into visible light by a phosphor layer deposited on the rear glass substrate. Visible light leaves the panel through the front glass. This so-called *reflective mode AC surface discharge design* has significant advantages [52] over DC discharges or vertical AC discharges where the plasma is sustained between electrodes located on the front and back plate (see also Sect. 3.2).

Figure 17: Range of Spindt-tip emitter-based mono- and polychrome FEDs developed by PixTech Inc. [10].

A three-dimensional schematic picture of the complete 3-electrode colour AC surface PDP as used in state-of-the-art products [52], [53] is shown in Figure 19 and is described in detail in Sect. 3.2. On the rear glass substrate channels are formed by barrier ribs. The depth of such channels is typically 100 – 200 μm and most commonly, the channels are fabricated by sandblasting in a low melting point soft glass layer. The channel width determines on the resolution of the display. For a 105 cm diagonal display with W(wide)-VGA resolution of 852 x 480 pixels, the channels are about 300 μm wide. On the bottom of each channel, a metal address electrode is positioned and covered by a dielectric. Sets of 3 channels are then covered with a red, green, and blue emitting phosphor layer, respectively form colour pixels. The front plate electrodes usually consist of Indium-Tin-Oxide (ITO), which is a conductive transparent material. However, since the peak currents in AC-PDPs are very high, the conductivity of the ITO is still insufficient

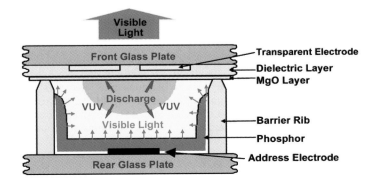

Figure 18: Schematic drawing of a discharge cell in a state-of-the-art AC-surface discharge type PDP. In this sketch the front plate electrodes are drawn in the same direction as the address electrode on the rear plate for visibiliy. In real panels the transparent discharge electrodes are perpendicular to the address electrode.

VIII Displays

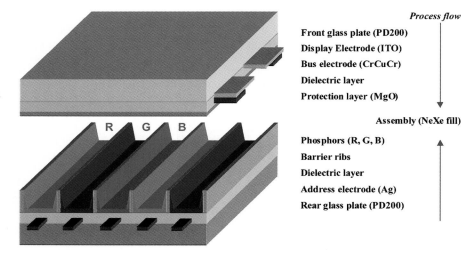

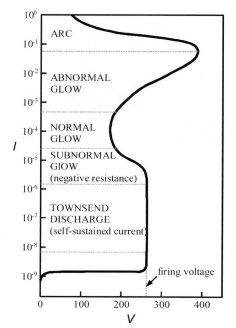

Figure 19: Schematic drawing of a 3-electrode AC-surface discharge type colour PDP. The sustain electrodes (two per pixel) are formed by the ITO display electrodes, with small metal layers (bus electrodes) on top.

Figure 20: Current-voltage (*I-V*) characteristics of a gas discharge. The different discharge modes are indicated. AC PDP discharges start out as Townsend discharges and convert into normal glow mode discharges.

and small extra (bus) electrodes made from copper or silver are applied on top of the ITO. On each intersection of a pair of display electrodes with an address electrode, an addressable pixel is formed. A pixel is selected by an appropriate voltage pulse applied between the address electrode and one display-electrode. Two display-electrodes form the so-called sustain electrodes for one pixel. After addressing, light pulses are generated by an AC-voltage of some hundred kHz applied between the sustain electrodes.

3.1 PDP Discharge

The basics of gas discharges are described in several textbooks and in the case of PDP-discharges, the book by Y. Raizer [11] is perhaps a good starting point. In Figure 20, the different current regions are shown as a function of the electrode voltages for a 2-electrode discharge. If the voltage at the electrodes surpasses a certain threshold, a discharge fires and a low current, self-sustaining carrier avalanche develops (Townsend regime). For higher currents, space charge in front of the cathode builds up and limits the discharge to the stable normal glow regime. For still higher voltages, the voltage across the cell varies strongly with time and all the different regimes shown in Figure 20 up to normal glow mode are traversed during the discharge.

In one-element gas discharges, e.g. in a discharge using just Ne gas, ions are typically generated by electron impact ionisation thus releasing a second free electron:

$$Ne + e^- \rightarrow Ne^+ + 2e^- \tag{16}$$

The second electron is also accelerated in the applied electric field and may generate further ions on its path to the anode, thus generating an avalanche process that results in a self-sustaining discharge. The generated ions are accelerated and neutralised at the cathode where they can release secondary electrons. Hence a high secondary electron emission coefficient of the anode surface material is very important for the firing of the discharge. Generally, generating secondary electrons is very important to obtain an efficient discharge. However, not all electron energy is used for ionisation processes. Part of the energy is consumed in order to generate excited (metastable) atoms, e.g.,

$$Ne + e^- \rightarrow Ne^* + e^- \tag{17}$$

These metastable atoms usually relax by emitting a photon. Furthermore, electrons from the surface may also be released by photo-effect or by de-excitation of metastables. If specific additional other species, e.g., argon atoms, are present in the gas phase, the so-called Penning ionisation process may play an important role. Energy may then be transferred from excited atoms to such species, resulting in ionisation of the additional gas. Mixing, e.g., Ne and Ar results in

$$Ne^* + Ar \rightarrow Ne + Ar^+ + e^- \tag{18}$$

Penning ionisation occurs only in certain gas mixtures (e.g. Ne/Ar) and leads to the efficient conversion of Ne metastables to Ar ions plus electrons, thus decreasing the required discharge firing voltage. In present day AC-PDPs, Ne/Xe mixtures are used in order to optimise the generation of VUV-radiation rather than minimizing the plasma firing voltage. VUV light is generated by excited Xe atoms and dimers in the Ne/Xe gas discharge. Figure 21 depicts the relevant energy levels and a typical emission spectrum for a low pressure Xe/Ne discharge.

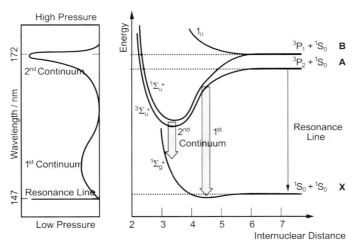

Figure 21: Energy levels of Xe and Xe_2 and typical emission spectrum of low pressure discharge [54].

For the excitation of the phosphors, which subsequently emit the visible light, the resonant radiation at 147 nm and the dimer radiation around 172 nm are most important. In the discharge, Xe is excited by e-impact into the resonant $^3P_1$, the metastable $^3P_2$ or a highly excited Xe** state

$$\begin{aligned} Xe + e^- &\rightarrow Xe(^3P_1) + e^- \\ &\rightarrow Xe(^3P_2) + e^- \\ &\rightarrow Xe^{**} + e^- \end{aligned} \qquad (19)$$

By emitting infrared photons, Xe** can relax

$$\begin{aligned} Xe^{**} &\rightarrow Xe(^3P_1) + h\nu(828 \text{ nm}) \\ &\rightarrow Xe(^3P_2) + h\nu(823 \text{ nm}) \end{aligned} \qquad (20)$$

The $^3P_1$ state can relax to the ground state by emitting a 147 nm photon

$$Xe(^3P_1) \rightarrow Xe + h\nu(147 \text{ nm}) \qquad (21)$$

or an excimer state is formed in the case of three body collisions with a Xe and any third atom (M):

$$Xe(^3P_1) + Xe + M \rightarrow Xe_2^* + M \qquad (22)$$

The excimer dissociates and emits a photon at about 150 nm (1st continuum) or 172 nm (2nd continuum).

3.2 Operation Principles of PDPs

In AC-PDPs, the electrodes are covered by a dielectric material and act like a capacitor. The capacity limits the discharge current in a very efficient way. Commonly, a square wave voltage pulse is applied as illustrated in Figure 22. If the voltage across the gap exceeds the firing voltage V_f, a fast transient discharge ignites. Subsequently, charge is collected on the dielectric layers, thus reducing the electrical field and the discharge extinguishes itself after about 100 ns. After reversing the voltage at the electrodes, the discharge process repeats itself, however, in the opposite direction.

A characteristic property of an AC discharge is a so-called memory effect, facilitating PDP addressing. With the first discharge, so-called wall charge is deposited on the dielectric layer. This wall charge reduces the voltage required for subsequent firing of the discharge and the cell is sustained at a voltage V_s lower than V_f. The difference between V_f and the minimum required sustain voltage $V_{s, min}$ is called operating margin. The reason for this name will become clear from the following.

As in any matrix display, a PDP pixel (or single discharge cell) can be addressed via a row and column electrode. Data signals for one row are connected to all columns of the display. When the row electrode is selected, light generation starts. I.e. light generation at one pixel is only possible while the pixel is addressed. Since a certain number of images have to be displayed per second, the time for addressing and light generation for one row is limited by the so-called frame time divided by the number of rows of the display.

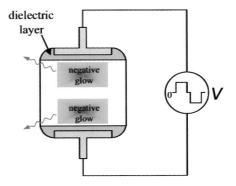

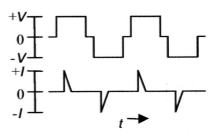

Figure 22: Operation principle of an AC-PDP discharge. Driven by a square wave voltage pulse a capacitive current is transported through the gas. The time dependence of voltage and current is shown.

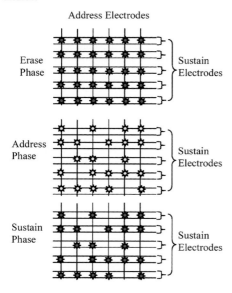

Figure 23: Erasing, addressing and light generation in a three-electrode surface discharge PDP. During erase, a strong discharge pulse is applied to all cells (between sustain electrodes), removing any wall charge from the dielectric. During addressing, one discharge from the address electrode to one sustain electrode is used to deposit wall charge onto the dielectric.

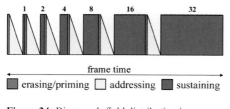

Figure 24: Binary sub-field distribution in an AC-PDP with separate periods for erasing, addressing and sustaining (producing light).

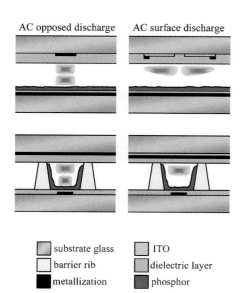

Figure 25: Opposing (left) and co-planar surface-type (right) discharge cell designs.

In order to increase the time for light generation PDPs employ a third electrode per pixel, with one address electrode on the back plate and two sustain electrodes on the front plate (see Figure 19). During an address phase, one single discharge from the address electrode to one of the sustain electrodes is used to deposit wall charge on the dielectric. In a second phase a sustain voltage with an amplitude between $V_{s,\,min}$ and V_f is applied to all sustain electrodes. Due to the memory effect of a discharge, only those cells will ignite which have received wall charge in the prior address phase. Before the next addressing phase is started, an erase pulse is applied to remove all wall charges [53]. Figure 23 displays the different phases of addressing and light generation of a three-electrode PDP matrix display.

During sustain, an AC voltage is supplied to all sustain electrodes. Due to the memory effect, only those pixels which have collected sufficient wall charge during the address period will ignite during the sustain period. The surface properties of the cell coating, e.g., MgO, are not only important for plasma firing but, based on their capability to store charge, have a fundamental influence on the operating margin of the display.

Because this addressing allows only to discriminate between cells to participate in the following sustain phase or not, several addressing-sustain- and erase phases have to be conducted within one frame time. Figure 24 displays the duration of the different erasing, addressing and sustain phases as a function of time, for a 6 bit binary addressing scheme. In that case 6 different sustain phases are combined within one frame time, varying binary in duration. The maximum time for light generation, i.e. the peak luminance, is much higher compared to a standard line at a time addressing.

Thus, for 6 sub-fields, 64 grey-levels can be realised by switching the pixels on or off for the appropriate sub-field. In real products, more sub-fields (up to 12) and other sub-field distributions are chosen in order to reduce motion artefacts (see, e.g., [55], [56]).

3.3 PDP Design and Manufacturing

Most, if not all, commercial PDPs employ a 3-electrode design similar to the one shown in Figure 19. The addressing discharge is applied between the address electrode and one of the sustain electrodes. Sustaining discharges are applied between the sustain electrodes only. This leads to a discharge very close to the surface of the front plates and prevents ions from hitting the phosphor layer. Damage by ion bombardment of phosphors was a major problem in the older, opposing electrodes PDP design, employing only two electrodes, one on the front and the second on the rear plate. Figure 25 compares opposing and co-planar, surface-type designs.

For the surface discharge, the dielectric layer on the front plate is covered by a thin protective layer. For this 500 – 1000 nm thick layer, e-beam evaporated MgO is commonly used. MgO is a very sputter resistant material and possesses a high ion-induced secondary electron emission coefficient (see Sect. 3.4).

One of the key technologies for PDPs is the precise and cost-effective manufacturing of back plates. Many different ways were proposed and demonstrated over the past years [59]. Today, most manufactures apply a sandblasting technique to structure channels and barrier ribs into a soft, low melting point glass. A typical process flow diagram for a complete PDP is shown in Figure 26 and Figure 27.

The cross section of a typical barrier rib structure is depicted in Figure 28.

3.4 High Gamma Coatings in PDP Cells

3.4.1 General

As mentioned earlier, in all state-of-the-art AC plasma display panels the electrodes are covered with glass that acts as a dielectric barrier. In addition, this glass layer is usually coated with a second thin film dielectric in order to prevent sputter damage during plasma operation. A 500 – 1000 nm thick layer of usually e-beam evaporated, polycrystalline MgO turns out to be sufficiently sputter resistant for this purpose.

Acting as the immediate interface between solid and the gas phase, this layer has to fulfil a second, even more important function. Upon interaction with ions created in the discharge, the surface layer needs to release as many electrons as possible in order to fire and sustain the plasma at low voltages. The ion-induced secondary electron emission coefficient γ_i of such a layer is defined as

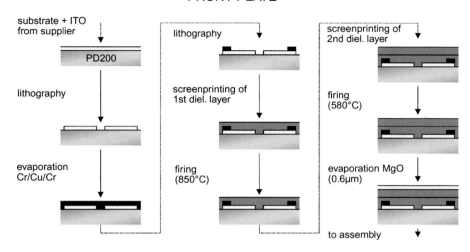

Figure 26: Typical process flow diagram for front plate manufacturing of PDPs. PD200 is a boro-silicate glass used in PDP manufacturing. The ITO display electrodes and the metal bus electrodes, which form the sustain electrodes of a display, are structured photo-lithographically. The dielectric layers covering the electrodes consist of low melting point glass powders, which become transparent after firing at 580 °C. The total thickness of the dielectric layers is about 30 – 60 μm. Finally a thin MgO layer is deposited, protecting the dielectric layers against bombardment of ionized particles from the discharge. The properties of the MgO layer have a fundamental impact on firing and display addressing margins.

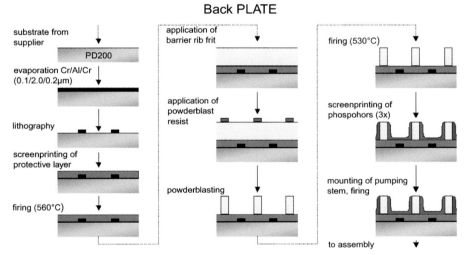

Figure 27: Typical process flow diagrams for back plate manufacturing of PDPs. Firstly, addressing electrodes are structured photo-lithographically and covered with a protective layer formed from low melting point glass powder. The barrier ribs are formed on top of the protection layer by sandblasting of a glass powder layer. The sandblast resist consists of a photo-lithographically structured Polyurethane layer. Finally, the red, green and blue emitting phosphor layers are screen printed between the barriers.

$$\gamma_i = \frac{N(emitted\ electrons)}{N(ions\ incident\ on\ surface)} \quad (23)$$

is related to the plasma firing voltage V_f according to Eq. (24)

$$V_f = \frac{D^2 \cdot p \cdot d}{\left(\ln \frac{C \cdot p \cdot d}{\ln(1/\gamma_i + 1)}\right)^2} \quad (24)$$

with p the pressure in the discharge cell, d the distance between the electrodes, and the known gas specific empirical fitting parameters C and D for the first ionisation coefficient α of the gas. Increasing γ_i increases the number of electrons available for the generation of a self-sustaining carrier avalanche, thus minimising the firing voltage V_f [11].

Measuring the γ_i-coefficients of insulators by means of ion-beam techniques ([16] – [21]) requires sophisticated equipment, and is usually done at very low pressure and high kinetic energy of the ions, i.e., not under panel operation conditions where the ions are basically thermal. Therefore, the relevance of the corresponding results for PDP operation is often doubtful. AC breakdown measurements under panel operation conditions, i.e. at around 500 mbar and electrode distances of 0.5 mm or less, are a fairly simple alternative ([22] – [24]). The breakdown voltage for a particular gas discharge set-up depends only on the product $p \cdot d$, with p the pressure in the system and d the distance between the electrodes (Paschen law; Eq. (24)). Determination of γ_i-coefficients is,

Figure 28: SEM cross section of a high resolution PDP back plate (micrograph from [58]).

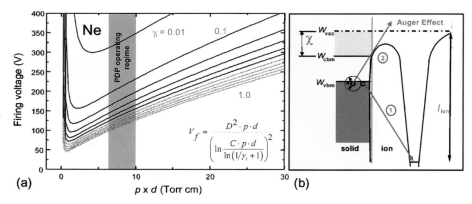

Figure 29: (a) Calculated Paschen curves for different γ_i-values and the tabulated constants C and D of Ne inserted into Eq. (24). (b) illustrates the Hagstrum model of ion-induced electron emission, where the ion arriving at the solid surface is neutralized by transfer of electron 1 and energy that is gained by this process is transferred to electron 2 which is then emitted into the vacuum, provided the condition $W_{ion} - 2(W_g + \chi) > 0$ is fulfilled [30].

therefore, possible by measuring breakdown voltages as a function of $p \cdot d$, i.e., by Paschen curve measurements. The procedure is described in detail in [25] and [26].

3.4.2 Ion-induced Electron Emission Mechanism and Data Analysis

Figure 29 (a) depicts Paschen curves that are calculated by inserting different γ_i-values into Eq. (24), using the values given in [11] for the coefficients C and D of neon. The operating regime of real PDPs is marked in Figure 29. Theoretically, varying γ_i from 0.01 to 1 leads to a reduction of panel firing voltage from approx. 320 V to less than 100 V. This emphasises the tremendous importance of optimizing the respective material properties for the fabrication of cost and performance competitive PDPs. Ion-induced emission processes alone, however, do not lead to γ_i-values that are substantially larger than 0.5, as delineated by red-dashing the respective calculated curves. The commonly accepted Hagstrum model [27] describes ion-induced electron emission from dielectric materials as dominated by Auger neutralisation of thermal (i.e., low kinetic energy) gas ions (Figure 29, b). Ions that arrive with thermal energies at a dielectric surface (like, e.g., in a PDP at pressures around 500 mbar) are neutralised by electron transfer from the valence band of the solid to the gas ion. The energy that is gained due to neutralisation is transferred to a second electron in the solid, which then can be emitted, from the surface. Ion-induced electron emission is feasible for

$$W_{ion} - 2(W_g + \chi) > 0 \tag{25}$$

with W_{ion} being the ionisation energy of the gas atoms, W_g the band gap energy, and χ the electron affinity of the solid. Within the framework of the Hagstrum model, for each noble gas ion arriving at the surface, a maximum of one electron may receive sufficient energy to be emitted. For isotropic emission, however, it is only possible to detect up to (approx.) 0.5 electrons per ion in the gas phase. Therefore, even for the best emitter surface, experimental γ_i-values are not expected to substantially exceed 0.5, as is confirmed by measurements on in-situ cleaned MgO single crystals [26] and high quality MgO films [25].

In Figure 30 (left), measured (symbols) Paschen curves for uncoated and MgO-coated glass are compared to Paschen curves calculated inserting $\gamma_i = 0.06$ (glass) and 0.5 (MgO film) into Eq. (24), thus illustrating the close match. The Paschen curves

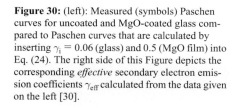

Figure 30: (left): Measured (symbols) Paschen curves for uncoated and MgO-coated glass compared to Paschen curves that are calculated by inserting $\gamma_i = 0.06$ (glass) and 0.5 (MgO film) into Eq. (24). The right side of this Figure depicts the corresponding *effective* secondary electron emission coefficients γ_{eff} calculated from the data given on the left [30].

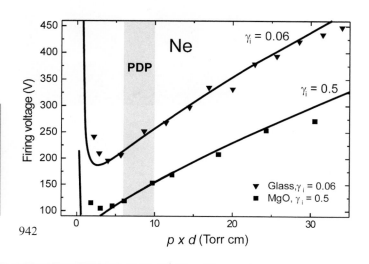
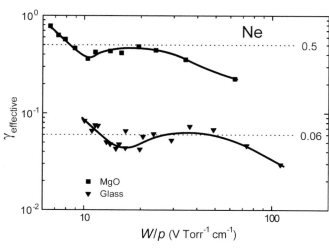

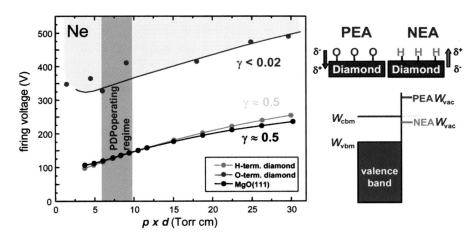

Figure 31: The influence of surface dipole-induced electron affinity changes on plasma firing voltage and γ_i-coefficient. Negative electron affinity (NEA) H-terminated CVD diamond shows a high γ_i-coefficient of 0.5, similar that of best quality single crystalline (111) oriented MgO. O-terminated diamond with a positive electron affinity (PEA) of +1.9 eV exhibits a γ_i-coefficient of < 0.02, which corresponds to an increase of the plasma firing voltage of more than 200 V in the $p \cdot d$-region of interest [30].

can also be converted into *effective* secondary electron emission coefficients γ_{eff} using the Townsend condition for discharge initiation,

$$\gamma_{\text{eff}} = \frac{1}{e^{\alpha d} - 1} \qquad (26)$$

and the tabulated first Townsend coefficients α [28] for Ne (see refs. [25] and [26] for details). The resulting γ_{eff}-curves are shown in Figure 30 (right). The dashed horizontal lines approach the measured values in the region were γ_{eff} (composed of contributions from photo emission and other processes) is almost totally dominated by ion-induced emission (see [13] for details). Although photo-induced processes play a role for emitting electrons from dielectric surfaces (especially for low values of field strength/pressure (E/p) (see [13] for regions and processes), ion-induced emission based on Auger neutralisation certainly dominates the emission process in a plasma panel as well as for breakdown measurement conditions. This is confirmed by the close match between calculated curves using Eq. (24) inserting a fixed, single γ_i-value and the measured Paschen curves. In the end, both, interpretation of Paschen curve data and γ_{eff}-curves give similar results for γ_i.

3.4.3 Parameters Influencing γ-Emission: Some Examples

As outlined by the Hagstrum model (Figure 29) and the cut-off condition given in Eq. (26), measured γ_i-data depend on properties of the bulk solid (e.g., the band gap energy). However, as Auger neutralisation takes place at the surface of the solid, surface-related properties (e.g., contamination or the (surface dipole-induced) electron affinity) play a dominant role. Therefore, proper coating preparation, contamination-free handling, and, where necessary, surface treatments that establish or maintain proper surface termination or remove unwanted contamination are essential for the practical performance of such coatings in assembled displays. It is well known that MgO reacts with water vapour and CO_2 when exposed to air. Bake-out of MgO at 300 – 500 °C is usually sufficient to remove absorbed water and even to decompose a thin layer of $MgCO_3$ that may form on the surface. Consequently, such layers reach the expected high γ-values of 0.5 after bake-out. Extended storage of such layers in air leads to surface reactions that lower the γ-coefficients considerably and bake-out at moderate temperatures (< 450 °C) is not sufficient to restore the proper surface. Consequently, γ drops to 0.2 or less and plasma firing occurs at substantially higher firing voltages.

The influence of surface dipole-induced electron affinity changes is demonstrated in Figure 31. Here, Paschen curves and γ-coefficients for a hydrogen-terminated vapour-deposited diamond thin film [29] are shown along with values obtained after replacing the surface-terminating hydrogen by oxygen using a 30 kHz oxygen discharge [30]. It is well known that H-terminated diamond exhibits a very low or even negative electron affinity, while O-terminated diamond possesses a positive electron affinity of +1.9 eV [31]. Consequently, H-terminated diamond is measured to have a high γ_i-coefficient of 0.5, while the same diamond layer after terminating its surface with oxygen (and reversing the surface dipole) exhibits a γ_i-coefficient of < 0.02 [30]. This corresponds to an increase of the plasma firing voltage of more than 200 V in the pd-region of interest.

In Figure 31, the measurements for diamond are not compared to those of randomly oriented MgO, but to values obtained for an MgO single crystal that is oriented into the (111) direction. This refers to the fact that crystal orientation plays an important role for γ. Differently oriented MgO single crystals terminate differently and for (111) oriented material, the surface consists either of a layer of Mg atoms or of O atoms, while other orientations lead to both Mg *and* O at the terminating surface. For (111)-oriented MgO, the surface dipole, similar to H-terminated diamond, reduces the electron affinity, thus leading to a higher γ_i-coefficient than for other crystal orientations [26]. For PDPs this implies that not only contamination-free deposition, proper handling, and thorough bake-out are required, but also oriented growth of the MgO layer with a strong (111) texture are needed to achieve optimum panel performance.

3.5 Phosphors for PDPs

In a Plasma Display Panel (PDP) phosphors have to convert vacuum UV (VUV) photons emitted by the Xe/Ne discharge between 147 nm to 190 nm into visible light. Materials applied today are closely related to those materials applied in fluorescent lighting. However, phosphors used in conventional cathode-ray tubes still outperform current PDP phosphors in TV display applications with respect to red and blue colour purity and life.

In fluorescent lighting UV-C light from a Hg discharge is penetrating the whole phosphor grain. VUV light in PDPs is absorbed via band gap absorption, with a penetration depth of about 100 nm, which compares with the electron range for voltages used in high voltage FEDs. Therefore, the phosphor surface and the limited excitation volume play an important role in the light emission process.

Apart from the VUV conversion, the phosphor layer in a PDP has not only to be geometrically dense to fully absorb the VUV radiation but also highly reflective for visible light, demanding a large optical thickness for the emitted light.

Table 1 shows the most relevant phosphor materials with their specific problem areas for PDP applications. Most PDPs nowadays use the so-called Willemite (Zn_2SiO_4:Mn) phosphor for green. Although this phosphor has an intrinsically long decay time due to principally strictly forbidden optical transitions within the d shell of the Mn^{2+} ions, phosphors can be fabricated with compositions that relax the selection rules sufficiently to achieve decay time constants below 10 ms suitable for display applications [43]. The so-called blue emitting BAM ($BaMgAl_{10}O_{17}$:Eu^{2+}) phosphor has the desired luminescence properties, however phosphor lifetime and also stability during the high temperature processing steps applied during PDP fabrication (> 400 °C) are subject to ongoing investigations. The red phosphor used mostly for efficiency reasons, (Y, Gd)BO_3:Eu^{3+}, does not have the deep red emission desired for emissive displays.

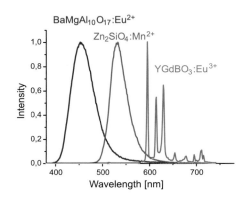

Figure 32: Emission spectra of the red, green and blue emitting PDP phosphors used by most PDP manufacturers.

Phosphor Composition	Emission Colour	Problem Area
$BaMgAl_{10}O_{17}$:Eu^{2+}	Blue	Lifetime, processing stability (efficiency, colour point)
Zn_2SiO_4:Mn^{2+}	Green	Decay time
$BaAl_{12}O_{19}$:Mn^{2+}	Green	Lifetime
(Y,Gd)BO_3:Eu^{3+}	Red	Colour point
Y_2O_3:Eu^{3+}	Red	Efficiency

Table 1: Basic characteristics of PDP phosphor materials. Luminescence originates from the cations for which the charge is indicated. Figure 32 displays emission spectra for those phosphors mostly applied in PDPs today.

FED phosphors or cathode-ray phosphors are not very efficient under VUV excitation.

3.5.1 Phosphor Efficiency

In photoluminescence, the so-called quantum efficiency $QE(\lambda)$ at the excitation wavelength λ is defined as the ratio of emitted quanta and the number of absorbed quanta at excitation wavelength λ. It is measured as light output LO with known reflectivity R according to Eq. (27). Absolute values for the quantum efficiency QE are obtained after determination of the reflectivity R in comparison with the light output for a given phosphor with a reference phosphor of known quantum efficiency.

$$LO(\lambda) = QE(\lambda) \cdot (1 - R(\lambda)) \qquad (27)$$

All PDP phosphors exhibit a high light output between 147 and 190 nm. In this wavelength range, the amount of reflected light is rather low, due to strong band gap absorption, and the quantum efficiency QE is close to the light output of the phosphor.

The quantum efficiencies of all phosphors listed in Table 1 are below one, although energetically, below 200 nm, the photon energy of the exciting photons would be high enough to produce two visible photons. In principle, such a *quantum cutting* process has been observed for special activator and host lattice systems in this wavelength regime, but it is not yet applicable practically [44].

Although the quantum efficiency of PDP phosphors is rather high, the energy efficiency of PDP phosphors reaches 25 – 30% only. However, CRT phosphors even have a lower energy efficiency, which on average is below 20 percent (cf. Sect. 2.3.1).

3.5.2 Electronic Transitions Involved in Europium Luminescence

Figure 33 displays the energy level diagram for Europium ions, which play an important role for phosphors in lighting and display applications. The understanding of the spectral properties associated with the emission spectra as described in detail in the legend of Figure 33, is crucial for phosphor development and the understanding of effects relevant to the application of PDP phosphors in PDPs (see below).

Eu^{3+} emission consists usually of narrow lines in the red spectral region. Most of the lines belong to transitions from the $^5D_0$ level to the ^7F_J ($J = 0, 1, 2, 3, 4, 5, 6$) levels, although transitions from other ^5D levels are observed frequently. The $^5D_0 \rightarrow {}^7F_J$ transitions are ideally suited for determination of the lattice site symmetry. For a position with inversion symmetry electric dipole transitions are strictly forbidden among the 4f levels. In that case only magnetic dipole transitions with the selection rules $\Delta J = 0, \pm 1$ are alowed ($J = 0$ to $J = 0$ forbidden). Without inversion symmetry electric dipole transition become allowed and some transitions with $\Delta J = 0, \pm 2$ appear and dominate the spectrum for even small deviations from inversion symmetry.

3.5.3 Colour Point and Efficiency of the Red Phosphors

Emissive colour displays (like PDPs or FEDs) aim for a high luminance (luminous efficacy) and a so-called large colour gamut. The colour gamut is the range of colours that may be displayed by the combination of the three primary colours. In the 1931 CIE diagram (cf. Introduction to Part VIII) the colour gamut is formed by the area covered by the triangle formed by the colour points of the red, green and blue emitting phosphors. For colour television it was predicted in 1955 already, that for red the above-mentioned requirements can only be met with a phosphor emitting by line around 610 nm. Meanwhile it is clear, that only the Eu^{3+} ion is able to satisfy this requirement [34].

The CIE colour point of the $(Y, Gd)BO_3$:Eu phosphor of $x = 0.640$ and $y = 0.360$ is not as red as the red emitting CRT (FED) phosphor Y_2O_2S:Eu with $x = 0.650$ and $y = 0.342$. The difference in colour co-ordinates is rather small. However, the numerical difference in this chromaticity system does not reflect the human perception in colour difference. The emission colour of the $(Y, Gd)BO_3$:Eu is perceived as somewhat orange compared to the deep red emission of the Y_2O_2S:Eu.

The orange appearance of the emission colour of the $YGdBO_3$:Eu^{3+} (YGB) phosphor is due to the relatively intense line at 594 nm which corresponds to a magnetic dipole transition $^5D_0 \rightarrow {}^7F_1$ while the emission lines for the electric dipole transitions $^5D_0 \rightarrow {}^7F_2$ at 612 and 627 nm are weaker (see Figure 34).

The spectrum can be understood from the site symmetry of the Eu^{3+} (see Figure 33) in the $(Y, Gd)BO_3$:Eu lattice. Eu^{3+} ions occupy two sites with C_3 symmetry with a very small deviation from a local S_6 symmetry with inversion symmetry. So electric dipole transitions ($^5D_0 \rightarrow {}^7F_2$) are almost forbidden. Figure 34 shows emission spectra of the $YGdBO_3$:Eu^{3+} phosphor together with alternative red emitting phosphors, for which the

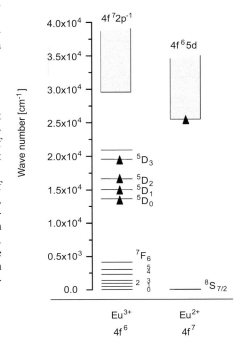

Figure 33: Energy level scheme of Eu^{3+} and Eu^{2+}. The energy axis (in cm^{-1}) is representative for ions in oxide lattices. Horizontal lines represent narrow energy states of 4f levels. Shaded areas represent broad charge transfer states in case of Eu^{3+} and $4f^65d$ states for Eu^{2+}. Triangles indicate levels from which radiative transitions were observed. For Eu^{2+} the 5d band covers almost all 4f levels, leading to a broad emission spectrum found in many lattices, covering the whole visible spectrum. The emission wavelength is determined by the position of the 5d levels which is a function of the crystal field.

Figure 34: Emission spectra of different Eu^{3+} phosphors. $YGdBO_3{:}Eu^{3+}$ (YGB) mainly used in PDPs has a colour point of $x = 0.640$ and $y = 0.360$. The colour point of $Y_2O_3{:}Eu$ is $x = 0.641$ and $y = 0.344$. The colour point of $YVO_4{:}Eu$ is $x = 0.645$ and $y = 0.343$. The colour point of $Y_2O_2S{:}Eu$ is $x = 0.650$ and $y = 0.342$. The $^5D_0 \rightarrow {}^7F_J$ levels to which the transitions can be mainly assigned to are indicated for the corresponding spectral regions in the top of the figure (compare also Figure 33).

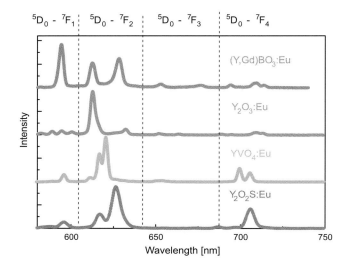

Figure 35: Excitation spectra (light output as a function of the excitation wavelength) of the red emitting phosphors depicted in Figure 34. In the wavelength regime of the Xe/Ne plasma emission (147 – 180 nm) the $YGdBO_3{:}Eu^{3+}$ (YGB) phosphor has a distinctly higher light output than all alternative phosphors.

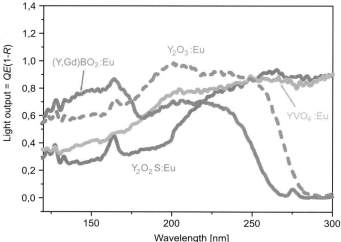

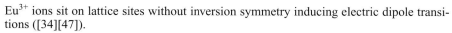

Eu^{3+} ions sit on lattice sites without inversion symmetry inducing electric dipole transitions ([34][47]).

Although differences in colour points are small, they are clearly perceived by the human eye.

The reason why $YGdBO_3{:}Eu^{3+}$ (YGB) phosphor is mainly used for PDPs is shown in Figure 35, where the light output (excitation spectra) for the phosphors discussed above are displayed as a function of the excitation wavelength. The light output of all alternative red emitting phosphors is inferior to the $YGdBO_3{:}Eu^{3+}$ (YGB) phosphor for excitation in the vacuum UV spectral range of 147 – 180 nm, the emission regime of the Ne/Xe gas discharge.

3.5.4 Stability and Colour Point of the Blue Emitting BAM Phosphor

The most vulnerable material with respect to panel processing and degradation is the blue emitting $BaMgAl_{10}O_{17}{:}Eu^{2+}$ (BAM) phosphor. Degradation effects are related to the BAM host lattice and the Eu^{2+} activator ion. Figure 36 displays the light output LO measured on BAM phosphor powders as a function of the annealing temperature for excitation at 147 nm and 254 nm, after annealing in ambient air. With increasing temperature the light output (and also the quantum efficiency) decreases significantly, starting at a lower temperature for the shorter wavelength (147 nm) excitation. The main reason for the drastic decrease in light output is the oxidation of the Eu^{2+} ion, also observed by Oshio et al. [48]. They confirmed the formation of Eu^{3+} containing $EuMgAl_{11}O_{19}$. Eu^{3+} emission is very inefficient in this lattice and is hardly visible.

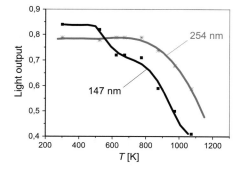

Figure 36: Light output of BAM phosphor powders as a function of annealing temperature for excitation at 147 and 254 nm as indicated in the figure. Annealing was done in ambient air.

In addition to the Eu^{2+} oxidation decreasing the light output of the phosphor, the Eu^{2+} emission itself can also be affected during PDP processing and panel operation. Figure 37 shows emission spectra of BAM phosphor layers after annealing in ambient air at 550 °C – such high temperatures are reached during PDP processing – for excitation at different wavelengths. In particular, when excited by short wavelength radiation, additional green emission occurs, resulting in a broadening of the emission spectra. As discussed in Sect. 3.5.2, the position of the Eu^{2+} emission band is determined by the crystal field. The spectra depicted in Figure 37 can be fitted quantitatively using an additional emission term centered at 495 nm which is associated with the formation of $Ba_{0.75}Al_{11}O_{17.25}$:Eu [46].

From these findings it is concluded that BAM particles have to be single phase with optimal stoichiometric composition. Excess of Al_2O_3 during BAM synthesis should be avoided, since new Al_2O_3 - rich phases that form, e.g., $Ba_{0.75}Al_{11}O_{17.25}$:Eu, are a source of additional emission at 495 nm. Recent experiments indicate that a small excess of Mg^{2+} used during BAM synthesis yields powders with enhanced photo and thermal stability [45][46].

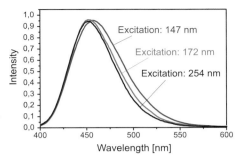

Figure 37: Luminescence spectra of a processed BAM phosphor layer for different excitation wavelength, indicated in the figure. The phosphor layers were baked in ambient air at 550°C for 1 hour. With decreasing excitation wavelength the green emission intensity increases. The green emission is due to the formation of $Ba_{0.75}Al_{11}O_{17.25}$:Eu which emits at 495 nm.

3.6 PDP Performance and Prospects

Today, AC-PDPs are the leading technology for large and flat displays and the dream of a flat TV hanging on the wall is now a reality. Although some high-resolution panels with diagonals of less than 70 cm are available on the market (e.g., a 64 cm diagonal SXGA from Fujitsu), PDPs appear to be best suited for display diagonals of 80 cm to 150 cm. An impressive prototype of a 60" PDP is depicted in Figure 38.

Table 2 shows a large number of PDP manufactures together with typical product specifications. Mostly, an aspect ratio of 16:9 is chosen, but also some 4:3 sets are available. Brightness and power consumption are important discriminators for a display technology and significant improvements were achieved for PDPs over the past years. A key parameter for high brightness and low power consumption is the luminous efficacy. Today, about 1 – 1.5 lm/W are standard for PDPs. To really surpass CRTs in terms of brightness and power consumption, 2 – 3 lm/W are required and for the ultimate low weight, low power flat television set of the 21st century, 5 lm/W are the final goal. Figure 39 delineates some important advances in performance over the past 20 years [60]. Major breakthroughs were the introduction of the surface discharge design, the transition from He+Xe- to Ne+Xe-discharges, the use of a light reflecting rather than of a light transmitting phosphor layer design, and the increase in the cell size (reduction of barrier width). Currently, improved discharge efficiencies and phosphor materials already allow the fabrication of research prototype PDPs with efficiencies of more than 3 lm/W ([61], [62], [45], [63], [64]).

Figure 38: A 60" diagonal HDTV-PDP from Plasmaco. Photograph from [50].

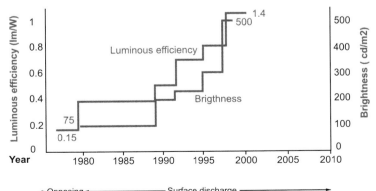

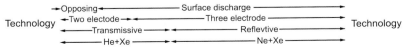

Figure 39: Improvements in luminous efficiency and brightness of PDPs over time (picture from [60]).

Manufacturer	Products
FHP, Fujitsu-Hitachi-Sony Joint Venture	32" (81 cm) – 42" (107 cm), 16:9 and 4:3, 852 * 480 to 1024 * 1024 (ALIS), 300 – 650 Cd/m^2
Pioneer	40" (102 cm) – 50" (127 cm), 16:9 and 4:3, 640 * 480 to 1280 * 1024 350 – 400 Cd/m^2
Matsushita (Panasonic)	37" (95 cm) – 42" (107 cm), 16:9, 852 * 480 to , 852 * 480 470 – 550 Cd/m^2
NEC	33" (83 cm) – 50" (127 cm, 4:3 and 16:9 640 * 480 to 1365 * 768 350 – 570 Cd/m^2
LG	40" (102 cm) – 50" (127 cm), 16:9 and 4:3, 640 * 480 to 1360 * 768, 280 – 350 Cd/m^2
Samsung	42" (107 cm), 16:9, 852 * 480, 350 Cd/m^2

Table 2: Manufactures and product specification range (screen diagonal, resolution and peak brightness) of actual PDP modules available on the market. ALIS refers to a new (interlace-like) addressing technology from Fujitsu [57].

Starting with displays for professional applications such as airport information displays or advertising panels, PDPs meanwhile also entered the consumer television market. Here they, of course, compete with CRTs and also with projection display systems. The two major issues for the PDP technology development today are cost and efficiency improvement. Compared to CRTs, PDPs are very expensive. It is generally assumed that a price level of 10000 Yen per inch (40 Euro per centimeter) is required, to make the flat PDP technology attractive for the consumer and to sell a significant number of PDP-based TV sets. This price level is expected for approx. the year 2003 and by that time a total world market of about 3 million units will be reached. In order to replace the large diagonal jumbo-CRTs the PDP set price needs to drop to that of a high performance CRT-based TV set. This is what all PDP development programs aim at today. In addition, many activities are directed towards the application of standardised materials and simplification of the manufacturing methods (see e.g. [59]).

4 Summary

Any display technology is compared to and will have to meet the performance data of state-of-the-art cathode-ray tubes (CRTs). Field emission displays (FEDs) as well as Plasma Display Panels (PDPs) share with CRTs the utilization of phosphors to generate visible light and different colours. The material requirements and performance of phosphors for FEDs and PDPs are summarized, including specific problems related to the light generation processes involved. The different phosphor materials used in FEDs and PDPs have in common that the penetration depth of excitation energy (provided by electrons in FEDs and vacuum UV photons in PDPs) is small. Performance and lifetime of both types of displays depend heavily on the quality and the proper excitation of the phosphor materials used. Although some smaller size (< 20 cm diagonal) FED products are commercially available, a break-through in performance and display size is only expected for cost efficient, easy-to-manufacture electron emitters and device structures. Carbon nanotubes (CNTs) are currently considered to be best suited to match these requirements.

One fundamental advantage of the CRT over all flat panel displays is its flexible-format-addressing technology. Increasing the display resolution requires only a modest increase in display complexity. In FEDs and PDPs any additional addressable display row or column requires additional electronics and drivers. Consequently, the complexity of the electronics to operate PDPs is a major reason for the still high price of

PDPs compared to CRTs. Driver electronics prices are determined to a large degree by the voltages and the power they can handle. Low switching voltages in PDPs and thus cheaper driver electronics are only possible by utilizing dielectric coatings with a high ion induced electron emission (gamma) coefficient inside the plasma discharge cells. High gamma values are achieved for low electron affinity materials, e.g. (111) textured MgO or H-terminated diamond, and proper, contamination-free processing conditions.

Table 3, compares important performance data of PDP and CRT displays. FEDs are commercially only available in small sizes (approx. 5" screen diagonal). Larger sizes of up to 17" diagonal were demonstrated, however, not (yet) commercialised. Reliable performance data are only available for small display formats.

	CRT Television Sets	PDP Television-Sets	FED displays (small formats)
Screen diagonals	max. 90 cm (max. 36")	80 cm – 150 cm (32" – 60")	13.2 cm (5.2")
Brightness (full white screen)	100 – 130 Cd/m^2	60 – 100 Cd/m^2	
Peak-Brightness (1 % screen area white)	≈ 500 Cd/m^2	≈ 500 Cd/m^2	Up to 1200 Cd/m^2 (monochrome with grey scale)
Luminous efficacy	2 – 3 lm/W	1 – 1.5 lm/W	
Power consumption for typical television signal	200 – 300 W	250 – 400 W	4 W at 25 % white 8 W at 50 % white
Lifetime	> 30000 hr	> 30000 hr	> 10000 hr
Weight	≈ 80 kg (36")	20 – 30 kg	300 g
Thickness	≥ 60 cm (36")	< 10 cm	< 1 cm (incl. Electronics)

Table 3: Typical performance numbers of PDP- and conventional CRT-based television sets and small FEDs in the year 2002.

Acknowledgements

The editor would like to thank Christian Ohly (FZ Jülich) for checking the symbols and formulas in this chapter.

References

[1] R.H. Fowler and L. Nordheim, Proc. Royal Soc. (London), A119, 173 (1928).

[2] I. Shah, Physics World **10**, 45 (1997).

[3] C.A. Spindt, I. Brodie, L. Humphrey, and E. Westerberg, J. Appl. Phys. **47**, 5248 (1976).

[4] W.A. de Heer, A. Chatelain, D. Ugarte, Science, **270**, 1179 (1995).

[5] F.J. Himpsel, J.A. Knapp, J.A. van Vechten, and D.E. Eastman, Phys. Rev. B**20**, 624 (1979).

[6] C. Bower, W. Zhu, S. Jin, and O. Zhou, Appl. Phys. Lett. **77**, 830 (2000).

[7] Y. C. Choi, Y.M. Shin, Y.H. Lee, B.S. Lee, G.S. Park, W.B. Choi, N.S. Lee, and J.M. Kim, Appl. Phys. Lett. **76**, 2367 (2000).

[8] J.M. Kim et al., ICNDST-7, July 23 – 28, 2000, City University Hong Kong, paper 9.1 (2000).

[9] J.E. Jaskie, MRS Bulletin, March 1996, 59 (1996).

[10] PixTech Inc. 3350 Scott Blvd. Bldg. 37, Santa Clara, CA 95054, sales brochure (2000).

[11] Yu. P. Raizer, in *Gas Discharge Physics,* Springer, Berlin, 1997.

[12] C.J. Curtin and Y. Iguchi, Conf. Proc. SID 00 Digest, 1263 (2000).

[13] A.V. Phelps and Z.L. Petrovic, Plasma Sources Sci. Technol. **8**, R21, (1999).

[14] P. K. Bachmann, H. Lade, K. Radermacher, D.U. Wiechert, H. Wilson, Diamond Films '95, paper 16.2, (1995).

[15] P.K. Bachmann, G.F. Zhong, D.U. Wiechert, J. Merikhi, unpublished (2000).

[16] H. Uchiike, K. Miura, N. Nakayama, T. Shinoda, Y. Fukushima, IEEE Trans. Electron Dev., ED-**23** 1211 (1976).

[17] N. J. Chou, J. Vac. Sci. Technol. **14**, 307 (1977).

[18] K. Yoshida, H. Uchiike and M. Sawa, Proc. IDW 98, 515 (1998).

[19] M. Ishimoto, S. Hidaka, K. Betsui and T. Shinoda, SID 99 Digest, 552 (1999).

[20] E.-H. Choi et al., Jpn. J. Appl. Phys. **37**, 7015 (1998).

[21] K. S. Moon, J. Lee and K.-W. Whang, J. Appl. Phys. **86**, 4049 (1999).

[22] G. Auday, Ph. Guillot, J. Galy, H. Brunet, Int. Conf. on Phen. in Ionized Gases (ICPIG XXIV) **4**, 69 (1999).

[23] O. Sahni and C. Lanza, J. Appl. Phys. **47**, 1337 (1976).

[24] M. O. Aboelfotoh and J. A. Lorenzen, J. Appl. Phys. **48**, 4754 (1977).

[25] V. v. Elsbergen, P.K. Bachmann, T. Juestel, Conf. Proc. SID 00 Digest, 220 (2000).

[26] V. v. Elsbergen, P.K. Bachmann, G. Zhong, Conf. Proc. IDW 00, 687 (2000).

[27] H. D. Hagstrum, Phys. Rev. **122**, 83 (1961).

[28] A.A. Kruithof, Physica **7**, 519 (1940).

[29] P.K. Bachmann, in *Industrial Handbook of Diamond* ed. by M. Prelas, G. Popovici, K. Bigelow, Marcel Dekker, 821 (1997).

[30] P. K. Bachmann, V. van Elsbergen, D.U. Wiechert, G. Zhong, Diamond 2000 Conf., Sept. 3 – 8, 2000, Porto, Portugal; Diamond and Rel. Mat. (accepted for publ.) (2001).

[31] P.K. Bachmann, W. Eberhardt, B. Kessler, H. Lade, K. Radermacher, D.U. Wiechert, H. Wilson, Diamond and Rel. Mater. **5**, 1378 (1996).

[32] L. Ozawa, Journal of the SID **6**, No. 4 (1998).

[33] H. Gläser, H. Bechtel, F. Busse, Journal of the SID **8**, 3, (2000).

[34] G. Blasse and B.C. Grabmaier, Luminescent Materials, Springer-Verlag Berlin, Heidelberg, New York (1994).

[35] D.J. Robbins, J. Electrochem. Soc. **127**, 2694 (1980).

[36] L.E. Shea, Electrochem. Soc. Interface **7**, 24 (1998).

[37] S.P. Shea, *Electron Beam Interactions with Solids*, Inc. AMF O'Hare, USA 145 (1982).

[38] D.B. Wittry, D.F. Kyser, J. Appl. Phys. **38**, 375 (1967).

[39] G. Gergely, J. Phys. Chem. Solids **17**, 112 (1960).

[40] H. Bechtel, W. Czarnojan, M. Haase, W. Mayr, H. Nikol, Philips Journal of Research **50**, 433 (1996).

[41] A.Bril, Physica **15**, 361 (1949).

[42] A. Bril, F.A. Kröger, Philips Techn. Rev. **12**, 120 (1950).
[43] E. van der Kolk, P. Dorenbos, C.W.E. van Eijk, H. Bechtel, T. Jüstel, H. Nikol, C.R. Ronda, D.U. Wiechert, J. Lumin. **87 – 89**, 1246 (2000).
[44] R.T. Wegh, H. Donker, K.D. Oskam, A. Meijerink, Science **283**, 663 (1999).
[45] T. Jüstel, H. Nikol, Adv. Materials **12**, 7, 527 (2000).
[46] M. Zachau, D. Schmidt, U. Müller, C.F. Chenot, WO 99/34389.
[47] S. Shionoya, W.M. Yen (editors), *Phosphor Handbook*, CRC Press, Boca Raton, Boston, New York (1998).
[48] S. Oshio, K. Kitamura, T. Nishiura, T. Shigeta, S. Horii, T. Matsuoka, Nat. Tech. Rep. **43,** 69 (1997).
[49] J.M. Kim et al., Conf. Proc. 20th IDRC, 386 (2000).
[50] L.F. Weber, *The Promise of Plasma Displays for HDTV,* Conf. Proc. SID 00 Digest,, 402 (2000).
[51] J. A. Castellano, *Handbook of Display Technology,* 111, Academic Press, San Diego (1992).
[52] Shinoda T. and Ninuma A., Conf. Proc. SID 86 Digest, 172 (1986).
[53] L.F. Weber, chapter 10, 332, in *Flat panel Displays and CRTs*, L.E. Tannas, ed., Van Nostrand Reinhold Company, New York, (1985).
[54] B. Gellert and U. Kogelschatz, Appl. Phys. B, **52**, 14 (1991).
[55] T. Holtslag, J. Hoppenbrouwers, R. v. Dijk, Proc. of the seventh Int. Display Workshops, IDW'99, 779 (1999).
[56] M.A. Klompenhouwer and G. de Haan, Conf. Proc. SID 00 Digest, 388 (2000).
[57] Y. Kanazawa, T. Ueda, S. Kuroki, K. Kariya, T. Hirose, Conf. Proc. SID 99 Digest, 154 (1999).
[58] M. Osawa, F. Asami, H. Inoue. M. Wakitani, *Overview of PDP*, in 2000 Nikkei Microdevices' Flat Panel Display Yearbook, 164, publ. by United States Display Consortium (http://www.usdc.org), ISBN 1-884730-21-3, (2000).
[59] M. Usui, T. Asano, *Standardizing Materials and Simplifying Manufacturing Methods and Equipment for PDP*, in 1999 Nikkei Microdevices' Flat Panel Display Yearbook, 131, published by United States Display Consortium (http://www.usdc.org), ISBN 1-884730-16-7, (1999).
[60] T. Shinoda, *PDP Principles and Structure - Surpassing CRTs to Enter the Home Market*, in 1999 Nikkei Microdevices' Flat Panel Display Yearbook, 117, published by United States Display Consortium (http://www.usdc.org), ISBN 1-884730-16-7, (1999).
[61] G. Oversluizen, S. de Zwart, S. van Heusden and T. Dekker, Conf. Proc. of the IDW 99, 591 (1999).
[62] G. Oversluizen, S. de Zwart, S. van Heusden and T. Dekker, Applied Physics Letters **77**, 948 (1999).
[63] H. Bechtel, T. Juestel, H. Glaeser, D. Wiechert, Conf. Proc. of the 20th IDRC '00, 366 (2000).
[64] G. Oversluizen, S. de Zwart, S. van Heusden and T. Dekker, Conf. Proc. of the IDW'00, 631 (2000).

Electronic Paper

Stefan Jung
Corporate Research, Infineon Technologies, Munich, Germany

Dietmar Theis
Corporate Technology, Siemens AG, Munich, Germany

Contents

1 Introduction	955
2 Microparticle-based Displays	956
2.1 Electrophoretic Displays	956
2.2 Rotating Ball Displays	960
2.3 Suspended-Particle Displays	961
3 Alternative Paper-like Display Technologies	961
3.1 Reflective Direct View Liquid Crystal Displays	961
3.2 Bistable Liquid Crystal Displays	962
3.3 Micromechanics-based Displays	963
4 Flexible Backplane Electronics	963
4.1 Printable Integrated Circuits	963
4.2 Fluidic Self-Assembly	964
5 Outlook and Vision	965

Electronic Paper

1 Introduction

There is a variety of physical effects that can be used to transform an electronic one to a visible signal. As presented before (see Figure 2 of Introduction to Part VIII), two basic display types can be categorized according to the question of how each picture element reacts on the electrical signal: either by **photon generation** or by **interaction with ambient photons**. After decades of research and development, display technology even today covers newly emerging technologies with many new concepts, particularly in the field of novel electro-optical materials.

The ultimate benchmark for *direct view* portable displays in terms of readability, ease-of-access and use, ruggedness and ultra-low power consumption is the electronic emulation of printed paper. Plain paper is a reflective 'device' with no energy consumption and accordingly, all electronic emulations are expected to be non-emitting reflective devices, modulating the ambient light rather than generating light. Direct-view reflective devices can be divided into *'matter moving'* or *chemical reaction types*. In this chapter we will disregard projection devices. In terms of direct view technologies for paper-like displays microparticle based displays, electrophoretic displays, liquid crystal technologies, interferometric modulation, and rotating ball displays and some micro-electromechanical technologies have been investigated reaching different levels of maturity. As a first approach one could probably accept black and white solutions, with a low frame rate. But ultimately the user would like to have a high speed device with fully saturated colour, and high resolution. Furthermore, it should offer a diffuse, Lambertian reflectance exceeding 80 %, appear paper-white, have a contrast better than 12:1, and be flexible and foldable like a newspaper magazine. An ideal electronic paper display operates under ultra-low power, is preferably bistable, robust, self-healing, light weight, user-friendly and environmentally compatible.

Achieving the full properties of printed paper in a *flexible electronic display device* manufacturable in print is a goal which up to now has not yet been achieved. There are, however, some exciting inroads into the solution of the problem and the aim of this article is to explain some of the promising approaches. The research field of electronic paper is broad and manifold and we do not attempt to cover every suggestion published, but rather present a selection of the more mature technologies.

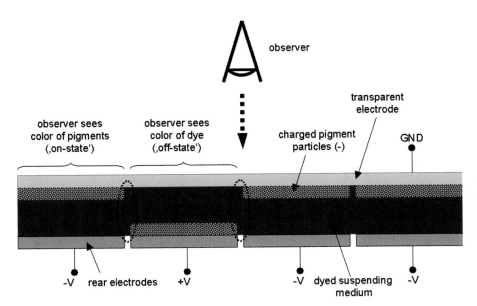

Figure 1: Schematic structure of an electrophoretic display. In this case, white negatively charged microparticles are suspended in a dark blue dyed liquid and sandwiched between two opposite electrodes. The white particles can be moved by an electrophoretic effect and thus modulate the reflectivity of the display cell.

2 Microparticle-based Displays

The common basis of all microparticle-based display techniques discussed in this section is the manipulation of small objects such as *pigments* or *rotating microscopic balls* by means of electrical forces. The mechanical manipulation of these objects allows for a spatial modulation of the optical properties of display cells such as their reflectance or transmission behaviour. Unlike micro-electro-mechanically machined display systems based on silicon chip technology (e.g. the digital mirror projection device [2]) these monolithic approaches do not apply to the requirements of paper-like direct-view displays on large substrates. For cost efficiency and manufacturability, hybrid realizations are favourable such as pigment-based printable electrophoretic materials combined with patterned electrode substrates. In the following, the most promising emerging techniques in this field are reviewed and several alternative approaches are discussed briefly.

2.1 Electrophoretic Displays

The electro-optical effect used in these types of display cells results from the movement of charged microparticles in a colloidal suspension. The migration in a liquid of small charged particles under influence of an electrical field is referred to as electrophoresis. Since the pigments are chosen to have reflectance properties different from the liquid, the optical characteristics varies depending on the distribution of the pigments in the bulk of the display cell. Research activities on these so-called electrophoretic displays (EPDs) have already been reported in the early 1970's [3]. The EPD concept as nonemissive display technology offers interesting aspects such as a very wide angular viewing range and a high contrast ratio over a wide range of illumination. Since then, much effort has been spent on the development of prototypes and their commercialization, since this display technology promises image quality and user convenience comparable to the one of classical printed paper. This analogy also led to the popular terms *electronic ink* or *electronic paper*. A good review on the historical roots and the basic operating principles of EPDs is given in [4].

Basic Structure and Operating Principle
As depicted in Figure 1, the above-mentioned suspension in EPD cells is commonly composed of a darkly dyed liquid with low reflectivity, whereas the microparticles are of a light color. Using Titanium dioxide particles, a white reflector can be achieved with pigment sizes of about 0.25 microns. When a thin layer of this suspension is sandwiched between two parallel electrodes (of which at least the one directed towards the observer is transparent) the particles either move to the top or to the bottom electrode, depending on the polarity of the applied field. Therefore, the observer either sees the opaque dark dye (referred to as 'off-state') or the reflective white particles ('on-state'). Since this effect is only controlled by an electric field perpendicular to the electrode planes, the addressing of many pixel elements can be realized in parallel by means of appropriate electrode structure. One of the most important advantages of EPDs with respect to energy efficiency is the bistability of the display cells, i.e. the cell remains in its original state (either 'on' or 'off') after the electric field is *removed*. This is achieved by matching the densities of the suspending medium and the microparticles, thus establishing *gravitational stability*. It is obvious that mechanical shocks or thermal influences will distort this stability.

In general, the resulting image *contrast* is determined by the dye concentration in the suspension, by the thickness of the layer, and by the effective electric field which is applied to the electrophoretic layer. A darker appearance of the off-state can be achieved by increasing the dye concentration and thus increasing the opacity of the display cell. However, when increasing the dye concentration above a distinct value, the brightness of the white particles in the on-state begins to diminish to a higher degree than it improves the opacity in off-state. An increase in dye concentration decreases the white state reflectivity because the dyed fluid between the interstices of the packed particles leads to increased absorbance within the pack. Both the thickness and the applied electric field do also have an impact on the contrast ratio between 'on' and 'off' states. Not only the switching speed but also the contrast rises with increasing electric field, because the stronger the particles are compressed behind the attracting electrode, the higher the particle density and reflectivity. A thicker EPD cell will show better contrast since the light absorption in off-state is improved. The on/off contrast ratio could even be improved by further increasing the thickness and decreasing the dye concentration. However, this may not be practical because the switching speed of the display which is determined in part by the distance between the electrodes may become too slow, or the required electrode voltages may become too large, respectively.

A further aspect is the interdependence of *switching speed* and applied *electric field*. For a fast display response, the electrophoretic mobility of the suspended particles should be large. This can be achieved by choosing a low viscosity liquid with a large dielectric constant. Since an EPD is envisioned for low-power applications, the suspension should be highly resistive in order to prevent current leakage.

As the particles are bunched together or even compressed in both display states, the danger of agglomeration and clustering is enlarged at high electric fields, which has a negative effect on lifetime and quality of the display cell. Therefore, the additional introduction of *repulsive forces* is necessary to prevent the individual microparticles from sticking together. Colloidal suspensions inherently tend to be unstable. The large surface free energy is reduced as soon as the particles bunch together, which in turn enlarges the attractive forces. Establishing a colloidal stability of the suspension is the most crucial aspect of EPDs and can be achieved by *electrostatic* or by *steric repulsion*. While electrostatic repulsion is based on a precise charging of the pigments, steric repulsion is accomplished by covering their surfaces with distinct polymer molecules. Finally, dye, solvent, and the microparticles must be prevented from interacting chemically, and also macroscopic electrohydrodynamics in the fluid must be considered. Another problem becomes obvious when paying attention to the interface regions between pixels as highlighted in Figure 1. Without the presence of barriers between two adjacent EPD cells in opposite states, not only vertical but also lateral diffusion effects of the microparticles might occur. The image contrast may be diminished, especially at higher image resolutions when the diameter of an EPD cell reaches the order of magnitude of the cell thickness. More detailed information on the extremely difficult composition of a stable and reliable suspension of EPDs is given in [4].

Microencapsulation of the Suspension

As described above, the drawbacks of electrophoretic materials include lack of long-term stability due to *pigment clustering*, *agglomeration*, and *lateral migration*. These problem get even larger when flexible substrates are used. Together with the colloidal stability, these problems present the hurdles for EPDs to develop into a mature technology.

There have been manifold attempts to solve these problems. Several proposals for modifications of cell design and electrode arrangements have been made, such as etched cavities and walls for each display cell for a compartmentalization of the particles. However, besides these architecture-related approaches, a material-related approach called *microencapsulation* is regarded as most relevant and will therefore be discussed in more detail. The above mentioned limiting factors can be diminished by confining the electrophoretic suspension locally in microcapsules which have sizes smaller than the scale on which these failures occur [5]. Another advantage of microencapsulation is that is allows for printing the display material like an ink with traditional techniques like

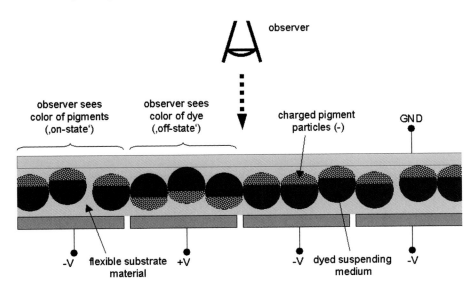

Figure 2: The microencapsulation of the suspension can be considered as the breakthrough for flexible EPDs. Agglomeration or clustering of the pigments cannot occur on larger scales than the capsule size, enhancing the lifetime of the display.

VIII Displays

Figure 3: Photograph of a microencapsulated electrophoretic suspension with an average capsule diameter of 70 microns. Several half-activated capsules are observed. This shows that the optical resolution is not limited to the capsule sizes [6].

screen printing. The schematic structure of the microencapsulated EPD is shown in Figure 2. The sizes of the capsules are approximately of the order of magnitude of the thickness of the display cell in Figure 1.

Similar to conventional EPDs, the dielectric suspension presented in [5] consists of titanium dioxide powder which is dispersed in a specific gravity matched mixture of tetrachloroethylene and an isoparaffinic solvent with a blue dye. Since tetrachloroethylene has severe environmental drawbacks, more benign solvents have been found and are meanwhile applied in the industry. The subsequent microencapsulation is caused by an in-situ polycondensation of urea and formaldehyde. The resulting capsules have mechanically strong, optically transparent walls with a relatively high dielectric constant. The latter property is important for maximizing the effective voltage drop across the internal suspension. After the microcapsules are sieved, washed, and dried, they are dispersed in a polymer binder. The contrast can be enhanced by optically index matching the binder to the microcapsule wall material.

The resulting dispersion is then ready for printing onto a substrate. Figure 3 shows a dielectric film with an average capsule diameter of 70 microns. As can be observed, the optical resolution of microparticles is not limited by the capsule size but rather by the electric field lines defined by the electrode structure [6]. The microcapsules have typical diameters between 30 and 300 microns.

Figure 4 shows measured results of the optical properties of a 200 microns thick film of electrophoretic ink [7]. The plot shows the trade-off between frame rate and achievable contrast. For increasing voltages, the curve begins to saturate, since all pigments traverse the whole capsule. Hence, increasing the voltage above a reasonable value does not increase contrast but rather enlarges the danger that the pigments stick together. The mobility of the charged particles $\mu = v/E$ with v being the particle velocity and E the electric field can be interpreted from the measured curves in Figure 4. As can be seen, the 1Hz curve begins to saturate at electrical fields above $E = 2$ V/μm. Thus, the transient time τ of a particle traversing through and back an EPD cell of thickness d is given by $\tau = 2d/v$, assuming a laminar flow of the pigments through the fluid. Thus, with $d = 50$ μm and $\tau = 1$s, a particle mobility of $\mu = 2d/(\tau E) = 0.5 \cdot 10^{-6}$ cm²/(Vs) results.

Electrophoretic ink offers many advantages in terms of readability and power consumption compared to other technological solutions for flexible paper-like displays. Figure 5 compares monochrome reflective LCDs, electronic ink displays, and a traditional piece of newspaper. The technology of EPD is still under development. Multicolor EPDs using color filters are reported in [9].

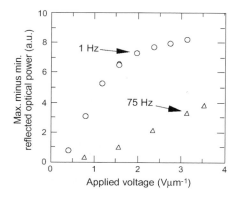

Figure 4: Properties of electronic ink consisting of white particles in a blue dye. The measurements were conducted on a 200-micron-thick film of microcapsules with a diameter of 30 to 50 microns. The plot shows the contrast versus applied electrical field for different switching frequencies [7].

Figure 5: Comparison of key figures of different reflective display technologies for electronic paper [1]. The power dissipation of a 320 x 240 pixel EPD matrix is given at 1 mW @ 0.1 Hz frame rate compared to 20 – 60 mW for a reflective monochrome LCD according to [8].

Display type Parameter	Paper	Cholesteric LCD	Gyricon	Electrophoretic	Reflective twisted nematic LCD
Characteristics of some highly readable yet low-power displays					
Contrast	20:1 laser print 7-10:1 newspaper	20-30:1	10:1	10-30:1	<5:1
Reflectivity	80% laser print 50% newspaper	40%	20%	40%	<5%
Viewing angle	All angles	All angles			Narrow
Flexibility	Yes	Moderately	Yes	Yes	No
Full color	Yes	Yes	No[a]	No[b]	No
Reflection type	Lambertian	Near Lambertian	Lambertian	Lambertian	Highly specular
Characteristics of electronic displays only					
Response time		30 – 100 ms	80 ms	100 ms	20 ms
Maximum voltage		40 V	90 V	90 V	5 V
Substrate		Plastic or glass	Plastic or glass	Plastic or glass	Glass
High-resolution drive scheme		Passive	Active	Active	Active
Multiplexing capability		High	None	None	Low

[a] Development is under way of cyan, magenta, and yellow cell stacks for subtractive color techniques
[b] Color filter arrays may be possible.
LCD = liquid-crystal display

Alternative EPD Concepts

As stated above, research and development in display technology does not follow a main stream path but rather pursues a broad range of different approaches. This trend can also be observed in the field of electrophoretic displays. A few selected alternative approaches are described in the following.

Most research is focused on improvements of the microparticles, on cell layout and electrode arrangements, cf. [10]. An EPD containing black and white pigments with opposite surface charges in a clear solution is presented in [11]. Since the particles attract themselves due to the opposite charging and tend to agglomerate, a polymeric stabilizer is needed for a stable suspension. Since no dyed solution is involved for achieving the image contrast, there is no need to separate the black and white particles completely for good image contrast, which results in relatively short switching times, cf. Figure 6. According to the authors, only a relatively small change of the relative locations of the black and white particles is necessary within the cells to produce a good contrast image. Grey scale becomes feasible by pulse amplitude control, pulse width modulations, or variations in the number of scans.

Yet another approach is described in [14], where positively charged toner particles are moved through clouds of white electrically neutral pigments.

Besides the above-mentioned microencapsulation, lateral particle movement can be prevented by using a pixelated foil interposed between front and bottom electrode. Figure 7 shows a sample foil with micropores as small cavities for dye and pigments [12]. The foil has been fabricated by replica molding of a polyurethane film on a PDMS master with lithographically structured fingers. Besides electrophoretic materials, the versatile foil can also be filled with other electro-optical materials such as *organic light emitting polymers*, or *polymer-dispersed liquid crystals*.

The pigments in reflective EPD cells do not necessarily move vertically between bottom and top electrodes. With an *in-plane* electrode arrangement as described in [13] it is possible to manufacture all driving electrodes onto the bottom substrate, and to move the pigments in lateral dimensions, as depicted in Figure 8, left. In this case, black-colored positively charged toner particles with 1-2 micron are used, and aluminum electrodes serve as light reflectors. The pigments are located in a reservoir area in 'on' state, while they occupy the entire cell area in 'off' state. By introducing an additional top control electrode together with additional barrier walls between reservoir and cell area (not shown in Figure 8) the bistability can be enhanced. Furthermore, the use of a control electrode is advantageous for passive matrix addressing. Since a clear fluid can be used, a good adaptation to a multicolor system is feasible by coloring the driving electrodes. However, driving voltages of several 100 volts are applied. A 5 x 5 pixel demonstrator with a cell aperture of 55% is reported to show image information bistability for one week without supply voltages connected.

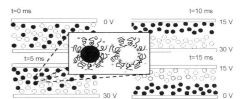

Figure 6: Schematic representation of an electrophoretic suspension containing black and white particles with opposite polarities in a clear medium. For details on the surface treatment cf. [11]. Both black and white images as well as grey scale formations for each individual pixel are possible.

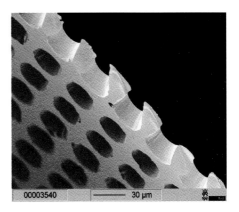

Figure 7: Microphotograph of a patterned pixelated foil with microholes for compartmentalization of the electrophoretic material [12]. The foil is placed between front and back electrode of the display.

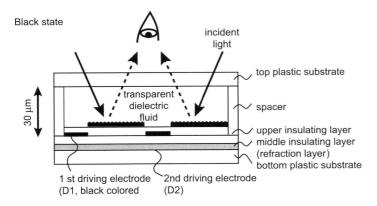

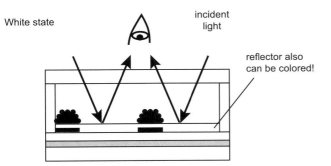

Figure 8: In-Plane electrode structure of an EPD with all electrodes on the bottom substrate [13]. The pigments are moved laterally either to a reservoir region, or cover a larger reflecting electrode. The disadvantage here is the reduced cell aperture and the resulting lower image contrast.

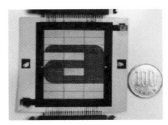

(a) Driven while flat

(b) Driven while deformed

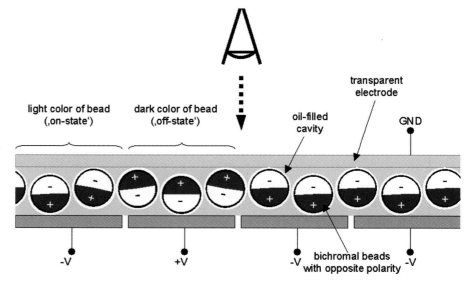

Figure 9: Principle of a Gyricon rotating ball display. Small hemispheric bichromal balls are dispersed in a thin elastomer sheet. Since each ball is placed in an oil-filled cavity, it is free to rotate when an electric field is applied. The observer either sees the dark or light hemispheres of the balls. In addition to the rotation, also a small migration of the balls takes place.

2.2 Rotating Ball Displays

The invention of the so-called 'Gyricon' rotating ball display principle in the late 1970's heads for reusable paper-like media which can be written and erased by means of electrostatic printing devices [15]. In fact, the Gyricon electronic paper has been the first approach towards flexible electronic paper displays. Since the driving voltages of Gyricon displays are higher than the ones for EPDs, they tend to use direct drive approaches instead of active matrix panels. This electronic paper media commonly applies some dedicated non-active external electrostatic device concept to be *reprogrammed*, such as a charged stylus or printer-like devices [16].

The basic operating principle of a Gyricon display is depicted in Figure 9. A clear and transparent elastomer sheet is sandwiched between two electrodes. Small balls or beads are randomly dispersed into this substrate sheet, each embedded into a liquid-filled cavity. The balls are composed of two different materials resulting in a structure of two hemispheres with contrasting color and opposite electrical charge, respectively. Thus, each ball has a macroscopic electrical dipole character and is free to rotate within its cavity when an electrical field is applied. The observer thus sees either the dark or the light side of the balls, which finally produces an image contrast. In the example shown in Figure 9, the dark side of the ball is charged positively while the white side is charged negatively. The difficulty is to start the rotation process of a fully vertically oriented ball, since the dipole moment is near zero under this condition. In case of an imperfect manufacturing process, especially at small ball sizes, the probability of immobility of a ball is large. For ball sizes of 100 μm and a thickness of the substrate sheet of 600 μm, minimum programming voltages of between 50 V and 150 V have been reported [17].

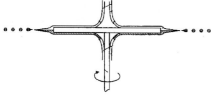

Figure 10: Schematic view of the fabrication technique of the tiny bichromal balls. Two molten plastics of different colors are introduced onto top and bottom of a spinning disk. The jets break up at the edge and the resulting balls solidify while moving through the air.

The bichromal balls are produced from low-cost molten pigmented plastic by means of a method shown in Figure 10 [18][19]. The wax-like plastics are introduced onto opposite sides of a spinning disk with a diameter of approximately 3 inches rotating at 45 rps. The plastics flow to the edges of the disk, and form a large number of small jets. These jets break up into balls and quickly solidify while moving through the air. The balls are then dispersed in an uncured elastomer. After a sheet is formed, the elastomer is cured and then soaked in a low viscosity oil. The oil causes the elastomer to swell and form oil-filled cavities around each single ball.

Figure 11 shows a microphotograph of a sample display with an average ball diameter of 100 μm. The response times for complete rotation are between 80 and 100 ms according to the authors. Finally, Figure 12 shows the complete demonstrator sheet which has been written by means of an external electrically charged writing devices. Overall displays thicknesses of between 120 to 800 μm have been reported.

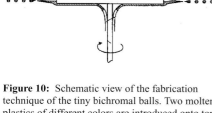

Figure 11: Photograph of a Gyricon display containing balls with an average diameter of 100 μm [18].

2.3 Suspended-Particle Displays

Although suspended microparticles are used here as described in the previous section, the operating principle of so-called suspended-particle displays (SPDs) [20] is different from EPDs or Gyricon displays. The roots of this light valve technology go back to the early 1930s [21]. The particles used here are not light-scattering, but light-absorbing, have shapes of rods or plates, and are suspended in a clear transparent liquid.

The operating principle is sketched in Figure 13. Without an electric field applied, the particles have randomized orientation due to Brownian movements and thus absorb incoming light. For an AC voltage applied to the electrodes, the particles align since they act like induced electrical dipoles and the light can pass through. Although the particles do not migrate from bottom to front electrode like in EPDs, they also suffer from agglomeration effects. The operating principle has a certain analogy to the one of PDLCDs. However, here the suspended particles operate independently, whereas a liquid crystal alignment has a large cooperative effect.

Figure 12: A demonstrator Gyricon display device, written by means of an external electrostatic scanning or printing device.

3 Alternative Paper-like Display Technologies

3.1 Reflective Direct View Liquid Crystal Displays

Although liquid crystal displays (cf. Chapter 37) already play a considerable role in the market, several drawbacks in optical performance have not been eliminated yet. Several mechanisms for modulating light have been implemented. These mechanisms include phase retardation, rotation of the polarisation, scattering of light, Bragg reflection and absorption. Polarisation and phase effects require the action of at least one *polariser*, whereas scattering, Bragg reflection and absorption do *not* require one.

In reflective displays employing either one or two polarisers a reflector is placed on the inner side of the rear glass plate. Due to the reflective character of the device, increasing ambient light should result in a more vivid image. However, the background ambient light is also reflected from the front surface, it overlaps with the displayed image and degrades contrast ratio. Low contrast ratio causes unsaturated colours. Two polariser twisted nematic (TN) LCDs show high contrast ratio but they achieve only low brightness and suffer from parallax problems (double image) which limit the resolution. In single polariser approaches this parallax is removed and reflectivity, i.e. brightness is enhanced. Contrast is lower than in the case of two polarisers but better than in a non-polariser LCD, which on the other hand shows better reflectivity. The single polariser LCD offering the compromise between brightness and contrast ratio can be regarded as the present mainstream approach. Several cell configurations exist for single polariser reflective devices. Since the single polariser behaves like two parallel

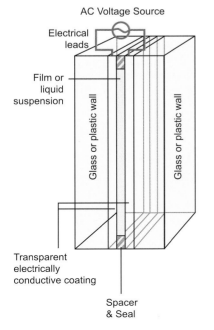

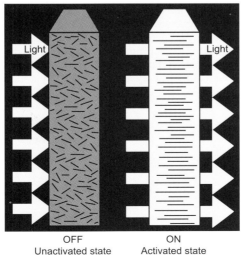

Figure 13: Structure and operating principle of a suspended-particle device. The particles are controlled by an AC voltage. In this light valve, non-oriented particles absorb incoming light (off-state), whereas oriented particles let the light pass through (on-state).

polarisers for the incident light, a quarter wave film is introduced beneath the polariser to emulate the crossed polariser configuration. Unfortunately the viewing angle of these cells is limited and has to be adjusted by the user for optimum contrast.

LCDs without polarisers – so called modified nematic liquid crystal displays - show high reflectivity and wide viewing angles, but limited contrast ratio [23]. The two main representatives are polymer dispersed liquid crystal displays (PDLCDs), and guest-host (GH) displays. PDLCDs operate on the basis of light scattering generated by the refractive index mismatch between micron sized liquid crystal droplets and the surrounding polymer matrices. In planar mode coloured light is reflected that meets the Bragg reflection condition. Three panels for red, green, and blue can be stacked together to achieve multiple colours with the trade-off of parallax problems [22]. A GH display employs the absorption anisotropy of 1–5 % dichroic dye molecules introduced into an aligned liquid crystal host. The host liquid crystal is transparent in the visible spectrum and the guest dyes absorb one polarisation of the incoming light and transmit the other. Special configurations with quarter wave films or chiral configurations have been designed to avoid a polariser. Certainly, reflector design is the most critical issue for all reflective direct-view displays.

3.2 Bistable Liquid Crystal Displays

A liquid crystal display is called *bistable* if it assumes two different configurations at zero electric field. The configuration displayed depends on the addressing history of the device. The respective two stable states represent minima in the total (electrical, dielectric and mechanical) energy stored. Bistable devices are able to maintain an image memory at zero field, like a printed page of a magazine. Obviously they consume substantially less energy than those requiring periodic refreshes. Bistability in LCDs occurs with ferroelectric, cholesteric and nematic liquid crystal materials.

Ferroelectric liquid (chiral smectic) crystal molecules are arranged in a helix form [25]. The pitch of this helix is roughly equal to the wavelength of visible light. Since the cell gap in these devices is smaller than the pitch, the ferroelectric liquid crystal is devoid of a helical structure. Under these conditions the liquid crystal assumes two stable states. If one of these states is made to coincide with the linear polarisation axis of the polariser film, a display effect can be produced, in which one of the bistable states transmits light and the other one does not. Unless a voltage is applied, the ferroelectric LCD remains stable in one of its two states (memory function). However, ferroelectric LCDs are less suitable for use in electronic paper application, due to the polarisers required.

Chiral nematic – or as they are often called – cholesteric liquid crystals are often composed of nematic material with the addition of chiral dopants. Chiral nematic-LCDs have two stable configurations, the reflective planar and the focal conic structure. As shown in Figure 14, the state of the liquid crystal is characterised by the direction of the helical axis. In the planar texture, the helical axis is perpendicular to the cell surface (top). The material reflects light if the Bragg reflection condition (wavelength is the index of refraction multiplied by the pitch length) is fulfilled. If the wavelength is in the visible light region the cell has a bright coloured appearance. In the focal conic texture the helical axis is more or less parallel to the cell surface (bottom). This is a multiple domain structure and the material is scattering. Finally, when the applied field is larger than a critical field the helical structure is destroyed and the liquid crystal molecules are aligned in the cell normal direction. With the appropriate surface anchoring condition both the planar texture and the focal conic texture can be stable at zero field. Bistability in nematic liquid crystals is imposed by a special surface alignment.

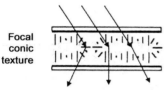

Figure 14: The structure of cholesteric textures (from [24]).

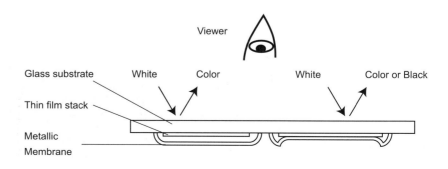

Figure 15: Display based on interferometric modulation. It is realized using microelectromechanical technology [26].

Apart from cholesterics, there are two competing approaches that are under development. These are the zenithally-bistable displays (ZBD) and the so-called 'Bi-Nem' displays. Both displays are bistable; the ZBDs have been demonstrated on glass and on plastic substrates. Both the cholesteric displays and the ZBD can be passively addressed. This is a potentially huge advantage in that high-resolution images can be achieved without an actice matrix panel. The update rate of the displays tends to be slow, however.

3.3 Micromechanics-based Displays

In order to finalize the review on display technology approaches suited for electronic paper, a concept based on micromechanics is briefly introduced. An example for a monolithically integrated display cell is the so-called interferometric modulator. A reflective self-supporting deformable membrane is fabricated by depositing a thin-film stack on a transparent glass substrate, as depicted in Figure 15.

Depending on the thickness of the stack and the thickness of the gap between stack and membrane, the optical resonance behaviour of the cavity can be controlled. Hence, light which enters through the glass substrate interferes depending on the membrane position which results in total or partial extinction of its spectrum. The membrane is controlled to collaps by applying a specific electrostatic force and is designed with a mechanical hysteresis, i.e. it remains in the collapsed state as long as a bias voltage is applied. Therefore, the power consumption of this quasi-bistable display is very low [26][27].

4 Flexible Backplane Electronics

Until now, various approaches of electro-optical materials well-suited for flexible paper-like displays have been discussed. These materials show a linear dependence between the applied electrical field and the optical effect. In order to avoid cross-talk and the activation of half-selected display cells in a row and column addressing scheme, *non-linear* elements are necessary in each cell in the display matrix. These non-linear elements are usually diodes or transistors operated as switches. They are integrated in the so-called display backplane. In conventional flat panel displays, the backplane is realized as an integrated circuit, e.g. thin film transistors on a glass substrate. In general, the backplane circuitry manages the addressing and the data routing to the display cells. However, for 'electronic paper' displays additional constraints become important. For economical reasons, especially when considering large-scale public displays or low-cost or even disposable applications, inexpensive production is mandatory. Furthermore, the backplane substrate should be thin and mechanically flexible, but at the same time robust against defects. A schematic view onto the backplane circuitry in a flat panel EPD is shown in Figure 16.

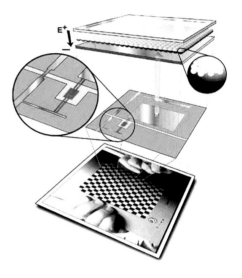

Figure 16: Concept of a flexible display backplane for electrophoretic ink displays with ink jet printed organic polymer transistors [32]. Each transistor passes through the programming voltage via the bit line (yellow) when addressed by the select line (grey) and induces the movement of the pigments by charging or discharging the display cell (top).

Various early approaches for active backplane circuitry are known from the literature. A silicon based backplane with NMOS transistors for electrophoretic displays has been presented in [28]. Since large scale integrated circuits deliver low manufacturing yields, this concept also offers some primitive redundancy strategies in order to cope with pixel defects. A first step towards the realisation of low-cost and flexible non-linear elements have been printable metal-insulator-metal diodes reported in [29]. The diodes have a sandwich layer structure that consists of Tantalum powder in polymer, Tantalum oxide, and Chrome powder in a polymer binder. Active matrix displays have been presented incorporating amorphous silicon (a-Si) transistors and microencapsulated electrophoretic materials on glass substrates [6], and on stainless steel foils [30].

4.1 Printable Integrated Circuits

Today, the manufacturing of integrated circuits is based on iterative structuring of functional layers by means of processing steps like lithography, deposition, and etching. Processing cost thus tends to rise with increasing substrate sizes. The paradigm of printable electronics is to achieve the structuring of active electronic devices and interconnects by means of *roll-to-roll* or *printing* processes directly on flexible substrates. Despite the relatively low performance achievable compared to silicon technologies, these technologies have the potential to allow for highly functional, mechanically flexible, and inexpensive display backplanes. Direct ink-jet printing of transistor circuits based on solution-processed polymer conductors, insulators, and self-organising semiconductors, including via-hole interconnections have been shown in [31]. Practical channel lengths of 5 micrometers, electron mobilities of 0.02 cm²/(Vs), and on-off current switching ratios of 10^5 were achieved.

The combination of an organic transistor active matrix backplane with electrophoretic ink has been reported in [33][34]. Figure 17 shows the 16 x 16 pixel dot matrix display with 5" x 5" dimension which can be operated even when flexed. The display is driven at 50 volts with an 80 Hz frame rate. The circuitry resembles the one known from select transistors in memory devices, such as a DRAM architecture. A programming voltage is passed through a bit line via a select transistor and charges the display cell. The gate of the select transistor is controlled by means of a word line. As stated above, an electrophoretic display cell has a small capacitance and a relatively high resistance. Therefore, only very small on-state currents I_{on} are needed to drive the display at reasonable frame rates. As a result, the low-performance properties of printed electronics match those of electrophoretic cells. However, the off-current I_{off} of the select transistors must be small enough to achieve reasonably large values for the matrix dimensions. A short calculation considering the target line address time, the required I_{on}, the maximum voltage drop across the transistor in its conducting state, and the parasitic charging current I_{off} of half-selected pixels result in $I_{on}/I_{off} > 40\,N$, where N is the number of pixel rows all connected to a single select line [35]. The transistor devices used in [33] show an $I_{on}/I_{off} = 1000$, thus allowing only a relatively small number of rows.

Transistor devices based on organic semiconducting polymers have properties which are disadvantageous for the application in paper-like displays. These materials are rather unstable and prone to the impact of oxygen and water and therefore require reliable sealing. It is obvious, that thin and large sized systems such as electronic paper used in rugged environments cannot be sealed easily without diminishing their flexibility, weight, and manufacturing cost. A solution to this problem might be printable electronics based on *suspended nanograins* of inorganic semiconducting materials. Since the melting temperature of nanocrystals is significantly lower than that of the respective bulk material, suspended nanograins can be structured by means of ink jet printing and subsequent melting or sintering [36]. Organic-Inorganic hybrid materials for low-cost manufacturing of thin-film transistors have been reported in [37]. For both organic and inorganic materials, printed electronics combines the advantages of monolithic integrated circuits and cost-efficient manufacturing using roll-to-roll processing.

Figure 17: Flexible display backplane with 16 x 16 select transistors on a rubber stamped Mylar substrate (left). When vertically combined with electrophoretic ink materials, a display results which can be flexed during operation (right) [33].

4.2 Fluidic Self-Assembly

An alternative method to integrate active electronic devices into flexible substrates is to sparsely disperse tiny silicon-based microchips on a flexible plastic substrate. Fluidic self-assembly (FSA) is a technology which is capable of accurately assembling large numbers of very small devices without the need for robotic pick-and-place and thus reducing the cost for handling and mounting millions of microchips on one substrate. Originally invented to combine the properties of silicon integrated circuits and optical III-V materials by mounting many small laser devices on wafer scale [38], this technology is now being commercialized for display backplanes and many more applications, such as RFID tags, sensors, and phased array antennas [39].

In the FSA process, silicon wafers are micromachined into trapezoidal elements from 30 microns to 1000 microns in size, e.g. using [111] facet selective etches. In the target substrate, corresponding holes are manufactured by laser drilling, stamping etc. All of the surfaces are treated to minimize surface forces. The tiny devices are suspended in a liquid and handled as a slurry. The suspension is flowed over the top of the substrate, filling the target sites. This leads to a planar assembly, aided by fluid flow and vibration. A schematic view of this self-organizing process is shown in Figure 18. After all holes in the target substrate are filled correctly, the tiny elements are fixed by a planarizing layer. Subsequent photolithographic steps process vias and a planar metallization layers for interconnecting the tiny chips, cf. Figure 18.

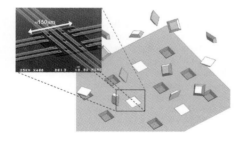

Figure 18: Novel mounting and packaging technology for very small pieces of silicon from 30 to 1000 microns in size. In the fluidic self-assembly process, thousands or millions of devices are suspended in a liquid and flowed over a substrate with receptor holes. After the holes are filled, they are interconnected by means of conventional metallization.

In the case of a display backplane, the tiny chips can contain the select transistors as shown in Figure 16. However, when embedding high-performance silicon circuits, the chips can manage tasks which are much more complex than simple pixel addressing. The wafer area needed for a given level of functionality is continuously decreasing as miniaturization of integrated circuits proceeds. Very powerful chip systems containing simple microcontroller cores and program and data memory are feasible on a few hundred microns wide chips. Hence it will be possible to distribute computational power for purposes such as image decompression, vector graphic calculations, or management of defective display regions directly into the backplane.

5 Outlook and Vision

Electronic Paper Displays will open a number of novel applications for the future information society. Although commercialization of electrophoretic displays is expected to start in the upcoming years, today this technology is still in its infancy. For all the envisioned applications, further efforts are necessary to develop paper-like systems with a real touch of paper.

One of the visions of the e-paper pioneers is the idea of the electronic book: "Such a book has hundreds of electronic page displays formed on real paper. On the spine are a small display and several buttons. The user may leaf through several thousand titles, select one he or she likes, wait a fraction of a second and open the book to read King Lear. When done with King Lear, another title may be selected ..." [40].

Yet another visionary goal is a digital newspaper that combines wireless communication devices with electronic paper. This so-called static radio paper could reconfigure itself over night and download the latest news content. Figure 19 illustrates a design study of an electronic newspaper reading device.

This chapter described selected technological approaches for electronic paper-like information displays. Obviously, the requirements for large-scale and high-resolution dot-matrix displays like the one depicted in Figure 19 are very compelling. Besides innovations in electro-optical materials, in the field of intelligent backplane electronics, novel concepts become mandatory to meet these challenges.

Figure 19: Visionary design study of an electronic newspaper reading device. The high-resolution multi-page display is connected to a control and wireless communication device on the bottom [41].

Acknowledgements

The authors would like to thank Paul Drzaic for carefully reviewing the manuscript and for his valuable comments. They also acknowledge the many helpful comments made by Jukka Hautanen and Jyrki Kimmel. The editor would like to thank Christian Ohly (FZ Jülich) for checking the symbols and formulas in this chapter.

References

[1] G. P. Crawford, IEEE Spectrum, **37**, 40 (2000).

[2] L. J. Hornbeck et al., DSC-Vol. 62 / HTD-Vol. 354, Micromechanical Systems (MEMS), ASME, 3-8 (1997).

[3] I. Ota, J. Ohnishi and M. Yoshiyama, Proceedings of the IEEE, **61**, 832 (1973).

[4] A. L. Dalisa, *Electrophoretic Displays*, in: Topics in Applied Physics, ed. J.I. Pankove, **40**, 213 (1980).

[5] B. Comiskey, J. D. Albert, and J. Jacobson, SID 1997 Digest, 75 (1997).

[6] P. Kazlas et al., SID 2001 Digest, 152 (2001).

[7] B. Comiskey, J. D. Albert, H. Yoshizawa, and J. Jacobson, Nature, **394**, 253 (1998).

[8] http://www.eink.com .

[9] G. M. Duthaler, M. Davis, E. Pratt, C. Gray, and K. Suzuki, in Proceedings of the International Display Research Conference (IDRC), 473 (2001).

[10] S. A. Swanson, M. W. Hart, and J. G. Gordon, SID 2000 Digest, 29 (2000).

[11] J. Hou, S. Shokhor, and S. Naar, SID 2001 Digest, 164 (2001).

[12] http://www.papyron.com .

[13] E. Kishi et al., SID 2000 Digest, 24 (2000).

[14] T. Kitamura, in Proceedings of the International Display Research Conference (IDRC), 1517 (2001).

[15] N. K. Sheridon and M. A. Berkovitz, SID 1977 Digest, 289 (1977).

[16] http://www.parc.xerox.com/dhl/projects/gyricon .

[17] M. E. Howard, E. A. Richley, R. Sprague, and N. K. Sheridon, SID 1998 Digest, 1010 (1998).

[18] N. K. Sheridon et al., International Display Research Conference, 82 (1997).

[19] J. M. Crowley et al., *Method and Apparatus for Fabricating Bichromal Balls for a Twisting Ball Display*, US Patent 5 262 098 (1993).

[20] R. L. Saxe, R. I. Thompson, and M. Forlini, in Proceedings of the International Display Research Conference, 175 (1982).

[21] E. H. Land, *Colloidal Suspensions and the Process of Making Same*, US Patent 1 951 664 (1934).

[22] P. Drzaic, World Scientific (1995).

[23] E. Lueder, *Liquid Crystal Displays – Addressing Schemes and Electro-Optical Effects*, Wiley (2001).

[24] S.-T. Wu and D.-K. Yang, *Reflective Liquid Crystal Displays*, Wiley (2001).

[25] M. Muecke, M. Randler, V. Frey, J. Brill, E. Lueder, and A. Karl, SID 2000 Digest, 1126 (2000).

[26] M. W. Miles, SID 2000 Digest, 32 (2000).

[27] http://www.iridigm.com .

[28] R. R. Shiffmann and R. H. Parker, Proceedings of the SID, **25**, 105-115 (1984).

[29] J. Park and J. M. Jacobson, Proceedings of the Material Research Society Symposium, **508**, 211 (1998).

[30] Y. Chen, K. Denis, P. Kazlas, P. Drzaic, SID 2001 Digest, 157 (2001).

[31] H. Sirringhaus et al., Science, **290**, 2123 (2000).

[32] J. A. Rogers, Science, **291**, 1502 (2001).

[33] K. Amundson et al., SID 2001 Digest, 160 (2001).

[34] J. A. Rogers et al., PNAS, **98**, 4835 (2001).

[35] K. Amundson and P. Drzaic, Proceedings of the International Display Research Conference, 2000.

[36] B. A. Ridley, B. Nivi, and J. M. Jacobson, Science, **286**, 746 (1999).

[37] C. R. Kagan, B. D. Mitzi, and C. D. Dimitrakopoulos, Science, **286**, 945 (1999).

[38] J. S. Smith, International Electron Devices Meeting (IEDM), Technical Digest, 201 (2000).

[39] http://www.alientechnology.com .

[40] J. Jacobson et al., IBM Systems Journal, **36**, 457 (1997).

[41] http://www.idsa.org/whatis/seewhat/idea99/winners/epaper.htm .

Abbreviations

1T/1C	Single Transistor / Single Capacitor Memory Cell
1T/1TMR	Single Transistor/ Single TMR Memory Cell
2DEG	Two Dimensional Electron Gas
2T/2TMR	Twin Transistor / Twin TMR Memory Cell
5-HT$_3$	5 – Hydroxytryptamin
A/D	Analog/Digital
ABS	Air Bearing Surface
AC	Alternating Current
Ach	Amine acetylcholine
ACPDP	AC Plasma Display Panel
ADC	Analog to Digital Converter
AES	Auger Electron Spectroscopy
AFC	Antiferromagnetically Coupled
AFM	Atomic Force Microscope
AFM	Antiferromagnetic
ALCVD	Atomic Layer CVD
ALD	Atomic Layer Deposition
ALE	Atomic Layer Epitaxy
ALIS	Alternate Lighting of Surfaces (addressing technology by Fujitsu)
Alq$_3$	tris (8-hydroxyquinoline) aluminum
ALU	Arithmetic-Logical Unit
AM	Active Matrix
AMR	Anisotropic Magnetoresistance
ARC	Anti Reflex Coatings
ARDE	Aspect Ratio Dependent Etch
a-Si	amorphous Silicon
ASIC	Application Specific Integrated Circuit
ATP	Adenosine Triphosphate
BAM	BaMgAl$_{10}$O$_{17}$:Eu^{2+} (phosphor material)
BARC	Bottom Anti Reflex Coatings
BAW	Bulk Acoustic Wave
BCS	Bardeen, Cooper and Schrieffer
BEOL	Back-End-of-Line
BER	Bit Error Rates
BGA	Ball-Grid-Array
BLT	(Bi,La)$_4$Ti$_3$O$_{12}$
BNBIQ	diButyl-Naphtylamine-Butylsulfonato-IsoQuinolinium
BPSG	Boron Phosphor Silicate Glass
BS	Block-Select
BSIM	Berkley Short-Channel IGFET Model
BST	(Ba,Sr)TiO$_3$
BTO	BaTiO$_3$
CAD	Central Aperture Detection
CAIBE	Chemical Assisted Ion Beam Etching
cAMP	Adenosine 3', 5' Cyclic Monophosphate
CAR	Chemically Amplified Resists
CB	Conduction Band
CBGA	Ceramic Ball-Grid-Array
CCD	Charge Coupled Device
CCFL	Cold Cathode Fluorescent Lamp
CD	Critical Dimension
CD	Compact Disc
CD-R	CD-Recordable
CD-ROM	CD – Read Only Memory
CD-W	CD- Rewritable
CET	Capacitor Equivalent Thickness
CFeRAM	Chain Ferroelectric Random Access Memory
cGMP	Guanosine 3', 5' Cyclic Monophosphate
Ch-LCD	Cholesteric Liquid Crystal Display
CIE	Commission International de l´Eclairage (International Commission on Illumination)
CIP	Current In Plane
CMOS	Complementary Metal Oxide Semiconductor
CMOSFET	Complementary Metal Oxide Semiconductor Field Effect Transistor
CMP	Chemical Mechanical Polishing
CNS	Central Nervous System
CNT	Carbon Nanotube
CNTFET	Carbon Nanotube Field Effect Transistor
CNT-SRAM	Carbon Nanotube Static Random Access Memory
COB	Chip on Board
Col	Columnar
CPP	Current Perpendicular to Plane
CPPD	Cell Plate Pulse Driven
CRT	Cathode Ray Tube
CSD	Chemical Solution Deposition
CTE	Coefficient of Thermal Expansion
Cub	Cubic

Abbreviations

CU	Control Unit	EL	Electroluminescence
CVD	Chemical Vapor Deposition	EML	Light Emission Layer in OLEDs
cw	Continuous Wave	EMS	Electromechanical Systems
D	Drain	EOSFET	Electrolyte Oxide Silicon Field Effect Transistor
DAC	Digital Analog Converter	EOT	Equivalent Oxide Thickness
DARPA	Defense Advanced Research Projects Agency	EPD	Electrophoretic Display
DAT	Digital Audio Type	EPR	Electron Paramagnetic Response
DBQW	Double Barrier Quantum Well	EPROM	Electrically Programmable Read Only Memory
DBR	Distributed Bragg Reflector	EPSP	Excitatory Postsynaptic Potential
DC	Direct Current	ESD	Electrostatic Discharge
DCFL	Direct Coupled FET Logic	ETL	Electron Transport Layer in OLEDs
EUV	Extreme Ultra-Violet	EUVL	Extreme UV Lithography
DFF	Delayed Flip Flop	EXAFS	Extended X-Ray Absorption Fine Structure
DG	Double Gate	EXOR	Exclusive OR
DI	De-Ionized Water	FAD	Front Aperture Detection
DIBL	Drain Induced Barrier Lowering	FED	Field Emission Display
DiI	1,1' –dioctadecyl-3,3,3' ,3' –tetramethyl-indocarbocyanine perchlorate	FeFET	Ferroelectric Field Effect Transistor
DNA	Deoxyribonucleic Acid	FeRAM	Ferroelectric Random Access Memory
DNQ	Diazonaphthoquinones System	FET	Field Effect Transistor
DoE	Design of Experiment	FFT	Fast Fourier Transform
DOF	Depth of Focus	FIB	Focused Ion Beam
DOS	Density of States	FIL	Fingerprint Sensor/Identifier LSI Chip
DOTFET	Nanoscaled Dot Field Effect Transistor	FIR	Finite Impulse Response
DOW	Direct Overwrite	FLIC	Fluorescence Interface Contrast
DQx	Data Line	FLOPS	Floating Point Operations per Second
DR	Dynamic Range	FM	Ferromagnetic
DRAM	Dynamic Random Access Memory	FM	Free Membrane
DRO	Destructive Read-Out	FPD	Field Programmable Device
DRT	Distribution of Relaxation Time	FPGA	Field Programmable Gate Array
DSP	Digital Signal Processor	FPN	Fixed-Pattern Noise
DSTN	Double Supertwisted Nematic	FSA	Fluidic Self Assembly
DUV	Deep Ultra-Violet	FSTN	Foil Supertwisted Nematic
DVD	Digital Versatile Disc	G	Gate
EBU	European Broadcasting Union	GABA	Amino Acid Glycine + γ-Amino Butyric Acid
ECDL	Exchanged Coupled Double Layer	GB	Gigabyte
ECR	Electron Cyclotron Resonance	GFP	Green Fluorescent Protein
EDX	Energy Dispersive X-ray Analysis	GH	Guest Host
EEDF	Electron Energy Distribution Function	GMR	Giant Magneto Resistance
EELS	Electron Energy Loss Spectroscopy	GND	ground
EEPROM	Electrically Erasable Programmable Read Only Memory	HB-DC	hexa-tert-butyl decacyclene
EFM	Electric Force Microscopy	HDD	Hard Disk Drives

Abbreviations

HDTV	High Definition Television	LCAO	Linear Combination of Atomic Orbital
HEK	Human Embryonic Kidney Cells	LCD	Liquid Crystal Display
HEMT	High Electron Mobility Transistor	LCOS	Liquid Crystal on Silicon
HGA	Head Gimbal Assembly	LDOS	Local Density of States
HMD	Head Mounted Display	LED	Light Emitting Diode
HME	Hybrid Molecular Devices	LEED	Low Electron Energy Diffraction
HOMO	Highest Occupied Molecular Orbital	LEEM	Low Energy Electron Microscopy
HREELS	High Resolution EELS	LIM-DOW	Laser Intensity Modulation DOW
HRTEM	High Resolution TEM	LO	Longitudinal Optical
hSlo	Calcium Activated Potassium Channel	LO	Light Output
HTL	Hole Transport Layer in OLEDs	LOCOS	Local Oxidation of Silicon
HTS	High Temperature Superconductor	low - K	low permittivity
IADF	Ion Angular Distribution Function	LPCVD	Low Pressure CVD
IBAD	Ion Beam Assisted Deposition	LPP	Laser Produced Plasma Sources
IBE	Ion Beam Etching	LSI	Large Scale Integration
IC	Integrated Circuit	LST	Lyddane Sachs Teller
ICP	Inductively Coupled Plasma	LTD	Long Term Depression
ID	Identification	LTM	Long Term Memory
IEC	Interlayer Exchange Coupling	LTP	Long Term Potentiation
IEDF	Ion Energy Distribution Function	LTS	Low Temperature Superconductor
IEDM	International Electron Device Meeting	LUMO	Lowest Unoccupied Molecular Orbital
III-V	Group III-Group V Semiconductor	LUT	Loop up Tables
ILB	Inner Lead Bonding	MAC	Multiply-Add Unit
ILD	Interlevel Dielectric	MAMOS	Magnetic Amplifying MO Systems
IP_3	inositol – 1, 4, 5 triphosphate	MBE	Molecular Beam Epitaxy
IPDS	Ion Projection Direct Structuring	MCB	Mechanically Controlled Break Junction
IPES	Inverse Photo Electron Spectroscopy	MCM	Multi-Chip-Module
IPL	Ion Projection Lithography	MD	Mini Disc
IPSP	Inhibitory Postsynaptic Potential	MEMS	Microelectromechanical System
$Ir(ppy)_3$	fac-tris(2-phenylpyridine) iridium (III)	$MF-ABO_3$	Metal-Ferroelectric
ISC	Intersystem Crossing	MFIS	Metal- Ferroelectric–Insulator–Semiconductor
ISFET	Ion-Sensitive Transistor		
IT	Information Technology	MFM	Magnetic Field Modulation Magnetic Force Microscopy
ITO	Indium-Tin Oxide		
ITRS	International Technology Roadmap for Semiconductors	MFMIS	Metal- Ferroelectric–Metal-Insulator–Semiconductor
		MFS	Metal-Ferroelectric-Semiconductor
J-FET	Junction Field Effect Transistor	MHL	Magnetization of Hard Layer
JM	Attached Membrane	MIGS	Metal Induced Gap States
JTL	Josephson Transmission Line	MIMD	Multiple Instruction - Multiple Data
KAMINA	Karlsruhe Micro Nose	MIPS	Million Instructions Per Second
KPD	KH_2PO_4	MISD	Multiple Instruction - Single Data
LAMPAR	Low-Noise Arrays of Mosfets for Pyroelectric Array Readout	MIT	Massachusetts Institute of Technology
		ML	Monolayer
LB	Langmuir-Blodgett		

Abbreviations

MLCT	Metal-to-Ligand Charge Transfer		NGL	Next Generation Lithography
mM	millimol/l		NIL	Nano Imprint Lithography
MME	Mono-Molecular Devices		NIR	Near-Infrared
MML	Monostable Multistable Logic		NMOS, n-MOS	N-Type Metal-Oxide Semiconductor
MO	Molecular Orbital		NMR	Nuclear Magnetic Resonance
MO	Magneto Optical		NOL	Nano Oxide Layer
MOBILE	Monostable-Bistable Transition Logic Element		NSIC	National Storage Industrial Consortium
MOCVD	Metal Organic CVD		NSOM	Near-Field Scanning Optical Microscopy
MOD	Metal Organic Decomposition		NW	Normally-White LCDs
MOKE	Magneto Optical Kerr Effect		OES	Optical Emission Spectroscopy
MOMBE	Metal Organic Molecular Beam Epitaxy		OFETs	Organic Thin-Film Transistors
MOS	Metal-Oxide-Semiconductor		OLB	Outer Lead Bonding
MOSFET	Metal Oxide Semiconductor Field Effect Transistor		OLED	Organic Light Emitting Diode / Device
MOVPE	Metal Organic Vapor Phase Epitaxy		OMCVD	see MOCVD
MPB	Morphotropic Phase Boundary		ONO	Oxide/Nitride/Oxide
MPU	Micro Processor Unit		PAC	Photoactive Compound
MQC	Macroscopic Quantum Coherence		PADOX	Pattern Dependent Oxidation
MQT	Macroscopic Quantum Tunneling		PAG	Photo Acid Generator
MR	Magnetoresistance		PAP	Photoaddressable Polymer
MRAM	Magnetic Random Access Memory		PARC	Pattern Recognition Unit
MSR	Magnetically Induced Superresolution		PBGA	Plastic Ball-Grid-Array
MTDATA	4,4',4''-tris(3-methylphenylphenylamino)triphenylamine		PC	Program Counter
			PC	Personal Computer
MTJ	Magnetic Tunnel Junctions		PCA	Principle Component Analysis
MWCNT	Multiwalled Carbon Nanotube		PD	Photo Diode
MWNT	Multi-Wall Nanotube		PDA	Personal Digital Assistant
N	Nematic		PDLC	Polymer-Dispersed Liquid Crystal
N^*	Chiral Nematic		PDLCD	Polymer Dispersed Liquid Crystal Display
NA	Numerical Aperture		PDP	Plasma Display Panel
NB	Normally-Black LCDs		PEA	Positive Electron Affinity
Nd:YAG	Neodym-doped Yttrium-Alluminium-Garnet		PEB	Post Exposure Bake
			PECVD	Plasma-Enhanced Chemical Vapor Deposition
NDP	Non-Driven Plate		PEEM	Photoemission Electron Microscope
NDR	Negative Differential Resistance		PET	Polyethylene Terephthalate
NDRO	Non-Destructive Read-Out		PFM	Piezoresponse Force Microscopy
NEA	Negative Electron Affinity		PFM	Pulse Frequency Modulation
NEMO	Nanoelectronic Modelling		PFPE	Perfluoropolyether
NEMS	Nano Electromechanical System		PFRAM	Polymeric Ferroelectric Random Access Memory
NEP	Noise Equivalent Power			
NETD	Noise Equivalent Temperature Difference		PGO	$Pb_3Ge_5O_{11}$
NEXI	Nonequilibrium Exchange Interaction		PL	Photoluminescence
NFR	Near Field Recording		PL	Plate Line

Abbreviations

PLD	Programmable Logic Devices	RSFQ	Rapid Single-Flux-Quantum Logic
PLD	Pulsed Laser Deposition	RTBT	Resonant Tunneling Bipolar Transistor
PM	Passive Matrix	RTD	Resonant Tunneling Diode, Two-Terminal Device
PMMA	Polymethylmethaacrylat		
p-MOS	p-type MOS	RTHEMT	Resonant Tunneling HEMT
POMBE	Pulsed Organo-Metallic Beam Epitaxy	RTP	Rapid Thermal Process
PROM	Programable Read Only Memory	RWL	Reference Word Line
PSG	Phosphorus Silicate Glass	RZ	Return to Zero
PSM	Phase Shift Method	S	Source
PT	$PbTiO_3$	S/N	Signal to Noise
PTCR	Positive Temperature Coefficient Resistors	SAC	Self-Aligned Contact
		SAED	Small Angle Electron Diffraction
PVDF	polyvinylidene di-fluoride	SAF	Synthetic Antiferromagnet
PVR	Peak-to-Valley (Current Density) Ratio	SAM	Self-Assembled Monolayer
PWB	Printed Wiring Board	SAW	Surface Acoustic Wave
PWM	Pulse Width Modulation	SBN	Strontium-Barium-Niobate
PXL	Proximity X-ray Lithography	SBT	$SrBi_2Ta_2O_9$
PZT	$PbZr_xTi_{1-x}O_3$, $Pb(Zr,Ti)O_3$	SC	Switched Capacitor
QC	Quantum Computing	SCALPEL	Scattering with Angular Limitation in Projection Electron Beam Lithography
QCA	Quantum Cellular Automata		
QE	Quantum Efficiency	SCM	Single Chip Module, Scanning Capacitance Microscopy
QWS	Quantum Well States		
R&D	Research and Development	SD	Sputter Deposition
R/W	Rewritable	SDR	Software Defined Radio
RAD	Rear Aperture Detection	SEM	Scanning Electron Microscopy
RAM	Random Access Memory	SET	Single Electron Transistor
RAMAC	Random Access Method of Accounting and Control	SFM	Scanning Force Microscope
		SFQ	Single-Flux-Quantum Logic
RBS	Rutherford Backscattering Spectroscopy	SID	Society for Information Display Spatially Immerse Display
RE	Rare Earth	SIL	Solid Immersion Lenses
redox	reduction-oxidation	SIMOX	Separation by Implantation of Oxygen
RF	Radio Frequency	SIMS	Secondary Ion Mass Spectroscopy
RHEED	Reflection High Energy Electron Diffraction	SiON	Silicon Oxynitrid
		SISD	Single Instruction - Single Data
RHET	Resonant Tunneling Hot Electron Transistor	SLM	Spatial Light Modulator
		Sm	Smectic
RIBE	Reactive Ion Beam Etching	SMD	Surface Mounted Device
RIE	Reactive Ion Etching	SMT	Surface Mount Technology
RISC	Reduced Instruction SET Computer	SNOM	Scanning Near-Field Optical Microscopy
R-LCD	Reflector Liquid Crystal Display	SNR	Signal-to-Noise Ratio
rms	root mean square	SoC	System-on-a-Chip
ROIC	Read Out Integrated Circuit	SOI	Silicon on Insulator
ROM	Read Only Memory	SON	Silicon on Nothing
RPL	Reference Plate Line	SPD	Suspended Particle Display

Abbreviations

SPEELS	Spin Polarized EELS		UHV-STM	Ultra High Vacuum Scanning Tunneling Microscopy
SPICE	Simulation Program with Integrated Circuit Emphasis		ULSI	Ultra Large Scale Integration
SPM	Scanning Probe Microscopy		US	Ultrasonic
SQUID	Superconducting Quantum Interference Device		UTB	Ultra Thin Body
			UTM	Turing Machine
SRAM	Static Random Access Memory		UTC-PD	Ultrahigh Speed Photodiode
SS	Stainless Steel		UV	Ultraviolet
SSFLC	Surface Stabilised Ferroelectric Liquid Crystal		VB	Valence Band
			VCR	Videocassette Recorder
STI	Shallow Trench Isolation		VFD	Vacuum Fluorescent Display
STM	Scanning Tunneling Microscopy		VGA	Video Graphics Adapter
STN	Supertwisted Nematic		VLIW	Very Large Instruction Word
STV	Protein Streptavidin		VLS	Vapor-Liquid-Solid
SURFACTANT	Surface Active Agent		VLSI	Very Large Scale Integration
SVGA	Super Video Graphics Adapter		VMD	Virtual Model Display
SWNT	Single-Wall Nanotube		VMRAM	Vertical MRAM
TAB	Tape Automated Bonding		V-PADOX	Vertical Pattern Dependent Oxidation
TAR	Top Anti Reflex Coatings		VRD	Virtual Retina Display
TC	Thermo-Compression		VSM	Vibrating Sample Magnetometer
TCP©	Transformer-Coupled Plasma		WDX	Wavelength Dispersive X-ray Analysis
TDEAZ	Tetradiethylaminozirconium		WE	Write Enable
TEM	Transmission Electron Microscopy		WORM	Write Once Read Many
TEOS	Tetra Ethyl Ortho Silane		WVGA	Wide Video Graphics Adapter
TEP	Transductive Extracellular Potential		XPS	X-ray Photoelectron Spectroscopy
Teramac	Tera (10^{12}) Multiple Architecture Computer		XRD	X-ray Diffraction
			XRF	X-ray Fluorescence
TFFE	Thin Film Ferroelectric		YGB	$YGdBO_3:E^{3+}$ (phosphor material)
TFT	Thin-Film Transistor		ZBD	Zenithally-Bistable Display
TGS	Tri-Glycerine Sulfate			
TIFF	Tagged Image File Format			
TM	Transition Metal			
TMR	Tunnel Magnetoresistance			
TN	Twisted Nematic			
TO	Transversal Optical			
TOLED	Transparent OLED			
TPD	N,N'-diphenyl-N,N'-bis(3-methylphenyl)1-1' biphenyl-4,4' diamine			
TS	Thermosonic			
TTF	Tetrathiafulvalene			
TTT	Time-Temperature-Transformation Diagram			
UBM	Under Bump Metallization			
UHV	Ultra-High Vacuum			

Symbols

$\boldsymbol{a}$	unit vector of the bravais lattice
$\boldsymbol{a}_1, \boldsymbol{a}_2$	unit vectors
$\boldsymbol{a}_i$	elemental translation vector
$a(k)$	gaussian distribution
a_i	activity of component I
A	acceleration, activation field
A	area
A	anisotropy of an etch process
A	Hamaker constant
A_{FM}	area of the free membrane
A_{JM}	area of the attached membrane
A_s	electron affinity
$\boldsymbol{B}$	magnetic induction
B	critical exponents
B	probability
B	barrier height
$\boldsymbol{C}_h$	chiral vector
C	van der Waals interaction constant
C	Curie constant
C	capacitance
C'	capacitance per unit length
C^*	channel capacity
C_0	capacitance in state '0'
C_1	capacitance in state '1'
C_{BL}	bitline capacitance
C_{FE}	capacitance of a ferroelectric capacitor
C_{JM}	global capacitance of the membrane
C_S	global capacitance of the chip
$CMTF_{resist}$	Critical Resist Modulation Transfer Function
c_{ij}	elastic tensor constants
c	sound velocity
c_0	light velocity
c_M	area specific capacitance of membrane
c_{ph}	velocity of sound
c_S	specific capacitance per unit area
c_{Sh}	shear sound velocity
$\boldsymbol{d}$	piezoelectric coefficient
$\boldsymbol{d}_{ij}$	tensor component in ij-direction
d	diameter
d	distance
d	thickness
d_J	distance of the cell-chip junction
d_{ox}	thickness of the oxide
$\boldsymbol{D}$	displacement vector
D	diffusion coefficient
D	thickness
D^*	detectivity
$D(W)$	density of states
DD	device density
E	effect of laser pulse
E_{bias}	internal bias field
E_C	critical electric field
$\boldsymbol{E}_C$	coercive electric field
E	illuminance
ER_L	lateral etch rate
ER_V	vertical etch rate
E_{th}	threshold field
e	elementary charge
$\mathfrak{F}$	Fourier Transform
f_{ij}	elastic depolarization coefficients
f	force constant
f	flow rate (gas or liquid)
f	Fermi-Dirac distribution function
f	frequency

Symbols

f_0	resonant frequency		I	electrical current
f_B	maximal signal frequency		I_C	capacitive current, maximum super-current
f_c	clock-frequency		I_D	drain current
f_g	cut-off frequency		I_l	leakage current
f_J	Josephson frequency		I_s	ionization energy
f_r	mechanical resonant frequency		j	growth rate of films
F	force		$\boldsymbol{J}$	magnetic dipole moment
F	feature size		J	magnetic coupling parameter
F	free energy		J	current
F_{el}	electrostatic force		$\boldsymbol{k}_\parallel$	parallel wave vector
F_r	reaction force		$\boldsymbol{k}_F$	Fermi wave vector
$\boldsymbol{g}_i$	primitive vector		k	bulk modulus
g	proximity gap		k	force constant, spring constant
g	conductance, transconductance		k	coupling coefficient
$\boldsymbol{G}$	reciprocal lattice vector		k	extinction coefficien
G	elastic shear modulus		k_B	Boltzmann constant
G	Landauer conductance		k_{eff}	effective inverse decay length
G	heat conductance		K	anisotropy constant
G'	conductance per unit length		K	rate constant
$\boldsymbol{H}$	Hamilton operator		K_p	Preston's coefficient
$\boldsymbol{H}$	magnetic field		l	lateral dimension, length
H	enthalpy		l	SQUID parameter
H	scaling factor		l_B	Bjerrum length
H	free energy		L	length
H	heat capacity		L	diffusion length
H	luminous exposure		L	luminance
H	average information		L'	inductance per unit length
H^*	information flux		L_D	Debye length
H_{ac}	transformation energy		$\boldsymbol{m}$	magnetisation
H_C	critical magentic field strenght		m	mass, decay rate
I	intensity		m_e	electron mass
I	luminous intensity		MFS	Minimum Feature Size
I	information		mM	millimol/l

Symbols

MTF	Modulation Transfer Function	P_d	power dissipation of a logic gate
M_w	molecular mass	P_x, P_y, P_z	components of the polarization vector
$\mathbf{n}$	normal vector	$\mathbf{q}$	wave vector
n_A	Avrami exponent	$\mathbf{q}$	reciprocal lattice vector
n_i	index of refraction	q	charge
n_o	ordinary index of refraction	$\mathbf{Q}$	spanning vector in k-space
n_{ox}	refractive index of silicon dioxide	Q	selectivity
$\mathbf{N}$	demagnetising tensor of particles	Q	charge
N	number	Q	luminous energy
N	density	Q_0	unloaded quality factor
N	noise signal power	QE	quantum efficiency
NA	numerical aperture	Q_{ext}	external quality factor
N_{ss}	surface state density	Q_L	loaded quality factor
OD	operation density	$\mathfrak{R}$	responsivity
$\mathbf{p}$	effective polarization	$\mathfrak{R}$	reluctance
p	pyroelectric coefficient	r	reflection factor
p	dipole moment	r	rate of information transfer
p	pressure	R	electrical resistance, surface resistant, tunnel resistance
p_{PY}	pyroelectric coefficient	R	radius
$\mathbf{P}$	polarization	R	reflectance, reflection coefficient, reflectivity
$\mathbf{P}_0$	spontaneous polarization		
$\mathbf{P}_h$	propagation vector	R	curvature at the apex of the tip
$\mathbf{P}_{irr}$	irreversible polarization	R	sensor responsivity
$\mathbf{P}_m$	maximal polarization	R'	resistance per unit length
$\mathbf{P}_R$	remanent polarization	s	spacing
$\mathbf{P}_S$	spontaneous polarization	$s(t), s_s(t)$	signal functions
$\mathbf{P}_{tot}$	total polarization	$\mathbf{S}$	mechanical strain tensor, elastical deformation
P	power, power density		
P	dielectric probability	$\mathbf{S}$	Pointing vector
P	spin polarization	S	area, squareness
P	total light intensity	S	signal power
PD	density of dissipated power	S	sensivity, sensing signal output
PD	path difference	S	order parameter
		S	spin quantum number

Symbols

S	entropy		V_{BL}	bitline voltage
S	weighted sum		V_{CC}	operation voltage
$S(f)$	signal function		V_{DD}	supply voltage
S_0	spontaneous strain		V_{DD}	power supply voltage
S_{21}	transmission element of scattering matrix		V_{DS}	voltage between drain and source
S_D	voltage swing		V_{GS}	voltage between metal gate and source
S_i	strain tensor component		$V_{in/out}$	potential inside/outside of the cell
$\mathbf{t}$	scattering matrix		V_J	transductive extracellular potential in the junction
t	thickness		V_{JE}	voltage between junction and bulk electrolyte
$tan\delta$	loss tangent		V_{REF}	reference voltage
t_{cycle}	cycling time		V_S	potential of the substrate
t_{diff}	diffusion time		V_T, V_{th}	threshold potential, threshold voltage
t_p	thickness		V_x	potential for any ion x
t_r	carrier transit time		W	energy, width
t_s	switching time		W_b	bandgap, band gap energy
$\mathbf{T}$	stress tensor		W_d	energy dissipation per logic operation
T	transmittance, transmission coefficient		W_{DW}	domain wall energie
T	temperature		W_E	electric energy
T	transparency		W_F	Fermi energy, Fermi level
T	pulse length		W_g	bandgap, band gap energy
T^{-1}	sampling rate		W_M	elastic energy
T_c	critical temperature, phase transition temperature		W_S	surface energy
T_c	coherent transmission coefficient		W_{tot}	total energy
T_g	glass-transition temperature		W_{vac}	vacuum energy level
T_M	melting point		$[x]_{o/i}$	concentration ion
T_{NI}	nematic-isotropic phase transition temperature		X	area factor
U	displacement, generalized field		Y	admittance, conductance
v	velocity		Z	impedance
v	potential difference		Z	tip-sample distance
V	voltage		Z_0	characteristic impedance of a line
V	volume		α	attenuation coefficient
V_b	bias voltage		α	critical parameter

Symbols

α	scaling factor	ε_r	relative permittivity
α	radial coordinate	ε_{tot}	total permittivity
β	exponential factor	η	efficiency
β	complex propagation constant	η	viscosity
γ	specific interface energy, surface tension, surface energy	θ	coverage per surface site
γ	damping coefficient	θ	angle of propagation
γ	emission coefficient	θ	Curie temperature
γ	critical exponent	θ	threshold value
γ_0	nearest-neighbor overlap integral	θ^*	effective charge
Γ	gamma function	θ_B	Brewster Angle
Γ	resonance half width	κ	filling factor
Γ	tunneling rate	κ	absorption index
δ	displacement	κ	thermal conductivity
δ	index contrast	κ	exponential constant
δ	skin depth	κ_D^{-1}	Debye length
$\delta(t)$	delta function	κ_F, κ_B	rate of exothermic (F), endothermic (B) energy transfers
Δ	gap energy	κ_g, κ_h	radiative decay rate of triplets on the guest and host molecules
Δ	density of the solid	κ_{nr}	rate constant for non-radiative decay of an exciton
Δf_{3dB}	resonant halfwidth frequency	κ_r	rate constant for radiative decay of an exciton
Δl	displacement	λ	mean free path, mean free length
$\Delta \mathbf{p}$	change of the electric dipole moment	λ	scaling factor
$\Delta \mathbf{P}_{ns}$	non-switching polarization	λ	wavelength
$\Delta \mathbf{P}_s$	switching polarization	λ_L	London penetration depth
ΔQ_{ns}	non-switching charge	λ_{sheet}	length constant
ΔQ_s	switching charge	μ	carrier mobility, hole mobility
$\Delta \mathbf{v}$	change of the elastic dipole moment	μ	reduced mass
ΔW	latent heat	μ	electrochemical potential
$\Delta \omega$	frequency shift	μ_{eff}	effective carrier mobility
ε	permittivity	ν	attempt frequency
ε_0	vacuum permittivity	ν	velocity
ε_∞	high frequency permittivity	ν_D	drift velocity
$\varepsilon_{extrinsic}$	extrinsic permittivity		
$\varepsilon_{intrinsic}$	intrinsic permittivity		

Symbols

ν_0	group velocity
ν_s	carrier velocity
π	out of plane orbital (C=C)
ρ	density
ρ_c	specific contact resistivity
σ	energy density
σ	in-plane molecular orbital (C-C)
Σ	surface charge density
τ	time constant, decay time, circuit delay time
τ^{-1}	scattering rate
τ_h	heat pulse duration
φ	phase
ϕ	scattering angle
ϕ	phase shift
ϕ	work function
ϕ	luminous flux
χ	susceptibility
χ	electron affinity
χ	volume fraction
Ψ	wave function
ω	angular frequency
ω_0	resonance frequency
Ω	number of possible configurations

Authors

Joerg Appenzeller (Chapter 19)
Research Staff Member
IBM Research Division
T.J. Watson Research Center, Route 134
Yorktown Heights, NY 10598, USA
joerga@us.ibm.com

Peter Atkins (Chapter 5)
Lincoln College
Oxford University
Oxford OX1 3DR, Great Britain
peter.atkins@lincoln.oxford.ac.uk

Peter K. Bachmann (Chapter 39)
Philips Research Laboratories
Weisshausstrasse 2
D-52066 Aachen, Germany
peter.bachmann@philips.com

Arnd Baumann (Chapter 6)
Institut für Biologische Informationsverarbeitung I
Forschungszentrum Jülich GmbH
D-52425 Jülich, Germany
a.baumann@fz-juelich.de

Helmut Bechtel (Chapter 39)
Philips Research Laboratories
Weisshausstrasse 2
D-52066 Aachen, Germany
helmut.bechtel@philips.com

Howard R. Beratan (Chapter 34)
Raytheon Commercial Infrared
13532 N. Central Expwy M/S 37
Dallas, Texas 75243, USA
h-beratan@raytheon.com

Thomas Bieringer (Chapter 27)
Bayer Polymers, IIS Innovation
Physics and Characterization
Building B 406
D-51368 Leverkusen, Germany
thomas.bieringer.tb@bayer-ag.de

Gerd K. Binnig (Chapter 28)
IBM Zürich Research Laboratory
Säumerstr. 4
CH-8803 Rüschlikon, Switzerland

Ulrich Böttger (Chapter 22)
Institut für Werkstoffe der Elektrotechnik II
RWTH Aachen
52056 Aachen, Germany
boettger@iwe.rwth-aachen.de

Werner Brockherde (Chapter 33)
Fraunhofer Institute of
Microelectronic Circuits and Systems
Finkenstraße 61
D-47057 Duisburg, Germany
werner.brockherde@ims.fraunhofer.de

Daniel E. Bürgler (Chapter 4)
Institut für Festkörperforschung
Forschungszentrum Jülich GmbH
D-52425 Jülich, Germany
d.buergler@fz-juelich.de

Christoph Buchal (Chapter 30)
Institut für Schichten und Grenzflächen
Forschungszentrum Jülich GmbH
D-52425 Jülich, Germany
c.buchal@fz-juelich.de

Mark DeHerrera (Chapter 23)
Motorola Semiconductor Products Sector,
Embedded Memory Center
7700 South River Parkway
Tempe, Arizona 85284, USA
mark.d@motorola.com

Michael Despont (Chapter 28)
IBM Zürich Research Laboratory
Säumerstr. 4
CH-8803 Rüschlikon, Switzerland
dpt@zurich.ibm.com

Andreas Dietzel (Chapter 24)
Micro- and Nano- Scale Engineering
TU Eindhoven
Den Dolech 2, Postbus 513
5600 MB Eindhoven, Netherlands
A.H.Dietzel@tue.nl

Michael Dolle (Chapter 7)
Infineon Technologies AG
Balanstr. 73
D-81541 München, Germany
michael.dolle@infineon.com

Urs Dürig (Chapter 28)
IBM Zürich Research Laboratory
Säumerstr. 4
CH-8803 Rüschlikon, Switzerland
drg@zurich.ibm.com

Hermann Dürr (Chapter 23)
BESSY
Albert-Einstein-Str. 15
D-12489 Berlin, Germany
hermann.duerr@bessy.de

Philipp Ebert (Chapter 12)
Institut für Festkörperforschung
Forschungszentrum Jülich GmbH
D-52425 Jülich, Germany
p.ebert@fz-juelich.de

Authors

Peter Ehrhart (Chapter 8)
Institut für Festkörperforschung
Forschungszentrum Jülich GmbH
D-52425 Jülich, Germany
p.ehrhart@fz-juelich.de

Stephen R. Forrest (Chapter 38)
Department of Electrical Engineering
and Princeton Materials Institute
EQUAD B-301
Princeton University
Princeton NJ 08544, USA
forrest@ee.princeton.edu

Arno Förster (Chapter 15)
Institut für Schichten und Grenzflächen
Forschungszentrum Jülich GmbH
D-52425 Jülich, Germany
a.foerster@fz-juelich.de

Peter Fromherz (Chapter 32)
Department of Membrane and Neurophysics
Max Planck Institute for Biochemistry
Am Klopferspitz 18a
D-82152 Martinsried, Germany
fromherz@biochem.mpg.de

Peter Griffin (Chapter 13)
CIS-X 301, Mail Code 4075
Stanford Nanofabrication Facility
Paul Allen Center for Integrated Systems Building
Stanford University, Stanford, CA 94305, USA
griffin@stanford.edu

Peter A. Grünberg (Chapter 4)
Institut für Festkörperforschung
Forschungszentrum Jülich GmbH
D-52425 Jülich, Germany
p.gruenberg@fz-juelich.de

Susanne Hoffmann-Eifert (Chapter 1)
Institut für Festkörperforschung
Forschungszentrum Jülich GmbH
D-52425 Jülich, Germany
su.hoffmann@fz-juelich.de

Bedrich J. Hosticka (Chapter 33)
Fraunhofer Institute of Microelectronic Circuits and Systems
Finkenstraße 61
D-47057 Duisburg, Germany
bedrich.hosticka@ims.fraunhofer.de

Wolfgang Hönlein (Chapter 19)
Infineon Technologies AG
Corporate Research CPR NP
Otto-Hahn-Ring 6
D-81739 München, Germany
wolfgang.hoenlein@infineon.com

Mirco Imlau (Chapter 27)
Physics Department
University of Osnabrück
Barbarastraße 7
D-49069 Osnabrück, Germany
mimlau@uos.de

Hiroshi Ishiwara (Chapter 14)
Frontier Collaborative Research Center
Tokyo Institute of Technology
4259 Nagatsuda, Midoriku, Yokohama 226-8503, Japan
ishiwara@pi.titech.ac.jp

Ralf Jeremias (Chapter 33)
Fraunhofer Institute of
Microelectronic Circuits and Systems
Finkenstraße 61
D-47057 Duisburg, Germany
ralf.jeremias@ims.fraunhofer.de

Ernesto Joselevich (Chapter 19)
Dept. of Materials and Interfaces
Weizmann Institute of Science
Rehovot, 76100, Israel
ernesto.joselevich@weizmann.ac.il

Stefan Jung (Chapter 40)
Emerging Technologies
Corporate Research
Infineon Technologies AG
Otto-Hahn-Ring 6, D-81730 München, Germany
stefan.jung@infineon.com

U. Benjamin Kaupp (Chapter 6)
Institut für Biologische Informationsverarbeitung I
Forschungszentrum Jülich GmbH
D-52425 Jülich, Germany
a.eckert@fz-juelich.de

Joel Kent (Chapter 36)
Elo TouchSystems, Inc.
6500 Kaiser Drive
Fremont, CA 94555-3613, USA
jkent@elotouch.com

Angus I. Kingon (Chapter 21)
Department of Material Science and Engineering
North Carolina State University
Raleigh, NC 27695-7919, USA
angus_kingon@ncsu.edu

Norbert Klein (Chapter 31)
Institut für Schichten und Grenzflächen
Forschungszentrum Jülich GmbH
D-52425 Jülich, Germany
n.klein@fz-juelich.de

Claus-Dieter Kohl (Chapter 35)
Physics Department
Justus Liebig University Giessen
Heinrich-Buff-Ring 16
D-35390 Giessen, Germany
kohl@ap.physik.uni-giessen.de

Authors

Hermann Kohlstedt (Chapter 14)
Institut für Festkörperforschung
Forschungszentrum Jülich GmbH
D-52425 Jülich, Germany
h.h.kohlstedt@fz-juelich.de

Mark A. Lantz (Chapter 28)
IBM Zürich Research Laboratory
Säumerstr. 4
CH-8803 Rüschlikon, Switzerland
mla@zurich.ibm.com

Hans Lüth (Chapter 3)
Institut für Schichten und Grenzflächen
Forschungszentrum Jülich GmbH
D-52425 Jülich, Germany
h.lueth@fz-juelich.de

Katsuyuki Machida (Chapter 36)
NTT Telecommunications Energy Labs.
3-1 Wakamiya Morinosato, Atsugi-shi, Kanagawa
243-0198, Japan
machi@aecl.ntt.co.jp

Koichi Maezawa (Chapter 15)
Graduate School of Engineering
Nagoya University Furo-cho, Chikusa-ku
Nagoya 464-8603, Japan
maezawa@nuee.nagoya-u.ac.jp

Siegfried Mantl (Chapter 13)
Institut für Schichten und Grenzflächen
Forschungszentrum Jülich GmbH
D-52425 Jülich, Germany
s.mantl@fz-juelich.de

Marcel Mayor (Chapter 20)
Institut für Nanotechnologie
Forschungszentrum Karlsruhe
D-76021 Karlsruhe, Germany
marcel.mayor@int.fzk.de

Simon McClatchie (Chapter 10)
Lam Research
4650 Cushing Parkway Mailstop CA4
Fremont CA 64538-6470, USA
simon.mcclatchie@lamrc.com

Jürgen Moers (Chapter 9)
Institut für Schichten und Grenzflächen
Forschungszentrum Jülich GmbH
D-52425 Jülich, Germany
j.moers@fz-juelich.de

Wilfried Mokwa (Chapter 29)
Institut für Werkstoffe der Elektrotechnik I
RWTH Aachen
D-52056 Aachen, Germany
mokwa@iwe.rwth-aachen.de

Paul Muralt (Chapter 34)
Ceramics Laboratory
Swiss Federal Institute of Technology EPFL
CH-1015 Lausanne, Switzerland
paul.muralt@epfl.ch

Serguey G. Odoulov (Chapter 27)
Institute of Physics
National Academy of Sciences
46, Science Ave
03 650 Kiev-39, Ukraine
odoulov@iop.kiev.ua

Shinji Okazaki (Chapter 9)
Association of Super-Advanced Electronics
Technologies(ASET)
EUV Process Technology Research Department
2-45 Aomi Kouto-ku Tokyo, 135-8073 Japan
okazaki@aset.tokyoinfo.or.jp

Martin Pfeiffer (Chapter 38)
Institute for Applied Photophysics (IAPP)
University of Technology Dresden
D-01069 Dresden, Germany
pfeiffer@iapp.de

Dieter Richter (Chapter 2)
Institut für Festkörperforschung
Forschungszentrum Jülich GmbH
D-52425 Jülich, Germany
d.richter@fz-juelich.de

Andreas Roelofs (Chapter 12)
Seagate Technology
1251 Waterfront Place
Pittsburgh, PA 15222, USA
Andreas.K.Roelofs@seagate.com

Klaus Röll (Chapter 25)
Experimental Physics Group IV
Kassel University
Heinrich-Plett-Str. 40
D-34132 Kassel, Germany
roell@physik.uni-kassel.de

Hugo E. Rothuizen (Chapter 28)
IBM Zürich Research Laboratory
Säumerstr. 4
CH-8803 Rüschlikon, Switzerland
rth@zurich.ibm.com

Stefan Schneider (Chapter 10)
Unaxis Balzers AG
P.O. Box 1000
FL-9496 Balzers, Liechtenstein
Stefan.Schneider@unaxis.com

Herbert Schroeder (Chapter 21)
Institut für Festkörperforschung
Forschungszentrum Jülich GmbH
D-52425 Jülich, Germany
he.schroeder@fz-juelich.de

Authors

Oliver H. Seeck (Chapter 11)
Hasylab am DESY
D-22603 Hamburg, Germany
Institut für Festkörperforschung,
Forschungszentrum Jülich GmbH
D-52425 Jülich, Germany
ohseeck@mail.desy.de

Michael Siegel (Chapter 17)
Institute of Micro- and Nanoelectronic Systems
University of Karlsruhe
Hertzstr. 16
76131 Karlsruhe, Germany
m.siegel@iegi01.etec.uni-karlsruhe.de

Jon M. Slaughter (Chapter 23)
Motorola Labs, Physical Sciences Research Labs
7700 South River Parkway
Tempe, Arizona 85284, USA
jon.slaughter@motorola.com

Gerd Spekowius (Chapter 39)
Philips Research Laboratories
Weisshausstrasse 2
D-52066 Aachen, Germany
gerd.spekowius@philips.com

Scott Summerfelt (Chapter 22)
Texas Instruments
Si Technology Development
13560 North Central Express, M/S 3736
Dallas, Texas 75243, USA, 972-995-2389
P.O. Box 650311, M/S 3736
Dallas, Texas 75265, USA
s-summerfelt@ti.com

Kristof Szot (Chapter 12)
Institut für Festkörperforschung
Forschungszentrum Jülich GmbH
D-52425 Jülich, Germany
k.szot@fz-juelich.de

Dietmar Theis (Chapter 40)
Siemens AG
Corporate Technology
CT SM CM
Otto-Hahn-Ring 6
D-81730 München
dietmar.theis@siemens.com

Susan Trolier-McKinstry (Chapter 2)
Department of Materials Science and Engineering
Pennsylvania State University
151 Materials Research Laboratory
University Park, PA 16802, USA
stmckinstry@psu.edu

Ken Uchida (Chapter 16)
Advanced LSI Technology
Laboratory Toshiba Corporation
8 Shinsugita-cho, Isogo-ku, Yokohama 235-8522, Japan
ken1.uchida@toshiba.co.jp

Alexey Ustinov (Chapter 18)
Institute of Physics III
University Erlangen-Nürnberg
Erwin-Rommel-Str. 1
D-91058 Erlangen, Germany
ustinov@physik.uni-erlangen.de

Peter Vettiger (Chapter 28)
IBM Zürich Research Laboratory
Säumerstr. 4
CH-8803 Rüschlikon, Switzerland
pv@zurich.ibm.com

Rainer Waser
Institut für Festkörperforschung
Forschungszentrum Jülich GmbH
D-52425 Jülich, Germany;
Institut für Werkstoffe der Elektronik II
RWTH Aachen
D-52056 Aachen
r.waser@fz-juelich.de
waser@iwe.rwth-aachen.de

Heiko Weber (Chapter 20)
Institut für Nanotechnologie
Forschungszentrum Karlsruhe
D-76021 Karlsruhe, Germany
heiko.weber@int.fzk.de

Theo Woike (Chapter 27)
Institute for Mineralogy and Geochemistry
University of Cologne
Zülpicher Str. 49b
D-50674 Köln, Germany
th.woike@uni-koeln.de

Shin-Tson Wu (Chapter 37)
University of Central Florida
School of Optics - CREOL
4000 Central Florida Boulevard
PO Box 162700
Orlando, Florida 32816-2700, USA
swu@mail.ucf.edu

Matthias Wuttig (Chapter 26)
I. Physikalisches Institut
RWTH Aachen
D-52056 Aachen, Germany
wuttig@physik.rwth-aachen.de

Edward W. A. Young (Chapter 13)
International SEMATECH (ISMT) assignee,
Philips Semiconductors, p/a IMEC
Kapeldreef 75
B-3001 Leuven, Belgium
edward.young@philips.com

Reiner Zorn (Chapter 37)
Institut für Festkörperforschung
Forschungszentrum Jülich GmbH
D-52425 Jülich, Germany
r.zorn@fz-juelich.de

Index

A

absorption 281, 735
ac Josephson effect 446
acceptor 921
access transistor 540
acoustic
    phonon 42
    wave device 851
acousto-optic 749
AC-surface discharge 937
action potential 147
activation energy 648
active matrix 528, 923
    display 900
adder 174
additive color mixing 882
address electrode 940
addressing 936, 939
adhesion layer 543
AFM thermomechanical recording 689
agglomeration 957
air-bearing surface 618, 625
algebra of sets 19
alkaline earth titanate 51
alkanes 129
alkoxide 210
alkynes 130
All-CNT nanoelectronics 494
alphabet 15
Alt-Pleshko scheme 897
amino acid 137
amorphous layer 913
amperometric cell 850
analog-to-digital converter (ADC) 419
analysis methods
    overview 195
anisotropic
    magnetoresistance (AMR) 118, 591, 623
    media 46
anisotropy 249, 891
anode voltage 936
antiferroelectric order 64
antiferromagnet 112, 597, 602
antiferromagnetic
    layer 592
aperture ratio 924
arctangent transition 624
areal density 617
arenes 130
arithmetic-logical unit (ALU) 178
aroma 854
aromatic hydrocarbons 130
artificial atom 429

ashing 251
aspect ratio dependent etch rate (ARDE) 260
aspherical surface 233
atomic
    layer deposition (ALD) 213, 373
    layer epitaxy 213
    polarizability 35
    resolution 303
attenuated PSM 227
attenuation 705
Auger
    electron 286
    electron spectroscopy (AES) 274, 289
automata 20
avalanche photodiode 744
axon 147, 148, 158
azobenzene 676

B

ballistic transport 476
band
    gap 739
    pass transmission 709
    structure 281
    width 737
barium strontium titanate 540
barium titanate 61, 63, 69
barrel
    reactor 253
    shifter 174
barrier 121
base band transmission 709
basic binary operation 343
$BaTiO_3$ 747
BCS theory 103
benchmark program 184, 345
bending magnet 275
BGA 723
bias voltage 253
bilinear coupling 115
binary
    addressing 940
    addressing scheme 940
binocular disparity 883
biometric 869
biquadratic coupling 115
birefringence 64, 678, 679, 892
bismuth layer structure 70
bistabile molecular structure 506
bistability 956
bit error rate (BER) 667
bitline 568

Index

capacitance 568, 572
capacity 533
Bloch wave 85
Boolean algebra 18
boosted voltage 568
Born approximation 275
Bosch Process 256
Bose-Chaudhuri-Hocquenghem-block code (BCH) 709
Bragg
    condition 277, 278
    mismatch 664
    peak 277
    reflection 284
    reflector 232, 741
bright field image 292
brightness 880
Brillouin zone 85
bucky
    ball 493
    shuttle device 494
buffer layer 387
bulk
    micromachining 840
    molecular system 502
bump 721
bus system 170, 177

C

capacitance 393
capacitive
    coupling 706
    touchscreen 864
carbon
    based electron emitter 931
carbon nanotube 473, 481, 530, 931
    circuit 492
    device 488
    field effect transistor (CNTFET) 487
    interconnect 485
    vias 486
carrier 15
catenane 507
cathode ray tubes (CRTs) 879
CD/DVD-Recordable (CD-R, DVD-R) 612
celebral cortex 349
cell
    body 147, 160
    plate 568
    plate line, local 571
cell plate
    line, global 571
cellular communication 756
central aperture detection (CAD) 639
channel capacity 707
characteristic
    impedance 706
    X-ray 285
charge
    based RAMs 531

carrier transport 916
coupled device (CCD) 811
injection 575, 920
polarization 36
qubit 464
separation 582
transport mechanism 670
chemical
    driving force 152, 154
    etching 249
    mechanical planarization (CMP) 716
    mechanical polishing (CMP) 249
    solution deposition (CSD) 208
    synthesis 138
    vapor deposition (CVD) 208, 371
chemically amplified resist (CAR) 240
chiral 891
    center 138
    nanotube 474
    vector 474
cholesteric 891
    liquid crystal material 962
CIE chromaticity diagram 882
circuit
    architecture 594
    imulation 175
circularly polarised wave 54
Clausius-Mossotti equation 35
clock frequency 445
CMOS 400, 401
    design phase 191
    device 445
    fabrication phase 191
    process 576
    technology overwiev 190
code 15
coercive
    field, frequency dependence 76, 575
    field, size effect 575
    voltage 573
coherent rotation 602
cold cathode fluorescent lamp 901
cold welding 924
Cole-Cole diagram 39
collision cascade 207
colloidal suspension 956
colorimetry 882
colour gamut 945
combinational logic circuits 19
common mode voltage 569
commutativity 18
compact disc (CD) 610
concatenability 324
conductive elastomer 869
conductivity 34
    quanta 100
conductor 34
cones 812
confinement of carrier 98
conformal deposition 213

Index

constitutive equation 45
contact
    lithography 224
    mode 308
contrast 286
controlled
    deposition 484
    growth 484
controlled NOT 21
conversion
    analog-to-digital (A/D) 704
Cooper pair 103, 445
copper interconnection 716
Coulomb
    blockade 426
    blockade device 332
    integral 132
    oscillation 428
critical
    dimension 255
    nucleus 648
    spanning vector 117
cross-contamination 580
cross-talk 963
crystallization rate 647
Curie temperature 62, 65
Curie-von Schweidler behavior 41, 574
Curie-Weiss law 51, 66
current
    perpendicular-to-plane 598
    response 573
    voltage characteristic 447
CVD growth 485
cycloaddition 141
cycloalkanes 129

D

damped
    oscillation 38
    oscillator 48
    oscillator model 48
damping 38
dark field image 292
data storage 387
dc sputtering 207
De Morgan's law 169
De Morgan's rules 19
dead layer 546
Debye relaxation 39
decoherence 468
    time 463
defect 917
    resonator 759
    trap 581
delta-sigma converter 419
demultiplexer 173
dendrites 147, 148, 160
density
    device 351
    of state 98
    operation 351
depolarization 148, 150
depolarizing field 582
design of experiment (DoE) 261
destructive read-out (DRO) 448
detectivity 835
devices 863
dexterous end effector 868
D-FF 418
D-flip-flop 171, 451
dielectric 34
    CMP 263
    coefficient 74
    constant 34
    displacement 34
    filter 770
    layer 940
    properties 543
    resonator 769
different carrier frequencies 711
diffraction mode 283
diffusion barrier 543
digit line 593
digital
    audion type (DAT) 609
    signal processor (DSP) 181, 339
    to-analog converter 455
    versatile disc 645
    video disc 645
dipole moment 35
dishing 267
disk media 125
dispersion 733
    equation 43
dispersive medium 46
display 923
    active matrix 883
    backplane 963
    field emission (FED) 929, 936
    flat panel 925
    flexible OLED 925
    guest-host 962
    head mounted (HMD) 884
    light controlling 879
    light generating 879
    microparticle-based 956
    passive matrix 883
    polymer dispersed liquid crystal 962
    projection 905
    spatially immerse (SID) 885
    TFT 900
    virtual model (VMD) 885
    virtual retinal (VRD) 884
distortion 705
distributivity 18
DNA computer 338
domain
    states 71
    wall 71

Index

    wall movement 74
doping 921
DOTFET 194
drift velocity 93
driveline 568
dry etching 249
DUV 239
DWDM 737, 738
dynamic addressing 896
dynamic RAM (DRAM) 529

E

EELS 288
EEPROMs 529
effective mass 93
efficiency saturation 933, 935, 936
elastic
    collision 252
    energy 647
electric
    arc discharge 480
    dipole transition 945
electric dipole transition 945
electrical resistance 92
electrochemical cell 850
electrode 155, 543
    material 545
    metal 577
    noble metal 577
    oxide 577
electroluminescence 913
electromagnetic resonator 755
electron
    affinity 915, 921, 931, 944
    beam evaporator 204
    cyclotron resonance (ECR) 254
    density 276, 280
    energy loss spectroscopy (EELS) 274, 282, 287
    microscopy 282
    penetration depth 933, 934
    phonon interaction 916
    transmission probability 409
    transport layer 913
    tunneling 407
electronegativity 136
electronic
    band structure 84
    ink 956
    nose 854
    paper 956
    polarization 35
    surface state 95
    transport phenomena 109
electronically controlled birefringence 898
electro-optical effect 56
electrophiles 138
electrophilic
    addition 141
    aromatic substitution 140

electrophoresis 956
electrophoretic display 956
electrostatic scanning force microscopy 308
electrotonic potential 150
ellipsometry 293
elliptically polarized wave 54
embedded
    memories 565
encoding
    channel 709
    line 709
endocytosis 160
endothermic energy transfer 920
endpoint 258
energy
    barrier 596, 597
    density 936
    dispersive X-ray analysis (EDX) 274
    dissipation 447
    domain wall 72
    efficiency 932, 945
    elastic 72
    electric 72
    free 65, 67, 647
    surface 72
entropy 17, 67, 68
equation of motion 37
equilibrium potential 152
equivalent thickness 541
erbium-doped optical fiber amplifier (EDFA) 729, 745
erosion 267
Esaki diode 414
ESD 871
etch rate 249
EUV 232
exchange
    anisotropy 112
    bias 112, 600, 601
exciton 915, 917
    blocking layer 917
    diffusion length 917
exocytosis 160
exothermic energy 647
    transfer 920
extended X-ray absorption fine structure (EXAFS) 274
extinction rule 278

F

FED structure 929
feedback prevention 324
FeFET 388
Fermi
    energy 89
    statistic 89
    surface 90
ferroelectric 747, 768
    capacitance 568
    capacitors, planar 573
    clamping in thin films 74

Index

gate 387
hysteresis, shift 582
liquid crystal 906
liquid crystal material 962
on a conductive oxide 389
polycrystalline ensemble 73
RAM (FeRAM) 529
single crystal 71
switching 76
thin films properties 70
ferromagnetic layer 592, 601
ferromagnets 114
field-programmable
    device (FPD) 339
    gate array (FPGA) 175, 339, 511
figure of merit 834
FIL-chip 872
filter 755
fingerprint 870
    sensor 863, 869
finite impulse response (FIR) 182
firing rate code model 347
flash memories 529
flat cathode-ray tube 929
flavor 856
flip-chip bonding 721
flip-flop 171
floating-point operation per second 345
floppy disc 610
fluidic self-assembly (FSA) 964
fluorescence 281, 282, 915
    efficiency 913
fly height 618, 620, 625
folded
    bitline 571
    3-D structure 584
food 856
force-based touchscreen 864
Förster transfer 918
Fourier transformation 37, 41, 276
Fowler-Nordheim equation 930
Franck-Condon factor 915
Fredkin gate 22
free enthalpy 647
free-programmable system 338
Frenkel exciton 915
frequency
    cut-off 705
frontier orbital 914
full-adder 174
fullerene 135
    recrystallization 482
functional group 135
fuzzy-logic 20

G

gain guiding 740
gas
    chromatographic analysis 856
    discharge source 230
    FET 851
    kinetic 200
    sensor 849
gate
    dielectric 359, 362, 368, 372, 374
    leakage 366
    stack 367, 379
    structure 391
Ga_2O_3 853
GeO_2 736
GeSbTe 651
Ge-Te compound 651
$GeTe-Sb_2Te_3$ 651
giant magnetoresistance (GMR) 109
    effect 598
    sensor 617
Goldman-Hodgkin-Katz (GHK) equation 153
graphene 473
ground-state depletion model 935
growth mode 201
gyricon 960

H

Hadamard transformation 462
hafnium
    oxide 372
    silicate 373
half-adder 174
Hamaker constant 304
Hamilton operator 83
    squid 466
Hamming distance 709
hard disk drive 610, 617
hard layer 600
hardwired system 338
harmonic oscillator 43
head-gimbal assembly 618
heteroepitaxy 202
heterointerface 917
heterojunction 917
heterostructure 95, 740
higher permittivity material 541
highest occupied orbital 134, 914
high-k material 370, 373
high-pressure Hg-lamp 229
high-resolution TEM (HRTEM) 285
high-speed operation 571
high-temperature
    superconducting film 760
    superconductor 447
    superconductor technology 453
Hillock formation 577
hole
    blocking 917
    transporting layer (HTL) 913
holographic video 885
homoepitaxy 202
HOMO-LUMO gap 134

Index

hopping 916
host 918
HRTEM 285
Hückel MO method 133
human nose 858
hybrid molecular electronic 502
hydrocarbon 129
hydrogen
    bonded systems 69
    diffusion 579
hyperpolarization 150
hysteresis loop 62, 600

I

identification 865
III-V-semiconductor 747
illuminance 880
illumination wavelength 224
imaging mode 283
immersion lens 653
incubation time 652
index guiding 741
indicatrix (index ellipsoid) 747
indium tin oxide (ITO) 895, 913
inductive 620
    effect 136
    write head 617, 622
inductively coupled plasma (ICP) 254
inelastic
    collision 252
    mean free path 291
information
    pragmatic level 13
    semantic level 13
    syntactic level 13
information flux 707
infrared touchscreen 865
InGaAlP laser diode 653
ink jet printing 914
inorganic storage material 667
integrate-and-fire model 347
interconnection 715
interface 95
    anisotropy 111
    capacitance 546
    layer 575
    state 95
interference mode 899
interferometric modulator 963
interlaced mode 884
interlayer exchange coupling (IEC) 109
internal quantum efficiency 919
International Technology Roadmap for Semiconductors (ITRS) 532
intersystem crossing (ISC) 919
intramolecular gate 492
inverse photo electron spectroscopy (IPES) 274
inversion center 62
inverter 168

ion
    beam etching (IBE) 251
    energy distribution function (IEDF) 261
    milling 284
    projection lithography 237
    pump 151
ionic
    crystal 69
    permeability 151, 153
    polarization 36
ionization energy 915, 921
irreversible polarization process 75
isomers 129
isothermal phase changes 649
ITRS roadmap 556

J

Johnson noise 834
Josephson
    effect 445
    junction 445, 465, 468
    transmission line 450

K

Karlqvist head 619
Kerr effect 56
ketonate 210
known good dies 724
Knudsen cell 204
Kramers-Kronig integral relationship 39

L

Landauer formula 102
Langmuir-Blodgett 208
laser produced plasma source 230
latent heat 68
lateral manipulation 313
lattice oscillation 42
Laue function 277
lead titanate zirconate (PZT) 69, 70
LiftMode 308
lift-off 924
ligand-gated channel 160, 162
light
    emitting layer 913
    generation 940
    output 945
$LiNbO_3$ 747
linear
    combination of atomic orbital 131, 915
    polisher 264
linearly polarised wave 54
liquid
    crystal 890
    crystal display 889
    crystal on silicon 905
$LiTaO_3$ 747
lithium-niobate 664, 674

Index

load capacitance 327
loading 260
local
    area network (LAN) 703
    density of states (LDOS) 410
    electric field 35
localized state 916
logic
    irreversible 20
    reversible 20
London equation 102
longitudinal
    E-wave 47
    phonon 43
    recording 617, 619, 626
long-term
    depression 163
    potentiation 163
loss tangent 36, 765
LO-TO splitting 50
low decoherence 463
low energy electron
    diffraction (LEED) 292
    microscopy (LEEM) 274
lowest unoccupied orbital 134, 914
low-k dielectric 718
low-temperature superconductor 447, 452
luminance 880, 945
luminous
    energy 880
    exposure 880
    flux 880
    intensity 880
    transfer function 881
Lyddane-Sachs-Teller relation 50, 69

M

Mach-Zehnder waveguide modulator 748
magnetic
    anisotropy 110, 599
    cartridge archive 610
    core memory 527
    dipole transition 945
    field modulation (MFM) 634
    flux 618
    flux quantization 448
    flux qubit 464
    potential 619
    random access memory (MRAM) 125
    recording 617
    scanning force microscopy 307
    stability 596, 598, 599
    transition 620
    tunnel junction (MTJ) 591
magnetization reversal 596, 600, 602
magneto
    crystalline anisotropy 110
    electronics 109
    optical disc 634

    optical Kerr effect (MOKE) 599, 601, 602, 633
    resistance (MR) ratio 592
    resistive (MR) sensor 617
    transport 125
mass storage device (MSD) 607
Mauguin limit 896
maximum clock frequency 327
Maxwell
    equation 44
    stress 309
    Wagner polarization 36
mean field theory 65
mechanically controlled break junction 512
mechanoreceptors 815
media
    exchangeable 609
    pre-written 608
    read/write 608
    write-once 608
medium 15
Meissner-Ochsenfeld effect 105
melt quenched amorphous bits 652
membrane potential 155, 158
memory
    capacitor 540
    cell 539, 593
    effect 939
    non-volatile 655
MEMS 869
mesomeric effect 136
metal 88
    CMP 263
    ferroelectric insulator semiconductor 389
    ferroelectric metal insulator semiconductor 389
    ferroelectric semiconductor 388
    gate 359, 360, 375
    induced gap state 96
    organic decomposition 214
    organic-CVD 209
    to-ligand charge transfer 919
metallization 579
microcalorimetric gas sensor 850
microcontroller 177
microelectrode 155
microelectromechanical system 747
microencapsulation 957
micromechanical switches 755
microprocessor 177
microraman spectroscopy 293
microstripline 760
microwave
    communication 761
    dielectric 764
    properties 755
Miller indices 277
Millipede 687, 688
minimum feature size 192, 556
misfit
    dislocation 203
    strain 73

Index

mode
- longitudinal 68
- softening 50, 68
- transverse 68

modulation 709
- differential pulse code (DPCM) 708
- pulse code (PCM) 708
- transfer function 226

modulator 746

molecular
- beam epitaxy 203
- crystal 914
- orbital 131
- wire 504

monomode waveguide 749
monomolecular electronic 502
monostable-bistable transition logic element 415
MOS capacitor 359, 373, 374
MOSFET 167, 363, 365
motion pictures expert group (MPEG) 708
MRAM 591
multi-chip-module 716
multilayer mirror 232
multi-level metallization 716
multiplexer 172

multiplexing
- addressing 883
- angular and wavelength 664
- holographic 680

multisensor array 858
multi-threading architecture 185
multi-wall nanotube 473
myelination 158
M#-Number 668

N

NAND gate 169, 448

nano
- dot memory 529
- oxide layer 120

nanodrive 697
nanomechanical three-terminal device 508
nanopore concept 512

nanotechnology
- bottom-up approach 192
- hybrid approach 192
- top-down approach 192

nanotubes for memory applications 493
narrow gap semiconductors 69
native MIPS 345
near-field optical microscopy (NSOM or SNOM) 653
Néel type - coupling 114
negative differential resistance 506

nematic
- liquid crystal material 962
- phase 891

Nernst equation 152, 154
neuron 147, 154, 155
neurotransmitter 148, 159, 160

new high permittivity material 540
n-MOS enhancement transistor 167
noise 705
noise equivalent power 835
non-contact mode 303
non-destructive read 448, 655
non-deterministic polynomial problem 337
non-equilibrium exchange interaction (NEXI) 109

non-linear
- characteristic 322
- dielectric 46, 543

non-radiative decay 917
non-switched charge 566
non-volatile 387, 591, 597
NOR gate 169
novolac/diazonaphthoquinones (DNQ) 239
nuclear magnetic resonance (NMR) 462
nucleation rate 648
nucleophiles 138

nucleophilic
- addition 141
- substitution 140

numerical aperture 225
Nyquist criterion 705

O

odorant receptor 814
off-axis illumination 228
Ohm's law 47

olfactory
- neuron 148, 149
- sensory neuron 814

olloidal stability 957
ON dielectric 541
open-bitline 571

optical
- amplifier 745
- diffraction 637
- electronic network switch 456
- phonon 42
- transition 944
- waveguide 733

orange peel type coupling 114
order parameter 65

organic
- light-emitting devices (OLEDs) 913
- vapor phase deposition (OVPD) 914

orientation polarization 36
orthogonal conductor 593
oscillation 38
overetch 250
overlap integral 132
oxynitride 367

P

Pacinian corpuscle 815
parallel plate capacitor 34
parallel processing 869
parameter 624, 625

Index

parasitic path 695
Paschen curve 942
passive
   device 769
   matrix 528, 923
   matrix memory 583
patch-clamp technique 156, 157
pattern recognition 854
π-bond 130
PCRAM 654
PEEM 600, 601, 602
pellistor 850
penetration depth 944
Penning ionisation 938
pentagon defect 482
penumbral blur 234
performance 345
pericyclic reaction 141
permalloy 618, 620
permanent dipole 582
permeability 151, 153, 154, 619, 620
permittivity 34, 35, 36
perovskite crystal structure 543
phase
   boundary energy 647
   change material 530
   change recording 645
   diagram 200
   diagram for barium titanate thin film 70
   shifting technique 226
   transition 61
   transition temperature 62, 65
phased array 738
phonon 287
   scattering 477
phosphor 932
   efficiency 934
phosphorescence 916
photo
   active compound 239
   addressable polymer 676
   chemistry 676, 677
   detector 743
   diode 821, 824
   electron spectroscopy 922
   emission electron microscopy (PEEM) 274, 599
   lithography 924
   receptors 812
   refractive crystal 675
   refractive effect 670
   voltaic cell 913
photoelectrical transduction process 813
photoelectron 290
photonic
   bandgap structure 755
   network 729
physical etching 249
piezoelectric
   coefficient 65, 70
   effect 64

response force microscopy (PFM) 309
pin photodiode 744
pipelining 179
pixel 821, 923
   parallel image processing 871
pixelated foil 959
π-junction 467
plasma
   density 252
   discharge 207
   display panels (PDPs) 929
   firing 943
   membrane 147, 160
   potential 253
plasmon 287
plateline 568
plug 551, 578
p-MOS enhancement transistor 168
Pockels effect 56
Poisson equation 34
polar
   optical phonon 44
   reaction 138
polariton 44
polarizability 38
polarization 34, 389
   electronic 62
   ionic 62
   self-polarized 74
   spontaneous 62
polishing pad 264
polycyclic aromatic molecules 130
polyenes 130
polymer 677, 914
   dispersed liquid crystal 908
   storage 689
polysilicon thin film transistor 923
post-exposure bake (PEB) 240
postsynaptic cell 148, 159
potentiometric cell 850
power
   amplification 322
   consumption 594
   dissipation 327
   dissipation limit 351
   efficiency 916
Poynting vector 53
p-polarization 55
Preston's law 263
presynaptic terminal 147, 160, 163
printable integrated circuit 963
programmable
   logic device (PLD) 175
   quantum interferometer 462
projection lithography 224
propagation
   delay time 325
   velocity 706
protective layer 940
protein 137

Index

proximity lithography 224
    X-ray (PXL) 234
pseudomorphic layer growth 278
PTE-diagram 652
pulsed laser deposition 206
punched card 609
pyroelectric coefficient 64
pyroelectricity 831

Q

quantum
    algorithm 463
    bit 461
    cellular automata 335
    coherent tunneling 464
    computing 337
    computing, criteria 463
    detector 831
    dot 97, 426, 429
    efficiency 945
    Fourier transformation 462
    gate 463
    measurement 463
    transport 100, 407
    well state 116
quaternary compound 739
qubit 461

R

radiative decay 917
radical
    addition 141
    reaction 138
random access memory 528, 654
    capacity 533
    device 527
    non-volatile 529
rapid single-flux quantum (RSFQ) 468
rare earth (RE) 635
Rayleigh
    criterion 225
    wave 867
RC delay 717
RC time 413
reactive ion
    beam etching (RIBE) 252, 579
    etching (RIE) 579
read
    after-pulse scheme 569
    failure 582
    heads 125
    only memory (ROM) 528
    only memory, programmable 528
    operation 595
    out scheme 531
    /write access times 591, 602
reading process 593
receptor potential 148, 158
reciprocal lattice 84

reconfigurable system 338
recrystallization 647
redundancy 18
reflection 913
    coefficient 100
    factor 706
    of light 55
reflective liquid crystal display 902
reflectivity 280
refraction of light 55
refractive index 34, 45, 46
refresh time 543
relative MIPS 345
relaxation 37
    current density 40
    phenomena 39
    process 38
    step 39
    time 39
relaxed layer growth 278
reluctance 619
remanent polarization 573
    loss 582
repolarization 155, 163
requirement catalogue 543
residual resistance 95
resistance-based RAM 532
resistive
    oxygen sensor 852
    semiconductor gas sensor 852
    touchscreen 864
resolution 227, 273, 283
resonance 37
    angular frequency 38
    frequency 43
    integral 132
    phenomena 37
resonant
    state lifetime 413
    tunneling 407
    tunneling transistor 414
response
    of human eye 881
    photopic 881
    scotopic 881
resting membrane potential 147, 155
retardation 50
retina 812
    implant 813
reversible polarization process 75
rhodopsin 813
roasting process 857
robot tactile sensor 868
rods 812
rotary polisher 264
RS-flip-flop 171
RSJ model 446
RTD/HEMT integration 416
Rutherford Backscattering Spectrometry (RBS) 274, 293

Index

S

sampling time 705
satellite communication 757
SAW 867
scaling
- rule 360, 365
- trend 532

scanning
- electron microscopy (SEM) 274, 282, 285
- near-field optical microscope (SNOM) 297
- probe microscopy (SPM) 297
- tunneling microscope (STM) 298

scattering at impurities 476
Schottky
- contact 96
- diode 744
- emitter 283

Schrödinger equation 83, 407
scooter mechanism 482
secondary
- electron 285
- electron emission coefficient 940, 943
- ion mass spectrometry (SIMS) 274, 289

selection rule 278
selectivity 250, 849
self-aligned contact (SAC) 257
self-assembly process 192
- chemically controlled 193
- physically controlled 194

self-organisation process 192
SEM 283
semiconductor 88
- integrated circuit 445
- laser 739

sense amplifier 568
sensor 125
- active 809
- array 811, 856
- chemical array 812
- modulating 809
- passive 809
- self-generating 809

sequential
- access device 527
- logic circuits 20

sequential access device 527
SFQ pulse 451
SFS junction 467
shadow-masking 924
shape anisotropy 110
sheet
- carrier density 392
- resistance 654

signal 15
- analog 704
- digital 704
- propagation in nanotubes 486
- transmission 704

signaling 147
silicon dioxide 360, 367, 368, 378

SIMS 290
single
- bit operation 462
- edge triggered flip-flop 172
- flux-quantum circuit 449
- flux-quantum logic 449
- mode fibers 736
- molecular system 502
- wall nanotube 473

single-electron
- box 426
- device 425, 431
- pump 431
- transistor 428, 430, 465
- turnstile 431

singlet exciton 916
SiO_2 736
Si-perovskite interface 388
skeletal rearrangement 143
slider 618, 625
slurry 262
small angle electron diffraction (SAED) 284
small molecule 914
smart card 566
Snell's law 55
SnO_2 853
soft
- layer 600
- mode 51
- phonon 68

software-defined radio 456
SOI (silicon on insulator) 375
SON (silicon on nothing) 379
source encoding 708
space
- charge limit 920
- domain 707

specific heat 67
speed of light 45
spike code model 347
spin
- accumulation 124
- polarization 121
- polarized current 335
- polarized EELS (SPEELS) 282
- transistor 334
- valve 112

spindependent
- reflectivity 116
- tunneling 121

Spindt-tip type FED structure 930
spintronic 333
s-polarization 55
spontaneous polarization 61
- temperature dependence 64

spring constant 304
sputter deposition 206, 634
sputtering 249, 251
- magnetron 207
- process 207

Index

RF 207
SRAM cell 170
stacked
    capacitor 541
    layer 725
    OLED display 924
stand-alone memories 565
standard observer 882
static RAM (SRAM) 529
stereoisomer 138
stereoscopy 883
Stewart-McCumber parameter 447
Stokes shift 915
storage
    capacity 528
    density 653
    medium 634
    requirement 649
strain relaxation 203
strained silicon 381
stress
    compressive 73
    tensile 73
strontium bismuth tantalate (SBT) 70
strontium titanate 51
structure
    factor 276
    orthorhombic 63
    rhombohedral 63
    tetragonal 63
subtractive color mixing 882
superconducting
    film 762
    quantum interference device (SQUID) 447, 464
superconductivity 102
superconductor 464
superheating 652
superscalar architecture 185
supertwisted nematic cell 899
surface
    acoustic wave (SAW) 738, 749
    discharge 940
    impedance 762
    micromachining 840
susceptibility 35, 36
suspended
    gate 852
    particle display 961
sustain electrode 938, 940
switched charge 566
switches 746, 771
switching
    algebra 19
    elements 19
    field distribution 597
    probability 596
    time 447
    time of ferroelectric 573
    variables 19
synaptic
    cleft 148, 159
    depression 163
    potential 148
    vesicle 148, 160
synchrotron radiation 275
system
    on a chip 372
    performance 184

T

tactile 864
    sensor 863
tape automated bonding 722
Ta_2O_5 540
technology
    additive methods 189
    modifying methods 189
    subtractive methods 189
    wafer-level 190
technology node 555
TEM 283
temporal resolution 602
terabit drive 697
Teramac concept 511
ternary compound 739
thermal
    budget 576
    detector 831
    evaporation 203, 914
    noise 834
    wavelength 836
thermo
    magnetic writing process 633
    mechanical sensing concept 689
    tropic liquid crystal 891
Thomson scattering 275
threshold 147, 149
    gate 325
    voltage 393
time
    domain 707
    of-flight 820
    temperature superposition principle 691
timing
    diagram 570
    scheme 531
    sequence 595
touchscreen 863
transducer 809
transfer matrix 408
    method 407
transflective display 905
transformer coupled plasma 254
transistor 387, 390
transition 617, 621, 622, 627
    dipole moment 917
    metal (TM) 635
transmission
    coefficient 100

electron microscopy (TEM) 274, 282
  line 706
  system 708
transparent OLED (TOLED) 924
transverse
  E-wave 47
  phonon 43
trench 260
  capacitor 541
TriCore microcontroller 181
triglycine sulfate 66
triplet
  exciton 916
  state 919
  -triplet annihilation 920
  -triplet quenching 920
tri-state buffer 170
TTT diagram 649
tunable SAW filter 751
tunnel
  effect 298
  junction 426, 529, 594
  magnetoresistance (TMR) 109, 591
  time 413
twisted nematic 961
  cell 894
two-bit quantum operation 462
two-terminal device 330
two-valued logic 19

U

ultimate computer 353
ultra-fast network switch 456
ultra-high vacuum 200
ultra-large scale integrated 543
ultrasonic touchscreen 865
undamped resonance 38
undulator 275
universal turing machine (UTM) 341

V

van der Pauw measurement 654
van der Waals force 303
vapor
  liquid-solid (VLS) macroscopic model 482
  pressure 200
variable-range hopping 916
variation principle 132
Venn diagram 19
vertical
  manipulation 315
  MRAM (VMRAM) 598
very long instruction word architectures (VLIW) 185
virtual reality 924
volatile 540
voltage
  clamp technique 156
  gated channel 156, 160
  shift 582

VSAL 654

W

Wannier exciton 915
wave
  character of the electron 409
  equation 45
  vector transfer 276
waveguide 761
wavelength
  intermediate 705
  large 705
  short 706
wavelength division multiplexing 737
WDM 737, 739
wet etching 249
wiggler 275, 282
wireless communication 703
wordline 568
work function 920, 930
WO_3 853
write
  failure 582
  field 617, 618
  operation 593

X

X-ray
  absorption 599, 600, 601
  diffraction (XRD) 274
  fluorescence (XRF) 274
  photoelectron spectroscopy (XPS) 274, 289
  scattering 274
  tube 275

Z

zirconium oxide 372

Numerics

1-D quantum wire 97
1T cell 567
1T/1C cell 567
1T/2C cell 567
1T-1C DRAM cell 539
2-D quantum well 97
2T/2C cell 567
3-D
  camera 820
  imager 822
  imaging 820
  integration 724